DICTIONARY
OF ECONOMIC PLANTS

by

J. C. TH. UPHOF
(Tampa, Fla.)

Emeritus Professor of Botany.
Formerly Faculty Member of the Michigan State Agricultural College and
State University of Arizona.
Economic Botanist to the Board of Economic Warfare, Washington, D. C.

WEINHEIM (BERGSTRASSE)
PUBLISHED BY H. R. ENGELMANN (J. CRAMER)
1959

NEW YORK
HAFNER PUBLISHING CO.

CODICOTE
WHELDON & WESLEY, LTD.

© BY J. CRAMER, PUBLISHER IN WEINHEIM

Printed in Germany by Richard Mayr in Würzburg

PREFACE

The purpose of this book is to present alphabetically a brief description of economic plants in their broadest sense, with their geographical distribution, their products and principal uses. Such a comprehensive work has long been needed. However, there are a number of useful publications which describe economic plants from certain areas, countries, plant families, genera or economic groups, such as those producing oils, rubber, drugs, spices, essential oils, gums, resins, and so on. These publications have paved the way to a considerable extent for the preparation of this book. Many of these special treatises are listed at the end of this work under „Bibliography".

Crops have been considered that are of importance to agriculture, forestry, fruit and vegetable growing, pharmacognosy; also those that are of importance to the world trade, as well as plants that are strictly of local value. A considerable number of species are given that are of ethnological interest in that they furnish food, medicine and other comodities of value in the daily life of primitive peoples. Many plants have been included that have proven to be of value in times of want during war time, or in certain countries during periods of famine.

The nomenclature of the adaptation of the scientific names is close to that of the Index Kewensis and its Supplements. Also the International Code of Botanical Nomenclature has been followed as far as it was feasible to do so.

With each species the name of the family to which it belongs has been given. It will be noticed that the names of practically all families end in -aceae, e. g. Rutaceae, Palmaceae, Zingiberaceae. After each species the name of the author who has given that particular name to the plant is indicated; for example L. stands for Carolus Linnaeus; Engl. for Adolf Engler; Lam. for Jean de Lamarck. In a number of instances other scientific names are given, they are synonyms. Because of the length of this work, it has only been possible to cross-reference the more important plants and their products.

Over 6000 different species have been considered, including the lower as well as the higher forms of plants. A number of species could not be included because the sources were not reliable or important data were uncertain.

A constructive criticism or advice which lead to the improvement of future editions will be gratefully received by the author.

Unfortunately, the publication of this book was delayed due to technical reasons.

Most of the manuscript was prepared in the libraries of the New York Botanical Garden, New York Public Library, U.S. Department of Agriculture, Cornell and Harvard Unitersity. In addition, much useful information was obtained through written inquiry to many parts of the world. The author ist most appreciative for the kind help he received from the Library Staff of the New York Botanical Garden and of the U.S. Department of Agriculture, Washington, D. C. Also an expression of appreciation is due his wife and his daughter and son-in-law, Mr. and Mrs. Harold W. Oliver, Jr. who helped in the preparation of the manuscript.

Tampa, Florida, March 1959.   J. C. Th. Uphof.

41015

ERRATA

Abies nobilis (Dougl.) Lindl. being the generally used name must be replaced by the new (1940) A. procera Rehd. as the valid name, according to the Int. Code of Bot. Nomencl., because A. nobilis (Dougl.) Lindl. is a newer homonym (1833) than A. nobilis Dietr. (1823) which after all is a never used and until 1940 forgotten synonym of A. pectinata DC.

Aristolochia brasiliensis Mart. and Zucc. Add the author names to the following species: A. galeta Mart. and Zucc., A. cymbifera Mart. and Zucc., A. rumicifolia Mart. and Zucc., A. odoratissima L., A. cordiflora Mutis ex H. B. K., A. turabacensis H. B. K., A. fragrantissima Ruiz and A. maxima Jacq.

Aphenanthe aspera Engl. read Aphananthe asepera Engl.

Cinchona sucirrubra Pavon read Cinchona succirubra Pavon.

Coprinus comatus (Fl. D.) Gray read blackish instead blackist.

Cupressus Benthami Endl. (1847) should be Cupressus lusitanica Mill. according the Int. Code of Bot. Nomencl. because the name is older (1768), though lusitanica (meaning from Portugal) is absurd for a Central American species.

Drug Lionsnear read Drug Lionsear.

Eucalyptus hemiphloia F. v. Muell. read Grey Box instead Gray Box.

Fraxinus oregana Nutt (1849) must be replaced by the older (1844), though not a much used name Fraxinus latifolia Benth. according to the Intern. Code of Bot. Nomencl.

Hyssopis officinalis L. read -pinene instead -pinense.

Oryza sativa L. Read culinary instead colinary.

Raphanus sativus L. read annual or biennial instead annual or perennial.

Simmondsia californica Nutt. (1844) must be Simmondsia chinensis (Link) Schneid. being an older (1822) name according the Int. Code of Bot. Nomencl., though chinensis (meaning from China) is a misplaced name for a well known plant from California and adj. Mexico.

A

Abaca → Musa textilis Née.

Abbevillea Fenzliana Berg., Guabiroba. (Myrtaceae). - Small tree. Brazil. Occasionally cultivated. Fruits edible, orange-yellow, flavor like a Guava; used for jelly.

Abbevilea Guaviroba Berg. → Campomanesia Guaviroba Benth. and Hook.

Abelmoschus moschatus Moench. → Hibiscus Abelmoschus L.

Aberemoa dioica Rodr. Borb. → Annona dioica St. Hil.

Aberia caffra Harv. and Sond. → Dovyalis caffra (Hook. f. and Harv.) Warb.

Aberia Gardneri Clos. → Dovyalis hebecarpa (Gardn.) Warb.

Abies alba Mill. →Abies pectinata DC.

Abies amabilis (Loud.) Forbes., Cascade Fir, Silver Fir, White Fir. (Pinaceae). - Tree. Pacific Coast of N. America. Wood hard, light, not strong, close-grained, light-brown; used for interior finish of buildings, source of wood pulp.

Abies balsamea (L.) Mill., Balsam Fir. (Pinaceae). Tree. N.America. Wood soft, light, coarse-grained, not strong, perishable, pale brown with yellow; used for crating, staves for fish and sugar barrels, butter boxes, food containers; pulpwood; made into wrapping paper and paperboard. Needles impart a pleasant scent and are used for stuffing cushions. Source of Canada Balsam, Canada Turpentine or Balsam of Fir, Balm of Gilead, a liquid oleoresin; used for mounting slides in microscopical technique, also as cement for lenses, manuf. of fine lacquers. Sp. Gr. 0.987 - 0.994. Has been used medicinally in catarrhal diseases of urogenital tracts; externally used for plasters.

Abies concolor Lindl. and Gord., White Fir. (Pinaceae). - Tree. Western N.America. Wood course-grained, very light, not strong, soft, not durable, pale brown to almost white; used for butter tubs, packing cases, doors, frames, general millwork, as building material and pulp wood.

Abies Delavayi Franch. (Pinaceae). - Tree. W. China. Wood light, soft, not very durable, white; used in China for house building, beams, interior finish, boards, general house-hold purposes, cheaper grades of coffins. Important timber in W.China.

Abies excelsa Lam. → Picea excelsa Link.

Abies grandis Lindl. Grand Fir, Lowland White Fir. (Pinaceae). - Tree. Western N. America. Wood coarse-grained, light, not strong, soft, not durable, light brown; used for interior finish of buildings, wooden-ware and packing cases. Source of pulpwood.

Abies lasiocarpa (Hook.) Nutt. Alpine Fir, Nutt. Rocky Mountain Fir. (Pinaceae). - Tree. Western N. America. Bark is source of a gum; used as antiseptic on wounds. Young twigs and leaves were burned as an incense by the Crow Indians. Wood ist source of wood pulp.

Abies magnifica Murr., Red Fir (Pinaceae). - Tree. Pacific Coast Area of N.America. Wood soft, not strong, light, fairly durable, light redbrown; used for fuel; manufactured into lumber for cheap buildings, and for packing cases; source of pulp-wood.

Abies nobilis (Dougl.) Lindl., Noble Fir. (Pinaceae). - Tree. Pacific Coast Region of N. America. Wood hard, strong, close-grained, light pale brown, red streaked; used for manuf. of lumber, often sold under the name of Larch; employed for interior finish of buildings, for packing cases, sash and door stock, Venetian blinds, ladder rails, aircraft-construction and to a limited extent for wood pulp.

Abies pectinata DC (syn. Abies alba Mill.), Silver Fir. (Pinaceae). - Tree, Europe, Asia Minor. Wood yellowish to reddish-white, soft, light, easily split, very elastic, not easy to bend, fairly durable; used for construction work, furniture, masts, matches, boxes, coopers work. Source of cellulose, pulpwood. Strassburg Turpentine is mainly produced from this tree in the Vosges. Principal Christmas tree in most countries of Europe. Leaves are source of Silver Pine Needle Oil, obtained by distillation; used for bath salts. bath tablets, bath balsams, room and theater sprays; in medical preparations used for healing and disinfecting. Can be blended with Eau de Cologne, lavender water and also with Chypre type perfumes. When inhaled is used for catarrh, colds and asthma; also sometimes used in cough drops.

Abies procera Rehd. → A. nobilis (Dougl.) Lindl.

Abies recurvata Mast. (Pinaceae). - Tree. W. Szechuan, in the Ming Valley (China). Wood close-grained, easily worked; used in China for construction of houses.

Abies religiosa (H. B. K.) Schlecht and Cham. (syn. A. hirtella Lindl.), Sacred Fir. (Pinaceae). - Tree. Mexico. Source of Aceite de Palo, an oleo resin; used in medicin for its balsamic properties. Also ingredient for paints. Wood used for building purposes and manuf. wood pulp.

Abies sachalinensis Mast. (Pinaceae). - Tree. Sachalin and Japan. Much used in Hokaidô (Jap.) for building purposes, manuf. wood pulp, ship building, water works, cooperage, household furniture, packing boxes, matches and wood shavings.

Abies squamata Mast., Flaky Fir. (Pinaceae). - Tree. Tibet, China esp., W. Szechuan. Wood of good quality; used in Tibet for construction of houses.

Abrasin Oil → Aleurites montana Willd.

Abrin → Abrus precatorius L.

Abroma angusta L. f., Cotton Abroma, Ramie sengat. (Sterculiaceae). Semiwoody plant. Malaysia. Bark source of a fibre, having some commercial value. Resembles jute, a good cordage fibre; used for clothes lines. Juice of the plant is said to be used in some parts of India for indysmenorrhea.

Abroma Mariae Mart. → Theobroma Mariae Schum.

Abronia fragrans Nutt. (Nyctaginaceae). Herbaceous plant. Western N. America, esp. California. Ground roots mixed with corn were consumed by the Indians.

Abronia latifolia Esch. (syn. A. arenaria Menz.), Yellow Sandverbena, Seacoast Abronia. (Nyctaginaceae). - Herbaceous perennial. California. Stout roots were used as food by the Chonook Indians.

Abronia, Seacoast → Abronia latifolia Esch.

Abrus precatorius L. (syn. Abrus Abrus [L.] Wight.) Crab's Eye, Indian Liquorice, Rosary Pea, Jequirity. (Leguminosaceae). - Vine. Tropics of Old and New World. Bright colored black and red seeds are used for beads. Roots called Wild or Indian Liquorice are used as substitute for Liquorice; they contain glycirrhizin. Seeds contain abrin, a toxalbumin, an exceedingly poisonous substance; very small amounts when brought into contact with wounds may be fatal.

Abrus pulchellus Wall. (Leguminosaceae). - Vine, India to S. Australia. Roots used in Malaysia as liquorice.

Absinthe → Artemisia Absinthium L.

Absinthe, Algerian → Artemisia Barrelieri Ben.

Abu Beka Resin → Gardenia lutea Fresen.

Abura → Mitragyna stipulosa Kuntze.

Abuta Candollei Triana and Planch. (syn. A. rufescens DC.). (Menispermaceae). - Vine. Guiana. The whitish roots are source of White Pereira. Used in medicine.

Abutilon asiaticum G. Don. (Malvaceae). Shrub. Tropics. Stem is source of a fibre; used for manuf. sacks and paper.

Abutilon Avicennae Gaertn. Chingma (Malvaceae). - Annual herb. Tropics. Much cultivated in China. Source of a fibre, called China Jute or Indian Mallow, being grayish-white, coarse and strong; used in the same way as jute; also employed for caulking boats.

Abutilon Bedfordianum St. Hil. (Malvaceae). - Shrub. Brazil. Stem is source of a fibre; used for cordage and manuf. paper. Introd. in Australia.

Abutilon indicum Sweet., India Abutilon, Country Mallow. (Malvaceae). - Shrub. Malaysia. Source of a fibre; easy to ret. Fibres are white, strong; made into cordage when derived from matured stems and woven into fabrics when derived from young stems. Takes dyes readily.

Abutilon intermedium Hochst. (syn. A. angulatum Mast.) (Malvaceae). - Shrub. Tropics. Bark source of a fibre; used in Madagascar for manuf. cloth.

Abutilon Jacquini G. Don. (syn. A. lignosum [Cav.] G. Don.). (Malvaceae). - Half shrub. Texas, Mexico and West Indies. In W. Mexico stems are source of a fibre; used for manuf. twine and rope.

Abutilon molle Sweet. (Malvaceae). - Shrub. S. America, esp. Brazil and Peru. Stem is source of a fibre; used for cordage.

Abutilon muticum Sweet. (syn. A. asiaticum Guill. and Pierre.) (Malvaceae). - Small tree. Tropics. Seeds are consumed by the Nomads in Sudan.

Abutilon octocarpum F. v. Muell. (Malvaceae). - Perennial plant. Australia. Bark is source of a fibre; used by the aborigines of Queensland for game netting.

Abutilon oxycarpum F. v. Muell. (Malvaceae). - Shrub. S. America, esp. Brazil also Australia. Occassionally cultivated. Source of a fibre, used for manuf. sails. Also A. venosum Lem.

Abutilon polyandrum G. Don. (Malvaceae). - Shrub. Trop. Asia, esp. India and Ceylon. Stem is source of a fibre; used for manuf. sails.

Abyssinian Myrrh Tree → Commiphora abyssinica (Berg.) Engl.

Acacia abyssinica Hochst. → A. glaucophylla Steud.

Acacia acuminata Benth., Raspberry Acacia. (Leguminosaceae). - Tree. W. Australia, Source of charcoal. Wood used by aborigines for weapons. Fragrant flowers have been suggested as a source of perfume.

Acacia albida Delile. (Leguminosaceae). - Tree. N. Africa. Twigs and leaves are used as camel food. Bark is employed in tanning. Stem is source of a Gum Arabic also called Gomme de Sénégal.

Acacia aneura F. v. Muell., Mulga Acacia. (Leguminosaceae). - Tree. Australia. Ground seeds are consumed as food by the aborigines of S. Australia. Wood was employed for making spear-throwers. Leaves are used as food for live-stock, also S. salicina Lindl.

Acacia Angico Mart.→ Piptadenia rigida Benth.

Acacia angustissima (Miller) Kuntz. (syn. Mimosa angustissima Miller), Prairie Acacia. (Leguminosaceae). - Large shrub. S. of United States, Mexico, Centr. America. Bark used for tanning in Oaxaca (Mex.). Pods consumed as food by Mazatecs, Chinantecs and other tribes in Mexico.

Acacia arabica Willd. (syn. A. nilotica Delile), Babul Acacia. (Leguminosaceae). - Tree. Trop. Africa. Wood heavy, hard, durable, close-grained, red-brown to deep red, resistant to water and white ants; used for sugar and oil presses, rice-pounders, agricultural implements, tool-handles, wheels, railroad-ties. Was used in Ancient Egypt since time immemorial for house-beams, statues, panelling and boats. Leaves and green pods are a food for camels, sheep and goats. Young bark is source of a fibre, Pods and bark are employed in tanning leather; called Bablah or Neb-neb; it is also used for dyeing clothes a yellow color. Stem is source of a soluble gum, called under various local names, Amrad, Amrawatti, Brown Barbary Gum. Tears are of moderate size, often vermiform, light dusky brown tint. Seed with dates are sometimes made into a fermented beverage.

Acacia, Babul → Acacia arabica Willd.

Acacia Bidwillii Benth.→A. pallida F. v. Muell.

Acacia binervata DC. (syn. A. umbrosa Cunningh.) Twinvein Wattle, Black Wattle. (Leguminosaceae). - Tree. New South Wales, Queensland. Bark source of valuable tanning material. Also **A. calamifolia** Sweet, **A. dealbata** Link., **A. harpophylla** F. v. Muell., **A. pycnantha** Benth., **A. podaliriaefolia** Cunningh. Wood of A. binervata is tough, light, close-grained; used for axe handles, bullock yokes.

Acacia calamifolia Sweet → A. binervata DC.

Acacia Catechu Willd., Catechu. (Leguminosaceae). - Tree. E. India. Source of Catechu, Black Catechu, Black Cutch, Pegu Catechu being an extract from the heartwood. Used for tanning, toilet preparations, dyeing fabrics of a black or brown color, also for staining of wood. Used medicinally as astringent, for diarrhea, sore throat. Contains gum catechin, catechu tannic acid, catechu red and quercin. Also from Acacia suma Kunz. It is sometimes called Terra Japonica a name also applied to Uncaria Gambier Roxb.

Acacia Cavenia Bert., Cavenia Acacia, Espino Cavan. (Leguminosaceae). - Small tree. S. America, esp. Chile. Allied to A. Farnesiana. Culti-vated. Source of a perfume, Cassie Romaine, obtained by petroleum ether extraction; about 200 kg. flowers produce 1 kg. concrete and 300 gr. absolute ess. oil. Application is the same as A. Farnesiana oil (Cassie Ancienne), though it is less fine and has a more spicy character. Much is derived from S. France.

Acacia cibaria F. v. Muell. (Leguminosaceae). - Tree. Australia esp. W. Australia, New South Wales. Seeds are eaten by the natives of Australia. Also A. longifolia Willd.

Acacia cineria Spreng. → Dichrostachys cinerea Wight and Arn.

Acacia concinna DC., Soap Pod. (Leguminosaceae). Prickly shrub. India. Pods are widely used as detergent for washing silks and woollen goods, washing hair and cleaning silver plates.

Acacia conferta Cunningh. (Leguminosaceae). - Australia. Flowers have been suggested as a source of perfume. Also A. harpophylla F. v. Muell. and A. pycnantha Benth.

Acacia Cunninghamii Hook., Kowarkul, Bastard Myal. (Leguminosaceae). - Tree. Australia, esp. New South Wales, Queensland. Wood close-grained, takes a good polish; used for cabinet work.

Acacia dealbata Link. Silver Wattle. (Leguminosaceae). - Shrub or tree. S. Australia, Produces a gum exceedingly viscose and as useful as Gum Arabic. Also A. binervata DC., A. decurrens Willd., A. elata Cunningh., A. glaucescens Willd., A. harpophylla F. v. Muell., A. pycnantha Benth.

Acacia decora Reichb., Graceful Wattle Acacia. (Leguminosaceae). - Tree. Australia. Gum from the stem is consumed by the aborigenes of N. Queensland. Also A. homalophylla Cunningh.

Acacia decurrens Willd., Greenwattle Acacia, Black Wattle. (Leguminosaceae). - Tree. Australia. Bark is used for tanning, contains 30 to 50⁰/o of tannin. Also A. decurrens var. mollisima Willd., A. dealbata Link, A. saligna Wendl.

Acacia dictyophleba F. v. Muell. (Leguminosaceae). - Tree. Australia. Seeds when pounded are used as food by the natives of Centr. Australia. Also A. stipuligera F. v. Muell.

Acacia doratoxylon Cunningh., Currawong Acacia. Spear Wood, Brigalow. (Leguminosaceae). - Tree. Australia. Wood tough, close grained, durable, heavy. Used for furniture, gates. Employed by aborigines for spears and boomerangs. Leaves are used as food for cattle. Scented flowers have been suggested as a source of perfume.

Acacia elata Cunningh. → A. dealbata Link.

Acacia Ehrenbergiana Hayne. (Leguminosaceae). - Shrub. Trop. Africa. Trunk is source of a gum.

Acacia etbaica Schweinf. → A. nubica Benth.

Acacia excelsa Benth., Ironwood, Brigalow. (Leguminosaceae). - Tree. Queensland. Wood close-grained, hard, very tough, elastic, violet scented. Used for cabinet-work.

Acacia falcata Willd. (syn. A. plagiophylla Spreng.), Burra Acacia. (Leguminosaceae). - Shrub or tree. Australia esp. New South Wales, Queensland. Used by aborigines to stupefy fish. Also A. penninervis Sieb., A. salicina Lindl.

Acacia Farnesiana (L.) Willd., Sweet Acacia. (Leguminosaceae). - Shrub or small tree. Tropics. Cultivated. Bark is used for tanning. Pods produce a black dye; used for ink and for giving leather a black color. Stems produce a second grade gum. Wood is very hard, tough; used in some parts of Africa for tentpegs, ploughs, ship's knees, cabinet work. The very fragrant Cassie Flowers are source of Cassie Ancienne, an ess. oil, used in perfumery. Much is derived from S. France. Flowers are laid between linen to impart a pleasent scent.

Acacia fistula Schweinf. → A. Seyal Delile.

Acacia Giraffae Willd. Giraffe Acacia. (Leguminosaceae). - Spiny shrub or small tree. S. Africa. Source of Cape Gum. Bark used for tanning. Pods are ground into flour and used as fodder for live-stock.

Acacia glaucescens Willd. → A. dealbata Link.

Acacia glaucophylla Steud. (Leguminosaceae). - Shrub or small tree. Trop. Africa. Stem is source of Somali Gum. Also A. abyssinica Hochst.

Acacia Greggi Gray., Cat's Claw. (Leguminosaceae). - Shrub or tree. Mexico and SW of United States. Source of a gum, similar to Gum Arabic, locally used. Pods produce a flour, made into cakes and consumed by the Indians.

Acacia gummifera Willd., Mogador Acacia. (Leguminosaceae). - Tree. N. Africa. Source of a resin, called Mogador or Morocco Gum. Much is derived from Morocco. Tears are usually dark colored and of a sweetish taste which do not dissolve completely in water.

Acacia harpophylla F. v. Muell., Sickeleleaf Acacia. Bigalow. (Leguminosaceae). - Tree. Queensland. Source of a reddish-brown dye; stains wool and cotton.

Acacia hemiteles Benth. → A. subcoerulea Lindl.

Acacia horrida Willd. (syn. A. Karroo Hayne). Allthorn Acacia. (Leguminosaceae). - Tree. S. Africa. Wood hard, tough; used for interior finish of buildings. Bark and pods are employed for tanning, imparting a reddish color to leather. Leaves and young pods are used as food for live-stock. Stems produce a gum, used locally; resembling Gum Arabic. Tears are pale yellow, hard, brittle, irregular; giving a rather dark mucilage. Sometimes called Cape Gum.

Acacia homalophylla Cunningh., Gidgee Acacia, Spear Wood, Curley Yarran. (Leguminosaceae). - Tree. Australia. Wood dark brown, fragrant; used for turner's work, tobacco pipes and fancy articles.

Acacia Jacquemontii Benth., Khumbut, Dhakki. (Leguminosaceae). - Small shrub. India. Source of a gum, inferior to Gum Arabic. Used for calico printing and in paper making.

Acacia Karroo Hayne → Acacia horrida Willd.

Acacia Koa Gray., Koa Acacia. (Leguminosaceae). - Tree. Hawaii. Wood hard; used by natives for spears and fancy paddles.

Acacia laeta R. Br. → A. Senegal (L.) Willd.

Acacia leucophloea Willd., Arinj. (Leguminosaceae). - Large tree. India, Burma. Source of a gum, used for adulterating gums of better quality.

Acacia leucophylla Lindl. → A. pendula Cunningh.

Acacia longifolia Willd. → A. cibaria F. v. Muell.

Acacia melanoxylon R. Br., Blackwood, Black Sally. (Leguminosaceae). - Tree. Australia esp. Tasmania, S. Australia, Victoria, New South Wales. Wood close-grained, hard. Used for furniture, cabinet-work, boat-building, tool-handles, oil-casks, naves of wheels, pianofortes, parts of organs, billiard-tables; cut into veneers. Takes a fine polish.

Acacia mellifera Benth. → A. Senegal (L.) Willd.

Acacia modesta Wall. (Leguminosaceae). - Medium sized tree. Sub-Himalaya, Plains of N. Pundjab. Source of a useful gum, Amritsar Gum, mainly derived from Pundjab. Used in India as medicine.

Acacia myriadena Bert. → Serianthes myriadena Planch.

Acacia Neboueb Baill. → A. stenocarpa Hochst.

Acacia nilotica Delile → A. arabica Willd.

Acacia nubica Benth. (Leguminosaceae). - Shrub. Red Sea area, Sudan. Bark is used for tanning. Also A. etbaica Schweinf.

Acacia Oswaldii F. v. Muell. Umbrella Acacia. (Leguminosaceae). - Tree. Australia. Seeds are consumed as food by the aborigenes of Australia.

Acacia pallida F. v. Muell. (Leguminosaceae). - Tree. Australia. Young roots when roasted are consumed as food by the aborigines of N. Queensland. Also A. Bidwillii Benth.

Acacia pendula Cunningh. (syn. A. leucophylla Lindl.), True Myall, Boree. (Leguminosaceae). - Tree. Australia esp. New South Wales, Queensland. Wood close-grained. hard dark, beautifully marked, scent of violets. Used for veneers, fancy boxes, tobacco pipes. Leaves are used as food for live-stock.

Acacia penninervis Sieb. → A. falcata Willd.

Acacia plagiophylla Spreng. → A. falcata Willd.

Acacia podaliriaefolia Cunningh. → A. binervata DC.

Acacia pycnantha Benth., Golden Wattle. (Leguminosaceae). - Shrub. Australia. Bark is used for tanning leather. Stem and branches are source of Australian or Wattle Gum, also A. decurrens Willd., A. homalophylla Cunningh.

Acacia Rehmanniana Shinz. → A. Senegal (L.) Willd.

Acacia Roemeriana Scheele. (Leguminosaceae). - Shrub. Mexico and SW United States. Flowers are source of a valuable honey in some parts of Texas.

Acacia rivalis J. M. Black. (Leguminosaceae). - Tree. Australia. Gum is consumed by the aborigines of S. Australia. Also the seeds are eaten as food when being ground.

Acacia salicina Lindl. → A. aneura F. v. Muell., A. falcata Willd.

Acacia saligna Wendl. → A. decurrens Willd.

Acacia samaryana Chev. → A. Senegal (L.) Willd.

Acacia Senegal (L.) Willd. (syn. A. Verek Guill and Perr.). Gumarabic Acacia. (Leguminosaceae). - Shrub or small tree. Trop. Africa. Source of a resin called Gum Arabic, known as Kordofan, Ghezirch Gum, Prickly Turkey, White Senaar, Senegal Gum, Gomme Blondes, Gomme Blanche. Is harvested from December until June. Much is derived from the Sudan. Gum is often graded in Middle, White, Middle Siftings, Very Large, Large and Small. Used for lustre to crape and silk, thickening colors, calico printing, manuf. ink, mucilage, confectionary. Important trading centers are Senegal and Kordofan. Less quantities are produced by A. laeta R. Br., A. mellifera Benth., A. Rehmanniana Shinz, A. samaryana Chev., A. campylocantha.

Acacia Seyal Delile (syn. A. Fistula Schweinf.), Seyal Acacia. (Leguminosaceae). - Small tree. Trop. Africa. Stem is source of a Gum Arabic, known in the trade as Suakim, Talca, Sennarr, Talha and Talki Gum. It is said that elephants feed on the pods.

Acacia Sieberiana DC. (Leguminosaceae). - Tree. Trop. Africa. Stem is source of a pale yellow to yellowish brown, soluble gum.

Acacia spirocarpa Hochst. (Leguminosaceae). - Tree. Trop. Africa. Bark is source of a strong fibre; used by the natives of Africa.

Acacia stenocarpa Hochst. Narrowwing Acacia. (Leguminosaceae). - Small tree. E. Africa. This is one of the species being a source of Gum Arabic of commerce, sometimes called Gomme Salobreda. Also A. Neboueb Baill.

Acacia stipuligera F. v. Muell. → A. dictyophleba F. v. Muell.

Acacia subcoerulea Lindl. (syn. A. hemiteles Benth.) Blue-leaved Acacia. (Leguminosaceae). - Tree. W. Australia. Source of a yellow dye.

Acacia subporosa F. v. Muell. (Leguminosaceae). - Tree. Australia esp. Victoria, New South Wales. Wood tough, elastic. Used for toolhandles, gun-stocks.

Acacia tortilis Hayne. (syn. A. fasciculata Guill and Perr.), Sejal, Talha. (Leguminosaceae). - Shrub or small tree. N. Africa, Arabia. Source of a resin, Gomme Rouge.

Acacia umbrosa Cunningh. → Acacia binervata DC.

Acacia Verek Guill and Perr. → A. Senegal (L.) Willd.

Acacia Visco Lorentz. (Leguminosaceae). - Tree. Argentina. Wood is used for house-building and for boxes.

Acaena pinnatifida Ruiz and Pav. (Rosaceae). - Perennial plant. Chile. Infusion is used by the natives as an astringent.

Acalypha indica L., Indian Nettle. (Euphorbiaceae). - Herb. Trop. Asia, Africa. Fresh or dried plants are used medicinally as gastro-intestinal irritant; large doses are emetic. Contains acalyphin, an ess. oil, resin, tannin and an alkaloid. Acanthaceae → Acanthus, Adhatoda, Andrographis, Barleria, Blepharis, Clinocanthus, Hypestes, Jacobinia, Justicia, Peristrophe, Rhinacanthus, Staurogyne, Strobilanthes, Tubiflora, Whitfieldia.

Acanthocereus pentagonus (L.) Britt. and Rose. (syn. Cereus pentagonus L.), Pitahaya, Pitahaya Naranjada, (Cactaceae). - Xerophytic tree. East coast of Mexico, Texas, Centr. America, Northern S. America. Fruits are red, edible.

Acanthopanax ricinifolium Seem. (syn. Kalopanax pictus [Thunb.] Nakai.) (Araliaceae). - Tree. China, Manchuria, Korea, Japan. Wood easily worked, pliable, resonant; used in China for drums, in boats and temples.

Acanthopanax spinosus Miq. (Araliaceae). - Shrub. Japan, China. Very young leaves are eaten during spring as a vegetable in Japan.

Acanthopeltis japonica Okam. (Gelidiaceae). - Red Alga. Japan. Used in Japan for the manufacture of agar. See under Gelidium Amansii (Lamour) Lamour.

Acanthophora specifera (Vahl.) Boeg. (Rhodomelaceae). - Red Alga. Indian Ocean. Consumed as food in the Kangean Islds., N.E. Java and Philipp. Islds.

Acanthorrhiza Warscewiczii Wendl. (Palmaceae). - Tall palm. Costa Rica, Panama. Leaves are used for thatching.

Acanthospermum hispidum DC. (Compositae). - Herbaceous plant. Brazil to Argentina. Plant is bitter and aromatic; used in some parts of S. America as diuretic and sudorific.

Acanthosphaera Ulei Warb., Balsamo (Moraceae). - Tree. Brazil, esp. Amazon region, E. Peru. Bitter latex is used locally as febrifuge.

Acanthosicyos horrida Welw., Narasplant. (Cucurbitaceae). - Thorny shrub. Trop. Africa. Fruits having the size of an orange; pleasant acid flavor; eaten fresh or preserved. Seeds called Butter Pits, are consumed as food by the Hottentots. Fresh fruit-pulp is used to separate casein from milk after being heated.

Acanthosyris falcata Griseb. (Santalaceae). - Shrub or tree. Argentina, Paraguay, Bolivia. Red fruits size of a cherry, are consumed by the natives; also used for making a liqueur. Wood is employed for making furniture.

Acanthus ebracteatus Vahl., Sea Holly. (Acanthaceae). - Saltbush. Trop. Asia. Decoction of boiled leaves is used as cough medicin in some parts of Malaya. It is taken with flowers of Averrhoa, black sugar-cane, cinnamon and crystalline sugar.

Accra Copal → Daniella Ogea Rolfe.

Aceite de Palo → Abies religiosa (H. B. K.) Schlecht and Cham.

Aceitunillo → Aextoxicon punctatum Ruiz and Pav.

Acer campestre L., Hedge Maple. (Aceraceae). - Tree. Europe, N. Iran, Asia Minor, Algeria. Wood light-brown to reddish-white, medium heavy, very hard, tough, very difficult to split, elastic; used for cutlery, turnery, mill-wheels, tobacco pipes (Ulmer pipes), agricultural implements, balls, veneer, Twigs are made into walking-sticks.

Acer circinatum Pursh., Vine Maple. (Aceraceae). - Tree. Pacific Coast region of N. America. Wood close-grained, hard, not strong, heavy, light brown; used for handles of axes and tools. Used by Indians for bows of fishing nets.

Acer crataegifolium Sieb. and Zucc. Hawthorn Maple. (Aceraceae). - Japan. Bark is used in Suruga province, Japan, as a paste in manuf. of paper.

Acer dasycarpum Ehrh. → Acer saccharinum L.

Acer eriocarpum Michx.→ Acer saccharinum L.

Acer Ginnala Max. (syn. A. tataricum L. var. Ginnala Max.). (Aceraceae). - Tree. Japan, China. Young leaves are used as a tea in Japan.

Acer macrophyllum Pursh., Broad-leaved Maple. (Aceraceae). - Tree. Pacific Coast region of N. America. Wood not strong, close-grained, soft, light, rich brown; used for interior finish of buildings, handles of axes and tools, also for furniture.

Acer mono Schum. → Acer pictum Thunb.

Acer Negundo L. (syn. Negundo aceroides Moench., N. fraxinifolium Nutt.). Box-Elder. (Aceraceae). - Tree. N. America. Wood not strong, close-grained, soft, light creamy white; used for wooden ware, paper pulp, cooperage, cheap furniture, interior finish of houses. Sap is occasionally source of sugar.

Acer pictum Thunb. (syn. A. mono Schum.) (Aceraceae). - Tree. Japan, Korea, Amur, China. Wood hard, close-grained, yellow-white. Used in Japan for furniture, cabinet-work, interior finish of buildings, railroad cars, ship building, important as fuel. Also A. Mayrii Schwer.

Acer platanoides L., Norway Maple. (Aceraceae). - Tree. Throughout Europe, Caucasia, Armenia. Wood reddish-white, fine-grained, very hard, heavy, elastic, easy to polish; used for turnery, wagons, furniture and manuf. of rifles.

Acer Pseudo-Platanus L. Plane Tree Maple, Scottish Maple. (Aceraceae). - Tree. Europe, Caucasia, Armenia, Asia Minor. Wood yellowish-white, fine structure, very hard, heavy, elastic, easy to work, polishes well, fairly resistant to insects; used for furniture, carriages, billiard-cues, shoe lasts, violins, wood carving, floors, and for rifles.

Acer rubrum L., Red Maple. (Aceraceae). - Tree. Eastern N. America to Florida and Texas. Wood not strong, close-grained, very heavy, light brown; used for furniture, turnery, gun-stocks, wooden ware, wood pulp.

Acer saccharinum L. (syn. A. dasycarpum Ehrh., A. eriocarpum Michx.), Silver Maple, White Maple (Aceraceae). - Tree. Eastern N. America to Florida. Wood close-grained, hard, strong, brittle, easily worked; used for veneer, cooperage, shoe-lasts, handles, spools, furniture, sporting goods, toys, motor vehicle parts, flooring, fuel, wood pulp. Sugar is occasionally obtained from the sap of the stem.

Acer saccharum Marsh., Sugar Maple. (Aceraceae). - Tree. Eastern N. America. Juice from stem is a source of maple syrup, maple sugar, used for flavoring and sweetening, different foods and candies. Much ist manuf. in the New England States. Wood close-grained, tough, hard, heavy, light brown, tinged with red; used for floors, interior finish of buildings, furniture, turnery, shoe-lasts; ship building, pegs, fuel, wood pulp.

Aceraceae → Acer.

Acetobacter aceti (Kützing) Beijerinck. (Bacteriaceae). - Bacil. Microorganism. Belongs to the acetic acid bacteria. The fermentation product is acetic acid, commercially known as vinegar, cider vinegar, apple vinegar, wine vinegar, malt vinegar, sugar vinegar, glucose vinegar, spirit vinegar, distilled vinegar, grain vinegar. To this group belong several other species of Acetobacter.

Acetobacter ascendens (Henneb.) Berg. et al. (Bacteriaceae). - Microorganism. Is able to produce ethanol under anaerobic conditions. Also A. suboxydans Kluyver and de Leeuw and A. Pasteurianum (Hansen) Beijerinck.

Acetobacter suboxidans Kluyver and de Leeuw. (Bacteriaceae). Bacil. Is important in the production of sorbose, a product of fermentation. Also A. cylinum (Brown) Beijerinck, A. aceti (Kützing) Beijerinck, A. melanogenum Beijerinck, A. xylinoides Heenneb. and A. rancens Beijerinck.

Acetone-athanol → Bacillus acetothyllicum Northtr.

Aceton-butanol → Clostridium acetobutylicum McCoy et al.

Achillea Millefolium L., Milfoil, Yarrow. (Compositae). - Perennial herb. Temp. zones. Leaves are source of a tea; used in home remedies for coughs, diseases of the bladder and kidneys, also for „cleaning" the blood; for sealing open wounds on hands, abcesses; ailments of the stomach. Herb is occacionally used in salads in soup (called Gründonnerstag-Suppe in Germany). Sometimes used in parts of Sweden as a substitute for tobacco. Leaves and flower tops are used medicinally, being a stimulant, tonic, mild diaphoretic and emmenagogue. A distilled oil, Oleum Millefolii is used as tonic, nervino-excitans and aromatic. Herb contains an ess. oil, and a bitter alkaloid, achilleine.

Achillea moschata Jacq., Musk Yarrow. (Compositae). - Perennial herb. Centr. Europe. Source in Italy of a liqueur, called Esprit d'Iva, Iva Liqueur, Iva Wine; in Oberengadin (Switzerl.) Iva Bitter. A tea is used as a home remedy, being tonic, stomachic, stimulant, diaphoretic. Herb contains alkaloids and an ess. oil cineol.

Achillea Ptarmica L., Sneezewort (Compositae). - Perennial herb. Temp. Europe, Asia. Cultivated. Roots formerly used as sneeze-pulver. Flower heads were employed in medicine.

Achillea Santolina L. (Compositae). - Herbaceous perennial. E. and N. Africa, Arabia, Iran. Used in Teheran, Iran for chest complaints; it is considered carminative and tonic.

Achillea sibirica Led. (Compositae). - Herbaceous perennial. Temp. N. Asia. Used in China medicinally.

Achras mammosa L. → Calocarpum mammosum Pierre.

Achras nigra Poir. → Dipholis nigra Gris.

Achras Sapota L., Sapodilla. (Sapotaceae). - Tree. S. Mexico, Guatemala to Honduras. Cultivated. Fruit edible, of fine quality, round to oval; pulp yellow-brown, soft, sweet, rich fine flavor. Stem is source of a gum; used as base for chewing gum, called Chicle.

Achyranthes aspera L. Prickly Chaff Flower. (Amaranthaceae). - Herbaceous plant. Trop. Africa, Asia, Australia. Used by the Shari Chad races (Afr.) for making salt. Macerated roots are used in India to relieve pain from scorpion stings. Leaves are eaten as vegetable in Java. Branches are source of tooth brushes in Hadramut (Arabia). Ashes from plant are employed as a source of alkali in dyeing.

Achyrocline flaccida DC. (Compositae). - Herbaceous plant. S. America. Plant is used in Brazil as a tonic and excitant, also as febrifuge, anthelmintic and antispasmodic.

Acid, citric → Aspergillus niger van Viegh., Citromyces glaber Wehmer, Citrus Limon Burman and Penicillium citrinum Thom.

Acid, d'lactic → Rhizopus oryzae Went and Prins geerl.

Acid, lactic → Lactobacillus Delbrueckii.

Acid, fumaric → Rhizopus nigricans Ehr.

Acid, propionic → Propionibacterium Freudenreichii von Niel.

Ackawai Nutmeg → Acrodiclidium Camara Schomb.

Acnistus arborescens (L.) Schlecht. Tree Wild Tobacco. (Solanaceae). - Shrub. Trop. America. Fruits are edible, made into jellies.

Acokanthera Ouabaio Cathel. (Apocynaceae). - Tree. Trop. Africa, esp. Somaliland. Source of Ouabain, a glucoside; used in medicine.

Acom → Dioscorea latifolia Benth.

Aconite, Indian → Aconitum ferox Wall.

Aconite, Japanese → Aconitum Fischeri Reichb.

Aconitum ferox Wall. Indian Aconite. (Ranunculaceae). - Perennial herb. Himalaya region. Roots are sold in bazaars of India for medicinal purposes.

Aconitum Fischeri Reichb. (Ranunculaceae). - Perennial herb. Kamschatka. Root commercially known as Japanese Aconite. Other asiatic species used among druggists are **A. chasmanthum** Stapf., **A. laciniatum** Stapf., **A. Balfouri** Stapf. and **A. deinorrhizum** Stapf.

Aconitum heterophyllum Wall. (Ranunculaceae). - Perennial herb. Himalayan region. Herb is said to be inert and is eaten by the hill people. It is claimed to be antiperiodic, aphrodisiac and tonic. It is used in native medicine as a mild and bitter tonic; sold in bazaars of the Orient.

Aconitum Napellus L., Monkshood. (Ranunculaceae). - Perennial herb. Centr. Europe. Cultivated. Dried roots are used medicinally as a heart and nerve sedative; locally as a tincture and as analgesic. Roots are collected in the autumn or during flowering time. Much derived from Germany and Switzerland. Also the herb, Herba Aconiti, composed of dried leaves and flowering tops is used for the same purpose. Contains aconitine, an alkaloid. In ancient times a decoction of the poisonous herb was given to criminals.

Acorus Calamus L., Calamus, Sweet Flag, Flag Root. (Araceae). - Perennial herb. Temp. zones of Old and New World. Rhizomes are used as flavoring agent, they are sometimes candied. Cultivated in Burma, Ceylon. Dried rootstocks come also from Russia, Netherlands, Germany, England and parts of the United States, among which Michigan, Indiana, Virginia, N. Carolina. Collected in autumn. Used medicinally as carminative, a stimulant and as an aromatic bitter tonic. Contains acorin, a bitter, viscid, aromatic glucosidal compound.

Acrocarpus fraxinifolius Wight and Arn., Pink Cedar, Red Cedar, Shingle Tree. (Leguminosaceae). - Tall tree. India to Mal. Archip. Wood used for tea-boxes, boards, furniture, shingles, building purposes; it is very hard, brown with even black stripes. Wood from some areas is more durable than that from others.

Acroceras sparsum Stapf. (Graminaceae). - Perennial grass. Malaysia. Robust herb; produces much foliage; used as food for livestock in Java.

Acrocomia mexicana Karw., Coyal. (Palmaceae). - Medium palm. Mexico to Guatemala. Fruit edible. Sold in markets.

Acrocomia sclerocarpa Mart. Paraguay Palm, Macaja, Micauba. (Palmaceae). - Tall palm. Trop. S. America, West Indies, Leewards Islds. Pulp oil and kernel oil are used for manuf. soap. When refined it is suitable for cooking.

Acrocomia vinifera Oerst. (Palmaceae). - Palm. Centr. America, esp. Nicaragua to Panama. Source of an intoxicating beverage, used since ancient times by the Indians. Fruits are eaten during times of scarcity.

Acrodiclidium Camara Schomb. (Lauraceae). - Tree. Guiana to Brazil. Fruits called Ackawai Nutmegs, are excitant, aromatic, carminative, antispasmodic, anti-diarrhetic, and antidysenteric. Wood is used for general carpentry.

Acronychia laurifolia Blume. (syn. Cyminosma resinosa DC., Jambolifera resinosa Lour.), (Rutaceae). - Small tree. Trop. Asia. Leaves when young are used as a condiment. In Cochin China leaves are put in stimulating baths, probably due to the action of an ess. oil. Bark is employed in S. Annam by fishermen for making boats water-tight. Root is supposed to be a fish poison used in Cochin China.

Acronychia odorata Baill. (syn. Cyminosma odorata DC., Jambolifera odorata Lour.) (Rutaceae). - Small shrub. Trop. Asia. Used in Annam for a fever, called benh-cum or dingue. Leaves are used as condiment in Malaysia.

Acronychia resinosa Forst. (Rutaceae). - Tree. Indo-China. Bark used for caulking boats, toughening fishing nets. Supposed to stupefy fish.

Acrostichum aureum L. Marsh Fern. (Polypodiaceae). - Tall Fern. Marshes throughout tropics. Very young leaves are eaten by natives of Borneo, Celebes, Timor. Among the Annamites the firm parchment-like fronds are dried, strung up on rods, and used in stead of straw thatch.

Actephila excelsa Muell. Arg. (Euphorbiaceae). - Small tree or shrub. Trop. Asia. Plant is source of a beverage.

Actinella biennis Gray (syn. Actinea Richardsonii Nutt.) (Compositae). - Herbaceous perennial. S.W. of United States. Bark of the roots was made into a chewing gum by Indians of New Mexico.

Actinella odorata Gray. (Compositae). - Perennial herb. S.W. of the United States and adj. Mexico. Decoction of flowertops was used as beverage by Indians of Texas.

Actinidia callosa Lindl. (syn. A. arguta Franch. and Sav., A. Kolomikta Maxim.). Kolomikta Vine. (Actinidiaceae). - Woody Vine. N. China, Manchuria, N. Korea, E. Siberia. Fruits, called Tara, are edible, having the size of a cherry to that of a plum. Much esteemed in some parts of China. In Siberia fruits are dried for winter use, called Kismis; they are baked in bread and pastry.

Actinidia chinensis Planch. (Actinidiaceae). - Woody vine. China, esp. Szechuan, Hupeh. Fruits are edible, of good quality, being of the size of a hen's egg; used in preserves and for jams.

Actinidia polygama Franch., Silver Vine. (Actinidiaceae). - Woody vine. China, Manchuria, Japan. Leaves boiled are eaten in some parts of Japan. Salted fruits are consumed by the Japanese. It is said that cats are fond of this plant.

Actinidiaceae → Actinidia.

Actinomycin → Streptomyces antibiotica (Waksman and Woodruff) Berg. et al.

Adam's Needle → Yucca filamentosa L.

Adansonia digitata L., Baobab, Monkey Bread, Ethiopian Sour Gourd, Cream of Tartar Tree. (Bombaceae). - Tall tree. Trop. Africa. Pulp of fruit is used as a source of food; used as food seasoner. Acid pulp is used as a rubber coagulant. It is used by pastoral tribes in Africa to curdle milk, resulting into Kwatakwari. Burnt pulp is used as a fumigant to combate biting insects on domestic animals. Young leaves are eaten in soup, also as pot-herb. Pulp is widely used in Africa as diaphoretic for fever and dysentery. In the Cameroons the bark is used for tanning. Seeds are eaten as food, they are also mixed with meal of millet. It is made into a thin gruel; also used as beverage. Pounded seeds are used in some regions as famine food. Inner bark is source of a strong fibre; used for horsegirths, cordage, tethering rope, strings of musical instruments, also source of paper material. Seeds are source of Baobab, Fony or Reniala Oil, a non-drying, golden yellow oil of pleasant taste, suitable for manuf. soap. Sp. Gr. O. 9198; Sap. Val. 190.5; Iod. No. 67.5.

Adansonia Gregorii F. v. Muell. (Bombaceae). - Tree. Trop. Australia. Seeds are eaten raw or roasted by the aborigenes of Queensland.

Adansonia madagascariensis Baill., Madagascar Baobab. (Bombaceae). - Tree. Madagascar. Bark is source of a fibre. Fruits are consumed by the natives.

Aden Senna → Cassia holosericea Fresen.

Adenanthera microsperma Teijsm. and Binn· (Leguminosaceae). - Tall tree. Java. Cultivated. Wood much esteemed, durable, very strong, hard, heavy, yellow-brown to brown-red; used for house building, furniture, bridges, rolls in sugar mills; is resistant to insects and decay. Bark is recommended for tanning.

Adenanthera pavonina L., Sandal Beadtree, Zumbic Tree, Circassian Tree. (Leguminosaceae). - Tree. India, S.E. China, Moluccas, etc. Wood called Red Sandal Wood is hard, close grained, red; used in India for house building, cabinet work, a source of red dye. Decoction of leaves is used in India for rheumatism and gout; its wood as a tonic. Bark used in Billiton for washing hair and clothes. Goldsmiths use seeds in soldering; also used for necklaces and other ornaments.

Adenia cissampeloides (Planch.) Harms. (syn. Ophiocaulon cissampeloides Hook. f.). (Passifloraceae). - Vine. Guinea. Plant when crushed is used as fish poison.

Adenium Honkel A.DC. (Apocynaceae). Trop. Africa. Powdered roots are used in Adamaiwa to stupefy fish. In E. Sudan juice from the plant is employed for poisoning arrows. Is used in the French Sudan as ordeal poison. Contains deniine, a cardiac poison, similar to digitalin.

Adenium multiflorum Klotzsch. (Apocynaceae). - Perennial herb. Trop. Africa. Used in some parts of Transvaal to stupefy fish.

Adenium speciosum Fenzl. (Apocynaceae). - Perennial herb. Trop. Africa. Juice of the plant is used by the natives of Africa as arrow poison.

Adenophora communis Fischer. (Campanulaceae). - Herbaceous perennial. Temp. Europe, Asia. Cultivated in Japan. Roots are eaten boiled, also used in soup.

Adenophora latifolia Fish. Broadleaf Ladybell. (Campanulaceae). - Perennial herb. Siberia, Roots are consumed as food in some parts of Siberia.

Adenophora polymorpha Ledeb. (Campanulaceae). - Perennial herb. Russia, Siberia, China. Used medicinally in China. Contains saponin.

Adenophora verticillata Fisch. (Campanulaceae). - Herbaceous perennial. Japan, Dahuria. Roots are used as food among the Ainu.

Adenopus breviflorus Benth. (Cucurbitaceae). - Vine. Trop. Africa. Fruits are used by tanners in Hausa and Yoruba (Afr.) for removing hair from the hides.

Adenostemma viscosum Forst. (Compositae). - Herbaceous plant. Trop. Africa, Asia. Herb is source of a blue dye.

Adenostoma sparsifolium Torr. Redshank Chamise. (Rosaceae). - Shrub or tree. Baja California (Mex.) to California. Used by the Coahuilla Indians of California for arrowheads and rabbit sticks.

Adhatoda Vasica Nees., Malabar Nut Tree. (Acanthaceae). - Small sub-herbaceous shrub. Trop. India. Herb said to have definite expectorant action, gives relief in bronchitis. Leaves when boiled with sawdust of Jack Wood are source of a yellow dye. Charcoal is used for gunpowder. Wood ist made into beads.

Adiantum aethiopicum L. (Polypodiaceae). - Perennial herb. Fern. Trop. Africa. Decoction is used in some parts of Cape Peninsula for coughs. Decoction of rhizomes is used by Basuto Kaffirs as an abortive.

Adiantum Capillus - Veneris. L., Maidenhair Fern. (Polypodiaceae). - Perennial herb. N. temp. zone. Used medicinally as emmenagogue; as tea in chronic respiratory affections.

Adiantum pedatum L., Maidenhair Fern. (Polypodiaceae). - Perennial fern. N. America, E. Asia. Rhizome is used medicinally as a stimulant, expectorant, demulcent; contains a bitter principal, volatile oil and tannin. Was used by the Indians as aromatic, bitter, demulcent, in pectoral affections and catarrhs.

Adina rubella Hance. (syn. A. rubescens Hemsl.) (Rubiaceae). - Tree. India, Mal. Archip. Wood hard, heavy, yellow; used for building purposes; lasts 20 to 30 years in the ground.

Adinandra integerrima Anders. (Theaceae). - Tree. Burma, Indo-China, Malacca. The red wood is used in Indo-China for handles of tools and for wagon making.

Adinobotrys atropurpureus Dunn. → Whitfordiodendron pubescens Burkill.

Adinobotrys erianthus Dunn. → Withofordiodendron eranthum Dunn.

Adju Mahogany → Canarium Mansfeldianum Engl.

Adonis aestivalis L., Summer Adonis, Pheasant's Eye. (Ranunculaceae). - Annual herb. Europe, Asia. Dried herb is used medicinally as cardiac tonic, indirect diuretic. Contains adonine, a glucoside.

Adonis vernalis L., Spring Adonis. (Ranunculaceae). - Perennial herb. Europe, N. Asia. Dried herb collected in spring is used medicinally as a cardiac stimulant. Contains adonidin and several glucosides. Has the same physiological action as Digitalis. Also are used **A. aestivalis** L. and **A. microcarpa.**

Adzuki Bean → Phaseolus angularis Wight.

Aechmea Magdalenae André. (syn. Ananas Magdalenae (André). Standl. (Bromeliaceae). - Perennial herb. Centr. and S. America. Leaves produce a tough fibre of excellent quality; used for rope and twine.

Aegle Marmelos Correa. Bael Fruit (Rubiaceae). Small tree. E. India. Cultivated. Fruits are size of an orange, hard shelled, soft aromatic, pleasant flavor; used for drinks and sherbets. Used medicinally for dysentry and dyspepsia.

Aegoceras majus Gaertn. (syn. A. corniculatum Blanco.) (Myrsinaceae). - Shrub. India, Malaysia to New Guinea and Australia. Bark is used as fish poison. Contains aegiceras-saponin. Wood is hard, heavy, ranging from red-brown to almost black; used for handles of knives.

Aegopodium Podagraria L., Goutweed, Bishop's Elder. (Umbelliferaceae). - Perennial herb. Europe, Asia; introd. in N. America. Leaves are sometimes eaten boiled, also used in salads and in soup.

Aeolanthus heliotropioides Oliv. (Labiaceae). - Perennial herb. Trop. W. Africa. Herb is used for flavoring soups by the natives.

Aeolanthus pubescens Benth. (Labiaceae). - Perennial herb. Trop. Africa. Cultivated. Herb is used for flavoring soup; also employed as febrifuge.

Aerobacillus polymyxa (Prazm.) Migula. (Bacteriaceae). Bacil. Microorganism. Is important in the production of 2,3-Butanediol which can be converted into 1,3-Butadiene; used in the production of rubber of the Buna type.

Aerobacter aerogenes (Prazm.) Migula. (Bacteriaceae). - Bacil. Microorganism. Is important in the production of 2,3-Butanediol. (→ Aerobacillus polymyxa).

Aerva lanata Juss. (Amarantaceae). - Shrub. Trop. Asia, Africa. Eaten as a spinach in E. Africa. Shoots are used in India in curry. Also a famine food in India.

Aerva tomentosa Lam. (Amarantaceae). - Shrub. Red Sea area, Trop. Africa. Woolly spikes are used in some parts of the Sudan in donkey saddles and for stuffing pillows. Roots are employed as tooth-brushes. Is used medicinally for camels and horses.

Aeschynomene aspera L., Sola Pith Plant. (Leguminosaceae). - Shrub. Trop. Asia. Very light pith is used in India for manuf. sun-helmets or sola-topis. Much is derived from Calcutta.

Aeschynomene grandiflora → Sesbania grandiflora (L.) Poir.

Aeschynomene spinulosa Roxb. → Sesbania aculeta Poir.

Aeschynomene uniflora E. Mey. (Leguminosaceae). - Perennial herb. S. Africa, India. Soft wood is used in India for manuf. helmets, toys, floats and similar products. Used in surgery for manuf. of tampons.

Aesculus arguta Buckl., Ohio Buckeye, Western Buckeye. (Hippocastanaceae). - Tree. Missouri to S.W. of United States. Ground seeds mixed with flour were used to stupefy fish. Decoction of inner part of fruit was employed by the Kiowa Indians (New Mex.) as a powerful emetic.

Aesculus californica Nutt., California Buckeye. (Hippocastanaceae). - Shrub or small tree. California. Seeds were eaten in large quantities by Indians of California. Seeds were boiled with much water to remove unpleasant taste.

Aesculus glabra Willd., Ohio Buckeye, Fetid Buckeye. (Hippocastanaceae). - Tree. E. United States. Wood not strong, close-grained, light, soft, whitish; used for wooden ware, artificial limbs, wooden hats and paper pulp.

Aesculus Hippocastanum L. Horse Chestnut. (Hippocastanaceae). - Tree. Balkan Pen. Caucasia, W. Iran, Himalaya. Cultivated. Wood very uniform, light, soft, easy to split, little elastic, not very durable; used for cutlery, furniture, for boxes, wagons, pushcarts; source of charcoal, used for gunpowder. Flowers are occasionally used as tincture for rheumatism. Seeds after removal of tannin and a glucosid, are made into flour; used for breadmaking when mixed with wheat or ryeflour, consumed in time of food shortage.

Aesculus octandra Marsh., Yellow Buckeye, Sweet Buckeye. (Hippocastanaceae). - Tree. Eastern United States to Texas. Wood close grained, soft, light, difficult to split; used for artificial limbs, wooden hats, wooden ware, paper pulp.

Aesculus Pavia L. (Pavia rubra Poir.), Red Buckeye. (Hippocastanaceae). - Shrub or tree. E. United States. Fresh seeds were macerated in water, mixed with wheat flour and used to stupefy fish by the Indians.

Aesculus turbinata Blume (syn. A. chinensis Engl.) Japanese Horse Chestnut. (Hippocastanaceae). - Tree. Japan, esp. Hokkaidô, Honshū; China. Wood soft, light, close grained, pale yellow. Used in Japan for furniture, inlaid works, statues, spools, boxes, interior finish of buildings and cars. Decoction of seeds is used by the Ainu for washing eyes of horses when discharging matter.

Aethusa Mutellina St. Lag. → Ligusticum Mutellina (L.) Crantz.

Aextoxicon punctatum Ruiz. and Pav., Palo Muerto, Aceitunillo. (Euphorbiaceae). - Tree. Chile. Wood pale brown, with a reddish hue, straight grained, easy to work, finishes smoothly, durable; used in Chile for general carpentry.

Afara Torminalia → Torminalia superba Engl. and Diels.

Affun Yam → Dioscorea cayennensis Lam.

Aframomum angustifolium Schum. (Zingiberaceae). - Perennial herb. Madagascar, Sychelles, Mauritius, E. Africa. Source of Madagascar Cardamon. Cultivated. Seeds used as condiment.

Aframomum Hanburyi Schum. (Zingiberaceae). - Perennial herb. Trop. Africa, Cameroon. Cultivated. Source of Cameroon Cardamon; used as condiment.

Aframomum mala Schum., East African Cardamon. (Zingiberaceae). - Herbaceous perennial. Trop. Africa. Seeds are used as a condiment.

Aframomum Melegueta Schum. (Zingiberaceae). - Perennial herb. Trop. Africa. Cultivated in French Guinea, Ivory Coast, Sierra Leone, Dahomey, etc. Seeds known as Grains of Paradise, are used as a condiment. Pungency of seeds is due to paradol related to gingerol.

African Blackwood → Dalbergia melanoxylon Guill and Perr.

African Bowstring Hemp → Sansevieria senegambica Baker.

African Breadfruit Tree → Treculia africana Decne.

African Ebony Tree → Diospyros mespiliformis Hochst.

African Elemi → Boswellia Frereana Birdw. and Canarium Schweinfurtii Engl.

African Grenadille Wood → Dalbergia melanoxylon Guill. and Pierre.

African Kino → Pterocarpus erinaceus Lam.

African Locust → Parkia africana R. Br.

African Mahogany → Afzelia africana Smith, Detarium senegalense Gmel., Dumoria Heckeli Chev. and Khaya senegalense Juss.

African Millet → Eleusine coracana (L.) Gaertn.

African Myrrh Tree → Commiphora africana Endl.

African Oak → Chlorophora excelsa (Welw.) Benth and Hook. f. and Oldfieldia africana Benth. and Hook.

African Padauk → Pterocarpus Soyauxii Taub.

African Peach Bitter → Sarcocephalus esculentus Afzel.

African Pepper → Xylopia aethiopica A. Rich.

African Rosewood → Pterocarpus erinaceus Lam.

African Sandalwood → Baphia nitida Lodd.

African Starapple → Chrysophyllum africanum A.DC.

African Teak → Oldfieldia africana Benth. and Hook.

African Tragacanth → Sterculia Tragacantha Lindl.

African Valerian → Fedia Cornucopiae DC.

African Walnut → Lovoa Klaineana Pierre and Sprague.

Afrolicania elaeosperma Mildbr. (Rosaceae). - Tree. Trop. W. Africa. Oil from fruits, called Mahogany Nuts, is recommended for manuf. of paints and varnishes. Product is sometimes called Po-Yoak Oil.

Afzelia africana Smith., African Mahogany. (Leguminosaceae). - Tree. Trop. Africa. Wood hard, durable, exported to Europe as African Mahogany; used for cabinet work, turnery, naval construction on the Congo. Burnt pods used in Sudan for manuf. a native soap.

Afzelia bijuga (Colebr.) Gray. (syn. Intsia amboinensis Thours). Fiji Afzelia, Molucca Ironwood. (Leguminosaceae). - Tree. Pacific Islds. Wood strong, durable; used for building houses, ships, telegraph poles, furniture, bridges, packing houses; employed by the Somoans for Kavabowls. Not suitable for marine construction.

Afzelia palembanica Baker. (syn. Intsia Bakeri Prain.) Malacca Teak, Ironwood. (Leguminosaceae). Malaysia. Wood very durable, hard, and strong; close-grained, light-brown; used for high grade construction work, furniture, railroad ties, beams. Source of brown and yellow dyes; used for coloring mats and clothes.

Agapanthus umbellatus L. Hér. (Amaryllidaceae). - Perennial herb. S. Africa. Roots are used in some parts of Africa for heart troubles and intestinal pain.

Agapetes saligna Benth. and Hook. (Ericaceae). - Shrub. Temp. India. Leaves are used as a tea in some parts of India.

Agar or Agar-Agar → Gelidium, Ahnfeltia, Graciliaria, Pterocladia. As supplementary agar → Acanthopeltis, Ceramium, Eucheuma, Gelidium, Graciliaria, Pterocladia.

Agar, Macassar → Eucheuma muricatum (Gmel.) Web. v. Bosse.

Agar, Sachalin → Ahnfeltia plicata (Huds.) Fries.

Agaric acid → Polyporus officinalis (Vill.) Fr.

Agaricine → Polyporus (Vill.) Fr.

Agaricaceae (Fungi) → Agaricus, Amanita, Armillaria, Cantharellus, Clitocybe, Clitopilus, Collybia, Conchomyces, Coprinus, Crepidotus, Flammulina, Gomphidius, Gymnopus, Hygrocybe, Hygrophorus, Hypholoma, Hyporhodius, Inocybe, Lactarius, Lentinus, Lepiota, Mycena, Oudemansiella, Paneolus, Panus, Pholiota, Pleurotus, Pluteus, Psalliota, Psathyrella, Rajapa, Rhodopaxillus, Rozites, Russula, Tricholoma, Volvaria.

Agaricus appendiculatus Bull. → Psathyrella Candolleana (Fr.) A. H. Smith.

Agaricus arvensis Schäff. → Psalliota arvensis (Schäff.) Fr.

Agaricus campestris L. → Psalliota campestris (L.) Fr.

Agaricus Canarii Jungh. → Oudemansiella Canari (Jungh.) v. Hoehm.

Agaricus decastes Fr. (syn. Lyophyllum aggregatus Schäff., Clitocybe multiceps Peck) (Agaricaceae). - Basidiomycete. Fungus. Fruitbodies are consumed as a food in Europe and Japan. Sold in markets.

Agaricus Djamor Fr. → Crepidotus Djamor (Fr.) v. Overeem.

Agaricus luzonicus Graff. (Agaricaceae). - Fungus. Trop. Asia. Fruitbodies are consumed as food in the Philipp. Islds. Also A. Boltoni Copel, A. argyrostectus Copel, A. manilensis Copel, A. Merrillii Copel.

Agaricus marginatus Batch → Pholiota marginata (Batch) Quél.

Agaricus ostreatus Jacq. → Pleurotus ostreatus (Jacq.) Quél.

Agaricus pratensis Scop. → Psalliota arvensis (Schäff.) Fr.

Agaricus procerus Scop. → Lepiota procera (Scop.) Quél.

Agarobilli → Caesalpinia brevifolium Baill.

Agarophyte → Gelidium cartilagineum (L.) Gaill.

Agarweed → Gelidium cartilagineum (L.) Gaill. var. robustum Gardner.

Agastache anethiodora (Nutt.) Britt., Giant Hyssop. (Labiaceae). - Perennial Herb. Missouri and Westw. of United States. Decoction of leaves was used as a beverage by several Indian tribes.

Agastache neomexicana (Briq.). Stand. (Labiaceae). - Perennial herb. New Mexico. Leaves were used for flavoring food by the Indians.

Agathis alba Foxw. (syn. Dammara alba Lam.). White Dammar Pine. (Pinaceae). - Tree. Mal. Archip. Source of Manila Copal, much of the quality depends upon the age of the product, being derived from living trees or obtained from the soil in semi-fossil condition. Grades are divided into hard, semi-hard, and soft. Used for varnishing enamals and interior work. Much is derived from the Indonesia, also from the Philipp. Islds. Product is also called Singapore, Pontuik, Macasar Manilla and Boed Copal.

Agathis australis Steud., Kauri. (Pinaceae). - Tall tree. New Zealand. Wood used for houses, bridges, wharves, boats, deck-planking, masts. Stem is source of a resin, called Kauri Gum, or Kauri Copal. Much is derived from the soil in a semi-fossilized condition. Used as substitute for amber, made into mouthpieces of tobacco-pipes, small ornaments and for manuf. of varnishes, suitable for outside work. Important grades are Range Gum composed of the best copal, followed by Swamp Gum and Bush Gum; the latter being obtained by tapping living stems. The gum resin was also used as masticatory by the aborigenes of New Zealand.

Agathis microstachys Warb. (Pinaceae). - Tree. Mal Archip. esp. Java. Wood used for lumber, esp. construction work.

Agathis lanceolata Panch. (Pinaceae). - Tree. New Caledonia. Stem is source of a Kauri or Dammar Resin. Also **A. ovata** (Moore) Warb. and **A. Moorei** (Lindl.) Warb.

Agathis loranthifolia Salish. (Pinaceae). - Tree. Malaysia, Burma. Stem is source of a copal. used in varnishes, sometimes called Manila Co-

pal; also known as Borneo, Celebes, Batjan, Molucca, Macassar, Menedo, Oli, Singapore, Labuan, Fidji, Sambas, Malengthet and Boed Copal or Manilla.

Agathis Palmerstoni F. v. Muell., Kauri. (Pinaceae). - Tree. Australia. Wood is source of lumber; used for different purposes.

Agathis vitiensis (Seem.) Warb. (syn. Dammara vitiensis Seem. (Pinaceae). - Tree. Fiji Islds. Source of Resin of Fiji; resembles Manila and Macassar Copals.

Agati Sesbania → Sesbania grandiflora Poir.

Agatophyllum aromaticum Willd. → Ravensara aromatica Gmel.

Agave atrovirens Karw. (syn. A. latissima Jacobi). (Amaryllidaceae). - Mexico. Acaulescens herbaceous tall plant. Mexico. Cultivated. Source of Pulque and Mezcal de Pulque. Among the varieties are Maguey Manso and Maguey Manso Fino.

Agave cantala Roxb. (Amaryllidaceae). - Origin unknown. Cultivated in tropics of Old and New World. Maguey or Manila Aloë. Source of a fibre known in India as Bombay Hemp, Bombay Aloë Fibre.

Agave compluviata Trel. (Amaryllidaceae). - Acaulescent perennial herb. Source of Aguamiel and Pulque. Near Comitan, Mex. a fine liqueur, Comiteca is distilled from fermented sugar cane juice and mash.

Agave Deserti Engelm. (Amaryllidaceae). - Perennial plant. S.W. of United States and adj. Mexico. Base of leaves were roasted in mescal pits, forming a sweet juicy food, eaten by the Indians. Also A. utahensis Engelm.

Agave falcata Engelm. (Amaryllidaceae). - Acaulescent, perennial tall herb. Mexico. Leaves are source of an Ixtle Fibre, Tampico Fibre.

Agave fourcroydes Lemaire. Henequen Agave. (Amaryllidaceae). - Acaulescent, tall herbaceous plant. Mexico, esp. Yucatan. Cultivated. Source of an excellent commercial fibre, called Henequen or Yucatan Sisal. Used extensively for binder twine in harvesting grain crops, also employed for guy ropes and general purpose ropes. Is not suitable for pulley blocks owing to swelling when wet and due to its harshness.

Agave Funkiana Koch and Bouché (Amaryllidaceae). Nuevo León and Tamaulipas (Mexico). Source of a fibre, Ixtle de Jaumava.

Agave gracilispina Engelm. (Amaryllidaceae). - Acaulescent perennial herb. Mexico, esp. San Luis Potosi. Source of a Pulque, an alcoholic beverage. Leaves source of a fibre, called Ixtle.

Agave heteracantha Zucc. (syn. A. Lecheguilla Torr.) Lecheguilla. (Amaryllidaceae). - Acaulescent perennial herb. Texas and Mexico. A fibre derived from the leaves, is exported as Ixtle; used for brushes, bagging and cordage. It is also called Jaumave Ixtle.

Agave Kirchneriana Berger (Amaryllidaceae). - Acaulescent tall perennial. Mexico esp. Geurrero. Produces an excellent fibre, Maguey delgado and Mezcal.

Agave lecheguilla Torr. → Agave heteracantha Zucc.

Agave lophanta Schiede. (related to A. Funkiana Koch and Bouché). (Amaryllidaceae). - Ascaulescent herbaceous plant. Mexico. Leaves are source of a fibre, called Ixtle de Jamauve, being of good quality, also called Rula Ixtle.

Agave mapisaga Trel. (Amaryllidaceae). , Acaulescent herb. Mexico, esp. San Luis Potosi. Source of Pulque.

Agave melliflua Trel. (Amaryllidaceae). - Acaulescent tall herb. Mexico, esp. Nuevo León. Source of Aquamiel and Pulque.

Agave rigida Mill. → Agave sisalana Perrine.

Agave Schottii Engelm. Amole. (Amaryllidaceae). Acaulescent perennial herb. Arizona, Sonora. Source of a substitute for soap.

Agave sisalana Perrine. (syn. A. rigida Mill.). Sisal Agave. (Amaryllidaceae). - Acaulescent, herbaceous plant. Yucatan (Mex.). Grown in tropics of Old and New world. Source of Sisal Hemp, Bahama Hemp, Yaxci. Fibre is used for general purpose ropes, hammocks; somtimes used in marine cordage. Used alone or mixed with Manila Hemp. When wet it swells and therefore can not be used for pulley blocks.

Agave tequillana Trel. Chino Azul. (Amaryllidaceae). - Acaulescent herbaceous plant. Mexico, esp. Jalisco. Source of Mezcal de Tequilo, a distilled alcoholic beverage.

Agave utahensis Engelm. (Amaryllidaceae). - Perennial. Western N. America, esp. Utah, Arizona, Nevada, California. Bulbous roots were roasted by the Indians and eaten as food.

Agave Victoriae-Reginae Moore. (Amaryllidaceae). - Acaulescent. Mexico, esp. Nuevo León, Coahuila, Durango. Source of Noa, a short, strong fibre.

Agave virginica L., False Aloe. (Amaryllidaceae). - Perennial herb. E. United States to Florida and Texas. Bitter root was used by the Indians for colic, stomachic and antispasmodic.

Agave zapupe Trel. (Amaryllidaceae). - Acaulescent herbaceous plant. Mexico, esp. Veracruz. Source of Zapupe Fibre.

Aglaia Diepenhorstii Miq. (syn. A. odoratissima Blume). (Meliaceae). - Shrub or small tree. Malaysia to Java. The very fragrant flowers are used in perfumery.

Aglaia Harmsiana Perk. (Meliaceae). - Tree. Throughout Philippines. Pulp of seeds is edible; resembling cranberries.

Aglaia odorata Lour. (Meliaceae). - Shrub. Indo-China, S. China. Cultivated. Flowers used by Chinese for scenting tea. Article of trade. In Java flowers are used to perfume clothes. Pro-

duces an ingredient applied to the body after childbirth. Much sold in bazaars.

Aglaia oligantha C.DC. Manatan. (Meliaceae). Trop. Asia, esp. Philippine Islands. Fruits are formed in bunches, roundish, 2 cm. across, acid, pulp edible.

Aglaia polystachya Wall. → Amoora Rohituka Wight and Arn.

Aglaia samoënsis Gray. (Meliaceae). - Small tree. Pacific Islands. Wood is used for houseposts. Fragrant flowers are used for scenting coconut oil, also used by the Samoans to decorate the hair and necklaces.

Agonandra brasiliensis Benth. and Hook. (Olacaceae). - Tree. Brazil. Seeds source of Ivory Wood Seed Oil or Paō Manfin Pil; semi-drying; after sulfonation gives a similar product as Turkey Red; Oil; by partial hydrogenation source of a rubber substitute. Sp. Gr. 0.9602; Sap. Val. 207.3; Iod. No. 112.3; Unsap. 0.6 %.

Agoseris aurantiaca Hook.) Greene. (syn. Macrorhynchus troximoides Torr. and Gray.) Orange Agoseris. (Compositae). - Herbaceous plant. Western N. America. Leaves were eaten as food by Indians of Utah and Nevada.

Agoseris villosa Rydb. (Compositae). - Perennial herb. N. W. of United States. Latex of the stem has been used for chewing by the Thompson Indians in Brit. Columbia.

Agrimonia Eupatoria L., Agrimony. (Rosaceae). - Perennial herb. Temp. Northern Hemisphere. Herb used since ancient times as a home remedy for chronic liver ailments. Used as diuretic, astringent, haemorrhagic, emmenagogue and anthelmintic. Stems and leaves are a source of a golden-yellow dye.

Agrimony → Agrimonia Eupatoria L.

Agriophyllum gobicum Bunge., Soulkhir. (Chenopodiaceae). - Annual herb. Mongolia, Siberia. Seeds form an important food among the natives of Mongolia. Plant is also used as a forage crop.

Agrocybe aegerita (Bruganti) Kühner → Pholiota cylindrica D.C.

Agropyron pauciflorum (Schwein.) Hitchc. (syn. A. tenerum Vasey). Slender Wheatgrass. (Graminaceae). - Perennial grass. North America and N.W. Mexico. Grown as a pasture grass.

Agropyron repens (L.) Beauv., Quackgrass. (Graminaceae). - Herbaceous perennial. Troughout northern temp. zones. Dried rhizome used medicinally. Collected in spring. Majority derived from Centr. Europe. A demulcent, also diuretic. For home remedy used as a tea for coughing. Poor people roast rootstocks as a substitute for coffee. Rootstock has been source of bread during times of famine.

Agropyron spicatum (Pursh.) Scribn., Bluebunch Wheatgrass. (Graminaceae). - Perennial grass. Dry places. North America. An important pasture grass.

Agrostemma Githago L. Corn Cuckle. (Caryophyllaceae). - Annual herb. Europe, introd. in N. America. Young leaves have been used as vegetable with vinegar and bacon for emergency food during times of want. Seeds are considered poisonous.

Agrostis alba L. Redtop. (Graminaceae). - Perennial herb. Europe, Asia, N. America. Cultivated as pasture grass, as a haycrop and for binding soil. Adaptable to wet and sour land.

Agrostis vulgaris With. (syn. A. tenuis Sibth.), Rhode Island Bent. (Graminaceae). - Perennial grass. Cultivated in the Old and New World. Important pasture grass.

Agrostistachys borneensis Bece. (Euphorbiaceae). - Shrub. Westcoast of Malaya, N. Borneo. Stem used by Dyaks of Singhi for blackening teeth.

Ahnfeltia concinna J. Ag. (Phyllophoraceae). - Red Alga, Sea Weed. Pac. Ocean. Used as food by the Hawaiians, called Limu Akiaki. Eaten boiled with squid and octopus, producing a jelly-like food.

Ahnfeltia plicata (Huds.) Fries. var. tobuchiensis Kanno and Matsub. (Phyllophoraceae). - Red Alga. Northeast Coast of Asia. Source of Sachalin agar; of commercial importance in Russia and Japan.

Ailanthus glandulosa Desf. (syn. A. altissima [Mill.] Swingle). - Tree of Heaven. (Simarubaceae). - Tree. China. Wood yellowish gray, fairly hard and heavy, difficult to split, easy to work and to polish; used for building and general construction, furniture, fancy articles.

Ailanthus malabarica DC. (Simarubaceae). - Tree. S.W. India, Ceylon. Resin of bark furnishes in India an incense; used in Hindu temples. Leaves and bark are much in demand as tonic, febrifuge; useful for dysentery complaints. Leaves source of a black dye used for coloring satin.

Ailanthus Vilmoriana Dode. (Simarubaceae). - Tree. W.China, esp. N.W. Fangsien. Cultivated. Source of food of a silk worm, Attacus cynthia. Also A. glandulosa Desf.

Aipi → Manihot dulce (J. F. Gmel.) Pax.

Aira brasiliensis Spreng. → Arundinella brasiliensis Radde.

Aira gitantea Steud → Gynerium sagittatum (Subl.) Beauv.

Airan → Panus rudis Fr.

Air Potato → Dioscorea bulbifera L.

Aizoaceae → Aizoon, Gisekia, Mesembryanthemum, Mollugo, Tetragonia, Trianthema.

Aizoon canariense L. (Aizooaceae). - Herbaceous plant. N. Africa, Can. Islds. Herb is used as food by the Tuareg (N. Afr.).

Ajowa → Carum copticum (L.) Benth. and Hook.

Ajowan Seed Oil → Carum copticum (L.) Benth. and Hook.

Akam Yam → Dioscorea latifolia Benth.

Akebia lobata (Houtt.) Decne. (Lardizabaliaceae). - Woody vine. Japan. Bleached young vines are made into baskets in Midsuguchi, Omi, Japan. Fruits are eaten in some parts of Japan where they are sold in markets. Dried, young leaves are used as a tea in Japan.

Akee Apple → Blighia sapida Koenig.

Ako Ogrea Gum → Paradaniella Oliveri Rolfe.

Akoke → Euphorbia lorifera Stapf.

Alang-Alang → Imperatia arundinaceae Cyrill.

Alangium Lamarckii Thwait. (syn. A. decapetalum Lam.). Cornaceae). - Shrub or small tree. Trop. Asia esp. S. India, Ceylon, Mal. Penins., Philipp. Islands. Wood close grained, heavy, brown. Used in Madras for pestles and oil mills. Juice of roots is said to be anthelmintic, purgative; used for dropsical cases.

Alangium salviifolium Wangerin. (Cornaceae). - Shrub or tree. Widely distrib. Trop. Asia. esp. Comoro Islands, India, Philipp. Islands, New Guinea. Much used in Indian medicine. Bark very bitter, emetic; substitute for ipecacuanha.

Alaria esculenta (Lynb.) Grev. Murlins, Bladder Locks. (Alariaceae). - Brown Alga. Cold areas of the North Atlantic. Used as food in some parts of Ireland and Scotland, esp. the „fingers" and the sweet midrib when stripped of its membrane.

Alariaceae (Brown Algae). → Alaria, Ecklonia, Undaria.

Alaska Cedar → Chamaecyparis Nootkatensis (Lamb.) Sudw.

Albarco → Cariniana pyriformis Miers.

Albertinia Candolleana Gardn. → Vanillosmopsis erythropappa Schultz-Bip.

Albizzia amara Boiv. (Leguminosaceae). - Tree. E. Africa. Trunk is source of a gum.

Albizzia anthelmintica Brongn. Mucenna Albizzia. (Leguminosaceae). - Shrub or small tree. E. Africa. Bark is used as an anthelmintic in Abyssinia.

Albizzia Brownii Walp. (Leguminosaceae). - Tree. Trop. Africa. Stem is source of a resin, called Nongo Gum. Wood is durable, very hard, dark brown; made into planks; used for doors, beams, general building purposes.

Albizzia fastigiata Oliv., Flatcrown Albizzia. (Leguminosaceae). - Tree. Madagascar, Trop. Africa. Tree. Source of a gum, similar to Gum Arabic. A native sauce is made from the seeds. Wood is soft; used for yokes, wheels, doors and general carpentry. Is also source of a dye; used in Madagascar.

Albizzia moluccana Miq. (Leguminosaceae). - Tall tree. Malaysia. Used for shading tea shrubs, however, is subject to wind damage. Wood is made into matches, tea boxes. In Amboina bark is employed for tanning nets. Wood is suitable for paper pulp.

Albizzia odoratissima Benth. (syn. A. leblecoides Benth.). Fragrant Albizzia. (Leguminosaceae). - Tree. Philipp. Islds. Bark is source of a fermented drink, called Basi, used in the Philippines.

Albizzia procera Benth. Tall Albizzia, Tee-coma. (Leguminosaceae). - Tree. Trop. Asia, N. Australia. Wood close grained, dark, easily worked. Used for building, cabinet-work, sugar cane crushers, wheels, rice pounders, agricultural implements, matches, house posts, bridges, charcoal.

Albizzia saponaria (Lour.) Blume. (Leguminosaceae). - Medium sized tree. Throughout Philipp. Islds. to New Guinea. Saponaceous bark is used locally in Philipp. Islds. for washing; fresh wood lathers freely with water.

Albizzia Welwitschii Oliv. (Leguminosaceae). - Tree. Trop. Africa. Wood rose-red, very heavy, hard, dense, very tough, strong, easy to work; used for furniture, turnery, wagons, flooring, walking sticks.

Alcarnoque → Mora megitosperma (Pitt.) Britt. and Rose.

Alchemilla vulgaris L., Lady's Mantle. (Rosaceae). - Perennial herb. Throughout Europe, temp. Asia, N. America. Is used in some countries as astringent, diuretic, emmenagogue; also for healing wounds.

Alchornea cordifolia Muell. Arg. (syn. A. cordata Benth.). (Euphorbiaceae). - Shrub or small tree. Trop. Africa. Plant is source of a black dye.

Alchornea villosa Muell. Arg., Ramie Bukit. (Euphorbiaceae). - Shrub. Malaysia. Bark very tough, source of a fibre, considered a substitute for Rami.

Alcornoco → Bowdichia virgillioides H.B.K.

Alder, Black → Alnus glutinosa Medic.

Alder Buckthorn → Rhamnus Frangula L.

Alder, European → Alnus glutinosa Medic.

Alder, Hazel → Alnus rugosa (Du Roi) Spreng.

Alder, Mountain → Alnus viridis DC.

Alder, Nodding → Alnus pendula Matsum.

Alder, Red → Alnus rubra Bong.

Alder, Smooth → Alnus rugosa (Du Roi) Spreng.

Alder, Spickled → Alnus incana (L.) Moench.

Alder, Western → Alnus rubra Bong.

Alder, White → Platylophus trifoliatus D. Don.

Alectoria Fremontii Tuckerm. (Usneaceae). - Lichen. Temp. zone, N. America. Plants were used as a famine food by Indians of California, Oregon and Montana.

Alectoria jujuba (L.) Ach. Horsehair Lichen. (Usneaceae). - Lichen. Temp. Zone. On twigs of trees. Has been used in England to stain woolens a pale green to brown-red color. Also used in perfumery though less than Evernia. Boiled with Camas roots, fermented and baked was consumed by the Indians of the Pacific Region of N. America.

Alectra melampyroides Benth. (Scrophulariaceae). - Herbaceous plant. S. Africa. Roots and stems are source of a golden-yellow dye.

Aletris farinosa L., Star Grass, Colic Root, Unicorn Root. (Liliaceae). - Perennial herb. East of N. America. Used medicinally. Dried rootstock and root are diuretic, uterine tonic. Much is supplied from Virginia, Tenessee and N. Carolina. Infusion of leaves in cold water was used by the Catawba Indians for colic and stomach ailments.

Alectryon excelsum Gaertn. (Sapindaceae). - Tree. New Zealand. Wood strong, durable, elastic, straight grained, easily worked, suitable where great strength and elasticity are required. Used for axe handles, cabinet work, panels, hubs, bent ware, agricultural implements. Seeds source of an oil used for making a native perfume.

Alectryon macrococcus Radlk. (Sapindaceae). - Tree. Hawaii. Scarlet fruits are used as food by the natives.

Alepidia amatymbica E. and Z. (Umbelliferaceae). - Perennial herb. S. Africa. A resin and the very bitter rhizome are used by the farmers and Kaffirs of S. Africa for stomach pain, as tonic and for colic. Is a laxative when taken in large quantities.

Aleppo Pine → Pinus halepensis Mill.

Alerce → Fitzroya cupressoides (Molina) Johnst.

Aleurites cordata, Steud., Japanese Tung Oil Tree. (Euphorbiaceae). - Tree. Japan. Cultivated. Seeds source of Japanese Tung Oil, a drying oil, used for waterprooving fabrics, paper; in paints and varnishes. Sp. G. 0.943-0.940; Sap. Val. 189-196; Iod. No. 148-160; Unsap. 0.4-0.8 %.

Aleurites Fordii Hemsl., Tung Oil Tree (Euphorbiaceae). - Tree. China. Cultivated in Old and New World. Seeds source of Tung Oil or Chinese Wood Oils, a drying oil. Has been used by Chinese for many centuries for water-proofing fabrics, wood, paper. Manuf. of paints, enamels, varnish, automobile brake lining, linoleum, lacquers, tiles, pressed fiber boards, printing ink. Used as cobalt, manganese, lead and tungate driers. Press cakes used as fertilizer. Sp. Gr. 0.939-0.943; Sap. Val. 189-195; Iod. No. 157-172; Unsap. 0.4-0.8 %. Source of India. Ink.

Aleurites triloba Forst. (syn. Aleurites moluccana Willd.) Candle Nut Oil Tree. (Euphorbiaceae). - Tree. Mal. Archip., Philipp. Islds, Polynesia, India Australia. Source of Lumbang Oil,

Candlenut Oil, a drying oil. Used as wood preservative, for varnishes, paints, burning oil, soap; protecting bottoms of small craft from marine borers. Sp. Gr. 0.920-0.927; Sap. Val. 190-193; Iod. No. 140-164; Unsap. 0.9 %. The product is also called Bagilumbang Oil.

Aleurites montana Willd. (Euphorbiaceae). - Tree. China, Indo-China. Cultivated. Source of Abrasin Oil or Mu Oil, a drying oil. Sp. Gr. 0.9360-0.9460; Sap. Val. 190.2-195; Iod. Val. 156-157; Unsap. 0.4-0.5 %.

Alexandrian Laurel → Calophyllum inophyllum L.

Alexandrian Senna → Cassia acutifolia Del.

Alfalfa → Medicago sativa L.

Alfalfa, Yellow Flowered → Medicago falfata L.

Algae → Economic Algae.

Algin → Laminaria digitata (L.) Edmonson., L. stenophylla (Kuetz.) J. Ag., L. saccharina (L.) Lamour., Macrocystis pyrifera (L.) C. Ag.

Alhagi camelorum Fisch., Camel's Thorn. (Leguminosaceae). - Shrub. Deserts of Iran, Syria, Egypt. A sugary exudation is produced by the plant which is used as a sweet-meat and in medicine in Iran, Syria, etc. A Manna from the pods, containing mainly cane sugar, is used as a laxative and expectorant.

Alhagi graecorum Boiss. (Leguminosaceae). - Shrub. Greece, etc. Plants used as fodder for camels.

Alhagi maurorum Medic. (Leguminosaceae). - Shrub. Desert regions, Syria, Egypt. Important food for camels. Twigs exude during flowering time a sweet substance or Manna, which is consumed by Beduins. Roots are eaten by some of the nomadic desert tribes during times of want.

Algaroba → Caesalpinia brevifolia Baill.

Algerian Absinthe → Artemisia Barrelieri Ben.

Alisma Plantago L., Water Plantain. (Alismaceae). - Perennial herb, Cosmopolit. Fresh rhizomes considered poisonous; when properly prepared they become harmless; used as food by the Kalmucks.

Alismaceae → Alisma, Sagittaria.

Aliziergeist → Sorbus torminalis (L.) Crantz.

Alkali Grass, Nuttall → Pucciniella Nuttaliana (Schult.) Hitchc.

Alkali Seepweed → Suaeda fruticosa Forsk.

Alkanet → Alkanna tinctoria (L.) Tausch.

Alkanna Root → Alkanna tinctoria (L.) Tausch.

Alkanna tinctoria (L.) Tausch., Alkanna (Boraginaceae). - Perennial herb. S. and E. Europe, Asiatic Turkey. Cultivated. Source of Alkanna, Alkanet, a red coloring agent derived from the roots. Used to detect oils, now replaced by Sudan III and IV. Alkanna root contains alkannin or anchusin. Sometimes used for coloring liqueurs, salves, fats, oils, cotton and silk.

Allamanda Blanchetii DC. (syn. Plumeria Blanchetii DC.) (Apocynaceae). - Shrub. Brazil. Sap is emetic and cathartic when used in small doses, poisonous in large quantities.

Allamanda Doniana Muell. Arg. (Apocynaceae). - Woody plant. Brazil. Bark is considered emetic-cathartic, used in Brazil.

Allanblackia oleifera Oliv. (Guttiferaceae). - Tree. E. Africa, esp. Congo, Cameroon. Seeds source of Kagné-Butter; used locally in food. Sap. Val. 197.9; Iod. No. 42. 3; Unsap. 1.60% Melt. Pt. 30-40° C.

Allanblackia Stuhlmanni Engl. (Guttiferaceae). - Tree. E. Africa. Bark is source of a red dye. Seeds produce the non-drying Mkani Fat; used locally in food. Sp. Gr. O. 8736; Sap. Val. 188.6; Iod. No. 37,5; Melt. Pt. 43-46° C.

Alleghany Blackberry → Rubus allegheniensis Porter.

Allgood → Chenopodium Bonus-Hendricus L.

Alliaria officinalis And. → Sisymbrium Alliaria (L.) Scop.

Alligator Juniper → Juniperus pachyphlaea Torr.

Alligator Pear → Persea americana Mill.

Allium Akaka Gmel. (Liliaceae according to some Amaryllidaceae). - Bulbous perennial. Temp. Asia. Bulbs are used as food in Iran. They are sold in bazaars of Teheran.

Allium Ampeloprasum L., Levant Garlic. (Liliaceae). - Bulbous perennial. Mediterranean, Orient. Bulbs are eaten raw by peasants in some parts of S. Europe.

Allium angulare Pall. (Liliaceae). - Bulbous perennial. Siberia. Bulbs are eaten as winter food in some parts of Siberia.

Allium ascalonicum L., Shallot. (Liliaceae). - Bulbous perennial. Probably from W. Asia. Known under cultivated condition. Seldom produces seeds. Important market vegetable. Used for flavoring; more delicate than onions.

Allium canadense L. Canada Garlic, Meadow Leek, Wild Garlic, Rose Leek. (Liliaceae). - Perennial bulbous herb. Eastern N. America, southward to Florida and Texas. Bulbs when boiled or pickled were consumed by the Indians.

Allium Cepa L., Onion. (Liliaceae). - Bulbous perennial. Probably from W. Asia. Widely cultivated as an annual vegetable in Old and New World. Bulbs are consumed raw or boiled, used as flavoring in pickles. They are dehydrated, also made into onion salt. Numerous varieties. Onions are sometimes divided into: I. American Group. Small bulbs, dense texture; red, white or yellow. Yellow varieties are: Yellow Globe, Yellow Globe Danvers, Ebeneger. Red Varieties: Southport Red Globe, Red Wetherfield.

Whit Varieties: Southport White Globe, White Pearl, Silverskin, White Queen. II. Foreign Group. Bermuda, Yellow Bermuda, Crystal White Wax, Red Bermuda, Sweet Spanish (Denia), Red Italian, Creole, Yellow Multipier, Egyptian (Perennial Tree), Characteristic scent of the onions is caused by allyl sulfide.

Allium cernuum Roth. Nodding Onion, Lady's Leek. (Liliaceae). - Bulbous perennial. Strongly flavored bulbs are used in soups, they are sometimes pickled.

Allium fistulosum L., Welsh Onion, Cibol, Stone Leek. (Liliaceae). - Bulbous perennial. Siberia. Cultivated as annual or perennial in Old and New World. Mainly grown for its leaves; used in salads.

Allium Ledebourianum Schult. (Liliaceae). - Perennial herb. Russia to Japan. Cultivated in Japan. Leaves and bulbs fresh or boiled are eaten in Japan.

Allium Macleanii Baker., Royal Salep. (Liliaceae). - Bulbous perennial. Iran and Afghanistan. Bulbs are consumed in some parts of Centr. Asia.

Allium nipponicum Franch. and Sav. (Liliaceae). - Bulbous plant. Japan. Bulbs used in salads by the Ainu.

Allium obliquum L., Twisledleaf Garlic. (Liliaceae). - Bulbous perennial. Siberia. Occasionally cultivated in some parts of Siberia. Bulbs are used as substitute for garlic.

Allium odorum L., Chinese Chive, Fragrant Onion. (Liliaceae). - Perennial bulbous plant. Siberia to China and Japan. Cultivated in Asia. Resemble chives. Used for flavoring. Chinese use the seeds for „purifying" the blood, also employed as a cordial and tonic.

Allium Porrum L., Leek. (Liliaceae). - Biennial herb. Mediterranean region. Culticated as a vegetable in the Old and New World. Bleached stems and leaves are eaten boiled, also used in soups. Varieties are: Giant Carentan, London Flag, Lyon, Large Musselburgh.

Allium rubellum Bieb. (Liliaceae). - Bulbous perennial. Centr. Asia. Bulbs are consumed among the hill-people in E. India.

Allium sativum L., Garlic. (Liliaceae). - Bulbous herb. Orient, S. Europe. Cultivated in the Old and New World. Widely used for flavoring food, meats, sausages. Bacteriocidal and antiseptic; used for caughs and colds. Rubefaciant, expectorant, diaphoretic in bronchitis. Home remedy for tooth and earache; used for intestinal worms. Juice is used in mending glass and china. Source of Oil of Garlic, derived from the bulb or entire plant; used medicinally as expectorant, stimulant; in nervous affections, hysteria, in coughs and colds. Contains allylpropyl disulfide, diallyl disulfide.

Allium Schoenoprasum L., Chive. (Liliaceae). - Bulbous perennial. Temp. Europe, Caucasia, Si-

beria, E. Asia, N. America. Cultivated. Leaves are used for flavoring.

Allium scorodoprasum L. Giant Garlic. (Liliaceae). - Bulbous perennial, Centr. and S. Europe, Asia Minor. Cultivated. Bulbs are used for flavoring foods.

Allium senescens L. (Liliaceae). - Bulbous perennial, Europe, temp. Asia. Bulbs are eaten as a vegetable in some parts of Japan.

Allium sphaerocephalum L., Ballhead Onion. (Liliaceae). - Bulbous perennial. Europe, temp. Asia. Bulbs are eaten among the people near Lake Baikal, Siberia.

Allium splendens Willd. (Liliaceae). - Bulbous perennial. Japan. Small bulbs are eaten boiled in Japan; also preserved as pickles in a boiled mixture of sake (rice beer), vinegar and soy.

Allium tricoccum Ait. Wild Leek. (Liliaceae). - Perennial herb. N. America. Bulbs are used for flavoring by country people.

Allium Victorialis L., Longroot Onion. (Liliaceae). - Bulbous perennial. Europe, Asia to Japan. Herb is used for colds by the Ainu.

Allophylus zeylanicus L. (syn. Schmidelia africana DC. (Sapindaceae). - Tree. Trop. Africa. Fruits are used in Abyssinia for tape worm. Dried fruits when pounded and mixed with flour are made into cakes.

Allspice → Pimenta officinalis Lindl.

Allspice (substitute) → Lindera Benzoin (L.) Blume.

Allspice, Carolina → Calycanthus floridus L.

Allthorn Acacia → Acacia horrida Willd.

Almeida alba St. Hil → Raputia alba Engl.

Almendero → Laplacea Curtyana A. Rich.

Almendro → Terminalia Catappa L.

Almond → Amygdalus communis L.

Almond, Bitter → Amygdalus communis L.

Almond, Dwarf → Amygdalus nana L.

Almond, Earth → Cyperus esculentus L.

Almond, Indian → Terminalia Catappa L.

Almond, Java → Canarium commune L.

Almond Leaved Willow → Salix triandra L.

Almond Shorea → Shorea eximea Scheffer.

Almond, Sweet → Amygdalus communis L.

Almond, Tropic → Terminalia Catappa L.

Almondette Tree → Buchanania Lanzan Spreng.

Alnus arguta (Schlecht.) Spach. (Betulaceae). - Tree. Mexico. Bark used for tanning leather, by Chinantecs, Zapotecs and Mazatecs (Mex.).

Alnus glutinosa Medic., European Alder, Black Alder. (Betulaceae). - Shrub or tree. Throughout Europe, Siberia, Caucasia, Asia Minor, N. Iran, N. Africa. Wood reddish, soft, fairly light, easily split, elastic; used for wood carving, cigar boxes, pumps, wooden shoes and slippers, water works, for moulds in manuf. glass. Bark is used for tanning, giving leather a red, hard appearance. Bark and leaves are used medicinally as astringent and alterative.

Alnus incana (L.) Moench., Spickled Alder. (Betulaceae). - Shrub or tree, Europe, Caucasia, N. America. Wood coarse-grained, light, soft, easy to split, fairly elastic. Used in the same way as A. glutinosa.

Alnus japonica Sieb. and Zucc., Japanese Alder. (Betulaceae). - Tree. Japan, Korea, Manchuria. Wood reddish, close-grained; used in Japan for turnery, charcoal for gun powder. Bark is source of a dye. Also A. hirsuta Turcz. and A. Maximowiczii Call.

Alnus pendula Matsum., Nodding Alder. (Betulaceae). - Small tree. Japan, esp. Hokkaidô. Wood used in Japan for turnery, combs. Decoction of strobulis is used for dyeing.

Alnus rubra Bong. (syn. Alnus oregana Nutt.), Red Alder, Western Alder. (Betulaceae). - Tree. Alaska to California. Wood soft, brittle, not strong, light, close grained, light brown tinged with red; used in Washington and Oregon for furniture. Indians of Alaska used the hallowed trunks for canoes.

Alnus rugosa (Du Roi) Spreng. (syn. A. serrulata Willd.) Hazel Alder, Smooth Alder. (Betulaceae). - Shrub or small tree. Eastern N. America to Florida and Texas. Decoction of boiled inner bark is used by the Cherokee Indians to induce vomiting. Bark is tonic, astringent, emetic; also used for indigestion due to general debility of the stomach.

Alnus tenuifolia Nutt. (Betulaceae). - Tree. Western N. America. Inner bark is the source of an orange dye, used by the Indians.

Alnus viridis DC. (syn. A. alnobetula Koch.), Mountain Alder. (Betulaceae). - Shrub or small tree. N. temp. Zone. Inner bark is used by some Eskimo tribes to stain the inner side of tanned skins a red color.

Alocasia cucullata Schott., Giant Taro. (Araceae). - Perennial herb. E. India. Corms are consumed in some parts of India.

Alocasia denudata Engl. (Araceae). - Herbaceous perennial. Malaysia. Iritant used in the dart poison among the Mantera and other tribes.

Alocasia indica Schott., Indo Malayan Alocasia. (Araceae). - Tall perennial herb. Throughout S.E. Asia and Pacific Islands. Cultivated. In India stems are used in curries. In Bengal stems are eaten when properly boiled and prepared.

Alocasia odora C. Koch. (Araceae). - Perennial herb. E. India, China. Corms are consumed in different parts of E. India.

Alocasia macrorhiza Schott. Giant Alocasia. (Araceae). - Perennial herb. Trop. Asia, Pacific Islds, Australia. Corms are consumed by the natives in different parts of tropical Asia.

Aloë abyssinica Lam. (Liliaceae). - Succulent shrub. E. Africa. Source of Mecca or Moka-Aloë from Red Sea region; also prepared in Jaferabad, Kathiawar. Sold in black circular cakes, forms yellow powder, strong aloë aroma.

Aloë candelabrum Tod. (Liliaceae). - Succulent shrub. S. Africa. Source of Natal Aloë.

Aloë ferox Mill. (syn. A. horrida Haw.). Cape Aloë. (Liliaceae). - Succulent perennial. S. Africa. Used medicinally in vet. medic. Evaporated juice from leaves is greenish black. Exported from Cape Peninsula. Also **A. africana** Mill. and **A. spicata** Baker. Uganda Aloë is considered an hepatic form of Cape Aloë.

Aloë Perryi Baker., Socotrine Aloe. (Liliaceae). - Perennial succulent plant. Island Socotra. Used medicinally. Evaporated juice from fleshy leaves is source of Socotra Aloë and probably also of Zanzibar Aloë. Is usually packed in skins of small animals. Has the same use as A. vera L.

Aloë tenuior Haw. (Liliaceae). - Perennial succulent plant. S. Africa. Root is used by the Kaffirs for tape-worm; recommended for children.

Aloë vera L., Curaçao Aloë, Barbados Aloë. (Liliaceae). - Succulent plant. Mediterranean region. Introd. in tropics of New World. Commercial product derived from Curaçao, Aruba, Bonaire. Used medicinally. Condensed juice from fleshy leaves becomes a blackish-brown, orange to dark reddish-brown powder. Contains aloin (barbaloin). Cathartic, acting on large intestine; vermifuge, emmenagogue. Product is called commercially Curaçao or Barbados Aloë. Certain varieties of this species are source of an Indian and Jaffarabad Aloë.

Aloë, Barbados → Aloë vera L.

Aloë, Cape → Aloë ferox Mill.

Aloë, Curaçao → Aloë vera L.

Aloë, False → Agave virginica L.

Aloë, Manila → Agave cantala Roxb.

Aloë, Moka → Aloë abyssinica Lam.

Aloë, Natal → Aloë Candelabrum Tod.

Aloë, Socotra → Aloë Perryi Baker.

Aloë, Uganda → Aloë ferox Mill.

Aloë, Zanzibar → Aloë Perryi Baker.

Aloës Wood → Aquilaria Agallocha Roxb.

Aloës Wood (substitute) → Gonostylus Miquelianus Teijsm and Binn.

Alopecurus pratensis L., Meadow Fox Tail. (Graminaceae). - Perennial grass. Europe, Asia, Caucasia. Used as meadow grass.

Aloysia citriodora Ort. → Lippia citriodora Kunth.

Alphitonia zizyphoides Gray. (syn. Pomaderris zizyphoides Hook. and Arn.). (Rhamnaceae). - Tree. New Caledonia. Wood greyish to violet, durable, easily worked; when it is polished it has the appearance of a light mahogany; used for interior work and cabinet making.

Alpine Dock → Rumex alpinus L.

Alpine Fir → Abies lasiocarpa (Hook.) Nutt.

Alpine Timothy → Phleum alpinus L.

Alpinia aromatica Aubl. → Renealma domingensis Horan.

Alpinia conchigera Griff. (syn. Languas conchigera Burkill.). (Zingiberaceae). - Perennial herb. Malaysia. Rhizome is used as condiment, esp. in Cochin China. It is also made into a rice spirit. Fruits are very pungent.

Alpinia Galanga Willd. (Languas Galanga [L.] Stuntz.) Langwas, Greater Galangal. (Zingiberaceae). - Perennial herb. Trop. Asia. Cultivated. Source of an ess. oil, called Essence d'Amali. Rhizomes are employed as condiment. Flowers are eaten raw, with other vegetables or in pickles in some parts of Java.

Alpinia malacensis Rosc. (Zingiberaceae). - Perennial herb. Mal. Archipel., E. India. Cultivated. Source of an ess. oil, called Essence d'Amali. Rhizome is used for chewing by the Amboinese.

Alpinia officinarum Hance. (syn. Languas officinarum [Hance] Farw.) Lesser Galangal, Small Galangal. (Zingiberaceae). - Perennial herb. East and S.E. Asia. Cultivated in China. Rhizome is source of Galangal. It is cut into small pieces and dried. Much is derived from Hainan, China. Used as a condiment. Employed in medicine, as aromatic, stimulant and carminative. Contains a cineol, eugenol, sesquiterpenes and a pungent acrid resin, galangol.

Alpinia pyramidata Blume. (Zingiberaceae). - Perennial herb. Trop. Asia esp. Philipp. Islds to Java. Root is used as condiment; similar to ginger. Sap cooked with sugar or honey, with water is source of an intoxicating drink; used in the Philipp. Islds. Rhizome is considered carminative and stimulant. Decoction of leaves is put in stimulant and aromatic baths in Philipp. Islds.

Alriba Resin → Canarium strictum Roxb.

Alsidium Helminthochorton (de la Tour.) Kuetz. Corsican Moss. (Rhodomelaceae). Red Alga. Mediterranean. Used as vermifuge.

Alsike Clover → Trifolium hybridum L.

Alsodeia physiphora Mart. (Violaceae). - Perennial plant. Brazil. Leaves are eaten as spinach in some parts of Brazil, esp. among the negro population.

Alsophila australis R. B. (Cyatheaceae). - Tree fern. Australia. Tissues in top of trunk contain starch which is eaten raw or roasted by aborigines. Hard part of stem is made into walking sticks and fancy furniture.

Alsophila myosuroides Liebm. (Cyatheaceae). - Tree-fern. S. Mexico and Centr. America. Spores occasionally used as styptic by Chinantecs and Zapotecs (Mex.).

Alsophila rufa Fée. (Cyatheaceae). - Tree-fern. Centr. and S. America. Pith used as food in the Choco Intendancy of Colombia during times of emergency.

Alstonia congensis Engl. (Apocynaceae). - Tree. Trop. Africa. Wood soft, white, light; used for furniture, war-drums, boats, native tools. Latex is used to adulterate rubber.

Alstonia constricta F. v. Muell. Fever Bark, Bitter Bark.)Apocynaceae). - Tree. Australia esp. New South Wales, Queensland. Bark very bitter, valued as febrifuge and tonic. Decoction sold in Australia as bitters. Contains a bitter principle alstonin.

Alstonia eximea Miq. (Apocynaceae). - Tree. Sumatra, Bangka. Source of Gutta Djelutung, used for adulterating Gutta Percha.

Alstonia grandifolia Miq. (Apocynaceae). - Tree. Sumatra. Source of Gutta Malaboeai, resembling Djelutung.

Alstonia scholaris R.Br. Palmira Alstonia, Pulai, Jelutong. (Apocynaceae). - Tall. tree. Ceylon to Australia. Used for centuries medicinally. Sold in most oriental drug shops. Tonic, febrifuge, vermifuge; used for intestinal troubles. Increases appetite. Contains alkloids: ditamine, echitamine, alstonine. Wood used for coffins.

Alstroemeria ligtu L. (Amaryllidaceae). - Perennial herb. Chile. Roots are source of a starch which is sold in some markets of Chile under the name of Chuño de Concepción. It is said the roots are consumed as food among some of the Indian tribes.

Alstroemeria haemantha Ruiz and Pav. (Amaryllidaceae). - Herbaceous perennial. Chile. Roots are source of a starch, eaten by the Indians. Also **A. revoluta** Ruiz and Pav., **A. versicolor** Ruiz. and Pav.

Alternanthera sessilis R. Br. Racaba (Amarantaceae). - Small herb. Tropics. Herb is eaten with fish among the natives of Belg. Congo. Leaves are consumed as spinach and in soup among the natives of Malaya. It is eaten with rice in Indonesia. Also **A. amoena** (Lem.) Voss., **A. philoxeroides** (Mart.) Griseb.

Althaea officinalis L., Marsh Mallow. (Malvaceae). - Perennial herb. Europe. Introd. in N. and S. America. Cultivated. Root, Radix Althaea, is used in medicine for coughs, intestinal catarrh, excipient in making pills, demulcent. Contains 25 to 35 % mucilage. Roots are used in tea and salads. Much is supplied by Holland, France and Germany. Decoction of root with sugar, arabic gum and white of egg is source of Pasta Althaea. Leaves collected during flowering time are used as an emollient and demulcent. Fibres from stems and roots have been occasionally used for manuf. paper.

Altingia excelsa Noron. (syn. Liquidambar Altingia Blume). (Hamamelidaceae). - Tall tree. Mal. Archip. to Yunnan. Wood heavy, fine grained, red to blackish brown; used for beams of houses, flooring. Stem source of a yellow scented resin, Getah Mala, seldom found in international trade. It is sometimes called Rasamala Resin or Rasamala Wood Oil.

Alui (house-hold medicin) → Andrographis paniculata Nees.

Alva Marina → Zostera marina L.

Alysicarpus rugosus DC. (Leguminosaceae). - Herbaceous. Trop. Africa, Asia, Australia, West Indies. In Katagum (Afr.) used as food for domestic animals.

Alysicarpus vaginalis DC. (syn. A. nummularifolius DC.). (Leguminosaceae). - Perennial herb. Java. Crop is used for terrassing land.

Alyxia lucida Wall. (Apocynaceae). - Woody plant. Malaysia, Madagascar. Infusion of leaves is used in Madagascar as a vermifuge. Bark and leaves are employed in manuf. rum.

Amanita caesarea (Scop.) Pers. (Agaricaceae). - Basidiomycete, Fungus. Warm-temperate zone. Fruitbodies edible, being a time-old food in Southern Europe; was considered a delicacy among the Romans.

Amanita muscaria (L.) Pers. (syn. Agaricus muscarius [L.] Pers.) (Agaricaceae). - Basidiomycete. Fungus. Temperate zone. Used in many countries to kill flies, after fruitbodies have been boiled and sugar added. Used in Northern Siberia as an intoxicating drug. Formerly used against ringworm. Fruitbodies are very poisonous. Sometimes called Fly Mushroom and Fly Amanita.

Amanita ovoidea (Scop.) Quel. (Agaricaceae). - Basidiomycete fungus. Mediterranean area. The white fruitbodies are used as food and are of fine quality. Sold in markets of Italy, Portugal, Catalonia and Southern France.

Amanita rubescens Pers. (Agaricaceae). - Basidiomycete. Fungus, Temp. and subtrop. zone. Fruitbodies are consumed as food, esp. in some parts of England. Used in ketchup. Also A. gemmata (Fr.) Gill., A. nivalis Grev. and A. spissa (Fr.) Quél.

Amanori → Porphyra.

Amaranthaceae → Achyranthes, Alternanthera, Amaranthus, Celosia, Deeringia, Digera, Iresine, Telanthera.

Amaranth, Green → Amaranthus retroflexus L.

Amaranth, Prostrate → Amaranthus blitoides Wats.

Amaranth, Redroot → Amaranthus retroflexus L.

Amaranth, Slim → Amaranthus hybridus L.

Amaranthus blitoides S. Wats. Prostrate Amaranth. (Amaranthaceae). - Herbaceous plant. Western N. America. Leaves were eaten as a potherb. Seeds were used for pinole, by different Indian tribes. Also A. diacanthus Raf., A. graecicans L., A. hybridus L., A. Palmeri S. Wats. and A. Torreyi Benth.

Amaranthus caudatus L., Inca Wheat, Quihuicha. (Amaranthaceae). - Annual herb. Trop. Africa, America, E. India. Culivated. Seeds are used in S. America as food; made into bread by the Indians. Leaves are consumed as potherb and in salads. Used since the times of the Incas. Plant is sometimes called Quinoa, a name also given to Chenopodium Quinoa Willd.

Amaranthus gangeticus L. (Amaranthacaeae). - Herbaceous plant. Tropics. Cultivated in India, Malaya, China, Japan. Plants are eaten during the young seedling stage, cooked and eaten like spinach.

Amaranthus grandiflorus J. M. Black. (Amaranthaceae). - Herbaceous plant. Australia. Seeds are used as food by the aborigines of Centr. Australia.

Amaranthus hybridus L., Slim Amaranth. (Amaranthaceae). - Annual herb. Temp. zone. Seeds are consumed by the Indians. Leaves are eaten as potherb in different countries. Cultivated in India for its seeds.

Amaranthus mangostanus L. (Amaranthaceae). - Annual herb. India. Cultivated, esp. in Japan. A hot weather vegetable. Leaves eaten boiled or after being salted, in Japan.

Amaranthus retroflexus L., Redroot Amaranth, Green Amaranth, Pigweed. (Amaranthaceae). - Annual herb. Temp. zone. Young parts of the plant are consumed as potherb. Seeds were used as food by the Indians.

Amaranthus viridis L. (Amaranthaceae). - Herbaceous plant. Tropics. Cultivated. Tops of the shoots are consumed as potherb in different parts of the tropics.

Amaryllidaceae → Agapanthus, Agave, Allium, Alstroemeria, Bomara, Buphane, Crinum, Curculigo, Furcraea, Haemanthus, Hymenocallis, Narcissus, Polianthes, Tulbaghia.

Amatou → Fomes fomentarius (L.) Gill.

Amatungula → Carissa Arduina Lam.

Amazon Poison Nut → Strychnos Castelnaei Wedd.

Ambarella → Spondias dulcis Forst. f.

Ambari Hemp → Hibiscus cannabinus L.

Amber → Pinus succinifera Conw.

Amber de Cuspinole → Hymenaea Courbaril L.

Amboina Sandalwood → Osmoxylon umbelliferum Merr.

Ambretta Seeds → Hibiscus Abelmoschus L.

Amburana acreana (Ducke) Sm., Cumarú, Imburana de Cheiro. (Leguminosaceae). - Tree. Brazil, esp. Acre Territory. Wood yellow to light brown, with some orange hue, scent and taste of vanilla, coarse grained. Used for general construction. Seeds are source of a volatile oil; used in perfumery.

Amburana cearensis (Allem.) Sm. (syn. Torresea cearensis Allem., A. Claudii Schw. and Taub.) (Leguminosaceae). - Tree. Brazil, N. Argentina. Resin of bark is source of a volatile oil; used in medicine. Wood used in Argent. for furniture; in N. E. Braz. for window frames, carpentry, cooperage, crating and general construction work.

Amelanchier alnifolia Nutt. Saskatoon Serviceberry, Western Service Berry. (Rosaceae). - Small tree. N. America. Fruits were used as food by Indians of different tribes.

Amelanchier canadensis (L.) Medic. Shadblow Serviceberry, Shad Bush, Service Berry. (Rosaceae). - Small tree. Eastern N. America. Wood very hard, strong, close grained, heavy, dark brown, sometimes tinged with red; used for small implements, handles of tools. Under the name of Lance Wood it is used for manuf. of fishing rods. Fruits of good flavor; used for pies, also canned for winter use.

American Barberry → Berberis canadensis Mill.

American Basswood → Tilia americana L.

American Beech → Fagus ferruginea Dryand.

American Blue Vervain → Verbena hastata L.

American Centaury → Sabatia angularis (L.) Pursh.

American Chestnut → Castanea dentata (Marsh.) Borkh.

American Elder → Sambucus canadensis L.

American Ephedra → Ephedra americana Humb. and Bonpl.

American False Hellebore → Veratrum viride Ait.

American Hornbeam → Carpinus caroliniana Walt.

American Holly → Ilex opaca Ait.

American Mastic → Schinus Molle L.

American Mistletoe → Phorodendron flavescens (Pursh) Nutt.

American Nutgalls → Quercus imbricaria Micjx.

American Pennyroyal → Hedeoma pulegioides (L.) Pers.

American Raspberry → Rubus strigosus Michx.

American Red Currant → Ribes triste Pall.

American Sea Rocket → Cakile edentula (Bigel) Hook.

American Senna → Cassia marilandica L.

American Smoketree → Rhus cotinoides (Nutt.) Britt.

American Spikenard → Aralia racemosa L.

American Upland Cotton → Gossypium hirsutum L.

American White Hellebore → Veratrum viride Ait.

American Wild Gooseberry → Ribes Cynosbati L.

American Wild Mint → Mentha canadensis L.

American Wormseed → Chenopodium ambrosioides L. var. anthelminticum L.

American Yew → Taxus canadensis Willd.

Amerimnun granadillo Standl. (Leguminosaceae). - Tree. Mexico, esp. Oaxaca to Michoacán. Wood used for cabinet work in Mexico.

Amannia senegalensis Lam. (Lythraceae). - Herbaceous plant. Trop. Africa. Used in some parts of Africa as a blistering agent. Also A. baccifera L.

Ammi Visnaga Lam. (Umbelliferaceae). - Perennial herb. Mediterranean region. Fruiting pedicels sold in Egyptian markets as tooth picks. Seeds used as diuretic in colic; also promoting the passage of small kidney stones.

Ammobroma sonorae Torr. (Lennoaceae). - Chlorophylless plant. - S. W. of the United States to N. W. Sonora (Mex.) Large succulent subterrenean stems are used as food raw, roasted or made into flour by the Papago Indians of Arizona.

Ammodaucus leucotrichus Coss. and Dur. (Umbelliferaceae). - Herbaceous plant. N. Africa. Cultivated in Sahara Oases, Mauritania, Upper Nile Valley. Seeds used as condiment in sauces. Infusion of leaves is used for chest complaints. Often sold in markets.

Ammodenia peploides (L.) Hook. → Arenaria peploides L.

Ammoniac, Moroccan → Ferula tingitana L.

Ammoniac, North African → Ferula tingitana L.

Ammoniac of Morocco Gum → Ferula communis L.

Ammoniac of Cyrenaica → Ferula marmarica Asch. and Taub.

Ammoniacum → Dorema Ammoniacum D. Don.

Ammophila arundinacea Host. (syn. A. arenaria Link., Psamma arenaria Roem. and Schult.) European Beachgrass. (Graminaceae). - Perennial grass. Coastal zone of Europe. Grown as a sand binder in sand dunes in different countries.

Amole → Agave Schottii Engelm.

Amomum aromaticum Roxb., Bengal Cardamon, Nepal Cardamon. (Zingiberaceae). - Herbaceous perennial. E. India. Cultivated in E. India. Plant is used as a condiment.

Amomum Cevuga Seemann. (Zingiberaceae). - Perennial herb. Pacific Islds., esp. Society Islands, Tahiti, Marquesas. Used in Fiji for scenting coconut oil. Leaves used for thatch and native beds in Tahiti.

Amomum Cardamomum Willd. Cardamon. (Zingiberaceae). - Perennial herb. Malaysia. Cultivated. Seeds are used as condiment, often used for flavoring cakes. Ground rhizomes are used medicinally for colds.

Amomum dealbatum Roxb. (Zingiberaceae). - Perennial herb. E. India, Malaysia. Seeds are used as a substitute for Cardamons.

Amomum kepulaga Sprague and Burkill., Round Cardamom. (Zingiberaceae). - Perennial herb. Cultivated in Malaysia. Fruits are used as spice and are also chewed by the natives to sweeten their breath. Employed medicinally for coughs and colds.

Amomum Krervanh Pierre. Krervanh. (Zingiberaceae). - Perennial herb. Cambodia, Mt. Krervanh. Cultivated. Fruits are used as a condiment, also in curries. Employed in Chinese pharmacy. Used in Europe in cordials and in sausages.

Amomum maximum Roxb., Java Cardamon. (Zingiberaceae). - Herbaceous perrenial. Malaysia. Cultivated in Java. Plant is used as a condiment by the natives.

Amomum subulatum Roxb. (Zingiberaceae). - Perennial herb. Trop. Asia. Source of Nepal Cardamon, used by the natives.

Amomum thyrsoideum Gagn., Krakorso. (Zingiberaceae). - Perennial herb. Trop. Asia, esp. Indo-China. Fruits used as a condiment in food and curries.

Amomum xanthioides Wall. (Zingiberaceae). - Perennial herb. Burma. Cultivated in India. Used as a condiment. Source of Bastard Siamese Cardamon or Wild Siamese Cardamon.

Amoora nitida Benth. (Meliaceae). - Tree. Australia. Wood is known as Incense Wood and is used for cabinet work.

Amoora Rohituka Wight and Arn. (syn. Aglaia polystachya Wall.) (Meliaceae). - Tree. Trop. Asia. Seeds are source of Rohituka Oil, having a camphor-like scent; used for illumination and in industries, esp. in Bengal, S. China und Indo-China.

Amoora Wallichii King. (Meliaceae). - Tall tree. Bengal. Excellent timber, red; used for furniture and decorative work.

Amorphophallus campanulatus Blume. Whitespot Giant Arum. (Araceae). - Perennial herb. Trop. Asia. Large roundish corms are consumed as food when boiled.

Amorphophallus Harmandii Engl. and Gehr. (Araceae). - Perennial herb. Tonkin. Corms are consumed boiled by the natives. Plant is occasionally cultivated.

Amorphophallus Rivieri Dur. (Araceae). - Perennial tall herbaceous plant. Trop. Asia. Var. Konjac is cultivated for its tubers which are eaten boiled, esp. in China and Japan. Source of Konjaku Flour or Konnjaku Powder.

Ampelocissus Martini Pl. (Vitaceae). - Woody vine. Philippéne Islands. Fruits edible, occur in large bunches, dark maroon red to black, fleshy, acid; excellent for jelly.

Ampelopsis quinquefolia Michx. → Parthenocissus quinquefolia (L.) Planch.

Ampelozizyphus amazonicus Ducke., Saracuramira. (Rhamnaceae). - Tree. Brazil, esp. Lower Amazon region, Guiana. Fragrant bark is used by Waiwai Indians of Brit. Guiana in place of soap.

Amphicarpa monoica (L.) Ell. (syn. Falcata comosa [L.] Kuntze.) Hog Peanut, Ground Peanut. (Leguminosaceae). - Herbaceous vine. Eastern and southern United States. Seeds were a source of food among the Indians.

Amphicarpa Edgeworthii Benth. (Leguminosaceae). - Herbaceous plant. China, Himalayas, Japan. Subterranean seeds used as food by the Ainu.

Amphilophis intermedia Stapf. (Graminaceae). - Perennial grass. Angola, Gold Coast, Cape Verde Islands, Barbados, Guiana. Recommended as fodder plant in dry lime-stone soils, as hay crop and for grazing.

Amphilophis pertusa Stapf., Sour. Grass. (Graminaceae). - Perennial grass. Trop. Africa, Arabia, India, Ceylon, Mauritius. Drought resistant fodder grass.

Amphimas pterocarpoides Harms. (Leguminosaceae). - Tree. Trop. Africa. Bark is source of a red sticky resin; a decoction is used in Liberia for dysentery.

Amphipterygium adstringens (Schlecht) Schiede → Juliana adstringens Schlecht.

Amrad → Acacia arabica Willd.

Amrawatti → Acacia arabica Willd.

Amritsar Gum → Acacia modesta Wal.

Amur Cork Tree → Phellodendron amurense Rupr.

Amygdalus communis L. (syn. Prunus Amygdalus Batsch.) Almond Tree. (Rosaceae). - Small tree. W. Asia, E. Mediterranean region. Cultivated since antiquity. Extensively grown in S. Europe and California. Sweet Almond, var. dulce DC. is source of Sweet Almonds, a valuable food, eaten green, ripe and dried; also salted, roasted, in bread and in pastry. Almond meal is made into macaroons and cakes. Source of almond extract. Varieties are: Drake, Ne Plus Ultra, Nonpareil, Texas Prolific, Languedoc. Many are thin shelled. Jordan, a fine flavored thick shelled variety, is derived from Malaya, used in confectionary. Kernels are source of Sweet Almond Oil, Oleum Amygdalae Expressum, nondrying; used medicinally as emollient, demulcent. Almond meal is used as a cleansing agent. Oil of Bitter Almond, Oleum Amygdalae Amarae, non-drying, is derived from kernels of var. amara DC., used as sedative, for cough remedies; much is derived from France, Morocco, Canary Islds, California, Iran, Portugal, Spain, Sicily, Syria. Sp. Gr. 0.9175; Sap. Val. 183-196; Iod. No. 95-102. A glucoside in the kernels gives rise to prussic acid.

Amygdalus leiocarpus Boiss. (Rosaceae). - Tree. Semi-arid regions of Iran, Arabia, Palestine. Source of a gum, called Cherry Gum, Kirschgummi, Persian Gum, Djedk-i-Ardjin, consumed by the natives. The material is also exported. Sold in bazaars.

Amygdalus nana L. (syn. Prunus nana [L.] Stokes) Dwarf Almond. (Rosaceae). - Eastern Europe, Siberia. Seeds are source of an oil being inferior to bitter almond oil. Used in S. Russia and Hungary. Produces also a bitter almond water. Fruits are eaten in Siberia.

Amygdalus Persica L. (syn. Prunus Persica [L.] Batsch.). Peach Tree. (Rosaceae). - Small tree. China. Cultivated since ancient times. Grown from temp. to subtr. zones of Old and New World. Important commercial fruit tree. Dessert fruit, consumed fresh, dried, preserved, canned, dehydrated, also as sweet pickles and peach butter. Source of peach cordials and peach brandy. Waste pits are used for manuf. of oil and paste. Stem is source of a gum; used in some parts of Asia. Var. scleropersica Reichb. Clingstone peaches, flesh adhares to the stone, e. g. Orange Clingstone, Amsden, Triumph, Tuskina, Heath; var. aganopersica Reichb. Freestone peach flesh easily separated from the stone, e. g. Hale's Early, Crawfords Early, Elberta, Alexander, Susquehanna, Lovell. Var. Honey Peach with oval long pointed fruits. Var. nectarina Ait. Nectarines with smooth skin, e. g. Humboldt, Fowton, Advance. Var. compressa Lout. (syn. var. platycarpa Decne.) Flat Peach, Peento; fruits flattened from above.

Amygdalus spartoides Spach. (syn. Prunus spartoides [Spach.] Schneid.). (Rosaceae). - Small tree. W. Asia. Source of a gum, sold in some parts of Iraq and Iran.

Amyris balsamifera L., West Indian Sandalwood, Rosewood, Candlewood, Touchwood. (Rutaceae). - Tree. Trop. America, esp. Cuba, Jamaica, Puerto Rico and Northern S. America. Source of West Indian Sandalwood Oil, Cayenne Linaloe Oil. Wood ist used for building purposes. Also used as incense.

Amyris Plumieri DC. (Rutaceae). - Tree. West Indies, Mexico. Source of Elemi of Mexico or Elemi of Yucatan; used in lacquers.

Amyris sylvatica Jacq. (Rutaceae). - Tree. Trop. America, esp. Jamaica, S. Domingo and Colombia. Wood is source of a resin; used for technical purposes.

Anabasis articulata (Forsk.) Moq. (Chenopodiaceae). - Shrub. Sahara region. Plant is used as food for camels. Stems produce a gum or Manna.

Anacamptis, pyramidalis (L.) Rich. (Orchidaceae). - Perennial herb. Centr. and S. Europe, Asia Minor, Orient, Iran and N. Africa. Tubers are source of salep, used in medicine.

Anacardiaceae → Anacardium, Astronium, Bouea, Buchanania, Corynocarpus, Cyrtocarpa, Dracontomelum, Gluta, Harpephyllum, Heeria, Kokkia, Lannea, Loxopterygium, Mangifera, Melanorhoea, Microstemon, Odina, Pachycormus, Pistacia, Pleiogynium, Poupartia, Quebrachia, Rhodosphaera, Rhus, Schinopsis, Schinus, Sclerocarya, Semecarpus, Sorindeia, Spondias.

Anacardium giganteum Loud. (Anacardiaceae). - Tree. Guiana. Fleshy fruit-stalks are said to be eaten in some parts of Dutch Guiana.

Anacardium occidentale L. Cashew. (Anacardiaceae). - Tree. Trop. America. Cultivated in tropics of Old and New World. Of commercial importance. Kidney shaped kernels, called Cashew Nuts are consumed roasted and salted, much is exported to N. America and Europe. Fruits contain 45 to 47% oil, has high heat resistance, excellent lubricant in magneto armatures in airplanes; also used in varnishes, inks, termite proofing lumber, insulating coatings. Kernels are said to give a flavor to Madeira Wine; is also mixed with chocolates. The Cashew Apple which is not a fruit, but the thickened fleshy stem below the Cashew Nut, is used when soft as jam, in sweet meats and made into a wine and beverage, called in Brazil Cajuado. The stem produces a gum used in varnishes, as protection for books, and woodwork, against ants and other insects. Called Cashawa Gum, Gomme d'Acajou. Much is derived from West Indies, Portug. E. Africa, Tanganyika and Kenya.

Anacardium humile St. Hil. (syn A. subterraneum Liais). (Anacardiaceae). - Shrub. Fruits are rich in oil, consumed in some parts of Brazil much like A. occidentale L. Also A. nanum St. Hil.

Anacardium orientale L. → Semecarpus Anacardium L. f.

Anacardium rhinocarpus DC. (Anacardiaceae). - Tree. Trop. America. Wood known as Esparvie, Espave Mahogany, being yellowish to dark brown or reddish; used instead of true Mahogany. Bark is source of a fibre, called Mijugua Fibre.

Anacyclus officinarum Hayne., Bertram. (Compositae). - Annual herb. Mediterranean region. Used in home remedies as a mouth wash and as a tincture for toothache.

Anacyclus Pyrethrum DC., Pellitory. (Compositae). - Perennial herb. Mediterranean region, Arabia, Syria. Produced commercially in Algeria. Source of an ess. oil, used in liqueurs, also made into mouthwashes, used for toothache. Roots are sometimes used for chewing. Dried roots, collected in autumn are used medicinally, being stimulant, rubefacient, irritant, sialogogue; contains a very pungent alkaloid pyrethrine.

Anadendrum montanum Schott. (Araceae). - Herbaceous perennial. Epiphyte. Trop. Asia. esp. Tonkin to Celebes. Leaves are used in curries.

Anagallis arvensis L., Cure All. Scarlet Pimpernel. (Primulaceae). - Small annual herb. Troughout Europe, Temp. Asia, N. Africa, N. America, Australia. Herb was formerly used in medicine, at the present employed as a home remedy; diaphoretic, diuretic, resolvent; for gallstones, liver cirrhosa, lung ailments.

Anamirta paniculata Colebr. (syn. A. Cocculus [L.] Wight and Arn.), Fish Berry, Indian Berry. (Menispermaceae). - Woody climber. Coast of Malabar, S.E. Asia. Fruits are used by the natives to stupefy fish. Dried ripe berries are used medicinally as parasiticide; a convulsant poison. Contains coculin. Much is shipped from Bombay and Calcutta.

Ananas comosus (L.) Merr. (syn. Ananas sativus [Lindl.]. Schult.), Pineapple. (Bromeliaceae). - Perennial herb. Brazil. Cultivated commercially in the tropics and parts of subtr. of Old and New World. Fruits are eaten raw, preserved, canned, in salads, candied, in jams, as fruit juice, in sherbets. Grated or crushed it is used for pies, ice ream, cakes, in confectionary. Cores are made into candies; used by confectionaries. Juice is source of alcohol and alc. beverage, Vin d'Ananas. Denatured alcohol with ether is used in automobile engines. Pressed peels and cores are used as food for livestock. Pineapple waste is made into vinegar. Var. sativus (Lindl.) Mez. (syn. Ananassa sativus Lindl., Bromelia Ananas L.) widely cultivated in the tropics. Var. lucidua (Mill.) Mez. (syn. Ananassa lucida Lindl.) of which many forms are grown in hothouses of Europe. Cultivated are: Red Spanish, Abachi, Smooth Cayenne (for canning), Sugar Loaf, Black Jamaica, Black Prince, White Antigua, Prince Albert, Porto Rico. Leaves are source of a hard fibre, called Piña fibre.

Ananas Magdalenae André → Aechmea Magdalenae (André.) Standl.

Anaptychia ciliaris (L.) Kbr. (syn. Parmelia ciliaris Ach.) (Physciaceae). - Lichen. Temp. zone, esp. Europe. On stems of broadleafed trees. Was during the 17th Century source of Cyprus Powder, a toilet article, being scented with musk, ambergris, oil of jasmine, roses or orange blossom. It was used to scent, cleanse and whiten the hair.

Anatto → Bixa Orellana L.

Anchomanes Hookeri Schott. (syn. Anchomanes difformis Engl.) (Araceae). - Perennial herb. Trop. W. Africa. Tubers are consumed as food by the natives during times of scarcity.

Ancistrocladus extensus Wall. (Ancistrocladaceae). - Vine. Burma to S. China und Singapore to Banka. Roots when boiled are used by the Malays for dysentery. In Siam very young leaves are used for flavoring.

Ancistrophyllum opacum Drude. (Palmaceae). - Palm. Trop. W. Africa. Stems are used for building houses and making baskets.

Ancistrophyllum secundiflorum Mann. and Wendl. (Palmaceae). - Climbing palm. Trop. Africa, esp. Mouth of Niger, Cameroons, Sierra Leone, Angola, Belg. and Fr. Congo. Stems are used in hut building, fish traps and baskets.

Anda-assy Oil → Joannesia princeps Vell.

Anda Gomesii Juss. → Joannesia princeps Vel.

Andaman Canary Tree → Canarium euphyllum Kurz.

Andaman Crape Myrtle → Lagerstroemia hypoleuca Kurz.

Andaman Zebrawood → Diospyros Kurzii Hiern.

Andean Blueberry → Vaccinium floribundum H. B. K.

Anderson Wolfberry → Lycium Andersonii Gray.

Andira Araroba Aguiar., Angelin Tree. (Leguminosaceae). - Tree. Trop. America. Source of Chrysarobin, derived from Goa or Bahia Powder being a substance deposited in the wood; used medicinally as an irritant and antiparasitic.

Andira excelsa H. B. K. (syn. Vouacapoua americana Aubl.) Patridge Wood, Brownheart, (Leguminosaceae). - Tree. Trop. America. Wood durable, dark brown or reddish brown, heavy, very hard, tough, strong, not difficult to work. Used for heavy construction work, furniture, wheelwright work, carpentry, cabinet work.

Andira Galeothiana Standl. (Leguminosaceae). - Tree. S. Mexico. Used for shade in coffee-plantations. Bark is an effictious vermifuge, used by Chinantecs (Mex.).

Andira inermis (Swartz) H. B. K. (Leguminosaceae). - Tree. S. Mexico, West Indies, Centr. and S. America, W. Africa. Wood durable, hard, yellowish to dark brown and black, takes high polish. Used in Tabasco (Mex.) for construction purposes. Shade tree in coffee plantations. Seeds and bark are a vermifuge, febrifuge, purgative and anthelmintic.

Andira stipulacea Benth. (Leguminosaceae). - Tree. Brazil. Wood is used for construction of boats, general carpentry. Seeds are vermifuge.

Andira vermifuga Mart. Brazilian Angelin Tree. (Leguminosaceae). - Tree. Brazil. Bark is used as a vermifuge, purgative, emetic, narcotic; poisonous in large doses.

Andiroba → Carapa guineensis Aubl.

Androcymbium punctatum Baker. (syn. Erythrostictus punctatus Schlecht.) (Liliaceae). - Herbaceous plant. N. Africa. Used as a condiment by the Tuareg (Afr.).

Andrographis paniculata Nees., Kariyat. (Acanthaceae). - Annual herb. India and Ceylon. Cultivated. Source of a household medicin, called Alui; much used in India, being a powerful bitter tonic, used for dysentery and diarrhea. Roots and leaves are stomachic, tonic, alterative, febrifuge and anthelmintic. Known in England as Halviva.

Andromeda polifolia L., Bog Rosemary. (Ericaceae). - Low shrub. Arctic regions, temp. Europe, Asia, N. America. Leaves and twigs are used in some parts of Russia for tanning.

Andropogon aciculatus Retz. (syn. Chrysopogon aciculatus Trin.), (Graminaceae). - Perennial grass. Tropics. Used in Philipp. Islands and Guam as food for cattle. Also for making hats and mats.

Andropogon candidus Trin. → Elionurus candidus Hack.

Andropogon caricosus L. (Graminaceae). - Perennial grass. Trop. Asia, Ins. Mascar. Source of an excellent hay.

Andropogon contortus L. → Heteropogon hirtus Pers.

Andropogon hispidus Willd. → Arundinella brasiliensis Raddi.

Andropogon laniger Desf. (Graminaceae). - Perennial herb. W. Africa, E. India. Rootstocks when dried, are sold in small sachets, imparting an agreeable scent. It is probably the Nara Syriagae of the Ancients.

Andropogon muricatus Retz. → Vetiveria zizanoides Stapf.

Andropogon rufus Kunth., Yaragua Grass. (Graminaceae). - Perennial grass. Trop. Africa, Brazil. Plant has been recommended as a fodder for livestock.

Andropogon saccharoides Sw. (syn. A. perforatus Trin., A. emersus Fourn.). (Graminaceae). - Plant is considered a good pasture grass for cattle.

Andropogon Sorghum (L.) Brot. → Sorghum vulgare Pers.

Aneilema beninense Kunth. (syn. Commelina beninensis Beauv.). (Commelinaceae). - Perennial herb. Throughout Trop. Africa. Decoction of plants is used by the natives as a laxative for children.

Anemone cernua Thuab. Nodding Anemone. (Ranunculaceae). - Perennial herb. Manchuria, China, esp. Honan; Japan. Used medicinally in China. Contains anemonin.

Anemone flaccida Fr. Schmidt. (Ranunculaceae). - Perennial herb. Japan, China. Leaves and stems are an article of diet among the Ainu.

Anemone Hepatica L. (syn. Hepatica triloba DC.). (Ranunculaceae). - Europe, N. America. Plant is used medicinally for its tonic properties.

Anemone narcissiflora L. (Ranunculaceae). - Perennial herb. N. temp. Hemisphere. Leaves are eaten as cress, soured or prepared in oil by the Eskimos of King Island (Alaska).

Anemone patens L. Spreading Pasqueflower. (Ranunculaceae). - Perennial herb. Europe, Siberia, Turkestan, Mongolia, Amur region. Herb used medicinally in the same way as A. Pulsatilla L.

Anemone pratensis L. Meadow Pasqueflower. (Ranunculaceae). - Herbaceous perennial. Europe, N. Asia. Used medicinally for the same purposes as A. Pulsatilla L.

Anemone Pulsatilla L., European Pasqueflower. Pulsatilla. (Ranunculaceae). - Perennial herb. Europe. Dried herb collected during flowering stage is used medicinally as emmengogue, alternative, diuretic and expectorant. Much is derived from Czechoslovakia. Contains an acrid ess. oil.

Anemonella thalictroides (L.) Spach. (syn. Syndesmon thalictrodes [L.] Hoffm.). Wild Rue, Rue Anemone. (Ranunculaceae). - Perennial herb. Eastern N. America. Starchy roots are used as food in mountainous districts of Pennsylvania where it is called Wild Potato.

Anemonopsis californica Hook. and Arn. (syn. Houttuynia californica Benth. and Hook.) (Saururaceae). - Perennial herb. S.W. of United States and adj. Mexico. Herb is used by the Mexican population to relieve colds, indigestion and to purify blood.

Anethum graveolens L., Dill. (Umbelliferaceae). - Annual herb. S. Europe, Caucasia, Iran, India. Cultivated in England, Germany, Rumania, United States. Fruits used for flavoring foods, pickles. Oil of Dill Seed, an ess. oil is used widely in food industry. Dried, ripe seeds used medicinally, being a stimulant and carminative.

Angado Mastiche → Echinops viscosus DC.

Angelica Archangelica L., Angelica. (Umbelliferaceae). - Biennial herb. Temp. Europe, Siberia, Kamtschatka, Himalaya. Cultivated in Belgium, Hungary etc. Candied leafstalks are used in pastries, cakes. Source of Oil of Angelica Root, ess. oil, used in cordials, Benedictine, Chartreuse, gins; in perfumes of the Oriental type. Dried rhizome, Angelicae Radix; dried leaves and flowering tops, Herba Angelicae, are used medicinally, being stomachic, carminative, stimulant. Ripe fruits are a carminative, stimulant; used in many home-remedies.

Angelica edulis Miyabe. (Umbelliferaceae). - Perennial herb. Japan. Stalks are used as food by the Ainu.

Angelica Levisticum All. → Levisticum officinale (Baill.) Koch.

Angelica sylvestris L., Woodland Angelica, Wild Angelica. (Umbelliferaceae). - Biennial to perennial herb. Europe, Asia, introd. in N. America. Young stems and leaves when boiled are sometimes eaten as a vegetable. Pulverized fruits are used to kill head parasites.

Angelica villosa (Walt.) B. S. P., Hairy Angelica. (Umbelliferaceae). - Perennial Herb. N. America to Florida. Leaves are recommended as a means to discourage the use of tobacco.

Angelica → Angelica Archangelica L.

Angelica, Hairy → Angelica villosa (Walt.) B. S. P.

Angelica, Wild → Angelica sylvestris L.

Angelim amarillo → Vataireopsis speciosa Ducke.

Angelin Tree, Brazilian → Andira vermifuga Mart.

Angelin Tree, Goa → Andira Araroba Aguiar.

Angelonia salicariaefolia H. B. (Scrophulariaceae). - Herbaceous. S. America. Used in some parts of Venezuela as diaphoretic.

Angico Gum → Piptadenia rigida Benth.

Angiopteris evecta Hoffm. (Marattiaceae). - Tall fern. Queensland. Pith rich in starch, eaten by aborigines. Produces an aromatic oil, used in Polynesia to perfume coconut oil.

Angola Weed → Roccella fuciformis (L.) Lam.

Angophora intermedia DC. (syn. Metrosideros floribunda Smith.) (Myrtaceae). - Tree. Victoria, New South Wales. Source of a reddish-brown brittle kino.

Angraecum fragrans Thouars. (Orchidaceae). - Perennial herb. Masc. Islds. Leaves are source of a tea, called Thé de Bourbon.

Angso Gutta Percha → Palaquium leiocarpon Boerl.

Aniba Canelilla (H. B. K.) Mez. (syn. Cryptocarya Canelilla H. B. K.). (Lauraceae). - Tree. Trop. America, esp. Peru. The cinnamon flavored bark is used as stimulating tea in Peru. Powdered bark is used for perfuming linen. Wood is source of an oil, called Sandalo Brasileiro.

Aniba perutilis Hemsl., Comino. (Lauraceae). - Tree. Andes of Columbia, Bolivia. Wood has satiny luster; used for fine furniture, interior finish of buildings, veneers for cabinet work.

Aniba rosaeodora Ducke. Bois de Rose Femelle. (Lauraceae). - Tree. French Guiana, E. Surinam, Brazil esp. the Lower Amazone region. Wood lustrous yellow, becomes lateron brownish, contains a volatile, fragrant oil; used for chests and drawers. Source of Huile de Linaloe, Essence de Bois Rose; used in perfumery.

Aniba terminalis Ducke. Pao Rosa (Lauraceae). - Tree. Amazone Basin, Guiana. One of the sources of Rose Wood Oil. Occasionally used in perfumery.

Anime Copal → Trachylobium Hornemannianum Hayne.

Anise, Chinese → Illicium verum Hook. f.

Anise, Star → Illicium verum Hook, f.

Anise Plant → Pimpinella Anisum L.

Anisette Liqueur → Pimpinella Anisatum L.

Anisomeles indica Kuntze. (Labiaceae). - Strong scented herb. Trop. Asia. Used in China, Ceylon and India medicinally as carminative, astringent, tonic; also for flavoring sago cakes. Made into a syrup by the Tamils.

Anisomeles salviaefolia R.Br. (Labiaceae). - Perennial herb. Australia. Flowers and herb have been suggested as a source of perfume.

Anisoptera laevis Ridl. (Dipterocarpaceae). - Tall tree. Trop. Asia. Wood fairly hard, heavy, pale yellow; used for boards, cupboards, construction work.

Anisoptera marginata Korth. (Dipterocarpaceae). - Tall tree. Borneo. Wood yellow, hard, durable, difficult to saw; used for construction of houses. Source of a Dammar.

Anisoptera thurifera Blanco. Palosapis Mersawa. (Dipterocarpaceae). - Philipp. Islands. Trunk source of Palosápis Oil. Wood yellowish, used for construction.

Anisosperma Passiflora Manso. (Cucurbitaceae). - Perennial vine. Brazil. Oil from seeds is used in Brazil as drastic and emetic; an ess. oil is employed as anthelmintic.

Anisum officinarum Moench. → Pimpinella Anisum L.

Anisum vulgare Gaertn. → Pimpinella Anisum L.

Anneslea fragrans Wall. (Theaceae). - Tree. Trop. Asia. Wood grey-brown to brown, hard, beautifully marked; used for fancy articles and furniture.

Annona aurantiaca Rodr. Borb. (Annonaceae). - Shrub. Brazil. Fruits are consumed in some parts of Brazil.

Annona cacans Warm. (Annonaceae). - Tree. Brazil. Fruits edible, consumed in some parts of Brazil.

Annona Cherimola Mill., Cherimoya. (Annonaceae). - Small tree. Trop. America. Grown in most parts of the tropics, though essentially subtropical. Fruit has good dessert qualities, subacid, delicate, eaten raw, in cooling drinks, sherbets. Several varieties, e. g., Anona de dedos Pentados, Cherimoya Lisa, Cherimoya de la Tetillas, Cherimoya de Puas. Highly flavored fruits are known from Madeira.

Annona delabripetala Raddi → Rollinia longifolia St. Hil.

Annona dioica St. Hil. (syn. Aberemoa dioica Rodr. Borb.). (Annonaceae). - Shrub. Brazil. Leaves are considered in some parts of Brazil antirheumatic; fruits are a emollient.

Annona diversifolia Safford., Ilama. (Annonaceae). - Small tree. Central America, esp. S. Mexico, Guatemala, El Salvador. Cultivated at low elevations. Introd. in S. Florida. Fine annonaceous fruit, eaten raw.

Annona glabra L., Pond Apple (Annonaceae). - Small tree. S. Florida, West Indies. Insipid fruits are occasionally used for jellies.

Annona longiflora S. Wats., Wild Cheremoya of Jalisco. (Annonaceae). - Shrub. Mexico, esp. Jalisco. Fruits edible. Sweetmeat prepared when boiled with sugar together with the fruit of Tejocote, Crataegus mexicana Moc. and Sessé. Fruits resemble Cherimoya.

Annona Mannii Oliv. (syn. Annonidium Mannii [Oliv.] Engl. and Diels). (Annonaceae). - Small tree. Trop. Africa. Fruits large, of good flavor, consumed by the natives.

Annona montana Macf., Mountain Soursop, Cimarrona. (Annonaceae). - Small tree. West Indies. Seldom cultivated. Fruits of a good flavor, refreshing juicy, subacid.

Annona muricata L., Soursop, Guanábana, Corossol. (Annonaceae). - Small tree. Trop. America. Cultivated in tropics of Old and New World. Fruits largest among Annona species; eaten raw, used in sherbets and other drinks; pulp also consumed with wine or cognac. In Java young fruits are used in sajoer, a soup. Several varieties in existance.

Annona purpurea Moc and Sassé. Soncoya. (Annonaceae). - Small tree. S. Mexico, Centr. America. Occasionally cultivated. Fruits agreeable taste. Sometimes sold in markets. Has been recommended for improvement by breeding.

Annona reticulata L., Bullock's Heart, Common Custard Apple, Corazon. (Annonaceae). - Small tree. Trop. America. Cultivated in tropics of Old and New World. Fruit inferior to the Cherimoya or Sugar Apple, flavorless; of fair quality.

Annona scleroderma Safford. Posh Té. (Annonaceae). - Small tree. Centr. America, esp. S. Mexico to Guatemala. Occasionally cultivated. Fruits aromatic, delicious, richer in flavour than the Soursop; thick peel; recommended for breeding purposes, to obtain thick-skinned hybrids suitable for shipping purposes.

Annona senegalensis Pers. (Annonaceae). - Small tree. W. Trop. Africa, esp. Senegal, Congo, Zambesie. Fruits edible. Not cultivated. Flowers when boiled are said to be used as a source of perfume by the Golo women, (Afr.).

Annona squamosa L., Sugar Apple, Custard Apple, Sweet Sop, Anon. (Annonaceae). - Small tree. Trop. America. Introduced in the tropics of the Old and New World. Popular fruit, excellent dessert qualities.

Annona squamosa L. x **A. Cherimola Mill.** Sugar Apple x Cherimoya. Atemoya. (Annonaceae). - A hybrid. Fruit of excellent quality. Has good shipping qualities. Resembles Cherimoya more than it does the Sugar Apple. Cultivated in tropics.

Annona testudina Safford. Anona del Monte. (Annonaceae). - Small tree. Honduras, Guatemala. Fruit resembles the Posh-Té, has a juicy pulp, much like that of the Cherimoya. Not cultivated, deserves more attention.

Annonaceae → Annona, Asimina, Canangium, Cyathocalyx, Cymbopetalum, Duguetia, Enantia, Guatteria, Hexalobus, Melodorum, Monodora, Oxandra, Polyalthia, Rollinia, Rollinopsis, Saccopetalum, Stelechocarpus, Unona, Uvaria, Xylopia.

Annonidium Mannii (Oliv.) Engl. and Diels. → Annona Mannii Oliv.

Annunciation Lily → Lilium candidum L.

Anodendron Candolleanum Wight. (Apocynaceae). - Woody vine. W. Malayan Penin. Borneo. Was once source of an important fibre.

Anodendron tenuiflorum Miq. (Apocynaceae). - Woody vine. Sumatra. Bark is source of a strong fibre; used for making fishing nets.

Anoectochilus Reinwardtii Blume (Orchidaceae). - Perennial herb. Mountains of Malaysia. Used as potherb. Sold in markets where abundant. Also A. geniculatus Ridley, A. albo-lineatus Par. and Reichenb.

Anogeissus latifolia Wall., Ghatti Tree. (Combretaceae). - Tree. E. India, Ceylon. Source of Ghatti Gum or India Gum, an exudation of the wood, composed of yellowish white tears. Soluble in water, giving a very sticky mucilage. Used in sweet meats and employed by dyers in cloth printing. Leaves produce a black dye. Wood is tough and durable.

Anogeissus leiocarpus Guill and Perr. (Combretaceae). - Tree. Trop. Africa, esp. West Africa to Abyssinia, Central Africa. Source of an insoluble gum, eaten in Kordofan. Ashes of wood used by natives as mordant for dying. Roots are made into chewsticks.

Anogeissus Schimperi Hochst. (Combretaceae). - Tree. Trop. Africa. Wood very hard, durable, termite proof, dark, dull brown, polishes well; used for house posts, house building, hoe-handles, logs for frame-work of wells. Ashes of burnt wood is used in dehairing goat-skins for the tanning bath. By the Yorubas, Afr. it is used as a mordant for Lonchocarpus indigo dye. Leaves are a source of a yellow dye. Plant is vermifuge for tape-worm of horses and donkeys. Roots are used as chewsticks, sold in local markets.

Anon → Annona squamosa L.

Anona del Monte → Annona testudina Safford.

Ant Tree, Surinam → Triplaris surinamensis Cham.

Anthemis nobilis L., Roman Chamomille, English Chamomille. (Compositae). - Perennial herb. S. and W. Europe, Mediterranean region. Cultivated. Source of Oil of Roman Chamomille, an ess. oil, light blue when fresh, of pleasant odor; used for flavoring fine liqueurs, especially of the French type; also used in perfumery. Dried, expanded flower-heads, Flores Chamomillae Romanae, are used medicinally; carminative, stimulant, tonic, nervine. Contains an ess. oil and a bitter glucoside.

Anthemis tinctoria L. (syn. Chamaemelum tinctorium All.). - Golden Chamomille, Dyer's Chamomille. (Compositae). - Perennial herb. Europe, W. Asia, introd. in N. America. Cultivated. Flowers are source of a yellow dye.

Anthemis Wiedemanniana Fisch. and Mey. (Compositae). - Herbaceous herb. Asia Minor, Iran, Afghanistan. Flowerheads are used in Iran as febrifuge and carminative.

Anthistiria ciliata L. Common Kangeroo Grass. (Graminaceae). - Perennial herb. E. India, Africa. Used as fodder grass in Australia.

Anthistiria imberbis Wood → Themeda triandra Forsk.

Anthistiria membranacea Lindl. Barcoo Grass. (Graminaceae). - Perennial herb. Australia. One of the best pasture grasses in Queensland.

Anthocephalus macrophyllus Havil. (Rubiaceae). - Tall tree. Amboina. Wood not very durable, yellowish; used in parts of Indonesia for construction of houses and boats.

Anthocleista nobilis G. Don. (Sapotaceae). - Tree. Trop. Africa. Ash derived from burned leaves is used for manuf. soap by the natives.

Antholyza paniculata Klatt. Aethiopian Madflower. (Iridaceae). - Perennial herb. S. Africa. Rootstock is used by the Zulus for dysentery and diarrhaea.

Anthoxanthum odoratum L., Spring Grass, Sweet Vernalgrass. (Graminaceae). - Perennial grass. Europe, Asia, Western N. Africa, Caucasia, N. America, Australia. Grown as forage grass, usually in mixture with other species. Gives a fragrance to the hay. Has little food value.

Anthriscus Cerefolium Hoffm. (syn. Chaerefolium Cerefolium (L.) Schinz., Scandix Cerefolium L.). Garden Chervil. (Umbelliferaceae). - Annual. Europe, Asia, Minor, Caucasia, Iran, E. Asia. Introd. in America, Australia and New Zealand. Cultivated. Aromatic leaves are used for flavoring. There is a curled-leaved variety.

Anthyllis Vulneraria L. Kidney Vetch. (Leguminosaceae). - Annual, biennial or perennial. Europe, Caucasia, Medit. Region. Cultivated; used as food for sheep and goats.

Antiaris toxicaria Lesch., Upas Tree. (Moraceae). - Tall tree. Trop. Asia, esp. S. India, Ceylon, Burma, S. China to E. Malaysia, Mal. Archip. Juice very toxic; principal poison used by different tribes for arrows and darts. Poison composed of two glucosides: antiarin-x and z. Act on the heart more powerfully than digitalin.

Antibacterials → Pencillium notatum Westling Streptomyces antibiotica (Waksman and Woodruff) Berg. et al and S. aureofaciens.

Antibiotics → Penicillium notatum Westling, Streptomyces antibiotica (Waksman and Woodruff) Berg et al and S. aureofaciens.

Antidesma Bunius (L.) Spreng., Bigney, China Laurel, Chinese Laurel. (Euphorbiaceae). - Tall tree. S.E. Asia and Australia. Berries red to black, small, in long racemes or clusters; used in syrups, jellies and in brandy; seldom consumed fresh. Also used in sauce with fish. Sold in local markets of Malaysia.

Antidesma platyphyllum H. Mann Bigney. (Euphorbiaceae). - Tree. Hawaii. Fruits pink; used in syrup and jellies, also made into wine. Wood close-grained, hard, reddish-brown; used by Hawaiians for Olona anvile.

Antidesma venosum E. Mey. (Euphorbiaceae). - Shrub. S. Africa. Fruits are eaten by the natives.

Antigonum leptopus Hook. and Arn., Mountainrose, Coralvine. (Polygonaceae). - Woody vine. Mexico, Centr. America, S. America. Tubers edible, nut-like flavor; eaten by natives.

Antirrhoea coriacea Urb. (Rubiaceae). - Tree. West Indies. Wood very strong, durable; used in carpentry work, furniture, house construction.

Antrocaryum Nannani De Wild. (Anacardiaceae). - Tree. Trop. Africa, esp. Belgian Congo. Bark contains tannin, has been recommended for tanning. Seeds are source of a non-drying, oil, called Congo Oil, said to be similar to Olive Oil, becoming slowly rancid. Pressed oil cakes are recommended as food for livestock.

Apama tomentosa Kuntze → Bragantia corymbosa Griff.

Apera crinita Palis. → Dichelachne crinita Hook. f.

Aphenanthe aspera Engl. (Urticaceae). - Tree. Japan. Hard and rough leaves are used as sandpaper to polish articles of wood. Leaves are gathered in autumn. Those from Terakuma, province Tamba, Japan, are most esteemed.

Aphloia mauritiana Baker. (syn. A. teaeformis Benn.) (Flacourtiaceae). - Tree. Ins. Mascar. Leaves used as tea in the Mascarine Islands.

Apio Arracaria → Arracaria xanthorrhiza Bancr.

Apios tuberosa Moench. (syn. Glycine Apios L.) Potato Bean, Ground Nut. (Leguminosaceae). - Vine. Eastern N. America to Florida and Texas. Sweet tubers are edible, they are boiled or roasted, being one of the best wild foods. Occasionally cultivated in different countries. Important food of the Indians.

Apitong Gurjung Oil Tree → Dipterocarpus grandiflorus Blanco.

Apitong Resin → Dipterocarpus grandiflorus Blanco.

Apium Bulbocastanum Caruel → Bunium Bulbocastanum L.

Apium Carvi Crantz → Carum Carvi L.

Apium graveolens L., Celery. (Umbelliferaceae). - Biennial herb. Europe, Asia to E. India,
N. America, S. America, New Zealand. Widely
cultivated as a vegetable in the Old and New
World. Var. silvestre Presl. Its leaves are used
for flavoring soups and vegetables. Var. rapa-
ceum (Miller) DC. Celeriac, Turnip-rooted Celery. Has a tuberous base, cooked as vegetable
or used in salads. Cultivated are: Early Prague
and Early Paris. Var. dulce (Miller) Pers.
Blanching Celery. Eaten when bleached, as a
cooked vegetable, in salads etc. Cultivated are:
Golden Plume, Early Blanching, Summer Pas-
cal. Seeds are used medicinally as carminative, stomachic, stimulant and emmenagogue.
Anodyne poultice, nerve tonic, has tonic effect
upon central nervous system. Oil of Celery,
an ess. oil from the seed, contains d-limonene, selinene, susquiterpene; used for manuf.
celery-salt, culinary sauces. Oleoresin of Celery
has ever increasing popularity among food manufacturers.

Apium Petroselinum L. → Petroselinum hortense Hoffm.

Aplopappus Nuttallii Torr. and Gray. (syn. Eriocarpum grindelioides Nutt.). (Compositae). -
Herbaceous plant. Western N. America. Tea derived from the roots is used as beverage by the
Hopi Indians; is also used against coughs.

Aplopappus spinulosus DC. (syn. Sideranthus
spinulosus [Pursh.]. Sweet.). (Compositae). - Perennial herb. Montana and Minnesota to Arizona and Texas. Leaves and roots used by the
Navajo Indians to cure tooth ache.

Apocynaceae → Acokanthera, Adenium, Allamanda, Alstonia, Alyxia, Anodendron, Apocynum, Aspidosperma, Carissa, Carpodinus, Cerbera, Chonemorpha, Clitandra, Conopharyngia,
Couma, Diplorhynchus, Dyera, Erystamia, Fosteronia, Funtumia, Geissospermum, Gonioma,
Hancornia, Haplophyton, Holarrhena, Landolphia, Mascarenhasia, Melodinus, Micrechites,
Ochrosia, Oncinotis, Parabarium, Parameria,
Picralima, Plectaneia, Plumeria, Polyadoa, Rauwolfia, Stemmadenia, Strophanthus, Tabernaemontana, Tabernanthe, Thevetia, Trachelospermum, Urceola, Urechites, Voacanga, Willughbeia, Xylinabaria, Zschokkea.

Apocynum androsaemifolium L., Spreading
Dogbane, Wild Ipecac. (Apocynaceae). - Perennial herb. N. America. Root is used medicinally,
being diaphoretic, cathartic, diuretic; contains
apocynein, apocynin and an ess. oil. Plant has
been suggested as a source of rubber during
times of scarcity.

Apocynum cannabinum L., Hemp Dogbane, Indian Hemp. (Apocynaceae). - Perennial herb.
N. America. Bark is used by the Indians as a
source of fibre made into cordage, sails, ropes,
fishing nets. Latex was used by the Kiowas and
other Indians as a source of chewing gum. Plant
has been suggested as a source of rubber dur-

ing times of scarcity. Dried rhizome and roots
are used medicinally, being cardiac stimulant,
expectorant, diaphoretic, emetic; contains apo-
cynamarin. Was used by the early settlers for
dropsy and as a diuretic.

Apocynum venetum L. (syn. A. sibiricum Pall.).
(Apocynaceae). - Perennial herb. S. Europe to
Siberia, Turkestan to China. Fibres called in
Turkestan Kendir or Turka; used for manuf.
cloth, string, sails and fishing nets.

Apodanthera smilacifolia Cogn. (Cucurbitaceae). - Vine. Brazil. Leaves and roots are considered antisiphylitic.

Apodytes dimitiata E. Mey. White Pear. (Icacinaceae). - Tree. S. Africa to Angola and
Abyssinia. Wood whitish, solid, close grained,
hard, strong, elastic; used by wagon makers.

Aponogeton distachyus Thunb., Cape Asparagus. (Aponogetonaceae). - Herbaceous water
plant. S. Africa. Flowering spikes are eaten as
pickles, also consumed as spinach in some parts
of S. Africa.

Aponogetonaceae → Aponogeton.

Aporosa frutescens Blume. (Euphorbiaceae). -
Shrub or small tree. Java. Wood used in Java
for house building and implements. Bark called
Sasah, employed as mordant of a dye derived
from Morinda, used in Batik work.

Apple → Pyrus Malus L.

Apple, African Mammee → Ochrocarpus africana (Don.) Oliv.

Apple, Akee → Blighia sapida Koenig.

Apple, Balsam → Momordica Balsamina L.

Apple Bell → Passiflora laurifolia L.

Apple, Bush → Heinsia pulchella Schum.

Apple, Chinese → Pyrus prunifolia Willd.

Apple, Chinese Crab → Pyrus baccata L.

Apple, Common → Pyrus Malus L.

Apple Common Custard →Annona reticulata L.

Apple, Crab → Pyrus angustifolius Ait. and
P. coronaria L.

Apple, Crab China → Pyrus hupehensis Pamp.

Apple, Custard → Annona squarosa L.

Apple, Dead Sea → Quercus tauricola Kotschy.

Apple, Elephant → Feronia limonia Swingle.

Apple, Emu → Owenia acidula F. v. Muell.

Apple, Kangeroo → Solanum aviculare Forst.

Apple, May → Azalea nudiflorum L. and Podophyllum peltatum L.

Apple, Monkey → Licania platypus (Hemsl.)
Pitt.

Apple, Mountain → Eugenia malaccensis L.

Apple, Oregon Crab → see Pyrus rivularis
Dougl.

Apple, Otaheite → Spondias dulcis Forst. f.

Apple, Pond → Annona glabra L.

Apple, Prairie → Psoralia esculenta Pursh.

Apple, Sugar → Annona squamosa L.

Apple Tree, Common → Pyrus Malus L.

Apple Wood → Feronoa lemonia Swingle and Pyris Malus L.

Apricot Briançon → Armeniaca brigantina Pers.

Apricot, Japanese → Armeniaca Mume Sieb.

Apricot Kernel Oil → Armeniaca vulgaris Lam.

Apricot Plum → Prunus Simonii Carr.

Apricot Tree → Armeniaca vulgaris Lam.

Apricot Vine → Passiflora incarnata L.

Aptandra Spruceana Miers. (Olacaceae). - Tree. Brazil. Bark of root is source of a greenish Sandal Oil, called Sando de Maranhão.

Apuleia leiocarpa (Vog.) Macbride. (syn. A. praecox Mart.) (Leguminosaceae). - Tree. S. America, esp. Argentina, Brazil, E. Peru, Venezuela. Wood golden-yellow to yellowish-brown, coppery hue on exposure, tough, hard, heavy, strong, durable; used for flooring, wheelwright work, door frames, shafts of vehicles, heavy construction work.

Aqua Aurantii Florum → Citrus Aurantium L.

Aquae Rosae Fortior. → Rosa centifolia L.

Aquamiel → Agave atrovirens Karw., A. compluviata Trel. and A. meliflua Trel.

Aquiboquil → Lardizabala biternata Ruiz and Pav.

Aquifoliaceae → Ilex.

Aquilaria Agallocha Roxb. (Thymelaeaceae). - Tree. India, China, Malaysia. Source of a fragrant wood, formed under pathological conditions, called Aloës Wood. Used as incense before Mahommedan prayers. Put into funeral pyre by the Siamese. Used by the Chinese as joss sticks. Wood ist distilled in Assam, giving an ess. oil called Chuwa and Aggar Atta.; used in perfumery.

Arabian Coffee → Coffea arabica L.

Arabian Dragon's Blood → Dracaena Cinnabari Balf.

Arabic Gum → Acacia Senegal (L.) Willd.

Arabic Gum (substitute) → Albizzia fastigiata Oliv.

Arabian Horse-Radish Tree → Moringa aptera (Forsk.) Gaertn.

Arabian Jasmine → Jasminum Sambac Ait.

Arabian Senna → Cassia angustifolia Vahl.

Arabian Tea → Catha edulis Forsk.

Arabian Wolfberry → Lycium arabicum Schweinf.

Araça → Psidium Araça Raddi.

Araceae → Acorus, Alocasia, Amorphophallus, Anchomanes, Arisaema, Arum, Caladium, Colocasia, Cyrosperma, Epipremnum, Homalomena, Lasia, Monstera, Orontium, Peltandra, Philo-

dendron, Pothoidium, Schizoglossum, Spathiphyllum, Spathyema, Typhonium, Typhonodorum, Xanthosoma.

Arachis hypogaea L., Pea Nut, Ground Nut, Goober. (Leguminosaceae). - Annual herb. Probably from S. America. Widely cultivated in warm countries. Fruits ripen below surface of ground. Seeds are important food when roasted, salted, in confectionary; when ground, made into peanut butter. Source of Peanut Oil, used in cooking, in salads, shortening in pastry and bread, oleomargarine, nutmargarine, canning sardines, deep-fat frying of potato chips and doughnuts; also made into soap. Peanut oil is able to take much heat. Young pods are consumed as vegetable. Peanut crop is used as food for livestock. Important for fattening hogs. Source of hay. Peanut meal is used as stock feed and as fertilizer. Some new commercial products are: peanut milk, peanut ice cream, peanut massage oil for infantile paralysis treatment, peanut paints, peanut dyes. Peanut hulls are used as fuel, filler for fertilizer or ground into meal for insulation blocks, bedding of stables, floor sweeping compounds. Large podded varieties are: Virginia Runner, African, North Carolina, Virginia Bunch. Small podded varieties are: Spanish, Small Spanish, Improved Spanish, Georgia Red, Valencia, Tennessee. Large podded varieties are mainly used for roasting and confectionary.

Arachis prostrata Benth. (Leguminosaceae). - Annual herb. S. America. This species has been recommended as green manure in some parts of Brazil.

Aralia chinensis L. Chinese Aralia. (Araliaceae). - Shrub. Japan, China. Very young leaves are consumed as a vegetable in Japan.

Aralia cordata Thunb. Udo. (Araliaceae). - Perennial herb. Japan. Cultivated in Japan. Young and soft stalks are eaten in Japan as a vegetable. Cultivated varieties are: Malt-udo, Kanūdo and Nenjui-udo.

Aralia racemosa L. American Spikenard. (Araliaceae). - Perennial herb. N. America. Rhizome and roots are used medicinally, being alterative, purgative, diaphoretic. Contain a bitter compound and an ess. oil. Is collected in the autumn.

Aralia spinosa L., Prickly Ash, Prickley Elder, Devil's Walking Stick, Hercules Club. (Araliaceae). - Shrub or small tree. Eastern United States to Florida and Texas; Manchuria, Japan. Bark is used as a stimulant, diaphoretic; contains a yellow ess. oil. Also used as a substitute for Zanthoxylum americanum Mill. Boiled, mashed roots were used by the Choctow Indians for boils. Roots kept in cold water were employed by the Koasati Indians for sore eyes.

Araliaceae → Acanthopanax, Aralia, Boerlagiodendron, Cuphocarpus, Dendropanax, Didymopanax, Fatsia, Hedera, Nothopanax, Osmoxylon, Panax, Reynoldsia, Schefflera.

Aramina Fibre → Urena lobata L.

Arangan Oil → Ganophyllum falcatum Blume.

Arar Tree → Tetraclinis articulata (Vahl.) Mast.

Araracanga → Aspidosperma desmanthum Muell. Arg.

Arariba → Centrolobium robustum Mart.

Arariba viridiflora Allem. → Sickingia viridiflora Schum.

Araroba Powder → Vataireopsis speciosa Ducke.

Araucaria Bidwillii Hook. Bunya-Bunya. (Araucariaceae). - Tall tree. E. Australia. Seeds edible, much esteemed by the natives.

Araucaria brasiliensis A. Rich., Parana Pine, Brazilian Pine. (Araucariaceae). - Tree. Brazil, Paraguay, N. Argentina. Important timber tree. Wood brown, often with bright red streaks, straight grained, easy to work, not resistant to decay; used for general building purposes; in United States it is used as slats for Venetian blinds, also as backing for electrotypes. Seeds are edible, sold in markets of Brazil.

Araucaria Cookii R. Br. (Araucariaceae). - Tree. New Caledonia. Wood ligth colored, fine-grained, keeps well under water, attacked by borers; used for general carpentry in New Caledonia.

Araucaria Cunninghamii Sweet., Moreton Bay Pine, Hoop Pine. (Araucariaceae). - Tree. New South Wales, Queensland. Wood strong, durable. Used for general work.

Araucaria imbricata Pav. (syn. A. araucana [Molina] Koch.) (Araucariaceae). - Tree. Coastal Cordillera, Arauco (Chile). Wood pale yellowish, of good quality; used for furniture, general construction work, doors, flooring, boxes, ceiling, window sash. Has been recommended for paper pulp. Seeds, piñones, are eaten roasted. An article of commers in Chile. Resin is used medicinally by the Araucanian Indians.

Araucariaceae → Araucaria.

Arauco → Araucaria imbricata Pav.

Arauijia sericifera Brot. (Asclepiadaceae). - Herbaceous plant. Argentina, Peru and adj. territory. Plant is source of a strong fibre which has been recommended for textile work.

Arbol del Hule → Castilla elastica Cerv.

Arbor Vitae → Thuya occidentalis L.

Arbor Vitae, Giant → Thuja plicata D. Donn.

Arbutus Andrachne Lam. (Ericaceae). - Shrub or small tree. Mediterranean region. Asia Minor. Fruits are eaten in some parts of Asia Minor.

Arbutus canariensis Duham. Canary Madrone. (Ericaceae). - Shrub. Canary Islands. Berries are made into a sweet-meat in some parts of the Canary Islands.

Arbutus Menziesii Pursh., Madrona. (Ericaceae). - Tree. Pacific Coast area of N. America and adj. Mexico. Wood close grained, hard, heavy, strong, light brown; used for furniture and charcoal. Bark is sometimes employed for tanning.

Arbutus Unedo L., Strawberry Tree. (Ericaceae). - Shrub or small tree. Mediterranean region. Berries edible, mealy, sweet; made into preserves, alcoholic beverage, wine, brandy and liqueur. Occasionally sold in markets of Europe. Bark contains andromedotoxin; used for diarrhea. Leaves, bark and fruits are used for tanning in some parts of the Mediterranean region. Bark contains 45 % tannin.

Arbutus xalapensis H.B.K. (Ericaceae). - Shrub. Mexico. Leaves are in Mexico source of food of a silkworm, Eucheiria socialis Westw. Also A. macrophylla Mart. and Gal.

Arcel → Parmelia caperata (L.) Ach.

Archangelica Gmelini DC. (syn. Coelopleurum Gmelini [DC] Ledeb.). (Umbelliferaceae). - Perennial herb. N. America, N. E. Asia. Herb is used by the Eskimos of Alaska much like the white man uses celery. Plant is considered an aromatic tonic.

Archil → Roccella phycopsis Ach.

Arctic Bramble → Rubus arcticus L.

Arctium Lappa L. (syn. Lappa major Gaertn.). Great Burdock. (Compositae). - Perennial herb. Temp. Europe, Asia, introd. in N. America. Roots eaten in Japan as a vegetable. Dried first year's root used medicinally, being diuretic, alterative, diaphoretic. Usually collected in autumn. Contains inuline, a bitter compound and an ess. oil.

Arctium minus (St. Hil.) Bern. (syn. Lappa minor Hill.). Common Burdock. (Compositae). - Perennial herb. Europe, Asia; introd. in N. America. Used medicinally as A. Lappa L.

Arctopus echinatus L. (Umbelliferaceae). - Perennial herb. S. Africa. Decoction of roots is used by the farmers and Kaffirs of S. Africa for „cleaning" the blood, as diuretic and for skin diseases.

Arctostaphylos arguta Zucc. (syn. Comarstaphylos arguta Zucc., Arbutus discolor Hook.) (Ericaceae). - Shrub. S. Mexico. Fruits and leaves possess narcotic properties. Extract from fruits is used in parts of Mexico as a hypnotic.

Arctostaphylos Manzanita Parry., Manzanita. (Ericaceae). - Oregon to California. Berries are used for jelly. Indians made a cider from the fruits.

Arctostaphylos nevadensis Gray. (Ericaceae). - Small shrub. California, Oregon, Nevada. Dried leaves were mixed with tobacco; used for smoking by the Klamath Indians in Oregon.

Arctostaphylos pungens H.B.K., Pointleaf Manzanita. (Ericaceae). - Shrub. Mexico, California. Fruits are used as food by the Zapotecs and other tribes in Mexico.

Arctostaphylos Uva-ursi (L.) Spreng. (syn. Arbutus Uva-ursi L.), Bearberry. (Ericaceae). - Small shrub. N. Hemisphere to Arctics. Fruits edible. Leaves are used for tanning in some parts of Sweden and Russia for preparation of „Russian Leather". Dried leaves, collected in autumn are used medicinally as astringent, tonic, diuretic; contain volatile oil, arbutin and quercetin. They are also used as tea in some parts of Russia, known as Kutai and Caucasian Tea.

Ardisia fuliginosa Blume. (Myrsinaceae). - Shrubby tree. Java. Sap from the stem, boiled with coconut oil, is used by natives for scurfy.

Arduina bispinosa L. → Carissa Arduina Lam.

Areca Caliso Bec. (Palmaceae). - Tall palm. Philippine Islands. Fruits are used by the Manobos (Philipp. Islds.) as a substitute for Arec Nuts. Also A. ipot Bec.

Areca Catechu L. Betelnut Palm. (Palmaceae). - Tall palm. Trop. Asia, esp. Malaysia, New Guinea. Source of Betel-Nuts; used as masticatory in betelchewing. Very young unfolded leaves are eaten as vegetable. Seeds contain alkaloids: arecoline and arecaidine. Seeds used medicinally, being diaphoretic, miotic, teniacide, astringent. Used in vet. medic. as laxative, teniafuge. Ground seeds are said to be used in Europe for certain tooth powders.

Areca sapida Soland. → Rhopalostylis sapida Wendl. and Drude.

Arenaria peploides L. (syn. Ammodenia peploides [L.)] Hook.) Seabeach Sandwort. (Caryophyllaceae). - Herbaceous plant. N. temp. and arct. regions. Leaves are consumed fresh, soured or in oil by the Eskimos of Alaska.

Arenga pinnata (Wurmb.). Merr. (syn. A. saccharifera Lab.). Sugar Palm. (Palmaceae). - Tall palm., Malaysia. Stem is source of a palm-sugar, widely used by the natives. Sap is made into Palm Wine or Toddy, upon distillation it yields an Arrack. Pith from stem is occasionally source of a sago. Young leaf-sheats produce a fibre. Wood is used for water-pipes.

Argan Gum → Argania Sideroxylon Roem.

Argania Sideroxylon Roem. (Sapotaceae). - Tree. Sahara region, Morocco. Source of Argan, a gum. Seeds are the source of a good oil. Seedcoats are given as food to cattle. Wood is hard and much esteemed by the natives.

Argemone intermedia Sweet. (Papaveraceae). - Annual herb. N. America. Ashes from the leaves are an ingredient for tattooing among the Kiowa Indians (New Mex.).

Argemone mexicana L., Mexican Prickly Poppy. (Papaveraceae). - Annual herb. S. W. United States, Mexico. Seeds source of Mexican or Prickly Poppy Seed Oil; used in Mexico as illuminant; in medicine as purgative; manuf. soap. Sp. Gr. 0.9220; Iod. No. 1237; Sap. Val. 192.7; Unsap. 1. 4%.

Argentine Jujub → Zizyphus Mistol Griseb.

Argyreia speciosa Sweet., Woolly Asiaglory. (Convolculaceae). - Vine. Malaysia. Leaves are used by the natives for poultices.

Arillastrum gummiferum Panch. → Spermolepis gummifera Brongn. and Cris.

Arinj → Acacia leucophloea Willd.

Arisaema japonicum Blume. (Araceae). - Herbaceous perennial. Japan. Corms used as food by the Ainu.; plant parts are put in hot ashes to bake.

Arisaema triphyllum (L.) Torr., Jack-in-the-Pulpit, Indian Turnip. (Araceae). - Perennial herb. Eastern N. America to Florida and Texas. Dried corms are used medicinally as stimulant, expectorant, diaphoretic and irritant. Tubers were used as food by the Indians.

Aristea alata Baker. (Iridaceae). - Herbaceous perennial. Trop. Africa. Sap of the plant is used by the Washambaa with juice of Pteridium aquilinum as a dye for tattooing. Also A. polycephala Harms.

Aristida pungens Desf. (syn. Arthratherum pungens Beauv.). (Graminaceae). - Herbaceous grass. Sahara region. Seeds are consumed as food by different N. African tribes during times of distress.

Aristolochia albida Duch. (Aristolochiaceae). - Vine. Trop. W. Africa. Bitter root is used as tonic and for Guinea worm in parts of Africa.

Aristolochia antihysterica Mart. (Aristolochiaceae). - Woody vine. Trop. America. Roots are used in some parts of Brazil as excitant, sudorific and emetic, also as emmenagogue.

Aristolochia barbata Jacq. (syn. Howardia barbata Kl.). (Aristolochiaceae). - Woody vine. S. America. Rhizome is used by the natives for snake-bites in some parts of Brazil.

Aristolochia bracteata Retz. (Aristolochiaceae). - Trailing shrub. India to Trop. Africa. Very bitter, has anthelmintic and purgative properties: Leaves and roots are used in Sudan for Guineaworm. Roots are used in Sudan for scorpion bites.

Aristolochia brasiliensis Mart. and Zucc. (Aristolochiaceae). - Brasil. The rhizome, Raiz de José Domingo, Papo de Peru is used against snakebites in Brazil, also A. galeta, A. cymbifera, A. rumicifolia, Also are used for the same purpose: A. anguicida from Central America; A. odoratissima from the West Indies and Central America; A. cordiflora, A. turbacensis from Colombia, A. fragrantissima from Peru and A. maxima from Guiana and Venezuela.

Aristolochia chilensis Miers. (Aristolochiaceae). - Woody vine. Chile. Roots are used in some parts of Chile as emmenagogue.

Aristolachia cordigera Willd. (syn. Howardia cordigera K.). (Aristolochiaceae). - Vine. Brazil. Roots are used in some parts of Brazil as emmenagogue, excitant and for intermittent fevers.

Aristolochia cymbifera Mart. and Zucc. (Aristolochiaceae). - Vine. Brazil, Uruguay, Paraguay and Argentina. Roots are used in some parts of Paraguay and Argentina as emmenagogue; in Brazil for diarrhea, asthma, as diaphoretic and diuretic.

Aristolochia glaucescens H. B. K. (syn. Abuta amara Aubl.). (Aristolochiaceae). - Vine. Colombia and adj. territory. The bitter roots are source of Yellow Pareira, containing alkaloids that resemble berberine and bebeerine; used in medicine as a bitter tonic and diuretic.

Aristolochia grandiflora Swartz. (Aristolochiaceae). - Woody vine. Mexico, esp. Chiapas and Yucatan; West Indies, Centr. America. Roots used for snake bites. Sudorific, emmenagogue, abortive properties. Leaves are used for colds, chills; also as sudorific among the Chinantecs (Mex.) and other tribes.

Aristolochia Kaempferi Willd. (Aristolochiaceae). - Eastern Asia. Is probably source of the Chinese drug ma-tou-ling, used against asthma, coughing, lung troubles and haemorrhoids. Probably also A. contorta Bunge.

Aristolochia longa L. (Aristolochiaceae). - Vine. Mediterranean region, Iran and Iraq. Roots used by the natives as a tonic, stomachic; also employed for snake bites.

Aristolochia maurorum L. (Aristolochiaceae). - Vine. Iraq. Iran. Root is used by some native tribes for healing wounds; also for scab of sheep.

Aristolochia pallida L. (Aristolochiaceae). - Asia Minor. Rhizome has been used against snakebites for centuries. Theophrast has already recommended the species for this purpose.

Aristolochia reticulata Nutt., Texas Snakeroot. (Aristolochiaceae). - Herbaceous perennial. Southern United States. Dried rhizome and roots used medicinally. See A. serpentaria L.

Aristolochia rotunda L. (Aristolochiaceae). - Vine. Mediterranean region, Iran. Tuberous roots are used in Iran as a tonic, vermifuge, diuretic and emmenagogue.

Aristolochia sericea Blanco. (Aristolochiaceae). - Woody vine. Philipp. Islds. Fresh parts of plants are used as emmenagogue, carminative and febrifuge. Root is said to be a violent abortive.

Aristolochia Serpentaria L. Virginia Snakeroot. (Aristolochiaceae). - Herbaceous perennial. Eastern N. America to Florida and Texas. Dried rhizome and roots are used medicinally; they are collected in autumn, being aromatic, tonic

and bitter stimulant, promote perspiration. Contain borneol, serpentarin, aristolochine. Boiled roots were used by the Chactaw Indians to alleviate pain of the stomach. Used against snakebites.

Aristolochia taliscana Hook. and Arn. (Aristolochiaceae). - Woody vine. Mexico, esp. Sinaloa and Jalisco. Used in Sinaloa for snakebites.

Aristolochia Tagala Cham. (Aristolochiaceae). - Woody vine. Philipp. Islds., esp. Luzon to Mindanao. Roots used as tonic, carminative, emmenagogue; effective for infantile tympanites when applied to the abdomen.

Aristolochia theriaca Mart. (Aristolochiaceae). - Woody vine. Brazil. Roots are used among the rural population of some parts of Brazil for snakebites.

Aristolochia Warmingii Mart. (Aristolochiaceae). - Woody vine. Brazil. Roots are considered antiseptic, digestive, febrifuge, tonic, diuretic, emmenagogue, abortive. Also A. Burcheli Mart., A. crenata Mart., A. elegans Mart., A. amazonica Ule., A. lutescens Duck.

Aristolochiaceae → Aristolochia, Asarum, Bragantia.

Aristotelia Macqui L'Hér., Maqui., Chilean Wineberry. (Elaeocarpaceae). - Shrub. Chile. Fruits small, edible. Much esteemed in some parts of Chile.

Aristotelia racemosa Hook. f. (Elaeacarpaceae). - Shrub or tree. Endemic, New Zealand. Wood white, straight-grained, often streaked. Used for cabinet-work, general turnery, for inlaying. Charcoal is used for certain gun-powders. Fruits are consumed by the natives.

Arizona Poplar → Populus arizonica Sarg.

Arjan Terminalia → Terminalia Arjuna Wight and Arn.

Arjona tuberosa Cav. (Santalaceae). - Herbeceous perennial. Patagonia. Tubers called Macachi, are sweet and are consumed by the natives.

Armand Pine → Pinus Armandi Franch.

Armeniaca brigantina Pers. (syn. Prunus brigantina Vill.). Briançon Apricot. (Rosaceae). - Tree. S. France. Cultivated. Seeds are source of a perfumed oil, called Huile de Marmotte.

Armeniaca Mume Sieb. (syn. Prunus Mume Sieb. and Zucc.), Japanese Apricot. (Rosaceae). - Small tree. Japan. Cultivated in Japan and China. Fruits eaten raw, preserved in sugar, salted and boiled. Many varieties, among which are Yatsubusa-no-mume, Bungo-mume, Komume and Tokomume.

Armeniaca vulgaris Lam. (syn. Prunus Armeniaca L.). Apricot Tree. (Rosaceae). - Small tree. W. Asia. Cultivated for centuries. Grown in temp. zones of the Old and New World. Important commercial fruit. Fruits eaten raw, dried, canned, candied, made into paste; used in pastries, pies, cakes. Over ripe fruits are made into canned pulp, used for ice cream. Source of apricot juice, brandy, liqueur, cordials. Varieties are: Blenheim, Royal, Moorpark, Tilton. Seeds are source of Apricot Kernel Oil, semi-drying, edible, used in toilet creams, pharmaceutical preparations. Sp. Gr. 0.9158; Sap. Val. 190,2; Iod. No. 108.7 Unsap. 0.7 %.

Armillaria bulbigera (A. and S.) Quél. (Agaricaceae). - Basidiomycete. Fungus. Fruitbodies are consumed as food. Sold in markets of Sweden.

Armillaria matsutake S. Ito and Imai. (Agaricaceae). - Basidiomycete. Fungus. Japan. Matsu Take. Fruitbodies are consumed as food in Japan. Sold in markets.

Armillaria mellea (Vahl.) Quél. (Agaricaceae). - Basidiomycete. Fungus. Cosmopolite. Fruitbodies wich have, when raw, an unpleasent taste, are eaten after being cooked, pickled or salted. Are consumed in some parts of Austria, Czechoslovakia and of Eastern Europe.

Armillaria ventricosa Peck. (Agaricaceae). - Basidiomycete. Fungus. N. America and Japan. Fruitbodies are consumed as food. Sold in markets.

Armoracea lapathifolia Gilib. → Cochlearia Armoracea L.

Arnebia hispidissima DC. (syn. Lithospermum hispidissimum Lehm.). (Boraginaceae). - Annual. Trop. Africa, Egypt, Arabia, N. India. Roots are source of a red dye.

Arnica → Arnica montana L.

Arnica montana L., Arnica. (Compositae). - Perennial herb. Europe, Asia, Western N. America. Dried flower heads are used medicinally, as vulnerary, irritant, tonic. Contains a bitter compound, arnicin and an ess. oil. Arnica Root is composed of the dried rhizomes and roots; used for the same purpose as the dried flower heads. Used in many home remedies. A tincture is used for sprains and bruises.

Arracacia xanthorrhiza Baner., Arracha, Apio Arracaria. (Umbelliferaceae). - Herbaceous perennial. Venezuela, Colombia, Ecuador. Cultivated in S. American countries. Important plant food among the péons. Roots resembling parsnip are eaten as vegetable; used in stews and soups, also boiled and fried, has a strong flavor.

Arracacha → Oxalis crenata Jacq.

Arracha → Arracacia xanthorrhiza Baner.

Arrack → Arenga pinnata (Wurmb.) Merr.

Arrayán → Eugenia foliosa DC.

Arrhenaterum avenaceum Beauv. (syn. A. elatior Beauv.), Tall Meadow Oat Grass. (Graminacae). - Perennial grass. Europe, Cultivated. Valuable pasture grass, produces a heavy yield of hay.

Arrowleaf → Sagittaria latifolia Willd.

Arrowleaf Balsamroot → Balsamorhiza sagittata (Pursh) Nutt.

Arrow poisons. Plants source of → Adenium, Antiaris, Buphane, Callicarpa, Chondodendron, Erythrophleum, Euphorbia, Hippomane, Indigo, Lophopetalum, Paullinia, Sapium, Sebastiana, Strophanthes, Strychnos, Urechites.

Arrowroot → Maranta arundinacea L.

Arrowroot, Bermuda → Maranta arundinacea L.

Arrowroot, Brazilian → Ipomoea Batatas Poir.

Arrowroot, East Indian → Curcuma angustifolia Roxb., C. leucorhiza Roxb.

Arrowroot, False → Curcuma Pierreana Gagn.

Arrowroot, Florida → Zamia integrifolia Ait.

Arrowroot, Guiana → Dioscorea Batatas Decne.

Arrowroot, Hawaiian → Tacca hawaiiensis Limpr. f.

Arrowroot, Pará → Manihot esculenta Crantz.

Arrowroot, Portland → Colocasia antiquorum Schott.

Arrowroot, Queensland → Canna edulis Ker-Gawl.

Arrowroot, St. Vincent → Maranta arundinacea L.

Arrowroot, Tahiti → Tacca pinnatifida Forst.

Arsenic, presence of → Penicillium brevicaule Sacc.

Artemisia Abrotanum L., Southern Wood. (Compositae). - Shrub. S. Europe, temp. Asia. Cultivated. Leaves are occasionally used as a tea.

Artemisia Absinthium L., Absinthe. (Compositae). - Perennial herb. Europe, S. Siberia, Kashmir, Mediteranean region, introd. in N. America. Cultivated in S. Europe, N. Africa, United States; among which New York, Michigan, Nebraska, Wisconsin. Used for the preparation of Absinthe, a liqueur, and in Vermouth Wine, Muse Verte. Harmful effect is caused by thujon and absinthin, giving rise to absinthism. Dried leaves and flowering tips are used medicinally, being a stimulant, stomachic, tonic, anthelmintic. Contains a bitter ess. oil, absinthol.

Artemisia afra Jacq. (Compositae). - Shrub. S. Africa. Herb used in some parts of Africa in the same way as A. Absinthium is used.

Artemisia Barrelieri Ben. (Compositae). - Perennial herb. S. Europe, esp. Spain. Source of an ess. oil; used for manuf. of Algerian Absinthe, has a powerful aromatic odor.

Artemisia Cina Berg. Levant Wormseed Plant. (Compositae). - Small shrub. Orient, Russian Turkestan. Cultivated in Russia and western United States. Dried flower-heads, Flores Cinae, are used medicinally, being anthelmintic; contain santonin.

Artemisia dracunculoides Pursh., False Tarragon. (Compositae). - Perennial herb. N. America. Oily seeds were eaten by the Indians of

Utah and Nevada. Leaves were consumed when baked between hot stones. Also seeds of A. tridentata Nutt., A. tripartita Rydb., A. Wrightii Gray and other spp.

Artemisia Dracunculus L., Tarragon, Estragon. (Compositae). - Perennial herb. Russia, W. Asia to Himalaya. Cultivated. Widely used as a condiment in salads, vinegar, pickled onions and cucumbers. Used to stimulate appetite; diuretic, anthelmintic, emmenagogue; used for toothache. Contains Oleum Dracunculi composed of esdragol, l-methylchavicol.

Artemisia filifolia Torr. (Compositae). - Perennial herb. N. America. Plants were chewed by the Tewa Indians of New Mexico, as a remedy for indigestion, biliousness and flatulence.

Artemisia frigida. Willd. (Compositae). - Shrub. N. America, Siberia. Leaves are used by the Hopi Indians for flavoring sweet-corn. It is said to be source of a comphor-like product.

Artemisia glutinosa J. Gray. (Compositae). Perennial herb. S. Europe, esp. France and Italy. Has been suggested as a source of ess. oil. Has the scent of a mixture of lavender, sage and rosemary.

Artemisia Herba-alba Asso. (Compositae). - Perennial herb., N. Africa. Herb is used by the natives for perfumery; as emmengagogue and vermifuge.

Artemisia judaica L. (Compositae). - Herbaceous plant. Syria, Egypt, Arabia etc. Herb is source of an aromatic bitter, used medicinally in Egypt and Arabia.

Artemisia Maciverae Hutch. and Dalz. (Compositae). - Perennial herb. Trop. Africa, esp. Sudan. Powdered leaves and stems, sold by Arabs of Sokoto, are used as fragrant powder, called Tazargade, used as vermifuge.

Artemisia mendozana DC. (Compositae). - Perennial plant. S. America. Infusion of plant is used in some parts of Argentina as stomachic and carminative.

Artemisia mexicana Willd. Mexican Mugwort. (Compositae). - Perennial herb. Western United States and Mexico. Used by Indians of S. California for intestinal worms, as a stimulant and emmenagogue. Leaves were chewed by Kiowa Indians for sore throat. Source of an ess. oil.

Artemisia pontica L. Roman Wormwood. (Compositae). - Shrub. S.E.Europe, W.Asiatic Plains, Mediterranean region. Source of Oil of Wormwood, similar to that of A. Absinthium L.

Artemisia tridentata Nutt., Sagebrush, Black Sage. (Compositae). - Shrub or small tree. N. America, W. of the Missouri. Leaves were chewed by the Tewa Indians for dispelling flatulence.

Artemisia vulgaris L., Mugwort. (Compositae). - Perennial herb. Temp. N. Hemisphpere. Young shoots and leaves are used as a condiment for geese and pork, in some parts of Europe. In the Far East they serve for flavoring festival ricecakes. Also used as a surrogate of tobacco. Young plants are eaten in Japan during the spring after being boiled. They are also used for flavoring mochi or dango (dumplings).

Arthraterum pungens Beauv. → Aristida pungens Desf.

Arthrocnemum glaucum (Del.) Ung-Sternb. (Chenopodiaceae). - Shrub. Mediterranean region. Plant is used in some parts of Greece in a garlic porridge.

Arthrocnemum indicum (Willd.) Moq. (Chenopodiaceae). - Shrub. Trop. Asia. Herb is consumed as a salad in some parts of India.

Arthrothamnus bifidus (Gmel.) Rupr. (Laminariaceae). - Brown Alga. Coasts of Japan and China. Used for the manufacturing of Kombu, esp. in Japan. Also A. kurilensis Rupr.

Artichoke, Chinese → Stachys Sieboldi Miq.

Artichoke, Globe → Cynara Scolymus L.

Artichoke, Japanese → Stachys Sieboldi Miq.

Artichoke, Jerusalem →Helianthus tuberosus L.

Artocarpus altissima J. J. Smith. (Moraceae). - Large tree. Mal. Archip. Wood brownish, streaked, close-grained, easily worked; a valuable timber for shipbuilding.

Artocarpus anisophylla Miq. (Moraceae). - Tall tree. Mal. Archipel. Wood soft, coarse-grained, durable, easily worked, yellow-brown; used for building houses and bridges. Thick, very sticky latex is used as bird-lime.

Artocarpus brasiliensis Gomez. (Morazeae). - Tree. Brazil. Fruits occasionally eaten in Brazil. Pulp of an agreeable taste. Seeds are consumed like chestnuts.

Artocarpus communis Forst. (syn. A. incisa L. f.). Breadfruit. (Moraceae). - Tall tree. Malay. Archip. Widely cultivated in the tropics. Fruit edible, often seedless; staple food among natives in many warm countries. Consumed boiled, cooked or baked; also cut in slices and fried like potatoes; occasionally eaten raw.

Artocarpus Cumingiana Trée. (Moraceae). - Philipp. Islds., esp. Luzon to Mindanao. Latex has been recommended as a source of chewing gum base, called Anubîng Gum.

Artocarpus Dadak Miq. (Moraceae). - Tall tree. Sumatra. Wood yellowish, durable, not attacked by insects; used for bridges, supports of floors; is too hard for manuf of boards. Fruits as large as chicken eggs, acid, edible.

Artocarpus elastica Reinw. Wild Breadfruit Tree. (Moraceae). - Tree. W. Mal. Arch. Wood soft, durable, coarse-grained, light yellow to brown, quality variable; used for boards, posts of houses, native boats, general carpentry. Is attacked by termites. Bark used for ropes, fancy materials, flower-baskets. Seeds are eaten roasted. Latex has been used as a chewing gum base, called Gumihan Gum; locally used as a bird lime.

Artocarpus glauca Blume. (Moraceae). - Medium sized tree. Java. Wood is used for house construction. Fruits are as large as plums, acid; locally used for jelly.

Artocarpus integrifolia L. f. Jackfruit. (Moraceae). - Tree. India and Malaysia. Widely cultivated in tropics of Old and New World. Fruits are 30 to 50 cm long, pulp soft, whitish, eaten fresh, cooked or fried, much eaten by the poorer classes. Pulp is somtimes boiled with milk. Also used as preserves. Seeds when roasted are considered a good food, much consumed in the tropics. Varieties may be divided into two classes, e. g. those with soft and those with hard rinds. Young flower clusters are eaten in Java as roedrak with syrup and agar-agar. Very young fruits are eaten in soup.

Artocarpus Kemando Miq. (Moraceae). - Small tree. W. Mal. Archipel. Latex has taste of coconutmilk; used by natives in sauces; too much latex is harmful, causing stomach-ache.

Artocarpus Lakoocha Roxb. Lakoocha. (Moraceae). - Tree. E. India, Malaysia. Cultivated in tropics of Old and New World. Fruits are edible, being of a pleasant, sub-acid flavor. Sold in bazaars. Wood durable, hard, yellow, suitable for polishing; used for furniture and for boat building.

Artocarpus lanceaefolia Roxb. (Moraceae). - Tall tree. Malaysia. Wood fairly hard, close-grained, durable, dark yellow, striped; takes polish; one of the best lumber species in the Mal. Peninsul. Suitable for general construction, furniture, chinese coffins.

Artocarpus Limpato Miq. (syn. Prainea Limpato Beumee). (Moraceae). - Tree. Sumatra. Wood hard, very durable, orange-red, not attacked by insects; used for houses, bridges; keeps well under water.

Artocarpus odoratissima Blanco. Marang. (Moraceae). - Medium sized tree. S. Philipp. Islds., Sulu Archipel. Cultivated. Introduced in tropics of the New World. Fruits roundish to oblong, 15 cm long, resemble Jackfruit, though of better quality; aromatic, fleshy, sweet, juicy, of a pleasant flavor.

Artocarpus polyphema Pers. (syn. A. Champeden (Lour.) Spreng.) Champedak. (Moraceae). - Medium sized tree. Trop. Asia. esp. W. Malaysia, Indo-China. Cultivated. Wood hard, durable, dark yellow to brown; used for furniture and boat building. Young fruits are eaten in Java in soup.

Artocarpus rigidus Blume. (syn. A. dimorphophylla Miq.) Monkey Jack (Moraceae). - Medium sized tree. Mal. Archip. Cultivated for its fruit. Wood yellow-brown to orange-red, fairly heavy, durable, resistant to termites; used for furniture, house-building, native boats. Sticky

latex is used in Batik-work. Fruits are roundish, flesh is of a pleasant flavor.

Artocarpus Teijsmannii Miq. (Moraceae). - Tall tree. Mal. Archipel. Wood used by natives for ship-building. Latex employed in Celebes as a bird-lime.

Arum maculatum L. (Araceae). - Perennial herb. Europe. Starchy tubers are edible roasted or properly boiled and prepared, after which the poisonous properties have been removed. Used during times of want.

Arundinaria falcata Nees. Himalayan Bamboo. (Graminaceae). - Woody grass. Malaysia. Stems used for hookah tubes, fishing rods, arrows, basket work, lining of roofs of houses.

Arundinaria gigantea (Walt.) Chapm. Large Cane, Southern Cane. (Graminaceae). - Perennial herb. Eastern and southern part of N. America. Seeds used by Indians and early settlers as a substitute for wheat. Young shoots are consumed as a potherb. Plant is also food for livestock.

Arundinaria Prainii Gamble. (Graminaceae). - Woody grass. E.India, esp. Naga and Jainita Hills. Stems used for basket work and hut building.

Arundinaria racemosa Munro. (Graminaceae). - Woody grass. Himalayan region. Culms used extensively for mat making, roofs of native houses.

Arundinaria Wightiana Nees. (Graminaceae). - Woody grass. Trop. Asia, esp. S. and W. India. Used for baskets, mats, walking-sticks. Much exported.

Arundinella brasiliensis Raddi. (syn. Aira brasiliensis Spreng., Andropogon hispidus Willd.). (Graminaceae). - Perennial herb. S. America. Used as forage grass in different parts of Brazil.

Arundo Donax L., Giant Reed. (Graminaceae). - Tall perennial grass. Mediterranean region, Trop. Asia. Cultivated in the Old and New World. Stems used in Europe for clarinets and bag pipes; employed in some parts of N. America for mats, lattices, screens in adobe houses.

Asafetida → Ferula Assa-foetida L.

Asakusanori → Porphyra.

Asarum arifolium Mich. (Aristolochiaceae). - Perennial herb. E. of United States to Florida and Louisiana. Tea from leaves is used by the Catawba Indians for pain in the stomach.

Asarum canadense L., American Wild. Ginger. (Aristochiaceae). - Perennial herb. Eastern N. America. Dried rootstock is sometimes used as spice, a substitute for ginger. Dried rhizomes and roots are used medicinally, they are collected in spring being a tonic, carminative, aromatic, stimulant; containing an ess. oil, pungent resin; a fragrant principle, asarol. Decoction of the herb with that of Anemone Hepatica and Epigaea repens was used by the Cherokee Indians for abdominal pains.

Asarum europaeum L., European Wild Ginger. (Aristolochiaceae). - Perennial herb. Centr. and S. Europe, Temp. Asia, Asia Minor. Dried rhizome is used in some countries as stimulant, diaphoretic and emmenagogue; used in France as emetic, after use of too much wine. Is an ingredient of Schneeberger Snuff. Contains asaron.

Asclepiadaceae → Arauijia, Asclepias, Asclepiodophora, Brachystelma, Calotropis, Cryptostegia, Fockea, Glossonema, Gomphocarpus, Gonocrypta, Gymnema, Hemidesmus, Hoya, Leptadenia, Marsdenia, Metaplexis, Morrenia, Omphalogonus, Pergularia, Periploca, Raphionacme, Sarcolobus, Secamone, Secamonopsis, Solenostemma, Telosma, Tylophora, Xysmalobium.

Asclepias eriocarpa Benth. Woollypod Milkweed. (Asclepiadaceae). - Perennial herb. S.W. of United States. Plant is source of a fibre which was used by the Concord Indians of Sacramento Valley, Calif. for manuf. ropes and string. Latex has been used as a source of chewing gum, used by the Luiseño Indians.

Asclepias erosa Torr. (Asclepiadaceae). - Perennial herb. S.W. of United States. Has been recommended as a source of rubber, during times of emergency.

Asclepias fruticosa L. → Gomphocarpus fruticosus R. Br.

Asclepias gigantea L. → Calotropis procera Ait.

Asclepias incarnata L., Swamp Milkweed. (Asclepiadaceae). - Perennial herb. N. America. Buds were eaten in soup and with deer. by Menominee Indians. Bark is source of Ozone Fibre, a good material, though thus far little exploited commercially.

Asclepias involucrata Engelm. Dwarf Milkweed. (Asclepiadaceae). - Perennial herb. S.W. of United States. Latex has been used as a source of chewing gum by the Hopi Indians.

Asclepias speciosa Showy Milkweed. Torr. (Asclepiadaceae). - Perennial herb. N. America. Buds when boiled were eaten with meat by the Indians. Also young shoots and leaves were consumed by the Hopi Indians. Latex was used by the Gosuite Indians as a chewing gum.

Asclepias subulata Decne. Desert Milkweed. (Asclepiadaceae). - Xerophytic perennial herb. S. W. of United States and adj. Mexico. Has been recommended as a source of rubber during times of emergency.

Asclepias syriaca L. Common Milkweed. (Asclepiadaceae). Perennial herb. N. America. Young bud sprouts and young green fruits were eaten by Indians of different tribes. Dried latex was used for chewing. Dried root is used medicinally, being alterative anodyne; containing asclepion- a bitter principale, asclepiadin and an ess. oil.

Asclepias tuberosa L. Butterfly-Weed. (Asclepiadaceae). - Perennial herb. N. America. Pods when boiled eaten with buffalo meat; tender shoots eaten as greens; roots were consumed boiled by Indians of different tribes. Source of Pleurisy Root, Orange Milkweed Root. Dried root is used medicinally, being expectorant, diaphoretic; contains a yellow glucoside, asclepiadin.

Asclepiodophora decumbens (Nutt.) Spider Antelope Horn. (Asclepiadaceae). - Western United States. Latex has been used as a source of chewing gum by the Indians in Utah and Nevada.

Ascophyllum nodosum (L.) Le Jolis. (syn. Fucus nodosus L.). Yellow Wrack. Knobbed Wrack, Sea Whistles. (Fucaceae). - Brown Alga. Temp. Atlantic Coasts. Kelp is used for manufacturing of iodine. Also employed as a green manure.

Alh, Black → Fraxinus nigra Marsh.

Ash, Blue → Fraxinus quadrangulata Michx.

Ash, Cape → Eckebergia capensis Sparm.

Ash, European → Fraxinus excelsior L.

Ash, Flowering → Fraxinus Ornus L.

Ash, Hercules Club Prickly → Zanthoxylum Clava Herculis L.

Ash, Japanese Prickly → Zanthoxylum piperitum DC.

Ash, Manna → Fraxinus Ornus L.

Ash, Mountain → Eucalyptus amygdalina Lab.

Ash, Northern Prickly → Zanthoxylum americanum Mill.

Ash, Oregon → Fraxinus oregana Nutt.

Ash, Prickly → Aralia spinosa L.

Ash, Prickly → Zanthoxylum Clava-Herculis L.

Ash, Red → Fraxinus pubescens Lam.

Ash, Senegal Prickly → Zanthoxylum senegalense DC.

Ash Twig Lichen →Ramalina fraxinea (L.) Ach.

Ash, White → Fraxinus americana L.

Ash, Wingleaf Princkly → Zanthoxylum alatum Steud.

Ash, Yellowheart Prickly → Zanthoxylum flavum Vahl.

Ashen Tephrosia → Tephrosia cinerea Pers.

Ashleaf Star Tree → Astronium fraxinifolium Schott.

Asimina costaricensis Don. Smith → Cymbopetalum costaricense (Don. Smith.) Fries.

Asimina reticulata Chapm. (syn. Pictyothamnus reticulatus (Chapm.) Small.). Seminole Tea. (Annonaceae). - Florida. A tea is made from th flowers which was used by the Seminole Indians for kidney troubles. Also A. speciosa Nash.

Asimina triloba (L.) Dunal. Pawpaw. (Annonaceae). - Shrub or small tree. Eastern N. America to Florida and Texas. Fruits greenish to yellowish, oval, of a sweet flavor; sold in markets. Has been recommended for improvement by scientific breeding. Fishermen used the inner bark from branches in spring to string fish.

Aspalathus cedarbergensis Bolus. (Leguminosaceae). - Small shrub. Cape Peninsula. Dried leaves are used as a tea in some parts of Cape Peninsula.

Asparagopsis Sanfordiana Harv. (Bonnemaisoniaceae). - Red Alga. Pacific Ocean. Called by Hawaiians Limu kohu. Consumed as food. The material is pounded, soaked in fresh water to remove the bitter flavor, afterward it is salted and eaten as a relish or salad with fish, meat and poi. It is also mixed with other sea algae and meat stews. It is made into inomena with other ingredients, as a condiment. The sea weed in powdered, pressed into balls, packed with salt in barrels and shipped to Honolulu.

Asparagus, Garden → Asparagus officinalis L.

Asparagus Bean → Dolichos sesquipedalis L.

Asparagus Bush → Dracaena Mannii Baker.

Asparagus, Cape → Aponogeton distachyus Thunb.

Asparagus Pea → Psochocarpus tetragonolobus DC.

Asparagus abyssinicus Hochst. (Liliaceae). - Perennial herb. Eritrea, Abyssinia. Large roots are fried and eaten in some parts of Abyssinia.

Asparagus acutifolia L. (Liliaceae). - Perennial herb. Mediterranean region. Stems are eaten as a vegetable, from Spain to Greece and the Orient.

Asparagus adscendens Roxb. (Liliaceae). - Perennial herb. Iran, Afghanistan, Turkestan, Himalayan region. Roots used in Iran as stimulant, diaphoretic. Also A. sarmentosus Willd. and A. racemosus Willd.

Asparagus africanus Lam. (Liliaceae). - Perennial herb. S. Africa. Leaves used in ointments by native women to stimulate growth of hair.

Asparagus albus L. White Asparagus (Liliaceae). - Perennial herb. High plateaux, N. Africa. Sold in markets of Algeria, as Asperge Sauvage; eaten as vegetable.

Asparagus lucidus Lindl. Shiny Asparagus. (Liliaceae). - Perennial herb. China, Japan. Tubers when boiled are eaten in Japan as a vegetable, also preserved in sugar.

Asparagus officinalis L. Garden Asparagus. (Liliaceae). - Perennial herb. Temp. Europe, Asia, N. Africa, N. America. Widely cultivated as a vegetable in temp. zones of Old and New World. Young stems when green or bleached are eaten boiled, also used in soups. Contains asparagine. Seeds can be used as substitute for coffee. Stems are said to be diuretic. Varieties are: Mary Washington, Martha Washington, Argentueil, Reading Giant.

Asparagus Pauli-Guilelmi Solms Laub. (Liliaceae). Perennial herb. Trop. Africa. Boiled tubers are consumed as food by natives of W.Africa. The plant is capable of being improved by breeding and cultivation.

Asparagus racemosus Willd. (Liliaceae). - Perennial herb. Red Sea area, India, Australia. Fruits are consumed by the Golos (Sudan). Roots are demulcent, anti-dysenteric and diuretic. In vet. medic. it is considered a demulcent.

Asparagus sarmentosus L. (Liliaceae). - Perennial herb. E. India, Ceylon. Fleshy roots are consumed as food in Ceylon; they are candied in some parts of China.

Aspen, Largetooth → Populus grandidentata Michx.

Aspen → Populus tremula L.

Aspen, Trembling → Populus tremuloides Michz.

Aspergillaceae → Aspergillus, Penicillium.

Aspergillus flavus Link. (Aspergillaceae). - Microorganism. Mold. Is used in Japan as a Koji, a starter which is used in several fermentation processes, among which that of Sake (Rice Wine) and Shoya (Soya Sauce). The strain used for Shoya should possess high proteolytic and amyloclatic action because there is little starch in the beans that can bu used for fermentation. In China Aspergillus oryzae, Zygosaccharomyces (a yeast) and Lactobacillus Delbruckii are used.

Aspergillus niger van Tiegh. (Aspergillaceae). - Fungus. A mold which is being used for the formation of citric acid as a result of fermentation. Also A. clavatus Desm. A. Wentii Wehmer.

Aspergillus oryzae (Ahlb.) Cohn. (Aspergillaceae). - Fungus. This mold is used in Japan in the production of Sake. (See Saccharomyces Sake). Is occasionarlly used in the production of beer.

Asperula odorata L. Sweet Woodruff. (Rubiaceae). Perennial herb. Throughout most of Europe, Siberia, N. Africa. Herb is used for flavoring a beverage, known as Maiwein, Maitrank or Maybowl; much used in Germany and Austria, Contains cumarin. Dried herb is put between clothes to impart a pleasant scent.

Asperula tinctoria L. Dyer's Woodruff. (Rubiaceae). - Perennial herb. S. Europe. Roots are the source of a red dye.

Asphodelus fistulosus L. (Liliaceae). - Bulbous perennial. Arabia, Mediterranean region. Bulbs are eaten boiled by the Beduins in Arabia.

Asphodelus luteus Reichb. (Liliaceae). - Bulbous perennial. S. Europe, Asia Minor. Bulbs were used by the ancient Greeks and Romans as potatoes are used at the present among our civilization.

Asphodelus microcarpus Viû. (Liliaceae). - Bulbous perennial. Mediterranean region, Canary Islds. Bulbs are source of a yellow dye; used in Mariût Egypt, for dyeing carpets.

Asphodelus pendulinus Coss. and Dur. (Liliaceae). - Bulbous perennial herb. Algeria to Arabia. Used in some parts of N. Africa and Arabia as diuretic.

Asphodelus racemosus L. (Liliaceae). - Bulbous perennial. S. Europe, Asia Minor. Roots are source of a gum; used in Turkey as a glue, esp. in book binding.

Aspidosperma Curranii Standl. Carreto. (Apocynaceae). - Tree. Trop. America, esp. Magdalena region, Colombia to Venezuela. Wood heavy, hard, very durable when in contact with soil; used for heavy bridges, railroad ties and houseposts.

Aspidosperma Cuspa (H. B. K.) Blake. (Apocynaceae). - Tree. Trop. America, esp. Venezuela, Guiana. Leaves and bark used in some parts of Venezuela as febrifuge.

Aspidosperma demanthum Muell. Arg. Araracanga. (Apocynaceae). - Tree. Brazil, esp. Lower Amazon. Wood orange brown, waxy feeling; used in Brazil for heavy construction work.

Aspidosperma eburneum Allem. Peguia Marfin, Pau Setim. (Apocynaceae). - Tree. Brazil, esp. Sao Paulo, Victoria. Wood of variable density; used for cabinet work and turnery.

Aspidosperma excelsum Benth. Yaruru, Paddle Wood. (Apocynaceae). - Tree. Trop. America. Wood used for paddles, tool-handles; cores of large trees for millrollers; it is durable, compact, strong and elastic. Bark is carminative, stomachic.

Aspidosperma nitidum Benth. (Apocynaceae). - Tree. Trop. America esp. Guiana, Brazil. Wood relatively hard, heavy; used for handles of tools and where much elasticity is required.

Aspidosperma polyneuron Muell. Arg. Peroba Rosa. (Apocynaceae). - Large tree. S. Brazil. Wood rose-red to yellowish, heavy, hard, uniform, fine-grained, fairly durable; used for general construction, flooring, framing, furniture, interior finish, cabinet work, sash and doors.

Aspidosperma quebracho blanco Schlecht. Quebracho Blanco. (Apocynaceae). - Tree. Argentina and adj. territory. Bark source of tannin, used for manuf. leather. Called Quebracho Blanco, White Quebracho.

Aspidosperma tomentosum Mart., Woolly White Quebracho. (Apocynaceae). - Tree. Brazil. Wood is known in the lumber trade as Piquia Peroba; used for carpentry.

Aspidosperma Vargasii DC. (Apocynaceae). - Tree. Venezuela. Wood yellowish, hard relatively heavy, fine-grained, easy to split and to polish; used for rulers, backs of brushes and similar material.

Aspidium Filix Mas. L. Male Fern, Male Shield Fern. (Polypodiaceae). - Perennial herb. Fern. Dried rhizome and stipes are used medicinally, being teniacide; contain filicic acid, albaspidin, filicin, filix red, filmaron and a resin.

Aspidium spinulosum (O. F. Muell.) Sw. var. dilatatum (Hoffm.) Hook. (Polypodiaceae). - Perennial herb. Fern. Temp. and Arct. N. Hemisphere. Rootstocks after being cooked on hot stones are consumed as food by Eskimos of Alaska and of other regions.

Aspilia latifolia Oliv. and Hiern., Hemorrhage Plant. (Compositae). - Perennial herb. Trop. Africa. Fresh plant has extraordinary properties of stopping bleeding.

Asplenium macrophyllum Sw. (Polypodiaceae). - Perennial fern. Philipp. Islands, esp. N. Luzon to S. Mindanao. Decoction of the fronds is a powerful diuretic; used for defective urinary secretion, esp. when induced by beri-beri.

Assam-Rubber → Ficus elastica Roxb.

Astelia grandis Hook. f. (Liliaceae). - Perennial plant. New Zealand. Source of a soft, strong, brown fibre; used for different purposes.

Astelia nervosa Banks and Soland. (Liliaceae). - Large herb. New Zealand. Berries eaten by Maoris.

Astelia trinervia Kirk. Kauriegrass. (Liliaceae). - Tall herb. New Zealand. Leaves are source of a fibre; used for various purposes, especially by the early settlers.

Aster cantoniensis Blume → Boltonia cantoniensis French. and Sav.

Aster macrophyllus L. Rough Tongues. (Compositae). - Perennial herb. Eastern N. America. Very young leaves are occasionally consumed as potherb.

Astilbe philippensis Henry. (Saxifragaceae). - Perennial herb. Philipp. Islands. Leaves are used for smoking by the Igorots (Philipp. Isld.).

Astragalus baëticus L. (Leguminosaceae). - Perennial herb. Spain, Sicily. Seeds are source of a substitute for coffee.

Astragalus canadensis L. Canada Milk Vetch. (Leguminosaceae). - Herbaceous perennial. N. America. Roots harvested in the fall or spring were eaten raw or boiled by the Blackfoot Indians.

Astragalus caryocarpus Ker.-Gawl. (Leguminosaceae). - Herbaceous plant. Western N. America. Pods raw or boiled were eaten by Indians of Montana.

Astragalus cylleneus Boiss and Heldr. (Leguminosaceae). - Shrub. Greece. Source of Morea Tragacanth.

Astragalus edulis Dur. (Leguminosaceae). - Perennial herb. Iran to N. Africa. Seeds are consumed as food in some parts of Iran.

Astragalus fasciculifolius Boiss. (Leguminosaceae). - Perennial plant. S.W. Asia. Source of a sweet, yellowish-brown, brittle exudation, Sarcocolla, is exported from Kurdistan. It is used as plasters by Parsi bone-setters; applied locally to face and ears for neuralgic pains. Used by harem-ladies to give a glossy appearance of the skin.

Astragalus Garbancillo Cav. (syn. A. Mandoni Rusby.) Garbancillo (Leguminosaceae). - Perennial herb or small shrub. Peru, Argentina, Chile. Used in soap for washing fine clothes, among the Indians of Peru. Supposed to have insecticidal properties.

Astragalus glycyphyllos L. Milk Vetch. (Leguminosaceae). - Perennial herb. Europe, Siberia to Altai. Sometimes cultivated as a fodder for livestock. Herb is occasionally used as tea.

Astragalus gummifer Lab. Tragacanth. (Leguminosaceae). - Herbaceous plant. Asia Minor, Iran, Syria, Greece, Armenia, Kurdistan. Source of Gum Tragacanth originating from injuries in the stem. Irregularly shaped tears are called Tragacanth Sorts; tears that are twisted and worm-like are called Vermiform Tragacanth. Has already been used by the Ancient Greeks and during the Middle Ages. Much is exported from Smyrna and ports along the Gulf of Persia. Used in cosmetics, calico printing, confectionary. Used medicinally as a suspending medium for insoluble powders, emulsifying oils and resins; adhesive in pill and troche masses. Contains bassorin which swells but does not dissolve in water. Material is also called Gomme Adragante. **A. microcephalus** Willd. from Asia Minor; **A. kurdicus** Boiss. from Kurdistan and Asia Minor and **A. stromatodes** Bunge from Iran and Asia Minor are also source of a Tragacanth.

Astragalus Henryi Oliv. (Leguminosaceae). - Perennial herb. China, esp. Hupeh, Szechuan. Widely used as a medicine in China. Also **A. Hoantchy** Franch and **A. mongholicus** Bunge. **Astragalus heratensis** Bunge. Indian Tragacanth. (Leguminosaceae). Centr. Asia. Source of a gum, Katira Gabina, much exported. Used for glazing and stiffening of local fabrics.

Astragalus pictus filifolius Gray. (Leguminosaceae). - Perennial herb. Western N. America, esp. Idaho to New Mexico. Roots were consumed by the Hopi Indians of Arizona.

Astragalus prolixus Sieb. (Leguminosaceae). - Herbaceous perennial. N. India, Red Sea area. Plant is probably one of the sources of Gum Tragacanth.

Astragalus pycnocladus Boiss. and Haussk. (Leguminosaceae). - Shrub. W. Russia, Iran. Source of Persian Tragacanth. Probably also **A. adscendens** Boiss. and Haussk. from S. W. Persia; **A. strobiliferus** Royle from Persia; **A. leiacladus** Boiss. from Centr. and S. Iran; **A. brachycalyx** Fischer from Iran and Kurdistan and **A. eriostylus** Boiss and Haussk. from S. W. Iran.

Astrakan Wheat → Triticum polonicum L.

Astrebla pectinata F. v. Muell. (syn. Danthonia pectinata Lindl.). Mitchell Grass. (Graminaceae). - Perennial herb. Australia esp. S. Australia, New Sorth Wales, Queensland. Drought resistant grass, a good pasture grass for cattle in Australia.

Astrebla triticoides F. v. Muell. (syn. Danthonia triticoides Lindl.) (Graminaceae). - Perennial herb, Australia, esp. S. Australia, New South Wales and Queensland. Strong growing grass, has highly fattening qualities, readily eaten by livestock.

Astrocaryum Ayri Mart. (syn. Toxophoenix aculeatissima Schott.). (Palmaceae). - Palm. Brazil. Fruits are used in some parts of Brazil for erysipelas.

Astrocaryum Jauari Mart. Jauary Palm. (Palmaceae). - Palm. Brazil. Fruits are used in some parts of Brazil as angle-bait to catch Tambaqui fishes. Kernels are source of an oil.

Astroloma humifusum R. Br. (Epacridaceae). - Small shrub. Australia. Fruits have a sweet pulp; much esteemed by the natives.

Astroloma pinifolia Benth. (Epacridaceae). - Small shrub. Australia. Fruits are edible, consumed by the natives.

Astronium fraxinifolium Schott. Ashleaf Star Tree. (Anacardiaceae). - Tree. Trop. America, esp. Colombia to Brazil. Wood much esteemed in Colombia; used for fine furniture and cabinet work.

Astronium graveolens Rich. (Anacardiaceae). - Tree. E. Brazil. Wood, called Gonçalo Alves, is moderately hard, heavy, fairly loose in structure, durable; suitable for veneers; takes an excellent polish.

Astronium Urundeuva (Allem.) Engl. (Anacardiaceae). - S. America. Wood cherry-red to dark brownish red; used for railroad ties, bridge timbers, posts, piling and for durable and heavy construction work.

Astrostylidium Pittieri Hack. (Graminaceae). - Perennial grass. Centr. America. Stems were frequently used by the ancient Mayas for making fish-spears. Also **A. spinosum** Swall.

Atalantia glauca Benth. and Hook. f. (Triphasia glauca Lindl.) Desert Lemon. (Rutaceae). - Tree. Australia, esp. New South Wales, Queensland. Fruits are used in drinks or as preserves in some parts of Australia.

Atalantia monophylla DC. (syn. Limonia monophylla Roxb.). (Rutaceae). - Shrub or small tree. E. India. Fruits source of an oil; used for chronic rheumatism.

Atalaya hemiglauca F. v. Muell. Cattle Bush, White Wood. (Sapindaceae). - Tree. S. Australia, New South Wales. Leaves eaten by stock.

Atamasco Lily →Zephyranthes Atamasco Herb.

Atemoya → Annona squamosa L. x A. Cherimola Mill.

Athamanta cretensis L. Candy Carrot. (Umbelliferaceae). - Perennial herb. S. Europe. Mediterranean region. Plant is used for flavoring liqueurs.

Athamanta Roxburghiana (Benth.) Wall. → Carum Roxburghianum Benth.

Atherosperma moschatum Labill. Sassafras. (Monimiaceae). - Tree. Australia, esp. Tasmania, New South Wales, Victoria. Fragrant bark used as tea in Tasmania.

Athrixia phylicoides DC. (Compositae). - Shrub. S. Africa. Infusion of the leaves is used in some parts of S. Africa as a tea.

Athyrium esculentum Copel (Polypodiaceae). - Perennial herb. Fern. Young fronds are eaten in Philipp. Islds.; used in salads or cooked. Sold in markets.

Atlantic Mountain Mint → Pycnanthemum incanum (L.) Michx.

Atlantic Yam → Dioscorea villosa L.

Atol de Piña → Bromelia Karatas L.

Atractylis gummifera L. (Compositae). - Perennial herb. Mediterranean region. Source of a gum; used to adulterate Mastic.

Atriplex campanulata Benth. Small Salt-Bush. (Chenopodiaceae). - Shrub. Australia, esp. S. Australia. New South Wales, Queensland. Used as food for stock. Also **A. halimoides** Lindl., **A. nummularia** Lindl., **A. semibaccata** R. Br., **A. spongiosa** F. v. Muell., **A. vesicaria** Heward.

Atriplex canescens (Pursh) Nutt., Fourwing Saltbush. (Chenopodiaceae). - Shrub. Western N. America to Mexico. Seeds are used as food by the Gosuite Indians of Utah.

Atriplex Halimus L. Mediterranean Saltbush. (Chenopodiaceae). - Shrub. N. and S. Africa. On alkali soils. Considered an important food for livestock in S. Africa. Used as food by the Tuareg (N. Afr.) during times of emergency. Ash from plants is used in manuf. soap.

Atriplex hortensis L., Mountain Spinach, Garden Orach. (Chenopodiaceae). - Annual herb. Temp. Europe, Asia. Introduced in N. America. Occasionally cultivated. Leaves are consumed as spinach.

Atropa Belladonna L. Belladonna. (Solanaceae). - Perennial herb. Centr. and S. Europe, Balkans, Caucasia, Iran etc.; introd. in N. America. Berries were formerly used by ladies as a cosmetic, to dilate their eye-pupils, giving them a striking appearance. Dried roots and leaves used medicinally as a sedative, anodyne, antispasmodic; decrease flow of many secretions. Used in ophthalmology to dilate pupils (mydriatic). Leaves stimulate central nervous system, also a stimulant and narcotic. Contains atropine, hyoscyamine, being extremely toxic alkaloids.

Attalea Cohune Mart. Cohune Palm. (Palmaceae). - Tall palm. Mexico, Central America. Seeds source of Cahune Oil; non-drying, used in food, as illuminant, manuf soap. Sp. Gr. 0.868-0.971; Iod No. 10-14; Sap. Val. 252-256; Melt. Pt. 18-20° C.; Unsap. 0.2 0.5%. Trunk is used for building. Leaves are employed for thatching. Very young buds are consumed as a vegetable. Young leaves are made into hats. Fruits are made into sweetmeats also used as fodder for livestock.

Attalea funifera Mart., Bahia Piassave. (Palmaceae). - Palm. Brazil. Leaves are source of a long fibre; used for different purposes. Seeds produce a good oil.

Aubergine → Solanum Melongana L.

Aubrya gabonensis Baill. (Humiraceae). - Woody plant. Trop. Africa. Fruits are source of a fermented drink, called Stouton; used by natives of Gabon. Wood is white, easy to work.

Aulospermum longipes (S. Wats.) Coult. and Rose. (syn. Cymopterus longipes S. Wats.) Ribseed. (Umbelliferaceae). - Herbaceous plant. Western N. America. Leaves were consumed boiled by Indians of Nevada and Utah.

Aulospermum purpureum (S. Wats.) Coult. and Rose. (syn. Cymopterus purpureus S. Wats.). (Umbelliferaceae). - Herbaceous plant. Western N. America and adj. Mexico. Herb was used to season mush and soup and was also eaten as potherb by Navajo Indians.

Aurantii Dulcis Cortex → Citrus sinensis Osbeck.

Aureomycin → Streptomyces aureofaciens.

Auricula Tree → Calotropis procera Ait.

Auricularia Auricula Judae (L.) Schroet. (syn. Hirneola Auricula-Judae (L.) Berk.) Jew's Ears. (Auriculariaceae). - Basidiomycete. Fungus. Temp. zone. Fruitbodies are extensively consumed as food in China.

Auricularia cornea Ehrenb. (Auriculariaceae). - Basidiomycete. Fungus. Tropics. Fruitbodies are used as food in the Philipp. Islds. Also **A. tenuis** Lev., **A. brasiliensis** Fr. and **A. Moellerii** Lloyd.

Auricularia delicata (Fr.) Henn. (Auriculariaceae). - Basidiomycete. Fungus. Malaysia. Fruitbodies are used as food in different parts of the Mal. Archipelago.

Auricularia polytricha (Mont.) Sacc. → Hirneola polytricha Mont.

Auriculariaceae (Fungi) → Auricularia, Hirneola.

Australian Desert Kumquat → Eremocitrus glaucca Swingle.

Australian Grass Tree → Xanthorrhoea australis R. Br.

Australian Gum → Acacia pycnantha Benth.

Australian Kino → Eucalyptus rostrata Schlecht.

Australian Sandalwood → Santalum lanceolatum R. Br.

Austrian Pine → Pinus nigra Arn.

Aveledoa nucifera Pitt. (Opiliaceae). - Tree. Venezuela. Fruits are consumed in some parts of Venezuela.

Avellano → Gevuina Avellana Molina.

Avena fatua L. Wild. Oat. (Graminaceae). - Perennial grass. Europe, Asia, introd. in N. America. Fruits were made into flour and consumed by several Indian tribes.

Avena nuda L. Naked Oats. (Graminaceae). - Annual grass. Origin unknown, probably from S. Europe. Occasionally cultivated as a cereal.

Avena orientalis Schreb. Hungarian Oat, Turkish Oat (Graminaceae). - Annual grass. Origin uncertain. S. Europe. Occasionally cultivated as cereal, esp. in S. E. Europe.

Avena sativa L., Common Oat. (Graminaceae). - Annual grass. Origin uncertain. Grown by the Lakedwellers of Europe. Raised in the temp. zone of the Old and New World. Cereal, source of oatmeal, very nutricious; used as rolled oats, breakfast foods, porridge, in cakes and biscuits; not suitable for bread. In Belgium source of white beer. Excellent fodder for horses, not suitable for hogs. Varieties are grouped into Winter and Springs Ooats. Color of hull may be white, yellow, gray, red or black. Most hull-less oats are of Asiatic origin. Numerous varieties, many of local importance, among which Fulham, Culberson, Red Rustproof, Burt, Albion, Kherson, Sixty Day, Golden Rain. Grains are source of Oat Oil, used for preparation of cereals. Sp. Gr. 0.925; Sap. Val. 185-192; Iod. No. 100-114; Unsap. 1.3 - 2.6%.

Avens Root → Genum urbanum L.

Averrhoa Bilimbi L., Bilimbi, Cucumber Tree. (Oxalidaceae). - Tree. Probably from Malayan region. Only known in cultivation. Fruits very acid, 5 to 7 cm long, light yellow; used in drinks, marmelade, jelly, syrup, candied or pickled.

Averrhoa Carambola L., Carambola. (Oxalidaceae). - Small tree. Trop. Asia, Cultivated in tropics of Old and New World. Fruits deeply angled, star-shaped on cross-section, yellow to pale-brown, 7 to 12 cm long, of a pleasant acid, quince-like flavor, eaten raw, in compots, jelly, jam, drinks, sliced in salads. Many varieties, among which Demak Blimbing, with large, sweet fruits.

Avicennia officinalis L. (syn. A. tomentosa Jacq., A. resinifera Forst.). (Verbenaceae). - Tree. Coastal area. Tropics. Fruits when baked or steamed are consumed as food by the aborigenes of N. Queensland. Bark employed for tan-ning, produces a harsh, pale brown, firm leather. Wood used for boat building and piles.

Avicennia marina Stapf. (Verbenaceae). - Shrub or small tree. Coast of E. Africa, Red Sea area. Bark is used for tanning.

Avicenna nitida Jacq. Black Mangrove (Verbenaceae). - Shrub or tree. Sea shores. Trop. America. Bark is used for tanning.

Avocado → Persea americana Mill.

Avocado, Coyo → Persea Schiedeana Nees.

Avocado, Mexican → Persea drymifolia Cham. and Schlecht.

Avocado Oil → Persea americana Mill.

Awarra Palm → Astrocaryum Jauari Mart.

Axonopus compressus (Swarz) Beauv. Carpet Grass. (Graminaceae). - Perennial grass. From N. America to Argentina. Recommended as pasture grass in alluvial and mucky soils.

Ayahuasa → Banisteria Caapi Griseb.

Aydendron Cujumary Mart. → Ocotea Cujumary Mart.

Aydendron firmulum Nees., Pichrim Bean. (Lauraceae). - Tree. Brazil. Aromatic seeds are used in some parts of Brazil as a condiment.

Aydendron floribundum Meisn. (syn. Cryptocarya minima Mez.). (Lauraceae). - Tree. Brazil. Wood is used for general pursoses, construction work. Decoction of bark is used for washing ulcers.

Aydendron panurense Meisn. (syn. Aniba panurensis Mez.). (Lauraceae). - Tree. S. America, esp. Brazil. Source of Cayenne Linaloe or Bois de Rose.

Azalea nudiflora L. (syn. Rhododendron nudiflorum Nutt.). Poinxter Flower. (Ericaceae). - Shrub. Eastern and Southern N. America. Galls called "May Apples", found on the leaves and tems, are pickled with spices and vinager, which were consumed since the days of the Pilgrims.

Azerolier → Crataegus Azarolus L.

Azolla pinnata R. Br. (Salviniaceae). - Small floating fern-ally. Trop. Asia. Used as green-manure, also as food for pigs and ducks in Indochina.

Azotobacter chroococcum Beijerinck. (Bacteriaceae). - Bacil. Microorganism. A non symbiotic soil organism being capable of fixing the nitrogen from the atmosphere. Also A. agile Beijerinck, A. Vinelandii Lopm., A. Beijerinckii Lipm., A. Woodstownii Lipm. and A. vitreum Loehnis and Westerm.

Aztec Clover → Trifolium amabile H. B. K.

Aztec Pine → Pinus Teocote Cham. and Schlecht.

Aztec Tobacco → Nicotiana rustica L.

B

Babaco → Carica pentagona Heilb.

Babassu Oil → Orbignya speciosa Barb.

Babassu Palm → Orbignya speciosa Barb.

Babiana plicata Ker-Gawl. (Iridaceae). - Perennial herb. S. Africa. Roots, called Baboon Root, were eaten by settlers in Cape Peninsula.

Bablah → Acacia arabica Willd.

Baboonroot → Babiana plicata Ker-Gawl.

Babul Acacia → Acacia arabica Willd.

Bacaba → Oenocarpus distichus Mart.

Bacaba Branca → Oenocarpus distichus Mort.

Bacidia muscorum (Sw.) Mudd. (Lecideaceae). - Lichen. Temp. zone. Source of a dye, to give woolens a red color, esp. used in Europe.

Baccaurea malayana King. (Euphorbiaceae). - Tall Tree. Mal. Penin., Sumatra. White flesh around seeds is edible; used for preparation of a liquor among the Mantera and Jakuns.

Baccaurea Motleyana Mull Arg. (Euphorbiaceae). - Tree. Malaysia. Cultivated. Fruits edible, 3 cm long, salmon buff; sweet to acid; eaten raw, cooked or preserved. Juice when fermented is made into a liqueur. Bark is source of a dye.

Baccaurea racemosa (Blume). Muell. Arg. (Euphorbiaceae). - Medium sized tree. Mal. Archipelago, esp. Sumatra and Java. Often cultivated by natives. Fruits roundish, yellow-green, subacid; 2 to 2.5 cm in diam.; pulp pink; much esteemed by the natives.

Baccaurea sapida Muell. Arg. (syn. Pierardia sapida Roxb.) (Euphorbiaceae). - Tree. Malaysia, India, China. Berries edible, 25 to 35 mm across; pulpy arillus, yellow, somewhat acid, agreeable taste. Cultivated among the Hindus.

Baccharis calliprinos Griseb. (Compositae). - Shrub. Argentina. Infusion of herb is used in some parts of Argentina for colic. Plant is source of a yellow dye.

Baccharis genistelloides Pers. (syn. B. triptera Mart.) (Compositae). - S. America, esp. Argentina, Paraguay, Uruguay and Peru. Herb is used in some parts of S. America as a tonic and febrifuge.

Baccharis tridentata Vahl. (Compositae). - Shrub. Brazil. Plant is used as febrifuge and diuretic in some parts of Brazil.

Bachang Mango → Mangifera foetida Lour.

Bachelor's Button, Yellow → Polygala Rugelii Shuttlw.

Bacillus acetoethylicum Northtr. (Bacteriaceae). - Bacil. Microorganism. Has been recommended in the production of acetone-ethanol, as result of fermentation.

Bacillus butylicus Fitz. (Bacteriaceae). - Bacil. Is able to produce butyric acid and butyl alcohol from carbohydrates.

Bacillus Freudenreichii (Migula) Chester. (Bacteriaceae). - Bacil. Microorganism. Agriculturally a useful soil microorganism, being capable of converting urea to ammonium carbonate.

Bacillus mecerans Schard. (Bacteriaceae). - Bacil. Is important in the production of acetone-ethanol as a result of fermentation.

Backhousia citriodora F. v. Muell. Citron Backhausia, Scrub Myrtle, Native Myrtle. (Myrtaceae). - Tree. Australia, esp. Queensland. Has been suggested as a source for perfum, also for manuf. soap.

Bacove → Musa sapientum L.

Bacteria, Useful → Microorganism.

Bacteriaceae →Acetobacter, Aerobacillus, Aerobacter, Azotobacter, Bacillus, Bacterium, Clostridium, Escherichia, Micrococcus, Nitrobacter, Pectinobacter, Plectridium, Proprionibacterium, Pseudomonas, Rhizobium, Sarcina.

Bacterium acidi-propriocia Scherm. (Bacteriaceae). - Bacil. Microorganism. Is important in the production of highgrade Emmenthaler cheese with characteristic flavor, having a normal eye development. Belongs to the propionic acid bacteria.

Bacterium mycoides (Grotenf.) Migula. (Bacteriaceae). - A soil bacteria that is capable of reducing nitrates to nitrites and finally to ammonia. Some species are able to reduce nitrate to nitrites. Also B. vulgatus Eggerth and Gagnon.

Bacterium rhamnosifermentans Castell. (Bacteriaceae). - Bacil. Is of importance in the fermentation of rhamnose. Propyle glycol being one of the end products.

Bactris Gasipaës H.B.K. (syn. Guilielma speciosa Mart.) Peach Palm. (Palmaceae). - Tall palm. S. America. Cultivated since pre-Columbian times. Peach-like fruits are important food among the natives in some regions of Brazil. Eaten when boiled or roasted; also source of a fermented drink. Fruits of certain varieties contain no kernels. Wood very hard; used for building purposes. Spines are used in tattooing.

Bactris minor Jacq. (Palmaceae). - Colombia. Fruits are edible and are sold in markets of Colombia.

Bacu → Cariniana pyriformis Miers.

Bacury → Platonia insignis Mart.

Bacury Kernel Oil → Platonia insignis Mart.

Baeckia frutescens L. (Myrtaceae). - Shrub or small tree. Mal. Penin., S. China, Sumatra to Australia. Leaves used for a refreshing tea; for fevers and lassitude. Often sold in Javanese bazaars. In Annam flowers are used in tea. Produces an ess. oil, Essence de Bruyère de Tonkin; used in France for perfuming soaps.

Bael Fruit → Aegle Marmelos Correa.

Bagasse → Saccharum officinale L.

Bagilumbang Oil → Aleurites triloba Forst.

Bahama Hemp → Agave sisalana Perrine.

Bahama White Wood → Canella alba Murr.

Bahia Grass → Paspalum notatum Fluegge.

Bahia Piassave → Attalea funifera Mart.

Bahia Rosewood → Dalbergia nigra Allem.

Bahia Wood → Caesalpinia brasiliensis Sw.

Baikal Shullcup →Scutellaria baicalensis Georg.

Baillonella ovata Pierre (syn. Mimusops Pierreana Engl.) (Sapotaceae). - Tree. Africa, esp. Gabon. Seeds are eaten by the natives with meat and fish. Source of Beurre d'Orère, Orère Butter.

Balanophoraceae → Balanophora, Langsdorffia, Thonningia.

Ballot — Exocarpus cupressiformis Lab.

Bakupari → Rheedia brasiliensis Planch. and Triana.

Balanites aegyptiaca Delile., Betu. (Zygophyllaceae). - Shrub. N. Trop. Africa, Arabia, Palestine. Seeds are source of Betu Oil, recommended for manuf. soap. Fruits are edible, source of an alcoholic beverage in Congo and Nupé. Leaves are eaten as vegetable in Bormu (W. Afr.). Roots and bark are considered purgative and vermifuge. Wood ist compact, fine-grained, hard; used in Abyssinia for clubs, ploughs, walking-sticks, turnery, general carpentry work, joists, mortars, pestles. Bark is source of a strong fibre.

Balanites Maughamii Sprague. (Zygophyllaceae). - Woody plant. Portug. E. Africa. Nuts are source of a clear oil, similar to Olive Oil.

Balanites orbicularis Sprague. (Zygophyllaceae). - Tree. Maritime region of Somaliland etc. Source of Hanjigoad, a gum resin. Tears are globular pieces from the size of a cherry to a pigeon's egg; dark greenish yellow to deep orange red.

Balanocarpus Heimii King. (syn. B. Wrayi King, B. maximum Ridley.) Chengal Tree. (Dipterocarpaceae). - Large tree. Malaysia, Wood heavy, hard, medium texture, straight-grained, pale yellow. A standard lumber, for bridges, railroad ties, heavy construction work, resistant to white ants and to decay. Source of a resin, Damar Penak, one of best dammar gums; used for certain varnishes. Criminally, thieves stupefy victims by burning dammar with Datura.

Balanocarpus maximus King. (Dipterocarpaceae). - Tree. Malaysis, esp. Perak. Source of a Damar Penak.

Balanophora elongata Blume. (Balanophoraceae). - Tree. Malaya, esp. Perak. Source of a wax, used by the natives for illumination and candles of small torches. Also B. globosa Jungh. and B. Ungeriana Val.

Balata → Manilkara bidentata (DC.) Chev.

Balata Blanca → Micropholis Melinoniana Pierre.

Balata Rosada → Sideroxylon cyrtobotryum Miq., S. resiniferum Ducke.

Balata Tree →Manilkara bidentata (DC.) Chev.

Baláu Resin → Dipterocarpus grandiflorus Blanco.

Baláustre → Centrolobium orinocense (Benth.) Pitt.

Balche (beverage) → Lonchocarpus longistylus Pitt.

Bald Cypress → Taxodium distichum (L.) Rich.

Baleric Myrobalans → Terminalia Bellerica Roxb.

Balfourodendron Riedelianum Engl., Pau-liso, Guatambú. (Rutaceae). - Tree. Subtrop., S.America, esp. Brazil, Paraguay, Argentina. Wood of excellent quality, much esteemed in Argentina; used for agricultural implements, interior finish of buildings, turnery, tool handles and furniture.

Baliospermum axillare Blume. (syn. B. montanum Muell. Arg.) (Euphorbiaceae). - Stout subherbaceous shrub. Trop. Asia, esp. India to Java. Decoction of leaves is used as purgative. Seeds are drastic and purgative.

Ballhead Onion → Allium sphaerocephalus L.

Balm → Melissa officinalis L.

Balm, Bee → Monarda didyma L.

Balm, Horse → Collinsonia canadensis L.

Balm of Gilead → Abies balsamea (L.) Mill. and Populus balsamifeea L.

Balmony → Chelone glabra L.

Baloghia lucida Endl. (syn. Codiaeum lucidum Muell. Arg.) Brush Bloodwood. (Euphorbiaceae). - Tree. Australia, esp. New South Wales, Queensland, Norfolk Islds. Source of a beautiful red pigment.

Baloghia Pancheri Baill. (Euphorbiaceae). - Tree. Australia and New Caledonia. Bark is used for tanning in some parts of New Caledonia.

Balsa → Ochroma lagopus Swartz.

Balsa Wood → Ochroma lagopus Swartz.

Balsam Apple → Momordica Balsamina L.

Balsam, Canada → Abies balsamea (L.) Mill.

Balsam, Copaiba → Copaifera coriacea Mart. and C. reticulata Ducke.

Balsam Capaiba Tree → Paradaniella Olivieri Rolfe.

Balsam, Duhnual → Commiphora Opobalsam Engl.

Balsam Fir → Abies balsamea (L.) Mill.

Balsam, Garden → Impatiens Balsamina L.

Balsam, Gurjum → Dipterocarpus alatus Roxb.

Balsam, Maracaibo → Copaifera officinalis L.

Balsam, Mecca → Commiphora Opobalsam Engl.

Balsam of Fir → Abies balsamea (M.) Mill.

Balsam of Gilead → Commiphora Opobalsam Engl.

Balsam, Oregon → Pseudotsuga taxifolia (Lam.) Britt.

Balsam Pear → Momordica Charantia L.

Balsam, Peru → Myrocylon Pereirae Klotzsch.

Balsam Poplar → Populus balsamifera Moench.

Balsam Poplar Buds → Populus balsamifera Moench.

Balsam, Tolu → Myroxylon toluiferum H.B.K.

Balsam Tree, Peru → Myroxylon Pereirae Klotzsch.

Balsam Tree, Tolu → Myroxylon toluiferum H.B.K.

Balsamo → Acanthosphaera Ulei Warb.

Balsamocarpum brevifolium (Baill.) Clos. → Caesalpinia brevifolium Baill.

Balsamodendron gileadense Kunth. → Commiphora Opobalsam Engl.

Balsamodendron Kataf Kunth. → Commiphora Kataf Engl.

Balsamodendron Mukul Hook. → Commiphora Mukul Engl.

Balsamorhiza deltoidea Nutt. Puget Balsamroot. (Compositae). - Perennial herb. Western N. America. Young sprouts were eaten raw; seeds were made into bread and roots eaten boiled by the Indians of Oregon, Washington and Brit. Columbia.

Balsamorhiza Hookeri Nutt., Hooker's Balsamroot. (Compositae). - Herbaceous perennial. Western N. America, esp. Washington to California, Utah. Roots when cooked were used as food by the Indians.

Balsamorhiza sagittata (Pursh.) Nutt., Balsam Root, Oregon Sunflower, Arrowleaf Balsam Root. (Compositae). - Herbaceous perennial. Western N. America. Fleshy roots were eaten after being cooked on hot stones; used as food by the Nez Percé Indians.

Balsamroot → Balsamorhiza sagittata (Pursh.) Nutt.

Balsamroot, Arrowleaf → Balsamorhiza sagittata (Pursh) Nutt.

Balsamroot, Hooker's → Balsamorhiza Hookeri Nutt.

Balsamroot, Puget → Balsamorhiza deltoidea Nutt.

Balsaminaceae → Impatiens.

Bambara Groundnut → Voandzeia subterranea Thon.

Bamboo, Berry bearing → Melocanna bambusoides Trin.

Bamboo, Black → Phyllostachys nigra Blanco.

Bamboo, Giant Timber → Phyllostachys bambusoides Sieb. and Zucc.

Bamboo, Hairy Sheeth Edible. → Phyllostachys edulis Riv.

Bamboo, Himalayan → Arundinaria falcata Nees.

Bamboo, Plains → Bambusa Balcooa Roxb.

Bamboo, Slender → Oxytenanthera Stocksii Munro.

Bamboo, Spiny → Bambusa arundinacea Willd.

Bamboo, Terai → Melocanna bambusoides Trin.

Bamboo, Tufted → Oxynanthera nigrociliata Munro.

Bambuk Butter → Butyrospermum Parkii (Don.) Kotschy.

Bambusa arundinacea Willd., Spiny Bamboo. (Graminaceae). - Woody grass. Trop Asia, esp. E. India. Very young buds are consumed as food in some parts of India.

Bambusa Balcooa Roxb. Plains Bamboo. (Graminaceae). - Tall woody grass. Plains E. India. One of the best and strongest bamboos for building purposes. Much found around Calcutta.

Bambusa cornuta Munro. (Graminaceae). - Woody grass. Java. Cultivated. Young tender shoots are consumed as vegetable.

Bambusa multiplex Raeusch. (Graminaceae). - Shrubby, woody grass. Cochin China, Japan. Cultivated in trop. Asia. Used for fishing rods. Young buds eaten as vegetable. Culms are source of paper.

Bambusa nana Roxb. (syn. B. glaucescens Sieb.) (Graminaceae). - Woody grass. China. Cultivated. Stems are used as fishing rods.

Bambusa Oldhami Munro. (Graminaceae). - Woody grass. China. Young buds are eaten as a vegetable in some parts of Trop. Asia.

Bambusa polymorpha Munro. (Graminaceae). - Tall woody grass. S. Asia, esp. in mixed forests of the Pegu Yomah and Martaban; westw. to Assam and E. Bengal. Excellent bamboo for roofs of houses, floors and walls. Has been recommended as a source of paper.

Bambusa spinosa Roxb. (Graminaceae). - Woody, tall grass. E. India. Cultivated. A timber bamboo, used for construction. Young tender shoots when boiled are consumed as a vegetable by the natives.

Bambusa Tulda Roxb. (Graminaceae). - Perennial woody grass. S. Asia, esp. Burma. Young buds are eaten in some parts of E. India.

Bambusa vulgaris Schrad. (Graminaceae). - Woody grass. Trop. Asia. Cultivated. Young buds are eaten as a vegetable.

Banabu → Lagerstroemia piriformis Koehne.

Banana, Casa → Sicana odorifera Naud.

Banana, Chinese Dwarf → Musa Cavendishii Lam.

Banana, Common → Musa sapientum L.

Banana, Dwarf Jamaica → Musa Cavendishii Lam·

Banana Fig → Musa sapientum L.

Banana, Japanese → Musa Basjoo Sieb.

Banana, Plantain → Musa paradisiaca L.

Banana, Red → Musa sapientum L.

Bangalay Eucalyptus → Eucalyptus botryoides Smith.

Bangiaceae (Red Algae) → Porphyra.

Banisteria Caapi Griseb., Yage (Malpighiaceae). - Liane, S. America, esp. Brazil, Columbia, Ecuador. Leaves and twigs are source of a beverage, used by the Indians of the n. w. of the Amazon region. It is known as Yage, Ayahuasca, Caapi. Contains telepathine, an alkaloid, said to cause hallucinations and gives a peculiar telepathic effect.

Banisteria lupuloides L. → Gouania lupuloides (L.) Urban.

Bank's Rose → Rosa Banksiae Ait.

Banksia integrifolia L.f. (syn. B. spicata Gaertn.) Beef Wood. (Proteaceae). — Tree. Australia, esp. Victoria, New South Wales and Queensland. Bark is used occasionally for tanning.

Banksia serrata L. f., Wattung-Urree. (Proteaceae). — Tree. Australia. Wood purplish, coarse, strong. Used for window frames, boat and ship building, furniture.

Baphia nitida Lodd. African Sandalwood. (Leguminosaceae). — Tree. Trop. Africa. Wood is source of Camwood of commerce; used for beams, pillars, walking sticks, pestles. Source of a red and brown dye; used for coloring wool. In Sierra Leone the herb ist used for skin diseases.

Baphia pubescens Hook. f. (Leguminosaceae). - Tree. Trop. Africa. Wood fine grained, heavier than water, turns red when exposed; has odor of violets; is source of a dye.

Baptisia leucantha Torr. and Gray. (Leguminosaceae). - Eastern N. America, esp. Florida to Louisiana. Plant is used among the Indians as a cathartic, strong emetic and in small doses as a laxative.

Baptisia tinctoria (L.) R. Br., Yellow Wild-Indigo, Rattle Weed. (Leguminosaceae). - Perennial herb. United States. Used by southern mountain people in the United States as a dye and flybrush. Dried root used medicinally, being emetic and cathartic in large doses. Contains cytisene, an alkaloid.

Barbados Aloë → Aloë vera L.

Barbados Cherry → Malpighia glabra L.

Barbados Gooseberry → Pereskia aculeata Mill.

Barbarea praecox R. Br. (syn. B. verna (Mill.) Asch.) Scurvy Grass, Winter Cress. (Cruciferaceae). - Biennial herb. Europe, naturalized in America. Occasionally cultivated as a winter salad. Used as pot herb in some parts of the United States, England and Germany. Seeds are used as a source of oil.

Barbarea vulgaris R. Br., Yellow Rocket. (Cruciferaceae). - Biennial or perennial herb. Temp. zone. Occasionally used in salads. Herba Barbariaea was formerly used in a balsam for healing wounds.

Barbasco → Paullinia pinnata L.

Barbatimao → Stryphnodendron Barbatinam Mart.

Barberry, American → Berberis canadensis Mill.

Barberry, European → Berberis vulgaris L.

Barberry, Indian → Berberis aristata DC.

Barberry, Magellan → Berberis buxifolia Lam.

Barcelona Nut → Corylus Avellana L.

Barcoo Grass → Anthistiria membranacea Lindl.

Bard Vetch → Vicia monantha Desf.

Barilla → Salsola Soda L.

Barister Gum → Mezoneuron Scortechinii F. v. Muell.

Barleria Prionitis L. (Acanthaceae). - Small prickly plant. Trop. Africa, Asia. Cultivated. Bitter juice is given in India to children for catarrh; in Java leaves are chewed for toothache; in Siam, a febrifuge is prepared from roots.

Barley, Common → Hordeum vulgare L.

Barley, Egyptian → Hordeum trifurcatum Ser.

Barley, Six-rowed → Hordeum hexastichon L.

Barley, Two-rowed → Hordeum distichon L.

Barnadine → Passiflora quadrangularis L.

Barosma betulina (Thunb.) Bartl. and Wendle, Bucco. (Rutaceae). - Shrub. S. Africa. Dried leaves used medicinally; a source of Short Buchu; diuretic, carminative. Contains an ess. oil, buchu camphor or diosphenol; also diosmin, an alkaloid. Leaves were used by the Hottentots before the coming of the White Man.

Barosma crenulata (L.) Hook., Buchu. (Rutaceae). - Low shrub. S. Africa. Dried leaves are source of Short Buchu, used medicinally. See B. betulina (Thunb.) Bartl.

Barosma serratifolia (Curtis) Willd., Buchu. (Rutaceae). - Shrub. S. Africa. Dried leaves are a source of Long Buchu, used medicinally as a diuretic and carminative.

Barringtonia acutangula Gaertn. (Lecythidaceae). - Tree. India to Moluccas. Bark used as fish poison. Wood is employed in local carpentry.

Barringtonia calyptrata R.Br. (Lecythidaceae). - Mangrove tree. Australia. Bark is used among the natives in N.Queensland to stupefy fish.

Barringtonia Careya F. v. Muell. (Lecythidaceae). - Woody plant. Australia. Bark of the stem is used to stupefy fish.

Barringtonia racemosa Roxb. (Lecythidaceae). - Tree. Malaysia, Pacific Islands. Seeds are source of oil, used as an illuminant. Roots and crushed fruits are used to stupefy fish. Young leaves are eaten raw or cooked by the natives.

Barringtonia speciosa Forst. (Lecythidaceae). - Medium sized tree. Trop. Asia, esp. India to Malaysia. Pounded fruits are used to stupefy fish. Pods when cooked are eaten in Indo China, regardless of being fish poison. Wood yellowish to red, light, soft, easy to split; used for furniture.

Bartsch (beverage) → Heracleum Sphondylium L.

Basanacantha armata Hook. f. (syn. Randia armata (Swartz) DC.). Crucito, Tintero (Rubiaceae). - Woody plant. Mexico to Panama. Fruits are edible, consumed by the natives.

Basella rubra L. (syn. B. alba L., B. cordifolia Lam.). Vine Spinach. (Basellaceae). - Herbaceous vine. Tropics. Cultivated. Leaves are consumed as pot-herb. Sap from fruits is used for coloring food and edible agar. Some varieties have green, other red leaves.

Basellaceae → Basella, Boussingaultia, Ullucus.

Basi (fermented drink) → Albizzia odoratissima Benth.

Basil, Holy → Ocimum sanctum L.

Basil, Sweet → Ocimum Basilicum L.

Basil, Wild → Calamintha Clinopodium Benth.

Basket Oak → Quercus Prinus L.

Basket Willow → Salix viminalis L.

Basketry. Some plants used for. → Akebia, Ancistrophyllum, Artocarpus, Arundinaria, Calamus, Calathea, Coccothrinax, Cocculus, Cocos, Costus, Daemonorops, Dasylirion, Dendrobium, Desmoncus, Eleocharis, Fimbrystilis, Flagellaria, Freycinetia, Hierochloë, Hyphaene, Jardinea, Jatropha, Juncus, Korthalsia, Lepironia, Ligustrum, Lygodium, Nepenthes, Nipa, Nolina, Oxytenanthera, Philodendron, Phrynium, Plectocomia, Pseudostachyum, Rhus, Rostkovia, Sabal, Salix, Spartium, Teinostachyum, Triticum, Vitex, Xerophyllum.

Bassia betis (Blanco) Merr. (Sapotaceae). - Tree. Philipp. Islands esp. Luzon to Mindanao. Fruits are source of Bétis Oil, locally used as illuminant.

Bassia latifolia Roxb. (Sapotaceae). - Tree. Trop. Asia, esp. Bengal, Indo-China, Malaysia. Flowers are source of an alcoholic beverage, called Mahua Spirit. The creamy white petals of the flowers are eaten fresh of after being dried. Seeds are source of an edible oil.

Bassia longifolia L. → Madhuca longifolia Gm.

Bassia malabarica Bedd. (Sapotaceae). - Tree. E. India. Flowers are eaten by the natives. Seeds are source of an oil.

Bassia Mottleyana Clarke (syn. Ganua Mottleyana Pierre.) (Sapotaceae). - Tree. Malaysia. Source of an inferior Gutta Percha. Seeds produce Katio Oil, used for pastries and foods; exported to Europe. Sp. Gr. O. 9173; Sap. Val. 189-193; Iod. No. 53-66; Unsap. 0.4-0.5%. Wood coarse, easily worked, reddish-brown; used for construction of houses and for boards.

Basswood → Tilia spp.

Basswood, American → Tilia americana L.

Bastard Bullet → Humiria floribunda Mart.

Bastard Cedar → Soymida febrifuga Juss.

Bastard Mahogany → Eucalyptus botryoides Smith.

Bastard Myal → Acacia Cunninghamii Hook.

Bastard Peppermint → Tristania suaveolens Smith.

Bastard Sandalwood → Eremophila Mitchelli Benth. and Myoporum platycarpum R. Br.

Bastard Siamese Cardamon → Amomum xanthioides Wall.

Bastard Teak → Butea superba Roxb.

Bastard Vervain → Stachytarpheta jamaicensis Vahl.

Batate → Ipomoea Batatas Poir.

Batavia Cinnamon → Cinnamomum Burmanni Blume.

Bati Oil → Gomphia parviflora DC.

Batidaceae → Batis.

Batiputa Oil → Gomphia parviflora DC.

Batis maritima L., Saltwort. (Batidaceae). - Small shrub. Sea-coast, Florida, West Indies, Mexico, Trop. America, Hawaii Islds. Leaves have a salty flavor, used occasionally in salads. In early days ashes were used for manuf. soap and glass.

Batjan Manila → Agathis loranthifolia Salisb.

Buttinan Crape Myrtle → Lagerstroemia piriformis Koehne.

Bauhinia esculenta Burch. (Leguminosaceae). - Woody plant. Trop. Africa. Pods are an important food among some of the natives of S. Africa.

Bauhinia malabarica Roxb. (Leguminosaceae). - Small tree. Trop. Asia. Leaves are used for flavoring meats and fish in the Philipp. Islds. and other parts of trop. Asia.

Bauhinia reticulata DC. (Leguminosaceae). - Small tree. Trop. Africa. Roots are source of a mahogany colored pigment. Bark produces a fibre, made into clothes and ropes by the Golo women (Afr.), Seeds are consumed by the natives. Pods and seeds are source of a black and blue dye. Ashes from the plant are used in manuf. soap.

Bauhinia retusa Roxb. (Leguminosaceae). - Woody plant. Himalayan region. Source of an inferior gum. Sold in N. India.

Bauhinia rufescens Lam. (Leguminosaceae). - Tree. W. Africa. Bark is used in French W. Africa for manuf. ropes, also for tanning leather; remedy for dysentery. Root is employed for intermittent fever. Decoction of leaves is used by natives for eye diseases. Wood used in carpentry.

Bauhinia Vahlii Wight and Arn. (Leguminosaceae). - Tree. India. Bark is source of a fibre, used for manuf. sails.

Bauhinia variegata L. (Leguminosaceae). - Tree. Burma, China, East Indies. Leaves are used in India for cigarette covers. Leaves and pods are eaten as vegetable. Bark is tonic, alterative, astringent; used for tanning and dyeing. Source of gum, Senn or Semba gona. Wood used for agricultural implements.

Bay Fat → Laurus nobilus L.

Bay, Loblolly → Gordonia Lasianthus (L.) Ellis.

Bay, Red → Persea Borbonia (L.) Raf.

Bay, Swamp → Magnolia glauca L.

Bay, Sweet → see Magnolia glauca L. and Persea Borbonia (L.) Raf.

Bay, True → Laurus nobilus L.

Bayberry → Myrica cerifera L.

Bdellium → Commiphora africana Endl.

Bdellium, Guban Myrrh → Commiphora Hildebrandtii Engl.

Bdellium, Habbak Dundas → Commiphora Hildebrandtii Engl.

Bdellium, Habbak Daseino → Commiphora Hildebrandtii Engl.

Bdellium, Habbak Dunkal → Commiphora Hildebrandtii Engl.

Bdellium, Habbak Harr → Commiphora Hildebrandtii Engl.

Bdellium, Habbak Alka Adacai → Commiphora Hildebrandtii Engl.

Bdellium, Habbak Tubuk → Commiphora Hildebrandtii Engl.

Bdellium, Harobil Myrrh → Commiphora Myrrha (Nees) Engl.

Bdellium, Indian → Commiphora Mukul Engl.

Bdellium, Mukul-i-Azrak → Commiphora Mukul Engl.

Bdellium, Mukul-i-Yahud → Commiphora Mukul Engl.

Bdellium of Bombay → Commiphora erythraea Engl.

Bdellium, Opaque→ Commiphora spp.

Bdellium, Perfumed → Commiphora erythraea Engl. var. glabrescens Engl.

Bdellium, Sakulali →Commiphora Mukul Engl.

Bea Gum → Caesalpinia praecox Ruiz and Pav.

Beach Plum → Prunus maritima Wang.

Beam Tree, Sandal → Adenanthera pavonina L.

Beaked Hazelnut → Corylus rostrata Ait.

Bead Tree → Elaeocarpus Ganitrus Roxb.

Beakpod Eucalyptus → Eucalyptus robusta Smith.

Beam Tree, White → Sorbus Aria Crantz.

Bean, Adzuki → Phaseolus angularis Wight.

Bean, Asparagus → Dolichos sesquipedalis L.

Bean, Bengal → Mucuna alterrima (Pier and Tracy) Holland.

Bean, Black Mauritius → Mucuna atterrima (Piper and Tracy) Holland.

Bean, Broad → Vicia Faba L.

Bean, Calabar → Physostigma venenosum Balf.

Bean, Castor → Ricinus communis L.

Bean, Common Coral → Erythrina Corallodendron L.

Bean, Eastern Coral → Erythrina herbacea L.

Bean, Florida Velvet → Mucuna Deeringiana (Bort.) Small.

Bean, Horse → Vicia Faba L.

Bean, Hyacinth → Dolichos Lablab L.

Bean, Ignatius → Strychnos Ignatii Berg.

Bean, Indian → Catalpa bignonioides Walt.

Bean, Indian Coral → Erythrina indica Lam.

Bean, Jack → Ganavalia ensiformis DC.

Bean, Jumping → Sapium biloculare (S. Wats.) Pax.

Bean, Jumping → Sebastiania Pavoniana Muell. Arg.

Bean, Jumpy → Leucaena glauca (L.) Benth.

Bean, Kidney → Phaseolus vulgaris L.

Bean, Lima → Phaseolus lunatus L.

Bean, Lucky → Thevetia nereifolia Juss.

Bean, Mescal → Sophora secundiflora (Orteg.) Lag.

Bean, Metcalf → Phaseolus retusus Benth.

Bean, Molucca→ Caesalpinia Bonducella Flem.

Bean, Moth → Phaseolus aconitifolius Jacq.

Bean, Mung → Phaseolus Mungo L.

Bean, Ordeal → Physostigma venenosum Balf.

Bean, Pichrim → Aydendron firmulum Nees. and Nectandra Puchury-minor Nees and Mart.

Bean, Pigeon → Vicia Faba L.

Bean, Potato → Apios tuberosa Moench.

Bean, Rice → Phaseolus calcaratus Roxb.

Bean, Screw →Strombocarpa odorata (Torr. and Frem.) Torr.

Bean, Sea → Mucuna gigantea DC.

Bean, Soy → Glycine soja Sieb. and Zucc.

Bean, St. Thomas → Entada scandens Benth.

Bean, Sweet → Gleditsia triacanthos L.

Bean, Tepary → Phaseolus acutifolius Gray var. latifolius Freem.

Bean, Tonka → Dipteryx odorata Willd.

Bean, Vanilla → Vanilla planifolia Andr.

Bean, Velvet → Mucuna nivea DC.

Bean, Wayaka Yam → Pachyrrhizus angulatus Rich.

Bean, West African Locust → Parkia filicoides Welw.

Bean, Wild → Phaseolus polytachyus (L.) B.S.P.

Bean, Yam → Pachyrrhizus erosus Urb., P. tuberosus Spreng.

Bearberry → Arctostaphylos Uva-ursi (L.) Spreng. Lonicera involucrata (Rich.) Banks. and Rhamnus Purshiana DC.

Bear Grass → Xerophyllum tenax (Pursh) Nutt.

Bear Grass → Yucca glauca Nutt.

Beardless Wild Rye →Elymus tricticoides Buckl.

Bearwood → Rhamnus Purshiana DC.

Bearded Useno → Usnea barbata Hoffm.

Beautyleaf, Ceylon → Calophyllum Calaba L.

Beaverhead Scurfpea → Psoralea castorea S. Wats.

Beaver Wood → Celtis occidentalis L.

Bebeeru → Nectandra Rodioei (Schomb,) Hook.

Beccabunga → Veronica Beccabunga L.

Beckmannia erucaeformis Host. (Graminaceae). - Herbeaceous plant. Temp. Asia, N.America. Grains are consumed in Japan.

Beckmannia syzigachne (Steud.) Fern. American Sloughgrass. (Graminaceae). - Annual grass. N. America, Asia. A forage grass for live-stock, also made into hay.

Bee Balm → Monarda didyma L.

Beebalm, Lemon → Monarda citriodora Cerv.

Beebalm, Pony → Monarda pectinata Nutt.

Beech, American → Fagus ferruginea Dryand.

Beech, Blue → Carpinus caroliniana Walt.

Beech, European → Fagus sylvatica L.

Beech, Malay Bush → Gmellina arborea Roxb.

Beech, Mountain → Nothofagus Cliffortoides Oerst.

Beech Nut Seed Oil → Fagus sylvatica L.

Beech, Oriental → Fagus orientalis Lipsky.

Beech, Red → Nothofagus fusca Oerst.

Beech, Silver → Nothofagus Menziesii Oerst.

Beefsteak Fungus → Fistulina hepatica (Huds.) Bull.

Beef Tongue → Fistulina hepatica (Huds.) Bull.

Beef Wood → Banksia integrifolia L. f., Grevillea striata R. Br. and Stenocarpus salignum R. Br.

Beer, Ginger → Saccharomyces pyriformis.

Beer, Spruce → Picea Mariana (Mill.) B.S.P.

Beet → Beta vulgaris L.

Beet, Red → Beta vulgaris L.

Beet, Sugar → Beta vulgaris L.

Beggarweed →Desmodium tortuosum (Sw.) DC.

Begonia bahiensis DC. (Begoniaceae). - Perennial herb. Brazil. Herb is considered in Brazil a febrifuge.

Begonia cucullata Willd. (syn. B. spathulata Lodd.). (Begoniaceae). - Perennial herb. E. Brazil. Herb is used as diuretic in some parts of Brazil.

Begonia luxurians Scheidw. (syn. Scheidweileria luxurians Kl.). (Begoniaceae). - Perennial herb. Brazil. Leaves are used in decoctions for fevers, in Brazil.

Begonia sanguinea Raddi. (syn. Pritzelia sanguinea Kl.). (Begoniaceae). - Perennial herb. Brazil. Herb is used in Brazil as diuretic.

Begonia tuberosa Lam. (Begoniaceae). - Perennial herb. Moluccas. Leaves are eaten among the natives of the Moluccas as sorrel, also cooked with fish and in sauce.

Begoniaceae → Begonia.

Behaimia cubensis Griseb. (Leguminosaceae). - Tree. Cuba. Wood strong, dark ash colored; used for construction work and general carpentry.

Beilschmiedia Baillonii Planch and Sébert. (Lauraceae). - Tree. New Caledonia. Wood durable, strong, easily worked; suitable for interior work and cabinet making in New Caledonia.

Beilschmiedia lanceolata Planch and Sébert. (Lauraceae). - Tree. New Caledonia. Wood grayish, long-grained; used for interior work.

Beilschmiedia odorata Baill. (Lauraceae). - Tree. New Caledonia. Wood is used for cabinet work. Also B. grandifolia Planch and Séb.

Beilschmiedia Tawa Hook f. (syn. Nesodaphne Tawa Hook f.). (Lauraceae). - Woody plant. New Zealand. Fruits are consumed by the aborigenes of North Island (New Zeal.).

Bejuco Blanco → Exogonium bracteatum (Cav.) Choisy.

Belamcanda chinensis DC., Blackberry Lily, Leopard Flower. (Iridaceae). - Perennial herb. China, Japan. Cultivated in Old and New World. Used in Chinese pharmacy for complaints of chest and liver; added in tonics and purgatives.

Belangera tomentosa Camb. (Cunoniaceae). - Tree. Brazil. The bitter, styptic bark is used as tonic in some parts of Brazil. Also B. speciosa Camb.

Belembe →Xanthosoma brasiliense (Desf.) Engl.

Bell Apple → Passiflora laurifolia L.

Belladonna → Atropa Belladonna L.

Belleric Terminalia → Terminalia Bellerica Roxb.

Bellwort → Uvularia sessiliflora L.

Benedictine (liqueur) → Angelica Archangelica L.

Bengal Bean → Mucuna alterima (Piper and Tracy) Holland.

Bengal Cardamon → Amomum aromaticum Roxb.

Bengal Kino → Butea superba Roxb.

Bengor Nut → Caesalpinia Bonducella Flem.

Benin Apata Wood → Microdesmis puberula Hook. f.

Benincasa cerifera Savi. (syn. B. hispida Cogn.), Wax Gourd, White Gourd, Chinese Preserving Melon. (Cucurbitaceae). - Annual vine. Trop. Asia, Africa. Cultivated in the tropics. Fruits used for sweet pickles, preserves, also put into curries; consumed as vegetable in Japan. Varieties are: Chosen-uri and Kiushiu.

Ben Oil → Moringa oleifera Juss.

Ben Seed Oil → Moringa oleifera Juss.

Bene → Sesamum indicum L.

Bengal Gambir Plant → Uncaria Gambir Roxb.

Benguet Pine → Pinus insularis Endl.

Benin Ebony → Diospyros crassiflora Hiern. and D. suaveolens Gurke.

Benin Mahogany → Guarea Thompsonii Sprague and Hutch., Khaya grandis Stapf., and K. Puchii Stapf.

Bent, Rhode Island → Agrostis vulgaris With.

Benzoin → Styrax Benzoin Dryander.

Benzoin, Siam → Styrax tonkinense Craib.

Benzoin, Sumatra → Styrax Benzoin Dryander.

Berberidaceae → Berberis, Caulophyllum, Jeffersonia, Mahonia, Podophyllum.

Berberis Aquifolium Pursh. → Mahonia Aquifolium (Lindl.) Don.

Berberis aristata DC., Indian Barberry. (Berberidaceae). - Shrub. N.W. Himalaya. Dried stem is used medicinally; a bitter tonic for intermittent fevers. Contains berberine an alkoloid; resin and tannin. A common household remedy among Hindus. Roots and stems are source of a yellow dye.

Berberis buxiifolia Lam., Magellan Barberry. (Berberidaceae). - Shrub. Chile to Strait of Magellan. Fruits edible, much consumed in some parts of Chile. Berries are made into preserves.

Berberis canadensis Mill., Alleghany Barberry, American Barberry. (Berberidaceae). - Shrub. Eastern N. America. Fruits are suitable for jellies.

Berberis Darwinii Hook. Michay. (Berberidaceae). - Shrub. Chile. The root and bark are used as tonic in some parts of Chile. Plant is also source of a yellow dye; used for dyeing clothes.

Berberis flexuosa Ruiz and Pav. (Berberidaceae). - Shrub. Argentina, Peru and Chile. Roots are source of a yellow dye; used for dyeing woolen goods.

Berberis haematocarpa Woot., Barberry. (Berberidaceae). - Shrub. New Mexico, Colorado. Fruits are used for jellies.

Berberis laurina Thunb. (Berberidaceae). - Shrub. Brazil. Root is source of a yellow dye; used for dyeing cotton.

Berberis lutea Ruiz. and Pav. (Berberidaceae). - Shrub. Peru, Ecuador. Wood is source of a yellow dye, used by Indians of Peru for coloring cottons.

Berberis Lycium Royle. (Berberidaceae). - Shrub. Himalayan region. Extract from roots, called Rasant, is much esteemed in native medicine. Considered febrifuge, gentle aperient, carminative; used in haemorrhoides.

Berberis repens Lindl. → Mahonia repens (Lindl.) G. Don.

Berberis rusciifolia Lam. (Berberidaceae). - Shrub. Brazil. Source of a yellow dye, used in some parts of Brazil.

Berberis vulgaris L. European Barberry. (Berberidaceae). - Shrub. Europe, Caucasia. Wood very hard, fine-grained, yellow; used for mosaic-work, turnery and tooth-picks. Wood and bark are source of a yellow dye, containing berberin, used for dyeing wool and leather. Small red berries are used as preserves. Confiture d'épine vinette, a French jam, is made from a seedless variety. Bark of stem and root is used medicinally, being tonic and alterative.

Bergamot → Citrus bergamia Risso and Poit.

Bergamot Mint → Mentha aquatica L.

Bergamot Oil → Citrus bergamia Risso and Poit.

Bergamot, Wild → Monarda fistulosa L.

Berlinia acuminata Soland. (Leguminosaceae). - Tree. Throughout Trop. W. Africa. Wood hard, white to light-red, streaked; used for carpentry, turnery, cabinet-work, canoes, drums, naval construction.

Bermuda Arrowroot → Maranta arundinacea L.

Bermuda Grass → Cynodon Dactylon (L.) Pers.

Berracos → Cyrtocarpa procera H. B. K.

Berria Ammonilla Roxb. (Tiliaceae). - Large tree. Ceylon, India, Mal. Archip. to Queensland. Source of a valuable lumber. Wood is hard, durable, flexible, very tough, dark red; used for house building, boats, wheelwork and shafts.

Berry bearing Bamboo → Melocanna bambusoides Trin.

Bertholletia excelsa H. B. K., Brazil Nut, Pará Nut. (Myrtaceae). - Tree. Brazil, Guiana, Venezuela. Seeds are consumed in large quantities, known as Brazil Nuts, used in international trade. Kernels are source of an oil, Brazil Nut Oil, used for edible purposes, manuf. of soap. Sp. Gr. O. 9170 - O. 9180; Sap. Val. 192-200; Iod. No. 98-106; Unsap. 0.5%. Also other species furnish Brazil Nuts. Bark is used in Pará for caulking ships.

Bertholletia nobilis H. B. K. (Myrtaceae). - Brazil. Seeds are edible, similar to B. excelsa H. B. K.

Betony, Common → Stachys officinalis (L.) Trev.

Bertram → Anacyclus officinarum Hayne.

Bertya Cunninghamii Planch. (Euphorbiaceae). - Tree. Australia, esp. Victoria, New South Wales. Source of a gum resin, abundantly produced by the plant.

Beta vulgaris L. Beet. (Chenopodiaceae). Annual to perennial herb. Cultigen. Derived from B. maritima L. Europe. Different races are grown as root or leaf crops in temp. zone of Old and New World. I. Group. Sugar Beet. Important commercial source of beet-sugar. Grown since modern times. Improved varieties contain 17 to 19% sugar. Waste pulp and tops are used as fodder for live-stock. Filter cakes employed as manure. Molasses made into alcohol and food for cattle. Varieties are: Klein Wanzleben, Rimpau. Dippe W. II. Group. Mangel-Wurzel. Roots used as fodder for cattle. Much grown in Europe and Canada. Varieties: Oberndorfer, Red Mammoth, Eckendorfer. III. Group. Red Beet. Red roots consumed as vegetable when cooked, in salads, also canned. Leaves sometimes eaten as pot-herb. Varieties are: Crosby's Egyptian, Eclipse, Early Model, Early Blood, Crimson Globe, Half Long Blood, Detroit Dark Red. IV. Group. Swiss Chard, Silver Beet. (var. Cicla L.) Leaves eaten as spinach, leaf-stalks as asparagus. Grown are: Fordhook Giant, Giant Lucullus, a red variety is Rhubarb Chard. Leaves of beets are sometimes used as substitute for tobacco.

Betelnut Palm → Areca Catechu L.

Betula Ermani Cham., Erman's Birch. Betulaceae. Tree. Japan. Bark used by the Ainu for healing wounds, said to prevent inflamation.

Betula japonica Sieb. (Betulaceae). Tree. Temp. Asia, esp. Hokkaidô (Jap.), Sachalin, Manchuria, China, Kamtchatka. Wood close-grained, hard, heavy, yellow-white, takes a good polish. Used in Japan for cheap domestic furniture, spools, wooden pipes. Bark used by the Ainu for domestic utensils and torches.

Betula lenta L., Cherry Birch, Black Birch. (Betulaceae). - Tree. Eastern N. America to Florida. Wood very strong, hard, close grained, heavy, dark brown; used for furniture, agricultural implements, wooden ware, fuel; sometimes for ship and boat building. Bark produces upon steam distillation an ess. oil, Oil of Sweet Birch, resembling Oil of Wintergreen; is chiefly composed of methyl salicylate; used medicinally for flavoring. An alcoholic beverage is made by fermenting the sugary sap from the stem in spring.

Betula lutea Michx., Yellow Birch (Betulaceae. - Tree. Eastern N. America. Wood very strong, close-grained, hard, heavy, light brown, tinged with red; used for furniture, button and tassel moulds, tubs of wheels, boxes and fuel.

Betula nigra L., Red Birch, River Birch. (Betulaceae). - Tree. Eastern part of N. America to Florida and Texas. Wood hard, strong, light, close-grained, light brown; used for woodenware, furniture, wooden shoes and turnery.

Betula papyrifera Marsh., Paper Birch, Canoe Birch. (Betulaceae). - Tree. N. America. Wood hard, strong, light, very close-grained, light brown tinged with red; used for wooden-ware, spools, shoe-lasts, in turnery, for pegs, also for wood pulp and fuel. Durable, tough, resinous bark is impervious to water and was used by northern Indians for canoes and baskets, drinking cups, bags and for covering their wigwams in winter.

Betula populifolia Marsh., Gray Birch. (Betulaceae). - Tree. Eastern N. America. Wood soft, not strong sight, not durable, close-grained, light brown; used for wooden-wares spools, clothespins, hoops for casks, wood-pulp, charcoal and fuel.

Betula pubescens Ehrh. (syn. B. alba L. pro parte) Birch. (Betulaceae). - Tree. Europe, Greenland, Temp. Asia to Kamtschatka. Wood yellowish to reddish-white, long grained, soft, difficult to split, elastic, not durable; suitable for wagons, cars, turnery, snow shoes, wooden nails, spoons. Stems are used for rustic garden furniture, bridges, etc. Juice from stem is used in some countries for making the hair attractive. Powdered bark is eaten by the natives of Kamtschatka with eggs of the sturgeon. Bark is eaten in Lapland during time of famine.

Betula utilis D. Don. (syn. B. Bhojpathra Wall.). Himalaya Birch, Indian Paper Birch (Betulaceae). - Tree. Kashmir, Himalayan region. Periderm from bark was used in former days as writing material in Kashmir. Also used as packing paper. Likewise B. alnoides Hamilt. (syn. B. acuminata Wall.).

Betula verrucosa Ehrh. (syn. B. pendula Roth., B. alba L. ex parte). Birch. (Betulaceae). - Tree. Europe, Caucacia, Temp. Asia, Mongolia. Wood somewhat similar to that of B. pubescens, is not easy to split, elastic, tough; used for spoons, pipes, nails, woodens hoes. Bark used for shoes, boxes, baskets, mats, clothes, tabacco boxes, also as covering of roofs of huts and houses. In some parts of Norway and Finland a bread is made from the bark, eaten during times of distress. Tar from Birch is an excellent preservative for leather and wood. Leaves with alum are a source of green dye; used with chalk, it gives a yellow color. Its soot is made into a black paint. Sugary sap in spring is made into a home-made wine, Birch Wine. Stems with the white bark are made into rustic garden furniture, bridges, etc. Bark and wood is source of an oil, derived from dry distillation; used medicinally in lotions or ointments, counter irritant, parasiticide, antiseptic in skin diseases. Much is derived from Russia, Poland and Finland.

Betulaceae → Alnus, Betula, Carpinus, Corylus, Ostrya.

Beurre d'Orère → Baillonella ovata Pierre.

Beverages. Source of some alcoholic. → Acacia, Acanthosyris, Achillea, Acrocomia, Agave, Albizzia, Amygdalus, Anacardium, Ananas, Annona, Antidesma, Arbutus, Arctostaphylos, Arenga, Armeniaca, Artemisia, Avena, Balanites, Bassia, Beta, Betula, Borassus, Buddleia, Calotropis, Carnegiea, Caryota, Catha, Ceratonia, Citrus, Coccolobis, Cocculus, Cocos, Cordyline, Cornus, Corypha, Daphne, Diospyros, Elaeagnus, Eleusine, Eugenia, Ficus, Genipa, Heracleum, Hordeum, Humulus, Hymenaea, Hyphaene, Imperata, Lolium, Lonchocarpus, Macaranga, Mahonia, Mammea, Manihot, Mauritia, Menyanthes, Musa, Myrciaria, Oenocarpus, Opuntia, Oryza, Panicum, Paullinia, Phoenix, Piper, Prosopis, Prunus, Pyrus, Raphia, Rheum, Ribes, Rubus, Saccharum, Sambucus, Schinus, Sclerocarya, Secale, Solanum, Sorbus, Spondias, Taraxacum, Trichocereus, Triticum, Vitis, Yucca, Zea, Zingiber, Zizyphus. See also under Cordials.

Beverages. Some sources of non-alcoholic. → Actephila, Actinella, Adansonia, Agastache, Alpinia, Alyxia, Ananas, Annonia, Armeniaca, Asperula, Atalantia, Aubrya, Averrhoa, Bactris, Banisteria, Bidens, Brosimum, Buettneria, Camellia, Cannabis, Carica, Catha, Cercidium, Chenopodium, Citrus, Cocos, Coffea, Cola, Coreopsis, Coriaria, Cymbidium, Cyperus, Dasylirion, Empetrum, Genipa, Gladiolus, Hydrangea, Lichtensteinia, Madhuca, Mahonia, Marumia, Mesona, Monstera, Mussaenda, Myrica, Myrtus, Ocimum, Oxalis, Passiflora, Picea, Pimpinella, Pithecolobium, Poncirus, Prunus, Punica, Pyrus, Quercus, Rhus, Ribes, Rosa, Rubus, Saccharum, Salvia, Scoparia, Securidaca, Sisymbrium, Smilax, Spondias, Sterculia, Tamarindus, Thelesperma, Theobroma, Vitis, Zingiber.

Bhang → Cannabis sativa L.

Bharbur Grass → Ischaemum angustifolium Hack.

Bible Frankincense → Boswellia Carterii Birdw.

Bidens Bigelovii Gray. (Compositae). - Herbaceous perennial. S.W. of United States and adj. Mexico. Decoction of flowertops were used as beverage by Indians of Texas.

Bidens pilosa L. (syn. B. chinensis Willd.) (Compositae). - Herbaceous. China, Japan, Philip. Islds. The Igorots of Benguet (Philipp.) mix the herb with half-boiled grains of rice for making a ricewine, called Tafei.

Bidens scandens L. → Salmea Eupatoria DC.

Bigarade → Citrus Aurantium L.

Big-bud Hickory → Carya tomentosa (Lam.) Nutt.

Big Shellbark Hickory → Carya laciniosa (Michx. f.) Loud.

Big Tree → Sequioa gigantea Lindl. and Gord.

Bigareaus (Cherries) → Prunus avium L.

Bigleaf Eucalyptus → Eucalyptus goniocalyx F. v. Muell.

Bignay China Laurel → Antidesma Bunias (L.) Spreng.

Bignonia alliacea Lam. (syn. Adenocalymna alliaceum Miers.) (Bignoniaceae). - Vine. Trop. America. Plant has been recommended as a vermifuge.

Bignonia Chica Humb. and Bonpl. (Bignoniaceae). - Tree. Trop. America. Flowers and leaves when boiled with other plant species are source of a red dye, used by the Indians to color their body.

Bignonia Copaia Aubl. → Jacaranda (Aubl.) Copaia D. Don.

Bignonia pentaphylla L. → Tabebuia pentaphylla (L.) Hemsl.

Bignonia stans L. → Tecoma stans (L.) H. B. K.

Bignoniaceae. → Bignonia, Catalpa, Crescentia, Cybistax, Diplanthera, Dolichandrone, Jacaranda, Kigelia, Markhamia, Newbouldia, Oroxylum, Parmentiera, Stereospermum, Tabebuia, Tecoma.

Bigroot Geranium → Geranium macrorrhizum L.

Big String Nettle → Urtica dioica L.

Bikkia mariannensis Brogn. (syn. Cormogonus mariennensis Raf.). Torchwood. (Rubiaceae). - Small tree. South Sea Islds. Wood ignites easily; used for torches by the natives.

Bilimbi → Averrhoa Bilimbi L.

Bilberry → Vaccinium meridionale Sw.

Billet Wood → Diospyros Dendo Welw.

Billian, Borneo → Eusideroxylon Zwageri Teijsm. and Binn.

Billyweb Sweetia → Sweetia panamensis Benth.

Bimlipatam Jute → Hibiscus cannabinus L.

Bimplipatam Tree → Hibiscus cannabinus L.

Binggas Terminalia → Terminalia comintana (Blanco) Merr.

Binjai Mango → Mangifera caesia Jacq.

Binukao → Garcinia Binucao Choisy.

Birch, Black → Betula lenta L.

Birch, Canoe → Betula papyrifera Marsh.

Birch, Cherry → Betula lenta L.

Birch, Erman's → Betula Ermani Cham.

Birch, Gray → Betula populifolia Marsh.

Birch, Himalayan → Betula utilis D. Don.

Birch, Indian Paper → Betula utilis D. Don.

Birch, Paper → Betula papyrifera Marsh.

Birch, Red → Betula nigra L.

Birch, River → Betula nigra L.

Birch Wine → Betula verrucosa Ehrh.

Birch, Yellow → Betula lutea Michx.

Birdlime, Plants source of → Artocarpus, Chrysophyllum, Conopharyngia, Cordia, Ficus, Ilex, Loranthus, Sapium, Trochodendron, Viscum.

Bird Rape → Brassica campestre L.

Birds Foot Trefoil → Lotus corniculatus L.

Biriba → Rollinia deliciosa Safford.

Bisobol Myrrh → Commiphora erythraea Engl. var. glabrescens Engl. and Opopanax Chironium (L.) Koch.

Bisabol Myrrh Tree → Commiphora erythraea Engl.

Biscayne Palm → Coccothrinax argentea (Lodd.) Sarg.

Bischoffia javanica Blume. (Euphorbiaceae). - Tree. S.E. Asia to Trop. Australia. Wood heavy, very hard, fine structure, dark reddish-brown; used for construction of houses, bridges, mining timber, durable when in contact with soil. Bark is source of a red dye, used to stain rattan baskets.

Bishop's Elder → Aegopodium Podagraria L.

Bistort → Polygonum Bistorta L.

Bitter Almond Oil → Amygdalus communis L.

Bitter Bark → Alstonia constricta F. v. Muell.

Bitter Casava → Manihot esculenta Crantz.

Bitter Cola → Cola acuminata Scott. and Engl.

Bitter Cress → Cardamine pennsylvinca Muhl.

Bitter Dock → Rumex obtusifolius L.

Bitter Fennel → Foeniculum vulgare Mill.

Bitter Leaf → Vernonia amygdalina Delile.

Bitter Lettuce → Lactuca virosa L.

Bitternut → Carya cordiformis (Wangh.) Koch.

Bitter Root → Lewisia rediviva Pursh.

Bitter Vetch → Vicia Ervilia (L.) Willd.

Bitter Wood → Quassia amara L.

Bittersweet → Solanum Dulcamara L.

Bitterwood, Mafura → Trichilia emetica Vahl.

Bitterwood, White → Trichilia spondioides Jacq.

Bixa Orellana L., Annatto Tree. (Bixaceae). - Tall shrub or small tree. Trop. America. Cultivated in tropics of Old and New World. Seed coat source of a yellow pigment used for coloring butter, cheese and ointments. Formerly used by Indians to paint their bodies when going to war.

Bixaceae → Bixa.

Black Alder → Alnus glutinosa Medic.

Black Ash → Fraxinus nigra Marsh.

Black Bamboo → Phyllostachys nigra Blanco.

Black Birch → Betula lenta L.

Black Butt → Eucalyptus pilularis Sm.

Black Cap → Rubus occidentalis L.

Black Carroway → Pimpinella Saxifraga L.

Black Catechu → Acacia Catechu Willd.

Black Cedar → Nectandra Pisi Miq.

Black Cohosh → Cimicifuga racemosa (L.) Nutt.

Black Cottonwood → Populus heterophylla L.

Black Crottle → Parmelia omphalodes (L.) Ach.

Black Cummin → Nigella sativa L.

Black Cutch → Acacia Catechu Willd.

Black Dammar Resin → Canarium bengalense Roxb.

Black Ebony → Diospyros Dendo Welw.

Black Gambir → Uncaria Gambir Roxb.

Black Gram → Phaseolus Mungo L.

Black Grama → Bouteloua eriopoda Torr.

Black Gum → Nyssa multiflora Wang.

Black Haw → Crataegus Douglasii Lindl. and Viburnum prunifolium L.

Black Hazel → Ostrya virginiana (Mill.) Willd.

Black Huckleberry → Gaylussacia baccata (Wang.) Koch.

Black Ironwood → Olea laurifolia Lam.

Black Lecanora → Haematomma ventosum (L.) Mass.

Black Locust → Robinia Pseudacacia L.

Black Mangrove → Avicennia nitida Jacq.

Black Mauritius Bean → Mucuna alterima (Piper and Tracy) Holland.

Black Medic → Medicago lupulina L.

Black Mulberry → Morus nigra L.

Black Mustard → Brassica nigra (L.) Koch.

Black Mustard Seed Oil → Brassica nigra (L.) Koch.

Black Nightshade → Solanum nigrum L.

Black Oak → Quercus velutina Lam.

Black Olive Tree → Laguncularia racemosa Gaertn. f.

Black Pepper → Piper nigrum L.

Black Persimon → Diospyros texana Scheele.

Black Pine → Callitris calcarata R. Br. and Podocarpus spicata R. Br.

Black Poplar → Populus nigra L.

Black Raspberry → Rubus occidentalis L.

Black Rosewood → Dalbergia latifolia Roxb.

Black Sage → Salvia mellifera Greene.

Black Sally → Acacia melanoxylon R. Br.

Black Salsify → Scorzonera hispanica L.

Black Sapote → Diospyros Ebenum Koen.

Black Sassafras → Cinnamomum Oliveri Bailey.

Black Sloe → Prunus umbellata Ell.

Black Snakeroot → Cimicifuga racemosa (L.) Nutt.

Black Spruce → Picea Mariana (Mill.) B.S.P.

Black Stinkwood → Ocotea bullata E. Mey.

Black Titi → Cyrilla racemiflora L.

Black Walnut → Juglans nigra L.

Black Wattle → Acacia binervata DC. and A. decurrens Willd.

Black Willow → Salix nigra Marsh.

Blackberry → Rubus.

Blackberry Lily → Belamcanda chinensis DC.

Blackseed Juniper → Juniperus saltuaria Rehd. and Wils.

Blackthorn Sloe → Prunus spinosa L.

Blackwood → Acacia melanoxylon R. Br.

Blackwood, African → Dalbergia melanoxylon Guill and Perr.

Bladder Campion → Cucubalus baccifer L.

Bladder Locks → Alaria esculenta (Lyngb.) Grev.

Bladder Wrack → Fucus.

Bladderseed → Levisticum officinale (Baill.) Koch.

Blackea trinervis Ruiz. and Pav., Mess Apple. (Melastomaceae). - Tree. Brit. Guiana. Bark is used for tanning. Also B. quinquenervis Aubl.

Blancmange → Chondrus crispus (L.) Stackh.

Blazing Star → Chamaelirium luteum (L.) Gray and Laciniaria punctata (Hook.) Kuntze.

Blepharis edulis Pers. (Acanthaceae). - Perennial herb. Sahara region, Iran, Arabia. Seeds are eaten by Nomads.

Blepharis linariaefolia Pers. (Acanthaceae). - Herbaceous plant. Trop. Africa. Plant is used as food for camels. Also B. edulis Pers. Seeds are eaten by the Nomads.

Blepharocalyx cisplatensis Griseb. (Myrtaceae). - Tree. Argentina, Uruguay and Chile. Herb is used in some countries of S. America as a tonic and astringent.

Blighia sapida Koenig Akee. (Sapindaceae). - Trop. Africa, West Indies. Fragrant flowers are source of a perfumed water; used as cosmetic by the natives of different parts of Africa. Fruit, Akee Apple is 7,5 cm long; fleshy arillus, has a nutty flavor, is firm and oily; becomes soon rancid; eaten when fried in butter or with salted fish. Plant has been introduced in the New World during the early slave trade. Cultivated in the West Indies.

Blind-Your-Eyes Tree → Excoecaria Agallocha L.

Blister Buttercup → Ranunculus sceleratus L.

Blistered Umbilicaria → Umbilicaria pustulata (L.) Hoffm.

Blood Root → Sanguinaria canadensis L.

Bloodwood → Eucalyptus corymbosa Smith.

Bloodwood, Brush → Baloghia lucida Endl.

Bloodwood, Brush → Synoum glandulosum Juss.

Bloodwood, Kutcha → Eucalyptus terminalis F. v. Muell.

Bloody Spotted Lecanora → Haematomma ventosum (L.) Mass.

Blue Ash → Fraxinus quadrangulata Michx.

Blue Beech → Carpinus caroliniana Walt.

Blue Cohosh → Caulophyllum thalictroides (L.) Michx.

Blue Ebony → Copaifera bracteata Benth.

Blue Flag → Iris versicolor L.

Blue Grama → Bouteloua gracilis (H.B.K.) Lag.

Blue Gum → Eucalyptus botryoides Smith and E. globulus L.

Blue Japanese Oak → Quercus glauca Thunb.

Blue-leaved Acacia → Acacia subcoerulea Lindl.

Blue Mahue → Hibiscus elatus Swartz.

Blue Sage → Salvia mellifera Greene.

Bluebeard → Salvia viridis L.

Blueberry → Vaccinium nitidum Andr.

Blueberry, Andean → Vaccinium floribundum H. B. K.

Blueberry, Colombia → Vaccinium floribundum H. B. K.

Blueberry Elder → Sambucus coerulea Raf.

Blueberry, Ground → Vaccinium Myrsinites Lam.

Blueberry, Swamp → Vaccinium corymbosum L.

Blueberry, Sweet → Vaccinium pennsylvanicum Lam.

Bluegrass, Canada → Poa compressa L.

Bluegrass, English → Festuca elatior L.

Bluegrass, Fowl → Poa palustris L.

Bluegrass, Kentucky → Poa pratensis L.

Bluegrass, Pine → Poa scabrella (Thurb.) Benth.

Bluegrass, Timberline → Poa rupicola Nash.

Bluegrass, Wood → Poa nemoralis L.

Bluestem Willow → Salix irrorata Anders.

Bluewood Condalia → Condalia obovata Hook.

Blumea balsamifera DC. (syn. Conyza balsamifera L.) Ngai Camphor. (Compositae). - Subshrub. Himalaya, Nepal to Philipp. Islands, Moluccas, Mal. Archip. Strongly camphor scented. Used medicinally by Malays Sold in bazaars. Leaves eaten with food as aromatic. Decoction of leaves and roots used for fever. Ngai Camphor is obtained by distillation.

Blumea myriocephala DC. (Compositae). - Tall herb. E. Himalaya to Mal. Penin. Cultivated in French Indo-China. Leaves used to season fish. Used medicinally as sudorific for bronchitis.

Boabab → Adansonia digitata L.

Boabab, Madagascar → Adansonia madagascariensis Baill.

Boabab Oil → Adansonia digitata L.

Bobartia indica L. (syn. Bobartia spathaca Ker-Gawl.). (Iridaceae). - Perennial herb. S. Africa. Tough rush-like leaves are used in S. Africa for fruit baskets.

Bocconia cordata S. Wats. (Papaveraceae). - Tree. Centr. America, esp. Mexico to Guatemala. Bark source of a yellow dye. Produces an alkaloid, used as local anesthetic by surgeons in Mexico.

Bocconia frutescens L. (Papaveraceae). - Shrub or tree. Mexico to Peru. Used in Mexico for skin eruptions, chronic ophthalmia; to remove warts. In Colombia infusion of roots is used for jaundice and dropsy. Yellow sap from the bark is used by certain Indian tribes to dye feathers.

Boea Copal → Agathis alba Foxw.

Boehmeria caudata Swartz. (B. flagelliformis Liebm.) Urticaceae. - Perennial herb. Trop. America. Bark source of a fibre of poor quality; used by Chinantes, (Mex.).

Boehmeria macrophylla D. Don. (Urticaceae). - Perennial herb. Nepal, Sikkim, Khasia Mts. Fibres from stem are manuf. into fishing-nets.

Boehmeria nivea (L.) Gaud., Ramie, China Grass. (Urticaceae). - Perennial herb. China. Cultivated in warm countries of Old and New World. Source of an excellent, strong fibre; very durable, has much luster. Important article of commerce is China, India etc. Used for fabrics, Chinese Linen, Canton Linen, Grass Linen, Swatow Grass, for table cloths, plushes, napkins, curtains, dress goods, knit material etc.

Boerhaavia plumbaginea Cav. (Nyctaginaceae). - Herbaceous plant. Trop. Africa, Mediterranean region, Arabia. Decoction from leaves used for jaundice in some parts of Africa, also for healing wounds.

Boerhaavia repens L. (syn. B. diffusa L.). (Nyctaginaceae). - Herbaceous plant. Pantropic. Roots are eaten as food by the natives of Centr. Australia. Leaves are consumed as pot herb and in soups. In Angola decoction of leaves is used for jaundice. Roots and leaves are considered an expectorant, emetic; used for asthma.

Boerhaavia tuberosa Lam. (Nyctagynaceae). - Perennial herb. Peru. Roots are consumed as a vegetable by the natives of Peru.

Boerlagiodendron palmatum Harms. (syn. Trevesia moluccana Miq.) (Araliaceae). - Shrub. Mal. Archip. Leaves and stems are eaten boiled in water or in coconut milk by the natives.

Bog Myrtle → Myrica Gale L.

Bog Rosemary → Andromeda polifolia L.

Bog Spruce → Picea Mariana (Mill.) B. S. P.

Bois de Rose Femelle → Aniba rosaedora Ducke.

Bokalahy Rubber → Marsdenia verrucosa Decne.

Bokhara Galls → Pistacia vera L.

Boldia fragrans Gay → Peumus Boldus Molina.

Boletus edulis Bull Polish Mushroom, Cepe (Polyporaceae). - Basidiomycete. Fungus. The fruitbodies are consumed as food and are much esteemed. Sold fresh and dried in markets of several countries in Europe. Also pickled. Also B. badius Fr., B. appendiculatus Shäff., B. chrysenteron Bull., B. scaber Bull. and others.

Boletus castaneus Bull. → Gyroporus castaneus (Bull.) Quél.

Boletus Chirurchorum → Polyporus officinalis (Vill.) Fr.

Boletus Laricis Jacq. → Polyporus officinalis (Vill.) Fr.

Boletus subtomentosus L. (Polyporaceae). - Basidiomycete. Fungus. Temp. zone. Fruitbodies are consumed as food in several countries of Europe and Asia.

Bolivian Black Walnut → Juglans boliviana Dode.

Bolongeta Ebony → Diospyros pilosanthera Blanco.

Boltonia cantoniensis Franch. and Sav. (syn. Aster cantoniensis Blume). (Compositae). - Perennial herb. China, Japan. Young leaves when boiled are consumed in Japan as a vegetable.

Bomarea acutifolius (Link and Otto). Herb. (Amaryllidaceae). - Herbaceous vine. S. Mexico and Centr. America. Starchy tubers are used as food by Mazatecs, Chinantecs and other Indian tribes in Mexico. Also B. ovata Mirb. B. edulis Herb. in Santo Domingo; U. salsilla Herb. in Chile; B. glaucescens Baker. in Ecuador.

Bombaceae → Bombacopsis, Bombax, Cavanillesia, Ceiba, Chorisia, Cullenia, Durio, Eriodendron, Matisia, Ochroma, Pachira, Quararibea.

Bombacopsis Fendleri (Seem.) Pitt. (Bombaceae). - Tree. Centr. America. Wood light, used for construction work.

Bombacopsis sepium Pitt., Saquen Saqui (Bombaceae). - Tree. Venezuela. Wood used for manuf. tanning vats and rum storage vats.

Bombax buonopozence Beauv., Silk Cotton Tree. (Bombaceae). - Tree. Trop. Africa. Source of a silky fibre, similar to Kapok. Decoction of bark used as emmenagogue by Yorubos. Bark mixed with tobacco flowers is used in Kontagora (Afr.) to improve appearance of teeth. Wood light, soft; used in French Guinea for canoes, tom-toms, household utensils, doors, boards.

Bombax campestris Schum. (Bombaceae). - Shrub. Brazil. Fruits are source of a yellow floss, being of slight value; used for stuffing pillows etc.

Bombax ellipticum H. B. K. (syn. B. mexicanum Hemsl.). (Bombaceae). - Large tree. Mexico to Centr. America. Decoction of bark and root is used for tooth ache and hardening of gums.

Bombax longiflorum Schum. (Bombaceae). - Tree. Brazil. Fruits are source of a yellow-brown floss; employed in Brazil for home-use in stuffing pillows etc. Has little commercial value. Bark is source of a tough fibre.

Bombax malabarica DC. (syn. Gossampinus heptaphylla Bakh.) (Bombaceae). - Tree. Malaysia. Wood has been recommended as a source of cellulose. Stem is source of a resin, called Gum of Malabar.

Bombax Manguba Mart. (Bombaceae). - Tree. Brazil. Seeds are source of a floss of inferior quality, sometimes used for stuffing pillows.

Bombay Aloe Figbre → Agave cantala Roxb.

Bombay Ebony → Diospyros montana Roxb., D. Gardneri Thw.

Bombay Hemp → Agave cantala Roxb.

Bombay Mace → Myristica malabarica Lam.

Bombay Mastic → Pistacia mutica Fisch and Mey.

Bombay Rosewood → Dalbergia latifolia Roxb.

Bonduc Nut → Caesalpinia Bonducella Flem.

Boneset → Eupatorium perfoliatum L.

Bonnaya antipoda Druce (syn. Ilysanthes antipoda Merr.). (Scrophulariaceae). - Herbaceous plant. Trop. Asia. Decoction of roots and leaves is used as vermifuge.

Bonnemaisoniaceae (red algae) → Asparagopsis

Bootlace Tree → Eperua falcata Aubl.

Borage → Borago officinalis L.

Borago officinalis L., Borage. (Boraginaceae). - Annual herb. Mediterrenean region. Cultivated. Blue flowers are used as garnish. Plant is used for flavoring of beverages, among which negus and claret cup. Flowers and dried shoots, Herba et Flores Borroginis, were used since the Middle Ages as tea or syrup, as refreshing drink, also as diurectic; in recent times also as diaphoretic.

Borassus flabellifer L., Palmyra Palm (Palmaceae). - Tall palm. Trop. Asia. Of much importance in India. Leaves used for writing material since time immemorial, also for palm-leaf books, fans, thatching, mats, hats, bags, buckets, green-manure, manuf. potash, primitive flutes. Source of different fibres, from leaf-stalk, from inner part of stem, from pericarp of fruit. Inflorescens source of sugar wine and vinegar. Pericarp is edible. Kernels are consumed when young. Salt is prepared from the leaves.

Bordeaux Turpentine → Pinus Pinaster Soland.

Boraginaceae → Alkanna, Arnebia, Borago, Cordia, Cynoglossum, Ehretia, Heliotrpium, Lithospermum, Macrotomia, Mertensia, Onosma, Patagonula, Pulmonaria, Symphytum.

Borecole → Brassica oleracea L. var. acephala DC.

Boree → Acacia pendula Cunningh.

Boridschah → Ferula galbaniflua Boiss. and Buhse.

Borneo Billian → Eusideroxylon Zwageri Teijsm. and Binn.

Borneo Camphor → Dryobalanops aromatica Gaertn. f.

Borneo Ironwood → Eusideroxylon Zwageri Teysm. and Binn.

Borneo Mahogany → Calophyllum inophyllum L.

Borneo Manila → Agathis loranthifolia Salisb.

Borneo Shorea → Shorea aptera Burck.

Borneo Tallow → Shorea aptera Burck. and S. stenocarpa Burck.

Boronia megastigma Nees. (Rutaceae). - Woody plant. S. W. of Western Australia. Source of Oil of Boronia, an ess. oil, used in perfumes, resembling violet and cassie.

Boronia Oil → Boronia megastigma Nees.

Borrera, Yellow → Theloschistes flavicans (Swartz) Muell. Arg.

Borreria articularis Williams. (Rubiaceae). - Herbaceous. Trop. Asia. Used for poultices. Sold by Chinese herbalists.

Borreria capitata (Ruiz and Pav.), DC, Poya. (Rubiaceae). - Perennial. Throughout S. America. Supposed to be used as a substitute for ipecacuanha.

Borzicactus sepium (H. B. K.) Britt. and Rose. (syn. Cactus sepium H. B. K., Clustocactus sepium Weber.). (Cactaceae). - Xerophytic shrub. Ecuador Fruits are consumed in some parts of Ecuador.

Boscia angustifolia Rich. (Capparidaceae). - Shrub or small tree. Sahara region. Seeds are eaten by the natives.

Boscia caffra Sond. → Maerua pedunculata Vahl.

Boscia octandra Hochst. (Capparidaceae). - Shrub or small tree. Trop. Africa. Fruits, called Kursan by the Arabs in Sudan, are edible. Emulsion of leaves is used as eyewash.

Boscia senegalensis Lam. (Capparidaceae). - Tree. Trop. Africa. Leaves and berries are used as food, in soup and mixed with cereals. In Sudan fruits are eaten as emergency food. Seeds are a substitute for coffee. Sold in native markets.

Bosé Wood → Staudtia kamerunensis Warb. and Xylopia striata Engl.

Bosquea angolensis Ficalho. (Moraceae). - Tree. Trop. W. Africa. Bark is source of a latex that turns orange to blood-red; used for adulterating rubber. Employed in S. Nigeria for water-proofing bags, used by hunters in which to keep powder dry.

Bosquea Phoberus Baill. (Moraceae). - Tree. Trop. Africa. Milky sap is source of a red dye.

Boswellia Ameero Balf. (Burseraceae). - Tree. Socotra. Source of a resin, used as incense.

Boswellia Carterii Bird. Bible Frankincense. (Burseraceae). - Wood plant. Iran, Iraq, Somaliland. Source of Olibanum, an incense. Zakana „male frankincense" is deep yellow or reddish, having circular tears; Kundura Unsa „female frankincense" is reddish white, translucent, has pale tears. Gum is used medicinally as expectorant, stimulant, in plasters, for fumigation. Contains gum, arabin, 6 to 8⁰/o bassarin and a bitter principle. The incense is much used during ceremonies in Roman Catholic Churches.

Boswellia Frereana Birdw. Elemi Frankincense. (Burseraceae). - Tree. Trop. Africa. Stem is source of a resin, called African Elemi, Loban Maidi Luban Meti, a superior frankincense; a lemon scented balsam, sold in pale topaz-yellow tears or flat irregular pieces.

Boswellia papyrifera Hochst., Elephant. Tree. (Burseraceae). - Tree. E. Africa. Source of a fragrant incense, used in churches. It is said that elephants feed on the tree. Also B. Bhaw-Dajiana Birdw. and B. neglecta Moore.

Boswellia serrata Roxb. Indian Frankincense, Indian Olibanum Tree. (Burseraceae). - Tree. N. W. India. Source of a gum, Salai-gugul, transparent, golden-yellow, hardens slowly; used in medicine. Wood used for tea-boxes, charcoal, bowls, dishes.

Botany Bay Gum → Xanthorrhoea hastilis R. Br.

Botaosinho Root → Croton humilis L.

Botoko Plum → Flacourtia Ramontchi L'Hér.

Botor tetragonolobus Adans → Psophocarpus tetragonolobus DC.

Bot Tree → Ficus religiosa L.

Botrychium ternatum Sw. (Ophioglossacae). - Perennial herb. Fern-ally. Japan, China. Leaves are eaten in Japan as a vegetable, said to have a soft delicious taste.

Bottle Gourd → Lagenaria vulgaris Ser.

Bouea burmanica Griff. (Anacardiaceae). - Tree. Cochin-China, Laos, Burma, Malaysia. Fruits edible, resemble a small mango.

Bouea macrophylla Griff. (Anacardiaceae). - Tree. Malaysia. Fruits resembling a yellow plum, are eaten by the natives raw or cooked, also as pickles. Young leaves are eaten with rice in Java.

Boulder Lichen → Parmelia conspersa (Ehrh.) Ach.

Boussingaultia baselloides H. B. K., Madeira Vine, Mignonette Vin. (Basellaceae). - Perennial vine. Mexico to Chile. Occasionally cultivated. Leaves are eaten as spinach. Tubers are consumed as food by the natives.

Bouteloua curtipendula (Michx.) Torr. Side-Oats Grama. (Graminaceae). - Perennial Grass. Prairies of North America. Important pasture grass.

Bouteloua eriopoda Torr., Black Grama. (Graminaceae). - Perennial grass. S. W. of United States and Northern Mexico. An important pasture grass.

Bouteloua filiformis (Fourn.) Griff., Grama Grass. (Graminaceae). - Perennial grass. S. W. of United States, Northern Mexico. Used as pasture grass.

Bouteloua gracilis (H. B. K.) Lag., Blue Grama. (Graminaceae). - Perennial grass. Plains. N. America. Used as pasture grass.

Bovista nigrescens Pers. (Lycoperdaceae). - Basidiomycete. Fungus. Fruitbodies known as puffballs are consumed as food. Also B. plumbea Pers., and B. pila Berk and Curt. The fruitbodies of B. plumbea were used as food by the Omaha Indians.

Bow Wood → Maclura aurantiaca Nutt.

Bowdichia virgilioides H. B. K. Alcornoco. (Leguminosaceae). - Tree. S. America, esp. Venezuela, Guianas, Brazil. Wood brown to reddish-brown, tough, very heavy, strong, difficult to work; used in Brazil for hubs, felloes of cartwheels.

Bowenia spectabilis Hook. (Cycadaceae). - Woody plant. Australia, esp. Queensland. Large yam-like rhizones are consumed by the natives.

Bowman's Root → Gillenia trifoliata (L.) Moench.

Box → Buxus sempervirens L.

Box, Brush → Tristania conferta R. Br.

Box, Cape → Buxus Macowani Oliv.

Box Elder → Acer Negundo L.

Box, Red → Eucalyptus polyanthemos Scham. and Tristania conferta R. Br.

Box, White → Eucalyptus hemiphloia F. v. Muell. and Tristania conferta R. B.

Box, Yellow → Eucalyptus hemiphloia F. v. Muell.

Boxwood → Gossypiospermum praecox (Griseb.) Wils., Jacaranda Copaia (Aubl.) D. Don. and Schaefferia frutescens Jacq.

Boxwood, Brazilian → Euxylophora paraensis Huber.

Boxwood, Ceylon → Plectronia didyma Bedd.

Boxwood, San Domingo → Phyllostylon brasiliensis Cap.

Boxwood, West African → Sarcocephalus Diderrichii De Wild.

Boxwood, West Indien → Tabebuia pentaphylla (L.) Hemsl.

Brachiaria distichophylla Stapf. (Graminaceae). - Annual grass. Trop. Africa, esp. Nigeria, Belg. Congo, Angola, Gold Coast, Senegamba. Excellent pasture-grass, much liked by Hippopotami.

Brachiaria disticha (L.) Stapf. (syn. Panicum distachum L.). (Graminaceae). - Herbaceous grass. Trop. Asia, from Punjab to Australia. Good fodder for cattle.

Brachylaena elliptica Less. (Compositae). - Arborescent shrub. S. Africa. Leaves are used in Cape Peninsula for diabetes.

Brachylaena Hutchinsii Hutch. (Compositae). - Herbaceous plant. Trop. Africa, esp. Kenya. Source of Muhugu Oil; recommended for soap perfumery and as a fixative in perfumery.

Brachystegia spicaeformis Benth. (Leguminosaceae). - Tree. Trop. Africa. Wood very hard, heavy; used in the Congo for construction and different kind of woodwork; employed in Zanzibar for kilts, band boxes, grain stores, matches, roofing for huts. Bark is source of a fibre, used in E. Africa for sacking material.

Brachystelma Bingeri Chev., Fikongo. (Asclepiadaceae). - Perennial herb. Trop. Africa, esp. French Sudan. Tubers are much esteemed as food by the natives of the Niger region.

Brachystelma lineare. Rich. (Asclepiadaceae). - Perennial plant. E. Africa. Tubers are eaten by the natives of Abyssinia.

Brachytrichia Quoyi (C. Ag.) Born. & Flah. (Rivulariaceae). - Blue-green alga. Warmer parts of the Pacific and Indian oceans. Used for food in China and Japan.

Bracken → Pteridium aquilinum (L.) Kuhn.

Brackenbridgea zanguebarica Oliv. (syn. Ochna alboserrata Engl.) (Ochnaceae). - Tree. Trop. Africa. Bark is source of a yellow dye.

Bragantia corymbosa Griff. (syn. Apama tomentosa Kuntze) (Aristolochiaceae). - Shrub. Java. Stems and leaves are used in Java for washing clothes. Also used for snake bites.

Brahea dulcis (H. B. K.) Mart. (Palmaceae). - Palm. Mexico. Trunks are used for frames of native houses. Leaves are employed for thatching. Fruits, called Michire or Miche, are sweet and edible.

Bramble, Arctic → Rubus arcticus L.

Bramble of the Cape → Rubus rosaefolius Smith.

Brasenia Schreberi J. F. Gmel. (syn. B. peltata Pursh). (Nymphaeaceae). - Perennial waterplant. Asia, Africa, Australia, N. America. Fresh young leaves when seasoned with vinegar are eaten in Japan during spring.

Brassica adpressa Bois. (syn. Sinapis incana L.) (Cruciferaceae). - Herbaceous plant. Greece, Turkey. Young plants are consumed in spring with oil and lemon juice in some parts of Greece.

Brassica alba (L.) Boiss. → Sinapis alba L.

Brassica arvensis (L.) Ktze., Charlock. (Cruciferaceae). - Europe, Introd. in N. America. It is said that the seeds are source of Dakota Mustard.

Brassica Besseriana Adrz. (syn. B. juncea (L.) Coss. var. eu-juncea Thell.). (Cruciferaceae). - Annual herb. Russia, Trop. Asia. Cultivated. Seeds are source of Sarepta Mustard.

Brassica campestris L. Bird Rape. (Cruciferaceae). - Annual herb. Temp. Europe, Asia, naturalized in N. America. Seeds source of Ravinson-Oil; grown in Black Sea area. Semi-drying. Used locally for manuf. soap, lubricant, burning oil. Sp. Gr. O. 9175 - O. 9217; Sap. Val. 173—181; Ion. No. 109-122; Unsap. 1.4 - 1.8%.

Brassica campestris L. var. chinoleifera. Chinese Golza (Cruciferaceae). - Annual herb. Cultigen. Cultivated in Asia. Source of Chinese Colza Oil; semi-drying; used in food, for technical purposes, similar to Rape-Oil. Sp. Gr. 0.9097; Sap. Val. 273.8; Iod. No. 100.3; Unsap. 96.1%.

Brassia chinensis L., Pak-choi, Chinese Cabbage. (Cruciferaceae). - Perennial to biennial herb. Originally from China. Cultivated in temp. zones of Old and New World. Consumed as a vegetable. Has somewhat the appearance of Swiss Chard. Numerous asiatic varieties among which: Chihli, Wong Bok, Hagaromo, Giant Shuntang, Santo-Sai.

Brassica integrifolia O. E. Schultz, Indian Brown Mustard, Rai. (Cruciferaceae). - Annual herb. India. Cultivated. Source of Indian Brown Mustard. Related to B. juncea (L.) Cosson.

Brassica japonica Sieb. Japanese Mustard, (Cruciferaceae). - Annual to perennial herb. Cultivated in Japan. Soft, thin leaves are consumed as greens.

Brassica juncea (L.) Coss. (syn. Synapis juncea L.) India Mustard. Leaf Mustard. (Cruciferaceae). - Annual to perennial herb. Europe and Asia. Has been cultivated for centuries. Leaves are consumed when cooked. Varieties are: Chinese Brown Leaved, Florida Broad Leaved, Komatsuma.

Brassica napiformis Bailey., Tuberous-rooted Chinese Mustard. (Cruciferaceae). - Annual to perennial herb. Cultigen. Origin uncertain. Varieties are chiefly grown for their fleshy, turniplike, edible roots.

Brassica Napo-Brassica Mill., Rutabaga. Swedish Turnip. (Cruciferaceae). - Annual to perennial herb. Cultigen. Has been cultivated for centuries. Roots which are more elongated and oval and larger than Turnips are consumed as a vegetable, especially for winter-use. Usually yellow fleshed, some are white. Also called Russian, Winter or Yellow Turnip. Varieties are: Golden Neckless, Sweet German, Long Island Improved. Most varieties need a few weeks longer to mature than Turnips.

Brassica Napus L., Rape, Colza (Cruciferaceae). - Annual to biennial herb. Cultigen. Origin uncertain. Has been cultivated since ancient times. Oil is obtained from the pressed seeds of annual varieties. Biennial forms are cultivated as a late summer and autumn forage for live-stock. Varieties are: Dwarf Essex and Victoria. Var. oleifera D. C. Seeds source of Rape Oil, Colza Oil, classified between non-drying and semi-drying. Cold pressed oil used in foods in Europe and India, oiling of loaves before baking, for lubricating delicate machineries, illumation, used as wool oil; for soft soap, tempering steel plates. Sp. Gr. 0.9139-0. 9160; Sap. Val. 170 - 180; Iod. No. 98 - 106; Unsap. 0.5 - 1.5%. Oils derived from B. Napus L. B. Rapa L. and B. campestris L. are often classified as Rape and Colza Oils. Similar oils are derived in E. India from B. glauca Roxb., B. dichotoma Roxb., B. ramosa Roxb. and B. juncea (L.) Coss.

Brassica oleracea L., Cabbage (Cruciferaceae). - Annual or biennial herb. Europe. Contains a large group of vegetable races. Origin uncertain. Grown since early centuries. Cultivated in temp. zone of Old and New World. Important commercial crops. I. Group. Var. acephala DC. Source of Kale and Collard, sometimes termed Marrow Cabbage or Borecole. Leaves are consumed as food. Cultivated are: Dwarf Curled Scotch, Early Curled Siberian, Dwarf Green Curled. Fall, winter and early spring vegetable. II. Group. Var. gemmifera Zenker. Brussels Sprouts. Side-buds or side-heads on stem used as vegetable when boiled. Variety: Long Island Improved. III. Group. Var. capitata L. Typical cabbage. Some varieties are grown for Sauerkraut. Divided into A. Wakefield and Winningstadt. Small pointed heads. Mature early, grown for an early crop. Grown are: Charleston Wakefield, Jersey Wakefield, Early Winningstadt. B. Copenhagen Market. Round heads, compact, light green, heads much larger than the former among which: Copenhagen Market, Golden Acre. C. Flat Dutch or Drumhead. Flat heads, large, fairly solid, among which: Glory of Enkhuizen, Early Summer, Succession. D. Savoy. Leaves of heads much wrinkled, foliage dark green among which: Savoy, Perfection Savoy, Drumhead Savoy. E. Danish Ballhead. Heads of medium size, very solid, excellent keeping qualities, late varieties, among which Danish Roundhead, Hollander. F. Alpha. Heads smaller than Wakefield, round solid. Grown are St. John Day, Miniature Marrow. G. Volga. Plants large, few outer leaves. Heads round, somewhat flattened, medium size. Among which Volga. H. Red Cabbage. Cabbages deep purple color, among which: Rock Red, Red Dutch, Red Danish, Mommoth Red Rock. IV. Group. Var. botrytis L. Cauliflower, Broccoli composed of undeveloped, crowded, abnormal flower-buds, among which are grown: Early Snowball, Snowdrift. V. Group. Var. italica Plenck. Asparagus or Sprouting Broccoli, among which Italian Green Sprouting. VI. Group. Var. gongyloides L. (syn. B. caulocarpa Pasq.). Kohlrabi. Turnip-like enlarged stem, sourrounded by leaves. Developed above ground, among which are grown: Green Vienna, White Vienna, Earliest Erfurt, Purple Vienna. Used as vegetable and as fodder for live-stock. Early spring or fall crop.

Brassica nigra (L.) Koch., Black Mustard. (Cruciferaceae). - Annual herb. Europe, Introd. in N. America. Cultivated as an agric. crop. Seeds are source of Black Mustard. They produce Black Mustard Seed Oil, a slow drying oil, used for edible purposes, as lubricant and for soft soaps. Sp. Gr. 0.917-0. 922; Sap. Val. 176-184; Iod. No. 114-124. Medicinally dried ripe seed is used as emetic, externally as a rubefacient. Used for mustard plasters. Contains a fixed oil, and sinigrin a glucoside, accompanied by myrosin, an enzyme. When the seeds are crushed with water, the sinigrin is hydrolyzed forming allyl-isothiocyanate (mustard oil).

Brassica pekinensis Rupr., Pe-tsai, Chinese Cabbage. (Cruciferaceae). - Annual to biennial herb. Cultivated. Originated from China. Grown in temperate zones of Old and New World. Used as a vegetable. Has the appearance of Cos-Lettuce.

Brassica Rapa L. Turnip. (Cruciferaceae). - Annual to perennial herb. Cultigen. Origin uncertain. Has been cultivated for centuries. A cool season-crop. Broad thick roots are consumed as a vegetable, also used as fodder for live-stock. Varieties: White Milan, Whit Flat Dutch, Yellow Aberdeen. Seeds source of Oil, see B. Napus L.

Brauneria angustifolia (DC). Heller. (syn. Echinacea angustifolia DC.). Rattlesnake Weed. (Compositae). - Perennial herb. Eastern N. America to Texas. Ground root was chewed and the juice swallowed by the Kiowa Indians for coughs and sore throats. Causes also a profusion of saliva.

Brauneria pallida (Nutt.) Britt. (syn. Echinacea pallida Nutt.). Purple Cone Flower (Compositae). - Perennial herb. Eastern N. America to Texas. Dried rhizome is used medicinally as alterative and diaphoretic; contains inuline, inuloid and resins.

Brayera → Hagenia abyssinica J. F. Gmel.

Brayera anthelminthica Kunth. → Hagenia abyssinica J. F. Gmel.

Brazil Copal → Hymenaea Courbaril L.

Brazil Krameria → Krameria argentea Mart.

Brazil Nut → Bertholletia excelsa H. B. K.

Brazil Nut Oil → Bertholletia excelsa H. B. K.

Brazil Redwood → Brosimum paraense Hub.

Brazil Tulipwood → Dalbergia cearensis Ducke.

Brazilian Angelin Tree → Andira vermifuga Mart.

Brazilian Arrowroot → Ipomoea Batatas Poir. and Manihot esculenta Crantz.

Brazilian Boxwood → Euxylophora paraensis Huber.

Brazilian Guava → Psidium Araça Raddi.

Brazilian Jalap → Piptostegia Pisonis Mart.

Brazilian Pine → Araucaria brasiliensis A. Rich.

Brazilian Redwood → Caesalpinia brasiliensis Sw.

Brazilian Rhatany → Krameria argentea Mart.

Brazilian Rosewood → Dalbergia nigra Allem. and Physocalymma scaberinum Pohl.

Brazilian Sassafras → Mespilodaphne Sassafras Meisn.

Brazilian Satinwood → Euxylophora paraensis Buber.

Brazilian Tea → Stachytarpheta jamaicensis Vahl.

Brazilwood → Caesalpinia echinata Lam. and Haematoxylon Brasilette Karst.

Breadroot, Common → Psoralea esculenta Pursh.

Bread, guarana → Paullinia Cupana Kunth.

Breadfruit → Artocarpus communis Forst.

Breadfruit Tree, African → Treculia africana Decne.

Breadfruit, Wild → Artocarpus elastica Reinw.

Breadnut Tree → Brosimum Alacastrum Sw.

Breadnut Tree, Pará → Brosimum paraense Hub.

Breadnut Tree, Ramon →Brosimum Alacastrum Swartz.

Breadroot, Indian → Psoralia esculenta Pursh.

Brehmia spinosa Harv. → Strychnos spinosa Lam.

Breynia rhamnoides (Retz) Muell. Arg. (Euphorbiaceae). - Trop. Asia. Bark astringent; used in the Philipp. Islands. to prevent hemorrhage.

Briançon Apricot → Armeniaca brigantina Pers.

Briar Root → Erica arborea L.

Brickelia Cavanillesii Gray. (syn. Coleosanthus squarrosus (Cav.) Blake, Eupatorium squarrosus Cav.) (Compositae). - Shrub. Mexico. Used as febrifuge, vermifuge; for diarrhoa.

Bridelia ferruginosa Benth. (Euphorbiaceae). - Shrub or small tree. Trop. Africa. A food plant of the African Silkworm, Anaphe infracta in Uganda. Also B. micrantha Baill.

Bridelia micrantha Baill. (Euphorbiaceae). - Woody plant. Trop. Africa. Leaves are in Natal, Transvaal, S. Cameroon source of a food of a silkworm Anaphe reticulata Walker and A. panda Boisd. Bark is source of a tannin, also of a red and black dye.

Bridelia minutiflora Hook. (Euphorbiaceae). - Tree. Trop. Asia esp. Indo-China, S. China, Philipp. Islds., Mal. Archipelago. Fruits ovoid, 10 mm. in diam. reddish; eaten in some countries by the natives.

Bridelia Moonii Thw. (syn. B. retusa Baill.) (Euphorbiaceae). - Tree. Ceylon. Wood hard, durable, very close-grained, brownish or olive, resistant to termites. Used for house-building, agricultural implements, carts.

Brigalow → Acacia doratoxylon Cunningh., A. excelsa Benth.

Brimstone Colored Lepraria → Lepraria chlorina (DC). Ach.

Brimstone Tree → Morinda critrifolia L.

Brindonnes → Garcinia indica Choisy.

Bristle Cone Pine → Pinus aristata Engelm.

Bristly Greenbrier → Smilax Bona-nox L.

Britoa acida Berg., Para Guava. (Myrtaceae). - Shrub or small tree. Brazil. Occasionally cultivated. Fruits edible, of good quality; used for jellies.

Brittle Willow → Salix fragilis L.

Broad Bean → Vicia Faba L.

Broadleaf Fig → Ficus platyphylla Delile.

Broadleaf Ironbark → Eucalyptus siderophloia Benth.

Broad leaved Dock → Rumex obtusifolius L.

Broadleaved Lavender →Lavandula Spica Cav.

Broadleaved Maple → Acer macrophyllum Pursh.

Broadleaved Water Gum → Tristania suaveolens Smith.

Broccoli → Brassica oleracea L. var. botrytis L.

Broccoli, Asparagus → Brassica oleracea L. var. italica Plenck.

Broccoli, Sprouting → Brassica oleracea L. var. italica Plenck.

Brodiaea capitata Benth. (syn. Dichelostemma capitatum (Benth.) Wood.). (Liliaceae). - Herbaceous bulbous plant. Western United States. Sweet bulbs were eaten as food by Indians of Arizona and California.

Bromelia Ananas L. → Ananas comosus (L.) Merr.

Bromelia Karatas L. (Bromeliaceae). - Perennial plant. Trop. America. Very young inflorecenses are sold in markets of El Salvador as vegetable. Ripe fruits are used for the preparation of Atol de Piña.

Bromelia Pinguin L., Piñuela. (Bromeliaceae). - Perennial herb. West Indies, Centr. America, Venezuela. Fruits edible, very acid. Sold in markets.

Bromelia serra Griseb. (syn. B. argentina Baker, Rhodostachys argentina Baker.) (Bromeliaceae). - Perennial herb., S. America, esp. Argentina, Brazil, Bolivia, Paraguay. Source of a Caraguata Fibre, used in S. America for sacks, sails and is recommended for manuf. paper.

Bromeliaceae → Aechmea, Ananas, Bromelia, Greigia, Karatas, Neoglaziovia, Puya, Tillandsia.

Bromus anomalus Rupr., Nodding Brome. (Graminaceae). - Perennial grass. Western United States, adj. Mexico. Excellent food for livestock.

Bromus catharticus Vahl., Rescue Grass. (Graminaceae). - Annual or biennial grass. Cultivated in the Southern part of the United States. A winter forage grass.

Bromus inermis Leyss., Smooth Broome. (Graminaceae). - Perennial grass. Europe, Asia. Cultivated as a hay crop in Old and New World.

Bromus unioloides H. B. K. (Graminaceae). - Annual or biennial grass. Probably from Argentina. Cultivated as a fodded plant, also used for pasturage. Identical or related to B. catharticus vahl.

Bronze Shield Lichen → Parmelia olivacea (L.) Ach.

Brook Evonymus → Evonymus americanus L.

Brook Lime → Veronica Beccapunga L.

Brookweed → Samolus Valerandi L.

Broomcorn → Sorghum vulgare Pers.

Broom, Scotch → Sarothamnus scoparius (L.) Wimmer.

Broom, Spanish → Spartium junceum L.

Broom, Sweet → Scoparia dulcis L.

Broom Tea Tree → Leptospermum scoparium Forst.

Broomjute Sida → Sida rhombifolia L.

Brosimum Alacastrum Swartz., Ramon Breadnut Tree. (Moraceae). - Tree. Tropical America. Seeds are edible when roasted. Leaves are fodder for live stock. Wood is hard, compact, fine-grained, whitish; used for carpentry work. Seeds are sometimes used as substitute for coffee.

Brosimum Aubletii Peopp. and Endl. (syn. B. guianensis (Aubl.) Huber. Piratinera guianensis Aubl.) Letterwood. (Moraceae). - Guiana. Tree. Wood reddish brown or brown with black markings; resembling letters or hieroglyphs. One of the most expensive woods; often sold by weight. Wood hard, heavy, strong, straight-grained, takes very smooth finish. Used for drum sticks, umbrella handles, fishing rods, butts, fancy articles, occasionally violin bows and cabinet work.

Brosimum costaricanum Liebm. (Moraceae). - Tree. Costa Rica. Leaves and branches are used as food for live-stock. Seeds are eaten boiled.

Brosimum paraense Hub. (syn. Ferolia guianensis Aubl.). Para Breadnut Tree. (Moraceae). - Tree. Trop. S. America. Source of Brazil Redwood, Cardinal Wood. Exported. Used for furniture and in fancy carpentry.

Brosimum utile (H. B. K.) Pitt., Cow Tree (Moraceae). - Tree. Trop. America. Latex from trunk is used as a beverage, apparently without disagreeable results. Latex has been employed as a base for chewing gum. Bark has been used by Indians for manuf. cloth, blankets and sails.

Broussonetia Kaempferi Sieb. (Moraceae). - Woody plant. Japan. Bark is used in Japan for manuf. of a paper.

Broussonetia papyrifera Vent., Paper Mulberry. (Moraceae). - Tree. China, Japan. Strips of bark are used in China for manuf. paper and clothing, called Tapa Cloth, which can easily be dyed. Fibre is also made into a kind of rope.

Brown Barbary Gum → Acacia arabica Willd.

Brown Heart → Andira excelsa H. B. K.

Brown Mahogany → Entandrophragma utile (Dawe and Sprague) Sprague., Lovoa Klaineana Pierre and Sprague and L. Swynnertonii Bak. f.

Brown Padauk → Pterocarpus macrocarpus Kurz.

Brown Sarsaparilla → Smilax Regelii Killip and Morton.

Brown Strophanthus → Strophanthus hispidus DC.

Brucea antidysenterica J. F. Mill. (Simarubaceae). - Tree. E. Africa, especially Abyssinia. The bitter bark and fruits are used in Abyssinia for diarrhea and fever.

Brucea sumatrana Roxb. (syn. B. amarissima Desv.) (Simarubaceae). - Small shrub. India to N. Australia. Bitter root is used against attacks of insects. Fruits, called in Java Makassaarsche Pitjes are sold for dysentery. They contain brucamarin, an alkaloid.

Bruguiera conjugata Merr. (syn. B. gymnorrhiza Lam.) (Rhizophoraceae). - Tree. Coastal zone. E. Africa to Pacif. Islds. Bark is of moderate tanning vallue. Wood used for charcoal. Phlobaphene in the tan is used as a black dye. Bark used in Indonesia to flavor a preparation of fish. Fruit sometimes used as astringent in betel quid.

Bruguiera parviflora Wight and Arn. (Rhizophoraceae). - Tree. Along coast, India, Malaya. Germinating embryo used by Malays as vegetable.

Brunfelsia Hopeana Benth. (syn. Francisea uniflora Pohl.) Manaca Raintree. (Solanaceae). - Shrub. Trop. America. Dried root is used as alterative in rheumatism and for syphilis. Contains a very poisonous alkaloid, called manacine, resembling strychnine in action. In Brazil all parts of the plant are being used.

Brush Bloodwood → Baloghia lucida Endl. and Synoum gladulosum Juss.

Brush Box → Tristania conferta R. Br.

Brussel Witloof → Cichorium Intibus L.

Brussels Sprouts → Brassica oleracea L. var. gemmifera Zenker.

Brya Ebenus DC. Ebony Cocuwood. (Leguminosaceae). - Tree. West Indies. Wood strong, compact, heavy; used for carpentry; one of the principal lumber trees in some parts of the West Indies.

Bryonia alba L. White Bryony. (Cucurbitaceae). - Perennial vine. Centr. Europe, Russia, Balkans, N. Iran. Dried roots, collected in summer and autumn, are used medicinally as cathartic; used in dropsy.

Bryonia dioica Jacq., Redberry Bryony. (Cucurbitaceae). - Perennial vine. Europe. Root, Radix Bryoniae, formerly used as drastic purgative, diuretic, cathartic, emetic; for dropsy, hemorrhagia. Contains bryoresin, bryonin and bryonidin.

Bryony, Redberry → Bryonia dioica Jacq.

Buaze Fibre → Securidaca longipedunculata Fres.

Bucco → Barosma betulina (Thunb.) Bartl.

Buchanania latifolia Roxb. (Anacardiaceae). - Tree. E. India, Burma. Seeds are consumed by the natives, being of a nutty, delicate flavor; also used in sweet-meets. Seeds are also source of an excellent oil. Source of Chironji-ki-gond, a gum, sold in bazaars of India; it has adhesive properties.

Buchanania Lanzan Spreng., Almondette Tree. (Anacardiaceae). - Tree. India, Malaysia. Seeds are rich in oil, sold in England under the name of Almondettes, used in sweet meats.

Buchnera leptostachya Benth. (Scrophulariaceae). - Herbaceous. Trop. Africa, Madagascar. Herb used by the native of Madagascar for staining the teeth.

Buchu, long → Barosma serratifolia (Curtis) Willd.

Buchu, Short → Barosma betulina (Thunb.) Bartl. and Wendl. and B. crenulata (L.) Hook.

Bucida Buceras L. → Laguncularia racemosa Gaertn. f.

Buckeye, California → Aesculus california Nutt.

Buckeye, Ohio → Aesculus arguta Buckl.

Buckeye, Red → Aesculus Pavia L.

Buckeye, Sweet → Aesculus octandra Marsh.

Buckeye, Western — Aesculus arguta Buckl.

Bucklandia populnea R.Br. (Hamamelidaceae). - Tree. Trop. Asia, esp. India to S. China, Mal. Penin. Wood is of good quality; used for window frames. Bark has been recommended for tanning.

Buckthorn, Alder → Rhamnus Frangula L.

Buckthorn, Cascara → Rhamnus Purshiana DC.

Buckthorn, Common → Rhamnus cathartica L.

Buckthorn, Dahurian →Rhamnus dahurica Pall.

Buckthorn, Gloss → Rhamnus Frangula L.

Buckthorn, Japanese → Rhamnus japonica Maxim.

Buckthorn Plantain → Plantago Coronopus L.

Buckthorn, Redberry → Rhamnus crocea Nutt.

Buckthorn, Sea → Hippophae rhamnoides L.

Buckthorn, Woolly → Bumelia lanuginosa (Michx.) Persoon.

Buckwheat → Fagopyrum esculentum Moench.

Buckwheat Bush →Cliftonia monophylla (Lam.) Sarg.

Buckwheat, Japanese → Fagopyrum esculentum Moench.

Buckwheat, Tatary → Fagopyrum tataricum Gaertn.

Buddleia brasiliensis Jacq. (syn. B. australis Vell., B. thapsoides Desf.). (Loganiaceae). - Shrub. Brazil. Plant is used to stupefy fish.

Buddleia cambara Arech. (Loganiaceae). - Small shrub. Uruguay to Brazil. Herb is used in Brazil in dometic medicine as pectoral.

Buddleia curviflora Hook. and Arn. (Loganiaceae). - Shrub. China. Branches and leaves are used in Japan to stupefy fish.

Buddleia madagascariensis Lam. (Loganiaceae). - Shrub. China. Branches and leaves are used in the manuf. of a native rum. Flowers are employed for dyeing cloth, called Ziafotsy.

Buddleia marrubiifolia Benth. (Loganiaceae). - Shrub. Mexico. Decoction of leaves is used in Coahuiila for giving a yellow to orange color to butter and vermicelli. Plant is used as diuretic and aperitive.

Buddleia salvifolia Lam. (Loganiaceae). - Shrub or small tree. S. Africa. Wood hard, tough, heavy; suitable for ramrods, yokes and rural utensils.

Buddleia scordioides H. B. K. (Loganiaceae). - Aromatic shrub. S.W. of the United States and Mexico. Leaves are used as a tea for indigestion in some parts of Mexico.

Buddleia tucumanensis Griseb. (Loganiaceae). - Shrub. Argentina. Twigs are used in some parts of Argentina as astringent and stimulant.

Buettneria carthagenensis Jacq. (syn. B. aculeata Jacq.). (Sterculiaceae). - Creeping shrub. Mexico to S. America. Roots are used in Venezuela as a substitute for sarsaparilla.

Buffalo Currant → Ribes aureum Pursh.

Buffalo Berry, Russet → Shepherdia canadensis (L.) Nutt.

Buffalo Berry, Silver → Shepherdia argentea Nutt.

Buffalo Thorn → Zizyphus mucronata Willd.

Bugle Weed → Lycopus virginicus L.

Bull Oak → Casuarina equisetifolia L.

Bull Pine → Pinus Sabiniana Doughl.

Bullace Plum → Prunus instititia L.

Bullet → Manilkara bidentata (DC.) Chev.

Bullet, Bastard → Humiria floribunda Mart.

Bullock's Heart → Annona reticulata L.

Bulnesia arborea (Jacq.) Engl., Maracaibo Lignum Vitae. (Zygophyllaceae). - Tree. Coastal zone of Colombia to Venezuela. Wood used for the same purpose as Lignum Vitae (Guaiacum) also made into collars of water turbines and brush backs.

Bulnesia Sarmienti Lorentz., Pao Santo, Paraguay Lignum. (Zygophyllaceae). - Tree. S. America esp. Gran Chaco of Argentina and Paraguay. Source of Oil of Guaiac Wood, ess. oil, obtained by steam distillation; soft rose-like scent; used in perfumery; conceals harsh notes of synthetic aromatics. Manuf. into soaps. Contains guaiol and bulnesol. Oil is also called Champaca Wood Oil, Oleum Ligni Guaiaci and Essence de Bois Gaiac. Timber is known as Palo Balsamo.

Bulrush, Great → Scirpus lacustris L.

Bulungu Resin → Canarium edule Hook. f.

Bumelia laetevirens Hemsl. (Sapotaceae). - Mexico. Fruits produce a kind of chicle. Sold in markets. Immature fruits are pickled in salt or vinegar.

Bumelia lanuginosa (Michx.) Persoon., Woolly Buckthorn, Gum Elastic. (Sapotaceae). - Shrub or small tree. Eastern part of United States to Florida and New Mexico. Black fruits are consumed by the Indians. Ground up bark is source of a mucilaginous substance which hardens quickly in the air, used as a chewing gum by the Kiowa Indians (New Mex.).

Bumelia nigra Sw. → Dipholis nigra Gris.

Buna → Platanus orientalis L.

Bunchosia armeniaca (Cav.) DC. Malpighiaceae). - Shrub or shrubby tree. Throughout Andean region. Cultivated in some parts of Ecuador. Fruits round, greenish, 2.5 to 4 cm. diam., pulp cream colored, sweet; occacionally consumed by the natives.

Bunias Erucago L. (Cruciferaceae). - Biennial herb. Mediterranean region, Asia Minor. Leaves are occasionally eaten in salads and as spinach.

Bunium Bulbocastanum L. (syn. Ligusticum Bulbocastanum Crantz., Apium Bulbocastanum Caruel.). (Umbelliferaceae). - Perennial herb. Europe. Tubers are consumed as a vegetable, sometimes sold in markets; also used as fodder for live-stock. Leaves are eaten as parsley; fruits as cumin. The plant was much cultivated in earlier centuries. Tubers are used medicinally as astringent.

Buntal Fibre → Corypha elata Roxb.

Bunya-Bunya → Araucaria Bidwillii Hook.

Buphane disticha Herb. (syn. B. toxacaria Herb.). (Amaryllidaceae). - Bulbous perennial. S. Africa. Juice from bulbs is used as arrow-poison by the natives. Used in vet. medicine for pyroplasmosis of cattle.

Buphthalmum oleraceum Lour. (Compositae). - Shrub. Trop. Asia. Young aromatic leaves are used in Annam as a condiment, esp. for fish.

Bur Clover → Medicago hispida Gaertn.

Bur Clover, Spotted → Medicago arabica (L.) All.

Bur Oak → Quercus macrocarpa Michx.

Burdekin Plum → Pleiogynium Solandri Engl.

Burdock, Common → Arctium minus (St. Hil.) Bern.

Burdock, Great → Arctium Lappa L.

Burgundy Pitch → Picea excelsa Link.

Burkea africana Hook. (Leguminosaceae). - Tree. S. Africa. Stem is source of a soluble gum, being dark colored and of fair quality. Bark is employed for tanning in S.W. Tanganyika.

Burma Gugertree → Schima Noronhae Reinw.

Burmese Lacquer → Melanorrhoea usitata Wall.

Burmese Varnish Tree → Melanorrhoea usitata Wall.

Burnet → Sanguisorba officinalis L.

Burning Bush → Evonymus atropurpureus Jacq.

Burra Acacia → Acacia falcata Willd.

Burrawang Nut → Macrozamia spiralis Miq.

Burrofat, Tree → Isomeris arborea L.

Bursa Opopanax → Commiphora Kataf (Forsk.) Engl.

Bursera Aloëxylon (Schiede) Engl. (syn. Elaphrium Aloëxylon Schiede.) Linaloé Tree. (Burseraceae). Shrub or small tree. Morelos, Puebla, Oaxaca (Mex.). Source of Lianloé Oil, obtained from distillation of the wood. Ripe fruits are source of Linaloé Seed Oil, used in perfumes.

Bursera Delpechianum Poisson. (Burseraceae). - Tree. Mexico. Source of Mexican Linaloé wood. It becomes scented when wounded, due to the formation of an ess. oil.

Bursera gummifera L. (syn. Elaphrium simaruba (L.) Rose.) (Burseraceae). - Tree. Trop. America. Bark is source of a resin, called West Indian Elemi, Elqueme or Tacamahaca. Used in varnishes and as substitute for Gum Arabic. Employed is mending broken china and glass. Has been used by the Mayas as incense since ancient times. Employed for painting canoes, as preserve from attacks of worms. Medicinally it is considered diaphoretic, diuretic, purgative; used for dropsy, dysentery and yellow fever. Much is derived from the West Indies.

Bursera jorullensis (H. B. K.) Engl. (syn. Elaphrium jorullensis H. B. K.). (Burseraceae). - Shrub or tree. Mexico. Source of a copal, derived from incissions in the trunk. It is called Copal de Penca. Produces a very glossy varnish. Used for uterine diseases. Bark is employed in tanning and dyeing.

Bursera odorata Brandeg. (syn. Elaphrium odoratum (Brandeg.) Rose.). (Burseraceae). - Shrub or small tree. Mexico. Source of a gum, being expectorant and purgative; considered cure for scorpion stings. Used for mending dishes. Bark is employed for tanning hides.

Burseraceae → Boswellia, Bursera, Canarium, Commiphora, Dacryodes, Icica, Protium, Santira.

Bush Apple → Heinsia pulchella Schum.

Bush Cherry → Eugenia myrtifolia Sims.

Bush Cinquefoil → Potentilla fruticosa L.

Bush Morning Glory → Ipomoea leptophylla Torr.

Bustic → Dipholis salicifolia (L.) A. DC.

Butanediol → Aerobacillus polymyxa (Prazm.) Migula and Aerogenes (Prazm.) Migula.

Butea Gum → Butea superba Roxb.

Butea superba Roxb. (syn. B. frondosa Roxb.). Bastard Teak, Bengal Kino. (Leguminoseaceae). - Tree. India, Malaysia. Valuable tree for recovering salt lands. Food of a lac insect. Source of Bengal Kino or Butea Gum, formed from scars in the bark. Gum is ruby colored, substitute for Kino. Pterocarpium Marsupium. It is astringent, mild in operation, adapted to children. Flowers and leaves are supposed to be diuretic, astringent and aphrodisiac. Flowers produce an inferior yellow dye. Pounded seeds are a powerful rubefacient, purgative, vermifuge. Wood is very durable under water, it is claimed more so than above the ground; used for well curbs, water scoops of wells, also source of a charcoal used for gun powder. Bark is employed for cordage, called Pala Fibre, used for manuf. sails.

Butomaceae → Butomus, Limnocharis.

Butomus umbellatus L. Flowering Rush. (Butomaceae). - Perennial herb. Grows in swamps. Europe, Asia. Rhizones are consumed as food by some races in Russia.

Butter, Bambuk → Butyrospermum Parkii (Don.) Kotschy.

Butter Bush → Pittosporum phillyraeoides DC.

Butter, Cacao → Theobroma Cacao L.

Butter, Dika → Irvingia gabonensis Baill.

Butter Fruit → Diospyros discolor Willd.

Butter, Galam → Butyrospermum Parkii (Don.) Kotschy.

Butter, Goa → Garcinia indica Choisy.

Butter, Illipé → Madhuca longifolia Gm.

Butter, Jaboty → Erisma calcaratum Warm.

Butter, Kagné → Allanblackia oleifera Oliv.

Butter, Kanga → Pentadesma butyracea Sabine.

Butter, Kokam → Garcinia indica Choisy.

Butter, Lamy → Pentadesma butyracea Sabine.

Butter, Indian → Madhuca butyracea Gm.

Butter, Orère → Baillonella ovata Pierre.

Butter, Otoba → Myristica otoba Humb. and Bompl.

Butter, Phulwara → Madhuca butyracea Gm.

Butter Pits → Acanthosicyos horrida Welw.

Butter, Shea → Butyrospermum Parkii (Don.) Kotschy.

Butter, Sierra Leone → Pentadesma butyracea Sabine.

Butter, Tallow Mowrah → Madhuca longifolia Gm.

Butter Tree → Pentadesma butyracea Sabine.

Butter, Ucuhuba→ Myristica surinemensis Roland.

Butterbur, Palmate → Petasites palmata Gray.

Buttercup, Blister → Ranunculus sceleratus L.

Butterfly Weed → Asclepias tuberosa L.

Butternut → Caryocar nuciferum L.

Butternut → Juglans cinerea L.

Butterwort → Pinguicula vulgaris Sm.

Button Bush → Cephalanthus occidentalis L.

Button Clover → Medicago orbucularis (L.) All.

Button Snakeroot → Eryngium yuccifolium Michx.

Buttonwood → Conocarpus erectus L. and Platanus occidentalis L.

Buttonwood, White → Languncularia racemosa Gaertn. f.

Butyrospermum Parkii (Don.) Kotschy., Shea Butter Tree. (Sapotaceae). - Tree. Trop. Africa, esp. W. Coast and Sudan. Seeds are source of Shea Butter, Bambuk Butter, Galam Butter. Used by natives as food and illuminant. Used in Europe as cooking fat; olein is employed in margarine; stearine and hydrogeneted fat as cacao butter substitute. Sp. Gr. 0.917-0.918; Sap. Val. 178-189; Iod. No. 56-65; Unsap. 2.2-11⁰/o.

Buxaceae → Buxus, Simmondsia, Styloceras.

Buxus japonica Muell. Arg. (Buxaceae). - Shrub. Japan. Wood yellow, fine-grained; used for combs, engraving blocks, stamps.

Buxus Macowani Oliv., Cape Box (Buxaceae). - Small tree. S. Africa. Wood light yellow, very close-grained, hard, durable; used for engravers' and turners' work; also for mathematical and musical instruments.

Buxus sempervirens L., Box. (Buxaceae). - Shrub or tree. South and Centr. Europe, N. Africa, Caucasia, Asia Minor to Caspic Sea. Wood heavy, very hard, difficult to split, durable, fine texture, light to dark yellow; used for statues, wood engraving combbacks, flutes, clarinets, scientific instruments, writing tables, tobacco-pipes, planes, furniture. Leaves are used as an adulterant of Folia Uvae Ursi.

Byrsonima crassifolia H. B. K. (Malpighiaceae). - Small tree. Trop. America. Berries are edible, yellow, size of a cherry, slightly acid. Sold in local markets of Mexico. Wood is source of a good charcoal.

Byrsonima cubensis Juss. (Malpighiaceae). - Tree. Cuba. Wood red, strong, compact, heavy; used for wooden ware.

Byrsonima spicata. Rich. Maricao (Malpighiaceae). - Tree. Trop. America, Porto Rico. Wood strong, fine-grained, compact, cinnamon brown; used for boards, framework of houses.

C

Caapi → Banisteria Caapi Griseb.

Cabbage → Brassica oleracea L.

Cabbage, Chinese → Brassica chinensis L. and B. pekinensis Rupr.

Cabbage, Kerguelen → Pringlea antiscorbutica R. Br.

Gabbage, Marrow → Brassica oleracea L. var. acephala DC.

Cabbage, Pak-Choi → Brasice chinensis L.

Cabbage, Palmetto → Sabal Palmetto (Walt.) Todd.

Cabbage, Pe-tsai → Brassica pekinensis Rupr.

Cabbage, Red → Brassica oleracea L.

Cabbage, Savoy → Brassica oleracea L. var. capitata L.

Cabbage, Skunk → Spathyema foetida (L.) Raf.

Cabbage, Swamp → Spathyema foetida (L.) Raf.

Cabelluda → Eugenia tomentosa Camb.

Cabralea Cangerana Sald., Pau de Santo, Cancharana. (Meliaceae). - Tree. Trop. S. America. Wood dull red or maroon, rather brittle, firm, strong, easy to work; used locally for construction work, interior and exterior of houses, furniture, joinery, sculpture, immages of saints. When sawdust is soaked in water, it is source of a red dye. Bark is called Cangerana; used in Brazil for swamp fevers.

Caburá → Myrocarpus frondosus Allem.

Cacao Blanco → Theobroma bicolor Humb. and Bonpl.

Cacao Butter → Theobroma Cacao L.

Cacao Calabacillo → Theobroma leiocarpa Bern.

Cacao de Mico → Theobrama purpureum Pitt. and T. speciosa Willd.

Cacao de Sonusco → Theobroma speciosa Willd.

Cacao Lagarto → Theobroma pentagona Bern.

Cacao, Madre de → Gliricidia sepium (Jacq.) Steud.

Cacao Montaras → Theobroma albiflora Steud.

Cacao Silvestre → Theobroma speciosa Willd.

Cacao Simarron → Theobroma albiflora Goud.

Cacao Tree → Theobroma Cacao L.

Cacaoti → Theobroma Mariae Schum.

Cacara erosa Thour. → Pachyrrhizus erosus Rich.

Cachibou → Calathea discolor Mey.

Cachiman → Rollinia Sieberi DC.

Cactaceae → Acanthocereus, Borzicactus, Carnegiea, Echinocactus, Echinocereus, Escontria, Espostoa, Ferocactus, Hylocereus, Lemaireocereus, Lophocereus, Lophophora, Machaerocereus, Myrtillocactus, Nopalea, Opuntia, Pachycereus, Pelecyphora, Pereskia, Pereskiopsis, Selenicereus, Trichocereus.

Cacti. Useful → Cactaceae.

Cactus Candy →Ferocactus Wislizeni (Engelm.) Britt. and Rose.

Cactus, Giant → Carnegiea gigantea (Engelm.) Britt. and Rose.

Cactus sepium H. B. K. → Borzicactus sepium (H. B. K.) Britt und Rose.

Cadaba farinosa Forsk. (Capparidaceae). - Woody plant. Trop. Africa, Arabia. Pounded leaves and twigs prepared with cereals are made into a cake or pudding and used as food by the natives, called Farsa or Balambo. Sold in native markets.

Caesalpinia arborea Zoll. →Peltophorum pterocarpum Backer.

Caesalpinia Bonducella Flem., Molucca Bean, Bonduc Nut, Physic Nut, Bengor Nut. (Leguminosaceae). - Tropics of Old and New World. Seeds are used for necklaces, bracelets, rosaries. In India they are mixed with black pepper, used as tonic and febrifuge. Bark is employed as tonic.

Caesalpinia Bonduc Roxb. (Leguminosaceae). - Woody climber. Tropics. Cultivated. Properties similar to C. crista L.

Caesalpinia brasiliensis Sw., Bahia Wood, Brazilian Redwood. (Leguminosaceae). - Tree. S. America. Wood is source of a red dye.

Caesalpinia brevifolium Baill. (syn. Balsamocarpum brevifolium (Baill.) Clos.). (Leguminosaceae). - Shrub. Chile. Fruits are used in Chile for tanning. They are called Agarobilli or Algaroba, containing 60% tannin.

Caesalpinia coriaria (Jacq.) Willd., Divi - Divi. (Leguminosaceae). - Shrub or tree. Mexico, West Indies, Centr. America, N. of S. America. Pods contain 25 to 30% tannin; used for tanning; much is exported. Also yield a black dye; used in Mexico for ink. Wood very solid, dark colored; used for general carpentry work.

Caesalpinia Crista L. (Leguminosaceae). - Woody climber. Tropics. Cultivated. Leaves, roots and fruits are used as tonic, antiperiodic. Seeds are employed in India for colic. Contains bonducin. Fat from the seeds is used in India for cosmetic preparations; supposed to soften the skin.

Caesalpinia digyna Rottl. (Leguminosaceae). - Woody plant. E. India. Roots and fruits, called Tari, are used for tanning.

Caesalpinia echinata Lam., St. Martha Wood, Peach Wood. (Leguminosaceae). - Tree. Colombia to Mexico. Wood is source of a red dye, becoming purplish with alkalies and yellow with acids. Also called Brazil Wood, Nicaragua Wood, Ymira Piranga, Pernambuco Wood.

Wood is red brown to dark brown, fine-grained, hard, heavy, dense, sinks in water, easy to split and to polish; used for turnery and ship building.

Caesalpinia eriostachys Benth. Iguanero, Palo Alejo. (Leguminosaceae). - Shrub or small tree. Mexico to Costa Rica. Bark used in Colima (Mex.) to stupefy fish.

Caesalpinia Gardneriana Benth. (Leguminosaceae). - Tree. Brazil. Bark is source of a yellow dye, used in Brazil.

Caesalpinia melanocarpa Griseb. (Leguminosaceae). - Tree. S. America. Leaves are used for tanning, contain 21% tannin. Called Guajacan.

Caesalpinia microphylla Mart. (Leguminosaceae). - Tree Brazil. Wood is used in Brazil for general carpentry.

Caesalpinia Paipae Ruiz and Pav. (syn. C. glabrata H. B. K., C. corymbosa Benth.). (Leguminosaceae). - Small tree or shrub. Peru, Ecuador. Pods source of a black dye, produces a good ink.

Caesalpinia praecox Ruiz and Pav. (Leguminosaceae). - Tree. S. America esp. Argentina, Chile. Stem is source of a gum, called Bea, containing 80% arabin.

Caesalpinia pulcherrima Swartz., Paradise Flower, Peacock Flower. (Leguminosaceae). - Tree. Tropics. All parts of plant are a powerful emmenagogue. Pods and leaves are used in East India as substitute for senna. Decoction of roots is used in Angola for intermittent fever.

Caesalpinia Rugeliana Urb. (Leguminosaceae). - Tree. Cuba. Wood reddish; used for inlay-work and turnery.

Caesalpinia Sappan L., Sappan Wood Tree. Japan Wood Tree. (Leguminosaceae). - Tree. Tropics. Wood is source of a red dye, used with alum as a mordant, for coloring matting and cakes. Wood is employed for cabinet work. Bark contains tannin; used in India with addition of iron to make a black dye.

Caesalpinia Volkensii Harms. (Leguminosaceae). - Liane. Trop. Africa. Roots are source of a red dye.

Caesalpiniaceae → Leguminosaceae.

Café du Sudan → Parkia africana R. Br.

Caffir Marvola Nut → Sclerocarya caffra Sond.

Cailliea callistachys Hassk. → Dichrostachys cinerea Wight and Arn.

Cainito → Chrysophyllum Cainito L.

Cajanus indicus Spreng., Pigeon Pea. (Leguminosaceae). - Shrub. Probably from E. India. Cultivated in Tropics of Old and New World. Peas are used as a nutritious and wholesome food, also as fodder for livestock. Very young seeds are eaten as green peas, when dry and ripe used in soups, eaten with rice etc. Also employed as hay and cover crop and as green-manure. Several varieties among which: Morgan Congo, Cuban Congo, No-eye Pea. Leaves are used in Madagascar as food of a silk-worm.

Cajaput Oil → Melaleuca Leucadendron L.

Cajaput Tree → Melaleuca Leucadendron L.

Cajuado (beverage) → Anacardium occidentale L.

Caka Bark → Erythrophleum guineense G. Don.

Cakile edentula (Bigel.) Hook., American Sea Rocket. (Cruciferaceae). - Herbaceous plant. Along beaches of N. America. During scarcity, the fleshy roots are dried, ground, mixed with flour and used for making bread. Leaves are occasionally used in salads and as a pot herb.

Calabar Bean → Physostigma venenosus Balf.

Calabar Ebony → Diospyros Dendo Welw.

Calabash → Lagenaria vulgaris Ser.

Calabash Gourd → Lagenaria vulgaris Ser.

Calabash Nutmeg → Monodora Myristica (Gaertn.) Dunal.

Calabash Tree → Crescentia Cujete L.

Calamondin → Citrus mitis Blanco.

Calamovilfa longifolia (Hook.) Hack. (Graminaceae). - Perennial grass. North America. Important for winter grasing in Nebraska and Dakota. Source of a fair, coarse hay.

Calamus → Acorus Calamus M.

Calderon Coyotillo → Karwinskia Calderoni Standl.

Caladium sororium Schott. (Araceae). - Perennial herb. Brazil. Fruits are consumed by the aborigenes of Amazonia (Braz.).

Caladium striatipes Schott. (Araceae). - Perennial herb. Brazil. Spadix and tubers are consumed as food in some parts of Brazil.

Calamintha Clinopodium Benth. (syn. Clinopodium vulgare L., Melissa vulgaris Trev., M. Clinopodium Benth.), Wild Basil. (Labiaceae). - Perennial herb. Europe, Asia, N. America. Sometimes used medicinally as stomachic, styptic, emmenagogue; also as excitans taken in white wine. Herb is source of a yellow and brown dye.

Calamintha graveolens Benth. (Labiaceae). - Perennial herb. S. W. Asia, Mediterranean region, Transcaucasia. Seeds are used as stimulant and as aphrodisiac.

Calamus aquatilis Ridl. Rattan, Bakan. (Palmaceae). - Long, slender, climbing palm. Malaysia. Stems are made into rough baskets.

Calamus asperrimus Blume. (Palmaceae). - Thin Rattan, W. Java, S. Sumatra. Stems easy to split, suitable for manuf. of bow-nets.

Calamus Barteri Becc. (Palmaceae). - Rattan. Niger Territory, Sierra Leone. Stems are used along Lower Niger, as rope and handles of native Kola baskets.

Calamus caesius Blume (Palmaceae). - Climbing palm. Mal. Penin., Sumatra, Borneo. Cultivated. Stems strong, supple, easy to split. Valuable for basketry, mats, seats of chairs. Much exported.

Calamus deeratus Mann. and Wendl. (Palmaceae). - Trop. Africa. Slender palm. Split stems are used for thatch, fences, house rafters, chairs, fishing weirs, baskets. Entire stems are used for suspension bridges.

Calamus inops Becc. (Palmaceae). - Rattan. Celebes. Stems used for rattan-furniture. Much is exported.

Calamus javensis Blume. (syn. C. equestris Blume). (Palmaceae). - Rattan palm. Mal. Archip., Sumatra, Borneo. Stems easy to split, strong, supple; used for fine basketry, mats; material is exported.

Calamus luridus Becc. (Palmaceae). - Climbing palm. So. Malaya Penin. to Sumatra. Canes are very durable; used for baskets.

Calamus Manan Miq. (Palmaceae). - Climbing palm or rattan. Malay Penin., Sumatra. Canes used for walking-sticks and polo-sticks, rattan furniture.

Calamus minahassae Warb. (Palmaceae). - Rattan palm. Celebes. Stems supple, easy to split, commercially known as Rottan Datoo; used for basketry, mats etc.; used in export-trade.

Calamus oblongus Reinw. (syn. Daemonorops oblongus Blume). (Palmaceae). - Mal. Archip. Stems used for manuf. light furniture and rattan mats.

Calamus optimus Becc. (Palmaceae). - Tall slender rattan. Borneo. Stems suitable for light furniture.

Calamus ornatus Blume. (Palmaceae). - A rattan. Sumatra to Philipp. Islds. Canes used for coarse tables. Culms do not split easily. Fruit edible, acid, sold locally in markets of Philipp. Islds.

Calamus ovoideus Thw. (syn. C. zeylanicum Thw.) (Palmaceae). - Rattan. Ceylon. Very young unfolded leaves are eaten raw or cooked as vegetable by the natives.

Calamus radiatus Thw. (Palmaceae). - Rattan. Ceylon. Stem used for basket-work, chair-bottoms.

Calamus retroflexua Becc. (Palmaceae). - Tall climbing palm. Sumatra, Borneo. Stem very tough, made into course ropes.

Calamus Scipionum Lour. (Palmaceae). - Rattan. Sumatra to Mal. Penin. Borneo. Canes used for walking-sticks, basket-handles. Buds eaten as food by Berembun tribes.

Calandrina balonensis Lindl. (syn. Claytonia balonensis F. v. Muell.). (Portulacaceae). - Peennial herb. Australia. Plants were used as food by early settlers.

Calandrina Menziesii Torr and Gray. (Portulacaceae). - Herbaceous perennial. California. Twigs and leaves are used as a pot-herb or as a garnish.

Calandrina polyandra Benth. (syn. Claytonia polyandra F. v. Muell.) (Portulacaceae). - Perennial herb. Australia. Used as food by the aborigines of W. Australia.

Calanthe mexicana Reichb. f. (Orchidaceae). - Terrestr. orchid. Mexico, Centr. America, West Indies. Powdered petals and sepals used by Mazatecs of the Cerro de los Frailes (Mex.) to stop nose-bleeding.

Calathea alluia (Aubl.) Lindl., Sweet Corn Root. (Marantaceae). - Herbaceous perennial. Probably from West Indies. Was source of the principal food of the aborigines. Tubers are eaten cooked.

Calathea discolor Mey., Cachibou. (Marantaceae). - Herbaceous perennial. West Indies, trop. America. Leaves very tough and durable when dry, made into waterproof baskets called Arima baskets; used for carrying clothes.

Calathea macrosepala Schum., Chufle. (Marantaceae). - Perennial herb. S. Mexico to S. America. Young flowers are cooked and consumed as a vegetable by the natives.

Calathea violacea (Rose) Lindl. (Marantaceae). - Perennial herb. Mexico, Centr. America to Brazil. Young flower clusters are cooked and eaten as a vegetable. Sold in local markets.

Calceolaria arachnoidea Grah. (Scrophulariaceae). - Herbaceous plant. Chile. Root is astringent and is used by country people in Chile. Root is also source of a red dye.

Calceolaria thyrsiflora Grah. (Scrophulariaceae). - Small shrub. Chile. Leaves are used in some parts of Chile for healing sore lips, affections of the tongue, throat and gums.

Calendula officinalis L., Marigold. (Compositae). - Annual herb. S. Europe and Levant. Cultivated in Old and New World. Dried, ligulate flowers are used medicinally, being carminative, stimulant; contain an ess. oil, an amorphous bitter compound and a gummi substance, calendulin. Used as a home remedy for jaundice, as stomachic and anthelmintic. Adulterant of saffron.

Caliatur Wood → Pterocarpus santalinus L. f.

California Buckeye → Aesculus californica Nutt.

Californian Goldenrod → Solidago californica Nutt.

California Holly → Photinia arbutifolia Lindl. and Rhamnus ilicifolium Kellogg.

California Juniper → Juniperus californica Carr.

Californian Laurel → Umbellularia californica Nutt.

Californian Pepper Tree → Schinus Molle L.

Californian Poppy → Eschscholtzia californica Cham.

Californian Rose → Rosa californica Cham. and Schlecht.

California Soap Plant → Chenopodium californicum Wats.

Californian Soap Root → Chlorogalum pomeridianum (Ker-Gawl) Kunth.

California White Oak → Quercus lobata Née.

Calisaya Bark → Cinchona Calisaya Wedd.

Calliandra anomala (Kunth) Macbr. (syn. Inga anomala (Kunth) Macbr.). Cabeza de Angel. (Leguminosaceae). - Shrub. Mexico to Guatemala. Used for tanning. Root used to retard fermentation of Tepache, a beverage made from pulque and sugar.

Calliandra formosa Benth. (syn. Lysiloma formosa Hitchc.) Sabicu. (Leguminosaceae). - Tree. Trop. America. Wood heavy, strong, close-grained; used for carpentry.

Calliandra Houstoniana (Mill.) Standl. (syn. Mimosa Houstonii L'Hér.). (Leguminosaceae). - Bark, called Pambotana Bark, is used in Europe as antiperiodic.

Callicarpa cana L. (Verbenaceae). - Shrub. Malaysia to Australia. Plant is used as arrow poison. In the Philipp. Islds. also to stupefy fish.

Callicarpa longifolia Lam. (Verbenaceae). - Malaysia to Australia. Used for poulticing in fever and colic among the Malays.

Callicoma Billardieri D. Don. (syn. Codia montana Labill.). (Saxifragaceae). - Tree. Australia to New Caledonia. Wood reddish, fine-grained, easily worked; used for turnery and handles of tools.

Callirhoe digitata Nutt., Finger Poppy Mallow. (Malvaceae). - Perennial herb. S. of United States. Roots being of a pleasant taste, were consumed by the Indians.

Callirhoe pedata Gray. Pimple Mallow. (Malvaceae). - Perennial herb. Western N. America. Roots were used as food by the Indians from Nebraska to Idaho.

Callitris arborea Schrad. (syn. Widdringtonia juniperioides Endl.) (Cupressaceae). - Tree. S. Africa. Source of a gum, used by the natives as diuretic. Wood suitable for furniture and cabinet work.

Callitris calcarata R. Br. (Frenela Endlicheri Porlat.) Black Pine, Murray Pine. (Cupressaceae). - Tree. Australia, esp. N. Victoria, Centr. Queensland. Source of Australian Sandarac, a yellow resin.

Callitris cupressoides Schrad. (Sm. Widdringtonia cupressoides Endl.) Sapreewood. (Cupressaceae). - Shrub. Cape Peninsula to Natal. Wood light; suitable for coopers' work and manuf. of pails.

Callitris quadrivalvis Vent. → Tetraclinis articulata (Vahl.) Mast.

Callitris robusta R. Br. (syn. C. glauca R. Br.). Black Pine, Common Pine, Marung. (Cupressaceae). - Tree. Australia, esp. N. Australia, Queensland to N. W. Australia. Wood durable, light to dark brown, excellent timber, easily worked. Used for flooring, weatherboards, tables. Resists teredo and white ants.

Callitropsis araucarioides Compton. (Cupressaceae). - Tree. New Caledonia. Source of Oil of Araucaria, ess. oil, viscid, used as neutral fixative for scenting soaps, cosmetics, lotions. Has a rose-like odor.

Calluna vulgaris Salisb., Common Heather. (Ericaceae). - Small shrub. Throughout Europe, N. Asia Minor, W. Siberia, N. America. Branches are made into brooms, much used in some parts of Europe. Herb. occasionally used in beer, also used as a tea. Bark has been recommended for tanning. Flowers are source of a commercial honey. Was source of a yellow dye, used for dyeing wool.

Calocarpum mammosum Pierre. (syn. Achras mammosa L., Lucuma mammosa Gaertn.) Sapote, Marmelade-Plum, Mamey Colorado. (Sapotaceae). - Tree. Centr. America. Cultivated in tropics. Fruits edible, elliptic, russet-brown, coarse skin; flesh salmon red to reddish-brown, firm, with one large seed. Eaten raw, in sherbets and as preserves. Used in Cuba in Guave-Cheese also made into a thick jam called Crema de Mamey Colorado.

Calocarpum viride Pitt., Green Sapote. (Sapotaceae). - Tree. Guatemala to Costa Rico. Cultivated. Fruits are more delicate than C. mammosum, Pulp pale red-brown, sweet, juicy; eaten fresh or preserved. Much esteemed among Indians. Frequently found in markets of Guatemala.

Calochortus aureus S. Wats. (Liliaceae). - Perennial bulbous plant. Western United States. Bulbs were consumed as food among the Navajo and Hopi Indians.

Calochortus elegans Pursh. (Liliaceae). - Perennial bulbous plant. Western N. America. Bulbs were consumed among the Indians of the N. W. of the United States.

Calochortus Gunnisonii S. Wats., Sagebrush Mariposa. (Liliaceae). - Perennial bulbous herb. Western United States esp. Montana to New Mexico and Arizona. Bulbs when boiled were consumed as food. Dry bulbs were also pounded into flour for making a porridge or mush by the Cheyenne Indians. Also other Calochortus species were used for this purpose.

Calochortus Nuttallii Torr. and Gray, Sego Lily, Mariposa Lily. (Liliaceae). - Bulbous perennial. Western N. America, esp. Montana to New Mexico, California, Oregon. Corms were boiled or roasted and used as food by the Indians.

Calochortus pulchellus Dougl. (Liliaceae). - Perennial bulbous herb. Western N. America, esp. California. Bulbs were consumed raw and roasted by the Indians of California.

Calodendrum capense Thunb. Cape Chestnut. (Rutaceae). - Tree. S. Africa. Seeds are source of a non-drying oil in Cape Peninsula; used for manuf. of soap. Wood is whitish, soft, tough; used for wagons, tent bows, yokes and planking.

Calonyction aculeatum (L.) House. (syn. C. speciosum Choisy, Ipomaea aculeata Kuntze) Moonflower. (Convolvulaceae). - Vine. Trop. of Old and New World. Milky juice is used to coagulate latex of Castilla elastica Cav.

Calophyllum brasiliense Camb. (Guttiferaceae). - Tree. Brazil. Wood is source of a yellowish-green Sandel Oil, called Sandalo Inglez.

Calophyllum Calaba L. (syn. C. Burmanni Wight.) Ceylon Beautyleaf. (Guttiferaceae). - Small tree. Ceylon. Oil from seeds is supposed to be a specific for itch. Source of a Tacamahaca-like resin, called Resina Ocuje. Wood is soft, reddish, rather light, durable; used for general carpentry.

Calophyllum inophyllum L. Alexandrian Laurel (Guttiferaceae). - Tree. E. Africa, East India, Polynesia. Source of Dilo or Domba-Oil; non-drying, unpleasant, aromatic taste. Used by natives for skin-diseases and rheumatism. Refined oil, intramusculary injected, releaves pain of leper patients. Sp. Gr. O. 9415; Sap. Val. 191-202; Iod. No. 82-98; Acid Val. 27-78; Unsap. 0.25-1.4%. Source of Takamahaca Resin. Wood hard, close-grained, elastic, durable, takes beautiful polish; used for boat building, cabinet work, tubs of cart wheels, rail-road ties. Sold as Borneo Mahogany.

Calophyllum longifolium Willd. (Guttiferaceae). - Tree. S. America. Source of a resin, called Maina.

Calophyllum saigonense Pierre. (Guttiferaceae). - Tree. Trop. Asia. Wood hard, durable; used for manuf. furniture and for boat building. Also C. dryobalanops Pierre and C. Thorelli Pierre.

Calophyllum spectabile Willd. (syn. C. Soulatti Burm.). (Guttiferaceae).- Tree. Trop. Asia, Pacif. Islds. Wood durable, elastic, ealisy worked; used for construction of houses, roofs, panels, masts. Bark is used among the natives in medicine.

Calophyllum tacamahaca Willd. (Guttiferaceae). - Tree. Madagascar. It is supposed that the stem is source of a Tacamahaca Resin.

Calophyllum Walkeri Wight. (Guttiferaceae). - Large tree. Ceylon, E. India. Wood durable, light red-brown; used for beams, posts, doorframes, rafters, ornamental panels, dadoes.

Caloplaca murorum (Hoffm.) Th. Fr. (Caloplacaceae). - Lichen. Temp. zone. On rocks and walls. Has been used in Sweden as source of a yellow dye, to color woolens.

Calopogonium coeruleum Hemsl. Jicuma. (Leguminosaceae). - Large vine. Centr. America. Used in El Salvador by laundresses to take out dirt, while rubbing the clothes.

Calopogonium mucunoides Desv. (Leguminosaceae). - Perennial herb. Guiana. Grown in some parts of tropics as green-manure.

Calotropis gigantea Ait. Tembega. (Asclepiadaceae). - Shrub. Trop. Asia, esp. India to Ceylon, China, Mal. Penin., Java, Lesser Sunda Islands, Moluccas. Leaves used by natives for poulticing soures. Chinese in Java candy central part of flower for sweatmeat. Leaves rubbed on skin of elephants are supposed to heal kesarayer disease. Fibre is of great strength. Flos used in India for stuffing quilts, upholdstery; also spun into fishing lines and nets.

Calotropis procera Ait. (syn. Asclepias gigantea L.). Auricula Tree. (Asclepiadaceae). - Shrub or small tree. Trop. Africa, India. Stems source of a strong fibre, used for ropes, fishing nets, lines, halters; durable under water. Floss from seeds used in matresses. Milky juice mixed with salt is used to remove hair from hides. Wood source of charcoal used for gun-powder. Leaves produce Merissa, a native beer in W. Africa. Juice from the plant is used in Sudan as an infanticide. Bark of roots is used for leprosy in India. Plant is source of a rubber-like product, called Mudar Gummi.

Caltha palustris L., Marsh Marigold. (Ranunculaceae). - Perennial herb. Europe, temp. Asia, North America. Stems and leaves have been used as a potherb. Flowerbuds are kept in vinegar and used as capers. Roots are consumed as food among the Ainu.

Caltrop, Water → Trapa natans L.

Calvatia cakavu (Zippel) v. Overeem. (Lycoperdaceae). - Basidiomycete. Fungus. Fruitbodies are consumed as food in some parts of Asia. Also for medical purposes. Likewise C. umbrinum Pers., C. natalense Cook and Mass and others.

Calvatia cyathiformis (Bosc.) Mor. (Lycoperdaceae). - Basidiomycete. Fungus. Temp. zone. Fruitbodies are consumed as food by the Omaha Indians.

Calvatia lilacina (Berk. and Mont.) Lloyd. (Lycoperdaceae). - Basidiomycete. Fungus. Fruitbodies are used to stop bleeding. Formerly used as a surgical dressing and as a haemostat known as Fungus Chirurgorum, Fungus Bovista and Crepitus Lupi. Used in beekeeping to intoxicate bees, by burning dried fruitbodies near the hives. Spores when mixed with honey are used against throat infections by the Chinese. Young fresh fruitbodies are consumed as food by the Iroqios Indians. The following puffballs are used for the same purpose: C. gigantea (Batch) Lloyd, C. candida (Rorsk) Holl. and others.

Calycanthaceae → Calycanthus, Chimonanthus

Calycanthus fertilis Walter (syn. C. glaucus Willd.). (Calycanthaceae). - Shrub. E. of United States. Bark, leaves and roots are used by certain Indian tribes as antiperiodic.

Calycanthus floridus L., Carolina Allspice. (Calycanthaceae). - Shrub. E. of United States. Aromatic bark was used by the Indians as spice.

Calycophyllum candidissimum (Vahl.) DC. (Rubiaceae). - Tree. Cuba, S. Mexico to Colombia, Venezuela. Wood brownish, strong, hard, heavy, tough, not difficult to work, resistant to the teredo. Source of Lemonwood of the bow-makers in the U.S.A. Used locally for agricultural implements, vehicles, tool handles, frames and turnery. Wood is made into charcoal. Wood is sometimes called Lancewood.

Calycophyllum multiflorum Griseb., Palo Blanco (Rubiaceae). - Tree. S. America, esp. Argentina, Paraguay; Matto Grosso (Braz.). Wood very fine, uniform, hard, straight-grained; used for implements, shoe-lasts, cogs of wheels, turnery; recommended for rulers, shuttles, flooring, wood pulleys.

Calypso borealis Salisb. (syn. C. bulbosa (L.) Okes). (Orchidaceae). - Perennial herb. Western N. America. Bulbs were eaten among the Creek Indians.

Calyptranthes aromatica St. Hil. (Myrtaceae). - Tree. S. Brazil. Plant is used in Brazil as a spice, resembling cloves.

Calyptranthes Schiediana Beng. (Myrtaceae). - Tree. Mexico. Leaves are used as a spice in Mexico.

Calyptrogyna sarapiquensis Wendl. (Palmaceae). - Palm. Costa Rica. Leaves are used by natives for covering roofs and walls of houses.

Calystegia sepium R. Br. (Convolvulaceae). - Perennial herb. Tropics and subtropics. Cultivated. Roots are boiled and consumed as a vegetable in China. Young shoots are eaten in some parts of India.

Camagoon Ebony → Diospyros pilosanthera Blanco.

Camarophyllus pratensis (Pers.) Karst. → Hygrophorus pratensis (Pers.) Fr.

Camass, Common → Camassia Quamash Greene.

Camassia esculenta Linal. Wild Hyacinth, Indigo Squil. (Liliaceae). - Bulbous perennial. Eastern and southern part of the United States. Bulbs are edible, used as food by the Indians.

Camassia Quamash Greene, Common Camass. (Liliaceae). - Bulbous herb. Western N. America. Bulb used as food boiled and roasted, by the Indians and early settlers.

Camel's Thorn → Alhagi camelorum Fisch.

Camellia japonica L., Camellia. (Theaceae). - Tree. Japan. Cultivated, also ornamentally. Seeds source of Tsubaki-Oil, non-drying, used as hair-oil. Sap. Val. 187.2; Iod. No. 78; Aic. Val. 1.05; Unsap. 0.2%.

Camellia drupifera Lour. (Theaceae). - Tree. Himalayan region, Burma, China. Seeds are source of an oil.

Camellia sinensis (L.) Kuntze. (syn. C. Thea Link, Thea sinensis L.). Tea. Chinse Tea. (Theaceae). - Tree. Assam, China. Cultivated since early times in China, Japan, Java and other asiatic countries. Cured leaves are used as a beverage. Green Tea is derived from dried green leaves to which belong Pekoes, Souchong and Congous. Black Tea has undergone fermentation or a chemical process, among which belong: Young Hyson, Hyson, Gunpowder and Imperial. Oolong Tea has been partly fermented, among which are: Oolong of Formosa, Oolong of Fou-tcheou and Oolong of Amoy. Tea is sometimes flavored with flowers of Jasminum Sambac, J. paniculata, Gardenia florida, Osmanthus fragrans and Aglaia odorata. Brick Tea is made of steamed leaves and twigs used in Russia. Tea contains 3 to 5% theine. Tea is considered astringent, stimulant and a nervine. Seeds are a source of oil.

Camellia Sasanqua Thunb., Sasanqua Camellia. (Theaceae). - Tree. Trop. Asia. Cultivated in Assam, China, Japan. Seeds are source of Tea Seed Oil; non-drying, textile oil, used in silk industry and manuf. soap. When refined it is suitable in food. Sp. Gr. 0.915-0.919; Sap. Val. 190-195; Iod. No. 80-87.

Cameroon Cardamon → Aframomum Hanburyi Schum.

Cameroon Copal → Copaifera Demeusii Harms.

Cameroon Mahogany → Entandrophragma Rederi Harms, Khaya euryphylla Harms and Entandrophragma Candollei Harms.

Camogan Ebony → Diospyros discolor Willd.

Campanula Rapunculus L., Rampion. (Campanulaceae). - Biennial herb. Europe, N. Africa, S. W. Asia, Siberia, Occasionally cultivated. Roots are consumed in salads.

Campanula versicolor Sibth. (Campanulaceae). - Perennial herb. Italy, Greece, Turkey. Herb is eaten as a vegetable and salad in some parts of Greece.

Campanulaceae → Adenophora, Campanula, Clermontia, Codonopsis, Isotoma, Lobelia.

Camphire → Lawsonia alba Lam.

Camphor →Cinnamomum Camphora (L.) Nees. and Eberm.

Camphor, Borneo → Dryobalanops aromatica Gaertn. f.

Camphor, Ngai → Blumea balsamifera DC.

Camphora officinarum Nees. → Cinnamomum Camphora (L.) Nees. and Eberm.

Campomanesia aromatica Griseb. (Myrtaceae). - Small shrub. Guiana and West Indies. Fruit edible, pleasant flavor, resembling that of strawberry.

Campomonesia Guaviroba Benth. and Hook. (syn. Abbevillea Guaviroba Berg.). Guabiraba. (Myrtaceae). - Small shrub. S. Brazil. Occasionally cultivated. Fruits edible, globular, 2 to 3 cm in diam., yellow-orange.

Campsiandra laurifolia Benth. (Leguminosaceae). - Tree. Brazil. Leaves and roots are used in Brazil as febrifuge, tonic, excitant, for tertian fevers, also for washing ulcers.

Camwood → Baphia nitida Lodd.

Canada Balsam → Abies balsamea (L.) Mill.

Canada Bluegrass → Poa compressa L.

Canada Crookneck → Cucurbita moschata Duch.

Canada Garlic → Allium canadensis L.

Canada Milk Vetch → Astragalus canadensis L.

Canada Pitch → Tsuga canadensis (L.) Carr.

Canada Plum → Prunus americana Marsh.

Canada Thistle → Cirsium arvense (L.) Scop.

Canada Turpentine → Abies balsamea (L.) Mill.

Canadian Golden Rod Oil → Solidago canadensis L.

Canadian Goldenrod → Solidago canadensis L.

Canaigre → Rumex hymenosepalus Torr.

Cañafistula → Peltophorum Vogelianum Walp.

Canangium odoratum L., Ylang Ylang. (Annonaceae). - Large tree. Trop. Asia. Cultivated in tropics of Old and New World. Flowers are source of an ess. oil; used in perfumery, soaps, cosmetics. They are also laid between cloth to impart an agreeable scent. Siamese use an infusion of the flowers after bathing. Coconut Oil is scented with Ylang Ylang and with other aromatics known as Macassar Oil.

Canarium bengalense Roxb. (Burseraceae). - Tree. E. India. Stem is source of a gum called Black Dammar Resin, recommended as a hard drying varnish for enamel painting, also known as East Indian Copal.

Canarium commune L. Java Almond. (Burseraceae). - Tree. Moluccas. Cultivated in Mal. Archipelago. Seeds edible, of a pleasant taste; consumed with rice in pastries; much esteemed by the natives. Seeds source of an oil, used for illumination and in food. Said to be source of an Manila Elemi, a resin, incense.

Canarium decumanum Gaertn. (Burseraceae). - Tree. Malaya. Seeds are edible and consumed by the natives. Also C. luzonicum Miq.

Canarium edule Hook. f. (Burseraceae). - Tree. Trop. Africa. Fruits edible, violet, size of a plum; consumed by the natives. Occasionally cultivated. Bark is source of a resin, Bulungu Resin, used by the natives for smoking and perfuming; it is also used for mending earthen ware and pitching inner surface of calabashes; employed for skin diseases and jiggers.

Canarium euphyllum Kurz., Andaman Canary Tree. (Burseraceae). - Tree. Andam. Islds. Wood is used for general carpentry.

Canarium hirsutum Willd. (Burseraceae). - Tree. Moluccas. Source of Damar Sengai, one of the hardest Malayan resins; used for varnishes.

Canarium legitinum Miq., Damar Itam. (Burseraceae). - Tree. Amborna; Ceram. Source of a resin, Damar Itam; used as torches by the natives.

Canarium luzonicum Miq. Elemi Canary Tree. (Burseraceae). - Tree. Philipp. Islds. Source of Manila Elemi, a resin product.; used in varnishes, for caulking boats and torches. Much is exported. Elemi Oil is used for the same purpose as turpentine.

Canarium Mansfeldianum Engl. (Burseraceae). - Tree. Trop. Africa. Source of Adju Mahogany or Gabon Mahogany.

Canarium ovatum Engl., Pili (Philipp.). (Burseraceae). - Tree. S. Luzon (Philipp.) Nuts are source of Pili-Nut Oil. Seeds when roasted have excellent flavor; are used for confectionary. Used in Camarines as an adulterant of chocolate. Uncooked nuts are purgative. Sp. Gr. of Pili Nut Oil is 0.9069; Sap. Val. 197.6; Iod. No. 56; Ac. Val. 1.92. Fruits are eaten cooked.

Canarium paniculatum (Lam.) Benth. (Burseraceae). - Tree. Mauritius. Stem is source of a light colored resin, also of Colophan Wood.

Canarium polyphyllum Schum. (Burseraceae). - Mal. Archip., New Guinea. Seeds are consumed by the natives in some parts of New Guinea.

Canarium pseudo-decumanum Hoehr. (Burseraceae). - Tall tree. Sumatra. Borneo. Resin is used by the natives for making vessels watertight.

Canarium rufum Benn. (Burseraceae). - Tree. Malacca, Philipp. Islds. Source of a dammar, recommended for varhishes.

Canarium samoaënse Engl. (Burseraceae). - Tree. Samoa. Trunks are used as canoes by the natives. A fragrant gum is used for perfuming coconut oil.

Canarium Schweinfurthii Engl., Papo Canary Tree. (Burseraceae). - Trop. Africa. Source of African Elemi or Elemi of Uganda, derived from incisions of the trunk; used for preparation of ointment and plasters; manuf. of printing inks, varnishes. Used in Uganda as incense in churches. Powdered bark in Angola is used for scorbutic ulcers. Fruit is used in Angola as condiment.

Canarium strictum Roxb. (Burseraceae). - Tree. Malaysia. Stem is source of Alriba Resin.

Canarium villosum Benth. and Hook. (Burseraceae). - Tree. Philipp. Islds. Trunk is source of Sahing or Pagsahingin Resin; used in Philipp Islds. as fuel, illumination, caulking material for bancas.

Canarium zeylanicum (Retz.) Blume. (Burseraceae). - Tree. Ceylon. Wood is used for manuf. of tea boxes.

Canary Grass → Phalaris canariensis L.

Canary Grass, Red → Phalaris arundinacea L.

Canary Madrone → Arbutus canariensis Duham.

Canary Tree, Andaman → Canarium euphyllum Kurz.

Canary Tree, Elemy → Canarium luzonicum Miq.

Canary Tree, Papo → Canarium Schweinfurthii Engl.

Canary Wood → Euxylophora paraensis Huber. and Morinda citrifolia L.

Canavalia ensiformis DC., Jack Bean, Chickasaw Lima. (Leguminosaceae). - Perennial vine. Probably native to West Indies. Cultivated in tropics of Old and New World. Used as food for live-stock. Pods when very young are eaten as snap-beans. Roasted seeds are used as substitute for coffee. Fresh, unripe seeds are considered poisonous. Also C. obtusifolia DC. In Japan pods of white varieties are eaten preserved in salt, those of black varieties are eaten fresh or boiled.

Candelariella vitellina (Ehrh.) Muell. Arg. (Lecanoraceae). - Lichen. Temp. zone. On stones and walls. Has been used in Sweden to dye woolens a yellow color.

Candelilla → Euphorbia antisyphilitica Zucc.

Candelilla Wax → Pedilanthus Pavonis (Klotzsch and Garcke) Boiss.

Candida Guillermondii (Cast.) Lang and Guerra. (Cryptococcaceae). - An asporogenous yeast. Has been found to produce ruboflavon as result of fermentation.

Candillo → Urena lobata L.

Candied Peel → Citrus Medica L.

Canella alba Murr. (Sym. Canella Winteriana (L.) Gaertn.). Wild Cinnamon (Canellaceae). - Tree. West Indies to S. Florida. Source of Canella, Canella Bark, Whit Cinnamon, Bahama White Wood, Wild Cinnamon. Contains 8% manitol, an ess. oil, eugenol, cineol, pinene and caryophyllene; being aromatic, stimulant, stomachic; used as spice and for flavoring tobacco.

Canellaceae → Canella, Cinnamodendron.

Candleberry → Myrica cerifera L.

Candlenut Oil → Aleurites triloba Forst.

Candlenut Oil Tree → Aleurites triloba Forst.

Candlewood → Amyris balsamifera L. and Dodonaea viscosa Jacp.

Cane, Sugar → Saccharum officinarum L.

Cane, Wild → Gynerium sagittatum (Aubl.) Beauv.

Canistel → Lucuma Rivicoa Gaertn. f.

Camota Rubber → Sapium taburu Ule.

Candy Carrot → Athamanta cretensis L.

Canella Bark → Canella alba Murr.

Cangerana Bark → Cabralea Cangerana Sald.

Canna bidentata Bertol. (Cannaceae). - Perennial herb. Trop. Africa. Seeds are used for rosaries and necklaces. Leaves are employed for wrapping food. Starch from rhizomes is consumed in times of scarcity.

Canna edulis Ker-Gawl, Queensland Arrowroot. (Cannaceae). - Trop. America. Cultivated in Tropics, among which especially West Indies and Hawaii. Very young tubers are consumed when cooked. Source of a starch known as Queensland Arrowroot, used for invalids and infants. Leaves and tubers are used as fodder for dairy cows.

Canna gigantea Desf. (syn. C. latifolia Rosc.). (Cannaceae). - Perennial herb. Brazil. Rhizome is considered diuretic and diaphoretic in some parts of Brazil.

Canna glauca L. (Cannaceae), - Perennial herb. Brazil to Mexico. Rootstocks are consumed in some parts of Brazil and Antilles.

Canna Warszewiczii Dietr. (syn. C. sanguinea Warsz.). (Cannaceae). - Perennial herb. Trop. America. Rootstock is considered diuretic, antiblenorrhagic; leaves are emollient.

Cannaceae → Canna.

Cannabis sativa L., Hemp. (Urticaceae). - Annual herb. Asia. Cultivated in the Old and New World. Grown since ancient times. Stem is source of Hemp Fibre, used for twine, cordage, bags, sacks, sailcloth, carpets, for caulking vessels, packing for pumps, engines, used in cooperage. Seeds are source of Hemp Seed Oil, used in manuf. of paints, varnishes, green soft soap and for edible purposes. Sp. Gr. 0.9285; Sap. Val. 90-193; Iod. No. 150-166; Unsap. 1-1.3%. Seeds are used as food for birds. Dried flowering tops of pistillate plants are used medicinally, being sedative, analgestic, narcotic. Contain 15 to 20% of a resin, cannabin and an ess. oil. Also source of Marihuana, used for smoking, unlawful in many countries, much is used by Mohammedan fakirs and the poorer classes in parts of Asia and Africa. Used by criminals to drug people, for this purpose it is mixed with Black Datura and sugar. Excessive smoking is physically and mentally injurious, producing moral weekness and depravity. Bhang, Siddhi or Patti are the dried leaves used for the preparation of a drink.

Canoe Birch → Betula papyrifera Marsh.

Cañon Live Oak → Quercus chrysolepis Liebm.

Cañon Palm → Washingtonia filifera (Lindl.) Wall.

Canteloupe → Cucumis Melo L.

Canterelle → Cantharellus cibarius Fr.

Cantharellus cibarius Fr. Canterelle. (Agricaceae). - Basidiomycete. Fungus. Temp. zone. The fruitbodies are consumed as food in several countries. They are of excellent quality. They are also dried for winter use. Sold in markets. Also C. cinnabarina Schw. from N. America.

Cantharellus clavatus Fr. (Agaricaceae). - Basidiomycete. Fungus. Fruitbodies are consumed as food. Often sold in markets in Europe. Also C. floccosus Schw.

Cantharellus glutinosus Pat. (Agaricaceae). - Basidiomycete. Fungus. In Indochina and Annam the fruitbodies are consumed as food.

Canthium longiflorum Hiern. (Rubiaceae). - Shrub or tree. Trop. Africa. Fruits edible, said to be of excellent quality.

Canthium glabriflorum Hiern. → Plectronia glabriflora Holland.

Canthium lanciflorum (Benth. and Hook.) Hiern. → Plectronia lanciflora Benth. and Hook.

Canton Lemon → Citrus limonia Osbeck.

Canton Linen → Boehmeria nivea (L.) Gaud.

Saobe Mahogany → Swietenia Mahogany Jacq.

Cape Aloë → Aloë ferox Mill.

Cape Ash → Eckebergia capensis Sparm.

Cape Asparagus → Aponogeton distachyus Thunb.

Cape Barren Tea → Correa alba Andr.

Cape Box → Buxus Macowani Oliv.

Cape Chestnut → Calodendron capense Thunb.

Cape Ebony → Euclea Pseud-ebenus E, Mey.

Cape Gum → Acacia giraffae Willd. and A. horrida Willd.

Cape Mahogany → Ptaeroxylon Obliquum (Thunb.) Radlk. and Trichillia emetica Vahl.

Cape Thorn → Zizyphus mucronata Willd.

Capellenia moluccana T. and B. → Endospermum muluccanum Becc.

Caper Bush → Capparis spinosa L.

Capers → Capparis spinosa L.

Capparidaceae → Boscia, Cadaba, Capparis, Cleome, Courbonia, Crataeva, Dipterygium, Gynandropsis, Isomeris, Maerua.

Capparis aphylla Roth. (syn. C. decidua Pax.). (Capparidaceae). - Shrub. Sahara region. Fruits are consumed by native tribes in N. Africa.

Capparis corymbifera E. Mey. (Capparidaceae). - Woody plant. S. Africa. Flowerbuds are eaten preserved in salted vinegar.

Capparis Mitchellii Lindl. (Capparidaceae). - Woody plant. Australia, esp. Tasmania, W. Australia. Fruits eaten by natives. Also C. nobilis F. v. Muell., C. canescens Banks.

Capparis spinosa L., Caper Bush. (Capparidaceae). - Spiny shrub. Mediterranean region. Cultivated. Flower buds ar consumed pickled, called Capers, an important commercial product, esp. in France.

Capparis tomentosa Lam. (Capparidaceae). - Small tree. Trop. Africa. In N. Nigeria the leaves are used as fodder for camels.

Capsella Bursa pastoris (L.) Medic. Shepherd's Purse. (Cruciferaceae). - Annual herb. Temp. zone. Herba Bursae Pastoris was formerly used as anti-dysenteric, diuretic, haemostatic, febrifuge. Contains bursin, an alkaloid.

Capsicum frutescens L. (syn. C. annuum L., C. baccatum L.). Green Pepper, Chilli, Cayenne Pepper. (Solanaceae). - Annual to perennial plant. Trop. America. A very variable species, sometimes divided into many species. Used for salads, vegetable, condiment, in sauces. The following groups or subspecies are distinguished: I. Group. subsp. cerasiforme Bailey. Cherry Peppers, very pungent, red, yellow or purplish. var. Oxheart, Cherry, Yellow Cherry. II. Group. subsp. fasciculatum Irish. Red Cluster Pepper, very pungent. III. Group. subsp. conoides Bailey. Cone Peppers. Very acrid; resembling cherry peppers, fruits conical, var. Tabasco, Coral Gem. IV. Group. subsp. longum Sendt. Long Peppers, var. Mild Black Nubian, County Fair, Ivory Tusk. V. Group, subsp. grossum Sendt. Sweet or Bell Peppers. Apple to tomatoshaped, mild flavor, used in salads or cooked filled with chopped meat, var. Ruby King, Sweet Mountain, Monstrous, California Wonder. VI. Group, subsp. acuminatum Fingh. Fruits larger than fasciculatum, var. Long Cayenne. Pimiento is used for canning. Paprika is used as a powdered spice, much is derived from Hungary and Spain. Rosenpaprika is Hungarian Paprika, derived by grinding selected pods from which stalks, stems and placenta have been removed. Koenigspaprika contains entire ground pods, stems and seeds. Pimenton is Spanish Paprika. Of Cayenne Pepper the dried, ripe fruit is used as a stimulant, rubefacient. Contains the extremely pungent capsaicin; 1 part in 11 000 000 parts of water has still a distinct pungent taste.

Capucine Pea → Pisum arvense L.

Capulin → Prunus serotina Ehr.

Caracu → Xanthosoma Caracu Koch.

Caragana arborescens Lam. Siberian Pea Shrub, Pea Tree. (Leguminosaceae). - Small tree. Siberia. Bark is source of a fibre. Young pods are eaten as a vegetable in some parts of Siberia.

Caragana Chamlagu Lam. (Leguminosaceae). - Shrub or small tree. N. China. Yellow flowers are eaten in some parts of N. China.

Caraguata Fibre → Bromelia serra Griseb.

Caranda → Carissa Carandas L.

Caraipa fasciculata Camb. (Guttiferaceae). - Tree. Guiana, Brazil. Trunk is source of a balsam, used for healing wounds.

Caraipa psidifolia Ducke. (Guttiferaceae). - Tree Brazil. Seeds are source of an oil, used in Brazil for skin diseases. Also C. Lacerdaei Barb. Rodr., C. minor Hub. and C. palustris Barb. Rodr.

Carallia calycina Benth. (Rhizophoraceae). - Large tree. Ceylon. Wood durable, yellowish-red to dark red-brown, beautifully marbled; takes a fine polish. Used for building purposes, panels, for cabinets.

Carambola → Averrhoa Carambola L.

Carapa procera DC. (Meliaceae). - Tropics. Seeds are probably source of Touloucouna Oil; used for curing yaws, burns, mosquito bites. Used in Marseille for manuf. soap. Bark is tonic and a febrifuge. Wood is fairly heavy, dark red brown, strong, durable; used for general carpentry and furniture.

Carapa guineensis Aubl. Andiroba, Crabwood. (Meliaceae). - Tree. S. America, West Indies, W. Africa. Seeds source of Andiroba or Craw-Wood Oil; non-drying, supposed to be poisonous. Used as illuminant, for soap; ointment as protection from insect bites. Sp. Gr. 0.927; Sap. Val. 197.3; Iod. No. 65.8; Unsap. 1%. Wood is durable, easily worked, reddish; used for furniture, house-building, instrument cases, construction work, recommended for hardwood joinery, display cabinets and cupboards.

Carapa moluccensis Lam. (syn. Xylocarpus moluccensis Roem. X. Granatum Koenig). (Meliaceae). - Tree. Trop. Asia, Africa, Pacif. Islds.. Wood very hard, brown, strong, durable; used for small objects, wooden pins, handles of tools, weapons, house posts. Bark is used in India, Indo China and Perak as tan, also used for toughening fish nets. Peel from fruits is eaten in soup, when dried it is used as an appetiser.

Caraway, Black → Pimpinella Saxifraga L.

Caraway, Common → Carum Carvi L.

Carbo Ligni Depuratus → Fagus sylvatica L.

Carbo Ligni Pulveratus → Fagus sylvatica L.

Cardamine nasturtioides Bert. (Cruciferaceae). - Herbaceous perennial. Chile. Herb is consumed as a salad in some parts of Chile.

Cardamine pennsylvanica Muhl., Bitter Cress. (Cruciferaceae). - Perennial herb. Eastern part of N. America. Is sometimes used as a substitute of Water Cress. Has been sugested as an emergency foodplant.

Cardamine pratensis L. Cuckoo Flower. (Cruciferaceae). - Perennial herb. N. America, Europe, temp. Asia. Sometimes used as a substitute for Cress. Emergency foodplant.

Cardamine yesoënsis Max. (Cruciferaceae). - Herbaceous perennial. Ainu (Jap.). Young leaves and rhizomes are eaten in the spring by the natives.

Cardamon → Elettaria Cardamomum (L.) Maton.

Cardamon (substitute) → Amomum dealbatum Roxb.

Cardamon, Bastard Siamese → Amomum xanthioides Wall.

Cardamon, Bengal → Amomum aromaticum Roxb.

Cardamon, Cameroon → Aframomum Hanburyi Schum.

Cardamon, Ceylon → Elettaria Cardamomum (L.) Maton. and E. major Sm.

Cardamon, Cluster → Elettaria Cardamum (L.) Maton.

Cardamon, East African → Aframomum mala Schum.

Cardamon, Java → Amomum maximum Roxb.

Cardamon, Madagascar → Aframomum angustifolium Schum.

Cardamon, Nepal → Amomum aromaticum Roxb. and A. subulatum Roxb.

Cardamon Oil → Elettaria Cardamomum (L.) Maton.

Cardamon, Round → Amomum kapulaga Sprague.

Cardamon, Siam → Elettaria Cardamomum (L.) Maton.

Cardiaca vulgaris Moench. → Leonurus cardiaca L.

Cardinal, Red → Erythrina arborea (Chapm.) Small.

Cardinal Wood → Brosimum paraense Hub.

Cardiogyne africana Bur. (Moraceae). - Shrub or small tree. Trop. Africa. Wood is source of a yellow dye.

Cardón Hecho → Pachycereus pecten- aboriginum (Engelm.) Britt. and Rose.

Cardoon → Cynara Cardunculus L.

Carduus edulis (Nutt.) Greene → Cirsium edule Nutt.

Carduus Marianum L. → Silybum Marianum Gaertn.

Carduus oleraceus Vill. → Cirsium oleraceum (L.) Scop.

Carduus undulatus Nutt. → Cirsium undulatum (Nutt.) Spreng.

Carex dispalatha Boott. (Cyperaceae). - Perennial herb. Japan. Cultivated in rice fields. Leaves are made into hats in Japan.

Carex Pierotii Miq. (Cyperaceae). - Perennial herb. Japan, China. Dried leaves are made into ropes in Japan.

Carex riparia Curtis (syn. C. acuta All., C. versicaria Leers.) (Cyperaceae). - Perennial herb. Europe, Caucasia, E. Asia, N. Africa. Plants are used in stables for cattle to stand upon.

Careya arborea Roxb. Tummy Wood. (Myrtaceae). - Tree. Asia, esp. India. Fibrous bark is used for cord in Indo-China. Leaves used as food for Tuscan Silkworm. Seeds are roasted and consumed by the natives.

Carib Grass → Eriochloa polystachya H. B. K.

Caribbean Princewood → Exostemma caribaeum Roem. and Schult.

Carica candamarcensis Hook. f. Mountain Papaya. (Caricaceae). - Tree. Colombia to Ecuador. Fruits eaten as preserves or candied, seldom consumed raw. Recommended for dyspeptics. Sold in inter-Andean markets.

Carica candicans Gray. (Caricaceae). - Tree. Peru. Dry, fibrous fruits are of a pleasant taste, eaten in some parts of Peru.

Carica cestriflora Solms. (syn. Vasconcella cestriflora A. DC.) Papaya de Terra Fria. (Caricaceae). - Small tree. New Granada, Colombia. Fruits edible, eaten as preserves.

Carica chrysophylla Heilb., Higicho. (Caricaceae). - Tree. Ecuador. Cultivated. Fruits have an aromatic flavor; used in preserves.

Carica Papaya L., Papaya, Melon Tree, Pawpaw. (Caricaceae). - Sparingly branched, somewhat succulent tree. Tropical America, probably from Mexico and Centr. America. Cultivated in tropics and subtropics of the Old and New World. Fruits of high dessert quality; also used in salads, pies, jellies, sherbets, preserves, pickles and candied. When green, eaten as vegetable. The plant produces a latex, containing papain, resembling animal pepsin. Tough meat becomes tender when wrapped and cooked with leaves; is sometimes used as tenderizer. Dessicated latex, a digestant, is of commercial importance. Fruitpulp is sometimes a ingredient in face creams and hair shampoos. There are many varieties, some of local importance, e. g. Fairchild, Graham, Betty (dioecious types), Solo, Blue Stem (hermaphrodites).

Carica peltata Hook. and Arn., Papaya de Mico. (Caricaceae). - Tree. Nicaragua to Costa Rica. Fruits small, are consumed by the natives.

Carica pentagona Heilb. Babaco. (Caricaceae). - Small tree. Ecuador and adjacent territory. Cultivated in Ecuador. Flesh of fruits nearly white, acid; eaten cooked, also as dulce or preserves, and in sauce.

Carica quercifolia (St. Hil.) Solms Laub., Higuera del Monte. (Caricaceae). - Tree. S. America, esp. Bolivia, Ecuador, Uruguay and N. Argentina. Fruits are eaten candied or as preserves.

Caricaceae → Carica, Jaracatia.

Cariniana pyriformis Miers. Bacu, Albarco (Lecythidaceae). - Tree. Colombia, Venezuela, Guiana. Wood was formerly known as Colombian Mahogany, has a high silica content. Manuf. into sliced or rotary veneers. Bark made into cordage by the Indians.

Caripé → Moquilea utilis Hook.

Carissa Arduina Lam. (syn. C. bispinosa (L.) Desf., Arduina bispinosa L.). Amatungula. (Apocynaceae). - Spiny shrub. S. Africa. Fruits red, size of a cherry, edible.

Carissa Brownii F. v. Muell. (Apocynaceae). - Shrub. Australia. Reddish, small fruits are consumed by the aborigenes of N. Queensland.

Carissa Carandas L. Caranda. (Apocynaceae). - Large shrub or small tree. India, Malaysia. Berries 2 cm in diam.; eaten raw; also made into jelly; ripe fruits are used for pies.

Carissa grandiflora A. DC., Natal Plum (Apocynaceae). - Large spreading, spiny shrub. Natal (S. Afr.). Cultivated in tropics and subtropics of Old and New World. Berries bright-red, 2 to 5 cm. long; half-ripe fruits are made into jelly; ripe fruits are used for pies. Sold in markets of Natal.

Carissa ovata R. Br. (Apocynaceae). - Small shrub. S. Australia. Fruits ovoid, as large as a small plum, pleasant taste; consumed by the natives.

Carline Thistle → Carlina acaulis L.

Carlina acaulis L., Carline Thistle. (Compositae). - Perennial herb. Mountainous regions. Europe. Root formerly used as emmenagogue, diaphoretic, stomachic; when used in strong doses also as a purgative.

Carludovica angustifolia Ruiz and Pav. (Cyclanthaceae). - Low palm-like plant. Peru. Leaves are used for thatching huts among the natives of Peru.

Carludovica divergens Drude. (Cyclanthaceae). - Palm-like plant. Peru, Brazil. Source of a fibre used for cordage in Peru.

Carludovica Labela Schult. (Cyclanthaceae). - Low palm-like plant. Mexico. Leaves are used for thatching roofs in parts of Mexico.

Carludovica palmata Ruiz and Pav., Panama Hat Palm, Palmita. (Cyclanthaceae). - Low palm-like plant. Centr. America to N.W. of S. America. Young leaves are made into Panama Hats. Important industry in Ecuador.

Carludovica sarmentosa Sagot. (Cyclanthaceae).- Low palm-like plant. Guiana. Used for manuf. of brooms.

Carnation → Dianthus Caryophyllus L.

Carnation Baboonroot → Babiana plicata Ker-Gawl.

Carnauba Wax → Copernica cerifera Mart.

Carnegiea gigantea (Engelm.) Britt. and Rose. (syn. Cereus giganteus Engelm.) Sahuaro, Giant Cactus (Captaceae). - Tall xerophytic tree. S. Arizona, California also, Sonora (Mex.). Dried woody ribs of stems used as lances by Indians, frame-work of huts. Fruits eaten raw, cooked, dried or preserved; source of a syrup, intoxicating beverage. Seeds made into a paste used as a butter on tortillas, or made into pinole.

Caroa Verdadeira Fibre → Neoglaziovia variegata (Arruda da Cam.) Mez.

Carob → Ceratonia Siliqua L.

Carob Gum → Ceratonia siliqua L.

Carobbe di Guidea → Pistacia Terebinthus L.

Carolina Allspice → Calycanthus floridus L.

Carosella → Foeniculum vulgare Mill.

Carpathian Turpentine → Pinus Cembra L.

Carpet Grass → Axonopus compressus (Swarz) Beauv.

Carpinus Betulus L. European Hornbeam. (Betulaceae). - Tree. Europe, Caucasia, Asia Minor, N. Iran. Wood grayish to yellowish, hard, heavy, difficult to split, not very durable; used for wagons, mills, tools and turnery.

Carpinus caroliniana Walt. American Hornbeam, Blue Beech. (Betulaceae). - N. America to Mexico and Centr. America. Wood light brown; used for handles of tools, small articles, levers and fuel.

Carpinus cordata Blume, Heartleaf Hornbeam. (Betulaceae). - Tree. Japan. Wood compact, hard, fine-grained, takes good polish; used in Japan for turnery, handles of umbrellas.

Carpinus laxiflora Blume. (Betulaceae). - Tree. Japan, Korea. Wood heavy, strong, hard, somewhat elastic, close-grained, yellowish-white; used in Japan for ski's, handles of agric. implements, sculpture, turnery and cabinet work.

Carpodinus chylorrhiza Schum. (Apocynaceae). - Half-shrub. Trop. Africa. Source of a rubber, derived from the root.

Carpodinus hirsuta Hua. (Apocynaceae). - Woody plant. W. Africa. Source of a fair rubber, known as La Glu and Accra Paste.

Carpodinus uniflora Stapf. (Apocynaceae). - Woody plant. Cameroon. Source of a rubber.

Carpolobia lutea G. Don. (Polygalaceae). - Tree. Trop. Africa esp. Nigeria. Wood very hard, resists termites; used for house-posts and walking-sticks.

Carpotroche brasiliensis Endl. (Flacourtiaceae). - Tree. Brazil. Source of Oleo de Sapucainha, Carpotroche Oil; parasiticide, specific for dermatosis, leprosy; also in veterinary medicine. Sp. Gr. 0.9545; Sap. Val. 202.5; Iod. No. 10.2—13.5; Unsap. 0.6%. Recommended for the same purpose are C. amazonica Benth, C. denticula Benth, C. grandiflora Spruce. C. laxiflora Benth and C. longifolia Benth.

Carpotroche Oil → Carpotroche brasiliensis Endl.

Carrageen → Chondrus crispus (L.) Stackh.

Carageenin → Chondrus crispus (L.) Stackh.

Carreto → Aspidosperma Curranii Standl.

Carrot → Daucus Carota L.

Carrot, Candy → Athamanta cretensis L.

Carrot, Gargan Death → Thapsia garganica L.

Cartagena Ipecac → Cephaëlis acuminata Karsten.

Cartán → Centrolobium orinocense (Benth.) Pitt.

Carthamus tinctoria L., Safflower. (Conpositae). - Annual herb. Mediterranean region, Asia Minor. Cultivated since ancient times. Flowers are source of a red and yellow dye, called Safflower. Fruits produce a drying oil, used in paints, varnish, linoleum, also for edible purposes. Sp. Gr. 0.925—0.928; Iod No. 140—150;

Sap. Val. 188—194; Unsap. 0.5—1,3%. Also source of Roghan or Afridi Wax, used in Afridi Wax Cloth. Fruits are edible; considered good food for poultry. Fried seeds are used in chutney. Pressed seed-cake is an excellent food for cattle. It is said that the fruits have the property of coagulating milk.

Caruba de Castilla → Passiflora mollissima (H. B.K.) Bailey.

Carum Bulbocastanum Koch. Earth Nut. (Umbelliferaceae). - Perennial herb. Europe, temp. Asia. Starchy tubers are eaten as vegetable and in salads; sometimes sold in markets. Has been recommended as an emergency food plant.

Carum Carvi L. (syn. Apium Carvi Crantz, Seseli Carvi Lam.). Caraway. (Umbelliferaceae). - Biennial herb. Mediterranean region, S. Russia, N. Iran, Himalaya. Cultivated in Holland, England, Russia, Germany, N. Africa, United States. Seeds are used for flavoring, in bread, meat, sausage, vegetables, in Sauerkraut, in cheese, brandy (Kümmelbranntwein), liqueur (Kümmel). Roots are eaten as a vegetable. Dried fruits used medicinally, being stimulant, aromatic, diaphoretic, diuretic, carminative. Contains an ess. oil, Oil of Caroway, Oleum Carvi.

Carum copticum (L.) Benth. and Hook. (syn. Ammi copticum L.). Ammi. (Umbelliferaceae). - Annual herb. India. Cultivated. Used as condiment and for the extraction of thymol. Source of Ajowan Oil. Antiseptic, carminative, aromatic. Omum Water is distilled from seeds, widely used in India as a medicine. Ajowan Seed Oil is semi-drying, used for technical purposes. Sp. Gr. 0.9267; Sap. Val. 176.9; Iod. No. 108.8; Unsap. 1.14%.

Carum Gairdneri (Hook and Arn.) Gray, Yampa. (Umbelliferaceae). - Western N. America. A favorite food among the Indians of many western States. Roots were eaten raw or preserved for winter use, having a pleasant nutty flavor. Also C. Kelloggi Gray and C. oreganum S. Wats.

Carum nigrum Baill → Pimpinella Saxifraga L.

Carum Roxburghianum Benth. (syn. Athamantha Roxburghiana (Benth.) Wall.) Ajmud (India). (Umbelliferaceae). - Annual herb. Trop. Asia. Cultivated in India, Ceylon, Cochin-China, Annam and Laos. Seeds used as a condiment, stimulant, carminative; also used in spices and curries.

Carumbium amboinicum Miq. → Pimelodendron amboinicum Hassk.

Carya cathayensis Sarg., Chinese Hickory. (Juglandaceae). - Tree. E. China. Nuts are consumed in some parts of China. They are made into sweet meat. Wood is strong, tough; used in China for handles.

Carya cordiformis (Wangh.) Koch. (syn. Hicoria minima Brit., C. amara Nutt.). Bitternut. (Juglandaceae). - Tree. N. America. Wood strong, tough, very hard, heavy, dark brown, close-grained; used for ox-yokes and hoops.

Carya glabra (Mill.) Sweet. (syn. Hicoria glabra (Mill.) Brit., C. porcina (Michx. f.) Nutt.). Pignut Hickory. (Juglandaceae). - Tree. Eastern N. America to Florida. Wood hard, tough, very strong, heavy, flexible, light to dark brown; used for manuf. wagons, agricultural implements, handles of tools and as fuel. Nuts are very variable in quality, some are astringent, others have a pleasant flavor.

Carya laciniosa (Michx. f.) Loud. (syn. Hicoria laciniosa (Michx. f.) Sarg., C. sulcata Nutt.). Big Shellbark Hickory. (Juglandaceae). - Tree. N. America. Wood used for tool-handles and agricultural implements. Nuts are used as food, sold in markets.

Carya ovata (Mill.) Koch. (syn. Hicoria ovata (Mill.) Brit., Carya alba Nutt.). Shagbark Hickory, Shellbark Hickory. (Juglandaceae). - Tree. N. America. Wood very hard tough, strong, heavy, closed-grained, light brown; used for manuf. of carriages, wagons, agricultural implements, axe-handles, baskets and fuel. Nuts are used as food, they are the common Hickory Nuts of commerce.

Carya Pecan (Marsh.) Engl. and Graebn. (syn. Hicoria Pecan Britt. C. olivaeformis Nutt.) Pecan. (Junglandaceae). - Tree. Eastern and southern United States, into Mexico. Commercially cultivated as a nut tree in the southern part of the United States. Paper-shelled or thin-shelled varieties are being grown among which: Curtis, Frotscher, Moneymaker, Pabst, Schley, Stuart and others, many being suitable to certain regions. Nuts eaten raw, in candies, cakes, salted etc. Wood hard, brittle, not strong, coarse-grained, heavy, light brown, red tinged; occasionally used for agricultural implements, manuf. of wagons, also as fuel.

Carya tomentosa (Lam.) Nutt. (syn. C. alba Koch not Nutt.). Mockernut, Big-bud Hickory. (Juglandaceae). - Tree. Eastern N. America to Florida. Wood used for the same purposes as C. ovata (Mill.) Koch.

Caryocar amygdaliferum Mutis. (Caryocaraceae). - Tree. S. America. Seeds are source of Sawarri or Suari Fat, being of pleasant flavor, used in food. Sp. Gr. 0.8981; Sap. Val. 197-199.5; Iod. No. 41-50; Melt. Pt. 30-37° C. Seeds are consumed among the Indians of Ecuador.

Garyocar brasiliense Camb. (Caryocaraceae). - Tree. Brazil. Yellow fruitpulp is eaten with meat by the Indians of the Mattogrosso region, Brazil.

Caryocar butyrospermum Willd. (Caryocaraceae). - Tree. Trop. America. Oily seeds are edible, consumed by the natives.

Caryocar glabrum Perr., Soapwood. (Caryocaraceae). - Tree. Guiana. Inner bark used by natives for washing hair, clothes etc.

Caryocar nuciferum L. (Caryocaraceae). - Tall tree. Brazil and Guiana. Cultivated in West Indies. Source of Suari Nuts or Butter Nuts, edible; produce an oil which is exported. Wood of good quality, durable; used for shipbuilding.

Caryocar tomentosum Willd., Suari Tree. (Caryocaraceae). - Large tree. Guiana. Source of Suari Nuts, composition similar to that of Brazil Nuts.

Caryocaraceae → Caryocar.

Caryophyllaceae → Agrostemma, Arenaria, Corrigiola, Cucubalus, Dianthus, Melandrium, Paronychia, Saponaria, Silene, Spergularia, Stellaria, Tunica.

Caryophyllus aromaticum L. → Eugenia caryophyllata Thunb.

Caryota mitis Lour. (Palmaceae). - Tall palm. India, Malaysia. Source of a sago. Very young unfolded leaves are eaten as vegetable. Fibre from the base of the leaves is used by wild tribes for the wad behind the dart in blowpipes.

Caryota urens L., Fish-tail Palm (Palmaceae). - Tall palm. India, Sikkim to Ceylon. Source of Kittul-Fibre or Ceylon-Piassava; very fine, like horse-hair; used for brushes. Source of a sago. Very young unfolding leaves edible. Juice from trunk made into sugar, also made into alcohol.

Casa Banana → Sicana odorifera Naud.

Casaba → Cucumis Melo L.

Cascade Fir → Abies amabile (Loud.) Fores.

Cascara Amarga → Sweetia panamensis Benth.

Cascara Buckthorn → Rhamnus Purshiana DC.

Cascara de Lingue → Persea Lingue Nees.

Cascara Sagrada → Rhamnus Purshiana DC.

Cascarilla Crespilla → Cinchona officinalis L.

Cascarilla Delgada → Cinchona pubescens Vahl.

Cascarilla Provinciana → Cinchona micrantha Ruiz and Pav.

Cascarilla Verde → Cinchona micrantha Ruiz and Pav., C. officinalis L.

Cascarille Bark → Croton Eluteria Benn.

Cascarille, Industrial → Croton echinocarpus Muell. Arg.

Casearia alba Rich. (Flacourtiaceae). - Tree. Cuba. Wood yellow, strong, compact; used for cabinet work.

Casearia hirsuta Sw. (Flacourtiaceae). - Tree. West Indies. Wood soft; used for furniture.

Casearia lucida Tul. (Flacourtiaceae). - Tree. Madagascar. Wood is used for making drums in Madagascar.

Casearia praecox Griseb. → Gossypiospermum praecox (Griseb.) Wils.

Casearia sylvestris Sw. (Flocourtiaceae). - Tree. Trop. America. Decotion or macerated roots, called Guassatonga, are used in Brazil for wounds and leprosy. Seeds are source of an oil, used as a substitute for Chaulmoogra Oil. Also C. singularis Eichl.

Cashawa Gum → Anacardium occidentale L.

Cashew → Anacardium occidentale L.

Cashew Apple → Anacardium occidentale L.

Cashew Marking Nut Tree → Semecarpus Anacardium L. f.

Cashew Nut → Anacardium occidentale L.

Casimiroa edulis La Llave., White Sapote, Zapote Blanco. (Rutaceae). - Medium sized tree. Mexico, Centr. America. Cultivated. Fruit edible, excellent for desserts, size of an orange, 8 cm. diam., yellowish; pulp yellow, soft, melting in the mouth, somewhat bitter sweet, of a peculiar aromatic flavor; eaten fresh, used in sherbets. There is much variation among the individual seedlings. C. tetrameria from Guatemala may be the same, or a related species.

Casis (Liqueur) → Ribus nigrum L.

Caspa Tea → Cyclopia subterenata Vog.

Caspic Willow → Salix acutifolia Willd.

Cassada → Dipholis salicifolia (L.) A. DC.

Cassava → Manihot esculenta Crantz.

Cassava, Bitter → Manihot esculenta Crantz.

Cassava, Sweet → Manihot esculenta Crantz.

Cassia Absus L. (Leguminosaceae). - Annual herb. Pantropic. Seeds are used in S. Africa for ringworm and ophthalmia, and as a cathartic in India.

Cassia acutifolia Del. (Leguminosaceae). - Shrub. Egypt, Sudan, Red Sea area. Leaves and pods are source of Alexandrian Senna of commerce.

Cassia alata L. Ringworm Senna (Leguminosaceae). - Shrub. Pantropic. Infusion of root is used in Guatemala for rheumatics and as a powerful drastic.

Cassia angustifolia Vahl. (syn. C. medicinalis Bisch.). Tennevelley Senna. (Leguminosaceae). - Shrub. Trop. Asia. Cultivated. Source of Tennevelley, Mecca or Arabia Senna. Dried leaves are used medicinally as laxative.

Cassia aphylla Cav. (Leguminosaceae). - Shrub. Argentina. Stems are used in some parts of Argentina for manuf. brooms.

Cassia crassiramea Benth. (Leguminosaceae). - Shrub. Argentina. Twigs are used for making brooms.

Cassia emarginata L. (Leguminosaceae). - Trop. America. Used in Jamaica as a dyewood.

Cassia Chamaecrista Chapm. (syn. Chamaecrista fasciculata (Michx.) Greene.). Partridge Pea. (Leguminosaceae). - Annual herb. Eastern United States to Florida and Texas. Flowers are source of a honey.

Cassia Fistula L., Purging Cassia. (Leguminosaceae). - Tree. Trop. Asia. Cultivated in tropics of Old and New World. Source of Cassia Fistula, Purging Cassia, Indian Laburnum, Cassia Pods. Dried fruit used medicinally as purgative, laxative, for habitual constipation.

Cassia holosericea Fresen. (Leguminosaceae). - Shrub-like plant. Abyssinia. Source of Aden Senna, used in medicine.

Cassia laevigata Willd. Smooth Senna. (Leguminosaceae). - Herbaceous plant. Pantropic. Seeds are used as a substitute for coffee in some parts of Guatemala. Used in Mexico as an emmenagogue.

Cassia marilandica L., American Senna. (Leguminosaceae). - E. of N. America. Perennial plant. Leaves or leaflets when collected at flowering time are sometimes used for their cathartic properties.

Cassia mimosoides L. (Leguminosaceae). - Tree. Trop. Asia, Africa. Decoction of leaves used as tea in Japan.

Cassia moschata H. B. K. (Leguminosaceae). - Perennial plant. S. America. Leaves are used as a purgative in some parts of Colombia.

Cassia obovata Collad. (Leguminosaceae). - Shrub. Tropics. Source of Dog Senna or Italian Senna; used in medicine.

Cassia occidentalis L., Coffee Senna. (Leguminosaceae). - Shrubby plant. Tropics. Seeds are used as substitute for coffee, called Mogdad or Negro Coffee. Medicinelly used as tonic, stomachic, diuretic, febrifuge; used for dropsy, fevers, rheumatism, ringworm, eczema.

Cassia oxyphylla Kunth. (Leguminosaceae). - Shrub. Mexico, Centr. America, N. W. of America. In Sinaloa used as an emetic.

Cassia sericea Sw. (syn. C. ornithopoides Lam.). (Leguminosaceae). - Trop. America. Seeds are used as substitute for coffee in Brazil. Leaves are employed as poultices for wounds; roots for combating dropsy.

Cassia Sieberiana DC. (Leguminosaceae). - Woody plant. Throughout trop. Africa. Decoction of roots is used as diuretic in Gambia. Pods are used to stupefy fish.

Cassia Sophera L. (Leguminosaceae). - Woody plant. Throughout tropics. Leaves, bark and seeds are cathartic. Juice of leaves is a specific for ringworm.

Cassia Tora L. Sickle Senna (Leguminosaceae). - Shrub. Tropics of Old and New World. Leaves purgative. Used in India for ring-worm and cutaneous diseases. Cultivated for its seeds to be used as mordant in dying cloth blue. In Mexico seeds used as substitute for coffee.

Cassia Buds → Cinnamomum Cassia Blume.

Cassia Cinnamon → Cinnamomum Cassia Blume.

Cassia, Indian → Cinnamomum Tamala (Buck. and Ham.) Nees. and Eberm.

Cassia Pods → Cassia Fistula L.

Cassie Ancienne → Acacia Farnesiana (L.) Willd.

Cassie Romaine → Acacia Cavenia Bert.

Cassiope tetragona (L.) D. Don. (Ericaceae). - Low shrub. Arctic. subartic regions. Plant is much used as fuel during summer time by the Eskimos.

Cassumar Ginger → Zingiber Cassumar Roxb.

Cassytha filiformis L. (Lauraceae). - Semiparasite. Tropics. Stems mashed in water are source of a brown dye; used in E. Africa.

Castanea crenata Sieb. and Zucc., Japanese Chestnut. (Fagaceae). - Tree. Japan. Wood strong, very hard, heavy, durable in soil, yellow brown; used in Japan for furniture, cabinet work, rail-road ties and ship building. Nuts are consumed by the Chinese and Japanese. Several varieties, among which Tamba and Shiba.

Castanea dentata (Marsh.) Borkh., American Chestnut. (Fagaceae). - Tree. Eastern North America. Wood strong, grayish brown or brown, soft, splits easily, often checks and warps during drying, easy to work, resistant to decay, contains from 6 to 11% tannin; used for furniture, caskets, boxes, crates, interior finish of houses, railroad ties, fence-posts; parts of tables, desks and pianos, paperpulp. Wood is important source of tanning extract derived by soaking chipped wood in hot water, the resulted liquor after evaporation at a certain concentration being used as extract. The edible sweet nuts are sold in local markets. Dried leaves contain 9% tannin and a mucilage; used as tonic and astringent. Some varieties grown for the nuts are: Ketchem, Watson and Griffin.

Castanea Henryi (Skan) Rehd. and Wils. Henry Chinquapin. (Fagaceae). - Tree. Centr. China. Wood much esteemed in China for building purposes.

Castanea javanica Blume → Castanopsis javanica A. DC.

Castanea mollissima Blume, Chinese Chestnut. (Fagaceae). - Tree. N. and W. China. Cultivated in China. Introd. in N. America. Nuts are much consumed by the Chinese. Several varieties, some with large nuts.

Castanea pumila (L.) Mill. Chinquapin. (Fagaceae). - Tree. Eastern part of the United States, southw. to Florida and Texas. Wood hard, strong, light, dark brown, coarse-grained; used for rail-road ties and fence posts. The sweat, edible nuts are sometimes sold in southern markets. Root is considered tonic, astringent, has been used for intermittent fevers.

Castanea sativa Mill. (syn. C. vesca Gaertn.). Spanish Chestnut. (Fagaceae). - Mediterranean region, Asia Minor, Caucasia. Cultivated. Wood relatively hard, easy to split, not easy to bend, durable, fine-grained; used for general carpentry, rail-road ties, manuf. of cellulose. Bark is used for tanning. Much cultivated for its fruits, consumed when roasted, boiled in water, substitute for coffee, made into flour, called Farine de Châtaignes; used in bread making, for thickening soups, fried in oil, (Paltenta, Nicoi,

Pattoni), also used in brandy, in confectionary (Marron Glacé), in different desserts; source of oil. Many commercial varieties, among which: Baurrue de Juillac, Gentile, Gentile Colombo, Batârd de Gourdon, Grosse Rouge, Serdonne, Longuipie, Luisant Pointu, Castania Lombarda, Martino.

Castanea sumatrana Oerst. → Castanopsis sumatrana A. DC.

Castanha de Maranhão → Sterculia Chicha St. Hil.

Castanha Oil → Telfairia pedata Hook.

Castanopsis argentea A. DC. (syn. Castanea argentea Blume) (Fagaceae). - Tree. Burma, Mal. Archipel. Wood durable, strong; used for building purposes. Decoction of bark used for dyeing, giving rattan-work a black color.

Castanopsis Boisii Hick and Camus. (Fagaceae). - Tree. Tonkin. Nuts are consumed as food by the natives of Tonkin. Sold in local markets of Hanoi, Tonkin etc.

Castanopsis chrysophylla A. DC., Golden Chinquapin, Golden-leaved Chestnut. (Fagaceae). - Tree. Washington to California. Wood closegrained, not strong, light, soft, light brown with red; sometimes used for agricultural implements, ploughs. The sweet fruits are eaten by the Indians.

Castanopsis sumatrana A. DC. (syn. C. inermis Benth. and Hook., Castanea sumatrana Oerst.) (Fagaceae). - Tree. Mal. Penins., Sumatra. Nuts are consumed as food boiled, parched or roasted. Sold in local markets of Sumatra.

Castanopsis javanica A. DC. (syn. Castanea javanica Blume (Fagaceae). - Tree. W. Mal. Archipel. Wood fairly heavy; used for building purposes.

Castanopsis philippensis Vid. (Fagaceae). - Tree. Philipp. Islds. Fruits are consumed among the natives, having the flavor of chestnuts.

Castanopsis sclerophylla Schottky. (Fagaceae). - Tree. East and Centr. China. Nuts used as food, made into paste, eaten by peasants in China.

Castanopsis tibetiana Hance. (Fagaceae). - Tree. China, Tibet. Nuts are consumed as food in some parts of China.

Castanospermum australis Cuningh. Moreth Bay Chestnut. (Leguminosaceae). - Tree. Australia esp. Queensland, New South Wales. Beans are roasted or made into a coarse flour, consumed by the natives.

Castilla costaricana Liebm. Hule (Moraceae). - Tree. Costa Rica. Latex from stem is source of a good rubber.

Castilla elastica Cerv., Arbol del Hule. Caucho (Moraceae). - Tree. S. Mexico to Central America. Source of a good rubber, Caucho Negro. Occasionally cultivated in tropics of Old and New World. Bark used by some Indians as a source for clothing and blankets.

Castilla lactiflora Pittier. (Moraceae). - Tree. Chiapas (Mex.). Source of rubber.

Castilla nicoyensis Pittier (Moraceae). - Tree. Near Nicoya, Costa Rica. Source of a rubber.

Castilla Ulei Warb. Cauchu. (Moraceae). - Tree. Amazone Region. Brazil. Source of a good rubber.

Castilloa → Castilia.

Castor Bean → Ricinus communis L.

Castor Oil → Ricinus communis L.

Casuarina collina Poisson. (Casuarinaceae). - New Caledonia. Wood very hard, dense, fine-grained, easily worked; used in N. Caledonia for wagon-making and turnery.

Casuarina Cunninghamiana Miq. (Casuarinaceae). - Tree. Australia, esp. New South Wales, Queensland. Wood hard, close-grained, used for shingles, staves.

Casuarina Deplancheana Miq. (Casurarinaceae). - Tree. New Caledonia and adj. islds. Wood heavy, fine-grained, very hard, yellowish, easily worked; used for wagon-making.

Casuarina equisetifolia L., Swamp Oak, Bull Oak (Casuarinaceae). - Tree. Australia, esp. New South Wales, Queensland, N. Australia, Pacif. Islds. Wood beautifully marked, coarse-grained. Used for shingles, gates, fencing.

Casuarina Junghuhniana Miq. (syn. C. montana Jungh.). (Casuarinaceae). - Tree. Java. Wood heavy, hard, fine-grained, light to dark brown, not easily worked; used for carpentry.

Casuarina Rumphiana Miq. (Casuarinaceae). - Tree. Moluccas. Wood has similar properties as C. equisetifolia L.

Casuarina stricta Ait. Shingle Oak, River Oak. (Casuarinaceae). - Tree. Australia. Wood not durable, close-grained, tough, reddish; used for furniture, turnery, bullock-yokes, wheel spokes, axe-handles, shingles.

Casuarina suberosa Otto and Dietr. Swamp Oak, River Black Oak. (Casuarinaceae). - Tree. Australia, esp. Tasmania, New South Wales, Queensland. Bark sometimes used for tanning.

Casuarina sumatrana Jungh. (Casuarinaceae). - Tree. Mal. Archipel. Wood extremely hard, very heavy, tough, durable; used for handles of tools.

Casuarina torulosa Ait. Forest Oak, River Oak. (Casuarinaceae). - Tree. New South Wales, Queensland, Wood attractively marked, close. Used for veneers, cabinet-work, shingles.

Casuarinaceae → Casuarina.

Catalonian Jasmine → Jasmimum grandiflorum L.

Catalpa bignonioides Walt. (syn. C. Catalpa (L.) Karst.) Common Catalpa, Indian Bean. (Bignoniaceae). - Tree. Eastern N. America to Florida and Texas. Wood course-grained, not strong, soft, very durable when in contact with soil, light brown; used for railroad-ties and fence posts.

Catalpa longissima Sims. (syn. Bignonia Quercus Lam.). (Bignoniaceae). - Tree. Antilles. Bark used for tanning in some regions of the West Indies.

Catalpa speciosa Warder., Catawba Tree, Western Catalpa. (Bignoniaceae). - Tree. N. America. Wood not strong, soft, coarse-grained, light, very durable when in contact with soil; used for rail-road ties, interior finish of houses, furniture, telegraph pols, fence-posts.

Catalpa, Common → Catalpa bignonioides Walt.

Catalpa, Western → Catalpa speciosa Warder.

Catenella impudica J. Ag. (Rhabdoniaceae). - Red Alga. Coast of Tenasserim etc. Consumed as food, collected in large quantities and sold in bazaars.

Cat Mint → Nepeta Cetaria L.

Cataria vulgaris Moench. → Nepeta Cataria L.

Catchweed → Galium Aparine L.

Catechu → Acacia Catechu Willd.

Catha edulis Forsk. (syn. Celastrus edulis Vahl.) Kât (Arab.). (Celastraceae). - Small tree. Arabia. Trop. Africa. Cultivated in Arabia, Yemen and Abyssinia. Leaves are used as a tea, much esteemed in Arabia and adj. territory. They are also used for chewing. Leaves form an ingredient of honey wine, used in Abyssinia.

Cathay Walnut → Juglans cathayensis Maxim.

Cativo → Prioria copaifera Griseb.

Catjang → Vigna Catjang Walp.

Catnip → Nepeta Cataria L.

Cato Seed Oil → Chisocheton Cumingianus (C. DC.) Harms.

Cat's Claw → Acacia Greggii Gray. and Pithecolobium Unguis-Cati Benth.

Cat's Ear → Hypochoeris maculata L.

Cattail Millet → Setaria glauca Beauv.

Cattail, Narrowleaf → Typha angustifolia L.

Cattle Bush → Atalaya hemiglauca F. v. Muell.

Cattley Guava → Psidium Cattleianum Sab.

Caucho → Castilla elastica Cerv.

Caucho Blanco → Sapium stylare Muell. Arg.

Caucho Blanco Rubber → Sapium pavonianum Muell. Arg. and S. Thomsoni God.

Caucho Mirado → Sapium stylare Muell. Arg.

Caucho Virgin Rubber → Sapium Thomsoni God.

Caulerpa Freycinetti I. Ag. (Caulerpaceae). - Green Alga. Pac. Ocean, Indian Ocean and adj. waters. Consumed in Malaysia. Also C. lactivirens Weber in Bali; C. racemosa Agardh. in E. Malaysia and C. clavifera Agardh. in Guam.

Caulerpa laetivirens W. V. B. (Caulerpaceae). - Green Alga. Sea Weed. Pacif. Ocean, and adj. seas. Is consumed as food in some parts of the Mal. Archipelago, esp. Bali.

Caulerpa peltata Lamour. (Caulerpaceae). - Green Alga. Sea Weed. Pac. Ocean. Is consumed as food in some parts of the Mal. Archipelago.

Cauliflower → Brasica oleracea L. var. botrytis L.

Caulophyllum thalictroides (L.) Michx. Blue Cohosh, Papoose Root, Squaw Root. (Berberidaceae). - Perennial herb. N. America. Dried roots and rootstock used medicinally as diuretic, emmenagogue and antispasmotic.

Caucasian Wing Nut → Pterocarya fraxinifolia Spach.

Cautivo → Prioria copaifera Griseb.

Cavanillesia planifolia H. B. K. (Bombaceae). - Tree. Panama, Colombia. Wood is used for canoes, floating rafts.

Cavenia Acacia → Acacia Cavenia Bert.

Cay-Cay Fat → Irvingia Oliveri Pierre.

Cayenne Linaloe Oil → Amyris balsamifera L.

Cayenne Pepper → Capsicum frutescens L.

Cayaponia Espelina Cogn. (syn. Parianthopodus Carijo Manso, P. Espelina Manso). (Cucurbitaceae). - Perennial plant. Brazil. Root is considered in some parts of Brazil as diuretic, tonic, antisiphylitic antidiarhhetic, purgative and drastic.

Cayaponia pedata Cogn. (Cucurbitaceae). - Vine. Brazil. Roots contain a violent drastic.

Ceanothus americanus L., New Jersey Tea, Red Root. (Rhamnaceae). - Shrub. Eastern N. America to Florida and Texas. Dried bark is hemostatic and blood coagulent: Leaves were used during revolutionary periods as a tea. It is said that the roots have also the property of hastening blood coagulation. Has been suggested as a remedy for lung hemorrhage.

Ceanothus reclinatus L. Hér. (syn. Colubrina reclinata (L'Hér.) Brogn.) Mabi. (Rhamnaceae). - Tree. West Indies, Porto Rico. Bark is used in the preparation of a drink, used in Porto Rico.

Ceara Rubber → Manihot Glaziovii Muell. Arg.

Cebil Colorado → Piptadenia macrocarpa Benth.

Cebil Gum → Piptadenia Cebil Griseb.

Cecropia pachystachya Tréc. (Moraceae). - Tree. S. America. Bark is source of a fibre; used in some parts of Brazil for manuf. sails. Also C. peltata L.

Cecropia peltata L. (Moraceae). - Tree. Trop. America. Trunk is used as trough to conduct water. Spongy wood is employed as tinder; used in Brazil for paper pulp. Juice is used in Mexico as a caustic for removal of warts; in S. America and West Indies for dysentery. Decoction of young leaves is considered by some Indian tribes for liver ailments and dropsy. Young buds are occasionally eaten as pot herb.

Cedar, Bastard → Soymida febrifuga Juss.

Cedar, Black → Nectandra Pisi Miq.

Cedar, Eastern Red. → Juniperus virginiana L.

Cedar, Himalayan → Cedrus Libani Barrel.

Cedar, Incense → Libocedrus decurrens Torr.

Cedar Lichen → Cetraria juniperina (L.) Ach.

Cedar Mahogany → Entandrophragma Candollei Harms. and E. cylindricum Sprague.

Cedar, Pencil → Dysoxylon Fraseranum Benth.

Cedar, Pink → Acrocarpus fraxinifolius Wight. and Arn.

Cedar, Red → Acrocarpus fraxinifolius Wight and Arn. and Cedrela Toona Roxb.

Cedar, Southern Red → Juniperus barbadensis L.

Cedar, Southern White → Chamaecyparis thyoides (L.) B. S. P.

Cedar, Spicy → Tylostemon Mannii Stapf.

Cedar, Stinking → Torreya taxifolia Arn.

Cedar, West Indian → Cedrela odorata L.

Cedar, Western White → Thuja plicata Donn.

Cedar, White → Thuja occidentalis L.

Cedar Wood Oil → Juniperus virginiana L.

Cedar, Yellow → Rhodosphaera rhodanthema Engl.

Cedrela fissilis Vell. (syn. C. brasiliensis Juss.) (Meliaceae). - Tree. Brazil. Wood is used for railroad ties, canoes, general carpentry, pencils. Bark is astringent and emetic, has been recommended for leucorrhea.

Cedrela Glaziovii DC. (Meliaceae). - Tree. S. America. Wood aromatic, resinous, of excellent quality; used for general construction, carpentry.

Cedrela Huberi Ducke. (Meliaceae). - Tree. Brazil. Wood aromatic, durable; used for construction work.

Cedrela longipes Blake. (Meliaceae). - Tree. Centr. America. Wood is used for house building and furniture in Guatemala.

Cedrela odorata L. Cigarbox Cedrela, West Indian Cedar, Cedro. (Meliaceae). - Large tree. Trop. America. Wood coarse, scented; used for cigar-boxes, also for furniture, moth-proof chests. Root bark is bitter; used as febrifuge. Seeds have vermifuge properties.

Cedrela Toona Roxb. (syn. C. febrifuga King.) Cedrela Tree. (Meliaceae). - Tall tree. India, Malaya to Australia. Wood one of the best timbers; exported to Europe; red, prettily marked, very durable, light, easily worked; used for furniture, house building, tea boxes in Assam, cigar boxes in Madras; oil casks in Travancore, also for musical instruments. Flowers produce a yellow and red dye; used in India. Also C. sureni Burkill from Malaysia.

Cedrela Tree → Cedrela Toona Roxb.

Cedro → Cedrela odorata L.

Cedro blanco → Cupressus Benthamii Endl.

Cedro Bordado → Panopsis rubescens (Pohl) Pitt.

Cedro Macho → Guarea Guara (Jacq.) Wils.

Cedro Oil → Citrus Limon Burman.

Celandine → Chelidonium majus L.

Cedronella triphylla Moench. (Labiaceae). - Perennial herb. Canary Islds. Leaves are used as a tea, called Thé des Canaries.

Cedrus atlantica Manetti. (Pinaceae). - Tree. N. Africa, esp. Atlas Mountains. Wood used locally as timber.

Cedrus deodara (Roxb.) Loud., Deodar. (Pinaceae). - Tree. Centr. Asia, esp. N. W. Himalaya, Afghanistan and N. Balutchistan. Important lumber tree, esp. in N. India; used for railroad ties, buildings and bridges.

Cedrus Libani Barrel, Hilamayan Cedar. (Pinaceae). - Large tree. W. Himalayas to Afghanistan, N. Africa. Principal timber in N. India. Wood moderately hard, durable, scented, resistant to white ants, yellowish-brown; used for building material, railroad ties. Suitable for Mohammedan and Sikh work. Wood source of oil, Kelon-ka-tel, resembling turpentine. Source of an ess. oil, used on a small scale for perfumery.

Ceiba acuminata (S. Wats.) Rose. (Bombaceae). - Tree. Mexico. Flos from seeds is used in Mexico for stuffing pillows, also made into candlewicks.

Ceiba pentandra (L.) Gaertn. (syn. Eriodendron anfractuosum DC. Bombax pentandrum L.) Kapok Tree, Silk Cotton Tree. (Bombaceae). - Large tree. Trop. America, Asia, Africa. Silky fibre from the pods, called Kapok, is widely used for stuffing pillows, mattresses, life preservers; also used as insulating material, among which in airoplanes against cold and sound; equipping automobile trucks for refrigeration. The fibre is resilient, buoyant, water resisting and highly moisture proof. Much is cultivated in Java. Seeds are source of an oil; used for illumination and for manuf. soap. Sp. Gr. 0.920-0.933; Sap. Val. 189-195; Iod. No. 86-100; Unsap. 0.8-1.6%. Wood is easily worked; used for packing-boxes and matches.

Ceiba sumauma Schum. (Bombaceae). - Tree. Brazil. Seeds are source of a floss or Paina, sometimes used for stuffing pillows. Of slight commercial value.

Celandrina, Lesser → Ranunculus Ficaria L.

Celastraceae → Catha, Celastrus, Elaeodendron, Evonymus, Goupia, Gymnosporia, Kokoona, Lophopetalum, Maytenus, Perrottetia, Pterocelastrus, Schaeffera, Tripterygium.

Celastrus edulis Vahl. → Catha edulis Forsk.

Celastrus senegalensis Lam. → Gymnosporia montana Benh.

Celastrus serratus Hochst. (syn. Celastrus obscurus Rich.) (Celastraceae). - Vine. Abyssinia. Used in Abyssinia in a medicine called Add-Add.

Celebes Manila → Agathis loranthifolia Salisb.

Celeriac → Apium graveolens L.

Celery → Apium graveolens L.

Celery, Blanching → Apium graveolens L.

Celery Leaved Crowfoot → Ranunculus sceleratus L.

Celery Pine → Phyllocladus trichomanoides D. Don.

Celery, Turnip-rooted → Apium graveolens L.

Celtuce → Lactuca sativa L.

Clavariella formosa (Pers.) Karst. → Clavaria formosa Pers.

Cellonia → Posidonia australis Hook. f.

Celosia argentea L. (Amarantaceae). - Annual herb. Tropics. Young shoots and leaves are consumed as vegetable among the natives of India.

Celosia laxa Schur. and Thonn. (Amarantaceae). - Herbaceous plant. Trop. Africa and Asia. Leaves are eaten as a vegetable by the inhabitants of Khasia.

Celosia scabra Schinz. (Amarantaceae). - Herbaceous plant. Trop. Africa. Decoction of the herb is used by the Aajamba (S. E. Afr.) for carcinoma.

Celosia trigyna L. (Amarantaceae). - Annual herb. Trop. Africa, Madagascar, Arabia. Used in some parts of Africa for pustular skin eruptions; in Nigeria and Uganda it is used for tapeworm.

Celtis australis L. (syn. C. lutea Pers.) European Hackberry. (Ulmaceae). - Tree. Mediterranean region. Wood yellow-brown to grayish, heavy, fairly hard, very tough, durable, easy to bend; used for wagon-building, blow-instuments, turnery, household-utensils. Stems are used for walking-sticks, fishing-rods and whip-rods. Seeds are edible, sometimes sold in markets of of the Balkans.

Celtis brasiliensis Planch. (syn. C. flagellaris Casar.) (Ulmaceae). - Tree. Brazil. Wood porous, elastic, soft, not durable; used for agricultural tools, carpentry, paper pulp. Bark is used as febrifuge.

Celtis Kraussiana Bernh., Camdeboo, Stinkwood. (Ulmaceae). - Tree. Cape Peninsula to Abyssinia. Wood heavy, hard, tough, strong, yellowish-white; used for planks, yokes, triggers, axe-handles, laths, fences, coopers' work, and rail-road ties.

Celtis occidentalis L. Common Hackberry, Beaver Wood. (Ulmaceae). - Tree. N. America. Wood not strong, soft, coarse-grained, heavy, light yellow; used for cheap furniture, agricultural implements, fence-posts, fuel. Fruits are sometimes eaten. Also C. mississippiensis Bosc. Wood was used by the Kiowa Indians as fuel on the altar during the Peyote Ceremony.

Celtis reticulata Torr. Western Hackberry. (Ulmaceae). - S. W. part of the United States. Fruits are consumed by the Indians.

Celtis Selloviana Niq. (Ulmaceae). - Tree. S. America. Fruits are consumed in some parts of S. America. Roots are source of a brown dye.

Celtis Tala Gill. (Ulmaceae). - Tree. S. America. An infusion of the leaves is used in some parts of Argentina and Chile for indigestion.

Centaurea acaulis L. (Compositae). - Perennial herb. N. Africa, Algeria. Source of a yellow dye.

Centaurea Calcitrapa L. (Compositae). - Herbaceous plant. Europe, Asia, Mediterranean region. Young leafy stems are consumed in some parts of Egypt.

Centaurea eryngioides Lam. (Compositae). - Herbaceous plant. E. Nile region, Arabia. Leaves are eaten as a salad by the Beduins of Arabia.

Centaurea Perrottetii DC. (Compositae). - Annual or biennial herb. Desert regions. Katagum, Senegambia, Nile region. Much used as food for camels.

Centaurea Rhaponticum L. (syn. Rhaponticum scariosum Lam.). (Compositae). - Perennial herb. Europe. Roots occasionally used in place of Chinese Rhubarb. They are considered purgative.

Centaurium umbellatum Gilib. → Erythraea Centaurium Pers.

Centaury → Erythraea Centaurium Pers.

Centaury, American → Sabbatia angularis (L.) Pursh.

Centipeda orbicularis Lour. (syn. C. minima Kuntze). (Compositae). - Herbaceous plant. Asia to Australia and Pacif. Islds. Used medicinally in China. Contains an ess. oil, myriogynic acid and a bitter principal.

Central American Sarsaparilla → Smilax Regelii Killip and Morton.

Centrolobium orinocense (Benth.) Pitt. Baláustre, Cartán. (Leguminosaceae). - Tree. Colombia, Venezuela. Wood reddish-orange; used for cabinet-work, fine furniture; in Venezuela also for frames of buildings and furniture.

Centrolobium robustum Mart. (Leguminosaceae). - Tree. Guiana, Brazil. Source of Zebrawood, Arariba, being uneven colored, light brown to dark brown, fairly heavy, hard, difficult to split, easy to polish; used for furniture, de luxe articles and boat-building.

Centrosema Plumierii Benth. (Leguminosaceae). - Climbing herb. Trop. America. In some tropical countries used as green-manure.

Centrosema pubescens Benth. (Leguminosaceae). - Herbaceous plant. Trop. America. Plant is used as a green manure in some tropical regions.

Cephaëlis acuminata Karsten, Cartagena Ipecac. (Rubiaceae). - Shrub. N. and Centr. Colombia to Nicaragua. Source of Cartagena, Panama and Nicaragua Ipecac. Much is exported from Cartagena and Savanilla. See C. Ipecacuanha (Brotero) Rich.

Cephaëlis barcellana Muell. Arg. (Rubiaceae). - Shrub. Trop. America, esp. Peru, Brazil, Venezuela, Colombia. Is employed by the natives for burns.

Cephaelis emetica Pers. (syn. Psychotria emetica L.). (Rubiaceae). - Shrub. Guatemala to Bolivia. Roots are source of Striated Ipecac, which is occasionally exported.

Cephaelis Ipecacuanha (Brotero) Rich. (syn. Uragoga Ipecacuanha Baill; Psychotria Ipecacuanha Stokes). Ipecac, Ipecacuanha. (Rubiaceae) - Shrub. S. America, esp. S. Brazil. Cultivated in Malaya. Dried rhizome and roots are used medicinally, being emetic, expectorant, astringent. Contains alkaloids, e. g. emetine, cephaeline, psychotrine. Much is exported from Matto Grossa, Braz. as Pará Ipecac. Roots are collected during the dry season.

Cephalandra indica Naud. → Coccinea indica Wight and Arn.

Cephalanthus occidentalis L., Button Bush. (Rubiaceae). - Shrub or small tree. N. America, Mexico, West Indies, E. Asia. Tea from inner bark used by the Meskwaki Indians as emetic. Bark has been used by the Indians as astringent, laxative and tonic.

Cephalanthus spathelliferus Baker. (Rubiaceae). - Tree. W. Madagascar. Wood is used in Madagascar for house-building.

Ceramiaceae (Red Algae) → Ceramium.

Ceramium hypnoides (J. Ag.) Okam. (syn. Campylophora hypnoides J. Ag.). (Ceraminaceae). - Red Alga. China, Japan. Used in the manufacture of agar in Japan. Also C. Boydenii Gepp.

Ceramium rubrum (Huds.) J. Ag. (Ceraminaceae). - Red Alga. Sea Weed. Atl., Pac., Ocean and adj. waters. Used for the production of Kanten in Japan, being of inferior quality.

Ceratonia Siliqua L. Carob, St. John's Bread. (Leguminosaceae). - Tree. Mediteranean region, Syria and adj. countries. Cultivated in the Old and New World. Fruits edible, tough, sweet, dark brown, rich in sugar and protein. Source of alcohol. Also important forage crop, much fed to horses and cattle. Seeds used as substitute for coffee, called Carob Coffee, especially in some parts of Austria. Seeds were used in former days as weight by jewelers and druggists. Decoction of the fruits is used for catarrhal infections. Several varieties are in existance. In Cyprus a molasses is made from the beans called Pasteli. Wood is easily worked, used for carts and furniture. Source of Locust Gum, Locus Bean Gum, Carob Gum.

Ceratopetalum gummiferum Sm. Christmas Bush. (Saxifragaceae). - Tree. Australia. Produces a beautiful kino, of a rich ruby color, transparent, very tough.

Ceratopteridaceae → Ceratopteris.

Ceratopteris thalictroides Brong. (Ceratoperidaceae). - Fern. Tropics. Young leaves are consumed in the spring as a vegetable in Japan.

Ceratotheca sesamoides Endl. (Pedaliaceae). - Herbaceous plant. Trop. Africa. Is consumed as a vegetable in some parts of Sudan. Often sold in markets. Seeds are eaten like those of Sesamum indicum. Cultivated.

Cerbera Manghas L. (Apocynaceae). - Tree. S. W. Asia, Malaysia, Polynesia, Trop. Australia. Wood source of a fine charcoal, Seeds used in Philipp. Islds. to stupefy fish.

Cercidiphyllum japonicum Sieb. and Zucc. (Cercidiphyllaceae). - Tree. Japan, China. Wood not strong, light, soft, straight and fine grained; used in Japan for interior finish of houses, furniture, cabinet work, boxes and engraving.

Cercidium Torreyanum (S. Wats.) Sarg. Palo Verde. (Leguminosaceae). - Small tree. S. W. of United States and Baja California (Mex.). Seeds when ground and made into cakes were consumed by the Indians of Arizona and California. Seeds were also prepared for a beverage.

Cercis canadensis L., Red-Bud. (Leguminosaceae). - Small tree. N. America. Flowers were used in salads and pickles, by the early French Canadians. Bark has been used as an active astringent by the Indians.

Cercocarpus latifolius Nutt. Mountain Mahogany. (Rosaceae). - Shrub or small tree. Western N. America. Wood is used for bows by Cosiute Indians of Utah.

Cercocarpus montanus Raf. (Rosaceae). - Shrub. New Mexico. Decoction of leaves used by Tewa Indians of New Mexico as a laxative.

Cercocarpus parvifolius Nutt. (syn. C. betuloides Nutt.) (Rosaceae). - Shrub or small tree. Baja Calif. (Mex.), California. Wood used for tool handles.

Cereus chiotilla Weber → Escontria chiotilla (Weber) Rose.

Cereus conglomeratus Berger → Echinocereus conglomeratus Forst.

Cereus enneacanthus Engelm. → Echinocereus enneacanthus Engelm.

Cereus giganteus Engelm. → Carnegiea gigantea (Engelm.) Britt. and Rose.

Cereus grandiflorus Mill. → Selenicereus grandiflorus (L.) Britt. and Rose.

Cereus griseus Haw. → Lemaireocereus griseus (Haw.) Britt. and Rose.

Cereus gummosus Engelm. → Machaerocereus gummosus (Engelm.) Britt. and Rose.

Cereus lanatus DC. → Epostoa lanata (H. B. K.) Britt. and Rose.

Cereus pecten-aboriginum Engelm. → Pachycereus pecten-aboriginum (Engelm.) Rose.

Cereus pentagonus L. → Acanthocereus pentagonus (L.) Britt. and Rose.

Cereus queretaroensis Weber → Lemaireocereus queretaroensis (Weber) Safford.

Cereus Thurberi Engelm. → Lemaireocereus Thurberi (Engelm.) Britt. and Rose.

Cereus undata Haw. → Hylocereus undatus (Haw.) Britt. and Rose.

Ceriman → Monstera deliciosa Liebm.

Ceriops Candolleanum Arn. (syn. C. tagal Rob.). (Rhizophoraceae). - Shrub or tree. Tropics. Wood most durable among the mangroves, yellow to orange; used for knees of boats, trunks for house building. Bark is used for tanning, contains 45% tannin.

Ceropteris calomelanos (L.) Und. (Polypodiaceae). - Perennial herb. Tropics. Leaves are used by the Creoles to stop bleeding of wounds.

Ceroxylon andicolum Humb and Bonpl., Wax Palm of the Andes. (Palmaceae). - Tall palm. Trop. S. America. Leaves source of a wax, used for manuf. of candles and wax matches.

Ceroxylon Klopstockiae Mart. (syn. Klopstockia cerifera Humb. and Bonpl.) (Palmaceae). Tall Palm. Northern S. America. Used in Colombia for manuf. of wax.

Ceruana pratensis Forsk. (Compositae). - Herbaceous plant. Egypt, Sudan. Used in Upper Egypt for manuf. of brooms. Has been found in ancient Egyptian Tombs.

Cervina Truffle → Hydnotria carnea (Corda) Zebel.

Cestrum Pseudoquina Mart. (Solanaceae). - Shrub. Brazil. Bark, called Quina do Matto, is used in Brazil for swamp fevers; used in tablets and pills. Also employed as a stomachic, and for digestive disturbances.

Cetraria aculeata (Schreb.) E. Fr. (Parmeliaceae). - Lichen. Temp. zone. On sterile soils. Has been used in Scotland and Canary Islds. to dye woolens a red-brown color.

Cetraria fahlunensis (L.) Schaer. (syn. Parmelia fahlunensis Ach.). Swedish Shield Lichen. (Parmeliaceae). - Lichen. Temp. zone. On rocks. Has been used in different parts of Europe to dye woolens a red-brown color.

Cetraria glauca (L.) Ach., Pale Shield Lichen. (Parmeliaceae). - Lichen. Temp. zone. On stems of trees. Has been used in some parts of Europe to dye woolens a chamois color.

Cetraria islandica (L.) Ach., Iceland Moss. (Parmeliaceae). - Lichen. Temp. and subarctic. zones. Source of Iceland Moss Jelly, being demulcent and nutrient. Plant is used as food, made into bread, after a bitter principle has been removed. Used in the tanning industry. Has been used in Iceland to dye woolens a brown color.

Much of the commercial moss is derived from Scandinavia, Germany, Switzerland and Austria. Employed as a substitute for salve bases; reduction of bitter taste in some drugs; as laxative; also as culture medium for bacteria. Source of alcohol.

Cetraria juniperina (L.) Ach., Cedar Lichen. (Parmeliaceae). - Lichen. Temp. zone. Source of a dye which has been used in Scandinavia for dyeing woolens a yellow color. Is employed to poison wolves.

Cetraria nivalis (L.) Ach., Snow Lichen. (Parmeliaceae). - Lichen. Tem. zone, esp. Europe. Source of a violet pigment; used for dying woolens.

Cetraria pinastri (Scop.) S. Gray, Pine Lichen. (Parmeliaceae). - Lichen. Temp. zone. Has been used in some parts of Europe as source of a dye to stain woolens a green color.

Ceylon Beautyleaf → Calophyllum Calaba L.

Ceylon Bowstring Hemp → Sansevieria zeylancia Willd.

Ceylon Boxwood → Plectronia didyma Bedd.

Ceylon Cardamon → Elettaria Cardamomum (L.) Maton and E. major Sm.

Ceylon Cinnamon → Cinnamomum zeylanicum Nees.

Ceylon Gooseberry → Dovyalis hebecarpa (Gardn.) Warb.

Ceylon Mango → Mangifera zeylanica Hook.

Ceylon Moss → Gracilaria lichenoides (L.) Harv.

Ceylon Piassava → Caryota urens L.

Chacoli (beverage) → Myrciaria cauliflora Berg.

Chaectospermum glutinosum (Blanco) Swingle. (Rutaceae). - Tree. Philipp. Islds. Juice from fruit is used to cure itch on skin of dogs. Also used as hair-tonic in Philipp. Islds.

Chaenomeles sinensis (Thouin) Koehne → Cydonia sinensis Thouin.

Chaerefolium Cerefolium (L.) Schinz. → Anthriscus Cerefolium Hoffm.

Chaerefolium sylvestre (L.) Shinz and Thell. (syn. Chaerophyllum sylvestre L., Myrrhis sylvestris Spreng.) Cow Parsley. (Umbelliferaceae). - Biennial or perennial herb. Europe, Caucasia, Siberia, N. Africa. Juice from leaves used as a home-remedy for skin-diseases.

Chaerophyllum bulbosum L. Turnip-Rooted Chervil. (Umbelliferaceae). - Herbaceous biennial to perennial. Europe, Balkans, Caucasia. Thick roots are sometimes boiled and eaten as vegetable; excellent for winter use. Occasionally cultivated.

Chaerophyllum sylvestre L. → Chaerefolium sylvestre (L.) Shinz and Thell.

Chaetocarpus castanocarpus Thw. (Euphorbiaceae). - Tree. Trop. Asia, esp. Malaya, Ceylon. Wood close-grained, hard, dense, durable, red-dish brown; used for building, beams, joists and posts.

Chaetochloa italica (L.) Scribn. → Setaria italica L.

Chaetomorpha crassa Kuetz. (Cladophoraceae). - Green Alga. Pac. Ocean and adj. waters. Consumed as food, eaten in salads. Of commercial importance for manuf. a sweet meat. Also C. antennica Kuetz. esp. in Hawaii Islds.

Chaetophoraceae (Green Algae) → Stigeoclonium.

Chaetoptelea mexicana Liebm. (syn. Ulmus mexicana (Liebm.) Planch. (Ulmaceae). - Large tree. Mexico to Panama. Wood used for lumber, Bark astringent, used for coughs by the natives.

Chailletia timorienses DC. → Dichapetalum timoriense Engl.

Chailletia toxicaria G. Don. (Chailletiaceae). - Woody plant. W. Africa. Seeds are used in W. Africa for destroying rats, stupefying fish, also for criminal poisoning. Seeds are sold in native markets.

Chamaecyparis Lawsoniana (Murr.) Parl. Lawson Cypress. (Cupressaceae). - Tree. Pacific Coast Region of N. America. Wood hard, strong, durable, easily worked, light, fragrant, light yellow to whitish; used for flooring, interior finish of buildings, fenceposts, ship and boat building, railroad ties, matches. Resin is a strong diuretic.

Chamaecyparis Nootkatensis (Lamb.) Sudw., Alaska Cedar, Yellow Cypress, Sitka Cypress. (Cupressaceae). - Alaska, British Columbia to Oregon. Wood very close-grained, somewhat brittle, hard, very durable, fragrant, yellow; used for furniture, interior finish of buildings, boat and ship-building.

Chamaecyparis thyoides (L.) B. S. P., Southern White Cear. (Cupressaceae). - Tree. Eastern N. America. Wood soft, light, not strong, somewhat fragrant, close-grained, light brown; used for woodenware, cooperage, boat-building, interior finish of houses, shingles, railroad ties, fence-posts.

Chamaedaphne calyculata (L.) Moench. (Ericaceae). - Shrub. N. temp. Zone. A tea was made from the fresh and dried leaves by the Ojibway Indians.

Chamaedorea Sartorii Liebm. (syn. Eleutheropetalum Sartorii (Liebm.) Liebm.) (Palmaceae). - Palm. S. Mexico. Young flower-clusters eaten as greens or cooked as vegetable by the Mazatecs, Chinantecs, Zapotecs and other tribes in Mexico.

Chamaedorea elegans Mart. (syn. Collina elegans (Mart.) Liebm.) (Palmaceae). - Palm. S. Mexico, Centr. America. Fruits are used as food by the Chinantecs, Zapotecs and other Indian tribes in Mexico and Centr. America. Very young spaths are cooked and eaten like asparagus.

Chamaedorea Tepejilote Liebm. (Palmaceae). - Small palm. Mexico. Unopened spaths are cooked, eaten as asparagus.

Chamaedrys, Germander → Teucrium Chamaedrys L.

Chamaedrys officinalis Moench. → Teucrium Chamaedrys L.

Chamaedrys Scordium Moench. → Teucrium Scordium L.

Chamaelirium luteum (L.) Gray., Blazing Star, Unicorn Root. (Liliaceae). - Perennial herb. Eastern N. America. Used medicinally. Dried rootstocks and roots, called Helonias, contain chamaelirin, a bitter glucoside. Diuretic and uterine tonic.

Chamaenerion angustifolium (L.) Scop. → Epilobium angustifolium L.

Chamaerops humilis L. (Palmaceae.) - Small palm. Mediterranean region. Leaves are source of a cordage which has much commercial value. Fibre is called Crin Végétale. Very young leaf-buds are consumed as vegetable.

Chamaesaracha Coronopus (Dunal) Gray. (syn. Solanum Coronopus Dunal.) (Solanaceae). - Herbaceous plant. Western United States and adj. Mexico. Berries are eaten by the Hopi and Navajo Indians.

Chamise, Redshank → Adenostoma sparsifolium Torr.

Chamomille → Matricaria Chamomilla L.

Chamomille, Dyer's → Anthemis tinctoria L.

Chamomille, English → Anthemis nobilis L.

Chamomille, Golden → Anthemis tinctoria L.

Chamomille, Roman → Anthemis nobilis L.

Champaca Wood Oil → Bulnesia Sarmienti Lorentz.

Champignon, Common Edible → Psalliota campestriis (L.) Fr.

Chapote → Diospyros texana Scheele.

Champedak → Artocarpus polyphema Pers.

Charcherquem → Visnea Mocanera L.

Charcoal. Plants sources of → Adhatoda, Aesculus, Albizzia, Alnus, Arbutus, Aristotelea, Bruguiera, Butea, Byrsonima, Calotropis, Calycophyllum, Cerbera, Cocos, Conocarpus, Corylus, Curatella, Euvonymus, Fagus, Guaruma, Heritiera, Ligustrum, Melicytus, Miconia, Parinarium, Paulownia, Pinus, Populus, Prosopis, Quercus, Rhus, Salix, Sesbania, Sweetia, Tilia.

Charlock → Brassica arvensis (L.) Ktze.

Chartreuse (liqueur) → Angelica Archangelica L.

Chasalia rostrata Miq. (syn. Psychotria rostrata Blume). (Rubiaceae). - Shrub. N. Malay Penin., Sumatra, Borneo. Decoction of leaves is used in Penang for constipation.

Chaste Tree → Vitex Agnus Castus L.

Chaste Tree, Guiana → Vitex divaricata Sw.

Chaste Tree, Molave → Vitex parviflora. Juss.

Chaste Tree, Timor → Vitex littoralis Cunningh.

Chaulmoogra → Taraktogenos Kurzii (King.) Pou.

Chaulmoogra Oil → Hydnocarpus anthelmintica Pierre and Taraktogenos Kurzii (King) Pou.

Chaulmoogra Oil (substitute) → Casearia sylvestris Sw.

Chaulmoogra Tree, Common → Hydnocarpus anthelmintica Pierre.

Chayote → Sechium edule Swartz.

Checker Tree → Sorbus torminalis (L.) Crantz.

Cheese, Camembert → Penicillium camemberti Thom.

Cheese, Emmenthaler → Bacterium acidi-propriocid Scherm.

Cheese, Roquefort → Penicillium roqueforti Thom.

Cheesemaker → Withamia coagulans Dunal.

Cheiranthus Cheiri L. Wallflower. (Cruciferaceae). - Biennial or perennial herb. S. and Centr. Europe. Cultivated. Flowers were formerly used as emmenagogue, anti-spasmodic, purgative, resolvens. Contains cheiranthin, a glucosid which is a heart-poison; also nerol, geraniol, benzylalcohol, and cheirinin, an alkaloid. Flowers are source of an ess. oil, however, seldom used in perfumery.

Chelidonium majus L., Celandine. (Papaveraceae). - Perennial herb. Europe, temp. Asia, introd. in N. America. Fresh yellow latex is used as home remedy for skin diseases, especially for summer freckles. Powdered drug is considered medicinally purgative, sedative, diuretic, diaphoretic and expectorant. Contains several alkaloids, among which chelidonine, chelerythrine and protopine.

Chelone glabra L., Balmony. (Scrophulariaceae). - Perennial herb. Eastern N. America. Herb is used medically as a tonic, has cathartic properties and used for expelling worms.

Chempaka → Michelia Champaca L.

Chene Gum → Spermolepis gummifera Brongn.

Chengal Tree → Balanocarpus Heimii King.

Chenopodiaceae → Agriophyllum, Anabasis, Arthrocnemum, Altriplex, Beta, Chenopodium, Cornulaca, Dondia, Haloxylon, Kochia, Monolepis, Salicornia, Salsola, Spinacea, Suaeda.

Chenopodium album L., Lamb's Quarters (Chenopodiaceae). - Annual herb. Temp. zone. Frequently a weed. Occasionally cultivated. Leaves and young tops when boiled are consumed as a pot-herb. Seeds are made into flour for cakes and gruel by the Indians.

Chenopodium ambrosioides L. Wormwood. (Chenopodiaceae). - Annual herb. Temp. to trop. Zone. Introd. in N. America. Leaves are used as condiment in soups by the Chinatects and Mezatects in Mexico, and other tribes. Var.

anthelminticus L. American Wormseed is cultivated. Source of Oil of Chenopodium, Oil of Wormseed, an ess. oil, obtained by steam distillation of the herb. Much is derived from Maryland. Used medicinally as anthelmintic for round worms, hook worms, intestinal amoeba.

Chenopodium auricomum Lindl. Salt-bush. (Chenopodiaceae). - Shrub. Tasmania, W. Australia. Used as feed for cattle, in Australia.

Chenopodium Bonus-Hendricus L. (syn. Ch. esculentus Salisb.). Allgood, Good King Henry. (Chenopodiaceae). - Herbaceous plant. Temp. zone. Occasionally cultivated. Consumed as a pot-herb. Also C. album L., C. rubra L., C. murale L., C. amaranticolor Coste and Reyn.

Chenopodium californicum Wats., California Soap Plant. (Chenopodiaceae). - Herbaceous plant. California. Roots have saponaceous properties; used for washing.

Chenopodium chilense L. (Chenopodiaceae). - Herbaceous plant. Chile and adj. territory. Herb is used as a vermifuge.

Chenopodium foetidum Schrader. (Chenopodiaceae). - Herbaceous plant. E. Africa. Herb is used by the natives to eradicate ants and other vermin.

Chenopodium Fremontii Wats. (Chenopodiaceae). - Western N. America. Roasted seeds were used when ground as food by the Klamat Indians in Oregon.

Chenopodium leptophyllum Nutt. (Chenopodiaceae). - Herbaceous plant. N. America. Seeds mixed with corn meal and salt; also raw or cooked plants were consumed by the Indians of Utah, Nevada and New Mexico.

Chenopodium Quinoa Willd. Quinoa. (Chenopodiaceae). - Annual herb. Chile, Peru and adj. territory. Cultivated since pre-Columbian times. Seeds are source of a food much esteemed by the native population; used in bread, porridge, soup. Also source of a beverage called Tschitscha.

Chenopodium rhadinostachyum F. v. Muell. (Chenopodiaceae). - Herbaceous plant. Australia. Seeds when ground are used as food by native tribes in Centr. Australia.

Chenopodium virgatum (L.) Aschers. (Chenopodiaceae). - Herbaceous plant. Europe. The red berries are sometimes used in cosmetics by country people in S. E. Europe. Also Ch. capitatum (L.) Aschers.

Chenopodium Vulneraria L. (Chenopodiaceae). - Herbaceous plant. Europe, Mediterranean region, Caucasia. Herb. is source of a yellow dye.

Cherbet Tokhum → Ocimum Basilicum L.

Cherimoya → Annona Cherimola Mill.

Cherimoya of Jalisco, Wild → Annona longiflora Wats.

Cherry, Common → Prunus avium L.

Cherry, Barbados → Malpighia glabra L.

Cherry Birch → Betula lenta L.

Cherry, Bush → Eugenia myrtifolia Sims.

Cherry, Chinese Sour → Prunus cantabrigiensis Stapf.

Cherry, Choke → Prunus virginiana L.

Cherry, Cornelian → Cornus Mas L.

Cherry Elaeagnus → Elaeagnus multiflora Thunb.

Cherry, European Bird. → Prunus Padus L.

Cherry, Ground → Physalis heterophylla Nees.

Cherry, Jamaica → Muntingia Calabura L.

Cherry Kernel Oil → Prunus Cerasus L.

Cherry Laurel → Prunus Laurocerasus L.

Cherry, Mahaleb → Prunus Mahaleb L.

Cherry, Manchu → Prunus tomentosa Thunb.

Cherry, Mountain → Prunus angustifolia Marsh.

Cherry Plum → Prunus cerasifera Ehrh.

Cherry, Rocky Mountain → Prunus melanocarpa (A. Nels.) Rydb.

Cherry, Rum → Prunus serotina Ehrh.

Cherry, Sour → Prunus Cerasus L.

Cherry, St. Lucie → Prunus Mahaleb L.

Cherry, Surinam → Eugenia uniflora L.

Cherry, Sweet → Prunus avium L.

Cherry, West Indian → Malpighia punicifolia L.

Cherry, Western Sand → Prunus Besseyi Bailey.

Cherry, Wild Black → Prunus serotina Ehrh.

Cherry, Winter → Physalis Alkekengi L.

Chervil, Garden → Anthriscus Cerefolium Hoffm.

Chervil, Turnip-rooted → Chaerophyllum bulbosum L.

Chervin → Sium Sisarum L.

Chestnut, American → Castanea dentata (Marsh.) Borkh.

Chestnut Cape → Calodendron capense Thunb.

Chestnut, Chinese → Castanea mollissima Blume.

Chestnut, Earth → Lathyrus tuberosus L.

Chestnut, Japanese → Castanea crenata Sieb. and Zucc.

Chestnut Oak → Quercus Prinus L.

Chestnut, Polynesian → Inocarpus edulis Forster.

Chestnut, Tahiti → Inocarpus edulis Forster.

Chestnut, Spanish → Castanea sativa Mill.

Chestnut Tongue → Fistulina hepatica (Huds.) Bull.

Chestnut, Water → Eleocharis tuberosa Schultes.

Chestnut, Water → Trapa natans L.

Chewing gum. Sources of → Achras, Actinella, Agathis, Agoseris, Apocynum, Artocarpus, Asclepias, Asclepiodophora, Bumelia, Chrysothamnus, Combretum, Couma, Dyera, Echinops, Encelia, Euphorbia, Ficus, Lygodesmia, Manilkara, Melicope, Pistacia, Schinus, Sideroxylon, Stemmadenia.

Chia→ Salvia chia Fern., S. Columbariae Benth.

Chicha → Ziziphus Mistol Griseb.

Chick Pea → Cicer arietinum L.

Chickasaw Lima → Canavalia ensiformis DC.

Chickasaw Plum → Prunus angustifolia Marsh.

Chickling Vetch → Lathyrus sativus L.

Chickweed, Common → Stellaria media (L.) Vill.

Chicle Gum → Achras Sapota L.

Chicmu → Trifolium amabile H. B. K.

Chicory → Cichorium Intibus L.

Chikrassia tabularis Juss., Chickrassia (Bengal) (Meliaceae). - Tree. Centr. Asia, Malaysia. Source of a gum; 10 gram gum with 100 gram of water gives an excellent mucilage.

Children's Tomato → Solanum anomalum Thonn.

Chile Hazel → Gevuina Avellana Molina.

Chile Laurel → Laurelia aromatica Juss.

Chilean Guava → Myrtus Ugni Mol.

Chilean Peppertree → Schinus latifolius Engl.

Chilean Trevoa → Trevoa trinervia Miers.

Chilean Wineberry → Aristotelia Maqui L'Hér.

Chilghosa Pine → Pinus Gerardiana Wall.

Chilli → Capsicum frutescens L.

Chiloe Strawberry → Fragaria chiloensis (L.) Duch.

Chilte → Cnidoscolus elasticus Lundell.

Chilte Blanco → Cnidoscolus elasticus Lundell.

Chilte, Highland → Cnidoscolus elasticus Lundell.

Chimaphila umbellata Nutt., Spotted Wintergreen. (Pirolaceae). - Half-shrub. Europe, temp. Asia, N. America. Leaves, Folia Chimaphilae, used in medicine for stones in the bladder and retarding excretion of urine. Contains ericolin, arbutin, chimaphilin, urson, tannin and gallic acid.

Chimonanthus fragrans Lindl. (Calicanthaceae). - Tree. Japan, China. Fragrant flowers are used in Japan in scent-bags and perfumed water.

China Jute → Abutilon Avicennae Gaertn.

China Linnen → Boehmeria nivea (L.) Gaud.

Chinaberry Tree → Melia Azedarach L.

Chinese Apple → Pyrus prunifolia Willd.

Chinese Anise → Illicium verum Hook. f.

Chinese Aralia → Aralia chinensis L.

Chinese Artichoke → Stachys Sieboldi Miq.

Chinese Cabbage → Brassica chinensis L.

Chinese Cabbage → Brassica pekinensis Rupr.

Chinese Chestnut → Castanea molissima Blume.

Chinese Chive → Allium odorum L.

Chinese Colza Oil → Brassica campestris L. var. chinoleifera.

Chinese Cotton → Gossypium Nanking Meyen.

Chinese Grabapple → Pyrus baccata L.

China Crab Apple → Pyrus hupehensis Pamp.

Chinese Dates → Ziziphus mauritiana Lam.

Chinese Dwarf Banana → Musa Cavendishii Lam.

Chinese Ephedra → Ephedra sinica Stapf.

Chinese Galls → Rhus semialata Murr.

China Grass → Boehmeria nivea (L.) Gaud.

Chinese Green → Rhamnus dahurica Pall.

Chinese Haw → Crataegus pentagyna Waldst. and Kit.

Chinese Hemlock → Tsuga chinensis Pritz.

Chinese Hickory → Carya cathayensis Sarg.

Chinese Jujub → Ziziphus Jujuba Mill.

Chinese Laurel → Antidesma Bunius (L.) Spreng.

Chinese Liquorice → Glycyrrhiza uralensis Fisch.

Chinese Matgrass → Cyperus tegetiformis Roxb.

Chinese Mustard → Brassica juncea L.

Chinese Pear → Pyrus chinensis Lindl.

Chinese Peony → Paeonia albiflora Pall.

Chinese Pistache → Pistacia chinensis Bunge.

Chinese Plum → Prunus salicina Lind.

Chinese Potato → Dioscorea Batatas Descne.

Chinese Preserving Melon → Benincasa cerifera Savi.

Chinese Quince → Cydonia sinensis Thouin.

Chinese Rhubarb (substitute). → Centaurea Rhaponticum L.

Chinese Sassafras → Sassafras Tzumu Hemsl.

Chinese Soapberry → Sapindus Mukorossi Gaertn.

Chinese Sour Cherry → Prunus cantabrigiensis Stapf.

Chinese Sumach → Rhus semialata Murr.

Chinese Sweet Gum → Liquidambar formosana Hance.

Chinese Torrey → Torreya grandis Fort.

Chinese White Pine → Pinus Armandi Franch.

Chinese Wolfberry → Lycium chinense Mill.

Chinese Wood Oil → Aleurites Fordii Hemsl.

Chinese Yam → Dioscorea Batatas Decne.

Chingma → Abutilon Avicennae Gaertn.

Chinquapin → Castanea pumila (L.) Mill.

Chinquapin, Golden → Castanopsis chrysophylla A. DC.

Chinquapin, Henry → Castania Henryi (Skan) Rhed. and Wils.

Chinquapin, Water → Nelumbium luteum (Willd.) Pers.

Chiococca alba (L.) Hitch. (syn. G. racemosa L.) (Rubiaceae). - Shrub. Subtr. and trop. America. Herb is used in some regions for snake bites.

Chiogenes hispidula (L.) Torr. and Gray. (Ericaceae). - Small shrub. Eastern N. America. A beverage was made from the leaves and sweetened with maple sugar by the Indians of Maine, Wisconsin and Minnesota.

Chionanthus virginica L., Fringe Tree, Old Man's Beard. (Oleaceae). - Tree. Eastern N. America to Florida and Texas. Bark of root used medicinally; contains a bitter glucoside chionanthin; alterative, tonic; for catarrhal jaundice. Bark is collected in the autumn, much is derived from Virginia and N. Carolina.

Chios Mastic → Pistacia lentiscus L.

Chiotilla → Escontria chiotilla (Weber) Rose.

Chirata → Swertia Chirata Buch-Ham.

Chirinda Medlar → Vanguieria esculenta S. Moore.

Chisocheton cuminginaus (C. DC.) Hamrs. (Meliaceae). - Tree. Philipp. Islds. Nuts are source of Catoseed Oil or Balukanag Oil, a non drying oil; used as purgative and for illumination.

Chisocheton pentadrum (Blanco) Merr. (Meliaceae). - Tree. Philipp. Islds. Oil is used locally in the Philippines as a hair cosmetic.

Chive → Allium Schoenoprasum L.

Chive, Chinese → Allium odorum L.

Chloranthaceae → Chloranthus, Hedyosum.

Chloranthus inconspicuus Swartz (Chloranthaceae). - Half-woody plant. N. China. Cultivated by Chinese. In Java flowers and leaves are placed in contact with tea while being dried, to impart scent.

Chloranthus officinalis Blume. (Chloranthaceae). - Woody plant. Malaysia. Decoction from leaves is used as beverage. Already used in Java before tea became known. Has sudorific action.

Chloris Gayana Kunth., Rhodes Grass. (Graminaceae). - Perennial grass. S. Africa. Cultivated in warm countries as pasture grass and for hay.

Chlorogalum pomeridianum (Ker-Gawl) Kunth., Californian Soaproot. (Liliaceae). - Bulbous perennial herb. W. of United States. Used locally for washing clothes, source of a good lather.

Chlorophora excelsa (Welw.) Benth. and Hook. f. (syn. Milicia africana Sim.). Iroko Fustic Tree, Odum, Mbang, African Oak. (Moraceae). - Tree. Trop. Africa. Wood brown to blackbrown, long and fine-grained, heavy, tough, difficult to split, not attacked by termites; used as substitute for oak and teak, for ship building, furniture, wagons. Latex is sometimes mixed with that of Funtumia.

Chlorophora tinctoria (L.) Gaud. (syn. Morus tinctoria L.). (Moraceae). - Shrub or tree. Trop. America. Wood light green, hard, close-grained, heavy, strong, tough, takes polish; used for furniture, wheels, interior finish. Wood furnishes a yellow, brown and green dye, known as Fustic. Bark is tonic, astringent; in large doses it is purgative.

Chlorosplenium aeruginosum (Oed.) De Not. (Helotiaceae). - Ascomycete. Fungus. Temp. zone. The mycelium is the cause of a deep and attractive color of the wood due to the formation of xylochloric acid. Oak wood when well stained is very expensive and known as Tunbridge Ware.

Chloroxylon Swietenia DC. Satinwood Tree. (Meliaceae). - Central India to Ceylon. Satinwood of Commerce; exported; used for cabinet work, furniture; locally for agricultural implements, carts, railroad ties. Satinwood may give irritation to the skin, caused by chloroxylonine, an alcaloid.

Chochoco → Mahonia chochoco Fedde.

Chocolate → Theobroma Cacao L.

Chocolate Colored Nephroma → Nephroma parilis Ach.

Chocolate, Gaboon → Irvingia Barteri Hook. f.

Chocolate Tree, Nicaraguan → Theobroma bicolor Humb. and Bonpl.

Choiromyces Magnusi (Matt.) Paol. (syn. Terfeza Magnusii Matt.) (Eutuberaceae). - Ascomycete. Fungus. S. Europe, esp. Sardinia and Portugal. Fruitbodies are consumed as food in some parts of southern Europe.

Choiromyces venosus Fr. (syn. Tuber album With.). (Eutuberaceae). - Ascomycete. Fungus. Europe. Fruitbodies are consumed as food, esp. in Russia Czechoslovakia and Moravia. Sold in markets. Known as Troitskie Truffle and White Truffle. Some cases of poisoning have been reported.

Choke Cherry → Prunus virginiana L.

Chokecherry, Gray → Prunus Grayana Max.

Chondodendron platyphyllum Miers. (Menispermaceae). Woody vine. Brazil. Roots are used in Brazil for dyspepsia, uterine colic.

Chondodendron tomentosum Ruiz and Pav. Pareira (Menispermaceae). - Vine. Brazil and Peru. Dried root, Pereira root, Radix pareira brava is used medicinally. Much is exported from Rio de Janeiro. Used as bitter tonic and diuretic. Contains alkaloids e. g. chondoinine, beheerin, misobeerine. Root is also used in the Rio Huallaga region (Peru) and several parts of Brazil and Venezuela for the preparation of curare, an arrowpoison. Also C. iquitanum Diels, C. candicans (L. C. Rich) Sandw., C. limacifolium (Diels) Mold. and C. polyanthemum Diels.

Chondrus crispus (L.) Stackh. Carrageen, Irish Moss. (Gigartinaceae). - Red Alga. Temp. N. Atlantic. Produces commercial carrageenin or Irish moss extract, a seaweed colloid. The purified extract is extensively manufactured in the U. S. A. Many people use the whole plants which are boiled in water in order to make the gelose. Source of blancmange, a jelly dessert. Used in the U. S. A. for stabilizing chocolate milk; in ice creams, ices, fruit juices, chocolate syrup, salad dressing, cheese, confectionaries and soups; emulsifier in pharmaceutical preparations, cough medicines, ointments, shaving creams, calico printing, emulsifying casein paints, in leathers.

Chondrus ocellatus Holm. (Cigartinaceae). - Red Alga. Japan. Used for sizings of cloth, calcimines and plaster for house walls in Japan. Also other species are used for the same purpose.

Chondrus plotynus J. Ag. (Gigartinaceae). - Red Alga. Coast of Japan, esp. Island of Yezo. In Japan the plants are made into a paste.

Chonemorpha elastica Merr. (Apocynaceae). - Woody vine. Minandao (Philipp. Islds.) Has been suggested as a source of rubber.

Chonemorpha macrophylla G. Don. (Apocynaceae). - Woody vine. E. India, Malaya. Bark is source of a fibre; is resistant to fresh and salt water; made into fishing nets.

Chorisia speciosa St. Hil. (Bombaceae). - Tree. Argentina to Brazil. Seeds are source of a floss, a Kapok-like material, called Paineira or Painaliferin, being of good quality, used for stuffing pillows, and cushions.

Ch'pei-tzu Galls → Rhus Potanini Maxim.

Christmas Berry → Photinia arbutifolia Lindl.

Christ Thorn → Zizyphus Spina-Christi Willd.

Christmas Rose → Helleborus niger L.

Chrozophora plicata Juss. var. obliquifolia Prain. (Euphorbiaceae). - Herbaceous plant. N. Africa, E. India. Fruits are source of a blueish dye. Seeds and leaves are used by the natives as purgative.

Chrozophora tinctoria Juss. (Euphorbiaceae). - Annual herb. Mediterranean region, Arabia, E. India. Source of a blue and green dye; used in France for coloring linen cloth, giving with ammonia vapor a red color, also used for coloring confectionary and the outside of certain cheeses.

Chrozophora verbascifolia Juss. (Euphorbiaceae). - Herbaceous plant. India, Iran, Iraq. Mediterranean region. Leaves, stalks and fruits are used in Iran locally for whooping cough, as alterative. Seeds are used by the Beduins of Arabia as substitute for clarified butter.

Chrosactinia mexicana Gray., Damianita. (Compositae). - Half shrub. Mexico. Source of a domestic medicine; considered antispasmodic, aphrodisiac, diuretic and sudorific.

Chrysanthemum Balsamita L. (syn. Pyrethrum Balsamita (L.) Willd.) Costmary, Mintgeranium. (Compositae). - Herbaceous plant. W. Asia. Herb used for flavoring ales.

Chrysanthemum cinerariaefolium (Trév.) Vis. (syn. Pyrethrum cinerariaefolium Trév.). (Compositae). - Perennial herb. Dalmatia. Cultivated in S. Europe, Croatia, Dalmatia, Herzegovina, United States. Source of Dalmatian Insect Powder, being composed of dried, unexpanded flower-heads; contains an ess. oil and resins. Used in different insecticides. Toxic principle is pyrethron.

Chrysanthemum coccineum Willd. (syn. C. roseum Adam., Pyrethrum roseum Lindl.). (Compositae). - Perennial herb. Caucasia, Armenia, N. Iran. Dried flower heads are source of Persian Insect Powder.

Chrysanthemum coronarium. L., Garland Chrysanthemum. (Compositae). - Annual herb. Mediterranean region. Cultivated as an ornamental in Old and New World. Eaten as a vegetable in China und Japan. Very young seedlings are cooked and eaten as a potherb.

Chrysanthemum indicum L. (Compositae). - Woody herb. China, Japan. Flowerheads are consumed when preserved in vinegar in some parts of Japan.

Chrysanthemum Leucanthemum L. Oxeyedaisy. (Compositae). - Perennial herb. Europe, temp. Asia, introd. in N. America. Used as a home remedy for catarrh; herb is made into syrup, essence and pastilles. Young leaves are occasionally eaten as pot herb, also in salads. Has been recommended as an emergency plant in time of want.

Chrysanthemum Marschallii Ascher. (Compositae). - Iran and adj. countries. Dried flowerheads have been suggested as a source of insect powder.

Chrysanthemum Parthenium (L.) Bernh. (syn. Matricaria Parthenium L.). Feverfew Chrysanthemum. (Compositae). - Perennial herb. Mediterranean region, Balkans, Asia Minor, Transcaucasia, Caucasia. Dried flowers are used in some parts of Europe in home remedies as tonic, for digestion, cleaning the blood, vermifuge, emmenagogue, abortive, insecticide. Also used as tea, in wine and certain pastries.

Chrysanthemum sinense Sab. (syn. Pyrethrum sinense DC.) (Compositae). - Perennial herb. Japan, China. Cultivated in Japan and China. Boiled flowerheads are eaten in Japan as vegetable. They are also dried and preserved. Leaves are also consumed as vegetable.

Chrysanthemum Tanacetum Karsch. → Tanacetum vulgare L.

Chrysarobin → Andira Araroba Aguiar.

Chrysobalanus Icaco L. Icaco Plum. (Rosaceae). - Shrub or small tree. Trop. and subtr. America. Fruits are eaten as preserves.

Chrysobalanus orbicularis Schum. and Thonn. (Rosaceae). - Tree. Trop. W. Africa. Dried fruits are consumed by the natives of Angola. Oil from seeds is used for candles in Gambia.

Chrysophyllum africanum A. DC., African Star-apple. (Sapotaceae). - Tree. Trop. Africa. Cultivated. Fruits are much esteemed by the natives, they are of an apricot color and have an pleasant acid pulp. They are called Odara Pears. Sold in markets. Oily seeds are used for manuf. soap. Wood is used for turnery, cabinet work, moulding, railway carriages; along the Ivory Coast for images and in S. Nigeria for fancy work.

Chrysophyllum albidum G. Don. (Sapotaceae). - Tree. Trop. Africa. Juice from leaves is used by natives instead of lime-juice, for coagulating flake rubber. Extract from tree used as bird-lime.

Chrysophyllum argenteum Jacq. (Sapotaceae). - Tree. West Indies. Fruits edible, consumed in some parts of the Antilles. Also C. glabrum Jacq.

Chrysophyllum Buranhem Riedel. → Lucuma glycyphloea Mart. and Eichl.

Chrysophyllum Cainito L. (syn. C. olivaeforme L.) Cainito., Star-Apple. (Sapotaceae). - Tree. West Indies, Centr. America. Cultivated in tropics. Fruits size of an apple, white to purple, hard; pulp of good flavor; eaten raw or as a preserve. Caimito Blanco is white; Caimito Morado is purple. Wood ist close-grained, hard, strong, very heavy, light brown; used for cabinet work.

Chrysophyllum lucumifolium Griseb. (Sapotaceae). - Tree. Argentina to Brazil. Wood compact, elastic, durable; used for agricultural implements, boxes, recommended for paper making. Fruit edible, insipid.

Chrysophyllum Macoucou Aubl. (syn. Aubletella Macoucou Pierre). (Sapotaceae). - Tree. Guiana. Fruits are edible, being yellow orange, size of a pear.

Chrysophyllum magalis-montana Sond. (Sapotaceae). - Small tree. S. Africa. Fruits are used as preserves, they are ellipsoid, 2.5 cm. across, bright scarlet, flesh of a pleasant flavor.

Chrysophyllum maytenoides Mart. (Sapotaceae). - Tree. Brazil. Wood elastic, compact; used for carpentry work.

Chrysophyllum Michino H. B. K. (Sapotaceae). - Colombia to Peru. Fruit yellow with whitish pulp, edible.

Chrysophyllum microcarpum Swartz. (Sapotaceae). - Tree. West Indies. Fruits size of a gooseberry, sweet, of a pleasant flavor.

Chrysophyllum rugosum Sw. → Micropholis Melinoniana Pierre.

Chrysophyllum sessilifolium Panch and Sebert. (Sapotaceae). - Tree. New Caledonia. Wood yellowish-red, hard, flexible, easily worked, attractive when varnished; used in New Caledonia for general carpentry and wagons.

Chrysophyllum Wakeri Planch and Sebert. (Sapotaceae). - Tree. New Caledonia. Wood yellowish, fine-grained, very hard and resistant; used for interior work, and turnery; one of the best lumber species in N. Caledonia.

Chrysophyllum Welwitschii Engl. (Sapotaceae). - Tree. W. Africa. Seeds used as ornaments among natives of Angola.

Chrysopogon aciculata Trin. → Andropogon aciculatus Retz.

Chrysosplene → Chrysosplenium alternifolium L.

Chrysoplenium alternifolium L., Chrysosplene, Golden Saxifrage. (Saxifragaceae). - Perennial herb. Europe, Caucasia, temp. Asia, N. America. Herb occasionally used in salads. Also C. oppositifolius L. Recommended as an emergency food plant.

Chrysothamnus confinis Greene. (Compositae). - Half-shrub. New Mexico. Flower-buds with salt were eaten by the Indians of New Mexico.

Chrysothamnus nauseosus Britt., Rabbit Bush. (Compositae). - Shrub. Western N. America. Source of a rubber. May become of importance as an emergency rubber plant. A gum was used by the Indians for chewing.

Chrysothamnus viscidiflorus (Hook.) Nutt. Rabbit Bush. (Compositae). - Shrub. Western N. America. Roots produce a chewing gum, used by the Indians of Utah and Nevada.

Chufa → Cyperus esculentus L.

Chuflé → Calathea macrosepala Schum.

Ch'ung tsao → Cordyceps sinensis (Berk.) Sacc.

Chuño de Concepción → Alstroemeria ligtu L.

Chupire Rubber → Euphorbia calyculata H.B.K.

Chupones → Greigia sphacelata Regel.

Churco Bark → Oxalis gigantea Barn.

Chuwah → Aquilaria Agallocha Roxb.

Chytranthus Mannii Hook. f. (Sapindaceae). - Tree. Trop. Africa. Fruits are consumed by the natives.

Cibol → Allium fistulosum L.

Cibotium Barometz Smith (Cyatheaceae). - Treefern. Assam, S. China to Sumatra, Borneo, Philip. Islds. Soft hairs on the buds toward the apex of the stem are widely used as stypic for dressing wounds. Used by Chinese for centuries.

Cibotium Menziesii Hook. (Cyatheaceae). - Tree-fern. Hawaii. Scales or pulu from stems used in Hawaii for stuffing pillows. Also C. Chamissoi Kaulf.

Cicer arietinum L., Chick Pea, Garbanzos. (Leguminosaceae). - Annual herb. Mediterrenean region. E. India. Cultivated in warm countries of Old and New World. Grown since antiquity. Seeds are consumed fresh or dried in various dishes and soups; also made into a flour for bread making. Used as substitute for coffee. Also fodder for live-stock. Many varieties. Roasted roots are used as substitute for coffee.

Cichorum Intibus L., Chicory, Succory. (Compositae). - Perennial herb. Temp. zone. Roots when roasted are a well known substitute for coffee. Grown as a vegetable, known when bleached as Brussel Witloof. Grown as a fodder for live-stock. Dried root used medicinally, being diuretic, tonic. Contains inulin. Much is cultivated in Holland, Belgium, France and Germany.

Cichorium Endivia L. Endive, Escarolle. (Compositae). - Annual to perennial herb. Probably from S. Europe and India. Cultivated in the Old and New World as a vegetable. Varieties of two groups are being grown. I. Curled Fringed Leaved group; used in salads: Giant Fringed, Green Curled and White Curled. II. Broad-leaved, its varieties are sometimes partly bleached, are eaten boiled: Batavian Full Hearted, Florida Deep Hearted. This group is also called Escarolle.

Cigarbox Cedrela → Cedrela odorata L.

Cilician Tulip → Tulipa montana Lindl.

Cimarrona → Annona montana Macf.

Cimicifuga foetida L. (Ranunculaceae). - Perennial herb. Europe, Siberia, China. Used medicinally in China. Contains cimicifugin.

Cimicifuga racemosa (L.) Nutt., Black Cohosh, Black Snakeroot. (Ranunculaceae). - Perennial herb. Eastern N. America. Dried rhizome and roots used medicinally; collected in autumn, especially in the Blue Ridge Mts. Used as a sedative, alterative and emmenagogue.

Cinchona Galisaya Wedd. Yellowbark Cinchona. (Rubiaceae). - Tree. Bolivia. Source of Yellow Bark or Calisaya Bark. Bark contains a number of alkaloids, among which quinine, quinidine, cinchonine and cinchonidine, from which different salts are derived, used medicinally as a febrifuge, antiperiodic and tonic.

Cinchona Ledgeriana Moens. (syn. C. Calisaya var. Ledgeriana How.). Ledgerbark Cinchona. (Rubiaceae). - Tree. Bolivia. Cultivated in Java, Ceylon and India. Source of Ledger Bark or Yellow Bark. Bark of stem is source of galenicals, that of the root produces alkaloids, especially quinine; used as febrifuge, tonic and antiperiodic. Excellent bark is obtained from the hybrid C. Ledgeriana Moens. x C. Calisaya Wedd.

Cinchona micrantha Ruiz and Pav. (syn. C. affinis Wedd.) Cascarilla Provinciana, Cascarilla Verde. (Rubiaceae). - Medium sized tree. Peru, Bolivia, Ecuador. Bark was formerly a more or less important source of quinine.

Cinchona officinalis L. (syn. C. lancifolia Mut., C. nitida Ruiz and Pav.) Cascarilla Verde, Ichu Cascarilla, Cascarilla Crespilla. (Rubiaceae). -

Tree. Peru, Ecuador, Bolivia, Colombia. Bark is an important source of quinine. Cultivated. Source of Peruvian, Crown, Jesuit's, Loxa, Pale Bark and Countes's Powder. Bark yields 1 to 4% alkaloids of which one half to two third is quinine. Used medicinally as C. Calisaya.

Cinchona pubescens Vahl. (syn. C. Morado Ruiz and Pav., C. grandifolia Mut.) Cascarilla, Cascarilla Delgada. (Rubiaceae). - Medium sized tree. Peru, Bolivia to Colombia. Bark is an important source of Wild Cinchona, used in medicine.

Cinchona succirubra Pavon. Redbark Cinchona. (Rubiaceae). - Tree. Peru and adj. territory. Source of Red Bark used for pharmaceutical preparations; used as febrifuge. Also called Druggist's Bark. Cultivated in Indonesia.

Cinnamodendron corticosum Miers. (Canellaceae). - Small tree. West Indies. Bark is used as a spice.

Cinnamomum Burmanni Blume, Batavia Cinnamon. (Lauraceae). - Tree. Java. Cultivated in Java and Sumatra. Dried bark source of important spice. Much used in The Netherlands. Wood heavy, soft, fine-grained; used in W. Java for building purposes.

Cinnamomum Camphora (L.) Nees. and Eberm. (syn. Camphora officinarum Nees.) Camphor Tree. (Lauraceae). - Tree. China and Japan. Cultivated commercially in China, Japan and Formosa. Introduced in the tropics and subtropics of the Old and New World. Source of Camphor derived from chipped wood of stems and roots by steam distillation. Contains a fixed volatile oil. Used medicinally as stimulant, antispasmodic, antiseptic and rubefacient. Employed for manuf. celluloid. Wood is beautifully grained, light brownish, takes polish; used in China and Japan for furniture, cabinets, interior finish of buildings.

Cinnamomum Cassia Blume (syn. C. obtusifolium Nees. var. Cassia Perrot and Eberm.). (Lauraceae). - Tree. S. E. China. Cultivated in Kwangsi and Kwangtung (China). Dried bark is source of an important spice. Shipped from Canton and Hong-Kong. Resembles Saigon Cinnamon. Is more astringent. Immature fruits are source of Cassia Buds, yield an ess. oil called Oil of Cassia, Oil of Cinnamon, Oleum Cinnamomi, is obtained by steam distillation from leaves, and twigs; rectified by distillation; used as a flavoring agent. Medicinally used as carminative. Contains cinnamic aldehyde.

Cinnamomum Culilawan Blume. (Lauraceae). - Tree. Mal. Archip., China. Very aromatic bark is used as condiment. Calyx around fruits is used in sauces by the natives.

Cinnamomum Loureirii Nees. (syn. C. obtusifolium Nees. var Loureirii Perrot and Eberm.) Saigon Cinnamon. (Lauraceae). - Tree. Cochin-China, China, Java, Annam, Japan. Cultivated. Spice from bark much esteemed in China and

Japan. Dried bark used medicinally as aromatic, carminative, astringent, flavoring agent. Contains an ess. oil.

Cinnamomum Massoia Schw. (syn. Massoia aromatica Becc.) (Lauraceae). - Tree. New Guinea. Source of a cinnamon substitute, called Massoia Bark.

Cinnamomum Mercadoi Vid. Kalingag. (Lauraceae). - Small tree. Philipp. Islds. Bark source of of Kalingag Oil, locally used as medicine. Has been recommended for root beers, has strong sassafras odor.

Cinnamomum mindanaense Elm., Mindanao Cinnamon. (Lauraceae). - Small tree. Mindanao (Philipp. Islands). Bark is sold as a commercial cinnamon.

Cinnamomum Oliveri Bailey, Oliver's Bark, Black Sassafras. (Lauraceae). - Tree. Australia, esp. Queensland. Used locally as a substitute for cinnamon.

Cinnamomum parthenoxylon Meissn. (Lauraceae). - Tall tree. Malaysia. Aromatic bark is used for flavoring.

Cinnamomum pedunculatum J. S. Presl. (Lauraceae). - Tree. S. Japan. Seeds are source of a wax; used in Japan for manuf. candles.

Cinnamomum Sintok Blume. Sintok. (Lauraceae). - Tree. Malaysia. Source of Sintok Bark; used in Indonesia for diarrhoea, is also considerred a vermifuge.

Cinnamomum Tamala (Buck. and Ham.) Nees and Eberm., Indian Cassia. (Lauraceae). - Tree. India. A source of cinnamon substitute.

Cinnamomum zeylanicum Nees. Ceylon Cinnamon. (Lauraceae). - Tree. Cultivated, kept as a shrub. India, Ceylon, Malaysia. Extensively cultivated in several parts of the tropics of the Old and New World. Dried bark is important spice. Several essential oils are found in the leaves, bark, stem and roots. Used for flavoring food, gum, candies; in dentifrices, perfumes, incense. Used in medicine as stomachic, carminative, astringent.

Cinnamon, Batavia → Cinnamomum Burmanni Blume.

Cinnamon, Ceylon → Cinnamomum zeylanicum Nees.

Cinnamon, Mindanao → Cinnamomum mindanaense Elm

Cinnamon, Wild. → Canella alba Murr.

Cinquefoil → Potentilla fruticosa L.

Cipres de las Guaytecas → Pilgerodendron uviferum (Don.) Florin.

Circassian Tree → Adenanthera pavonina L.

Cirouaballi, Yellow → Nectandra Pisi Miq.

Cirsium arvense (L.) Scop. (syn. Cnicus arvensis Hoffm.) Canada Thistle. (Compositae). - Herbaceous plant. Temp. Europe, Asia. Introd. in N. America. Herb has been used to coagulate milk.

Cirsium Drummondii Torr. and Gray. (syn. Cnicus Drummondii Gray.) (Compositae). - Perennial herb. Western N. America. Roots are consumed by the Indians of Wyoming, Montana, Utah and Nevada.

Cirsium edule Nutt. (syn. Carduus edulis (Nutt.) Greene, Cnicus edulis Gray.) (Compositae). - Perennial herb. Western N. America. Soft, sweet stems when peeled were eaten by the Cheyenne and other Indians.

Cirsium oleraceum (L.) Scop. (syn. Cnicus oleraceus L., Carduus oleraceus Vill.) (Compositae). - Perennial herb. Europe to Siberia. Very young succulent parts of the plant are used as vegetable in some regions of Russia and Siberia.

Cirsium tuberosum (L.) All. (syn. Carduus bulbosum DC., Cnicus tuberosus Willd.). (Compositae). - Perennial herb. Europe. Roots harvested in autumn are sometimes consumed as vegetable; they may be stored throughout the winter. Recommended during times of food emergency.

Cirsium undulatum (Nutt.) Spreng. (syn. Carduus undulatus Nutt., Cnicus undulatus Gray.) (Compositae). - Perennial herb. N. America. Roots when cooked are consumed by several Indian tribes. Also C. scopulorum (Greene) Cook, C. occidentale (Nutt.) Jeps., C. virginianum (L.) Michx. Recommended as emergency food during times of want.

Círuela → Spondias purpurea L.

Cissampelos capensis Thunb. (Menispermaceae). - Vine. Cape Peninsula. Used among the farmers as an emetic and laxative. Leaves are used by the Kaffirs for snake-bites.

Cissampelos fasciculata Benth. (syn. C. Caapeba Vell., C. denuata Miers.) (Menispermaceae). - Woody vine. Brazil. Root is considered astringent, tonic and febrifuge.

Cissampelos Pareira L. (syn. C. acuminata Benth.) (Menispermaceae). - Scandent shrub. Tropics. Poultice of leaves used by Chinantecs and other tribes in trop. America for snakebites. Contains pelosine, an alkaloid. Plant is supposed to be diuretic, expectorant, emmenagogue, febrifuge. Roots are used to prevent a threatened abortion, relieves menorrhagia; arrest uterine hemorrhage.

Cissus populnea Guill. and Perr. → Vitis pallida Wight and Arn.

Cissus quadrangularis L. → Vitis quadrangularis Wall.

Cissus sicyoides L. → Vitis discolor Dalz.

Cistanche lutea Hoffmzg. and Link (syn. Phelypaea lutea Desf.) (Orobanchaceae). - Parasitic herb. N. Africa. Plants are eaten by the Tourecs as asparagus.

Cistaceae → Cistus, Helianthemum.

Cistanthera papaverifera Chev. (Tiliaceae). - Tree. Trop. W. Africa. Wood is one of the best Ironwoods in Africa. Very hard, fairly durable.

light-red to brown-red; used for rail-road ties, bulkheads in ships. Along Gold Coast used for handles of native axes, hoes and mortars.

Cistus albidus L. (Cistaceae). - Shrub. Mediterranean region. Leaves are used among the Arabs in Algeria as a tea.

Cistus ladaniferus L. (Cistaceae). - Shrub. Mediterranean region. Source of Gum Labdanum or Droga de Jara. Most of the resin is produced during times of the greatest heat in July and August. Source of Labdanum derived by distillation of the balsamic exudation. Used for perfumes, soaps, powders, creams, cosmetics; recommended for deodorant or insecticide industries. Much is derived from S. France, Spain, Morocco, Corsica and Greece. Gum is dark, plastic, has heavy odor, resembles ambergris. Odor is agreeable and heavy, acts as excellent fixative. Used as substitute of natural ambergris. Oil blends well with lavender, pine, as well as all types of Oriental bouquets.

Cistus salvifolius L. (Cistaceae). - Shrub. Mediterranean region. Herbaceous parts of plant with fruit-coat of Pome Granate are used for tanning. The root is used by Arabs for hemorrhages and bronchitis.

Cistus villosus L. (Cistaceae). - Shrub. Spain. Source of a Labdanum Balsam once much esteemed in medicine.

Citheroxylum quadrangulare Jacq. Pedulo Colorado. (Verbenaceae). - Tree. West Indies. Wood strong, reddish; used for general building purposes, windows, doors, beams; for guitars used by natives.

Citrange → Poncirus trifoliata Raf. x Citrus sinensis Osbeck.

Citrangequat → (Poncirus trifoliata Raf. x Citrus sinensis Osbeck). x Fortunella margarita Swingle.

Citromyces glaber Wehmer. (Moniliaceae). - Microorganism. Is able to produce citric acid from nutrient sucrose solutions, containing calcium carbonate. Also C. Pfeifferianus Wehmer.

Citron → Citrullus vulgaris Schrad. and Citrus Medica L.

Citron, Jewish → Citrus Medica L.

Citronella → Cymbopogon Nardus (L.) Rendle.

Citrullus Colocynthis (L.) Schrad., Colocynth. (Cucurbitaceae). - Perennial herbaceous vine. Dry areas of Africa, trop. Asia. Cultivated in India and Mediterranean region. Dried pulp from fruits is used medicinally as a drastic purgative, hydrogogue, cathartic. Contains citrullin.

Citrullus vulgaris Schrad., Watermelon. (Cucurbitaceae). - Annual vine. Trop. Africa. Cultivated in warm countries of Old and New World. Fruits eaten ripe. Rind of fruits sometimes preserved in sugar. Numerous varieties, e. g., Florida Favorite, Black Diamond, Tom Watson, Klondyke, Angelo, New Hampshire Midget, Stone Mountain. The Citron having

inedible raw flesh, is used in pickles and preserved in syrup. Varieties Red Citron, Green Giant.

Citrus aurantifolia Swingle. (syn. C. Medica L. var. acida Brandis). Lime. (Rutaceae). - Small tree. India. Cultivated in tropics and subtropics of Old and New World. Often escaped of cultivation. Fruits used for drinks; carbonated and non-carbonated beverages, food products, candies, confectionary, source of commercial lime-juice. Cold-pressed oil has very fine odor and flavor, different from distilled oil. Contains citral, methyl anthranilate, sesquiterpene. Oil of Lime is a volatile oil derived from the fruit. Varieties are: Tahiti, Mexican Lime, Kusaie; also seedlings are much grown. The Rangpur Lime being oval and orange can hardly be classified in this group. Seeds are source of Lime Seed Oil, a semi-drying oil, suitable for manuf. soap., Sp. Gr. 0.92636; Iod No. 109.5; Sap. Val. 197.7; Acid Val. 13.6.

Citrus aurantifolia Swingle x Fortunella margarita Swingle. Lime x Nagami Kunquat. Limequat. (Rutaceae). - Small tree. Artificial hybrid. Fruits oval, yellow, pleasantly acid. Of no economic value.

Citrus Aurantium L. Sour Orange, Bigarade, Seville Orange. (Rutaceae). - Tree. S. E. Asia, Cochin-China. Grown in subtropics of the Old and New World. Used as stock of Orange, Grapefruit varieties etc. to bud upon. Fruits appearance of Orange, very sour and bitter; used for drinks and in marmelades. Peel is used in Curaçao, a liqueur. Source of Oil of Bitter Orange, volatile, pale yellow, bitter, carminative; used for flavoring and in perfumery. Contains d-lemonen, citral, methyl anthranilate, decyl aldehyde. Sp. Gr. 0.842-0.848. Oil of Neroli is derived from distilled flowers; used in perfumery, for flavoring. Contains limonenes, l-linalol and geraniol. Oil of Petigrain is derived from distilled leaves; much comes from S. France and Paraguay; used in perfumery - Eau de Cologne, creams, toilet waters and powders. Contains pyrrol, furfural, dipentine, nerol, camphene, l-linalol and geraniol. Dried peel is used medicinally as tonic, stomachic, carminative and stimulant. Flowers are source of Orange Flower Water or Aqua Aurantii Florum.

Citrus bergamia Risso and Poit., Bergamot (Rutaceae). - Tree. Trop. Asia. Cultivated in Calabria (S. Italy). Peel is source of Bergamot Oil, a yellowish-green ess. oil of pleasant odor; used in perfumery, pomades, hair-oils, Eau de Cologne. Contains l-linalyl, acetate, l-linalol, d-limonene, bergaptene and diptene.

Citrus grandis Osbeck. (syn. C. maxima (Burm.) Merr.) Pummelo, Shaddock. (Rutaceae). - Tree. Malaya. Grown in the tropics and subtropics of the New and Old World. Fruits large, yellow, resembling grapefruit. Has little commercial value. Vartieties: Mammoth, Pink and Tresca.

Citrus Hystrix DC. (Rutaceae). - Small thorny tree. Philipp. Islds. Fruits lemonscented, used by Filipinos with fish, also in drinks. Peel is used in some parts of Java for flavoring meat. It is also sometimes candied.

Citrus japonica Thunb. → Fortunella japonica (Thunb.) Swingle.

Citrus limetta Risso. Sweet Lemon. (Rutaceae). - Small tree. Trop. Asia. Cultivated in some countries. Fruits yellow, sweet, insipid; with slight lemon flavor.

Citrus Limon Burman. Lemon. (Rubaceae). - Tree. Probably from subtr. Asia. Cultigen. Cultivated in S. Europe, Italy, California, Florida etc. Fruits used for beverages, flavoring, confectionary, source of citric acid, canned lemon juice. Numerous varieties among which Lisbon, Eureka, Genoa, Villafranca, Florida Everbearing. Oil of Lemon, Cedro Oil, a volatile oil is derived from the peel, is carminative; used for flavoring liqueurs, foods,beverages, pastry, candies; used in perfumes. Contains limonene, terpinene, phellandrine, pinene, citronellal, sesquiterpenes. Seeds are source of Lemon Seed Oil, Lemon Pip Oil, a semi-drying oil, used for manuf. soap. Sp. Gr. 0.921-0.923; Sap. Val. 188-196; Iod. No. 103-109. Outer yellow peel is used medicinally as stimulant, stomachic, chiefly with other drugs. Oil of Lemon is carminative, stimulant and stomachic.

Citrus limonia Osbeck, Ninmeng, Canton Lemon. (Rutaceae). - Tree. Subtr. China. Fruits having properties of ordinary Lemon. C. limonia var. otaitensis Tanaka Otahite Orange; var. Khatta Tanaka, Khatta Orange.

Citrus longispina West. Tamisan, Kamisan. (Rutaceae). - Small tree. Cebu, Bohol (Philipp. Islds.) Fruits very juicy, good flavor, mildly acid; sometimes used as break-fast fruit; used in ades.

Citrus margarita Lour. → Fortunella margarita (Lour.) Swingle.

Citrus Medica L., Citron. (Rutaceae). - Tree. Cultigen from subtr. Asia. Cultivated in Mediteranean region. S. Italy, Corsica. Fruits yellow, very large, with very thick peel. Source of commercial Candied Peel. Used in confectionary etc. The Etrog or Jewish Citron is used by the Jews during the Feast of the Tabernacles. Some varieties are Cédratier Ordinaire, A Gros Fruits, Cédratier Poncire, Tubéreux.

Citrus microcarpa Bunge, Musk Lime. (Rutaceae). - Small tree. Malaysia. Plant produces a very small orange with peculiar musky fragrance. By some considered different from C. mitis Blanco, Calamondin.

Citrus mitis Blancoo Calamondin (Rutaceae). - Tree. Philipp. Islds. Occasionally cultivated in subtropics and tropics of the Old and New World. Fruits have the appearance of a small manderine or tangerine, very acid; used in drinks, tea, marmelade or jelly.

Citrus paradisi Macf., Grapefruit (Rutaceae). - Small tree. Origin uncertain, probably a cultigen. May have originated as a seedling in West Indies. Important fruit tree. Cultivated in many parts of subtropics of Old and New World, among which Florida, Puerto Rico, S. Texas etc. Fruits eaten raw, or canned. Source of canned or bottled grapefruit juice; occasionally a candied peel is sold in the local markets. Important varieties are: Duncan, Hall, Marsh's Seedless, Mc. Carty, Walters, Foster, Pink Marsh's Seedless. Seeds are source of Grapefruit Seed Oil, has been recommended to be used after sulfonation in dyeing cotton goods; source of a medium hard soap with good lathering properties. Sp. Gr. 0.9170; Sap. Val. 194; Unsap. 0.7%. Waste products derived from peels of different citrus products are ground and used as fodder for live-stock.

Citrus paradisi Macf. x C. reticulata Blanco var. deliciosa Andr. Grapefruit x Tangerine. Tangelo. (Rutaceae). - Of hybrid origin. Fruit has appearance of an orange, pleasant flavor, of local commercial value. Varieties are: Sampson, Thornton, Nocatee, Seminole, Minneola and Umatilla.

Citrus reticulata Blanco (syn. C. nobilis Andrws non Lour.). King Orange. (Rutaceae). - Small tree. Cochin China. Cultivated in subtropics and tropics of the Old and New World. Fruits size of an orange, coarse peel; of excellent dessert qualities. var. deliciosa Mandarine, Tangerine; var. Unshiu, Satsuma Orange; var. papillaris, Tizon Orange. Source of Mandarine and Tangerine Oil, derived from Sicily where it is produced by the old sponge expressing method; 1000 fruits yield 400 gr. of ess. oil. Much is derived by moderne methods from Florida and Brazil. Contains d-limonene, octyl aldehyde, decyl aldehyde, citral, linalol etc.

Citrus sinensis Osbeck., Sweet Orange. (Rutaceae). - Tree. Cultigen, probably from China and Cochin-China. Widely cultivated in subtropics and parts of tropics of Old and New World, among which Mediterranean region, California, Arizona, S. Texas, Florida, S. America, Australia, S. Africa. Dessert fruit; juice used as beverage, commercial canned and bottled orange juice; for flavoring in sherberts, ice creams, marmelades, jellies, etc. Peel is source of Sweet Orange Peel, Aurantii Dulcis Cortex. Numerous varieties, e. g., I. Spanish Oranges: Homosassa, Indian River, Parson Brown. II. Mediterranean Oranges: Hamlin, Jaffa, Lue-Gim Gong, Valencia, Pineapple. III. Blood Oranges: Ruby, Maltese, St. Michael. IV. Navel Oranges: Washington Navel, Australian. Oil of Sweet Orange is a volatile derived from fresh peel; is carminative; used for flavoring, perfumes. Contains d-limonene, linalol, citral, decyl aldehyde, terpinol. Sweet Orange Peel is used for flavoring medicines. Seeds are source of Orange Seed Oil, Orange Pip Oil, used in manuf. of soap. Sp. Gr. 0.921-0.925; Sap. Val. 194-197; Iod. Val. 98-104. Flowers are source of an excellent honey.

Citrus sinensis Osbeck x C. reticulata Blaneo var. Sweet Orange x a Manderine. Temple Orange. Temple Manderine. Tree. Probably of hybrid origin. Fruits appearance of orange, very fine flavor; of commercial importance. Originated in Florida. Bittersweet is a hybrid of Sweet x Sour Orange.

Citrus Webberi West., Kalpi. (Rutaceae). - Small tree. Philipp. Islds. Fruits are used like the lemon.

Cladium effusum (Sw.) Torr. (syn. Mariscus jamaicensis (Crantz) Britt. Saw Grass. (Cyperaceae). - Perennial herb. Southern United States, Trop. America. Plant is sometimes used for manuf. of a cheap paper.

Cladonia coccifera (L.) Willd. (Cladoniaceae). - Lichen. Temp. zone. Sterile soils. Has been used in some parts of Europe to dye woolens a red purple color.

Cladonia fimbriata (L.) Willd., Trumpet Lichen. (Cladoniaceae). - Lichen. Temp. zone. Sterile soils. Has been used to give woolens a red color.

Cladonia gracilis (L.) Willd. (Cladoniaceae). - Lichen. Temp. zone. Has been source of a dye to color woolens an ash-green color.

Cladonia pyxidata (L.) Hoffm. (Cladoniaceae). - Lichen. Temp. zone. Has been used to give an ash-green color to woolens.

Cladonia rangiferina (L.) Web., Reindeer Moss. (Cladoniaceae). - Lichen. Temp. and subarctic zones. An important food of reindeer in arctic zones. Also C. alpestris (L.) Rabenh. Contains 30% mannose. Has been used in some parts of Europe to dye woolens an iron-red color. An ess. oil has been suggested for use in perfumery. Source of alcohol.

Cladonia sylvatica (L.) Hoffm. (syn. C. tenuis (Flke) Harm.). (Cladoniaceae). - Lichen. Temp. zone. Source of an ess. oil which has been suggested for use in perfumery.

Cladophora nitida Kuetz. (Cladophoraceae). - Green Alga. Fresh water. Eaten among the Hawaiians with fresh water shrimps.

Cladophoraceae (Green Algae) → Cladophora, Pithophora.

Cladosiphon decipiens (Sut.) Okam. (Mesogloiaceae). - Brown Alga. Sea Weed. Consumed in Japan.

Cladrastis amurensis Benth. (syn. Maackia amurensis Rupr. and Maxim.) (Leguminosaceae). - Japan, Korea, Manchuria, Amur region. Wood heavy, strong, very hard, close-grained, dark brown; used in Japan for interior finish of houses, furniture, utensils, gun stocks, handles of implements.

Cladrastis lutea (Michx.) Koch. Yellow Ash, Yellow Wood. (Leguminosaceae). - Tree. Southern and Southeastern United States. Wood

relatively heavy, strong, close-grained, yellow; used for gunstocks. Heartwood is source of a yellow dye.

Clamy Horseseed Bush → Dodonaea viscosa Jacq.

Claoxylon indicum Hassk. (syn. C. Polot Merr.). (Euphorbiaceae). - Tree. Trop. Asia. Young leaves are eaten as lalab with rice among the Javanese.

Clappertonia ficifolia Hook. → Honckenya ficifolia Willd.

Clary Sage → Salvia Sclarea L.

Clasping Mullein → Verbascum phlomoides L.

Clausenia anisata Hook. f. (Rutaceae). - Shrub. Trop. Africa. Anise scented leaves are used by the natives to combate mosquitos.

Clausenia anisum-olens (Blano) Merr. (Rutaceae). - Tree. Philipp. Islds. Leaves when stuffed in pillows give a soporific effect. Also employed in Philipp. Islds. in baths for rheumatism.

Clausenia inaequalis Benth. (Rutaceae). - Woody plant. Trop. Africa. Decoction of plant is used by the Zulus for intestinal worms.

Clausenia Lansium Skeels, Wampi. (Rutaceae). - Small tree. S. China. Cultivated. Fruits have the size of a grape, yellowish, of an agreeable flavor. Introd. in the New World.

Clausenia Willdenowii Wight and Arn. (Rutaceae). - Small tree. Trop. Asia. Berries edible, size of a cherry, fleshy, pleasant flavor.

Clavaria aurea Schäff. (syn. Clavariella aurea (Schäff.) Karst.) (Clavariaceae). - Basidiomycete. Fungus. Fruitbcdies are consumed as food. Frequently sold in markets of Europe. Also C. Wettsteienii Sing., C. flava Schäff. and C. botrytis Pers.

Clavaria formosa Pers. (syn. Clavariella formosa Pers.) Karst.) (Clavariaceae). - Basidiomycete. Fungus. Fruitbodies are sometiemes used as a mild anthelmintic for children. C. Mairei Donk has a purgative effect.

Clavaria Zippeli Lex. (Clavariaceae). - Basidiomycete. Fungus. Trop. Asia, esp. Malaya. Fruitbodies, known as Majang, are consumed as food by the natives. Sold in markets.

Clavariaceae (Fungi) → Clavaria, Sparassis.

Clavariella aurea (Schäff.) Karst. → Clavaria aurea Schäff.

Claviceps purpurea (Fr.) Tul. Ergot. (Hypocreaceae). - Ascomycete. Fungus. Fungus parasites in a number of grasses among which rye. The dried sclerotium is known as Ergot, Ergot of Rye and Secale Cornuti. It is a source of ergosterine known in vitamine D synthesis. Much is derived from Russia and Spain. Has characteristic action on the contraction of the smooth muscle, the uterus and the bladder. Controls bleeding.

Claytonia acutifolia Pal. (Portulacaceae). - Herbaceous plant. Temp. and arctic zones, E. Siberia, Alaska. Fleshy taproots are consumed as food by the Eskimos of King Island (Alaska).

Claytonia balonensis F. v. Muell → Calandrina balonensis Lindl.

Claytonia caroliniana Michx. (Portulacaceae). - Herbaceous perennial. Eastern N. America. Roots are edible and are recommended in times of scarcity.

Claytonia perfoliata Don. (syn. Montia perfoliata (Willd.) How. Winter Purslane. (Portulacaceae). - Annual. N. America, introd. in Europe. Occasionally cultivated. Vegetable, eaten like spinach.

Claytonia virginiana L., Rose Elf, Spring Beauty. (Portulacaceae). Herbaceous perennial. Eastern N. America. Starchy bulbs were much valued as food by some Indian tribes.

Cleistanthus collinus Benth. (Euphorbiaceae). - Small tree. E. India. Dried fruit is used in criminal poisoning. Seeds are employed for stupefying fish.

Clematis dioica L. (Ranunculaceae). - Woody vine. Mexico, West Indies, Centr. America, S. America. An ointment is made from the leaves used for cutaneous diseases. Infusion from flowers and leaves is used as a cosmetic.

Clematis Thunbergii Steud. (syn. C. hirsuta Guill. and Poir.) (Ranunculaceae). - Vine. S. Africa. Leaves used by the natives of Senegal for skin diseases.

Clematis Vitalbla L. (Ranunculaceae). - Herbaceous vine. Centr. and So. Europe, Caucasia. Young sprouts are sometimes consumed as food in some countries of Europe.

Cleome integrifolia Torr. (syn. Peritoma serrulatum DC.). (Capparidaceae). - Herbaceous plant. N. America. Plants were gathered in spring by the Tewa Indians of New Mexico and after removing the alkaline taste, were eaten with cornmeal porridge. Source of a black dye; used by the Pueblo Indians for decorating their pottery.

Cleome viscosa L. (syn. Polanisia icosandra Wight and Arn.). Capparidaceae). - Herbaceous plant. Tropics. Sold in native herb shops in the Far East. Used as stimulant in food for giving appetite. Pods are made into pickles.

Clermontia Gaudichaudii (Gaud.) Hbd. (Campanulaceae). - Shrub or small tree. Hawaii. Sweet yellow berries are eaten by the natives.

Clerodendron glabrum E. Mey. (Verbenaceae). - Small tree. Cape Peninsula. Bark is used by the natives as a purgative for calves.

Clerodendron serratum Spreng. (Verbenaceae). - Shrub. E. India. Mal. Archip. Young leaves and flower-clusters are eaten as lalab with rice, among the Javanese.

Clethra barbinersis Sieb. and Zucc. (Ericaceae). - Shrub. Japan. Young leaves are consumed with rice among Japanese peasants of remote mountain villages.

Clibadium surinamense L. (Compositae). - Herbaceous plant. Guiana. Leaves and stems are used to stupefy fish.

Clidemia blepharoides DC. (syn. Melastoma coccineum Vell.). (Melastomaceae). - Epiphyte. Brazil. Leaves are used in some parts of Brazil for treating ulcers.

Clidemia Deppeana Steud. (syn. Melastoma petiolare Schlecht.) (Melastomaceae). - Woody plant. S. Mexico, Centr. America. Fruits are used as food by the Mazatecs, Chinatecs and Zapotecs (Mex.). Also C. chinantlana (Naud.) Triana, C. dependens D. Don., C. hirta (L.) D. Don., and C. Naudiniana Cogn.

Cliffortia ilicifolia L. (Rosaceae). - Woody plant. S. Africa. Decoction of leaves known as Thorntea is used in S. Africa as emollient and expectorant.

Cliftonia monophylla (Lam.) Sarg., Titi, Buckwheat Bush. (Cyrillaceae). - Shrub or small tree. Coastal plain to Florida and Louisiana. Flowers are source of a commercial honey.

Climbing Entada → Entada scandens Benth.

Clinacanthus Burmani Nees. (syn. Justicia nutans Burm., J. fulgida Blume). (Acanthaceae). - Shrub. Trop. Asia. Young leaves are eaten as vegetable in Annam; used in a dish called banh manh cong. Plant also used for eye diseases.

Clinopodium laevigatum Standl. (Labiaceae). - Shrub. Mexico. Leaves used with sugar are a popular tea along West Coast of Mexico.

Clinopodium macrostemum (Benth.) Kuntze. Te del Monte. (Labiaceae). - Shrub. Mexico. Leaves are used as a tea.

Clinopodium vulgare L. → Calamintha Clinopodium Benth.

Clinostigma oncorhyncha Beccari. (Palmaceae). - Tall Palm. Pacif. Islds. Flexible wood is used by the natives of Samoa for construction of roofs.

Clitandra flavidiflora Hall. (Apocynaceae). - Woody plant. W. Africa. Source of a rubber.

Clitandra nzunde De Wild. (Apocynaceae). - Woody plant. Belgian Congo. Source of a good black rubber.

Clitandra orientalis Schum. (C. Arnoldiana De Wild.) Kappa. (Apocynaceae). - Woody plant. Trop. Africa, esp. Belgian Congo, Uganda. Source of a good rubber, known as Noire du Congo, Kappa. Sap of Costus Lukasiana Schum. is used to coagulate the latex.

Clitandra Simoni Gilg. (Apocynaceae). - Woody plant. N. W. Cameroon. Source of a good rubber.

Clitocybe hypocalamus v. Overeem. (Agaricaceae). - Basidiomycete. Fungus. Trop. Asia. Grows on Calamus spp. Fruitbodies are much esteemed as food among the natives of the Mal. Archipelago.

Clitocybe multiceps Peck. → Agaricus decastes Fr.

Clitocybe tessulata (Bull.) Sing. (syn. Pleurotus ulmarius Bull.) (Agaricaceae). - Basidiomycete Fungus. Temp. zone. Fruitbodies grow on branches of trees. They are consumed by North American Indians and occasionally by the white population.

Clitopilus prunulus (Scop.) Quél. (Agaricaceae). - Basidiomycete Fungus. Cosmopolite, known as Mousseron. Fruitbodies are eaten in Europe and Asia, where they are often sold in markets.

Clitoria cajanifolia Barth. (Leguminosaceae). - Herbaceous plant. Tropics. Is sometimes used as a green manure in warm countries.

Clitoria Ternata L. (Leguminosaceae). - Herbaceous vine. Tropics. Roots are a powerful cathartic. Seeds are used as purgative in Sudan.

Clostridium acetobutylicum McCoy et al. (Bacteriaceae). - Bacil. This microorganism is of importance in the acetone-butanol fermentation.

Clostridium butyricum Prazm. (Bacteriaceae). - Bacil. This organism takes an important part in the retting process of flax and hemp fibres. Also C. felsineum (Carbone and Tomb.) Bergey et al.

Clostridium Pasteurianum Winogr. (Bacteriaceae). - Bacil. A non symbiotic soil microorganism being capable of fixing the nitrogen from the atmosphere.

Cloudberry → Rubus Chamaemorus L.

Clove → Eugenia caryophyllata Thunb.

Clove Bark Tree → Dicypellum caryophyllatum Nees.

Clove Oil → Eugenia caryophyllata Thunb.

Clove Pink → Dianthus Caryophyllus L.

Clove Tree → Eugenia caryophyllata Thunb.

Clover, Button → Medicago orbicularis (L.) All.

Clover, Alsike → Trifolium hybridum L.

Clover, Aztec → Trifolium amabile H. B. K.

Clover, Crimson → Trifolium incarnatum L.

Clover, Egyptian → Trifolium alexandrinum L.

Clover, Holy → Onobrychus viciaefolia Scop.

Clover, Hop → Medicago lupulina L.

Clover, Hungarian → Trifolium pannonicum Jacq.

Clover, Japan → Lespedeza striata Hook.

Clover, Persian → Trifolium resupinatum L.

Clover, Purple Prairie → Petalostemon purpureum (Vent.) Rydb.

Clover, Red → Trifolium pratense L.

Clover, Slender Prairie → Petalostemon oligophyllum (Torr.) Rydb.

Clover, Small Flowered → Trifolium parviflorum Ehrh.

Clover, Snail → Medicago scutellata (L.) Willd.

Clover, Teasel → Trifolium parviflorum Ehrh.

Clover, White → Trifolium repens L.

Club Moss → Lycopodium spec.

Club Wheat → Triticum compactum Host.

Clubmoss, Common → Lycopodium clavatum L.

Clusia flava L., Monkey Apple, Fat Pork. (Guttiferaceae). - Tree. West Indies. Stem is source of Hog Gummi, used for healing wounds.

Clusia fluminensis Planch and Triana. (Guttiferaceae). - Shrub. Brazil. Stem is source of a resin, used in some countries in vet. medicine.

Clusia insignis Mart. (Guttiferaceae). - Tree. Brazil. Flowers are said to be source of a resin; used in Brazil for healing wounds.

Clusia minor L. (Clusiaceae). - Small tree. S. Mexico to Panama, and S. America. Source of an elastic gum; used for bandaging of hernia among children.

Clusia palmicida Rich. (syn. Clusia alba Choisy.) (Guttiferaceae). - Tree. S. America. Plant is source of an incense; used in Colombia in churches.

Cluster Cardamon → Elettaria Cardamomum (L.) Maton.

Cluster Pine → Pinus Pinaster Soland.

Clustocactus sepium Weber → Borzicactus sepium H. B. K.

Cluytia hirsuta Muell. Arg. (Euphorbiaceae). - Woody plant. S. Africa. Used by the Kaffirs in S. Africa for splenic fever. Decoction in brandy is used by farmers of S. Africa as stomachic.

Cluytia similis Muell. Arg. (Euphorbiaceae). - Tree. S. Africa. Herb is used as antidote for anthrax and the root for snake-bites by the natives of S. Africa.

Cnestis corniculata Lam. Oboqui, Furuluga. (Connaraceae). - Woody plant. Trop. Africa, esp. Senegambia, Sierra Leone, Gaboon. Leaves are a powerful astringent.

Cnestis ferruginea DC. (Connaraceae). - Shrub. Trop. Africa. Decoction of leaves used as laxative in Yoruba (Afr.). Bitter fruits are used in Sierra Leone to clean teeth.

Cnestis polyphylla Lam. (Connaraceae). - Woody plant. Madagascar. Used in Madagascar for poisoning dogs.

Cnicus arvensis Hoffm. → Cirsium arvense (L.) Scop.

Cnicus Drummondii Gray → Cirsium Drummondii Torr. and Gray.

Cnicus japonicus Max. (Compositae). - Perennial herb. Japan. Young leaves are consumed as a vegetable in Japan.

Cnicus oleraceus L. → Cirsium oleraceum (L.) Scop.

Cnicus tuberosus Willd. → Cirsium tuberosum (L.) All.

Cnidoscolus elasticus Lundell. (Euphorbiaceae).- Small xerophytic tree. Mexico, esp. Sinaloa, Durango. Source of Highland Chilte Rubber, latex is composed of 44 to 50% rubber; is coagulated by water. C. tepiquensis Lundell is source of Chilte Blanco.

Coagulants of Milk → Milk, Coagulants of

Cob Nut → Corylus Avellana L.

Cob Nut → Omphalea triandra L.

Coca Huanuca → Erythroxylum Coca Lam.

Coca Tree → Erythroxylon Coca Lam.

Coca, Truxillo → Erythroxylon nova-granatense (Morris) Hier.

Coccaceae → Micrococcus, Sarcina.

Coccinea indica Wight and Arn. (syn. Cephalandra indica Naud.). (Cucurbitaceae). - Perennial vine. Trop. Asia, Red Sea area to Sudan. Fruits are eaten raw, candied or cooked in some parts of Sudan. In Java and Indo China young shoots and fruits are eaten with rice. Plant is sometimes cultivated.

Coccolobis grandifolia Jacq. (syn. C. pubescens L.). (Polygonaceae). - Tree. Trop. America. Wood strong, heavy, reddish, resistant; used for general carpentry.

Coccolobis laurifolia Jacq., Pigeon Plum. (Polygonaceae). - Florida, Antilles, Bahamas, S. America. Wood dark brown, very hard, heavy, brittle, strong, close-grained, sometimes used for cabinet making.

Coccolobis uvifera (L.) Jacq., Sea-Grape. (Polygonaceae). - Shrub or tree. Trop. America. Wood hard, close-grained, very heavy, dark brown to violet; sometimes used for furniture. Fruits are edible; used for jelly; made in the West Indies into a fermented drink. Extract from the bark is known as Jamaica Kino.

Coccothrinax argentea (Lodd.) Sarg. (syn. C. jucunda Sarg.). Biscayne Palm. (Palmaceae). - Small palm. S. Florida. Young, tough leaves are used for making hats, baskets and similar material.

Cocculus Bakis Guill. and Perr. → Tinospora Bakis Miers.

Cocculus Cebatha DC. (syn. Menispermum edule Vahl.). (Menispermaceae). - Woody vine. Arabia. Fruits edible; mixed in Arabia with rasins etc. to make an alcoholic beverage.

Cocculus Ferrandianus Gaudich. (Menispermaceae). - Woody vine. Hawaii Islands. Fruits used to capture fish.

Cocculus filipendula Mart. (Menispermaceae). - Woody plant. Brazil. Bitter and poisonous; used as emmenagogue and for snake-bites.

Cocculus laurifolius DC. (Menispermaceae). - Woody vine. Japan to Java. Bark source of alkaloids, cocculine and coclaurine, having properties of curare.

Coculus Leaeba DC. (Menispermaceae). - Woody plant. Trop. W. Africa. Roots used in Senegal and Fr. Sudan for periodic fevers. Contains sangoline, an alkaloid.

Cocculus palmatus Hook. → Jateorhiza Miersii Oliv.

Cocculus pendulus (Forst.) Diels (Menispermaceae). - Arabia, Iran, Afghanistan. Vine. Juice of the plant is used by the Arabs in the preparation of a fermented beverage; in Afghanistan it is employed against fever.

Cocculus sarmentosum (Lour.) Diels. (Menispermaceae). - Malaysia. Vine. Roots are used for the preparation of a poison.

Cocculus Thunbergii DC. (Menispermaceae). - Woody vine. Japan. Cultivated. Bleached tendrils are used for making baskets, manuf. in Midsuguchi (Omi), Japan.

Cochlearia Armoracea L. (syn. Armoracea lapathifolia Gilib.). Horse Radish. (Cruciferaceae). - Perennial herb. Europe, W. Asia. Fresh roots are a source of Horse Radish, used as a condiment for flavoring meat, vegetables; in pickles; promotes appetite; invigorates digestion. Cultivated. Contains sinigrin and myorosin; when crushed and moistened it gives rise to an ess. oil allyl isothiocyanate. Roots are made into Horse Radish Powder, Horse Radish Vinegar and Horse Radish Sauce.

Cochlearia officinalis L., Spoonwort, Scorbute Grass, Scurvy-Grass. (Cruciferaceae). — Perennial herb. Europe and temp. Asia. Is occasionally grown for its antiscorbutic properties.

Cochlospermaceae → Cochlospermum.

Cochlospermum angolense Welw. and Oliv. (Cochlospermaceae). - Woody plant. Angola. Seeds are source of a red dye; used by the natives.

Cochlospermum Gossypium DC. (Cochlospermaceae). - Small tree. E. India. Source of an insoluble gum; used as substitute for Tragacanth. It occurs in irregular, rounded, translucent clumps of a pale, bluff color. Cultivated in India near temples for its yellow flowers.

Cochlospermum niloticum Oliv. (Cochlospermaceae). - Shrubby plant. Trop. Africa. Swollen subterranean part of the plant is chewed in the Nuba Mountain Province as a tonic.

Cochlospermum Planchoni Hook. (Cochlospermaceae). - Woody plant. Trop. Africa. Tubers are source of a yellow dye. Also C. niloticum Oliv.

Cochlospermum tinctorum Perr. (Cochlospermaceae). - Perennial plant. Trop. Africa. In Sudan source of a yellow dye. Used by the Hausas with Indigo for making a green sacred dye.

Bark is used as fibre, made into rope; occasionally exported. Plant is used medicinally as emmenagogue.

Cochlospermum vitifolium Spreng. (syn. Maximiliana vitifolia Krug. and Urb.). (Cochlospermaceae). - Tree. Mexico. Bark is used in Sinaloa (Mex.) for making ropes.

Cocillana → Guarea Rusbyi (Britt.) Rusby.

Cock Spur Thorn → Crataegus Crus-galli L.

Cockur → Ochrolechia tartarea (L.) Mass.

Coco de Chili → Jubaea chilensis (Molina) Baill.

Coco Grass → Cyperus rotundus L.

Cocobola Wood → Lecythis costaricensis Pitt. and Dalbergia retusa Hemsl.

Coconut, Double → Lodoicea callipyge Comm.

Coco Nut Palm → Cocos nucifera L.

Coconut Oil → Cocos nucifera L.

Coco Palm → Cocos nucifera L.

Cocora → Grias peruviana Miers

Cocorite Palm → Maximiliana caribaea Griseb.

Cocos coronata Mart. (Palmaceae). - Tall palm. Trop. S. America. Seeds source of Ouricur Palm Kernel Oil, non-drying. Exported to U. S. A. and Europe. Used for manuf. of margarine. Sp. Gr. 0.9221; Sap. Val. 236.9; Iod. No. 14.7; Unsap. 0.27%.

Cocos nucifera L., Coco Palm (Palmaceae). - Tall palm. Malaya. Cultivated in tropics of Old and New World. Important commercial crop. Meat of seeds eaten raw, shredded and dried which is used in pastries and in confectionary; also very much used as dried copra for preparation of coconut oil, employed in food products, especially margarines, in soaps, cosmetics, salves, shampoos, shaving creams, as illuminant, in toilet preparations. Pressed oil cake is an excellent food for livestock. Coconut milk is a refreshing drink in many parts of the tropics. Young inflorescences are source of palm sugar, palm wine, arrack and vinegar. Coir or rough fibre composing husk of fruit is used for mats, ropes, baskets, brushes etc. Husks are used for cups, bowls, spoons, as fuel; source of an excellent charcoal used for gas masks. Young buds are consumed as vegetable and in salads. Leaves are made into mats, baskets, used for thatching. Wood is employed for cabinet work, construction, building and firewood. Numerous varieties, among which suitable for copra are: Bukay, Daakan, Galimba, Hongola, Laguna, Limba, Malaya, Pati. Suitable for sweetmeats is Kapuno.

Cocos Romanzoffiana Cham. (Palmaceae). - Palm. Trop. America. The very young buds are consumed as food, especially when preserved in oil or in vinegar.

Cocos Yatay Mart. (Palmaceae). - Palm. Argentina and adj. territory. Fruits are used by the natives for making brandy. The very young buds are consumed as food.

Codia montana Labill. → Callicoma Billardieri D. Don.

Codiaeum lucidum Muell. Arg. → Baloghia lucida Endl.

Codiaceae (Green Algae) → Codium.

Codium fragile (Sm.) Huit. (Codiaceae). - Green Alga. Plants are consumed in Japan as food.

Codium Lindenbergii Binder (Codiaceae). - Green Alga. Plants are consumed in Japan when fresh; also after being dried or salted.

Codium Muelleri Kuetz. (Codiaceae). - Green Alga. Pac. Ocean. Consumed as food among the Hawaiians. Plants are pounded, mixed with salted squid with addition of chili peppers.

Codium tenus Kurz. (Codiaceae). - Green Alga. Seaweed. Eaten by the natives in Philipp. Islds. Also C. tomentosum Stackh. in Malaya. Sold in native markets.

Codium tomentosum (Huds.) Stackh. (Codiaceae). - Green Alga. Plants are used in Japan as food. They are eaten in soup, also with soya sauce and vineger.

Codonopsis Targshen Oliv. (Campanulaceae). - Herbaceous herb. Hupeh, Szechuan, Shensi (China). Root is source of an important Chinese drug, supposed to have tonic and aphrodisiac properties. C. lanceolata Benth. produces a similar, though inferior drug.

Codonopsis ussuriensis Hemsl. (Campanulaceae). - Herbaceous perennial. Japan, China. Roots used raw or roasted, eaten as food by the Ainu.

Coelococcus amicarum Warb. Polynesian Ivory-Nut Palm. (Palmaceae). - Tall palm. Carolina Islands. Cultivated in Philipp. Islds. Large, very hard, ivory-like seeds are used for manuf. of buttons.

Coelopleurum Gmelini (DC) Ledeb. → Archangelica Gmelini DC.

Coelorhopalon obovatum (Berk.) v. Overeem. → Xylaria obovatum Berk.

Coffea arabica L., Arabian Coffee. (Rubiaceae). - Tree. E. Africa. Cultivated in the tropics of the Old and New World, among which in Java, Malaysia, Centr. America, Colombia, Brazil etc. Roasted coffee beans are source of a beverage, also used for flavoring ice cream, candies and pastries. Source of coffein. Dried ripe seeds are used medicinally as stimulant, nervine, diuretic; act on the central nervous system, kidneys, heart and muscles. Some varieties are Bourbon, Maragogipe, San Ramón, Mocha, Columnaris and Murta.

Coffea Arnoldiana De Wild. (Rubiaceae). - Tree. Belgian Congo. Related to C. excelsa. Has been recommended for cultivation. Trees do not need shade and are adapted to a dry climate.

Coffea breviceps Hiern. (Rubiaceae). - Tree. Trop. Africa. Seeds have been recommended as a source of coffee.

Coffea canephora Pierre (Rubiaceae). - Small tree. French and Belgian Congo. Occasionally cultivated. Is related to C. arabica L. Susceptable to Hemileia vastarix. Beans produce a pleasant beverage.

Coffea congensis Froehner. (Rubiaceae). - Tree. French and Belgian Congo. Related to C. arabica L. Var. Chaloti Pierre is occasionally cultivated. Said to be immune from Hemileia vastatrix.

Coffea Dewevrei De Wild. (Rubiaceae). - Tree. Trop Africa, esp. Congo region. Has been recommended for trial as a coffee crop.

Coffea excelsa Chev. (Rubiaceae). - Tree. French Africa. Cultivated in tropics of Old and New World. Source of a good coffee, said to be superior to that of C. liberica and C. robusta, though inferior to C. arabica. Trees are excellent yielders.

Coffea Humblotiana Baill. (Rubiaceae). - Tree. French Africa. Ins. Comor. Has been recommended for cultivation. Seeds do not contain caffeine, but contain a bitter substance cofamarine; has been suggested for persons who can not use caffeine. Also C. Bonnieri Dub., C. Gallienii Dub. and C. Mogeneti Dub. from Madagascar.

Coffea liberica Hiern., Liberian Coffee. (Rubiaceae). - Small tree. Trop. Africa, esp. Sierra Leone, Angola, Congo, Gabon, Liberia. Cultivated in Trop. Africa, India, Ceylon, East Indies, Java, Madagascar, West Indies, Guiana, Surinam. Considered inferior to Arabian, Robusta and Sierra Leone Coffee. Adapted to warmer and moister climates than C. arabica.

Coffea Maclaudii Chev. (Rubiaceae). - Tree. French Guinea. Seeds are source of a good coffee, resembling that of C. excelsa.

Coffea quillon Wester. Rubiaceae. - Tree. Trop. Africa, esp. French Congo. Trees are excellent yielders and have been recommended for trial as coffee producers.

Coffea robusta Linden. (syn. C. Laurentii De Wild.). Robusta Coffee, Rio Nunez Coffee, Congo Coffee. (Rubiaceae). - Small tree. Congo. Cultivated in Java, Sumatra, Trinidad etc. Quality of berries nearly equal to A. arabica.

Coffea stenophylla G. Don. Highland Coffee of Sierra Leone. (Rubiaceae). - Small tree. Sierra Leone. Occasionally cultivated in preference to C. liberica. Grown in Sierra Leone, West Indies, India and Ceylon. Berries small, beans of superior quality, are claimed to be equal to Mocha.

Coffee Tree, Kentucky → Gymnocladus dioica (L.) Koch.

Coffee, Arabian → Coffea arabica L.

Coffee, Highland → Coffea stenophylla G. Don.

Coffee, Liberian → Coffea liberia Hiern.

Coffee, Mocha → Coffea arabica L.

Coffee, Robusta → Coffea robusta Linden.

Coffee Senna → Cassia occidentalis L.

Coffee substitutes → Agropyron, Asparagus, Astragalus, Boscia, Brosimum, Canavalia, Cassia, Castanea, Ceratonia, Cicer, Cichorium, Crataegus, Daucus, Dioscorea, Galium, Glycine, Gymnocladus, Hordeum, Iris, Lathyrus, Leontodon, Ligustrum, Parkia, Phorodendron, Pithecolobium, Scorzonera, Secale, Silybum, Simmondsia, Sium, Sorbus, Taraxacum.

Cogswellia ambigua (Nutt.) Jones. (syn. Eulophus ambiguus Nutt.) (Umbelliferaceae). - Perennial herb. Western N. America. Roots when ground into flour and made into cakes, were eaten by Indians of the Northwest. Also C. Canbyi (Coult. and Rose) Jones, C. cous (S. Wats.) Jones, C. farinosa (Hook.) Jones, C. foeniculacea (Nutt.) Coult and Rose and others.

Cohoba → Piptadenia peregrina (L.) Benth.

Cohoba Tree → Piptadenia peregrina (L.) Benth.

Cohosh, Blue → Caulophyllum thalictroides (L.) Michx.

Cohosh, Black → Cimicifuga racemosa (L.) Nutt.

Cohune Oil → Attalea Cohune Mart.

Cohune Palm → Attalea Cohune Mart.

Coir → Cocos nucifera L.

Coix Lachryma-Joby L. Job Tears. (Graminaceae). - Annual grass. Africa, Asia. Cultivated in warm countries of Old and New World. Parched seeds used as tea in Japan; also in N. Tonkin. Source of „been" by the Nagas. Grains widely used in bead-chains, roseries etc. Foliage used as fodder.

Cola → Cola acuminata Schott and Endl. and C. nitida Schott and Endl.

Cola, Bitter → Cola acuminata Schott. and Endl.

Cola, False → Cola acuminata Schott. and Endl.

Cola, Male → Cola acuminata Schott. and Endl.

Cola acuminata Schott and Endl. (syn. Garcinia Kola Heckel). Cola. (Sterculiaceae). - Tree. Trop. Africa. Cola nuts are used for chewing among the natives. Contain 2.4 to 2.6% caffeine. Much exported to Europe and America for preparation of a beverage, pills, bonbons, pastills, etc. Cultivated in tropics of the Old and New World. Product is called Male Cola, Bester Cola, False Cola.

Cola Ballayi Cornu. (Sterculiaceae). - Tree. Trop. E. Africa. Seeds are sometimes mixed with those of C. nitida. They are said to be deficient in alkaloids.

Cola cordifolia R. Br. (Sterculiaceae). - Tree. Trop. Africa. Seeds are eaten by the natives.

Cola diversifolia De Wild. (Sterculiaceae). - Trop. Africa. Seeds edible, consumed by the natives.

Cola nitida Schott. and Endl. Cola. (Sterculiaceae). - Tree. Trop. Africa, esp. Sierra Leone to Congo. Cultivated in Africa, Asia, West Indies and Brazil. Dried cotyledons are used medicinally, acting on the central nervous system, being nervine, stimulant, tonic and astringent. Closely related to C. acuminata.

Colibah → Eucalyptus microtheca F. v. Muell.

Colic Root → Aletris farinosa L.

Colchicum autumnale L., Meadow Saffron. (Liliaceae). - Perennial tuberous herb. Europe. Seeds used medicinally as antirheumatic in gout and rheumatism, alterative, sedative, diuretic. Dried corms have the same properties. Much is supplied by Germany and Italy. Contains colchicin, a toxic alkaloid. Used in experimental genetics as a means of causing permanent characteristics in plant and animals.

Colchicum luteum Baker. (Liliaceae). - Bulbous perennial. Temp. Himalaya to Iran. Plant used in Iran for rheumatism. Contains colchicine.

Colchicum montanum L. (syn. C. Ritchii R. Br.) (Liliaceae). - Bulbous perennial. Egypt. Corms are said to be used in Egypt for fattening, commonly used in Mariût.

Coleus aromaticus Benth. (syn. C. amboinicus Lour.). (Labiaceae). - Perennial herb. Origin uncertain, probably from Malaya. Herb is used for seasoning meat dishes. Decoction of herb is employed for washing clothes and hair. Cultivated.

Coleus Dazo Chev. (Labiaceae). - Perennial herb. Congo, Sudan. Cultivated. The tubers are eaten as food by natives in parts of Africa.

Coleus edulis Vatke. (Labiaceae). - Herbaceous plant. E. Africa. Cultivated in the plateaux of Abyssinia. Tubers are consumed as food by the natives.

Coleus rotundifolius Chev. and Perrot. (syn. Plectranthus rotundifolius Spreng., P. tuberosus Blume.). Hausa Potato. (Labiaceae). - Perennial herb. Trop. Africa, Madagascar, Malaysia, Mauritius, East Indies etc. Cultivated. Tubers are consumed as food, like potatoes.

Coli Bacil → Eschirichia coli (Migula) Castell and Cham.

Collard → Brassica oleracea L. var. acephala DC.

Colletia cruciata Gill. and Hook. (Rhamnaceae). - Tree. Argentina to Chile. Wood used for manuf. wagons, carts, handles of tools and for making houses.

Colletia ferox Gill. and Hook. (Rhamnaceae). - Tree. S. America. Decoction of plant is used in some parts of Argentina, Uruguay and adj. territory as febrifuge.

Collina elegans (Mart.) Liebm. → Chamaedorea elegans Mart.

Collinsonia canadensis L., Stoneroot, Horse Balm. (Labiaceae). - Perennial herb. Eastern N. America to Florida. Dried rhizome and roots are used medicinally as diaphoretic, diuretic; for infantile and biliary colic and dropsy. Contains mucilage, a saponin-like glucoside and a resin.

Collybia Boryana (Mont.) Sacc. → Lentinus cubensis Berk and Curt.

Collybia butyracea (Bull.) Quél. (Agaricaceae). - Basidiomycete. Fungus. Temp. zone. Fruitbodies occasionally consumed in America and Europa; they are esp. appreciated and sold in Japan. Also: C. acercata (Fr.) Gill. and C. distorta (Fr.) Quél.

Collybia eurhiza Berk → Rajapa eurhiza (Berk.) Sing.

Collybia microcarpa (Berk. and Br.) Hoehn. (Entoloma microcarpum [Berk. and Br.] Sacc.) (Agaricaceae). - Basidiomycete. Fungus. Trop. Asia, esp. Malaya. The mycelium occurs in termite nests. Fruitbodies are consumed as food by the natives of Malaya.

Callybia velutipes (Curt.) Quél. → Flammulina velutipes (Curt.) Karst.

Colocasia antiquorum Schott. Taro. (Araceae). - Large perennial herb. Trop. Asia, China, Japan. Tubers are consumed in many tropical countries. Source of Portland Arrowroot.

Colocasia esculenta Schott., Dasheen. (Araceae). - Perennial herb. Trop. Asia. Cultivated in the tropics of the Old and New World. Tubers are consumed boiled or fried. Young leaves are eaten like spinach and are also consumed when bleached.

Cococynth → Citrullus Colocynthis (L.) Schrad.

Colombia Copal → Hymenaea Courbaril L.

Colombian Blueberry → Vaccinium floribundum H.B.K.

Colombian Mahagony → Cariniana pyriformis Miers.

Colombia Virgin Rubber → Sapium Thomsoni God.

Colombian Berry → Rubus macrocarpus Benth.

Colophan Wood → Canarium paniculatum (Lam.) Benth.

Colophony → Pinus palustris Mill., P. Merkusii Jungh. and De Vr.

Colorado Grass → Panicum texanum Buckl.

Colpoon compressus Berg. (Santalaceae). - Shrub or small tree. S. Africa. Bark is used for tanning.

Coltsfoot → Tussilago Farfara L.

Coltsfoot, Sweet → Petasites palmata Gray.

Colubrina ferruginosa Brogn. (Rhamnaceae). - Tree. West Indies. Wood strong, compact, red; used for general carpentry.

Colubrina oppositifolia Brongn., Kauila. (Rhamnaceae). - Tree. Hawaii Islds. Wood hard, was formerly used for building Hawaiian grass-huts, also used for manuf. of the ie kuku or kapa beater and for making spears. Rounded and polished rods were used as hairpins by Hawaiian women.

Colubrina reclinata (L'Hér.) Brogn. → Ceanothus reclinatus L'Hér.

Colubrina rufa Reiss. (Rhamnaceae). - Shrub or tree. Brazil. Occasionally cultivated. Bark called Saguaragy, used in Brazil for fevers. Bark is exported. Wood is strong, fine structure, durable; used for cabinet work, ship building, bridges, vehicles, rail-road ties, fence posts.

Columbia javanica Blume. (Tiliaceae). - Tree. Java. Bark source of a fibre; used by natives for fishing-nets, cordage and ropes.

Columbine Meadow Rue → Thalictrum aquifolium L.

Colza → Brassica Napus L.

Colza, Chinese → Brassica campestris var. chinoleifera.

Colza Oil → Brassica Napus L.

Comarum palustre L. (syn. Potentilla palustris Scop., P. comarum Nestl.). (Rosaceae). - Perennial herb. Europe, temp. Asia, introd. in N. America. Flowers are source of a red dye; used in some parts of Europe for coloring wool. Rhizome is used as home remedy; occasionally for tanning.

Combee Resin → Gardenia gummifera L. f.

Combee Resin → Gardenia lucida Roxb.

Combretaceae → Anogeissus, Combretum, Conocarpus, Laguncularia, Lumnitzera, Quisqualis, Terminalia.

Combretum alatum Guill and Perr. (syn. C. micranthum G. Don. (Combretaceae). - Tree. W. Africa. A decoction of the leaves is used by the natives for black fever. It is claimed that a tea is used by Europeans to get accustomed to the climate of W. Africa.

Combretum bracteatum Engl. and Diels. (Combretaceae). - Tree. W. Africa. Decoction of leaves is used as febrifuge and tonic.

Combretum coccineum Lam. (Combretaceae). - Vine, Madagascar. Bark is source of a fibre; used in Madagascar.

Combretum erythrophyllum Sond. (Combretaceae). - Shrub or small tree. S. Africa. Stem is source of a gum; used in S. Africa as a substitute for Gum of Tragacanth.

Combretum ghasalense Engl. and Diels. (Combretaceae). - Tree. Sudan. Wood is used as an incense by women of the Sudan.

Combretum glutinosum Guill and Perr. Simba Bali. (Combretaceae). - Tree. Trop. Africa. Root and bark are source of a yellow dye in Senegambia; Alkaline ash is used for fixing Indigo Blue in cotton. Decoction of leaves is used for washing wounds.

Combretum Hartemannianum Schweinf. (Combretaceae). - Tree. Trop. Africa. Stems are source of a gum. Used by the Nuba as a perfume, also as a substitute for Gum Arabic., called Mumuye Gum.

Combretum hypotilinum Diels. (Combretaceae). - Tree. Trop. W. Africa. Is source of a gum in Hause, called Taramniiya, used to cure toothache. In Gambia leaves are used as a purgative.

Combretum leonense Engl. and Diels. (Combretaceae). - Tree. Trop. Africa. Source of Mumuye Gum.

Combretum lecananthum Engl. and Diels. (Combretaceae). - Small tree. Trop. W. Africa. Source of a white, yellow to redish gum, chewed by the natives.

Combretum sokodense Engl. (Combretaceae). - Tree. Trop. W. Africa. One of the sources of Mumuye Gum derived from Bornu and Adamwa. Decoction of roots is used for dysentery.

Combretum sundaicum Miq. (Combretaceae). - Tall climber. Malaysia. It has been claimed that leaves used as a tea are a cure for opium-craving. Contain no alkaloids; there is apparently a resin and tannic acid.

Combretum trifoliatum Vent. (Combretaceae). - Tree. Trop. Africa. Wood is source of a scented material; used by the native women of Sudan in perfumery.

Comfrey, Prickly → Symphytum asperrinum Don.

Commelinaceae → Aneilema, Floscopa.

Commersonia echinata Forsts. (Sterculiaceae). - Tree. Malaya, New South Wales, Queensland. Bark produces a fibre. Used by aborigines of Australia for Kangeroo or fishing nets.

Commiphora abyssinica (Berg.) Engl. Abyssinian Myrrh. Tree. (Burseraceae). - Shrub or small tree. Coast of Somaliland, Abyssinia, Nubia, Arabia. A gum-resin is derived from wounds in the stems. Sometimes called Hotai. Used medicinally as a stimulant, stomachic, astringent; in mouth washes. Contains a green thickis ess. oil, cuminol and eugenol. Oil of Myrrh is used in perfumes of the Oriental type; blends well with vetiver, geranium and sandal.

Commiphora africana Endl. African Myrrh Tree. (Burseraceae). - Small tree. Sudan, Abyssinia. Stem is source of a Bdellium, a resin, resembling Myrrh.

Commiphora Berryi Engl. (Burseraceae). - Small tree. E. India. Stem is source of a gummi exudation, called Mulu Kilavary.

Commiphora erythraea Engl. Bisabol Myrrh Tree. (Burseraceae). - Tree. Trop. Africa. Source of Bisabol Myrrha, Coarse Myrrha from Aden; Perfumed Bdellium of Bombay.

Commiphora erythraea Engl. var. glabrescens Engl. Haddi Tree. (Burseraceae). - Tree. Somaliland. Source of an important Bdellium, somtimes called Perfumed Bdellium. Sold in irregular clumps, having the color of true Myrrh. Has a powerful scent. Bark is burned in huts of the natives to throw off a pleasant odor. Fruits are eaten by camels. Source of Bisabol Myrrh and of Gafal Wood used as incense and perfumes. Was used for embalming among the Ancients.

Commiphora Hildebrandtii Engl. (Burseraceae). - Tree. Maritime hills, Somaliland. Source of a rare Bdellium, resembling that of Guban Myrrh with which it is sometimes mixed. Tears are described of being of a „calf's-foot-jelly-like" color. In Somaliland are other Bdelliums that need careful studies; thus far the plants have not been properly identified. Under Somali names, are known the Hodai Tree producing an opaque bdellium, known as Habbak Hodai, found in irregular masses, being dirty milky white to dull reddish. Dunkal Tree is source of the rare Habbak Dunkal, an opaque bdellium, being dull, yellowish red. Hagar Ad Tree is source of Habbak Harr which is sometimes mixed with Guban Myrrh. Garon Gurun. Tree produces Habbak Garon Gurun, a dark red bdellium, composed of irregular clumps. Other unidentified bdelliums are: Habbak Daseino, Habbak Ilka Adaxai, Habbak Dundas and Habbak Tubuk.

Commiphora Kataf Engl. (syn. Balsamodendron Kataf Kunth.). Opopanax Myrrh Tree. (Burseraceae). - Tree. Trop. Africa. Stem is probably a source of Bursa Opopanax, a resin used in perfumery.

Commiphora Mukul Engl. Mukul Myrrh Tree. Bdellium. (Burseraceae). - Tree. Arabia, to E. India. Source of a resin. Indian Bdellium. Mukul-i-Azrak has a reddish tinge; Mukul-i-Yahud has a yellow tinge; Sakulali has a brown color. Used in Iran in muscular rheumatism and as stomachic.

Commiphora Myrrha (Nees.) Engl. Common Myrrh Tree. Harobol Myrrh. (Burseraceae). - Shrub or small tree. Arabia, Somaliland, Abyssinia. Source of Harobol Myrrh. Derived from wounds in the stem. The gum-resin is brown-black. Used for perfumes and as incense during religious ceremonies. Was employed by the Ancients for embalming.

Commiphora opobalsam Engl. (syn. Balsamodendron gileadense Kunth.). Mecca Myrrh. (Burseraceae). - Small tree. Arabia. Source of Balsam of Gilead, Duhnual Balsam, Mecca Balsam, derived from the wounds of the stems. Product is brown to reddish. Used by the natives as stomachic, for healing wounds, diaphoretic. Employed in perfumes of the Oriental type and for incense. Product is relatively rare.

Commiphora pedunculata Engl. (Burseraceae). - Small tree. Trop. Africa. Stem is source of a resin, similar to Olibanum.

Commiphora pilosa Engl. (Burseraceae). - Tree. Trop. Africa. Bark is source of a reddish-brown dye; used for dyeing cloth.

Commiphora Schimperi Engl. (Burseraceae). - Small thorny shrub. Abyssinia, SW.-Arabia. Source of Myrrh.

Common Apple Tree → Pyrus Malus L.

Common Banana → Musa sapientum L.

Common Barley → Hordeum vulgare L.

Common Betony →Stachys officinalis (L.) Trev.

Common Borneo Camphor → Dryobalanops aromatica Gaertn.

Common Breadroot →Psoralea esculenta Pursh.

Common Buckthorn → Rhamnus cathartica L.

Common Buckwheat → Fagopyrum esculentum Moench.

Common Burdock → Arctium minus (St. Hil.) Bern.

Common Camas → Camassia Quamash Greene.

Common Catalpa →Catalpa bignonioides Walt.

Common Chaulmoogra Tree — Hydnocarpus anthelmintica Pierre.

Common Chickweed →Stellaria media (L.) Vill.

Common Clubmoss → Lycopodium clavatum L.

Common Confrey → Symphytum officinale L.

Common Coral Bean → Erythrina Corallodendron L.

Common Cow Parsnip → Heracleum Sphondylium L.

Common Custard Apple →Annona reticulata L.

Common Everlasting → Gnaphalium polycephalum Michx.

Common Gorse → Ulex europaeus L.

Common Gromwell → Lithospermum officinale L.

Common Heather → Calluna vulgaris Salisb.

Common Hyacinth → Hyacinthus orientalis L.

Common Jasmine → Jasminum officinale L.

Common Jasmin Orange → Murraya exotica L.

Common Jujub → Zizyphus Jujuba Mill.

Common Kangeroo Grass →Anthistiria ciliata L.

Common Lespedeza → Lespedeza striata Hook.

Common Licorice → Glycyrrhiza glabra L.

Common Lizardtail → Saururus cernuus L.

Common Mallow → Malva rotundifolia L.

Common Mignonette → Reseda odorata L.

Common Milkweed → Asclepias syriaca L.

Common Motherwort → Leonurus cardiaca L.

Common Myrrh Tree — Commiphora Myrrha (Nees.) Engl.

Common Oat — Avena sativa L.

Common Persimon → Diospyros virginiana L.

Common Pokeberry → Phytolacca decandra L.

Common Rue → Ruta graveolens L.

Common Smoketree → Rhus Cotinus L.

Common Tigerflower → Tigridia Pavonia (L. f.) Ker-Gawl.

Common Tobacco → Nicotiana Tabacum L.

Common Twig Lichen → Ramalina calicaris (L.) Röhling.

Common Valerian → Valeriana officinalis L.

Common Vetch → Vicia sativa L.

Common Wheat → Triticum vulgare Vill.

Common Willow → Salix Caprea L.

Common Whip Tree → Luehea divaricata Mart.

Compositae. → Acanthospermum, Achillea, Achyrocline, Actinella, Adenostemma, Agoseris, Anacyclus, Athemis, Aplopappus, Arctium, Arnica, Artemisia, Aspilia, Aster, Athrixia, Atractylis, Baccharis, Balsamorhiza, Bidens, Blumea, Boltonia, Brachylaena, Brauneria, Brickelia, Buphtalmum, Caldendula, Carlina, Carthamus, Centaurea, Centipeda, Ceruana, Chrysactinia, Chrysanthemum, Cichorium, Cirsium, Clibadium, Cnicus, Conyza, Coreops, Cynara, Dicoma, Diotis, Echinops, Eclipta, Elephantopus, Emilia, Encelia, Enhydra, Erigeron, Eriocephalus, Eupatorium, Gnaphalium, Grangea, Helenium, Grindelia, Guizotia, Gundelia, Gynura, Helianthus, Helichrysum, Heterothalamus, Hymenopappus, Hypochoeris, Inula, Kleinia, Laciniaria, Lactuca, Leontodon, Leyssera, Libothamnus, Lygodesmia, Madia, Matricaria, Microseris, Mutisia, Olearia, Onoperdon, Parthenium, Pectis, Pentzia, Perezia, Petasites, Pluchea, Polymnia, Proustia, Psiadia, Pulicaria, Pyrhopappus, Salmea, Saussurea, Scolymus, Scorzonera, Senecio, Silybum, Solidago, Sparganophorus, Sphaeranthus, Spilanthus, Stenocline, Stevia, Tagetes, Tanacetum, Taraxacum, Tarchonanthes, Thelesperma, Tragopogon, Trilisia, Tussilago, Vanillosmopsis, Venidium, Vernonia, Wyethia.

Conanthera bifolia Ruiz. and Pav. (Liliaceae). - Bulbous herb. Chile. Bulbs are source of a food, eaten by the Indians.

Conanthera Simsii Sweet. (syn. Conanthera campanulata Lindl., Cumingia campanulata D. Don.). (Liliaceae). - Perennial herb. Chile. Tubers are consumed as food in some parts of Chile.

Conchomyces verrucisporus Van Overveen. (Agaricaceae). - Basidiomycete Fungus. Malayan Archipelago. Fruitbodies are consumed by the natives as a food.

Condalia lycioides Gray (syn. Zyziphus lycioides Gray). Southwestern Condalia, Crucillo. (Rhamnaceae). - Shrub. S.W. of United States and Mexico. Bark used as soap. Pimas in Arizona use decoction of roots for sore eyes.

Condalia obovata Hook. Bluewood Condalia. (Rhamnaceae). - Shrub or small tree. Nuevo Leon and Texas. Wood produces a blue dye. Fruits edible, made into good jelly.

Condiments → Acrodiclidium, Acronychia, Adansonia, Aeolanthus, Aframomum, Agastache, Allium, Alpinia, Ammodaucus, Amomum, Androcymbium, Anethum, Anisomeles, Archangelica, Artemisia, Aydendron, Blumea, Buphtalmum, Capparis, Capsicum, Carum, Cochlearia, Coleus, Coriandrum, Crithmum, Curcuma, Cymbopogon, Cynomorium, Dicypellium, Dracontomelum, Drimys, Elettaria, Eryngium, Eupatorium, Eustrema, Fraxinus, Glochidion, Gynandropsis, Hemerocallis, Heracleum, Horsfieldia, Hyptis, Kaemfera, Laurus, Lepidium, Levisticum, Lysimachia, Majorana, Marismius, Melissa, Mentha, Monarda, Moringa, Myristica, Ochocoa, Ocimum, Odyendyea, Origanum, Parkia, Perilla, Persea, Petasites, Petroselinum, Phaeomeria, Phellopterus, Pimenta, Pimpinella, Poliomintha, Pycnanthemum, Rosmarinus, Ruta, Salvia, Sarothamnus, Satureja, Sparganophorus, Tagetes, Thymus, Unona, Zingiber. See also Spices, Essential Oils.

Condorvine → Marsdenia Reichenbachii Triana.

Condurango Bark → Marsdenia Reichenbachii Triana.

Condurango Wine → Marsdenia Reichenbachii Triana.

Conehead Thyme → Thymus capitatus Hoffm. and Link.

Conessi Bark → Holarrhena antidysenterica Wall.

Confectionary. Plants used in → Acacia, Acer, Acorus, Alhagi, Amygdalus, Anacardium, Ananas, Angelica, Anageissus, Arachis, Asparagus, Astragalus, Buchanania, Calodropis, Canarium, Carica, Carya, Castanea, Chondrus, Chrysanthemum, Citrus, Cocos, Coffea, Cola, Corylus, Corypha, Craniolaria, Echinocactus, Elettaria, Eryngium, Eugenia, Euphorbia, Ferocactus, Fagara, Hyphaene, Jacaratia, Juglans, Levisticum, Macadamia, Maranta, Marrubium, Nelumbium, Nipa, Opuntia, Passiflora, Phaseolus, Pistacia, Quercus, Rosa, Saccharum, Sesamum, Telosma, Theobroma, Torreya, Trapa, Triticum, Vanilla, Vitis, Zea, Zingiber, Zizyphus. See also under Condiments, Spices, Essential Oils, Flour.

Confiture d'Épine Vinette → Berberis vulgaris L.

Cone Flower, Purple → Brauneria angustifolia (DC.) Heller.

Confrey, Common → Symphytum officinalis L.

Congo Copal Tree → Copaifera Demeusii Harms.

Congo Goober → Voandzeia subterranea Thon.

Congo Copal → Copaifera Demeusii Harms.

Congo Mallee Eucalyptus → Eucalyptus dumosa Cunningh.

Conium maculatum L., Poison Hemlock. (Umbelliferaceae). - Annual or biennial herb. Temp. N. Hemisphere. Very poisonous. Dried, full-grown unripe fruits are used medicinally, being antispamodic, anodyne, sedative. Contains coniine, an alkaloid. Decoction of unripe fruits was used as a hemlock poison for putting to death criminals among the Ancient Greeks. It is supposed that Socrates was put to death by drinking this poison. Entire fresh plant is the source of Succus Conii.

Connaraceae → Cnestis, Connarus, Rourea.

Conarum africanus Lam. (Connaraceae). - Tree. Trop. Africa. Seeds are made into a flour; used as purge in Sierra Leone. Efficient tenefuge; used as anthelmintic along the West Coast of Africa.

Conobea scoparioides Benth. (Scrophulariaceae). - Perennial plant. S. America. Used for toothache in the Chaco Intendency of Colombia.

Conocarpus erectus L., Buttonwood. (Combretaceae). - Tree. Coastal zone, Trop. America. Wood suitable for charcoal. Bark bitter, astringent; used for tanning leather.

Conopharyngia Chippii Stapf. (Apocynaceae). - Woody plant. Trop. W. Africa. Latex is an ingrediant of paste-rubber, called Bede-Bede.

Conopharyngia crassa Stapf. (Apocynaceae). - Woody plant. Trop. W. Africa. Juice is used to coagulate rubber.

Conopharyngia pachysiphon Stapf. (syn. Tabernaemontana pachysiphon Stapf.) (Apocynaceae). - Small tree. Trop. Africa. Latex used for adulterating rubber. Bark source of a fibre; used for making cloth. Juice from tree is used as bird-lime. Leaves are source of a black substance; used by native women for coloring hair.

Conostegia mexicana Cogn. (Melastomaceae). - Woody plant. S. Mexico. Berries used as food by Chinantecs and Mazatecs (Mex.). Also C. arborea (Schlecht.), Schauer, C. subhirsuta DC., C. xalapensis (Bonpl.) D. Don.

Conostegia xalapensis D. Don. (syn. Melastoma xalapense Bonpl.) Capiroto. (Melastomaceae). - Shrub. Trop. America. Fruits are edible, of good flavor; sold in markets of Centr. America.

Convallaria majalis L. Lily-of-the-Valley. (Liliaceae). - Perennial herb. Europe, temp. Asia, Japan. Cultivated in the Old and New World. Dried rhizome and roots are used medicinally. They are collected late in summer. Considered a heart tonic and diuretic. Contains convallamarin, a glycoside. Racemes and herb are used for the same purpose. Flowers are occasionally source of an ess. oil; used as concrete or absolute.

Convolvulaceae → Argyreia, Calonyction, Calystegia, Convolvulus, Cressa, Exogonium, Ipomoea, Piptostegia, Rivea.

Convolvulus floridus L. f. (Convolvulaceae). - Herbaceous plant. Canarian Islds. Teneriffa. Roots and rootstocks are source of an ess. oil, Oil of Rhodium, Guadil. Also C. soparius L. f.

Convolvulus Scammonia L., Scammony. (Convolvulaceae). - Perennial vine. Mediterranean region. Asia Minor. Source of Levant Scammony, Levant Scammony Resin, Resina Scammoniae, being an exudation from the roots, a brownish-black to greenish-gray substance, brittle. It is a hydragogue cathartic. Dried root is used medicinally for the same purpose. Much is exported from Aleppo and Smyrna.

Convolvulus Soldanella L. (Convolvulaceae). - Perennial creeping plant. Throughout Europe, temp. Asia, N. Africa, N. and S. America. Herb, Herba Soldanellae is used as diuretic, also as a purgative.

Conyza balsamifera L. → Blumea balsamifera DC.

Conyza persicaefolia Oliv. and Hiern. (Compositae). - Herbaceous plant. Trop. Africa. Sap from leaves is source of a black dye; used in the Belgian Congo.

Copaiba Balsam → Copaifera coriacea Mart.

Copaifera bracteata Benth. (Leguminosaceae). - Tree. Northern S. America. Source of Blue Ebony, Violet Wood, Marawayana, being reddish violet, heavy, fairly hard, easy to split and to work, polishes well; used for furniture and fancy articles.

Copaifera coriacea Mart., Copaiba. (Leguminosaceae). - Tree. Brazil, Colombia, banks of the Orinocco. Source of Copaiba Balsam. Used in varnishes, for photographic paper, removing old varnish from oil paintings. Used medicinally, diuretic, stimulant, expectorant, genito-urinary disinfectant; for diarrhea, chronic dysentery.

Copaifera Demeusii Harms. (Leguminosaceae). - Tree. Belg. Congo. A source of Congo Copal; used for varnishes and lacqeurs; also called Cameroon Copal.

Copaifera Gorskiana Benth. (Leguminosaceae). - Tree. Trop. Africa. Source of Inhambane Copal, a good hard resin, not completely fossilized, very little on the market.

Copaifera Guibourtiana Benth. (syn. Guibourtia copallifera Benn.) (Leguminosaceae). - Tree. Trop. Africa. Source of Sierra Leone Copal. Used in varnishes.

Copaifera guyanensis (Desf.) Benth. (Leguminosaceae). - Tree. Amazon basin, Brasil, Giuana. Similar to C. coriacea Mart.

Copaifera hymenaefolia Moric. (Leguminosaceae). - Tree. West Indies. Wood close-grained, durable; used for railroad-ties, poles etc.

Copaifera Lansdorffii Desf. (Leguminosaceae). - Tree. Brazil. Wood employed locally for turnery, carriages, furniture, ship building. Bark is used for tanning. Source of a Copaiba Balsam. Used for skin diseases, eczema, gonorrhoea; also used in perfumery, for manuf. of pomades.

Copaifera mopane J. Kirk. (Leguminosaceae). - Tree. Trop. Africa. Source of a Congo Copal.

Copaifera officinalis L. (syn. C. Jacquini Desf.) (Leguminosaceae). - Venezuela, Guiana. Source of a resin, called Maracaibo Resin.

Copaifera reticulata Ducke. (Leguminosaceae). - Tree. Brazil. Source of Copaiba Balsam, Oleo Vermelho; used medicinally for healing wounds, for skin diseases and gonorrhea. Also C. guanensis (Desf.) Benth., C. multijuga Hayne, C. Martii Hayne, C. glycicarpa Ducke.

Copaifera Salikounda Heckel. (Leguminosaceae). - Tree. W. Africa. Source of a brittle, light-yellow, hard, Sierra Leone Copal derived from exudations of the stem.

Coontie → Zamia integrifolia Ait.

Copaiba Balsam → Copaifera reticulata Ducke.

Copal → Agathis loranthifolia Salisb.

Copal-Accra → Daniella Ogea Rolfe.

Copal- Anime → Trachylobium Hornemannianum Hayne.

Copal, Boea → Agathis alba Foxw.

Copal, Brazil → Hymenaea Courbaril L.

Copal, Cameroon → Copaifera Demeusii Harms.

Copal, Colombia → Hymenaea Courbaril L.

Copal, Congo → Copaifera Demeusii Harms.

Copal de Penca — Bursera jorullensis Engl.

Copal, East Indian → Canarium bengalense Roxb.

Copal-Gum of the Gold Coast → Daniella similis Craib.

Copal, Kauri → Agathis australis Steud.

Copal, Illorin → Daniella thurifera J. J. Benn.

Copal, Indian → Vateria indica L.

Copal-Inhambane → Copaifera Gorskiana Benth.

Copal, Madagascar → Trachylobium Hornemannianum Hayne and T. verrucosum Oliv.

Copal, Mozambique → Trachylobium Hornemannianum Hayne.

Copal, Niger → Daniella oblonga Oliv.

Copal, Ogea → Daniella Ogea Rolfe.

Copal, Sierra Leone — Copaifera Guibourtiana Benth.

Copal Tree → Daniella similis Craib.

Copal Tree, Congo → Copaifera Demeusii Harms.

Copal, Zanzibar → Trachylobium Hornemannianum Hayne.

Copernicia australis Becc. (Palmaceae). - Tall palm. Trop. S. America. Trunk resistant and durable; used for telegraph poles, construction work. Leaves used for covering huts.

Copernicia cerifera Mart. (Palmaceae). - Tall palm. S. America, esp. Brazil. Leaves are source of Carnauba Wax, has a high melting point, very hard; used for phonograph records, high luster wax varnishes, polishing pastes, candles. Grades of the wax are divided into yellow, yellow prime, yellow medium, grey fatty and chalky. Much is exported from Ceara, Para and Pernambuco, Brazil.

Copernicia tectorum Mart. (Palmaceae). - Tall Palm. Colombia. Leaves are used for manuf. hats. Trunk is employed for beams of houses in the country. Also C. Sancta Martae Becc.

Coprinus ater Copel. (Agaricaceae). - Basidiomycete. Fungus. Malayan region. Fruitbodies are consumed as food by the natives of the Philippine Islands. Also C. Bryanti Copel, C. concolor Copel, C. confertus Copel and C. rimosus Copel.

Coprinus atramentarius (Bull.) Fr. (Agaricaceae). - Basidiomycete. Fungus. Temp. zone. Old blackish fruitbodies are manufactured into an ink, having the appearance of Chinese ink. Has been recommended for signing important documents, as it can be microscopically easily identified on account of the presence of the spores. Has been advised for police work.

Coprinus comatus (Fn.) Gray. (Agaricaceae). - Basidiomycete Fungus. Temp. zone. The fruitbodies, called Shaggy Mane, have a fine flavor, though they must be eaten soon after picking, as they would turn into a blackist material. This material is sometimes used for manuf. a black ink, though it is of a rather poor quality. It is used for retouching.

Coprinus fimetarius (L.) Fr. (Agaricaceae). - Basidiomycete. Fungus. N. America. Fruitbodies are edible.

Coprinus macrorhizus (Pers.) Rea. (Agaricaceae). - Basidiomycete. Fungus. Trop. Asia. Fruitbodies are much esteemed as a delicious food by the natives of different parts of the Malayan Archipelago.

Coprinus micaceus (Bull.) Fr. (Agaricaceae). - Basiliomycete. Fungus. Temp. zone. Fruitbodies are edible, having a delicate flavor.

Coprinus microsporus Berk. and Broome. (Agaricaceae). - Basidiomycete. Fungus. Trop. Asia. Fruitbodies are consumed as food by the natives of Mal. Archipelago.

Coptis anemonaefolia Sieb. and Zucc. (Ranunculaceae). - Herbaceous perennial. Japan. Herb is used medicinally in Japan.

Coptis brachypetala Sieb. and Zucc. (Ranunculaceae). - Perennial herb. Japan. Roots are source of a yellow dye; used in Japan. Also C. occidentalis Nutt.

Coptis chinensis Wils. (Ranunculaceae). - Herbaceous perennial. China. Used in Chinese medicine as tonic and blood-purifier; infusion for dyspepsia.

Coptis Teeta Wall. (Ranunculaceae). - Small herbaceous herb. Mountains, Upper Assam. Used as eye-salve. Bark of root contains berberine. Used in native medicine as bitter tonic. Sold in bazaars.

Coptis trifolia (L.) Salisb., Goldthread. (Ranunculaceae). - Small herbaceous perennial. N. America. Entire dried plant is used medicinally as stomachic, tonic, has similar action as Hydrastis. Used locally for ulcerated mouth. Contains berberine and coptin, being alkaloids.

Coquitos → Jubaea chilensis (Molina) Baill.

Coral Peony → Paeonia corallina Retz.

Coral Tree → Erythrina indica Lam.

Corchorus capsularis L., Jute Plant. (Tiliaceae). - Annual plant. Tropics. Cultivated in warm countries of the Old and New World. Bark is source of a fibre, called Jute. Of much commercial importance. It is obtained by retting in pools and tanks. Jute is used for gunny and potato sacks, burlap, coarse cloth, twine, curtains, carpets. Short fibres from lower ends of stems form jute butts used in manuf. of paper. Much is derived from E. India. It is an upland species.

Corchorus olitorius L. (Tiliaceae). - Herbaceous plant. Tropics. Cultivated. Source of a fibre, similar to that of C. capsularis. Leaves used in W. Africa as a pot-herb, eaten as substitute of spinach by Europeans. It is a lowland species.

Corchorus siliquosus L. (Tiliaceae). - Herbaceous to woody shrub. Texas, West Indies, Centr. America, N. of S. America. Leaves are used as substitute for Chinese tea.

Corchorus tridens L. (Tiliaceae). - Woody plant. Tropics. Leaves when boiled are consumed by the natives in trop. Africa.

Cordia alba (Jacq.) Roem. (Boraginaceae). - Shrub or small tree. Mexico, West Indies to Colombia, Venezuela. Wood strong, hard, yellow; used for carpentry. Concoction of flowers is used to induce perspiration. Fruits are used in Oaxaca to coagulate indigo.

Cordia alliodora (Ruiz and Pav.) Cham. (Boraginaceae). - Tree. Mexico, West Indies, Centr. and S. America. Wood used for cabinet work, ceiling, flooring, beams. Decoction of leaves is stomachic, tonic; used for catarrh. Ointment of pulverized seeds used in West Indies for cutaneous diseases. Fruits are used as food by a number of Indian tribes in Mexico.

Cordia Boissieri DC. (Boraginaceae). - Shrub or small tree. S.W. of United States and Mexico. Jelly of fruits is used for coughs and colds. Decoction of leaves is a domestic remedy for the same ailments.

Cordia dodecandra DC. (Boraginaceae). - Tree. Mexico, Guatemala. Cultivated for its edible fruits. Wood used for making furniture.

Cordia gerascanthus L. Capá Prieta. (Boraginaceae). - Tree. West Indies, Porto Rico. Wood rose colored; used for doors, Venetian blinds, sweeps of sugarmills, furniture, boats.

Cordia Irwingii Baker. (Boraginaceae). - Tree. W. Africa. Wood very durable; used for shingles.

Cordia Myxa L. (Boraginaceae). - Tree. Greece, Turkey. Viscous pulp from fruits is used for catching birds; used in S. Iran as demulcent; in Orient for coughs and chest complaints. Wood yellow brown, easy to polish; used for cabinet work. Known as Karthoum or Sudan Teak.

Cordia Rothii Roem. and Schultes. (Boraginaceae.) - Shrub or tree. Trop. Africa, Arabia, India. Fruits are pickled and eaten in India. Bark is source of a fibre; used for ropes. Wood is used in India for agricultural implements and general building material. A gum from the plant is used to adulterate Gum Arabic.

Cordia subcordata Lam. (syn. C. Sebestiana Forst.) (Boraginaceae). - Tree. Tropics. Wood is used by the Hawaiians for rafts. Fruits serve for pasting their lapa clothing.

Cordials. Plants used in → Acanthosyris, Achillea, Agave, Anacyclus, Angelica, Anthemis, Arbutus, Armeniaca, Artemisia, Athamanta, Baccaurea, Carum, Citrus, Cuminum, Daucus, Elettaria, Erythraea, Foeniculum, Geum, Hyssopus, Inula, Laserpitium, Levisticum, Mammea, Marrubium, Matricaria, Melissa, Mentha, Pimpinella, Prunus, Salvia, Sorbus, Zipyphora. See also under Flavoring, Spices, Essential Oils.

Cordyceps sinensis (Berk.) Sacc. Ch'ung-tsao. (Chin.) (Hypocreaceae). - Ascomycete. Fungus. The mycelium parasites in a caterpillar, Hepialus sp., found in the western uplands of China. Used in China as antidote for opium poisoning and cure for opium haling; also as tonic and stimulant for convalescent people.

Cordylanthus Wrightii Gray., Club Flower. (Scrophulariaceae). - Annual herb. W. Texas to Arizona. Is used by Hopi Indians for bleaching.

Cordyline australis Hook. f. Palm Lily. (Liliaceae). - Tree. New Zealand. Leaves source of a fibre; used by settlers in place of twine. Leaves used for paper.

Cordyline indivisa Steud. (Liliaceae). - Tree. New Zealand. A fibre derived from the leaves is used by the Maoris for making garments.

Cordyline terminalis Kuenth. (syn. C. fruticosa Goep., Taetsia ferrea Medic.). Palm Lily. (Liliaceae). - Shrub or small tree. India, Pacific Islds., Australia. Sweet roots were a source of a fermented drink among the aboriginal Hawaiians. In Samao leaves are made into fringed skirts. Leaves are used in Polynesian Islds. for wrapping fish, before being put in the ovens to bake.

Coreopsis cardaminefolia (DC.) Torr and Gray. (Compositae). - Annual herb. Louisiana to New Mexico and Kansas. Decoction of plant was used as beverage by the Zuni Indians.

Coriander → Coriandrum sativum L.

Coriander Seed Oil → Coriandrum sativum L.

Coriandrum sativum L., Coriander. (Umbelliferaceae). - Annual herb. Mediterranean region, Caucasia. Cultivated. Fruits used for flavoring, in sausages, soups, cakes, liqueurs, gin essences, spicy sauces. Source of Coriander Seed Oil, ess. oil used for flavoring canned soups and foods, confectionary. Also used in perfumes, harmonizing well with jasmine; also in manuf. soap. Dried ripe fruits used medicinally; being carminative, aromatic. Oil of Coriander or Oleum Coriandri contains coriandrol, d-x-pinene, z-pinene.

Coriaria myrtifolia L. (Coriariaceae). - Tree. Mediterranean region. Bark and leaves are used for tanning.

Coriaria ruscifolia L. (Coriariaceae). - Shrub. New Zealand, Chile, Peru. Plant is source of a black dye.

Coriaria sarmentosa Forst. Tuhu. (Coriariaceae). - Small tree. New Zealand. Shoots and seeds are poisonous to livestock. Berries are source of a beverage prepared by the Maoris.

Coriaria thymifolia Humb. and Bonpl. (Coriariaceae). - Woody plant. S. America. Berries are source of a black ink; used in some parts of Colombia.

Coriariaceae → Coriaria.

Coridothymus capitatus Reichb. → Thymus capitatus Hoffm. and Link.

Cork Oak → Quercus Suber L.

Cork Tree, Amur → Phellodendron amurense Rupr.

Cork Tree, Sachalin → Phellodendron sachalinense Sarg.

Corks → Parmelia omphalodes (L.) Ach.

Corkwood → Duboisia myoporoides R.Br., Leitneria floridana Chapm. and Myrianthum arboreum Beauv.

Cork Wood Tree → Entelea arborescens R.Br.

Cormigonum mariannensis Raf → Bikkia mariannensis Brogn.

Corn → Triticum and Zea mais L.

Corn Cuckle → Agrostemme Githago L.

Corn, Indian → Zea Mays L.

Corn, Pop → Zea Mays L.

Corn Poppy → Papaver Rhoeas L.

Corn Salad → Valerianella olitoria Pollich.

Corn Salad, Italian → Valerianella erioparpa Desv.

Corn, Squirrel → Dicentra canadensis (Gold.) Walp.

Corn, Turkey → Dicentra canadensis (Gold.) Walp.

Cornaceae → Alangium, Cornus, Curtisia, Griselinia Helwingia.

Cornelian Cherry → Cornus Mas L.

Cornurlaca monacantha Del. (Chenopodiaceae). - Shrubby plant. Desert regions of W. Africa. A valuable fodder for camels in desert areas.

Cornus alternifolia L. f., Pagoda Dogwood. (Cornaceae). - Shrub or small tree. N. America. Wood hard, close-grained, heavy, brown; adapted for turnery.

Cornus florida L., Flowering Dogwood. (Cornaceae). - Small tree. Eastern N. America. Wood hard, close-grained, heavy, strong, brown; used for hubs of small wheels, handles of tools, barrel-hoops, bearings of machinery, turnery, occasionally for engraver's blocks. Root of bark is astringent, tonic, antiperiodic. Contains cornin, a bitter principle.

Cornus Mas. L. (syn. C. mascula Hort.) Cornelian Cherry. (Cornaceae). - Shrub or small tree. Europe, Asia Minor. Fruits edible slightly sweet, agreeable taste; eaten fresh, also as preserves or marmelade. Fruits are also source of an alcoholic beverage, known in France as Vin de Cornoulle.

Cornus Nuttallii Aud. Western Flowering Dogwood. (Cornaceae). - Tree. Pacific Coast region of N. America. Wood very hard, close-grained, heavy, strong, light brown; used for handles of tools, mauls and cabinet work. Bark is used in place of quinine as a home remedy.

Cornus stolonifera Michx. (Cornaceae). - Shrub. N. America. Inner bark used by Indians for smoking; with or without tobacco.

Cornus suecica L. (Cornáceae). - Low herbaceous plant. N. America, N. Europe, Asia. Fruits small, red; consumed by the Eskimos.

Coromandel Ebony → Diospyros hirsuta L. f.

Corossol → Annona muricata L.

Correa alba Andr. (syn. C. cotinifolia Salisb.) (Rutaceae). - Tree. Tasmania. Leaves occasionally used as tea. Called Cape Barren Tea.

Corrigiola littoralis L. (syn. Corrigiola telephiifolia Pourr.) (Caryophyllaceae). - Herbaceous plant. Europe, Mediterranean region. Pulverized root is used in Morocco in perfumes. Is also a tonic and used for gastralgia.

Corsican Moss → Alsidium Helminthochorton (de la Tour). Kuetz.

Cortaderia argentea Stapf. (syn. Gynerium argenteum Nees.). Pampas Grass. (Graminaceae). - Perennial grass. S. America. Cultivated. Plant is used in some parts of S. America for manuf. paper.

Corteza de Ojé → Ficus anthelmintica Mart.

Cortinarius emodensis Berk. (Agaricaceae). - Basidiomycete. Fungus. Himalayas. Fruitbodies called Onglau are consumed as food. Also C. ararmillatus Fr., C. elatus (Pers.) Fr., C. fulgens (A. and S.) Fr., C. latus (Pers.) Fr., C. multiformis Fr. and C. violaceus (L.) Fr.

Cortinellus shiitake Henn. → Lentinus edodes (Bernk.) Sing.

Corydalis ambigua Cham. and Schlecht. (Fumariaceae). - Herbaceous perennial. Kamschatka. Japan. Small tubers are eaten by the inhabitants of Ainu.

Corydalis solida Swartz. (Fumariaceae). - Perennial herb. Europe, temp. Asia. The small tubers are consumed as food in some parts of Siberia.

Corydalis bulbosa DC. (Fumariaceae). - Perennial herb. Temp. Europe, Asia. Tubers are boiled and eaten by the Kalmuks.

Corylus Avellana L., European Hazel (Betulaceae). - Shrub. Europe, Temp. Asia. Widely cultivated in temperate zones of Old and New World. Hazel Nut, Cob Nuts, Barcelona Nuts of much commercial value, eaten raw, in confectionary, as „noces" in oil. Source of a clear, yellow oil used in food, for painting, perfumes, fuel, in machineries, for manuf. soap. Wood reddish-white, soft, easy to split, not very durable; used for handles, sieves, parts of aeroplanes, walking sticks; source of charcoal made into gun powder. The principal varieties are: White Filbert, Cosford, Dowton Large Square, Grandis. Hazel Nut Oil, Filbert Oil, a non-drying oil has Sp. gr. 0.917; Sap. Val. 190-197; Iod. No. 84-90; Unsap. 0.3-0.5%. Leaves are sometimes used for smoking during times of lack of tobacco.

Corylus Colurna L., Turkish Hazel. (Betulaceae). - Tree. S. E. Europe, to S.W. Siberia. Occasionally cultivated for its nuts, called Constantinople Nut or Filbert of Constantinople.

Corylus heterophylla Fisch. Siberian Hazelnut. (Betulaceae). - Shrub or small tree. S.W. China. Cultivated in China. Fruits are consumed by the natives.

Corylus rostrata Ait., Beaked Hazelnut. (Betulaceae). - Shrub. Eastern N. America. Nuts are used as food. Also C. americana Walt.

Corylus tubulosa Willd. Lambert's Filbert. (Betulaceae). - Shrub or small tree. S. Europe. Occasionally cultivated for its nuts. Some varieties with large fruits are: Ceret, Large Long Spanish.

Corynanthe paniculata Welw. (Rubiaceae). - Large tree. Trop. Africa. Wood durable, hard, white, very dense, fine-grained; used for building purposes.

Corynanthe Yohimbe Schum. (Rubiaceae). - Tree. W. Africa. Source of Yohimbe Bark, containing Yohimbine, an alkaloid.

Corynocarpus laevigata Forst. (Anacardiaceae). - Tree. New Zealand. Fruits are consumed by the aborigenes of New Zealand. Trunks are used by the Maoris for canoes. Seeds are an important staple food among the natives.

Corynostylis arborea (L.) Blake. (syn. Viola arborea L., C. Hybanthus Mart. and Zucc.) (Violaceae). - Shrub. Mexico, esp. Veracruz to Yucatan. into S. America. Root used in S. America as an emetic.

Corypha elata Roxb. (Palmaceae). - Tall palm. E. India to Philippines. Sap is source of a palmwine (Tuba), alcohol, vinegar, sugar and syrup.

The trunk produces a starch. Buds are used as salad and as vegetable. Kernels of young fruits are made into sweet-meats. Mature seeds are made into buttons and rosaries. Petioles are source of Buntal-Fibre from which Lucban Hats are made. Leaves are used for covering tobacco-bales. Unfolded leaves produce a very fine fibre, used for cloth, fancy articles and as string. Fibres from ribs of unfolded leaves are used for manuf. Calasiao or Pototan Hats.

Corypha umbraculifera L., Talipot Palm. (Palmaceae). - Tall palm. India, Malaya, Ceylon. Leaves are used in Ceylon for thatching, umbrellas and tents. Sacred Buddist Books were written on strips of the leaves. Seeds are made into buttons and similar comodities. Pounded young fruits are said to be used to stupefy fish.

Coscinium fenestratum Colebr. (Menispermaceae). - Vine. E. India. Wood source of a yellow dye, resembling tumeric. Root considered bitter tonic, stomachic.

Cosmetics. Some plants used in →Adenanthera, Asparagus, Astragalus, Betula, Blighia, Caesalpinia, Caryocar, Chaetospermum, Chenopodium, Cistus, Clematis, Cocos, Coleus, Conopharyngia, Copaifera, Cremaspora, Crossopteryx, Cymbopogon, Ecliptia, Entada, Euphorbia, Euvonymus, Feretia, Flagellaria, Gardenia, Gaultheria, Hibiscus, Humiria, Hyssopus, Iris, Juniperus, Lawsonia, Leea, Limnophila, Lysimachia, Machilus, Mallotus, Melilotus, Mesua, Michelia, Mirabilis, Monarda, Myroxylon, Nardostachys, Ocimum, Oryza, Parinarium, Picea, Pimenta, Podaxis, Psittacanthus, Salvia, Santalum, Saurauia, Sesamum, Sorindeia, Sphaeranthus, Urginea, Zea, Zygophyllum. See also under Essential Oils, Perfums.

Costa Rican Guava → Psidium Friedrichsthalianum Ndz.

Costa Rica Sarsaparilla → Smilax Regelii Killip and Morton.

Costmary → Chrysanthemum Balsamita L.

Costus → Saussurea Lappa C. B. Clarke.

Costus afer Ker-Gawl., Ginger Lily. (Zingiberaceae). - Perennial herb. Trop. Africa. Decoction of powdered fruits is used as cough medicin among the natives. Stem is used to alleviate nausea. Strips made from the outer stem are made into baskets.

Costus Lucanusianus Braun. and Schum. (Zingiberaceae). - Perennial herb. Trop. Africa, esp. S. Nigeria, Cameroons, Lower Congo. Decoction of plant is sometimes used by the natives to coagulate latex of Landolphia owariensis.

Costus Root Oil →Saussurea Lappa C. B. Clarke.

Costus spicatus (Jacq.) Swartz. (Zingiberaceae). - Perennial herb. S. Mexico, Central America to N. of S. America. Acid sap is used in a home-remedy as a diuretic.

Cotinus americanus Nutt. → Rhus cotinoides (Nutt.) Britt.

Cotinus coggygria Scop. → Rhus Cotinus L.

Cotoneaster racemiflora (Desf.) Koch. (syn. C. nummularia Fisch and Mey). (Rosaceae). - Shrub. N. Africa, Centr. Asia, esp. India, Afghanistan, Iran, Turkestan. Plant is source of Shir-Khist, a Manna-like substance, being whitish and sweet; used in India and Iran. It contains 13% saccharin and 37.5% dextrose.

Cotton Abroma → Abroma angusta L. f.

Cotton, American Upland → Gossypium hirsutum L.

Cotton, Arabian → Gossypium herbaceum L.

Cotton, Bahia → Gossypium brasiliense Macf.

Cotton, Chinese → Gossypium Nanking Meyen.

Cotton, Gallini → Gossypium barbadense L.

Cotton Gum → Nyssa aquatica L.

Cotton- Kathiawar → Gossypium obtusifolium Roxb.

Cotton, Khaki → Gossypium Nanking Meyen.

Cotton, Kidney → Gossypium brasiliense Macf.

Cotton, Kumpta → Gossypium obtusifolium Roxb.

Cotton, Levant → Gossypium herbaceum L.

Cotton, Maltese → Gossypium herbaceum L.

Cotton, Nanking →Gossypium Nanking Meyen.

Cotton, Pernambuco → Gossypium brasiliense Macf.

Cotton, Peruvian → Gossypium peruvianum Cav.

Cotton, Red Peruvian → Gossypium microcarpum Tod.

Cotton, Sea Island → Gossypium barbadense L.

Cotton Seed Oil → Gossypium herbaceum L. etc.

Cotton, Short Staple American → Gossypium herbaceum L.

Cotton, Siam → Gossypium Nanking Meyen.

Cotton, Surat → Gossypium obtusifolium Roxb.

Cotton, Syrian → Gossypium herbaceum L.

Cotton, Tree → Gossypium arboreum L.

Cotton Tree→Ochroma limonensis Rowlee.

Cottonwood, Swamp → Populus heteroophylla L.

Cotyledon edulis (Nutt.) Brewer. (Crassulaceae). - Herbaceous Plant. California. Tender leaves were eaten raw by Indians of California. Also C. lanceolata (Nutt.) Brewer and Wats., C. pulverulenta (Nutt.) Brewer and Wats.

Couepia polyandra (H.B.K.) Rose. (syn. Hirtella polyandra H.B.K.) (Rosaceae). - Shrub or tree. Mexico. Fruits are consumed by the natives, especially of the w. coast of Mexico. They are called Zapote Amarillo.

Coula edulis Baill. (Olacaceae). - Tree. Trop. Africa. Seeds are edible when fresh or cooked, they are also fermented by the natives to serve as a condiment with meat. Wood is brownish, easy to work, polishes well; used for turnery, as substitute for Mahogany.

Coulter Pine → Pinus Coulteri D. Don.

Couma guianensis Aubl. (Apocynaceae). - Tree. Guiana. Fruits edible, spherical, size of a Guava, pulp sweet, of pleasant flavor.

Couma macrocarpa Barb. (Apocynaceae). - Tree. Brazil. A source of a chewing-gum base. also C. guatemalensis Standl.

Couma utilis Muell. Arg. (Apocynaceae). - Tree. Brazil. Pulverized fruits are used in Brazil for skin parasites of humans and of animals.

Country Ipecacuanha →Naregamia alata Wight and Arn.

Country Mallow → Abutilon indicum Sweet.

Couranira → Humiria balsamifera St. Hil.

Couratari Tauari Berg., Tauary. (Lecythidaceae). - Tree. S. America. Bark is source of a cloth used by Indians in S. America.

Courbaril → Hymenaea Courbaril L.

Courbonia virgata Brongn. (Capparidaceae). - Shrub. Red Sea area, Trop. Africa. Fruits are edible. Ashes from stems and leaves are used as source of salt in parts of Sudan and Arabia.

Couroupita peruviana Berg., Aiauma (Peru). (Lecythidaceae). - Tree. Peru. Wood soft; its pulp is used after exposure by Indians for skin-diseases of animals.

Coutoubea Apicata Aubl. (syn. C. alba Lam.). (Gentianaceae). - Annual herb. Trop. America. Bitter root is used as stomachic, tonic, anthelmintic and febrifuge.

Covillea tridentata (DC.) Vaill → Larrea mexicana Moric.

Cowania mexicana D. Don. (syn. C. Stansburiana Torr.) (Rosaceae). - Shrub. Mexico to S. California. Thin, silky inner bark is used by the Indians of Utah and Nevada for clothing, also made into ropes, mats and sandals.

Cowberry → Vaccinium Vitis-Idaea L.

Cow Parsley → Chaerefolium sylvestre (L.) Shinz and Thell.

Cow Parsnip, Common → Heracleum Sphondylium L.

Cowpea → Vigna sinensis Endl.

Cowpea, Hindu → Vigna Catjang Walp.

Cow Tamarind → Pithecolobium Saman Benth.

Cow Tree → Brosimum utile (H.B.K.) Pitt.

Cowslip → Primula veris L.

Coxoba Snuff → Piptadenia peregrina Benth.

Coyal → Acrocomia mexicana Karw.

Coyo Avocado → Persea Schiedeana Nees.

Coyotillo, Calderon → Karwinskia Calderoni Standl.

Coyotillo, Humboldt → Karwinskia Humboldtiana (Roem. and Schult.) Zucc.

Crab Apple → Pyrus angustifolia Ait. and P. coronaria L.

Crabeye Lichen → Lecanora parella Mass.

Crab's Claw → Stratiotes aloides L.

Crab's Eye → Abrus precatorius L.

Crabwood → Carapa guineensis Aubl.

Crambe maritima L., Sea-Kale. (Cruciferaceae). Perennial herb. Seacost. Europe. Bleached leaf-stalks are consumed as a vegetable. Cultivated, esp. in England. Varieties: Purple-Tipped, Lily White.

Crambe tatarica Jacq., Tatarian Sea-Kale. (Cruciferaceae). - Perennial herb. S.E. Europe, S.O. Russia, W. Siberia. Bleached stems are occasionally consumed like Crambe maritima L.

Cranberry → Vaccinium macrocarpon Ait.

Cranberry, Mountain → Vaccinium erythrocarpum Michx.

Cranberry, Small → Vaccinium Oxycoccus L.

Cranberry Tree → Viburnum Opulus L.

Crane's Bill, Wood → Geranium sylvaticum L.

Craniolaria annua L. (Pedaliaceae). - Annual. Northern S. America. Decoction of the roots is used as a mild laxative among the natives. The fleshy roots when preserved in sugar were consumed as a delicacy among the Creoles in some parts of the West Indies.

Crape Myrtle Andaman → Lagerstroemia hypoleuca Kurz.

Crape Myrtle, Battinan → Lagerstroemia piriformis Koehne.

Crape Myrtle, Queen → Lagerstroemia Flos-Reginae Retz.

Crassulaceae → Cotyledon, Sedum, Sempervivum.

Crataegus Azarolus L. (syn. C. Aronica Bosc.). Azerolier (Rosaceae). - Shrub or small tree. S. Europe, Orient, N. Africa. Often cultivated in S. Europe and N. Africa. Fruits eaten fresh, also preserved. There are varieties with yellow, red and white fruits. Sold in markets.

Crataegus Crus-galli L., Cock-spur Thorn. (Rosaceae). - Small tree. Eastern N. America to Florida and westward. Wood heavy, very fine grained, hard; used for tool-handles.

Crataegus Douglasii Lindl. Black Haw. (Rosaceae). - Tree. Pacific Coast Region of N. America. Fruits pleasant sweet and juicy; used for jellies.

Crataegus crenulata Roxb. (syn. Pyracantha crenulata [Roxb.] Roem.) (Rosaceae). - Shrub or small tree. Leaves are used in China as a tea.

Crataegus flava Ait., Summer Haw. (Rosaceae). - Small tree. Southern United States. Yellowish, somewhat pear-shaped fruits are esteemed for jellies.

Crataegus hupehensis Wils. (Rosaceae). - Tree. Hupeh (China). Cultivated in Hupeh. Fruits esteemed in some parts of China, esp. near

Hsïngshan Hsien. Fruits are scarlet, $2^{1}/_{2}$ cm, diam. insipid.

Crataegus mexicana Moc. and Sessé. (syn. C. subserrata Benth., C. hyplosia Koch.). (Rosaceae). - Tree. Mexico. Fruits are used in Mexico as preserves and jellies.

Crataegus mollis (T. and G.) Scheele. (Rosaceae). - Tree. Centr. United States. Fruits used for jellies.

Crataegus oxyacantha L. (syn. Mespilus oxyacantha Crantz). May, Hawthorn. (Rosaceae). - Tree. Europe. Young leaves used as substitute for tea, used for „cleaning" the blood; also as substitute of tobacco. Seeds employed in World War as substitute for coffee. Dried fruit pulp was formerly added to flour in some countries of Europe. Decoction from leaves when frequently used, is supposed to diminish high blood pressure and has been recommended for arteriosclerosis. Wood reddish-white, hard, heavy, tough, difficult to split; used in turnery, wagons; straight twigs are made into walking-sticks.

Crataegus pentagyna Waldst. and Kit. (syn. C. pinnatifida Bunge). Chinese Haw. (Rosaceae). - Tree. N. China, Korea, Siberia. Cultivated in China as a fruit tree. Fruits are consumed when stewed, candied, preserved, in sweet meats or as jelly. Often sold in markets of China.

Crataegus stipulosa (H.B.K.) Steud., Manzanilla. (Rosaceae). - Tree. Guatemala to Ecuador. Cultivated. Fruits mealy, pleasant flavor; used for jellies and preserves; also stewed and made into syrup. Sold in markets.

Crataeva macrocarpa Kurz. (syn. Capparis magna Lour.) (Capparidaceae). - Small shrub. Trop. Asia esp. Cochin-China. Young leaves are eaten in Cochin-China as a vegetable.

Crataeva religiosa Forst. (syn. C. Nurvala Buch.). - (Capparidaceae). - Shrub or small tree. Trop. Africa, Asia. Bark stimulates appetite and is also a laxative, much used in India. Burmese consume pickled flowers as stomachic.

Craterispermum laurinum Benth. (Rubiaceae). - Shrub. Trop. Africa. Bark when beaten with grass is source of a yellow dye; used in S. Keone for dying clothes.

Crawfood Plantain → Plantago Coronopus. L.

Cream of Tartar Tree → Adansonia digitata L.

Creeper, Virginia → Parthenocissus quinquefolia (L.) Planch.

Creeping Thyme → Thymus Serpyllum L.

Crema de Caruba → Passiflora mollissima (H. B.K.) Bailey.

Crema de Mamay Colorado → Calocarpum mammosum Pierre.

Cremaspora africana Benth. (Rubiaceae). - Shrub. Trop. Africa. Fruits are source of a blue-black dye; used in W. Africa as a cosmetic for face and body.

Cremastra Wallichiana Lindl. Hakkuri. (Orchidaceae). - Perennial herb. Japan, China, Himalayas. Roots used by the Ainu for toothache.

Creosote → Fagus ferruginea Dryand., F. sylvatica L.

Creosote Bush → Larrea mexicana Moric.

Crepidotus Djamor (Fr.) v. Overeem. (syn. Agaricus Djamor Fr.). (Agaricaceae). - Basidiomycete. Fungus. Trop. Asia. Fruitbodies are much prized as food among the natives of the Malayan Archipelago.

Crepitus Lupi → Calvatia lilacina (Berk. and Mont.) Lloyd.

Crescentia Cujete L., Calabash Tree, Cujete. (Bignoniaceae). - Tree. Trop. America. Cultivated in Old World. Seeds eaten cooked. Pulp used in domestic medicin, emollient, laxative, astringent, expectorant. Shell of fruits used for drinking cups, often being decorated. Young fruits used as pickles, considered in Jamaica equal to pickled wallnuts. Three varieties are recognized in Guatemala; e. g.: Morro, Jicara and Guacal.

Cress, Bitter → Cardamine pennsylvanica Muhl.

Cress, Garden → Lepidium sativum L.

Cress, Penny → Thlaspi arvense L.

Cress, Water → Nasturtium officinale R. Br.

Cress, Winter → Barbarea praecox R. Br.

Cressa cretica L. (Convolvulaceae). - Perennial herb. Tropics. Plant is used as a tonic in some parts of the Sudan.

Crested Dog's Tail → Cynosurus cristatus L.

Cretian Dost → Origanum Onites L.

Crimson Clover → Trifolium incarnatum L.

Crinkleroot → Dentaria diphylla Michx.

Crinum flaccidum Herb. (Amaryllidaceae). - Bulbous perennial. Australia. Produces a kind of arrowroot; eaten by the natives.

Crin Végétale → Chamaerops humilis L. and Tillandsïa usneoides, L.

Crithmum maritimum L., Samphire. (Umbelliferaceae). - Perennial herb. Temp. zone. Salty leaves are eaten in salads and as condiment; also used as capers and pickles in vinegar.

Crocosmia aurea Planch. (syn. Tritonia aurea Papp.). (Iridaceae). - Perennial herb. Trop. Africa. Flowers are source of a yellow dye that can be used as a substitute for saffron.

Crocus sativus L. Saffron Crocus. (Iridaceae). - Perennial bulbous herb. Centr. and S. Europe, Asia Minor, Iran. Cultivated in France, Spain and other countries. Styles of flowers are source of saffron, 100 000 flowers produce 1 kg saffron. Used as a coloring agent, contains crocin, a yellow glucoside. Used medicinally to promote eruptions in measles; emmenagogue and diaphoretic. Employed as a home-remedy.

Cromwell, Hoary → Lithospermum canescens Lehm.

Crookneck, Canada → Cucurbita moschata Duch.

Crookneck, Winter → Cucurbita moschata Duch.

Crossopteryx Kotschyana Fenzl. (Rubiaceae). - Small tree. Trop. Africa. Bark called African Bark; used in S. Leone as febrifuge. In Mahi seeds are used by natives to fumigate their bark-clothes. Seeds when powdered are made into a pomade; used for rubbing the body. Wood light brown, very hard, fine structure; employed by the natives for tables containing expressions of the Koran.

Crotal → Ochrolechia tartarea (L.) Mass.

Crotalaria anagyroides H. B. K. (Leguminosaceae). Annual herb. Venezuela. Grown as a cover crop and green manure in warm countries.

Crotalaria Burhia Buch.-Ham. (Leguminosaceae). - Herbaceous plant. E. India. Cultivated in Shind. Stems source of a fibre; used for manuf. sails.

Crotalaria cannabina Schweinf. (Leguminosaceae). - Herbaceous plant. Sudan. Cultivated. Bark is source of a fibre; used for manuf. of ropes by the Bongos (Afr.).

Crotalaria glauca Willd. (Leguminosaceae). - Herbaceous. Trop. Africa. Leaves, flowers and pods are eaten by the natives.

Crotalaria guatemalensis Benth. (syn. C. Carmioli Polak). (Leguminosaceae). - Herbaceous plant. Panama to Mexico. Leaves and young branches are occasionally used in Centr. America as pot-herb.

Crotalaria intermedia Kotschy (Leguminosaceae). - Annual herb. Trop. Africa. Bark is source of a strong fibre; used for manuf. string by the Bangos (Afr.).

Crotalaria juncea L. Sunn or San Hemp. (Leguminosaceae). - Annual herb. Origin uncertain. Cultivated since ancient times. Bark source of a strong fibre, more enduring than Jute; used for cordage, canvas, fishing-nets. Also employed as green-manure.

Crotalaria retusa L. (Leguminosaceae). - Herbaceous herb. Tropics. Source of a fibre; used for cordage and canvas. Cultivated.

Crotalaria Saltiana Andr. (Leguminosaceae). - Herbaceous plant. E. Africa. Bark is source of a fibre; used in Bahr-el-Ghazal, Afr. for manuf. nets to catch big game.

Crotalaria striata DC. (Leguminosaceae). - Annual herb. Tropics. Used as a cover crop and green manure in tropical and subtropical countries.

Crotalaria usaramoensis Baker. (Leguminosaceae). - Annual herb. Trop. Africa. Grown as a cover crop and green manure in warm regions.

Croton alanosanus Rosa. (Euphorbiaceae). - Shrub. Mexico, esp. Sonora to Oaxaca. Resin is used as remedy for toothache.

Croton caudatus Geisel. (Euphorbiaceae). - Stragling shrub. India, Malaysia, Concoction of roots is used to relieve constipation.

Croton Cortesianus H.B.K. (syn. Croton trichocarpus Torr.) (Euphorbiaceae). - Shrub. Mexico. Juice is used as caustic for treatment of skin diseases.

Croton corymbulosus Rottr. (Euphorbiaceae). - Herbaceous plant. S.W. of United States. Infusion of flowering tops are made into a beverage, called Chaparral Tea.

Croton Draco Schleicht. Sangre de Drago. (Euphorbiaceae). - Tree. Mexico, esp. San Luis Potosi and Veracruz to Chiapas. Bitter bloodred sap is source of a red dye; used as remedy for hoof diseases of horses.

Croton echinocarpus Muell. Arg. (Euphorbiaceae). - Shrub. Brazil. Bark is used in Brazil in bitters, called Industrial Cascarille; strong drastic, anthelmintic. Decoction is used by Indians of Brazil for ailments of the eyes incl. of trachoma. Also C. floribundus Spr.

Croton Eluteria Benn. (Euphorbiaceae). - Treelike shrub. West Indies. Source of Cascarilla Bark, usually supplied by the Bahamas. Dried bark is used as a tonic and aromatic bitter. Contains a volatile oil.

Croton humilis L. (Euphorbiaceae). - Small shrub. Trop. America. Sometimes cultivated. Roots called Botaosinho; used in Brazil for skin and urinary diseases. Also C. campestre St. Hil. and some unidentified species.

Croton linearis Jacq. (Euphorbiaceae). - Shrub. Florida, West Indies. Leaves are used for tea.

Croton macrostachys Hochst. (Euphorbiaceae). - Shrub. Abyssinia. Juice from leaves is used by the natives as an anthelmintic.

Croton niveus Jacq. (syn. C. pseudo-china Schlecht.) (Euphorbiaceae). - Shrub or tree. Mexico, Centr. America, Northern S. America, West Indies. Bark substitute for Cascarilla Bark, known as Copalchi or Quina Blanca; used for intermittent fevers.

Croton reflexifolius H.B.K. (Euphorbiaceae). - Small tree. Mexico to Colombia. Used as febrifuge and as a tonic in some parts of Centr. America.

Croton setigerus Hook., Turkey Mullein. (Euphorbiaceae). - Herbaceous plant. S.W. of United States. Macerated stems and leaves are used to stupefy fish.

Croton Tiglium L. Purging Croton. Euphorbiaceae. Small tree. Trop. Asia. Cultivated. Seeds source of Croton Oil, being drastic purgative, causes often pustular eruptions on the skin. Seeds are used to stupefy fish.

Croton xalapensis H. B. K. (Euphorbiaceae). - Shrub or small tree. Trop. America. Trunk exudes a gum which is used in some parts of Mexico for cleening the teeth.

Croton Oil → Croton Tiglium L.

Crottle → Ochrolechia tartarea (L.) Mass.

Crottle, Black → Parmelia omphalodes (L.) Ach.

Crottle, Dark → Parmelia physodes (L.) Ach.

Crottle, White → Pertusaria corallina (L.) Th. Fr.

Crowberry → Empetrum nigrum L.

Crowfoot, Celery Leaved → Ranunculus sceleratus L.

Crucillo → Condalia lycioides Gray.

Crucito → Basanacantha armata Hook. f.

Cryptocarya Canelilla H.B.K. → Aniba Canelilla (H.B.K.) Mez.

Cryptocarya membranacea Thw. Gal-mora. (Lauraceae). - Large tree. Ceylon. Wood close-grained, easily worked, straw yellow. Used for sas bars, picture frames.

Cryptocarya minima Mez. → Aydendron floribundum Meisn.

Cryptocarya nitida Phil. (syn. Bellota Miersii C. Gay.). (Lauraceae). - Tree. Chile. In some parts of Chile where the plant is abundant, it is source of a fibre.

Cryptocarya Peumus Nees. (syn. Laurus Peumus Domb.). (Lauraceae). - Tree. Chile. Bark is used for tanning in some parts of Chile.

Cryptococcaceae → Candida.

Cryptonemia decumbens Weber. (Grateloupiaceae). - Red Alga. Pac. Ocean. Used as food among the natives of the Polynesian Islands.

Cryptomeria japonica D. Don. (Taxodiaceae). - Large tree. Japan. Wood fine-grained, light, reddish, brown; used for house-building, shipbuilding, bridges, boxes, tubs etc. Leaves are source of incense-sticks. Old wood when buried in soil, becomes dark green, which is much esteemed, and known as Jindai-sugi.

Cryptoporus volvatus (Peck) Shear. (Polyporaceae). - Basidiomycete. Fungus. North America and Eastern Asia. Fruitbodies growing on wood of Conifers are eaten when raw by the Tibetans.

Cryptostegia grandiflora R. Br. (Asclepiadaceae). - Woody vine. Trop. Africa or India. Often cultivated as ornamental. Source of Palay Rubber. Occasionally cultivated commercially. Cultivated as source of rubber in World War II.

Cryptostegia madagascariensis Bojer., Lombiro. (Asclepiadaceae). - Shrubby vine. Madagascar. Bark is used by the Sakalava (Madag.) for manuf. of rum; A fibre is employed for making fishing nets. Latex contains 7 to 10% rubber which is of slight commercial value.

Cryptotaenia canadensis DC. (Umbelliferaceae). - Perennial herb. N. America. Cultivated in Japan. Young leaves, are eaten in the spring when boiled; roots are eaten in Japan when fried.

Crystal Tea Ledum → Ledum palustre L.

Ctenolophon parvifolium Oliv. (Linaceae). - Tree. Malaya. Wood durable, hard; used for house construction.

Cuba Hemp → Furcraea cubensis Vent.

Cube Gambir → Uncaria Gambir Roxb.

Cubeba → Piper Cubeba L.

Cubebs → Piper Cubeba L.

Cruciferaceae → Barbarea, Brassica, Bunias, Cakile, Capsella, Cardamine, Cheiranthus, Cochlearia, Crambe, Dentaria, Diplotaxis, Eruca, Erucaria, Eustrema, Hesperis, Isatis, Lepidium, Lunaria, Morettia, Nasturtium, Pringlea, Pugionum, Raphanus, Senebiera, Sinapis, Sisynbrium, Sophia, Stanleya, Stanleyella, Thlaspi, Zilla.

Cuckoo Flower → Cardamine pratensis L.

Cucúa → Poulsenia armata (Miq.) Standl.

Cucubalus baccifer. L. (syn. C. baccatus Gueld.) Bladder Campion. (Caryophyllaceae). - Perennial herb. Centr. and S. Europe, Caucasia, S. Siberia. Herb was formerly used as an astringent.

Cucumber → Cucumis sativus L.

Cucumber, Mandera → Cucumis Sacleuxii Paill. and Bois.

Cucumber Root, Indian → Medeola virginiana L.

Cucumber, Squirting → Ecballium Elaterium A. Rich.

Cucumber tree → Magnolia acuminata L.

Cucumis Anguria L., West Indian Gherkin, Gooseberry Gourd. (Cucurbitaceae). - Annual vine. Probably from American tropics. Young fruits are eaten boiled; are also frequently used in pickles.

Cucumis conomon Roxb., Oriental Pickling Melon. (Cucurbitaceae). - Annual vine. China, Japan. Very young fruits are used in soup; they are also put into pickles.

Cucumis Melo L., Melon, Muskmelon, Canteloupe. (Cucurbitaceae). - Annual vine. Probably from S. Asia, trop. Africa. Cultivated in Old and New World. Fruits eaten for breakfast or dessert. Numerous varieties which are divided in the following groups. I. Jenny Lind group: Jenny Lind, Fordhook, Champlain. II. Netted Gem group: Netted Gem, Early Watters, Sugar Sweet. III. Pollock group: Golden Pollock, Netted Rock, Eden Gem, Salmon Tint. IV. Hackensack group: Long Island Beauty, Baltimore Market, Large Hackensack. V. Tip Top group: Tip Top, Irondequoit, Bender's Surprise, Ohio Sugar. VI. Osage group: Extra Early Osage, Miller's Cream, Emerald Gem. VII. Hoodoo group: Hearts of Gold, Defender. VIII. Burrell's Gem Group: Ordway Pink Meat, Pink Queen, Burrell's Gem. IX. Persian Group: Persian. Botanically this species is devided into var. reticulatus Naud. Netted and nutmeg melons; var. cantasoupensis Naud. European canteloupe me-

lons; var. saccharinus Naud. Pineapple melons, sucrins: var. inodorus Winter melons, Honey Dew, Casaba; var. chito Naud. Small fruits. Orange Melon, used for making preserves; var. dudaine Naud. Pomegranate Melon, Queen Anne's Pocket Melon; var. cocomon Naud. Oriental Pickling Melon.

Cucumis metuliferus E. Mey. (Cucurbitaceae). - Vine. Trop. Africa. Cultivated. Fruits are used in salads in some parts of Africa.

Cucumis prophetarum L. (Cucurbitaceae). - Annual vine. Trop. Africa. Bitter fruit is used in Sudan as an emetic.

Cucumis Sacleuxii Paill. and Bois., Mandera Cucumber. (Cucurbitaceae). - Annual vine. Zanzibar. Young fruits are preserved as pickles.

Cucumis sativus L., Cucumber. (Cucurbitaceae). - Annual vine. S. Asia. Cultivated in Old and New World. Used as salad. Small fruits called Gherkins are much used as pickles. Numerous varieties, some are green, others yellow to almost white. I. Black Spine varieties: Netted Russian, Everbearing, New Siberian. Early Cluster type: Early Flame, Early Cluster, Medium Green type: Chicago Pickling Nichols Medium. Long Green type: Japanese Climbing, Long Green. II. White Spine varieties: White spine type: White Wonder, Davis Perfect, Fordhook Famous. English Forcing type: Telegraph, Lorne, Edinburgh.

Cucumis trigonus Roxb. (Cucurbitaceae). - Perennial vine. Australia. Fruits are consumed by the natives.

Cucurbita ficifolia Bouché. (syn. C. melanosperma A. Br.). (Cucurbitaceae). - Perennial vine. E. Asia. Cultivated in some parts of the tropics of the Old and New World. Fruits are consumed boiled and preserved.

Cucurbita foetidissima H. B. K., Wild Gourd. (Cucurbitaceae). - Creeping vine. S.W. of United States and adj. Mexico. Ground roots in cold water are used as laxative among the Teiwa Indians of New Mexico. Seeds are used as food by the Luiseño Indians (Calif.). Ripe fruits are employed as substitute for soap.

Cucurbita maxima Duch. Squash. (Cucurbitaceae). - Annual vine. Origin uncertain, probably from America. Cultivated, already known among the Indians before pre-Columbian times. Fruits are consumed. Varieties are divided into I. Banana Group to which belong: Winnebago, Banana, Gilmore, Alligator. II. Hubbard Group containing: Blue Hubbard, Chicago Warted Hubbard, Delicious, Ironclad, Marblehead. III. Turban Group: Essex Hybrid, Victor, Warren, American Turban. IV. Mammouth Group: Estampes, Genuine Mammoth, Mammoth Chili, Atlas.

Cucurbita moschata Duch., Cushaw, Canada Crookneck, Winter Crookneck. (Cucurbitaceae). - Annual vine. Probably from Trop. Asia. Cultivated. Fruits are consumed boiled. They are divided in I. Cheese Group: Large Cheese, Quaker Pie, Calhoun. II. Cushaw Group: Tennessee White Potato, Mammoth Golden Cushaw, Small Garden Cushaw, Japanese Pie, White Cushaw.

Cucurbita Pepo L., Pumpkin. (Cucurbitaceae). - Annual vine. Origin uncertain, probably from the Himalayan region. Cultivated. Fruits are eaten boiled as a vegetable, also used in pies. They are divided in I. Connecticut Field Group: Sugar Pumpkin, Pie Pumpkin, Connecticut Field. II. Fordhook Group: Perfect Gem, Winter Nut, Fordhook, Table Queen. III. Patty Pan Group: White Bush, Scallop, Golden Custard. IV. Crookneck Group: Yellow Summer, Crookneck, White Summer Crookneck. V. Vegetable Marrow Group: Cocozella, Zucchini, Long White Marrow. Seeds are source of Pumpkin Seed Oil, a semi-drying oil, used for illumination and in food. Sp.Gr. 0.9159; Sap. Val. 174.2; Iod No. 116.8; Unsap. 1.58%. Dried, ripe seeds are used medicinally as anthelmintic, taenifuge.

Cucurbitaceae → Acanthosicyos, Adenopus, Anisoperma, Apodanthera, Benincasa, Bryonia, Cayaponia, Citrullus, Coccinea, Cucumis, Cucurbita, Cyclanthera, Ecballium, Echinocystis, Elaterium, Lagenaria, Luffa, Melothria, Momordica, Polakowskia, Sechium, Sicana, Telfairia, Trichosanthes.

Cudbear → Ochrolechia tatarea (L.) Mass.

Cudrania javanensis Trecul (syn. Vaniera cochinchinensis Lour.). (Moraceae). - Spiny herb. S. Japan, Himalayas to Australia. Young leaves are consumed in the Moluccas when raw. Fruits are eaten fresh or preserved in different parts of Japan. Wood is source of a yellow dye, after being treated with alum; it produces a green dye after treatment with indigo.

Cudrania triloba Hance. (syn. C. tricuspidata [Carr.] Bureau, Maclura tricuspidata Carr.) (Moraceae). - Shrub or tree. China, Korea, Japan. Fruits are edible, sweet, round, 25 mm. across, sub-acid, consumed in some parts of China.

Cujin → Inga Rensoni Pitt.

Cullenia excelsa Wight., Wild Durian. (Bombaceae). - Tree. Ceylon, India. Wood easily split, pale yellowish brown; used for cases and cigar boxes.

Cumarú → Amburana acreana (Ducke) Sm.

Cumin → Cuminum Cyminum L.

Cumin, Black → Nigella sativa L.

Cuminum Cyminum L., Cumin. (Umbelliferaceae). - Annual herb. Mediterannean region to Turkestan. Cultivated in Old and New World. Seeds used for flavoring dishes, in curry powders, cordials, in Leiden cheese. Used in home remedies, antispasmodic, antihysteric, for stomach ache. Source of Oil of Cumin. Used in perfumery, imparts distinctive tones with violet, jasmine, acacia, lily-of-the-valley and for orien-

tal bouquets. Contains cominic aldehyde, pinene, dipentene.

Cunila origanoides (L.) Butt. (syn. C. Mariana L., Mappia origanoides (L.) House. Stone Mint, Dittany.) (Labiaceae). - Herbaceous plant. Eastern N. America to Florida and Texas. Herb was used as tea by the Indians and early settlers for colds and fevers.

Cunninghamia sinensis R. Br. (syn. C. lanceolata Hook.) (Taxodiaceae). - Tree. Hupeh, Szechuan, Yunnan, Kwantung (China). Wood light, fragrant, easily worked, durable; used in China for house building, general carpentry, tea-chests, pillars, boats andw here great strength is required. The leading coniferous species for reforestation in China.

Cunoniaceae → Belangera, Cunonia, Davidsonia, Platylophus, Spiraenthemum, Weinmannia.

Cunonia capensis L., Red Alder. (Cunoniaceae). - Tree. South Africa. Wood moderately light, soft, compact, durable in water, takes polish; suitable for furniture, construction of mills and wagons.

Cupania Cunninghamiana Hook f. → Diploglottis Cunninghamiana Hook. f.

Cupania glabra Sw. (Sapindaceae). - Tree. West Indies. Wood strong, compact, red; used for general carpentry.

Cupania gracilis Panch and Sebert. (Sapindaceae). - Small tree. New Caledonia. Wood is recommended in New Caledonia for turnery and cabinet-work.

Cupania pseudorhus A. Rich. (syn. Jagera pseudorhus Radlk.) (Sapindaceae). - Small tree. Australia. Used by natives in Queensland as fish-poison. Bark was used during World War I as substitute for Quillaja Bark, causing a head or foam in cordials.

Cupania stipitata Panch and Sebirt. (Sapindaceae). Small tree. New Caledonia. Wood fine-grained, pale red; recommended in New Caledonia for cabinet making.

Cuphea glutinosa Cham. and Schlecht. (Lythraceae). - Herbaceous plant. Brazil to Argentina. Infusion of leaves is used in S. America as diuretic, purgative and to „clean" the blood.

Cuphocarpus inermis Baker. (Araliaceae). - Tree. E. Madagascar. Wood is used in Madagascar for manuf. musical instruments.

Cuprea Bark → Remijia pedunculata Flueck.

Cupressaceae → Callitris, Callitropsis, Chamaecyparis, Cupressus, Fitzroya, Juniprus, Libocedrus, Pilgerodendron, Tetraclinis, Thuja, Thujopsis.

Cupressus Benthamii Endl. (syn. C. lusitanica Mill.). Cedro blanco. (Cupressacae). - From Tepic and Veracruz (Mexico) to Costa Rica. Tree. Wood is used locally in Mexico for various purposes. Bark is considered an astringent in some parts of Mexico.

Cupressus funebris Endl. (Cupressaceae). - Tree. Hupeh, Szechuan, Chekiang (China). Cultivated in Centr. China. Wood fairly hard, close-grained, white, durable, tough; used in China for general carpentry, agricultural implements, coffins, boat-building, construction of houses. Tree is planted in China over tombs, near shrines and temples.

Cupressus sempervirens L., Italian Cypress. (Cupressaceae). - Tree. S. Europe and W. Asia. Source of Oil of Cypress, derived from the leaf material. Recommended for perfumery. Oil has an amber-like note, due to sesquiterpene alcohol; blends well with Chypre types of perfumes; is also suitable in manuf. of soaps. Used medicinally for whooping cough. Much is derived from Bouches du Rhône region, France.

Cupressus torulosa Don. (Cupressaceae). - Tree. Himalayas to W. China. Wood durable; used in China for various purposes.

Curaçao Aloë → Aloë vera L.

Curaçao Liqueur → Citrus Aurantium L.

Curanga fel-terrae Merr. (syn. C. amara Juss.) (Scrophulariaceae). - Perennial herb. Indo-China to Philipp. Islds., Moluccas. Leaves stimulant to intestines, sudorific, diuretic, emmenagogue; for early stages of dropsy; in colic.

Curare poison → Chondodendron, Cocculus, Strychnos, Telitoxicum.

Curatella americana L., Sandpaper Tree. (Dilleniaceae). - Small tree. Centr. America to Colombia and Venezuela. Wood coarse, heavy, hard, reddish-brown; used for cabinetwork; also made into charcoal. Leaves contain much silica and are used as substitute for sandpaper. Bark is used for tanning.

Curcas Oil → Jatropha Curcas L.

Curculigo latifolia Dryand. (Amaryllidaceae). - Perennial herb. India to Malaya. Leaves are source of a fibre; used for manuf. fishing nets by the Dyaks of Borneo.

Curculigo orchioides Gaertn. (Amaryllidaceae). - Perennial herb. Trop. Asia. Plant used in the Philipp. Islds. for skin-diseases.

Curculigo recurvata Dryand. (Amaryllidaceae). - Perennial herb. India to Australia. Leaves are source of a fibre; used to make false hair by the hill people of Luzon.

Curcuma → Curcuma longa L.

Curcuma aeruginosa Roxb. (Zingiberaceae). - Perennial herb. Burma. Rhizome is source of a starch.

Curcuma angustifolia Roxb., East Indian Arrowroot. (Zingiberaceae). - Perennial herb. Himalayan region. Cultivated. Rhizomes are source of a starch, called East Indian Arrowroot, Tik, Tikur, Tikor Starch or Travancore Starch.

Curcuma leucorhiza Roxb. (Zingiberaceae). - Perennial herb. E. India. Cultivated. Rhizomes are source of a starch, known as East Indian Arrowroot.

Curcuma longa L. (syn. C. domestica Loir.) Curcuma, Turmeric. (Zingiberaceae). - Perennial herb. E. India. Cultivated, esp. in India, Cochin-China, China and Java. Source of Turmeric or Curcuma, derived from powdered rhizomes. Used as coloring agent and as a condiment, in mixed spice-powders, food preparations, prepared mustards etc. Employed medicinally as carminative and aromatic stimulant. Contains an orange-yellow ess. oil and curcumin. Has been used as adulterant of ginger and mustard. Curcuma Paper used as indicator.

Curcuma Pierreana Gagn. (Zingiberaceae). - Perennial herb. Malaya. Cultivated in Annam. Source of False Arrow-Root.

Curcuma xanthorrhiza Roxb. (Zingiberaceae). - Perennial herb. Amboina. Occasionally cultivated in Java. Rhizomes are source of a starch; used as porrage and pudding.

Curcuma Zeodoaria Roscoe., Zedoary. (Zingiberaceae). - Perennial herb. E. India. Cultivated in S.E. Asia, Ceylon and Madagascar. Cut and dried rhizome is source of Zedoary or Zedoaria, used as a condiment. Is employed medicinally as aromatic stimulant and carminative. Contains a-cienol and a resin. Also used in cosmetics.

Curled Dock — Rumex crispus L.

Curly Yarran →Acacia homalophylla Cunningh.

Curlycup Grindelia → Grindelia squarosa (Pursh.) Dun.

Currant, American Red → Ribes triste Pall.

Currant, Buffalo → Ribes aureum Pursh.

Currant, European Black → Ribes nigrum L.

Currant, Golden → Ribes aureum Pursh.

Currant, Missouri → Ribes aureum Pursh.

Currant, Native → Leptomeria acida R. Br.

Currant, Prickly → Ribes lacustre Poir.

Currant, Red → Ribes rubrum L.

Currant Rhubarb → Rheum Ribes L.

Currant, Swamp Red → Ribes triste Pall.

Currant, Wild Black → Ribes americanum Mill.

Currants, Dried → Vitis vinifera L. var. apyrena Risso.

Currawong Acacia → Acacia doratoxylon Cunningh.

Curtisia faginea Ait., Assagay Tree. Cape Lance Wood. (Cornaceae). - Tree. S. Africa. Wood durable, close-grained, takes good finish, red to brown; used for furniture, spokes and felloes.

Curuba → Passiflora maliformis L. and Sicana odorifera Naud.

Curupag → Piptadenia macrocarpa Benth.

Cushaw → Cucurbita moschata Duch.

Cusparia febrifuga Humb. (syn. C. trifoliata Engl.). (Rutaceae). - Tree. Brazil. Bitter bark is used as tonic, antidysenteric. In Brazil the plant has a reputation of being febrifuge.

Cusso → Hagenia abyssinica J. F. Gmel.

Custard Apple → Annona squarosa L.

Custard Apple, Common →Annona reticulata L.

Cutch, Black → Acacia Catechu Willd.

Cuxiniquil → Inga Preussii Harms.

Cyamopsis psoralioides DC. Cluster Bean. (Leguminosaceae). - Perennial plant. E. India. Unripe seeds and pods are consumed like string beans in E. India. It is also grown as green manure and as fodder for live-stock.

Cyathea dealbata Swartz. (Cyathaceae). - Tree fern. New Zealand. Pith of the stem is eaten by the natives.

Cyathea medullaris Swartz. Black-stemmed Tree Fern. (Cyatheaceae). - Tall treefern. Australia esp. Tasmania, Victoria, New South Wales. Pith contains much starch, consumed by the aborigines.

Cyathea mexicana Schlecht. and Cham. Ocopetate, Cola de Mono. (Cyatheaceae). - Treefern. S. Mexico. to Centr. America. Scales from the fronds are used in Veracruz to arrest hemorrhage.

Cyathea usambariensis Hiern. (Cyatheaceae). - Tree fern. Trop. E. Africa. Pith and young leaves used by the natives in E. Africa for tapeworm.

Cyathea Vieillardii Mett. (Cyatheaceae). - Treefern. New Caledonia. A mucilaginous substance derived from the base of the fronds is consumed as food by the inhabitants of New Caledonia.

Cyatheaceae → Alsophila, Cibotium, Cyathea, Dicksonia, Hemitelia.

Cyathocalyx globusus Merr. (Annonaceae). - Tree. Philipp. Islands. Seeds used by Negrittos as a substitute for Arac Nuts for chewing.

Cyathocalyx zeylanicus Champ. (Annonaceae). - Tree. Ceylon. Wood quickly decomposes; used for tea boxes, small cases. Sticks supply Kandyan Chief's staffs. Flowers are eaten as addition to betel leaf in parts of Sabaragamuwa (Ceylon).

Cybistax Donnell-Smithii (Rose) Seib. (syn. Tabebuia Donnell-Smithii Rose). (Bignoniaceae). - Tree. S. Mexico, Centr. America. Wood yellowish white to light yellow brown, light firm, medium texture, rather coarse, straight-grained, easy to work, not resistant to decay; used for furniture. Is known as Primavera, formerly as White Mahogany.

Cybistax Sprucei Schum. (Bignoniaceae). - Tree. Andean region of Peru. Cultivated. Leaves are source of a blue dye.

Cycadaceae → Bowenia, Cycas, Dioon, Encephalartos, Macrozamia, Zamia.

Cycas circinalis L. (Cycadaceae). - Palm-like plant. Moluccas. Stem is source of a sago, of commercial importance.

Cycas media R. Br. (Cycadaceae). - Palm-like plant. Australia, Kernels are considered poisonous when fresh. When roasted and pounded

several times, they are consumed as food by the aborigenes of N. Queensland.

Cycas revoluta Thunb. (Cycadaceae). - Palm-like tree. Japan. Vermilion-red, 2.2 cm. long seeds are eaten in Japan when fresh or roasted.

Cycas Rumphii Miq. (Cycadaceae). - Palmlike plant. Mal. Archip., Australia. Young leaves before being unfolded are eaten as vegetable.

Cyclamen europaeum L., Snowbread (Primulaceae). - Small perennial herb. Parts of Centr. Europe. Root, was formerly used as a purgative. Used at present as a home-remedy for pain in the bladder, as purgative, for colic, tooth-ache and as emmenagogue.

Cyclanthaceae → Carludovica.

Cyclanthera edulis Naud., Pepino de Comer (Cucurbitaceae). - Annual vine. Peru and Bolivia. Fruits are used in pickles, also consumed as a vegetable.

Cyclanthera pedata Schrad. Achocha. (Cucurbitaceae). - Vine. Trop. America. Fruits are consumed as a vegetable in some parts of S. America, esp. in Peru and Bolivia. Related or identical with the previous species.

Cyclea peltata Hook. (syn. C. barbata Miers.) (Menispermaceae). - Malaya. Tubers are an important medicine in Java and Cochin China; used for fevers, as tonic and stomachic. Employed in Chinese drugs. Leaves are used in Tjintjaoo Idjo, a delicacy in Java.

Cyclophorus adnascens Desv. (Polypodiaceae). - Perennial herb. Africa, trop. Asia to Polynesia. Used by Malays for dysentery.

Cyclopia genistoides Vent. (Leguminosaceae). - Woody plant. S. Africa. Leaves are used as a substitute for tea.

Cyclopia subterenata Vog. (syn. C. brachypoda Benth. (Leguminosaceae). - Woody plant. S. Africa. Leaves are source of the commercial Caspa Tea.

Cyclopia tenuifolia Lehm. (Leguminosaceae). - Small shrub. Cape Peninsula. Leaves after having been subject to fermentation are used as tea.

Cydonia oblonga Mill. (syn. C. vulgaris Persoon. Pyrus Cydonia L.). Quince. (Rosaceae). - Small tree. Transcaucasia, Iran, Turkestan, S. E. Arabia. Cultivated in temperate regions of Old and New World. Fruits pear or apple-shaped; used in jams, marmelades; preserved with pears. The mucilaginous seed is used as demulcent vehicle for other remedies, especially for skin lotions. Most seeds are imported from Russia, Iran, Spain, Portugal and S. France.

Cydonia sinensis Thouin. (syn. Chaenomeles sinensis (Thouin) Koehne. Chinese Quince. (Rosaceae). - Small tree. N. China. Cultivated. Fruits very large. Used in China for perfuming rooms after being placed in bowls, spreading a delightful spicy scent. Used by foreigners in sweetmeats.

Cymbidium canaliculatum R. Br. (Orchidaceae). - Perennial herb. Australia. Pseudobulbs are consumed as food by the aborigenes of N. Queensland.

Cymbidium virescens Lindl. (Orchidaceae). - Perennial herb. Japan. Flowers are salted, put in hot water and used as a beverage in Japan. They are also preserved in plum vinegar.

Cymbopetalum costaricense (Don. Smith.). Fries. (syn. Asimina costaricensis Don. Smith.). (Annonaceae). - Tree. Costa Rica. Petals from flowers are used by Indians to give a distinct flavor to chocolate.

Cymbopetalum penduliflorum Baill. Flor de la Oreja. (Annonaceae). - Shrub or small tree. Mexico, Guatemala. Cultivated. Spicy dried petals with vanilla were used by Aztecs in pre-Columbian times to flavor chocolate. Also used medicinally.

Cymbopogon citratus (DC.) Stapf. (Graminaceae). - Perennial grass. Trop. Asia. Cultivated. Source of Lemongrass Oil, an ess. oil derived from steam distillation. Used for perfuming soaps, cosmetics, for flavoring and in many technical preparations. Has powerful fresh lemon scent. Oil is also source of ionone which is made into a synthetic violet.

Cymbopogon Martini Stapf. Rosha Grass. (Graminaceae). - Perennial grass. E. India. Cultivated in India, Java and Sumatra. Source of Palmarosa Oil, an ess. oil obtained by steam distillation; used in perfumery. It produces a high grade geraniol. Much used in cosmetics and soap, giving a lasting and pronounced rose note. Was formerly used as adulterant of Bulgarian Rose Oil. Oils have also been manufactured from C. caesius Stapf. and C. flexuosa Stapf. from India, source of Oil of Inchi Grass.; C. senaarensis Chiov, produces Oil of Mahareb Grass, in Sudan; C. caesius Stapf. produces Oil of Kachi Grass from Bengalore and Mysore; C. coloratus Stapf from Fiji produces an oil; C. giganteus Chiov. gives Oil of Tsauri Grass from Nigeria. Composition and quality of Cymbopogon oils are much dependent upon locality.

Cymbopogon Nardus (L.) Rendle. (syn. Andropogon Nardus L.) (Graminaceae). - Perennial grass. Trop. Asia. Cultivated in Java, Ceylon and other parts of the tropics. Source of Citronella, obtained by steam distillation. Used in perfumery, for scenting soaps, varnishes, insecticides, spraying liquids for households and theaters, disinfectants, shoe polishes and other preparations. Leaves are used by the natives for flavoring soups, in cooking fish, preparing curry. They are also employed as a tea.

Cymbopogon praximus Stapf. (Graminaceae). - Perennial, frangrant grass. Nupe, Nubia, Sudan, Abyssinia. Used for thatching.

Cymbopogon Schoenanthus Spreng., Camel Grass. (Graminaceae). - Perennial grass. Eritrea, Nile Land, Tunis, Marocco, Arabia, Iraq, Iran,

Panjab etc. Source of an ess. oil, sold in bazaars of Panjab; used for medicinal purposes. Sometimes employed in perfumery. Was much esteemed by the ancient Greeks and Romans.

Cyminosma odorata DC. → Acronychia odorata Baill.

Cyminosma resinosa DC. → Acronychia laurifolia Blume.

Cymopterus Fendleri Gray. (Umbelliferaceae). - Herbaceous plant. Colorado, Utah, New Mexico. Aromatic roots were used for flavoring meat by Indians of New Mexico. Leaves were also consumed as greens.

Cymopterus globosus S. Wats. (Umbelliferaceae). - Herbaceous plant. Nevada to California. Roots being of a pleasant taste, were eaten by the Indians. Leaves were also consumed as greens. Likewise C. acaulis (Pursh.) Rydb.

Cymopterus longipes S. Wats → Aulospermum longipes. (S. Wats) Coult. and Rose.

Cymopterus montaus Nutt. Gamote. (Umbelliferaceae). - Herbaceous plant. Western N. America esp. Kansas and westward to N. Mexico. Parsnip-like roots are eatenb y the Indians and Mexicans.

Cymopterus purpureus S. Wats. → Aulospermum purpureum (S. Wats.). Coult. and Rose.

Cynara Cardunculus L., Cardoon. (Compositae). - Large perennial herb. Mediterranean region. Cultivated. Bleached leaf stalks are eaten as a vegetable when boiled. Varieties: Tours, Spanish.

Cynara Scolymus L. Globe Artichoke. (Compositae). - Large perennial herb. S. Europe. Cultivated. Soft fleshy receptacle of flowerhead bud and thick base of the scales around flowerhead are eaten as vegetable when boiled. Varieties ar: Large Green Paris, Laon, Violet Hatif.

Cynodon Dactylon (L.) Pers., Bermuda Grass. (Graminaceae). - Perennial grass. Southern United States. Used as pasture grass.

Cynoglossum officinale L., Hounds Tongue. (Boraginaceae). - Perennial herb. Temp. Europe, Asia, N. America. Roots, collected in early summer, also leaves and herb are used medicinally as a sedative, demulcent. Has paralyzing effect on central motor system. Contain a resin, tannin and inulin. Young leaves are used in some localities, e. g. Tessin (Switzerl.) as a salad and vegetable.

Cynometra cauliflora L. (Leguminosaceae). - Small tree. E. India, Malaya. Fleshy, wrinkled, sub-acid pods are of a pleasant flavor, suitable for preserves. Young fruits are pickled, eaten with Spanish pepper, also prepared with fish, soja and other ingredients.

Cynometra cubensis Rich. (Leguminosaceae). - Tree. Cuba. Wood dark reddish; used for carpentry.

Cynometra sessiliflora Harms. (Leguminosaceae). - Tree. Congo region. Source of a Copal.

Cynomoriaceae → Cynomorium.

Cynomorium coccineum L. (Cynomoriaceae). - Shrub. N. Africa. Pulverized roots are used as a condiment by the Tuareg (N. Afr.).

Cynosurus cristatus L., Crested Dogtail, Dog's Tail Grass. (Graminaceae). - Annual grass. Europe, Caucasia, Asia Minor, N. America. Grown as a pasture grass and for hay production.

Cyperaceae → Carex, Cladium, Cyperus, Eleocharis, Fimbristylis, Kyllingia, Lepironia, Mariscus, Scirpodendron, Scirpus.

Cyperus alopecuroides Rottb. (Cyperaceae). - Perennial herb. Tropics. Stems used in several parts of Egypt for manuf. of mats.

Cyperus aristatus Rottb. (syn. C. inflexus Muhl.) (Cyperaceae). - Perennial herb. N. and S. America. Tuberous roots were used as food among the Indians of New Mexico.

Cyperus articulatus L. (Cyperaceae). - Perennial herb. Warm countries of the Old and the New World. Occasionally cultivated. Fragrant, sweet scented tuberous roots are used for perfuming clothing etc. Stems are made into mats.

Cyperus Cephalotes Vahl. (syn. C. natans Buch-Ham.) (Cyperaceae). - Perennial herb. Trop. Asia, Australia. Cultivated in rice fields in Japan. Stems are made into green mats, called Riukiu-omote.

Cyperus esculentum L. Chufa, Earth Almond, Yellow Nut Grass. (Cyperaceae). - Perennial herb. Cosmopolit .. Tuberous rootstocks are consumed as food, also used as food for hogs. Juice from pressed tubers are source of a drink, Horchata de Chufas. Cultivated since early times.

Cyperus Haspan L. (Cyperaceae). - Perennial herb. Trop. Africa, India. Salt is prepared from the ash of the plant, made on a small scale by some tribes in E. Africa.

Cyperus longus L., English Galingale. (Cyperaceae). - Perennial herb. Europe, Orient. Rhizomes have scent of violets, used in perfumery; sometimes added to Lavender Water.

Cyperus maculatus Boeck. (Cyperaceae). Perennial herb. Trop. Africa. Tubers are source of a perfume, sold in markets of Nupe (Afr.).

Cyperus malaccensis Lam. (Cyperaceae). - Perennial herb. Trop. Asia to Australia. Used for matting.

Cyperus nudicaulis Poir. (Cyperaceae). - Perennial herb. Madagascar. Stems are used for making hats, called Panama Hats of Madagascar.

Cyperus papyrus L., Papyrus Plant. (Cyperaceae). - Tall perennial herb. Nile region, Syria, Siciliy. Stems were source of a paper material, used by the Ancient Egyptians.

Cyperus rotundus L. (syn. C. hexastachyos Rottb.) Nut Grass, Cocograss. (Cyperaceae). - Perennial herb. Tropics. Dried tuberous roots, known as Souchet, have aromatic properties; used in perfumes. Used by the natives of India

for perfuming hair and clothes, also employed in perfumery. It was the Radix Junci of the Romans. Was employed by the Scythians for embalming. Also C. scariosus.

Cyperus tegetiformis Roxb., Chinese Matgrass. (Cyperaceae). - Perennial grass. Asia. Cultivated, especially in China. Source of a fibre, made into mats and similar material.

Cyperus vaginatus R.Br. (Cyperaceae). - Perennial herb. Australia. Source of a fibre, made into fishing nets and cordage; used by the natives.

Cypholophus macrophyllus Wedd. (Urticaceae). - Woody plant. Philipp. Ilds., Samoa. Bark is source of a fibre; used for manuf. mats by the natives.

Cyphomandra betacea Sendt. Tree-Tomato. (Solanaceae). - Tree. Peru, Brazil. Cultivated for centuries among Peruvian Indians. Introduced in tropics and subtropics of Old and New World. Fruits egg-shaped, smooth, reddish-yellow, fleshy, agreeable flavor; eaten raw; also suitable for jams and preserves.

Cyphomandra Hartwegi Sendt. (Solanaceae). - Perennial plant. S. America. Berries reddish, eaten cooked by the natives of Colombia, Chile, Argentina. Fruits are sold in markets of Chile and Argentina.

Cypress, Bald → Taxodium distichum (L.) Rich.

Cypress, Italian — Cupressus sempervurens L.

Cypress, Lawson → Chamaecyparis Lawsoniana (Murr.) Parl.

Cypress, Sitka → Chamaecyparis Nootkatensis (Lamb.) Sudw.

Cypress, Southern → Taraxacum distichum (L.) Rich.

Cypress, Yellow → Chamaecyparis Nootkatensis (Lamb) Sudw.

Cypripedium parviflorum Salisb., Small Yellow Lady's Slipper. (Orchidaceae). - Perennial herb. N. America. Dried rhizome used medicinally as a nerve stimulant and antispasmodic.

Cypripedium pubescens Willd. (Orchidaceae). - Perennial herb. N. America. Dried rhizome and roots are used medicinally as a nerve stimulant and antispasmodic.

Cyprus Powder → Anaptychia ciliaris (L.) Kbr.

Cyrillaceae → Cliftonia, Cyrilla.

Cyrilla racemiflora L., Ironwood, Black Titi. (Cyrillaceae). - Shrub or small tree. S. Virginia to Florida and Texas. Flowers are source of a commercial honey.

Cyrtocarpa procera H.B.K. (Anacardiaceae). - Tree. Mexico. Bark used as substitute for soap. Fruit edible, known as Berracos. Wood soft, strong scent; used for trays, small images.

Cyrtosperma Merkusii (Hassk.) Schott. (Araceae). - Herbaceous plant. Philipp. Islds. to Sumatra. The large starchy rootsstocks are consumed as food. Decoction of the flower cluster is used as emmenagogue and embolic.

Cystoclonium armatum Harv. (Rhodophyllaceae). - Red Alga. Sea Weed. Pac. Ocean. Consumed as food in Japan.

Cytinus Hypocistis L. (Rafflesiaceae). - Mediterranean region. Parasitic on roots, esp. on Cistus. A decoction of the plant and the fruits (Succus Hypocistidis) is used against dysentry. The young plants are eaten like asparagus.

Cytisus pallidus Poir. (Leguminosaceae). - Shrub. Canary Islands. Cultivated in Canary Islands as a forage crop. Also C. palmensis Hutch. and C. stenopetalus Christ.

Cyttaria Darwinii Berk. (Cyttariaceae). - Ascomycete (Fungus). Southern temperate and antarctic South America. The stromata form the principal vegetable food among the Fuegians and Patagonians. In the same region occurs C. Berteroi Berk.

Cyttariaceae (Fungi) → Cyttaria.

D

Dacrydium cupressinum Soland. Red Pine. (Podocarpaceae). - Tree. New Zealand. Wood red or yellow, beautifully marked. Used for panels, ceilings, dadoes, general house-building, bridges, keels of ships, cabinet-work. Bark used for tanning, imparts a red color to the skin.

Dacrydium elatum Wall. (Podocarpaceae). - Tree. Malaya. Wood fairly hard, heavy, fine-grained light-brown; made into boards, posts and boxes.

Dacrydium Franklinii Hook., Huon Pine. (Podocarpaceae). - Tree. Tasmania, New Zealand, New Caledonia, Mal. Archip. Source of Huon Pine Wood Oil; used for medicinal soaps, toilet waters; preservative in casein paints; recommended as a source for manuf. vanillin. Contains methyl eugenol.

Dacrydium intermedium T. Kirk. Yellow Silver Pine. (Podocarpaceae). - Tree. New Zealand. Wood firm, compact, straight grained, resinous, reddish yellow; used for boatbuilding.

Dacrydium westlandicum T. Kirk. Westland Pine. (Podocarpaceae). - Tree. New Zealand. Wood very valuable, firm, compact, even grain. Used for cabinet-work, agricultural implements, bridges, wharves, general house building.

Dacryodes hexandra Griseb. (Burseraceae). - Tree. S. Domingo. Source of a resin, called Tabonico.

Dactylis glomerata L. Orchard Grass. (Graminaceae). - Perennial grass. Europe. Cultivated in most temp. countries; used as pasture and as a hay-crop for fattening cattle for the market.

Daemonorops Draco Blume. Dragon's Blood Palm. (Palmaceae). - Rattan palm. Malaya. Source of a Dragon's Blood, Resina Draconis, Sanguis Draconis, a resin derived from the scales of the fruits. It is of a red color; composed of rounded or flattened masses weighing up to several kg. or of sticks being 20 to 30 cm long. The resin is a stimulant and astringent; was formerly used in dentifrices and for mouth washes. Also used in varnishes and lacquers, giving a mahogany stain. Technically it is used in photo engraving on zinc, protecting metal parts not to be etched. Also D. didymophyllus Becc. from Malacca; D. micranthus Becc. from Penang, Perak and Malacca; D. Motleyi Becc. and D. sparsiflorus Becc. from Borneo.

Daemonorops fissus Blume. (Palmaceae). - Rattan. Borneo. Stems are used for mats.

Daemonorops Forbesii Becc. (Palmaceae). - Rattan-palm. Sumatra. Stems easy to split; used for basket-work.

Daemonorops longipes Mart. (Palmaceae). - Rattan. Mal. Archipel. Used in Billiton for rattan-furniture.

Daemonorops trichrous Miq. (Palmaceae). - Rattan-palm. Sumatra, Bangka. Stems easy to split; used for basket-work, mats etc.

Dahomey Rubber → Ficus Vogelii Miq.

Dahoon Holly → Ilex Cassine L.

Dahurian Buckthorn → Rhamnus dahurica Pall.

Dahurian Lily → Lilium dauricum Ker-Gawl.

Daimyo Oak → Quercus dentata Thunb.

Dais glaucescens Decne. (Thymelaeaceae). - Shrub. Madagascar. Bark is source of a fibre; used in Madagascar for manuf. string.

Daisybush, Travers → Olearia Traversii F. v. Muell.

Dakota Mustard → Brassica arvensis (L.) Ktze.

Dalbergia Baroni Baker. (Leguminosaceae). - Tree. Madagascar. Wood is source of Rosewood of Madagascar. Also D. Greveana Baill.

Dalbergia cochinchinensis Pierre. (Leguminosaceae). - Tree. Trop. Asia, esp. Siam. Wood is source of Rosewood of Siam.

Dalbergia cearensis Ducke. Brazil Tulipwood, Kingwood, Violeta. (Leguminosaceae). - Tree. Brazil. Wood finly striped, fragrant, uniform, not difficult to work, has violet-brown and black or black violet alternating concentric layers; takes high, waxy, natural polish. Used for marquetry, inlay-work, turnery, fancy articles.

Dalbergia cubilquitzensis (D. Sm.) Pitt., Rosewood of Guatemala. (Leguminosaceae). - Tree. Guatemala, Honduras. Wood orange with violet stripes, waxy appearance, heavy, hard, tough, strong, takes lustrous finish, not difficult to work; used for cabinet-work, brush-backs, interior finish of buildings, axles, tongues of wagons, spokes of truck wheels.

Dalbergia hupeana Nance. (Leguminosaceae). - Tree. Szechuan, Hupeh, Chekiang (China). Wood hard, durable; used in China for oil-presses, tool-handles, wheel-spokes, blocks, pulleys and where great strength is required, for making wheelborrows used in Chengtu Plain.

Dalbergia Junghuhnii Benth. (syn. D. parviflora Roxb., D. zollingeriana Miq.). (Leguminosaceae). - Woody vine. India, Malaya. Heartwood of thick stems and roots is scented; used in China for joss-sticks, called 'hsiang-chen-hsiang. Used as incense in Hindoo and Chinese Temples. Wood is called Lacca Lignum.

Dalbergia latifolia Roxb. Rosewood of Southern India, East Indian Rosewood, Black Rosewood, Malabar Rosewood, Bombay Rosewood, Rosetta Rosewood, Javanese Palissander. (Leguminosaceae). - Tree. E. India. Wood light, red to deep rich purple, streaked with golden-yellow to black, close, firm. Used for turnery, flooring, etc.

Dalbergia melanoxylon Guill. and Perr., African Blackwood, African Grenadille Wood, Senegal Ebony, Mozambique Ebony, Unyoro Ebony (Leguminosaceae). - Tree. Trop. Africa. Valuable timber, hard, heavy, substitute for True Ebony; used for furniture, carving, musical instruments. Made into arrows and wooden hammers by the natives. Root is considered a cure for tooth ache.

Dalbergia Miscolobium Benth. (syn. Miscolobium nigrum Mart.). (Leguminosaceae). - Tree. Brazil. Source of a wood of fine quality.

Dalbergia nigra Allem. (Leguminosaceae). - Tree. Brazil, esp. Bahia. Source of Brazilian Rosewood, Rio Rosewood or Bahia Rosewood, occasionally called White Rosewood. Wood variable, brown to chocolate brown or violet, streaked with black, rose-scented, straight grained, medium texture, excellent working properties; used for cabinet work, pianos, furniture, handles of fine tools, brush backs, radio cabinets, marquetry, plane handles, butcher knife handles, billiard tables etc.

Dalbergia retusa Hemsl., Nicaragua Rosewood, Cocobolo, Palo negro. (Leguminosaceae). - Tree. Central America. Wood variable in color, rainbowhued to deep red. Used in cutlery, knife-handles, small tool-handles, brush-backs, Tunbridgeware, expensive machinery, musical and scientific instruments, steering wheels, inlaying, small attachments of automobiles, launches, jewelry boxes, rosary-beads, forks, spoons, canes, chessmen and buttons.

Dalbergia Sissoo Roxb., Sissoo of India. (Leguminosaceae). - Tree. Trop. Asia, esp. India, Centr. Asia. Wood brown, firm, hard, durable, does not split, strong, seasons well, takes a beautiful finish; used for fine carving, parquet flooring and furniture.

Dalbergia Spruceana Benth., Jacarandá do Pará. (Leguminosaceae). - Tree. S. America, esp. Brazil. Wood fine uniform, straight-grained, hard, compact, heavy, strong. Source of Rosewood of Lower Amazon. Used for cabinet work.

Dalbergia Stevensoni Standl. (Leguminosaceae). - Small tree. Brit. Honduras. Wood very durable, pinkish-brown or purplish. Source of Rosewood of British Honduras. Used for musical instruments, marimbas and xylophones. See also Platymiscium sp. which is used for marimbas in Guatemala.

Dalbergia stipulaceae Roxb. (syn. D. ferruginea Roxb.) (Leguminosaceae). - Tree. Philipp. Islds. Infusion of wood and roots is considered an emmenagogue; used as abortive taken in moderate amounts.

Daldinia concentrica (Bolt.) Ces. and De Not. (Xylariaceae). ¬ Ascomycete fungus. The blackish stromata are powdered and used among the Moluccans for skin diseases and ulcers.

Dalea Emoryi Gray. (Leguminosaceae). - Shrub. Western United States and adj. Mexico. Branches are source of a yellowish-brown dye, used by the desert Indians of S. California for coloring deer skins.

Dallis Grass → Paspalum dilatatum Poir.

Dalmatian Insect Powder → Chrysanthemum cinerariaefolium (Trév.) Vis.

Damar → Dammar.

Damask → Hesperis matronalis L.

Damask Rose → Rosa damascena Mill.

Damiana → Turnera diffusa Willd.

Damianita → Chrysactinia mexicana Gray.

Dammara alba Lam. → Agathis alba Foxw.

Dammar Batu → Hopea Maranti Miq.

Dammar Gaging → Shorea leprosula Miq.

Dammar Hiroe → Vatica papuana Dyer.

Dammar Kedemut → Hopea fagifolia Miq.

Dammar Itam → Canarium legitinum Miq.

Dammar Kloepoep → Shorea eximea Scheffer.

Dammar Kumus → Shorea glauca King.

Dammar, Njating Matapoesa → Hopea Sangal Korth.

Dammar, Njating Matpleppek → Hopea Sangal Korth.

Dammar Penak → Balanocarpus Heimii King.

Dammar Pine, White → Agathis alba Foxw.

Dammar Rasak → Vatica Rassak Blume.

Dammar Resin → Agathis lanceolata Panch. and Shorea Wiesneri Schiff.

Dammar, Rock → Hopea odorata Roxb.

Dammar, Rose → Vatica Rassak Blume.

Dammar Sengai → Canarium hirsutum Willd.

Dammar Temak → Shorea hypochra Hance.

Dammar Tubang → Shorea eximea Scheffer.

Dammar Tenang → Shorea Koordersii Brandis.

Dammar, White → Vateria indica L.

Damson Plum → Prunus instititia L.

Danais Gerrardi Baker. (Rubiaceae). - Vine. Madagascar. Roots are source of a dye; used by the Sihanaka of Madagascar. Bark produces a fibre.

Dandelion → Taraxacum officinale Weber.

Dandelion Wine → Taraxacum officinale Weber.

Dangleberry → Gaylussacia frondosa Torr. and Gray.

Daniella oblonga Oliv. (Leguminosaceae). - Tree. Trop. Africa. Stem is source of a resin, called Niger Copal.

Daniella Ogea Rolfe. (Leguminosaceae). - Tree. W. Africa. Stems are source of Ogea or Accra Copal, of importance in manuf. varnishes.

Daniella similis Craib. Copal Tree. (Leguminosaceae). - Tree. Trop. Africa, esp. Nigeria, Gold Coast. Source of Gum Opal of Gold Coast; derived from the stems and roots of old trees. When resin is buried in the soil, it produces a semi-fossilised opal.

Daniella thurifera Bennett. (Leguminosaceae). - Tree. Trop. Africa. Wood moderately hard, light, coarse-grained, streaked brown, easily attacked by insects; used for knife handles, light furniture, packing cases. Stem is source of Ogea Gum of Sierra Leone, a Frankincense; sold as a body perfume. It is also called Illorin Gum and Balsam of Copaiba.

Danthonia pectinata Lindl. → Astrebla pectinata F. v. Muell.

Danthonia triticoides Lindl → Astrebla triticoides F. v. Muell.

Daphne cannabina Wall. (Thymelaeaceae). - Shrub. Temp. Himalaya, from Chamba to Bhutan. Bark is source of Nepal paper.

Daphne Cneorum L. (Thymelaeaceae). - Shrub. S. Europe. Bark is used in Spain to stupefy fish.

Daphne Gnidium L. (Thymelaeaceae). - Shrub. Mediterranean region. Crushed roots are used in Sardinia to stupefy fish. Powdered bark is used as abortive.

Daphne Mezereum L., Mezereon. (Thymelaeaceae). - Small shrub. Europe, Siberia, Caucasia, Asia Minor, N. Africa, Altai. Dried bark used medicinally, much being derived from Thuringia, S. France, Algeria. Diuretic, stimulant, sialogogue. Contains mezerin, an acrid resin; daphnin, a bitter glucoside. All parts of the plants are poisonous.

Daphne odora Thunb. (Thymelaeaceae). - Shrub. Japan. Cultivated. The fragrant flowers are used in Japan in sachets and for perfumed water.

Daphne oleoides Schreb. (Thymalaeaceae). - Small shrub. Europe to Centr. Asia. Decoction of roots is used by the natives of Centr. Asia as purgative. A spirit is supposed to be distilled from the berries in Sutlej Valley (Asia).

Daphne pseudo-mezereum Gray. (Thymelaeaceae). - Shrub. Japan. Bark is source of a fibre; used in Japan for manuf. paper.

Daphniphyllum humile Maxim. (Euphorbiaceae). - Shrub. Japan. Leaves smoked in place of tobacco by the Ainu.

Daphnopsis brasiliensis Mart. and Zucc. (Thymelaeaceae). - Shrub. Brazil. Bark from young plants is used in Brazil as a drastic, also for erysipelas and psoriasis.

Daphnopsis Swartzii Meissn. (Thymelaeaceae). - Shrub. West Indies. This species is supposed to have similar medicinal properties as Daphne Mezereum L.

Dark Catechu → Acacia Catechu Willd.

Dark Crottle → Parmelia physodes (L.) Ach.

Darkleaf Malanga → Xanthosoma atrovirens C. Koch.

Darling Plum → Reynosia septentrionalis Urb.

Darnel → Lolium temulentum L.

Daru Urandra → Urandra corniculata Foxw.

Darwinia fascicularis Rudge. (Myrtaceae). - Woody plant. Australia. Source of an ess. oil; has been recommended for perfumery. Also D. grandiflora.

Dasheen → Colocasia antiquorum Schott. var. esculenta Scott.

Dasylirion simplex. Trel. Sotol. (Liliaceae). - Low, shrub-like plant. Durango (Mex.). Leaves used for making baskets and „sopladores" for fanning charcoal fires.

Dasylirion texanum Scheele., Texas Sotol. (Liliaceae). - Perennial. Texas. Central part of bud, containing a sugary pulp, was used as food by the Indians. It was also prepared for Sotol, a beverage.

Dasylirion Wheeleri S. Wats., Wheeler Sotol. (Liliaceae). - Perennial herb. New Mexico and Arizona. Central part of plant was roasted in mescal pits and consumed by Indians. Also made into a beverage.

Date, Chinese → Zizyphus mauritiana Lam.

Date, Common → Phoenix dactylifera L.

Date Palm → Phoenix dactylifera L.

Dátil → Yucca baccata Torr.

Datisca cannabina L. (Datiscaceae). - Perennial herb. Centr. Asia. Cultivated. Leaves, roots and stems are source of a yellow dye. Was formerly extensively used in S. France, the Orient and India for dyeing silk after being treated with alum. Root is source of a bitter substance, called datiscin.

Datiscaceae → Datisca.

Datura fastuosa L. (syn. D. alba Nees. and Esenb.). (Solanaceae). - Herbaceous plant. Trop. Africa. Source of a blue and green dye, used in Zanzibar. Leaves are used in cigarettes for asthma.

Datura Metel L. Hindu Datura. (Solanaceae). - Herbaceous plant. India. Cultivated. Introduced in many parts of the tropics and subtropics of Old and New World. Dried leaves and seeds are used medicinally; see Belladonna. Leaves are used in cigarettes for asthma.

Datura meteloides DC., Sacred Datura. (Solanaceae). - Perennial herb. S.W. of United States and adj. Mexico. Leaves and roots are the source of a stupefying beverage; used by the Indians of New Mexico, Arizona and California.

Datura Stramonium L., Thornapple, Jimson Weed, Stramonium. (Solanaceae). - Annual herb. Centr. and S. Europe, Balkans, Syria, Egypt, Asia, Africa, N. America. Cultivated. Dried leaves and flower tops used medicinally; having action much like Belladonna. Narcotic, mydriatic, Contains daturine.

Daucus Carota L., Carrot. (Umbelliferaceae). - Annual or biennial herb. Temp. zone. Widely cultivated as vegetable and as food for domestic animals. Deep orange colored varieties are used as a vegetable; white and yellow varieties and very large red varieties are used as fodder for animals. Numerous varieties.   I. Group. Roots distinctly pointed, long: Long Orange, Purple, White Belgian, Yellow Belgian. Long: Daver's Half Long, Vosges White. II. Group. Roots blunt toward lower end. Half long: Nantes, Chantenay. Very short: French Forcing. Pigment from juice is sometimes used for coloring butter. Syrup used for sweetening. Alcoholic tincture of carrot-seed is used in French liqueurs. Oil of Carrot from seeds is volatile; used for flavoring, in liqueurs, perfumery. Contains daucol, carotol, pinene, l-limonene, asarone. Carrots sometimes cause allergic reactions in humans. Roasted roots have been used as substitute for coffee; recommended in times of want.

Daucus pussilus Michx., Rattlesnake Weed. (Umbelliferaceae). - Biennial herb. N. America and Mexico. Roots were consumed raw or cooked by the Navajo and Nez Percé Indians.

Davidsonia pruriens Muell. Arg., Do-rog. (Cunoniaceae). - Tree. Queensland. Berry plum-shaped, pulp red; used for jelly and preserves.

Davilla Kunthii St. Hil. (Dilleniaceae). - Woody vine. Mexico. Centr. America, Colombia. Stems tough, flexible; used for tying framework of huts of natives. Bark source of a black dye.

Daylily, Grassleaf → Hemerocallis minor L.

Daylily, Tawny → Hemerocallis fulva L.

Dead Nettle, Purple → Lamium purpureum L.

Dead Nettle, White → Lamium album L.

Dead Sea Apple → Quercus tauricola Kotschy.

Debregeasia edulis Wedd. (Urticaceae). - Shrub. Japan. Fruits, yellow, resemble strawberries; eaten in Japan.

Debregeasia hypoleuca Wedd. (Urticaceae). - Shrub. W. temp. Himalaya. Source of a fibre; used by the hill people; made into ropes and cordage.

Decan Hemp → Hibiscus cannabinus L.

Deeringia amaranthoides (Lam.) Merr. (Amaranthaceae). - Herbaceous plant. Trop. Asia. Young leafy shoots are eaten with rice in Indonesia.

Deer's Tongue → Saxifraga erosa Bursh.

Deer's Tongue → Trilisa odoratissima (Walt.) Cass.

Delima sarmentosa L. → Tetracera sarmentosa Vahl.

Delphinium Ajacis L. Rocket Larkspur. (Ranunculaceae). - Annual herb. Europe, Mediterranean region. Dried, ripe seeds used medicinally as tincture employed as a parasiticide. Contains delphinine, an alkaloid.

Delphinium Consolida L., Forking Larkspur. (Ranunculaceae). - Annual herb. Europe, temp. Asia. Flowers after having been treated with alum were formerly used in some parts of Europe as a source of a green dye.

Delphinium Staphisagria L., Stavisacre. (Ranunculaceae). - Annual or biennial herb. S. Europe, Asia Minor. Poisonous seeds are source of an insecticide. Used in medicine, antineuralgic, antispasmodic, in neuralgia, toothache, asthma.

Delphinium Zalil Aitch. and Hemsl. (Ranunculaceae). - Perennial herb. Iran, Afghanistan. Flowers are source of a yellow or orange-yellow dye called Asbarg, used for coloring silk and cotton. Much is used in Punjab. Dye is often used with that of Datisca canabina L. with alum as a mordant.

Demarara Greenheart → Acotea Rodiaei (Schomb.) Mez.

Demarara Mahogany → Caraoa procera DC.

Dendrobium purpureum Roxb. (Orchidaceae). - Perennial herb. E. Malaya. Used for poultices by natives.

Dendrobium salaccense Lindl. (Orchidaceae). - Perennial herb. Trop. Asia, esp. Malayan Penin., Java. Used by natives for flavoring rice.

Dendrobium utile Smith. (Orchidaceae). - Herbaceous perennial, epiphyte. Mal. Penin. Stems that have been split length-wise are used for different kinds of basket-work, having a yellowish color.

Dendrocalamus giganteus Munro. (Graminaceae). - Tall, woody grass, bamboo. Burma, Siam and southward. Stem used by natives for water-buckets.

Dendrocalamus Hamiltonii Nees and Arn. (Graminaceae). - Tall, woody grass, bamboo. Himalayas to Burma. Used for baskets. Young shoots of var. edulis are eaten as food when boiled.

Dendrocalamus Hookeri Munro. (Graminaceae). - Tall woody grass. Khasia, Jaintia hills. Stems used by natives for water and milk pails.

Dendrocalamus latifolius Munro. (Graminaceae). - Tall woody grass. China. Very young shoots are consumed as a vegetable in some parts of China and Japan.

Dendropanax arborea Decne and Planch. (Araliaceae). - Tree. West Indies to Northern S. America. Wood strong, yellow with dark streaks; used for general carpentry work.

Dendropogon usnoides (L.) Raf. → Tillandsia usneoides L.

Dendrosma Deplanchei Planch and Sebert. (Rutaceae). - Tree. New Caledonia. Wood is used for interior finish of houses and for general carpentry.

Dentaria diphylla Michx., Crinkleroot. (Cruciferaceae). - Perennial herb. Eastern N. America. Roots were eaten raw or boiled by Iroquois Indians. Also D. laciniata Muhl.

Deodar → Cedrus deodara (Roxb.) Loud.

Dermatocarpon miniatum (L.) Mann., Sprout Lichen. (Dermatocarpaceae). - Lichen. Temp. zone. On rocks. A dye has been used in some parts of Europe to give an ash-green color to woolens.

Derris elliptica Benth., Derris. (Leguminosaceae). - Woody vine. E. India, Malaya to New Guinea. Cultivated in the tropics of the Old and New World. Powdered root is in recent years much used as an effective insecticide. Contains rotenon, deguelin, tephrosin and toxicarol. Used in some parts of trop. Asia to stupefy fish, and for poisoning arrows.

Derris Koolgibberah F. M. Bailey. (Leguminosaceae). - Vine. Australia. Crushed plant is used in Queensland for killing fish; 1 : 1000 in water is fatal in one hour.

Derris malaccensis Prain. (Leguminosaceae). - Woody plant. Malaya. Has been mentioned as source of insecticide. Contains 0.1 to 0.7% rotenon.

Derris polyantha Miq. (Leguminosaceae). - Vine. Sumatra. Root has been suggested as a source of rotenon.

Derris uligonosa Benth. (Leguminosaceae). - Climbing plant. Australia, Asia, Africa, Madagascar. Stems when powdered are used to stupefy fish. Has been recommended as an insecticide.

Derris Root → Derris elliptica Benth.

Desbordesia insignis Pierre. (Simarubaceae). - Tree. Gabon (Afr.). Seeds are used in sauces by the natives.

Deschampsia caespitosa (L.) Beauv., Tussock Grass, Tufted Hairgrass. (Graminaceae). - Annual grass. Western N. America to Baja California (Mex.). Grass used for hay and pasture.

Desert Lemon → Atalantia glauca Benth. and Hook. f.

Desert Milkweed → Asclepias subulata Decaisne.

Desert Pepperweed → Lepidium Fremontii S. Wats.

Desert Seepweed →Suaeda suffrutescens Wats.

Desert Trumpet → Eriogonum inflatum Torr.

Desmodium discolor Vog. (Leguminosaceae). - Shrub. Brazil. Grown as forage plant in warm countries.

Desmodium gangeticum DC. (Leguminosaceae). - Woody plant. Tropics of Old World. Used in India for catarrh and as febrifuge; green-manure.

Desmodium gyroides DC. (Leguminosaceae). - Shrub like. Trop. Asia. Used as green-manure in different parts of the tropics.

Desmodium heterophyllum DC. (Leguminosaceae). - Trailing shrub. S.E. Asia to Philipp. Islds. Recommended as fodder for cattle.

Desmodium latifolium DC. (syn. D. lasiocarpum DC.) (Leguminosaceae). - Herbaceous plant. Trop. Africa, Asia. Plant is used in Nigeria as a food for horses.

Desmodium Oldhami Oliv. (Leguminosaceae). - Perennial herb. Japan. Leaves are used as a tea in some parts of Japan.

Desmodium tortuosum (Sw.) DC. (syn. Meibomia purpurea [Mill.] Vail.) Beggarweed. (Leguminosaceae). - Perennial herb. Tropic and subtropics. Used as green manure in warm countries.

Desmodium triflorum DC. (Leguminosaceae). - Herbaceous. Tropics of Old and New World. Used for dysentery. Recommended as cover-crop and green-manure.

Desmoncus chinantlensis Liebm. (Palmaceae). - Palm. S.E. Mexico. Leaves used by natives of Mexico for basketry.

Desmoncus major Crueg, Picmoc (Palmaceae). - Climbing palm. Trinidad, S. America. Mature stems used for basket-work.

Detarium senegalense Gmel. (syn. D. microcarpum Guill. and Perr.). (Leguminosaceae). - Tree. Trop. Africa. Wood grey, fine, regular grained, hard, works well, resistant to insects; a source of African Mahogany; used for carpentry, joinery etc. Produces a fragrant resin. Sweet pulp from fruits is used as substitute for sugar. Sold in markets of Senegal.

Determa → Ocotea rubra Mez.

Deutzia scabra Thunb. (Saxifragacea) - Shrub or small tree Japan. Wood fine-grained, whitish; used in Japan for mosaic work and wooden nails.

Devilpepper, Java → Rauwolfia serpentina Benth.

Devilpepper, Trinidad → Rauwolfia canescens L.

Dewberry,European → Rubus caesius L.

Dewberry, Northern → Rubus flagellaris Ait.

Dewberry, Southern → Rubus trivialis Michx.

Dhakki → Acacia Jacquemontii Benth.

Dkam-ka-gond Gum → Woodfordia floribunda Salisb.

Dhaura Gum → Woodfordia floribunda Salisb.

Dhupa Fat → Vateria indica L.

Dialium indum L., Tamarind Plum. (Leguminosaceae). - Tree. Mal. Archipel. Fruits small, black; consumed by native children and women as a delicacy; sold in markets. Wood very hard, very heavy, strong, dark yellow-brown to chestnut-brown; important lumber for construction of houses, boats and made into rulers.

Dialium Maingayi Baker. (Leguminosaceae). - Tree. Mal. Penins. Fruits edible, sold in native markets.

Dialium ovoideum Thwaites. (Leguminosaceae). - Tree. Ceylon. Fruits eaten by the natives. Sold in bazaars of Ceylon.

Dialium platysepalum Baker. (Leguminosaceae). Large tree. Malaya. Fruits are brownish black, pulp brown, acid; consumed as a delicacy among the natives.

Dialyanthera otoba (H. and B.) Warb. (Myristicaceae). - Tree. Costa Rica, Panama, Colombia, Venezuela, Peru. Fat containing seeds used in Colombia for combating parasites of animals.

Dianella nemorosa Lam. (syn. D. sandwicensis Hook. and Arn.). (Liliaceae). - Perennial herb. Hawaii Islds. Juice from the berries is source of a pale blue dye; used by the Hawaiians.

Dianthus Caryophyllus L., Carnation, Clove, Pink, Picotee. (Caryophyllaceae). - Perennial herb. Mediterranean region. Cultivated as an ornamental. Flowers are source of an ess. oil, used in perfumery, which is manuf. with volatile solvents; used for high grade perfumes, having excellent fixation value. About 400 to 500 kg flowers produce 1 kg concrete being source of about 100 gr. absolute. Oil contains eugenol, benzyl benzoate, benzyl salicylate, methyl salicylate, phenyl ethyl alcohol. Much is derived from the French and Italian Riviera. There is much variation as to odor.

Diatoms (Diatomaceae). - Microscopic one-celled algae, composed of numerous species, living in fresh and salt water of different parts of the world, prefering the cooler regions. F o s-s i l s p e c i e s are a source of Diatomaceous Earth, also called Infusorial Earth, Silicious Earth, Fossil Flour, Kieselgur being composed of fragments of the outer silicious walls. Vast deposits are known from various parts of the world. A white, gray to light buff powder. Used

in clarifying pharmaceuticals. Has the capacity of taking up and of holding of about four times its weight of water, therefore much used as an absorbant for liquids and for dispensing fluid extracts in powderform. Clarifying of oils, varnishes, filtering liquids, heat insulations, fire brick, fire- and acidproof packing materials. Filler for paper, paints. Important in the manuf. of dynamite. Putzpomades, various metal polishes, nail polishes, dentifices. Recent species. Pleurosigma angulatum W. Sm. is used to test the accuracy of microscope lenses. Also some other species are used for this purpose, e. g. Triceratium favus Ehrbg., Navicula lyra Ehrbg., Stauromeis phoenicentaneis Ehrbg. Surirella gemma Ehrbg. (Diatom test plates). Diatom type plates are microscopic slides on which 20 or 100 different species are neatly arranged. Under each species occurs the sci. name in microphotography. Also commercial microscopic slides are made, showing a number of different species that are arranged artistically into designs.

Diatomaceae → Diatoms.

Diatomaceous Earth → Diatoms.

Dicalyx tinctorius Blume → Symplocos fasciculata Zoll.

Dicentra canadensis (Gold.) Walp., Squirrel Corn, Turkey Corn. (Fumariaceae). - Small perennial herb. Eastern N. America. Dried tubers used medicinally as a tonic, alterative. Contains several alkaloids among which protopine and corydaline; also a bitter principle.

Dicentra Cucullaria (L.) Bernh., Dutchman's Breeches. (Fumariaceae). - Small perennial herb. Eastern N. America. Dried tubers are used medicinally as a tonic and alterative.

Dichapetalaceae → Dichapetalum.

Dichapetalum timoriense Engl. (syn. Chailletia timoriensis DC.). (Dichapetalaceae). - Creeping shrub. Java. Bark used for making fishing nets by the natives. Young leaves are eaten cooked. Fruits are cooked with fish.

Dichelachne crinita Hook. f. (syn. D. Hookeriana Trin., Stipa Dichelachne Steud.) Shorthair Grass. (Graminaceae). - Perennial herb. Australia, New Zealand. Important pasture grass in Australia.

Dichelachne sciurea Hook. f. (syn. D. Sieberiana Trin., Stipa Dichelachne Steud.) Shortair Plumegrass. (Graminaceae). - Perennial herb. New South Wales, Queensland, New Zealand. A wintergrass, valuable as fodder plant in Australia.

Dichelostemma capitatum (Benth.) Wood. → Brodiaea capitata Benth.

Dichrostachys cinerea Wight and Arn. (syn. Acacia cinerea Spreng., Cailliea callistachys Hassk.) (Leguminosaceae). - Shrub. Trop. Africa. Bark is employed for tanning in some parts of Africa. Root is used by the natives for intestinal worms. Wood is very heavy, extremely hard, strong, durable, red-brown; used for cogwheels, pegs and de luxe articles.

Dicksonia antartica Lab. (Cyatheaceae). - Tall tree-fern. Australia. Pulp of upper part of trunk roasted in ashes, consumed by aboriginals. Is source of a starch.

Dicksonia Blumei Moore. (Cyatheaceae). - Tree fern. Mal. Archip. Scales from the leaves are used to stop bleeding.

Dicoma anomala Sonder. (Compositae). Herbaceous plant. Trop. Africa. Root used by natives for dysentery and diarrhea.

Dicrastylis ochrostricta F. v. Muell. (Verbenaceae). - Tree. Australia. Tomentose mass at base of the stem is used by the Wilpirri (Centr. Australia) to decorate the body.

Dictamnus albus L. (syn. Fraxinella Dictamnus Moench.), Dithany. (Rutaceae). - Perennial herb. S. and temp. Europe, Siberia, N. China. Young leaves used as a tea in some parts of Siberia. Root is used medicinally, being an uterine stimulant; source of dictamine, an alkaloid.

Dictyophora indusiata (Vent.) Ad. Fischer. (Phallaceae). - Basidiomycete Fungus. Tropics. The fruitbodies when pulverized are mixed with flowers of Hibiscus rosa-sinensis which is put on tumors.

Dictyopteris plagiogramma Mont. Limu Lipoa (Hawaii). (Dictyotaceae). - Brown Alga. Plants are a favorite food among the Hawaiians.

Dictyota acutiloba J. Ag. (Dictyotaceae). - Brown Alga. Sea Weed. Pac. Ocean. Consumed as food among the Hawaiians. Sold in markets of Hawaii Islds. Called Limu Lipoa.

Dictyota apiculata J. Ag. (Dictyotaceae). - Brown Alga. Pac. Ocean. Plants are used as food among the natives of the Pacific Islands. Dictyotaceae (Brown Algae) → Dictyopteris, Dictyota.

Dicypellium caryophyllatum (Mart.) Nees. (Lauraceae). - Tree. Trop. America. Bark is source of Clove Bark; sold in long quills; used for flavoring. Produces Clove Bark oil, obtained from distillation; has a strong clove-like scent.

Didymocarpus reptans Jack. (Gesneriaceae). - Creeping herb. Sumatra, Mal. Penin. Decoction of leaves used for dysentery, colic, constipation.

Didymopanax longipetiolatum March., Mandioqueira. (Araliaceae). - Tree. S. Brazil. Wood pale brownish, fine structure, straight-grained, easy to work, not resistant to decay; used in Brazil for boxes that need rough handling.

Didymopanax Morototoni (Aubl.) Decne. Matchwood. (Araliaceae). - Tree. West Indies, Centr. America, Venezuela. Wood soft, light, brittle, not durable, close-grained; used for general carpentry, interior construction of buildings, boxes, paper pulp.

Digenea simplex (Wulf.) J. Ag. (Rhodomelaceae). - Red Alga. Sea Weed. Tropical Seas. Is used as food in Japan. Employed in Chinese and Japanese pharmacy for worm diseases, has similar properties as santonin. Anthelmintic and laxative for infants. Source of macnin.

Digera arvensis Forsk. (Amaranthaceae). Herbaceous plant. Trop. Africa, Asia. Plant is eaten as potherb in Sudan, also as fodder for livestock.

Digger Pine → Pinus Sabiniana Dougl.

Digitalis purpurea L., Foxglove. (Scrophulariaceae). - Herbaceous perennial. Centr. and S. Europe. Cultivated. Dried leaves collected from the first and second year old plants, late in summer or autumn, are used medicinally. Action on central nervous system, heart and vessels; cardic tonic and stimulant. Contains several glucosides, e. g. digitaline, digitoxin, digitonin and digitalein. Occasionally are used D. lanata Ehrh., D. lutea L., D. ferruginea L., and D. grandiflora Jacq.

Digitaria exilis Stapf. (syn. Paspalum exile Kipp., P. longiflorum Chev.). (Graminaceae). - Annual grass. Trop. Africa. Cultivated. Grains are used as cereal by the natives of Hausa.

Digitaria Iburua Stapf. (Graminaceae). - Annual grass. Hausaland (Afr.). Cultivated. Cereal, used as a food by the natives.

Digitaria longiflora Pers. (syn. Paspalum brevifolium Flueg. (Graminaceae). - Perennial grass. Mal. Archipel. Recommended as fodder for cattle in Mal. Archipel.

Dika Butter → Irvingia gabonensis Baill.

Dika Bread → Irvingia Barteri Hook. f.

Dika Nut → Irvingia Barteri Hook. f.

Dikkamaly Resin → Gardenia gummifera L. f.

Dill → Anethum graveolens L.

Dilo Oil → Calophyllum inophyllum L.

Dillenia megalantha M. Bayani. (Dilleniaceae). - Tree. Philipp. Islds. Fruits edible, green, fleshy, juicy, acid; used for preserves.

Dillenia mindanaense Elm. (Dilleniaceae). - Tree. Philipp. Islds. Fleshy petals and fruits are edible; recommended for preserves.

Dillenia pentagyna Roxb. (Dilleniaceae). - Tree. India, Malaya. Leaves when old contain much silica; used as „sand-paper".

Dillenia philippensis Rolfe. (Dilleniaceae). - Medium sized tree. Philipp. Islds. Fruits are smooth, green, somewhat flattened; recommended for preserves. Bark is source of a red dye.

Dillenia Reifferscheidia F. Vil. (Dilleniaceae). - Tree. Philipp. Islds. Fruits are used in the Philipp. Islds. in sauces and jam.

Dilleniaceae → Curatella, Davilla, Dillenia, Saurauia, Tetracera.

Dilsea edulis Stackh. (syn. Iridaea edulis [Stackh.] Harv.) (Dumontiaceae). - Red Alga. N. temp. Atlantic. Sometimes eaten among fishermen in Ireland and south-west of England.

Dimorphandra Gonggrijpii Sandw. (syn. Mora Gonggrijpii [Kleinh.] Sandw. Morabukea. (Leguminosaceae). - Tree. Guiana, Venezuela. Wood reddish brown with slight purplish hue, heavy, hard, tough, durable, not difficult to work; used for general construction work and ship building.

Dimorphandra megistosperma Pitt. (Leguminosaceae). - Tree. Centr. America. Seeds are source of a red dye.

Dimorphandra Mora Benth. (syn. Mora excelsa Benth.) Mora. (Leguminosaceae). - Tree. Guiana, Orinoco Delta (Venez.), Trinidad. Wood resistant to decay, durable resistant to Teredo and other marine borers. Used for ship-building, rail-road ties, paving blocks, house frames. Also M. paraensis Ducke, Brazil.

Dingleberry → Vaccinium erythrocarpum Michx.

Dionycha Bojeri Naud. (Melastomaceae). - Shrub or small tree. Madagascar. Source of a black dye; used in Madagascar for coloring silk.

Dioon edule Lindl. Chamal, Palma de la Virgin, Palma de Macetas. (Cycadaceae). - Palm like plant. Mexico. Seeds are eaten in some parts of Mexico boiled or roasted. Decoction of seeds is used for neuralgia.

Dioscorea alata L. White Yam, Water Yam, Greater Yam, Ten-Months Yam. Name de Agua. (Dioscoreaceae). - Perennial vine. South Pacif. Islands. Cultivated in tropics of Old and New World. Tubers are consumed as food. Some weigh as much as 50 kg. Several varieties; white ones are usually prefered to those that are red. Plant produces aerial tubers. Roots are eaten baked, boiled, mashed, fried, as yam salad or roasted with meat. Good varieties are suitable for marketing.

Dioscorea altissima Lam. Name Dunguey. (Dioscoreaceae). - Perennial vine. West Indies Roots coarse, very irregular, has been used as food for many centuries in Trop.

Dioscorea atropurpurea Roxb. (Dioscoreaceae). - Perennial vine. Burma, Malaya. Tubers are consumed as food; they are sometimes called Malacca Yams.

Dioscorea Batatas Decne. (syn. D. divaricata Blanco.). Chinese Potato, Chinese Yam, Name de China. (Dioscoreaceae). - Perennial herb. Trop. Asia, Malaya. Widely cultivated in Tropics of Old and New World. Tubers much consumed in different countries, having high nutritious value. A starch derived from the tubers is called Guiana Arrowroot.

Dioscorea Bemandry Jum. and Perr. (Dioscoreaceae). - Herbaceous vine. Madagascar. Tubers are consumed as food by the natives of Madagascar.

Dioscorea bulbifera L., Air Potato. (Dioscoreaceae). - Trop. Asia. Perennial vine. Tubers are consumed as food in different parts of the tropics. Cultivated.

Dioscorea cayenensis Lam. (syn. D. rotundata Poir.) Guinea Yam, Yellow Yam, Affun Yam, Negro Yam, Name Amarillo, Name Guinea. (Dioscoreaceae). - Perennial vine. Probably from trop. America. Cultivated in tropics of Old and New World. Roots edible, weighing from 2 to 20 kg, having a roughened and barklike exterior, with pale jellow flesh. One of the best yams for the home and market in Porto Rico and the West Indian Islands.

Dioscorea cinnamomifolia Hook. (syn. D. teretiuscula Klotzsch., D. tuberosa Vell.). (Dioscoreaceae). - Herbaceous vine. Brazil. Cultivated. Large, starchy tubers are consumed as food among the rural population in some parts of Brazil.

Dioscorea dodecaneura Vell. (syn. D. hebantha Mart.). (Dioscoreaceae). - Vine. Brazil. Tubers are consumed as food in some parts of Brazil.

Dioscorea dumetorum (Kunth) Pax. (Dioscoreaceae). - Perennial vine. E. Africa. Tubers are an important food among the natives of Usumbara (Afr.). Cultivated varieties are Kiwa with gold-yellow tubers; Sinquekano with white, mealy tubers; Vigonjo with yellow roots.

Dioscorea esculenta (Lour.) Burk., Potato Yam, Lesser Asiatic Yam, Fancy Yam, Name Papa. (Dioscoreaceae). - Perennial vine. Probably from Africa. Tubers are consumed as food. Several are produced by one plant, having the appearance of potatoes. Flesh is snowy white, mealy and slightly sweet.

Dioscorea glandulosa Klotzsch. (Dioscoreaceae). - Vine. Brazil. Tubers are consumed in some parts of Brazil.

Dioscorea globosa Roxb. (Dioscoreaceae). - Perennial vine. E. India. Roots are consumed as food among the natives of India.

Dioscorea hastata Vell. (Dioscoreaceae). - Perennial vine. S. America. Tubers are of good quality, being consumed as food in some parts of Brazil.

Dioscorea hastifolia Nees. (Dioscoreaceae). - Perennial vine. W. Australia. Source of a hard yam; tubers consumed by aborigins. Cultivated in a primitive way.

Dioscorea hylophila Harms. (Dioscoreaceae). - Perennial vine. Usambara (Afr.). Stem is source of a fibre; used by the natives of Usumbara for different purposes.

Dioscorea latifolia Benth., Akam Yam, Acom, Name Akam. (Dioscoreaceae). - Perennial vine. Trop. W. Africa. Grown in the tropics of Old and New World. Tubers are consumed as food.

Dioscorea lutea G.F.W. Mey. (syn. D. heptoneura Vell.). Dioscoreaceae). - Herbaceous vine. Trop. America. Tubers are consumed as food, they are also roasted and ground; used as substitute for coffee.

Dioscorea luzonensis Schau. (Dioscoreaceae). - Perennial vine. Luzon (Philipp. Islds.). Tubers of wild plants are eaten as potatoes in Philippines.

Dioscorea Maciba Jum. and Perr. (Dioscoreaceae). - Perennial vine. Madagascar. Tubers are source of a food in some parts of Madagascar.

Dioscorea papuana Rich. (Dioscoreaceae). - Perennial vine. Pacif. Islds., Papuasia. Cultivated by natives. Tuber used as food.

Dioscorea pyrifolia Kunth. (Dioscoreaceae). - Climbing perennial herb. Mal. Penin. Wild food plant; tubers eaten by the Sakai.

Dioscorea rhipogonoides Oliv. Dye yam. (Dioscoreaceae). - Vine. Kwangtung, Kwangsi (China), Formosa. Roots used by fishermen of Formosa for tanning and dyeing nets; in Pakhoi for coloring coarse native cotton-cloth and in Canton for dyeing grass-cloth.

Dioscorea spinosa Roxb., Spiny Yam, Wild Yam. (Dioscoreaceae). - Perennial spiny vine. E. India, Polynesia. Tubers of wild plants are eaten boiled by the natives of the Pacif. Islands.

Dioscorea transversa R. Br. (Dioscoreaceae). - Perennial vine. Australia. Tubers when raw or roasted are consumed by the aborigenes of N. Queensland.

Dioscorea trifida L. f., Yampi, Name Mapicey. (Dioscoreaceae). - Perennial vine. West Indies, Guiana. Cultivated in tropical America. Roots edible, excellent for table use. One plant produces several tubers. Varieties are: Morado, Blanco and Cush-Cush. Has been suggested as a source of cortisone, also D. cayensis Lam., D. esculenta (Lour.) Burk., D. latifolia Benth., D. rotunda Poir., D. alata L.

Dioscorea trifoliata H. B. K. (syn. D. triloba Kast.). (Dioscoreaceae). - Vine. S. America. Sweet tubers and their starch are consumed in some parts of Brazil.

Dioscorea villosa L. Atlantic Yam, Wild Yam. (Dioscoreaceae). - Herbaceous Perennial. N. America to Florida and Texas. Dried rhizome is used medicinally, being an expectorant and diaphoretic; contains an acrid resin and a compound related to saponin. Much of the supply comes from N. Carolina, Virginia, Indiana and Michigan.

Dioscoreaceae → Dioscorea.

Diospyros atropurpurea Gurke. (Ebenaceae). - Tree. Trop. W. Africa. Woody very hard, a brown ebony, having black streaks. Exported from Cameroon. Fruit is edible.

Diospyros bipindensis Gürke. (Ebenaceae). - Tree. Trop. Africa. Source of South Cameroon Ebony.

Diospyros camerunensis Gurke. (Ebenaceae). - Tree. Trop. W. Africa. Wood tough, hard, a black ebony type, resilient, heavy, not floating; used in Liberia and along Gold Coast for handles of axes and implements, pestles and house poles.

Diospyros Conzattii Standl. (Ebenaceae). - Tree. Oaxaca (Mex.). Fruits edible, being of fine flavor.

Diospyros crassiflora Hiern., Benin Ebony. (Ebenaceae). - Tree. Trop. Africa. Wood a black ebony; used for combs. Exported from Benin.

Diospyros Dendo Welw., Black Ebony, Billet Wood, Gaboon Ebony, Calabar Ebony. (Ebenaceae). - Tree. Trop. Africa, esp. Angola, Cameroon. Wood strong, dense, durable, black in the center of the trunk; used for turnery, inlaying etc.

Diospyros discolor Willd. (D. Mabola Roxb.). Mabolo. (Ebenaceae). - Tree. Malaya, Philipp. Islds. Occasionally cultivated in Old and New World. Fruits called Butter Fruit, roundish, velvety with dry flesh of a dark cream color; consumed by the natives. Source of Camogan Ebony. Also D. multiflora Blanco.

Diospyros Ebenaster Retz., Zapote Negro, Black Sapote. (Ebenaceae). - Tree. Mexico. Cultivated. Fruits consumed fresh; pulp blackish; pulp eaten by adding some orange or lemon juice.

Diospyros Ebenum Koenig. Ebony Tree, Sapote Negro, Black Sapote. (Ebenaceae). - Tree. India. Wood very hard, very closely and evenly grained; used for turnery, keys of pianos, rulers, backs of brushes, stands for ornaments. Source of Ebony Wood.

Diospyros embryopteris Pers. (Ebenaceae). - Tree. Trop. Asia, Africa. Used by natives as dye, for tanning, toughening fishing nets. Fruits edible. Gum derived from unripe fruits is used in India to paint bottoms of boats as a preservative.

Diospyros flavescens Gürke. (Ebenaceae). - Tree. Trop. Africa. Wood is source of a Gabon Ebony. Also D. evila Pierre.

Diospyros Gardneri Thw. (syn. D. pyregrina [Gaertn.] Gurke). (Ebenaceae). - Tree. E. India, Ceylon. Source of Bombay Ebony. Also D. ramiflora Roxb. D. sylvatica Roxb., D. tupru Buch. and D. insignis L. f.

Diospyros haplostylis Boiv. (Ebenaceae). - Tree. Madagascar. Source of a Madagascar Ebony. Also D. microrhombus Hiern. and D. Perrieri Jum.

Diospyros hirsuta L. f. (Ebenaceae). - Tree. E. India and Ceylon. Source of Coromandel Ebony. Also D. oppositifolia Thw. and D. quaesita Thw.

Diospyros Kaki L. f. (syn. D. chinensis Blume). Kaki, Japanese Persimon. (Ebenaceae). - Small tree. China. Commercially grown in China, Japan, Mediterranean region, Southern United States. Fruits have size and appearance of tomatoes, orange to red, orange yellow; pulp of the same color, eaten raw. In China also consumed when dried, called Ki-kwe. Numerous varieties among which are: Fuyuaki, Hackiya, Tamopan, Tane-nashi, Tsuru.

Diospyros Kurzii Hiern. (Ebenaceae). - Tree. Nicobar and adjacent islands. Source of a Zebra Wood or Andaman Zebra Wood, being black and gray striped, seldom entirely black, very heavy, hard, durable, easy to work; used for furniture, luxe articles and inlay work.

Diospyros Lotus L. (Ebenaceae). - Tree. W.Asia to China. Cultivated. Fruits are consumed fresh, dried or over-ripe (bletted), like Medlar. Has been suggested as a stock for the Persimmon.

Diospyros Loureiriana G. Don. (Ebenaceae). - Small tree. Trop. Afr. Roots are used by the natives of Mozambique for imparting a red color to the teeth.

Diospyros Malacapai A. DC. (Ebenaceae). - Tree. Philipp. Islds. Source of White Ebony. Also D. chrysophyllos A. DC., D. melanida Poir. from Mascar. Islds., D. foliosa Wall. from India.

Diospyros marmorata Parker. (Ebenaceae). - Tree. Malaya. Source of the beautiful, handsome Andamanese Marble Wood, streaked grey with black; used for cabinet work.

Diospyros melanoxylon Roxb. (Ebenaceae). - Tree. India. Ceylon. Source of Coromandel Ebony, resembling True Ebony. Leaves used for wrapping cigarettes.

Diospyros mespiliformis Hochst. Zanzibar Ebony, West African Ebony, Swamp Ebony. (Ebenaceae). - Tree. Trop. Africa. Wood dark brown to black, very hard, heavy, gives a fine polish; used by the natives for tool handles, combs, walking sticks, cudgels. Is exported to Europe and America where it is made into furniture, rulers, backs of combs and similar comodities. Pulp of fruit is made into a fermented drink. In Hausa (Afr.) it is used for making a soft toffee, called Ma'di.

Diospyros microphyllum Bedd. (syn. D. buxifolia Hiern.). (Ebenaceae). - Tree. S. India, Mal. Penins. Wood very hard, brittle, heavy, deep black; source of an Ebony Wood.

Diospyros mombuttensis Guerke. (Ebenaceae). - Tree. Trop. Africa, esp. Lagos, Nigeria etc. Wood used for walking-sticks, tool-handles etc. Source of Yoruba Ebony.

Diospyros montana Roxb. (Ebenaceae). - Tree. E. India, Ceylon. Source of Bombay Ebony.

Diospyros pilosanthera Blanco. (Ebenaceae). - Tree. Philipp. Islds. Source of Camagoon Ebony or Bolongeta Ebony.

Diospyros rubra Gaertn. (Ebenaceae). - Tree. Mauritius. Source of Red Mauritius Ebony.

Diospyros suaveolens Guerke. (Ebenacae). - Tree. Trop. Africa. Wood is source of Benin Ebony.

Diospyros tesselaria Poir. (Ebenaceae). - Tree. Mauritius. Source of Mauritius Ebony.

Diospyros texana Scheele, Black Persimon, Chapote. (Ebenaceae). - Tree. Texas and adj. Mexico. Wood heavy, black heartwood; used for handles of tools, turnery. Fruits are used by Mexicans for dyeing sheepskins.

Diospyros utilis Koord. and Val., Macassar Ebbeny. (Ebenaceae). - Tree. Moluccas, Ternate, Celebes. Source of a lumber. Wood hard, dark; used for furniture, handles of tools, turnery.

Diospyros virginiana L., Common Persimmon. (Ebenaceae). - Tree. Eastern N. America to Florida and Texas. Wood strong, heavy, dark brown almost black; used for wooden-ware, turnery, shoe-lasts, shuttles, plane-stocks. Fruits are edible, variable in taste and quality. Some varieties are in existence. Seedligs used in the United States as stock of the Japanese Persimmon.

Diospyros Wallichii King. (Ebenaceae). - Tree. Siam. Fruit used for fish-poison.

Diotis candidissima Desf. (Compositae). - Herbaceous plant. Mediterranean region. Herb is used by the Arabs as febrifuge and emmenagogue.

Dips → Vitis vinifera L.

Dipholis nigra Gris. (syn. Achras nigra Poir., Bumelia nigra Sw.).(Sapotaceae).- Tree. West Indies. Source of a substitute balata, Balata Batârd, Acomat-Batârd.

Dipholis salicifolia (L.) A. DC. Bustic, Cassada. (Sapotaceae). - Shrub or tree. S. Florida, West Indies, Mexico. Wood extremely hard, close grained, strong very heavy, dark brown; used for cabinet work.

Diphysa robinioides Benth. (Leguminosaceae). - Shrub or small tree. Mexico, Centr. America. Wood is source of a yellow dye.

Diplanthera speciosa Vieil. (syn. D. Deplanchei F. v. Muell.). (Bignoniaceae). - Tree. New Caledonia. Wood hard, fine-grained; suitable for turnery and cabinet work.

Diplazium asperum Blume. (Polypodiaceae). - Perennial herb. Fern. Trop. Asia. Young leaves are eaten raw or boiled by the natives.

Diplazium esculentum Swartz. (Polypodiaceae). - Tall plant. Mal. Archip. Trop. Asia. Very young leaves are eaten as lalab with rice among the Javanese.

Diplodiscus paniculatus Turcz., Barobo. (Tiliaceae). - Tree. Philipp. Islds. Starchy seeds are eaten boiled by the natives.

Diploglottis Cunninghamii Hook. f. (syn. Diploglottis Cunninghamiana Hook. f.) (Sapindaceae). - Small tree. Australia, esp. Queensland. Fruits edible; arillus orange-red, sweet, subacid, pleasant flavor; used in Australia for preserves.

Diplorhynchus mossambicensis Benth. (Apocynaceae). - Tree. Trop. Africa, esp. Zambesi, Congo. Claimed to be a source of a Guta Percha.

Diplospora malaccense Hook. f. (Rubiaceae). - Tree. Siam to Singapore. Infusion of leaves used as tea in Perak and Malacca.

Diplotaxis Duveyrierana Coss. (Cruciferaceae). - Herbaceous plant. Sahara region. Used as food by the Tuareg (N. Afr.).

Diplotaxis pendula DC. (Cruciferaceae). - Herbaceous plant. Algrian Sahara. Used as food by the Tuareg (N. Afr.).

Diplotaxis Sieberi Presl. (syn. D. acris Boiss.). (Cruciferaceae). - Herbaceous plant. Nile region, Arabia. Leaves are consumed as a salad by the Beduins of Arabia.

Diplotropis brachypetalum Tul. (Leguminosaceae). - Tree. Guiana. Wood hard, takes fine polish; used for furniture, boat building and house frames. Decoction of bark is used by the Indians to destroy vermin.

Dipsaceae → Dipsacus.

Dipsacus fullonus L. (syn. D. sativus Honck.) Teasel. (Dipsaceae). - Tall perennial herb. Occasionally cultivated. Europe, Mediterranean region. Asia Minor. Old flower-heads are a source of fuller's teasels, used in the wool industry.

Dipterocarpaceae → Anisoptera, Balanocarpus, Dipterocarpus, Doona, Dryobalanops, Hopea, Isoptera, Shorea, Vateria, Vatica.

Dipterocarpus alatus Roxb. (syn. D. costatus Gaertn.). Gurjum. (Dipterocarpaceae). - Tree. E. India, Burma. Source of Gurjum Balsam used by the natives of S. India for gonorrhoea; also employed to adulterate Copaiba Balsam and Maracaibo Balsam. Used in varnishes and for caulking boats; also D. turbinatus Gaertn. f.

Dipterocarpus gracilis Blume. (Dipterocarpaceae). - Tall tree. Mal. Archip. Wood heavy, hard, close-grained; used for house-building. Source of a balsam, used in paint-oil.

Dipterocarpus grandiflorus Blanco. (Dipterocarpaceae). - Philipp. Islds. Trunk source of Baláu or Apitong Resin; used in Philippines for varnishing; giving a tough, brilliant, durable coating; dries slowly. Wood hard, strong, stiff, red to reddish-brown; used for joists, rafters, beams, flooring, wharf and bridge-construction, furniture and for paving blocks.

Dipterocarpus hispidus Thw. (Dipterocarpaceae). - Large tree. Ceylon. Wood durable when seasoned, pale reddish-brown. Used for boat-building, loft floors.

Dipterocarpus Kerrii King (Dipterocarpaceae). - Tree. Malaya. Source of Minyak Keruong, a liquid oleo-resin; used locally in medicine and for caulking boats. Also D. Crinitus.

Dipterocarpus Kunstleri King. (Dipterocarpaceae). - Tree. Trop. Asia. Wood fairly hard and heavy, grey-brown to reddish brown, strong; recommended for house-building.

Dipterocarpu s vernicifluus Blanco. (Dipterocarpaceae). - Tree. Philipp. Islds. Trunk source of Baláu or Panau Resin; used like that of D. grandiflorus Blanco.

Dipterygium glaucum Decn. (Capparidaceae). - Shrubby plant. Nubia, Sudan. Arabia. Plant is used as food for camels.

Dipteryx odorata Willd. (syn. Coumarouna odorata [Willd.] Aubl.) Tonka Tree. (Leguminosaceae). - Tree. Guiana, Venezuela. Seeds source of Tonka Beans, often called Dutch Tonka, producing after a process of fermentation a coumarin, used for flavoring; also employed in perfumery. Supposed to be narcotic and a stimulant. C. oppositifolia is source of English Tonka. During the process of curing the beans are often steeped in rum, which takes place in Trinidad.

Dipteryx panamensis (Pitt.) Rec. (Leguminosaceae). - Tree. Nicaragua to Panama. Roasted seeds are eaten in some parts of Panama.

Dirca palustris L., Leatherwood, Swampwood. (Thymelaeaceae). - Shrub. Eastern N. America to Florida and Louisiana. The tough bark is source of a fibre which was used by the Indians.

Disoon platycarpus F. v. Muell. → Myoporum platycarpum R. Br.

Diserneston gummiferum Jaub. and Spach. → Dorema Ammoniacum D. Don.

Dishcloth Gourd. → Luffa acutangula Roxb.

Dissotis rotundifolia Triana. (Melastomaceae). - Woody plant. Trop. Africa. Leaves are used by the natives as an anthelmintic.

Disterigma margaricoccum Blake, Chirimote. (Ericaceae). - Shrub. Ecuador. Fruits edible, tender, crisp, juicy, sub-acid, very refreshing; occasionally sold in Ecuador.

Disterigma Popenoei Blake. (Ericaceae). - Shrub. Ecuador. Fruits edible, resembling the Chirimote; soft, juicy, pulp, acidulous. Occasionally sold in markets of Ecuador.

Distylium racemosum Sieb. and Zucc. (Hamamelidaceae). - Tree. China, Japan. Wood hard, fine-grained, dark brown; used in Japan for comb-backs, chop-sticks, musical instruments, boxes. Ash from the wood is employed as glazing material for porcelain.

Dithany → Dictamnus albus L.

Dittany → Cunila origanoides (L.) Butt.

Dittelasma Rarak Hook. f. → Sapindus Rarak DC.

Divi-Divi → Caesalpinia coriaria (Jacq.) Willd.

Djamoer rajap → Rajapa eurhiza (Berk.) Sing.

Djati → Tectona grandis L.

Djeluton → Dyera costulata (Miq.) Hook. f.

Dobera Roxburghii Planch. (Salvadoraceae). - Tree. Trop. Africa. India. Flowers are source of an ess. oil; used in Sudan by the women as a perfume.

Dock, Alpine → Rumex alpinus L.

Dock, Broad Leaved → Rumex obtusifolius L.

Dock, Curled → Rumex crispus L.

Dock, Patience → Rumex Patientia L.

Dock, Spanish Rhubarb → Rumex abyssinica Jacq.

Dockowar Grass Tree → Xanthorrhoea arborea R. Br.

Doctor's Gum → Symphonia globulifera L.

Dodecatheon Hendersonii A. Gray. Henderson Shooting Star. (Primulaceae). - Perennial herb. Western N. America. Roasted roots and leaves were consumed as food by the Indians of the western states.

Dodol → Garcinia Mangostana L.

Dodonaea lobulata F. v. Muell., Hop Bush. (Sapindaceae). - Shrub. Australia. Shrub is used as fodder for live-stock.

Dodonaea madagascariensis Radlk. (Sapindaceae). - Woody plant. Madagascar. Leaves are a source of food of a silkworm, Libethra cajani Vins.

Dodonaea Thunbergiana Eckl. and Zeyh. (Sapindaceae). - Shrub or small tree. S. Africa. Decoction of plant is used by the farmers as a mild purgative.

Dodonaea viscosa Jacq., Switchsorrel, Candlewood. (Sapindaceae). - Shrub. Tropics. Bark used for preparations of astringent baths and fomentations. In Australia fruits occasionally used as hops. Wood hard, brown, close-grained; used in India for tool-handles, turnery, walking-sticks, engraving.

Dogbane, Hemp → Apocynum cannabinum L.

Dogbane, Spreading → Apocynum androsaemifolium L.

Doghip → Rosa canina L.

Dog Lichen → Peltigera canina (L.) Willd.

Dog Nettle → Urtica urens L.

Dog Rose → Rosa canina L.

Dog Senna → Cassia obovata Collad.

Dog's Tail, Crested → Cynosurus cristatus L.

Dog Tooth → Erythronium Dens canis L.

Dogwood, Flowering → Cornus florida L.

Dogwood, Jamaica → Piscidia Erythrina L.

Dogwood, Pagoda → Cornus alternifolia L. f.

Dogwood, Western Flowering → Cornus Nuttallii Aud.

Dolichandrone spathacea Schum. (Bignoniaceae). - Tree. Trop. Asia. Wood easily worked, not durable; used for small household goods and toys, also for floats of fishing nets.

Dolichos biflorus L. (Leguminosaceae). - Vine. Tropics. Plant is used as food for live-stock and as greenmanure in some warm countries.

Dolichos Lablab L., Hyacinth Bean, Lablab, Monavist. (Leguminosaceae). - Annual herb. Tropics. Cultivated in warm countries. Pods are eaten as snap-beans; also seeds are consumed as food. Plants also grown as forage for live-stock and as greenmanure.

Dolichos pruriens Jacq →Mucuna pruriens DC.

Dolichos pseudopachyrrhizus Harms. (Leguminosaceae). - Vine. Trop. Africa. Used by the natives as an insecticide.

Dolichos sesquipedalis L., Asparagus Bean. (Leguminosaceae). - Herbaceous vine. Trop. Asia. Young green pods are consumed as snap beans. Also used a forage for live-stock.

Dolichos sinensis L. → Vigna sinensis Endl.

Dolichos sphaerospermus DC. (Leguminosaceae). - Annual vine. Jamaica. Seeds are consumed as a vegetable.

Domba Oil → Calophyllium inophyllum L.

Dombeya Burgessiae Gerr. (Sterculiceae). - Shrub. S. Africa. Wood is used by the natives of E. Africa for making fire by friction.

Dombeya spectabilis Bojer. (Sterculiaceae). - Tree. Trop. Africa. Bark is used in some parts of Africa for making a rough cordage.

Doom Bark →Erythropheum guineense G. Don.

Donax cuspidata Schum. (Zingiberaceae). - Perennial herb. Trop. Africa. Leaves are much used by the natives of N. Nigeria for wrapping Kola Nuts and food. Fibre is used for making fishing-nets in Gold Coast.

Dondia suffrutescens (S. Wats.) Heller. (Chenopodiaceae). - Half-shrub. Arizona, California. Leaves were eaten boiled or as greens by Indians of Arizona and California.

Doona zeylanica Thw. (Dipterocarpaceae). - Tree. Ceylon. Wood hard, easy to split, durable; used for housebuilding, shingles, flooring, boards, joists.

Dorema Ammoniacum D. Don. (syn. Disereneston gummiferum Jaub. and Spach.). (Umbelliferaceae). - Perennial herb. Arabia to Iran and India. Source of Ammoniakum, a gum-resinous exudation, caused by punctures from insects made in the stems and leaf-stalks. Tears are pale yellowish brown, irregular to round. Gum is used medicinally as expectorant, carminative, antiseptic. Contains an acid resin, an ester of ammoresinotannol and an indifferent resin. Much is shipped from Bombay.

Dorstenia brasiliensis Lam. (Moraceae). - Perennial herb. S. America. Rhizome is used in some parts of Brazil as a stimulant, diaphoretic, tonic, purgative, diuretic, emetic and emmenagogue.

Dorstenia Contrajerva L. (Moraceae). - Herbaceous perennial. Trop. America. Dried rhizomes are used for flavoring cigarettes. Infusions from root are used in Costa Rica as febrifuge.

Dorstenia Drakena L. (Moraceae). - Perennial herb. Mexico. Tea from leaves used by Mazatecs (Mex.) for relieving alcoholic indulgences.

Doryphora Sassafras Endl., Sassafras. (Monimiaceae). - Tree. Australia, esp. New South Wales, Queensland. Wood fragrant. Used for interior finish of houses, furniture, packing boxes for keeping out insects.

Dotted Saxifrage → Saxifraga punctata L.

Double Claw → Martynia probocidea Glox.

Double Coconut → Lodoicea callipyge Comm.

Douglas Fir → Pseudotsuga taxifolia (Lam.) Britt.

Douglas Knotweed → Polygonum Douglasii Greene.

Douglas Spruce → Pseudotsuga taxifolia (Lam.) Britt.

Doum Palm → Hyphaene thebaica Mart.

Doum Palm, Egyptian → Hyphaene thebaica Mart.

Doum Palm, South African → Hyphaene crinita Gaertn.

Doundaké → Sarcocephalus esculentus Afz.

Dovyalis caffra (Hook. f. and Harv.) Warb. (syn. Aberia caffra Harv. and Sond.) Umkokolo, Kei Apple. (Flacourtiaceae). - Small tree. S. Africa. Introduced in the tropics of the New World. Fruits globose, fleshy, shape of a small apple, 2.5 to 3 cm across, greenish-yellow; pulp very juicy, pleasant flavor, aromatic, acid; used for jam, compote and marmelade; unripe fruits are made into jelly.

Dovyalis hebecarpa (Gardn.) Warb. (syn. Aberia Gardneri Clos.) Ceylon Gooseberry, Ketembilla. (Flacourtiaceae). - Small spiny tree or shrub. Trop. Asia, esp. India and Ceylon. Introd. in tropics of Old and New World. Cultivated. Berries dark purple to black, velvety, juicy, acid, 2 to 2.5 cm across, resembling a gooseberry; recommended for jelly and preserves, also served with fish and meat.

Downy Myrtle → Rhodomyrtus tomentosa Wight.

Dracaena Cinnabari Balf. Socotra Dragon Blood Dracaena. (Liliaceae). - Tree. Socotra. Source of a resin, called Dragon's Blood Resin, occurs in tears of a beautiful garnet color when broken. Used in varnishes. Employed in medicin as an astringent and to stop hemorrhage. Resin is sometimes called Socotra Dragon Blood.

Dracaena congesta Ridl. (Liliaceae). - Herbaceous plant. Mal. Archip. Decoction of root used as vermifuge and for rheumatism.

Dracaena Draco L. Dragon Dracaena, Dragon Blood Tree. (Liliaceae). - Tree. Canary Islds.

Source of a red resin, Dragon's Blood, Sanguis Draconis, Sang de Dragon. Dragon Blood is used in the varnish industry, for pigment paper; in pharmacy for red plasters.

Dracaena Mannii Baker., Asparagus Bush. (Liliaceae). - Woody plant. Trop. Africa. Source of a light colored dye; used by natives in S. Nigeria. Young shoots are consumed by the natives and Europeans as Asparagus. Chopped young leaves are eaten with rice by the natives.

Dracaena schizantha Baker. (Liliaceae). - Treelike. E. Africa, S. Arabia. Source of a red resin, Arabian Dragon's Blood, also called Socotra Dragon Blood.

Dracontomelum mangiferum Blume. (Anacardiaceae). - Tree. India to Malaya. Fruit edible, sold in markets of Tonkin and Borneo. Used by Malays as a sour ralish with fish. Flowers are used to flavor food.

Dragon Dracaena → Dracaena Draco L.

Dragon's Blood → Daemonorops Draco Blume, Dracaena Draco L. and Pergularia africana N. E. Br.

Dragon's Blood, Arabian → Dracaena Cinnabari Balf. f.

Dragon's Blood Palm → Daemonorops Draco Blume.

Dragon's Blood Padauk →Pterocarpus Draco L.

Dragon's Blood Resin → Dracaena Cinnabari Balf.

Dragon Blood, Socotra → Dracaena schizantha Baker.

Dragon Blood Tree → Dracaena Draco L.

Dragon Spruce → Picea asperata Mast.

Drimia ciliaris Jacq. (syn. Idothea ciliaris Kth.). (Liliaceae). - Bulbous perennial. S. Africa. Juice from the bulbs is used by the farmers as emetic, diuretic and expectorant.

Drimia Cowanii Ridl. (Liliaceae). - Bulbous herb. Madagascar. Bulbs are used by the Betsiles of Madagascar as rat-poison.

Drimys aromatica F. v. Muell. (Magnoliaceae). - Tree. Australia. Fruits are used as spice, substitute for pepper and allspice in Australia.

Drimys axillaris Forster. (Magnoliaceae). - Small tree. New Zealand. Wood reddish; used for inlay-work. Bark very aromatic, tonic and astringent.

Drimys Winteri Forst., Wintersbark Drimys. (Magnoliaceae). - Tree. Argentina and Chile to Mexico. Source of Winter's Bark used in domestic medicine as aromatic and pungent, having antiscorbutic and tonic properties. Used in Brazil for dysentery and gastric disturbances. Powdered bark is used as condiment in parts of Mexico and Brazil. Wood is locally used for interior finish of houses, boxes and cases.

Droga de Jara → Cistus ladaniferus L.

Drooping Leucothoe → Leucothoe Catesbaei (Walt.) Gray.

Dropseed, Mesa → Sporolobus flexuosus (Thurb) Rydb.

Dropwort, Water → Oenanthe stolonifera Wall.

Drosera ramentacea Burch. (syn. D. madagascariensis DC.). Droseraceae. Herbaceous plant. Madagascar. Herb is used in Madagascar to preserve teeth, also a remedy for dyspepsia and coughs.

Drosera rotundifolia L., Sundew. (Droseraceae). - Small perennial- insectiverous herb. Temp. Europe, Asia, N. America. Dried herb used medicinally as an antispasmodic; for whooping cough, bronchitis. Contains an enzyme that converts albumen into peptones. Also D. longifolia L. and D. anglica Huds.

Drosera Whittakerii Planch. (Droseraceae). - Perennial herb. Australia. Root-tubers are source of a red dye, suitable for dyeing silk.

Droseraceae → Drosera.

Drug Eyebright → Euphrasia officinalis L.

Drug Lionsear → Leonites Leonurus R. Br.   /

Drug Centaurium → Erythraea Centaur. Pers.

Drug Speedwell → Veronica officinalis L.

Druggist's Bark → Cinchona succirubra Pavon.

Drugsquill, India → Urginea indica Kunth.

Drummond's Ironweed → Vernonia missurica Raf.

Dryas octopetala L. (Rosaceae). - Low shrub. Alpine, subalpine, arctic regions of Europe, Asia, N. America. Leaves used in the East and West Alpes as a tea, called Schweizertee or Kaisertee.

Drynaria quercifolia (L.) J. Sm. (Polypodiaceae). - Fern. Malaya, Philipp. Islds. Decoction of rhizome is used as astringent.

Drynaria rigidula Bedd. (Polypodiaceae). - Perennial herb. Fern. Malaya. Used by natives for persistant diarrhoea and virulent gonorrhoea. Young leaves eaten in Celebes.

Dryobalanops aromatica Gaertn. (syn. D. Camphora Colebr.). Common Borneo Camphor Tree. (Dipterocarpaceae). - Tall tree. Mal. Archip. Wood coarse-grained, splits easily; used for house building, bridges, boards. Stem is source of a balsam, said to be comparable with Camphor, source of Borneol.

Dryobalanops oiocarpa v. Sl. (Dipterocarpaceae). - Tree. Borneo. Wood light red, easy worked, does not shrink; resistant to borers; used for native vessels, posts, house-building and household utensils.

Dryopteris anthamantica Ktze. (Polypodiaceae). - Perennial herb. Fern. Rhizome is used in Transvaal as vermifuge, especially tapeworm.

Dryopteris paleacea (Sw.) Christ. (Polypodiaceae). - Perennial herb. Fern. Trop. America. Is used as an anthelmintic in Colombia.

Dryopteris pteroides Kuntze. (Polypodiaceae). - Perennial herb, fern. Philipp. Islds. Stems used in Philipp. Islds. for decorating baskets.

Duabanga moluccana Blume. (Sonneratiaceae). - Tree. Mal. Archip. Wood strong, light; used in N. Celebes for boards. Decoction of chopped bark with that of Mallotus moluccana is source of a black dye; used to stain basketry derived from Pandanus.

Duabanga sonneratioides Buch-Ham. (Sonneratiaceae). - E. India. Wood soft, with light yellowish streaks, easily worked; used for canoes, tea boxes, house and boat building.

Duboisia Hopwoodii F. v. Muell., Pituri, Pitchery, (Solanaceae). - Woody plant. Australia. Leaves and finely broken twigs are used as mastigatory by the aborigenes of Centr. Australia, similar to Coca in Peru.

Duboisia myoporoides R. Br., Corkwood, Mgmeo. (Solanaceae). - Woody plant. Australia. Has intoxicating properties; used by the aborigines. Employed to stupefy fish. Contains duboisine, a mydriatic alkaloid.

Duck Potato → Sagittaria latifolia Willd.

Duckweed → Lemna minor L.

Duguetia vallicola Macbride. Yaya. (Annonaceae). - Tree. Colombia. Wood used in Colombia for handles of tools.

Duhnual Balsam → Commiphora Opobalsam Engl.

Dukes (Cherries) → Prunus avium L.

Dulse → Rhodymenia palmata (L.) Grev.

Dumontiaceae (Red Algae) → Dilsea.

Dumoria Heckeli Chev. (Sapotaceae). - Tree. Trop. W. Africa. Wood beautifully grained, fine red; used for cabinet work, parts of automobiles, railway carriages. Source of an African Mahogany; exported to Europe. Seeds source of Doumori Butter, used as food by the inhabitants of French W. Africa.

Dundee Fibre → Sesbania aculeata Poir.

Dunkai Tree → under Commiphora Hildebrandtii Engl.

Durian → Durio zibethinus Murr.

Durian, Wild → Cullenia excelsa Wight.

Durio zibethinus Murr., Durian. (Bombaceae). - Large Tree. Malaya, Philipp. Islds., Mal. Archipel. Cultivated. Introduced in tropics of the New World. Tropical fruit. Characteristic flavor, described as being a mixture of old cheese and onions, flavored with turpentine; weighs about 5 pounds. Much esteemed among natives. Eaten as a sauce with rice. Pulp of ripe fruits is fermented in jars and is eaten as a side dish. Unripe fruits are boiled and consumed as a vegetable or in soup. Sliced seeds when dried or baked in coconut oil with spices are eaten with the rijsttafel in Java. They are also eaten with sugar. Seeds are consumed when roasted.

Durra → Sorghum vulgare Pers.

Durum Wheat → Triticum durum Desf.

Durvillea antartica (Cham.) Hariot. (syn. D. utilis Bory). Bull Kelp. (Durvilleaceae). - Brown Alga. Antarctic region, coasts of New Zealand and Australia. Used by the natives of Australia and New Zealand as a food. Made by the Maoris into bags for packing and preserving mutton-birds. Used as manure. Sold on the markets of South Chile where it is used as a vegetable and in soups.

Durvilleaceae (Brown Algae) → Durvillea.

Dutchman's Breeches → Dicentra Cucullaria (L.) Bernh.

Dutch Tonka → Dipteryx odorata Willd.

Dwarf Cape Gooseberry → Physalis pubescens L.

Dwarf Elder → Sambucus Ebulus L.

Dwarf Jamaica Banana →Musa chinensis Sweet.

Dwarf Milkweed → Asclepias involucrata Engelm.

Dwarf Pine Needle Oil →Pinus pumilio Haenke.

Dwarf Sumach → Rhus copallina L.

Dye Bedstraw → Galium tinctorium L.

Dyera costulata (Miq.) Hook. f., Jelutong, Djelutong. (Apocynaceae). - Tree. Malaya. Source of a substitute for Chicle. Also used in manuf. asbestos, cellulose, linoleum. Formerly known as Dead Borneo and Pontianak Gutta.

Dyera laxiflora Hook. f. (Apocynaceae). - Tree. Mal. Archip. Source of a djelutong. Also D. Lowii Hook.

Dyer's Chamomille → Anthemis tinctoria L.

Dyer's Greenwood → Genista tinctoria L.

Dyer's Indian Mulberry → Morinda tinctoria Roxb.

Dyer's Woodruff → Asperula tinctoria L.

Dyestuffs. Lichens source of → Bacidia, Caloplaca, Candelariella, Cetraria, Cladonia, Dermatocarpon, Evernia, Gyrophora, Lecanora, Lepraria, Lobaria, Nephroma, Ochrolechia, Parmelia, ePrtusaria, Ramalina, Rhizocarpon, Roccella, Solorina, Stereocaulon, Sticta, Thelochistes, Umbilicaria, Usnea, Variolaria, Xanthoria.

Dyestuffs. Source of → Acacia, Adenanthera, Adenostemma, Adhatoda, Afzelia, Agrimonia, Ailanthus, Albizzia, Alchornea, Alectra, Alkanna, Allanblackia, Alnus, Anogeissus, Anthemis, Arachis, Aristea, Arnebia, Asperula, Asphodelus, Baccaurea, Baccharis, Baloghia, Baphia, Baptisia, Basella, Bauhinia, Berberis, Betula, Bignonia, Bischoffia, Bixa, Bocconia, Bosquea, Brackenbridgea, Bridelia, Bruguiera, Buddleia, Butea, Cabralea, Caesalpinia, Calamintha, Calceolaria, Calendula, Calluna, Cardiogyne, Carthamus, Cassytha, Cedrela, Celtis, Centaurea, Chenopodium, Chlorophora, Chrozophora, Cla-

drastis, Cleome, Cochlospermum, Comarum, Combretum, Condalia, Conyza, Coprinus, Coptis, Coriaria, Echinocvstis, Evonymus, Feronia, Fraxinus, Coscinium, Craterospermum, Cremaspora, Crocosma, Crocus, Croton, Cudrania, Curcuma, Cybistax, Dalea, Danais, Datisca, Datura, Daucus, Davilla, Delphinium, Dianella, Dillenia, Dimorphandra, Dionycha, Diospyros, Diphysa, Drosera, Duabanga, Echinodontium, Eclipta, Elaeocarpus, Elaeodendron, Elephantorrhiza, Enantia, Entada, Erica, Ervatamia, Erythroxylon, Euclea, Eugenia, Eupatorium, Ficus, Flemingia, Flindersia, Galium Garcinia, Gardenia, Genipa, Genista, Gentiana, Geranium, Gordonia, Gossypium, Grumilea, Haematoxylon, Haronga, Hedera, Helianthus, Heterothalamus, Homalanthus, Hydrastis, Hymenocardia, Hypestis, Ilex, Impatiens, Indigo, Inotus, Inula, Isatis, Jacobinia, Jatropha, Juglans, Juliana, Koelreuteria, Laetinovorus, Lafoensia, Lannea, Lecontea, Leonurus, Ligustrum, Lithospermum, Lonchocarpus, Maclura, Macrotomia, Mahonia, Mallotus, Mangifera, Marsdenia, Medicago, Melanorhoea, Mercurialis, Mirabilis, Mitragyna, Morinda, Myrica, Nolina, Nyctanthes, Ochrosia, Odina Onopordon, Onosma, Opuntia, Papaver, Parthenocissus, Peganum, Peltophorum, Peristrophe, Phellinus, Phellodendron, Phyllocladus, Phytolacca, Pipturus, Pisolithus Pithecolobium, Platycarya, Polygala, Polygonum, Polyporus, Pseudocedrela, Psoralea, Pterocarpus, Pterolobium, Pyrus, Quercus, Randia, Raphiolepis, Rauwolfia, Reseda, Rhamnus, Rhus, Rivina, Royena, Rubia, Rumex, Sadlera, Salix, Sambucus, Sapium, Sarcocephalus, Sarothamnus, Sickingia, Sisyrinchium, Sophora, Sorghum, Sorindeia, Stanleyellia, Strobilanthes, Suaeda, Symplocos, Tabebuia, Taxus, Tecoma, Terminalia, Teucrium, Toddalia, Trema, Trifolium, Uapaca, Urtica, Vauquelinia, Ventilago, Vernonia, Whitfieldia, Woolfordia, Xanthophyllum, Xanthorrhiza.

Dysoxylon acutangulum Miq. (Meliaceae). - Tree. Sumatra. Wood strong, very beautiful; used for furniture, chinese coffins and for building purposes.

Dysoxylon decandrum (Blanco) Merr. (Meliaceae). - Tree. Philippine Islds. Finely powdered bark is used as a safe emetic among the Philippinos.

Dysoxylum euphlebium Merr. (Meliaceae). - Tree. Amboina. Wood whitish; occasionally made into boards, used for house-building. All parts of the plant have a scent of onions. Young leaves are boiled with fish.

Dysoxylon Fraseranum Benth., Rosewood, Pencil Cedar. (Meliaceae). - Tree. Australia esp. New South Wales, Queensland. Wood fragrant, reddish, easily worked; used for cabinet-work, wood-engraving, ship-building, turnery. Wood is called Roho Mahogany or Australian Mahogany. Also D. Lessertianum Benth.

Dysoxylon Muelleri Benth. Turnip Wood, Pencil Cedar. (Meliaceae). - Tree. Australia, esp. New South Wales, Queensland. Wood rich red, easily worked; used for cabinet-work.

Dysoxylon spectabile Hook. f. (Meliaceae). - Tree. New Zealand. Wood light, strong, fairly durable, pale red; used for cabinet work. Decoction of the bitter leaves is used a tonic by the bushmen of New Zealand.

E

Eagle Fern → Pteridium aquilinum (L.) Kuhn.

Early Sweet Blueberry →Vaccinium pennsylvanicum Lam.

Earth Almond → Cyperus esculentus L.

Earth Chestnut → Lathyrus tuberosus L.

Earth Nut → Carum Bulbocastanum Koch.

East African Cardamon → Aframomum mala Schum.

East Indian Arrowroot → Curcuma angustifolia Roxb., C. leucorhiza Roxb.

East Indian Copal → Canarium bengalense Roxb.

East Indian Kino → Pterocarpus Marsupium Roxb.

East Indian Rosebay → Ervatamia coronaria Stapf.

East Indian Rosewood → Dalbergia latifolia Roxb.

East Indian Screw Tree → Helicteres Isora L.

Eastern Coral Bean → Erythrina herbacea L.

Eastern Hemlock → Tsuga canadensis (L.) Carr.

Eastern Lichen → Stereocaulon paschale (L.) Hoffm.

Eastern Red Cedar → Juniperus virginiana L.

Eau de Creole → Mammea americana L.

Ebenaceae → Diospyros, Euclea, Maba, Royena.

Ebony, Benin → Diospyros crassiflora Hiern. and D. suaveolens Gurke.

Ebony, Black → Diospyros Dendo Welw.

Ebony, Blue → Copaifera bracteata Benth.

Ebony, Bolongeta → Diospyros pilosanthera Blanco.

Ebony, Bombay → Diospyros montana Roxb., D. Gardneri Thw.

Ebony, Calabar → Diospyros Dendo Welw.

Ebony, Camagoon → Diospyros pilosanthera Blanco.

Ebony, Camogan → Diospyros discolor Willd.

Ebony, Cape → Euclea Pseudebenus E. Mey.

Ebony Cocus Wood → Brya Ebenus DC.

Ebony, Coromandel → Diospyros hirsuta L. f.

Ebony, Gabon → Diospyros Dendo Welw. and D. flavescens Gurke.

Ebony, German → Taxus baccata L.

Ebony, Green → Tecoma leucoxylon (L.) Mart.

Ebony, Greenheart → Tecoma leucoxylon (L.) Mart.

Ebony, Macassar → Diospyros utilis Koord. and Val.

Ebony, Madagascar → Diospyros haplostylis Boiv.

Ebony, Mauritius → Diospyros rubra Gaertn., P. tesselaria Poir.

Ebony, Mozambique → Dalbergia melanoxylon Guill. and Perr.

Ebony, South Cameroon → Diospyros bipidensis Gurke.

Ebony, Swamp → Diospyros mespiliformis Hochst.

Ebony Tree → Diospyros Ebenum Koen.

Ebony Tree, African → Diospyros mespiliformis Hochst.

Ebony, Unyoro → Dalbergia melanoxylon Guill and Perr.

Ebony, Zanzibar → Diospyros mespiliformis Hochst.

Ebony, White → Diospyros Malacapai A. DC.

Ecballium Elaterium A. Rich., Squirting Cucumber. (Cucurbitaceae). - Perennial vine. Mediterranean region. Cultivated. Juice from full-grown, unripe fruits, is used medicinally; has strong purgative properties. Contains elaterium and elaterin.

Ecclinusa balata Ducke. (Sapotaceae). — Medium sized tree. Brazil. Stem is source of a balata-like substance called Abiurana, of an inferior quality. Much is exported from Manaos.

Ecclinusa sanguinolenta Pierre. (syn. Ragala sanguinolenta Pierre). (Sapotaceae). - Tree. Guiana, Amazone Region. Source of a substitute balata, Balata saignant, Balata rouge, Wapo.

Echinacea angustifolia DC. → Brauneria angustifolia (DC.) Heller.

Echinocactus horizonthalonius Lem. (syn. E. equitans Scheidw.). (Cacataceae). - Xerophyte. S. W. of the United States. Pulp of plant is used for making sweetmeats.

EchinocactusWilliamsii Lem. → Lophophora Williamsii (Lem.) Coulter.

Echinocactus Wislizeni Engelm. → Ferocactus Wislizenia (Engelm.) Britt. and Rose.

Echinocereus conglomeratus Forst. (syn. Cereus conglomeratus Berger.) Pitahaya de Agosto. (Cactaceae). - Xerophytic shrub. Mexico. Fruits are edible.

Echinocereus dasyacanthus Engelm. (syn. Cereus dasyacanthus Engelm.) (Cactaceae). - Xerophytic shrub. Chihuahua (Mex.) W. Texas, New Mexico. Fruits edible, 2.5 to 3.5 cm across, purplish.

Echinocereus enneacanthus Engelm. (syn. Cereus enneacanthus Engelm.) Strawberry Cactus (Cactaceae). - Xerophytic shrub. N. Mexico. New Mexico and Texas. Fruits edible, delicious strawberry flavor, eaten raw or in preserves.

Echinocereus stramineus (Engelm.) Ruempl. Pitahaya. Mexican Strawberry. (Cactaceae). - Shrub. Chihuahua (Mex.), W. Texas, New Mexico. Fruits edible, of pleasant flavor.

Echinochloa colonum (L.) Link., Shama Millet, Jungle Rice. (Graminaceae). - Annual grass. Cultivated in warm countries of Old and New World, as fodder for livestock.

Echinochloa frumentacea (Roxb.) Link. (syn. Panicum frumentacea Roxb.). Japanese Millet, Sanwa Millet, Billion Dollar Grass. (Graminaceae). - Annual grass. E. Asia. Cultivated as fodder for livestock.

Echinocystis macrocarpa Nutt. (Cucurbitaceae). - Vine. California. Roots are used as purgative. Seeds are source of a red dye, used by the Luisiño Indians in California.

Echinodontium tinctorium (Ell. and W.) Ell. and Ev. (Polyporaceae). - Basidiomycete Fungus. Northwestern N. America. The fruitbodies are the source of a dye used by the American Indians.

Echinops dahuricus Fisch. (Compositae). - Perennial herb. Fukien, Shantung (China), Mongolia. Used in China medicinally.

Echinops viscosus DC. (Compositae). - Perennial herb. Mediterranean region, Greece. Source of a gum, called Angado Mastiche, used in some ports of Greece for chewing.

Eckebergia capensis Sparm. Cape Ash. (Meliaceae). - Tree. S. Africa. Wood white, soft, not durable, open-grained; used for furniture, beams, planks, wagon work.

Ecklonia bicyclis Kjellm. (syn. Capea elongata Mart.) (Alariaceae). - Brown Alga. Plants are used as food in Japan. They are also used in manuf. of iodine.

Ecklonia cava Kjellm. (Alariaceae). - Brown Alga. Pac. Ocean. Used in Japan as manure, also source of iodine and potash.

Ecklonia kurome Okam (Alariaceae). - Brown Alga. Seas around Japan and China. Plants are used as food in the Orient. esp. in China and Japan.

Ecklonia latifolia Kjellm. (Alariaceae). - Brown Alga. Plants are used as food among the Chinese and Japanese.

Ecklonia radicosa Okam. (syn. Laminaria radicosa Kjellm.) (Alariaceae). - Brown Alga. Plants are used as food in Japan.

Eclipta erecta L. (syn. E. alba Hassk.) (Compositae). - Herbaceous plant. Tropics. Decoction of the leaves used in India as laxative. Plant is source of a black stain used to color the hair. A black bluish dye is used by some races in India for tattooeing, after puncturing parts of the skin.

Economic Algae → Blue-Green Algae, Brown Algae, Diatoms, Green Algae and Red Algae.

Economic Blue-green algae → Brachytrichia, Nostoc.

Economic Brown Algae → Alariaceae, Dictyotaceae, Durvilleaceae, Fucaceae, Laminariaceae, Lessoniaceae, Mesogloiaceae, Scytosiphonaceae.

Economic Ferns and Fern Allies → Ceratopteridaceae, Cyatheaceae, Equisetaceae, Gleicheniaceae, Hymenophyllaceae, Isoetaceae, Lycopodiaceae, Marattiaceae, Ophioglossaceae, Polypodiaceae, Salviniaceae, Schizaceae, Selaginellaceae.

Economic Flowering Plants among Families of → Acanthaceae, Aceraceae, Actinidiaceae, Aizoaceae, Alismaceae, Amarantaceae, Amaryllidaceae, Anacardiaceae, Ancistrocladaceae, Annonaceae, Apocynaceae, Aponogetonaceae, Aquifoliaceae, Araceae, Araliaceae, Araucariaceae, Aristolochiaceae, Asclepiadaceae, Balanophoraceae, Balsaminaceae, Basellaceae, Batidaceae, Begoniaceae, Berberidaceae, Betulaceae, Bignoniaceae, Bixaceae, Bombaceae, Boraginaceae, Bromeliaceae, Burseraceae, Butomaceae, Buxaceae, Cactaceae, Calycanthaceae, Campanulaceae, Canellaceae, Cannaceae, Capparidaceae, Caricaceae, Caryocaraceae, Caryophyllaceae, Casuarinaceae, Celastraceae, Chenopodiaceae, Chloranthaceae, Cistaceae, Cochlospermaceae, Combretaceae, Commelinaceae, Compositae, Connaraceae, Convolvulaceae, Coriariaceae, Cornaceae, Crassulaceae, Cruciferaceae, Cucurbitaceae, Cunoniaceae, Cupressaceae, Cycadaceae, Cyclanthaceae, Cynomoriaceae, Cyperaceae, Cyrillaceae, Cytinaceae, Datiscaceae, Dichapetalaceae, Dilleniaceae, Dioscoreaceae, Dipsaceae, Dipterocarpaceae, Droseraceae, Ebenaceae, Elaeagnaceae, Empetraceae, Epacridaceae, Ephedraceae, Ericaceae, Erythroxylaceae, Eucommiaceae, Eucryphiaceae, Euphorbiaceae, Fagaceae, Flacourtiaceae, Flagillariaceae, Fouquieriaceae, Frankeniaceae, Fumariaceae, Garryaceae, Gentianaceae, Geraniaceae, Gesneriaceae, Ginkgoaceae, Globulariaceae, Gnetaceae, Goodeniaceae, Graminaceae, Grossulariaceae, Guttiferaceae, Halorrhagidaceae, Hamamelidaceae, Hernandiaceae, Hippocastanaceae, Hippocratea-

ceae, Humiraceae, Hydnaceae, Hydrophyllaceae, Hypericaceae, Icacinaceae, Iridaceae,
Juglandaceae, Juncaceae, Juncaginaceae, Julianaceae, Krameriaceae, Labiaceae, Lardizabalaceae, Lauraceae, Lecythidaceae, Leguminosaceae, Leitneriaceae, Lemnaceae, Lennoaceae,
Lentibulariaceae, Liliaceae, Linaceae, Loasaceae, Loganiaceae, Loranthaceae, Lythraceae,
Magnoliaceae, Malpighiaceae, Malvaceae, Marantaceae, Martyniaceae, Melastomaceae, Meliaceae, Melianthaceae, Menispermaceae, Monimiaceae, Moraceae, Moringiaceae, Myoporaceae, Myricaceae, Myristicaceae, Myrsinaceae,
Myrtaceae, Najadaceae, Nepenthaceae, Nyctaginaceae, Nymphaeaceae, Nyssaceae, Ochnaceae, Olacaceae, Oleaceae, Onagraceae, Opiliaceae, Orchidaceae, Orobanchaceae, Oxalidaceae, Palmaceae, Pandaceae, Pandanaceae, Papaveraceae, Passifloraceae, Pedaliaceae, Penaeaceae, Phytalaccaceae, Pinaceae, Piperaceae,
Pittosporaceae, Plantaginaceae, Platanaceae,
Plumbaginaceae, Podocarpaceae, Polemoniaceae, Polygalaceae, Polygonaceae, Pontederiaceae, Portulacaceae, Potamogetonaceae, Primulaceae, Proteaceae, Punicaceae, Pyrolaceae,
Rafflesiaceae, Ranunculaceae, Resedaceae, Restionaceae, Rhamnaceae, Rhizophoraceae, Rosaceae, Roxburghiaceae, Rubiaceae, Rutaceae,
Sabiaceae, Salicaceae, Salvadoraceae, Santalaceae, Sapindaceae, Sapotaceae, Saururaceae,
Saxifragaceae, Scrophulariaceae, Simarubaceae,
Solanaceae, Sonneratiaceae, Sparganiaceae,
Sterculiaceae, Styraceae, Symplocaceae, Taccaceae, Tamaricaceae, Taxodiaceae, Taxaceae,
Theaceae, Thelygonaceae, Theophrastaceae,
Thymelaeaceae, Tiliaceae, Trochodendraceae,
Tropaeolaceae, Turneraceae, Typhaceae, Ulmaceae, Umbelliferaceae, Urticaceae, Valerianaceae, Verbenaceae, Violaceae, Vitaceae, Vo-
chyaceae, Xyridaceae, Zingiberaceae, Zygophyllaceae.

Economic Fungi → Agariaceae, Auriculariaceae, Clavariaceae, Cyttariaceae, Elaphomycetaceae, Eutuberaceae, Helotiaceae, Helvellaceae, Hydnaceae, Hymenogastraceae, Hypo-
creataceae, Lycoperdaceae, Phallaceae, Podaxaceae, Polyporaceae, Sclerodermataceae, Terfeziaceae, Tremellaceae, Xylariaceae.

Economic Green Algae → Cladophoraceae,
Chaetophoraceae, Codiaceae, Hydrodyctiona-
ceae, Ulvaceae.

Economic Lichens → Alectoria, Anaptychia,
Bacidia, Caloplaca, Candelariella, Cetraria,
Cladonia, Dermatocarpon, Evernia, Gyrophora,
Haematomma, Lecanora, Lecidea, Lepraria,
Letharia, Lobaria, Nephroma, Ochrolechia,
Parmelia, Peltigera, Pertusaria, Physica, Rama-
lina, Rhizocarpon, Raccella, Solorina, Stereo-
caulon, Sticta, Theloschistes, Umbilicaria, Ur-
ceolaria, Usnea, Viriolaria, Xanthoria.

Economic Red Algae → Bonnemaisoniaceae,
Bangiaceae, Ceramiaceae, Dumontiaceae, Endocladiaceae, Gelidiaceae, Gracilariaceae, Gi-

gartinaceae, Grateloupiaceae, Phyllophoraceae,
Rhabdoniaceae, Rhodomelaceae, Rhodophyllaceae, Rhodymeniaceae, Solieriaceae, Sphaerococcaceae.

Edible Fungi → Agaricus, Amanita, Armillaria,
Auricularia, Boletus, Bovista, Calvatia, Cantharellus, Choiromyces, Clavaria, Clitocybe, Clitopilus, Collybia, Conchromyces, Coprinus, Cortinarius, Crepidotus, Cryptoporus, Cyttaria,
Favolus, Fistulina, Flammulina, Gomphidius,
Gymnopus, Gyroporus, Helvella, Hericium, Hirneola, Hydnotria, Hydnum, Hygrocybe, Hygrophorus, Hypholoma, Hyporhodius, Inocybe,
Lactarius, Lentinus, Lepiota, Lycoperdon, Morchella, Mycena, Panus, Pholiota, Pleurotus, Pluteus, Podaxis, Polyporus, Polystictus, Pompholyx, Poria, Psalliota, Plathyrella, Rajapa, Rhizopogon, Rhodopaxillus, Rozites, Russula, Sparassis, Terfezia, Tirmania, Tremella, Tricholoma,
Tuber.

Ecuador Walnut → Juglans Honorei Dode.

Edgeworthia Gardneri Meissn. (Thymelaea-
ceae). - Shrub. Centr. and E. Himalaya, Nepal,
Manipur, Burma, China to Japan. Cultivated.
Source of Nepal paper.

Edible Tulip → Tulipa edulis Baker.

Edible Valerian → Valeriana edulis Nutt.

Eel Grass → Vallisneria americana Michx.

Efwatakala Grass → Melinis minutiflora Beauv.

Eggplant → Solanum Melongana L.

Eglantine Rose → Rosa Eglanteria L.

Egyptian Barley → Hordeum trifurcatum Ser.

Egyptian Clover → Trifolium alexandrinum L.

Egyptian Doum Palm → Hyphaene thebaica
Mart.

Egyptian Lettuce Seed Oil → Lactuca Scariola L.

Egyptian Pistache → Pistacia Khinjuk Stocks.

Egyptian Lupine → Lupinus Termis Forsk.

Eheretia acuminata R. Br. (Boraginaceae). -
Tree. China, Korea, Japan to Australia. Wood
light, tough; used in China as carrying-poles.
Also E. Dicksoni Hance.

Eheretia buxifolia Roxb. (syn. E. microphylla
Lam.) (Boraginaceae). - Shrub. Philipp. Islds.
Leaves are used locally in Philipp. Islds. as a
substitute for tea.

Eheretia elliptica DC., Sugarberry, Knackaway. (Boraginaceae). - Shrub or tree. Mexico
and W. Texas. Wood is used for tool handles,
yokes, axles and wheel spokes.

Eichel Kaffee → Quercus Robur L.

Eichhornia crassipes Solms. (syn. Piaropus crassipes (Mart. Britt.). Water Hyacinth. (Pontederiaceae). - Floating waterplant. Tropics and
subtropics of Old and New World. Obstructive
to navigation where abundant. Has been used
as food for cattle; has little nutritive value. Is
recommended as a source of cellulose.

Einkorn → Triticum monococcum L.

Ekanda Rubber → Raphionacma utilis N. E. Br.

Ekonge → Sterculia oblonga Mart.

Elaeagnaceae → Elaeagnus, Hippophaë, Shepherdia.

Elaeagnus latifolia L. (Elaeagnaceae). - Shrub or small tree. Trop. Asia. Fruits are consumed among the natives of Nepal and Hindustan.

Elaeagnus multiflora Thunb. Cherry Elaeagnus. (Elaeagnaceae). - Tree. Japan. Fruits are eaten as preserves, also source of an alcoholic beverage; used in some parts of Japan.

Elaeagnus philippensis Perk. (Elaeagnaceae). - Shrub. Philipp. Islds. Fruits are used for a very fine jelly; they are pinkish to pale red, sub-acid to sour; 2 to 2.5 cm. long.

Elaeagnus umbellata Thunb. (Elaeagnaceae). - Shrub or small tree. Japan. Fruits when scalded are eaten by the Ainu.

Elaeis guineensis Jacq. Oil Palm (Palmaceae). - Tall palm. Trop. W. Africa. Cultivated in Malay. Penin., Sumatra. Seeds source of Palm Kernel Oil; nondrying, used for manuf. margarine, olein for soap making; stearin, substitute for cacao butter; sometimes used as lubricant. Sp. Gr. 0.873; Sap. Val. 244-255; Iod. No. 16-23; Melt. Pt. 24-30° C. Palm Oil used for soap; tin plate industry, protecting cleaned iron surfaces before tin is applied. Refined, it is suitable for margerine, vegetable shortenings. Sp. Gr. 0.9209-0.9250; Sap. Val 196-205; Iod. No. 48-60; Unsap. 0.2-0.5%.

Elaeis melanococca Gaertn. Noli Palm (Palmaceae). - Tall palm. Trop. America. Seeds source of an oil, non-drying, recommended for food manuf. soap., illumination and machineries.

Elaeocarpus Bancroftii F. v. Muell. (Elaeocarpaceae). - Tree. Queensland. Seeds have a pleasant flavor; consumed by the aborigines.

Elaeocarpus Baudouini Brongn. and Gris. (Elaeocarpaceae). - Tree. New Caledonia. Wood is grayish-green, durable, easily worked; suitable for interior work.

Elaeocarpus bifida Hook. and Arn. (Elaeocarpaceae). - Tree. Hawaii. Islds. Bark of tree is source of a fibre; made into cordage; used by Hawaiians.

Elaeocarpus dentatus Vahl. (Elaeocarpaceae). - Tree. New Zealand. Fruit resembles a damson, much esteemed among the Maoris. Bark source of a blue-black dye; used by the natives for dyeing their garments. Bark is also used for tanning.

Elaeocarpus Ganitrus Roxb. Bead Tree (Elaeocarpaceae). - Large tree. India, Malaya. Hard seeds are made into sacred beads; used by Brahmins and Sinyasis; 101 beads are made in each chain, representing the hundred and one eyes of Shiv. Also used for rosaries, hatpins, buttons and similar comodities.

Elaeocarpus Hookerianus Raoul. (Elaeocarpaceae). - Tree. New Zealand. Bark is source of a blue dye; used by the aborigines of New Zealand.

Elaeocarpus oppositifolius Miq. (Elaeocarpaceae). - Shrub or shrub-like tree. Amboina. Cultivated in Java. Berries are red, 5 cm. long, eaten raw or cooked by the natives.

Elaeocarpus persicifolius Brongn. and Gris. (Elaeocarpaceae). - Tree. New Caledonia. Wood light; used for making small vessels in New Caledonia.

Elaeocarpus serratus L. (Elaeocarpaceae). - Tree. N. E. Himalaya, E. Bengal, Malaysia. Fruits, called Wild Olives, are used in curries, also plicked as olives, eaten by notives.

Elaeodendron lycioides Baker. (Celastraceae). - Shrub. Madagascar. Source of a dye, used by the Sakalava of Madagascar to give a red color to the finger-nails.

Elaeophorbia drupifera Stapf. (Euphorbiaceae). - Tree. W. Africa. Juice from tree is used as a fish poison.

Elaphomyces cervinus (L.) Schrot. (Elaphomycetaceae). - Ascomycete Fungus. Temp. zone. Fruitbodies are used medicinally and sold in rural drugstores in Central Europe; being used as aphrodisiacum for cattle.

Elaphomycetaceae → Elaphomyces.

Elaphrium Aloëxylon Schiede → Bursera Aloëxylon Engl.

Elaphrium jorullensis H. B. K. → Bursera jorullensis (H. B. K.) Engl.

Elaphrium simaruba (L.) Rose → Bursera gummifera L.

Elateriospermum Tapos Blume. (Euphorbiaceae). - Medium sized tree. W. Mal. Archip. Wood hard, durable; used for general carpentry. Seeds are eaten roasted by natives in parts of Sumatra, when raw they are poisonous; they are sold in local markets or pasars.

Elaterium ciliatum Cogn. (Cucurbitaceae). - Vine. Centr. America. Young fruits are consumed as a vegetable when cooked.

Elder, American → Sambucus canadensis L.

Elder, Blueberry → Sambucus coerulea Raf.

Elder, Dwarf → Sambucus Ebulus L.

Elder, European → Sambucus nigra L.

Elder, Mexican → Sambucus mexicana Presl.

Elecampane → Inula Helenium L.

Elecampane Oil → Inula Helenium L.

Elemi, African → Boswellia Frereana Birdw. and Canarium Schweinfurthii Engl.

Elemi Canary Tree → Canarium luzonicum Miq.

Elemi Frankincense → Boswellia Frereana Birdw.

Elemi of Guiana → Icica viridiflora Lam.

Elemi, Manila → Canarium commune L.

Elemi of Mexico → Amyris Plumieri DC.

Elemi Occidental → Protium Icicariba March.

Elemi of Yucatan → Amyris Plumieri DC.

Elemi Oil → Canarium luzonicum Miq.

Elemi, West Indian → Bursera gummifera L.

Eleocharis austro-caledonica Vieill. (Cyperaceae). - Perennial herb. New Caledonia. Stems are used for manuf. of baskets in New Caledonia.

Eleocharis plantaginea R. Br. (Cyperaceae). - Perennial herb. Java, Celebes. Cultivated. Used for basket work in Indonesia.

Eleocharis sphacelata R. Br. (syn. E. plantaginea F. v. Muell.). Maya. (Cyperaceae). - Perennial herb. Australia. Spherical tubers are eaten raw by the aborigines of Australia.

Eleocharis tuberosa Schultes., Water Chestnut. (Cyperaceae). - Perennial herb. E. India, China, Japan. Cultivated in the Far-East. Tubers are consumed in several Chinese dishes.

Elephant Apple → Feronia limonia Swingle.

Elephant Tree → Boswellia papyrifera Hochst.

Elephant's Foot, Prickly Leaved → Elephantopus scaber L.

Elephantopus scaber L., Prickly Leaved Elephant's Foot. (Compositae). - Herbaceous plant. Pantropic. Has diuretic and febrifuge properties.

Elephantorrhiza Burchellii Benth. (Leguminosaceae). - Acaulescent plant. S. Africa. Plant is source of a dye.

Elettaria Cardamomum (L.) Maton., Cardamon. (Zingiberaceae). - Perennial herb. E. India, Malaysia. Source of Cardamon Seeds, Siam Cardamon, also known as Cluster Cardamon. Some grades are Mysore, Malabar, Mangalore, Allepy, Madras Cardamon. Cultivated in India, Malaya and Mal. Archip. Used as a condiment, in cordials, bitters, artificial fruit flavors, cakes, gingerbread, sausages, pickles. When used in perfumes, it is put in certain types of Eau de Cologne. Used medicinally as carminative and aromatic stimulant. E. cardamomum var. major is source of Ceylon Cardamon. Contains ess. oil, Oleum Cardamoni composed of limonene, d-a-terpineal, borneol, cineol, sabinene.

Elettaria speciosa Blume. (Zingiberaceae). - Perennial herb. Java. Leaves are cooked with rice. Flowers and fruits are used as a substitute for tamarind by the natives. Fruits are candied in some parts of Java.

Elettariopsis sumatrana Valet. (Zingiberaceae). - Herbaceous perennial. Siam, W. Centr. Malaysia. In Sumatra used for flavoring. Juice used for stings of scorpions.

Eleusine aegyptiaca Desf. (Graminaceae). - Perennial grass. Tropics, subtropics. Ground seeds cooked are consumed in different countries, also fodder for live-stock.

Eleusine coracana (L.) Gaertn., African Millet, Korakan. (Graminaceae). - Annual grass. Trop. Asia. Cultivated. Native to India. Source of Ragi Flour, used in cakes, puddings also made into an alcoholic beverage.

Eleusine indica Gaertn., Goose Grass. (Graminaceae). - Annual grass. Tropics. Young seedlings are consumed in Java with rice. Stems are used for making mats. A food for live-stock.

Eleusine Tocussa Fresen. (Graminaceae). - Annual herb. Abyssinia. Grains are source of food in some parts of Abyssinia.

Elionurus candidus Hack. (syn. Andropogon candidus Trin., Lycurus muticus Spreng.). (Graminaceae). - Perennial grass. S. America. The plant is considered an excellent sand-binder in dunes.

Elm, English → Ulmus campestris L.

Elm, European → Ulmus campestris L.

Elm, Japanese → Ulmus japonica Sarg.

Elm, Mountain → Ulmus montana With.

Elm, Red → Ulmus alata Michx.

Elm, Rock → Ulmus Thomasii Sarg.

Elm, Slippery → Ulmus fulva Michx.

Elm, Sweet → Ulmus fulva Michx.

Elm, White → Ulmus americana L.

Elm, Winged → Ulmus alata Michx.

Elqueme Gum → Bursera gummifera L.

Elssholtzia cristata Willd. (Labiaceae). - Aromatic herbaceous perennial. Europe, temp. Asia. Used by the Ainu for persons suffering from the after effects of intoxication.

Elymus arenarius L. (Graminaceae). - Perennial grass. Temp. Europe, Asia etc. Cultivated in Ugo (Japan), used for mats, ropes and made into paper.

Elymus condensatus Presl., Giant Wild Rye. (Graminaceae). - Herbaceous herb. Western part of N. America. Seeds were made into flour and consumed by the Indians in bread.

Elymus triticoides Buckl., Wild Wheat, Wild Bearless Rye, Squaw Grass. (Graminaceae). - Perennial herb. Western N. America, Mexico. Grains were made into a flour and consumed by the Indians as porridge or in cakes.

Elytraria tridentata Vahl. → Tubiflora squamosa (Jacq.) Kuntze.

Embelia philippensis A. DC. (Myrsinaceae). - Woody vine. N. Luzon to S. Minandao (Philippine Islands). Leaves ar eaten with fish by the Filipinos.

Embelia Ribes. Burm. f. (Myrsinaceae). - Woody climber. India, Malaysia. Dried fruits are used as adulterant of black pepper, also used as stomachic, tonic, astringent, anthelmintic, for tapeworm in children. Contains 2 to 3% embelin.

Emblic → Phyllanthus Emblica L.

Embothrium rubricaule Giord. → Stenocarpus salignus R. Br.

Emex spinosa Campd. (Polygonaceae). - Perennial herb. Europe, Africa. Decoction of leaves is used by the Kaffirs as a tonic and for indigestion. In large doses it is used for intestinal troubles of cattle and horses.

Emilia sonchifolia DC. (Compositae). - Annual. Trop. Asia, Africa. Cultivated. Young leaves are eaten as lalab with rice and in soup among the Javanese.

Emmer → Triticum dicoccum Schrank.

Emory Bush Mint → Hyptis Emory Torr.

Emory Oak → Quercus Emoryi Torr.

Empetraceae → Empetrum.

Empetrum nigrum L. Crowberry. (Empetraceae). - Small shrub. Temp. Europe, Asia, N. America. Berries edible, made into a beverage with sour milk, used in Iceland; consumed in Greenland with fat of seal.

Empetrum rubrum Vahl. (Empetraceae). - Chile, Patagonia, Falkland Islands. Berries are consumed in antarctic regions. Supposed to have tonic properties.

Emu Apple → Owenia acidula F. v. Muell.

Emu Bush → Eremophila oppositifolia R. Br.

Enalus acoroides Steud. (syn. E. Koenigii Rich.) (Hydrocharitaceae). - Submersed saltwater-plant. Malaya. Used in Ceram and New Guinea as a source of fibre made into fishing-nets. Seeds are eaten boiled or roasted by the natives.

Enantia chlorantha Oliv. (Annonaceae). - Tree. Trop. Africa. Wood fine-grained, uniform, soft yellow; turning brown; used for house-building, furniture, general carpentry. Is sometimes called African Whitewood.

Enantia Kummeriae Engl. and Diels. (Annonaceae). - Tree. Trop. Africa. Wood source of a red dye.

Enantia polycarpa Engl. (Annonaceae). - Tree. Trop. Africa, esp. Nigeria. Source of a yellow dye, used by natives for coloring mats and skins. Specific for ulcers. Extract of bark used in Sierra Leone as unguent for sores.

Encelia farinosa Gray., White Brittlebush. (Compositae). - Perennial bush. S. W. of United States and adj. Mexico. Gum derived from the plant was used for chewing by the Indians of Arizona. A resin is sometimes employed as incense in churches.

Encephalartos caffer Miq. (Cycadaceae). - Palm-like plant. S. Africa. Stems are source of a starch, used by Hottentots for making bread. Also E. Hildebrandtii A. Br.

Encephalartos Miquellii F. v. Muell. → Macrozamia Miquellii A. Dc.

Encephalartos spiralis Lehm. → Macrozamia spiralis Miq.

Endive → Cichorium Endivia L.

Endocladiaceae → Gloiopeltis.

Endospermum chinense Benth. (Euphorbiaceae). - Tree. Trop. China, Malaya. Wood yellow, used for matches, packing cases; suggested for manuf. paper.

Endospermum moluccanum Becc. (syn. Capellenia moluccana T. and B.) (Euphorbiaceae). - Tall tree. Malaya, Moluccas. Wood very light; used for floats of fishing-nets. Young leaves when cooked used as vegetable by the natives, mildly purgative; old leaves are strongly laxative.

Endospermum peltatum Merr. (Euphorbiaceae). - Tree. Philippine Islands. Wood is used for manuf. of matches.

Engelhardtia serrata Blume. (Juglandaceae). - Tall tree. Malaya. Wood light, soft, close-grained, yellowish-white to grayish-red, used for construction of houses.

Engelhardtia spicata Blume. (syn. Juglans pterococca Roxb.). (Juglandaceae). - Tree. China. The bark is used for tanning in some parts of China.

Engelmann Spruce → Picea Engelmanni (Parry) Engelm.

Englerodendron alata Horold., Manzanilla. (Ericaceae). - Shrub. Ecuador. Fruits globose, light red, soft, juicy; flavor of a pear.

Engleromyces Goetzii Henn. (Xylariaceae). - Ascomycete Fungus. Nyassa region, Africa. Dried fruitbodies are cooked and employed against stomach-aches by the natives.

English Bluegrass → Festuca elatior L.

English Chamomille → Anthemis nobilis L.

English Elm → Ulmus campestris L.

English Galingale → Cyperus longus L.

English Ivy → Hedera Helix L.

English Oak → Quercus Robur L.

English Walnut → Juglans regia L.

English Yew → Taxus baccata L.

Enhydra fluctuans Lour. (Compositae). - Herbaceous. Moist places. Trop. Asia. Young shoots eaten as salad in Malaya.

Enicostema littorale Blume. (Gentianaceae). - Herbaceous plant. Tropics. Bitter herb is used in Sudan as stomachic, tonic and laxative.

Entada polyphylla Benth., Pashaco (Leguminosaceae). - Tree. Peru, Brazil, Guiana. Source of a resin; used for dyeing leather black.

Entada scandens Benth. (syn. E. phaseloides (L.) Merr.) Climbing Entada, St. Thomas Bean. (Leguminosaceae). - Vine. Tropics. Cultivated. Produces the largest legumes. Bark contains saponin; used in oriental countries for washing the hair. Sold in bazaars. Young leaves are eaten as vegetable. Beans are consumed roasted. Plant is source of a fibre; used in some countries for fishing nets, sails and cords.

Entandrophragma Candollei Harms. Heavy Sapele. (Meliaceae). - Tree. Trop. Africa. Wood brown, not very hard and heavy, scented, difficult to split, easy to polish; used for furniture and de luxe articles. Called Sapele Mahogany, Scented Mahogany; E. macrophyllum Chev., source of Tiama Mahogany; E. Rederi Harms, produces Cameroon Mahogany; E. utile (Dawe and Sprague), Sprague source of Brown or Short-capsuled Mahogany or Heavy Mahogany; E. angolense (Welw.) A. DC. produces Ljebu Mahogany, which are all African species.

Entelea arborescens R. Br. Cork Wood Tree. (Tiliaceae). - Tree. New Zealand. Wood very light, half the weight of cork. Used by Maoris for floats of fishing-nets, small rafts.

Enterolobium cyclocarpum (Jacq.) Griseb. (syn. Mimosa cyclocarpa Jacq.) (Leguminosaceae). - Tree. Mexico, Centr. America, West Indies, N. of S. America. Wood very durable in water, used for canoes, water troughs, cabinet work. Young pods sometimes eaten when cooked. Bark and fruits used as a substitute for soap, for washing woolen goods. Syrup from bark used for colds. Fruits used as soap substitute in Venezuela. Is source of a gum called Goma de Caro.

Enterolobium Timboüva Mart. Timboüva. (Leguminosaceae). - Tree. S. America, esp. Argentina, Uruguay, Paraguay, Brazil. Parts of the plant are used in stead of soap. Contains saponin. Roots and seedcoats are source of a rapid working enthelmintic; used in Brazil.

Enteromorpha compressa Grev. (Ulvaceae). - Green Alga. Pac. and other Oceans. Plants are used as food by the natives of Japan, New Caledonia and other Pacif. Islands. Also E. ramulosa Kutz, E. complanata Kutz.

Enteromorpha prolifera (Muell.) J. Ag. (Ulvaceae). - Green Alga. Sea Weed. Pac. Ocean. Consumed as food among the Hawaiians. Called Limu Eleele. Eaten during native feasts. Plants are put in hot gravy or broth and put in meat stews. Sometimes the algae are consumed after undergoing a ripening process.

Entoloma clypeatus (L.) Quél. → Hyporhodius clypeatus (L.) Schroet.

Entoloma microcarpum (Berk. and Br.) Sacc. → Collybia microcarpa (Berk. and Br.) Hoehn.

Eon → Fillaeopsis discophora Harms.

Epacridaceae → Astroloma, Leucopogon, Monotoca, Styphelia.

Eperua falcata Aubl. Wallaba, Bootlace Tree. (Leguminosaceae). - Tree. Guiana. Wood hard, durable; made into shingles, telegraph-poles, construction of houses.

Ephedra americana Humb. and Bonpl. (syn. E. andina Poepp.). American Ephedra. (Ephedraceae). - Shrub. S. America. Infusion and decoction of branches and of the roots are used in S. America as diuretic and for „cleaning" the blood.

Ephedra antisyphilitica Berland. Syn. E. pedunculata Engelm. (Ephedraceae). - Slender shrub. Dry plains. Mexico, W. Texas. Decoction of stems is used as a cure for renal diseases.

Ephedra aspera Engelm. (Ephedraceae). - Shrub. Mexico. Decoction of stems used for renal diseases.

Ephedra nevadensis S. Wats., Nevada Joint Fr. (Ephedraceae). - Shrub. Nevada to California. Seeds when roasted were ground into flour and made into bread by Indians of several tribes. Used in urogenital diseases.

Ephedra ochreata Miers. (Ephedraceae). - Shrub. Argentina. An infusion of the branches is used in Argentina as diuretic and antiblenorrhagic.

Ephedra sinica Stapf. Chinese Ephedra, Ma-Huang. (Gnetaceae). - Shrub. Seacoast of S. China. Used medicinally. Source of ephedrine, an alkaloid. Excites sympathetic nervous system; causes rise of blood-pressure and mydriasis; diminishes hyperemnia. Drug is exported from Canton. Has been used in Chinese medicine for over 5000 years. Also E. equisetina Bunge.

Ephedra trifurca Torr. Longleaf Ephedra. Joint Fir. (Ephedraceae). - Shrub. S. W. of United States, Mexico. Tea from the branches was a popular drink among the Indians and Mexicans, called Desert Tea or Teamster's Tea. Also used medicinally.

Ephedraceae → Ephedra.

Epicampes macroura Benth., Zakaton Grass. (Graminaceae). - Perennial grass. Mexico. Cleaned roots are source of Zakaton, Rice Root, Mexican Whish, Dogtorth, used for manuf. brushes and brooms. Much exported to U. S. A., France and Germany. Also E. stricta Presl.

Epidendrum cochleatum L. (Orchidaceae). - Epiphyt. orchid. Mexico, Centr. America, West Indies N. of S. America. Pseudobulbs used by Mazatecs as a source of mucilage. Probably also E. vitellinum Lindl.

Epigaea repens L., Ground Laurel. (Ericaceae). - Small shrub. Eastern N. America to Florida. Infusion of herb is used by the Cherokee Indians for diarrhea in children. Plant is considered astringent and tonic.

Epilobium angustifolium L. (syn. Chamaenerion angustifolium (L.) Scop.) (Onagraceae). - Perennial herb. Temp. Europe, Asia, N. America. Leaves are used as a tea in some parts of Russia. Known as Kapporie tea, Kapor tea and Iwan-tschai.

Epiprenum giganteum Schott. (Araceae). - Fleshy climber. China, Indo-China, Malaya to N. Australia. Very poisonous, used criminally; a dart-poison.

Equisetaceae → Equisetum.

Equisetum arvense L. Common Horse Tail. (Equisetaceae). - Perennial herb. Fern-ally. Temp. zone. Strobili are eaten in Japan after being boiled, also salted and when kept in vinegar mixed with soy, after being boiled in water. The base of the plant was used as food by the Kiowa Indians (M. Mex.). Herba Equiseti is used as home remedy for kidney and bladder diseases in some parts of Europe.

Equisetum bogotense H. B. K. (syn. E. chilense Presl., E. quitense Fée.). (Equisetaceae). - Perennial herb. Fern ally. Colombia to Peru and Ecuador. Used in Colombia as an astringent, diuretic, antihemorragic, antidysenteric, antigonorrhoeic. Used in Venezuela for diabetis; in Peru as diuretic. Also E. giganteum H. B. K.

Equisetum hiemale L. Syowring Puish. (Equisetaceae). - Perennial herb. Fern-ally. Temp. zone. Stalks are used in Japan for polishing. Those from the village Waka-mori, province Tamba, are most esteemed. They are used by cabinet makers in Europe for polishing furniture and wooden floors.

Equisetum laevigatum A. Br. (Equisetaceae). - Perennial herb. Fern Ally. N. America. Dried plant was ground and made into a mush by the Indians of New Mexico.

Equisetum pratense Ehrh. (Equisetaceae). - Perennial herb. Fern Ally. Temp. zone. Tuberlike rhizomes were consumed by the Indians in Minnesota.

Equisetum robustum A. Br. (Equisetaceae). - Herbaceous plant. Fern-ally. N. America, Mexico. Used in parts of Mexico as a diuretic and antiblenorrhagic.

Eragrostis abyssinica Schrad. → Poa abyssinica Juss.

Eragrostis Brownei Nees. (Graminaceae). - Perennial herb. Australia. Excellent pasture grass in Australia.

Eragrostis cynosuroides Beauv. (syn. Poa cynosuroides Retz.) (Graminaceae). - Herbaceous grass. Nile region. Stems are used for manuf. mats in the Nile region.

Eragrostis lugens Nees. (syn. Poa microstachya Link.). (Graminaceae). - Perennial grass. Trop. America. Is considered an excellent forage grass.

Erdorseille → Ochrolechia tartarea (L.) Mass.

Eremocitrus glauca Swingle, Australian Desert Kumquat. (Rutaceae). - Spiny shrub. Australia. Fruit small, light yellow. Occasionally recommended for drinks and jam.

Eremophila Mitchelli Benth., Sandalwood, Bastard Sandalwood, Rosewood. (Myoporaceae). - Tree. S. Australia, New South Wales, Queensland. Wood very hard, fragrant, brown, beautifully grained. Used for cabinet-work, veneer.

Eremophila oppositifolia R. Br. (syn. E. arborescens Cunningh.) Emu Bush. (Myoporaceae). - Tree. Australia. Bark is used for tanning.

Eremospatha macrocarpa Mann and Wendl. (Palmaceae). - Rattan-palm. Trop. Africa. Canes are used for hammock bridges, baskets, binding of house- rafters. Rind is made into garden furniture and chairs in Cameroon.

Eremothecium Asbyii Schopf. (Saccharomycetaceae). - Yeast. Microorganism. Is a source of Riboflavin which is manufactured commercially.

Eremurus Aucherianus Boiss. (Liliaceae). - Perennial herb. Centr. Asia. Roots are source of a mucilage; used for cement's leather, employed in book binding. Leaves are consumed as food.

Eremurus spectabilis Bieb. (Liliaceae). - Tall perennial herb. Centr. Asia. Young shoots are consumed as a vegetable; they are sold in markets of Caucasia, Crimea and Kurdistan.

Erica arborea L. (Ericaceae). - Small tree. Mediterranean region, esp. Italy and S. France. Source of Briar Root; used for manuf. of bowls, tobacco pipes etc. Much is produced in Leghorn, Italy.

Erica Tetralix L. (Ericaceae). - Small shrub. Temp. Europe. Source of a yellow dye, used in some parts of the Scotch Highlands. Also E. cinerea L.

Ericaceae → Agapetes, Andromeda, Arbutus, Arctostaphylos, Azalea, Calluna, Chamaedaphne, Chiogenes, Clethra, Disterigma, Englerodendron, Epigaea, Erica, Gaultheria, Gaylussacia, Kalmia, Ledum, Leucothoë, Macleania, Rhododendron, Vaccinium.

Erigeron affinis DC. (Compositae). - Perennial herb. Mexico. Roots ar used among the rural population for toothache, as dentifrice and as a stimulant.

Erigeron canadensis L., Fleabane. (Compositae). - Annual herb. N. America, Europe. Leaves and tops used in dysentery, diarrhea, hemorrhage, gonorrhea and other urogenital diseases.

Eriobotrya japonica (syn. Photinia japonica Thunb.) Loquat, Japanese Plum, Japanese Medlar. (Rosaceae). - Tree. Probably Centr. E. China. Commercially cultivated in China, Japan, California. Grown in subtropics. Fruits with wooly skin, yellow or orange, containing one or more rather large seeds. Eaten raw, in preserves, jelly, compotes or stewed. Many varieties, among which are; Early Red, Tanaka, Victor, Champagne, Gold Nugget. Japanese varieties are: Shiroko-buda, Tobuda and Nagato-buda.

Eriocarpum grindelioides Nutt. → Aplopappus Nuttallii Torr, and Gray.

Eriocephalus spinescens Burch. (Compositae). - Small shrub. S. Africa. Used as fodder for livestock in Western S. Africa.

Eriochloa polystachya H.B.K. (syn. E. subglabra [Nash] Hitch.) Carib. Grass. (Graminaceae). - Annual grass. Trop. America. Cultivated as fodder for cattle.

Eriochloa subglabra Hitch. (Graminaceae). - Perennial grass. Java. Recommended as fodder for cattle in Java.

Eriodendron aesculifolium DC. (Bombaceae). - Tree. Mexico. Young fruits and ripe seeds are eaten in some parts of Yucatan (Mex.).

Eriodendron anfractuosum DC. → Ceiba pentandra (L.) Gaertn.

Eriodictyon californica (Hook. and Arn.) Torr., Yerba Santa. (Hydrophyllaceae). - Shrub. California to NO. Mexico. Dried leaves are used medicinally as a stimulating expectorant, to disguise bitterness in quinine. Contains a resin, pentatriacontane; cerotinic acid. Was used as medicine by the Indians. Dried leaves are source of a bitter tea, used in California as an expctorant, blood purifier and tonic.

Erioglossum edule Blume. (syn. E. rubigonosum Brand.) (Sapindaceae). - Small tree. Trop. Asia. Fruits are edible, 2 cm across. roundish-oblong.

Eriogonum corymbosum Benth. (Polygonaceae). - Herbaceous plant. Western N. America. Boiled leaves with corn meal were consumed by the Indians of Arizona.

Eriogonum inflatum Torr., Desert Trumpet. (Polygonaceae). - Perennial herb. Western United States. Tender stems before flowering are eaten raw.

Erisma calcaratum Warm. (Vochyaceae). - Tree. S. America. Seeds are source of Jaboty Butter or Jaboty Tallow; used for candles and soap. Sp. Gr. 0.8760; Sap. Val. 233; Iod. No. 54; Melt. Pt. 42.5° C.

Ergosterine → Claviceps purpurea (Fr.) Tul.

Ergot → Claviceps purpurea (Fr.) Tul.

Ergot of Rye → Claviceps purpurea (Fr.) Tul.

Erodium cicutarium (L.) L'Hér. Storks Bill. (Geraniaceae). - Annual to biennial. Temp. Zone. The remnants of the styles remaining on the fruits are very hygroscopic, curling in various degrees from moist to dry. Used for certain types of hygrometers.

Erodium hirtum Willd. (Geraniaceae). - Herbaceous plant. N. Africa, Algerian Sahara. Roots are consumed as food by the Tuareg (N. Afr.).

Erman's Birch → Betula Ermani Cham.

Eroteum theaoides Sw → Eurya theoides (Sw.) Blume.

Ertrog → Citrus Medica L.

Eruca sativa L. (syn. E. vesicaria [L.] Cav.) Rocket Salad, Roquette. (Cruciferaceae). - Annual. Centr. Europe, Mediterranean region, Asia. Introd. in America and Australia. Cultivated since ancient times. At present grown esp. in N. India, Punjab and Sind. Source of Jamba Oil, semi drying, has characteristic odor; used for burning. Sp. Gr. 0.915—0.917; Sap.Val. 172—175; Iod. No. 95—104; Unsap. 0.4—1.0%. Seeds are used in mustard. Plant is diuretic, antiscorbutic, stimulant, rubefacient and stomachic.

Erucaria aleppica Gaertn. (Cruciferaceae). - Herbaceous plant. Greece, Turkey, Arabia. Fleshy leaves and young shoots eaten boiled or in salad used in parts of Greece.

Ervatamia coronaria Stapf. (syn. Tabernaemontana coronaria Willd., T. divaricata R. Br.) Grape Jasmine, E. India Rosebay. (Apocynaceae). - Shrub. Probably N. India, Malaya. Cultivated. Pulp around seeds used as dye; wood medicinally used as refrigerant; used as incense; in perfuming. In Indonesia decoction of roots used for diarrhea.

Ervatamia cylindrocarpa King and Gamble. (Apocynaceae). - Small shrub. Malay. Penin. Leaves when pounded with rice and tumeric are used for eczema and itch.

Ervatamia hirta King and Gamble. (Apocynaceae). - Shrub or small tree. Trop. Asia, esp. Malacca, Negri Sembilan, N. E. Pahang. Used in Malacca for ulceration of nose. Likewise E. malaccensis King and Gamble.

Ervil → Vicia Ervilia (L.) Willd.

Ervilia sativa Link. → Vicia Ervilia (L.) Willd.

Ervum Ervilla L. → Vicia Ervilia (L.) Willd.

Ervum Lens L. (syn. Lens esculenta Moench. L. culinaris Med.). Lentil. (Leguminosaceae). - Annual herb. Mediterranean region, Syria. Cultivated since ancient times. Seeds consumed boiled, in soups, made into flour for bread. Fodder for domestic animals. Several varieties which are divided into Summer and Winter Lentils. A preparation called Revalenta Arabica is used as food for invalids.

Eryngium aquaticum L. (syn. E. virginianum Lam.). (Umbelliferaceae). - Perennial herb. Eastern United States to Florida and Louisiana. Decoction of boiled root is used by the Koasati Indians as emetic. Root is diaphoretic and expectorant; in large quantities an emetic.

Eryngium campestre L. (Umbelliferaceae). - Perennial herb. Throughout Europe. Candied roots are occasionally eaten in France and England. They are also consumed as food by the Kalmucks.

Eryngium Carlinae Delav. (Umbelliferaceae). - Perennial herb. Trop. America. Decoction of herb used in Centr. America for treating digestive disturbances in children.

Eryngium foetidum L. (Umbelliferaceae). - Herbaceous plant. Trop. America. Roots have offensive odor, when used as condiment in soups and meat dishes they impart a very agreeable flavor.

Eryngium maritimum L. (Umbelliferaceae). - Perennial herb. Along coast of Mediterranean, Atlantic and North Sea. Very young tops of roots are occasionally consumed as asparagus.

Eryngium synchaetum (A Gray) Rose. (Umbelliferaceae). - Perennial herb. Southern United States. Decoction of roots is used as ceremonial „black drink" by Seminole Indians.

Eryngium yuccifolium Michx., Rattlesnake Master, Button Snakeroot. (Umbelliferaceae). - Perennial herb. Eastern United States to Florida and Texas. Mashed roots in cold water are used by the Creek Indians for kidney troubles. Plant is diuretic, diaphoretic and expectorant; in large quantities it is emetic.

Erythea edulis (Wendl.) S. Wats. (Palmaceae). - Tall palm. Guadeloupe Isld.; Baja California (Mex.). Pulp of fruit is edible, sweet. Young leaf-buds from heart of tree are consumed as vegetable.

Erythraea Centaurium Pers. (syn. Centaurium umbellatum Gilib.). Drug Centaurium. Centaury. (Gentianaceae). - Annual or biennial herb. Europe, Caucasia, Iran, N. Africa, introd. into N. America. Dried, flowering herb, used medicinally as a tonic and has been claimed to be a febrifuge. Contains a bitter principle, ess. oil, a resin, and erytaurin, a glucoside. Tea or a liqueur is used as a home remedy in some countries of Europe, for chronic catarrh of the stomach.

Erythrina arborea (Chapm.) Small. (syn. E. herbacea var. arborea Chapm.). Red Cardinal. (Leguminosaceae). - Shrub or small tree. So. Florida. Brilliant red seeds are used for beads; called Coral Beans.

Erythrina Corallodendron L. Common Coralbean (Leguminosaceae). - Tree. Trop. America. Bark is used in Brazil for asthma as a tincture or extract. Bark derived from moist regions is said to be worthless.

Erythrina edulis Triana (Leguminosaceae). - Woody plant. Northern S. America, esp. Colombia. Seeds are consumed in some parts of Colombia when boiled.

Erythrina glauca Willd. (Leguminosaceae). - Tree. Guiana, Venezuela. Used as shade tree in cacao plantations in trop. America.

Erythrina herbacea L. Eastern Coralbean (Leguminosaceae). - Shrub or small tree. South Eastern United States, West Indies and Mexico. In San Luis Potosi (Mex.), seeds are used for poisoning rats. Roots were used among the Indians as diaphoretic.

Erythrina indica Lam. (syn. E. variegata L.) India Coralbean. (Leguminosaceae). - Trop. Asia, Australia. White wood is made in Siam into a powder; used for facepowder. Leaves and stems have been recommended as an excellent fodder for cattle, sheep and goats.

Erythrina micropteryx Poepp. Anauca. (Leguminosaceae). - Tree. Peru. Used as shade tree in cacao plantations.

Erythrina monosperma Gaud. (Leguminosaceae). - Shrub or small tree. Hawaii Islnds. The bark is source of a fibre; used in Hawaii for fishing-nets.

Erythrina mossambicensis Sim. (Leguminosaceae). - Small tree. S. Africa. Wood light; used for marimbas in Portugeese E. Africa; also for making fire by friction.

Erythrina rubrinervia H. B. K., Gallito (Leguminosaceae). - Shrub or small tree. Centr. America and adi. territory. In El Salvador and Guatemala the flowers and buds are eaten like string-beans. Young leaves are used in soup.

Erythrina Versperilio Benth. (Leguminosaceae). - Tree. Australia. Wood is used for making shields by the Wilparri in Centr. Australia.

Erythrina Zeyheri Haw. (Leguminosaceae). - Acaulescent plant. S. Africa. Red seeds are made into necklaces, bracelets etc.

Erythronium Dens-canis L. (Liliaceae). - Perennial herb. Temp. Europa, N. Asia. Roots are source of a starch, made in Japan; used for vermicelli and cakes. Leaves are eaten boiled. Bulbs are eaten, often with reindeer's or cow's milk in Mongolia and Siberia.

Erythronium grandiflorum Pursh. (Liliaceae). - Perennial bulbous herb. Western N. America. Bulbs were consumed os food by the Indians along the Pacific Coast of N. America.

Erythrophleum Fordii Oliv. (Leguminosaceae). - Tree. S. China, Indo-China. Wood tans leather, giving a fine structure with a chestnut color, sensitive to light.

Erythrophleum guineense G. Don. Sassy Bark Tree, Redwater Tree, Sasswood. (Leguminosaceae). - Tree. Trop. Africa. Decoction of bark is used as arrow poison and to stupefy fish among the natives of W. Africa. Bark known as Doom Bark, Caka Bark, Ordeal Bark is source of erythrophleine; an alkaloid, resembling digitalin, recommended as local anesthetic in dentistry; heart stimulant, astringent; for colic dysentery and diarrhea, also used for treatment of the eye. Wood is hard, durable; used for construction work.

Erythropsis Barteri Ridley. (Sterculiaceae). - Tree. Trop. W. Africa. Bark source of a fibre, made into cloth along the Gold Coast, also used by the natives for cordage. Fibre is harsh, fairly strong, suitable for ropes.

Erythrostictus punctatus Schlecht → Androcymbium punctatum Baker.

Erythroxylaceae → Erythroxylon.

Erythroxylon anguifugum Mart. (Erythroxylaceae). - Tree. Brazil. Decoction of bark is used in Brazil for snake-bites.

Erythroxylon citrifolium St. Hil. (Erythroxylaceae). - Tree. Brazil. Bark is used in Brazil for healing wounds.

Erythroxylon Coca Lam., Coca Tree. (Erythroxylaceae). - Small tree. Bolivia, Peru. Cultivated in Java and Ceylon. Source of Huanuca Coca from Bolivia. Dried leaves used medicinally. Source of several alkaloids, e. g. cocaine, cinnamyl-cocaine; cerebral stimulant, narcotic; used as local anesthetic. Leaves are chewed by the Indians as a stimulant.

Erythroxylon monogynum Roxb. (Erythroxylaceae). - Tree. E. Indies. Wood is source of a tar; used for vessels. Source of an ess. oil, of a Sandalwood odor; has been recommended for perfumery.

Erythroxylon myrtoides Bojer. (Erythroxylaceae). - Shrub. Madagascar. Wood almost black; used for ornamental purposes in Madagascar.

Erythroxylon nova-granatense (Morris) Hier., Truxillo-Coca. (Erythroxylaceae). - Small tree. Peru, Bolivia. Cultivated. Source of Truxillo Coca. See E. Coca Lam.

Erythroxylon obovatum Macf. (Erythroxylaceae). - Tree. West Indies. Wood compact, strong, red; used for telegraph poles and for support.

Erythroxylon ovatum Cav. (Erythroxylaceae). - Tree. West Indies, S. America. Wood is used for furniture in some parts of Argentina.

Erythroxylon suberosum St. Hil. (Erythroxylacete). - Tree. Brazil. Bark is source of a redbrown dye; used for dyeing woolen goods. Also E. tortuosum Mart.

Escarolle → Cichorium Endivia L.

Escherichia coli (Migula) Castell and Chalm. (Bacteriaceae). - Bacil. A harmless microorganism living in the intestinal tracts of man and animals. It is useful during the examination of drinking water which shows by the presence of this species that also pathogenic bacteria (cholera, typhus etc.) of man may be present in the water. With the aid of the presumtive and the confirming test of the coli bacil, its presence may be presumed that the water may be contaminated.

Eschscholzia californica Cham., Californian Poppy. (Papaveraceae). - Annual herb. S. E. of United States. Leaves were eaten as greens, boiled or roasted on hot stones by Indians of California.

Escontria chiotilla (Weber) Rose. (syn. Cereus chiotilla Weber). Chiotilla (Cactaceae). - Xerophyte. Mexico. Fruits edible, sold in markets.

Esenbeckia febrifuga Juss. (Rubiaceae). - Tree. S. E. Brazil, Paraguay, Argentina to Brazil. Bark used as a substitute of quinine. Wood made into spoons, dishes etc.

Esparcet → Onobrychus viciaefolia Scop.

Esparcet, Spanish → Hedysarum coronarium L.

Esparvie → Anacardium Rhinocarpus DC.

Espave Mahogany → Anacardium Rhinocarpus DC.

Espino Cavan → Acacia Cavenia Bert.

Espostoa lanata (H. B. K.) Britt. and Rose. (syn. Pilocereus lanatus Weber, Cereus lanatus DC.). (Cactaceae). - Small xerophytic tree. Ecuador, Peru. Fruits are sweet, edible, called Soroco in S. Ecuador.

Esprit d'Iva → Achillea moschata Jacq.

Essence d'Amali → Alpinia malaccensis Rosc.

Essence de Bois Gaiac → Bulnesia Sarmienti Lorentz.

Essence de Bruyère de Tonkin → Baeckea frutescens L.

Estragon → Artemisia Dracunculus L.

Ethiopian Madflower → Antholyza paniculata Klatt.

Ethiopian Sour Gourd → Adansonia digitata L.

Eucalyptus amygdalina Lab. Willowleaf Eucalyptus, Peppermint Tree, Mountain Ash. (Myrtaceae). - Tall tree. Australia, esp. New South Wales. Source of a port-wine colored kino, very friable, soluble in cold water, called Ribbon Gum Kino. Source of an essential oil. Wood light, splits easily, does not twist when dry. Used for carpentry, shingles, ship-building. Wood has been recommended as a source of cellulose and woodpulp.

Eucalyptus botryoides Smith. Blue Gum, Bangalay Eucalyptus, Bastard Mahogany, Swamp Mahogany (Myrtaceae). - Tree. Australia, esp. Victoria, New South Wales. Wood tough, hard, durable. Used for ship-building, knees of boats, felloes of wheels.

Eucalyptus citriodora Hook. (Myrtaceae). - Tree. Australia. Source of an ess. oil, rich in citronellal, also traces of geraniol and pinene.

Eucalyptus cneorifolia DC. (Myrtaceae). - Tree. Kangeroo Islnd., S. Australia. Source of a volatile oil, contains high percentage cineol, also pinene, cymene, cuminal, phellandral.

Eucalyptus corymbosa Smith. Bloodwood (Myrtaceae). - Tree. Australia, esp. New South Wales, Queensland. Source of an abundance of reddish kino.

Eucalyptus diversicolor F. v. Muell. Karri (Myrtaceae). - Tree. S. W. Australia. Wood light, tough, bends freely. Used for spokes, felloes, ship-building, masts, rudders.

Eucalyptus dumosa Cunningh. Cangoo Mallee Eucalyptus, Larap. (Myrtaceae). - Shrub. Australia, esp. Victoria, New South Wales. Plant produces a manna, starch-like substance, called Lerp. Formation is caused by an insect. Lerp is eaten by the natives. Also source of an ess. oil.

Eucalyptus globulus Lab. Fever Tree, Blue Gum, Tasmanian Blue Eucalyptus, Tasmanian Blue Gum, Victoria Blue Gum. (Myrtaceae). - Tree. Australia. Cultivated. Wood pale, heavy, strong, hard, durable. Used for ship-building, implements, spokes, rims, axle-beds, ploughbars, picks, forks, handles of axes, wood pulp. Source of Oil of Eucalyptus, Oleum Eucalypti, an ess. oil derived from the fresh leaves; being antiseptic, expectorant, febrifuge antiperiodic, diaphoretic. Contains eucalyptic acid, cineol, a bitter principle and veryl alcohol. Dried leaves are used medicinally as febrifuge, antiseptic, antimalarial, antiperiodic.

Eucalyptus gomphocephala DC. Tuart Eucalyptus, Touart, White Gum. (Myrtaceae). - Tree. Australia. Wood hard, very heavy, close-grained, durable, tough, strong, difficult to cleave, pale yellowish. Used for ship-building withstands action of water.

Eucalyptus goniocalyx F. v. Muell. Bigleaf Eucalyptus, Spotted gum, Grey Gum, Blue Gum. (Myrtaceae). - Tree. Australia, esp. Victoria, New South Wales. Wood tough, hard, durable, pale yellowish to brown, lasts long underground. Used for wheelwrights, spokes, ship and boat-building, rail-road ties.

Eucalyptus haemastoma Smith. Scribbly Gum. (Myrtaceae). - Tree. Australia, esp. New South Wales, Queensland. Bark occasionally used for tanning.

Eucalyptus hemiphloia F. v. Muell. Gray Box, Yellow Box, White Box. (Myrtaceae). - Tree. Australia. Wood tough, hard, durable, heavy, Excellent timber. Used for rail-road ties, bridge piles, mining slabs, plankings, handles, wheelwrights, shipbuilding.

Eucalyptus intertexta R. T. Baker (Myrtaceae). - Woody plant. Australia. Seeds when ground are consumed as food by the aborigines of S. Australia.

Eucalyptus leucoxylon F. v. Muell. (syn. E. conoidea Benth.) White Ironbark, White Gum. (Myrtaceae). - Tree. Victoria. Source of an essential oil. High in cineol content. Plant produces a kino, easily soluble in water. Wood durable, very strong, hard, tough, yellow-white to pale purplish white; used for rail-road ties, felloes of wheels, bridge piles, telephone poles, axe handles and bullock yokes.

Eucalyptus Macarthuri Deane and Maiden. (Myrtaceae). - Tree. Australia. Source of an ess. oil. Used in Australia as a denaturant of alcohol for perfumery. Contains much geranyl acetate and some geraniol.

Eucalyptus maculata Hook. (syn. E. variegata F. v. Muell.) Spotted Gum. (Myrtaceae). - Tree. Australia. Source of a yellowish-brown to olive colored kino, friable, can be easily crushed to a fine powder. Wood strong, durable, close grained; used for ship building, naves of wheels, bridges, cubes for street paving, shingles.

Eucalyptus macrorhynchia F. v. Muell., Stringybark. (Myrtaceae). - Tree. Australia, esp. Victoria, New South Wales. Source of a rich ruby colored kino, easily friable.

Eucalyptus mannifera Mudic, Manna Eucalyptus. (Myrtaceae). - Tree. Australia. Exudation from stems, called Eucalyptus Manna, is used as food by some of the natives of Australia. Also E. viminalis Labill., E. pulverulenta Sims and E. Gunnii Hook. f.

Eucalyptus marginata Smith. Jerrah (Myrtaceae). - Tree. Australia. Wood resistant to infl. of air. weather and moisture. Used for different purposes.

Eucalyptus microtheca F. v. Muell. Colibah, Floodbed Box (Myrtaceae). - Woody plant. Australia. Used in Queensland by the natives to stupefy fish.

Eucalyptus obliqua L'Hér. Messmate Stringy Bark, Stringy Bark. (Myrtaceae). - Tree. Australia. Wood light, easily worked. Used for different purposes, boards, shingles, posts, rafters etc. Bark employed for thatching roofs, for door mats, rope, bands for hay; very durable.

Eucalyptus pilularis Sm. Blackbutt. (Myrtaceae). - Tree. Australia. Wood strong, durable, yellowish. Used for ships' decks, bridge plankings, paving cubes, rail-road ties, telegraph poles.

Eucalyptus piperata Smith. Sydney Peppermint Eucalyptus, White Stringy Bark, Redwood, Blackbutt. (Myrtaceae). - Tree. Australia. Wood durable, resistant to damp soil. Used for shingles, house building, posts, rough indoor house work. Leaves and branches are source of pipertone; recommended as a raw material for thymol and menthol.

Eucalyptus polyanthemos Scham. Red Box, Grey Box, Bastard Box. (Myrtaceae). - Tree. Victoria, New South Wales. Wood hard, durable. Used for naves, felloes, cogs, slabs in mines.

Eucalyptus polybractea F. v. Muell. Blue Malle (Myrtaceae). - Shrub. Australia, esp. Victoria, New South Wales. Source of a medicinal oil; high in cineol content, also pinene, autralol, cuminal and phellandral.

Eucalyptus resinifera Smith. (syn. E. spectabilis F. v. Muell., E. hemilampra F. v. Muell.) Kino Eucalyptus, Red or Forests Mahogany. (Myrtaceae). - Tree. Australia, esp. New South Wales, Queensland. Source of a kino, sometimes mistaken for Botany Bay Kino. Clear ruby color, tough, horny, bright fracture, dissolves in water. Wood strong durable. Used for ships' knees, posts, shingles, rafters, general building purposes; lasts under ground.

Eucalyptus robusta Smith. Beakpod Eucalyptus, White Mahogany, Swamp Mahogany, Kimbarra (Austr.) (Myrtaceae). - Tree. Coastal region, New South Wales. Wood brittle, difficult to split, reddish. Used for shingles, ship-building, wheel wrights' work, general building.

Eucalyptus rostrata Schlecht. Longbeak Eucalyptus, Australian Kino, Red Gum (Myrtaceae). - Tree. Australia. Source of Red Gum being an exudation from the stem. Used medicinally for coughs, common colds, dysentery, relaxed throat; contains kinotannic acid, kino red and a glucoside. Wood very hard, difficult to work, very strong, durable; used for posts in damp soil, ship building, rail-road ties, bridges; resistant to teredo, chelura and termites. Wood has been recommended as a source of cellulose and woodpulp, also E. regnans

F. v. Muell, E. obliqua L'Hér., E. viminalis Lab., E. Sieberiana F. v. Muell., E. saligna Smith and other species.

Eucalyptus siderophloia Benth. Broadleaf Ironbark, Ironbark, Red Ironbark. (Myrtaceae). - Tree. Queensland. Wood very strong, durable, hard, difficult to work, heavy, light colored. Used for rail-road ties, shingles, fencing, beams, plough beams. Source of a Botany Bay Kino; tears horney, of a rich ruby color, difficult to crush.

Eucalyptus Sieberiana F. v. Muell. Cabbage Gum. (Myrtaceae). - Tree. Tasmania, Victoria, New South Wales. Source of a kino, easily soluble, rich garnet color, its powder is orange; appearance of E. amygdalina-kino.

Eucalyptus Staigeriana F. v. Muell. Lemon Scented Iron Bark. (Myrtaceae). - Tree. Australia. Source of an ess. oil, containing 60% laevolimonene, 16% citral, also some geraniol and geranyl acetate.

Eucalyptus Stuartiana F. v. Muell. (Myrtaceae). - Tree. Australia. Source of an ess. oil, having the scent of apples. Has been recommended for perfumery.

Eucalyptus tereticornis Smith. (syn. E. subulatum Cunningh.) Red Gum, Flooded Gum, Grey Gum, Blue Gum. (Myrtaceae). - Tree. Australia, esp. New South Wales, Queensland. Wood heavy, close-grained. Used for ship-building, fencing, general building purposes, plough beams, poles, rail-road ties.

Eucalyptus terminalis F. v. Muell. Kutcha Bloodwood Eucalyptus (Myrtaceae). - Tree. Australia. Leaves and stems are source of a gum or Manna; consumed by the natives of N. Queensland.

Eucalyptus viminalis Ldb. (syn. E. mannifera Cunning., E. persicifolia Lodd.) Ribbon Eucalyptus, White Gum, Swamp Gum, Manna Gum. (Myrtaceae). - Tree. Australia. Wood coursegrained, brown, durable for underground work. Used for shingles, general rough building, rails, wheel wrights' work. Bark exudes a kind of manna, a white substance of very pleasant sweet taste, much esteemed by the aborigenes of Australia.

Eucalyptus Manna → Eucalyptus mannifera Mudic.

Eucarya acuminata (R. Br.) Sprague. (Santalaceae). - Shrub or tree. Australia. Fleshy fruits are eaten raw or in preserves. Seeds are consumed as food.

Eucarya spicata Sprag. and Summ. (syn. Santalum spicatum DC.) West Australian Sandalwood. (Santalaceae). - Tree. S. W. Australia. Source of West Australian Sandalwood Oil substitute of Sandal Oil, obtained by distillation of the wood. A temple incense in China. Used for pharmaceutical preparations, urinary antiseptec. Contains fusanols, sesquiterpene alcohols, santalol.

Eucheuma gelatinae (Esp.) J. Ag. (Solieriaceae). - Red Alga. Indian Ocean, Trop. Pacific. Used for manufacturing agar used as food in China; for the manufacturing of plasters for house walls, calcimines and sizings of cloths in Japan, also for mounting chinese paintings, lanterns and maps.

Eucheuma muricatum (Gmel.) Web. v. Bosse. (syn. E. spinosum (L.) J. Ag.) (Solieriaceae). - Red Alga. Indian Ocean, Trop. Pacific. Source of agar-agar. Extensively used, much imported into China from Japan and East Indies. Consumed as food and used for making agar jelly. Source of Macassar Agar.

Eucheuma papulosa Cotton and Yedo. (syn. Callymenia papulosa Mont., Euhymenia papulosa Kuetz.) (Solieraceae). - Red Alga. Coast around China, Japan, Formosa, Hawaii, Red Sea, Somaliland and adj. territory; Dried and cured plants are used as food and employed in the preparation of isinglass. The majority is derived from China and occurs on the markets of Japan and China.

Eucheuma speciosa Ag. (syn. Gigartina speciosa Sind.) (Solieraceae). - Red Alga. Coast of W. Australia. A very gelatinous plant; used for making blancmange, and jelly, esp in Australia.

Euchlaena mexicana Schrad., Teosinte. (Graminaceae). - Tall annual grass. Mexico. Occasionally cultivated in the S. of United States and Mexico as green forage for live-stock. It is thought that from this species originated the Maize.

Euclea natalensis A. DC. (Ebenaceae). - Shrub. Natal and Transvaal. Bark of roots is source of a dye, supposed to be a good floor stain.

Euclea Pseudebenus E. Mey., Cape Ebony. (Ebenaceae). - Tree. S. Africa. Wood is used for furniture and general carpentry work.

Eucommia ulmoides Hook. Tu-chung. (Chin.) (Eucommiaceae). - Tree. Central China. Bark a tonic, acts on liver and kidneys used by Chinese. Leaves, bark of stem and roots, also the seeds contain a gum, which has been suggested as a source of rubber in temperate countries.

Eucommiaceae → Eucommia.

Eucryphia cordifolia Cav., Ulmo. (Eucryphiaceae). - Tree. Chile. Wood strong, hard, very close-grained, not difficult to work. Used for flooring, furniture, rail-road ties, telegraph poles, parts of vehicles, oars, yokes. Used by Indians for canoes. Bark is source of a commercial tannin. Flowers are a source of excellent honey.

Eucryphia Moorei F. v. Muell., White Sally. (Eucryphiaceae). - Tree. Australia, esp. Victoria, New South Wales. Bark is sometimes used for tanning.

Eucryphiaceae → Eucryphia.

Eugeissona triste Griff. (Palmaceae). - Low palm. Malaya, Midrib of leaves used by Benua, Pangan and other tribes for blow-pipe darts.

Roots used for flooring Besisi-houses. Midribs made into matting for inner and outer walls of Malay houses.

Eugenia aquea Burm. f., Watery Rose-Apple. (Myrtaceae). - Small tree. Ceylon, Malaya. Cultivated in Java. Fruits red, edible, insipid, eaten raw, also made into Roedjak, a syrup. Sold in local markets of Java.

Eugenia axillaris (Sw.) Willd., White Stopper. (Myrtaceae). - Shrub or small tree. Florida, West Indies. Wood hard, heavy, strong, very close-grained; locally used for cabinet work.

Eugenia Brackenridgei Gray. (Myrtaceae). - Tree. New Caledonia, Ins. Fiji. Wood red to almost black, very hard, easily worked and polished; used for interior work, cabinet making and turnery.

Eugenia caryophyllata Thunb. (syn. E. aromatica Baill., Caryophyllum aromaticum L.) Clove Tree. (Myrtaceae). - Moluccas. Cultivated in Sumatra, Java, Amboina, Madagascar, Mauritius, Zanzibar, West Indies etc. Dried flower-buds are the source of the commercial cloves, used as spice, for flavoring foods and candies. Produces Oil of Cloves, Oleum Caryophylli; used medicinally as stimulant, carminative, antiseptic, against toothache and for flavoring. Contains 84 to 90% eugenol. Also source of vanillin.

Eugenia clavimyrtus Koord. and Valet. (syn. Jambosa glabrata DC. (Myrtaceae). - Small tree. Java. Wood occasionally used for building. Bark in Java source of a black dye; used for staining cloth.

Eugenia confusa DC., Ironwood, Red Stopper. (Myrtaceae). - Tree. Florida, West Indies. Wood hard, strong, heavy, close-grained, red brown; locally used for cabinet work.

Eugenia Conzattii Standl. (Myrtaceae). - Tree. S. Mexico. Fruits eaten by several Indian tribes in Mexico.

Eugenia cordata Laws., Water Tree. (Myrtaceae). - Tree. Cape Peninsula. Wood dark red-brown, heavy, hard, strong, elastic, durable; used for beams, rafters, boat-planking, felloes, mill-work, water-wheels.

Eugenia cuprea Koord. and Valet. (Myrtaceae). - Large tree. Java, Wood heavy, close-grained, of fine construction; used in Java for building of houses.

Eugenia Curranii C.B.R., Lipoti. (Myrtaceae). - Medium sized tree. Philipp. Islds. Occasionally cultivated. Fruits edible, pleasant acid flavor, dark red to black, 2 cm. across; eaten raw, also made into wine, and an excellent jelly.

Eugenia cymosa Lam. (syn. Jambosa tenuicuspis Miq.) (Myrtaceae). - Tree. Mauritius, Mal. Archipel, esp. Java, Sumatra. Bark source of a black dye. Bark is pounded, soaked in water; boiled with coconut milk in which cloth is stained and dried.

Eugenia densiflora DC. (syn. Jambosa densiflora DC.) (Myrtaceae). - Tree. Mal. Archipel. Bark source of a brown dye. Flowers are eaten with sambal by the Javanese. Fruits edible, of a fair quality.

Eugenia Dombeyana DC. Grumichama. (Myrtaceae). - Small tree. Peru, S. Brazil. Occasionally cultivated. Fruits small, dark crimson, much variation as to quality, pleasant flavor when raw; suitable for jams, jellies and pies.

Eugenia edulis Benth. (Myrtaceae). - Tree. Brazil to Argentina. Fruits edible, made into vinegar in Argentina. Sold in markets.

Eugenia floribunda West. Rumberry, Murta. (Myrtaceae). - Tree. West Indies. Dark red to black fruits are used as a basis of a liqueur that is mainly exported to Denmark; also source of a good jam.

Eugenia foliosa DC. (syn. Myrtus foliosa H. B. K.) Arrayán. (Myrtaceae). - Shrub or small tree. Ecuador. Fruits dark purple, delicate texture; whitish juicy pulp, sub-acid, pleasant flavor; eaten raw.

Eugenia formosa Wall. (Myrtaceae). - Shrub or small tree. Trop. Asia. Cultivated in Cochin-China. Fruits small, round, insipid; sold in local markets.

Eugenia grandis Wight. (Myrtaceae). - Tree. Mal. Archipel. Australia. Wood relatively heavy, not hard, durable, brown; used in Java for building purposes, boat and house-construction.

Eugenia jamboloides Koord. and Valet. (syn. Syzygium racemosum DC.). (Myrtaceae). - Tree. Java. Bark source of a black dye. Wood occasionally employed for building purposes in Java.

Eugenia Jambolana Lam. (syn. Syzygium Jambolanum DC.). Java Plum. Jambolan. (Myrtaceae). - Tree. Trop. Asia to Australia. Small purple fruits are made into preserves; some varieties are seedless, being of a pleasant flavor.

Eugenia Jambos L. (syn. Jambosa vulgaris DC.). Rose Apple, Jambos. (Myrtaceae). - Tree. Tropics of the Old World, introduced in the Americas. Cultivated. Fruits reddish, juicy, sweet, crisp; used for jellies with addition of some other acid fruit; also in sauces. Flowers and fruits are candied. Bark is sometimes employed for tanning in Réunion.

Eugenia javanica Lam. (syn. E. mananquil Blanco.) Samarang Roseapple, Mankil. (Myrtaceae). - Tree. Malaya to Philipp. Islds. Fruits are edible, of a good flavor, ovoid, 4 cm. long, red, fleshy, acid. Much consumed in Java.

Eugenia Klotzschiana Berg., Pera do Campo. (Myrtaceae). - Small tree. Centr. Brazil. Occasionally cultivated. Fruits pear-shaped, edible, juicy, acid, very aromatic; used for jellies.

Eugenia lineata (DC) Duth. (syn. Jambosa lineata DC.) (Myrtaceae). - Tree. Mal. Archipel. Wood used for construction of houses. Bark employed for tanning fishing-nets,

Eugenia Luschnathiana Berg., Pitomba. (Myrtaceae). - Tree. Brazil, esp. Bahia. Sometimes cultivated. Fruit broadly ovate, orange-yellow, juicy, aromatic; used for jellies.

Eugenia Maire Cunningh. (Myrtaceae). - Small tree. New Zealand. Wood very straight, even-grained, heavy, dense, hard, durable, of great strength. Used for cabinet-work, jetty-piles, mooring-posts.

Eugenia malaccensis L. (syn. E. macrophylla Lam.) Mountain Apple. (Myrtaceae). - Medium sized tree. Trop. Asia, esp. Mal. Archipel. Cultivated. Fruits red, edible, not rich flavored. Several varieties.

Eugenia myrtifolia Sims. Bush Cherry, Red Myrtle. (Myrtaceae). - Tree. Australia, esp. New South Wales, Queensland. Wood elastic, seasons well, light; reddish or yellowish. Used for boat-building, oars. Employed by aborigines for making shields and boomerangs. Fruits are made into jellies in different parts of Australia.

Eugenia opaca Koord. and Valet. (Myrtaceae). - Tree. Java. Wood heavy, close-grained, light-brown; used for construction of houses.

Eugenia operculata Roxb. (syn. Syzygium nodosum Miq.) (Myrtaceae). - Tree. Trop. Asia. Bark is used for tanning by the natives.

Eugenia Pimenta DC. → Pimenta officinalis Lindl.

Eugenia polyantha Wight. (syn. E. lucidula Miq.) (Myrtaceae). - Tree. Mal. Archipel. Bark used for tanning fishing-nets; also for staining bamboo material. Decoction of leaves and bark is used in Java for diarrhea.

Eugenia polycephala Miq. (syn. Jambosa cauliflora Miq.) (Myrtaceae). - Tree. Borneo, India. Cultivated. Fruits edible, made into jelly, much esteemed by the natives. Has been recommended for improvement by scientific breeding.

Eugenia polycephaloides C. B. R., Maigang. (Myrtaceae). - Tree. Philippine Islands. Fruits are edible, 1 cm. across, dark red, acid, made into an excellent tart jelly.

Eugenia rhombea (Berg.) Urban., Red Stopper. (Myrtaceae). - Tree. Florida, West Indies. Wood close-grained, hard, strong, light brown; locally used for cabinet work.

Eugenia Selloi Berg. (syn. Phyllocalyx edulis Berg.). Pitanga tuba. (Myrtaceae). - Shrub or small tree. Brazil. Fruits ribbed, aromatic, acid; suitable for jam.

Eugenia Thumra Roxb. (syn. E. acuminatissima Kurz.) (Myrtaceae). - Mal. Archip. Wood strong, course-grained, red, attacked by white ants; used in Java for building purposes. Decoction of bark is used as a dye for fibre. Fruits are occasionally eaten.

Eugenia tomentosa Camb., Cabelluda. (Myrtaceae). - Shrub or small tree. Brazil. Fruits resemble gooseberries; they are juicy, of a pleasant sub-acid flavor, suitable for jellies.

Eugenia uniflora L., Pitanga, Surinam-Cherry. (Myrtaceae). - Shrub or small tree. Trop. America. Cultivated in tropics and subtropics. Fruits the size of a cherry, ribbed, orange, orange-red to almost black, edible, aromatic flavor, sweet, eaten raw, made into jam, jelly, compote, sherbet, syrup. Much esteemed in Brazil. The crushed pungent and scented leaves when spread on the floor of barns, will keep out insects.

Eugenia Uvalha Camb, Uvalha. (Myrtaceae). - Shrub or small tree. S. Brazil. Cultivated in Brazil. Fruits yellowish or orange, juicy, of a pleasant aroma; used for drinks.

Eugenia zeylanica Wight. (syn. E. spicata Lam.) (Myrtaceae). - Tree. Mal. Archip. Bark is source of a black dye.

Eulophus ambiguus Nutt. → Cogswellia ambigua (Nutt.) Jones.

Eupatorium Ayapana Vent. (syn. E. triplinerve Vahl.). (Compositae). - Perennial herb. Trop. America. Cultivated. Leaves are source of a tea, being digestive and stimulant.

Eupatorium cannabinum L., Hemp Eupatorium, Waterhemp. (Compositae). - Perennial herb. Europe, Asia, N. Africa. Roots and leaves, Radix et Herba Cannabis Aquatica were formerly used in medicine; at present employed as home-remedy in some countries for dropsy. Contains eupatorin, a glucosid.

Eupatorium chinense L. (Compositae). - Perennial herb. Chihli, Shantung, Kwantung (China), Japan. Used medicinally in China.

Eupatorium Dalea L. (Compositae). - Perennial herb. West Indies, S. America. Herb is strongly cumarin-scented; used as substitute for ranilla.

Eupatorium indigofera Perodi (Compositae). - Shrub. Argentina, Paraguay, and adj. territory. Leaves are source of a blue dye. Also E. laeve DC.

Eupatorium perfoliatum L., Boneset. (Compositae). - Perennial herb. N. America. Dried leaves and flower tops when collected in July and August, are used medicinally as emetic, diaphoretic, stimulant, cathartic; used for colds. Contains an ess. oil, a bitter glucoside, eupatorin and resin.

Eupatorium Rebaudianum Bertoni (Compositae). - Perennial herb. S. America, esp. Paraguay. Plant has been suggested as a sweetening agent. Contains estevin, a glucosid which is 150 times sweeter than sugar.

Eupatorium stoechadesmum Hance. (Compositae). - Perennial herb. China. Cultivated. Aromatic leaves are used for flavoring arec nuts. In Annam considered a tonic and aphrodisiac.

Eupatorium triplinerve Vahl. (syn. E. Ayapana Vent.) (Compositae). - Perennial. Trop. Asia. Sometimes cultivated by natives. Used as condiment and to stimulate digestion.

Euphorbia aegyptiaca Boiss. (Euphorbiaceae). - Prostrate shrub. Sudan. Egypt. Plant is used as a purgative by the natives.

Euphorbia antiquorum L. (Euphorbiaceae). - Xerophytic tree. S. W. Asia, Africa, Poisonous. Young shoots made into sweet-meat in Java by Chinese, after a long time boiling and re-boiling in sugar. Likewise E. neriifolia L., E. trigona Haw.

Euphorbia antisyphilitica Zucc. (syn. E. cerifera Alc.). Candellila. (Euphorbiaceae). - Shrub. Mexico and S. W. of United States. Source of Candelilla Wax, when refined used for polishes and creams, for foot and leather ware, floor and furniture varnishes. Mixed with rubber, gutta percha, it is used in electric insulating material and varnishes. Made into water proofing fabrics, dental moulding, paper sizes, paint removers, metal lacqeurs, lithographic colors, sealing waxes, gramophone records, chewing gum; in shoe polish, water proofing; when mixed with paraffin it is used for manuf. candles.

Euphorbia calyculata H.B.K., Chipire. (Euphorbiaceae). - Shrub or small tree. Mexico. Source of Chupire Rubber, a substance of minor importance.

Euphorbia Candelabrum Trém. (Euphorbiaceae). - Tree. Trop. Africa. Juice from the stem is used by the natives of the Sudan as arrowpoison.

Euphorbia Characias L. (Euphorbiaceae). - Herbaceous plant. Mediterranean region. Crushed plant is used in Greece to stupefy fish.

Euphorbia corollata L. (Euphorbiaceae). - Herbaceous plant. Eastern N. America to Florida and Louisiana. Plant was used by the Indians as an expectorant and diaphoretic; in larger amounts as an emetic. Source of Purging or Emetic Root.

Euphorbia cotinifolia L. (Euphorbiaceae). - Shrub. Mexico, Centr. America, N. of S. America. Milky sap has violent emetic-cathartic properties. Very poisonous. Used by some Indians for criminal poisoning, also as arrow poison; for stupefying fish.

Euphorbia dendroides L. (Euphorbiaceae). - Herbaceous plant. Mediterranean region. Crushed plant is used in S. Europe to stupefy fish.

Euphorbia fulva Stapf. (syn. E. elastica Alt. and Rose, Euphorbiodendron fulvum Millsp.). (Euphorbiaceae). - Tree. Mexico. Plant is a minor source of rubber, Palo Amarillo Rubber.

Euphorbia Intisy Drake del Cast., Intisy, Herotin. (Euphorbiaceae). - Large shrub or tree. Madagascar. Once source of the important Intisy Rubber. Species is almost exterminated.

Euphorbia Ipecacuanha L., Ipecac Spurge. (Euphorbiaceae). - Herbaceous plant. N. America. Source of Ipecac Spurge; used medicinally as an emetic. Contains euphorbon.

Euphorbia lorifera Stapf., Koko, Akoko. (Euphorbiaceae). - Small tree. Hawaii Islds. Latex has been recommended as a source of rubber during times of emergency, being of poor quality. Has also been advised as a chewing gum base.

Euphorbia marginata Pursh. Snow-in-the-Mountain. (Euphorbiaceae). - Annual herb. N. America. Latex is a source of chewing gum among the Kiowa Indians (New Mex.).

Euphorbia pilulifera L. (syn. E. hirta L.). (Euphorbiaceae). - Herbaceous plant. Tropics. Entire plant is considered sedative, haemostatic, soporific, decoction is very efficacious for allaying the dyspnoea of asthmatics.

Euphorbia Preslii Guss. (syn. E. hypericifolia Gray). (Euphorbiaceae). - Herbaceous plant. Eastern N. America. Was used by some Indian tribes for dysentery, diarrhea.

Euphorbia primulaefolia Baker. (Euphorbiaceae). - Small shrub. Madagascar. Herb. has violent purgative properties; used in Madagascar as rat-poison.

Euphorbia resinifera Berg. (Euphorbiaceae). - Xerophytic shrub. N. Africa. Source of a resin, called Gummi Resina Euphorbium; used in vet. medic. and in paints. Also E. canariensis L., E. officinalis L. and E. antiquroum L. These resins need careful investigations. Was formerly used as drastic purgative. Much is derived from the S. slopes of the Atlas Mountains.

Euphorbia venefica Trém. (Euphorbiaceae). - Bushy shrub. Trop. Africa. Juice is used by several native tribes as arrow-poison.

Euphorbiaceae → Acalypha, Actephila, Aetoxicon, Agrostistachys, Alchornea, Aleurites, Antidesma, Aporosa, Baccaurea, Baliospermum, Baloghia, Bertya, Bischoffia, Brynia, Bridelia, Chaetocarpus, Chrozophora, Claoxylon, Cleistanthus, Cluytia, Cnidosculus, Croton, Daphniphyllum, Elateriospermum, Endospermum, Euphorbia, Excoecaria. Glochidion, Hevea, Hieronima, Hippomane, Homalanthus, Homomoia, Hura, Hymenocardia, Jatropha, Joannesia, Macaranga, Mallotus, Manihot, Manniophyton, Mercurialis, Microdesmis, Omphalea, Palaquium, Pedilanthus, Phyllanthus, Pimelodendron, Plukenetia, Pycnoscoma, Ricinodendron, Ricinus, Rottlera, Sapium, Sauropus, Sebastiania, Spondianthus, Styllingia, Tetracarpidium, Toxicodendron, Uapaca.

Euphorbiodendron fulvum Millsp. → Euphorbia fulva Stapf.

Euphoria longana Lam. → Nephelium Longana (Lam.) Cam.

Euphoria Nephelium DC. → Nephelium lappaceum L.

Euphrasia officinalis L. (syn. E. Rostkoviana Hayne.) Drug Eyebright. (Scrophulariaceae). - Annual herb. Europe. The herb was formerly extensively used in medicin as a tonic-exitans and for diseases of the eye.

European Alder → Alnus glutinosa Medic.

European Ash → Fraxinus excelsior L.

European Barberry → Berberis vulgaris L.

European Beachgrass → Ammophila arundinacea Host.

European Beech → Fagus sylvatica L.

European Bird Cherry → Prunus Padus L.

European Black Currant → Ribes nigrum L.

European Dewberry → Rubus caesius L.

European Elder → Sambucus nigra L.

European Goat's Rue → Galega officinalis L.

European Goldenrod → Solidago virgaurea L.

European Gooseberry → Ribes Grossularia L.

European Hackberry → Celtis australis L.

European Hazel → Corylus Avellana L.

European Holly → Ilex Aquifolium L.

European Hornbeam → Carpinus Betulus L.

European Larch → Larix europaea DC.

European Mandrake → Mandragora officinarum L.

European Misletoe → Viscum album L.

European Mountain Ash → Sorbus Aucuparia L.

European Pasqueflower → Anemone Pulsatilla L.

European Pennyroyal → Mentha Pulegium L.

European Privet → Ligustrum vulgare L.

European Raspberry → Rubus Idaeus L.

European Turkey Oak → Quercus Cerris L.

Euonymus → Evonymus.

Eurya japonica Thunb. (Theaceae). - Tree. Indo Malayan region. Wood is used for wagon making and turnery.

Eurya ochnacea (DC.) Szysz. (Theaceae). - Tree. Trop. Asia. Wood is used by the natives for house and boat building.

Eurya theoides (Sw.) Blume. (syn. Eroteum theaoides Sw.). (Theaceae). - Tree. West Indies and Centr. America. Leaves are used in some parts of Cuba as a substitute for tea.

Euryale ferox Salisb. (Nymphaeaceae). - Large waterplant, Trop. Asia. Seeds are eaten in some parts of Japan. They are also dried and are a source of starch. Very young stalks and roots are edible.

Eusideroxylon Zwageri Teijsm. and Binn., Borneo Billian, Borneo Ironwood. (Lauraceae). - Tree. Borneo, Sumatra. Wood extremely heavy, very hard, durable, sinks in water, straight-grained, splits readily, yellow brown to dark brown; used for heavy construction work, har-

bours, rail-road ties, telephone poles, bridges, floors in factories, street paving, house posts, shingles; much is exported.

Euterpe oleracea Engel. (Palmaceae). - Tall palm. Trop. America. Fruits are source of a popular beverage in Pará. Very young center of the plant is consumed as vegetable, as palm-cabbage, said to resemble artichoke.

Eutrema Wasabi Max. Wasabi. (Cruciferaceae). - Perennial herb. Japan. Cultivated. Roots are used in Japan as a condiment.

Eutuberaceae (Fungi) → Choiromyces, Hydnotria, Tuber.

Euxylophora paraensis Huber. Brazilian Satinwood, Brazilian Boxwood, Canary Wood, Pau Amarello (Braz.). (Rubiaceae). - Tree. Lower Amazon region (Braz.), Wood clear yellow, hard, heavy, straight-grained, takes high polish; used for doors, tables, floors, furniture.

Evening Primrose → Oenothera biennis L.

Evergreen Sumach → Rhus sempervirens Schule.

Everlasting- Common → Gnaphalium polycephalum Michx.

Evernia furfuracea E. Fr. → Parmelia furfuracea (L.) Ach.

Evernia prunastri (L.) Ach. (syn. Lobaria prunastri Hoffm., Parmelia prunastri Ach.) Oak Moss. (Usneaceae). - Lichen. N. Temp. zone, esp. Europe. On stems of various species of trees. Source of an ess. oil, extracted by volatile solvents; used in perfumery; known as Mousse Chêne, Eichenmoss. Lichens derived from oak stems are reputed to yield the best oleo-resin for the perfumery industry. Much is derived from France, Czechoslovakia, Herzogovina and Piedmont (Italy). Used in soap making, as an impalpable powder, or in the form of a resin. Plant is also source of a color pigment; used for dyeing woolens a violet color. The lichen was used by the ancient Egyptians as leavening agent in making bread which is at the present continued by the Copts and Arabs. It is also used as an old tonic for intestinal weakness. Also considered in perfumery are E. furfuracea E. Fr. (syn. Parmelia furfuracea (L.) Ach., E. mesomorpha Nyl. (syn. Latheria thamnodes (Flot.) Hue.

Evernia vulpina (L.) Ach. (syn. Letharia vulpina (L.) Waino. (Usneaceae). - Lichen. Temp. zone. On stems of Conifers. Has been used in Norway and Sweden as as source of a yellow dye to color woolens.

Evia amara Comm. → Spondias mangifera Willd.

Evodia rutaecarpa Hook. (Rutaceae). - Woody plant. Kiangsi (China), Japan. Used medicinally in China. Contains evodine, evodiamine and rutaecarpine.

Evonymus americanus L., Strawberry Bush, Brook Evonymus. (Celastraceae). - Shrub. Eastern N. America to Florida and Texas. Bark was used by different Indian tribes as cathartic; also as irritant on gastro-intestinal mucous membrane.

Evonymus atropurpureus Jacq., Indian Arrow Wood, Burning Bush. (Celastraceae). - Shrub or small tree. Eastern N. America to Florida and Texas. Source of Wahoo Bark; used as a cathartic.

Evonymus europaeus L., Spindle Tree. (Celastraceae). - Shrub or small tree. Europe, W. Siberia, Asia Minor, Caucasia, Turkestan. Wood fine grained, difficult to split, not very durable; used for turnery, statues, instruments; source of charcoal, used for gun powder. Fruits and seeds are source of a yellow dye; used for coloring butter. Seeds produce an oil, sometimes used for manuf. soap.

Evonymus hians Koehne. (Celastraceae). - Small tree or shrub. Rich soils, forests. Hokkaidô, Honshū, Shikoku (Japan). Wood heavy, hard, very close-grained. Used in Japan for printing blocks, mosaic works, combs, in turnery.

Evonymus oxyphyllus Miq. (Celastraceae). - Small tree. Japan, Korea, China. Wood hard, elastic, very fine grained, takes polish. Used in Japan for stamp and printing blocks, mosaic work, small fancy wares. In early days used for bows.

Evonymus tingens Wall. (Celastraceae). - Shrub or tree. W. Himalayan region. Arillus and bark are used by the natives as cosmetic. Bark is purgative, also E. crenulatus Wall., E. pendulus Wall.

Excoecaria Agallocha L. Blind-Your-Eyes-Tree. (Euphorbiaceae). - Tree. S. E. Asia to trop. Australia. Certain parts of the wood are used for incense.

Exocarpus cupressiformis Lab., Ballot. (Santalaceae). - Tree. Australia. Wood close-grained. Used for turnery, cabinet-work, gun-stocks, tool-handles, map-rollers, and to a limited extent for engraving. Used by natives for spearthrowers.

Exocarpus latifolia R. Br. (syn. E. miniata Zipp.) Shrub Sandal - Wood. (Santalaceae). - Woody plant. Australia. Wood hard, coarsegrained, dark colored, fragrant. Used for cabinet-work, easily polished.

Excoecaria indica Muell. Arg. → Sapium indicum Willd.

Exogonium bracteatum (Cav.) Choisy. (syn. Ipomaea bracteata Cav.) Jicama, Bejuco Blanco. (Convolvulaceae). - Woody vine. Mexico. Large succulent roots are eaten raw or cooked among natives of Pacific Coast of Mexico.

Exogonium jalapa Nutt. and Coxe. → Ipomoea Purga Hayne.

Exostemma caribaeum Roem. and Schult. Caribbean Princewood. (Rubiaceae). - Tree. West Indies. Wood strong, saffron-like color; used for turnery and cabinet-work.

Eyebright, Drug → Euphrasia officinalis L.

Eysenhardtia polystachya (Ortega) Sarg. (syn. E. amorphoides H. B. K.). Palo Guate, Kidney Wood. (Leguminosaceae). - Shrub or tree. Mexico and S. W. of United States. Wood known for its peculiar properties. Infusion of heartwood in water is at first golden-yellow, afterward deepens to orange. Viewed against a black background is shows beautiful peacock blue fluorescence. Wood is used in Mexico as drinking trough for watering fowls. Flowers are source of a good honey.

F

Faba St. Ignatii → Strychnos Ignatii Berg.

Faba vulgaris → Vicia Faba L.

Fabiana imbricata Ruiz and Pav. (Solanaceae). - Woody plant. Peru and Chile. Infusion of leaves is used by the natives as diuretic.

Fagaceae → Castanea, Castanopsis, Fagus, Lithocarpus, Nothofagus, Quercus.

Fagara angolensis Engl. (Rutaceae). - Tree. Trop. W. Africa. Wood tough, medium coarsegrained, light, pale yellow; used for canoes, boxes, plywood, general carpentry and for paperpulp.

Fagara coriacea Kr. and Urb. → Zanthoxylum emarginatum Sw.

Fagara piperita Poureira → Zanthoxylum nitidum DC.

Fagonia cretica L. (syn. F. sinaica Boiss.). (Zygophyllaceae). - Shrub. N. and S. Africa. Herb is used as food for camels and mules.

Fagopyrum esculentum Moench., Common Buckwheat. (Polygonaceae). - Annual herb. Probably from Centr. and W. China. Cultivated in Old and New World. Source of Buckwheat Flour. Much employed in United States in pancake flour, made into buckwheat cakes. Also mixed with other flours. Groats are kernels with hulls removed, used as a breakfast food, thickening for soups, gravies, dressing, porridge. Also grown as green manure and orchard covercrop. Flowers are source of commercial honey. Japanese Buckwheat and Silver-hull are grouped under this species.

Fagopyrum tataricum Gaertn., Tatary Buckwheat, India Wheat, Siberian Buckwheat. (Polygonaceae). - Annual herb. Asia. Seeds are source of a flour used for different foods.

Fagraea fragrans Roxb. (syn. F. peregrina Blume.) (Loganiaceae). - Tree. Burma, Malaya. Wood hard, easily worked, does not split easily, durable, light yellow; used for bridges, house-building and furniture.

Fagraea grandis Panch and Sebert (Loganiaceae). - Tree. New Caledonia. Wood yellowish, fine-grained; esteemed by the natives of N. Caledonia for making sculptures.

Fagraea racemosa Jacq. (Loganiaceae). - Shrub. Malaya. Much used by Malayas in medicine. Roots used as tonic after fever.

Fagus betuloides Mirb. → Nothofagus betuloides Blume.

Fagus Cliffortoides Hook. f. → Nothofagus Cliffortoides Oerst.

Fagus ferruginea Dryand. (syn. F. grandifolia Ehrh., F. americana Sweet.). American Beech. (Fagaceae). - Tree. Eastern N. America to Florida and Texas. Wood strong, heavy, hard, no taste or odor; after steaming easy to bend, close-grained, reddish-brown to almost white; used for crossties, cooperage, pulpwood, boxes, crates, veneer, handles, flooring, woodenware, laundry appliances, general millwork, food containers, spools, toys and fuel. When steamed, wood is valuable for curved parts of chairs. The sweet nuts are edible; occasionally sold in northern markets. Wood tar is obtained by destructive distillation of the wood; source of creosote; used for medical purposes; is colorless to light yellow, stimulating expectorant, antiseptic, antipyretic; used for chronic bronchitis, vomiting sea-sickness, pulmonary tuberculosis. Wood is also source of methyl alcohol and acetic acid. Guaiacol is derived from beachwood creosote by fractional distillation, being expectorant, and intestinal antiseptic.

Fagus fusca Hook. f. → Nothofagus fusca Oerst.

Fagus Menziesii Hook. f. → Nothofagus Menziesii Oerst.

Fagus obliqua Mirb. → Nothofagus obligua Blume.

Fagus orientalis Lipsky. Oriental Beech. (Fagaceae). - Tree. Balkan Penin., Asia Minor, Caucasia, N. Iran. Wood is used for the same purposes as F. sylvatica L. to which it is related.

Fagus procera Poepp. and Endl. → Nothofagus procera Oerst.

Fagus Sieboldi Endl. Buna-no-ki. (Fagaceae). - Tree. Japan. Wood hard, tough, close-grained, takes polish. Used in Japan for manuf. chairs, furniture, farm implements.

Fagus sinensis Oliv. (syn. F. sylvatica var. longipes Oliv.) (Fagaceae). - Tree. Yunnan, Hupeh, Szechuan (China). Wood used in China for boat-building, plows, handles of tools.

Fagus Solandri Hook. f. → Nothofagus Solandri Oerst.

Fagus sylvatica L., European Beech. (Fagaceae). - Tree. Europe, Asia Minor, Centr. Asia. Wood reddish, heavy, hard, durable, withstands influence of water, easy to split, little elastic strong; used for wagons, furniture (Wiener Möbel), agric. implements, wooden shoes, spoons, plates, turnery, pianos, ship-building, railroad-ties, brush backs, meatchoppers, construction of dams, water-mills, excelsior, wood-pulp, excellent fuel. Formerly used for manuf. of wood alcohol and tar. Source of creosote being antiseptic, antipyretic, also of a charcoal called Carbo Ligni Pulveratus; purified charcoal Carbo Ligni Depuratus used for phosphorus and alkali poisoning. Seeds produce an oil, used for food, illumination, manuf. of soap. Sp. Gr. 0.9205-0.9225; Sap. Val. 191-196; Iod. No. 111-120; Unsap. 0.27%. Nuts are consumed roasted. Leaves have been used as substitute for tobacco.

Fahtsai → Nostoc commune Vauch. var flagelliforme (Berk. & Cart.) Born. and Flah.

Falcata comosa (L.) Kuntze → Amphicarpa monoica (L.) Ell.

False Aloe → Agave virginica L.

False Arrowroot → Curcuma Pierreana Gagn.

False Cola → Cola acuminata Schott. and Endl.

False Hellebore, White → Veratrum album L.

False Hellebore, American → Veratrum viride Ait.

False Tarragon → Artemisia dracunculoides Pursh.

Fame Flower, Potherb → Talinum triangulare Willd.

Fame Flower, Orange → Talinum aurantiacum Engelm.

Famine plants → Food plants used in emergency.

Fancy Yam → Dioscorea esculenta (Lour.) Burk.

Faramea odoratissima DC. (Rubiaceae). - Tree. West Indies, S. America. Wood strong, compact, pale yellow. used for general carpentry.

Farinaceous Ipecac → Richardsonia pilosa H. B. K.

Fat, Bay → Laurus nobilis L.

Fat, Cay-Cay → Irvingia Oliveri Pierre.

Fat, Dhupa → Vateria indica L.

Fat, Laurel Berry → Laurus nobilis L.

Fat, Malukang → Polygala butyraceum Heck.

Fat, Mkani → Allanblackia Stuhlmanii Engl.

Fat, Njatuo → Palaquium oleiferum Blanco.

Fat, Sawari → Caryocar amygdaliferum Mutis.

Fat, Siak → Palaquium oleiferum Blanco.

Fat, Suari → Caryocar amygdaliferum Mutis.

Fat, Taban Merak → Palaquium oleiferum Blanco.

Fatchoy → Nostoc commune Vauch. var. flagelliforme (Berk. & Cart.) Born. & Flah.

Fatsia horrida Benth. and Hook. (Araliaceae). - Shrub. Western N. America. Young succulent stems are consumed as food by the Eskimos of Alaska. Boiled bark is used as medicine.

Fatsia papyrifera Hook. (syn. Aralia papyrifera Hook., Tetrapanax papyrifera Koch.) (Araliaceae). - Irrigular shaped tree. China. Pith is source of ricepaper which is much manufactured in China. Used for surgical dressings and by chinese artists to paint upon; also made into toys and artificial flowers. Much is derived from Prov. Kweichou and manuf. in Chunkiang.

Faurea saligna Harv. (Proteaceae). - Tree. S. Africa, Angola, Mozambique. Wood hard, durable, resistant; used for building, and source of charcoal.

Faurea speciosa Welw. (Proteaceae). - Tree. Trop. Africa. Wood yellowish to reddish. Used in fine turnery and furniture.

Favolus spathulatus (Jungh.) Bresw. (Polyporaceae). - Basidiomycete. Fungus. Trop. Asia. Fruitbodies are consumed by the natives of the Sunda Islds.

Federita → Sorghum vulgare Pers.

Fedia Cornucopiae DC., African Valerian. (Valerianaceae). - Perennial herb. Mediterranean region. Leaves are consumed in some parts of Africa as a pot-herb.

Feijoa Sellowiana Berg., Feijoa (Myrtaceae). - Shrub or small tree. S. Brazil, Uruguay, Paraguay, N. Argentina. Cultivated in tropics and subtropics. Fruit edible raw, of a pleasant flavor, also eaten stewed, in jams or jelly. There is much difference in flavor between fruits of the different seedling trees.

Fendler Potato → Solanum Fendleri Gray.

Fennel → Foeniculum vulgare Mill.

Fennel, Bitter → Foeniculum vulgare Mill.

Fennel, Florence → Foeniculum vulgare Mill.

Fennel, Hog's → Peucedanum officinale L.

Fennel, Italian → Foeniculum vulgare Mill.

Fennel, Sweet → Foeniculum vulgare Mill.

Fenugrec → Trigonella Foenum Graecum L.

Fenugreek → Trigonella Foenum Graecum L.

Feretia canthioides Hiern. (Rubiaceae). - Woody plant. Trop. Africa. Fruits occasionally used in N. Nigeria as a cosmetic; with indigo it is used for making markings on the face. Decoction of roots is used for gonorrhoea.

Fern, Maidenhair → Adiantum.

Fern, Male Shield → Aspidium Filix Mas L.

Fern, Marsh → Acrostichum aureum L.

Fern, Ostrich → Onoclea Struthiopteris (L.) Hoffm.

Fernleaf Nitta Tree → Parkia filicoidea Welw.

Ferocactus hamatocanthus (Muehlenpf.) Britt. and Rose. (syn. Echinocactus hamatocanthus Muehlenpf.) (Cactaceae). - Xerophyte. N. of Mexico, Texas, New Mexico. In Nuevo León fruits are used in cooking, as a substitute for lemons.

Ferocactus Wislizeni (Engelm.) Britt. and Rose. (syn. Echinocactus Wislizeni Engelm.) (Cactaceae). - Xerophyte. Mexico, Texas to Arizona. Pulp of stems is source of a candy, eaten in Mexico and U.S.A. Often sold in stores.

Ferolia guyanensis Aubl., Satinwood. (Rosaceae). - Tree. Guiana. Wood yellowish red brown, beautifully marked; used for manuf. of furniture.

Feronia limonia Swingle (syn. F. elephantum Correa.) Elephant Apple, Wood Apple. (Rutaceae). - Small tree. India, Ceylon. Cultivated. Fruit edible, size of an orange, pulp aromatic and glutinous; eaten raw, in jellies and in sherbets; considered stomachic and stimulant. In India leaves are used for indigestion. Juice with orpiment makes a yellow ink; used by the Siamese for writing on palm leaves. Source of Feronia Gum or Velampisini, supposed to be equal to Gum Arabic. Dissolves easily in water. Used in water colors and as a glue, also employed in varnishes and certain paints.

Feroniella lucida Swingle. (Rutaceae). - Spiny tree. Java. Fruits globose, 8 cm. across; eaten in some parts of Java.

Ferraria purgans Mart. (Iridaceae). - Perennial herb. Trop. America. Roots are used in some S. American countries as a purgative, tonic and emmenagogue.

Fertilizers, Plants used for → Aleurites, Arachis, Asophyllum, Azolla, Beta, Cajanus, Calopogon, Centrosema, Clitoria, Crotalaria, Cyamopsis, Desmodium, Dolichos, Durvillea, Fagopyrum, Fucus, Gossypium, Indigo, Leucaenol, Lupinus, Lycopersicum, Medicago, Mimosa, Mucuna, Najas, Ornithopus, Pisum, Sargassum, Sesbania, Stratiotes, Tephrosia, Trifolium, Uraria, Vicia, Vigna.

Ferula Assa-foetida L., Asafetida. (Umbelliferaceae). - Perennial herb. Iran, W. Afghanistan. Source of Asafetida, a gum resin, exuded from incisions in living rhizomes and roots. Produces Oleum Asae Foetidae, an ess. oil, derived from steam distillation. Gum-resin is used medicinally as expectorant, laxative, antispasmodic, expectorant. It is composed of a reddish-brown amorphous resin.

Ferula communis L. (Umbelliferaceae). - Perennial herb. N. Africa. Var. brevifolia Marcz is source of Ammoniac of Morocco, a gum.

Ferula foetida (Bunge) Regel. (Umbellifera-ceae). - Perennial herb. Turkestan. Source of a resin having the same use as that from F. Assa-foetida L.

Ferula galbaniflua Boiss. and Buhse, Galba-num. (Umbelliferaceae). - Perennial herb. Iran, Turkestan. Source of Galbanum, a gum-resi-nous exudation from wounds in the stems. Composed of irregular masses or of tears, orange-brown to bluish green and brownish black. Known in Iran as Kasnih, Boridschah; in Afghanistan as Badra-Kema and Bi-ri-jih. Much is shipped from Bombay and Asia Minor. Gum used medicinally as carminative, expectorant, antispasmodic. Contains an ess. oil; a resin, um-belliferon and galbaresinotannol.

Ferula marmarica Asch. and Taub. (Umbelli-feraceae). - Perennial herb. N. Africa. Source of Ammoniac of Cyrenaica, a gum.

Ferula Narthex Boiss. Silphium of the Ancients. (Umbelliferaceae). - Perennial herb. Baltisthan. Probably used as a spice by the ancient Egyp-tians and Greeks about the 6 Cent. B. C. The Romans ascribed marvlous properties to this plant.

Ferula persica Willd. (Umbelliferaceae). - Per-ennial herb. Caucasia, Iran. Source of Saga-penum Gum; sold in tears or cakes; locally used for rheumatism and lumbago.

Ferula Schair Borszcz. (Umbelliferaceae). - Perennial herb. Turkestan. Plant is probably source of a Galbanum.

Ferula Sumbul Hook. f. (Umbelliferaceae). - Perennial herb. Centr. Asia. Root has a musk-like scent, source of Mush or Violet Root, used as stimulant, nervous tonic, antispasmodic; for hysteria and nervous disorders. Used in Iran as a perfume and as incense during religious ceremonies.

Ferula Szowitziana D. C. Sagapen. (Umbelli-feraceae). - Perennial herb. Centr. Asia. Source of Sagapen Resin, has scent of Galbanum. At present very little in use.

Ferula tingitana L. (syn. F. sancta Boiss.) (Um-belliferaceae). - Perennial herb. Syria and N. Africa. Source of North African or Moroc-can Ammoniac, Gummi Resina Ammoniacum.

Fescue, Greenleaf → Festuca viridula Vasey.

Fescue, Meadow → Festuca elatior L.

Fescue, Red → Festuca rubra L.

Fescue, Sheep's → Festuca ovina L.

Fescue, Thurber → Festuca Thurberi Vasey.

Festuca elatior L. Meadow Fescue, English Bluegrass. (Graminaceae). - Perennial grass. Europe. Cultivated in Old and New World. Va-luble for pastures.

Festuca ovina L., Sheep's Fescue (Gramina-ceae). - Perennial grass. Europe. Cultivated in Old and New World; valuable feed for sheep.

Festuca rubra L., Red Fescue. (Graminaceae). - Perennial grass. Temp. N. Hemisphere. Grown as a pasture grass.

Festuca Thurberi Vasey, Thurber Fescue. (Gra-minaceae). - Perennial herb. Dry slopes. We-stern United States. A pasture grass for sub-alpine regions.

Festuca viridula Vasey. Greenleaf Fescue. (Graminaceae). - Perennial grass. Mountain meadows. Western N. America. Outstanding pasture grass in sub alpine regions.

Fetia Star Tree → Astronium graveolens Jacq.

Fetid Buckeye → Aesculus glabra Willd.

Fetid Marigold → Pectis papposa Harv.

Fetter Bush → Leucothoë Catesbaei (Walt.) Gray.

Fever Bark → Alstonia constricta F. v. Muell.

Fever Bush → Garrya elliptica Dougl. and Ilex verticillata (L.) Gray.

Fever Tree → Eucalyptus globulus Lab.

Feverfew Chrysanthemum → Chrysanthemum Parthenium (L.) Bernh.

Fibre, Aramina → Urena lobata L.

Fibre, Buaze → Securidaca longipedunculata Fres.

Fibre, Buntal → Corypha elata Roxb.

Fibre, Caraguata → Bromelia serra Griseb.

Fibre, Caroa Verdadeira → Neoglaziovia varie-gata (Arruda da Cam.) Mez.

Fibre, Coir → Cocos nucifera L.

Fibre, Dundee → Sesbania aculeta Poir.

Fibre, Flax → Linum usitatissimum L.

Fibre, Ixtle → Agave falcata Engelm., A. graci-lispina Engelm., A. heteracantha Zucc., and A. lophantha Schiede.

Fibre, Kendir → Apocynum venetum L.

Fibre, Kittul → Caryota urens L.

Fibre, Mesakhi → Villebrunea integrifolia Gaud.

Fibre, Mijagua → Anacardium Rhinocar-pus DC.

Fibre, Ozone → Asclepias incarnata L.

Fibre, Paineira → Chorisia speciosa St. Hil.

Fibre, Pala → Butea superba Roxb.

Fibre, Piña → Ananas comosus (L.) Merr.

Fibre, Posidonia → Posidonia australis Hook. f.

Fibre, Raffia → Raphia gigantea Chev., R. pe-dunculata Beauv.

Fibre, Tampico → Agave falcata Engelm.

Fibre, Turka → Apocynum venetum L.

Fibre, Zapupe → Agave zapupe Trel.

Fibres. Plants source of → Abroma, Abutilon, Acacia, Adansonia, Aechmea, Agave, Alchor-nea, Anacardium, Ananas, Anodendron, Apocy-num, Arauijia, Arenga, Artocarpus, Asclepias, Astelia, Attalea, Balanites, Bauhinia, Boehmeria, Bombax, Borassus, Brachystegia, Bromelia, Broussonetia, Butea, Calotropis, Cannabis, Carex, Careya, Cariniana, Carludovica, Caryota, Castilla, Cecropia, Ceiba, Chamaecyparis, Chonemorpha, Chorisia, Cochlospermum, Cocos, Columbia, Combretum, Commersonia, Conopharyngia, Corchorus, Cordia, Cordyline, Corypha, Couratari, Cowania, Crotalaria, Cryptocarya, Curculigo, Cyperus, Cypholophus, Dais, Danais, Daphne, Debregeasia, Dichapetalum, Dioscorea, Dombeya, Donax, Elaeocarpus, Elymus, Enalus, Entada, Erythrina, Erythropsis, Ficus, Fimbristylis, Firmiana, Freycinetia, Furcraea, Girardinia, Gnetum, Gomphocarpus. Gossypium, Grewia, Guatteria, Helicteres, Hesperaloe, Heteropogon, Heteropteris, Hibiscus, Hoheria, Holarrhena, Honkenya, Humulus, Hyphaene, Inophloeum, Ischaemum, Kleinhovia, Kydia, Lagetta, Laportea, Leopoldina, Linum, Lygium, Malachra, Manicaria, Manniophyton, Mauritia, Melochia, Milletia, Morus, Muntingia, Musa, Neoglaziovia, Pandanus, Parameria, Passerina, Pavonia, Philodendron, Phoenix, Phormium, Pigafettia, Pipturus, Plagianthus, Polygala, Posidonia, Pothoidium, Poulzolziar, Psoralea, Puya, Raphia, Rourea, Rulingia, Samuela, Sansevieria, Sarcochlamys, Sarothamnus, Securidaca, Sesbania, Sida, Soymida, Sparmanina, Sterculia, Thespesia, Thuja, Tilia, Tillandsia, Touchardia, Trachycarpus, Trema, Trichospermum, Triumfetta, Ulmus, Urena, Urtica, Villebrunea, Washingtonia, Wissadula, Wistaria, Xylopia, Yucca.

Ficus Anomani Hutch. (Moraceae). - Tree. Trop. Africa, esp. Cameroons, S. Leone, Ivory Coast. Extract from plant is used in Gold Coast as a bird lime, mainly for catching small parrots.

Ficus anthelmintica Mart. (Moraceae). - Large tree. Peru, Brazil. Juice, Leche de Ojé and bark, Corteza de Ojé are used for tertian fever in the region of Iquitos (Peru).

Ficus aspera Forst. Rough-leaved Fig, Noomaie, Balemo. (Moraceae). - Tree. Australia, esp. Victoria, Queensland. Fruit black, eaten by aborigenes. Also F. glomerata Willd., F. platypoda Cunningh.

Ficus asperifolia Miq. (Moraceae). - Tree. Trop. Africa. The coarse leaves are sometimes used by the natives as sandpaper.

Ficus capensis Thunb. (Moraceae). - Tree. S. Africa. Wood is used by the natives to make fire by friction. Fruits are consumed by the natives, having the taste of ordinary figs.

Ficus Carica L. Common Fig Tree. (Moraceae). - Tree. Syria, Asia Minor, India, Greece, Mediterranean region. Cultivated since times immemorial mild temperate and subtropical zones. Of much economic importance. Fruit (pseudofruit) purple, greenish yellow; eaten raw, dried or preserved. When dried and ground used as substitute for coffee. In Majorca latex is used to coagulate milk.

Numerous varieties, e. g.: Deep purple varieties: Sultane, Brown Turkey, Brunswick, Fique de Nance, Greenish Varieties: Tapa Cartin, Celeste, Blanch d'Argenteuil, Blanchette, Gentile. Varieties used for drying: Smyrna, Adriatic, Varieties for canning: Magnolia. Caprifigs or male trees, produce the pollen necessary for fertilization.

Ficus consociata Blume. (Moraceae). - Tree. W. Malaysia. Bark source of bark-cloth used for binding books in tropics. Likewise F. gibbosa Blume.

Ficus dicranostyla Mildbr. (Moraceae). - Tree. Trop. Africa. In Togo leaves and bark are mixed with water; used for dehairing animal hides.

Ficus elastica Roxb., Karet Tree. (Moraceae). - Tall tree. Trop. Asia. Was formerly an impostant source of rubber, is now replaced by Hevea. It is called Assam or Indian Rubber.

Ficus exasperata Vahl. (Moraceae). - Tree. Trop. Africa, Arabia. Leaves are used by the natives as sandpaper for woodwork.

Ficus glumosa Delil. (Moraceae). - Tree. E. Africa. Bark is used in different parts of Africa for tanning. Latex is source of a bird lime. Succulent fruits are edible. A fibre is produced from the bark and made into cloth.

Ficus gnaphalocarpa Steud. (Moraceae). - Tree. Trop. Africa. Latex is source of a Guta Percha which is occasionally exported. Juice from fruits is made into an excellent spirit, having flavor of gin.

Ficus involuta Miq. (Moraceae). - Tree. Mexico, Centr. America. Figs are edible, consumed by the Indians.

Ficus microcarpa Vahl. (syn. F. Thonningii Blume). (Moraceae). - Medium sized tree. Trop. Africa. Fibre from the bark is made into cloth by the natives.

Ficus natalensis Hochst. (Moraceae). - Tree. Trop. Africa, esp. Gold Coast, Fernando Po, Cameroons, Uganda, Lower Guinea, Natal, Mozambique etc. Bark is source of a fibre, made by the natives of Uganda into an excellent cloth.

Ficus petiolaris H. B. K. (Moraceae). - Tree. Mexico. Bark was used by the ancient Indians for manuf. of paper. Strong elastic aereal roots are used by the Chinantecs (Mex.) for suspension bridges. Source of a gum acid to be used by surgeons for treating broken bones and hernia.

Ficus platyphylla Delile, Broadleaf Fig. (Moraceae). - Tree. Trop. Africa. Dried latex is used by the Dikas (Afr.) for cleaning brass instruments. It is also employed as a base for chewing gum. A rubber is called Red Cano Rubber, Niger Gutta, Ogbagha Rubber. Latex is also used as bird-lime. In Upper Nile region bark is used for tanning, and for manuf. of ropes, cordage and bark-cloth.

Ficus prolixa Forsk., Tahitian Fig. (Moraceae). - Tree. Pacif. Islds. A fine grade tapa is made from young stems and aereal roots, in the Marquesas. In Rapa aereal roots are source of a fibre, made into cordage.

Ficus pumila L., Okgue. (Moraceae). - Tree. Japan, Formosa, China. Fruits are used in jellies in Japan.

Ficus Radula Willd. (Moraceae). - Tree. Peru to S. Mexico. Bark used by Indians of Peru and Brazil for mats and cloth.

Ficus religiosa L. Peepul Tree, Bot Tree. (Moraceae). - Large tree. Trop. Asia. Much cultivated. Held sacred by Hindus and Buddhists. Gum occasionally source of a sealing-wax, used by artificers for filling up cavities of ornaments. Source of a bird-lime. Bark produces a fibre; formerly made in Burma into a paper. Figs small, used as famine food. Leaves are in Assam source of food of a silkworm, Theophila religiosae Helf. Lives also on F. indica.

Ficus Ribes Reinw. (Moraceae). - Tree. Java. Bark and leaves are sometimes used by the Javanese in Betel chewing. Extract is used for malaria. Sold in bazaars.

Ficus Roxburghii Wall. (Moraceae). - Tree. India, Indo-China. Cultivated. Figs are large, consumed by the natives.

Ficus Rumphii Blume. (Moraceae). - Tree. Mal. Archip. Fruits are consumed ripe by the natives of Bali.

Ficus Sakalavarum Baker. (Moraceae). - Large tree. Madagascar. Fruits edible, 8 cm. long, pear-shaped or roundish, yellow to red, sweet; taste resembling common figs. Eaten raw or in preserves.

Ficus Sycomorus L., Sycomore Fig. (Moraceae). - Tree. N. Africa. Cultivated. Figs edible, slightly aromatic; consumed in many parts of the Orient. Sold in markets. Wood used by the ancient Pharaos for making sarcophages. Latex is able to coagulate milk.

Ficus tinctoria Forster., Dye Fig. (Moraceae). - Small tree. Polynesia. Fruits eaten by natives of Rurutii. In Rapa bark is made into cordage. In Tahiti juice from fruit mixed with that of Cordia subcordata is source of a scarlet dye, to paint cloth; faces of warriors and actors among natives.

Ficus toxicaria L. (Moraceae). - Tree. Java. Plant is source of a wax which has been recommended in Java for batik work.

Ficus Tsjaleka Burm. f. (syn. F. venosa Dryand). (Moraceae). - Tree. E. India, Himalayan region. Leaves are in N. W. Himalaya source of a food of a silkworm, Ocinara signifera Walker and O. lida Moore.

Ficus Vallis-Choudae Delile. (Moraceae). - Tree. Trop. Africa. Bark is consumed with Cola by the natives of Sierra Leone.

Ficus variegata Blume. (Moraceae). - Tall tree. Trop. Asia. Plant is source of a wax which is used in Java for batik work. Is sold in markets of Java.

Ficus verrucolosa Warb. (Moraceae). - Tree. Trop. Africa. Fruits when ripe are consumed by the natives of Angola.

Ficus Vogelii Miq. Dob. (Moraceae). - Tree. Trop. Africa. Source of Dahomey Rubber, of good quality. It is said that the wood is used as soap by the natives.

Field Mint → Mentha arvensis L.

Field Pea → Pisum arvense L.

Fig → Ficus Carica L.

Fig, Banana → Musa sapientum L.

Fig, Hottentot → Mesembryanthemum edule L.

Fig, Sycomore → Ficus Sycomorus L.

Fig Tree → Ficus Carica L.

Fiji Afzelia → Afzelia bijuga (Colebr.) Gray.

Fiji Sandalwood → Santalum Yasi Seem.

Filbert, Lambert's → Coryllus tubulosa Willd.

Filbert, Siberian → Corylus heterophylla Fisch.

Filipendula Ulmaria (L.) Maxim. (syn. Spiraea Ulmaria L.). Queen-of-the-Meadow. (Rosaceae). - Perennial herb. Temp. Europe, Asia, introd. in N. America. Flowers and stems, used as home-remedy for „cleaning" the blood, as astringent, diuretic, antispasmodic, stomachic and styptic. Source of Oil of Meadowsweet, derived from flowerbuds by distillation. Oil has great fragrance. Contains salicylic aldehyde, methyl salicylate, vanillin and heliotropin.

Fillaeopsis discophora Harms., Eon, Totum. (Leguminosaceae). - Tree. Trop. W. Africa. Wood light brown, very dense, hard, heavy, coarse-grained, difficult to split, easy to polish; used for house construction, interior decoration, flooring, window-frames, furniture, street paving blocks.

Fimbristylis globulosa Kunth. (Cyperaceae). - Perennial herb. Trop. Asia. Cultivated. Used in different parts of Asia for manuf. of mats, basket work, tobacco cases and similar comodities. Also F. diphylla Vahl.

Fimbristylis spadicea Vahl. (Cyperaceae). - Perennial herb. Trop. America. Source of a fibre, known as Esparto Chino, Esparto Mulato or Gamelotte Fibre.

Finger Poppy Mallow → Callirhoe digitata Nutt.

Fingerleaf Morningglory → Ipomoea digitata L.

Fingotra Rubber → Landolphia crassipes Radl.

Fir, Alpine → Abies lasiocarpa (Hook.) Nutt.

Fir, Balsam → Abies balsamea (L.) Mill.

Fir, Cascade → Abies amabile (Loud.) Fores.

Fir, Douglas → Pseudotsuga taxifolia (Lam.) Britt.

Fir, Grand → Abies grandis Lindl.

Fir, Flaky → Abies squamata Mast.

Fir, Lowland White → Abies grandis Lindl.

Fir, Noble → Abies nobilis Lindl.

Fir, Red → Abies magnifica Murr.

Fir, Rocky Mountain → Abies lasiocarpa (Hook.) Nutt.

Fir, Silver → Abies amabilis (Loud.) Forbes. and A. pectinata DC.

Fir, White → Abies amabilis (Loud.) Forbes and A. concolor Lindl. and Gord.

Firmiana Barteri (Mast.) Schum. (Sterculiaceae). - Tree. Trop. Africa, esp. Togo. Wood whitish, very light; used by the natives for plates and dishes; also for floats of fishing-nets.

Firmiana simplex (L.) Wight. (syn. F. plantanifolia Brit., Sterculia plantanifolia L. f.) Phoenix Tree. (Sterculiaceae). - Tree. Hupeh, Shantung (China) to Formosa. Bark source of a fibre obtained by retting; used in China. Wood used in Szechuan (China) for furniture.

Fish Berry → Anamirta paniculata Colebr.

Fish Tail Palm → Caryota urens L.

Fishdeath Tephrosia → Tephrosia toxicaria (Sw.) Pers.

Fishfuddle Tree, Jamaica → Piscidia Erythrina L.

Fish Poisons. Plants used as → Acacia, Aeronychia, Adenia, Adenium, Aegiceras, Aesculus, Anamirta, Barringtonia, Buddleia, Caesalpinia, Callicarpa, Cassia Cerbera, Chailletia, Cleistanthus, Cocculus, Corypha, Croton, Cupania, Daphne, Derris, Duboisia, Euphorbia, Erythrophleum, Eucalyptus, Euphorbia, Frankenia, Gardenia, Jacquinia, Lonchocarpus, Machaerocereus, Milletia, Morelia, Oenanthe, Paullinia, Piscidia, Prosopis, Raputia, Salmea, Sapium, Schima, Tephrosia, Ternstroemia, Urechites, Verbascum.

Fistulina hepatica (Huds.) Bull. Vegetable Beefsteak, Beefsteak Fungus, Beef Tongue, Oak Tongue, Chestnut Tongue (Polyporaceae). - Basidiomycete. Fungus. Temp. zone. On trunks of trees. Fruitbodies are consumed as food raw or prepared in different ways. The mycelium gives a special stain to the wood, giving it a brownish color. It is known as Brown Oak and is valued by timber merchants.

Fitzroya cupressoides (Molina) Johnst. (syn. F. patagonica Hook.), Lahuán, Alerce. (Cupressaceae). - Tree. S. Chile. Wood of excellent quality, light, durable; used in Chile for general carpentry, musical instruments, honey barrels, pencils, cigar-boxes, troughs.

Flacourtia cataphracta Roxb., Runeala Plum. (Flacourtiaceae). - Spiny shrub, or small tree. India, Malaysia. Introduced in trop. America. Fruits round, 2 cm. across, russet-purple; pulp creamy white, slightly acid, pleasant flavor; used for compotes.

Flacourtia euphlebia Merr. (Flacourtiaceae). - Shrub. Philipp. Islds., esp. Mindanao. Fruits edible, 2 cm. across, roundish, dark purple, subacid; recommended for jellies.

Flacourtia inermis Roxb. Plum of Martinique. (Flacourtiaceae). - Small tree. Origin uncertain. Berries size of a small cherry, brilliant red, slightly acid; suitable for jelly, syrup and preserves; usually too acid to be eaten when fresh. Some varieties, however, have sweet fruits.

Flacourtia Ramontchi L'Hér. Botoko Plum, Governor's Plum, Ramontschi. (Flacourtiaceae). - Shrub. Malaysia, Madagascar. Cultivated in tropics of Old and New World. Fruits round, 3 cm. long, acid, dark purple to black; used for jam and preserves, also eaten with fish. Used in some parts of Trop. Africa for jaundice and enlarged spleen. Wood even-grained, hard, red, durable, does not warp, splits; used for agricultural implements and turnery.

Flacourtia Rukam Zoll. and Mor. Rukam. (Flacourtiaceae). - Spiny tree. Malaysia and Philipp. Islds. Cultivated. Fruit round, dark-red. A variety Rukum is subacid, eaten raw.

Flacourtia sepiaria Roxb. (Flacourtiaceae). - Spiny plant. Trop. Asia, esp. Cochin-China, Malaysia. Introduced in tropical America. Fruits round, dark purple, 1 to 2 cm. in diam., of a pleasant flavor, sweet, mildly acid, juicy; used in pies.

Flacourtiaceae → Aphloia, Carpotroche, Dovyalis, Flacourtia, Gossypiospermum, Homalium, Hydnocarpus, Oncoba, Ryparosa, Scolopia, Taraktogenos, Zuelania.

Flag Root → Acorus Calamus L.

Flagellaria indica L. (Flagellariaceae). - Semiwoody climber. E. India, Malaya. Mascarene Islds. to New Caledonia. Tough stems used in Siam and E. Malaysia for basket making and stitching ataps. Leaves used for hair-washes.

Flagellariaceae → Flagellaria.

Flaky Fir → Abies squamata Mast.

Flamulina velutipes (Curt.) Karst, (syn. Collybia velutipes (Curt.) Quél. (Agaricaceae). - Basidiomycete. Fungus. Temp. zone. The fruitbodies are consumed as a food, having a pleasant taste.

Flannel, Mullein → Verbascum Thapsus L.

Flat Crown Albizzia → Albizzia fastigiata Willd.

Flat Pea → Lathyrus sylvestris L.

Flatwoods Plum → Prunus umbellata Ell.

Flavoring. Plants used for → Abrus, Acer, Acorus, Aegle, Allium, Ancistrocladus, Anetum, Anisomeles, Anthemis, Anthriscus, Apium, Asperula, Athamanta, Bauhinia, Betula, Buettneria, Carum, Chrysanthemum, Citrus, Coffea, Coriandrum, Cuminum, Cymbopetalum, Cymopterus, Daucus, Dendrobium, Dicypellium, Dipteryx, Elettariopsis, Eugenia, Foeniculum,

Gaultheria, Horsfieldia, Humulus, Hyptis, Hyssopus, Illicium, Lagoecia, Laserpitium, Laurus, Levisticum, Litsea, Majorana, Melissa, Mentha, Monarda, Myristica, Myrrhis, Nigella, Parkia, Petroselinum, Peucedanum, Pogogyne, Ptelea, Pycnanthemum, Quararibea, Rosmarinus, Ruta, Salvia, Sarothamnus, Sassafras, Satureja, Selenipedium, Sisymbrium, Thymus, Trigonella, Trilisa, Vanilla, Zingiber. See also under Spices, Essential Oils and Condiments.

Flax Fibre → Linum usitatissimum L.

Flax, New Zealand → Phormium tenax Forst.

Flax Plant → Linum usitatissimum L.

Flax, Purging → Linum catharticum L.

Flax, Rocky Mountain → Linum Lewesii Pursh

Fleabane → Erigeron canadensis L.

Fleabane, Marsh → Pluchea sericea (Nutt.) Cov.

Flea Seed → Plantago Psyllium L.

Fleeceflower, Japanese → Polygonum cuspidatum Sieb. and Zucc.

Flemingia congesta Roxb. (syn. F. rhodocarpa Baker.) (Leguminosaceae). - Perennial herb. Trop. Asia. Source of Waras, a brilliant orange dye, used for dyeing silk.

Flindersia australis R. Br. Flindose, Rasp-Pad. (Rutaceae). - Tree. Australia esp. New South Wales, Queensland. Wood hard, strong, close-grained, durable; used for staves.

Flindersia maculosa F. v. Muell. (Rutaceae). - Tree. Australia, esp. New South Wales, Queensland. Plant is source of a gum, contains 80% arabin.

Flindersia Oxleyana F. v. Muell. (syn. Oxleya xanthoxyla Hook.) Light Yellow Wood. (Rutaceae). - Tree. Australia, esp. New South Wales, Queensland. Wood is source of a yellow dye.

Flindosa → Flindersia australis R. Br.

Floodbed Eucalyptus → Eucalptus microtheca F. v. Muell.

Flor de la Oreja → Cymbopetalum penduliflorum Baill.

Florence Fennel → Foeniculum vulgare Mill.

Florentine Iris → Iris florentina L.

Flores Cinae → Artemisia Cina Berg.

Flores Sambuci → Sambucus nigra L.

Florida Arrowroot → Zamia integrifolia Ait.

Florida Moss → Tillandsia usneoides L.

Florida Torreya → Torreya taxifolia Arn.

Florida Velvet Bean → Mucuna Deeringiana (Bort.) Small.

Florida Woodbox → Schaefferia frutescens Jacq.

Floscopa scandens Lour. (Commelinaceae). - Herbeceous herb. Trop. Asia, Australia. Juice of leaves used for sore eyes and ophthalmia.

Flour → Starch.

Flowering Ash → Fraxinus Ornus L.

Flowering Dogwood → Cornus florida L.

Flowering Rush → Butomus umbellatus L.

Flowering Spurge → Euphorbia corollata L.

Flowering Usnea → Usnea florida (L.) Hoffm.

Fly Amanita → Amanita muscaria (L.) Pers.

Fly Mushroom → Amanita muscaria (L.) Pers.

Fockea multiflora Schum. (Asclepiadaceae). - Woody vine. Angola. Latex used to adulterate rubber.

Foeniculum vulgare Mill. (syn. Foeniculum officinale All.) Fennel. (Umbelliferaceae). - Biennial plant. Centr. and W. Europe, Mediterranean region, Abyssinia, S. Africa, China, introd. in N. and S. America, New Zealand. Cultivated in France, Holland, Germany, Galicia, Bulgaria, Rumania, Russia, India and Japan. Used for flavoring foods, pickles, in Absinthe and other liqueurs. Fruits used medicinally and in home remedies, being stomachic, carminative, stimulant, prevents colic in infants. Source of Oil of Fennel, Oleum Foeniculi; contains 50 to 60% anethol, 20% fenchose, chavicol, anisic aldehyde. Differs in various localities. Var. piperitum (Ucr.) Cout. (syn. F. piperitum Ucr.). Source of Bitter Fennel. Var. dulce (Mill.) Thell. (syn. F. dulce Mill.) Florence Fennel, Sweet or Roman Fennel. Cultivated in S. France and Mediterranean region. Var. azoricum (Mill.) Thell. (syn. F. azoricum Mill.). Carosella, Italian Fennel. Cultigen, originated in Italy. Has very broad leaf-stalk bases, eaten raw, as a vegetable, in salads.

Fomes applanatus aut. → Ganoderma applanatus (Wallr.) Karst.

Fomes fomentarium (L.) Gill. (Polyporaceae). - Basidiomycete. Fungus. Temp. zone. The fruit-bodies were used as tinder or „amadou" until the beginning of this century. Fomes is used in pulverized condition in Siberia as ingredient of snuff. Recently it has been used for manuf. of buttons, bedroom slippers, flower pods, smoking caps etc. Used by dentists for absorption and compressing.

Fomes hemitephrus Berk. (Polyporaceae). - Basidiomycete. Fungus. Fruitbodies are made into razor strops in Australia.

Fomes igniarius (L.) Gill. (Polyporaceae). - Basidiomycete. Fungus. The fruitbodies are sometimes used as a source of a brown dye.

Fony Oil → Adansonia digitata L.

Food Candle Tree → Parmentaria edulis DC.

Food Inga → Inga edulis Mart.

Forest Mahogany → Eucalyptus resinifera Smith.

Food Plants → Condiments, Economic Algae, Edible Fungi, Emergency Food Plants, Flour, Food Plants of Primitive Peoples, Fruits, Seeds, Starch, Vegetables.

Food Plants, Emergency. → Abronia, Acacia, Achillea, Adansonia, Aegopodium, Aerva, Aesculus, Agoseris, Agropyron, Agrostemma, Alhagi, Alisma, Alsophila, Althaea, Ammobroma, Amphicarpa, Anchomanes, Arctium, Aristida, Arum, Atriplex, Betula, Butomus, Cakile, Camassia, Cardamine, Carum, Chenopodium, Chrsyosplenium, Cirsium, Claytonia, Cynoglossum, Ficus, Gisekia, Glaux, Haloxylon, Heritiera, Hydrophyllum, Hypochoeris, Icacina, Lactuca, Lamium, Leptochloa, Lycopus, Lythrum, Menyanthes, Nelumbium, Nuphar, Nymphaea, Oenothera, Plantago, Polygonatum, Potentilla, Pteridium, Quercus, Ranunculus, Reseda, Salicornia, Salsola, Samolus, Sanguisorba, Sedum, Stellaria, Symphytum, Themeda, Trianthema, Triglochin, Trillium, Tulipa, Tussilago, Urtica, Uvularia, Valeriana, Xysmalobium. See also: plants consumed by primitive peoples.

Food Plants for Live Stock. → Acacia, Acroceras, Agriophyllum, Agropyrum, Agrostis, Alhagi, Alysicarpus, Amphilophis, Anabasis, Ananas, Andropogon, Anthistiria, Anthoxanthum, Anthyllis, Arachis, Argania, Arrhenaterum, Arundinella, Astragalus, Astrebla, Atalaya, Atriplex, Avena, Axonopus, Azolla, Beckmannia, Beta, Blepharis, Bouteloua, Brachiaria, Bromus, Brosimum, Bunium, Cajanus, Calamovilfa, Canavalia, Canna, Capparis, Carthamus, Centaurea, Ceratonia, Chenopodium, Chloris, Cicer, Citrus, Cladonia, Cocos, Cornulaca, Cyamopsis, Cynodon, Cyonurus, Cytisus, Dactylis, Daucus, Deschampsia, Desmodium, Dichelachne, Digera, Dipterygium, Dodonaea, Echinochloa, Eichhornia, Eleusine, Eragrostis, Eriocephalus, Eriochloa, Erythrina, Euchlaena, Fagara, Festuca, Fucus, Galega, Glycine, Gosypium, Gourliea, Haemarthria, Haloxylon, Hedysarum, Helianthus, Heteropogon, Holcus, Hordeum, Hymenachne, Imperata, Indigo, Isachne, Ischemum, Kochia, Lathyrus, Lemna, Leptochloa, Lespedeza, Ligusticum, Lolium, Lotus, Ludwigia, Lupinus, Lycopersicum, Macrocystis, Manihot, Medicago, Melilotus, Melinis, Mesembryanthemum, Momordica, Morettia, Muehlenbergia, Neurada, Onobrychus, Ononis, Opuntia, Oryza, Panicum, Paspalum, Pentzia, Phalaris, Phaseolus, Phleum, Pistacia, Pisum, Poa, Polytoca, Portulacaria, Prosopis, Puccinellia, Pueraria, Quercus, Randonia, Rottboellia, Saccharum, Salicornia, Salsola, Sarothamnus, Secale, Sesbania, Setaria, Solanum, Sorghum, Spergularia, Stipa, Stylosanthes, Suaeda, Symphytum, Themeda, Tillandsia, Tricholaena, Trifolium, Trigonella, Tripsacum, Tristachya, Triticum, Ulex, Ulva, Vallisneria, Vicia, Vigna, Xanthosoma, Zea, Zizyphus, Zornia, Zygophyllum.

Food, Plants consumed by primitive peoples. → Abronia, Abutilon, Acacia, Acanthophora, Acanthosicyos, Acrostichum, Aeolanthus, Aesculus, Agoseris, Agriophyllum, Ahnfeltia, Aizoon, Alisma, Alstroemia, Amaranthus, Amelanchier, Ammobroma, Amphicarpa, Anemone, Angelica, Antigonum, Apios, Araucaria, Arenaria, Arisaema, Aristida, Arjona, Artemisia, Artocarpus, Asparagopsis, Asphodelus, Aspidium, Astragalus, Atriplex, Aulospermum, Avicennia, Bactris, Balsamorhiza, Barringtonia, Bauhinia, Begonia, Blepharis, Boerlagiodendron, Boscoa, Brachystelma, Brodiaea, Cadaba, Calamus, Calathea, Calirrhoe, Calochortus, Caltha, Calypso, Camassia, Caragana, Carapa, Cardamine, Carum, Caryocar, Castanospermum, Cecropia, Cetraria, Chrysothamnus, Cirsium, Cleome, Codonopsis, Cogwellia, Conanthera, Conostegia, Cordia, Corydalis, Cotyledon, Crotalaria, Cyathea, Cycas, Cymbidium, Cyperus, Dasylirion, Dentaria, Dicksonia, Digera, Digitaria, Dioon, Diplotaxis, Dodecatheon, Dondia, Dracaena, Durvillea, Eleocharis, Elymus, Eriogonum, Erodium, Eryngium, Erythronium, Eschscholzia, Eucalyptus, Fatsia, Fritillaria, Funkia, Gastrodia, Geonoma, Gladiolus, Gourliea, Gracilaria, Habenaria, Hedysarum, Helianthus, Helichrysum, Heracleum, Hippurus, Hoffmanseggia, Hyptis, Inocarpus, Ipomoea, Iris, Isomeris, Jatropha, Juniperus, Justicia, Kerstingiella, Laciniaria, Lallemantia, Laportea, Lathyrus, Lavatera, Lecanora, Lepidium, Leptadenia, Lewisia, Limnanthemum, Linum, Livistonia, Ludwigia, Lycium, Lycopus, Lygodesmia, Macrozamia, Maerua, Malabaila, Marattia, Mariscus, Medeola, Mentzelia, Mertensia, Metaplexis, Miconia, Microseris, Microtis, Mollugo, Monolepis, Monstera, Najas, Nelumbium, Nereocystis, Nopalea, Nothopanax, Nuphar, Nymphaea, Ochrocarpus, Onoclea, Opuntia, Orbignya, Orobanche, Orontium, Orthoceras, Oryzopsis, Osmunda, Oxalis, Pachira, Pachycereus, Paeonia, Pandanus, Pangium, Parosela, Peltandra, Pennisetum, Peperomia, Petasites, Phellopterus, Phorodendron, Phragmites, Phytolacca, Pisonia, Pithecolobium, Plectranthus, Pluchea, Poga, Polakowskia, Poliomintha, Polygonatum, Polygonum, Polymnia, Polyporus, Populus, Porphyra, Portulaca, Pringlea, Prosopis, Pseudocymopterus, Psophocarpus, Psoralea, Pugionum, Quercus, Ranunculus, Raphionacma, Rhus, Rinorea, Robinia, Rumex, Sabal, Sadlera, Sagittaria, Salicornia, Salix, Salvadora, Samuela, Sarcocephalus, Sargassum, Sauropus, Saxifraga, Scirpus, Sedum, Semecarpus, Senebiera, Sesamum, Simmondsia, Smilax, Socratea, Salonum, Solidago, Sophia, Sparganium, Spathyema, Sphenostylis, Spilanthes, Sporolobus, Stanleya, Sticta, Talinum, Telfairea, Telosma, Tetracarpidium, Tetrapleura, Thalictrum, Tigridia, Trichosanthes, Trifolium, Triglochin, Tsuga, Tulipa, Tylostemon, Typhonium, Typhonodorum, Ullucus, Ulmus, Valeriana, Vallisneria, Veronica, Viola, Washingtonia, Wyethia, Xanthorrhoea, Yucca, Zilla.

Food Plants for Silkworm Species. → Ailanthus, Arbutus, Bridelia, Cajanus, Careya, Dodonaea, Ficus, Liquidambar, Melastoma, Michelia, Morus, Symphonia, Terminalia, Zizyphus.

Forest Oak → Casuarina torulosa Ait.

Forestiera acuminata (Michx.) Poir., Swamp Privet. (Oleaceae). - Shrub or small tree. Eastern United States. Wood strong, hard, close-grained; suitable for turnery.

Forking Larkspur → Delphinium Consolida L.

Formosa Sweet Gum → Liquidambar formosana Hance.

Forsteronia floribunda Muell. Arg. (Apocynaceae). - Woody plant. Jamaica, Brazil. Source of a rubber which is available in small amounts.

Forsteronia gracilis Muell. Arg. (Apocynaceae). - Woody plant. Guiana. Source of a rubber which is of little commercial value.

Fortunella crassifolia Swingle, Meiwa Kumquat. (Rutaceae). - Small tree. China, Japan. Fruits round, orange, sweet; often consumed raw. Occasionally cultivated.

Fortunella japonica (Thunb.) Swingle. (syn. Citrus japonica Thunb.) Marumi Kumquat. (Rutaceae). - Small tree. Japan. Fruits round, orange, acid. Have the taste of the Nagami. Occasionally cultivated. Used for jams, jellies, preserves.

Fortunella margarita (Lour.) Swingle. (syn. Citrus margarita Lour.). Nagami Kumquat. (Rutaceae). - Small tree. Japan. Fruits oval, orange, acid. Commercially cultivated in Florida. Used for jams, jellies and preserves.

Fouquieria Macdougalii Nash. Jabonecillo, Chunari. (Fouquieriaceae). - Tree. Mexico, esp. Sonora, Sinaloa. Bark substitute for soap; used for washing woolen goods.

Fouquieria splendens Engelm., Ocotillo. (Fouquieriaceae). - Shrub. S. W. of United States and adj. Mexico. Has been recommended as a source of rubber during times of emergency.

Fouquiriaceae → Fouquieria.

Four o'clock → Mirabilis Jalapa L.

Fourwing Saltbush → Atriplex canescens (Pursh) Nutt.

Fourwing Sophora → Sophora tetraptera Ait.

Fowl Bluegrass → Poa palustris L.

Fox Grape → Vitis Labrusca L.

Fox Tail, Meadow → Alopecurus pratensis L.

Foxberry → Vaccinium Vitis-Idaea L.

Foxglove → Digitalis purpurea L.

Foxtail Millet → Setaria italica L.

Foxtail Pine → Pinus aristata Engelm.

Fragaria chiloensis (L.) Duch. Chiloe Strawberry. (Rosaceae). - Perennial herb. Pacific Coast region of N. and S. America. Cultivated in Andes region. Sold in markets. Has been used as one of the parent species of a number of large fruited strawberry varieties in N. America and Europe.

Fragaria vesca L., Strawberry. (Rosaceae). - Perennial herb. Europe, temp. Asia. Widely cultivated in temp. zones of the Old and New World, as a dessert fruit. Much used for jams, in ice cream industry, for pastries, drinks, flavoring, preserves, home-made wine, canning industry, packed in dry sugar in cold storage, frozen pack strawberries are used by ice-cream makers and bakers and to a limited extent in the retail trade. Numerous red and white varieties, many of local importance. American varieties are: Aroma, Dunlap, Clark, Progressive, Howard 17, Corvallis, Ettersberg 121, Nich Omer, Blakemore, Klondike. Blakemore is used for preserving; Marshall, Corvallis, Ettersby for freezing; Klondike is of commercial importance in the S. States. Some European varieties are: Laxton Noble, Royal Sovereign, Jucunda, Louis Gauthier, Madame Moutot. Most varieties are of hybrid origin, having been obtained from crosses with F. chiloensis Duch. Medicinally strawberries are used as syrup for pharmaceutical preparations. Leaves are used as substitute for tea.

Fragaria virginiana Duch., Virginian Strawberry. (Rosaceae). - Perennial herb. Eastern N. America to Texas and Arizona. Fruit red, occasionally white, pulp juicy, of pleasant flavor.

Fragrant Onion → Allium odorum L.

Fragrant Sandalwood → Fusanus spicatus R. Br.

Fragrant Sumach → Rhus aromatica Ait.

Franciscea uniflora Pohl → Brunfelsia Hopeana Benth.

Frangula Alnus Mill. → Rhamnus Frangula L.

Frankenia Berteroana Gay. (Frankeniaceae). - Shrub. Chile. Salt obtained from the ash of the plant is used by the natives.

Frankenia ericifolia C. Sm. (Frankeniaceae). - Shrub. Cape Verde Islds., Can. Islds. Used as fish poison in the Cape Verde Islds.

Frankenia portulacaefolia Spreng. (syn. Beatsonia portulacaefolia Roxb.). (Frankeniaceae). - Shrub. St. Helena. Herb is used as a tea among the inhabitants of St. Helena.

Frankeniaceae → Frankenia.

Frankincense → Boswellia Carterii Birdw.

Frankincense, Bible → Boswellia Carterii Birdw.

Frankincense, Elemi → Boswellia Frereana Birdw.

Frankincense, Indian → Boswellia serrata Roxb.

Frankincense, Male → Boswellia Carterii Birdw.

Fraxinella Dictamnus Moench → Dictamnus albus L.

Fraxinus americana L., White Ash. (Oleaceae). - Tree. Eastern N. America to Florida and Texas. Important lumber tree. Wood strong, hard, heavy, tough, close-grained, brown; used for interior finish of buildings, oars, furniture, carriages, agricultural implements, tool-handles. Bark from trunk, branches and roots are a bitter tonic, astringent; collected in spring.

Fraxinus excelsior L., European Ash. (Oleaceae). - Tree. Europe, Caucasia, Wood yellow-brown, fine and long-grained, heavy, fairly hard, elastic, difficult to split, easy to polish; used for general carpentry, handles of tools, agricultural machineries, furniture, in aeroplanes, rulers, sporting-materials.

Fraxinus Griffithii Clarke. (syn. F. Edenii Boerl. and Koord.) (Oleaceae). - Tree. Himalayan region to Java. Leaves are smoked as surrogate of opium, resembling its scent and taste, however, not its results.

Fraxinus Mariesii Hook. (Oleaceae). - Shrub or small tree. Centr. China. An insect wax, Pela, formed by Coccus pe-la is found on the tree. Pela is used in China for coating candles made from vegetable tallow; for imparting a glossy surface to silks and for coating pills. Also F. chinensis Roxb.

Fraxinus nigra Marsh. (syn. F. sambucifolia Lam.). Black Ash. (Oleaceae). - Tree. Eastern N. America. Wood not strong, soft, heavy, coarse-grained, tough, easily separated into thin layers, durable, dark brown; used for cabinet making, interior finish of houses, barrel-hoops, baskets. Ash Burl veneering are slices obtained from knots or burls from the trunk and large branches.

Fraxinus oregana Nutt. (F. latifolia Benth.) Oregon Ash. (Oleaceae). - Tree. Pacific Coast of N. America. Valuable timber tree. Wood brittle, coarse-grained, hard, light brown; used for wagons, carriages, interior finish of buildings and furniture.

Fraxinus Ornus L., Flowering Ash, Manna Ash. (Oleaceae). - Tree. S. Europe, Balkans, Asia Minor. Source of a Manna, an exudation from the stem, being a mild laxative nutrient; contains mannite, manneotetrose, manninotriose, glucose, levulose and resin.

Fraxinus oxyphylla Mar. (Oleaceae). - Tree. Morocco, Algeria, Fruits are used in Morocco as an condiment and aphrodisiac.

Fraxinus pubescens Lam. (syn. F. pensylvanica Marsh.) Red Ash. (Oleaceae). - Tree. Eastern N. America. Wood strong, brittle, hard, heavy, coarse-grained, light brown; used for the same purpose as F. americana L.

Fraxinus quadrangulata Michx. Blue Ash. (Olaceae). - Eastern N. America. Tree. Wood close-grained, hard, heavy, brittle, light yellow, brown, streaked. Used for construction work, flooring, agricultural implements, wagons. A blue dye is derived from the inner bark after maceration in water.

Fraxinus Sieboldiana Blume. (syn. F. longicuspis Fr. and Sav.). (Oleaceae). - Tree. Japan, Korea. Wood elastic, soft, light; used in Japan for furniture, utensils, base-ball bats, frames of tennis rackets.

Fremontia californica Torr., Slippery Elm. (Sterculiaceae). - Shrub or small tree. California to Baja California. Mucilaginous inner bark is sometimes used in poultices.

French Berries → Rhamnus infectorius L.

French Lavender → Lavandula stoechas L.

French Oil Turpentine → Pinus Pinaster Soland.

French Rose → Rosa gallica L.

Frenela Endlicheri Porlat → Callitris calcarata R. Br.

Frenela robusta Cunningh. → Callitris robusta R. Br.

Freycinetia arborea Gaud. (Pandanaceae). - Hawaii. Islds. Fibre from the stem is used in Hawaii for cordage to bind rafters, for baskets, funnelshaped traps for catching small fish and shrimp, also plaited into helmets. Cordage is frequently used by hula dancers.

Freycinetia Banksii Cunningh. (Pandanaceae). - Woody plant. New Zealand. Leaves are source of a fibre, made into articles of clothing and war-mats; used by aborigens of New Zealand.

Freycinetia funicularis Merr. (Pandanaceae). - Tall vine. Moluccas. Young inflorenscense is boiled and eaten as vegtable by natives.

Fringe Tree → Chionanthus virginica L.

Fringed Rue → Ruta chalepensis L.

Fritillaria camchatcensis Ker - Gawl. (Liliaceae). - Bulbous perennial. N. E. Asia, Alaska. Bulbs are consumed as food by the Eskimos of Alaska. Among the Ainu they are eaten with fat or rice.

Fritillaria pudica (Pursh) Spreng. (Liliaceae). - Perennial bulbous plant. Western N. America. Bulbs were consumed raw or boiled by the Indians of British Columbia, Montana and Utah. Bulbs were also dried for winter use.

Fritillaria Roylei Hook. (Liliaceae). - Bulbous perennial. Alpine regions. W. China. Used in Chinese pharmacy; much exported from Monking Ting and Tachien. Bulbs are pounded and when prepared are used for asthma.

Frost Grape → Vitis vulpina L.

Frostweed → Helianthemum canadense (L.) Michx.

Fuchsia macrostemma Ruiz and Pav. (Onagraceae). - Shrub. Chile. Wood is source of a black dve. Also F. magellanica Lam. Leaves and bark are used as diuretic and febrifuge.

Fructus Rhamni → Rhamnus infectorius L.

Fruits. Edible Species of → Abbevillea, Acanthocereus, Acanthosicyos, Achras, Acnistus, Acrocomia, Actinidia, Adansonia, Aglaia, Akebia, Alectryon, Allophylus, Amelanchier, Ampelocissus, Amygdalus, Ananas, Annona, Antidesma, Arbutus, Arctostaphylos, Aristotelia, Armeniaca, Artocarpus, Asimina, Astelia, Astroloma, Atalanta, Aveledoa, Averrhoa, Baccaurea, Bactris, Basanacantha, Beilschmiedia, Berberis, Blighia, Borsicactus, Bouea, Bridelia, Britoa, Bromelia, Bumelia, Bunchosia, Bysonima, Calamus, Calocarpum, Campomanesia, Canarium, Canthium, Carica, Carissa, Carnegiea, Casimiroa, Castanea, Castanopsis, Celtis, Ceratonia, Chamaesaracha, Chrysobalanus, Chrysophyllum, Citrullus, Citrus, Clausenia, Clermontia, Clidemia, Coccinea, Coccolobis, Cocculus, Cocos, Condalia, Cordia, Cornus, Corylus, Couepia, Couma, Crataegus, Cucumis, Cudrania, Cydonia, Dialium, Dillenia, Diospyros, Diploglottis, Disterigma, Dovyalis, Durio, Echinocereus, Elaeagnus, Elaeocarpus, Englerodendron, Eriobotrya, Eriodendron, Erioglossum, Escontria, Espostoa, Eucarya, Eugenia, Feijoa, Ferocactus, Feronia, Feroniella, Ficus, Flacourtia, Fortunella, Fagara, Fragarid, Garcinia, Gaultheria, Gaylussacia, Genipa, Geoffraea, Greigia, Guazuma, Guilielma, Hancornia, Haronga, Harpephyllum, Heinsia, Heterodendrum, Hippophaë, Hovenia, Hydnora, Hylocereus, Hymenaea, Hyphaene, Inga, Jacaratia, Jarilla, Jessenia, Juglans, Karatas, Lansium, Lardizabala, Lemaireocereus, Leptomeria, Licania, Lophocereus, Lucuma, Lycium, Machaerocereus, Macleania, Malpighia, Mammea, Mangifera, Matisia, Melicocca, Melodorum, Memecylon, Mesembryanthemum, Mespilus, Microcitrus, Monstera, Morinda, Morus, Mouriria, Muehlenbeckia, Musa, Myrciaria, Myrica, Myrtus, Nephelium, Nitraria, Nyssa, Opuntia, Owenia, Pandanus, Parinarium, Parkia, Parmentiera, Passiflora, Pereskia, Pereskiopsis, Persea, Philodendron, Phoenix, Phyllanthus, Physalis, Plectronia, Podocarpus, Podophyllum, Pometia, Prunus, Pyrus, Randia, Reptonia, Reynosia, Rheedia, Rhodomyrtus, Ribes, Rollinia, Rolliniopsis, Rosa, Rubus, Sageretia, Salacia, Salichroa, Sambucus, Sandoricum, Sarcocephalus, Saurauia, Scaevola, Sclerocarya, Shepherdia, Solanum, Sorbus, Sorindeia, Spondias, Stauntonia, Stelechocarpus, Strychnos, Terminalia, Tetrastigma, Timonius, Trichocereus, Triphasia, Uapaca, Uvaria, Vaccinium, Vangueria, Viburnum, Vitex, Vitis, Willighbeia, Ximenia, Zalacca, Zizyphus. See also under Alcoholic and Non Alcoholic Beverage, Seeds and Grains.

Fucaceae (Brown Algae) → Ascophyllum, Fucus, Turbinaria.

Fucus fuscatus C. Ag. (Fucaceae). - Brown Alga. Pacific Ocean. Consumed in some parts of Alaska.

Fucus vesiculosus L., Bladder Wrack, Lady Wrack, Sea Ware, Black Tang, Bladder Fucus. (Fucaceae). - Brown Alga. Temp. N. Atlantic. Considered antiscorbutic; used for goiter; for the treatment of obesity and as alterative. Formerly used for the manufacturing of kelp, for iodine and potash. Used in Ireland as manure; in Gothland for feeding of hogs; as a winter food for sheep and cattle in the Western Hebrides. Other species are used for the same purpose; eg. F. serratus L.

Fuller's Teasels → Dipsacus fullonus L.

Fungi → Economic Fungi and Economic Microorganism.

Fungi, edible → Edible Fungi.

Fungus Bovista → Calvatia lilacina (Berk. and Mont.) Lloyd.

Fungus Chirurgorum → Calvatia lilacina (Berk. and Mont.) Lloyd.

Fungus Chirurgorum → Polyporus officinalis (Vill.) Fr.

Fungus, Ink → Coprinus atramentarius (Bull.) Fr.

Funori → Gloiopeltis furcata (Post & Rupr.) J. Ag.

Fumaria parviflora Lam. (Fumariaceae). - Perennial herb. Iran, Afghanistan, Turkestan. Herb and fruits are used in Iran for „purifying" blood, also used as a laxative and diuretic.

Fumariaceae → Corydalis, Dicentra, Fumaria.

Funkia ovata Spreng. Shibo-shi. (Liliaceae). - Perennial herb. Japan. White parts of the leaf stalks are used as food when boiled by the Ainu.

Funtumia elastica Stapf. (syn. Kickxia elastica Preuss.) Silkrubber Tree. (Apocynaceae). - Tree. West Coast Africa to Belgian Congo and Uganda. Most important rubber tree of Africa. Cultivated. Source of Silkrubber or Lagos Silk Rubber.

Furcraea Cabuya Trel. (Amaryllidaceae). - Agave-like plant. Costa Rica to Panama. Leaves are source of a strong fibre; used for cordage.

Furcraea cubensis Vent. (Amaryllidaceae). - Agave-like plant. Trop. America. Leaves are source of a fibre, made into Cuba Hemp.

Furcraea gigantea Vent. (Amaryllidaceae). - Agave-like plant. Mexico. Cultivated in tropics of Old and New World. Leaves are source of a strong fibre, known as Mauritius Hemp.

Furcraea Selloa Koch. (syn. F. samalana Trel.). (Amaryllidaceae). - Agave-like plant. S. America. Leaves are source of an excellent fibre.

Fusanus acuminatus R. Br. (syn. Santalum acuminatum A. DC., S. Preissianum Miq.) South Australian Sandalwood Tree. (Santalaceae). - Tree. Australia. Wood hard, closegrained, pleasant scented; used for cabinet work. Source of South Australian Sandalwood Oil, having a scent of roses. Fleshy pericarp of the fruit is made into an excellent jelly, used in Australia.

Fusanus spicatus R. Br. (syn. Santalum spicatum A. DC.) Fragrant Sandal Wood. (Santalaceae). - Tree. S. and W. Australia. Wood is of much commercial value. Used for cabinetwork.

Fustic Tree, Iroco → Chlorophora excelsa (Welw.) Benth. and Hook. f.

G

Gabon Chocolate → Irvingia Barteri Hook. f.

Gabon Ebony → Diospyros Dendo Welw. and D. flavescens Gurke.

Gabon Mahogany → Khaya Klainei Pierre.

Gafal Wood → Commiphora erythraea Engl. var. glabrescens Engl.

Galam Butter → Butyrospermum Parkii (Don.) Kotschy.

Galangal, Greater → Alpinia Galanga Willd.

Galangal, Lesser → Alpinia officinarum Hance.

Galangal, Small → Alpinia officinarum Hance.

Galbanum → Ferula galbaniflua Boiss. and Buhse and F. Schair Borszcs.

Gale, Sweet → Myrica Gale L.

Galega officinalis L., Galega, European Goat's Rue. (Leguminosaceae). - Perennial herb. Centr. and S. Europe, Caucasia, Asia Minor, Iran. Cultivated. Used as food for livestock. Dried flowering tops are used medicinally, as a mild astringent, tonic; claimed to be a galactagogue. Contains a bitter principle and tannic acid.

Galega toxicaria Sw. → Tephrosia toxicaria (Sw.) Pers.

Galeopsis ochroleuca Lam. (syn. Galeopsis dubea Leers. Tetrahit longiflorum Moench). Hempnettle. (Labiaceae). - Annual herb. Europe. Herb, Herba Galeopsidis is used for ailments of lungs, intestines and spleen.

Galipea jasminiflora (St. Hil.) Engl. (Rutaceae). - Tree. S. America. Bark is used in place of quinine in some parts of Brazil. Decoction of roots is used for warts.

Galium Aparine L. Catchweed, Goose-Grass. (Rubiaceae). - Annual herb. Europe, Siberia, Centr. Asia, N. and S. America. Fruits are used in some parts of Ireland as substitute for coffee.

Galium orizabense Hemsl. (Rubiaceae). - Perennial herb. S. Mexico. Powdered leaves and stems are used by the Mazatecs (Mex.) for intestinal parasites; is considered an efficacious febrifuge.

Galium tinctorium L. Dye Bedstrow. (Rubiaceae). - Perennial herb. Europe, N. America. Roots are source of a red dye.

Gallesia Gorazema (Vell.) Moq. (Phytolaccaceae). - Tree. Brazil. Decoction of wood is used in Brazil for diseases of the lymphatic system and for intestinal worms. An ess. oil is used in S. America for genorrhea.

Gallesia integrifolia (Spreng.) Harms. Guararema. (Phytolaccaceae). - Tree. Trop. America, esp. Peru, Brazil. Leaves are used instead of soap in parts of Peru and Brazil.

Gallberry → Ilex glabra (L.) Gray.

Gallito → Erythrina rubrinerva H. B. K.

Galls, American Nutgalls → Quercus imbricaria Michx.

Galls, Bokhara → Pistacia vera L.

Galls, Chinese → Rhus semialata Murr.

Galls, Ch'-pei-tzu → Rhus Potanini Maxim.

Galls, Istrian → Quercus Ilex L.

Galls, Japanese → Rhus semialata Murr.

Galls, Morea → Quercus Cerris L.

Galls, Pistacia → Pistacia Terebinthus L.

Galls, Smyrna → Quercus lusitanica Lam.

Galls, Takut → Tamarix articulata Vahl.

Galls, Teggant → Tamarix articulata Vahl.

Galls, Wu-pei-tzu → Rhus semialata Murr.

Gambir → Uncaria Gambir Roxb.

Gambir, Black → Uncaria Gambir Roxb.

Gambir, Cube→ Uncaria Gambir Roxb.

Gambir Plant, Bengal → Uncaria Gambir Roxb.

Gamboge → Garcinia Hanburyi Hook. f. and G. Morella Desv.

Gamboge, Pipe → Garcinia Hanburyi Hook. f.

Gamboge, Siam → Garcinia Hanburyi Hook. f.

Gamboge Tree, Siam → Garcinia Hanburyi Hook. f.

Gamote → Cymopterus montanus Nutt.

Ganoderma applanatum (Wallr.) Karst. (syn. Fomes applanatus aut.). (Polyporaceae). - Basidiomycete Fungus. Cosmopolite. Used in the same way as Fomes fomentarius.

Ganophyllum falcatum Blume. (Sapindaceae). - Large tree. Malacca, Mal. Archip. Wood very strong, durable; used in Java for bridges, construction of houses. Seeds are source of an oil, called Arangan Oil, a solid fat, used for illumination.

Ganua Mottleyana Pierre → Bassia Mottleyana Clarke.

Garbancillo → Astragalus Garbancillo Cav.

Garbanzos → Cicer arietinum L.

Garcinia atroviridis Griffith. (Guttiferaceae). - Tree. Assam, Mal. Peninsula. Occasionally cultivated. Fruits large, yellow; pulp acid, must be eaten with much sugar; make excellent jelly and compote. Fruits when cut into pieces and dried are consumed in soup.

Garcinia bancana Miq. (Guttiferaceae). - Small tree. Mal. Archipel. Fruits are consumed by the natives.

Garcinia Benthami Pierre. (Guttiferaceae). - Tree. Cochin-China, Malaya. Fruits edible; pulp white of a pleasant flavor.

Garcinia Binucao Choisy. Binukao. (Guttiferaceae). - Small to medium sized tree. E. India to Philippine Islands. Fruits lemon-yellow, flattened, 4 or more cm. in diam., acid; eaten with fish by the Filipinos.

Garcinia celebica L. (syn. G. cornea Blume). (Guttiferaceae). - Medium sized tree. Mal. Archipel. Wood used for general building purposes.

Garcinia dulcis Kurz. (Guttiferaceae). - Tree. Philippine Islands to Java. Bark source of a green dye, used separately or with indigo, giving a brown color. Fruit orange to yellow, subacid; sometimes consumed by the natives. Occasionally cultivated.

Garcinia Hanburyi Hook. f. Gamboge. Siam Gamboge Tree. (Guttiferaceae). - Tree. E. India, Malaya. Bark is source of Pipe Gamboge, Siam Gamboge, a gum resin, being a powerful cathartic. Also used as pigment in water colors, lacquer varnishes and brass-work.

Garcinia Harmandii Pierre. (Guttiferaceae). - Tree. Cambodia. Fruits edible, of a pleasant flavor.

Garcinia Hombroniana Pierre. (Guttiferaceae). - Tree. Nicobar Islands, Mal. Penin. Fruit edible, has flavor of peaches, being of fine quality. Has been recommended for improvement by selection and hybridization with other species.

Garcinia indica Choisy. (Guttiferaceae). - Tree. Trop. Asia. Cultivated. Fruits called Brindonnes, resemble somewhat the Mangustan; they have a pleasant acid flavor when eaten raw; also used in jelly and syrup. Source of Goa or Kokam Butter. Sap. Val. 187-191.7; Iod. No. 25-36; Unsap. 2.3%.

Garcinia Kola Heckel → Cola acuminata Schott. and Endl.

Garcinia latissima Miq. (Guttiferaceae). - Tree. Malay Archip. Wood hard; used for tobacco pipes.

Garcinia Mangostana L. Mangosteen, Dodol. (Guttiferaceae). - Small tree. Moluccas. Cultivated in tropics. Outstanding delicious dessert fruit. Berries round, size of a mandarine, thick, smooth rind, red-purple; pulp appearance of a well-ripened plum; seeds 0 to 3. Eaten raw or in different preparations. Pulp is also cooked with rice, called Lempog in Java or is cooked with syrup, called Dodol.

Garcinia merguensis Wight. (Guttiferaceae). - Tree. Trop. Asia, esp. Cambodia, Tenasserim to Mal. Penin. Resin source of varnish.

Garcinia Mestoni Baill. (Guttiferaceae). - Tree. Queensland. Fruits size of an orange, edible, of a pleasant flavor.

Garcinia Morella Desv. (Guttiferaceae). - Tree. India. Source of commercial Gamboge; used in varnishes and in water-colors, in food, as ointment and illuminant. Sp. Gr. 0.900; Sap. Val. 195-198; Iod. No. 54-56; Melt. Pt. 34 to 37° C. Likewise G. cochinchinensis Chois., G. cambogia Desv., G. pictoria Roxb. and G. travancorica Hook. f.

Garcinia nigro-lineata Planch. (Guttiferaceae). - Tree. Malacca. Fruits edible, oval, 3 to 4 cm. long, orange; pulp sweet and agreeable; consumed by the natives.

Garcinia ovalifolia Oliv. (Guttiferaceae). - Tree. Trop. Africa, esp. Nigeria, Congo. Wood used in Congo for canoes.

Garcinia parvifolia Miq. (Guttiferaceae). - Small tree. Malaysia. Fruits yellow, size of a cherry, consumed by the natives with Spanish pepper, a fish product, soja and other material.

Garcinia Planchoni Pierre. (Guttiferaceae). - Tree. Cochin-China. Fruits edible, 7 to 8 cm. long, yellow-green; pulp sub-acid, agreeable taste. Also eaten dried.

Garcinia Prainiana Kz. (Guttiferaceae). - Small tree. Malaysia. Fruits are edible, sub-acid flavor.

Garcinia terpnophylla Thw. (Guttiferaceae). - Medium tree. Ceylon. Wood smooth, close-grained, very hard, durable, yellow brown; used for compression ressisting timber, beams, posts.

Garcinia venulosa Choisy. (Guttiferaceae). - Tree. E. India, Malaysia, Philippine Islands. Fruits eaten fresh.

Garden Balsam → Impatiens Balsamina L.

Garden Chervil → Anthriscus Cerefolium Hoffm.

Garden Cress → Lepidium sativum L.

Garden Orach → Atriplex hortensis L.

Garden Pea → Pisum sativum L.

Garden Rhubarb → Rheum Rhaponticum L.

Gardenia brasiliensis Spreng. (Rubiaceae). - Woody plant. Guiana, Brazil. Fruits are consumed by the aborigenes.

Gardenia Brighami Mann. (Rubiaceae). - Small tree. Hawaii. Islands. Yellow pulp from fruits is used by the Hawaiians for dyeing tapa clothing.

Gardenia erubescens Stapf. and Hutch. (Rubiaceae). - Tree. Trop. Africa. Yellow fruits are used by the natives in sauces and soup. Seeds are used in Hausa (Nigeria) in a black cosmetic.

Gardenia florida L. (syn. G. jasminoides Ellis). Rubiaceae. Shrub or small tree. Trop. China. Flowers are source of an ess. oil, used in perfumery, being a base of Gardenia Perfumes, blending with jasmin and tuberose; also with certain synthetics among which heliotrope, linalool, hydroxy-citronellal, styrolyl acetate. Oil is obtained with volatile solvents. About 3000 to 4000 kg. flowers produce 1 kg. concrete and about 500 gr. absolute. Contains benzyl acetate, linalool, terpineol, styrolyl acetate, linalyl acetate, methyl anthranilate. Flowers are also used for scenting tea. Fruits are source of a yellow dye sold in some parts of trop. Asia.

Gardenia gummifera L. f. (syn. G. arborea Roxb., G. inermis Dietr.). (Rubiaceae). - Tree. E. India. Source of a resin, called Dikkamaly or Combee.

Gardenia Jovis-tonantis Hiern. (Rubiaceae). - Tree. Trop. W. Africa. Decoction of roots is used in Fr. Guinea to restore failing strength. Fruit is used for stupefying fish. Seeds are source of a black stain used for the skin.

Gardenia Kalbreyeri Hiern. (Rubiaceae). - Woody plant. Trop. Africa. Fruits are source of a black cosmetic used by the natives.

Gardenia lucida Roxb. (Rubiaceae). - Tree. Burma. Source of Combee Resin.

Gardenia lucens Planch. and Sebert. (Rubiaceae). - Tree. New Caledonia. Wood light, hard, fine-grained; esteemed by the natives for making small wooden articles.

Gardenia lutea Fresen. (Rubiaceae). - Shrub. Abyssinia, Blue Nile region. Roots when boiled with Sorghum flour are used by natives for black water fever. Fruit is edible. Wood is dense, even-grained, hard, light yellow; used for knife-handles. Source of a fragrant resin, called by Arabs Abu Beka.

Gardenia Remyi Mann. (Rubiaceae). - Tree. Hawaii. Islands. The gelatinous leaf-buds are used by the Hawaiians as a cement. Pulp from fruits is a source of yellow dye.

Gardenia Rothmannia L. f., Candlewood S. Afr.) (Rubiaceae). - Medium sized tree. S. Africa, esp. Cape Peninsula. Wood very hard, close-grained; suitable for engraving work; used for tools, felloes, wagon work.

Gardenia Standleyana Pax. → Randia maculata DC.

Gardenia taitensis DC. (Rubiaceae). - Tree. Tahiti. Infusion of the flowers is used by the natives of Tahiti for certain headaches. Flowers are employed for scenting coconut oil, also leis, for hair ornaments and necklaces.

Gardenia Thunbergia L. f. (syn. G. ternifolia Schum. and Thonn. (Rubiaceae). - Woody plant. Trop. Africa. Fruit is source of a black cosmetic. Roots and stems with palm wine, Guinea grains and roots of Butyrospermum are used for severe constipation. Ash of wood is used for manuf. soap, also as lye in dyeing. Flowers produce an ess. oil, used by native women of the Sudan in perfumery. Wood is white; employed by the natives for spoons and other utensils.

Gardenia viscidissima S. Moore. (Rubiaceae). - Shrub. W. Africa, Ivory Coast. Wood is used for different purposes among the natives.

Gardenia Vogelii Hook, f. (Rubiaceae). - Shrub. Trop. Africa. Source of a dye, used by the natives of Djurland to paint their bodies.

Gargan Death Carrot → Thapsia garganica L.

Garland Chrysanhemum → Chrysanthemum coronarium L.

Garlic → Allium sativum L.

Garlic, Canada → Allium canadensis L.

Garlic, Giant → Allium Scorodoprasum L.

Garlic, Hedge → Sisymbrium Alliaria (L.) Scop.

Garlic, Levant → Allium Ampeloprasum L.

Garlic, Twistledleaf → Allium obliquum L.

Garlic, Wild → Allium canadense L.

Garrya elliptica Dougl., Quinine Bush, Fever Bush. (Garryaceae). - Shrub. Western N. America,esp. Oregon, California. Bitter decoction of bark and leaves is used to relieve intermittent fevers; employed as home remedy.

Garrya Fremontii Torr., Skunk Bush, Quinine Bush. (Garryaceae). - Shrub. Oregon, California. Leaves used in Calif. as antiperiodic and tonic. Contains garryine, an alkaloid.

Garryaceae → Garrya.

Gastrochilus panduratum Ridl. (Zingiberaceae). - Perennial herb. Malaysia, Java. Cultivated. Rhizomes and roots used as spice in food; stomachic, tonic.

Gastrodia Cunningamii Hook. f. (Orchidaceae). - Perennial herb. New Zealand. Tubers are consumed as food by the aborigenes of New Zealand.

Gastrodia sesamoides R. Br. (Orchidaceae). - Perennial herb. Tasmania. Tubers shape of kidney potatoes are consumed by natives of Tasmania.

Gatapa → Hibiscus tiliaceus L.

Gaultheria antipoda Foster. (Ericaceae). - Shrub. Tasmania. New Zealand. Fruits edible being of a good flavor.

Gautheria fragantissima Wall. Indian Winter Green. (Ericaceae). - Shrub. Travancore, Burma, Assam. Source of Indian Winter Green Oil.

Gautheria fragantissima Wall. var. **punctata** Smith. (Ericaceae). - Shrub. Java, Sumatra. Source of an essential oil, similar to Wintergreen. Used in Java for perfumery and as hair-oil. Decoction of leaves forms part of a medicinal tea, used by the natives.

Gaultheria hispida R. Br. (Ericaceae). - Shrub. Australia. Fruits are eaten by the natives.

Gaultheria leucocarpa Blume. (Ericaceae). - Small shrub. Java and Sumatra. Source of an essential oil.

Gaultheria Myrsinites Hook. (Ericaceae). - Shrub. Western N. America. Fruits are used in preserves.

Gaultheria procumbens L., Winter Green, Teaberry, Winter Berry. (Ericaceae). - Low shrub. Eastern N. America. Leaves yield an ess. oil, Oil of Gaultheria, Wintergreen Oil, obtained by steam distillation; used for flavoring, contains methyl salicylate, is also derived synthetically. Antiseptic, antirheumatic; leaves used as tea, called Mountain Tea. Berries are eaten in pies.

Gaultheria Shallon Pursh. Shallon. (Ericaceae). - Shrub. Brit. Columb. to California. Dried fruits are eaten in winter, much esteemed by the Indians.

Gaylusacia baccata (Wang.) Koch. Black Huckleberry. (Ericaceae). - Shrub. Eastern N. America. Fruits are edible; used for pies, also eaten raw. The large fruited improved varieties are canned. Fruits are also consumed of G. ursina Curtis, Bear Huckleberry; G. dumosa (And.) T. and G. Dwarf Huckleberry; G. brachycera (Michx.) Gray, Box Huckleberry and G. frondosa Torr. and Gray., Dangleberry.

Gaz → Quercus Cerris L.

Geissorhiza Bojeri Baker. (Iridaceae). - Perennial herb. Madagascar. Bulb is used by the natives as stomachic and digestive.

Geissospermum Vellosii Allem. (Apocynaceae). - Woody plant. Trop. America. Stem is source of Pāo Pereira, derived from Brazil, used as a tonic and febrifuge.

Gelidiaceae (Red Algae) → Acanthopeltis, Gelidium, Pterocladia.

Gelidiopsis rigida W.V.B. (Sphaerococcaceae). - Red Alga. Sea Weed. Pac. Ocean and adj. waters. Is consumed and much prized in some parts of the Mal. Archipelago.

Gelidium Amansii (Lamour.) Lamour. (Gelidiaceae). - Red Alga. Coasts of China, Japan, South Africa. Manufactured into agar, or kanten. Of much value in China and Japan. Used in jellies, food preparations, confectionaries, creams, medical emulsions, cosmetics; in the canning industry, for ices, malted milk, cream cheese; extensively used for bacteriological culture media and microscopic technic, coagulant for the precipitation of barium sulphate, electroplating of lead, wire drawing lubricants; manufacturing of submarine storage batteries; used in backings for films; insecticide activator, used in the standard test of Avena for growth hormones, sizing textiles and paper, for backings of gelatinous rolls for hectographs, for pH measuring equipment.

Gelidium cartilagineum (L.) Gaill. var, robustum Gardner. Agarweed. (Gelidiaceae). - Red Alga. Pacific Coast of N. America. Manufactured into agar in California. Also G. pacificum Okam., G. japonicum (Harv.) Okam., G. divaricatum Mart., G. arborescens Gard., G. nudifrons Gardn., G. latifolium Born., G. subcostatum Okam., G. linoides Kuetz. being used for the same purpose in Japan, China and other countries. They are also extensively used for the preparation of homemade agar gel. Product often known as Isinglass.

Gelidium divaricatum Mast. (Gelidiaceae). - Red Alga. Sea Weed. Pac. Ocean. Consumed as food in China. Plants are boiled with addition of vinegar or sweetened to taste. Collected seaweed is washed in water and dried in the sun.

Gelidium rigens Mast. (Gelidiaceae). - Red Alga. Sea Weed. Tropical waters. Consumed as food among the Japanese. Plants are usually dried in the sun. Sold commercially.

Gelsemium sempervirens (L.) Ait. f., Yellow Jessamine. (Spigeliaceae). - Woody vine. Eastern United States. Dried roots are medicinally used. They have a depressant action on central nervous system; are antispasmodic, nervine, sedative, mydriatic. Much is supplied from N. and S. Carolina, Tennessee and Georgia. Root contains strong toxic alkaloids, e. g.: gelsemine, gelsemidine, etc.

Geneps → Melicocca bijuga L.

Genipa americana L., Marmalade Box, Genipa. (Rubiaceae). - Large tree. Trop. America. Popular fruit in Brazil, Puerto Rico etc. Fruits size of an orange, round, russet-brown; pulp brownish, flavor very pronounced. Fruits must soften after being picked. Source of Genipapado, a refreshing drink and Licor de Genipado, an alcoholic beverage. Plant produces a dye, used by Braz. Indians in tattooing. Wood flexible, strong, light colored, resistant; used for general carpentry.

Genipa Caruto H. B. K. (Rubiaceae). - Tree. S. America. The fruits are used by the natives of S. America as a laxative.

Genipado → Genipa americana L.

Genista germanica L. (syn. G. villosa Lam.). (Leguminosaceae). - Half-shrub. Europe. Was source of a yellow dye.

Genista Raetam Forsk. (syn. Retama Raetam Webb. and Berth.) (Leguminosaceae). - Shrubby plant. Mediterranean region. Herb is used in large doses as a dangerous abortive; in small quantities as a purgative and vermifuge.

Genista Saharae Coss. (Leguminosaceae). - Shrub. Algeria. Used as a food for camels.

Genista tinctoria L. (syn. Spartium tinctorium Roth.) Dyer's Greenwood. (Leguminosaceae). - Small shrub or half-shrub. Europe, Ural, Caucasia, Asia Minor, S. W. Siberia. Twigs, leaves and flowers are source of a yellow dye, used for coloring linen and wool. A decoction of the plant is used as home-remedy, as a purgative and for excretion of urine.

Gentian Bitter → Gentiana lutea L.

Gentian, White → Laserpitium latifolium L.

Gentian, Yellow → Gentiana lutea L/

Gentiana adsurgens Cerc. (Gentianaceae). - Alpine perennial herb. S. Mexico. Decoction from roots used as stomachic and stimulant by Maya. Indians. With Ottoa oenanthioides H. B. K. it is given as a purgative.

Gentiana Kurroo Royle. (Gentianaceae). - Perennial herb. Kashmir, N. W. Himalaya. Has been suggested as substitute for roots of G. lutea L.

Gentiana lutea L., Yellow Gentian. (Gentianaceae). - Perennial herb. Mountain meadows, Centr. and S. Europe, Asia Minor. Extract of root used to improve appetite, stimulates gastric secretion; used in manuf. liqueurs. Made into Gentian Bitter. Employed in home-remedies. Dried rhizomes and roots, collected in the late summer and autumn and allowed to be cured, are used medicinally as a bitter tonic and stomachic. Contains a bitter glucoside, gentiopicrin. Also G. purpurea L., G. pannonica Scop. and G. punctata Gebh. are occasionally used for the same purposes.

Gentiana Pneumonanthe L. (Gentianaceae). - Perennial herb. Europe, Caucasia, temp. Asia. Flowers are source of a blue dye.

Gentianaceae → Coutoubea, Enicostema, Erythraea, Gentiana, Limnanthemum, Menyanthes, Sabatia, Sebaea, Swertia, Tachia.

Geodorum nutans (Presl.) Ames. (Orchidaceae). - Perennial herb. Philippine Islands to Formosa. Tuberous roots are source of a glue-like substance; used for cementing parts of guitars and mandolins. Glue has great tenacity.

Geoffraea superba Humb. and Bonpl. (Leguminosaceae). - Tree. S. America. Fruits, called Umari are used as food in some parts of Brazil.

Geonoma binervia Oerst. (Palmaceae). - Palm. Mexico, Centr. America. Young flowers-clusters used when cooked as food by the Chinantecs (Mex.). Large leaves are employed for thatching.

Geophila obvallata F. Didr. (Rubiaceae). - Woody plant. Trop. Africa. Leaves are cooked with food and used for diarrhea of children in Liberia.

Georgea Bark → Pinckneya pubens Michx.

Geraniaceae → Erodium, Geranium, Pelargonium, Wendtia.

Geranium macrorrhizum L., Bigroot Geranium. (Geraniaceae). - Perennial herb. Balkans to Central Europe. Is used as a aphrodisiac in Bulgaria.

Geranium mexicanum H. B. K. (Geraniaceae). - Perennial herb. Mexico, Centr. America. Decoction of leaves with that of Gaultheria is used by Mazatecs (Mex.) as tonic, specially for elderly patients.

Geranium sylvaticum L., Wood Crane's Bill. (Geraniaceae). - Perennial herb. Europe, Asia. Flowers are a source of a blue dye; used in parts of the Black Forests (Germany).

Geranium Oil → Pelargonium Radula L'Hér.

German Ebony → Taxus baccata L.

German Spearmint Oil → Mentha picata L.

Germander, Common → Teucrium Chamaedrys L.

Germander, Water → Teucrium Scordium L.

Germander, Wood→ Teucrium Scorodonia L.

Gerrard Vetch → Vicia Cracca L.

Gesneria allagorphylla Mart. (syn. G. grandis Hort., G. nitida Hort.). (Gesneriaceae). - Perennial herb. Brazil. Tuberous roots are considered an emollient and a tonic.

Gesneriaceae → Didymocarpus, Gesneria.

Getah Mala → Altingia excelsa Noron.

Getah Pootih → Palaquium Maingayi King and Gamb.

Geum urbanum L., Avens Root. (Rosaceae). - Perennial herb. Throughout temp. Europe, Asia, N. America, Australia. Rhizom occasionally used as a spice, having the scent of cloves with a trace of cinnamon. Used with orange peel and put into wine, it is supposed to produce a palatable Vermouth. It is sometimes used in liqueurs. Imparts an agreeable taste to beer.

Gevuina Avellana Molina, Chile Hazel. (Proteaceae). - Tree. Chile. Seeds are consumed in some parts of Chile, having a pleasant flavor, resembling hazelnuts. Much esteemed among the Chileans. Wood pale brown, with pinkish hue, light, firm, strong, medium texture, easily worked, not durable when in contact with the soil; used for turnery, picture frames, furniture and shingles.

Ghatta Tree → Anogeissus latifolius Wall.

Gherkin, West Indian → Cucumis Anguria L.

Ghezireh Gum → Acacia Senegal (L.) Willd.

Giant Alocasia → Alocasia macrorrhiza Schott.

Giant Arbor Vitae → Thuja plicata Donn.

Giant Bur Reed → Sparganium eurycarpum Engelm.

Giant Cactus → Carnegiea gigantea (Engelm.) Britt. and Rose.

Giant Garlic → Allium Scordoprasum L.

Giant Grenadilla → Passiflora quadrangularis L.

Giant Hyssop → Agastache anethiodora (Nutt.) Britt.

Giant Nettle → Laportea gigas Wedd.

Giant Raffia Palm → Raphia gigantea Chev.

Giant Reed → Arundo Donax L.

Giant Rye → Triticum polonicum L.

Giant Timber Bamboo → Phyllostachys bambusoides Sieb. and Zucc.

Giant Wild Rye → Elymus condensatus Presl.

Gidgee Acacia → Acacia homalophylla F. v. Muell.

Gigantochloa ater Kurz. (Graminaceae). - Tall woody bamboo. Trop. Asia. Young buds are used as vegetable in some parts of Trop. Asia.

Giganthochloa verticillata Munro. (Graminaceae). - Tall woody bamboo. Java. Cultivated. Young buds are eaten cooked, of delicate flavor; also consumed after being kept in vinegar. Several varieties are in cultivation.

Gigartina horrida Harv. (Gigartinaceae). - Red Alga. Sea Weed. Pac. Ocean and adj. waters. Plants are source of an Agar Agar, esp. manuf. in Malaysia.

Gigartina stellata (Stackh.) Batt. (Gigartinaceae). - Red Alga. Temp. North Atlantic. Used for the manufacturing of carrageenin, see Chondrus crispus (L.) Stackh. Consumed in some parts of Great Britain.

Gigartina Teedi (Roth.) Lam. (Gigartinaceae). - Red Alga. Sea Weed. Plants are consumed as food in Japan.

Gigartinaceae (Red Algae) → Chondrus, Gigartina, Mastocarpus.

Gillenia trifoliata (L.) Moench. Bowman's Root. (Rosaceae). - Perennial herb. Eastern United States. Plant was used by the Indians as a mild emetic.

Ginger → Zingiber officinale Rosco.

Ginger Bread Plum → Parinarium macrophylla Sab.

Ginger Bread Palm → Hyphaene coriacea Gaertn.

Ginger, Cassumar → Zingiber Cassumar Roxb.

Ginger, Japanese → Zingiber Mioga (Thunb.) Rosc.

Ginger, Mioga → Zingiber Mioga (Thunb.) Rosc.

Ginger, Wild → Asarum canadense L.

Ginger, Zerumbet → Zingiber Zerumbet (L.) Smith.

Ginkgo biloba L., Ginkgo (Ginkgoaceae). - Tree. Japan. Cultivated in Old and New World. Grown in Asia around Buddhist Temples. Seeds are roasted and sold as a delicacy in S. and Centr. China.

Ginkgoaceae → Ginkgo.

Ginseng → Panax quinquefolia L.

Giraffe Acacia → Acacia giraffae Willd.

Girardinia palmata Gaud. (syn. G. heterophylla Decne.) Nilgiri Nettle. (Urticaceae). - Woody plant. N. W. Himalaya to Malaysia. Stem is source of a fibre, made by the natives into a strong cloth.

Gironniera subaequalis Planch. (Ulmaceae). - Tree. S. E. Asia, Mal. Archip. Wood soft, light, easily worked; used for general construction.

Gisekia pharnaceoides L. (Aizoaceae). - Herbaceous. Trop. Africa, E. India. Leaves eaten as pot-herb during famines; also used for preparation of „dal".

Gladiolus edulis Burch. (Iridaceae). - Bulbous perennial. S. Africa. Corms when roasted are sometimes used as food by the natives.

Gladiolus quartinianus Rich. (Iridaceae). - Perennial herb. Nigeria, Portug. E. Africa. Cultivated by the Igara in Afr. Corm used as a food; it is pounded in water with guinea-corn flour. It is also made into a cooling beverage in Bassa, Afr.

Gladiolus spicatus Klatt. (Iridaceae). - Perennial herb. Trop. Africa. Corms are used as food by Lokoja tribes (Afr.)

Gladiolus zambesiacus Baker. (Iridaceae). - Perennial herb. E. Africa. Corms are consumed as food in Njelekwa (E. Afr.).

Glandbearing Oak → Quercus glandulifera Blume.

Glasswort, Leadbush → Salicornia fruticosa L.

Glaucium flavum Crantz. (syn. G. luteum Scop.) Horned Poppy. (Papaveraceae). - Biennial to perennial herb. Mediterranean region to Centr. Europe; introd. in N. America. Oil from pressed seeds is used as illuminant and in food; also manuf. into soap.

Glaux maritima L., Sea Milkwort. (Primulaceae). - Herbaceous plant. Europe, temp. Asia, introd. in N., America. Young shoots have been used in salads. Has been recommended as emergency food in times of want.

Gleditsia amorphoides Taub. (syn. Gorugandora amorphoides Griseb.) (Leguminosaceae). - Tree. Brazil. Wood is compact, strong, elastic, very durable; used for general carpentry work.

Gleditsia japonica Miq. (Leguminosaceae). - Tree. Japan. Cultivated. Juice from fruits is used in Japan for washing, also for cleaning furniture.

Gleditsia macrantha Desf. (Leguminosaceae). - Tree. Hupeh, Scechuan (China). Pods are used in China as a substitute for soap.

Gleditsia sinensis Lam. (syn. G. horrida Willd.) (Leguminosaceae). - Tree. E. China, Yangtze Valley. Wood used for general carpentry in China. Pods are broken up and used for laundry work. Make good lather in cold and hot water; used for tanning hides. Also G. Delavayi Franch.

Gleditsia triacanthos L., Honey Locust, Sweet Bean. (Leguminosaceae). - Tree. Eastern N. America to Florida and Texas. Wood coarse-grained, hard, very durable, strong, resists soil; used for railroad ties, tubs for wheels, fence posts.

Gleichenia dichotoma Hook. (syn. G. Hermanni R. Br.) (Gleicheniaceae). - Fern. Australia. Roots source of a starch, consumed by aborigines of Australia.

Gleichenia linearis C. B. Clarke. (Gleicheniaceae). - Fern. Throughout tropics. In Malay Penin. stems used for pens, also woven into mats, for partition of walls in houses, traps at fishing-stakes, stool and chair seats, pouches, caps.

Gleicheniaceae → Gleichenia.

Gli (beverage) → Liechtensteinia pyrethrifolia Cham. and Schlecht.

Gliricidia sepium (Jacq.) Steud. (syn. Lonchocarpus maculatus DC.) Madre de Cacao. (Leguminosaceae). - Tree. Mexico, Centr. America,

N. of S. America. Naturalized in Cuba and Philippine Islands. Shade tree in coffee plantations. Seeds or powdered bark with rice used as a poison for mice and rats.

Globe Artichoke → Cynara Scolymus L.

Globularia Alypum L. (Globulariaceae). - Perennial herb. Mediterranean region. Herb used as purgative, also used by natives for intermittent fevers and as aphrodisiac.

Globulariaceae → Globularia.

Glochidion Llanosi Muell. Arg. (syn. Phyllanthus Llanosi Muell. Arg.). (Euphorbiaceae). - Shrub. Indo-China, Philippine Islands. Young shoots are used as a spice with fish by the natives.

Glochidion marianum Muell. Arg. (Euphorbiaceae). - Shrub. Guam, Pacific Islands. Wood very strong, fine-grained; used by natives for cart-shafts.

Gloiopeltis furcata (Post. and Rupr.) J. Ag. (Endocladiaceae). - Red Alga. Pacific North America, China, Japan. Important source of hailo in China and of funori in Japan. Used for sizing silk and other textiles. Also used as a glue for binding of Chinese paintings. Also G. tenax (Turn.) J. Ag.

Gloss Buckthorn → Rhamnus Frangula L.

Glossonema Boveanum Desf. (Asclepiadaceae). - Woody plant. Arabia, N. Africa, E. Nile region. Fruits are consumed by the Beduins.

Glossy Privet → Ligustrum lucidum Ait.

Gluta rhengas L., Rengasz. (Anacardiaceae). - Tree. Malaysia. Wood reddish, not very hard, fairly light, easy to split; used for furniture, building material, fancy articles and inlay work.

Glyceria fluitans (L.) R. Br. (syn. Panicularia fluitans Kuntze.) Mammagrass, Sugar Grass. (Graminaceae). - N. America, Europe, Asia. Seeds were used as food by the Klamath and other Indians in N. America.

Glycine Apios L. → Apios tuberosa Moench.

Glycine soja Sieb. and Zucc. (syn. G. hispida Max., Soja hispida Moench., G. max Mer.). Soybean. (Leguminosaceae). - Annual herb. S. W. Asia. Cultivated since ancient times. Grown in many varieties in Old and New World. Used for many purposes. Beans are consumed as green vegetable, in salads and canned. Source of a meal, made into breakfast foods, infant food, crackers, bread, cakes, muffins, biscuits, macaroni; in ice creams and chocolate bars; substitute for coffee, source of casein, cheese. Soy sauce has been subject to fermentation. When mixed with wheat it is suitable for bread and cakes. Seeds are source of an oil, made into glycerine; used in enemals, varnishes, paints, waterproof goods, linoleum, hard soaps, liquid shampoo, paste soap for hospital use, oil cloth, used in metal moulding, foundry cores; used with rubber for manuf. mats, hose etc., rubber substitutes, lubricant,

in printing inks. When refined used for cooking, salads, margarine, shortening. It is also used for illumination. Sp. Gr. 0.922-0.925; Sap. Val. 189.9-194.3; Iod. No. 103-152; Unsap. 0.50-1.8%. An artificial milk from seeds is used in China and Japan. Seedlings are eaten in China in various dishes. Tempé, a food product is obtained by the aid of a fungus, Aspirgillus oryzae. Other preparations used in Asia are Teoufu, Tao-cho, Sho-yu. The crop is also suitable as forage, hay, silage and pasture for live-stock and as a green manure.

Glycyrrhiza asperrima L. f. (Leguminosaceae). - Perennial herb. Siberia, Centr. Asia. Leaves are used as a tea by the Kalmucks.

Glycyrrhiza glabra L., Common Licorice, Liquorice. (Leguminosaceae). - Perennial herb. Europe, Mediterranean region. Cultivated. Dried rhizome and roots are source of Licorice. Var. typica Regel and Herder produce Spanish or Italian Licorice; var. glandulifera Waldstein is source of Russian Licorice. Used medicinally as mild laxative, expectorant and demulcent; for masking taste of drugs, like quinine and aloë. Also used for manuf. of pills and troches. Sweet roots are used for chewing, also employed in confectionary and in chewing tobacco.

Glycyrrhiza lepidota (Nutt.) Pursh., Wild Liquorice. (Leguminosaceae). - Herbaceous perennial. N. America. Long fleshy roots were eaten by the Indians. Plants are sometimes grown near Indian villages. Roots contain 6% glycyrrhizin.

Glycyrrhiza uralensis Fisch., Chinese Liquorice. (Leguminosaceae). - Perennial herb. Siberia, China. Plant is used in China as emollient.

Gmelina arborea Roxb. (syn. Premna arborea Roth.) Malay Bushbeech. (Verbenaceae). - Tree. E. India, Malaya, Pacif. Islands. Parts of the plant form with the roots of Epipremnum pinnatum Yoro or Awalho of the Fiji islanders. Medicinally considered and anodyne. Wood is source of mining timber.

Gnaphalium polycephalum Michx. (syn. G. obtusifolium L.). Common Everlasting. (Compositae). - Annual. N. America. Herb was used by the Indians for intestinal and pulmonary catarrh; also for fomentation of bruises. Plant is supposed to be an anodyne.

Gnetaceae → Gnetum.

Gnetum Gnemon L. (Gnetaceae). - Woody plant. Malaysia, trop. Asia. Fruits eaten in Philippine Islands cooked or roasted. Young leaves eaten as a vegetable. Bark source of a fibre. Sometimes cultivated for its edible leaves. Also G. indicum (Lour.) Merr.

Gnetum scandens Roxb. (Gnetaceae). - Vine. E. Himalaya to S. Indo-China. Source of a fibre; used by the Andamese.

Goa Angelin Tree → Andira Araroba Aguiar.

Goa Bean → Psophocarpus tetragonolobus DC.

Goa Butter → Garcinia indica Choisy.

Goa Ipecacuanha → Naregamia alata Wight and Arn.

Goa Powder → Andira Araroba Aguiar.

Goatnut → Simmondsia californica Nutt.

Goat's Rue → Galega officinalis L.

Goat Willow → Salix Caprea L.

Goldband Lily → Lilium auratum Lindl.

Gold Edge Lichen → Sticta crocata (L.) Ach.

Golden Chain → Laburnum anagyroides Med.

Golden Chamomile → Anthemis tinctoria L.

Golden Chinquapin → Castanopsis chrysophylla A. DC.

Golden Currant → Ribes aureum Pursh.

Golden Gram → Phaseolus Mungo L.

Golden Ragwort → Senecio aureus L.

Goldenrod → Solidago.

Goldenrod, Californian → Solidago californica Nutt.

Goldenrod, Canadian → Solidago canadensis L.

Goldenrod, Sweet → Solidago odora Ait.

Golden Seal → Hydrastis canadensis L.

Golden Wattle → Acacia pycnantha Benth.

Goldenweed, Nuttall → Aplopappus Nuttallii Torr. and Gray.

Goldmoss Stonecrop → Sedum acre L.

Goldthread → Coptis trifolia (L.) Salisb.

Goma Anime de Mexico → Hymenaea Courbaril L.

Gom, Resina de Mamey → Mammea americana L.

Gombo → Hibiscus esculentus L.

Gomme d'Acajou → Anacardium occidentale L.

Gomme Blanche → Acacia Senegal (L.) Willd.

Gomme Blondes → Acacia Senegal (L.) Willd.

Gomme de Benaile → Moringa oleifera Juss.

Gomme de Sénégal → Acacia albida Delile and Sterculia tomentosa Guill. and Perr.

Gomme Rouge → Acacia tortilis Hayne.

Gomme Salobreda → Acacia stenocarpa Hochst.

Gomphia parviflora DC. (syn. Ourata parviflora Baill.). (Ochnaceae). - Shrub. Brazil. Seeds are source of an oil; used in Brazil for skin diseases and leprosy. The product is called Batiputa or Bati Oil, non drying. Sap. Val. 192-202; Iod. No. 51-70.

Gomphidius glutinosus (Schäff.) Fr. (Agaricaceae). - Basidiomycete Fungus. Temp. zone. The fruitbodies are consumed as food, having a pleasant flavor. Also G. subroseus Kauffm. from Western N. America; G. oregonensis Peck; from the same region; G. roseus (Fr.) Karts. from Europe and Western Siberia.

Gomphocarpus fruticosus R. Br. (syn. Asclepias fruticosa L.). (Asclepiadaceae). - Small shrub. Africa. Floss from seeds is used for stuffing pillows in different parts of Africa.

Gonioma Kamassi Mey, Kamassiwood. (Apocynaceae). - Tree. Cape Peninsula. Wood yellowish, dense, hard, close-grained, easily worked; used for engraving work, fancy turnery. Exported to Europe as Boxwood.

Gonocrypta Grevii Baill., Kompitro. (Asclepiadaceae). - Woody plant. Madagascar. Latex of stem is source of a good rubber.

Gonyo Oil → Antrocaryum Nannani de Willd. Anacardiaceae.

Gonystylus Miquelianus Teijsm. and Binn. (syn. Gonystylus bancanus Baill., Aquilaria bancana Miq.) (Thymelaeaceae). - Tree. Malaysia. Wood hard, fine-grained; used for boards and posts, also used as incense. Substitute for Aloes Wood. Sold in markets of Java. Oil of wood is used as incense, smoke is remedy for asthma.

Goober → Arachis hypogaea L.

Goober, Congo → Voandzeia subterranea Thon.

Good King Henry → Chenopodium Bonus-Hendricus L.

Goodeniaceae → Scaevola.

Gooseberry, American Wild → Ribes Cynosbati L.

Gooseberry, Barbados → Pereskia aculeata Mill.

Gooseberry, Ceylon → Dovyalis hebecarpa (Gardn.) Warb.

Gooseberry, Dwarf Cape → Physalis pubescens L.

Gooseberry, European → Ribes Grossularia L.

Gooseberry, Otaheite → Phyllanthus distichus (L.) Muell. Arg.

Gooseberry, Prickly → Ribes Cynosbati L.

Gooseberry, Smooth → Ribes oxycanthoides L.

Gooseberry, West Indian → Pereskia aculeta Mill.

Gooseberry Wine → Ribes Grossularia L.

Goose Grass → Eleusine indica Gaertn. and Galium Aparine L.

Gordonia Lasianthus (L.) Ellis., Loblolly Bay, Tan Bay. (Theaceae). - Small tree. S. Virginia to Florida and Louisiana. Wood close-grained, not durable, light, soft, light red; locally used for cabinet work. Bark contains tannin, sometimes used for tanning.

Gordonia excelsa Blume. (Theaceae). - Tree. Mal. Archipel., Java. Wood heavy, hard, close-grained, reddish to blackish-brown; used as building material, fine wood-work; rice-pestles. Bark is source of a black dye; used for staining basket-work.

Gorse, Common → Ulex europaeus L.

Gossweilerodendron balsamiferum Harms. (Leguminosaceae). - Tree. Trop. Africa, esp. Nigeria to Congo. Stem source of a copal-resin; used by natives for illumination. Wood strong moderately hard, easily worked, takes a good polish, light yellow;) used for tabletops, cheap furniture, light building material and construction work.

Gossypiospermum praecox (Griseb.) Wils. (syn. Casearia praecox Griseb.) Boxwood. (Flacourtiaceae). - Small to medium sized tree. Cuba, Dominican Rep., Venezuela. Wood lemon-yellow to almost white; used for engraving blocks, precision rulers, veneers, turnery, combs, shuttles, spindles for silk-mills, key-boards, piano keys, inlay, jewelers' burnishing wheels.

Gossypium arboreum L., Tree Cotton. (Malvaceae). - Tall herbaceous plant. India and Africa. Source of Tree Cotton. Cultivated in India and Africa. Var. neglecta Watt. produces very short staple, cultiv. in India and Burma; var. sanguinea Watt. staple strong of good quality.

Gossypium barbadense L., Sea Island Cotton. (Malvaceae). - Perennial, grown as annual. Var. maritima Watt. Source of Sea Island Cotton of commerce, Gallini Cotton of Egypt. Fibre longest of any cotton species, 3 to 5 cm. long. strong, of excellent quality.

Gossypium brasiliense Macf., Pernambuco Cotton (Malvaceae). - Herbaceous. Brazil. Source of Bahia, Pernambuca, Chain, Kidney, Stone Cotton. Lint very fine, silky; plentiful in pods.

Gossypium herbaceum L., Short Staple American Cotton(Malvaceae). - Annual. Origin unknown. Source of some Short Staple American Cotton varieties, also Syrian, Maltese, Levant and Arabian Cotton. Cotton are the hairy fibres from the epidermis of the seeds, used in numerous cotton goods, rubber-tire fabrics, stuffing cushions and pillows, manuf. of twine, ropes, carpets, mercerized cotton, rayon etc. Absorbent cotton has been cleaned from oily covering substance, being almost pure cellulose; used, when sterilized, in surgical dressings. Seeds are source of Cottonseed Oil, semi-drying; used for lard substitutes, vegetables shortening, oleomargarine, cooking oil, salad oil, manuf. of soap, soap powders. Sp. Gr. 0.9174; Sap. Val. 195; Iod. No. 108.2; Unsap. 0.9⁰/₀. Oil Cake is used as fertilizer, fodder for cattle and dyestuff. Stalks can be employed in manuf. paper also used as fuel. Hulls used as food for live-stock, fertilizer, lining oil-wells, production of xvlose which can be converted into alcohol, explosives. Root Bark is medicinally excellent oxytoxic, of value in arresting hemorrhage, also known as emmenagogue. Petals are in some parts of India source of a yellow and brown dye. Flowers are source of a mild honey.

Gossypium hirsutum L., American Upland Cotton (Malvaceae). - Herbaceous plant. Source of Short Staple American, American Upland Cotton. Staple is of good quality.

Gossypium microcarpum Tod. Red Peruvian Cotton. (Malvaceae). - Herbaceous. South America. Source of a fibre.

Gossypium obtusifolium Roxb., Kumpta Cotton. (Malvaceae). - Source of Kumpta, Kathiawar, Surat, Broach Cotton. Widely grown in India. Fibres coarse, reddish white.

Gossypium Nanking Meyen., Nanking Cotton, Chinese Cotton. (Malvaceae). - Herbeceous. Trop. Asia. Cultivated in Asia, Afr. Source of Chinese, Nanking, Khaki, Siam Cotton. Fibres silky, somewhat reddish; var. Bani Watt. Cultigen, high grade fibre; var. Roji Watt. harsh, short fibres.

Gossypium peruvianum Cav. (syn. G. barbadense Oliv.) Peruvian Cotton, Andes Cotton. (Malvaceae). - Herbeceous. S. America. Source of a fibre of good quality.

Gouania javanica Miq. (Rhamnaceae). - Woody climber. Indo-China to Philippine Islands, Malacca. Roots used for poultices on sores.

Gouania leptostachya DC. (Rhamnaceae). - Woody vine. Malaya. Decoction from bark is used by natives to wash hair. In Java pulped stems, roots and leaves are employed for skin complaints.

Gouania lupuloides (L.) Urban. (syn. Banisteria lupuloides L.) Toot Brush Tree, Chaw Stick. (Rhamnaceae). - Woody plant. Trop. America. Pieces of the stems are used for chewing and to heal and harden the gums. Dried powdered stems are made into dentifrices. Exported to Europe.

Gouania tiliaefolia Lam. (Rhamnaceae). - Woody vine. Philippine Islands. Roots used as substitute for soap.

Goupi Wood → Goupia glabra Aubl.

Goupia glabra Aubl. (Celastraceae). - Tree. Guiana. Source of Goupi Wood, being brownish-red, very hard, heavy, tough, easy to work and to polish; used for furniture, railroad ties, street-paving blocks, boat building.

Gourd. Bottle → Lagenaria vulgaris Ser.

Gourd, Calabash → Lagenaria vulgaris Ser.

Gourd, Dishcloth → Luffa acutangula Roxb.

Gourd, Sinkwa Towel → Luffa acutangula Roxb.

Gourd, Snake → Trichosanthes Anguinea L.

Gourd, Wax → Benincasa cerifera Savi.

Gourd, White → Benincasa cerifera Savi.

Gourd, Wild → Cucurbita foetidissima H. B. K.

Gourliea decorticans Gill. (Leguminosaceae). - Small Tree. Chile. Fruits are edible, resembling a jujube, fleshy, sub-acid. Source of an important food among the natives of the Chaco region. Also used as fodder for live-stock. Called Chanal or Chanar.

Goutweed → Aegopodium Podagraria L.

Governor's Plum → Flacourtia Ramontchi L'Hér.

Graceful Wattle Acacia → Acacia decora Reichb.

Gracilaria compressa (J. Ag.) Grev. (Gracilariaceae). - Red Alga. Plants are consumed as food in Japan.

Gracilaria confervoides (L.) Grev. (Gracilariaceae). - Red Alga. Cosmopolitan. Manufactured into agar in North Carolina, Australia and South Africa. When fresh, it is used as a food in Oriental countries.

Gracilaria coronopifolia J. Ag. (Gracilariaceae). - Red Alga. Sea Weed. Pac. Ocean and adj. waters. Consumed as food by the Hawaiians, who call it Limu Manauea. It is eaten with squid or octopus, producing a jelly much esteemed by the inhabitants. The sea weed is also consumed with chicken. It is sometimes grated with coconut or coconut milk. Also palatable in vegetable soups.

Gracilaria lichenoides (L.) Harv. Ceylon Moss. (Gracilariaceae). - Red Alga. Indian Ocean, Trop. Pacific. Used as food and for manufacturing of jelly in Asia. Much is exported to China, as Ceylon Moss. Is used in Chinese pharmacy where it is considered pectoral and antidysenteric.

Gracilariaceae (red algae) → Gracilaria.

Graines d'Avignon → Rhamnus infectorius L.

Grains → under Seeds and Grains.

Grains of Paradise → Aframomum Melehueta Schum.

Grains of Selim → Xylopia aethiopica A. Rich.

Gram, Black → Phaseolus Mungo L.

Grama, Black → Bouteloua eriopoda Torr.

Grama, Blue → Bouteloua gracilis (H. B. K.) Lag.

Grama Grass → Bouteloua filiformis (Fourn.) Griff.

Grama, Side-Oats → Bouteloua curtipendula (Michx.) Torr.

Graminaceae → Acroceras, Agropyron, Agrostis, Alopecurus, Ammophila, Amphilophis, Andropogon, Anthistiria, Anthoxanthum, Aristida, Arrhenaterum, Arundinaria, Arundinella, Arundo, Astrebla, Astrostylidium, Avena, Axonopus, Bambusa, Beckmannia, Bouteloua, Brachiaria, Bromus, Calamovilfa, Chloris, Coix, Cortaderia, Cymbopogon, Cynodon, Cynosurus, Dactylis, Dendrocalamus, Deschampia, Dichelachne, Digitaria, Echinochloa, Eleusine, Elionurus, Elymus, Epicampes, Eragrostis, Eriochloa, Euchlaena, Festuca, Giganthochloa, Glyceria, Gynerium, Haemarthria, Heteropogon, Hierochloe, Holcus, Hordeum, Hymenachne, Hyparrhenia, Imperata. Isachne, Ischaemum, Jardinea, Latipes, Leptochloa, Lolium, Lygium, Melinis, Melocalamus, Melocanna, Muehlenbergia, Ochlandra, Oryza, Oryzopsis, Panicum,

Pariana, Paspalum, Pennisetum, Phalaris, Phleum, Phragmitis, Phyllostachys, Poa, Polytoca, Pseudostachyum, Puccinellia, Rottboellia, Saccharum, Schizachyrium, Secale, Setaria, Sorghum, Sporolobus, Stipa, Teinostachyum, Thelepogon, Themeda, Thyrsostachys, Thysanolaena, Tricholaena, Tripsacum, Tristachya, Triticum, Uniola, Vetiveria, Zea, Zizania.

Grand Fir → Abies grandis Lindl.

Grangea maderaspatana Poir. (Compositae). - Herbaceous plant. Trop. Africa. Leaves are said to have anodyne properties.

Grape, Common → Vitis vinifera L.

Grape, Fox → Vitis Labrusca L.

Grape, Frost → Vitis vulpina L.

Grapehoney → Vitis vinifera L.

Grape Jasmine → Ervatamia coronaria Stapf.

Grape Mango → Sorindeia madagascariensis DC.

Grape, Muscadine → Vitis rotundifolia Michx.

Grape, Oregon → Mahonia Aquifolium (Lindl.) G. Don.

Grape, River Bank → Vitis vulpina L.

Grape, Sea → Coccolobis uvifera (L.) Jacq.

Grape, Summer → Vitis aestivalis Michx.

Grapefruit → Citrus paradisi Macf.

Grapefruit Seed Oil → Citrus paradisi Macf.

Grasses, Useful → Graminaceae.

Grass Linen → Boehmeria nivea (L.) Gaud.

Grass Pea → Lathyrus sativus L.

Grass Peavine → Lathyrus sativus L.

Grass, Swatow → Boehmeria nivea (L.) Gaud.

Grass Tree, Australian → Xanthorrhoea australis R. Br.

Grass Tree, Dockowar → Xanthorrhoea arborea R. Br.

Grass Tree Gum → Xanthorrhoea australis R. Br.

Grass Tree Gum → Xanthorrhoea hastilis R. Br.

Grass Tree, Spearleaf → Xanthorrhoea hastilis R. Br.

Grass Wack → Zostera marina L.

Grass Weed → Zostera marina L.

Grassleaf Daylily → Hemerocallis minor L.

Grateloupia atfinis (Harv.) Okam. (Grateloupiaceae). - Red Alga. Sea Weed. Consumed as food in Japan. Also G. filicina (Wulf.) J. Ag.

Grateloupia ligulata Holmes. (Grateloupiaceae). - Red Alga. Sea Weed. Pac. Ocean. Is consumed as food in China. Plants are collected late in spring or early summer.

Grateloupiaceae (Red Algae) → Cryptonemia, Grateloupia, Halymenia.

Gray Birch → Betula populifolia Marsh.

Gray Box Eucalyptus → Eucalyptus hemiphloia F. v. Muell.

Gray Sarsaparilla → Smilax aristolochiaefolia Mill.

Grayish Urceolaria → Urceolaria cinerea Ach.

Grays Chokecherry → Prunus Grayana Max.

Great Bulrush → Scirpus lacustris L.

Great Burdock → Arctium Lappa L.

Great Laurel → Rhododendron maximum L.

Greater Galangal → Alpinia Galanga Willd.

Greater Yam → Dioscorea alata L.

Greek Juniper → Juniperus excelsa Bieb.

Greek Turpentine → Pinus halepensis Mill.

Green Amaranth → Amaranthus retroflexus L.

Green, Chinese → Rhamnus dahurica Pall.

Green Ebony → Tecoma leucoxylon (L.) Mart.

Greenbrier, Bristly → Smilax Bona-nox L.

Greenbrier, Long Stalked → Smilax pseudo-China L.

Greenbrier, Saw → Smilax Bona nox L.

Greenheart → Nectandra Rodoei (Schomb.) Hook.

Greenheart, Demarara → Ocotea Rodiaei (Schomb.) Mez.

Greenheart Ebony → Tecoma leucoxylon (L.) Mart.

Greenheart, Surinam → Tecoma leucoxylon (L.) Mart.

Greenleaf Fescue → Festuca viridula Vasey.

Greenmanure → under Fertilizers.

Green Pepper → Capsicum frutescens L.

Green Sapote → Calocarpum viride Pitt.

Green Sprangletop → Leptochloa dubia (H. B. K.) Nees.

Green Strophanthus → Strophanthus Kombe Oliv.

Greenwattle Acacia → Acacia decurrens Willd.

Greigia sphacelata Regel. (Bromeliaceae). - Perennial hrb. Chile. Berries called Chupones are edible.

Grenadilla de Quijos → Passiflora Popenovii Killipp.

Grenadilla, Giant → Passiflora quadrangularis L.

Grenadilla, Purple → Passiflora edulis Sims.

Grenadilla Real → Passiflora quadrangularis L.

Grenadilla, Sweet → Passiflora ligularis Juss.

Grenadille Wood, African → Dalbergia melanoxylon Guill and Pierre.

Grenadine → Punica Granatum L.

Grevillea Gillivrayi Hook. (Proteaceae). - Tree. New Caledonia. Wood is reddish, fine-grained; used for cabinet work.

Grevillea robusta Cunningh, Silk Oak, Warragarra. (Proteaceae). - Tree. Australian, esp. New South Wales, Queensland. Wood light, hard, easily worked; used for cabinet-work, interior finish of houses. Tree used in some countries for shade in coffee plantations.

Grevillea striata R. Br. (syn. G. lineata R. Br.) Beef-wood, Turraie. (Proteaceae). - Tree. Australia. Wood hard, close-grained, takes a good polish; used for furniture, cabinet-work, fancy work.

Grewia occidentalis L. (Tiliaceae). - Shrub or small tree. Trop. Africa, esp. Cape Penins. Wood is used by the Bushmen for making bows.

Grewia populifolia Vahl. (syn. G. betulaefolia Juss.). (Tiliaceae). - Woody plant. Trop. Africa. Bark is source of a strong fibre which is used for different purposes by the natives.

Grewia excelsa Vahl. (Tiliaceae). - Woody plant. Trop. Africa, Asia. Stem is source of a fibre; used for different purposes.

Grewia macrophylla Baker. (Tiliaceae). - Shrub. Madagascar. Bark is source of a fibre; used by the Shihanoeka of Madagascar.

Grewia mollis Juss. (Tiliaceae). - Shrub or small tree. Trop. Africa. Bark is source of a fibre. Ashes from the wood are used for manuf. of salt. Wood is used in the Sudan for bows and arrows. Fruit is edible. Mucilaginous bark is eaten by the natives in soups.

Grias peruviana Miers., Cocora. (Lecythidaceae). - Trop. America, esp. Peru, Colombia and Venezuela. Scrapings from germinating seeds in tapid water are used by the Indians to cause vomiting.

Grindelia robusta Nutt., Shore Grindelia. (Compositae). - Western United States. Decoction of leaves and young flowering tops is used in home remedies, as mild stomachic, to „purify" the blood, also to relieve throad and lung troubles.

Grindelia squarosa (Pursh) Dun., Curlycup Grindelia. (Compositae). - Herbaceous plant. Western N. America. Dried leaves and flower tops are used medicinally, being antispasmodic, sedative, expectorant; used for burns, Ivy poisoning, whooping cough. Contains a resinous substance, ess. oil, grindelol.

Griselinia littoralis Raoul. (Cornaceae). - Tree. New Zealand. Wood firm, dense, compact, slightly brittle, reddish; used for rail-road ties, building of ships and boats.

Griselinia lucida Forst. (Cornaceae). - Tree. New Zealand. Wood compact, dense, very durable, brownish; used for millwright's work and posts.

Gromwell, Common → Lithospermum officinale L.

Grossulariaceae → Ribes.

Ground Blueberry → Vaccinium Myrsinites Lam.

Ground Cherry → Physalis heterophylla Nees.

Ground Hemlock → Taxus canadensis Willd.

Ground Laurel → Epigaea repens L.

Ground Nut → Arachis hypogaea L. and Apios tuberosa Moench.

Groundnut, Bambara → Voandzeia subterranea Thon.

Ground Peanut → Amphicarpa monoica (L.) Ell.

Groundnut Peavine — Lathyrus tuberosus L.

Grumichama → Eugenia Dombeyana DC.

Grumilea psychotrioides DC. (Rubiaceae). - Woody plant. Trop. Africa. Source of a red dye; used in S. Leone for dyeing cloth.

Guabiroba → Abbevillea Fenzliana Berg.

Guabiraba → Campomanesia Guaviroba Benth. and Hook.

Guadil → Convolvulus floridus L. f.

Guaiac → Guaiacum sanctum L.

Guaiacum officinale L., Lignum sactum, Guajacan Negro. (Zygophyllaceae). - Tree. Trop. America. Wood greenish brown, very hard and heavy, not easy to split, light, very durable, tough; used for different parts in ship construction, wooden hammers and different materials.

Guaiacum sanctum L., Lignum-Vitae. (Zygophyllaceae). - Shrub or tree. Trop. America. Heartwood is source of a resin, used for making small objects where strength, hardiness and weight are required. Guaiac was introd. in Europe in 1526 from San Domingo. It is brown to reddish brown, irregularl umps, mild laxative, diuretic, contains guaiaconic acid, 10% guaiaretic acid, 15% vanillin, guaiac yellow. It is obtained by boring a log longitudinally, heating it in a sloping position after which the melted resin escapes from the log. Wood is also cut into chips, boiled in water while the resin rises to the top. Also G. officinale L.

Guajacan Negro → Guaiacum officinale L.

Guanábana → Annona muricata L.

Guarana Bread → Paullinia Cupana Kunth.

Guararema → Gallesia integrifolia (Spreng.) Hamrs.

Guarea africana Welw. (Meliaceae). - Tree. Trop. Africa. Wood light, easy to work, whitish; used for general carpentry and interior finish of houses.

Guarea cedrata (Chev.) Pell. (syn. Trichelia cedrata Chev.). Pink Mahogany, Pink African Cedar. (Meliaceae). - Tree. Trop. W. Africa. Wood fine-grained, light pale mahogany; used for boats, canoes, cabinet work, furniture; is said to be termite-proof.

Guarea Guara (Jacq.) Wils. (syn. G. parva C. DC). Cedro Macho. (Meliaceae). - Tree. Centr. America. Wood hard, reddish brown, durable, strong, medium heavy, gives a fine finish; used for general carpentry.

Guarea Martiana C. DC. (syn. G. purgans St. Hil.) (Meliaceae). - Tree. Trop. America. Bark is a powerful depurative; used in Brazil in patent medicines.

Guarea Rusbyi (Britt.) Rusby, Cocillana. (Meliaceae). - Tree. E. slope of Andes in Bolivia. Source of Guapi Bark. Used by natives as an emetic. Employed in medicine as an expectorant; large doses as emetic. Contains rusbyine, an alkaloid.

Guarea Thompsoni Sprague and Hitch., Benin Mahogany. (Meliaceae). - Tree. Trop. Africa. Wood used for furniture; exported to Europe.

Guarea trichilioides L. (Meliaceae). - Tree. West Indies to Brazil. Wood is source of a reddish Sandal Oil, called Sandalo do Pará. Powdered bark is used in medicine as an emetic and hemostatic. Wood is purplish red; used for furniture, indoor construction and cabinet work.

Guassatonga → Casearia sylvestris Sw.

Guatemalan Walnut → Juglans mollis Engelm.

Guatteria Cargadero Triana and Planch. (Annonaceae). - Tree. S. America. Bark is source of a fibre; used for various purposes in some parts of Colombia.

Guava → Psdium Guajava L.

Guava, Brazilian → Psidium Araça Raddi.

Guava, Cattley → Psidium Cattleianum Sab.

Guava, Chilean → Myrtus Ugni Mol.

Guava, Costa Rican → Psidium Friedrichsthalianum Ndz.

Guavo Bejúco → Inga edulis Mart.

Guavo de Castilla → Inga spectabilis Willd.

Guavo de Mono → Inga Goldmanii Pittier.

Guavo Peludo → Inga Macuna Walpers.

Guavo Real → Inga radians Pitt.

Guayule → Parthenium argentatum Gray.

Guazuma grandiflora G. Don. (syn. Theobroma grandiflora Schum.) (Sterculiaceae). - Tree. Amazon (Braz.). Cultivated in some parts of Amazon basin. Fruits edible; used in sherbets.

Guazuma tomentosa H. B. K. (Sterculiaceae). - Tree. Trop. America. Wood light, resistant, very elastic, yellow-red; used for shoelasts.

Guazuma ulmifolia Lam. (syn. Theobroma Guazuma L.) Tablote, Guacima, Cablote. (Sterculiaceae). - Shrub or tree. Mexico, Trop. America. Wood fibrous, light, coarse-grained, grayish; used for ribs of small boats, barrel staves, furniture, shoe-lasts, paneling, charcoal for gunpowder. Juice is employed to clarify syrup in manif. of sugar. Bark was formerly used for asthma, at the present as a remedy for promoting hair growth.

Cuban Myrrh → Commiphora Hildebrandtii Engl.

Guettarda Angelica Mart. (Rubiaceae). - Tree. Brazil. Root is used in vet. medicine, being vulneric, astringent; used in Brazil for diarrhea of cattle and horses.

Guettarda speciosa L. (Rubiaceae). - Tree. Trop. Asia. Madagascar. Wood source of Zebra Wood. (Also Connarus guianensis Lam. from S. America).

Gugertree, Burma → Schima Noronhae Reinw.

Guiana Arrowroot → Dioscorea Batatas Decne.

Guiana Cashew → Anacardium Rhinocarpus DC.

Guiana Chaste Tree → Vitex divaricata Sw.

Guiana Symphonia → Symphonia glubulifera L.

Guibourtia copallifera Benn. → Copaifera Guibourtiana Benth.

Guioa Koelreuteria (Blanco) Merr. (syn. G. Perrottetii Radlk.) (Sapindaceae). - Tree. N. Luzon to S. Mindanao (Philippine Islands). Oil from seeds is used in Philipp. Islds. for certain skindiseases.

Guizotia abyssinica (L. f.) Cass. (syn. G. oleifera DC.) (Compositae). - Herbaceous plant. Trop. Africa. Cultivated in E. Africa. Fruits source of Niger Seed Oil, a drying oil; used in food, paint, for manuf, soap, as illuminant; adulterant of Rape Oil. Sp. Gr. 0.924-0.927; Sap. Val. 189-193; Iod. No. 126-134; Unsap. 0.5-1.2%. It is sometimes used as adulterant of Sesam Oil. The product is also called Ramtilla or Werinnua Oil. In Abyssinia pressed seeds with honey are made into cakes.

Gulancha Tinospora → Tinospora cordifolia Miers.

Gulangabin → Rosa Eglanteria L.

Gul-i-Pista → Pistacia vera L.

Gullan → Passiflora psilantha (Sodiro) Kilipp.

Gully Root → Petiveria alliacea L.

Guilielma insignis Mart. (Palmaceae). - Tall palm. S. America. Fruits are juicy; consumed raw or dried.

Guilielma speciosa Mart. Bactris Gasipaës H.B.K. Guilielma utilis Oerst. (Palmaceae). - Slender palm. Centr. America. Occasionally cultivated. Fruits when boiled are consumed in Costa Rica, they have the taste of sweet potatoes. Wood is made by the Indians into bows, arrows and clubs.

Guimba → Xylopia obtusifolia A. Rich.

Guinea Grass → Panicum maximum Jacq.

Guinea Pepper → Xylopia aethiopica A. Rich.

Guinea Yam → Dioscorea cayennensis Lam.

Guisaro → Psidium molle Bertol.

Güisquil → Sechium edule Swartz.

Gum, Ako Ogea → Paradaniella Oliveri Rolfe.

Gum Ammoniacum → Dorema Ammoniacum D. Don.

Gum Ammoniac of Cyrenaica → Ferula marmarica Asch. and Taub.

Gum, Ammoniac of Morocco → Ferula communis L.

Gum, Amritsar → Acacia modesta Wall.

Gum, Angado Mastiche → Echinops viscosus DC.

Gum, Angico → Piptadenia rigida Benth.

Gum Arabic → Acacia albida Delile, and A. Senegal (L.) Willd.

Gum Arabic Acacia → Acacia Senegal (L.) Willd.

Gum Arabic (substitutes) → Acacia dealbata Link., A. Greggi Gray, A. horrida Willd., A. Jacquimontii Benth., A. leucophloea Willd., A. Seyal Delile, A. Sieberiana DC., and A. stenocarpa Hochst.

Gum, Argan → Argania Sideroxylon Roem.

Gum, Australian → Acacia pycnantha Benth.

Gum, Barister → Mezoneuron Scortechinii F. v. Muell.

Gum, Bea → Caesalpinia praecox Ruiz and Pav.

Gum, Bisabol Myrrha → Commiphora erythraea Engl.

Gum, Black → Nyssa multiflora Wang.

Gum, Blue → Eucalyptus botryoides Smith. and E. globulus Lah.

Gum, Botany Bay → Xanthorrhoea hastalis R.Br.

Gum, Broadleaved Water → Tristania suaveolens Smith.

Gum, Brown Barbary → Acacia arabica Willd.

Gum, Butea → Butea superba Roxb.

Gum, Cape → Acacia giraffae Willd. and A. horrida Willd.

Gum Cashawa → Anacardium occidentale L.

Gum, Cebil → Piptadenia Cebil Griseb.

Gum, Chene → Spermolepis gummifera Brongn.

Gum, Chicle → Achras Sapota L.

Gum Copal of Gold Coast → Daniella similis Craib.

Gum, Dham-ka-gond → Woodfordia floribunda Salisb.

Gum, Dhaura → Woodfordia floribunda Salisb.

Gum, Doctor's → Symphonia globulifera L.

Gum, Elqueme → Bursera gummifera L.

Gum, Formosa Sweet → Liquidambar formosana Hance.

Gum, Galbanum → Ferula galbaniflua Boiss. and Buhse.

Gum, Gamboge → Garcinia Hanburyi Hook. f.

Gum Ghatti of Bombay → Anogeissus latifolia Wall.

Gum, Ghezireh → Acacia Senegal (L.) Willd.

Gum, Grass Tree → Xanthorrhoea australis R. Br. and X. hastilis R.Br.

Gum, Gumihan → Artocarpus elastica Reinw.

Gum, Hanjigoad → Balanites orbicularis Sprague.

Gum, Hog → Clusia flava L.

Gum, Icacia chiche → Sterculia tomentosa Guill. and Perr.

Gum, India → Anogeissus latifolia Wall.

Gum, Jingan → Odina Wodier Roxb.

Gum, Karaya → Sterculia urens Roxb.

Gum, Kateira → Sterculia urens Roxb.

Gum, Ketira → Cochlospermum Gossypium L.

Gum, Katira Gabina → Astragalus heratensis Bunge.

Gum, Kauri → Agathis australis Steud.

Gum, Mogador → Acacia gummifera Willd.

Gum, Morocco → Acacia gummifera Willd.

Gum, Mudar → Calotropis procera Ait.

Gum, Mulu Kilavary → Commiphora Berryi Engl.

Gum, Mumuye →Combretum Hartemannianum Schweinf. C. leonense Engl. and Diels. and C. sokodense Engl.

Gum, Nongo → Albizzia Brownii Walp.

Gum of Labdanum → Cistus ladaniferus L.

Gum of Opopanax → Opopanax Chironium (L.) Koch.

Gum, Oriental Sweet → Liquidambar orientale Mill.

Gum, Persian → Amygdalus leiocarpus Boiss.

Gum, Red → Liquidambar styraciflua L.

Gum, Resina de Cuapinole → Hymenaea Courharil L.

Gum, Sagapan → Ferula Szowitziana DC.

Gum, Sagapenum → Ferula persica Willd.

Gum, Salai-gugul → Boswellia serrata Roxb.

Gum, Sarcocolla → Penaea Sarcocolla L.

Gum, Scribbly →Eucalyptus haemastoma Smith.

Gum, Sembagona → Bauhinia variegata L.

Gum, Sennarr → Acacia Seyal Delile.

Gum, Senegal → Acacia Senegal (L.) Willd.

Gum, Senn → Bauhinia variegata L.

Gum, Somali → Acacia glaucophylla Steud.

Gum, Sour → Nyssa multiflora Wang.

Gum, Spotted → Eucalyptus goniocalyx F. v. Muell.

Gum, Spruce → Picea Mariana (Mill.) B.S.P.

Gum, Suakim → Acacia Seyal Delile.

Gum, Tacamahaca→Bursera gummifera L. and Populus balsamifera L.

Gum, Talca → Acacia Seyal Delile.

Gum, Talha → Acacia Seyal Delile.

Gum, Talki → Acacia Seyal Delile.

Gum, Taramniya → Combretum hypotilinum Diels.

Gum Tragacanth → Astragalus gummifera Lab.

Gum Tragacanth (substitute) → Astragalus prolixus Sieb.

Gum, Tupelo → Nyssa aquatica L.

Gum Turpentine → Pinus palustris Mill.

Gum, Velampisini → Feronia limonia Swingle.

Gum, Water → Nyssa biflora Walt.

Gum, Wattle → Acacia pycnantha Benth.

Gum, White →Eucalyptus gomphocephala DC.

Gum, Yakka → Xanthorrhoea australis R. Br.

Gummi Resina Ammoniacum → Ferula tingitana L.

Gummi Peucedani → Peucedanum officinale L.

Gummi Resina Euphorbium → Euphorbia resinifera Berg.

Gumíhan Gum → Artocarpus elastica Reinw.

Gums. Sources of → Abies, Acacia, Albizzia, Amygdalus, Anacardium, Anogeissus, Argania, Asphodelus, Astragalus, Atractylis, Bauhinia, Bombax, Buchanania, Burkea, Bursera, Butea, Caesalpinia, Callitris, Ceratonia, Chikrassia, Cistus, Clusia, Cochlospermum, Combretum, Commiphora, Cordia, Diospyros, Encelia, Enterolobium, Epidendrum, Eremurus, Feronia, Ferula, Ficus, Flindersia, Garcinia, Geodorum, Hymenaea, Macaranga, Mezoneuron, Moringa, Odina, Opopanax, Owenia, Panax, Penaea, Peucedanum, Phragmites, Picea, Pinus, Piptadenia, Pithecolobium, Populus, Prosopis, Prunus, Pseudocedrela, Pterocarpus, Puya, Spondias, Sterculia, Styrax, Woodfordia, Xanthorrhoea, Zuelania. See also under: kinos, frankincense, resins.

Gundelia Tournefortii L. (Compositae). - Perennial herb. Mediterranean region, Syria, Iran. Young leaves are eaten as a vegetable.

Gunnera chilensis Lam. (syn. G. scabra Ruiz and Pav.). (Halorrhagaceae). - Tall perennial plant. Chile. Root, Palo Pangue, is used in Chile for tanning. Contains 9⁰/o tannin. Young peeled leaf stalks are eaten as vegetable in some parts of Chile.

Gunnera perpensa L. (Halorrhagaceae). - Perennial herb. S. Africa. Decoction of roots is used by the farmers of S. Africa for dyspepsia; decoction in brandy is employed for kidney troubles.

Gunpowder. Charcoal source of → Adhatoda, Aesculus, Alnus, Aristotlea, Butea, Calotropis, Corylus, Evonymus, Guazuma, Melicytus, Paulownia, Populus, Salix, Tilia.

Gurjum → Dipterocarpus alatus Roxb.

Gurjum Balsam → Dipterocarpus alatus Roxb.

Gutta Djelutung → Alstonia eximea Miq.

Gutta Malaboeai → Alstonia grandifolia Miq.

Gutta Percha → Diplorhynchus mossambicensis Benth., Ficus graphalocarpa Steud., Palaquium spp., Payena Leerii Kurz and P. obscura Burck.

Gutta Percha (substitute) → Bassia Mottleyana Clarke.

Gutta Percha, Angso → Palaquium leiocarpon Boerl.

Gutta Percha, Philippine → Palaquium Aherninanum Merr.

Gutta Percha Tree, Malay → Palaquium Gutta Burck.

Gutta Pontinak → Dyera costulata (Miq.) Hook. f.

Gutta Pootih → Palaquium Maingayi King and Gamb.

Guttiferaceae → Allanblackia, Calophyllum, Caraipa, Clusia, Carcinia, Haploclathra, Haronga, Kielmeyera, Mammea, Mesua, Montrouziera, Ochrocarpus, Pentadesma, Platonia, Psorospermum, Rheedia, Symphonia, Vismia.

Gymnacranthera canarica Warb. (syn. Myristica canarica Bedd.) (Myristicaceae). - Tree. S. India. Seeds are source of an oil; used for manuf. candles.

Gymnartocarpus Woodii Merr. (Moraceae). - Tree. Centr. and S. Luzon, Mindanao (Philippine Islands). Seeds are eaten boiled or roasted in the Philippines.

Gymnema sylvestre R. Br. (Asclepiadaceae). - Vine. E. India, Trop. Africa, Australia. Leaves destroy timely the power of the tongue to distinguish between bitter and sweet.

Gymnocladus chinensis Baill. (Leguminosaceae). - Tree. Centr. China. Pods that have been swollen in water and crushed called. Fei-tsaotou, are used in laundry work; esp. for washing fine fabrics. Seeds are ground, mixed with cloves, putchuck, musk, camphor and sandalwood to manuf. a perfumed soap called P'ingshe fei-tsao.

Gymnocladus doica (L.) Koch. (syn. G. canadensis Lam.). Kentucky Coffee Tree. (Leguminosaceae). - Tree. Eastern United States. Wood strong, coarse-grained, not hard, heavy, durable when in contact with soil, light brown; used for railroad ties, cabinet work, fence posts. Roasted seeds have been used as a substitute for coffee.

Gymnogongrus flabelliformis Harv. (Phyllophoraceae). - Red Alga. Sea Weed. Pac. Ocean and adj. waters. Consumed as food in Japan.

Gymnogongrus pinnulata Harv. (Phyllophoraceae). - Red Alga. On rocks in the seas around Japan and China. Plants are made into jelly and used as food in Japan. They are also made into a paste which is used for hair shampoo.

Gymnopus microcarpus (Berk. and Broome) v. Overeem. (Agaricaceae). - Basidiomycete. Fungus. Trop. Asia. Fruitbodies are consumed as food among the natives of Malaysia and Ceylon.

Gymnosporia montana Benth. (syn. Celastrus senegalensis Lam.). (Celastraceae). - Shrub.

Trop. Africa. Decoction of stem is used in Senegal for colic, dysentery and diarrhea among children.

Gymnosporia senegalensis Loes. (Celastrinaceae). - Woody plant. Trop. Africa, India, Medit. Region. Bark of roots used in Senegambia and Senegal for chronic dysentery. Ashes from the plant are used in Sudan as substitute for salt.

Gynandropsis pentaphylla DC. (Capparidaceae). - Woody plant. Tropics. Seeds are anthelmintic. Source of an ess. oil, has properties of garlic or mustard oil, is considered antiscorbutic. Leaves are eaten as potherb in Nigeria and Brit. India. Also used for flavoring sauces.

Gynandropsis speciosa DC. (Capparidaceae). - Annual herb. Trop. America. Leaves are in some countries used as a vegetable.

Gynerium argenteum Nees. → Cortaderia argentea Stapf.

Gynerium sagittatum (Aubl.) Beauv. (Aira gigantea Steud.) Wild Cane. (Graminaceae). - Perennial grass. West Indies, S. Mexico to S. America. Stems are employed for light construction work and lattices.

Gynura cernua Benth (Compositae). - Annual herb. Trop. Africa. Used medicinally in Madagascar and as a vegetable in Layos.

Gyrinops Walla Gaertn. (Thymelaeaceae). - Tree. Ceylon. Wood soft, light, whitish; used for inlay work and fancy cabinet work

Gyrocarpus americanus Jacq. (Hernandiaceae). - Tree. Tropics. Wood soft, white, light; used for toys and boxes.

Gyrophora cylindrica (L.) Ach. (syn. Umbilicaria cylindrica Dub.). (Gyrophoraceae). - Lichen. Temp. and subarctic zone. Source of a dye, used in Iceland to dye woolens a greenbrown color.

Gyrophora deusta (L.) Ach. (syn. Umbilicaria flocculosa Hoffm.). Rock Tripe. (Gynophoraceae). - Lichen. Temp. and subartic. zone. On rocks. Used in Sweden to dye woolens a violet color. Was also used in paints. Linnaeus mentions that this paint was called Tousch, which was much used in Sweden.

Gyrophora esculenta Miyoshi (syn. Umbilicaria esculenta (Hoffm.). (Gyrophoraceae). - Lichen. Japan. Is consumed as a delicacy in Japan. known as Iwa-take or Rock Mushroom.

Gyrophora vellea (L.) Ach. (Gyrophoraceae). - Lichen. Temp. zone. On rocks. Was used in Sweden to dye woolens a violet color.

Gyroporus castaneus (Bull.) Quél. (syn. Boletus castaneus Bull.) (Polyporaceae). - Basidiomycete. Fungus. Temp. and subtrop. zone of Northern Hemisphere. Fruitbodies are consumed as food. Sold in markets of Europe and Asia. Also G. purpurinus (Snell.) Sing. and G. cyanescens (Bull.) Quél.

H

Habbak Daseino Bdellium → Commiphora Hildebrandtii Engl.

Habbak Dundas Bdellium → Commiphora Hildebrandtii Engl.

Habbak Dunkal Bdellium → Commiphora Hildebrandtii Engl.

Habbak Harr Bdellium → Commiphora Hildebrandtii Engl.

Habbak Ilka Adaxai Bdellium → Commiphora Hildebrandtii Engl.

Habbak Tubuk Bdellium → Commiphora Hildebrandtii Engl.

Habenaria Rumphii Lindl. (Orchidaceae). - Herbaceous perennial. Amboina. Root tubers are made into a preserve; used in Indonesia.

Habenaria sparciflora S. Wats. (Orchidaceae). - Perennial herb. Southwestern N. America and adj. Mexico. Plants were consumed as food by the Indians during times of want.

Habzelia obtusifolia A. DC. → Xylopia obtusifolia A. Rich.

Hackberry → Celtis occidentalis L.

Hackberry, European → Celtis australis L.

Hackberry, Western → Celtis reticulata Torr.

Haddi Tree → Commiphora erythraea Engl. var. glabrescens Engl.

Haemanthus coccineus L. (Amaryllidaceae). - Bulbous perennial. S. Africa. Bulbs kept in vinegar are used by the farmers of Cape Peninsula as expectorant and diuretic in asthma and dropsy.

Haemarthria fasciculata Kunth. (syn. Rottboellia fasciculata Desf.) (Graminaceae). - Perennial grass. Warm countries of Old and New World. Excellent pasture grass for cattle.

Haemotomma ventosum (L.) Mass. (syn. Lecanora ventosa Ach.). Black Lecanora, Bloody Spotted Lecanora. (Lecanoraceae). - Lichen. Temp. zone. On rocks. Used in Sweden to give a red-brown color to woolens.

Haematoxylin → Haematoxylon campechianum L.

Haematoxylon Brasiletto Karst., Brazilwood, Nicaraguawood. (Leguminosaceae). - Shrub or tree. Trop. America. Wood is bright orange, becoming red upon exposure, very hard, compact, medium to fine-grained. Source of Brasilin, a dye. Wood is exported from the W. Coast of Mexico.

Haematoxylon campechianum L., Logwood, Lignum Campechianum, Palo Campechio. (Leguminosaceae). - Tree. Trop. America. Wood brownish-red to blood-red, becoming brown violet upon exposure, hard, fairly easy to split; used for furniture. Heartwood is source of a dye, called Haematoxylin, becoming red upon exposure. Used in wool industry and in microscopial staining technique; source of an ink. Wood is medicinally a mild astringent used in diarrhea, also for dysentery. Wood is used for furniture and fancy articles. Flowers are source of a good honey.

Hagar Ad Tree → under Commiphora Hildebrandtii Engl.

Hagenia abyssinica J. F. Gmel. (syn. Brayera anthelmintica Kunth.) (Rosaceae). - Tree. Abyssinia and adj. territory. Cultivated in Abyssinia. Pistilate flowers, Flores Coso, are used medicinally as an anthelmintic and taenifuge. Contain cosotoxin.

Haiari, White → Lonchocarpus densiflorus Benth.

Hairy Angelica → Angelica villosa (Walt.) B. S. P.

Hairy Origanum → Origanum hirtum Link.

Hairy Sheeth Edible Bamboo → Phyllostachys edulis Riv.

Hairy Vetch → Vicia villosa Roth.

Hakea leucoptera R. Br., Pine Bush, Needle Bush. (Proteaceae). - Tree. Australia. Wood soft, close-grained, takes a good polish; used for veneers, tobacco pipes, cigaret holders. Roots are used by the aborigenes of S. Australia as a means to obtain water.

Hakea rubricaulis Colla → Stenocarpus salignus R. Br.

Hal-dummala Resin → Vateria acuminata Hayne.

Halfa → Stipa tenacissima L.

Halorrhagaceae → Gunnera, Hippuris.

Haloxylon persicum Bunge. (Chenopodiaceae). - Tree. Central Asia. Wood is used for general carpentry. Plants are grown for sand-binders. Also H. aphyllum (Minkw.) M. M. Iljin.

Haloxylon salicornicum Bunge. (Chenopodiaceae). - Herbaceous plant. Saline places. Iran, Afghanistan. Crop is used as forage. Young branches are sometimes eaten by natives in time of famine.

Haloxylon Schweinfurthii Aschers. (Chenopodiaceae). - Shrub. Arabian desert and adj. territory. Shrub is source of a Manna, caused by an unknown insect. The gum is consumed by the Beduins.

Halymenia formosa Harv. (Grateloupaceae). - Red Algae. Pacif. Ocean. Plants are used as food in the Philippine Islands.

Hamabo Hibiscus → Hibiscus Hamabo Sieb.

Hamamelidaceae → Altingia, Bucklandia, Distylium, Hamamelis, Liquidambar.

Hamamelis virginiana L. Witch-Hazel (Hamamelidaceae). - Shrub or small tree. Eastern N. America to Florida and Texas. Bark and leaves are used medicinally. Much is derived from Virginia, Tennessee and S. Carolina. Contains an ess. oil; a bitter principle called hamamelitannin, being astringent and hemostatic; used for hemorrhoids, varicose veins and bruises. Hamamelis Cortex or Witchhazel Bark contains an ess. oil, gallotannic acid, a glucosidal tannin and gallic acid, used for the preparation of Hamamelis Water or Extract of Witchhazel, being hemostatic. Decoction from bark was used by the Indians for healing wounds.

Hancornia speciosa Gomez, Mangabeira (Braz.). (Apocynaceae). - Small tree. Brazil, esp. Campos Cerrados, Matto Grosso, Minas Geraes, Bahia, Pernambuco. Source of Mangabeira Rubber; exported from Bahia and Pernambuco. Fruits edible, ovoid, size of a plum, yellow with red; much esteemed as a marmalade in some parts of Brazil.

Hanjigoad Gum → Balanites orbicularis Sprague.

Hannoa Klaineana Pierre and Engl. (Simarubaceae). - Tree. Trop. W. Africa. Wood light, soft, fibrous, white with satiny reflections; used for planks, canoes, small houses, ceilings. Decoction of bark is used for colic.

Hansonia apiculata (Rees.) Lindner. (Saccharomycetaceae). - Yeast. Microorganism. A yeast of which it has been reported that it adds during fermentation of grapejuice a particular taste and bouquet to the wine.

Hapalopilus nidulans (Fr.) Karst. (syn. Polyporus nidulans Fr.) (Polyporaceae). - Basidiomycete Fungus. Temp. zone. Fruitbodies are made into bottle corks in Russia and Scandinavia.

Haploclathra paniculata Benth. (Guttiferaceae). - Tree. N. Brazil. Wood beautiful red, called Mura Piranga, used for manuf. different instruments.

Haplophyton cimicidum A. DC. Hierba de la Cucaracha (Apocynaceae). - Woody plant. S. Arizona, Mexico, Centr. America, Cuba. Used in Mexico as insecticide. Decoction of plant with cornmeal kills cockroaches, and human parasites.

Hard Pear → Strychnos Henningsii Gilg.

Harobol Myrrh → Commiphora Myrrha (Nees) Engl.

Haronga madagascariensis Chois. (Guttiferaceae). - Shrub or tree. E. Africa, Madagascar. Fruits edible. Seeds are used in cookery in French Guinea. Inner bark is source of a yellow dye. Wood is very beautiful, easily worked; employed for different purposes. Stem is source of a resin, used by the natives for fastening their arrowpoints. Leaves are used in Madagascar for dysentery.

Harpephyllum caffrum Bernh. Kaffir Plum. (Anacardiaceae). - Tree. S. Africa. Fruits are red, 2.5 cm. across, acid, juicy pulp; is made in S. Africa into a jelly.

Harpullia arborea (Blanco) Radlk. (Sapindaceae). - Tree. N. Luzon to S. of Sulu Archip. (Philipp. Islds.) Pounded bark is used as a substitute for soap.

Harpullia pendula Planch. Tulip-Wood, Mogun-Mogun. (Sapindaceae). - Tree. New South Wales, Queensland. Wood firm, close-grained, beautifully marked, black to yellow. Used for cabinet-work.

Harrisonia perforata (Blanco) Merr. (Simarubaceae). - Tree. Philipp. Islds. Decoction from root-bark is a very efficacious remedy for dysentery and diarrhea.

Hart's Tongue → Scolopendrium vulgare Sm.

Hausa Potato → Coleus rotundifolius Chev. and Perrot.

Haw, Black → Crataegus Douglasii Lindl. and Viburnum prunifolium L.

Haw, Chinese → Crataegus pentagyna Waldst. and Kit.

Hawaiian Arrowroot → Tacca hawaiiensis Limpr. f.

Hawaian Sandalwood → Santalum Freycinetianum Gaud.

Hawthorn → Crataegus Oxyacantha L.

Hawthorn Maple → Acer crataegifolium Sieb. and Zucc.

Hazel Alder → Alnus rugosa (Du Roi) Spreng.

Hazel, Chile → Gevuina Avellana Molina.

Hazel, European → Corylus Avellana L.

Hazel, Turkish → Corylus Colurna L.

Hazelnut, Beaked → Coryllus rostrata Ait.

Hazelnut Oil → Corylus Avellana L.

Hazelraw → Lobaria pulmonaria (L.) Hoffm.

Hazelrottle → Lobaria pulmonaria (L.) Hoffm.

Headache Tree → Premnia integrifolia L.

Heartleaf Hornbeam → Carpinus cordata Blume.

Heath Tea Tree → Leptospermum ericoides Rich.

Heavy Mahogany → Entandrophragma Candollei Harms.

Heavy Sapele → Entandrophragma Candollei Harms.

Heckeria umbellata (L.) Kunth. → Piper umbellatum L.

Hedeoma pulegioides (L.) Pers., American Pennyroyal. (Labiaceae). - Annual herb. N. America. Dried leaves and flowering tops are used medicinally, being stimulant, carminative; used for colds. Contains an ess. oil, ketone pulegone and a bitter principle. Decoction of dried leaves is used as home remedy for flatulent colic, bowel complaints and stomach disorders.

Hedera Helix L., English Ivy. (Araliaceae). - Woody vine. Europe, Temp. Asia. Cultivated. Leaves boiled with soda are said to be suitable for washing clothes. Young twigs are source of a yellow and brown dye. The hardwood can be used as a substitute for Buxus in engraving.

Hedge Garlic → Sisymbrium Alliaria Scop.

Hedge Maple → Acer campestre L.

Hedycarya angustifolia Cunningh. (syn.H. Cunninghamiana Tul., H. australasica A. DC.) Native Mulberry, Smooth Holly. (Monimiaceae). - Tall shrub or tree. Australia, esp. Victoria, New South Wales, Queensland. Wood close-grained, very light, tough. Used for cabinet-work. Employed by aborigines of Australia to obtain fire by friction.

Hedychium coronarium Koenig., Garland-Flowers. (Zingiberaceae). - Perennial herb. India, Malaya. Has been recommended as a source of paper-pulp.

Hedychium longicornutum Griff. (Zingiberaceae). - Perennial herb. E. Indies, Malaysia. Roots are used for ear-ache.

Hedychium spicatum Ham. (Zingiberaceae). - Perennial herb. E. India. Rhizome is used for perfumery in some parts of Trop. Asia.

Hedyosmum brasiliense Mart. (syn. H. Bonplandianum Mart.). (Chloranthaceae). - Shrub. Brazil. Plant is aromatic, febrifuge. Decoction of leaves is used as sudorific, stomachic, diuretic, tonic and aphrodisiac.

Hedysarum coronarium L. Spanish Espercet. (Leguminosaceae). - Perennial herb. Mediterranean region. Occasionally cultivated as food for cattle.

Hedysarum Mackenzii Rich. Liquorice Root. (Leguminosaceae). - Herbaceous perennial. Central Canada to Alaska. Long sweet roots have the appearance of Liquorice, and are eaten in the spring by he Indians.

Heeria reticulata (Bak. f.) Engl. (Anacardiaceae). - Tree. Trop. Africa. Used among the natives of Trop. Africa as galactogogue and aphrodisiac.

Heimia salicifolia (H. B. K.) Link. (Lythraceae). - Shrub. Trop. America, Centr. America, Jamaica to S. America. Decoction of plant produces a pleasant and mild intoxication during which time everything appears to be yellow.

Heinsia pulchella Schum. (Rubiaceae). - Tree. Trop. W. Africa. Yellow to red fruits, called Bush Apples, are edible, pleasant flavor. Wood flexible, hard; used for spring traps and tool handles. Pulverized leaves have the scent of Anthoxanthum odoratum, a grass; made into pomade and used by native women.

Helenium tenuifolium Nutt. Bitter Sneezeweed. (Compositae). - Annual herb. N. America. Plant is used by some Indian tribes as errhine; promoting discharge of mucous from the nose.

Helianthemum canadense (L.) Michx., Frostweed. (Cistaceae). - Low shrub. N. America. Dried herb is used medicinally as tonic, astringent, alterative. Contains an ess. oil and a bitter principle.

Helianthus annuus L., Sunflower. (Compositae). - Annual herbaceous. Western N. America. Much grown for its oil in Russia, Bulgaria, India, China, Argentina and United States. Fruits source of Sunflower Seed Oil, semi-drying, pleasant flavor; used in foods, salads, butter substitutes, burning oil, Russian varnishes, Dutch enamel paint. Oil cake valuable as fodder for cattle. Sp. Gr. 0.992-0.926; Sap. Val. 189-194; Iod. No. 120-136; Unsap. 0.7-1.20%. Stems and leaves used as fodder for cattle, also as silage. Flowers are source of a honey. Varieties are: Black Seed, Black Giant, Giant Russian, Mammoth, White Russian. A yellow dye was made from the flowers; used by the Indians. Pith from stems is used in microscopical technique for making slides.

Helianthus doronicoides Lam. Oblonglead Sunflower. (Compositae). - Perennial herb. Western United States. Tubers were consumed as food by the Indians of the Central States of the U.S.A.

Helianthus giganteus L. (Compositae). - Perennial herb. N. America. Fruits were made into flour, mixed with corn meal, and used for making bread by the Choctaw Indians.

Helianthus Maximiliani Schrad. (Compositae). - Prairies United States. Perennial herb. Tubers were consumed by the Sioux Indians and other tribes.

Helianthus tuberosus L. Jerusalem Artichoke, Topinambour. (Compositae). - Perennial herb. N. America. Cultivated in Old and New World. Tubers contain much inulin. Consumed boiled as food. Varieties are Jerusalem White, Veitch's Improved Long White, Sutton's New White. In some countries the rhizoms are used a fodder for live-stock.

Helichrysum arenarium Moench. (Compositae). - Perennial herb. Europe, Caucasia. Flowerheads were formerly used for skin diseases, warts and as diuretic. Also H. Stoechas DC.

Helichrysum cochinchinensis Spreng. (Compositae). - Herbaceous. Cochin-China. Young twigs are used in Tonkin with rice flower in a dish called bârh khúc.

Helichrysum serpyllifolium Lessing. (Compositae). - Herbaceous plant. S. Africa, esp. Cape Peninsula, Transvaal to Rhodesia. Leaves are used as tea in some parts of S. Africa; called Hottentot tea.

Heliconia Bihai L. (svn. H. caribaea Lam.) (Musaceae). - Perennial herb. Trop. America. Young shoots are sometimes consumed as a vegetable in the West Indies. Herb has been recommended for manuf. paper material. Leaves are occasionally used for covering roofs.

Heliconia brasiliensis Hook. (Musaceae). - Perennial herb. Brazil. The roots are considered in some parts of Brazil as antigonorrheic; the seeds antidiarrhetic.

Heliconia Schiedeana Klotsch. (Musaceae). - Perennial herb. S. Mexico. Large leaves used by natives for thatching and as wrapping material.

Helicteres Isora L., East Indian Screw Tree. (Stereuliaceae). - Trop. Asia to Australia. Bark is source of a strong fibre, made into cordage. Pods are for sale in bazaars as a drug; used for intestinal complaints, colic diarrhea, chronic dysentery, flatulence, also to improve appetite.

Heliotropium anchusaefolium Poir. (Boraginaceae). - Herbaceous plant. Argentina. An infusion of the leaves is used by country people of Argentina as sudorific.

Heliotropium elongatum Willd. (Boraginaceae). - Shrub. Brazil. Herb is astringent, resolvent, antiasthmatic and diuretic.

Hellebore → Helleborus niger L.

Hellebore, American White → Veratrum viride Ait.

Hellebore, White → Veratrum album L.

Helleborus niger L. Hellebore, Christmas Rose. (Ranunculaceae). - Perennial herb. Europe. Dried rhizome and roots are used medicinally as a drastic hydragogue, cathartic, as emmenagogue, local anestetic, and heart stimulant. Contains helleborine and helleborain, Also H. viridis L.

Helminol → Digenea simplex (Wulf.) C. Ag.

Helminthocladaceae (Red Algae) → Nemalion.

Helminthostachys zeylanica Hook. f. (Ophioglossaceae). - Perennial herb, fern-ally. From Himalaya to Queensland and New Caledonia. Rhizome used for malaria; exported to China; tonic, also eaten with betel.

Helotiaceae (Fungi) → Chlorosplenium.

Helvella crispa (Scop.) Fr. (Helvellaceae). - Ascomycete. Fungus. Temp. zone. Fruitbodies are consumed as food in some parts of Asia. Also H. gigas Krombh., H. lacunosa Afzel and H. infulva Schaff.

Helvellaceae (Fungi) → Helvella, Morchella, Verpa.

Helwingia rusciflora Willd. (Cornacea). - Shrub. Japan, China. Young leaves when boiled are consumed as a vegetable in Japan.

Hemarthria compressa R. B. → Rottboellia compressa L. f.

Hemerocallis fulva L. Tawny Daylily. (Liliaceae). - Perennial herb. Europe, Asia. Dried flowers are used in China and Japan as a condiment. Much is exported to Shantung. Cultivated also as an ornamental.

Hemerocallis minor L. (syn. H. graminea Andr. Grassleaf Daylily). (Liliaceae). - Perennial herb. China, Japan. Flowers are consumed in some parts of China and Japan.

Hemidesmus indicus R. Br. (syn. H. Wallichii Miq.). (Asclepiadaceae). - Woody plant. E. India, Ceylon. Source of Indian Sarsaparilla; used in native medicin, being demulcent, diuretic, alterative.

Hemitelia lucida (Fée) Maxon. (Cyatheaceae). - Fern. Endemic in N.E. Oaxaca (Mex.). Spores used among Chinantecs as styptic for small wounds. Also H. apiculata Hook. and H. mexicana Liebm.

Hemlock, Chinese → Tsuga chinensis Pritz.

Hemlock, Eastern → Tsuga canadensis (L.) Carr.

Hemlock, Ground → Taxus canadensis Willd.

Hemlock, Oil → Tsuga canadensis (L.) Carr.

Hemlock Pitch → Tsuga canadensis (L.) Carr.

Hemlock, Poison → Conium maculatum L.

Hemlock, Western → Tsuga heterophylla (Raf.) Sarg.

Hemlock, Yunnan → Tsuga yunnanensis (Franch) Mast.

Hemorrhage Plant → Aspilia latifolia Oliv. and Hiern.

Hemp → Cannabis sativa L.

Hemp, African Bowstring → Sansevieria senegambica Baker.

Hemp, Ambari → Hibiscus cannabinus L.

Hemp, Bahama → Agave sisalana Perrine.

Hemp. Bombay → Agave cantala Roxb.

Hemp, Ceylon Bowstring → Sansevieria zeylanica Willd.

Hemp, Cuban → Furcraea cubensis Vent.

Hemp Dogbane → Apocynum cannabinum L.

Hemp Eupatorium → Eupatorium cannabinum L.

Hemp, Indian → Apocynum cannabinum L.

Hemp, Manila → Musa textilis Née.

Hemp, Mauritius → Furcraea gigantea Vent.

Hemp Nettle → Caleopsis ochroleuca Lam.

Hemp, Queensland → Sida rhombifolia L.

Hemp Seed Oil → Cannabis sativa L.

Hemp, Sisal → Agave sisalana Perrine.

Hemp, Sunn → Crotalaria juncea L.

Hen - and - Chickens → Sempervivum tectorum L.

Henbane → Hyoscyamus niger L.

Henderson Shooting Star. → Dodecatheon Hendersonii A. Gray.

Henequen → Agave fourcroydes Lemaire.

Henequen Agave → Agave fourcroydes Lemaire.

Henna → Lawsonia alba Lam.

Henna Shrub → Lawsonia alba Lam.

Henry Chinquapin → Castania Henryi (Skan) Rhed. and Wils.

Hepatica triloba DC. → Anemone Hepatica L.

Heracleum lanatum Michx. (Umbelliferaceae). - Perennial herb. N. America. Roots were eaten cooked, tasting like rutabaga. Young flowers and stems were eaten by many Indian tribes.

Heracleum persicum Desf. (Umbelliferaceae). - Perennial herb. Iran. Seeds are used as a condiment in pickles.

Heracleum Sphondylium L., Common Cow Parsnip. (Umbelliferaceae). - Biennial to perennial herb. Europe, Asia, N. America. From the boiled leaves and fruits an alcoholic beverage, called Bartsch, is prepared, used by some of the poorer classes in Slavic countries. Employed in France in liqueurs. Was formerly used medicinally.

Hercules Club Prickly Ash → Zanthoxylum Clava Herculis L.

Hericium coralloides (L.) Pers. (Hydnaceae). - Basidiomycete (Fungus). Temp. and subtrop. zones. The fruitbodies are consumed in some Asiatic countries.

Heritiera littoralis Dryand., Lookingglass Tree. (Sterculiaceae). - Tree. Trop. of Old World. Wood durable, strong; used for boat-building, masts, spokes of wheels, plows, knees of boats.

Heritiera minor Lam., Sundri Tree. (Sterculiaceae). - Tree. India, Mauritius. Wood source of charcoal; used in Calcutta. Seeds eaten as famine-food.

Herminiera elaphroxylon Guill. and Per. (Leguminosaceae). - Tree. Trop. Africa, Madagascar. Wood very light, strong, tough, durable; logs used by natives of Nile region for crossing rivers; made into rafts in the Congo, used in Mossamedes and Benguela for beds, stools, landing boats.

Hernandia nukuhivensis F. Br. (Hernandiaceae). - Tree. Pacif. Islds. Trunk is used for canoes by the Fatuhivans.

Hernandia peltata Meisen, Jack-in-the-Box. (Hernandiaceae). - Tree. Guam, Tahiti. Juice from the leaves is depilatory; destroys hair without pain.

Hernandiaceae → Cyrocarpus, Hernandia.

Herperoyucca Whipplei (Torr.) Baker → Yucca Whipplei Torr.

Herpestis Monniera H.B.K. (Scrophulariaceae). - Annual creeping herb. Tropics. Leaves, stems and roots used in Hindu medicin as nerve tonic, supposed to be a direct cardiac tonic; used by the Chinese as a tonic on the intestines.

Herrania albiflora Goud. → Theobroma albiflora Goud.

Hesperoloe funifera (Koch.) Trel Tree. (syn. Yucca funifera Koch.). (Liliaceae). - Low plant. Mexico. Cultivated in Nuevo León for its fibre, obtained from leaves; of excellent quality. Exported as Ixtli or Tampico Fibre.

Hesperis matronalis L., Damask. (Cruciferaceae). - Biennial or perennial herb. Europe, W.

and Centr. Asia. Cultivated. Pressed seeds are source of an oil, Honesty Oil, Huile de Julienne, Rotreps Oel.

Heterochordaria abietina (Rupr.) J. Ag. (Mesoglouiaceae). - Brown Alga. Pacif. and Atlant. Ocean and adj. waters. Consumed as food in Japan.

Heterodendrum oleifolium Desf. (Sapindaceae). - Tree. Australia. Red fruits are eaten when fresh bv the aborigenes in S. Australia.

Heteromorpha arborescens Cham. and Schlecht. (Umbelliferaceae). - Small shrub or herbaceous. Trop. Africa, esp. Cape Peninsula, Natal, Mozambique, Centr. Africa, Abyssinia. Tincture or infusion of inner bark of root used by Kafirs for colic.

Heteropogon hirtus Pers. (syn. H. contortus (L.) Beauv., Andropogon contortus L.). Spearhead, Tanglegrass. (Graminaceae). - Perennial grass. Tropics of Old and New World. Cultivated as fodder for live stock. Sourc of a fibre; used in India for coarse mats and for thatching. Used by the Old Hawaiians for making grass houses.

Heteropteris umbellata Juss. (syn. H. glabra Hook. and Arn.) (Malpighiaceae). - Liane. Argentina. Stems are source of a fibre; used for coarse ropes in some parts of Argentina.

Heterospatha elata Scheff. (Palmaceae). - Tall slender palm. Philipp. Islds., Malaya. In the Philipp. Islds. the nuts are used as masticatory with betel. Buds are used as vegetable. Petioles are source of splints; used for manuf. baskets. Leaflets are made in Bohol (Philipp. Islds.) into sun hats, called Salokots.

Heterothalamus brunoides Less. (Syn. Marshallia brunoides Less.) (Compositae). - Shrub. Argentina to Brazil. Plant is used as aromatic, excitant and febrifuge. It is also source of a yellow dye. The branches are used for making brooms.

Hevea Benthamiana Muell. Arg. (Euphorbiaceae). - Tree. Amazon Region (Brazil). Source of rubber. Occasionally cultivated. Likewise: H. discolor Muell. Arg., H. rigidifolia (Benth.) Muell. Arg., H. Foxii Hub., H. pauciflora Muell Arg., H. cuneata Hub.

Hevea brasiliensis Muell. Arg. Para Rubber-Tree. (Euphorbiaceae). - Tree. Amazone Region, Brazil, Peru, Bolivia. Cultivated in Indonesia, Malaya, Liberia, Brazil and other parts of the tropics. Source of Hevea or Para Rubber, the most important natural rubber. Many improved varieties or clones. Some of the commercial products are called Borracha Fina, Para Fina, Negro Heads, Sernambi and numerous other grades. Latex coagulates with the aid of acetic acid, formic acid and alum. Seeds are source of Para Rubber Seed Oil, recommended for manuf. soap. Sp. Gr. 0.924-0.930; Sap. Val. 186-195; Iod. No. 135-143; Unsap. 0.5-0.8%.

Hexagona Mori Poll. → Polyporus alveolaris (D. C.) Bond and Sing.

Hexalobus senegalensis A. DC. (syn. Uvaria monopetala Guill. and Perr.). (Annonaceae). - Woody plant. Trop. Africa. Roots, leaves and twigs are used in Senegal as expectorant and for diarrhea.

Hibiscus Abelmoschus L. (syn. Abelmoschus moschatus Moench.) Muskmallow. (Malvaceae). - Annual or perennial herb. Source of a fibre, said to be used for sails. Seeds, called Ambrette Seeds produce an oil; used in perfumery.

Hibiscus Bancroftianus Macf. (Malvaceae). - Herbaceous plant. West Indies. Plant is used medicinally like Althaea officinalis L.

Hibiscus cannabinus L., Kenaf Hibiscus, Bimplipatam Tree. (Malvaceae). - Annual to perennial. India and Africa. Cultivated in the tropics. Source of Ambari, Gambo, Decan Hemp or Bimplipatam Jute. Fibre competes with jute, though is somewhat coarser; used for the same purpose. In Africa seeds are source of an oil, used for burning.

Hibiscus elatus Swartz. Blue Mahue. (Malvaceae). - Tree. West Ind. Wood used for gunstocks, carriage-poles, fishing-rods, ship's knees, cabinet work.

Hibiscus esculentus L., Okra, Gombo. (Malvaceae). - Annual. Trop. Asia. Cultivated in warm countries of Old and New World. Young pods are eaten as a vegetable, in soups, catsup; also dried or canned for winter use. Several varieties among which are Dwarf Green, Louisiana Green Velvet, Louisiana White Velvet.

Hibiscus Hamabo Sieb. and Zucc. (Malvaceae). - Woody plant. S. Japan. Bark is source of a strong fibre; used in Japan for rope.

Hibiscus macrophyllus Roxb. (Malvaceae). - Tree. E. India to Java. Cultivated. Wood light, soft, fine-grained, light grayish brown; used in Java for construction of houses. Bark is source of a fibre; used in Siam, Burma and Malaya for rough ropes, mats and strings.

Hibiscus panduraeformis Burm. f. (Malvaceae). - Perennial herb. Trop. Australia. Bark is source of a fibre; used by aborigenes of Queensland for making twine, bags etc.

Hibiscus quinquelobus G. Don. (syn. Hibiscus sterculifolius Steud.) (Malvaceae). - Tree. Trop. Africa. Bark is source of a fibre, comparable with Jute, called West African Jute; used for cordage. Wood is hard, resembling walnut; used for general carpentry.

Hibiscus rosa-sinensis L. (Malvaceae). - Tall shrub or small tree. China, Japan. Much cultivated as ornamntal. Chinese women use juice of petals for blackening eyebrowes. In Malaya decoction of roots is used for sore eyes. Bark employed by Chinese and Annamese as emmenagogue.

Hibiscus Sabdariffa L., Roselle, Jamaica Sorrel (Malvaceae). - Herb or shrub. East Indies. Cultivated in tropics of Old and New World. Red, fleshy calyx is used for jellies and sauces. Stem source of a fibre; used for cordage. Varieties are: Victor, Rico and Archer.

Hibiscus tiliaceus L. (syn. H. guiensis DC., H. similis Blume, Paritium tiliaceum (L.) Britt. Gatapa, Majagua. (Malvaceae). - Tropics. Source of an important fibre; used for cordage, tow, sails, fishing nets, coarse bags. Exported to Europe.

Hibiscus unidas Lindl. (Malvaceae). - Woody plant. Brazil. The bark is source of a fibre; used as cordage and in textiles.

Hickory, Big-bud → Carya tomentosa (Lam.) Nutt.

Hickory, Big Shellbark → Carya laciniosa (Michx. f.) Loud.

Hickory, Chinese → Carya cathayensis Sarg.

Hickory, Pignut → Carya glabra (Mill.) Sweet.

Hickory Pine → Pinus pungens Michx.

Hickory, Shagbark → Carya ovata (Mill.) Koch.

Hicoria glabra (Mill.) Britt. → Carya glabra (Mill.) Sweet.

Hicoria laciniosa (Michx. f.) Sarg. → Carya laciniosa (Michx. f.) Loud.

Hicoria minima Britt. → Carya cordiformis (Wangh.) Koch.

Hicoria Pecan Britt. → Carya Pecan (Marsh.) Engl. and Graebn.

Hierba de la Cucaracha → Haplophyton cimicidum A. DC.

Hierochloë odorata (L.) Beauv. (syn. Holcus odoratus L. Torresia odorata Hitchc.). Sweet grass. (Graminaceae). - Perennial grass. Europe, Asia, N. America. Dried foliage is burned as an incense during ceremonies by the Kiowa Indians (New Mex.), also used as a perfume for clothing. Also employed for manuf. Sweet-Grass Baskets.

Hieronima alchorneoides Allem. Urucurana. (Euphorbiaceae). - Tree. Brazil. Roots an energic depurative; used in patent medicines in Brazil, Argentina and United States.

Hieronima caribaea Muell. Arg. Tapana. (Euphorbiaceae). - Tree. Trinidad, S. America. Wood durable, strong, easy to work, chocolate-brown; used for building purposes, furniture, wagons, wheelwrights.

Hieronima cubana Muell. Arg. (Euphorbiaceae). - Tree. Cuba. Wood saffron-colored, of excellent quality; used for cabinet work.

High Bush Blueberry → Vaccinium corymbosum L.

High Mallow → Malva sylvestris L.

Highland Coffee → Coffea stenophylla G. Don.

Higicho → Carica chrysophylla Heilb.

Higuera del Monte → Carica quercifolia (St. Hil.) Solms Laub.

Hilo Grass → Paspalum conjugatum Berg.

Himalaya Birch → Betula utilis D. Don.

Himalayan Bamboo → Arundinaria falcata Nees.

Himalayan Cedar → Cedrus Libani Barrel.

Himalayan May Apple → Podocarpus Emodi Wall.

Himalayan Rhubarb → Rheum Emodi Wall.

Hinahina → Melicytus ramiflorus Forster.

Hindu Cowpea → Vigna Catjang Walp.

Hindu Datura → Datura Metel L.

Hippobroma longiflora Don. → Isotoma longiflora Presl.

Hippocastanaceae → Aesculus.

Hippocratea acapulcensis H. B. K. (Hippocrateaceae). - Creeping shrub. Mexico to Northern S. America. Tincture of seeds is used to kill parasites on the human body.

Hippocratea senegalensis Lam. → Salacia senegalensis DC.

Hippocrateaceae → Hippocratea, Salacia.

Hips, Rose → Rosa pomifera Herrm.

Hippomane Mancinella L. (Euphorbiaceae). - Tree. Trop. America. Latex is used by the Caribs for poison arrows. Wood used in West Indies for cabinet work and interior finish.

Hippophae rhamnoides L., Sea Buckthorn. (Eleaegnaceae). - Tree. Europa, Asia. Berries are sometimes made into jelly. They are eaten with milk and cheese among the Siberians and Tartars. In some parts of France berries are made into a sauce and eaten with fish and meat. Wood is fine-grained, hard, fairly heavy; sometimes used for turnery.

Hippuris vulgaris L. (syn. H. tetraphylla L.). Marsetail. (Halorrhagaceae). - Perennial waterplant. Temp. N. Hemisphere. Leaves when young are consumed by the Eskimos of Alaska.

Hiptage benghalensis Kurz. (syn. H. madablota Gaertn.) (Malpighiaceae). - Vine. India to Burma and Malaya. Leaves used in India in cutaneous diseases; plant has insecticidal properties.

Hirneola Auricula Judae (L.) Berk. → Auricularia Auricula Judae (L.) Schroet.

Hirneola polytricha Mont. (syn. Auricularia polytricha (Mont.) Sacc.) (Auriculariaceae). - Basidiomycete. Fungus. Asia and Australia. Fruitbodies are consumed as food in several countries. Has commercial value in China.

Hoarhound → Marrubium vulgare L.

Hoary Gromwell → Lithospermum canescens Lehm.

Hodai Tree→ Commiphora Hildebrandtii Engl.

Hoffmanseggia densiflora Benth. (Leguminosaceae). - Herbaceous plant. S.W. of United States and adj. Mexico. Tubers were roasted and eaten as food by Indians of Texas, New Mexico and Arizona. Also H. falcaria Cav.

Hog' Fennel → Peucedanum officinale L.

Hog Gummi → Clusia fulva L.

Hog Peanut → Amphicarpa monoica (L.) Ell.

Hog Plum → Prunus umbellata Ell., Spondias lutea L., and Symphonia globulifera L.

Hoheria populnea Cunningh. (Malvaceae). - Small tree. New Zealand. Wood used for cabinet work, inlaying. Bark very strong, tough, used as substitute for rope and cord; also used as a demulcent drink by the Maoris.

Holarrhena africana A. DC. (Apocynaceae). - Shrub or small tree. Trop. Africa. Flos from seeds is used to stuff pillows, in S. Leone.

Holarrhena antidysenterica Wall., Tellicherry Bark. (Apocynaceae). - Small tree. Trop. Asia. Used in Arabic and Indian medicine. Seeds are considered a tonic, aphrodisiac; used for dysentery, diarrhea, fevers, flatulence and billious affections. Sold in bazaars. Source of Conessi or Kurchi Bark.

Holarrhena microteranthera Schum., Piripiri. (Apocynaceae). - Woody plant. Trop. Africa. Source of a rubber.

Holarrhena Wulfsbergii Stapf., Male Rubber Tree. (Apocynaceae). - Shrub or tree. Trop. W. Africa. Latex is used to adulterate rubber. Wood is used by the natives for making combs, immages, handles of matchetes. Bark kept in palm wine, is used by natives for dysentery.

Holcus halepensis L. → Sorghum halepense (L.) Pers.

Holcus lanatus L., Velvet Grass. (Graminaceae). - Perennial grass. Europe, temp. Asia; escaped in N. America. Grown for pasture and hay; recommended as horse feed.

Holleyhock → Althaea rosea Cav.

Holly → Ilex Aquifolium L.

Holly, California → Photinia arbutifolia Lindl.

Holly, California →Rhamnus ilicifolius Kellogg.

Holly, Dahoon → Ilex Cassine L.

Holly, European → Ilex Aquifolium L.

Holly, Longstalk → Ilex pedunculata Miq.

Holly Oak → Quercus Ilex L.

Holly, Smooth → Hedycarya angustifolia Cunningh.

Holocalyx Balansae Mich. (Leguminosaceae). - Tree. Paraguay to Brazil. Wood yellowish, compact, durable, heavy, not elastic; used for manuf. of de luxe articles.

Holy Basil → Ocimum sanctum L.

Holy Clover → Onobrychus viciaefolia Scop.

Holy Thistle → Silybum Marianum Gaertn.

Homalanthus populifolius R. Grah. (Euphorbiaceae). - Shrub or small tree. Malaya to Australia. Bark and leaves are source of a black dye; used by the natives for staining rattan goods, basketry of Corypha.

Homalium caryophyllaceum Benth. (syn. H. frutescens King, H. foetidum Ridl.). (Flacourtiaceae). - Small tree. Malaya. Wood hard, heavy, compact, light brown; used for building purposes.

Homalium foetidum Benth. (Flacourtiaceae). - Tree. E. Mal. Archipel, Celebes, Moluccas. Wood heavy, brown, becomes black in sea-water; used for building boats and tall buildings.

Homalium tomentosum Benth. (Flacourtiaceae). - Tree. Burma, Malaya. Source of Moulmein Lancewood.

Homalium vitiense Benth., Oueri. (Flacourtiaceae). - Tree. New Caledonia. Wood yellowish white, durable, fine-grained, easily worked; used for interior work.

Homalomena philippensis Engl. (Araceae). - Herbaceous. Philipp. Islds. Rhizomes are said to be antirheumatic when used as embrocation.

Home remedies. Plants used in →Achillea, Agrimonia, Agropyron, Althaea, Ammodaucus, Anagallis, Arnica, Calendula, Chrysactinia, Chrysanthemum, Comarum, Crescentia, Crocus, Cunila, Cyclamen, Ephedra, Eupatorium, Filipendula, Garrya, Genista, Grindelia, Haemanthus, Hedeoma, Heliotropium, Juglans, Lantana, Lepidium, Malva, Melissa, Nasturtium, Nigella, Ocotea, Papaver, Pimpinella, Piptadenia, Polygonatum, Populus, Primula, Protea, Pyrola, Ribes, Rosmarinus, Salix, Salvia, Sambucus, Sassafras, Sempervivum, Stachys, Tachia, Tanacetum, Tillandsia, Trevoa.

Homonia riparia Lour. (Euphorbiaceae). - Shrub or tree. Himalaya to Philipp. Islds., Mal. Archip. Leaves and fruits used by natives of N. Perak for skin diseases. In Java juice used for blackening teeth and making them firm when loose.

Homoranthus Oil → Homoranthus virgatus Cunningh.

Homoranthus virgatus Cunningh. (syn. Homoranthus flavescens Cunningh.). (Myrtaceae). - Shrub. New South Wales. Source of Homoranthus Oil, an ess. oil, contains 80% ocimene; has been recommended for perfumery.

Honckenya ficifolia Willd. (syn. Clappertonia ficifolia Hook.). (Tiliaceae). - Woody plant. Trop. Africa. Source of a valuable fibre resembles jute.

Honduras Lancewood → Lonchocarpus hondurensis Benth.

Honduras Mahogany → Swietenia macrophylla King.

Honduras Sarsaparilla → Smilax Regelii Killip and Morton.

Honesty Oil → Hesperis matronalis L.

Honey. Source of →Acacia, Calluna, Cassia, Citrus, Cliftonia, Cyrilla, Eucryphia, Eysenhardtia, Fagopyrum, Gossypium, Haematoxylon, Helianthus, Ilex, Lavandula, Medicago, Melilotus, Nyssa, Onobrychus, Oxydendrum, Phacelia, Piper, Pluchea, Prosopis, Protea, Rhizophora, Robinia, Sabal, Salvia, Serenoa, Tilia, Trifolium.

Honey Dew → Cucumis Melo L.

Honey Locust → Gleditsia triacanthos L.

Honeysuckle, Maori → Knightia excelsa R. Br.

Hongay Ooil → Pongamia pinnata (L.) Merr.

Hooboobali → Loxopterygium Sagotii Hook.

Hooker's Balsamroot → Balsamorhiza Hookeri Nutt.

Hoop Pine → Araucaria Cunninghamii Sweet.

Hop → Humulus Lupulus L.

Hop Clover → Medicago lupulina L.

Hop Hornbeam → Ostrya carpinifolia Scop.

Hop Tree → Ptelea trifoliata L.

Hopea Balageran Korth. (Diptoracarpaceae). - Tree. Mal. Archip. Source of a resin, called Njating Mahabong.

Hopea dealbata Hance. (Dipteracorpaceae). - Tree. Trop. Asia, esp. Indo-China, Siam, Lankawi Islands, Perak. Wood durable; used for heavy conotruction work. Likewise H. ferrea de Lans., H. globosa Brandis, H. multiflora Brandis.

Hopea fagifolia Miq. (Dipterocarpaceae). - Tree. Bangka. Source of a resin, called Dammar Kedemut.

Hopea Maranti Miq. (Dipterocarpaceae). - Tree. Mal. Archip. Source of a resin, called Dammar batu.

Hopea Mengarawan Miq. (Dipterocarpaceae). - Tree. Sumatra. Wood hard, close-grained, easily worked and to split, durable; used for general carpentry, furniture, boats, barrels. Source of a Dammar, used in varnishes, batik industry and illumination. Also H. dryobalanoides Miq., H. globosa Brandis, H. Griffithii Kurz, H. intermedia King, H. micrantha Hook and H. myrtifolia Miq.

Hopea nutans Ridl. (syn. H. Lowii Foxw.) Nodding Merawan, Giam. (Dipterocarpaceae). - Tree. Malay. Penin. Wood heavy, very hard, very durable; resists salt-water; used for buildings and construction.

Hopea odorata Roxb. (Dipterocarpaceae). - Tree. Burma, Brit. Malaya, Perak, Trengganu. Chief timber south of Tenasserim. Used for boats, cart-wheels, oil and sugar-cane presses, building, deck planks, bridges. Source of Rock Dammar, used by Burmese as a varnish over paint; for caulking boats; for sores and wounds. Sold in bazaars.

Hopea gangal Korth. (Dipterocarpaceae). - Tree. Borneo. Source of a resin, called Nyating Matapoesa when derived from old trees. It is called Njating Matpleppek when derived from young plants.

Horchata de Chufas (drink) → Cyperus esculentus L.

Hordeum distichon L., Two-rowed Barley. (Graminaceae). - Annual grass. Cultivated in temperate zones of Old and New World. Source of malt, used in manuf. of beer. Grains when roasted are a substitute for coffee. Varieties are: Chevalier, Hannchen, White Smyrna. Here belongs H. zeocriton L.

Hordeum hexastichon L. Six-rowed Barley. (Graminaceae). - Annual grass. Cultivated. Grown for the same purpose as Two-rowed barley. Varieties are: Manchuria, Oderbrucker, Nepal, Horsford, Mariout, Utah Winter.

Hordeum trifurcatum Ser, Egyptian Barley. (Graminaceae). - Annual herb. Asia Minor, N. Africa. Cultivated in N. Africa. Source of flour; used for bread, cakes etc.

Hordeum vulgare L., Common Barley. (Graminaceae). - Annual grass. Cultivated. Probably from S.W. Asia, N. Africa, S.E. Asia. Cultivated. Source of a bread flour. Unleavned barley cakes are esteemed in some parts of rural Scotland, and some other northern countries of Europe. Used as breakfast cereal. Pearled barley used in soups. Important food for children. Source of malt in manuf. of beer. Commercial malt extracts. Straw is used as fodder for livestock and as stable bedding.

Hornbeam, American → Carpinus caroliniana Walt.

Hornbeam, European → Carpinus Betulus L.

Hornbeam, Heartleaf → Carpinus cordata Blume.

Horned Poppy → Glaucium flavum Crantz.

Horse Balm → Collinsonia canadensis L.

Horse Bean → Vicia Faba L.

Horse Chestnut → Aesculus Hippocastanum L.

Horse Chestnut, Japanese → Aesculus turbinata Blume.

Horse Mint → Monarda fistulosa L.

Horse Nettle → Solanum carolinense L.

Horse Radish → Cochlearia Armoracea L.

Horse-Radish-Tree → Moringa oleifera Juss.

Horse-Radish Tree, Arabian → Moringa aptera (Forsk.) Gaertn.

Horse Seed Bush, Clamy → Dodonaea viscosa Jacq.

Horse Sugar → Symplocos tinctoria (L.) L. Her.

Horse Tail → Equisetum spp.

Horsehair Lichen → Alectoria jubata (L.) Ach.

Horsfieldia sylvestris Warb. (Myristicaceae). - Tree. E. Malaya. Fruit-coat pleasant taste, used for flavoring. Seeds source of oil; used for manuf. rough candles.

Hortulana Plum → Prunus hortulana Bailey.

Hoslundia opposita Vahl. (Labiacea). - Herbaceous plant. Trop. Africa. Source of Kamynye Oil; has an odor resembling that of vanillin; has been recommended in perfumery as a fixative.

Hottentot Fig → Mesenbryanthemum edule L.

Hottentot Tea → Helichrysum serpyllifolium Lessing.

Hottentot Tabacco → Tarchonanthes camphoratus L.

Hounds Tongue → Cynoglossum officinale L.

Houseleek → Sempervivum tectorum L.

Houttuynia californica Benth. and Hook. → Anemonopsis californica Hook. and Arn.

Hovenia dulcis Thunb., Japanese Raisin Tree. (Rhamnaceae). - Small tree. Himalayan region, China, Japan. Fruit dry, edible, reddish brown, sweet, sub-acid; much esteemed in some parts of Japan and China.

Howardia barbata Kl. → Aristolochia barbata Jacq.

Howardia brasiliensis Kl. → Aristolochia cymbifera Mart. and Zucc.

Howardia cordigera Kl. → Aristolochia cordigera Willd.

Hoya australis R. Br. (Asclepiadaceae). - Woody vine. Polynesia. Australia. White flowers are used as adornment, during ceremonial dances in Samoa.

Hoya coronaria Blume. (Asclepiadaceae). Epiphyte. Malaya. Used in Java with Capsicum leaves to stimulate digestion.

Hoya latifolia G. Don. (Asclepiadaceae). - Epiphyte. Mal. Penin. to Java. Latex used for ascites, diuretic.

Hoya Rumphii Blume. (Asclepiadaceae). - Woody climber. Mal. Archip. Decoction of plants used in Java for dangerous stings from poisonus fishes.

Huan Pine Wood Oil → Dacrydium Franklinii Hook.

Huanuca Coca → Erythroxylon Coca Lam.

Hubertia Ambavilla Bory → Senecio Ambavilla Pers.

Huckleberry → Gaylussacia baccata (Wang.) Koch.

Huckleberry, Black → Gaylussacia baccata (Wang.) Koch.

Hugonia Mystax L. (Linaceae). - Woody plant. E. India. Roots used in India for intestinal worms and snake-bites.

Huile de Juliene → Hesperis matronalis L.

Huile de Marmotte → Armeniaca brigantina Pers.

Huili huiste → Karwinskia Calderonii Standl.

Hule → Castilla costaricana Liebm.

Humboldt Coyotillo → Karwinskia Humboltiana (Roem. and Schult.) Zucc.

Humiria balsamifera St. Hil., Oloroso, Couranira. (Humiriaceae). - Tree. Trop. America. Wood used for general carpentry and spokes of wheels.

Humiria floribunda Mart., Bastard Bullet. (Humiriaceae). - Tree. Brazil, Guiana. Wood reddish brown, easily polished; used for furniture, house-framing, wheel-spokes. Bark when attacked by a certain fungus emits an agreeable perfume; used by the Indians for scenting hairoil.

Humiriaceae → Aubrya, Humiria.

Humulus Lupulus L., Hop. (Urticaceae). - Perennial vine. Europe, Asia, N. America. Cultivated in England, Germany, United States, S. America, Australia. Bitter substance found in glandular hairs of the strobilus is used in flavoring of beer. Stems are source of a fibre. Young bleached tops are sometimes eaten as vegetable in some parts of Belgium. Oil of Hops is used for certain perfumes of the chypre or fougère-type. Dried strobilus used medicinally as a bitter tonic, sedative, hypnotic. Contains lupuline, humuline, xanthohomul, cerotic acid and resins.

Hungarian Clover → Trifolium pannonicum Jacq.

Hungarian Oat → Avena orientalis Schreb.

Hungarian Turpentine → Pinus Cembra L.

Hungarian Vetch → Vicia pannonica Crantz.

Hungarian Water → Rosmarinus officinalis L.

Hura crepitans L., Sandbox Tree. (Euphorbiaceae). - Tree. Trop. America. Fruits are occasionally used as small boxes containing sand to dry ink-writing.

Hura polyandra Baill. (Euphorbiaceae). - Large tree. Mexico. Wood used for telegraph poles. Milky juice used for poisoning fish. Seeds violent purgative; used for poisoning coyotes.

Hyacinth Bean → Dolichos Lablab L.

Hyacinth, Common → Hyacinthus orientalis L.

Hyacinth, Tassel Grape → Muscari comosum Mill.

Hyacinth, Water → Eichhornia crassipes Solms.

Hyacinth, Wild → Camassia esculenta Lindl.

Hyacinthus orientalis L., Common Hyacinth. (Liliaceae). - Perennial herbaceous bulbous plant. Mediterranean region. Much cultivated in the Netherlands and S. France. Flowers are source of an ess. oil; used in perfumery; Has to compete with a synthetic product phenyl acetaldehyde. About 6000 kg. flowers produce 1 kg. absolute. Contains benzyl alcohol, cinnamic alcohol, phenyl ethyl alcohol, cinnamic aldehyde, benzaldehyde, hydrochinon dimethyl ether, eugenol and benzyl benzoate.

Hydnaceae (Fungi) → Hericium, Hydnum.

Hydnocarpus anthelmintica Pierre, Common Chaulmoogra Tree. (Flacourtiaceae). - Tree. Siam, Cochin China. Seeds source of a nondrying oil, used for leprosy. Sp. Gr. 0.943-0.952; Sap. Val. 191.4-226.5; Iod. No. 84.5-86.4; Ac. Val. 0.2.

Hydnocarpus heterophyllum Kurz. → Taraktogenos Kurzii Kind.

Hydnocarpus Wightiana Blume. (Flacourtiaceae). - Tree. Pennin. India. Seeds source of a non-drying oil; much used for leprosy in India. Sp. Gr. 0.9330; Sap. Val. 197.2; Iod. No. 103; Melt. Pt. 20.5° C.

Hydnora esculenta Jum. and Poir. (Hydnoraceae). - Herbaceous parasite on Acacia. Madagascar. Fruits edible, pleasant flavor, size of a pomegranate, with thick skin; much esteemed by the natives.

Hydnoraceae → Hydnora.

Hydnotria carnea (Corda) Zobel. (Eutuberaceae). - Ascomycete Fungus. Temp. zone. The fruitbodies, resembling truffles, are consumed and sold in Prague markets. They have the name of červena tartofle (red truffles).

Hydnum fragile Petch. (Hydnaceae). - Basidiomycete. Fungus. Trop. Asia. Fruitbodies are much esteemed as food in Java and other parts of Malaysia.

Hydrangea arborescens L., Smooth Hydrangea, Mountain Hydrangea, Seven-Bark. (Saxifraceae). - Tree-like shrub. Eastern United States. Dried rhizome and root is used medicinally as diaphoretic, cathartic and diuretic; contains hydrangin, an alkaloid. Is collected in autumn, much is derived from Virginia, N. Carolina, Michigan and Illinois.

Hydrangea paniculata Sieb. (Saxifragaceae). - Shrub or small tree. Japan, Sachalin. Wood very hard, fine-grained; used in Japan for walking sticks, umbrella handles, tobacco pipes, wooden nails. Bark is used in Japan for manuf. paper.

Hydrangea Thunbergii Sieb. (Saxifragaceae). - Shrub. Japan. Young leaves when gathered are steamed, rolled between hands, dried and used for a sweet beverage, called Amacha (Sweet Tea) in Japan. It is also mixed with soy.

Hydrastis canadensis L., Golden Seal. (Ranunculaceae). - Perennial herb. Eastern N. America. Cultivated in N. Carolina, Oregon and Washington. Dried rhizomes when collected in autumn used in medicine, hemostatic, antiperiodic; uterine hemorage, hemorrhoids, astringent, for inflamation of mucous membrane, as a bitter tonic. Contains hydrastine, berberine and canadine, being alkaloids. Plant was source of a yellow dye; used by the Indians.

Hydroclathrus cancellatus Bary. (Scytosiphonaceae). - Brown Alga. Pac. Ocean. Plants are used in China as fertilizer. They are collected late in spring and early in summer.

Hydrocotyle asiatica L. (syn. Centella asiatica Urb.) (Umbelliferaceae). - Perennial herb. Trop. Asia. Leaves are eaten raw or cooked with rice among the Japanese.

Hydrocharitaceae → Enalus, Ottelea, Stratiotes, Vallisneria.

Hydrodictyon reticulatum (L.) Lagerh. (Hydrodictyonaceae). -.Green Alga. Fresh Water. Eaten by the Hawaiians with fresh water shrimps and salt.

Hydrodictyonaceae (Green Algae) → Hydrodictyon.

Hydrolea zeylanica vahl. (Hydrophyllaceae). - Herbaceous plant. Trop. Asia. Cultivated. Young leaves are eaten as lalab with rice in Java.

Hydrophyllaceae → Eriodictyon, Hydrolea, Phacelia.

Hydrophyllum appendiculatum Michx., Waterleaf. (Hydrophyllaceae). - Herbaceous perennial. Eastern and southern United States. In some states the young shoots are eaten in salads, formerly used by the early settlers.

Hydrophyllum virginicum L., Virginia Waterleaf. (Hydrophyllaceae). - Herbaceous perennial. Eastern N. America. Shoots when young and tender were consumed by the Indians.

Hygrocybe punicea (Fr.) Karst. (syn. Hygrophorus puniceus Fr. (Agaricaceae). - Basidiomycete Fungus. Temp. zone. The red fruitbodies are consumed, having a very pleasant taste. Sold in markets of Sweden.

Hygrophorus marzuoles (Fr.) Bres. (Agaricaceae). - Basidiomycete, Fungus. Mountainous regions of Europe. Fruitbodies are consumed as food. Sold in markets. Also H. chrysodon (Batsch.) Fr.,; H. hypothejus Fr., H. Queletii Bres., H. lucorum Kalchbr. and others.

Hygrophorus pratensis (Pers.) Fr. (syn. Camarophyllus pratensis (Pers.) Karst.) (Agaricaceae). - Basidiomycete. Fungus. Temp. zone. Fruitbodies are consumed as food. Sold in markets of Europe.

Hygrophorus puniceus Fr. → Hygrocybe punicea (Fr.) Karst.

Hylocereus undatus (Haw.) Britt. and Rose. (syn. Cereus undatus Haw.). Pitahaya, Pitahaya Oregona (Mex.) (Cactaceae). - Xerophytic shrub. Mexico. Cultivated in Mexico and Centr. America. Fruits edible, oblong, 10 to 12 cm. diam. excellent quality; much sold in markets.

Hymenaea Courbaril L., Courbaril. (Leguminosaceae). - Tree. S. Mexico, Centr. America, West Indies, S. America. Wood heavy, tough, very hard, appearance of Mahogany; used for general construction work, ship building, furniture, sugar mills. Used by Indians in Brazil for canoes. Pulp around seeds is edible, mixed with water to make „atole", also an alcoholic beverage. A gum „Resina de Cuapinole", Colombia Copal, Brazil Copal, „Goma Anime de Me-

xico", „Amber de Cuapinole" used in Mexico as incense in churches, also used for varnish, patent. leather.

Hymenaea stilbocarpa Hayne. (Leguminosaceae). - Tree. Brazil. Source of a Copal, similar to that from H. Courbaril L.

Hymenaea verrucosa Gaertn. (Leguminosaceae). - Tree. Madagascar. Source of Madagascar Copal; used in varnishes, giving a darker color than Zanzibar Copal.

Hymenachne aurita Backer. (syn. Panicum javanicum Nees.). (Graminaceae). - Perennial grass. Mal. Archipel. Used as fodder for cattle.

Hymenachne interrupta Buese. (Graminaceae). - Perennial grass. Mal. Archipel. Much used as fodder for live-stock.

Hymenocardia acida Tul. (Euphorbiaceae). - Shrub. Trop. Africa. Bark is source of a red dye; used by the Bongos (Afr.) as antiseptic.

Hymenocallis tubiflora Salisb. (syn. Pancratium guianense Ker-Gawl, H. guianensis Herb.). (Amaryllidaceae). - Bulbous perennial. Guiana to Brazil. Bulbs are used in some parts of Brazil as astringent, expectorant, diuretic.

Hymenodictyon excelsum Wall. (Rubiaceae). - Tree. Trop. Asia. Wood soft, brownish-gray; used for tea boxes, scabboards, grain measures, toys, Burmese school slates. Bark is used as febrifuge and antiperiodic.

Hymenogastraceae (Fungi) → Rhizopogon.

Hymenopappus filifolius Hook. (Compositae). Perennial herb. Western N. America. Roots were used for chewing by Indians of New Mexico.

Hymenophyllaceae → Hymenophyllum.

Hymenophyllum plumosum Kaulf. (Hymenophyllaceae). - Perennial fern. S. America. An infusion of the plant is used as sudorific and diuretic in some parts of S. America.

Hypholoma appendiculatum (Bull.) Karst. Agaricaceae). - Basidiomycete. Fungus. Temp. zone. Fruitbodies are edible and often of good quality. They can be dried for winter use, retaining their good flavor.

Hypholoma Candolleanum (Fr.) Quél. → Psathyrella Candolleana (F.) A. H. Smith.

Hyoscyamus muticus L. (Solanaceae). - Herbaceous plant. Egyptian desert, Arabia, Iran to India. Medicinally used since ancient times. At present valuable as a sedative, for nerves, in cerebral and spinal troubles; for insomnia. Smoked by Bedouins to cure tooth-ache. Sometimes used instead of opium, as a narcotic.

Hyoscyamus niger L. Black Henbane. (Solanaceae). - Annual to perennial herb. Europe, Asia, N. Africa. Cultivated in Russia, Germany and England. Dried leaves, collected during flowering time are used medicinally; being anodyne, narcotic, hypnotic. Contains hyocyamine and hyoscine, being alkaloids.

Hyparrhenia Ruprechtii Fourn. (syn. Andropogon Ruprechtii Hack.) (Graminaceae). - Perennial grass. Trop. Africa. Used for thatching. Also H. rufa Stapf, H. soluta Stapf. and H. subplumosa Stapf.

Hypestes verticillaris R. Br. (Acanthaceae). - Herbaceous plant. Trop. Africa. Roots are used in Meshra El Zeraf (Sudan) for dyeing mats.

Hypericaceae → Hypericum.

Hypericum connatum Lam. (Hypericaceae). - Herbaceous plant. Argentina, Paraguay, Uruguay. The top of the flowergroup is used in S. America as a tonic and vulnerary.

Hypericum laxiusculum St. Hil. (Hypericaceae). - Shrub. Brazil. Herb is astringent, aromatic, excitant antispasmodic.

Hypericum teretiusculum St. Hil. (Hypericaceae). - Shrub. Brazil. Herb is used in some parts of Brazil as excitant, aromatic and emmenagogue.

Hyphaene Argun Mart. (syn. Medemia Argun Benth. and Hook.) (Palmaceae). - Palm. Nubia and adj. territory. Fruits are consumed by the native tribes of N. Africa.

Hyphaene coriacea Gaertn., Ginger Bread Palm. (Palmaceae). - Tall palm. Coastal region E. Africa. Pulp of fruits is eaten by the natives. Occasionally a source of palm-wine. Leaves are used for making mats, baskets and for covering huts.

Hyphaene crinita Gaertn. South African Doumpalm. (Palmaceae). - Tall palm. S. Africa. Young leaves are source of a fibre, made into strings, ropes and mats. Older leaves are made into hats, baskets and similar material. Kernels are source of a vegetable ivory. Sap of the stem is made into an alcoholic beverage by the natives.

Hyphaene thebaica Mart. Egyptian Doum Palm. (Palmaceae). - Tall, branching palm. Trop. Africa, esp. Nileland, Nubia, Eritrea, Kordofan, Somaliland, Abyssinia, Brit. E. Africa. Leaves are made into rope, mats, huts, tents, paper. Kernels are sometimes the source of cheap buttons; they are also made into small perfume boxes. Nuts when pounded are used for dressing wounds. When unripe are eaten raw. Rind of nut is used as food, also made into molasses, cakes and sweetmeats.

Hyphaene ventricosa Kirk. (Palmaceae). - Tall palm. Trop. Africa. The seeds are used for the manuf. of buttons and similar articles.

Hypnea cenomyce J. Ag. (Sphaerococcaceae). - Red Alga. Sea Weed. Pac. Ocean, Indian Ocean and adj. waters. Consumed as food among the natives of Indonesia.

Hypnea cervicornis J. Ag. (Sphaerococcaceae). - Red Alga. Sea Weed. Pac. Ocean. Is consumed as food in China. Plants are collected in summer.

Hypnea nidifica J. Ag. (Sphaerococcaceae). - Red Alga. Sea Weed. Pac. Ocean. Consumed as food by the Hawaiians, who call it Limu Huna. It is much prized when eaten with squid or octopus, forming a jelly. Also eaten with chicken, thickening the broth. Sometimes eaten with grated coconut and coconut milk. Also used in soups.

Hypnea musciformis (Wulf.) J. Ag. (Sphaerococcaceae). - Red Alga. Sea Weed. Pac. Ocean. Plants are consumed as food in China. They are cooked with meat and vegetables, forming a palatable jelly. Plants are collected in spring.

Hypochoeris maculata L., Cat's Ear. (Compositae). - Perennial herb. Europe, Siberia. Young leaves are sometimes consumed as salads. Recommended during times of emergency.

Hypochoeris scorzonerae F. v. Muell. (syn. Achyrophorus scorzonerae DC.). (Compositae). - Herbaceous plant. Chine. Plant is used by the inhabitants as diuretic. Is sold in the markets.

Hypocreaceae (Fungi) → Claviceps, Cordyceps.

Hyporhodius clypeatus (L.) Schroet. (syn. Entoloma clypeatus (L.) Quél. (Agaricacae) - Basidyomycete. Fungus. The fruitbodies are consumed as food in Europe and Asia. Are sold in markets of Sweden and Indochina.

Hyptis albida H. B. K. (Labiaceae). - Shrub. Sonora, Mexico. Leaves are used for flavoring food.

Hyptis Emoryi Torr. (syn. H. lanata Torr.) Emory Bushmint. (Labiaceae). - Shrub. Southwest of the United States. Seeds known as Chia (see also Salvia), used as food by Indians in Mexico.

Hyptis fasciculata Benth. (Labiaceae). - Herbaceous plant. S. America. Herb is used as carminative, anticatarrthic and sudorific.

Hyptis mutabilis (Rich.) Briq. (Labiaceae). - Herbaceous plant. Centr. America. Leaves are made into tea, as a remedy for pains in the stomach.

Hyptis pectinata Poit. (Labiaceae). - Annual herb. Pantropic. Source of a resin; sometimes used by natives of Africa as an incense.

Hyptis spicigera Lam. (Labiaceae). - Annual herb. Trop. Africa. Seeds when made into a pulp are used in stews and gravies by the Zandes (Sudan).

Hyssop → Hyssopus officinalis L.

Hyssop, Giant → Agastache anethiodora (Nutt.) Britt.

Hyssop, Wild → Verbena hastata L.

Hyssopus officinalis L. Hyssop. (Labiaceae). - Half-shrub. Medit. Region, Balkan, Asia Minor, Iran. Occasionally cultivated. Used for flavoring liqueurs; carminative. Contains b-pinene, x-pinense, camphene, l-pinocamphone, sesquiterpenes.

I

Ibota Privet → Ligustrum Ibota Sieb.

Icacia chiche Gum → Sterculia tometosa Guill. and Perr.

Icacina senegalensis Juss. (Icacinaceae). - Woody plant. Trop. Africa. Seeds and tubers are consumed by the natives of Shari, Afr. during shortage of food.

Icacina trichantha Oliv. (Icacinaceae). - Trop. Africa. Roots are used in soups and in other foods by the natives of W. Africa.

Icacinaceae → Apodytes, Icacina, Urandra.

Icaco Plum → Chrysobalanus Icaco L.

Ice Plant → Mesembryanthemum crystalinum L.

Iceland Moss → Cetraria islandica (L.) Ach.

Ichtyomethia piscipula (L.) Hitchc. → Piscidia Erythrina L.

Icica viridiflora Lam. (syn. Protium guianense March.) (Burseraceae). - Tree. Trop. America. Source of Elemi of Guiana; used for lacquers; also used as incense in churches.

Ignatia amara L. → Strychnos Ignatii Berg.

Ignatius Bean → Strychnos Ignatii Berg.

Ijebu Mahogany → Entandrophragma angolense (Welw.) A. DC. and E. Candoliei Harms.

Ilex amara Perodi. (Aquifoliaceae). - Tree. Argentina, Paraguay. Leaves are occasionally used as a tea or mixed with those of Yerba Maté.

Ilex Aquifolium L. European Holly. (Aquifoliaceae). - Tree. Centr. and S. Europe, W. Asia, Transcaucasia, N. Iran, N. Africa, Centr. China. Wood white, fairly strong, heavy, very difficult to split, easy to polish; used for inlay work, wood cuts, handles of kettles, veneer, walking sticks, mechanical instruments. Twigs are used in Europe for Christmas decoration.

Ilex capensis Harv. and Sond., Water Tree. (Aquifoliaceae). - Tree. S. Africa. Wood whitish, beautifully mottled, easily worked; used for spokes.

Ilex Cassine L., Dahoon Holly, Cassena. (Aquifoliaceae). - Tree. East and South east of N. America. Leaves are used as tea, occasionally they are sold in local markets along the South Atlantic Coast. A decoction was much used by the Creek Indians.

Ilex conocarpa Reiss. (Aquifoliaceae). - Shrub. Brazil. Leaves are used in Brazil as a tonic, diuretic and stomachic; used instead of Yerba Maté.

Ilex glabra (L.) Gray, Inkberry, Gallberry. (Aquifoliaceae). - Shrub. S. W. of the United States. Flowers are source of a honey.

Ilex integra Thunb. (Aquifoliaceae). - Tree. Japan. Cultivated. Pounded bark is used in Japan for bird-lime.

Ilex macropoda Miq. (Aquifoliaceae). - Tree. Japan, Korea, Wood light, tough, hard, close-grained, light blueish white; used in Japan for mosaic work, utensils, turnery, matches.

Ilex malabarica Bedd. (syn. I. Wightiana Dalz.) (Aquifoliaceae). - Tall tree. S. India, Ceylon. Wood close-grained, soft, light, creamy-white; used for tea boxes and packing material, esp. in Ceylon.

Ilex medica Reiss. (Aquifoliaceae). - Tree. Brazil. Infusion of leaves is used in Brazil as stomachic and diuretic.

Ilex Opaca Ait., American Holly. (Aquifoliaceae). - Tree. Eastern N. America to Florida and Texas. Wood not strong, light, tough, close-grained; used for cabinet making, turnery, interior finish of houses. Twigs used in United States for Christmas decoration.

Ilex paraguensis St. Hill., Paraguay Tea, Yerba Maté. (Aquifoliaceae). - Tree. S. America, esp. N. Argentina, Basin of the Parana, Uruguay, Paraguay to S. Matto Grosse (Braz.) Of commercial importance. Cultivated. Source of Maté, leaves generally used as tea in many countries of S. America. Derived from cultivated and wild plants. Product is usually graded into Maté grosso, Maté entrefino and Maté fino.

Ilex pedunculasa Miq. Longstalk Holly. (Aquifoliaceae). - Tree. Japan. Leaves when boiled are source of a brown dye, used in Japan.

Ilex pseudobuxus Reiss. (Aquifoliaceae). - Tree. Brazil. Leaves are used as a substitute for Yerba Maté. Wood is used for building and general carpentry.

Ilex Sebertii Panch. (Aquifoliaceae). - Small tree. New Caledonia. Wood yellowish white, dense, durable, easily worked; suitable for interior of houses and for planks.

Ilex theezans Mart. (syn. I. gigantea Bonpl.). (Aquifoliaceae). - Tree. Brazil. Leaves are diuretic, stomachic and stimulant. Also used as a tea.

Ilex verticillata (L.) Gray, Winterberry, Fever Bush, Black Alder. (Aquifoliaceae). - Shrub or small tree. Eastern N.America. Dried leaves are sometimes used as a substitute for tea.

Ilex vomitoria Ait., Yaupon (Aquifoliaceae). - Small tree. S. and S. E. of United States. Decoction of leaves is source of a drink, used by the Indians, during ceremonials, forming a part of the Bush Ceremonial. The tea is known as Black Drink. Plant is emetic. Leaves are also used as tea among the white populations, sometimes sold in local markets.

Illicium verum Hook. f., Star Anise, Chinese Anise. (Magnoliaceae). - Tree. S. E. Asia. Cultivated in S. China, French Indo-China, Phillipine Islands, Japan, Jamaica. Much is shipped from Tonkin. Source of an ess. oil; used for flavoring. Dried ripe fruit used medicinally as a stimulant carminative. Contains anethol. Fruits from I. religiosum, cultivated around Buddhist Temples, are very poisonous.

Illipé Butter → Madhuca longifolia Gm.

Illorin Gum → Daniella thurifera Bennett

Ilysanthes antipoda Merr. → Bonnaya antipoda Druce.

Imbu → Spondias tuberosa Arruda.

Impatiens Balsamina L., Garden Balsam. (Balsaminaceae). - Perennial herb. India. Cultivated as ornamental. Flowers used in parts of Asia instead of Henna for dyeing finger-nails.

Imperata arundinaceae Cyrill. Lalang, Alang-Alang (Graminaceae). - Perennial grass. Temp. region. Leaves are made into mats and farmer's raincoats in Japan. Used locally in China for manuf, paper; is often mixed with rice straw. Stems are used in Philipp. Islds. as braiding material for hats. Leaves are used for thatching. In Malaya beer has been made from the starch of the rhizomes.

Imperata brasiliensis Trin. (syn. I. caudata Chapm., I. Sape Anders., Saccharum Sape St. Hil.). (Graminaceae). - Perennial grass. Trop. America. Plant is considered an excellent forage grass for live-stock in Brazil. Rootstock is said to be diuretic.

Inca Wheat → Amaranthus caudatus L.

Incense. Source of → Abies, Ailanthus, Amyris, Aquilaria, Boswellia, Bursera, Canarium, Clusia, Combretum, Commiphora, Cryptomeria, Encelia, Ervatamia, Eucarya, Excoecaria, Ferula, Gonystylus, Hierochloë, Hymenaea, Hyptis, Icica, Lansium, Libothamnus, Meliosma, Pistacia, Protium, Santalum, Styrax, Tetraclinis.

Incense Cedar → Libocedrus decurrens Torr.

Incense Wood → Amoora nitida Benth.

Incienso → Myrocarpus frondosus Allem.

India Abutilon → Abutilon indicum Sweet.

India Coral Bean → Erythrina indica Lam.

India Gum → Anogeissus latifolia Wall.

India Drugsquill → Urginea indica Kunth.

India Madder → Rubia cordifolia L.

India Mustard → Brassica juncea (L.) Coss.

Indian Pokeberry → Phytolacca acinosa Roxb.

Indian Aloë → Aloë vera L.

Indian Aconite → Aconitum ferox Wall.

Indian Almond → Terminalia Catappa L.

Indian Arrow Wood → Evonymus atropurpureus Jacq.

Indian Barberry → Berberis aristata DC.

Indian Bean → Catalpa bignonioides Walt.

Indian Berry → Anamirta paniculata Colebr.

Indian Bdellium → Commiphora Mukul Engl.

Indian Blue Pine → Pinus excelsa Wall.

Indian Breadroot → Psoralia esculenta Push.

Indian Butter → Madhuca butyracea Gm.

Indian Cassia → Cinnamomum Tamala (Buck and Ham.) Nees. and Eberm.

Indian Copal → Vateria indica L.

Indian Corn → Zea Mays L.

Indian Cucumber Root → Medeola virginiana L.

Indian Frankincense → Boswellia serrata Roxb.

Indian Hemp → Apocynum cannabinum L.

Indian Jalap → Ipomoea Turpethum R. Br.

Indian Kale → Xanthosoma atrovirens C. Koch. and X. violaceum Schott.

Indian Laburnum → Cassia Fistula L.

Indian Liquorice → Abrus precatorius L.

Indian Long Pepper → Piper longum L.

Indian Mulberry → Morinda citrifolia L.

Indian Nettle → Acalypha indica L.

Indian Mallow → Abutilon Avicennae Gaertn.

Indian Olibanum Tree → Boswellia serrata Roxb.

Indian Olive → Olea cuspidata Wall.

Indian Paper Birch → Betula utilis D. Don.

Indian Poke → Veratrum viride Ait.

Indian Redwood → Soymida febrifuga Juss.

Indian Rennet → Withamia coagulans Dunal.

Indian Rhubarb → Saxifraga peltata Torr.

Indian Rice → Zizania aquatica L.

Indian Rice Grass → Oryzopsis hymenoides (Roem. and Schult.) Baker.

Indian Rubber → Ficus elastica Roxb.

Indian Sarsaparilla → Hemidesmus indicus R. Br.

Indian Tobacco → Lobelia inflata L.

Indian Tragacanth → Astragalus heratensis Bunge.

Indian Turpentine → Pinus longifolia Roxb.

Indian Yellow → Mangifera indica L.

Indian Winter Green → Gaultheria fragantissimum Wall.

Indigo → Indigofera Anil L. and other spp.

Indigo Plant → Indigofera Anil L.

Indigo Squill → Camassia esculenta Lindl.

Indigo, Yellow Wild → Baptisia tinctoria (L.) R. Br.

Indigo, Yoruba → Lonchocarpus cyanescens Benth.

Indigo, West African → Lonchocarpus cyanescens Benth.

Indigofera Anil L. (syn. I. suffruticosa Mill.), Indigo Plant. (Leguminosaceae). - Shrublike plant. Trop. America. Cultivated in the Old and New World. Was once a source of Indigo, a blue dye. Now largely replaced by coal-tar dyes.

Indigofera arrecta Hochst. (Leguminosaceae). - Herbaceous. E. Africa. Source of Degendeg; principal indigo producing species in Abyssinia. Also grown as green manure.

Indigofera diphylla Vent. (Leguminosaceae). - Herbaceous. Trop. W. Africa. Source of a blue dye, used by natives of Senegambia.

Indigofera pascuorum Benth. (Leguminosaceae). - Shrub-like plant. Trop. America. Infusion of the roots is used as a febrifuge and stimulant.

Indigofera pauciflora Delil. (Leguminosaceae). - Herbaceous plant. Trop. Asia, Africa. Plant is used as fodder for camels. Roots when boiled in milk are used as purgative in the Sudan.

Indigofera simplicifolia Lam. (Leguminosaceae). - Herbaceous plant. Trop. Africa. Roots used in Fulani (Afr.) as arrow-poison.

Indigofera tinctoria L. (syn. I. sumatrana Gaertn.) True Indigo Plant. (Leguminosaceae). - Shrubby plant. Mal. Archip. Cultivated. Source of Indigo, a blue dye.

Indo Malayan Alocasia → Alocasia indica Schott.

Industrial Cascarille → Croton echinocarpus Muell. Arg.

Infusorial Earth → Diatoms

Inga anomala Kunth. → Calliandra anomala (Kunth.) Macbride.

Inga Cipo → Inga edulis Mart.

Inga Goldmanii Pittier., Guavo de Mono. (Leguminosaceae). - Tree. Costa Rica and Panama. Used as shade in coffee plantations at lower altitudes.

Inga edulis Mart. Food Inga, Inga Cipo, Guavo-Bejúco. (Leguminosaceae). - Tree. Mexico to Panama and southward. Pods are consumed by the natives. Trees are used as shade for the coffee trees at lower altitude belts.

Inga leptoloba Schlecht., Quamo. (Leguminosaceae). - Tree. S. Mexico to Costa Rica and S. America. Used in Centr. America as shade tree of the coffee trees in upper altitude belts.

Inga Macuna Walpers and Duchas, Guavo Peludo. (Leguminosaceae). - Medium sized tree. Panama. Pulp from the fruits is consumed by the Chocó Indians of Panama.

Inga Paterno Harms. Paterno. (Leguminosaceae). - Medium sized tree. Mexico to Costa Rica. Pods are eaten among the natives in Centr. America. Offered in local markets.

Inga Pittieri Micheli. (Leguminosaceae). - Small tree. Costa Rico, Panama. Used in Centr. America as shade in coffee plantations in lower altitude regions.

Inga Preussii Harms. Cuxiniquil. (Leguminosaceae). - Tree. El Salvador. Used as shade tree in coffee plantations at lower altitude belts.

Inga punctata Willd., Guavo. (Leguminosaceae). - Tree. S. Mexico to Costa Rica. Used in Centr. America as a shade tree in coffee plantations in the upper altitude belt.

Inga radians Pitt., Guavo Real. (Leguminosaceae). - Tree. S. Mexico to S. America. Pods are consumed in Centr. America. Sold in native markets.

Inga Rensoni Pitt. Cujin. (Leguminosaceae). - Medium sized tree. El Salvador. Sweet pulp from pods is very much relished among the natives of El Salvador.

Inga Ruiziana G. Don., Toparejo. (Leguminosaceae). - Medium sized tree. Nicaragua to Panama. Fruits are consumed by the natives.

Inga spectabilis Willd. Guavo de Castilla. (Leguminosaceae). - Tree. Costa Rica to Panama. Pods are eaten in Centr. America.

Inga spuria Humb. and Bonpl., Nacaspilo. (Leguminosaceae). - Small tree. Mexico to Panama and S. America. Pods are consumed by the natives of Centr. America.

Inhambane Copal →Copaifera Gorskiana Benth.

Inkberry → Ilex glabra (L.) Gray.

Ink Caps → Coprinus atramentarius (Bull.) Fr.

Ink, Fungus → Coprinus atramentarius (Bull.) Fr.

Inocarpus edulis Forst. (syn. Bocoa edulis Baill.). Tahiti or Polynesian Chestnut. (Leguminosaceae). - Tree. Pacif. Islands. Seeds are eaten raw, boiled or toasted by the natives, having the flavor of chestnuts. Seeds are an important source of food among inhabitants of Samoa and neighboring islands.

Inocybe cutifracta Petch. (Agaricaceae). - Basidiomycete Fungus. Trop. Asia. Fruitbodies are consumed as food by the natives, also sold in markets. Some cases of poisoning have been noticed.

Inodes mexicana (Marsh.) Standl. → Sabal mexicana Mart.

Inodes texana Cook. → Sabal texana Becc.

Inomena → Asparagopsis Sanfordiana Harv.

Inophloeum aromaticum (Miq.) Pitt. (Moraceae). - Tree. Centr. and trop. America. Inner bark is made into cloth, used by women among the Cuna, Chocó and Guaymi Indians. Also used for blankets and hammocks; formerly made into sails for canoes.

Insect Powder, Dalmatian → Chrysanthemum cinerariaefolium (Trév.) Vis.

Insect Powder, Persian → Chrysanthemum coccineum Willd.

Ionotus hispidus (Bull.) Karst. → Polyporus hispidus Bull.

Irish Moss → Chondrus crispus (L.) Stackh.

Isinglass → Gelidium cartilagineum Gaill.

Isoetaceae → Isoetes.

Isoetes Martii A. Braun. (Isoetaceae). - Fern ally. Brazil. Stem bases are considered useful for treating snake-bites.

Intisy Rubber → Euphorbia Intisy Drake del Cast.

Intsia amboinensis Thouars → Afzelia bijuga Gray.

Intsia Bakeri Prain → Afzelia palembanica Baker.

Inula Helenium L., Elecampane. (Compositae). - Perennial herb. Probably from Centr. Asia. Cultivated. Occasionally used as diaphoretic, tonic, diuretic, expectorant, for chronic bronchitis, coughs, chronic diarrhoea. Source of a blue dye. Plant contains up to 44% inulin, also alantol. Was formerly used as an ingredient of Absinth. Root is source of Elecampane Oil, semi-solid, ess. oil.

Inula viscosa Ait. (Compositae). - Perennial herb. Mediterranean region. Source of a yellow dye, used by country people in Greece.

Ionidium album St. Hil. (Violaceae). - Perennial plant. Argentina to Brazil. Plant is considered as an emetic and purgative in some parts of S. America.

Ionidium Ipecacuanha Vent. (Violacae). - Herbaceous plant. Brazil. Source of White Ipecac. Dried roots are used medicinally as an emetic.

Ipecac → Cephaëlis Ipecacuanha (Brotero) Rich.

Ipecac, Cartagena → Cephaëlis acuminata Karsten.

Ipecac, Farinaceous → Richardsonia pilosa H.B.K.

Ipecac, Pará → Cephaëlis Ipecacuanha (Brotero) Rich.

Ipecac Spurge → Euphorbia Ipecacuanha L.

Ipecac, Undulated → Richardsonia pilosa H.B.K.

Ipecac, White → Ionidium Ipecacuanha Vent.

Ipecacuanha → Cephaëlis Ipecacuanha (Brotero) Rich.

Ipecacuanha, Country → Naregamia alata wight and Arn.

Ipecacuanha, Goa → Naregamia alata Wight and Arn.

Ipecacuanha, Portugueese → Naregamia alata Wight and Arn.

Ipomaea aculeata Kuntze → Calonyction aculeatum (L.) House.

Ipomoea altissima Mart. (syn. Operculia altissima Meisn.). (Convolvulaceae). - Vine. Brazil. Root is used as drastic; is poisonous in large doses.

Ipomoea angustifolia Jacq. (syn. Convolvulus japonicus Thunb.). (Convolvulaceae). - Vine. Japan. Roots roasted or boiled are used as food by the Ainu.

Ipomoea aquaticia Forsk. (Convolvulaceae). - Perennial creeping herb. Tropics. Cultivated in some parts of Asia. Leaves and shoots are cooked and eaten in trop. Asia. Sometimes called Chinese Cabbage.

Ipomoea Batatas Poir. (syn. Batatas edulis Choisy.) Sweet Potato, Batate. (Convolvulaceae). - Perennial creeping plant. Origin unknown. Tuberous roots are consumed as food, prepared in different ways. Cultivated. Of commercial importance. Tubers are sometimes canned; are also source of starch, called Brazilian Arrowroot; dextrine and alcohol. Yams are those that have a moist, soft texture. Numerous varieties are divided into different groups. I. Belmont Group: Eclipse Sugar Yam, Vineless Pumkin Yam, Belmont Old Time Yam. II. Ticotea Group: Tocotea, Koali. III. Spanish Group: Yellow Spanish, Bermuda, Red Spanish, Triumph, Porto Rico, Creola. IV. Shanghai Group: Shanghai, Minnet Yam. V. Florida Group: Florida, Nancy Hall, General Grand Vineless. VI. Southern Queen Group: White Yam, Southern Queen. VII. Pumpkin Group: Pumpkin Yam, White Gilke, Norton. VIII. Jersey Group: Red Jersey, Big-Stem Jersey, Yellow Jersey, Bush.

Ipomoea bracteta Cav. → Exogonium bracteatum (Cav.) Choisy.

Ipomoea digitata L. Fingerleaf Morningglory. (Convolvulaceae). - Tree vine. Tropics. India, Malaya, Ceylon. Root tonic, alterative, aphrodisiac.

Ipomoea leptophylla Torr. Bush Morning Glory. (Convolvulaceae). - Herbaceous perennial. Western N. America. Roots when roasted are used as food by several Indian tribes in times of famine.

Ipomoea mammosa Chois. (syn. Merremia mammosa Hall.) (Convolvulaceae). - Philipp. Islands. Roots are consumed as vegetable in some parts of the Philipp. Islds.

Ipomoea murocoides Roem. and Schult. (Convolvulaceae). - Tree. Mexico, Centr. America. Ashes from plants are used in Guatemala as soap for washing clothes.

Ipomoea orizabensis Ledenois., Mexican Scammony. (Convolvulaceae). - Perennial vine. Mexico. Source of Orizaba Jalap, Ipomea, Mexican Scammony. Much is derived from Orizaba (Vera Cruz, Mex.). Dried root is used medicinally as a hydragogue cathartic. Contains a glucosidal resin, pale yellow ess. oil and sconoletin.

Ipomoea palmato-pinnata Benth. and Hook. (syn. I. gigantea Choisy). (Convolvulaceae). - Herbaceous perennial vine. S. America, esp. Paraguay, Argentina and Brazil. Dried and pulverized root is used in S. America as purgative.

Ipomoea Purga Hayne. (syn. Exogonium jalapa Nutt. and Coxe. Jalap. (Convolvulaceae). - Perennial climber. Mexico. Cultivated in Mexico, West Indies and E. Indies. Dried tuberous root is used medicinally, being purgative, hydragogue and cathartic. Contains 8 to 12% of a resin and an ess. oil.

Ipomoea simulans Hanbury. (Convolvulaceae). - Vine. Mexican Andes. Plant is source of Tampico Jalap; used in medicine.

Ipomoea Turpethum R. Br. (syn. Operculina Turpethum S. Manso.). Turpeth, Indian Jalap. (Convolvulaceae). - Vine. Trop. Asia, Australia. Plant is source of Turpeth Root or Indian Jalap. Used as purgative by the natives. Employed in the Orient since early times. Is considered an excellent substitute for Ipomoea purga.

Iresine arbuscula Uline and Bray. (Amaranthaceae). - Tree. Centr. America. Ash from wood is used for manuf. of soap in some parts of Guatemala.

Iresine calea (Ibanez) Standl. (Amaranthaceae). - Shrub. Mexico. Plant is used as diuretic and diaphoretic.

Iresine Celosia L. Jubas Bush. (Amaranthaceae). - Herbaceous plant. West Indies. Decoction of herb is used in Cuba for stomach ailments.

Iresine paniculata (L.) O. Ktze. (Amaranthaceae). - Herbaceous plant. Trop. America. Decoction of herb is used in some parts of Mexico for fever.

Iriartea durissima Oerst. → Socratea durissima (Oerst.) Wendl.

Iridaceae → Antholyza, Aristea, Babiana, Belamcanda, Bobartia, Crocosma, Crocus, Ferraria, Geissorhiza, Gladiolus, Iris, Sisyrinchium, Tigridia, Trimeza.

Iripil Bark Tree → Pentaclethra filamentosa Benth.

Iris filifolia Boiss. (syn. I. juncea Brot.) (Iridaceae). - Perennial herb. Mediterranean region. Tubers are consumed as food by the Arabs.

Iris florentina L., Florentine Iris. (Iridaceae). - Perennial herb. S. Europe. Cultivated. Peeled and dried rhizome, known as Orris is, when powdered, source of toilet powders, dentifrices, sachet and dusting powders. An ess. oil is used in perfumery for violet combinations and as a fixative. Contains a volat. oil known as Orris Butter, in which is found iridin. Rootstocks are collected in autumn. Much is derived from France and Italy, some from Germany. To some extent are also used I. pallida Lam. and I. germanica L.

Iris setosa Pall. (Iridaceae). - Perennial herb. Alaska, Siberia. Seeds when roasted and ground are used as coffee by the Eskimos in some parts of Alaska.

Iris versicolor L., Blue Flag. (Iridaceae). - Herbaceous perennial. Eastern N. America. Dried rhizomes are used medicinally as emetic, diuretic and cathartic. Contains an acrid resin and ess. oil. Also I. caroliniana Wats.

Iroco Fustic Tree → Chlorophora excelsa (Welw). Benth. and Hook. f.

Iron Oak → Quercus obtusiloba Michx.

Ironbark → Eucalyptus siderophloia Benth.

Ironbark, Broadleaf → Eucalyptus siderophloia Benth.

Ironbark, Limonscented → Eucalyptus Staigeriana F. v. Muell.

Ironbark, Red → Eucalyptus leucoxylon F. v. Muell.

Ironbark, White → Eucalyptus leucoxylon F. v. Muell.

Ironweed, Kinka Oil → Vernonia anthelmintica Willd.

Ironwood → Acacia excelsa Benth., Afzelia palembanica Baker, Casuarina spp., Eugenia confusa DC., Olea paniculata R.Br., Ostrya virginiana (Mill.) Willd.

Ironwood, Black → Olea laurifolia Lam.

Ironwood, Borneo → Eusideroxylon Zwageri Teysm. and Binn.

Ironwood, Molucca → Afzelia bijuga Gray.

Ironwood, Nigerian → Vernonia nigritiana Oliv. Hiern.

Ironwood, Red → Reynosia septentrionalis Urb.

Ironwood, White → Toddalia lanceolata Lam.

Irvingia Barteri Hook. f. Dika Bread, Dika Nut, Gaboon Chocolate (Simarubaceae). - West Africa. Fruit edible. Seed used by the natives in Undike Bread. Fat from seeds has been recommended for manuf. of soap and candles.

Irvingia gabonensis Baill. Wild Manⁿo (Simarubaceae). - Tree. Trop. Africa. Seeds are source of Dikka Butter, Dikka Fatusedas food of natives. Sap. Val. 241—250; Iod. no. 1.8-9.8. Kern used by the natives in Dika Bread. Wood is light yellow to reddish yellow, tough, heavy, very hard, rather difficult to split; used for street paving.

Irvingia Oliveri Pierre. (Simarubaceae). Trop. Asia, esp. Cambodia, Cochin China. Seeds source of Cay-Cay Fat; non-drying, manuf. for candles. Sp. Gr. 0.9128—0,9130.

Irvingia Smithii Hook. f. (Simarubaceae). - Trop. Africa. Source of an ess. oil; used by the natives for perfumery.

Isachne australis R.Br. (Graminaceae). - Perennial Grass. Australia. Recommended as food for live-stock in tropical countries.

Isatan → Isatis tinctoria L.

Isatis japonica Miq. (Cruciferaceae). - Perennial herb. Japan. Leaves are source of a green dye, used in Japan.

Isatis tinctoria L. Woad. (Cruciferaceae). - Perennial or biennial. S.E. Europe to W. Asia. Cultivated. Source of a blue dye. Was used by early Britons to paint their bodies.

Ischaemum angustifolium Hack. Bharbur Grass. (Graminaceae). - Perennial Grass. India, Kashmir, Afghanistan. Plant is used for manuf. of paper, also made into mats. A fibre is used for sails, ropes and strings.

Ischaemum timorense Kunth. (Graminaceae). - Perennial. Mal. Archipel. Used as food for domestic animals.

Isertia Pittieri Standl. (Rubiaceae). - South America. Leaves are used as a substitute for soap in the Chaco Intendancy.

Isomeria arborea Nutt. (Capparidaceae). - Shrub. Green pods are eaten cooked by the Coahuilla Indians.

Isonandra Gutta → Palaguium Gutta Burck.

Isotoma longiflora Presl. (Campanulaceae). - Herbacous plant. Trop. America. Used in some-parts of Brazil as anti-asthmatic.

Isoptera borneensis Scheff. (Dipteraocarpaceae). - Borneo. Wood is resistant to moisture; used for buildings.

Istrian Galls. → Quercus Ilex L.

Italian Corn Salad. → Valerianella eriocarpa Desv.

Italian Cypress → Cupressus sempervirens L.

Italian Fennel → Foeniculum vulgare Mill.

Italian Rye → Lolium italicum R. Br.

Italian Senna → Cassia obovata Collad.

Iva Bitter→ Achillea moschata Jacq.

Iva Liqueur → Achillea moschata Jacq.

Iva Wine → Achillea moschata Jacq.

Ivory Nut Palm → Coelococcus amicarum Warb.

Ivy, English → Hedera Helix L.

Ixora concinna R. Br. (Rubiaceae). - Small tree. East Indies. Used by the natives as walking canes.

Ixora fulgens Roxb. (Rubiaceae). - Malaya. Small tree. Used by the natives against tooth ache.

Ixora longituba Boerl. (Rubiaceae). - Pulverized roots and twigs are used by the natives to give a red or reddish brown color to basket work.

Ixtle → Agave.

Ixtle de Jaumava → Agave Funkiana Koch and Bouché.

Ixtle de Palma → Samuela carnerosa Trel.

J

Jaborandi Pepper → Piper longum L.

Jaborandi, Pernambuco → Pilocarpus Jaborandi Holmes.

Jaboticaba → Myrciaria cauliflora Berg.

Jaboticaba Macia → Myrciaria tenella Berg.

Jaboticaba do Matto → Myrciaria Jaboticaba Berg.

Jaboty Butter → Erisma calcaratum Warm.

Jaboty Tallow → Erisma calcaratum Warm.

Jacaranda brasiliana Pers. (syn. Bignonia brasiliana Lam.) (Bignoniaceae). - Tree. Brasil. Wood very durable, used for carpentry work.

Jacaranda Copaia (Aubl.) D. Don. (syn. Bignonia Copaia Aubl.). Boxwood, Palo de Buba. (Bignoniaceae). - Tree. Brit. Honduras to Brazil. Wood whitish, light, firm, medium to coarse texture, straight grained, easily worked; used for cheap work in houses, manuf. matches, boxes, coffins, buck shells. Employed by Indians to add buoyancy to rafts of havier timbers. Wood is source of wood pulp. Bark is considered emetic and cathartic.

Jacaranda micrantha Cham. (Bignoniaceae). - Tree. Brazil. Wood is used for general carpentry and interior finish.

Jacaranda oxyphylla Cham. (syn. J. caroba DC. var. oxyphylla Bur.) (Bignoniaceae). - Shrub. Brazil. Bark is reputed in Brazil of being cathartic, diuretic and antisiphylitic.

Jacaranda mimosaefolia D. Don. (syn. J. ovalifolia R. Br.) (Bignoniaceae). - Trop. S. America. Wood durable, compact, fragile; used for general carpentry.

Jacaranda subrhombea DC. (syn. J. obovata Mart.). (Bignoniaceae). - Tree. Brazil. Bark is considered sudorific.

Jacaratia mexicana A. DC. Papaya Orejona. (Caricaceae). - Tree. Mexico. Centr. America. Fruit eaten cooked, as a salad or made into sweet meats.

Jack Bean → Canavalia ensiformis DC.

Jack-in-the-Pulpit. → Arisaema triphyllum (L.) Torr.

Jack Oak → Quercus marylandica Du Roi.

Jack Pine → Pinus Banksiana Lamb.

Jackfruit → Artocarpus integrifolia L. f.

Jacobinia spicigera (Schlecht). Bailey. (syn. Justicia spicigera Schlecht). (Acanthaceae). - Shrub. Mexico. Leaves when placed in hot water become blackish, lateron blue; used by Mexican laundresses for whitening clothes, similar to Indigo.

Jacobinia tinctoria (Oerst.) Hemsl. (Acanthaceae). - Shrub. Centr. America. Plants are source of a blue dye.

Jacquinia pungens Gray. (Myrsiniaceae).- Shrub or small tree. Trop. America. Fruits used along the W. Coast of Mexico to stupfy fish.

Jacquinia umbellata A. DC. (syn. J. aurantiaca Bart.) (Myrsiniaceae). - Shrub or small tree. Mexico. Centr. America, West Indies. Crushed fruits are used to stupefy fish.

Jacquinia Seleriana Urb. and Loes. (Myrsiniaceae). - Shrub. Oaxaca (Mex.). Fruits used to stupefy fish.

Jaffarabad Aloë → Aloë ferox Mill. and A. vera L.

Jagera pseudorhus Radlk. → Cupania pseudorhus A. Rich.

Jalap → Ipomoea Purga Hayne.

Jalap, Brazilian → Piptostegia Pisonis Mart.

Jalap, Indian → Ipomoea Turpethum R. Br.

Jalocote → Pinus Teocote Cham. and Schlecht.

Jamaica Cherry → Muntingia Calabura L.

Jamaica Dogwood → Piscidia Erythrina L.

Jamaica Kino → Coccolobis uvifera (L.) Jacq.

Jamaica Quassia Tree → Picrasma excelsa Planch.

Jamaica Sarsaparilla → Smilax Regelii Killip and Morton.

Jamaica Simuruba Bark → Simaruba amara Aubl.

Jamaica Sorrel → Hibiscus Sabdariffa L.

Jamba Oil → Eruca sativa L.

Jambolan → Eugenia Jambolana Lam.

Jambolifera odorata Lour. → Acronychia odorata Baill.

Jambolifera resinosa Lour. → Acronychia laurifolia Blume.

Jambos → Eugenia Jambos L.

Jambosa cauliflora Miq. → Eugenia polycephala Miq.

Jambosa densiflora DC. → Eugenia densiflora DC.

Jambosa glabrata DC. → Eugenia clavimyrtus Koord. and Valet.

Jambosa lineata DC. → Eugenia lineata (DC.) Duth.

Jambosa tenuicuspis Miq. → Eugenia cymosa Lam.

Jambosa vulgaris DC. → Eugenia Jambos L.

Jamrosa Bark → Terminalia mauritiana Lam.

Japan Clover → Lespedeza striata Hook.

Japan Tallow → Rhus succedanea L.

Japan Wood → Caesalpinia Sappan L.

Japanese Aconite → Aconitum Fischeri Reichb.

Japanese Alder → Alnus japonica Sieb. and Zucc.

Japanese Apricot → Armeniaca Mume Sieb.

Japanese Artichoke → Stachys Sieboldi Miq.

Japanese Banana → Musa Basjoo Sieb.

Japanese Buckthorn → Rhamnus japonica Maxim.

Japanese Buckwheat → Fagopyrum esculentum Moench.

Japanese Chestnut → Castanea crenata Sieb. and Zucc.

Japanese Elm → Ulmus japonica Sar.

Japanese Fleeceflower → Polygonum cuspidatum Sieb. and Zucc.

Japanese Galls → Rhus semialata Murr.

Japanese Ginger → Zingiber Mioga (Thunb.) Rosc.

Japanese Horse Chestnut → Aesculus turbinata Blume.

Japanese Lacquer → Rhus vernicifera DC.

Japanese Linden → Tilia japonica Simk.

Japanese Medlar → Eriobotrya japonica Lindl.

Japanese Millet → Echinochlea frumentacea (Roxb.) Link.

Japanese Mint Oil → Mentha arvensis L.

Japanese Mustard → Brassica japonica Sieb.

Japanese Persimmon → Diospyros Kaki L. f.

Japanese Plum → Eriobotrya japonica Lindl. and Prunus salicina Lindl.

Japanese Poplar → Populus Maximowiczii Henry.

Japanese Prickly Ash → Zanthoxylum piperitum DC.

Japanese Privet →Ligustrum japonicum Thunb.

Japanese Raisin Tree → Hovenia dulcis Thunb.

Japanese Snakegourd → Trichosanthes cucumeroides Max.

Japanese Staunton Vine → Stauntonia hexaphylla Decne.

Japanese Stone Pine → Pinus pumila Reg.

Japanese Torreya → Torreya nucifera Sieb. and Zucc.

Japanes Tung Oil → Aleurites cordata Steud.

Japanese Turpentine → Larix dahurica Turcz.

Japanese Wingnut → Pterocarya rhoifolia Sieb. and Zucc.

Japanese Zelkova → Zelkova serrata (Thunb.) Mak.

Jardinea congoensis Franch. (Graminaceae). - Perennial grass. Trop. Africa, esp. Nigeria, Togoland, Belg. and Fr. Congo. Stems and leaves are used for baskets, screens and mats.

Jarilla caudata (Brandeg.) Standl., Jarilla. (Caricaceae). - Small shrub. Mexico. Fruit used for preserves and sweet-meats.

Jaropa de Mora → Rubus glaucus Benth.

Jasmine, Arabian → Jasminum Sambac Ait.

Jasmine, Catalonian → Jasminum grandiflorum L.

Jasmine, Common → Jasminum officinale L.

Jasmine, Italian → Jasminum grandiflorum L.

Jasminum grandiflorum L., Catalonian Jasmine, Italian Jasmine. (Oleacae). - Shrub. Cultivated. Flowers are source of an essential oil; used in perfumery.

Jasminum niloticum Gilg. (Oleaceae). - Shrub. Trop. Africa. Flowers are source of an ess. oil, used as perfume by the women of Sudan.

Jasminum odoratissimum L. (Oleaceae). - Shrub. Madeira. Source of an essential oil; used in perfumery. Cultivated.

Jasminum officinale L., Common Jasmine. (Oleaceae). - Slender, vine-like plant. Centr. Asia. Cultivated. Very fragrant flowers are source of an ess. oil; used in perfumery.

Jasminum paniculatum Roxb. (Oleaceae). - Shrub. Temp. China. Flowers are used in China for scenting tea.

Jasminum Sambac Ait., Arabian Jasmine. (Oleaceae). - Shrub or shrubby vine. Trop. Asia. Cultivated. Flowers ar used for scenting tea.

Jatamansi → Nardostachys Jatamansi DC.

Jateorhiza Columba Miers. (syn. Jateorhiza palmata (Lam.) Miers.) Colombo. (Menispermaceae). - Woody climber. Trop. Africa. Dried root when dug in dry season, is used medicinally. Used as a bitter tonic, contains jateorrhizine, palmatine and columbamine, being alkaloids.

Jateorhiza Miersii Oliv. (syn. J. palmata Miers., Cocculus palmatus Hook.). (Menispermaceae). - Vine. Trop. Africa. Cultivated in Bourbon, Mauritius etc. Plant is used medicinally as stomachic and as digestive, also used for diarrhea, dysentery and colic.

Jatropha aconitifolia Mill. (Euphorbiaceae). - Tree. S. Mexico, Centr. America. In some parts of Mexico the young laves are boiled and eaten as a vegetable. Was also used for this purpose by the ancient Mayas.

Jatropha cinerea (Ortez) Muell. Arg. (Euphorbiaceae). - Shrub or small tree. Baja Calif. Sonora and Sinaloa (Mex.). Decoction is used as mordant in dyeing.

Jatropha Curcas L., Physic Nut, Purging Nut. (Euphorbiaceae). - Tree. Tropics. Cultivated in tropics. Source of Curcas Oil, powerful purgative; also used for manuf. candles, soap, illumination, lubricating; used in wool-industry.

Jatropha sphathulata (Oreg.) Muell. Arg. Sangre de Grado, Terote. (Euphorbiaceae). - Shrub. Mexico and S. W. of United States. Stems flexible and tough, used for whips and for making baskets. Bark used for tanning and is source of a dark red dye.

Jatropha Zeyheri Sond. (Euphorbiaceae). - Shrub Mexico. Stem contains 22.4% tannin, used for tanning; produces a pale-brown leather.

Jauary Palm → Astrocaryum Jauari Mart.

Java Almond → Canarium commune L.

Java Cardamon → Amomum maximum Roxb.

Java Devilpepper → Rauwolfia serpentina Benth.

Java Nato Tree → Palaquium javanense Burck.

Java Plum → Eugenia Jambolana Lam.

Javanese Long Pepper → Piper retrofractum Vahl.

Javanese Palissander → Dalbergia latifolia Roxb.

Jeffersonia diphylla (L.) Pers., Twinleaf. (Berberidaceae). - Eastern N. America. Herbaceous. Used as expectorant and tonic.

Jelly, Iceland Moss → Cetraria islandica (L.) Ach.

Jelutong → Alstonia scholaris R. Br. and Dyera costulata (Miq.) Hook. f.

Jenny Stone Crop → Sedum reflexum L.

Jequie Rubber → Manihot dichotoma Ule.

Jequirity → Abrus precatorius L.

Jerrah → Eucalyptus marginata Smith.

Jerusalen Artichoke → Helianthus tuberosus L.

Jerusalem Rye → Triticum polonicum L.

Jessamine, Yellow → Gelsemium sempervirens (L.) Ait. f.

Jessenia polycarpa Karst. (Palmaceae). - Tall palm. Trop. S. America. Pericarp of the fruit is sweet and edible. Seeds are the source of an oil.

Jesuit's Bark → Cinchona officinalis L.

Jew's Ears → Auricularia Auricula Judae (L.) Schroet.

Jew Plum → Spondias dulcis Forst. f.

Jewish Citron → Citrus Medica L.

Jicama → Calopogon coeruleum Hemsl. and Exogonium bracteatum (Cav.) Choisy.

Jícana → Pachyrrhizus palmatilobus Benth. and Hook.

Jimson Weed → Datura Stramonium L.

Jindai-Sugi → Cryptomeria japonica D. Don.

Jingan Gum → Odina Wodier Roxb.

Joannesia princeps Vell. (syn. Anda Gomesii Juss.). (Euphorbiaceae). - Tree. Brazil. Seeds are source of a thick oil, called Anda-assy Oil; used as purgative and for skin diseases. It has been claimed that the laxative effect is four times that of the Castor Oil. Sp. Gr. 0.927-0.9229; Sap. Val. 192; Iod. No. 115.7; Unsap. 1.2%.

Job Tears → Coix Lachryma-Joby L.

Jobo → Spondias lutea L.

Jocote → Spondias purpurea L.

Jodina rhombifolia Hook. and Arn. (Santalaceae). - Tree. Trop. America, esp. S. Brazil, Argentina, Uruguya. Bark is source of a commercial tanning material.

Johnson Grass → Sorghum halepense (L.) Pers.

Jointfir → Ephedra trifurca Torr.

Jojoba Oil → Simmondsia californica Nutt.

Jonquil → Narcissus Jonquilla L.

Joshua Tree → Yucca arborescens Trel.

Josswood → Mitragyna africana Korth.

Joyapa → Macleania Popenoei Blake.

Jubaea chilensis (Molina) Baill. (syn. J. spectabilis H. B. K.) Coco de Chile, Winepalm of Chile (Palmaceae). - Tall palm. Chile. Kernel of seeds edible; sometimes sold in local markets of Chile, called Coquitos. Seeds source of an edible oil, used in food. Source of Palm Honey, Palm Wine. Fruits are candied. Leaves are used for baskets etc.

Juglandaceae → Carya, Engelhardtia, Juglans, Platycarya, Pterocarya.

Juglans boliviana Dode. Bolivia Black Walnut. (Juglandaceae). - Tree. Bolivia. Nuts edible, of good quality.

Juglans cathayensis Maxim. Cathay Walnut. (Juglandaceae). - Tree. Centr. China. Kernels eaten by the natives of China.

Juglans cinerea L., Butternut. (Juglandaceae). - Tree. Eastern part of N. America. Wood coarse-grained, light brown, turning darker upon exposure, light, soft, not strong; used for manuf. of furniture, interior finish of houses. Nuts are used as food. Butternut Bark being the inner bark of the root; is used as mild cathartic and for fevers. Collected in autumn. Contains resinoid juglandin, juglone, juglandic acid and ess. oil. Sugar may be made from the sap. Green husks of the fruit may be used to dye cloth; giving it a yellow to orange color.

Juglans Duclouxiana Dode. (Juglandaceae). - Tree. Mountain regions of Asia. Cultivated in China. Fruits edible, consumed by the natives.

Juglans Honorei Dode. Ecuador Walnut, Nogal. (Juglandaceae). - Tree. Highlands of Ecuador. Occasionally cultivated. Wood strong, beautifully marked; used for different purposes. Source of a dye; used by the Indians of Imbabura. Nuts edible, thick-shelled; kernel rich and of good flavor; made into sweetmeats, called Nogada de Ibarra. Sold in local markets.

Juglans kamaonia Dode. (syn. J. regia L: var. kamaonia DC.) (Juglandaceae). - Tall tree. Centr. and W. Himalaya region. Nuts are source of a food among the native population of China.

Juglans mandschurica Maxim., Mandchurian Walnut. (Juglandaceae). - Tree. Mandchuria, Amur region. Wood brown, hard, heavy, easy to split, little elastic; used for general carpentry work.

Juglans mollis Engelm. (syn. J. mexicana S. Wats.) Guatemala Walnut. (Juglandaceae). - Tree. Mexico, Guatemala. Wood highly valued; used for tubs, bowls. Husks of fruits are a source of a coffee colored dye. Leaves used for rheumatism.

Juglans nigra L. Black Walnut. (Juglandaceae). - Tree. N. America. Important lumber tree. Wood hard, strong, heavy, coarse-grained, dark brown, very durable; used for cabinet making, boat and ship building, interior finish of buildings, gun-stocks. Nuts used as food; forming an important diet among the Indians.

Juglans pterococca Roxb. → Engelhardtia spicata Blume.

Juglans regia L., English Walnut. (Juglandaceae). - Tree. Probably from S. E. Europe to Himalaya and China. Cultivated in temp. zones of the Old and New World. Nuts are consumed fresh, salted; in various confectionaries and pastries. Young fruits are made into pickles. Source of a light yellow oil; when cold pressed it is used in foods; dry pressed for soap, paints. Sp. Gr. 0.925-0.927; Sap. Val. 189-197.3; Iod. No. 132-152. Wood hard, heavy, durable, close grained; used for furniture, gun-stocks. Ground nut shells used as adulterant of spices. Crushed leaves or a decoction, used as repellent of insects; also used as tea. Decoction of leaves, bark and husks with alum used for staining wool a brown color. Hulls of fruits are used medicinally as astringent; in home remedies employed for „cleaning" the blood, as astringent, antiscrophulosum and for intestinal worms. Some varieties are Mayette, Meylanaise, Gourlande, Corne, Marmot and Brantome.

Juglans rupestris Engelm. Texas Walnut. (Juglandaceae). - Tree. S. W. of United States and adj. Mexico. Kernels of the thickwalled nuts are eaten by the Indians of New Mexico and Arizona.

Juglans Sieboldiana Maxim. Siebold Walnut. (Juglandaceae). - Tree. Sachalin, Japan, China, Wood soft, light, not easily cracked or warped, dark brown. Used in Japan for gun-stocks, cabinet-work, various utensils. Bark and exocarp of fruit is used for dyeing.

Jujub, Argentine → Zipyphus Mistol Griseb.

Jujub, Chinese → Zizyphus Jujuba Mill.

Juliana adstringens Schlecht. (syn. Amphiyterygium adstringens (Schlecht) Schiede. (Julianiaceae). - Small tree. Mexico. Bark yields a red dye.

Julianaceae → Juliana.

Jumping Bean → Sapium biloculare (S. Wats.) Pax. and Sebastiana Pavoniana Muell. Arg.

Jumpy Bean → Leucaena glauca (L.) Benth.

Juncaceae → Juncus, Rostkovia.

Juncaginaceae → Lilaea, Triglochin.

Juncus acutus L. (Juncaceae). - Perennial herb. Temp. region. Stems are used for manuf. of mats.

Juncus effusus L. (syn. J. communis E. Mey.) (Juncaceae). - Perennial herb. Temp. zone. Cultivated. Stems are made into mats. Pith is used in Japan as a wick.

Juncus glaucus Sibth. Rush. (Juncaceae). - Perennial herb. Centr., S. Europe, temp. Asia, N. and S. Africa. Stems used for mats and baskets. Also J. conglomeratus L.

Juncus procerus E. Mey. (Juncaceae). - Perennial herb. Chile. Stem is used for manuf. of cord, in some parts of Chile.

Jungle Rice → Echinochloa colonum (L.) Link.

Juniper → Juniperus communis L.

Juniper, Alligator → Juniperus pachyphlaea Torr.

Juniper, Blackseed → Juniperus saltuaria Rehd. and Wils.

Juniper, California → Juniprus californica Carr.

Juniper, Greek → Juniperus excelsa Bieb.

Juniper, Mexican → Juniperus mexicana Schlecht.

Juniper Mistletoe → Phorodendron juniperinum Engelm.

Juniper, Sierra → Juniperus occidentalis Hook.

Juniper Wood Oil → Juniperus communis L.

Juniperus barbadensis L., Southern Red Cedar. (Cupressaceae). - Florida, West Indies. Wood was formerly used for lead-pencils. Supply has now been exhausted.

Juniperus californica Carr. California Juniper. (Cupressaceae). - Tree. California to Baja California. Fruits when fresh are ground into flour, were used as food by the Indians.

Juniperus communis L. Common Juniper. (Cupressaceae). - Tree. Temp. Europe, Asia, N. America. Sweet aromatic fruits are used for flavoring gin, liqueurs and cordials. Fruits require 2 years to mature. Contain 0.5 to 1.5% ess. oil, 10% resin, 15 to 30% dextrose. Fruits of highest quality come from Apennine Mts., Italy; much is also exported from Yugoslavia, Czechoslovakia and Hungary. Berries from many other regions are less flavored. Oil of Juniper Berries are obtained by steam distillation. Is diuretic, urogenital irritant. Wood is diuretic, diaphoretic. Tops of twigs are considered diuretic. Juniper Wood Oil, preferably from the roots is used in vet. medicine. Berries are also used for flavoring certain meats.

Juniperus excelsa Bieb. Greek Juniper. (Cupressaceae). — Tree, Asia minor, Afghanistan. Fruits and their oil are used as diuretic; for dysmenorrhoea and intestinal indigestion. Leaves are used in Khorasan (Asia) as an incense.

Juniperus mexicana Spreng. Mexican Juniper. (Cupressaceae). - Tree. W. Texas, Mexico, Guatemala. Wood hard, close-grained, brown; used for general construction, telegraph poles, railroad ties.

Juniperus occidentalis Hook. Sierra Juniper. (Cupressaceae). - Pacific Coast Area of N. America. Wood used for fencing and fuel. Fruits serve as food for Californian Indians.

Juniperus oxycedrus L., Prickly Juniper. (Cupressaceae). - Shrub. Mediterranean region. Caucasia. Source of Oil of Cade, a parasiticide and mild antiseptic, obtained from distillation of the heartwood.

Juniperus pachyphlaea Torr. Alligator Juniper. (Cupressaceae). - Tree. S. W. of United States and adj. Mexico. Fruits are consumed by the Indians.

Juniperus Sabina L. Savin. (Cupressaceae). - Shrub. S. Europe, Caucasia, S. Ural, Centr. N. Asia. Decoction of leaves is used in parts of Europe to combat lice. Leaves are kept between clothes during summer to combat moths. Young twigs, collected in spring are used medicinally, being a diuretic. They are source of Oil of Savin; used in pharmaceutical preparations and to some extent in perfuming of cosmetics. It is poisonous and occasionally used for abortion.

Juniperus saltuaria Rehd. and Wils. Black Seed Juniper. (Cupressaceae). - Tree. N. W. China, common near Sungpan. Wood is used for building purposes.

Juniperus utahensis (Engelm.) Lemm. Utah Juniper. (Cupressaceae). - Tree. Western United States. Fresh and ground fruits were consumed by the Indians, often put into cakes.

Juniperus uvifera Don. → Pilgerodendron uviferum (Don.) Florin.

Juniperus virginiana L., Eastern Red Cedar. (Cupressaceae). - N. America to Florida and Louisiana. Wood close-grained, not strong, brittle, light, dull red, easily worked, frangrant, durable; used for interior finish of buildings, sills, lining of closets, chests for preserving against insects, wooden-ware, pails, lead-pensils, posts. Red Cedar Wood Oil is derived from distillied heart-wood, containing 2.5 to 4.5% oil, composed of cedar camphor or cedrol and cedrene. Most oil is produced by cedar-pencil slat manufacturers. Oil is used as moth repellent, insecticide in dusting compounds and in sprays. Also employed as an efficacious fixative in perfumery and scenting of soap. Has been used for abortion, in some cases causing death. Source of Immersion-Oil, widely used for high powered objectives in microscopical examination, having a refract. index of 1.515 at 18° C. Leaves were used as incense during religious ceremonies by the Plains Indians.

Jura Turpentine → Picea excelsa Link.

Justicia insularis T. And. (Acanthaceae). - Shrub. Lower Congo. Young shoots and leaves are eaten as salad by the natives. Also the roots are consumed when boiled.

Justicia nutans Burm. → Clinocanthus Burmani Nees.

Justicia spicigera Schlecht. → Jacobinia spicigera (Schlecht) Bailey.

Jute → Corchorus capsularis L.

Jute, Bimlipatum → Hibiscus cannabinus L.

Jute, China → Abutilon Avicennae Gaertn.

Jute Plant → Corchorus capsularis L.

K

Kaempfera aethiopica (Solms). Benth. (Zingiberaceae). - Perenial herb. Trop. Africa esp. Nigeria, Gold Coast. Tubers are used as a ginger-like spice in E. Africa.

Kaempfera angustifolia Rosc. (Zingiberaceae). - Perennial herb. Trop. Asia. Rhizomes are used for coughs and as a mastigatory.

Kaempfera Galanga L. (Zingiberaceae). - Perennial herb. Trop. Asia. Cultivated. Rhizome, known as Galanga, is used as a condiment. Cultivated, esp. in India, Malaya and Cochin China.

Kafir → Sorghum vulgare Pers.

Kaffir Plum → Odina caffra Sond.

Kaffir Potato → Plectranthus esculentus N. E. Br.

Kagné Butter → Allanblackia oleifera Oliv.

Kahika → Podocarpus dacrydioides Rich.

Kaiser Tee → Dryas octopetala L.

Kalmia latifolia L., Laurel, Mountain Laurel. (Ericaceae). - Shrub or small tree. Eastern N. America. Wood strong, brittle, heavy, hard, close-grained, brown; used for handles of tools, turnery, fuel. Roots are made into tobacco-pipes.

Kallstroemia maxima (L.) Torr. and Gray. (Zygophyllaceae). - Herb. S. W. of United States to Centr. America. Young plants are sometimes used in Centr. America as a potherb.

Kaki → Diospyros Kaki L. f.

Kale → Brassica oleracea L. var. acephala DC.

Kale, Indian → Xanthosoma atrovirens C. Koch.

Kalingag Oil → Cinnamomum Mercadoi Vid.

Koloempang Oil → Sterculia foetida L.

Kalopanax pictus (Thunb.) Nakai → Acanthopanax ricinifolium Seem.

Kalpi → Citrus Webberi West.

Kamahi → Weinmannia racemosa L.

Kamala Powder → Mallotus philippensis Muell. Arg.

Kamala Tree → Mallotus philippinensis Muell. Arg.

Kamisan → Citrus longispina West.

Kandelia Rheedei Wight and Arn. (Rhizophoraceae). - Small tree. S. E. Asia, W. Malaysia. Bark used in Tonkin for tanning.

Kanga Butter → Pentadesma butyracea Sabine.

Kangeroo Apple → Solanum aviculare Forst.

Kangeroo Grass, Common → Anthistiria ciliata L.

Kanten → Agar.

Kapok Fibre → Ceiba pentandra (L.) Gaertn.

Kapok Tree → Ceiba pentandra (L.) Gaertn.

Kapor Tea → Epilobium angustifolium L.

Kappa → Clitandra orientalis Schum.

Kapporie Tea → Epiloboum angustifolium L.

Karamanni Wax → Symphonia globulifera L.

Karatas Plumieri Morr. (Bromeliaceae). - Perennial herb. Antilles, Martinique, Panama. Fruits edible and of a pleasant flavor.

Karaya Gum → Sterculia urens Roxb.

Karet Tree → Ficus elastica Roxb.

Kariyat → Andrographia paniculata Nees.

Karkalia → Mesembryanthemum aequilaterale Haw.

Karo Pittosporum → Pittosporum crassifolium Soland.

Karri → Eucalyptus diversicolor F. v. Muell.

Kartoum Teak → Cordia Myxa L.

Karwinskia Calderonii Standl. Calderon Coytillo, Huili huiste. (Rhamnaceae). - Tree. El Salvador. Wood of excellent quality, dull red to reddish brown, heavy, very hard, strong, durable; used for hubs of wheels, rail-road ties, weaver's shuttles, mortars pestles, and bowling balls.

Karwinskia Humboldtiana (Roem. and Schult.) Zucc. Humboldt Coytillo. (Rhamnaceae). - S. W. of United States, Mexico. Shrub or small tree. Pulp of fruit is edible; seeds have poisonous properties containing an oil, paralyzing to the motor nerves. Used in Mexico as anticonvulsive, esp. tetanus.

Kasnih → Ferula galbaniflua Bois. and Buhse.

Kât → Catha edulis Forsk.

Kateira Gum → Sterculia urens Roxb.

Kateiragum Sterculia → Sterculia urens Roxb.

Katio Oil → Bassia Mottleyana Clarke.

Katira Gabina Gum → Astragalus heratensis Bunge.

Kauila → Colubrina oppositifolia Brongn.

Kauki → Manilkara Kauki (L.) Dubard.

Kauri → Agathis australis Steud.

Kauri Copal → Agathis australis Steud.

Kauri Gum → Agathis australis Steud.

Kawa-Kawa → Piper methysticum Porst.

Kayu-galu Oil → Sindora inermis Merr.

Kei Apple → Dovyalis caffra (Hook. f. and Harv.) Warb.

Kelon-ka-tel Oil → Cedrus Libani Barrel.

Kelp, Bladder → Nereocystis Luetkeana (Mert.) Post & Rupr.

Kelp, Bull →Durvillea antartica (Cham.) Hariot.

Kelp, Broadleaf → Laminaria saccharina (L.) Lamour.

Kelp, California Giant. → Macrocystis pyrifera (L.) C. Ag.

Kelp, Horsetail → Laminaria digitata (L.) Edmonson.

Kelp, Japanese → Laminaria japonica Aresch.

Kemang → Mangifera caesia Jack.

Kenaf Hibiscus → Hibiscus cannabinus L.

Kendir Fibre → Apocynum venetum L.

Kentucky Bluegrass → Poa pratensis L.

Kentucky Coffee Tree→ Gymnocladus dioica (L.) Koch.

Kerguelen Cabbage → Pringlea antiscorbutica R. Br.

Kermadecia rotundifolia Brongn. and Gris. (Proteaceae). - Tree. New Caledonia. Wood is used in New Caledonia for general carpentry work and interior finish of houses.

Kerstingiella geocarpa Harms. (Leguminosaceae). - Herbaceous plant. Nigeria, Dahomey. Cultivated. Seeds are eaten by the natives of Nigeria and adj. territory.

Keteleria Davidiana Beissn. (Pinaceae). - Tree. Hupeh, Shensi, Yunnan, Szechuan (China). Wood soft, light, close-grained; used in China for construction of houses.

Ketembilla → Dovyalis hebecarpa (Gardn.) Warb.

Ketira Gum → Cochlospermum Gossypium L.

Khasia Pin → Pinus Khasya Royle.

Khasian Madder → Rubia khasiana Kurz.

Khatta Orange → Citrus limonia Osbeck.

Khaya anthotheca C. DC., White Mahogany, Smoth-barked African Mahogany. (Meliaceae). - Tree. Trop. W. Africa. Wood fine-grained, pinkish white, lateron becoming mahogany-brown. Exported to Europe. Bitter bark is used in Angola for fever.

Khaya ivorensis Chev. Ivory Coast Khaya, Red Mahogany. (Meliaceae). - Tree. Trop. W. Africa. Wood coarse-grained; used for building, canoes, chairs, handles of implements; exported to France. K. Klainei Pierri is source of Gabun Mahogany; K. euryphylla Harms of Cameroon Mahogany; K. grandis Stapf. and K. Punchii Stapf. of Benin Mahogany; K. madagascariensis J. and P. of Madagascar Mahogany.

Khaya senegalensis Juss., Senegal Khaya, African Mahogany. (Meliaceae). - Trop. Africa. Wood resembles that of Swietenia Mahogany; used for cabinet work, joinery; can be cut into fine planks, also used for canoes. Much exported.

Khumbut → Acacia Jacquemontii Benth.

Kickxia elastica Preuss. → Funtumia elastica Stapf.

Kidney Bean → Phaseolus vulgaris L.

Kidney Vetch → Anthyllis Vulneraria L.

Kidney Wood → Eysenhardtia polystachya (Ortega) Sarg.

Kielmeyera coriacea Mart., Pau Campo. (Gutti-feraceae). - Tree. S. America, esp. Brazil. Bark is used in Brazil as a source of cork. Has been suggested as a substitute of cork-oak bark during times of scarcity.

Kigelia acutifolia Engl. (Bignoniaceae). - Tree. Trop. Africa. Bark used for dysentery in some parts of Cameroon.

Kigelia africana Benth. African Sausage Tree. (Bignoniaceae). - Tree. S. Africa. Fruit is used as purgative and for dysentery. Sold in native markets.

Kiggelaria africana L., Natal Mahogany (Samydaceae). - Tree. S. Africa, esp. Cape Peninsula to Natal and Transvaal. Wood pink, compact, close-grained; used for boards, cabinet work and furniture.

Kingwood → Dalbergia cearensis Ducke.

Kinka Oil Ironweed → Vernonia anthelmintica Willd.

Kino, African → Pterocarpus erinaceus Lam.

Kino, Australian → Eucalyptus rostrata Schlecht.

Kino, Bengal → Butea superba Roxb.

Kino, East Indian → Pterocarpus Marsupium Roxb.

Kino Eucalyptus → Eucalyptus resinifera Smith.

Kino Gum, Red→ Eucalyptus rostrata Schlecht.

Kino, Jamaica → Coccolobis uvifera (L.) Jacq.

Kino, Malabar → Pterocarpus Marsupium Roxb.

Kino, Ribbon Gum → Eucalyptus amygdalina Lab.

Kirschwasser → Prunus Cerasus L.

Kismis → Actinidia callosa Lindl.

Kittul Fibre → Caryota urens L.

Klaineodoxia gabonensis Pierre. (Simaruba-ceae). - Tree. Trop. W. Africa. Seeds are eaten fresh, roasted or crushed into a paste. Wood open-grained, fairly coarse, heavy, very hard, red-brown to golden brown; used for planks of steamers on the Congo. In Liberia wood is used for poles and boards.

Klamath Plum → Prunus subcordata Benth.

Kleinhovia Hospita L. (Sterculiaceae). - Small tree. Trop. Asia, from Mascarene Islands to Polynesia. Juice of leaves is esteemed as eyewash. Bark source of cordage.

Kleinia pteroneura DC. (syn. Senecio pteroneurus Sch.) (Compositae). - Succulent herb N. Africa. Aromatic juice is used for intestinal troubles. Twigs are used by the natives for rheumatics.

Knackaway → Ehretia elliptica DC.

Knightia excelsa R. Br., Rewarewa, Maori Honey Suckle. (Proteaceae). - Tree. New Zealand. Wood is esteemed for cabinet work. It is deep red, straight-grained, beautifully marbled.

Knot Grass → Polygonum aviculare L.

Knotweed, Sachalin → Polygonum sachalinense F. Schmidt.

Koa Acacia → Acacia Koa Gray.

Kobus Magnolia → Magnolia Kobus DC.

Kochia aphylla R. Br., Salt-Bush. (Chenopodiaceae). - Shrub. Australia. Important as stock feed, during protracted droughts when nothing else is obtainable.

Koda Millet → Paspalum scorbiculatum L.

Koellia albescens Ktze. → Pycnanthmum albescens Torr. and Gray.

Koellia incanum Ktze. → Pycnanthemum incanum (L.) Michx.

Koellia virginiana L. → Pycnanthemum virginianum (L.) Durand and Jacks.

Koelreuteria apiculata Rehd. and Wils. (syn. K. paniculata var. apiculata Rehd. and Wils.) (Sapindaceae). - Tree. China, Korea, Japan. Flowers source of a yellow dye in China.

Koelreuteria paniculata Laxm. (Sapindaceae). - Tree. Kansu, Szechuan, Chekiang, Chihli (China). Flowers used medicinally in China. Source of a yellow dye. Black seeds are used for necklaces.

Kohlrabi → Brassica oleracea L. var. gongyloides L.

Kokam Butter → Garcinia indica Choisy.

Kokkia Rockii Lewton. (Anacardiaceae). - Tree. Haiwaii. Bark contains a reddishbrown juice; used by Hawaiians for dyeing fish-nets.

Koko → Euphorbia lorifera Stapf.

Kokoona zeylanica Thw. Kokum, Wana pota. (Celastraceae). - Tall tree. Ceylon. Seeds are source of an acrid oil, Pota-eta-tel, used as a preventative for leeches.

Kok-saghyz → Taraxacum kok-saghyz Rodin.

Kolomitka Vine → Actinidia callosa Lindl.

Kombu → Laminaria japonica Aresch., L. ochotensis Miyabe etc.

Konjac → Amorphophallus Rivieri Dur.

Konjaku Flour → Amorphophallus Rivieri Dur.

Konnyaku Powder → Amorphophallus Rivieri Dur.

Koolim → Scorodocarpus borneensis (Baill.) Becc.

Korakan → Eleusine coracana (L.) Gaertn.

Kordofan → Acacia Senegal (L.) Willd.

Korean Pine → Pinus koraiensis Sieb. and Zucc.

Korthalsia scaphigera Mart. (Palmaceae). - Rattan. Mal. Archipel. Stems are used for basketwork.

Kosteletzkya pentacarpa Led. (Malvaceae). - Perennial herb. Russia. Has medicinally similar properties as Althaea officinalis L.

Kowarkul → Acacia Cunninghamii Hook.

Krakorso → Amomum thyrsoideum Gagn.

Krameria argentea Mart. (Leguminosaceae). - Shrub. Brazil. Source of Pará or Brazilian Rhatany. Dried roots used medicinally as an astringent and tonic; also used for tanning. Contains 7.2% tannin.

Krameria tomentosa St. Hil. (Leguminosaceae). - Woody plant. Trop. America- esp. West Indies, Mexico, N. Brazil, Colombia. Source of Savanilla-Ratanhia Root, used for tanning.

Krameria triandra Ruiz. and Pav. Peruvian Krameria. (Leguminosaceae). - Low shrub. Bolivian and Peruvian Cordilleras. Source of Peruvian Rhatany. Dried root is used medicinally, astringent and tonic. Contains krameric acid. Was used in early days by native women of Lima as astringent and tooth preservative. Root is also used for tanning, containing 10% tannin. Powdered root is said to be used in certain tooth-powders.

Krervanh → Amomum Krervanh Pierre.

Krobonko → Telfairea occidentale Hook. f.

Kudzu Vine → Pueraria Thunbergiana Benth.

Kümmel → Carum Carvi L.

Kumpta Cotton → Gossypium obtusifolium Roxb.

Kumquat, Australian Desert → Eremocitrus glauca Swingle.

Kumquat, Marumi → Fortunella japonica (Thunb.) Swingle.

Kumquat, Meiwa → Fortunella crassifolia Swingle.

Kumquat, Nagami → Fortunella margarita (Lour.) Swingle.

Kundura Unsa → Boswellia Carterii Birdw.

Kurchi Bark → Holarrhena antidysenterica Wall.

Kurwini Mango → Mangifera odorata Grif.

Kussum Oil → Schleichera trijuga Willd.

Kussum Tree → Schleichera trijuga Willd.

Kutcha Bloodwood → Eucalyptus terminalis F. v. Muell.

Kutira Gum → Cochlospermum Gossypium L.

Kydia calycina Roxb. (Malvaceae). - Small tree. Dry regions of India and Burma. Fibrous bark used in India as source of course ropes. Mucilaginous material obtained from the stems is used for clarifying sugar.

Kyllingia triceps Rottb. (Cyperaceae). - Perennial herb. Tropics. Aromatic root is used in some countries as antispasmodic and for leucorrhea.

Kwatakwari → Adansonia digitata .

L

Labiaceae → Aeolanthus, Agastache, Anisome-les, Calamintha, Cedronella, Clinopodium, Coleus, Collinsonia, Cunila, Elsholtzia, Galeopsis, Hedeoma, Hoslundia, Hyptis, Hyssopus, Lallemantia, Lamium, Lavandula, Leonitis, Leunurus, Leucas, Lycopus, Majorana, Marrubium, Melissa, Mentha, Mesona, Micromeria, Microtanea, Mischosma, Monarda, Nepeta, Ocimum, Origanum, Orthosiphon, Perilla, Plectranthus, Pogogyne, Pogostemon, Poliomintha, Prunella, Pycnanthemum, Pycnonthamnus, Rosmarinus, Salvia, Satureja, Scutellaria, Sideritis, Stachys, Teucrium, Thymus, Zizyphora.

Labuan Manila → Agathis loranthifolia Salisb.

Lablab → Dolichos Lablab L.

Labrador Tea → Ledum groenlandicum Oeder.

Laburnum anagyroides Med. (syn. Laburnum vulgare Griseb.) Golden-Chain. (Leguminosaceae). - Tree. Centr. Europe, Balkans. Wood very hard; used for turnery, instruments. Leaves have been mentioned as a substitute for tobacco, although every part of the plant, esp. the seeds are poisonous.

Laburnum vulgare Griseb. → Laburnum anagyroides Med.

Laburnum, Indian → Cassia Fistula L.

Lacca Lignum → Dalbergia Junghuhnii Benth.

Lace Bark → Lagetta lintearia Lam.

Laciniaria punctata (Hook.) Kuntze. (syn. Liatris punctata Hook.) (Compositae). - Perennial herb. N. America. Roots are consumed as food by the Tewa Indians in New Mexico.

Lacktree, Malay → Schleichera oleosa Merr.

Lacquer, Burmese → Melanorrhoea usitata Wall.

Lacquer, Japanese → Rhus vernicifera DC.

Lactarius congolensis Beeli (Agaricaceae). - Basidiomycete Fungus. Trop. Africa. Fruitbodies are consumed by the natives of the Belgian Congo.

Lactarius deliciosus (L.) Gray. (Agaricaceae). - Basidiomycete. Fungus. Temp. zone. Fruitbodies are consumed as food. With L. sanguifluus, it is the most used mushroom in Catalonia. Sold in markets. Much consumed in Europe, Siberia and Japan. Also L. sanguiluus (Paul) Fr.; and L. subpurpureus Peck from North America.

Lactarius helvus (Fr.) Fr. (Agaricaceae). - Basidiomycete Fungus. Northern temp. zone. Fruitbodies when pulverized are used for flavoring soups and salads in some countries of Europe. Also L. camphoratus L.

Lactarius piperatus (Scop.) Gray. (Agaricaceae). - Basidiomycete. Fungus. Temp. zone. The acrid fruitbodies are consumed in several countries. Were used as drug by the Chinese in Yunnan. Considered by French physicians as a antiblennorhagic when taken internally.

Lactarius torminosus Schäff. (Agaricaceae). - Basidiomycete. Fungus. Temp. zone. The fruitbodies are much consumed as food, either fried or boiled, esp. in Sweden and Russia.

Lactarius scrobiculatus (Scop.) (Agaricaceae). - Basidiomycet. Fungus. Temp. zone. Fruitbodies are consumed when pickled or salted in various parts of Russia. Also L. repraesentaneus Britz., L. vellereus Fr. and L. rufus (Scop.) Fr.

Lactarius volemus (Fr.) Fr. (Agaricaceae). - Basidiomycete Fungus. Temp. and subtrop. zones. Fruitbodies are eaten, they have a fine flavor. Sold in markets of Europe. Also L. luteolus Peck from N. America; L. hygrophoroides B. and C., from Eastern Asia and America; L. flavidulus Imai from Japan and L. lignyotus Fr. from the temp. zone.

Lactobacillus acidophilus (Moro) Holland. (Bacteriaceae). - Bacil. Acidophilus milk is obtained by inoculating, sterile, fresh milk or partially skimmed milk with the aid of pure cultures. The product should be slightly sour in flavor and should have an odor like that of buttermilk.

Lactobacillus bulgaricus (Luerssen and Kühn) Holland. (Bacteriaceae). - Bacil. Is source of a fermented milk. It is the dominant organism in Yoghurt, originally used by Bulgarian tribes.

Lactobacillus casei Rogers. (Bacteriaceae). - Bacil. Takes part in the manufacturing process of Kumiss, a fermented milk; originally used as food among the Kumanes in Russia. Also a yeast species takes part in the fermentation process. It is prepared from mare's or cow's milk. Fermentation takes place in skin or leather bags or in open vessels.

Lactobacillus cucumeris Berg. et al. (Bacteriaceae). - This microorganism takes an important part in the manufacturing of Sauerkraut. Also L. plantarum Holland. After this process the fermentation is completed by L. pentoaceticus Fred. Peterss. and Davenp.

Lactobacillus Delbrueckii (Bacteriaceae). - Bacil. White calcium lactate is manufactured as a fermentation product. Is also important in the production of lactic acid.

Lactuca canadensis L. Wild. Canada. Lettuce. (Compositae). - Annual. N. America, West Indies. Leaves and stems when a few cm long make a good potherb. Considered an emergency foodplant.

Lactuca denticulata Maxim. (Compositae). - Herbaceous herb. China, Japan. Leaves eaten in China as a vegetable.

Lactuca perennis L. Perennial Lettuce. (Compositae). - Perennial herb. Centr. and S. Europe. Young or bleached leaves are used as salad in some parts of S. Europe.

Lactuca sativa L., Lettuce. (Compositae). - Annual. Origin uncertain, may have been derived from L. Scariola L. Cultivated as vegetable in Old and New World. The different varieties may be divided as follows: I. Butter, Cabbage Heads: Philadelphia Butter, California Cream Butter, Big Boston, Sugar Loaf. Bunching varietis: Golden Heart, Oak Leaved, Earliest Cutting. II. Crisp, Cabbage Heads: Brittle Ice, Hansen, Iceberg, Mammoth, Mignonette. Bunching varieties: Boston Curled, Grand Rapids, Chartier, Prize Head. III. Cos, Spathulate-leaved varieties: Dwarf White Cos, Giant White Cos, Paris White Cos. Lanceolate-leaved varieties: Asparagus, Lobed Leaved. Lettuce contains a bitter principal, called lactucin. Celtuce is a large-leaved thick-stemmed form of lettuce, which is grown for its thick stem. It is eaten raw or cooked.

Lactuca Scariola L., Prickly Lettuce. (Compositae). - Herbaceous annual. Europe, Asia, naturalized in North America etc. Grown in Upper-Egypt. Seeds source of Egyptian Lettuce Seed Oil; semi-drying; pleasant flavor, used in foods. Sp. Gr. 0.9247—0.9334; Sap. Val. 190; Iod. No. 122—136.

Lactuca Thunbergii Maxim. (Compositae). - Herbaceous. Temp. China, Mongolia, Japan. Used medicinally in China.

Lactuca virosa L., Bitter Lettuce, Lettuce-Opium. (Compositae). - Biennial herb. Centr. and S. Europe. Cultivated in Germany, France and England. Source of Lactucarium. Dried milky juice is used medicinally, being a mild hypnotic, sedative, expectorant, anodyne, diuretic. Contains the bitter principles lactuco-picrin, lactucin; and the very bitter lactucic acid; also lacturerol. Used as a home remedy in cough mixtures.

Lactucarium → Lactuca virosa L.

Lady's Comb → Scandix Pecten-Veneris L.

Lady's Leek → Allium cernuum Roth.

Lady's Mantle → Alchemilla vulgaris L.

Lady's Slipper, Yellow → Cypripedium parviflorum Salisb.

Lafoensia punicaefolia DC. (Lythraceae). - Tree. Trop. America. Tree produces a fine yellow dye.

Lagenaria vulgaris Ser., Calabash. (Cucurbitaceae). - Vine. Probably from Africa. Dry shell of fruit is used for bowls, ladles, bottles, floats, pipes, musical instruments, blowing horns etc. among natives in Africa. Used as cups for drinking Yerba Maté. Young fruits are eaten boiled in some parts of Africa and Asia. Pulp around seeds is purgative; used medicinally in India. Plant has been cultivated since ancient times.

Lagerstroemia Flos-reginae Retz. (syn. L. speciosa Pers.) Queen Crapemyrtle. (Lythraceae). - Tree. Mal. Archip. Wood close-grained, not attacked by insects, brown to reddish brown; used for construction of houses, bridges, railroad ties and boards.

Lagerstroemia hypoleuca Kurz., Andaman Crape Myrtle. (Lythraraceae). - Tree. Andam. Islands. Lumber is known as Pyinma Andaman; used for general carpentry work.

Lagerstroemia parviflora Roxb., Little Flower Lagerstroemia. (Lythraceae). - Tree. E. India. Wood is known as Sida in the lumber trade and is used for general carpentry work.

Lagerstroemia piriformis Koehne, Battinan Crape Myrtle. (Lythraraceae). - Tree. Philippine Islands. Wood is known as Banabu in the lumber trade and is used for general carpentry work.

Lagetta lintearia Lam. Lace Bark. Tree. (Thymelaeaceae). - Small tree. West Indies. Stretched inner bark is source of Lace Bark; made in Jamaica into lace-like material employed for ornamental purposes and as textile.

Lagoecia cuminoides L. (Umbelliferaceae). - Annual herb. Mediteranean region. Seeds are occasionally used in place of Cuminum Cyminum, Cumin for flavoring.

Laguncularia racemosa Gaertn. f. (syn. Bucida Buceras L.). White Mangrove, White Buttonwood, Black Olive Tree. (Combretaceae). - Shrub or tree. Coastal zone. Florida to Trop. America. Bark rich in tanin; used for tanning leather; is an astringent and a tonic.

Lahuán → Fitzroya cupressoides (Molina) Johnst.

Lalang → Imperatia arundinacea Cyrill.

Lallemantia ibirica. Fisch and Mey. (Labiaceae). - Perennial herb. Iran, Afghanistan, Caucasia. Leaves are used as a potherb in Iran. Source of Lallemantia Oil; used in varnishes. Sold in bazaars of Tabriz, Iran for medical use.

Lallemantia Royleana Benth. (Labiaceae). - Perennial herb. Iran to Himalayan region. Seeds are used in Iran for coughs, as aphrodisiac and cardial stimulant.

Lallemantia Oil → Lallemantia ibirica Fisch and Mey.

Lamb's Quarters → Chenopodium album L.

Lambert's Filbert → Coryllus tubulosa Willd.

Lamedor de Moca → Visnea Mocanera L.

Laminaria digitata (L.) Edmonson. Horsetail Kelp, Seaweed, Tangle, Sea Girdles, Sea-staff. (Laminariaceae). - Brown Alga. Temp. N. Atlantic. Manufactured into algin, a seaweed colloid, especially in U.S.A. and England. Used for stabilizing ices and ice creams; in milk pudding, icings, jellies, food preparations, chocolate milk; in pharmaceutical preparations, among which in pills, tablets, sulfanilimide; for hand lotions, cosmetic creams and jellies; in tooth pastes; treatment of boiler water; in casein emulsion paints; sizing of textiles; alginic texture fibres; used as dental impression material; for thickning of printing pastes. Young parts of the plants are used as food. Stems are made into handles of knives and similar materials. Was formerly an important source of iodine and potash. Also L. stenophylla (Kuetz.) J. Ag.

Laminaria japonica Aresch. Japanese Kelp. (Laminariaceae). - Brown Alga. Coast of Eastern Asia, especially Japan, Korea, Siberia. Source of kombu in Japan, being manufactured in a shredded and powdered form, consumed in sauces, soup, and sprinkled on rice; cooked with meat or eaten as a vegetable, and pickle; it is also used when candied and in confectionary, as a substitute for tea; an ingredient used by saké drinkers. Eaten in China stewed with pork. Also L. angustata Kjellm., L. cichorioides Miyabe, L. religiosa Miyabe and L. longipedalis Okam. are used for the same purpose.

Laminaria potatorum Lab. (Laminariceae). - Brown Alga. Sea Weed. Temp. parts of the seas. Plants are an important source of food among the aborigenes of Australia.

Laminaria saccharina (L.) Lamour. Sweet Tangle, Broadleaf Kelp. (Laminariaceae). - Brown Alga. Temp. North Atlantic. Manufactured into algin in the U.S.A. and England. Source of potash, iodine and mannite. See L. digitata (L.) Edmonson.

Laminariaceae (Brown Algae) → Arthrothamnus, Ecklonia, Laminaria.

Lamium album L., White Deadnettle. (Labiaceae). - Perennial herb. Europe, Asia; introd. in N. America. Tops of stems are sometimes eaten as vegetable. Flowers are occasionally used for leucorrhea, catarrh and dropsy. Roots in wine are recommended for stones in the kidney.

Lamium purpureum L., Purple Deadnettle. (Labiaceae). - Annual to perennial herb. Europe, Mediterranean region; introd. in N. America. Herb is sometimes used as styptic, diuretic and purgative.

Lamprothamnus Fosheri Hutch. → Morelia senegalensis A. Rich.

Lamy Butter → Pentadesma butyracea Sabine.

Lanamar → Posidonia australis Hook. f.

Lance Wood → Amelanchier canadensis (L.) Medic.

Lance Wood → Calycophyllum candissimum (Vahl.) DC.

Lancewood, Honduras → Lonchocarpus hondurensis Benth.

Lanceleaf Sandalwood → Santalum lanceolatum R. Br.

Landolphia comorensis Benth. (Apocynaceae). - Woody vine. Trop. Africa. Occasionally source of a good rubber.

Landolphia crassipes Radl. (Apocynaceae). - Woody vine. Madagascar. Source of a rubber, known as Fingotra.

Landolphia Dawei Stapf. (Apocynaceae). - Woody vine. Uganda. Source of a rubber.

Landolphia dondeensis Busse. (Apocynaceae). - Woody vine. E. Africa. Source of a good rubber.

Landolphia Droogmansiana de Wild. (Apocynaceae). - Woody vine. Congo. Source of a good rubber.

Landolphia Gentilii de Wild. (Apocynaceae). - Woody vine. Congo. Source of a good rubber. Known as Rouge de Kassai.

Landolphia Heudelotii A. DC. (Apocynaceae). - Woody vine. Guinea, Sudan, Senegal. Source of a good rubber. Occasionally cultivated.

Landolphia kilimandjarica Stapf. (syn. Clintandra kilimandjarica Warb.) (Apocynaceae). - Woody vine. Trop. Africa, esp. Kilimandscharo. Source of a rubber.

Landolphia Kirkii Dyer. (Apocynaceae). - Woody vine E. Africa. Important source of African wild rubber, known as Mozambique Rouge, Mozambique Blanc, Nyassa and Pine Rubber. Latex coagulates by sea-water, also by lime-juice.

Landolphia Klainei Pierre. (Apocynaceae). - Woody vine. Trop. W. Africa. Source of a rubber. Sold as spindles, balls and thimbles.

Landolphia lucida Schum. (Apocynaceae). - Woody vine. Trop. Africa, esp. Usambara. Source of a rubber.

Landolphia madagascariensis Benth. and Hook. (Apocynaceae). - Woody vine. Madagascar. Source of a rubber, known as Madagascar Rouge. Latex coagulated by 5% H_2SO_4. Rubber not very elastic.

Landolphia mandrianambo Pierre, Mandrianambo. (Apocynaceae). - Woody vine. Madagascar. Source of a rubber.

Landolphia owariensis Beauv. (Apocynaceae). - Woody climber. Trop. Africa, esp. W. Africa to Angola. Source of a rubber known as Rouge du Kassai, Rouge du Congo.

Landolphia Perieri Jum, Piralahi. (Apocynaceae). - Woody vine. Madagascar. Source of a good rubber, called Majunga rouge.

Landolphia senegalensis Kotschy and Peyr. (Apocynaceae). - Woody vine. Senegal, Nigeria, Sudan. Important source of rubber.

Landolphia Stolzii Busse. (Apocynaceae). - Woody vine. E. Africa. Source of a good rubber.

Landolphia Thollonii Dewevre. (Apocynaceae). - Woody climber. Congo, Angola. Source of a rubber obtained from the roots, sometimes called Caoutchouc des Herbes.

Landolphia ugandensis Stapf. (Apocynaceae). - Woody vine. Uganda. Source of a good rubber.

Langsat → Lansium domesticum Jack.

Langsdorffia hypogaea Mart. (Balanophoraceae). - Parasite. Mexico to S. America. Plants are source of a wax, called Siejas; sold in markets of some countries of S. America. Made into candles.

Languas → Alpinia Galanga Willd.

Languas conchigera Burkill → Alpinia conchigera Griff.

Languas Galanga (L.) Stuntz → Alpinia Galanga Willd.

Languas officinarum (Hance) Farw. → Alpinia officinarum Hance.

Lannea amaniensis Engl. and Krause. (Anacardiaceae). - Tree. Trop. Africa. Bark is source of a red dye; used by the Washamba for dyeing cloth. Also L. Stuhlmannii Engl.

Lannea grandis Engl. (Anacardiaceae). - Tree. Mal. Archip. Young leaves are eaten with rice among the Javanese.

Lansium domesticum Jack. Langsat. (Meliaceae). - Medium sized tree. Trop. Asia, esp. Malaya, Cochin-China. Cultivated in tropics of Old and New World. Fruits oval to round, 2 to 5 cm long, straw-colored; skin leathery; pulp whitish, juicy, aromatic; eaten raw, excellent dessert qualities. Dried peels of the fruits are used as incense in some parts of Java.

Lantana Camara L. (Verbenaceae). - Shrub. Trop. America. Infusion of aromatic leaves is sometimes used in home remedies as a tonic and stimulant.

Lantana microphylla Mart. (Verbenaceae). - Shrub. S. America. Leaves are used in some parts of Brazil as aromatic and anti-rheumatic. Fruits are tonic and stimulant.

Lapachillo → Sweetia elegans Benth.

Lapacho → Tabebuia ipe (Mart.) Standl.

Laplacea Curtyana A. Rich., Almendera. (Theaceae). - Tree. West Indies, esp. Cuba. Wood strong, hard; used for cabinet work and general carpentry.

Laplacea semiserrata Camb. (Theaceae). - Tree. Brazil. Decoction of seeds used in Brazil as aphrodisiac and diuretic.

Laportea bulbifera Wedd. (Urticaceae). - Herbaceous plant. Japan. Young shoots are boiled in spring; eaten by the Ainu.

Laportea gigas Wedd. (syn. Urtica gigas Cunningh). Giant Nettle. (Urticaceae). - Tree. New South Wales, Queenisland. Bark yields an excellent fibre.

Laportea Meyeniana (Walp.) Warb. (Urticaceae). - Woody plant. Philipp. Islands. Infusion of leaves and roots are used in Philipp. Islands as diuretic in case of urinary retention.

Laportea photiniphylla. (syn. Urtica photiniphylla Cunningh). Small-leaved Nettle. (Urticaceae). - Woody plant. Australia, esp. New South Wales, Queeensland. Inner bark yields a fibre; used by aborigines of Australia for cordage, dill-bags, fishing-nets.

Lappa major Gaertn. → Arctium Lappa L.

Lappa minor Hill. → Arctium minus (St. Hill.) Bern.

Larap → Eucalyptus dumosa Cunningh.

Larch → Larix americana Michx.

Larch (as lumber trade name) → Abies nobilis Lindl.

Larch, European → Larix europaea DC.

Larch, Western → Larix occidentalis Nutt.

Lardizabala biternata. Ruiz. and Pav. (Lardizabalaceae). - Woody plant. Chile, Peru. Fruits edible, called Aquiboquil; pulp sweet, pleasant flavor; sold in local markets in Chile and Peru.

Lardizabalaceae → Akebia, Lardizabala, Stauntonia.

Large Cane → Arundinaria gigantea (Walt.) Chapm.

Largetooth Aspen → Populus grandidentata Michx.

Larix americana Michx. (syn. L. laricina [Du Roi] Koch.) Tamarack, Eastern Larch. (Pinaceae). - Tree. Eastern N. America to Rocky Mountains. Wood very strong, hard, heavy, durable, close-grained, light brown; used for telegraph poles, upper knees of small vessels, rail-road ties, fence-posts.

Larix dahurica Turcz. Guimatsu, Kui. (Pinaceae). - Tree. Japan, E. Siberia. Wood hard, durable; used in Japan for pillars, foundation piles, beams, supports of mines, railroad ties, bridges, cooperage, pipes for waterworks, ship-building. Bark source of Japanese Terpentine and of tannin.

Larix europaea DC. (syn. L. decidua Mill.) European Larch. (Pinaceae). - Tree. Centr.Europe. Wood elastic, durable, easy to split, reddish-brown, slightly attacked by insects; used for construction work, general carpentry, water-pipes, shingles, furniture, masts, waterworks, A sugar-like product, Manna of Briançon, was formely used medicinally for chronic bladder and bronchial catarrh. It is also used in salves and plasters. Source of Venice Turpentine, Terebinthina Laricina or Terebinthina Veneta.

Larix occidentalis Nutt. Western Larch, Tamarack. (Pinaceae). - Tree. Western N. America. Wood very hard, strong, close-grained, very heavy, bright light red, very durable when in contact with soil; used for rail-road ties, interior finish of buildings, cabinet-making and fence-posts.

Larix Potaninii Batalin. Chinese Larch. (Pinaceae). - Tree. W. Szechuan (China). A valuable timber-tree in China.

Larkspur, Forking → Delphinium Consolida L.

Larkspur, Rocket → Delphinium Ajacis L.

Larrea mexicana Moric. (syn. Covillea tridenata [DC]. Vaill.) Creosote Bush, Greasewood. (Zygophyllaceae). - Shrub. S.W. of the United States and adj. Mexico. Twigs and leaves steeped in boiling water are source of an antiseptic lotion; used for sores and wounds of men and domestic animals. Flower buds are pickled in vinegar and eaten as capers. Plant is employed in some parts of Mexico for rheumatism. Decoction of leaves is used in baths and fomentations.

Larrea nitida Cav. (Zygophyllaceae). - Shrub. Argentina. Herb is used in Chile as excitant, vulneric, emmenagogue and for difficult digestion.

Laserpitium latifolium L. Laserwort, Woundwort. White Gentian. (Umbelliferaceae). - Perennial herb. Europe. Fruits are used by mountain people as stomach tonic; a decoction is used in beer. Employed in vet. medicine.

Laserpitium prutenicum L. (syn. L. selenioides Crantz.). (Umbelliferaceae). - Biennial to perennial herb. Europe. Root is used in home remedies for skin diseases. A resin from the roots is made in France etc. into Thaspia-Plasters.

Laserpitium Siler L. (syn. Siler montanum Crantz.) L. montanum Lam.). (Umbelliferaceae). - Perennial herb. Centr. and S. Europe. Roots and fruit are used by mountain people as a condiment, also for toothache. Fruits are used in Austria for preparation of a liqueur. Was formerly used medicinally.

Lasertwort → Laserpitium latifolium L.

Lasia aculeata Lour. (syn. L. spinosa Thw.). (Araceae). - Perennial herb. Trop. Asia, esp. Bengal, Ceylon and China, W. Malaya. Peeled leaf-stalks and young leaves are eaten as pot herb. Used in Ceylon in curries; in Java in rice.

Lasiophon Kraussii Meis. (Thymelaeaceae). - Woody plant. Trop. W. Africa. Leaves and roots are used as drastic purgative and as fish poison. Also employed for criminal poisoning.

Lastrea anthamantica Moore. (Polypodiaceae). - Perennial herb. Fern. S. Africa. Powdered rhizome used by the Zulus for intestinal worms, esp. tape-worm.

Lathraea Squamaria L., Toothwort. (Scrophulariaceae). - Perennial root-parasite. Centr. Europe, temp. Asia to Himalaya. Root was formerly used for colic, epilepsy, and for convulsions.

Lathyrus ochroleucus Hook. (Leguminosaceae). - Herbaceous plant. Western N. America. Seeds were consumed as food by the Chiptewa and Oijbway Indians.

Lathyrus Ochrus DC. (Leguminosaceae). - Herbaceous plant. S. Europe. Grown occasionally as fodder for live-stock in Greece.

Lathyrus odoratus L., Sweet Pea. (Leguminosaceae). - Annual vine. S. Europe, esp. Italy. Cultivated as ornamental. Flowers are source of an ess. oil, occasionally used in perfumery. Much is manuf. synthetically.

Lathyrus ornatus Nutt. (Leguminosaceae). - Herbaceous plant. Western N. America. Pods were consumed as food by the Indians of Nebraska.

Lathyrus polymorphus Nutt. (syn. L. decaphyllus Pursh.). (Leguminosaceae). - Herbaceous plant. Western N. America. Pods were consumed as food by the Indians of New Mexico.

Lathyrus sativus L., Grass Pearine. (Leguminosaceae). - Annual herb. Mediterranean region. Used as food for sheep. Seeds sometimes used in soup. Considered in some regions poisonous. Cultivated in Europe since ancient times.

Lathyrus sylvestris L., Plat Pea. (Leguminosaceae). - Perennial herb. Europe, W. Asia. Plant is sometimes used as food for live-stock.

Lathyrus tuberosus L., Groundnut Peavine, Earth Chestnut. (Leguminosaceae). - Perennial vine. Europe, W. Asia, Syria, Balkans. Introd. in N. America. Occasionally cultivated and sold in local markets. Tubers boiled eaten as vegetable. During the XVI Century flowers were distilled for perfume.

Lathyrus vestitus Nutt. (syn. L. maritimus Torr.). (Leguminosaceae). - Perennial herb. W. of N. America. Seeds when roasted are used as coffee by the Eskimos of Alaska.

Latipes senegalensis Kunth. (Graminaceae). - Annual grass. Trop. Africa. Seeds are eaten by desert tribes.

Lauraceae → Aniba, Aydendron, Beilschmiedia, Cassytha, Cinnamomum, Cryptocarya, Dicypellium, Eusideroxylon, Laurus, Lindera, Litsea, Macnilus, Massoia, Mespilodaphne, Miscanteca, Nectandra, Ocotea, Persea, Ravensara, Sassafras, Tylostemon, Umbellularia.

Laurel → Kalmia latifolia L. and Laurus nobilis L.

Laurel, Alexandrian → Calophyllum inophyllum L.

Laurel Berry Fat → Laurus nobilus L.

Laurel, Bignay China → Antidesma Bunias (L.) Spreng.

Laurel, Californian → Umbellularia california Nutt.

Laurel, Cherry → Prunus Laurocerasus L.

Laurel, Chile → Laurelia aromatica Juss.

Laurel, Chinese → Antidesma Bunius (L.) Spreng.

Laurel, Great → Rhododendron maximum L.

Laurel, Ground → Epigaea repens L.

Laurel Oak → Quercus imbricaria Michx.

Laurelia aromatica Juss., Chile Saurel. (Monimiaceae). - Woody plant. Chile, Peru. Aromatic seeds are used as spice in some parts of Peru.

Laurelia Novae-Zelandiae Cunningh. (Monimiaceae). - Tree. New Zealand. Wood soft, of great strength, very tough, does not split. Used for boat-building, furniture.

Laurencia papillosa (Forsk.) Grev. (Rhodomelaceae). - Red Alga. Pac. Ocean, Indian Ocean, Mediterranean and adj. seas. Consumed as food by the Hawaiians, who call it Limu Lipeepee. Sold in markets of Honolulu. Also L. obtusa (Huds.) Lamour.

Laurencia pinnatifida (Gmel.) Gamour. Pepper Dulce. (Rhodomelaceae). - Red Alga. Sea Weed. Atl. Ocean, North Sea and adj. waters. The pungent plants are used a condiment in Scotland.

Laurencia Wrightii Kuetz. (Rhodomelaceae). - Red Alga. Plants are used as food by the natives of New Caledonia and other Pacif. Islds.

Laurocerasus officinalis Roem. → Prunus Laurocerasus L.

Laurus nobilis L. Laurel, True Bay. (Lauraceae). - Tree. Mediterranean region. Leaves are used as condiment in different foods. Seeds are source of Laurel Berry Fat or Bay Fat; used for manuf. soap, also used in vet. medic. Sap. Val. 200.9; Iod. No. 82.4; Unsap. 1.0%; Acid. Val. 9.4.

Laurus pyrifolia Willd. → Ocotea squarroa Mart.

Lavandin → Lavandula officinalis Chaix. x L. latifolia Medic.

Lavandula dentata L. Toothed Lavender. (Labiaceae). - Small shrub. Mediterranean region to Iran, Afghanistan. Infusion of leaves is used in Iran for catarrh and for washing wounds.

Lavandula latifolia Medic. (syn. L. Spica Cav.) Broadleaved Lavender. (Labiaceae). - Half-shrub. Mediterranean region. Source of an ess. oil, Oleum Spicae, Oil of Spike, Essence d'Aspic, camphor and cineol scented. Cultivated in S. France and England. Used in vet. medic., in porcelain painting and perfumery. Leaves and flowers are used for scenting laundry. Used in home remedies as emmenagogue and abortive. Excellent source of honey.

Lavandula officinalis Chaix. (syn. L. vera DC., L. spica L.). Lavender. (Labiaceae). - Small shub. S. Europe. Cultivated. Source of Oil of Lavender; used for perfumery, soap industry, lavender water. Is used medicinally as mild stimulant, carminative, for flavoring pharmac. preparations. Contains l-linalyl acetate, gernaiol, linalol. Bunches of herb are used for imparting a pleasant scent in linen.

Lavandula officinalis Chaix. x L. latifolia Medic. Lavandin. (Labiaceae). - A hybrid being very hardy and produces a better yield of oil than its parent plants. Cultivated.

Lavandula stoechas L. (syn. Stoechas officinarum Moench). French Lavender. (Labiaceae). - Shrub. Mediterranean region. Source of Stoechas Oil; used in some parts of Spain for cramp, asthma and lung ailments. Is rich in d-camphen.

Lavatera plebeia Sims. (syn. L. Behriana Schl.) Tree Mallow. (Malvaceae). - Perennial herb.

Australia. Roots of a white-flowered variety are consumed by natives of Australia.

Lavender → Lavandula officinalis Chaix.

Lavender, Broadleaved → Lavandula latifolia Medic.

Lavender, French → Lavandula stoechas L.

Lavender, Toothed → Lavandula dentata L.

Laver → Porphyra.

Laver, Lettuce → Ulva Lactuca.

Lawson Cypress → Chamaecyparis Lawsoniana (Murr.) Parl.

Lawsonia alba Lam. (syn. Lawsonia inermis L.) Henna Shrub. Camphire. (Lythraceae). - Shrub. Trop. Africa, Asia. Much cultivated in Mahomedan countries. Dried leaves are the source of a green powder; used in cosmetics. Sold in markets and bazaars. Used for tinting hands, feet and nails. With Indigo it gives beard and hair a fine blue-black gloss. Hoof, tails, manes and forehead of whitehorses are colored coppery-red. A fragrant ess. oil, called Mehndi, is prepared from the flowers by distillation, used by Arabs for religious feasts. Flowers are also used by Afr. races to impart a fine scent to oils and pomades. Decoction of bark is used in Arabic medicine for jaundice and nervous symptoms. Caucasian Cossacks color their sheepskin bonnets with henna as a protective color. Mummies of Ancient Egypt were found wrapped in henna-dyed cloth.

Leadbush Glasswort → Salicornia fruticosa L.

Leaf Mustard → Brassica juncea L.

Leatherroot Scurfpea → Psoralea macrostachya DC.

Leatherwood → Dirca palustris L.

Leavenworth Goldenrod → Solidago Leavenworthii T. and G.

Lecanora calcarea (L.) Nyl. (Lecanoraceae). - Lichen. Temp. zone. On rocks. Used in Sweden to give a red-brown color to woolens.

Lecanora esculenta Everson., Manna Lichen. (Lecanoraceae). - Lichen. N. Africa, Asia Minor, Greece. Ground mixed with meal is consumed as food by several desert tribes. Is probably the biblical manna of the Israelites.

Lecanora parella Mass., Light Crottle, Crabeye Lichen. (Lecanoraceae). - Lichen. Temp. zone. On rocks. Was source of Orseille d'Auvergne; dye used in France and Great Britain to give woolens a violet color.

Lecanora tartarea Mass. (Lecanoraceae). - Lichen. Temp. zone. On rocks and dry moors. Source of a red or crimson dye. Used in Sweden and Scotland for dying yarn and cloth. Material is collected in May and June; is steeped in stale urine for three weeks; producing a blueish black mass, which is made into cakes of about ³/₄ pounds. They are hung up to dry in peat smoke. When dry the dye can be used for many years. A source of Lacmus, Turnshe, Lacca Musica.

Lecanora ventosa Ach. → Haematomma ventosum (L.) Mass.

Lecanora, Black → Haematomma ventosum (L.) Mass.

Lecanora, Bloody Spotted → Haematomma ventosum (L.) Mass.

Leccinum aurantiacum(Bull.) Gray → Polyporus versipellis Fr.

Leche de Ojé → Ficus anthelmintica Mart.

Lechuguilla → Agave heteracantha Zucc.

Lecidea, Red Fruited → Mycoblastus sanguinarius (L.) Norm.

Lecidea sanguinaria Ach. → Mycoblastus sanguinarius (L.) Norm.

Lecontea amazonica Ducke. (Rubiaceae). - Tree. Brazil. Root is source of a yellow aromatic Sandal Oil, called Oleo Sandalo do Norte.

Lecontea Bojeriana A. Rich. (Rubiaceae). - Woody vine. Madagascar. Stem is source of a black dye; used in Madagascar.

Lecythidaceae → Barringtonia, Cariniana, Couratari, Couroupita, Grias, Lecythis, Napoleona, Planchonia.

Lecythis costaricensis Pitt. Lecythidaceae. - Tall tree. S. and Centr. America. Wood yellowish, red to brownred, fine-grained, dense, very hard, heavy, difficult to split and to cut; used for fancy work, handles of knives, backs of brushes. Wood is called Cocobola Wood.

Lecythis grandiflora Aubl. (Lecythidaceae). - Tree. Guiana. Wood light red to orange red dense, close-grained, very strong, recommended for spokes. Seeds edible, contain a fine oil.

Lecythis longipes Poit. (syn. Eschweilera longipes (Poit.) Miers.) Toledo Wood. (Lecythidaceae). - Tree. Northern S. America, esp. Dutch Guiana. Wood rich in silica; olive brown to reddish brown; heavy, compact, extremely hard, very durable, straight-grained, difficult to work, easy to split, resistant to marine borers, not attacked by Teredo. Used for marine construction work in brackish waters. In importance the wood is followed by L. Sagotiana Miers, and L. subglandulosa (Steud.) Miers.

Lecythis ollaria L. Monkey Pod (Lecythidaceae). - Trop. America. Seeds are source of Sapucaja Oil; used for illumination and manuf. of soap. Seeds are also edible. Likewise L. grandiflora L., L. usitata Miers, L. urnigera L. Wood reddish to grayish brown, easy to split, polishes well, resistent to Teredo and barnacles; used for wharves, piles, sluices, house-framing. Bark is recommended for tanning.

Lecythis paraensis Ducke. (Lecythidaceae). - Tree. Brazil. Seeds are edible, sometimes sold as Para Nuts, Bertholletia excelsa H. B. K.

Lecythis usitata Miers. (Lecythidaceae). - Tree. Brazil. Nuts edible, of excellent quality and flavor.

Lecythis Zabucajo Aubl. (Lecythidaceae). - Tree. Trop. America. Nuts are edible, of excellent quality.

Ledum groenlandicum Oeder, Labrador Tea. (Ericaceae). - Small shrub. Arctic regions, Greenland, Labrador, Canada to N. of United States. Leaves were used as tea during the Revolutionary War in the U.S.A.

Ledum palustre L. Crystal Tea Ledum. (Ericaceae). - Shrub. N. Hemisphere. Leaves used as a tea by the Ainu and Eskimos.

Ledger Bark →Chinchona Ledgeriana Moens.

Leea Curtisii King. Mali-mali. (Leeaceae). - Shrub. Trop. Asia esp. Pera, Mal. Penins. Leaves pounded with Chinese tabacco are used on the head to preserve the hair.

Leea guinensis G. Don. (Leeaceae). - Vine. Trop. Africa. Leaves are used by the natives of Cameroon for colic.

Leea rubra Blume. (Leeaceae). - Shrub. Trop. Asia, esp. Indo-China to Java. In Indo-China used for tape-worm; Causes also intoxication.

Leeaceae → Leea.

Leek → Allium Porrum L.

Leek, Lady's → Allium cernuum Roth.

Leek, Meadow → Allium canadensis L.

Leek, Rose → Allium canadense L.

Leek, Stone → Allium fistulosum L.

Leek, Wild → Allium tricoccum Ait.

Leguminosaceae → Abrus, Acacia, Acrocarpus, Adenanthera, Aeschynomene, Afzelia, Albizzia, Alhagi, Alysicarpus, Amburana, Amerimnon, Amphicarpa, Amphimas, Andira, Anthyllis, Apios, Apuleia, Arachis, Aspalathus, Astragalus, Baphia, Baptisia, Bauhinia, Behaimia, Berlinia, Bowdichia, Brachystegia, Brya, Burkea, Butea, Caesalpinia, Cajanus, Calliandra, Calopogon, Campsiandra, Canavalia, Caragana, Cassia, Castanospermum, Centrolobium, Centrosoma, Ceratonia, Cercidium, Cercis, Cicer, Cladrastis, Clitoria, Copaifera, Crotalaria, Cyamopsis, Cyclopia, Cynometra, Cytisus, Dalbergia, Dalea, Daniella, Derris, Desmodium, Detarium, Dialium, Dichrostachys, Dimorphandra, Diphysa, Diplotropis, Dipteryx, Elephantorrhiza, Entada, Enterolobium, Eperua, Ervum, Erythrina, Erythrophleum, Eysenhardtia, Fillaeopsis, Flemingia, Galega, Genista, Geoffraea, Gleditsia, Gliricidia, Glycine, Glycyrrhiza, Gossweilerodendron, Gourliea, Gymnocladus, Haematoxylon, Hedysarum, Herminiera, Hoffmanseggia, Holocalyx, Hymenaea, Indigo, Inga, Inocarpus, Kerstingiella, Krameria, Laburnum, Lathyrus, Lespedeza, Leucaena, Lonchocarpus, Lotus, Lupinus, Lysiloma, Machaerium, Medicago, Melanoxylon, Melilotus, Mezoneurum, Milletia, Mimosa, Mucuna, Myroxylon, Neobaronia, Neptunia, Olneya, Onobrychus, Ononis, Ormosia, Ornithopus, Pachyrrhizus, Paradaniella, Parkia, Parkinsonia, Parosela, Peltogyne, Peltophorum, Pentaclethra, Periandra, Pericopsis, Petaloste-

mon, Phaseolus, Physostigma, Piptadenia, Piscidia, Pisum, Pithecolobium, Plathymenia, Platymiscium, Pongamia, Prioria, Prosopis, Psochocarpus, Psoralea, Pterocarpus, Pterogyne, Pterolobium, Pueraria, Rafnia, Recordoxylon, Robinia, Sarothamnus, Serianthes, Sesbania, Sindora,
Sophora, Spartium, Sphenostylis, Strombocarpa,
Stylosanthes, Styphnodendron, Swartzia, Sweetia, Tamarindus, Tephrosia, Tetrapleura, Trachylobium, Trifolium, Trigonella, Ulex, Uraria,
Vatairea, Vataireopsis, Vicia, Vigna, Virgilia,
Voandzeia, Whitfordiodendron, Wistaria, Xylia,
Zollernia, Zornia.

Leitneria floridana Chapm. Florida Cork-wood.
(Leitneriaceae). - Tree. Near Appalachicola River in Florida; Brazos River, Texas and S.E.
Missouri. Wood close-grained soft, extremely
light, pale yellow; used for floats of fishing
nets. Is lighter than cork, Sp. Gr. 0.207, while
that of cork is 0.240.

Leitneriaceae → Leitneria.

Lemaireocereus chichipe (Goss.) Britt. and
Rose. Chichipe (Cactaceae). - Xerophytic tree.
Mexico. Fruits edible; sold in markets.

Lemaireocereus griseus (Haw.) Britt. and Rose.
(syn. Cereus griseus Haw.), (Cactaceae). - Xerophytic tree. Venezuela, Curaçao. Cultivated.
Fruit edible, delicious flavor. Wood rich in
potash; its ash is used as fertilizer.

Lemaireocereus queretaroensis (Weber) Safford. (syn. Cereus queretaroensis Weber). Pitahaya (Cactaceae). - Xerophytic tree. Mexico.
Fruits edible.

Lemaireocereus Thurberi (Engelm.) Britt. and
Rose. (syn. Cereus Thurberi Engelm.) Pitahaya, Pitaya dulce (Mex.) (Cactaceae). - Xerophytic plant. Mexico, Arizona. Fruits edible.

Lemna minor L., Duckweed. (Lemnaceae). -
Small floating water plant. Cosmopolitic.
A food for ducks and geese. Also L. trisulca L.,
L. oligorrhiza Kurz. and other spp.

Lemnaceae → Lemna.

Lemon → Citrus Limon Burman.

Lemon, Beebalm → Monarda citriodora Cerv.

Lemon, Canton → Citrus limonia Osbeck.

Lemon, Desert → Atalantia glauca Benth. and
Hook. f.

Lemon Pip Oil → Citrus Limon Burman.

Lemon Seed Oil → Citrus Limon Burman.

Lemon, Sweet → Citrus limetta Risso.

Lemon Verbena → Lippia citriodora Kunth.

Lemon, Water → Passiflora laurifolia L.

Lemonade Sumach → Rhus trilobata Nutt.

Lemonade Tree → Rhus typhina L.

Lemongrass Oil → Cymbopogon citratus (DC.)
Stapf.

Lemonwood → Calycophyllum candidissimum
(Vahl.) DC.

Lennoaceae → Ammobroma.

Lens esculenta Moench. → Ervum Lens L.

Lentibulariaceae → Pinguicula.

Lentil → Ervum Lens L.

Lentinus cubensis Berk and Curt. (syn. Collybia Boryana (Mont.) Sacc.) (Agaricaceae). - Basidiomycete. Fungus. Neotropics. Young fruitbodies are consumed in different countries.

Lentinus edodes (Berk.) Sing. (syn. Cortinellus
shiitake Henn.) (Agaricaceae). - Basidiomycete
Fungus. Eastern Asia. Fruitbodies are much
consumed among the Japanese and Chinese.
Known among the Japanese as shiitake. Sold in
markets when fresh and dry. Canned in Japan.
Cultivated in various parts of China and Japan.
Also exported.

Lentinus Goossensiae Beeli (Agaricaceae). - Basidiomycete Fungus. Belgian Congo. Fruitbodies are consumed by the natives. Also L. li-
vidus Beeli and L. piperatus Beeli.

Lentinus exilis Klotz. (Agaricaceae). - Trop.
Asia, esp. Malaya. Fruitbodies are consumed as
food in the Philippine Islands.

Lentinus rudis (Fr.) Henn. → Panus rudis Fr.

Lentinus sajor-caju (Fr.) Fr. (Agaricaceae). -
Basidiomycete Fungus. Fruitbodies are much
consumed in China, Indochina, Philippines and
East Indies. Called sajor caju in Malayan. Also:
L. connatus Berk., L. djamor Fr. and L. Arau-
cariae Pat.

Lentinus tuber-regium Fr. (Agaricaceae). - Basidiomycete Fungus. The sclerotia, known as
Pachyma tuber-regium, are consumed by some
races in Africa. Used as medicine, esp. for fever,
in the Malay Archipelago.

Lentisk Pistache → Pistacia lentiscus L.

Leonitis Leonurus R. Brown, Lionsear.
(Labiaceae). - Perennial plant. S. Africa. Decoction of herb is used by farmers and Kaffirs
for snake-bites. Infusion of leaves and flowers
is employed for tapeworm and skindiseases; decoction of leaves is an emmenagogue.

Leontodon carolinianum Walt. → Pyrrhopappus carolinianus (Walt.) DC.

Leontodon hispidus L. (Compositae). Perennial
herb. Throughout Europe, Caucasia, N. Iran,
Asia Minor. Roasted roots have been recommended as a substitute for coffee.

Leonurus cardiaca L. (syn. Cardiaca vulgaris
Moench.). Common Motherwort. (Labiaceae). -
Perennial herb. Europe, Asia. Herb is source of
a dark olive green dye, used in some of the Danube countries. Herb was formerly used for stomach ailments and for children diseases.

Leonurus sibiricus L. Siberian Motherwort. (Labiaceae). - Perennial herb. Siberia, Manchuria;
Hupeh, Kwantung (China), Mongolia, Siberia.
Used medicinally in China. Contains an ess. oil
and leonurin.

Leopard Lily → Belamcanda chinensis DC.

Leopoldina major Wall. (Palmaceae). - Tall palm. Brazil. Ash from fruits is used by certain Indian tribes as a substitute for salt.

Leopoldina Piassaba Wall. Pará Piassaba. (Palmaceae). - Tall Palm. Trop. America. Leaves are source of Piassaba Fibre; used for making heavy ropes, taking the place of Manilla Hemp; it is also used for brooms and brushes.

Lepargyrea canadensis Greene → Shephardia canadensis (L.) Nutt.

Lepidium Fremontii S. Wats. Desert Pepperweed. (Cruciferaceae). - Semi-shrub. Southwest United States. Seeds were used by the Indians of Arizona as food and for flavoring.

Lepidium Meyenii Walpers, Maca. (Cruciferaceae). - Perennial herb. Andine region, Bolivia, Peru. Cultivated. Eaten as a vegetable in Peru, Bolivia and adj. territory.

Lepidium oleraceum Forsk. (Cruciferaceae). - Perennial herb. New Zealand. Plant is used as pot-herb by settlers of South Island (New Zealand).

Lepidium rotundum DC. (Cruciferaceae). - Perennial herb. Australia. Plant when cooked is consumed as vegetable by the aborigenes of S. Australia.

Lepidium sativum L., Garden-Cress, Pepper Grass. (Cruciferaceae). - Annual herb. Temp. zone. Cultivated as a salad plant. Roots are occasionally used as a condiment. In Abyssinia seeds are source of an edible oil. Used medicinally as antiscorbutic, for „blood cleansing".

Lepidopetalum Perrottetii (Camb.) Blume. (Sapindaceae). - Tree. Philipp. Islands. Powdered seeds ar used in some parts of Philipp. Islands for killing wild dogs.

Lepiota congolensis Beeli (Agaricaceae). - Basidiomycete, Fungus. Trop. Africa. The fruitbodies are consumed as food by the natives of the Belgian Congo.

Lepiota procerea (Scop.) Quél. (syn. Agaricus procerus Scop.) Parasol Mushroom. (Agaricaceae). - Basidiomycete. Fungus. Temp. zone. The large fruitbodies are consumed as food and are of excellent quality. Also L. mastoidea (Fr.) Quél., L. rachodes (Vitt.) Quél. and other species.

Lepironia mucronata Rich. (Cyperaceae). - Perennial herb. S. E. Asia. Mal. Archipel. Fiji. Cultivated in Indonesia. Used for basket work, mats for packing tobacco, rubber, kapok, cotton, gambir etc.

Lepisanthes cuneata Hiern. (Sapindaceae). - Small tree. Mal. Penin. Leaves and stems used for coughs.

Lepraria, Brimstone Colored → Lepraria chlorina (DC.) Ach.

Lepraria chlorina (DC.) Ach., Brimstone Colored Lepraria. Lichen. Temp. zone. Used in Scandinavia as a source of brown dye to color woolens.

Lepraria iolithus (L.) Ach. Lichen. Temp. zone. Used in Scandinavia as source of a brown dye to dye woolens.

Leptactina senegambica Hook. (Rubiaceae). - Shrub. Trop. Africa. Source of an ess. oil; used in perfumery; gardenia-scented. Also L. densiflora Hook.

Leptadenia Spartum Wight. (syn. L. pyrotechnica Decne.) (Asclepiadaceae). - Woody plant. Trop. Africa. Slimy fruits and young twigs are used as food by the Beduins. Wood is used as fuel.

Leptandra virginica (L.) Nutt. → Veronica virginica L.

Leptocarpus chilensis Mast. (syn. Schoenodum chilense Gray.) Canutillo. (Restionaceae). - Herbaceous plant. Chile. Used in Chile for manuf. mats.

Leptochlaena pauciflora Baker. (Schizochlaenaceae). - Tree. Madagascar. Wood used in Madagascar for house building.

Leptochloa capillacea Beauv. (syn. L. chinensis Nees.) (Graminaceae). - Perennial grass. Africa, Australia. Grass is used as fodder for live stock. Seeds are used as food in Africa during times of famine.

Leptochloa dubia (H. B. K.) Nees. Green Sprangletop. (Graminaceae). - Perennial grass. Southern United States to Argentina. Important for grazing and for hay.

Leptomeria acida R. Br., Native Currant. (Santalaceae). - Woody plant. Tasmania, New South Wales, Queensland. Berries of sub-acid flavor; used as jelly and preserves in Australia. Also L. aphylla R. Br., L. Billardieri R. Br.

Leptonychia chrysocarpa Schum. (Sterculiaceae). - Small tree. Sudan. Wood is used for carving various ornaments.

Leptospermum ericoides Rich. Heathtea Tree. Manuka. (Myrtaceae). - Tree. New Zealand. Wood hard, durable; used for jetty piles, spokes and wheels.

Leptospermum flavescens Sm. (syn. L. Thea Willd.) Yellow Tea Tree. (Myrtaceae). - Tree. Australia. Leaves are used as a tea in some parts of Australia.

Leptospermum scoparium Forst. Broom Tea Tree. (Myrtaceae). - Shrub or small tree. New Zealand. Wood deep red, straight-grained, compact, strong, elastic. Used for cabinet-work, inlaying. Pungent leaves sometimes used in place of tea in New Zeal. Wood used by Maoris for paddles and spears. Bark was used by the aborigenes of New Zealand for covering roofs of huts.

Lespedeza bicolor Turcz. Shrub Lespedeza. (Leguminosaceae). - Perennial herb. China, Korea. Plant has been recommended as fodder for live-stock and as a means for soil conservation.

Lespedeza cyrtobotrya Miq. (Leguminosaceae). - Perennial herb. Japan. The plant has been suggested for forage and for soil conservation.

Lespedeza sericea Benth. (Leguminosaceae). - Perennial herb. Himalayan region, Korea, Japan. Recommended for forage, soil conservation and erosion control.

Lespedeza striata Hook. Common Lespedeza, Japan Clover. (Leguminosaceae). - Herbaceous plant. Eastern Asia. Plant is being grown as pasturage for cattle and as a hay crop.

Lesser Asiatic Yam → Dioscorea esculenta (Lour.) Burk.

Lesser Celandrine → Ranunculus Ficaria L.

Lesser Galangal → Alpinia officinarum Hance.

Lessoniaceae (Brown Algae) → Macrocystis, Nereocystis.

Letharia vulpina (L.) Waino → Evernia vulpina (L.) Ach.

Letterwood → Brosimum Aubletii Poepp. and Endl.

Lettuce → Lactuca sativa L.

Lettuce, Bitter → Lactuca virosa L.

Lettuce, Mountain → Saxifraga erosa Pursh.

Lettuce, Opium → Lactuca virosa L.

Lettuce, Perennial → Lactuca perennis L.

Lettuce, Prickly → Lactuca Scariola L.

Lettuce Tree → Pisonia alba Span.

Lettuce, Wild → Lactuca canadensis L.

Leucadendron argenteum R. Br. Silver Tree. (Proteaceae). - Tree. Cape Peninsula. Soft, silky leaves are sold as curios, bookmarks, mats, fancy articles. Wood soft, spongy; occasionally used for boxes.

Leucaena esculenta (Moc. and Sessé) Benth. (syn. Mimosa esculenta Moc. and Sessé.) (Leguminosaceae). - Tree. Jalisco to Chapas (Mex.). In Mexico seeds are eaten with salt. Pods are sold in local markets.

Leucaena glauca Benth. (syn. Mimosa glauca L.). Jumpy Bean. (Leguminosaceae). - Woody plant. Tropics. Young fruits, leaves and flower-buds are used as a sidedish with rice in Java. Used as greenmanure in tropical countries. The darkbrown seeds are strung into necklaces; used by many people as an ornament.

Leucas martinicensis R. Br. Wild Tea Bush. (Labiaceae). - Perennial herb. Tropics. Herb has mint-like odor; is burned to expel mosquitos. Infusion of leaves is used for gastro-intestinal troubles.

Leucopogon Vieillardii Brongn. and Gris. (Epacridaceae). - Small tree. New Caledonia. Wood fine-grained, hard, close-grained; recommended for inlay-work.

Leucospermum conocarpum R. Br. (Proteaceae). - Shrub or small tree. S. Africa, esp. Cape Peninsula. Bark used for tanning leather. A decoction of bark is recommended as a powerful astringent.

Leucothoë Catesbaei (Walt.) Gray. Fetter Bush. Drooping Leucothoë. (Ericaceae). - Shrub. Eastern United States. Plant is used by some Indian tribes as errhine, promoting discharge of mucous from the nose.

Levant Garlic → Allium Ampeloprasum L.

Levan Madder → Rubia peregrina L.

Levant Scammony → Convolvulus Scammonia L.

Levant Scammony Resin → Convolvulus Scammonia L.

Levant Storax → Liquidambar orientalis Mill.

Levant Worm Seed Plant → Artemisia Cina Berg.

Levisticum officinale (Baill.) Koch. (syn. Angelica Levisticum Baill.). Garden Lovage, Bladder Seed, Lovage angelica. (Umbelliferaceae). - Herbaceous plant. S. Europe. Leaves are sometimes eaten bleached, like celery. Fruits are used for flavoring food, soup, confectionary, also liqueurs of the French type. Root is source of Oil of Lovage; used for flavoring. Infusion of roots is used medicinally, being diaphoretic, emmenagogue, carminative, stimulant, aromatic.

Lewisia rediviva Pursh., Bitter Root. (Portulacaceae). - Perennial herb. Western N. America. Root was consumed as a food by some of the Western Indians.

Lexias, Dried → Vitis vinifera L.

Leyssera gnaphaloides L. (Compositae). - Small shrub. S. Africa. Infusion of leaves is used in S. Africa as a tea.

Liatris punctata Hook. → Laciniaria punctata (Hook.) Kuntze.

Liberian Coffee → Coffea liberica Hiern.

Libocedrus Bidwillii Hook. f. (Cupressaceae). - New Zealand. Wood red, straight in grain, very durable. Used for bridges, piles, house-blocks, rail-road ties, shingles, telegraph-poles.

Libocedrus decurrens Torr. California Incense Cedar. (Cupressaceae). - Tree. Western United States. Wood close-grained, soft, light, very durable when in contact with the soil, light reddish brown; used for shingles, laths, fencing, furniture, construction of flumes, interior finish of buildings.

Libocedrus Doniana Endl. (Cupressaceae). - Tree. Endemic to New Zealand. Wood easily worked, oft of great beauty, dark red with dark streaks. Used for shingles, posts, rails, for general building purposes.

Libothamnus nerifolius Ernst. (syn. Trixis neriifolia Bonpl.) (Compositae). - Woody plant. Venezuela. Source of a resin, called Incienso de los Criollos; used as incense.

Licania arborea Seem. (Rosaceae). - Tree. S. Mexico to S. America. Seeds contain an oil; used for burning, also for manuf. candles, soap and greese.

Licania floribunda Benth. (Rosaceae). - Tree. Trop. America. Wood is strong; used for general construction and carpentry work. Also L. micrantha Miq., L. sclerophylla Mart. and L. Turiuva Cham. and Schlecht.

Licania microcarpa Hook. f. (Rosaceae). - Tree. Brazil. Wood is used for general construction and carpentry. Bark is considered astringent.

Licania platypus (Hemsl.) Pitt. Sansapote, Zapote Cabillo. (Rosaceae). - Tree. Centr. America and S. America. Fruits 15 cm. long, rough, brownish. Pulp is occasionally eaten. Sold in markets. Is considered by many unwholesome. Called in Brit. Honduras Monkey Apple.

Licania rigida Benth. (Rosaceae). - Tree. Brazil, esp. Bahia, Ceara, Perahyba, Piauhy. Seeds source of Oiticica Oil, a drying oil; used in varnish, protective coatings. Sp. Gr. 0.9630-0.9697; Sap. Val. 186-195; Iod. No. 140-152; Unsap. 0.4-0.9%.

Lichen, Ash Twig → Ramalina fraxinea (L.) Ach.

Lichen, Boulder → Parmelia conspersa (Ehrh.) Ach.

Lichen, Bronze Shield. → Parmelia olivacea (L.) Ach.

Lichen, Cedar → Cetraria juniperina (L.) Ach.

Lichen, Common Twig → Ramalina calicaris (L.) Röhling.

Lichen, Crabeye → Lecanora parella Mass.

Lichen, Dog → Peltigera canina (L.) Willd.

Lichen, Gold Edge → Sticta crocata (L.) Ach.

Lichen, Horsehair → Alectoria jubata (L.) Ach.

Lichen, Manna → Lecanora esculenta Everson.

Lichen, Map → Rhizocarpon geographicum (L.) DC. f. atrovirens (L.) Mass.

Lichen, Mealy Blister → Physcia pulverulenta (Schreb.) Nyl.

Lichen, Midnight → Parmelia stygia (L.) Ach.

Lichen, Pale Shield → Cetraria glauca (L.) Ach.

Lichen, Pine → Cetraria pinastri (Scop.) S. Gray.

Lichen, Powdered Swiss. → Nephroma parilis Ach.

Lichen, Puffed Shield → Parmelia physodes (L.) Ach.

Lichen, Ring → Parmelia centrifuga (L.) Ach.

Lichen, Rose and Gold → Sticta aurata Ach.

Lichen, Smoky Shield → Parmelia omphalodes (L.) Ach.

Lichen, Snow → Cetraria nivalis (L.) Ach.

Lichen, Sprout → Dermatocarpon miniatus (L.) Mann.

Lichen, Swedish Shield → Cetraria fahlunensis (L.) Schaer.

Lichen, Trumpet → Cladonia fimbriata (L.) Willd.

Lichen, Warty Leather → Lobaria scrobiculata (Scop.) DC.

Lichen, Wrinkled Shield → Parmelia caperata (L.) Ach.

Lichens source of dyestuffs → Dyestuffs. Lichens source of.

Lichens source of perfumery → Perfumery.

Lichtensteinia pyrethrifolia Cham. and Schlecht. (Umbelliferaceae). - Perennial herb. S. Africa. Source of an intoxicating beverage, called Gli, used by the Hottentots in S. Africa.

Licor de Genipado → Genipa americana L.

Licorice → Glycyrrhiza glabra L.

Licorice, Italian → Glycyrrhiza glabra L.

Licorice, Russian → Glycyrrhiza glabra L.

Licorice, Spanish → Glycyrrhiza glabra L.

Light Crottle → Lecanora parella Mass.

Lignum Campechianum → Haematoxylon campechianum L.

Lignum Quassiae Jamaicense → Picrasma excelsa Planch.

Lignum Sanctum →Guaiacum officinale L.

Lignum Santali Rubrum. → Pterocarpus santalinus L. f.

Lignum Vitae → Guaiacum sanctum L.

Lignum Vitae, Maracaibo → Bulnesia arborea (Jacq.) Engl.

Ligusticum Bulbocastanum Crantz. → Bunium Bulbocastanum L.

Ligusticum Hultenii Fern. (Umbelliferaceae). - Perennial herb. Alaska and W. Canada. Herb is eaten with fish by the Eskimos of Alaska.

Ligusticum Mutellina (L.) Crantz. (syn. Meum Mutellina Gaertn., Aethusa Mutellina St. Lag.). (Umbelliferaceae). - Perennial herb. Centr. and S. Europe. Excellent fodder plant for live-stock in mountainous areas. Leaves are used in place of parsly; dried herb employed as tea. Decoction of leaves is sometimes given as a stomachic.

Ligustrum Ibota Sieb., Ibota Privet. (Oleaceae). - Shrub or tree. Japan, China. Roasted seeds are used in Japan as a substiute for coffee. Stems produce a wax, accumulated by action of certain insects; used for various industrial purposes.

Ligustrum japonicum Thunb. Japanese Privet. (Oleaceae). - Shrub. Japan. Cultivated. Roasted seeds are used as a coffee in some parts of Japan. Also L. nepalense Wall.

Ligustrum lucidum Ait., Glossy Privet. (Oleaceae). - Shrub or tree. Hupeh, Chekiang, Szechuan (China). A commercial insect wax is produced on the branches; used in China.

Ligustrum vulgare L. European Privet. (Oleaceae). - Shrub or small tree. Europe, W. Asia, N. Africa, introd. in N. America. Wood hard, difficult to split; used for small tools; source of charcoal. Bark produces a yellow dye, employed for coloring wool. Fruits produce an ink, also a green and black dye. Young twigs are suitable for basketry.

Lilaea subulata H. B. K. (Juncaginaceae). - Perennial herb. Trop. America. Leaves are used by the natives for covering huts and for manuf. brooms.

Liliaceae → Agapanthus, Aletris, Allium, Aloë, Androcymbium, Asparagus, Asphodelus, Astelia, Brodiaea, Calochortus, Camassia, Chamaelirium, Chlorogalum, Colchicum, Convallaria, Cordyline, Dasylirion, Dianella, Dracaena, Drimia, Eremurus, Erythronium, Fritillaria, Funkia, Hemerocallis, Hesperaloe, Hyacinthus, Lilium, Medeola, Merendera, Molina, Muscari, Phormium, Polygonatum, Rhipogonum, Samuela, Sansevieria, Schoenocaulon, Smilax, Taetsia, Trillium, Tulbaghia, Tulipa, Urginea, Uvularia, Veratrum, Xanthorrhoea, Xerophyllum, Yucca.

Lilium auratum Lindl. Goldband Lily (Liliaceae). - Bulbous perennial. Japan. Bulbs grown to great size in Japan where they are consumed as a vegtable.

Lilium candidum L. Madonna, Annunciation Lily, Bourbon Lily. (Liliaceae). - Perennial, bulbous herbaceous plant. Mediterranean region, S. W. Asia. Cultivated as an ornamental. Flowers are source of an ess. oil, used in perfumery. About 480 to 500 kg. flowers produce 1 kg. concrete from which 280 to 300 gr. absolute is obtained. Much derived from S. France. Also L. longiflorum Thunb. Easter Lily has been suggested for the same purpose.

Lilium cordifolium Thunb. (Liliaceae). - Bulbous perennial. Japan. Cultivated. Bulbs are source of a starch made in Japan, where it is used as a food. Young leaves are eaten as a vegetable.

Lilium dauricum Ker-Gawl, Dahurian Lily. (Liliaceae). - Bulbous plant. Dahuria. Bulbs are used as food by the Ainu.

Lilium Glehni F. Schmidt. Turep, Oba-ubayuri. (Liliaceae). - Bulbous perennial. Japan. Bulbs when specially prepared are used as food by the Ainu.

Lilium lancifolium Thunb. (Liliaceae). — Bulbous perennial. Japan. Cultivated in Japan, where bulbs are consumed as food.

Lilium maculatum Thunb. (syn. L. avenaceum Fisch.). (Liliaceae). - Bulbous perennial. Japan. Bulbs used as food by the Ainu. eaten with rice or millets.

Lilium Martagon L., Martagon Lily. (Liliaceae). - Bulbous perennial. Temp. Europe, Asia. Bulbs eaten as food by natives of Mongolia and Siberia, also among the Cossacks along the Volga. Bulbs are often dried in the sun and eaten with reindeer's or cow's milk.

Lilium parviflorum (Hook.) Holz. (Liliaceae). - Bulbous perennial. Western N. America. Bulbs were consumed as food by the Indians of British Columbia.

Lilium philadelphicum L. Wood Lily. (Liliaceae). - Bulbous perennial. N. America. Bulbs are eaten like potatoes by the Indians of Wisconsin and Minnesota.

Lilium pomponicum L. Pompon Lily. (Liliaceae). - Bulbous perennial Siberia. Bulbs when boiled are used as food among the Tartars and the inhabitants of Kamchatka.

Lilium Sargentiae Wils., Sargent Lily. (Liliaceae). - Bulbous perennial. China. Flowers are consumed in some parts of China.

Lilium superbum L. Turkcap Lily. (Liliaceae). - Bulbous perennial herb. Eastern United States. Bulbs were consumed by the Indians, also used for thickening soups.

Lilium tigrinum Ker-Gawl., Tiger Lily. (Liliaceae). - Bulbous plant. China, Japan. Cultivated as foodplant in China and Japan. Bulbs are eaten boiled in China, Cochin China and Japan. Sold in markets. Also L. tenuifolium Fisch and L. spectabilis Link.

Lily, Annuciation → Lilium candidum L.

Lily, Atamasco → Zephyranthes Atamasco Herb.

Lily, Blackberry → Belamcanda chinensis DC.,

Lily, Dahurian → Lilium dauricum Ker-Gawl.

Lily, Goldband → Lilium auratum Lindl.

Lily, Leopard → Belamcanda chinensis DC.

Lily, Madonna → Lilium candidum L.

Lily-of-the-Valley → Convallaria majalis L.

Lily, Palm → Cordyline australis Hook. f.

Lily, Pompon → Lilium pomponicum L.

Lily, Sargent → Lilium Sargentiae Wils.

Lily, Turkcap → Lilium superbum L.

Lily, Wood → Lilium philadelphicum L.

Lima Bean → Phaseolus lunatus L.

Lima, Chickasaw → Canavalia ensiformis DC.

Lima Weed. → Roccella fuciformis (L.) Lam.

Limaō do Matto → Rheedia edulis Planch. and Triana.

Limber Pine → Pinus flexilis James.

Lime → Citrus aurantifolia Swingle.

Lime, Musk → Citrus microcarpa Bunge.

Lime, Ogeeche → Nyssa Ogeche Marsh.

Lime, Queensland Wild → Microcitrus inodora Swingle.

Lime, Russel River → Microcitrus inodora Swingle.

Lime Seed Oil → Citrus aurantifolia Swingle.

Limequat → Citrus aurantifolia Swingle x Fortunella margarita Swingle.

Limestone Urceolaria → Urceolaria calcarea Sommerf.

Limnanthemum crenatum F. v. Muell. (Gentianaceae). - Perennial waterplant. Australia. Small round tubers when roasted are consumed as food by the aborigenes of Queensland.

Limnanthemum cristatum Griseb. (Gentianaceae). - Perennial herbaceous waterplant Swamps, China, Malaysia. Used as food in China.

Limnocharis emarginata Humb. and Bonpl. (syn. L. flava Buch.) Yellow Velvetleaf. (Butomaceae). - Perennial herb. Mal. Archip. Young leaves and inflorenscenses are eaten with rice among the Javanese.

Limnophila aromatica Mer. (syn. L. villosa Blume). (Scrophulariaceae). - Small aromatic shrub. India to Australia. Used in Malaysia for poulticing sores on legs. In India decoction used as expectorant.

Limnophila Roxburghii G. Don. (Scrophulariaceae). - Herbaceous plant. Trop. Asia, Pacif. Islands. Aromatic leaves are used in cooking; also for perfuming the hair.

Limonia monophylla Roxb. → Atalantia monophylla DC.

Limonscented Iron Bark → Eucalyptus Staigeriana F. v. Muell.

Limu Akiaki → Ahnfeldtia concinna J. Ag.

Limu Eleele → Enteromorpha prolifera (Muell.) J. Ag.

Limu Huna → Hypnea nidifica J. Ag.

Limu Kala → Sargassum echinocarpum J. Ag.

Limu Kohu → Asparagopsis Sanfordiana Harv.

Limu Lipahapaha → Ulva fasciata Delile.

Limu Lipeepee → Laurencia papillosa (Forsk.) Grev.

Limu Lipoa → Dictyota acutiloba J. Ag.

Limu Manauea → Graciliaria coronopifolia J. Ag.

Linaceae → Ctenolophon, Hugonia, Ixonanthes, Linum.

Linaloé Oil → Bursera Aloëcylon Engl. Aniba rosaeodora Ducke.

Linaloé Seed Oil → Bursera Aloëxylon Engl.

Linaloé Wood, Mexican → Bursera Delpechianum Poisson.

Linden, Japanese → Tilia japonica Simk.

Linden, Tuan → Tilia Tuan Szyszyl.

Lindenleaf Sage → Salvia tilaefolia Vahl.

Lindera Benzoin (L.) Blume (syn. Benzoin aestivalis (L.) Nees., Benzoin Benzoin Coult. Spice Bush. (Lauraceae). - Shrub. Eastern N. America to Florida and Louisiana. Leaves are used as a substitute for tea. Berries when dried and powdered were sometimes used instead of Allspice during the revolutionary days.

Lindera praecox Blume. (Lauraceae). - Tree. Japan. Seeds are source of an oil; used in Japan for illumination.

Linen → Linum usitatissimum L.

Linen, Canton → Boehmeria nivea (L.) Gaud.

Linen, China → Boehmeria nivea (L.) Gaud.

Linen, Grass → Boehmeria nivea (L.) Gaud.

Linseed Oil → Linum usitatissimum L.

Linum catharticum L., Purging Flax. (Linaceae). - Annual herb. Europe, Caucasia, W. Asia, Iran, N. Africa. The herb was formerly used as a purgative and diuretic.

Linum Lewisii Pursh., Rocky Mountain Flax. (Linaceae). - Western N. America. Stems were source of a fibre; used by the Klamath Indians and other tribes; made into strings, cords, baskets, mats, meshes for snow-shoes and fishing-nets. Seeds are used in cooking by some Indian tribes.

Linum marginale Cunningh. (Linaceae). - Herbaceous. Australia. Muciloginous seeds consumed by aborigines of Australia. Stem yields a very good fibre; used by the natives for fishing nets and cordage.

Linum usitatissimum L. Common Flax Plant. (Linaceae). - Annual herb. Temp. Europe, Asia, introd. in other continents. Cultivated in temp. zones, since early times. Stem is source of Flax Fibre producing a finer fabric than cotton. Linen goods were known since praehistoric times. Fibre is obtained by retting; a curing of the stems in water. They are durable and have a great tensile strength. Used for cloth, thread, carpets, twine, canvas, fish and seine lines, insulating material, writing paper. Much is derived from Russia, Belgium, The Netherlands, France. Seeds are source of Linseed Oil, a drying oil; when cold-pressed used for eating purposes; when hot-pressed used in paints, varnishes, printing ink, water proofing, soft soap, linoleum, oil-cloth, acid-refined oil in wet plumbing of white lead, thick lithographic varnishes. Sp. Gr. 0.927-0.931; Sap. Val. 189-196; Iod. No. 170-204; Unsap. 0.5-1.6%. Dried ripe seeds used medicinally, being demulcent, emollient; used internally as demulcent, laxative; externally for scalds and burns. Contains a mucilage.

Lionsear, Drug → Leonitis Leonurus R. Br.

Lipoti → Eugenia Curranii C. B. R.

Lippia citriodora Kunth. (syn. Aloysia citriodora Ort.). Lemon-Verbena. (Verbenaceae). - Shrub. Uruguay, Chile, Argentina. Leaves are used as a tea in some parts of S. America.

Lippia dulcis Trév. (Verbenaceae). - Trailing shrub. Trop. America. Dried leaves are used medicinally as demulcent and expectorant; containing ess. oil, camphor and lippiol.

Lippia geminata H. B. K. (Verbenaceae). - Perennial plant. N. and S. America. Leaves are used in some parts of Brazil and Paraguay as a stomachic and nervin.

Lippia ligustrina (lag.) Britt. (Verbenaceae). - S. W. of United States and Mexico. In Coahuila (Mex.) used as remedy for bladder diseases. Emmenagogue and antispasmodic. Used in S. Europe in perfumery.

Lippia lycioides Steud. (Verbenaceae). - Herbaceous plant. N. and S. America. Infusion of flowers is used for catarrh and colds.

Lippia scaberrima Sond. (Verbenaceae). - Small shrub. S. Africa. Dried herb is used medicinally for its hemostatic properties; contains an ess. oil.

Lippia umbellata Cav. (Verbenaceae). - Shrub or tree. Mexico. Centr. America. Plant used for colic.

Lippia Pseudo-thea Schau. (syn. Lantana Pseudo-thea St. Hil.) (Verbenaceae). - Small shrub. S. America. Leaves are used as a tea in some parts of Brazil.

Liqueurs → Cordials.

Liquidambar Altingia Blume → Altingia excelsa Noron.

Liquidambar formosana Hance. Formosa Sweet Gum, Chinese Sweet Gum. (Hamamelidaceae). - Tree. W. and E. China. Wood used as teachests for the higher grade teas. In Kwantung leaves are used as food for a silk-worm.

Liquidambar orientalis Mill. Oriental Sweet Gum, Levant Storax. (Hamamelidaceae). - Tree. W. Asia, Asia Minor. Source of Levant Storax, being an greyish, opaque, sticky mass; sometimes solid or semi-solid, typical taste and color. Parasiticide; for scabies, skin diseases. Used in fumigating powders and pastilles. Stimulating and expectorant. Used for scenting soaps, in various perfumes of the Oriental type. Application in insolation of cinnamic alcohol. Contains α storesin and β storesin, being resins, also cinnamic acid.

Liquidambar styraciflua L. American Sweet Gum, Red Gum. (Hamamelidaceae). - Tree. N. America to Centr. America. Wood hard, heavy, close-grained, not strong, brown, tinged with red; used for cabinet work, interior and exterior finish of buildings, furniture, veneer, wooden dishes, fruit boxes, street pavement, excelsior, radio, phonograph and kitchen cabinets, pulpwood. Sometimes marketed as Satin Walnut and California Red Gum. Trunk is source of American Storax or Styrax, a balsam; much is derived from Honduras and Guatemala. It is a stimulant, expectorant, antiseptic; applied externally for scabies and other skin diseases; in fumigating powders and pastilles. Oil of Styrax is used for scenting soap and perfumery, esp. of the Oriental type.

Liquorice → Glycyrrhiza glabra L.

Liquorice (substitute) → Periandra dulcis Mart.

Liquorice, Chinese → Glycyrrhiza uralensis Fisch.

Liquorice, Indian → Abrus precatorius L.

Liquorice Root → Hedysarum Mackenzii Rich.

Liquorice, Wild → Glycyrrhiza lepidota (Nutt.) Pursh.

Liriodendron tulipifera L., Tulip Tree, Yellow Poplar. (Magnoliaceae). - Tree. Eastern N. America. Wood brittle, soft, not strong, light, easily worked, light yellow to brown; used for wooden-ware, shingles, interior finish, cabinet-work, boxes, fixtures, radio and phonograph cabinets. Inner bark especially of the root is acrid; used as stimulant in native medicine. Source of tulipiferine, an alkaloid which acts violently on the heart and nervous system.

Litchi → Nephelium Litchi Camb.

Lithocarpus densiflora (Hook. and Arn.) Rehd. (syn. Pasania densiflora Oerst., Quercus densiflora Sarg,) Tanbark Oak, Chestnut Oak. (Fagaceae). - Tree. Oregon and California. Bark rich in tannin; used for tanning leather.

Lithospermum canescens Lehm. Hoary Gromwell. (Boraginaceae). - Herbaceous plant. Western N. America. Roots are source of a red dye; used by several Indian tribes.

Lithospermum erythrorhizon Sieb. and Zucc. (Boraginaceae). - Perennial herb. Japan. Cultivated. Roots are source of a purple dye; used in Japan.

Lithospermum hispidissimum Lehm → Arnebia hispidissima DC.

Lithospermum officinale Common Gromwell. L. (Boraginaceae). - Perennial herb. Europe, W. Asia, Caucasia, Iran, introd. in N. America. Leaves source of a tea, called Bohamian or Croatian Tea.

Lithospermum carolinense (Walt.) G. F. Gmel., Carolina Gromwell. (Boraginaceae). - Herbaceous plant. N. America. Source of a dye; used as face paint by the Chippewa Indians.

Litmus → Ochrolechia tartarea (L.) Mass., Rocella fuciformis (L.) Lam. and R. tinctoria DC.

Litrea molle Griseb. → Schinus latifolius Engl.

Litsea calicaris Kir. (Lauraceae). - Tree. New Zealand. Wood firm, very elastic. Suitable where strength, thoughness and elasticity are required; used for cooper's ware, bullock-eyes, ships' blocks, coach panels, shafts, wheelwrights.

Litsea Cubeba Pers. (syn. L. citrata Blume, Tetranthera polyantha Wall.) (Lauraceae). - Shrub or tree. China to Java. Used by natives in medicine. Source of an essential oil. Scented young fruits are used in Java as Sambal; for flavoring goats meat.

Litsea glauca Sieb. (Lauraceae). - Tree. China, Japan. In Japan source of a fixed oil; used for burning and manuf. of soap.

Litsea Neesiana (Schauer) Hemsl., Laurel de la Sierra. (Lauraceae). - Shrub or tree. Mexico. Leaves used in Sinaloa for colic pains.

Litsea novoleontes Bartlett. (Lauraceae). - Shrub. Mexico. Leaves used as a tea with milk and sugar. Induces perspiration.

Litsea tetranthera Mirb. (syn. L. sebifera Blume). (Lauraceae). - Tree. Trop. Asia, esp. India, Malaya. Cultivated. Source of an oil; manuf. into soap; used by the Chinese. Bark is used in the Philipp. Islands for intestinal catarrh.

Little Crowfoot → Eleusine aegyptiaca Desf.

Little Leaf Tunicflower → Tunica prolifera (L.) Scop.

Live Oak → Quercus chrysolepis Liebm. and Q. virginiana Mill.

Livistonia australis Mart. (syn. Corypha australis R. Br., L. inermis Wendl.). Cabbage Palm. (Palmaceae). - Tall palm. Australia. Young unfolded leaves formed in the center of the palm are consumed raw, cooked or baked by the aborigenes.

Livistonia cochinchinensis Blume. (Palmaceae). - Tall palm. Cochin-China. Ripe fruits are eaten by natives of N. Annam and Tonkin. Leaves are used for building roofs of native buildings.

Livistonia Jenkinsiana Griff. (Palmaceae). - Tall palm. N.E. India. Leaves used for thatch and hats. Likewise L. speciosa Kurz in Chittagong; L. chinensis Mart. in Philipp. Islds.; L. australis Mart. in Australia.

Livistonia sariba Merr. (Palmaceae). - Tall palm. Trop. Asia, esp. China, Philipp. Islands to Penang. Leaves used for thatch, matting, hats, rain-coats. Midrib of leaves used for blow-pipe darts by Jakuns of Langat. Endosperm eaten in Indo-China after maceration in vinager or in salt solution.

Loasaceae → Mentzelia.

Loban Maidi → Boswellia Frereana Birdw.

Lobaria calicaris Hoffm. → Ramalina calicaris (L.) Röhling.

Lobaria prunastri Hoffm. → Evernia prunastri (L.) Ach.

Lobaria pulmonaria (L.) Hoffm., Lungwort, Hazelraw, Hazelrottle. Rage. (Stictaceae). - Lichen, Subalpine woods, Europe. On stems of trees. Source of an ess. oil, used in perfumery. Species of tree has much influence upon the quality of the oil. Is said also to be used as tanning material. Produces an orange brown dye, employed in Scandinavia and Great Britain to stain woolens. Plant is also used as Muscus pulmonarius in home remedies.

Lobaria scrobiculata (Scop.). DC., Oak Rage, Warty Leather Lichen. (Stictaceae). - Lichen. Temp. zone. On stems and rocks. Used in Scotland and England as source of a brown dye to color woolens.

Lobelia inflata L., Indian Tobacco. (Campanulaceae). - Annual herb. N. America. Cultivated in New York andMassachusetts. Dried leaves and stem tops are used medicinally; being emetic, expectorant, nauseant. Contains an emetic, acrid alkaloid, lobeline and lobelanidine and a purgent ess. oil lobelianin. Plant is poisonous. Also used are L. cardinalis L., L. syphilitica L. and others.

Lobelia laxiflora H.B.K. (Lobeliaceae). - Perennial herb. Mexico and adj. territory. Roots are considered in Mexico as emetic, expectorant, and antiasthmatic.

Lobelia longiflora L. → Isotoma longiflora Presl.

Lobelia pinifolia L. (Campanulaceae). - Small shrub. S. Africa. Resinous root is used as stimulant, diaphoretic; for skin diseases and chronic rheumatism.

Lobelia succulenta Blume. (Campanulaceae). - Perennial. Java. Young leaves are used with rice by the Javanese.

Loblolly Bay → Gordonia Lasianthus (L.) Ellis.

Loblolly Pine → Pinus Taeda L.

Locust → Robinia Pseudacacia L.

Locust, African → Parkia africana R. Br.

Locust Bean Gum → Ceratonia siliqua L.

Locust, Black → Robinia Pseudacacia L.

Locust Gum → Ceratonia siliqua L.

Locust, Honey → Gleditsia triacanthos L.

Locust, West African →Parkia filicoidea Welw.

Lodgepole Pine → Pinus Murrayana Balf.

Lodoicea callipyge Comm. (syn. L. maldivica Pers., L. seychellarum Pabill.) Double Coconut. (Palmaceae). - Tall palm. Seychelles. Old leaves are used as thatch; young leaves are made into hats. Shells of the large fruits are used for water vessels, or when cut up, also used for platters.

Loeselia mexicana (Lam.) Brand. (Polemoniaceae). — Shrub. Mexico. Decoction of leaves is used in Mexico for fevers; is emetic, diuretic, purgative, sudorific. As a wash prevents falling of hair. Contains loeseline, an alkaloid.

Loganiaceae → Buddleia, Fagraea, Gelsemium, Nuxia, Spigelia, Strychnos.

Logwood → Haematoxylon campechianaum L.

Lolium italicum R. Br., Italian Rye-Grass. (Graminaceae). - Annual to perennial grass. Europe. Cultivated in Old and New World as pasture grass and for hay.

Lolium perenne L., Perennial Rye-Grass. (Graminaceae). - Perennial grass. Europe. Cultivated in Old and New World. Used in permanent pasture mixtures with other grasses.

Lolium temulentum L. Darnel. (Graminaceae). - Annual grass. Temp. zone. The poisonous seeds which contain the mycelium of a fungus, are in some countries occasionally mixed with barley, to give beer an intoxicating effect.

Lomatia hirsuta (Lam.) Diels. Palo Negro. (Proteaceae). - Tree. S. America, esp. Chile, Patagonia, S. Peru. Dark brown sap from bark is used by rural settlers to dye their ponchos. Wood is brown, much esteemed in Chile for furniture.

Lonchocarpus cyanescens Benth. West African Indigo, Yoruba Indigo. (Leguminosaceae). - Woody vine. Trop. Africa. Leaves and young roots are source of a blue dye. Root has been suggested for leprosy.

Lonchocarpus densiflorus Benth., White Haiari. (Leguminosaceae). - Vine. Guiana. Used as a fish poison in Guiana.

Lonchocarpus hondurensis Benth., Honduras Lancewood. (Leguminosaceae). - Tree. Centr. America. Wood is known as Rosa Morada in the lumbertrade and is used for carpentry and general purposes.

Lonchocarpus longistylus Pitt. (Leguminosaceae). - Tree. Mexico, Centr. America. Beverage is made from the bark in Yucatan (Mex.). By the ancient Mayas the bark was soaked in water with honey, which was fermented, being a source of a drink called Balche.

Lonchocarpus latifolius H.B.K. (syn. L. oxycarpus DC.) (Leguminosaceae). - Tree. West Indies. Wood is used for construction work, marquetry and wagons.

Lonchocarpus maculatus DC. → Gliricidia sepium (Jacq.) Steud.

Lonchocarpus Nicou DC. (syn. L. floribundus [Benth.] Killip). (Leguminosaceae). - Liane. Brazil, Peru. Roots have been suggested as a source of rotenone; used as an insecticide. They contain 0.75 to 1% rotenone, also deguelin, tephrosin and toxicarol.

Lonchocarpus sericeus H. B. K. (Leguminosaceae). - Tree. Trop. Africa, America. Wood very hard, durable, keeps under water; used for construction work, turnery. Bark is considered a laxative.

Lonchocarpus urucu Killip. (Leguminosaceae). - Shrubby liane. Trop. S. America. Cultivated. Roots have been suggested as a source of rotenone; used as an insecticide.

Lonchocarpus utilis Smith. (Leguminosaceae). - Woody plant. Peru. Has recently been cultivated as a source of rotenon; an important insecticide.

Long Buchu → Barosma serratifolia (Curtis) Willd.

Long Leaved Pine → Pinus palustris Mill.

Long Pair Plume Grass → Dichelachne crinita Hook. f.

Long Pepper → Piper longum L.

Long Stalked Greenbrier → Smilax pseudo-China L.

Long Yam → Dioscorea transversa R. Br.

Longan berry → Nephelium Longana (Lam.) Cam.

Longbeak Eucalyptus → Eucalyptus rostrata Schlecht.

Longleaf Ephedra → Ephedra trifurca Torr.

Longstalk Holly → Ilex pedunculatus Miq.

Lonicera angustifolia Wall. (Caprifoliaceae). - Shrub. Himalayan region. Berries sweet, size of a pea; eaten by the natives.

Lonicera coerulea L. var. edulis Reg. (Caprifoliaceae). - Small tree. Siberia, N. Caucasia. Berries ovoid, blue; eaten raw or preserved in some parts of Siberia.

Lonicera involucrata (Rich.) Banks. Bearberry. Twinbery. (Caprifoliaceae). - Shrub. N. America. Fruits were consumed by the Indians, hunters and miners.

Lookingglass Tree →Heritiera littoralis Dryand.

Loosestrife, Purple → Lythrum Salicaria L.

Loostrife, Spiked → Lythrum Salicaria L.

Lopez Root → Toddalia aculeata Pers.

Lophira alata Banks. Niam Tree, African Oak, Dwarf Ironwood. (Ochnaceae). - Tree Trop. Africa. Seeds are source of Niam Fat, Niam or Meni Oil; used by the natives in food and as a hair oil, also suitable for soap making. Sp. Gr. 0.9016—0.9044; Sap. Val. 182—195; Iod. No. 70—72.5; Insap. 0.5—0.9%. Wood red brown, very dense and hard, heavy, sinks in water, easy to split and to work; used for flooring, steps, bridges, rail-road ties, paving blocks, furniture and turnery.

Lophopetalum toxicum Loher. (Celastraceae). - Tree. Philippine Islands. Bark is source of an arrow-poison, contains lophopetalin.

Lophocereus Schottii (Engelm.) Britt. and Rose. (syn. Cereus Schottii Engelm.). Cina, Cabesa de Viejo. (Cactaceae). - Xerophytic tree. Sonora, Baja California (Mex.), Arizona. Fruits are edible.

Lophophora Williamsii (Lem.) Coulter. (syn. Echinocactus Williamsii Lem.). Peyote. (Cactaceae). - Low xerophytic, succulent cactus. S. Texas and adj. Mexico. The dried crowns, called Mescal Buttons are chewed by Indians, of several tribes, during religious ceremonies. Contains a narcotic substance, anhalonin, an alkaloid, causing remarkable visions, hallucinations and a feeling of well-being. Has been used since pre-Columbian times.

Lophostemon arborescens Schott → Tristania conferta R. Br.

Loquat → Eriobotrya japonica Lindl.

Loranthaceae → Loranthus, Phorodendron, Phthyrusa, Psittacanthus, Struthanthus, Viscum.

Loranthus calyculata DC. → Psittacanthus calyculata (DC.) Don.

Loranthus divaricatus H.B.K. (Loranthaceae). - Parasitic shrub, living on trees. S. America. A glue is produced from the plant; used for catching birds.

Loranthus grandiflorus King. (Loranthaceae). - Woody parasite. Trop. Asia, esp. Lower Siam, Sumatra, Mal. Penin. In Mal. Penin. leaves are used with turmeric and rice as poultice for ringworm.

Loranthus nummulariaefolius Chev. (Loranthaceae). - Parasite, woody plant. Trop. Africa, Somaliland etc. Bark used by natives for tanning.

Loranthus Zeyheri Harv. (Loranthaceae). - Shrub, parasite on Acacia caffra and A. Karroo. S. Africa. A rubber-like substance is used locally for preparation of a bird-lime.

Lotos → Nelumbium speciosum Willd.

Lotus corniculatus L., Birds-foot Trefoil. (Leguminosaceae). - Perennial herb. Europe, Asia, India, N. and E. Africa. Grown as food for livestock.

Lotus Fruit → Zizyphus Lotus Desf.

Louisiana Moss → Tillandsia usneoides L.

Lovage → Levisticum officinale (Baill.) Koch.

Lovoa Klaineana Pierre and Sprague. African Walnut, Brown Mahogany, Tiger Wood Lovoa. (Meliaceae). - Tree. Trop. W. Africa. Wood light, soft resembling Mahogany in grain, walnut brown; used for furniture, linings, panelling, decorative work. Supposed to be fire resistant.

Lovoa Swynnertonii Bak. fil. (Meliaceae). - Tree. Trop. Africa. Wood durable, heavy, difficult to work; source of Brown Mahogony; used for furniture and fancy articles.

Lowland Ribbon Wood → Plagianthes betulinus Cunningh.

Lowland White Fir. → Abies grandis Lindl.

Loxopterygium Sagotii Hook., Hooboobali Tree. (Anacardiaceae). - Tree. Guiana. Wood is known as Hooboobali in the lumber trade and is used for general carpentry work.

Luban Meti → Boswellia Frereana Birdw.

Lucerne → Medicago sativa L.

Lucerne, Wild →Stylosanthes mucronata Willd.

Lucky Beans → Thevetia nereifolia Juss.

Lucuma Arguacoënsium Karst. (Sapotaceae). - Tree. Colombia. Fruits pear shaped, yellow edible; much esteemed by the inhabitants.

Lucuma bifera Molina. (Sapotaceae). - Tree. Chile, Peru. Grown by the Indians for centuries. Fruits roundish, size of an apple, must be eaten when very ripe, of excellent flavor. Sold in markets of Peru and Chile.

Lucuma Caimito Roem. (syn. Pouteria Caimito Radlk.). Abui. (Sapotaceae). - Tree. Brazil, Peru. Cultivated. Fruits edible, sold in some local markets, esp. Brazil.

Lucuma glycyphloea Mart. and Eichl. (syn. Pradosia lactescens Radlk., Chrysophyllum Buran-

hem Riedel). (Sapotaceae). - Tree. Brazil. Bark is used as tonic, astringent, hemostatic.

Lucuma mammosa Gaertn. →Calocarpum mammosum Pierre.

Lucuma obovata H.B.K., Lucmo. (Sapotaceae). - Small tree. Peru, Chile. Cultivated in Peru. Fruits round to oval; pulp yellow, mealy. Fruits are kept in straw for a few days before being eaten.

Lucuma procera Mart. (syn. Urbanella procera Pierre). Macarandiba. (Sapotaceae). - Large tree. Brazil. Cultivated. Much esteemed dessert fruit in Brazil, also eaten preserved.

Lucuma Rivicoa Gaertn.f.(syn.L.nervosa A.DC.), Canistel. (Sapotaceae). - Tree. Centr. America, West Indies. Fruits ovoid, of good flavor, bright orange with mealy soft pulp, eaten fresh, in custards and sherbets. Plant is also source of a spice, called Canistel; used in some parts of Brazil.

Lucuma salicifolia H.B.K., Yellow Sapote, Zapote Amarillo. (Sapotaceae). - Small tree. Mexico to Costa Rica and Panama. Fruits eaten raw; sold in local markets.

Ludwigia repens Sw. (Onagraceae). - Perennial herb. Tropics. Tops of stems are eaten by natives of Cochin-China as vegetable. Also used as fodder for hogs.

Luehea divaricata Mart. Common Whiptree. (Tiliaceae). - Tree. S. Brazil, Argentina. Wood brown or brownish, fairly hard, heavy, strong, tough, not very durable; used for brush-backs, shoe-soles, pack-saddles, saddle frames, interior construction, wooden ware, furniture and general carpentry.

Luffa acutangula Roxb., Sinkwa Towel Gourd, Dishcloth Gourd. (Cucurbitaceae). - Annual vine. Trop. Asia. Cultivated in some parts of China, Japan. Very young fruits are consumed as vegetable; used with meats or in soup in China and Japan.

Luffa aegyptiaca Mill. (syn. L. cylindrica Roem.), Suakwa. Vegetable Spunge. (Cucurbitaceae). - Annual vine. Trop. Africa and Asia. Cultivated in the tropics of the Old and New World. Young fruits are eaten like squash, in soups and stews. The bleached skeleton of the fruits is used as a sponge for scrubbing.

Luffa purgans Mart. (syn. L. operculata Cogn., Momordica operculata L.). (Cucurbitaceae). - Vine. Trop. America. Fruits after removal of the soft tissues, may be used as a sponge.

Lumbang Oil → Aleurites triloba Forst.

Lumber, Economic and Commercial → Wood, Economic and Commercial.

Lumnitzera coccinea Wight and Arn. (L. littorea Voight). Red-Flowered Mangrove. (Combretaceae). - Small tree. Salt water swamps. Malacca to Polynesia. Wood heavy, fine-grained, yellow-brown; used for boat-building in Polynesia.

Lunaria annua L., Penny Flower. (Cruciferaceae). - Annual to biennial. Europe. Cultivated. Roots are edible before the development of the flowers.

Lungwort → Lobaria pulmonaria (L.) Hoffm., Pulmonaria officinalis L., Sticta aurata Ach. and S. pulmonacea Ach.

Lupine, Egyptian → Lupinus Termis Forsk.

Lupine, Yellow → Lupinus luteus L.

Lupinus albus L. (L. sativus Gaertn.), White Lupine.). (Leguminosaceae). - Annual or biennial plant. Mediterranean region. Grown as food for cattle. Was grown as foodcrop among the Romans. Also grown as green manure.

Lupinus littoralis Dougl., Shore Lupine. (Leguminosaceae). - Herbaceous plant. Western N. America. Roots were consumed as food by the Indians of Oregon and Washington.

Lupinus luteus L. European Yellow Lupine. (Leguminosaceae). - Annual. S. Europe. Grown as green manure in very poor soils, especially in Europe. Roasted seeds have been used as a substitute for coffee for centuries. Fresh seeds contain lupinotoxin, a poison, that is removed by a special process.

Lupinus Termis Forsk., Egyptian Lupine. (Leguminosaceae). - Annual herb. Mediterranean region. Cultivated in Egypt. Seeds sold in markets; eaten by the poorer classes.

Lusitanian Oak → Quercus lusitanica Lam.

Luzon Pine → Pinus insularis Endl.

Lychnis diurna Sibth. → Melandrium album (Mill.) Garcke.

Lycium Andersonii Gray. Anderson Wolfberry. (Solanaceae). - Shrub. Arizona to California. Berries are eaten fresh or dried; used in soup or mush by Indians of Arizona and California. Also L. Berlandieri Gray, L. Fremontii Gray, L. pallidum Miers.

Lycium arabicum Schweinf. Arabian Wolfberry. (Solanaceae). - Shrub. N. Africa, Arabia. Fruits are consumed by the Arabs and Beduins.

Lycium chinense Mill., Chinese Wolfberry. (Solanaceae). - Small shrub-like vine. Eastern Asia. Young leaves are consumed as a vegetable.

Lycium sandwicense Gray. (Solanaceae). - Shrub. Hawaiian Islands. Berries are consumed by the natives.

Lycoperdaceae → Bovista, Calvatia, Lycoperdon.

Lycoperdon perlatum Pers. (Lycoperdaceae). - Basidiomycete. Fungus. Fruitbodies, called puffballs, are consumed as food.

Lycoperdon fuligineum Berk and Curtis. (Lycoperdaceae). - Basidiomycete. Fungus. Trop. Asia. Young fruitbodies are consumed by the natives in different parts of Asia. Also L. piriforme Schaff., L. pratense Pers. L. umbrinum Pers. and other species.

Lycoperdon gemmatum Batsch. (Lycoperdaceae). - Basidiomycete. Fungus. Temp. zone. Fruitbodies were eaten fresh or roasted by the Omaha Indians.

Lycoperdon umbrinum Pers. (Lycoperdaceae). - Basidiomycete. Fungus. Cosmopolit. Juice from fruitbodies is used by the women of Ternate to make their hair glossy and black.

Lycopersicum esculentum Mill. (syn. Solanum Lycopersicum L.). Common Tomato. (Solanaceae). - Annual to perennial plant. S. America. Widely cultivated in the Old and New World. Salad fruit, also used in soups, purée, catsup, paste, chili sauces; prepared with meat and vegetables. Canned soups, canned juice. Green fruits are used in pickles Several varieties. I. Standard, not potato-leaved red: Stone, Globe, Ponderosa, Marglobe. Yellow: Honor Bright, Golden Queen. Potato-leaved red: Mikado, Magnus, Ox Heart. II. Dwarf red: Dwarf Stone, Yellow: Yellow Prince. III. Small fruited, red: Red Currant, Red Cherry. Yellow: Plum, Yellow Cherry. Seeds are source of Tomato Seed Oil, a semi-drying oil, suitable for edible purposes and manuf. soap. Press cake is suitable as fodder for cattle and as fertilizer.

Lycopodiaceae → Lycopodium.

Lycopodium cernuum L. (Lycopodiaceae). - Perennial herb. Fern ally. Tropics. Herb when dried is used for stuffing pillows.

Lycopodium clavatum L., Common Clubmoss. (Lycopodiaceae). - Perennial herb. Fern-ally, Temp. N. Hemisphere. Spores were formerly used to cause artificial lightning on the stage, explosives, pyrotechnics. Used medicinally as a dusting powder for protecting tender surfaces, as absorbent; preventing adhering of pills and for suppositories. Occasionally used are L. Selago L., S. annotinum L., and L. complanatum L. Stems are made into matting in some parts of Sweden.

Lycopus asper Greene. (syn. L. lucidus Turcz. var. americanus Gray.). (Labiaceae). - Perennial herb. N. America. Rootstock when boiled or dried was eaten by the Indians of Minnesota and Wisconsin. Considered an emergency foodplant.

Lycopus lucidus Turcz. (Labiaceae). - Perennial herb. Japan, China. Rhizomes are eaten in Japan as a vegetable when boiled, also when salted.

Lycopus virginicus L. (syn. L. uniflorus Michx. Virginian Bugle Weed. (Labiaceae). - Perennial herb. N. America. Rootstock when cooked is eaten by the Indians of Brit. Columbia.

Lygeum Spartum L. (Graminaceae). - Grass. N. Africa, esp. Tripolis, Tunis, Algeria. Source of a fibre; used for manuf. of mats, ropes, sails etc. See also Stipa tenacissima L.

Lygodesmia grandiflora (Nutt.) Torr. (Compositae). - Perennial herb. Western N. America. Leaves when boiled with meat were consumed by the Hopi Indians in Arizona.

Lygodesmia juncea (Pursh.) D. Don. Rush Skeleton Weed. (Compositae). - Herbaceous perennial. Plains of N. America. Juice used by the Indians of Missouri Valley for chewing. Stems are cut into pieces, from which the latex was collected.

Lygodium circinatum Sw. (Schizaceae). - Climbing fern. Trop. Asia, esp. Mal. Archip. Philippine Islands. Stems are used for basket-work.

Lygodium scandens Swartz. (Schizaceae). - Climbing fern. Tropics of Old and New World. Stems made into baskets in Bismarck Archipelago. In Philipp. Islands made into hats; also bags used for keeping betel-nuts of natives.

Lyophyllum aggregatus Schäff. → Agaricus decastes Fr.

Lyre Leaved Sage → Salvia lyrata L.

Lysiloma divaricata (Jacq.) Macbride. (Leguminosaceae). - Shrub or tree. Mexico. Bark is used for tanning.

Lysiloma latisiliqua Benth. (syn. L. bahamensis Benth.). (Leguminosaceae). - Tree. West Indies, Mexico. Wood is used in the West Indies for boat building.

Lysiloma Sabicu Benth. (Leguminosaceae). - Tree. Cuba, Bahamas, Yucatán (Mex.). Wood heavy, hard, fine-grained, durable in water. Used for ship-building and cabinet work.

Lysimachia clethroides Duby. (Primulaceae). - Herbaceous perennial. China, Japan. Leaves are used as a condiment in Tonkin.

Lysimachia Foenum-graecum Hance. (Primulaceae). - Perennial herb. China. Scent of fenugreek. Usd by Chinese women to scent hair. Roots are chewed to correct fetid breath.

Lysimachia Fortunei Max. (Primilaceae). - Perennial herb. China, Japan, Leaves are used as a condiment in Tonkin.

Lysimachia Nummularia L., Moneywort. (Primulaceae). - Low perennial herb. Europe, introd. in N. America. Herb was formerly used to heal wounds. Leaves and flowers are occasionally used as a tea.

Lysimachia obovata Buch-Ham. (syn. L. candida Lindl.) (Primulaceae). - Perennial herb. India. Used as potherb, esp. in Manipur.

Lysimachia quadrifolia L. Fourleaf Loosestrife. (Primulaceae). - Perennial herb. Eastern N. America. Plant is used by the Indians as an astringent and stomachic.

Lythraceae→Ammannia, Cuphea, Heimia, Lafoensia, Lagerstroemia, Lawsonia, Lythrum, Nesaea, Pemphis, Physocalymna, Woodfordia.

Lythrum Salicaria L. Purple Loosestrife, Spiked Loosestrife. (Lythraceae). - Perennial herb. Europe, intr. in N. America. Young leavy shoots have been used as vegetable; recommended as an emergency foodplant.

M

Maba abyssinica Hiern. (Ebenaceae). - Small tree. Abyssinia. Wood dark, easily polished; used for gun-stocks.

Maba buxifolia Pers. (syn. M. Ebenus Wight) (Ebenaceae). - Tree. Trop. Asia, Africa. Source of an Ebeny Wood; used for furniture and handles of tools.

Maba sandwicensis A. DC. (Ebenaceae). - Tree. Hawaii. Islands. Wood very hard, close-grained, reddish-brown; used by Hawaiians for building houses devoted to the gods.

Mabi → Ceanothus reclinatus L'Her.

Mabola → Diospyros discolor Willd.

Maca → Lepidium Meyenii Walpers.

Macachi → Arjona tuberosa Cav.

Macadamia ternifolia F. v. Muell., Queensland Nut. (Protaceae). - Medium or tall tree. Australia, esp. Queensland, New South Wales. Introduced in different parts of the tropics. Commercially grown in Hawaii. Seeds edible, excellent when roasted and salted.

Macaja → Acrocomia sclerocarpa Mart.

Macaranga gigantea Muell. Arg. (Euphorbiaceae). - Small tree. Trop. Asia, esp. Mal. Penin., Sumatra, Java. Decoction of root-bark is used in Malaya for diarrhea-dysentery. A reddish brown sap from the tree is used by the natives for glueing wooden objects. Also M. Diepenhorstii Muell. Arg.

Macaranga grandifolia (Blanco) Merr. (Euphorbiaceae). - Woody plant. Philippine Islands. Resin from the stems is used in the Philipp. Islds. as an astringent gargle for ulcerated mouth.

Macaranga hypoleuca Muell. Arg. (Euphorbiaceae). - Tree. Sumatra, Borneo, Mal. Penin. Used as febrifuge, expectorant, anti-spasmodic.

Macaranga involucrata Baill. (Euphorbiaceae). - Shrub or small tree. Moluccas. Decoction of bark with cinnamon and licorice, cubebs and cummin is used for gargling for swollen tonsils.

Macaranga Tanarius Muell. Arg. (Euphorbiaceae). - Small tree. Trop. Asia, esp. Tenasserim, S. China, Malaysia. Bark used by Ilocanos in Philippine Islands for manuf. of an alcoholic drink. Source of Binunga Gum; used in the Philipp. Islds. for fastening parts of musical instruments, violins and guitars.

Macaroni Wheat → Triticum durum Desf.

Macassar Agar → Eucheuma muricatum (Gmel.) Web. v. Bosse.

Macassar Ebony → Diospyros utilis Koord. and Val.

Macassar Manila → Agathis alba Foxw.

Macassar Oil → Schleichera trijuga Willd.

Mace → Myristica fragrans Houttuijn.

Mace Bombay → Myristica malabarica Lam.

Maceron → Smyrnium Olusatrum L.

Machaerium angustifolium Vog. (syn. (Drepanocarpus isadelphus E. Mey.). (Leguminosaceae). - Tree. Brazil. Bark is source of a gum-resin; considered in some parts of Brazil as antidote for snake-bites.

Machaerium macrocarpum Ducke. (Leguminosaceae). - Tree. Brazil. Fruits are used in Brazil as diuretic. Also M. acutifolium Vog., M. caudatum Ducke, M. macrophyllum Benth.

Machaerocereus gummosus (Engelm.) Britt. and Rose. (syn. Cereus gummosus Engelm.) Pitahaya, Pitahaya agria. (Cactaceae). - Xerophytic tree. Baja California (Mex.). Fruit edible, agreeable, acid, much esteemed. Crushed stems used to stupefy fish.

Machilus nanmu Hemsl. (Lauraceae). - Tree. Trop. Asia, esp. India, China. Wood used for Chinese coffins.

Machilus odoratissima Nees. (Lauraceae). - Tree. India, China. Wood used for building native houses and tea-boxes.

Machilus pauhoi Kanch. (Lauraceae). - Tree. S. China, India. Source of Pau-Hoi, wood-shavings; used by Chinese women to extract a mucilage for plastering down their hair.

Macleania ecuadorienses Horold. (Ericaceae). - Slender shrub. Ecuador. Fruit globose, 1 cm. in diam. purplish black, soft; delicate, juicy pulp, sweet, agreeable taste; eaten by the natives of Ecuador.

Macleania Popenoei Blake, Joyapa. (Ericaceae). - Shrub. Ecuador and adjacent territory. Fruits edible, round, 1 cm. across, soft, juicy, sweet; occasionally sold in local markets of Ecuador.

Maclura aurantiaca Nutt. (syn. Toxylon pomiferum Raf.). Osage Orange, Bow Wood. (Moraceae). - Southern and southeastern United States. Cultivated. Wood very strong, durable, hard, heavy, coarse-grained, flexible, bright orange, becoming brown on exposure; used for railroad ties, wheel-stock, fence-posts; in demand where great strength and durability are required. Bark of root is source of a yellow dye; used for dyeing fabrics. Wood was used by Osage and other Indian tribes for war-clubs and bows.

Macnin → Digenea simplex (Wulf.) J. Ag.

Macqui → Aristotelia Macqui L'Hér.

Macrochloa tenacissima (L.) Kunth. → Stipa tenacissima L.

Macrocystis pyrifera (L.) C. Ag. California Giant Kelp. (Lessoniaceae). - Brown Alga. Pacific. N. America to South Atlantic. Manufactured into algin in the U. S. A.; see also Laminaria digitata (L.) Edmonson. Was formerly a source of iodine, potash, calcium acetate and acetone. Occasionally used in animal feeds.

Macrorhynchus troximoides Torr. and Gray. → Agoseris aurantiaca (Hook.) Greene.

Macrotomia Cephalotes Bois. (Boraginaceae). - Perennial herb. Greece, Asia Minor. Source of Syrian Alkanet, used as adulterant of Alkanna tinctoria.

Macrozamia Miquelii A. DC. (syn. Encephalartos Miquelii F. v. Muell.). (Cycadaceae). - Palm-like plant. Australia. Seeds when baked in ashes or when kernels are soaked in water for several days and than pounded, are consumed as food by the aborigenes of Queensland.

Macrozamia spiralis Miq. (Zamia spiralis R. Br., Encephalartos spiralis Lehm.) Burrawang Nut. (Cycadaceae). - Palm-like plant. New South Wales, Queensland. Kernels source of an arrow-root of good quality. Seeds consumed by aborigines.

Madagascar Boabab → Adansonia madagascariensis Baill.

Madagascar Cardamon → Aframomum angustifolium Schum.

Madagascar Copal → Trachylobium Hornemannianum Hayne and T. verrucosum Oliv.

Madagascar Ebony → Diospyros haplostylis Boiv.

Madagascar Mahogany → Khaya madagascariensis J. and P.

Madagascar Nutmeg → Ravensara aromatica Gmel.

Madagascar Raffia Palm → Raphia pedunculata Beauv.

Madder → Rubia tinctorium L.

Madder, India → Rubia cordifolia L.

Madder, Khasian → Rubia khasiana Kurz.

Madder, Levant → Rubia peregrina L.

Madder, Sikkim → Rubia sikkimensis Kurz.

Madeira Vine → Boussingaultia baselloides H. B. K.

Madflower, Ethiopian → Antholyza paniculata Klatt.

Madhuca butyracea Gm. Illipe Butter Tree. (Sapotaceae). - Tree. Trop. Asia, esp. India. Seeds source of Phulwara or Indian Butter; non-drying; has little flavor, good keeping qualities. Used by natives in food. Very little exported. Sp. Gr. 0.856-0.862; Sap. Val. 191-200; Iod. No. 40-51.

Madhuca latifolia Macbr. (Sapotaceae). - Tree. N. India. Flowers rich in nectar; used for sweetening food, source of sugar, fermented liquors and acetone. Of some commercial value.

Madhuca longifolia Gm. (syn. Bassia longifolia L.) Mowra Butter Tree. (Sapotaceae). - Tree. S. India, W. Ghats, Kanaria to Travancora. Seeds source of Illipe Butter or Tallow Mowra Butter; used for soaps and candles; treatment for skin diseases; when refined used in butter. Seeds exported to Europe. Press cake suitable as fertilizer. Sp. Gr. 0.856-0.864; Sap. Val. 188-202; Iod. No. 58-63; Unsap. 1.4-2.6%. Also M. latifolia Macbr.

Madia sativa Molina. (Compositae). - Annual herb. Chile. Seeds are a source of oil; used in Chile.

Madonna Lily → Lilium candidum L.

Madre de Cacao → Gliricidia sepium (Jacq.) Steud.

Madrona → Arbutus Menziesii Pursh.

Madrone, Canary → Arbutus canariensis Duham.

Maerua angolensis DC. (Capparidaceae). - Tree. Trop. Africa, esp. Nigeria, Angola etc. Wood very hard, heavy, close-grained, yellowish; recommended for joinery work. Leaves are eaten as vegetable by the natives, also used as a purgative.

Maerua pedunculosa Vahl. (syn. Niebuhria pedunculosa Hochst., Boscia caffra Sond.). (Capparidaceae). - Tree. Cape Peninsula to Natal. Roots eaten by natives during times of scarcity.

Maesa tetrandra A. DC. (Myrsinaceae). - Shrub. Mal. Archip. Crushed roots are used by natives for fever.

Mafura Bitterwood → Trichilia emetica Vahl.

Mafura Oil → Trichilia emetica Vahl.

Mafura Tallow → Trichilia emetica Vahl.

Magellan Barberry → Berberis buxifolia Lam.

Magnolia acuminata L. (syn. Tulipastrum acuminata (L.) Small.). Cucumber Tree, Mountain Magnolia. (Magnoliaceae). - Eastern United States. Wood soft, light, not strong, durable, close-grained, light yellow-brown; used for flooring, cabinet-work, valuable for pump-logs. Wood of M. macrophylla Michx. is used for the same purpose.

Magnolia Blumei Prantl. (syn. Manglietia glauca Blume). (Magnoliaceae). - Tree. Java. Wood fine-grained, strong, durable, brown, easily worked; used in Java for houses, bridges, furniture and tools.

Magnolia glauca L. (syn. M. virginiana L.) Sweet-Bay, Swamp-Bay. (Magnoliaceae). - Tree. Eastern United States to Texas. Wood light brown tinged with red, soft; sometimes used for manuf. of broom handles and other woodware.

Magnolia grandiflora L., Southern Magnolia. (Magnoliaceae). - Tree. Southeast and south of the United States. Bark is used as a stimulant tonic and diaphoretic.

Magnolia javanica Koord. and Val. (Magnoliaceae). - Tree. Java, Sumatra. Wood durable, beautifully grained; used in Java for handles of weapons.

Magnolia Kobus DC., Kobus Magnolia. (Magnoliaceae). - Tree. Japan, Korea. Wood is soft, light, close-grained, light yellow; used in Japan for utensils, engraving, matches. Bark is used for colds by the Ainu.

Magnolia mexicana DC. (syn. Talauma mexicana (DC.) Don. (Magnoliaceae). - Tree. Mexico. Bark is used in Mexico as domestic medicine, fevers, heart affections, paralysis, epilepsy.

Magnolia obovata Thunb. (Magnoliaceae). - Tree. Japan, Centr. China. Wood soft, easily worked, close-grained, light-yellowish brown-gray. Used in Japan for furniture, utensils, cabinet-work, engraving.

Magnolia officinalis Rehd. and Wils. (Magnoliaceae). - Tree. Centr. China. Extract from bark used in Chinese pharmacie as tonic, aphrodisiac and for colds.

Magnoliaceae → Drimys, Illicium, Liriodendron, Magnolia, Michelia.

Maguey → Agave cantala Roxb.

Maguey Delgado → Agave Kirchneriana Berger.

Maguey Manso Fino → Agave atrovirens Karw.

Mahaleb Cherry → Prunus Mahaleb L.

Mahogany → Swietenia Mahagoni Jacq.

Mahogany, Adju → Canarium Mansfeldianum Engl.

Mahogany, African → Detarium senegalense Gmel., Khaya senegalensis Juss. and Kiggelaria africana L.

Mahogany, Bastard → Eucalyptus botryoides Smith.

Mahogany, Benin → Guarea Thompsonii Sprague and Hutch. and Khaya Punchii Stapf.

Mahogany, Borneo → Calophyllum inophyllum L.

Mahogany, Brown → Entandrophragma utile (Dawe and Sprague) Sprague, Lovoa Klaineana Pierre and Sprague and L. Swynnertonii Bak. f.

Mahogany, Cameroon → Entandrophragma Gandollei Harms, E. Rederi Harms and Khaya euryphylla Harms.

Mahogany, Coaba → Swietenia Mahagony Jacq.

Mahogany, Cape → Ptaeroxylon obliquum (Thunb.) Radlk.

Mahogany, Cedar → Entandrophragma Candollei Harms. and E. cylindricus Sprague.

Mahogany, Colombian → Cariniana pyriformis Miers.

Mahogany, Demarara → Carapa procera DC.

Mahogany, Espave → Anacardium Rhinocarpus DC.

Mahogany, Forest → Eucalyptus resinifere Smith.

Mahogany, Gabon → Khaya Klainei Pierre.

Mahogany, Heavy → Entandrophragma Candollei Harms.

Mahogany, Ijebu → Entandrophragma angolense (Welw.) A. DC.

Mahogany, Ijebu → Entandrophragma Candollei Harms.

Mahogany, Madagascar → Khaya madagascariensis J. and P.

Mahogany, Mexican → Swietenia humilis Zucc.

Mahogany, Mountain → Cercocarpus ledifolium Nutt.

Mahogany, Red → Eucalyptus resinifera Smith. and Khaya ivorensis Chev.

Mahogany, Sandomingo → Swietenia Mahagony Jacq.

Mahogany, Sapelé → Entandrophragma Candollei Harms.

Mahogany, Scented → Carapa procera DC. and Entadrophragma Candollei Harms.

Mahogany, Smooth barked African → Khaya anthotheca A. DC.

Mahogany, Swamp → Eucalyptus botryoides Smith.

Mahogany, Tiama → Entandrophragma Candollei Harms. and E. macrophyllum Chev.

Mahogany, Venezuela → Swietenia Candollei Pitt.

Mahogany, West African → Mitragyna macrophylla Hiern.

Mahogany, West Indies → Swietenia Mahagony Jacq.

Mahogany, White → Cybistax Donnell-Smithii (Rose) Seib., Eucalyptus robusta Smith and Khaya anthotheca C. DC.

Mahonia Aquifolium (Lindl.) Don. (syn. Berberis Aquifolium Pursh.). Oregon Grape. (Berberidaceae). - Shrub. Western N. America. Dried rhizome and roots used medicinally. Much is derived from Washington, Oregon and Califorenia. Used as bitter tonic and alterative. Contains berberine, oxyacanthine and berbadine being alkaloids.

Mahonia repens (Lindl.) G. Don. (syn. Berberis repens Lindl., G. Don., Odostemon Aquifolium Rydb.). Oregon Grape. (Berberidaceae). - Shrub. Western N. America. Decoction of

bark was formerly used by the early settlers for mountain fever and kidney troubles, as a tonic and febrifuge. Ripe fruits are used as food and in jellies, for wine and lemonades.

Mahonia trifoliolatus (Moric) Heller. (syn. Berberis ilicifolia Scheele). (Berberidaceae). - Shrub. Mexico. W. Texas, New Mexico. Acid fruits used as preserves and in tarts. Wood suitable for tanning and making ink; source of a yellow dye.

Mahonia chochoco Fedde, Chochoco. (Berberidaceae). - Shrub or tree. Mexico. Wood used as a dye and for tanning.

Mahonia philippensis Takida. (syn. M. nepalensis DC.) (Berberidaceae). - Shrub. Luzon (Philipp. Islds.) Plant is source of a yellow dye.

Ma-Huang → Ephedra sinica Stapf.

Maidenhair Fern → Adiantum spec.

Maigang → Eugenia polycephaloides C. B. R.

Maina Resin → Calophyllum longifolium Willd.

Maize → Zea Mays L.

Majagua → Hibiscus tiliaceus L.

Majang → Clavaria Zippeli Lev.

Majorana hortensis Moench. (syn. Origanum Majorana L.). Sweet Marjoram. (Labiaceae). - Annual herb. S. Europe, Mediterranean region, Arabia. Cultivated in Old and New World. Herb used for flavoring meat, soups, stews, meat-pies, dressings, geese, sausages, canned meats. Oil of Marjoram, an ess. oil derived from steam distillation, is used for flavoring sausage, canned meats and soups. Dried leaves and flowering tops are used medicinally, being a carminative, stimulant. Contains a greenish ess. oil, a mixture of borneol and camphor.

Majunga Noir Rubber → Mascarenhasia arborescens A. DC.

Mala Insana → Quercus tauricola Kotschy.

Mala Sodomitica → Quercus tauricola Kotschy.

Malabaila pumila Bois. (Umbelliferaceae). - Herbaceous plant. E. Nile region, Sahara. Roots are used as food by the Beduins.

Malabar Kino → Pterocarpus Marsupium Roxb.

Malabar Nut Tree → Adhatoda Vasica Nees.

Malabar Rosewood → Dalbergia latifolia Roxb.

Malabar Tallow → Vateria indica L.

Malacca Teak → Afzelia palembanica Baker.

Malachra capitata L. (Malvaceae). - Annual herb. Tropics. Cultivated. Source of excellent fibre, 20 to 25 dm. long. Similar to jute.

Malanga, Darkleaf → Xanthosoma atrovirens C. Koch.

Malanga, Primrose → Xanthosoma violaceum Schott.

Malay Bush Beech → Gmelina arborea Roxb.

Malay Gutta Percha Tree → Palaquium Gutta Burck.

Malay Lacktree → Schleichera oleosa Merr.

Male Cola → Cola acuminata Schott. and Endl.

Male Fern → Aspidium Filix Mas L.

Male Rubber Tree → Holarrhena Wulfsbergii Stapf.

Male Shield Ferz → Aspidium Folix Mas. L.

Mallee → Eucalyptus dumosa Cunningh.

Malenghet Manila → Agathis loranthifolia Salisb.

Malotus cochinchinensis Lour. (syn. M. paniculatus Muell. Arg.). (Euphorbiaceae). - Tree. S. China to N. Australia. Wood very light; used in Indo-China for matches, packing cases.

Mallotus discolor F. v. Muell. (syn. Rottlera discolor F. v. Muell.) (Euphorbiaceae). - Tree. Australia, esp. New South Wales, Queensland. Capsules produce a bright yellow dye.

Mallotus floribunda Muell. Arg. (Euphorbiaceae). - Small tree. Trop. Asia, esp. Annam and Tenasserim to New Guinea. Male flowers used with rice-powder for toilet-powder and scented powders, also used medicinally.

Mallotus philippensis Muell. Arg. Kamala Tree. (Euphorbiaceae). - Small tree. N.W. Himalaya to E. Australia. Glandular hairs from fruits are source of a beautiful, expensive dye, called Kamala Powder, giving a rich golden red color on silk. An alkali is necessary during the manuf. process. Contains rottlerin, yellow homo-rottlerin and salmon iso-rottlerin. Has been used medicinally for centuries as anthelmintic, also for skin complaints.

Mallow, Common → Malva rotundifolia L.

Mallow, Country → Abutilon indicum Sweet.

Mallow, Finger Poppy → Callirhoe digitata Nutt.

Mallow, High → Malva sylvestris L.

Mallow, Indian → Abutilon Avicennae Gaertn.

Mallow, Narrowleaf Globe → Sphaeralcea angustifolia (Cav.) G. Don.

Mallow, Pimple → Callirrhoe pedata Gray.

Mallow, Tree → Lavatera plebeia Sims.

Malpighia glabra L., Barbados Cherry. (Malpighiaceae). - Shrub or small tree. Trop. America, West Indies and S. Texas. Cultivated in the tropics. Fruits are scarlet, juicy pulp, acid; used in jellies and preserves.

Malpighia punicifolia L., West Indian Cherry Tree. (Malpighiaceae). - Tree. Trop. America. Fruits are red, acid of fair quality; used for jams, preserves and sauces.

Malpighia urens L. (Malpighiaceae). - Shrub or small tree. Trop. America. Fruits reddish, edible, refreshing, juicy. Called Grosse Cerise in Martinique.

Malpighiaceae → Banisteria, Bunchosia, Byrsonima, Heteropteris, Hiptage, Malpighia.

Malukang Fat → Polygala butyraceum Heck.

Maluko → Pisonia alba Span.

Malus angustifolia (Ait.) Michx. → Pyrus angustifolia Ait.

Malus baccata (L.) Borkh. → Pyrus baccata L.

Malus communis DC. → Pyrus Malus L.

Malus coronaria (L.) Mill. → Pyrus coronaria L.

Malus hupehensis (Pamp.) Rehd. → Pyrus hupehensis Pamp.

Malus prunifolia (Willd.) Borkh. →Pyrus prunifolia Willd.

Malus pumila Mill. → Pyrus Malus L.

Malus rivularis Roem. → Pyrus rivularis Dougl.

Malva rotundifolia L., Common Mallow. (Malvaceae). - Perennial herb. Temp. zone. Dried leaves used medicinally; demulcent, emollient. Contains a mucilage and tannin.

Malva sylvestris L. High Mallow. (Malvaceae). - Biennial or perennial herb. Europe, Asia, introd. in N. and S. America, Australia. Leaves are used as substitute for tea; those that are collected during flowering time are used as expectorant, contain a slimy substance. Flowers, Flores Malvae, are used for gargling and for mouth washes.

Malvaceae → Abutilon, Althaea, Calirrhoe, Gossypium, Hibiscus, Hoheria, Kosteletzkya, Kydia, Lavatera, Malachra, Malva, Maxwellia, Pavonia, Plagianthus, Sida, Sphaeralcea, Thespesia, Urena, Wissadula.

Mammea americana L., Mamey. (Guttiferaceae). - Tree. West Indies, Northern S. America. Cultivated in trop. America. Fruits roundish, rough leathery skin; flesh pale-yellow, 15 to 20 cm long. Eaten fresh, with wine, sugar or cream; as sauce, preserves and jam. Mamey preserve is of some commercial importance in Cuba. Mature green fruits are made into jelly, contain much pectin. Stem is source of a resin, called Resina de Mamey. Fruit pulp produces a wine. Scented flowers are used for manuf. a liqueur, Eau de Creole.

Mamey → Mammea americana L.

Mamey Colorado → Calocarpum mammosum Pierre.

Mammoth Tree → Sequioa gigantea Lindl. and Gord.

Mamoncillo → Melicocca bijuga L.

Manaca Rain Tree →Brunfelsia Hopeana Benth.

Manchu Cherry → Prunus tomentosa Thunb.

Manchurian Walnut → Juglans mandschurica Maxim.

Mandarine Oil → Citrus reticulata Blanco.

Mandarine Orange → Citrus reticulata Blanco.

Mandera Cucumber → Cucumis Sacleuxii Paill. and Bois.

Mandioca doce → Manihot dulce (J. F. Gmel.) Pax.

Mandioqueira → Didymopanax longipetiolatum March.

Mandragora officinarum L., European Mandrake, Mandragora. (Solanaceae). - Perennial herb. Mediterranean region. Root was much esteemed among the ancients for its narcotic properties. Is at the present used for colic, asthma, coughs, hayfever. Root contains two alkaloids e. g. mandragorine and one resembling hyoscyamine.

Mandrake, European → Mandragora officinarum L.

Mangabeira Rubber → Hancornia speciosa Gomez.

Mangel Wurzel → Beta vulgaris L.

Mangifera altissima Blanco. Paho. Pahutan Mango. (Anacardiaceae). - Large tree. Philippine Islands. Fruits 5 to 8 cm long, green to yellowish, smooth; used in the Philippines in pickles.

Mangifera caesia Jack. (Anacardiaceae). - Tree. Malaya. Fruits are eaten preserved in Malaya. Cultivated.

Mangifera foetida Lour. Bachang Mango. (Anacardiaceae). - Tree, Malaya. Fruit acid, terpentine flavor; before being eaten the Javanese place the peeled fruits in lime-water or for some time in syrup. Young fruits are eaten in Java as Sambal, with Cayenne Pepper, Soja and a fish product.

Mangifera indica L. Common Mango. (Anacardiaceae). - Tree. India, Malaya, Cochin-China. Widely cultivated in tropics of Old and New World. Important fruit species. Fruits eaten raw, preserved, pickled when green, as a sauce and in drinks. Numerous varieties, many of local importance, e. g. I. Mulgoba Group: Mulgoba, Haden. II. Alpohnse Group: Bennett, Amini, Rajpuri, Alphonse, Pairi. III. Sandersha Group: Sandersha. IV. Combodiana Group: Mulgoba, Haden. V. Alphonse Group: Wootten, Whitney, Roberts, Steward. VI. West Indian Group: Lewis, Victoria. Most fruits from seedling trees contain much fibre, those of improved varieties are almost fibreless. A yellow dye is obtained from urine of cows that have been fed with leaves of the plant. Much is derived from Monghyr, Bengal. Dye is called Indian Yellow, Jaune Indien, Monghyr Piuri, used in aquarel and oil painting, resistant to light.

Mangifera laurina Blume, Monjet Mango. (Anacardiaceae). - Tree. Mal. Archip. Fruits are consumed ripe by the natives.

Mangifera odorata Grif., Kurwini Mango. (Anacardiaceae). - Tree. Mal. Archip. Fruits are consumed ripe by the natives and Europeans; they are said to be of a good flavor.

Mangifera Rumphii Pierre, (Anacardiaceae). - Tree. Mal. Acrhip. Fruits are consumed ripe and are also made into preserves by the natives.

Mangifera zeylanica Hook., Ceylon Mango Tree. (Anacardiaceae). - Tree. Ceylon. The fruits are consumed in some parts of Ceylon.

Manglietia glauca Blume → Magnolia Blumei Prantl.

Mango → Mangifera indica L.

Mango, Bachang → Mangifera foetida Lour.

Mango, Binjai → Mangifera caesia Jacq.

Mango, Ceylon → Mangifera zeylanica Hook.

Mango, Grape → Sorindeia madagascariensis DC.

Mango, Kurwini → Mangifera odorata Grif.

Mango, Monjet. → Mangifera laurina Blume.

Mango, Pahutan → Mangifera altissima Blanco.

Mango, Wild → Irvingia gabonensis Baill.

Mangosteen → Garcinia Mangostana L.

Mangrove, Red → Rhizophora Mangle L.

Mangrove, Red Flowered → Lumnitzera coccinea Wight. and Arn.

Mangrove, White → Laguncularia racemosa Gaertn. f.

Manicaria saccifera Gaertn., Ubussu. (Palmaceae). - Large palm. Brazil and Guiana. Leaves are widely used in Brazil for covering roofs and are an article of commerce. Fibre from the spath is made into mats.

Manihot, Brazilian →Manihot esculenta Crantz.

Manihot dichotoma Ule, Manicoba de Jequie. (Euphorbiaceae). - Tree. Bahia, Brazil. Source of Jequie Rubber, Remanso or Mule Gum.

Manihot dulcis (J. F. Gmel.) Pax. Aipi, Mandioca doce, Yuca dulce. (Euphorbiaceae). - South America, esp. Brasil. Roots are source of a starch; used as food. Several vartieties are in existance.

Manihot esculenta Crantz. (syn. M. utilissima Pohl). Common Cassava, Manioc. (Euphorbiaceae). - Shrub or small tree. S. America. Cultivated in the tropics and sometimes subtropics of the Old and New World. Roots are source of a starch, forming an important food in S. and Centr. America, also in other countries. Known as Cassava, Manioc, Pará or Brazilian Arrowroot. Used for cassava bread, in West Indian pepperpot, cassareep, in sauces of the Worcester group, for sizing cotton fabrics. Starch is made into tapioca, used in soups, puddings etc. Plant is also source of an alcoholic beverage. Fresh roots contain esp. in the bark prussic acid. Varieties are divided into two groups. I. Group. Bitter Cassava. Roots contain 0.02 to 0.03% prussic acid. They are usually consumed boiled as vegetable. II. Group. Sweet lassava. Roots contain 0.007% prussic acid. They are used in the preparation of starch. Numerous varieties, among which: Yellow Bell, Blue Beard White, Pacho III, White Top, Icery, Singapore, Constantin, Pitter, Florida Sweet, Cenaguen. Roots are also used as fodder for livestock.

Manihot Glaziovii Muell. Arg. Manicoba, Mandihoba. (Euphorbiaceae). - Tree. Rio Grande do Norte, Paradyha, Ceara (Braz.). Cultivated in Trop. Africa, Asia etc. Source of Ceara Rubber.

Manihot heptaphylla Ule. Manicoba da São Francisco (Euphorbiaceae). - Tree. Along Rio de São Francisco (Braz.). Source of São Francisco Rubber.

Manihot piauhyensis Ule, Manicoba de Piauhy. (Euphorbiaceae). - Small tree. Piauhy (Braz.). Source of Piauhy Rubber.

Manila Aloe → Agave cantala Roxb.

Manila, Batjan → Agathis loranthifolia Salisb.

Manila, Borneo → Agathis loranthifolia Salisb.

Manila Copal → Agathis alba Foxw.

Manila Elemi → Canarium commune L. and C. luzonicum Miq.

Manila Hemp → Musa textilis Née.

Manila, Labuan → Agathis loranthifolia Salisb.

Manila, Macassar → Agathis alba Foxw.

Manila, Malenghet → Agathis loranthifolia Salisb.

Manila, Menado → Agathis loranthifolia Salisb.

Manila, Molucca → Agathis loranthifolia Salisb.

Manila, Pontiak → Agathis alba Foxw.

Manila, Singapore → Agathis alba Foxw.

Manila Tamarind → Pithecolobium dulce Benth.

Manilkara dariensis (Pitt.) Standl. (syn. Mimusops dariensis Pitt.) (Sapotaceae). - Tree. Panama. Source of Panama Balata.

Manilkara Elengi (L.) Chev. (Sapotaceae). - Tree. Origin uncertain. Cultivated in Mal. Archip. Decoction of roots is used for fever and as a mouth wash. Flowers are placed between clothes to impart a pleasant scent, source of an ess. oil obtained by steam distillation. Bark is tonic and febrifuge.

Manilkara Huberi (Ducke) Chev. (Sapotaceae). - Tree. Amazone Region (Brazil). Stem is source of a balata.

Manilkara bidentata (DC.) Chev. (syn. Mimusops globosa Gaertn. f., M. bidentata DC.) Balata Tree, Bully, Bullet, Purgio, Quinilla. (Sapotaceae). - S. America. Latex is source of commercial Balata, a non-elastic rubber; used for machine belts, boot soles. Substitute of chicle. Wood hard, dense. heavy, dark red; used for building purposes, windmill arms, shingles, wheel spokes, bridge piles in fresh water, posts, rail-road ties.

Manilkara Kauki (L.) Dubard. (syn. Mimusops Kauki L.) Kauki. (Sapotaceae). - Tree. Mal. Archip. Wood close-grained, heavy, durable, reddish, marbled, resistant to damp soil; used for beams, handles of tools and furniture.

Manilkara Sieberi (A. DC.) Dubard. (syn. Mimusops Sieberi A. DC.) (Sapotaceae). - Tree. Trinidad. Source of a substitute for balata.

Manilkara spectabilis (Pitt.) Standl. (syn. Mimusops spectabilis Pitt.) (Sapotaceae). - Tree. Costa Rica. Wood hard, heavy; used for railroad ties; resistant to water and moist soil.

Manketti Nut Oil → Ricinodendron Rautanenii Schinz.

Mankil → Eugenia javanica Lam.

Manicoba → Manihot Glaziovii Muell. Arg.

Manicoba de Jequie → Manihot dichotoma Ule.

Manicoba de Piauhy → Manihot piauhyensis Ule.

Manioc → Manihot esculenta Crantz.

Manna → Alhagi camelorum Fisch., A. maurorum Medic., Fraxinus Ornis L., Haloxylon Schweinfurthii Anders, Lecanora esculenta Ev.

Manna Ash → Fraxinus Ornus L.

Manna, Eucalyptus → Eucalyptus mannifera Mudic.

Manna Gum → Eucalyptus viminalis Lab.

Manna Lichen. → Lecanora esculenta Everson.

Manna Oak → Quercus Cerris L. and Q. persica Jaub. and Spach.

Manna of Briançon → Larix europaea DC.

Manna. Source of → Alhagi, Anabasis, Cotoneaster, Eucalyptus, Fraxinus, Haloxylon, Larix, Lecanora, Quercus.

Manniophyton africanus Muell. Arg. (Euphorbiaceae). - Liane. Trop. Africa. Bark source of a fibre; used for ropes, fishing-nets in the Congo.

Manuba Mahogany → Trichilia emetica Vahl.

Manures → under Fertilizers.

Manzanilla → Crataegus stipulosa (H. B. K.) Steud. and Englerodendron alata Horold.

Manzanita → Arctostaphylos Manzanita Parry.

Manzanita, Pointleaf → Arctostaphylos pungens H. B. K.

Maori Honeysuckle → Knightia excelsa R. Br.

Maple, Broad-leaved → Acer macrophyllum Pursh.

Maple, Hawthorn → Acer crataegifolium Sieb. and Zucc.

Maple, Hedge → Acer campestre L.

Maple, Norway → Acer platanoides L.

Maple, Red → Acer rubrum L.

Maple, Scottish → Acer Pseudo-Platanus L.

Maple, Silver → Acer saccharinum L.

Maple Sugar → Acer saccharum Marsh.

Maple Syrup → Acer saccharum Marsh.

Maple, Vine → Acer circinatum Pursh.

Maple, White → Acer saccharinum L.

Mappia oliganoides (L.) House → Cunila origanoides (L.) Butt.

Maracaibo Balsam → Copaifera officinalis L.

Maracaibo Lignum Vitae → Bulnesia arborea (Jacq.) Engl.

Marachino → Prunus Cerasus L.

Maranháo Nuts → Sterculia Chicha St. Hil.

Maranta arundinacea L., Arrowroot. (Marantaceae). - Herbaceous perennial. West Indies, North of S. America. Cultivated in tropics of Old and New World. Source of Arrowroot Starch, Bermuda Arrowroot, St. Vincent Arrowroot, derived from the rootstocks. Grown commercially in St. Vincent, Jamaica and Bermuda. Used for pastries and biscuits.

Marantaceae → Calathea, Maranta, Phrynium.

Marattia fraxinea Smith. (syn. M. salicina Smith.) (Marattiaceae). - Tree-fern. New South Wales, Queensland. Pith of stem is source of starch, consumed by aborigines.

Marattiaceae → Angiopteris, Marattia.

Marawayana → Copaifera braceata Benth.

Marble Wood → Olea paniculata R. Br.

Maretail → Hippurus vulgaris L.

Margosa → Melia Azadirachta L.

Margosa Oil → Melia Azadirachta L.

Margyricarpus setosus Ruiz and Pav. Pearlfruit. (Rosaceae). - Half-shrub. S. America. Infusion of the herb is used in some parts of Chile as diuretic.

Marianne Yellow Wood → Ochrosia mariannensis A. DC.

Marigold → Calendula officinalis L.

Marigold, Fatid → Pectis papposa Harv.

Marigold, Marsh → Caltha palustris L.

Marigold, Sweet → Tagetes lucida Cav.

Marihuana → Cannabis sativa L.

Mariposa, Sagebrush → Calochortus Gunnisonii S. Wats.

Mariscus jamaicensis (Crantz) Britt. → Cladium effusum (Sw.) Torr.

Mariscus umbellatus Vahl. (Cyperaceae). - Perennial herb. Tropics of Old World. Cultivated. Rhizomes are roasted and eaten by the natives in some parts of Africa.

Marjoram, Pot → Origanum vulgare L.

Marjoram, Sweet → Majorana hortensis Moench.

Markhamia lanata Schum. (Bignoniaceae). - Tree. Trop. Africa. Wood yellowish; used for buildings, construction work, bridges, wagons and cheap furniture.

Marmelade Box → Genipa americana L.

Marmelade Plum → Calocarpum mammosum Pierre.

Marrow, Vegetable → Cucurbita Pepo L.

Marrubium vulgare L., Hoarhound. (Labiaceae). - Perennial herb. Europe, introd. in N. America. A tea from the dried, bitter herb is used as home remedy for debility and coulds; is expectorant and promotes perspiration; in large doses it is laxative. Much used in certain candies; used for coughs and to relieve sore throat. Source of an ess. oil; used in liqueurs.

Marsdenia tinctoria R. Br. (Asclepiadaceae). - Climbing plant. Mal. Archipel. Cultivated. Source of an indigo-like dye.

Marsdenia Reichenbachii Triana. (syn. M. Cundurango Reich. f.). Condorvine. (Asclepiadaceae). - Ecuador to Colombia. Dried bark is a source of Cundurango Bark; used in Cundurango Wine as a stomachic. Contains cundurangin and cundurit.

Marsdenia tomentosa Morr. and Decne. (Asclepiadaceae). - Vine. Japan. Vines are used in Japan for making ropes and bow-strings.

Marsdenia verrucosa Decne. Bokalahy. (Asclepiadaceae). - Woody plant. Madagascar. Source of a good rubber. Latex coagulates by lemonjuice, sea-water, tamarind juice and upon heating.

Marsdenia zimapanica Hemsl. Tequampatli. (Asclepiadaceae). - Vine. Mexico. Root mixed with meat is used to poison coyotes.

Marsh Buckbean → Menyanthes trifoliata L.

Marsh Fern → Acrostichum aureum L.

Marsh Fleabane → Pluchea sericea (Nutt.) Cov.

Marsh Mallow → Althaea officinalis L.

Marsh Marigold → Caltha palustris L.

Marsh Samphire → Salicornia herbacea L.

Marsh Woodwort → Stachys palustris L.

Marshallia aliena Spreng. → Heterothalamus brunoides Less.

Marsippospermum grandiflorum Hook. → Rostkovia grandiflora Hook. f.

Martagon Lily → Lilium Martagon L.

Martynia montevidensis Cham. (Pedaliaceae). - Annual herb. Argentina to Brazil. The seeds are used in some parts of Argentina as an emolient, resolutive and for cataplasm.

Martynia probocidea Glox. (syn. M. louisiana Mill.). Unicorn Plant, Ram's Horn, Double Claw. (Martyniaceae). - Annual herb. N. America and Mexico. Very young fruits are eaten as pickles. Occasionally cultivated.

Martyniaceae → Martynia.

Marumia muscosa Blume. (Melastomaceae). - Woody vine. W. Java. Sap from stems is used by natives for infected eyes. Young shoots are eaten with rice. Berries are made into marmelade, preserves or drinks.

Marvey → Olea paniculata R. Br.

Mascarenhasia arborescens A. DC. (Apocynaceae). - Tree. Madagascar. Source of a rubber, called Majunga noir.

Mascarenhasia elastica Schum. (Apocynaceae). - Tree. E. Africa. Source of a rubber. Latex coagulates on the stem or by being laid on the arms of the collectors. Known as Mgoa Rubber.

Mascarenhasia lisianthiflora A. DC. (Asclepiadaceae). - Small tree. Madagascar. Source of a rubber. Latex coagulates quickly in the air, also by alcohol.

Mascarenhasia longifolia Jum. (Apocynaceae). - Tree. Madagascar. Source of a good rubber.

Mascarenhasia mangorensis Jum. and Pierre. (Apocynaceae). - Tree. Madagascar. Old parts of plants are source of a rubber.

Massoia aromatica Becc. (syn. Sassafras Goesianum Teysm. and Binn.) (Lauraceae). - Tree. New Guinea. Bark is used by natives of Java, Sumatra and Borneo as lotion.

Massoia Bark → Cinnamomum Massoia Schew.

Masson Pine → Pinus Massoniana D. Don.

Mastic → Sideroxylon Mastichodendron Jacq

Mastic, American → Schinus Molle L.

Mastic, Bombay → Pistacia mutica Fisch and Mey.

Mastic, Chios → Pistacia lentiscus L.

Mastic Thyme → Thymus Mastichina L.

Masticatory Plants. → Achras, Alpinia, Anacyclus, Areca, Catha, Cochlospermum, Cola, Cyathocalyx, Duboisia, Ficus, Glycyrrhiza, Gouana, Heterospathe, Hymenopappus, Kaempfera, Lophophora, Lysimachia, Mesembryanthemum, Napoleona, Neea, Nicotiana, Petalostemon, Piper, Premna, Saccharum, Schradera, Shorea, Sphaeralcea, Sphenocentrum, Tarchonanthes, Toddalia, Uncaria, Vernonia, Xylopia.

Mastocarpus Klenzinaus Kuetz. (Gigartinaceae). - Red Alga. Malayan Seas. Plants are used as a source of agar. They are also used as food by the natives.

Matricaria Parthenium L. → Chrysanthemum Parthenium (L.) Bernh.

Matayba apetala Radlk. → Ratomia apetala Griseb.

Matchwood → Didymopanax Morototoni (Aubl.) Decne.

Matgrass, Chinese → Cyperus tegetiformis Roxb.

Matisia cordata Humb. and Bonpl. Sapote. (Bombaceae). - Tree. S. America. Fruits edible, brownish green, skin leathery, pulp orange-yellow, sweet, of a pleasant flavor.

Matricaria Chamomilla L., Chamomille. (Compositae). - Annual herb. Europe ,Asia, Iran, Afghanistan. Cultivated in Old and New World. Source of an aromatic and bitter tea, used in home remedies as anthelmintic and antispasmodic. Dried flowerheads are used medicinally as stimulant, diaphoretic, carminative, nervine. Contains an ess. oil, a bitter compound. Some grades are known as Hungarian and German Chamomille. Oil of Chamomille an ess. oil is used for flavoring fine liqueurs of the French type; in perfumes and for scenting chamomille shampoos.

Matricaria discoidea DC. (Compositae). - Annual herb. Asia, introd. in Europe and N. America. Herb used as a tea instead of M. Chamomilla. Source of Oleum and Aqua Chamomilla.

Matricaria multiflora Fenzl. (syn. Tanacetum multiflorum Thunb.) (Compositae). - Perennial herb. S. Africa. Powder or infusion of plant is used in S. Africa for intestinal worms, also used as tonic.

Matsu Take → Armillaria matsutake S. Ito and Imai.

Maul Oak → Quercus chrysolepis Liebm.

Mauritia Carana Wall. (Palmaceae). - Palm. Trop. America. Young leaves are source of a fibre. Fruits produce a beverage. Stems are employed in building.

Mauritia flexuosa L. f. (Palmaceae). - Tall palm. Trop. S. America. Leaves are used for covering roofs and are also source of a fibre. Leaf-sheats are made into sandals. Fruits produce a beverage and an edible oil. From the stem a wine and a sago is prepared. Stems are used for posts and floating bridges. Hard seeds are made into buttons and other small objects.

Mauritia vinifera Mart. (Palmaceae). - Tall palm. Brazil. Fruits are source of an alcoholic beverage. Leaves produce a fibre. Pulp of the fruits is used with food, which is much esteemed in some parts of Brazil.

Mauritius Ebony → Diospyros rubra Gaertn., D. tesselaria Poir.

Mauritius Hemp → Furcraea gigantea Vent.

Mauritius Raspberry → Rubus rosaefolius Smith.

Mauritius Terminalia → Terminalia mauritiana Lam.

Maximiliana caribaea Griseb., Cocorite Palm. (Palmaceae). - Tall palm. West Indies, North of S. America. Leaves used for thatching.

Maximiliana vitifolia Krug. and Urb. → Cochlospermum vitifolium Spreng.

Maxwellia lepidota Baill. (Malvaceae). - Tree. New Caledonia. Wood yellow, easy to bend, flexible, easy to work; recommended for handles of tools.

May → Crataegus Oxyacantha L.

May Apple → Azalea nudiflora L. and Podophyllum peltatum L.

May Apple, Himalayan → Podocarpus Emodi Wall.

May Pop → Passiflora incarnata L.

Mayberry of Burbank → Rubus microphyllus L. f.

Maytenus Boaria Mol. (syn. M. chilensis DC.). (Celastraceae). - Tree. Chile. Leaves are used in Chile as febrifuge.

Maytenus obtusifolia Mart. (Celastraceae). - Tree. Brazil, esp. Bahia. Wood hard, heavy, not durable when in contact with soil; used for rural construction work and carpentry.

Mazzard → Prunus avium L.

Mbang → Chlorophra excelsa (Welw.) Benth. and Hook. f.

Meadow Fescue → Festuca elatior L.

Meadow Fox Tail → Alopecurus pratensis L.

Meadow Leak → Allium canadense L.

Meadow Oatgrass, Tall → Arrhenaterum avenaceum Beauv.

Meadow Pasqueflower → Anemone pratensis L.

Meadow Saffron → Colchicum autumnale L.

Meadow Sage → Salvia pratensis L.

Mealy Blister Lichen → Physica pulverulenta (Schreb.) Nyl.

Mealy Ramalina → Ramalina farinacea (L.) Ach.

Mecca Aloe → Aloë abyssinica Lam.

Mecca Balsam → Commiphora Opobalsam Engl.

Mecca Myrrh Tree → Commiphora Opobalsam Engl.

Mecca Senna → Cassia angustifolia Vahl.

Medeola virginiana L., Indian Cucumber-Root. (Liliaceae). - Perennial herb. Eastern N. America to Florida. Tuberous rhizomes were used as food by the Indians.

Medic, Black → Medicago lupulina L.

Medic, Tree → Medicago arborea L.

Medicago arabica (L.) All., Spotted Bur Clover. (Leguminosaceae). - Perennial herb. S. Europe, Algeria to Asia. Sometimes grown as pasture, also used as hay.

Medicago arborea L. Tree Medic. (Leguminosaceae). - Perennial herb. Mediterranean region, Asia, N. Africa, Greek Archipelago, S. Italy. Cultivated as food for cattle, sheep and goats.

Medicago falcata L., Yellow-flowered Alfalfa. (Leguminosaceae). - Perennial herb. Europe, Asia, N. Africa. Cultivated in both hemispheres. Green fodder for live-stock, also grown for hay.

Medicago hispida Gaertn., Bur Clover. (Leguminosaceae). - Annual herb. Mediterranean region. Cultivated as a green manure, for pasture and as a hay crop.

Medicago lupulina L., Hop Clover, Black Medic. (Leguminosaceae). - Annual herb. Temp. Europe, Asia, N. Africa. Cultivated in Old and New World as green fodder for live stock, green manure an as hay crop.

Medicago orbicularis (L.) All., Button Clover. (Leguminosaceae). - Perennial herb. Mediterranean region. Cultivated as a hay crop, green manure and for pasture.

Medicago ruthenica Trautv. (Leguminosaceae). - Perennial herb. Lake Baikal to E. Siberia, Manchuria, Mongolia, Korea, N. China. Grown in Asia as a fodder for cattle, horses, sheep and goats.

Medicago sativa L., Alfalfa, Lucerne. (Leguminosaceae). - Perennial herb. Temp. Europe, Asia, N. Africa. Widely cultivated in Old and New World, as fodder for live-stock, for pasturage, as hay crop, silage, soiling; also made

into meal. Several varieties and strains among which: Peruvian, Turkestan, Arabian. Seeds are source of a drying oil, has been suggested for manuf. of paints. Sp. Gr. 0.9147; Sap. Val. 172,3; Iod. No. 154.2; Unsap. 4.4%. Flowers are source of a pleasant, slightly mint flavored honey. Leaves and stems are source of a commercial chlorophyll.

Medicago scutellata (L.) Willd., Snail Clover. (Leguminosaceae). - Perennial herb. Algeria to Asia Minor. Cultivated as a pasture crop.

Medicinal plants used among different peoples. → Abies, Abroma, Abrus, Abuta, Acacia, Acalypha, Acanthospermum, Acanthosphaera, Acanthus, Achillea, Achyranthes, Achyrocline, Acokanthera, Aconitum, Acorus, Acrodiclidium, Adansonia, Adenanthera, Adenophora, Adhatoda, Adiantum, Adonis, Aegle, Aeolanthus, Aesculus, Agapanthus, Agave, Agrimonia, Agropyron, Ailanthus, Alangium, Albizzia, Alchemella, Aletris, Alepidia, Alhagi, Allamanda, Allium, Alnus, Aloë, Alpinia, Alsidium, Alsophila, Alstonia, Althaea, Alyxia, Amanita, Amburana, Ammania, Ammi, Ammodaucus, Amomum, Amphimas, Anacyclus, Anamirta, Ancistrocladus, Andira, Andrographis, Aneilema, Anemone, Anemonopsis, Anethum, Angelica, Angelonia, Anisomeles, Anisoperma, Annona, Anogeisus, Anthemis, Antholyza, Apium, Apocynum, Apodanthera, Aralia, Arbutus, Arctium, Arctopus, Arctostaphylos, Ardisia, Argemone, Argyreia Arisaema, Aristolochia, Arnica, Artemisia, Asarum, Asclepias, Asimina, Asparagus, Asphodelus, Aspidium, Aspidosperma, Aspilia, Astragalus, Astrocaryum, Atalantia, Atropa, Baccharis, Balanites, Baliospermum, Baphia, Baptisia, Barleria, Barosma, Bauhinia, Begonia, Balamcanda, Belangera, Berberis, Betula, Blepharocalyx, Blumea, Bocconia, Boerhaavia, Borago, Borreria, Boswellia, Brachylaena, Brassica, Brauneria, Breynia, Bickelia, Brucea, Brunfelsia, Bryonia, Buddleia, Bunium, Buphane, Bursera, Butea, Cabralea, Caesalpinia, Calamintha, Calanthe, Calendula, Calliandra, Calophyllum, Calycanthus, Calvatia, Campsiandra, Capsella, Capsicum, Carapa, Carica, Carlina, Carum, Casearia, Cassia, Castanea, Caulophyllum, Cayaponia, Ceanothus, Cecropia, Cedrela, Celastrus, Celosia, Celtis, Centipeda, Cephaëlis, Cephalanthus, Ceratonia, Cercis, Cercocarpus, Ceropteris, Cestrum, Cetraria, Chaetospermum, Chaerefolium, Chamaelirium, Chasalia, Cheiranthus, Chelidonium, Chelone, Chenopodium, Chimaphila, Chiococca, Chionanthus, Chonapodendron Chrozophora, Chrysanthemum, Cibotium, Cichorium, Cimicifuga, Cinchona, Cinnamomum, Cissampelos, Cistus, Citrullus, Citrus, Clavariella, Claviceps, Clematis, Clerodendron, Clidemia, Cliffortia, Clitoria, Clusia, Cluytia, Cnestis, Cocculus, Cochlearia, Cochlospermum, Codonopsis, Coffea, Cola, Colchicum, Collinsonia, Colubrina, Combretum, Commiphora, Coniam, Connarum, Convallaria, Convolvulus, Copaifera, Coptis, Cordia, Cordyceps, Coriandrum, Cornus, Corrigiola, Corynanthe, Corynostylis, Coscinium, Costus, Couma, Couroupita, Coutoubea, Crocus, Croton, Cucubalus, Cuphea, Cucumis, Cupressus, Curanga, Curcuma, Cusparia, Cyathea, Cyclamen, Cyclea, Cydonia, Cymbopogon, Cynoglossum, Cypripedium, Daldinia, Daemonorops, Dalbergia, Daphne, Daphnopsis, Datura, Delphinium, Dicentra, Dicksonia, Dicoma, Dictamnus, Dictyophora, Didymocarpus, Digenia, Digitalis, Dioscorea, Diotis, Dissotis, Dorema, Dorstenia, Drimia, Drimys, Drosera, Drynaria, Dryopteris, Dysoxylon, Ecballium, Echinops, Elaphomyces, Elephantopus, Elettaria, Engleromyces, Embelia, Emex, Enicostema, Enterolobium, Ephedra, Epigaea, Equisetum, Erigeron, Eriodictyon, Eruca, Ervatamia, Eryngium, Erythraea, Erythrina, Erythrophleum, Erythroxylon, Esenbeckia, Eucalyptus, Eucarya, Eucommia, Eugenia, Eupatorium, Euphorbia, Euphrasia, Evodia, Evonymus, Exogonium, Fabiana, Fagus, Feraria, Ferula, Ficus, Filipendula, Floscopa, Foeniculum, Fomitopsis, Fragaria, Fraxinus, Fritillaria, Fuchsia, Fucus, Fumaria, Galega, Galeopsis, Galipea, Gallesia, Ganoderma, Garernia, Geissospermum, Gelsemium, Genista, Gentiana, Gesneria, Glycyrrhiza, Gomphia, Gossypium, Gouania, Grangea, Grindelia, Guarea, Guettarda, Guioa, Gymnosporia, Haematoxylon, Hagenia, Hamamelis, Harrisonia, Hedeoma, Hedyosum, Heeria, Helianthemum, Helicteres, Helleborus, Hyptis, Helminthostachys, Hemidesmus, Hemitelia, Herpestis, Heterothalamus, Hexalobus, Hieronima, Holarrhena, Homalomena, Humulus, Hydnocarpus, Hydrastis, Hymenodictyon, Hymenophyllum, Hyoscyamus, Hypericum, Hypochoeris, Ilex, Illicium, Inula, Ionidium, Ipomoea, Iresine, Iris, Isotoma, Ixora, Jacaranda, Jateorhiza, Jatropha Jefersonia Joannesia, Juglans, Juniperus, Karwinskia, Kigelia, Kosteletzkya, Krameria, Kyllingia, Lactuca, Lagenaria, Laminaria, Lamium, Larrea, Laserpitium, Lasiophon, Lastrea, Lathraea, Lavandula, Leea, Lentinus, Leonurus, Leucas, Levisticum, Ligusticum, Linum, Lippia, Liquidambar, Liriodendron, Lobelia, Loeselia, Lonchocarpus, Lucuma, Lycopodium, Macaranga, Magnolia, Mahonia, Majorana, Mallotus, Malva, Mandragora, Marrubium, Marsdenia, Matricaria, Melaleuca, Melastoma, Melia, Melianthus, Melicocca, Melothria, Menispermum, Mentha, Menyanthes, Mespilodaphne, Metrosideros, Monodora, Morinda, Myristics, Myroxylon, Myrrhis, Myrtus, Naregamia, Nasturtium, Nectandra, Nepeta, Newbouldia, Olea, Omphalea, Oncoba, Orchis, Origanum, Paederia, Paeonia, Panaeolus, Panax, Pnicum, Papaver, Parthenium, Parthenocissus, Passiflora, Pausinystalia, Petiveria, Petroselinum, Peucedanum, Peumus, Phyllanthus, Phyrostigma, Phytolacca, Picea, Picralima, Picrasma, Picrorhiza, Pilocarpus, Pimenta, Pimpinella, Pinckneya, Pinus, Piper, Piptadenia, Piptostegia, Pistacia, Plantago, Plumbago, Plumeria, Podaxis, Podophyllum, Polygala, Polygonum, Polypodium, Polyporus, Polystictus, Pongamia, Populus, Poris , Potentilla, Prunus, Pseudocinchona, Psidium, Psororspermum, Psychotria, Pterocarpus, Renealima, Ptychopetalum, Pulmonaria, Punica, Pycnanthemum, Pycnarhena, Pycnocarpus, Quassia Quercus, Rauwolfia, Remijia, Rhamnus, Rheum, Rhododendron, Rhus, Richardsonia, Ricinus, Rogersia, Rosa, Rosmarinus, Rubia, Rumex, Ruta, Sabatia, Salix, Salvia, Samadera, Sambucus, Sanguinaria, Santalum, Saponaria, Sarcocephalus, Sargassum, Sarothamnus, Sassafras, Satureja, Saururus, Schinus, Schizoglossum, Scopolia, Scutellaria, Selaginella, Selenicereus, Senecio, Serenoa, Sesamum, Simaba, Siparuna, Sisymbrium, Smilax, Solanum, Solidago, Sorbus, Spathyema, Spigelia, Stenocline, Streblus, Strophanthus, Strychnos, Styphnodendron, Styllingia, Styrax, Sweetia, Swertia, Symphonia, Symplocos, Tabebuia, Tabernanthe, Tagetes, Tamarindus, Tamarix, Tanacetum, Taraktogenos, Taraxacum, Tarchonanthus, Taxodium, Tecoma, Telanthera, Tephrosia, Terminalia, Teucrium, Thalictrum, Thapsia, Theobroma, Thevetia, Thlaspi, Thuja, Thymus, Tilia, Tinospora, Trifolium, Trigonella, Trillium, Trimeza, Turnera, Tussilago, Tylophora, Ulmus, Ulva, Uraria, Urechites, Urginea, Urtica, Valeriana, Vanilla, Vataireopsis, Vatica, Veratrum, Verbascum, Verbena, Vernonia, Viburnum, Viola, Vismia, Visnea, Wendtia, Xylaria, Xylopia, Xyris, Zanthocylum, Zea, Zingiber.

Medicinal Rhubarb → Rheum officinale Baill.

Medinilla Hasseltii Blume. (Melastomaceae). - Woody vine. Mal. Archip. Young leaves are eaten by natives of S. Sumatra with rice. Crushed fruits are eaten with fish.

Mediterranean Saltbrush → Atriplex Halimus L.

Medlar → Mespilus germanica L.

Medlar, Chirinda → Vangueria esculenta S. Moore.

Medlar, Japanese → Eriobotrya japonica Lindl.

Medlar, White → Vangueria esculenta S. Moore.

Megabarea Trillesii Pierre → Spondianthus Preusii Engl.

Mehndi Oil → Lawsonia alba Lam.

Melaleuca ericifolia Smith. (svn. M. nodosa Link., M. heliophila F. v. Muell.) (Myrtaceae). - Shrub. Australia. Used in Australia for consolidating muddy shores; lives in very salty soil.

Melaleuca Leucadendron L. Cajaput Tree. (Myrtaceae). - Tree. Mal. Archip. to Australia. Fresh twigs and leaves are source of Cajaput Oil, having an agreeable camphor-like odor, yellowish, volatile. Used medicinally, being antiseptic, anthelmintic, carminative, mild counter-irritant; contains eucalyptol, l-pinene, aldehydes and terpinol. Wood is hard, heavy, close-grained, light with dark shades; used for ship-building, posts; withstands moist soil.

Melaleuca linariifolia Sm. (Myrtaceae). - Tree. Australia. Leaves are source of an ess. oil being antiseptic; has been recommended for antiseptic soaps.

Melaleuca Preissiana Schan. var. leiostachya Schan. (syn. M. parviflora Lindl.) (Myrtaceae). - Shrub. Australia. Cultivated, useful in preventing moving of coastal sands. Also M. linariifolia var. trichostachya Smith.

Melaleuca suaveolens Gaertn. → Tristania suaveolens Smith.

Melaleuca uncinata R. Br. Tea-Tree. (Myrtaceae). - Tree. Australia. Leaves when chewed, are used in curing ordinary catarrh among the natives.

Melaleuca viridiflora Brogn. and Gris. (Myrtaceae). - Tree. New Caledonia. Leaves are source of Niaouli Oil, derived from steam distillation. Oil is antiseptic; used as substitute for Oil of Eucalyptus or Oil of Cajaput, used for coughs, neuralgia and rheumatism. About 50 to 75 kg. fresh leaves produce half a kg. of oil, containing about 67% cineol. Most is derived from New Caledonia. A small amount is derived from French Indo-China.

Melaleuca Wilsonii F. v. Muell. (Myrtaceae). - Tree. Victoria, S. Australia. Source of an essential oil, resembling that of Melaleuca Leucodendron L.

Melandrium album (Mil.) Garcke (syn. M. dioicum Schinz and Thell., M. diurnum Fries, Lychnis diurna Sibth.) White Campion. (Caryophyllaceae). - Annual or biennial herb. Europe, N. Africa, Asia Minor to Armenia, Siberia. Roots were formerly used for washing clothes. Also M. rubrum Garcke.

Melanorhoea inappendiculata King. (Anacardiaceae). - Tree. Mal. Penin., India. Resin source of a lasting black ink. Plant sometimes used for criminal poisoning.

Melanorhoea laccifera Pierre. (Anacardiaceae). - Tree. Cochin-China. Source of a lacquer in French Indo-China.

Melanorhoea usitata Wall. Burmese Varnish-Tree. (Anacardiaceae). - Tree. Trop. Asia, esp. Manipur, Burma, Siam. Source of a valuable Burmese Lacquer; varnish exported from Burma. Used as natural varnish for preserving woodwork. As a cement used in Burmese glass-mosaics, also in Burmese lacquer ware.

Melanoxylon amazonicum Ducke → Record-oxylon amazonicum Ducke.

Melanoxylon Brauna Schott., Brauna, Grauna. (Leguminosaceae). - Tree. Brazil, esp. Bahia to Sao Paulo. Wood dark brown to blackish, heavy, tough, very hard, strong; used for spokes of wheels, bridge timbers, sills, beams, posts, rail-road ties. Bark is used for tanning.

Melastoma malabathricum L. Singapore Rhododendron. (Melastomaceae). - Small shrub. E. India Malay to Australia. In Mal. Penin.

powdered leaves are sprinkled over healing pocks of small pocks preventing marking. Wood-tar used for blackening teeth. Foliage source of food of the Atlas silk-worm.

Melastoma petiolare Schlecht. → Clidemia Doppeana Steud.

Melastoma xalapense Bonpl. → Conostegia xalapensis D. Don.

Melastomaceae → Blakea, Clidemia, Conostegia, Dionycha, Dissotis, Marumia, Medinella, Melastoma, Memecylon, Miconia, Mouriria.

Melia Azedarach L. Chinaberry Tree. Umbrella Tree. (Meliaceae). - Tree. S. W. Asia, Iran, Asia Minor. Cultivated in tropics and subtropics of Old and New World. Seeds used for beads and rosaries. Bark of roots used medicinally, has anthelmintic properties; contains a light-yellow resin. Arabs and Persians use juice from leaves as vermifuge, diuretic and emmenagogue. Fruits are source of flea powder and insecticide. Wood is used for cabinet making.

Melia Azadirachta L. Margosa. (Meliaceae). - Tree. E. India, Ceylon. Seeds source of Margosa Oil; non-drying; used by natives of India for skin diseases. Recently used for manuf. medicated soap. Sp. Gr. 0.9159-0.9182; Sap. Val. 198.5-204.1; Iod. No. 69.3-71.9; Unsap. 0.7-1.1%. It is claimed that pressed leaves kept in books in libraries in India keep out book-mites and other insects. Powdered rootbark is considered astringent and febrifuge.

Melia dubia Cav. (Meliaceae). - Tree. Malaysia. Wood reddish brown; used for building purposes, furniture and tea-boxes.

Meliaceae → Aglaia, Amoora, Cabralea, Carapa, Cedrela, Chickrassia, Chisocheton, Chloroxylon, Dysoxylon, Eckebergia, Entandrophragma, Guarea, Khaya, Lansium, Lovoa, Melia, Naregamia, Pseudocedrela, Ptaeroxylon, Pterorhachis, Sandoricum, Soymida, Suretenia, Synoum, Trichilia.

Melianthaceae → Melianthus, Owenia.

Melianthus comosus Vahl. (Melianthaceae). - Woody plant. S. Africa. Bark of roots and also leaves used in S. Africa for snake-bites.

Melianthus major L. (Melianthaceae). - Tree. S. Africa. Decoction of leaves is used by farmers of S. Africa for healing wounds.

Meliococca bijuga L., Geneps. Spanish Lime, Mamoneillo. (Sapindaceae). - Tree. Trop. America. Decoction of bark is used in Trop. America for dysentery. Fruits are green, leathery; pulp yellowish, juicy, sweet, pleasant flavor, somewhat acid, edible; sold in local markets.

Melicope ternata Forst. (Rutaceae). - Small tree. New Zealand. Source of a gum, chewed by the natives.

Melicytus ramiflorus Forster. Hinahina, Mahoe. (Violaceae). - Tree. New Zealand. Wood source of a charcoal, suitable for certain gun-powders.

Melilotus alba Desr., White Sweet Clover. (Leguminosaceae). - Biennial herb. Europe, Asia. Occasionally cultivated; used as food for livestock. Source of a good honey.

Melilotus altissima Thuill. (Leguminosaceae). - Biennial herb. Europe, temp. Asia. Sometimes cultivated as a fodder for horses. Herb is used like Trigonella coerulea for flavoring green cheese in Switzerland. It is also mixed with tobacco snuff and in cosmetics.

Melilotus elegans Sal. (Leguminosaceae). - Biennial herb. Mediterranean region, E. Africa, Abyssinia. Herb is mixed with butter, also used in hair-pomade by women in some parts of Abyssinia.

Melilotus officinalis (L.) Lam., Yellow Sweetclover. (Leguminosaceae). - Biennial herb, sometimes annual. Europe, temp. Asia, N. Africa, introd. in N. America. Herb is used for flavoring certain cheeses, put in tobacco snuff, a substitute for Tonka Beans. Used against moths. Roots are consumed as food among the Kalmuks. Flowers are a source of honey.

Melilotus ruthenica Ser. (syn. M. wolgica Poir.) (Leguminosaceae). - Perennial herb. S. Russia. Roots are consumed as food by the Kalmucks.

Melinis minutiflora Beauv. (Graminaceae). - Perennial grass. Trop. America. Source of Oil of Efwatakala Grass. Has odor of Cumin, said to be repellant for mosquito and tsetse fly. Also cultivated as a fodderplant in Australia and Rhodesia.

Meliosma buchananiifolia Merr. (Sabiaceae). - Tree. S. China. Bark is used by the Chinese as incense.

Meliosa Clinopodium Benth. → Calamintha Clinopodium Benth.

Melissa officinalis L. Common Balm. (Labiaceae). - Perennial herb. Mediterranean region, Caucasia, Syria, Iran, Turkestan, S. W. Siberia. Cultivated. Used for flavoring salads and soups, also in liquers. Oil of Balm is employed in perfumery. Decoction, known as Balm Tea, a home-remedy is sometimes used for head and toothache.

Melissa vulgaris Trev. → Calamintha Clinopodium Benth.

Melocalamus compactiflorus Benth. and Hook. f. (Graminaceae). - Woody grass. Trop. Asia, esp. E. Bengal, Burma. Culms are made into shoes; used by Shan Kachin and Chinese traders. Seeds are large, mealy, edible; resembling chestnuts.

Melocanna bambusoides Trin., Terai Bamboo, Berry-bearing Bamboo. (Graminaceae). - Woody grass. E. Bengal, Burma. Valuable for building, mat-making; much exported to Lower Bengal, Resistant to white ants.

Melochia corchorifolia L. (Sterculiaceae) - Shrub. Pantropic. Stem is source of a fibre. Bark of root is chewed in Kordofan (Afr.) for sore lips.

Melodinus ovalis Boerl. (Apocynaceae). - Tree. Borneo. Source of a rubber.

Melodinus pulchrinervius Boerl. (Apocynaceae). - Tree. Borneo. Used as an adulterant of Guta Percha.

Melodorum Leichhardtii Benth. (syn. Unona Leichhardtii F. v. Muell.). Merangara. (Annonaceae). - Tree. Australia. Fruits are consumed raw by the aborigenes of Australia.

Melon → Cucumis Melo L.

Melon, Chinese Preserving → Benincasa cerifera Savi.

Melon, Oriental Pickling → Cucumis Conomon Roxb.

Melon Tree → Carica Papaya L.

Melothria fluminensis Gardn. (Cucurbitaceae). - Perennial vine. Brazil. Seeds are used in Brazil for colic of cattle.

Melothria maderaspatana Cogn. (syn. Mukia scabrella Arn.). (Cucurbitaceae). - Herbaceous vine. Trop. Africa, Asia. Used along the westcoast of Africa as aperient and sudorific, also for flatulency.

Memecylon costatum Miq. (Melastomaceae). - Tree. Mal. Archip. Wood tough, durable, fine-grained; used for furniture, housegold articles and building of houses.

Memecylon edule Roxb. (syn. M. tinctorium Koenig). (Melastomaceae). - Small tree. Coast of Indian Ocean, Malabar to Sumatra. Wood very hard, heavy, light brown; used for rafters and house-posts. Leaves source of a yellow dye. Fruits edible, pulp astringent; eaten in different parts of trop. Asia.

Memecylon polyanthemos Hook. f. (Melastomaceae). - Tree. Trop. Africa. Stems are used in Liberia for stools, furniture, beds and housepoles.

Menado Manila → Agathis loranthifolia Salisb.

Meni Oil → Lophira alata Banks.

Menispermaceae → Abuta, Anamirta, Chonadendron, Cissampelos, Cocculus, Coscinium, Cyclea, Jateorhiza, Menispermum, Pycnarrhena, Sphenocentrum, Tinospora.

Menispermum canadense L. Common Moonseed. (Menispermaceae). - Woody vine. Eastern N. America. Dried rhizome used medicinally as a tonic, diuretic and alterative; contains a bitter alkaloid, menispine, also berberine and oxyacanthine.

Menispermum edule Vahl. → Cocculus Cebatha DC.

Mentha arvensis L., Field Mint. (Labiaceae). - Perennial herb. Temp. Europe, Asia, America. Var. piperacens Malinv. is source of Japanese Mint Oil, containing 80 to 90% of menthol. Much is derived from Japan, China and is recently cultivated in Brazil. Used in the preparations of certain kinds of cigarettes, pharmaceuticals and oral preparations.

Mentha aquatica L. (syn. Mentha citrata Ehrh.) Bergamot Mint. (Labiaceae). - Perennial herb. S. Europe, Asia, N. Africa. Source of a lemon scented ess. oil; used in perfumery.

Mentha canadensis L., American Wild Mint. (Labiaceae). - Perennial herb. N. America. Leaves are used for flavoring. Occasionally cultivated. Roasted leaves were eaten by the Indians in Maine. Herb produces an ess. oil from which pulegone and thymol are isolated.

Mentha piperita L., Peppermint. (Labiaceae). - Perennial herb. Supposedly a hybrid. M. aquatica L. x M. spicata L. Cultivated in England, France, Italy; United States, among which Michigan, Oregon, New York, Indiana; also in Japan and Manchuria. Source of Peppermint, Oil, an ess. oil, used for flavoring confectionary, candies, liqueur, chewing gum, peppermints. Dried herb used medicinally, being carminative, nervine, stimulant. Menthol is used for nasal sprays.

Mentha Pulegium L., European Pennyroyal. (Labiaceae).- Perennial herb, Europe, Mediterranean region to Iran. Cultivated. Source of Oil of Pennyroyal; used for manuf. of soaps and synthetic menthol. Used medicinally for flatulent colic, stomach ailments; as warm infusion it promotes perspiration.

Mentha spicata L. (syn. M. viridis L.) Spearmint. (Labiaceae). - Perennial herb. Europe, N. America. Cultivated in Old and New World. Source of Spearmint Oil, an ess. oil; used for flavoring candies, tooth paste, chewing gum. Dried tops and leaves are used medicinally, being a stimulant, carminative, nervine. Contains carvone. Mentha spicata var. crispa Schrad. (syn. Mentha crispa auct. non L.) is source of German Spearmint Oil. Cultivated.

Mentzelia albicaulis Dougl. (Loasaceae). - Annual to perennial herb. Western and south eastern part of the United States. Seeds are made into a parched meal; used as food by the Indians of Arizona, California and Oregon.

Menyanthes trifoliata L., Marsh Buckbean. (Gentianaceae). - Perennial herb, Europe, Asia, N. America. Used in Russia as emergency food. In Scandinavia it is added to beer. Also used as a substitute for tea. Dried leaves used medicinally as a tonic and febrifuge. Contains a bitter glucoside, menyanthin.

Merawan, Nodding → Hopea nutans Ridl.

Merawan, Tingan → Hopea odorata Roxb.

Mercurialis leiocarpa Sieb. and Zucc. (Euphorbiaceae). - Perennial herb. Japan. Source of a blue dye, formerly used in Japan for printing on clothes.

Merendera persica Boiss. and Kotschy. (Liliaceae). - Perennial herb. N. Iran, Afghanistan, and adj. territory. Corms are used in Iran for rheumatism.

Merendon Virola → Virola merendonis Pitt.

Merissa → Calotropis procera Ait.

Meristotheca papulosa (Mont.) J. Ag. (Solieriaceae). - Red Alga. Indian and Pacif. Ocean. Consumed as food in Japan and China. The latter country imports considerable amounts from the East Indies and Japan.

Merkus Pine → Pinus Merkusii Jungh. and De Vr.

Mertensia maritima (L.) S. F. Gray. (Boraginaceae). - Perennial herb. Temp. and arct. N. Hemisphere. Rhizome is consumed as food by the Eskimos of Alaska.

Mesea Dropseed → Sporolobus flexuosus (Thurb.) Rydb.

Mesakhi Fibre → Villebrunea integrifolia Gaud.

Mescal Beans → Sophora secundiflora (Orteg.) Lag.

Mescalbean Sophora → Sophora sericea Nutt.

Mescal Buttons → Lophophora Williamsii (Lem.) Coulter

Mesembryanthemum acinaciforme L. (Aizoaceae). - Herbaceous perennial. S. Africa, introduced along the Mediterranean. Fruits have the size of a gooseberry, edible, of inferior quality, insipid; consumed by the Hottentots.

Mesembryanthemum aequilaterale Haw. (syn. M. glaucescens Haw.). Karkalia (Aizoaceae). - Perennial herb. Australia, S. W. America. Fruits eaten raw by aborigines. Leaves eaten when baked.

Mesembryanthemum chilense Mol. (Aizoaceae). - Small shrub. Chile. Fruits are consumed in some parts of Chile.

Mesembryanthemum crystalinum L., Ice Plant. (Aizoaceae). - Annual herb. S. Africa, Mediterranean region, Canary Islands. Cultivated. Leaves are eaten as spinach, also used in salads. Also M. angulatum Thunb.

Mesembryanthemum edule L., Hottentot Fig. (Aizoaceae). - Herbaceous perennial. S. Africa. Fruits edible, eaten by Hottentots.

Mesembryanthemum floribundum Haw. (Aizoaceae). - Prostrate succulent shrub. S. Africa. Recommended in S. Africa as an excellent pasture for small stock.

Mesembryanthemum Forskalei Hochst. (Aizoaceae). - Succulent shrub. E. Africa, Nile region. Crushed fruits are used by Beduins in making bread.

Mesembryanthemum micranthum Haw. (Aizoaceae). - Perennial herb. S. Africa. Ash of the plant is used as a source of soda, employed by the Hottentots for washing.

Mesembryanthemum tortuosum L. (Aizoaceae). - Perennial herb. S. Africa. Leaves, having slight narcotic properties; are used for chewing by the Hottentots.

Mesogloia crassa Suring. (Mesogloiaceae). - Brown Alga. Sea Weed. Consumed when fresh in Japan.

Mesogloiaceae (Brown Algae). → Cladosiphor, Heterochordaria, Mesogloia.

Mesona palustris Blume. (Labiaceae). - Perennial herb. Java. Decoction of leaves is used in Java as a cooling drink.

Mespilodaphne indecora Meissn. → Ocotea indecora Schott.

Mespilodaphne Sassafras Meissn. (syn. Ocotea Sassafras Mez.). (Lauraceae). - Tree. Brazil. Aromatic root and bark are sudorific, antirheumatic, diuretic. Leaves are considered in Brazil as diuretic. Source of Brazilian Sassafras Oil, obtained by distillation of the wood; used for isolation of safrol and conversion into heliotropine. Employed in soaps, disinfectants, deodorants, sprays, and other technical preparations. The oil replaces the Japanese artificial sassafras oil being a fraction of camphor oil. It has been suggested that Ocotea pretiosa Benth. is identical.

Mespilus Aria Scop. → Sorbus Aria Crantz.

Mespilus Aucuparia All. → Sorbus Aucuparia L.

Mespilus germanica L. (syn. Pyrus germanica Hook f.) Medlar. (Rosaceae). - Small tree. Europe, Asia. Cultivated. Fruits edible, especially after having been touched by the frost, when they become soft and overripe. Some varieties are sweet, others are acid. Occasionally made into cider. Varieties are Dutch and Nottingham.

Mespilus Oxyacantha Crantz → Crataegus Oxyacantha L.

Mesquite → Prosopis juliflora DC.

Mesquite Mistletoe → Phorodendron californicum Nutt.

Messmate Stringybark Eucalyptus → Eucalyptus obliqua L'Hér.

Mesua ferrea L. Ironwod. (Guttiferaceae). - Tree. Trop. Asia, India, Mal. Archipel. Wood very hard; used for walking sticks and handles of lances. Flowers and flower-buds used by natives in medicine and in cosmetics. In Madura pillows are filled with the stamens to import a pleasant scent. Tree is often grown near temples.

Metaplexis Stauntoni Schult. (Asclepiadaceae). - Vine. China, Japan. Roots are used as food by the Ainu.

Metcalf Bean → Phaseolus retusus Benth.

Metrosideros glomulifera Smith → Syncarpia laurifolia Ten.

Metrosideros lucida Rich. (Myrtaceae). - Tree. New Zealand. Wood pale red, compact tough, of great strength. Used for ship building, general carpentry work.

Metrosideros robusta Cunningh. Northern Rata. (Myrtaceae). - Tree. Epiphytic or terrestrical. New Zealand. Wood red, straight-grained, dense, hard, heavy, durable. Used for rail-road wagons, carriages, telephone-poles, bridges, wharves, wheel-wrights material, spokes, hubs, felloes.

Metrosideros scandens Soland. (syn. M. perforata Rich.). Rata Vine. (Myrtaceae). - Shrub or woody vine. New Zealand. Inner bark is used to heal sores and to stop bleeding.

Metrosideros tomentosa Rich. (Myrtaceae). - Tree. New Zealand. Wood deep red, dense, compact, heavy, durable, great strength. Used for ship timber, for boards, machine-beds and bearings; for framing, sills of dock-gates, resists teredo. Decoction of bark used by bushmen for dysentery.

Metrosideros vera Lindl. (syn. Nania vera Miq.) (Myrtaceae). - Tree. Mal. Arch., esp. Celebes. Wood durable, hard, resistant to weather, soil and water, also to marine borers; used for rudders and parts of anchors. Is too hard for building purposes.

Metrosideros villosa Sm. (syn. M. polymorpha Gaud., M. collina Gray.) (Myrtaceae). - Tree. Hawaii Islds. Wood used by the Hawaiians for carving idols, spears, mallets and the outrigger canoe.

Metroxylum elatum Mart. → Pigafettia elata Wendl.

Metroxylon laeve Mart. (Palmaceae). - Tall palm. Siam. Stem is source of a sago.

Metroxylon Rumphii Mart. (Palmaceae). - Tall palm. Malaysia. Cultivated in the tropics of the Old and New World. Trunk is source of sago; an important commercial product.

Metroxylon Sagu Rottb., Sago Palm. (Palmaceae). - Palm. Moluccas. Sago is manufactured from the trunk by maceration before the palm is in flower.

Meum Mutellina Gaertn. → Ligusticum Mutellina (L.) Crantz.

Mexican Avocado → Persea drymifolia Cham. and Schlecht.

Mexican Elder → Sambucus mexicana Presl.

Mexican Juniper → Juniperus mexicana Spreng.

Mexican Linoloe Wood. → Bursera Delpechianum Poisson.

Mexican Mahogany → Swietenia humilis Zucc.

Mexican Mock Orange → Philadelphus mexicanus Schlecht.

Mexican Mugwort → Artemisia mexicana Willd.

Mexican Prickly Poppy → Argemone mexicana L.

Mexican Sarsaparilla → Smilax aristolochiaefolia Mill.

Mexican Scammony → Ipomoea orizabensis Ledenois.

Mexican Strewberry → Echinocereus stramineus (Engelm.) Ruempl.

Mexican Sycomore → Platanus mexicana Moric.

Mezcal → Agave Kirchneriana Berger.

Mezcal de Pulque → Agave atrovirens Karw.

Mezcal de Tequilo → Agave tequilana Trel.

Mezereon → Daphne Mesereum L.

Mezoneuron Kauaiense (Mann.) Hbd. (syn. Caesalpinia Kauaiensis Mann.). (Leguminosaceae). — Tree. Hawaii Islds. Wood very hard, close-grained, very durable, almost black; used by Hawaiians for spears and the Laau melo-melo, an implement used for fishing.

Mezoneuron Scortechinii F. v. Muell. (Leguminosaceae). - Tree. Australia. Source of Barister Gum, resembling Cherry Gum and Tragacanth.

Mgmeo → Duboisia myoporoides R. Br.

Micauba → Acrocomia sclerocarpa Mart.

Michelia celebica Koord. (Magnoliaceae). - Tall tree. Celebes. Wood much esteemed, easily worked, excellent qualitiy, light, durable, yellow-brown; used for general carpentry, inside of buildings, coffins.

Michelia Champaca L., Chêmpaka (Mal.) (Magnoliaceae). - Tree. India, Malaysia. Flowers used in Siam as cosmetic; used after bathing, also to scent hair-oils; put between clothes; source of a volatile oil used in perfumes. Wood durable; used for boards, furniture, door panels, housebuilding, tea-boxes, canoes, decorative work. Bark used as febrifuge. Leaves are source of food of a silk-worm, Antheraea assamensis Helfer.

Michelia fuscata Blume. (syn. M. Figo Spreng.) Banana-Shrub, (Magnoliaceae). - Small tree. China. Cultivated in warm countries. Flowers banana-scented. Used by Chinese for scenting hair-oil.

Michelia montana Blume. (Magnoliaceae). - Tree. Mal. Archip. Wood dark yellow-brown to dark-brown, light, fine-grained, durable; used for building houses and bridges.

Michelia Tsiampaca L. (syn. M. velutina Blume). (Magnoliaceae). - Tall tree. Java, Amboina. Wood lemon-yellow, close-grained, with light brown streaks, durable, strong, easily worked; used for construction of houses and furniture.

Miconia albicans (Swartz.) Triana. (Syn. Melastoma albicans Swartz.). (Melastomaceae). - Shrub. Trop. America. Fruits are used as food by the Chinantecs and Zapotecs (Mex.).

Miconia argentea (Swartz.) DC. (syn. Melastoma argentea Swartz.). (Melastomaceae). - Shrub or tree. S. Mexico, Guatemala to Panama. Wood source of charcoal much esteemed by the Chinantecs (Mex.).

Miconia Liebmannii Cogn. (Melastomaceae). - Tree or shrub. Endemic in Oaxaca (Mex.). Fruits are used as food by Indians of N. E. Oaxaca (Mex.).

Miconia Willdenowii Klotzsch. (Melastomaceae). - Shrub. Brazil. Bark is used in Brazil for swamp fevers. Leaves are substitute for tea, contain 0.22 % caffein.

Micrechites napensis. Quint. (Apocynaceae). - Tree Cochin-China. Source of good rubber. Latex coagulates by lemon-juice.

Microcitrus inodora Swingle. Queensland Wild. Lime. (Rutaceae). Shrub. Queensland. Fruits called Russell River Limes can be used as ordinary limes.

Micrococcus urea Cohn. (Coccaceae). - Bacil. Agriculturally an important soil microorganism being capable of converting urea into ammonium carbonate.

Microdesmis puberula Hook. f. (Euphorbiaceae). - Tree. Trop. Africa. Wood called Benin Apata Wood, is flexible, hard, very fine texture, easily worked, brown; used for handles of hoes, knives, spring traps, combs, spoons and walking-sticks.

Micromeria Chamissonis (Benth.) Greene. (Thymus Chamissonis Benth.) Yerba Buena. (Labiaceae). - Perennial herb. Western N. America. Dried leaves and stems are used as as tea among the Indians of California.

Micromeria Douglasii Benth. (Labiaceae). - Herbaceous plant. Western N. America. Herb is source of a beverage; used by the Luiseño Indians of California.

Microorganism, Useful → Bacteria: Coccaceae and Bacteriaceae. Molds → Actinomycetaceae, Aspergillaceae, Moniliaceae, Mucoraceae. Yeasts: → Saccharomycetaceae. Cryptococcaceae.

Micropholis Melinoniana Pierre. (syn. Stephanoluma rugosa Bn. Chrysophyllum rugosum Sw.) (Sapotaceae). - Tree Guiana. Source of a substitute balata, Balata Blanc.

Microseris Forsteri Hook. (syn. Scorzonera scapigera Forst.) (Compositae). Herbaceous perennial. Australia. Tubers are used as food by the aborigins.

Microstemon velutina Engl. (Anacardiaceae). - Tree. Trop. Asia, esp. So. Peninsular Siam to Malacca. Latex of bark used for ring-worm.

Microtanea cymosa Prain., Khasia Patchouli. (Labiaceae). Woody plant. Trop. Asia, esp. India, Khasia and Assam. Source of Khasia Patchouli; used in Kwangtung (China) and in Indo-China. Also M. robusta Hemsl. Employed in the same way as Patchouli.

Microtis porrifolia R. Br. (Orchidaceae). - Perennial herb. New Zealand. Tubers are consumed as food by the aborigenes of New Zealand.

Midnight Lichen → Parmelia stygia (L.) Ach.

Miel de Tuna → Opuntia megacantha Salm-Dyck.

Mignonette, Common → Reseda odorata L.

Mignonette Vine → Boussingaultia baselloides H. B. K.

Milfoil → Achillea Millefolium L.

Milicia africana Sim → Chlorophora excelsa (Welw.) Benth. and Hook. f.

Mijagua Fibre → Anacardium Rhinocarpus DC.

Milk, Acidophilus → Lactobacillus acidophilus (Moro) Holland.

Milk, Kefir → Saccharomyces Kefir Beijerinck.

Milk, Kumiss → Lactobacillus casei Rogers

Milk, Yoghurt → Lactobacillus bulgaricus (Luersen and Kuhn) Holland.

Milk. Coagulants of → Acanthosicyos, Adansonia, Carthamus, Cirsium, Ficus, Morrenia, Panus. Pavonia, Pinguicula, Sideroxylon, Solanum, Withania.

Milk Parsley → Peucedanum palustre (L.) Moench.

Milk Tree → Sapium Aucuparium Jacq.

Milk Vetch → Astragalus Glycyphyllos L.

Milkweed, Common → Asclepias syriaca L.

Milkweed, Desert → Asclepias subulata Decaisne.

Milkweed, Dwarf → Asclepias involucrata Engelm.

Milkweed, Showy → Asclepias speciosa Torr.

Milkweed, Swamp → Asclepias incarnata L.

Milkweed, Woolly → Asclepias eriocarpa Benth.

Millet, African → Eleusine coracana (L.) Gaertn.

Millet, Cattail → Setaria glauca (L.) R. Br.

Millet, Foxtail → Setaria italica L.

Millet, Japanese → Echinochloa frumentacea (Roxb.) Link.

Millet, Koda → Paspalum scrorbiculatum L.

Millet, Native → Panicum decompositum R. Br.

Millet, Pearl → Setaria glauca Beauv.

Millet, Proso → Panicum miliaceum L.

Millet, Sanwa → Echinochloa frumentacea (Roxb.) Link.

Millet, Shama → Echinochloa colonum (L.) Link.

Millet, Texas → Panicum texanum Buckl.

Milletia atropurpurea Benth. (Leguminosaceae). Tree. Malaysia. Twigs and roots have been recommended as source of an insecticide. Contains rotenon. Is used by the natives to stupefy fish. Also M. auriculata Baker from India and M. Taiwaniana Hayata from Formosa.

Milletia auriculata Baker. (Leguminosaceae). - Woody climber. Himalayan region. India. In India source of rough cordage.

Millettia caffra Meisn. (Leguminosaceae). - Tree. S. Africa. Wood very heavy, hard, not elastic, close-grained, compact, durable in contact with soil; used for spokes and bearings of light machinery. Fruits used by Kaffirs as vermicide.

Millettia eriantha Benth. → Whitfordiodendron erianthum Dunn.

Millettia sericea. Wight and Arn. (syn. Pongamia sericea Vent.) (Leguminosaceae). Woody vine. Burma and Malaysia. Roots used by natives to stupefy fish.

Milo → Sorghum vulgare Pers.

Mimetes lyrigera Knight. (Proteaceae). - Tree. S. Africa. Bark is rich in tannin, used in S. Africa for tanning leather.

Mimosa Barbatinam Vell. → Stryphnodendron Barbatinam Mart.

Mimosa cyclocarpa Jacq. → Enterolobium cyclocarpum (Jacq.) Griseb.

Mimosa esculenta Moc and Sessé → Leucaena esculenta (Moc. and Sessé) Benth.

Mimosa flexicaulis Benth. → Pithecolobium flexicaule Coult.

Mimosa glauca L. → Leucaena glauca (L.) Benth.

Mimosa guianensis Aubl. → Stryphnodendron guianense Benth.

Mimosa Houstonii L'Her. → Calliandra Houstoniana (Mill.) Standl.

Mimosa invisa Mart. (Leguminosaceae). - Herbaceous plant. Trop. America. Has been recommended as green manure in warm countries.

Mimosaceae → Leguminosaceae.

Mimusops bidentata DC. → Manilkara bidentata (DC.) Chev.

Mimusops djave (Lam.) Engl. (Sapotaceae). - Tree. Gold-Coast, Cameroon, Nigeria, Gabon. Seeds source of Djave or Adjab-Butter. Used by natives in food. In Europe used for manuf. soap. Press-cakes are poisonous. Sp. Gr. 0.8979; Sap. Val. 182—188; Iod. No. 56—65; Unsap. 2—3 %.

Mimusops Pierreana Engl. → Baillonella ovata Pierre.

Mindanao Cinnamon → Cinnamomum mindanaense Elm.

Minjak Tagkawang Tallow → Isoptera borneensis Scheff.

Mint, American Wild → Mentha canadensis L.

Mint, Atlantin Mountain → Pycnanthemum incanum (L.) Michx.

Mint, Cat → Nepeta Cetaria L.

Mint, Emory Bush → Hyptis Emory Torr.

Mint Field → Mentha arvensis L.

Mint, Horse → Monarda fistulosa L.

Mint, Pepper → Mentha piperita L.

Mint, Spear → Mentha piperita L.

Mint, Stone → Cunila origanoides (L.) Butt.

Mint, Virginia Mountain → Pycnanthemum virginianum (L.) Durand and Jacks.

Mint Geranium → Chrysanthemum Balsamita L.

Minyak Keruong Resin → Dipterocarpus Kerii King.

Mioga Ginger → Zingiber Mioga (Thunb.) Rosc.

Mirabilis Jalapa L., Four o'clock. (Nyctaginaceae). - Perennial herb. Tropics. Powdered seeds are used in Japan as cosmetic. Roots are used for dropsy. A crimson dye is obtained from steeped flowers in water; used in China to color jellies made from seaweeds, also for cakes.

Miscanteca anacardioides Benth. (Lauraceae). - Tree. Brazil. Bark used as an abortive. Contains an ess. oil resembling apiol. Also M. Duckei Samp.

Mischosma riparium Hochst. (Labiaceae). - Perennial plant. S. Africa. Infusion of leaves and roots is used by the Zulus as emetic.

Miscolobium nigrum Mart. → Dalbergia Miscolobium Benth.

Missouri Currant → Ribes aureum Pursh.

Mistletoe, American → Phoradendron flavescens (Pursh) Nutt.

Mistletoe, European → Viscum album L.

Mistletoe, Juniper → Phoradendron juniperinum Engelm.

Mistletoe, Mesquile → Phoradendron californicum Nutt.

Mitchell Grass → Astrebla pectinata F. v. Muell.

Mitracarpum scabrum Zucc. (Rubiaceae). - Annual herb. Trop. Africa. Dried leaves are supposed to heal old ulcers rapidly. Antidote for arrow poison.

Mitragyna africana Korth. Josswood. (Rubiaceae). - Shrub or tree. Trop. Africa. Wood used for carving, tools, planes, and Musselman writing boards. Bark is source of a yellow dye.

Mitragyna macrophylla Hiern. (Rubiaceae). - Tree. Trop. Africa, esp. S. Leone, Gold Coast, Angola, Trop. W. Africa. Wood cross-grained, light brown; used by Kroomen for canoes; in Angola for house-building, furniture, carpentry, cabinet work. Wood is sometimes exported as West African Mahogany. Roots when boiled are used in S. Leone for colic.

Mitragyna speciosa Korth. (Rubiaceae). - Tree. Mal. Peninsul., Lower Siam. Cultivated. Leaves are chewed or smoked like opium; substitute for opium. Supposed to be more harmful than opium.

Mitragyna stipulosa Kuntze. Abura, African Linden. (Rubiaceae). - Tree. Trop. W. Africa. Wood light, soft, brownish-yellow, durable in water, not immune from borers, easily worked; used for canoes, boards, roofing, doors, furniture, drums, boxes, paddles, barrels. Leaves are used for coughs, as febrifuge.

Miyama Cherry → Prunus Maximowiczii Rupr.

Mkani Fat → Allanblackia Stuhlmanii Engl.

Mobola Plum → Parinarium Mobola Oliv.

Mocha Coffee → Coffea arabica L.

Mock Oyster → Pleurotus ostreatus (Jacq.) Quél.

Mockernut → Carya tomentosa (Lam.) Nutt.

Mogador Acacia → Acacia gummifera Willd.

Mogador Gum → Acacia gummifera Willd.

Mogun-Mogun → Harpullia pendula Planch.

Mohria thurifraga Sw. (Polypodiaceae). - Perennial herb. Fern. Cape Peninsula. Dried, powdered leaves with fat are used by the natives for healing burns.

Moka Aloë → Aloë abyssinica Lam.

Molasses → Saccharum officinale L.

Molave Chaste Tree → Vitex parviflora Juss.

Molds, Useful → Microorganism.

Molito → Rhus Cotinus L.

Mollugo hirta Thunb. (syn. Glinus lotoides Loefl.). (Aizoaceae). - Annual herb. Tropics and subtropics. Tender leaves are eaten by the natives of the Sudan.

Molucca Bean → Caesalpinia Bonducella Flem.

Molucca Ironwood → Afzelia bijuga Gray.

Molucca Manila → Agathis loranthifolia Salisb.

Mombin, Yellow → Spondias lutea L.

Momordica Balsamina L. Balsam Apple. (Cucurbitaceae). - Herbaceous vine. Tropics. Cultivated. Fruits are eaten in the Sudan. Used in Syria for curing wounds. Leaves and stems are used as camel fodder.

Momordica Charantia L. Balsampear. (Cucurbitaceae). - Vine. Tropics. Cultivated. Leaves when par-boiled are consumed as a vegetable in some parts of Peru. Unripe fruits are consumed when boiled or fried, also eaten in salads. Sap from leaves and fruits is used for colic and worms.

Momordica cochinchinensis Spreng. (Cucurbitaceae). - Woody vine. Trop. Asia. Roots produce a lather with water; used for washing clothes.

Momordica Schimperiana Naud. (Cucurbitaceae). - Perennial vine. Trop. Africa. Used by the natives as an insecticide.

Monantha Vetch → Vicia monanthos Desf.

Monarda citriodora Cerv. Lemon Bee Balm. (Labiaceae). - Herbaceous perennial. New Mexico, Arizona and adj. Mexico. Herb was eaten with meat by Hopi Indians of Arizona.

Monarda didyma L. Oswego Tea, Oswego Bee Balm. (Labiaceae). - Perennial herb. Eastern N. America. Leaves sometimes used as a tea; also flavoring of food. Has mint flavor. Source of thymol, Oil of Thyme. M. fistulosa L. is sometimes used for the same purpose.

Monarda fistulosa L. (syn. M. menthaefolia Grah.) Wild Bergamot, Horse Mint. Bee Balm. (Labiaceae). - Perennial herb. N. America. Leaves eaten boiled with meat, by the Tewa Indians. Decoction made into hair-pomade; used by the Omeha and Ponca Indians. Herb is considered an active diaphoretic.

Monarda pectinata Nutt. Pony Bee Balm. (Labiaceae). Perennial herb. Western N. America, Utah. Herb was used for flavoring food by the Indians of New Mexico.

Monavist → Dolichos Lablab L.

Moneywort → Lysimachia Nummularia L.

Monghyr Piuri → Mangifera indica L.

Mongolia Oak → Quercus mongolia Fischer.

Monilia candida Bon. (Candida tropicalis (Cast.) Berkh.). (Moniliaceae). - Fungus. This microorganism has been used in the production of animal fodder from surplus and waste material etc. see Torula pulcherrima.

Moniliaceae → Citromyces and Monilia.

Monimiaceae → Atherosperma, Dorypha, Hedycarya, Laurelia, Peumus, Siparuna.

Monkshood → Aconitum Napellus L.

Monochoria vaginalis Presl. (Pontederiaceae). - Herbaceous waterplant. Indonesia. Roots are used by the Javanese for stomach troubles, toothache and diseases of the liver. Leaves are used for fever.

Monochoria hastata Solms. (Pontederiaceae). - Herbaceous waterplant. Indonesia, Malaysia. Rootstocks are used as food for pigs in Minahassa.

Monodora angolensis Welw. Angola Calabash. (Annonaceae). - Tree. Trop. Africa. Seeds are used in some parts of Africa as a condiment.

Monodora Myristica Dunal. Calabash Nutmeg. (Annonaceae). - Tree. Africa esp. Sierra Leone, Upper Guinea, Cameroon, Gabon. Endosperm of the seeds is used as a condiment, is highly appreciated by the natives of some parts of Africa. Seeds are made into rosaries. Also used medicinally. Sold in markets.

Monolepis Nuttaliana (Schultes) Greene. (Chenopodiaceae). - Herbaceous plant. Western N. America. Washed roots cooked with fat and salt were consumed as food by the Indians of Arizona. Seeds were used as pinole.

Monstera deliciosa Liebm. (syn. Philodendron pertusum Kunth). Ceriman. (Araceae). - Climbing plant. Mexico, Guatemala. Fruits are consumed; they are of pleasant flavor, having an aroma of a pineapple combined with that of a banana. They contain small needle shaped crystals, raphids, entering the tongue which makes eating often unpleasant. Pulp is used in ices and drinks.

Montanoa tomentosa Cervantes. (syn. Eriocomo fragrans D. Don). Hierba de la Perida. (Compositae). - Shrub. San Luis Potosi to Oaxaca (Mexico). Plant has stomachic, diuretic and pectoral properties. Commonly used in Mexico as an aid to women in childbirth. Ground roots mixed with lukewarm water are said to be used for dysentry.

Montia fontana L. (Portulacaceae). - Herbaceous plant. Centr. Europe. Leaves are sometimes used in salads, especially in some mountainous areas of France.

Montrouziera sphaeraeflora Planch. Houp. (Guttiferaceae). - Tree. New Caledonia. Wood reddish yellow veined, durable, easily worked; used for different purposes.

Moonseed → Menispermum.

Moquinia polymorpha DC. Tatané-moroli. (Caprifoliaceae). - Tree. Brazil. Wood compact, durable; used for general carpentry.

Mora → Dimorphandra Mora B. and H.

Moraceae. → Antiaris, Artocarpus, Brosimum, Broussonetia, Canabis, Castilla, Conocephalus, Dorstenia, Ficus, Humulus, Maclura, Morus, Musanga, Myriantus, Poulsenia, Treculia.

Morel → Morchella esculenta L.

Morchella esculenta L. Morel. (Helvellaceae). - Fungus. Ascomycete. Fruitbodies are consumed as food. There are many other species eaten in various countries of the Old and New World.

Morinda trifolia L. Indian Mulberry. (Rubiaceae). - Small tree. Indonesia, Malaya. Leaves are source of a red; the roots of a yellow dye. The red dye is used for batik work in Java. Plants are cultivated in Java. Young leaves are sometimes consumed as vegetable. A decoction of fruits, bark and roots is used by the Javanese to clean wounds. Wood (Togari wood of Madras) is compact, not attacked by insects.

Morinda bracteata Roxb. (Rubiaceae). - Tree. Malaysia. Roots are used to give a red color to linen and basketwork.

Morinda tinctoria Roxb. (Rubiaceae). - Tree. Southeastern Asia, India. Roots are source of a dye, used for coloring linen and woolen goods.

Morinda speciosa Roxb. (Rubiaceae). - Small shrub. Malaysia. A decoction of the roots is used for rheumatism by the natives.

Moringa oleifera Juss. (syn. M. pterygosperma Gaertn.) Horseradish Tree. (Moringaceae). - Tree. India. Grown in the tropics of the Old and New World. Source of Oil of Bene, which is used in cosmetics; for enfleurage; lubricating watches and delicate instruments. Employed in India as purgative and for rheumatics. Also M. arabica Pers. (syn. M. aptera Gaertn.).

Moringaceae → Moringa.

Morning Glory → Convolvulus and Ipomoea.

Moronobea coccinea Aubl. (Guttiferaceae). - Tree. Trop. America. Source of Bois Cochon, easy to split; used for building purposes. Anani is a resin to make vessels watertight.

Morphine → Papaver somniferum L.

Morus alba L. White Mulberrry. (Moraceae). - Tree. China. Widely cultivated. Has been used since time immemoreal as the food plant of the silkworm, Bombyx mori. Fruits are of little value.

Morus nigra L. Black Mulberry. (Moraceae). - Tree. Probably native to the Orient. Cultivated in Europe and the Near East for its fruit which has an agreeable taste. The Black Persian Mulberry is probably a variety.

Morus rubra L. Red Mulberry. (Moraceae). - Tree. Eastern N. America. Wood light, not strong, tough, coarse grained, very durable; used for fancing, cooperage, ship and boatbuilding.

Moschosma polystachyum Benth. (Labiaceae). - Herbaceous plant. Malaysia. Bruised parts of plant used externally and internally for persons that have been frightened, as it has a calming effect on the nervous system, used by the Javanese.

Mouriria domingensis Spach. (Melastomaceae). - Small tree. West Indies. Fruits of the size of a small plum are edible and are of a pleasant flavor.

Mouriria Pusa Gardn. (Melastomaceae). - Small tree. Brazil. The black fruits are edible and of a fine flavor. They are sold in the markets of Ceara, under the name of Pusa.

Mucor piriformis A. Fisch. (Mucoraceae). - Fungus. Microorganism. A mold that has been suggested in the formation of citris acid as a result of a fermentation process.

Mucoraceae → Mucor, Rhizopus.

Mucuna alterrima (Piper and Tracy) Holland. (Leguminosaceae). - Herbaceous vine. Trop. Asia. Is used in warm countries as a green manure and cover crop.

Mucuna Deeringiana (Bort) Small. (Stizolobium Deeringianum Bort.) Florida Velvet Bean. (Leguminaceae). - Annual vine. Probably from So. Asia or Malaysia. Grown in warm countries as a covercrop, greenmanure and forage plant.

Mucuna gigantea DC. (Leguminosaceae). - Herbaceous vine. Malaysia. The seeds are consumed in some parts of Malaya.

Mucuna Jonghuniana Backer. (Leguminosaceae). - Herbaceous vine. Beans are used as beads in some parts of Java, especially by children. It is supposed that they will keep them immune from diseases.

Mucuna nivea DC. Lyon Bean. (Leguminosaceae). - Herbaceous vine. East India. The pods are eaten as a vegetable in some parts of India. Plant is also grown as a covercrop and for greenmanure.

Mucuna pruriens DC. (Leguminosaceae). - Perennial herb. East Indies. Is sometimes used as an anthelmintic.

Muehlenbeckia adpressa Meissn. (syn. M. Gunii Hock. f.). (Polygonaceae). - Small climbing shrub. Australia. The white berries resemble gooseberries, are used in pies, puddings and confectionary.

Muehlenbergia Huegelii Tren. (Graminaceae). - Perennial grass. Java. The plant is used as food for livestock in some parts of Java. It is much liked by cattle.

Mulberry, Black → Morus nigra L.

Mulberry, Indian → Morinda citrifolia L.

Mulberry, Paper → Broussonetia papyrifera (L.) Vent.

Mulberry, Persian Black → Morus nigra L.

Mulberry, Red → Morus rubra L.

Mulberry, White → Morus alba L.

Mullein → Verbascum.

Mung Bean → Phaseolus aureus Roxb.

Muntingia calabura L. Capulin. (Elaeocarpaceae). - Tree. Trop. America. The sweet, red or pale-yellow fruits are edible and are of a pleasant flavor.

Murlins → Alaria esculenta (L.) Grev.

Murraya Koenigii Kurz. (Rutaceae). - Small tree. Himalaya region. Bark, leaves and roots are used by the natives as a tonic.

Murraya paniculata (L.) Jacq. (syn. Chalcas paniculata L.) Cosmetic Barktree. (Rutaceae). - Tree. Malaysia. Wood called Satinwood is fine, hard, light yellow, becoming later brown; used in Java for cutlery, especially handles of knives and walking sticks. It is said that the bark is used as a cosmetic.

Musa acuminata Colla. (Musaceae). - Tall perennial plant. Java to New Guinea. Fruits are edible, consumed by the natives, known as Pisang Jacki.

Musa Basjo Sieb. and Zucc. (Musaceae). - Tall herbaceous plant. Japan. Source of a fibre in some parts of Japan; used for the manuf. of cloth and for making sails.

Musa Cavendishii Lam. (syn. M. nana Lour.) Dwarf Banana, Chinese Banana, Cavendish Banana. (Musaceae). - Relative low plant. Eastern trop. Asia. Grown in the tropics and subtropics of the Old and New World. Hardiest among the fruiting bananas. Fruits relatively small, of a good flavor, shipping qualities fair. Grown extensively in the Canary Islands.

Musa corniculata Rumph. (Musaceae). - Tall perennial herb. Réunion. Fruits very large, consumed as food by the natives of Réunion.

Musa discolor Horan. (Musaceae). - Tall perennial herb. New Caledonia. A fibre is obtained from the leaves; used locally. Fruits are consumed by the natives.

Musa Fehi Bert. Aiori. (Musaceae). - Tall perennial herb. New Caledonia, Tahiti. The fruits are consumed by the natives as a food, especially in Tahiti and other islands of Polynesia.

Musa Holstii K. Schum. (Musaceae). - Tall perennial herb. East Africa. Source of a good fibre which has been recommended in cases of emergency. Also M. ulugurensis Warb.

Musa oleracea Vieillard. Banana Poreté. (Musaceae). - Tall perennial herb. New Caledonia. Fruits are consumed as food by the natives of New Caledonia.

Musa paradisiaca L. Plantain. (Musaceae). - Tall perennial herb. Trop. Asia. Known as Cooking Banana, Adam's Banana. Boiled fruits are fried and served in place of bread in many tropical countries. Ripe fruits are roasted or boiled in their skins and peeled. Flour is considered excellent for invalids and children. There are numerous varieties, some of local interest, among which Macho is the best; Hembra is smaller and similar in flavor.

Musa sapientum L. (syn. M. paradisiaca L. var. sapientum Kuntze). Banana. (Musaceae). - Tall perennial herbaceous plant. Trop. Asia. Grown extensively on a commercial scale in the Old and New World. Eaten as a dessert fruit. Fruits are also dried (Banana Figs) and ground into a flour. A mixture with wheat flour is excellent. Source of a coffee surogat, alcohol, oil, banana preserves (Bananes cristallisées). There are numerous varieties. Gros Michael is widely grown for export, esp. in Trop. America. Ripe bananas are eaten fried and are delicious. Var. rubra Baker is the Red Banana, Colorada or Morada. Its fruits are thick, with dark red or purple skin.

Musa silvestris Lamarie. Layason. (Musaceae). - Tall perennial herb. Tonkin, Philippine Islands, Moluccas. Source of a fibre being of local interest in some parts of the Philippines. Is probably related to M. textilis Née.

Musa textilis Née. Abaca. Manila Hemp. (Musaceae). - Tall perennial herb. Philippine Islands. Grown in the tropics of the Old and New World. Source of an excellent fibre, called Manila Hemp; used for the manufacturing of cordage, especially marine cables, binder twine, bagging, papier mâché, wrapping paper, and a lustrous cloth known as sinamay. The fibres are white to reddish-yellow, elastic and not injured by salt and fresh water. Much is derived from the Philippine Islands, Honduras and Panama.

Musa tikap Warb. (Musaceae). - Tall perennial herb. Carol. Islands. A fibre is obtained from the leaves; used by the natives for manuf. of cloth.

Musaceae → Heliconia, Musa, Ravenala.

Musanga Smithii R. Br. Kombo-Kombo. (Moraceae). - Small tree. Trop. Africa. A beverage is prepared from the plant by the natives.

Muscadine Grape → Vitis rotundifolia Michx.

Muscari comosum Mill. (Liliaceae). - Bulbous plant. Central and southern Europe. Bulbs are consumed in some parts of Greece, especially in early spring.

Mushrooms, Edible → Edible Fungi.

Mussaenda frondosa L. (Rubiaceae). - Shrub. Malaysia. Juice of the plant is used by the Japanese for infection of the eyes. Decoction of the leaves against intestinal worms.

Mussaenda variabilis Hemsl. (Rubiaceae). - Climbing shrub. Malaysia. Decoction of the roots is used by the natives against coughing and that of the leaves against fever.

Mussaeindopsis Beccariania Beill. (Rubiaceae). - Tree. Western Malayan Penins. Wood brownish-yellow, durable, heavy, resistant to weather; used for bridges and house building.

Muskmelon → Cucumis Melo L.

Mustard, Black → Brassica nigra (L.) Koch.

Mustard, Dakota → Brassica arvensis (L.) Ktze.

Mustard, Indian → Brassica juncea (L.) Coss.

Mustard, Indian Brown → Brassica integrifolia O. K. Schultz.

Mustard, Japanese → Brassica japonica Sieb.

Mustard, Sarepta → Brassica Besseriana Adrz.

Mustard, Tuberous-rooted Chinese → Brassica napiformis Bailey.

Mustard, White → Sinapis alba L.

Myoporum debile R. Br. (Myoporaceae). - Tree. Australia. Fruits are consumed by the natives in some parts of Australia. Also M. platycarpum R. Br. and M. serratum R. Br.

Myoporum platycarpum R. Br. (Myoporaceae). - Small tree. Australia. The tree forms an excudation, a Manna which is sold in the markts. Much is derived from Fowler Bay, S. Australia.

Myoporum sandwicensis Gray. Bastard Sandalwood. (Myoporaceae) . -Tree. Hawaii Islands. Wood is sold as a substitute for Sandalwood.

Myrciaria cauliflora Berg. (Myrtaceae). - Tree. So. Brazil. Fruits are edible and are of a good flavor. Is closely related to M. Jaboticaba Berg.

Myrciaria Jaboticaba Berg. Jaboticaba. (Myrtaceae). - Tree. So Brazil. A highly esteemed fruit in Brazil and other parts of the tropics and subtropics. Eaten raw. Fruits are source of a jelly and an alcoholic beverage. Sold in markets.

Myrciaria tenella Ber. (Myrtaceae). - Tree. So. Brazil. Fruits are of a pleasant flavor, eaten in different parts of Brazil.

Myrciaria trunciflora Berg. (Myrtaceae). - Tree. So. Brazil. A much esteemed fruit, eaten in Brazil. Sold in markets.

Myriocarpa longipes Liebm. (Urticaceae). - Shrub. Mexico. Used in Oaxaca (Mex.) as a remedy against malaria.

Myrianthus arborea Palisot. (Moraceae). - Tree. Congo. Fruits are edible and are much esteemed by the natives.

Myrica carolinensis Mill. Bayberry. (Myricaceae). - Small tree. Eastern and Southern United States. Is source of a wax, derived from the surface of the fruits. Used for manuf. of candles.

Myrica cerifera L. Wax Myrtle. (Myricaceae). - Tree. Eastern and Southern United States. A wax is derived that has developed around the fruits. It is removed by boiling the fruits in water. Used for manuf. of candles. Wax Myrtle Bark is derived from the roots; it is sold in strips and quills, is astringent and tonic. Also the leaves and bark of the Sweet Fern, Myrica (Comptonia) asplenifolia L.

Myrica gale L. Sweet Gale. (Myricaceae). - Small shrub. Temp. Europe, Asia and N. America. The leaves are used in medicine, they are considered aromatic and astringent.

Myrica javanica Bluem. (Myricaceae). - Tree. Java. Kernels of the fruits are consumed by the Javanese.

Myrica mexicana Willd. Arbol de la Cera. (Myricaceae). - Shrub small tree. Mexico. Fruits are source of a wax; used for candles. Is also a popular remedy, taken internally for jaundice and diarrhoea. Sold in markets. Decoction of rootbark is supposed to be acrid, astringent, in large doses emetic.

Myrica Nagi Thunb. (syn. M. rubra Sieb. and Zucc.) Ioobai, Yama Momo. (Myricaceae). - Tree. Trop. and subtr. Asia. Cultivated in China for centuries. The kernels of the fruits are edible and much esteemed in some parts of China and surrounding countries.

Myrica Pringlei Greenm. (Myricaceae). - Small shrub. Mexico. Fruits are source of a wax.

Myricaceae → Myrica.

Myristica argentea Warb. (Myristicaceae). - Tree. New Guinea. Furnishes the Macassar or Papua Nutmegs; they are very pungent and have the scent of methyl silicylate.

Myristica fatua Houtt. Moutain Nutmeg. (Myristicaceae). - Tree. Moluccas, Amboina, Banda. The nuts are occasionally used as a condiment. They have little commercial value.

Myristica fragrans Houtt. Nutmeg. (Myristicaceae). - Tree. Moluccas. Cultivated in the tropics of the Old and New World. Source of Nutmeg and Mace, used as condiments. Mace is derived from the arillus that surrounds the seeds. Used for savory dishes, pickles, katchup, sauces, puddings, and in beverages. The use of too much nutmeg is toxic. Nutmeg butter is derived from nuts that are unfit for the spice trade; it is used for ointments and for candles. Mace furnishes a similar butter. Fresh husks of the ripe fruits are source of a jelly. An essential oil, Oleum Myristicae is used in medicine as aromatic and carminative; used for dentifices; is also used in perfumes and the tobacco industry.

Myristica iners Blume (Myristicaceae). - Tree. Western Java. The wood is used by the Javanese for fumigating and scenting cloth.

Myristica Komba Baill. (Myristicaceae). - Tree. West Africa, esp. Gabon, Senegal. Fruits are used as a medicine by the natives.

Myristica malabarica Warb. (Myristicaceae). - Tree. So. India. Source of Bombay or Wild Nutmeg, is slighty aromatic and has little value as a spice. It is used as an adulterant of the true mace.

Myristica succedanea Blume. (Myristicaceae). - Tree. Moluccas. Seeds and mace are occasionally used as condiment.

Myristicaceae → Horsfieldia, Myristica, Ochocoa, Virola.

Myrobalan → Phyllanthus Emblica L., Spondias lutea L. and Terminalia Catappa L.

Myroxylon Balsamum (L.) Harms. (syn. M. toluiferum HBK). Balsam of Tolu. (Leguminosaceae). - Tree. Venezuela, Colombia, Peru. Source of a balsam, Opobalsam or Balsam of Tolu. It is collected in gourds, has a brown to yellowbrown color and is of a plastic nature. Used for ointments and salves, as an expectorant and antiseptic, in bronchitis, coughs and colds. Sometimes used to flavor cough syrups. Much is used as fixative in perfumery.

Myroxylon orbiculatum Forst. (Leguminosaceae). - Tree. Polynesian Islands. Wood is used for perfuming coconut oil by the natives. Also M. suaveolens Forst. M. hawaiense. (Seem.) Kuntze and M. Hillebrandii (Wawra) Kuntze.

Myroxylon Pereira (Royle) Klotzch. Balsam of Peru. (Leguminosaceae). - Tree. Central America. Cultivated in trop. countries. Wood valuable, resembles Mahogany. Balsam of Peru is derived from the stems; it is a thick, syrupy, dark reddish brown substance, being a pathogenic product, as a result of wounding the stem. Used in medicine for healing slow wounds and skin diseases, also used against coughs, bronchitis. Employed as a fixative in perfumery. Much is derived from El Salvador.

Myrrh, Bisabol → Commiphora erythraea (Ehrenb.) Engl.

Myrrh, Garden → Myrrhis odorata (L.) Scop.

Myrrh, Herabol → Commiphora Myrrha (Nees) Engl.

Myrrh, Sweet → Commiphora erythraea (Ehrenb.) Engl.

Myrrh, Sweet Scented → Myrrhis odorata (L.) Scop.

Myrrhis odorata (L.) Scop. Garden Myrrh., Sweet Scented Myrrh. (Unbelliferaceae). - Perennial herb. Europe, Caucasia. Used as a home medicine for „cleansing" the blood, expectorant. It is supposed that it increases the milkproduction of cattle. Roots and fruits are used for flavoring brandy. Is occasionally cultivated.

Myrsinaceae → Jacquinia, Maesa, Myrsine, Reptonia, Theophrasta.

Myrsine capitellata Wallich. (Myrsinaceae). - Tree. Trop. Asia, Himalaya region. Fruits are consumed by the natives. Also M. semiserrata Wallich.

Myrsine Grisebachii Hieron. Polo Blanco, Langa Marca. (Myrsinaceae). - Tree. Argentina. Wood is good qualitiy, used for furniture, turnery and wagons, known as Polo Blanco.

Myrsine melanophloea R. Br. Cape Beech. (Myrsinaceae). - Tree. So. Africa. Leaves are used as an astringent. Wood hard, tough, brownish red, of attractive appearance, easily worked; used for wagon building.

Myrsine urvillea DC. Matsam. (Myrsinaceae). - Tree. New Zealand. Bark is used for tanning.

Myrtaceae → Callistemon, Eucalyptus, Eugenia, Feijoa, Leptospermum, Melaleuca, Metrosideros, Myrciaria, Myrtus, Pimenta, Psidium, Rhodomyrtus.

Myrtle → Myrtus communis L.

Myrtle, Wax → Myrica cerifera L.

Myrtus communis L. Myrtle. (Myrtaceae). - Small tree. Mediterranean region. Green and dried fruits are occasionally used as a condiment, also as a stomachic. Myrtle oil, Essence de Myrte, Oleum Myrti is obtained in small quantities by distillation from the leaves in So. Europe, Syria, Asia Minor, India and other countries; used as aromatic and tonic. Wood is hard, elastic, close-grained; used for walking-sticks, furniture and handles of tools.

Mysore Gardamon → Elattaria Gardamomum (L.) Maton.

Myxopyrum nervosum Blume (Oleaceae). - Climbing shrub. Malaysia. Decoction of the roots is used by the Javanese for pain in the joints. Bark is used in Eastern Java as a binding material.

N

Nacaspilo → Inga spuria Humb. and Bonpl.

Nailwood, Silver → Paronychia argentea Lam.

Najadaceae → Najas.

Najas flexilis (Willd.) Rostk. and Schmidt. (Najadaceae). - Submersed herbaceous waterplant. N. America, West Indies, Old World. Where it grows abundantly, plants are sometimes used as fertilizer. They are also dried and used as packing material. Also other species are used.

Najas major All. (Najadaceae). - Perennial water plant. Temp. and trop. zones. Plant is eaten as food by the Hawaiians with salt, like watercress. Is considered appetizing with raw fresh water shrimps, opai and crabs. Sold in markets of Honolulu.

Naked Oat → Avena nuda L.

Nania vera Miq. → Metrosideros vera Lindl.

Nanking Cotton → Gossypium Nanking Meyen.

Nanny Berry → Viburnum cassinoides L.

Napier Grass → Pennisetum purpureum Schum.

Napoleona Heudelotii Juss. (Lecythidaceae). - Tree. Trop. Africa. Macerated fruits are used by the natives of Sierra Leone for inguinal hernia.

Napoleona leonensis Hutch. and Dalz. (Lecythidaceae). - Tree. Trop. Africa, esp. Liberia, Sierra Leone, Ivory Coast. Chopped bark is used in Liberia when mixed with cola nuts for chewing. It is also mixed with rice used as food.

Naranjillo → Solanum quitoense Lam.

Narasplant → Acanthosicyos horrida Welw.

Narbonne Vetch → Vicia narbonensis L.

Narcissus Jonquilla L., Jonquil. (Amaryllidaceae). - Perennial bulbous herb. Mediterranean region. Cultivated as an ornamental. Source of an ess. oil, obtained by enfleurage, also by the process of maceration. Used in perfumery, in compounds of the heavy type, in Oriental and floral bouquets, especially in heavy narcissus perfumes. About 450 - 500 kg. flowers produce 1 kg. concrete, giving 450 to 500 gr. of alcohol soluble absolute. Contains benzyl benzoate, methyl benzoate, methyl cinnamate, indol, linalol.

Narcissus poeticus L., Poets's Narcissus. (Amaryllidaceae). - Perennial bulbous herb. Europe. Widely cultivated as an ornamental. Flowers. are source of an ess. oil; used in perfumery. Is extracted by volatile solvents. About 500 kg. flowers produce 1 kg. concrete giving 270 to 300 gr. alcohol soluble absolute.

Nardostachys Jatamansi DC., Jatamansi. (Valerianaceae). - Perennial herb. Alpine Himalayas, Garhwa to Sikkim. Source of an essential oil, supposed to improve hair growth and causing blackness of hair. Recommended in perfumery. It is the Nardus Root or Spikenard of the Ancients; used for nervous disorders.

Nardus Root → Nardostachys Jatamansi DC.

Naregamia alata Wight and Arn. (Meliaceae). - Shrub. E. India. Root used for rheumatism, dysentery and as emetic; known as Country Ipecacuanha, Goa and Portuguese Ipecacuanha.

Nargusta Terminalia → Terminalia obovata Steud.

Narrowleaf Cattail → Typha angustifolia L.

Narrowleaf Globe Mallow → Sphaeralcea angustifolia (Cav.) G. Don.

Nasturtium → Tropaeolum majus L.

Nasturtium officinale R. Br. (syn. Roripa Nasturtium-aquaticum (L.) v. Hayek). Water Cress. (Cruciferaceae). - Perennial herb. Europe, temp. Asia, N. America. Cultivated. Eaten as a salad. Pungent flavor is caused by gluconasturtin. Seeds are occasionally used in mustard. Was formerly used medicinaly as antiscorbutic, febrifuge, antiscrophulasic. Employed in home remedies as „bloodcleanser" and for kidney troubles.

Nasturtium palustre D. C. (Cruciferaceae). - Perennial herb. S. Europe. Leaves are consumed in some parts of France like Water Cress.

Nasturtium, Tuber → Tropaeolum tuberosum Ruiz. and Pav.

Natal Aloë → Aloë Candelabrum Tod.

Natal Grass → Tricholaena rosea Nees.

Natal Mahogany → Kiggelaria africana L.

Natal Plum → Carissa grandiflora A. DC.

Native Currant → Leptomeria acida R. Br.

Native Millet → Panicum decompositum R. Br.

Nato → Mora megitosperma (Pitt.) Britt. and Rose.

Nato Tree, Java → Palaquium javanense Burck.

Neb-neb → Acacia arabica Willd.

Nectandra cinnamomoides Nees. (Lauraceae). - Tree. Equatorial Andes. Cultivated in Ecuador. Calyx of flowers and bark is used as spice in some parts of Ecuador.

Nectandra Pisi Miq., Black Cedar, Yellow Cirouaballi. (Lauraceae). - Tree. Trop. America, esp. Guiana. Wood. light to dark brown, fine-grained; used for ship building, harbor construction work.

Nectandra Puchury-minor Nees. and Mart. (syn. Ocotea Puchury-minor Mart.). (Lauraceae). - Tree. Brazil. Source of Pichurim or Puchury Beans; used in medicine.

Nectandra Rodioei (Schomb.) Hook. (Lauraceae). - Tree. Trop. America. Wood known as Greenheart, Bebeeru, Itauba Branca; being very heavy, very elastic, hard, tough, dense, durable, sinks in water, easy to split and to polish, resistant to termites and borers; used for ship buildings, docks, harbor works.

Nectarine → Amygdalus Persica L.

Neea parviflora Poepp. and Endl. (Nyctaginaceae). - Woody plant. Peru and adj. territory. Parts of the plant are chewed by the Indians of the Putumaye and Caqueta regions (Peru), as a dental preservative; it gives a black stain to the teeth.

Neea theifera Oest. (Nyctaginaceae). - Woody plant. Brazil. Leaves are used as a tea in some parts of Brazil.

Needle Bush → Hakea leucoptera R. Br.

Needle Oil, Silver Pine → Abies pectinata DC.

Negro Pepper → Xylopia aethiopica A. Rich.

Negro Yam → Dioscorea cayennensis Lam.

Negundo aceroides Moench. → Acer Negundo L.

Negundo fraxinifolium Nutt. → Acer Negundo L.

Nelumbium luteum (Willd.) Pers., Water Chinquapin, Water-Nut, Duck Acorn, Lotus. (Nymphaeaceae). - Perennial herbaceous waterplant. Eastern United States to Florida and Texas. Seeds were used as food by the Indians. Entire plant, especially the large, starchy rhizomes are edible.

Nelumbium speciosum Willd., Lotos. (Nymphaeaceae). - Perennial water plant. Trop. Asia. Cultivated. Rhizomes are consumed in China, Japan and other Asiatic countries when boiled, also preserved in sugar. Source of a starch, Lotus Meal. Boiled young leaves are eaten as vegetable. Kernels without bitter embryos, are dried; used in bakery, also made into starch. Seeds are also consumed when dry. In Indo China stamens are used for flavoring tea.

Nemalion lubricum Duby. (Helminthocladiaceae). - Red Alga. Pac. Ocean and adj. waters. Source of food in Japan. Also N. vermiculare Suring.

Neobaronia phyllanthoides Baker. (Leguminosaceae). - Tree. Madagascar. Wood very hard; used for handles of spades in Madagascar. Also N. xiphoclada Baker.

Neoglaziovia variegata (Arruda da Cam.) Mez. (syn. Dyckia Glaziovii Baker.). Caroa Verdadeira. (Bromeliaceae). - Perennial herb. Brazil, Argentina. Leaves are source of a fibre; used by natives for manuf. nets, also for packing tobacco. Has been recommended for manuf. paper and artificial silk.

Neosia Pine → Pinus Gerardiana Wall.

Neowashingtonia filifera Sudw. → Washingtonia filifera (Lindl.) Wendl.

Nepal Cardamon → Amomum aromaticum Roxb. and A. subulatum Roxb.

Nepenthaceae → Nepenthes.

Nepenthes distillatoria L., Pitcher Plant. (Nepenthaceae). - Woody climber. Ceylon. Stems used in Ceylon for small durable fancy baskets, tiffin baskets, teapot-holders.

Nepenthes Reinwardtiana Miq. (Nepenthaceae). - Woody vine. Mal. Archipelago. Stems are used for manuf. coarse ropes.

Nepeta Cataria L. (syn. Cataria vulgaris Moench.). Catnip, Catmint. (Labiaceae). - Perennial herb. Europe, introd. in N. America. Cultivated in Old and New World. Dried leaves and flowering tops are used medicinally, being stimulant, tonic, carminative, diaphoretic; for infantile colic. Contains a bitter principle. Cats are fond of this plant.

Nephelium chryseum Blume (Sapindaceae). - Small tree. Malaya. Fruit edible; consumed by the natives.

Nephelium lappaceum L. (syn. Euphoria Nephelium DC.) Rambutan. (Sapindaceae). - Small tree. Mal. Archipelago. Cultivated in trop. Asia. Excellent dessert fruit, also used as compote. Seeds occasionally eaten roasted. Numerous varieties, e. g. Rambutan Sinjonju, Rambutan Silengkeng, Rambutan Simatjan. Seeds are source of Rambutan Tallow. Sp. Gr. 0.86; Sap. Val. 193; Iod. No. 43.8; Melt. Pt. 40°—42°C.

Nephelium Litchi Camb. Litchi. (Sapindaceae). - Tree. S. China. Widely cultivated in warm countries of Old and New World. Favorite fruit in China; used since times immemoreal. Fruits are surrounded by a rough brittle shell, inside is a light gelatinous, agreeable, sweet, mildly acid pulp which is eaten fresh or dried. Dried fruits, Litchi Nuts, are much exported. Numerous varieties, among which Bedana, Chuimachi, Chumfung, Homshiuchi, Katjatkwo, Kwaluk, Sheungshuwal, Sweetcliff and Tongpok.

Nephelium Longana (Lam.) Cam. (syn. Euphoria Longana Lam.). Longan. (Sapindaceae). - Tree. Centr. and S. China. Source of a popular fruit in China, called Long-an fruits, Longyen or Lenkeng. Much esteemed in China. Fruits are eaten raw, preserved or dried. Cultivated.

Nephelium mutabile Blume., Pulasan. (Sapindaceae). - Medium sized tree. Malaya, probably from Java and Borneo. One of the best tropical fruits. Much cultivated in China, Malaysia, Indonesia etc. Fruits are eaten, fresh or as compote. Sometimes confused with the Rambutan. Some varieties are sweet, juicy or sub-acid, of excellent quality.

Nephroma, Chocolate Colored → Nephroma parilis Ach.

Nephroma parilis Ach., Chocolate Colored Nephroma, Powdered Swiss Lichen. (Peltigeraceae). - Lichen. Temp., subarctic zone. On stems, soil and rocks. Used in Scotland as source of a blue dye, used for coloring woolens.

Neptunia oleracea Lour. (syn. N. prostrata Baill.). (Leguminosaceae). - Waterplant. Tropics. Sprouts are eaten in some countries as a potherb.

Nereocystis Luetkeana (Mert.) Post. and Rupr. Bladder Kelp. (Lessoniaceae). - Brown Alga. Coast of Pacific North America. Succulent parts of the plant are used as food by Orientals in America. Stalks made into fishing lines by the Indians. Fronds when dried and powdered used in pharmacy in pills as supplement of mineral salts necessary for humans. „Seatron" was a food preparation, resembling candied citron.

Nerium Oleander L., Oleander. (Apocynaceae). Shrub or small tree. Medit. Region. Cultivated in Old and New World. Cardiac poison. Used to destroy rats. Used in criminal poisoning.

Nesodaphne Tawa Hook. f. → Beilschmiedia Tawa Hook f.

Nettle, Big String → Urtica dioica L.

Nettle, Dog → Urtica urens L.

Nettle, Giant → Laportea gigas Wedd.

Nettle Hemp → Galeopsis ochroleuca Lam.

Nettle, Horse → Solanum carolinense L.

Nettle, Indian → Acalypha indica L.

Nettle, Nilgiri → Girardinia palmata Gaud.

Nettle, Small → Urtica urens L.

Nettle, Stinging → Urtica dioica L.

Netleaf Oak → Quercus reticulata H. B. K.

Neurada procumbens L. (Rosaceae). - Shrub. E. India, Orient. Plant is used as a source of fodder for camels.

Nevada Jointfir → Ephedra nevadensis S. Wats.

New Caledonian Sandalwood → Santalum austro-caledonicum Vieil.

New Jersey Tea — Ceanothus americanus L.

New Zealand Flax → Phormium tenax Forst.

New Zealand Passionflower → Tetraspathaea australis Raoul.

New Zealand Spinach → Tetragonia expansa Murr.

Newbouldia laevis Seem. (Bignoniaceae). - Tree. Trop. Africa, esp. Sierra Leone to the Congo. Used along the Gold Coast for fever and dysentry. Decoction of leaves used for curing sore eyes in S. Nigeria. Bark used along the Gold Coast as a stomachic.

Ngai Camphor → Blumea balsamifera DC.

N'Ghat Oil → Plukenetia conophora Muell. Arg.

Nicaragua Chocolate Tree → Theobroma bicolor Humb. and Bonpl.

Nicaragua Rosewood → Dalbergia retusa Hemsl.

Nicaragua Wood → Haematoxylon Brasiletto Karst.

Niam Oil → Lophira alata Banks.

Niaouli Oil → Melaleuca viridiflora Brogn.

Nicotiana alata Link and Otto. (Solanaceae). - Herbaceous plant. S. America. Leaves are used for smoking and as a masticatory in some parts of S. America.

Nicotiana attenuata Torr. (Solanaceae). - Herbaceous plant. S. W. of United States and adj. Mexico. Dried leaves are smoked in pipes and in cigarettes of corn-husks by the Tewa Indians of New Mexiko.

Nicotiana glauca Graham., Tree Tobacco. (Solanaceae). - Shrub. S. America introd. in Mexico and S. W. of United States. Contains anabasine, an alkaloid, said to be more efficacious than nicotine for killing aphids.

Nicotiana quadrivalvis Pursh. (Solanaceae). - Annual herb. Western United States. Was formerly cultivated by the Indians and used for smoking, esp. when mixed with leaves of Arctostaphylos and Cornus stolonifera.

Nicotiana rustica L. Aztec Tobacco. (Solanaceae). - Annual herb. Probably from Mexico. Leaves were extensively used for smoking by the Indians of the Eastern United States. Grown on a small scale in Centr. Europe, Asia and East Indies. Used for smoking and manuf. of insecticides.

Nicotiana Tabacum L. Common Tobacco Plant. (Solanaceae). - Annual herb. Trop. America. Originally grown by the Indians. Cultivated in the Old and New World. Dried and cured leaves are used for manuf. cigars, as wrapper leaf, binder leaf, filler leaf, further for cigarettes, pipe tobacco, chewing tobacco, snuff; source of nicotine, nicotine sulfate, insecticides. Numerous varieties, suitable for certain particular areas and special purposes, e. g. I. Group. brasiliensis Comes. Broad-leaf tobacco. Cultivated commercially, var.: Marygold, Tennessee Red, Granville County Yellow, Ruffled Leaf. II. Group. havanensis Comes. Havanna Tobacco.

Native to Mexico. Cultivated, var.: Wilson's Hybrid, General Grand, Comstock, Little Dutch. III. Group. fruticosa Comes. Narrow-leaved or shrubby tobacco. Native to Mexico, Brazil. Cultivated. IV. Group. lancifolia. Comes. Large-leaved tobacco. Mexico. V. Group. virginica. Comes. Virginia Tobacco. Probably from Orinoco region. Commerc. grown, var.: Yellow Mammoth, Kentucky, Yellow, Golden Leaf, Golden Finder, One Sucker. Cavendish is composed of tobacco leaves that are moistened with syrup, being pressed into cakes, used for chewing.

Nicotiana trigonophylla Dunal. (syn. N. Palmeri Gray). (Solanaceae). Annual to perennial herb. Southwest of the United States and Mexico. Leaves are smoked by the Indians; esp. durin ceremonial occasions.

Nicou → Lonchocarpus Nicou DC.

Niebuhria pedunculosa Hochst. → Maerua pedunculosa Vahl.

Niepa Bark Tree → Samadera indica Gaertn.

Nigella sativa L., Black Cummin, Black Caraway. (Ranunculaceae). - Annual. Centr. and S. Europe, N. Africa, W. Asia. Cultivated. Pungent seeds are used for seasoning, also mixed with bread in different parts of Europe. In Greece the seeds are mixed with those of Sesam and spread over the bread. As a home-remedy they are used for intestinal worms and jaundice. Also N. damascena L.

Niger Copal → Daniella oblongo Oliv.

Niger Seed Oil → Guizotia abyssinica (L.) Cass.

Nigerian Ironweed → Vernonia nigritiana Oliv. and. Hiern.

Nightblooming Cereus → Selenicereus grandiflorus (L.) Britt. and Rose.

Nightshade, Black → Solanum nigrum L.

Nightshade, Silverleaf → Solanum elaeagnifolium Cav.

Nikau Palm → Rhopalostylis sapida Wendl. and Drude.

Nilgiri Nettle → Girardinia palmata Gaud.

Nima quassioides Buch-Ham. → Picrasma quassioides Benn.

Ninmeng → Citrus limonia Osbeck.

Nipa fruticans Thunb., Nipa Palm. (Palmaceae). - Palm. E. India to Australia. Leaves used for thatching, umbrellas, hats, baskets, mats, cigarette wrappers. Petioles are made into arrows. Inflorescense source of sugar (ancient industry), vinegar, alcohol. Seeds used by chinese for sweet-meat. Immature fruits are boiled with sugar and used as preserves.

Nipa Palm → Nipa fruticans Thunb.

Nistamal → Paullinia pinnata L.

Nitta Tree, Fernleaf → Parkia filicoidea Welw.

Nitraria Schoberi L. (Zygophyllaceae). - Shrub. Russia, N. Asia to Australia. Fruits edible, size of an olive, red, of a pleasant flavor, much esteemed by the natives of Australia.

Nitraria tridentata Desf. (Zygophyllaceae). - Shrub. N. Africa, Sahara region. Berries are consumed by several native N. Afr. tribes.

Nitrobacter Winogradskyi Buchanan. (Bacteriaceae). - Bacil. Useful soil microorganism being capable of oxydizing nitrites to nitrates, a process known as nitrification. It is unable to oxydize ammonia into nitrites. Also N. punctatum Sack, N. flavum Sack and N. opacum Sack.

Njatuo Fat → Palaquium oleiferum Blanco.

Noa Fibre → Agave Victoriae-Reginae Moore.

Noble Fir → Abies nobilis Lindl.

Nodding Alder → Alnus pendula Matsum.

Nodding Anemone → Anemone cernua Thunb.

Nodding Brome → Bromus anomalus Rupr.

Nodding Merawan → Hopea nutans Ridl.

Nogal → Juglans Honorei Dode.

Noir du Congo (rubber) → Clitandra orientalis Schum.

Noli Palm → Elaeis melanocarpa Gaertn.

Nolina longifolia (Schult.) Hemsl., Zacate. (Liliaceae). - Trunk to 3 meter high. Mexico. Leaves very tough, used for brooms, baskets, thatching, mats, coarse hats.

Nolina recurvata Hemsl. (syn. Beaucarnea recurvata Lem., Dasylirion recurvatum Macbride) (Liliaceae). - Trunk loosely branched. Mexico, esp. Veracruz. Leaves used for baskets, mats, etc. Also N. inermis (S. Wats) Rose, N. stricta Lem. and others.

Nonda Tree → Parinarium nonda F. v. Muell.

Nongo Gum → Albizzia Brownii Walp.

Nopal → Nopalea cochenillifera (L.) Salm-Dyck.

Nopal Chamacuero → Nopalea dejecta Sam-Dyck.

Nopal Cordón → Opuntia streptacantha Lem.

Nopal de Castilla → Opuntia megacantha Salm-Dyck.

Nopalea cochenillifera (L.) Salm-Dyck. (syn. Opuntia cochenillifera Mill.) Nopal. (Cactaceae). - Shrub-like xerophyte. Mexico. Cultivated. Used as food-plant for the cochineal insect, source of a red dye, cochineal. Fruit edible. Joints used for poultices.

Nopalea dejecta Salm-Dyck., Nopal chamacuero. (Cactaceae). - Shrub-like, xerophyte. Mexico. Cultivated. Fruits edible. Joints made into strips; eaten as a vegetable by Indians.

Nopalea Karwinskiana (Salm-Dyck) Schum., Nopalillo. (Cactaceae). - Shrub-like xerophyte. Mexico. Roots used for dysentery.

Nopalillo → Nopalea Karwinskiana (Salm-Dyck) Schum.

Norino-tsukudani → Porphyra tenera Kjellm.

North African Ammoniac → Ferula tingitana L.

Northern Dewberry → Rubus flagellaris Ait.

Northern Prickly Ash → Zanthoxylum americanum Mill.

Northern Rata → Metrosideros robusta Cunningh.

Norway Maple → Acer platanoides L.

Norway Spruce → Picea excelsa Link.

Nostoc commune Vauch. var. flagelliforme (Berk. and Cart.) Born and Flah. (Nostocaceae). - Blue-green Alga. Used as food in China, under the name of fatchoy and fahtsai.

Nostocaceae (Blue-green Algae) → Nostoc.

Nothofagus betuloides Blume. (syn. Fagus betuloides Mirb.) (Fagaceae). - Tree. S. Chile. Wood much used locally for the same purpose as the Northern Beech.

Nothofagus cliffortoides Oerst. (syn. Fagus cliffortoides Hook f.) Mountain Beech. (Fagaceae). - Small tree. New Zealand. Wood easily worked. Used for telephone poles, wharf-piles, general carpentry work.

Nothofagus fusca Oerst. (syn. Fagus fusca Hook. f.). Red Beech. (Fagaceae). - Tree. New Zealand. Wood of fine quality. Used for indoor work, rail-road ties, piles for wharves and bridges, mine props, framing, flooring-joists and joiners' work.

Nothofagus Menziesii Oerst. (syn. Fagus Menziesii Hook. f.). Silver Beech. (Fagaceae). - Tree. Australia, New Zealand. Wood very tough, strong, elastic, deep red, very straight, not durable when exposed to weather. Used for house-blocks, railroad ties, weather boarding, tubs, buckets, cooper's wares, wine casks.

Nothofagus obliqua Blume. (syn. Fagus obliqua Mirb.) Roble Pellin. (Fagaceae). - Tree. Chile. Wood darkred, heavy, very hard, resistant to decay; used for rail-road ties and construction-work.

Nothofagus procera Oerst. (syn. Fagus procera Poepp. and Endl.). Rauli. (Fagaceae). - Tree. Chile. Important hardwood tree in Chile. Wood red to cherry color, easy to work; used for cabinet work, furniture, cooperage, mouldings.

Notnofagus Solandri Oerst. (syn. Fagus Solandri Hook. f.) (Fagaceae). - Tree. New Zealand. Wood pale red, heartwood black, heavy, tough, strong, durable; used for bridges, construction work, rail-road ties, gate posts.

Nothopanax Edgerleyi Hook. f. → Panax Edgerleyi Hook. f.

Nothopanax fruticosum Miq. (syn. Panax fruticosum L., Polyscias fruticosa Harms.) (Araliaceae). - Shrub. Ternate. Cultivated in Java. Roots and leaves have scent of parsley, diuretic. In Ternate leaves are eaten by natives with fish, meat or in soup.

Nothopanax pinnatum Miq. (syn. Polyscias Rumphiana Harms.) (Araliaceae). - Shrub. Moluccas. Cultivated. Young leaves are much used by the natives as vegetable, cooked with coconutmilk or with fish.

Nsense → Clitocybe nebularis (Batch) Quel.

Nuphar advena Ait. f. (syn. Nymphaea advena Ait.). Yellow Pondlily, Cow Lily. (Nymphaeaceae). - Perennial water plant. Eastern N. America. Thick, fleshy rootstocks are eaten raw, roasted, boiled with meat. Seeds were ground to thicken soup. Used by many Indian tribes. Recommended as an emergency food plant.

Nuphar luteum Sm. (Nymphaeaceae). - Perennial waterplant. Europe, temp. Asia. Rhizome is source of starch, recommended for times of food emergency.

Nuphar polysepala Engelm. (Nymphaeaceae). - Perennial waterplant. Western N. America. Roasted seeds are much esteemed as food by the Klamath and other Indians.

Nut, Barcelona → Corylus Avellana L.

Nut, Bengor → Caesalpinia Bonducella Flem.

Nut, Bonduc → Caesalpinia Bonducella Flem.

Nut, Brazil → Bertholletia excelsa H. B. K.

Nut, Burrawang → Macrozamia spiralis Miq.

Nut, Cashew → Anacardium occidentale L.

Nut, Cob → Corylus Avellana L. and Omphalia triandra L.

Nut, Dika → Irvingia Barteri Hook. f.

Nut, Earth → Carum Bulbocastanum Koch.

Nut Grass → Cyperus rotundus L.

Nut Grass, Yellow → Cyperus esculentus L.

Nut, Kaffir Marvola → Sclerocarya caffra Sond.

Nut, Mahagony → Afrolicania elaeosperma Mildbr.

Nut, Maranhão → Sterculia Chicha St. Hil.

Nut, Pará → Bertholletia excelsa H. B. K.

Nut, Physic → Caesalpinia Bonducella Flem. and Jatropha Curcas L.

Nut, Pig → Omphalea triandra L.

Nut, Pine → Pinus cembroides Zucc. and P. edulis Engelm.

Nut, Pistachio → Pinstacia vera L.

Nut, Purging → Jatropha Curcas L.

Nut, Queensland → Macadamia ternifolia F. v. Muell.

Nut, Singhara → Trapa bispinosa Roxb.

Nut, Suari → Caryocar nuciferum L.

Nut Tree, Malabar → Adhatoda Vasica Nees.

Nutmeg → Myristica fragrans Houttuijn.

Nutmeg, Ackawai → Acrodiclidium Camara Schomb.

Nutmeg, Calabash → Monodora Myristica (Gaertn.) Dunal.

Nutmeg, Madagascar → Ravensara aromatica Gmel.

Nuttall Alkali Grass → Pucciniella Nuttaliana (Schult.) Hitchc.

Nuttall Goldenweed → Aplopappus Nuttallii Torr. and Gray.

Nux Vomica → Strychnos nux-vomica L.

Nuxia sphaerocarpa Baker. (Loganiaceae). - Tree. Madagascar. Wood. Used in Madagascar for building houses. Also N. terminaloides Baker and N. carpitata Baker.

Nyassa Rubber → Landolphia Kirkii Dyer.

Nyctaginaceae → Abronia, Boerhaavia, Mirabilis, Neea, Pisonia, Vieillardia.

Nyctanthes Arbor-tristis L. (Oleaceae). - Large shrub. India. Cultivated near Temples. Source of a saffron-yellow dye.

Nymphaea alba L., White Waterlily. (Nymphaeaceae). - Perennial water plant. Europe, temp. Asia. Rootstocks are source of a starch; recommended for times of food emergency.

Nymphaea calliantha Con. (Nymphaeaceae). - Perennial waterplant. S. Africa. Seeds are consumed by the natives of Transvaal.

Nymphaea gigantea Hook. (Nymphaeaceae). - Perennial waterplant. Australia. Porous seedstalk and fruit is peeled and eaten raw or roasted by the aborigenes of Australia. Also tubers are consumed.

Nymphaea Lotus L. (Nymphaeaceae). - Perennial waterplant. Trop. Africa, Asia. Powdered roots are used for dyspepsia, dysentery and piles. Powdered seeds are employed for skin diseases in Sudan. Seeds are consumed as famine food in India.

Nymphaea stellata Willd. (Nymphaeaceae). - Perennial waterplant. Trop. Africa, Asia. Rootstocks and seeds are eaten as food during times of famine in different parts of Asia and Africa.

Nymphaeaceae → Brasenia, Euryale, Nelumbium, Nuphar, Nymphaea.

Nyssa aquatica L. Water Tupelo. Tupelo-Gum, Cotton-Gum. (Nyssaceae). - Tree. Eastern United States to Florida and Texas. Wood not strong, light, soft, tough, difficult to split, close-grained, light brown to almost white; used for fruit and vegetable boxes, wooden ware, veneers, pulp, wooden shoes, broom handles. Wood of roots is sometimes used instead of cork for floats of nets. Flowers are source of honey.

Nyssa biflora Walt., Water Gum, Water Tupelo. (Nyssaceae). - Tree. New Jersey to Florida and Texas. Wood is used for the same purposes as N. multiflora Wang.

Nyssa multiflora Wang. (syn. sylvatica Marsh.) Black Gum, Pepperidge, Sour Gum. (Nyssaceae). - Tree. Eastern N. America. Wood strong, very tough, soft, difficult to split, heavy, not durable, light yellow. Used for rollers in glass factories, wharf-piles, ox-yokes, veneer, wooden pipes, hubs of wheels, soles of shoes, fruit boxes, crates, woodpulp.

Nyssa Ogeche Marsh., Ogeeche Lime, Sour Tupelo, Tupelo Gum. (Nyssaceae). - Tree. S. Carolina to Florida. Fruits are consumed preserved, having the name of Ogeche Limes. Flowers source of commercial honey.

Nyssaceae → Nyssa.

O

Oak, African → Oldfieldia africana Benth. and Hook.

Oak, Basket → Quercus Prinus L.

Oak, Blue Japanese → Quercus glauca Thunb.

Oak, Bull → Casuarina equisetifolia L.

Oak, Bur → Quercus macrocarpa Michx.

Oak, California White → Quercus lobata Née.

Oak, Canyon Life → Quercus chrysolepis Liebm.

Oak, Chestnut → Quercus Prinus L.

Oak, Cork → Quercus Suber L.

Oak, Daimyo → Quercus dentata Thunb.

Oak, Emory → Quercus Emoryi Torr.

Oak, European Turkey → Quercus Cerris L.

Oak, Forest → Casuarina torulosa Ait.

Oak, Glandbearing → Quercus glandulifera Blume.

Oak, Holly → Quercus Ilex L.

Oak, Iron → Quercus obtusiloba Michx.

Oak, Jack → Quercus marylandica Du Roi.

Oak, Laurel → Quercus imbricaria Michx.

Oak, Live → Quercus virginiana Mill.

Oak Lung → Sticta pulmonacea Ach.

Oak, Lusitanian → Quercus lusitanica Lam.

Oak Manna → Quercus Cerris L. and Q. persica Jaub. and Spach.

Oak, Maul → Quercus chrysolepis Liebm.

Oak, Mongolia → Quercus mongolica Fischer.

Oak Moss → Evernia prunastri (L.) Ach.

Oak, Mossy Cup → Quercus macrocarpa Michx.

Oak, Netleaf → Quercus reticulata H. B. K.

Oak, Oregon White → Quercus Garryana Hook.

Oak, Post → Quercus obtusiloba Michx.

Oak, Oriental → Quercus variabilis Blume.

Oak, Overcup → Quercus lyrata Walt.

Oak, Portuguese → Quercus lusitanica Lam.

Oak Rage → Lobaria scrobiculata (Scop.) DC.

Oak, Red → Quercus rubra L.

Oak, Red Southern → Quercus texana Buckl.

Oak, River → Casuarina stricta Ait.

Oak, River Black → Casuarina suberosa Otto and Dietr.

Oak, Scarlet → Quercus coccinea Wangenb.

Oak, Shin → Quercus undulata Torr.

Oak, Shingle → Casuarina stricta Ait. and Q. imbricaria Michx.

Oak, Silk → Grevillea robusta Cunningh.

Oak, Silky → Stenocarpus salignus R. Br.

Oak, Spanish → Quercus digitata (Marsh) Sudw.

Oak, Swamp → Casuarina equisetifolia L. and C. suberosa Otto and Dietr.

Oak, Swamp Spanish → Quercus Pagoda Raf. and Q. palustris Du Roi.

Oak, Swamp, White → Quercus bicolor Willd.

Oak, Texas → Quercus texana Buckl.

Oak, Valley → Quercus lobata Née.

Oak, Wavyleaf → Quercus undulata Torr.

Oak, Western Black → Quercus Emoryi Torr.

Oak, White → Quercus alba L.

Oak, Willow → Quercus Phellos L.

Oak, Yellow Bark → Quercus velutina Lam.

Oakesia sessiliflora (L.) Wats → Uvularia sessiliflora L.

Oat, Common → Avena sativa L.

Oat, Hungarian → Avena orientalis Schreb.

Oat, Naked → Avena nuda L.

Oat Oil → Avena sativa L.

Oat, Wild → Avena fatúa L.

Oca → Oxalis crenata Jacq.

Ochanostachys amentacea Masters. (Olacaceae). - Tree. Trop. Asia, Malay Penin., Borneo. Wood hard, heavy, close-grained, durable, strong, resistant to white ants; used for furniture, house posts, indoor work, floor beams, telephone poles.

Ochlandra travancorica Benth. (Graminaceae). - Shrubby, woody bamboo. S. India. Used in Travancore for manuf. paper.

Ochna alboserrata Engl. → Brackenridgea zanguebarica Oliv.

Ochnaceae → Brackenridgea, Gomphia, Lophira.

Ochocoa gaboni Pierre. (Myristicaceae). - Tree. Gabon. Oily seeds are used as a condiment by the natives of Gabon.

Ochrocarpus africana (Don.) Oliv., African Mammee Apple. (Guttiferaceae). - Tree. Trop. Africa. Seeds are consumed as food among the natives in some parts of Africa.

Ochrolechia tartarea (L.) Mass., Crotal, Crottle, Cockur. (Lecanoraceae). - Lichen. Temp. zone On soils and rocks. Source of a purple-crimson and blue dye. Cudbaer, Tincture of Cudbear. Produces a litmus, Erdorseile.

Ochroma lagopus Swartz. Balsa. (Bombaceae). - Tree. Trop. America. Source of Balsa Wood of commerce. One of the lightest woods. Used for manuf. of lifepreservers, insulating refrigerators. Decays quickly, which is prevented by impregnation with hot parafin.

Ochroma limonensis Rowlee., Cotton Tree. (Bombaceae). - Tree. Centr. America. Wood pale reddish to white, spongy, very soft, lighter than cork; used by Indians for rafts and balsas.

Ochrosia mariannensis A. DC. Marianne Yellow Wood. (Apocynaceae). - Tree. Guam, Polynesia, Marian. Islands. Wood fine grained, takes a fine polish, yellow; used in Guam for furniture.

Ochrosia sandwicensis A. DC. (Apocynaceae). - Small tree. Hawaii Islands. Bark and roots are source of a yellow dye; used for staining the tapa or paper clothing of the Hawaiians.

Ocimum Basilicum L., Sweet Basil. (Labiaceae). - Annual herb. Probably from India, S. E. Asia, N. E. Africa. Cultivated. Herb used as condiment. Source of Oil of Sweet Basil, an ess. oil; used in condiment mixtures, catsups, mustards, vinegars. Also in cosmetics and perfumes; blends well with jasmine types. Contains methylchavicol, eucalyptol, estragol, linalol. There are many varieties, some having different characteristics as to flavoring and composition among which var. anisatum Hort., var. thyrsiflora Wight, var. comosum Hort. Seeds are source of a beverage, Cherbet Tokhum, used in Mediterranean countries.

Ocimum canum Sims., Hoary Basil. (Labiaceae). - Perennial herb. Tropics of Old World. Herb is used as pot-herb; sold in markets of India. Leaves are made into a paste which is used in Sudan for skin diseases.

Ocimum grandiflorum Blume → Orthosiphon stamineum Benth.

Ocimum graveolens A. Br. (Labiaceae). - Perennial herb. Abyssinia. Has been recommended as a condiment for flavoring foods.

Ocimum sanctum L. Holy Basil. (Labiaceae). - Half shrub. Tropics of Old World. Cultivated. Sacred plant among the Hindus. Leaves used as condiment, in salads and other foods.

Ocote → Pinus Teocote Cham. and Schlecht.

Ocotea bullata E. Mey., Black Stinkwood. (Lauraceae). - Tree. S. Africa. Wood hard, dark brown, takes polish; used for building purposes, cabinet and carpenters work, gunstocks, wagons, planks, beams, doors, furniture, window frames. Bark is an astringent, used for tanning.

Ocotea caudata Mez. (Lauraceae). - Tree. Trop. S. America, esp. Guiana. Source of a highly odorous ess. oil, contains myristic aldehyde, cineal, terpinol and a sesquiterpene. Also O. usambarensis Engl.

Ocotea Cujumary Mart. (syn. Aydendron Cujumary Mart., Oreodaphne floribunda Benth.). (Lauraceae). - Tree. Brazil. Wood is used for construction of boats and for general carpentry.

Ocotea indecora Schott. (syn. Mespilodaphne indecora Meissn.). (Lauraceae). - Tree. Guiana to Brazil. Bark of the root is considered in Brazil antirheumatic.

Ocotea Rodiaei (Schomb.) Mex., Demarara Greenheart. (Lauraceae). - Large tree. Guiana. Wood strong, hard, durable, lustrous light to dark olive or blackish. Used for marine construction, piers, piling, lock and sluice gates, ship-building, board-walks, paving blocks. Sometimes confused with Surinam Groenhart, Tabebuia spec.

Ocotea rubra Mez., Determa, Wane. (Lauraceae). - Tree. Guianas; Lower Amazon region. (Braz.). Wood light reddish-brown, coarse-graine, strong, hard, excellent working properties, resistant to insects. Used for interior construction, boat planking, sugarboxes, furniture and canoes.

Ocotea Sassafras Mez. → Mespilodaphne Sassafras Meissn.

Ocotea squarrosa Mart. (syn. Laurus pyrifolia Willd.). (Lauraceae). - Shrub. Brazil. Bark is used as a tonic. Leaves are employed in domestic medicine in different parts of Brazil.

Ocotillo → Fouquieria splendens Engelm.

Odara Pear → Chrysophyllum africanum A. DC.

Odina acida Walp. (syn. Lannea acida A. Rich.) (Anacardiaceae). - Shrub or small tree. Trop. Africa. Source of an edible gum. Fruits are edible, sub-acid. Powdered bark mixed with other material is used by the natives as a paint for the face.

Odina antiscorbutica Rich. (Anacardiaceae). - Tree. Trop. Africa. Decoction of bark is used by natives of Angola for scorbutic ulcers of mouth, also for scurvy. Cultivated near villages.

Odina caffra Sond., Kafir Plum. (Anacardiaceae). - Tree. S. Africa. Wood heavy, hard, strong, elastic; used for furniture, planking, beams, rafters etc.

Odina fruticosa Hochst. (Anacardiaceae). - Tree. Abyssinia. Source of a gum; used for adulterating Gum Arabic.

Odina gummifera Blume (Anacardiaceae). - Tree. Java. Source of a gum, has been recommended for technical purposes.

Odina Wodier Roxb. (Anacardiaceae). - Tree. E. India, Burma. Stem is source of a soluble resin, called Jingan Gum; used in India for calico printing, size in white washing, for protecting fishing nets. It is composed of arabic galactan.

Odostemon Aquifolium Rydb. → Mahonia repens (Lindl.) G. Don.

Odum → Chlorophora excelsa (Welw.) Benth. and Hook. f.

Odyendyea gabonensis Pierre. (Simarubaceae). - Trop. Africa, esp. Gabon. Fruits called Nzeng; used as condiment by the natives.

Oenanthe crocata L. (Umbelliferaceae). - Perennial herb. S. Europe. Crushed roots are used in Sardinia for stupefying fish.

Oenanthe sarmentosa Presl. Water Parsley. (Umbelliferaceae). - Perennial herb. Pacific N. America. Black tubers of a sweet, cream-like taste were boiled and consumed by Indians of N. W. of United States.

Oenanthe stolonifera Wall., Water Dropwort. (Umbelliferaceae). - Perennial herb. Indo-China, China, Japan. Cultivated. Young shoots and leaves are used for flavoring fowl and fish-soup among the Chinese.

Oenocarpus Bacaba Mart. (Palmaceae). - Palm. Guinea to Brazil. Fruits are source of an oil; used similarly as Olive Oil.

Oenocarpus Bataua Mart., Tooroo. (Palmaceae). - Tall palm. Brazil, Guiana. Fleshy outside part of the fruit is used for making a beverage.

Oenocarpus distichus Mart. Bacaba. (Palmaceae). - Palm. S. America. Fruits are source of a beverage, called Bacaba Branca, is a white much esteemed drink; Bacaba Vernelha is a yellow-red beverage which is used pure, with sugar or with Manioc. Fruits are also source of a greenish oil; used in cooking.

Oenothera biennis L. (syn. Onagra biennis Scop.) Common Evening Primrose. (Onagraceae). - Bienniel herb. Europe and N. America. Roots when boiled are consumed as a vegetable, occasionally cultivated as a food plant. Young shoots are eaten in salads.

Ogeeche Lime → Nyssa Ogeche Marsh.

Ohio Buckeye → Aesculus arguta Buckl.

Oil of Abrasin → Aleurites montana Willd.

Oil of Ajowan Seed. → Carum copticum (L.) Benth. and Hook.

Oil of Allspice → Pimenta officinalis Lindl.

Oil of Ambrette → Hibiscus Abelmoschus L.

Oil of American Wormseed → Chenopodium ambrosioides L. var. anthelminthicum (L.) Gray.

Oil of Anda-assy → Joannesia princeps Vell.

Oil of Angelica Root → Angelica Archangelica L.

Oil of Anise → Pimpinella Anisum L.

Oil of Apricot Kernel → Armeniaca vulgaris Lam.

Oil of Arangan → Ganophyllum falcatum Blume.

Oil of Araucaria → Callitropsis araucarioides Compton.

Oil of Arbor Vitae → Thuya occidentalis L.

Oil of Asae foetida → Ferula Asa-foetida L.

Oil of Avocado → Persea americana Mill.

Oil of Babassu → Orbignya speciosa Bark.

Oil of Bacury Kernel → Platonia insignis Mart.

Oil of Bagilumbang → Aleurites triloba Forst.

Oil of Balm → Melissa officinalis L.

Oil of Banana → Musa sapientum L.

Oil of Baobab → Adansonia digitata L.

Oil of Bati → Gomphia parviflora DC.

Oil of Batiputa → Gomphia parviflora DC.

Oil of Bay → Pimenta acris (Swartz) Kostel.

Oil of Beech Nut Seed → Fagus sylvatica L.

Oil of Ben → Moringa oleifera Juss.

Oil Ben Seed → Moringa oleifera Juss.

Oil of Bene → Sesamum indicum L.

Oil of Bergamot → Citrus bergamia Risso and Poit.

Oil of Black Mustard Seed → Brassica nigra (L.) Koch.

Oil of Bitter Almond → Amygdalus communis L.

Oil of Bitter Orange → Citrus Aurantium L.

Oil of Boronia → Boronia megastigma Nees.

Oil of Brazil Nut → Bartholletia excelsa H. B. K.

Oil of Brazilian Sassafras → Mespilodaphne Sassafras Meissn.

Oil of Cade → Juniperus Oxycedrus L.

Oil of Cajaput → Melaleuca Leucadendron L.

Oil of Camphor → Cinnamomum Camphora

Oil of Canadian Golden Rod → Solidago canadensis L.

Oil of Candlenut → Aleurites triloba Forst.

Oil of Cardamon → Elettaria Cardamomum (L.) Maton.

Oil of Caroway → Carum Carvi L.

Oil of Carpotroche → Carpotroche brasiliensis Endl.

Oil of Carrot Seeds → Daucus Carota L.

Oil of Cashew Nuts → Anacardium occidentale L.

Oil of Cassia → Cinnamomum Cassia Blume.

Oil of Castanha → Telfeirea pedata Hook.

Oil of Cato Seed → Chisocheton Cumingianus (C. DC.) Harms.

Oil of Castor Seed → Ricinus communis L.

Oil of Cayenne Linaloe → Amyris balsamifera L.

Oil Cedar Wood → Juniperus virginiana L.

Oil of Cedro → Citrus Limon Burman.

Oil of Celery → Apium graveolens L.

Oil of Chamomille → Matricaria Chamomilla L.

Oil of Champaca Wood → Bulnesia Sarmienti Lorentz.

Oil of Chaulmoogra → Hydnocarpus anthelmintica Pierre, and Taraktogenos Kurzii (King) Pou.

Oil of Chêmpaka → Michelia Champaca L.

Oil of Chenopodium → Chenopodium ambrosioides L. var. anthelminthicum (L.) Gray.

Oil of Cherry Kernel → Prunus Cerasus L.

Oil of Chinese Colza → Brassica campestris L. var. chinoleifera.

Oil of Chinese Wood → Aleurites Fordii Hemsl.

Oil of Chuwah → Aquilaria Agallocha Roxb.

Oil of Cinnamon → Cinnamomum Cassia Blume, C. zeylanica Nees.

Oil of Citronella → Cymbopogon Nardus (L.) Rendle.

Oil of Clary → Salvia Sclarea L.

Oil of Clove → Eugenia Caryophyllata Thunb.

Oil of Coconut → Cocos nucifera L.

Oil of Cohune → Attalea Cohune Mart.

Oil of Colza → Brassica Napus L.

Oil of Copaiba → Copaiba spp.

Oil of Coriander → Coriandrum sativum L.

Oil of Coriander Seed → Coriandrum sativum L.

Oil of Corn → Zea Mays L.

Oil of Costus Root → Saussurea Lappa C. B. Clarke.

Oil of Cotton Seed → Gossypium spp.

Oil of Croton → Croton Tiglium L.

Oil of Cubeb → Piper Cubeba L.

Oil of Cumin → Cuminum Cyminum L.

Oil of Curcas → Jatropha Curcas L.

Oil of Cypress → Cupressus sempervirens L.

Oil of Dill → Anethum graveolens L.

Oil of Dilo → Calophyllum inophyllum L.

Oil of Domba → Calophyllum inophyllum L.

Oil of Dwarf Pine Needle → Pinus pumilio Haenke.

Oil of East African Sandalwood → Osyris tenuifolia Engl.

Oil of Efwatakala Grass → Melinis minutiflora Beauv.

Oil of Egyptian Lettuce Seed → Lactuca Scariola L.

Oil of Elecampane → Inula Helenium L.

Oil of Elemi → Canarium luzonicum Miq.

Oil of Eucalyptus → Eucalyptus globulus Lab.

Oil of Fennel → Foeniculum vulgare Mill.

Oil of Flax Seed → Linum usitatissimum L.

Oil of Fony → Adansonia digitata L.

Oil of Gardenia Flower → Gardenia florida L.

Oil of Garlic → Allium sativum L.

Oil of Gaultheria → Gaultheria procumbens L.

Oil of Geranium → Pelargonium Radula L'Hér.

Oil of German Spearmint → Mentha spicata L.

Oil of Gingli → Sesamim indicum L.

Oil of Gonyo → Antrocaryum Nannai de Wild.

Oil of Grapefruit Seed → Citrus paradisi Macf.

Oil of Gudiac Wood → Bulnesia Sarmienti Lorentz.

Oil of Hazel Nut → Coryllus Avellana L.

Oil of Hemlock → Tsuga canadensis (L.) Carr.

Oil of Hemp Seed → Cannabis sativa L.

Oil of Homoranthus → Homoranthus virgatus Cunningh.

Oil of Honesty → Hesperis matronalis L.

Oil of Hongay → Pongamia pinnata (L.) Merr.

Oil of Hops → Humulus Lupulus L.

Oil of Huon Pine Wood → Dacrydium Franklinii Hook.

Oil of Hyacinth Flower → Hyacinthus orientalis L.

Oil of Inchi → Cymbopogon flexuosa Stapf. and C. caesius Stapf.

Oil of Ivory Wood Seed → Agonandra brasiliensis Benth. and Hook.

Oil of Jamba → Eruca sativa L.

Oil of Japanese Mint. → Mentha arvensis L.

Oil of Japanese Tung → Aleurites cordata Steud.

Oil of Jatamansi → Nardostachys Jatamansi DC.

Oil of Jojoba → Simmondsia californica Nutt.

Oil of Jonquil → Narcissus Jonquilla L.

Oil of Juniper Berries → Juniperus communis L.

Oil of Juniper Wood → Juniperus communis L.

Oil of Kachi → Cymbopogon caesius Stapf.

Oil of Kalingag → Cinnamomum Mercadoi Vid.

Oil of Kaloempang → Sterculia foetida L.

Oil of Katio → Bassia Mottelyana Clarke.

Oil of Kayu-galu → Sindoria inermis Merr.

Oil of Kelon-ka-tel. → Cedrus Libani Barrel.

Oil of Kussum → Schleichera trijuga Willd.

Oil of Lallemantia → Lallemantia ibirica Fisch and Mey.

Oil of Lavender → Lavandula officinalis Chaix.

Oil of Lemon → Citrus Limon Burman.

Oil of Lemon Pip → Citrus Limon Burman.

Oil of Lemon Seed → Citrus Limon Burman.

Oil of Lemongrass → Cymbopogon citratus (DC.) Stapf.

Oil of Lime Seed → Citrus aurantifolia Swingle.

Oil of Linseed → Linum usitatissimum L.

Oil of Linoloé → Bursera Aloëxylon Engl.

Oil of Linaloé Seed → Bursera Aloëxylon Engl.

Oil of Lumbang → Aleurites triloba Forst.

Oil of Macassar → Schleichera trijuga Willd.

Oil of Mace → Myristica fragrans Houtt.

Oil of Mafura → Trichilia emetica Vahl.

Oil of Mahareb Grass → Cymbopogon senaarensis Chiov.

Oil of Manketti Nut → Ricinodendron Rautanenii Schinz.

Oil of Marjoram → Marjoram hortensis Moench. and Thymus Mastichina L.

Oil of Mandarine → Citrus reticulata Blanco.

Oil of Margosa → Melia Azadirachta L.

Oil of Meadowsweet → Filipendula Ulmaria (L.) Maxim.

Oil of Mehndi → Lawsonia alba Lam.

Oil of Meni → Lophira alata Banks.

Oil of Mignonette → Reseda odorata L.

Oil of Mu → Aleurites montana Willd.

Oil of Muhumgu → Brachylaena Hutchinsii Hutch.

Oil of Myrcia → Pimenta acris (Swartz) Kostel.

Oil of Myristica → Myristica fragrans Houtt.

Oil of Myrrh → Commiphora abyssinica (Berg.) Engl.

Oil of Myrtle → Myrtus communis L.

Oil of Narcissus → Narcissus poeticus L.

Oil of Neroli → Citrus Aurantium L.

Oil of N'Ghat → Plukenetia conophora Muell. Arg.

Oil of Niam → Lophira alata Banks.

Oil of Niaouli → Melaleuca viridiflora Brongn.

Oil of Niger Seed → Guizotia abyssinica (L.) Cass.

Oil of Nutmeg → Myristica fragrans Houtt.

Oil of Olive → Olea europaea L.

Oil of Oiticica → Licania rigida Benth.

Oil of Orange → Citrus sinensis Osbeck.

Oil of Orange Flower → Citrus Aurantium L.

Oil of Ouricury Palm Kernel → Cocos coronata Mart.

Oil of Owala → Pentaclethra macrophylla Benth.

Oil of Palm Kernel → Elaeis guineensis Jacq.

Oil of Palmarosa → Cymbopogon Martini Stapf.

Oil of Palosápis → Anisoptera thurifera Blanco.

Oil of Pão Manfin → Agonandra brasiliensis Benth. and Hook.

Oil of Para Rubber Seed → Hevea brasiliensis Muell. Arg.

Oil of Parsley → Petroselinum hortense Hoffm.

Oil of Patchouly → Pogostemon Cablin Benth.

Oil of Pea Nut → Arachis hypogaea L.

Oil of Peach Kernel → Amygdalus Persica L.

Oil of Pennyroyal → Mentha Pulegium L.

Oil of Peppermint → Mentha piperita L.

Oil of Perilla → Perilla frutescens (L.) Britt.

Oil of Petitgrain → Citrus Aurantium L.

Oil of Pili Nut → Canarium ovatum Engl.

Oil of Pimenta → Pimenta officinalis Lindl.

Oil of Pine Needle → Pinus sylvestris L.

Oil of Pistachio Nut → Pistacia vera L.

Oil of Po-Yak → Afrolicania elaeosperma Mildbr.

Oil of Pongam → Pongamia pinnata (L.) Merr.

Oil of Poppy Seed → Papaver somniferum L.

Oil of Prickly Poppy Seed → Argemone mexicana L.

Oil of Pumpkin Seed → Cucurbita Pepo L.

Oil of Ramtilla → Guizotia abyssinica (L. f.) Cass.

Oil of Rape → Brassica Napus L.

Oil of Ravinson → Brassica campestris L.

Oil of Reniala → Adansonia digitata L.

Oil of Reseda → Reseda odorata L.

Oil of Rhodium → Convolvulus floridus L. f.

Oil of Rice → Oryza sativa L.

Oil of Ricinus → Ricinus communis L.

Oil of Rohituka → Amoora Rohituka Wight and Arn.

Oil of Roman Chamomille → Anthemis nobilis L.

Oil of Rosemary → Rosmarinus officinalis L.

Oil of Roses → Rosa damascena Mill.

Oil of Rosin → Pinus palustris Mill.

Oil of Rue → Ruta graveolens L.

Oil of Sage → Salvia officinalis L.

Oil of Santal → Santalum album L.

Oil of Sapucaja → Lecythis Ollara L.

Oil of Sassafras → Sassafras albidum (Nutt.) Nees.

Oil of Savin → Juniperus Sabina L.

Oil of Sesame → Sesamum indicum L.

Oil of Simson → Trianthema salsoloides Fenzl.

Oil of Spanish Origanum → Thymus capitatus Hoffm. and Link.

Oil of Spearmint → Mentha spicata L.

Oil of Spike → Lavandula Spica Cav.

Oil of Star Anise → Illicium verum Hook f.

Oil Stoechas → Lavandula stoechas L.

Oil of Sunflower Seed → Helianthus annuus L.

Oil of Supa → Sindora Supa Merr.

Oil of South Australian Sandalwood → Fusanus acuminatus R. Br.

Oil of Sweet Almona → Amygdalus communis L.

Oil of Sweet Basil → Ocimum Basilicum L.

Oil of Sweet Birch → Betula lenta L.

Oil of Sweet Orange → Citrus sinensis Osbeck.

Oil of Sweet Orange Pip → Citrus sinensis Osbeck.

Oil of Tancy → Tanacetum vulgare L.

Oil of Tangerine → Citrus reticulata Blanco.

Oil of Tea Seed → Camellia Sasanagua Thunb.

Oil of Teel → Sesamim indicum L.

Oil of Theobroma → Theobroma cacao L.

Oil of Thyme → Thymus vulgaris L.

Oil of Tigli → Croton Tiglium L.

Oil of Touloucouna → Carapa procera DC.

Oil of Tsauri → Cymbopogon giganteus Ciov.

Oil of Tsubaki → Camellia japonica L.

Oil of Tuberose Flower → Polianthes tuberosa L.

Oil of Tung → Aleurites Fordii Hemsl.

Oil of Turkey Red → Agonandra brasiliensis Benth. and Hook. and Ricinus communis L.

Oil of Turpentine → Pinus palustris Mill.

Oil of Turpentine, French → Pinus Pinaster Soland.

Oil of Warinnum → Guizotia abyssinica (L. f.) Cass.

Oil of West Australian Sandalwood → Eucalyptus spicata Sprague and Sum.

Oil of White Cedar → Thuya occidentalis L.

Oil of White Mustard Seed. → Sinapis alba L.

Oil of Wild Marjoram → Thymus Mastichina L.

Oil of Wild Olive Seed → Ximenia americana L.

Oil of Winter Green → Gaultheria procumbens L.

Oil of Wormseed → Chenopodium ambrosioides L. var. anthelminticum L.

Oil of Zit-el-Harmel → Peganum Harmala L.

Oil Palm → Elaeis guineensis Jacq.

Oils. Sources of essential → Abies, Acacia, Achillea, Allium, Alpinia, Amburana, Amyris, Anacyclus, Anethum, Angelica, Angiopteris, Aniba Anisoptera, Anthemis, Apium, Aptandra, Aquilaria, Artemisia, Aydendron, Baeckea, Betula, Boronia, Bulnesia, Bursera, Calophyllum, Canangium, Carum, Cedrus, Cheiranthus, Chenopodium, Cinnamomum, Cistis, Citrus, Commiphora, Convallaria, Convolvulus, Coriandrum, Cuminum, Cupressus, Cymbopogon, Darwinia, Daucus, Dianthus, Dobera, Erythroxylon, Eucalyptus, Eucarya, Eugenia, Ferula, Filipendula, Foeniculum, Fusanus, Gallesia, Gardenia, Gaultheria, Gynandropsis, Homoranthus, Hoslundia, Humulus, Hyacinthus, Illicium, Inula, Iris, Irvingia, Jasminum, Juniperus, Lallemantia, Lathyrus, Lavandula, Lawsonia, Lecontea, Levisticum, Lilium, Litsea, Madhuca, Majorana, Manilkara, Marrubium, Matricaria, Melaleuca, Melissa, Mentha, Mespilodaphne, Michelia, Microtanea, Miscanteca, Monarda, Myristica, Myroxylon, Myrtus, Narcissus, Ocotea, Opopanax, Osyris, Pandanus, Pelargonium, Petroselinum, Picea, Pimenta, Pimpinella, Pinus, Piper, Pogostemon, Polianthes, Reseda, Rosa, Rosmarinus, Ruta, Salvia, Santalum, Sassafras, Satureja, Solidago, Spartium, Tagetes, Tanacetum, Thuja, Thymus, Tsuga, Umbellularia, Unona, Valeriana, Vetiveria, Viola, Vitex.

Oils, Sources of Fatty → Acrocomia, Adansonia, Afrolicania, Agonandra, Aleurites, Allanblackia, Amoora, Amygdalus, Anacardium, Anisosperma, Antrocaryum, Arachis, Argania, Argemone, Armeniaca, Astrocaryum, Attalea, Avena, Baillonella, Balanites, Barbarea, Barringtonia, Bassia, Bertholletia, Brachylaena, Brassica, Buchanania, Butyrospermum, Calodendrum, Calophyllum, Camellia, Canarium, Canabis, Caraipa, Carapa, Carpotroche, Carthamus, Caryocar, Casearia, Castanea, Ceiba, Chisocheton, Chrysobalanus, Chrysophyllum, Citrus, Cocos, Corylus, Croton, Cucurbita, Dumoria, Elaeis, Eruca, Evonymus, Fagus, Ganophyllum, Garcinia, Glaucium, Glycine, Gomphia, Gossypium, Guioa, Guizotia, Helianthus, Hesperis, Hevea, Hibiscus, Horsfieldia, Hydnocarpus, Irvingia, Isoptera, Jatropha, Jessenia, Joannesia, Jubaea, Juglans, Kokoona, Lactuca, Laurus, Lepidium, Leptactina, Licania, Lindera, Linum, Lophira, Lycopersicum, Madhuca, Madia, Mauritia, Medicago, Melia, Mimusops, Moringa, Nephelium, Oenocarpus, Olea, Omphalea, Oncoba, Orbignya, Palaquium, Panda, Pangium, Papaver, Pappea, Peganum, Pentaclethra, Pentadesma, Perilla, Persea, Pistacia, Platonia, Plukenetia, Pongamia, Prunus, Ricinodendron, Ricinus, Salvadora, Sapindus, Schleichera, Sesamum, Shorea, Simmondsia, Sinapis, Sindora, Solidago, Sterculia, Symphonia, Taraktogenos, Telfairia, Tetracarpidium, Theobroma, Thevetia, Torreya, Trichilia, Vitis, Xanthophyllum, Ximenia, Xvlopia, Zea.

Oilnut → Pyrularia pubera Michx.

Oka Oxalis → Oxalis crenata Jacq.

Okra → Hibiscus esculentus L.

Okgue → Ficus pumila L.

Olacaceae → Aptandra, Coula, Ochanostachys, Olax, Ptychopetalum, Scorodocarpus, Strombosia, Ximenia.

Olax Wightiana Wall. (Syn. O. zeylanica Wall.) (Olacaceae). - Small tree. Ceylon. Leaves are consumed as vegetable among the natives of Ceylon.

Old Man's Beard → Chionanthus virginica L.

Old Man's Beard → Usnea barbata Hoffm.

Oldenlandia umbellata L. Indian Madder, Chirval, Saya. (Rubiaceae). - Biennial herb. Trop. Asia, esp. Bengal to Ceylon, N. Burma, Abyssinia. Bark of root with alum is source of a red dye. Was formerly much used in Madras for dyeing bandana handkerchiefs.

Oldfieldia africana Benth. and Hook., African Teak, African Oak. (Euphorbiaceae). - Tree. Trop. Africa. Wood strong, heavy, sinks in water; used for boat-building.

Olea Cunninghamiana Hook. f. (Oleaceae). - Tree. New Zealand. Wood very strong, deep brown, heavy, dense, compact, easily worked, takes a finish. Used for rail-road carriages, wagons, cabinet work, turnery, bowls, bridges, wharves.

Olea cuspidata Wall (syn. O. ferruginea Royle). Indian Olive. (Oleaceae). - Tree. Himalayan region. Wood is known as Indian Olive in the lumber trade and is used for general carpentry work.

Olea europaea L., Common Olive Tree. (Oleaceae). - Tree. Mediterannean region. Cultivated since ancient times. Grown in subtropic, usually semi-arid regions of the Old and New World. Of much commercial importance. Cured ripe, near ripe and stuffed olives are used as pickles. Fruits are source of Olive Oil, nondrying, chiefly used for food, in salads, in cooking, for canning sardines; for medicial purposes as laxative, emollient and demulcent.; also used as lubricant and for manuf. soap. Inedible oil is „denatured" by rosemary oil. Sp. Gr. 0.9145—0.9191; Sap. Val. 185—200; Iod. No. 77—94; Unsap. 0.6—1.3 %. Varieties are: Mission, Mazanillo, Sevillano or Queen Olive and Ascolano. Pomace from pressed olives is used in California as fuel; in Europe pomace is dried and remaining oil extracted with trichloroethylene or carbon disulfide; the resulting oil is blended with edible olive oil or used for manuf. soap. Wood very hard, heavy; used for turnery, canes and brushes.

Olea fragrans Thunb. → Osmanthus fragrans Lour.

Olea laurifolia Lam., Black Ironwood. (Oleaceae). - Tree. South Africa, esp. Cape Peninsula. Wood hard, heavy, close-grained, brownish; used for axles, poles, wagon-work, tools, ploughs and other agricultural implements.

Olea paniculata R. Br., Ironwood, Marblewood, Marvey. (Oleaceae). - Tree. Australia, esp. New South Wales, Queensland. Wood hard, tough, durable, close-grained. Used for staves and turnery.

Olea sandwicensis Gray → Osmanthus sandwicensis Benth. and Hook.

Olea Thozetii Panch and Sebert. (Oleaceae). - Tree. New Caledonia, Australia. Wood is strong, close-grained; used for cabinet-work.

Olea verrucosa Link., Warty Olive. (Oleaceae). - Tree. South Africa, esp. Cape Peninsula, Natal and adj. territory. Wood very compact, heavy, very hard; used for furniture, tools, wagon work, construction of mills, millwheels.

Oleaceae → Chionanthus, Forestiera, Fraxinus, Jasminum, Ligustrum, Nyctanthes, Olea, Osmanthus, Phillyrea.

Oleander → Nerium Oleander L.

Olearia Colensoi Hook. f. (Compositae). - Shrub or fmall tree. New Zealand. Wood hard, compact, satiny lustre, streaked or clouded, light-brown. Used for different ornamental work.

Olearia Traversii F. v. Muell. Travers Daisybush. (Compositae). - Small tree. New Zealand. Wood dense, heavy, compact, firm, satiny lustre. Used for cabinet work.

Olibanum → Boswellia Carterii Birdw.

Oleum → Oil.

Oligoceras Eberhardtii Gagnep. (Euphorbiaceae). - Tree. Annam. The fruits are edible and are esteemed by the natives of Annam. Sold in local markets.

Olive → Olea europea L.

Olneya Tesota Geay. Tesota. Sonora Ironwood. (Leguminosaceae). - Tree. California to New Mexico and adj. Mexico. Seeds are consumed by the Mohave and other Indian tribes. They are often stored for the winter.

Omphalia diandra L. (Euphorbiaceae). - Climbing plant. West India. Fruits are said to be edible and consumed by the natives.

Omphalia triandra L. (Euphorbiaceae). - Tree. Trop. America. Latex may be of value as a source of rubber during times of emergency. Fruits are edible and by some much esteemed. Occasionally cultivated.

Omphalocarpus anocentrum Pierre. (Sapotaceae). - Tree. Trop. Africa. Seeds are source of an oil; used by the natives.

Omphalogonus calophyllus Bail. (Asclepiadaceae). - Tree. East Africa, Zanzibar. Latex is source of a rubber which is difficult to gather. Of probable use in times of emergency.

Oncinotis hirta Oliv. (Apocynaceae). - Tree. Congo region. Source of a latex which is source of a rubber. Is recommended in case of emergency.

Oncoba spinosa Forsk. Kpoe, Krutu. (Flacourtiaceae). - Tree. Trop. Africa. The pulverized large fruits are used as a snuff by the natives. Wood is easily split; used for inlay work.

Onocarpus vitiensis Gray. (Palmaceae). - Tall palm. Fiji Islands. Fruits are source of a beverage, used by the natives.

Oncosperma filamentosa Blume (Palmaceae) - Tall palm. Malaysia. The very young, not unfolded leaves are used as a vegetable in some parts of Trop. Asia. Wood of the stem is durable, especially in water; used for floors of native homes. Truncs are used for pile (lake) dwellings. Also O. horridum Scheff.

Oncosperma horridum Scheff. (Palmaceae). - Tall palm. Malaysia. Source of Palmite, which is supposed to have an agreeable taste.

Ongokea kamerunensis Engl. Ingo. (Olacaceae). - Tree. West Africa, esp. Cameroon. Wood yellow; used for building purposes.

Onion, Garden → Allium Cepa L.

Onion, Multiplier → Allium Cepa L. var. solaninum Alef.

Onion, Potato → Allium Cepa L. var. solaninum Alef.

Onion, Sea → Urginea Scilla Steinh.

Onion, Spring → Allium fistulosum L.

Onion, Top → Allium Cepa L. var. viviparum Metz.

Onion, Welsh → Allium fistulosum L.

Onobrychis viciifolia Scop. (syn. O. sativa Lam.) Esparcette. (Leguminosaceae). - Perennial herb. Europe. W. Asia, introd. in N. America. Cultivated. Grown as a fodder for livestock, esp. in some parts of Europe.

Ononis spinosa L. Restharrow. (Leguminosaceae). - Perennial herbaceous. Europe to Turkestan. Used in some parts of Central Europe for home remedies against catarrh of the bladder, gall stones and gout; a decoction of the stem is used for eczema, irritation of the skin and scorbut; a decoction of the flowers used for „cleansing" the blood.

Onopordon Acanthium L. (Compositae). - Perennial herb. Floral parts are used in the adulteration of Saffron.

Onosma echinoides (Boraginaceae). - Perennial herb. Southern and Central Europe to W. Himalaya. Roots are source of a red dye, used in different parts of India for dyeing woolens, oil and fat in place of Alkanna.

Operculia terpethum. Indian Jalap. (Convolvulaceae). - Herbaceous plant. East India. Source of a resin, derived from the root.

Ophioglossum reticulatum L. (Ophioglossaceae). - Herbaceous perennial. Fern-ally. Malaysia. The young, sweet leaves are consumed as a potherb in some parts of the Moluccas.

Opium → Papaver somniferum L.

Opobalsam → Myroxylon Balsamum (L.) Harms.

Opopanax Chironium (L.) Koch. Opopanax. (Umbelliferaceae). - Herbaceous plant. Mediterranean region. A source of Opopanax; used in perfumery. Was formerly used in medicine.

Opopanax → Opopanax Chrironium (L.) Koch.

Opuntia azurea Rose. Nopalillo, Nopal cyotillo. (Cactaceae). - Compact xerophytic shrub. Zacatecas and Durango (Mex.). Fruits are juicy and edible, consumed by the Mexicans.

Opuntia basiliana Engelm. and Bigel. (Cactaceae). - Xerophytic shrub. Southw. N. America. In spring, joints, buds and flowers are prepared as food by steaming in a pit in the ground by the Indians of California and New Mexico.

Opuntia Bigelowii Engelm. (Cactaceae). - Small xerophytic tree. Southwest. N. America and adj. Mexico. Stems are used as food for livestock. To make them more palatable, the spines are burned off with torches. Also O. cholla Weber, O. fulgida Engelm., O spinosior Engelm. and others.

Opuntia camanchica Engelm. (Cactaceae). - Xerophytic shrub. Southw. of N. America and adj. Mexico. Fruits are consumed as food by the Indians of California, Arizona and New Mexico.

Opuntia clavata Engelm. (Cactaceae). - Xerophytic shrub. New Mexico. Stems and fruits were roasted and used as food by the Indians of New Mexico.

Opuntia Engelmanii Salm-Dyck. (Cactaceae). - Xerophytic shrub. Southwest of N. America. Fruits are eaten raw or cooked; stems are sometimes fried by the Indians of California, Arizona and New Mexico.

Opuntia ficus-indica Mill. Nopal, Indian Fig, Tuna de Castilla. (Cactaceae). - Xerophytic shrub. Mexico. Cultivated in tropical and subtropical countries for their fruits, known as Tunas, Indian Figs or Prickly Pears. The fruits are held in high esteem among the Mexicans. Considerable quantities are shipped to the United States. They are of rather high nutrative value, and are eaten fresh, dried or prepared in various ways.

Opuntia fulgida Engelm. Cholla. (Cactaceae). - Succulent, xerophytic shrub. Southwest N. America and adj. Mexico. Stem is source of a gum; used in Mexico as a size or stiffening. Is sold locally in markets.

Opuntia humifusa Raf. (Cactaceae). - Xerophytic shrub. Southwestern United States. Fruits are eaten fresh or stewed, also dried for winter; used by the Indians of Nebraska, North and South Dakota.

Opuntia imbricata (Haw.) DC. Tuna, Joconoxtla. (Cactaceae). - Treelike xerophyte. Southwestern United States and adj. Mexico. Since early times a decoction of the fruit was prepared to set cochineal dyes, and is also used for this purpose at the present. Canes are made from the frame work of the stems after the soft tissues have been removed.

Opuntia leucotricha DC. Nopal Durazillo. (Cactaceae). - Shrubby xerophytic tree. Central Mexico. Fruits are aromatic and edible.

Opuntia megacantha Salm-Dyck (syn. O. castillae Griffith). Tuna, Nopal. (Cactaceae). - Xerophytic shrub. Mexico. Widely cultivated in Mexico for its fruits from which the best edible tunas are derived. Many of the varieties have local names. Some are large fruited and of excellent quality and are during their season the principal food of the people. They are consumed raw, cooked in various ways and as sweetmeats. Queso de Tuna is composed of the dried fruit pressed into large cakes; they are widely sold in the markets. Miel de Tuna is a syrup prepared from the fruits. Melcoacha is a thick paste made by boiling down the juice. Coloncha is the boiled and fermented juice. Nochote is a fermented beverage obtained from tuna juice, pulque and water. The tender young joints are often cooked as vegetable. They are also used as poultices to reduce imflamation. Juice of the joints is sometimes boiled with tallow in the manuf. of candles.

Opuntia streptacantha Lem. Tuna Cardona, Nopal Cordón. (Cactaceae). - Much branched xerophytic tree. Centr. Mexico. One of the most important Opuntias in Mexico. Grown for its fruits.

Opuntia versicolor Engelm. (Cactaceae). - Small xerophytic tree. Southwestern United States and adj. Mexico. Fruits are eaten by the Pima Indians of Arizona; either raw or preserved.

Opuntia spp. Spineless Prickly Pears. (Cactaceae). - Spineless xerophytic shrubs. A number of spineless Platopuntias, belonging to different species are grown in the southwestern United States and Mexico as food for livestock. They lack the large spines peculiar to most of the species.

Orach, Garden → Atriplex hortensis L.

Orange, Bitter → Citrus Aurantium L.

Orange, Blood → Citrus sinensis (L.) Osbeck.

Orange, Kid-Glove → Citrus nobilis Lour. and var. deliciosa (Ten.) Swingle, var. unshiu (Mak.) Swingle.

Orange, King → Citrus nobilis Lour.

Orange, Mandarin → Citrus nobilis Lour. var. deliciosa (Ten.) Swingle.

Orange, Mexican Mock. → Philadelphus mexicanus Schlecht.

Orange, Osage→ Maclura aurantiaca Nutt.

Orange, Satsuma → Citrus nobilis Lour. var. unshiu (Mak.) Swingle.

Orange, Sevilla → Citrus Aurantium L.

Orange, Sour → Citrus Aurantium L.

Orange, Sweet → Citrus sinensis (L.) Osbeck.

Orange, Tangerine → Citrus nobilis Lour. var. deliciosa (Ten.) Swingle.

Orbigya Cohune (Mart.) Dahlgr. → Attalea Cohune Mart.

Orbignya Martiana Barb. (Palmaceae). - Tall Palm. South America. The smoke of the burning seeds is sometimes used as a coagulant of the latex of Hevea brasiliensis.

Orbigya speciosa Berk. Babassu Palm. (Palmaceae). - Tall palm. Trop. So. America, esp. Brazil. Fruits are source of an excellent oil, called Babussu Oil; used for manuf. of soap and margerine.

Orcein → Rocella.

Orcello → Rocella fuciformis (L.) Lam.

Orchidaceae → Anacamptis, Anaectochilus, Angraecum, Calypso, Cremastra, Cymbidium, Cypripedium, Dendrobium, Epidendrum, Gastrodia, Geodorum, Habenaria, Micretis, Orchis, Platanthera, Vanilla.

Orchids, Useful → Orchidaceae.

Orchil → Rocella fusiformis (L.) Lam.

Orchis latifolia L. Orchis. (Orchidaceae). - Perennial herb. Europe, Western Asia. Root tubers are source of Salep, a fine white to yellowish white powder; used as food in oriental countries, also given to convalescent children. Is demulcent and nutrient. Consumed in Greece and Turkey with honey. Sometimes used medicinally for intestinal catarrh. A home remedie in the Orient as a aphrodisiacum. Used in India in sweetmets and chocolates. Much is derived from Germany, Levant, Anatolia, Iran and India. Also O. militaris L., O. mascula L., O. Morio L., O. ustulata L. and others.

Orcinol → Rocella.

Ordeal Bean → Physostigma venenosus Ralph.

Oregon Balsam → Pseudosuga taxifolia Lam.

Oreodoxa oleracea Mart. → Roystonia oleracea (Mart.) Cook.

Oreodoxa regia HBK. → Roystonia regia (HBK.) Cook.

Oriental Cashew Nut → Semecarpus Anacardium L. f.

Oriental Persimmon → Diospyros Kaki L.

Origanum creticum Sut. (Labiaceae). - Perennial herb. So. Europe. Known as Spanish Hops, sold for Thyme; used in medicine.

Origanum Dictamnus L. (Labiaceae). - Perennial herb. Eastern Mediterranean region. Occasionally sold for Thyme; used in medicine.

Origanum glandulosum Desf. (Labiaceae). - Herbaceous perennial. N. Africa. Used in some parts of Algeria as a condiment, also O. compactum Benth.

Origanum Majorana L. → Majorana hortensis Moench.

Origanum Onites L. (Labiaceae). - Perennial herb. So. Europe, Asia Minor. Used as a condiment, is of inferior quality.

Origanum vulgare L. Pot Marjoram. (Labiaceae). - Herbaceous perennial. Europe. Used as a condiment and in home remedies, for intestinal pains, diseases of the air passages and toothache. It is said that the plant drives away ants.

Orinoco Simaruba → Simaruba amara Aubl.

Orites excelsa R. Br. Sily Oak, Red Ash. (Proteaceae). - Tree. N. Australia. Wood gray, of attractive appearance, durable, easy to polish; used for agricultural implements.

Orizaba → Ipomoea orizabensis Ledenois.

Orlean → Bixa orellana L.

Ormosia calavensis Azaol. (Leguminosaceae). - Tree. Malaysia. Decoction of the bruised leaves is used by the natives for stomach ache.

Ormosia sumatrana Prain. (Leguminosaceae). - Tree. Sumatra, Java, Moluccas. Wood hard, not durable; used for floors and beams of native houses.

Ornithogalum pyrenaicum L. (Liliaceae). - Bulbous plant. So. France and Spain. Flowerclusters are occasionally consumed like asparagus, early in spring.

Ornithogalum umbellataum L. (Liliaceae). - Bulbous plant. Europe. Bulbs are occasionally consumed as food. Also O. narbonensis L.

Ornithopus sativus Link. Seradella. (Leguminosaceae). - Annual to perennial herb. Europe. Grown as a cover crop and green manure in some countries of Europe. Suitable for sandy soils.

Orobanchaceae → Orobanche, Phelypaea.

Orobanche californica Cham. and Schlecht. (Orobanchaceae). - Herbaceous parasite. Western N. America. Succulent underground stems are used as food by the Indians of Nevada and California.

Orobanche fasciculata Nutt. (Oribanchaceae). - Herbaceous parasite. Western N. America. Entire plant is used as food by the Indians of Utah and Nevada.

Orobanche ludoviciana Nutt. (Orobanchaceae). Herbaceous parasite. Western N. America. Yellow, tender roots are used as food after roasting on coals by the Indians of Utah, Nevada and California.

Orobanche tuberosa (Gray) Heller. (Orobanchaceae). - Herbaceous parasite. Roots are consumed as food by the Indians of California.

Orontium aquaticum L. Goldenclub. (Araceae). Perennial herb. Eastern N. America. Rootstocks and seeds were consumed by the Indians of New York and Virginia. Repeated boiling or washing, were required to remove the acrid taste.

Oroxylum indicum Vent. (syn. Calosanthes indica Blume). (Bignoniaceae). - Tree. India, Mal. Archip. The bitter bark is used by the Javanese for stomach ailments; sold in markets.

Orris → Iris florentina L.

Orris Root → Iris florentina L.

Orseille → Rocella tinctoria DC.

Orthosiphon grandiflorus Bold. (Labiaceae). - Herbaceous perennial. Trop. Asia, Mal. Archip. Leaves are diuretic; used by the natives for catarrh of the bladder. Cultivated.

Orthosiphon rubicundus Benth. (syn. Plectranthus tuberosus Roxb.). (Labiaceae). - Herbaceous plant. Trop. Africa. Tubers are source of a starch; consumed by the natives. Cultivated.

Oryza sativa L. Rice. (Graminaceae). - Annual grass. Important graincrop, grown in warm countries of the Old and New World. Important diet in many Asiatic countries. Very nutritious when not polished. Polished rice has lost much of its value, especially in proteins. Rice hulls and rice straw are used as stockfeed. Straw is made into hats, shoes and other material. Source of Rice Oil, Sake, Rice Wine much consumed in Japan. (→ also Saccharomyces Sake Yabe). Numerous varieties, divided into two groups: Upland, grown without submersion and Lowland grown in flooded land during the growing season. There are several thousand of varieties, divided into common (starchy) and glutinous types; the latter containing a sugary material instead of starch. The glutinous varieties are used in the Orient for special purposes among which sweetmeats. The starchy varieties are used in various dishes, cakes, soups, pastry, breakfast foods, starch, paste etc. They are divided in short-, medium- and long grain groups. In some countries short grained are often prefered. The long grained have the highest market value. Short grain varieties are: Colusa, Unsen, Calore, Shiriki. Medium grain varieties: Early Prolific, Shoemed, Zenith, Calady, Blue Rose. Long grain varieties: Honduras, Edith, Fortuna, Iola. Varieties differ in culinary properties. The glutinous and starchy types in some countries are divided into white and colored rice. The white are used for human consumption, the colored are raised as food for livestock and are used as human food only in time of scarcity. Paddy is rice in the husk.

Oryzopsis hymenoides (Roem. and Schult). (Graminaceae). - Perennial grass. Western N. America. Seeds are used as a source of food by the Indians of Montana, Utah and Arizona.

Osier → Salix. spp.

Osmanthus americana (L.) Gray. Native Olive., Devil Wood. (Oleaceae). - Tree. Eastern N. America. Wood heavy, very hard, strong, close-grained, dark brown, difficult to work, durable, occasionally used for different purposes.

Osmanthus aquifolium (Sieb. and Zucc.) Benth. and Hook. Hiiragi. (Oleaceae). - Wood is used for manuf. small furniture, combs and toys in various parts of Japan.

Osmanthus fragrans Lour. (Oleaceae). - Tree. China and adj. countries. The very scented flowers are used by the Chinese to impart a pleasant aroma to the tea.

Osmelia celebica Koord. (Flacourtiaceae). - Tree. Celebes. Wood very strong; used for construction of houses.

Osmorhiza Claytoni (Michx.) Clarke. (Umbelliferaceae). - Perennial herb. Western N. America. Roots were used by the Indians of Wisconsin to gain weight.

Osmaronia cerasiformis (Torr. and Gray) Greene. Osoberry. (Rosaceae). - Shrub. Western N. America. Fruits are consumed by the Indians of British Columbia.

Osmunda cinnamomea L. Cimmamon Fern. (Osmundaceae). - Perennial herb. Fern. Eastern N. America. Young fronds boiled were consumed as food in soup by the Menominee Indians.

Osmunda claytoniana L. (Osmundaceae). - Perennial herb. Fern. The rhizomes are sometimes used as an adulterant of Aspidium filix mas (L.) Sw.

Osmunda regalis L. (Osmundaceae). - Perennial herb. Fern. Trichomes (hairs) of the young leaves are used in Japan with wool for making a cloth and a material for raincoats.

Osmundaceae → Osmunda.

Ostrya carpinifola Scop. European Hop Hornbeam. (Betulaceae). - Tree. Central and So. Europe, Asia Minor and adj. countries. Wood yellowishred, close-grained, hard, very tough; used for different purposes, also made into charcoal.

Ostrya virginiana Koch. American Hop Hornbeam. (Betulaceae). - Tree. Eastern N. America. Wood hard, strong, durable, tough, light brown; used for fance posts, handles of tools, mallets and many small articles.

Osyris tenuifolia Engl. (Santalaceae). - Tree. East Africa, esp. Kilimandscharo, Usambara. Source of East African Sandalwood. Wood brown to brown, similar to Santalum album. East African Sandalwood Oil is used in perfumery.

Otaheite Apple → Spondias cytherea Sonner.

Otaheite Gooseberry → Phyllanthus distichus (L.) Muell. Arg.

Otophora alata Blume (Sapindaceae). - Tree. Borneo, Java. The dark purple fruits are consumed by the natives.

Otophora spectabilis Blume (Sapindaceae). - Shrub or small tree. Malaysia. Occasionally cultivated. The orange brown fruits are consumed by the natives.

Ottar of Roses → Rosa damascena Mill.

Otto of Roses → Rosa damascena Mill.

Ougeinia dalbergioides Benth. Sannan, Tinnus. (Leguminosaceae). - Tree. N. India. Wood light to red brown, hard, tough, durable, easy to polish; used for agricultural implements, furniture and building purposes.

Ouratea angustifolia Gilg. (Ochnaceae). - Tree. Ceylon. Wood is used for building purposes in Ceylon, known as Bokaara-gass.

Ouratea sumatrana Gilg. (Ochnaceae). - Tree. Malaysia, Indonesia. Wood hard, heavy, light to dark brown, strong, not easy to split; used for boats, pumps. Decoction of roots and leaves is used for dysentry.

Ourouparia Gambir Baill. → Uncaria Gambir Roxb.

Owenia acidula F. Muell. Emu Apple, Bulloo. (Meliaceae). - Tree. Australia. The reddish fruits possess a red refreshing pulp, eaten in some parts of Australia.

Owenia cerasifera F. v Muell. (Meliaceae). - Tree. Queensland. Source of an excellent wood; used for turnery and fine table work.

Oxalis Acetosella L. Wood Sorrel. (Oxalidaceae). - Herbaceous plant. Europe, N. America. The leaves are sometimes used instead of Sorrel, Rumex acetosa L. Also O. corniculata L. and O. stricta L.

Oxalis cernua Thunb. (Oxalidaceae). - Perennial herb. So. Africa. The small bulbs are sometimes consumed as food in So. France and N. Africa.

Oxalis crenata Jacq. (Oxalidaceae). - Herbaceous perennial. Peru. The tubers have been consumed since time immemorial as a vegetable by the early inhabitants of Peru. The plant was lateron introduced into Europe where it was grown as a vegetable mainly by amateurs. It is called Oca.

Oxalis Deppei Lodd. (Oxalidaceae). - Herbaceous perennial. Mexico. Tubers are consumed as food for which they were once cultivated in some parts of Europe, esp. France and Belgium.

Oxyanthus speciosum DC. Msala. (Rubiaceae). - Tree. East Africa, esp. Usambara. Source of a wood; used for different purposes.

Oxycoccus → Vaccinium.

Oxyria digyna (L.) Hill. Mountain Sorrel. (Polygonaceae). - Small perennial herb. Northern Hemisphare. The leaves are consumed like Sorrel and mixed with other herbs by some Indian tribes in Alaska.

Oysterplant → Tragopogon porrifolius L.

P

Pachira aquatica Aubl. Provision Tree. (Bombaceae). - Tree. S. Mexico, Centr. and S. America. Seeds eaten roasted by natives.

Pachycereus pecten-aboriginum (Engelm.) Britt. and Rose (syn. Cereus pecten-aboriginum Engelm.) Cardón hecho, Hecho. (Cactaceae). - Xerophytic tree. Mexico. Seeds ground into a meal; used in cakes; eaten by Indians and Mexicans.

Pachycereus Pringlei (S. Wats.) Britt. and Rose. (syn. Cereus Pringlei S. Wats.). Cardón. (Cactaceae). - Xerophytic tree. Mexico. Stems used for huts. Pulp of fruits and seeds made into a flour; used for tomales, eaten by Indians.

Pachycormus discolor (Benth.) Cov. (syn. Rhus Veitchiana Kellogg). (Anacardiaceae). - Mexico esp. Baja California. Bark used for tanning; exported to Europe.

Pachylobus edulis G. Don → Canarium edule Hook. f.

Pachyma cocos (Wolf) Fr. → Poria cocos Wolf.

Pachyma tuber-regium (Wolf) Fr. → Lentinus tuber-regium Fr.

Pachyrrhizus angulatus Rich. Wayaka Yambean. (Leguminosaceae). - Herbaceous vine. Malaya. Tubers are consumed as a vegetable in different parts of trop. Asia. Sold in markets.

Pachyrrhizus bulbosus Spreng. Yambean. (Leguminosaceae). - Perennial herb. Origin uncertain, probably from Trop. Asia. Cultivated in tropics of Old and New World. Young pods are consumed as a vegetable when boiled. Tuberous rhizomes are consumed raw or boiled as a substitute for yams.

Pachyrrhizus erosus Rich. (syn. Cacara erosa Thour). (Leguminosaceae). - Perennial herb. Probably native to Mexico. Cultivated in tropics of Old and New World. Young roots are eaten when raw; cooked or in soup.

Pachyrrhizus palmatilobus Benth. and Hook. Jicana. (Leguminosaceae). - Herbaceous vine. Trop. America. Cultivated in Mexico and Centr. America. Large turnip-like roots are consumed by the natives.

Pachyrrhizus tuberosus Spreng. Yam Bean. (Leguminosaceae). - Herbaceous vine. West Indies, S. America. Cultivated in Tropics. Young pods are eaten as a vegetable, like French beans. Tubers are a substitute for Yam. Source of a pure white starch, resembling arrowroot, palatable; used for custards and puddings.

Pacific Plum → Prunus subcordata Benth.

Pacific Yew → Taxus brevifolia Nutt.

Padauk, African → Pterocarpus Soyauxii Taub.

Padauk, Brown → Pterocarpus macrocarpus Kurz.

Padauk, Dragonsblood → Pterocarpus Draco L.

Padauk, Red → Pterocarpus macrocarpus Kurz.

Padauk, Sandalwood → Pterocarpus santalinus L. f.

Padauk, Vengai → Pterocarpus Marsupium Roxb.

Padbruggea pubescens Craib → Whitfordiodendron pubescens Burkill.

Paddlewood → Aspidosperma excelsum Benth. and Uvaria Nuesgenii Diels.

Paddy → Oryza sativa L.

Paederia foetida L. (Rubiaceae). - Woody climber. India, China, Phillipp. Islands, Malaya. Powdered leaves in water used for intestinal complaints, widely employed in Malaya and Java.

Paeonia albiflora Pall., Chinese Peony. (Ranunculaceae). - Perennial herb. Manchuria, China, Siberia. Plant is used medicinally in some parts of China. Roots when boiled are used as food by Mongolians and Tatars.

Paeonia corallina Retz., Coral Peony. (Ranunculaceae). - Perennial herb. N. W. Africa. Plant is used by Moroccans as antispasmodic.

Paeonia Moutan Sims., Tree-Peony. (Ranunculaceae). - Perennial. Kansu (China). Cultivated. Used medicinally in China. Contains phaeonol, a glucosid. Flowers are said to be eaten as a vegetable in some parts of Japan. Also P. officinalis L.

Pagoda Dogwood → Cornus alternifolia L. f.

Pagsahingin Resin → Canarium villosum Benth. and Hook.

Paho → Mangifera altissima Blanco.

Pahoorie → Platonia grandiflora Plach. and Triana.

Pahutan Mango → Mangifera altissima Blanco.

Paineira Fibre → Chorisia speciosa St. Hil.

Pak-choi → Brassica chinensis L.

Pala Fibre → Butea superba Roxb.

Palaquium Aherninanum Merr. (Sapotaceae). - Tree. Mindanao (Philipp. Islds.). The best source of Philippine Gutta-Percha; used for insulation of submarine and underground electrical cables, surgical appliances, funnels, bottles, outer covering of golf-balls.

Palaquium Burckii Lam. (Sapotaceae). - Tree. Sumatra. Wood whitish, brown; used for boards and boats. Fruit pulp edible. Seeds are used as food and are source of a fat, Vegetable Tallow Siak. Seeds are known as Siak Illipe Nuts.

Palaquium Gutta Burck. (syn. Isonandra Gutta Hook. f.) Malay Gutta Percha Tree. (Sapotaceae). - Tree. Mal. Archipelago. Cultivated. Source of Gutta Percha, of commercial value. Used for insolation of cables kept under water.

Palaquium javanense Burck. Java Nato Tree. (Sapotaceae). - Tree. E. Java. Seeds source of a butter; used by natives for illumination and as food.

Palaquium leiocarpon Boerl. (Sapotaceae). - Tree. Borneo. Source of Angso, a Gutta Percha.

Palaquium macrocarpon Burck. (Sapotaceae). - Tree. Sumatra, N. Celebes, Moluccas. Wood heavy, elastic, hard, durable, dark brown; used for construction of houses.

Palaquium Maingayi King and Gamb. (Sapotaceae). - Tree. Malacca. Source of Getah Pootih a good Gutta Percha, is often mixed with Djelutong.

Palaquium obovatum Engler, var. occidentale. Lam. (Sapotaceae). - Tree. Mal. Archipelago. Wood hard, not easily split, red; little attacked by termites, keeps well under water. Source of a good Gutta Percha.

Palaquium oleiferum Blanco. (Sapotaceae). - Tree. Malaya. Seeds source of Siak, Njatuo or Taban Merak Fat or Tallow. Used by natives in food; when refined used as substitute for cacaobutter. Sap. Val. 186; Iod. No. 38; Unsap. 1.2 %.

Palaquium Oxleyanum Pierre. (Sapotaceae). - Tree. Mal. Penins. Source of a Gutta Percha which is sometimes mixed with Djelutong.

Palaquium rostratum Burck. (Sapotaceae). - Tree. Sumatra, Bangka, W. Borneo. Wood light, durable, reddish brown, easily worked; used for floors, boards, household goods, beams, construction of houses and canoes.

Palay Rubber → Cryptostegia grandiflora R. Br.

Pale Shield Lichen → Cetraria glauca (L.) Ach.

Palissander, Javanese → Dalbergia latifolia Roxb.

Palm, Awara → Astrocaryum Jauari Mart.

Palm, Babassu → Orbignya speciosa Bark.

Palm, Betelnut → Areca Catechu L.

Palm, Biscayne → Coccothrinax argentea (Lodd.) Sarg.

Palm, Cocorite → Maximiliana caribaea Griseb.

Palm, Cohune → Attalea Cohune Mart.

Palm, Cañon → Washingtonia filifera (Lindl.) Wendl.

Palm, Coco Nut → Cocos nucifera L.

Palm, Date → Phoenix dactylifera L.

Palm, Doum → Hyphaene thebaica Mart.

Palm, Fish Tail → Caryota urens L.

Palm, Ginger Bread → Hyphaene coriacea Gaertn.

Palm, Ivory → Phytelephas macrocarpa Ruiz and Pav. and P. Seemanni Cook.

Palm, Ivory Nut → Coelococcus amicarum Warb.

Palm, Jauary → Astrocaryum Jauari Mart.

Palm Kernel Oil → Elaeis guineensis Jacq.

Palm Lily → Cordyline australis Hook. f.

Palm, Nikau → Rhopalostylis sapida Wendl. and Drude.

Palm, Nipa → Nipa fruticans Thunb.

Palm, Oil → Elaeis guineesis Jacq.

Palm, Palmyra → Borassus flabellifer L.

Palm, Panama Hat → Carludovica palmata Ruiz. and Pav.

Palm, Paraguay → Acrocomia sclerocarpa Mart.

Palm, Peach → Bactris Gasipaës H. B. K.

Palm, Sugar → Arenga pinnata (Wurmb.) Merr.

Palm, Tagua → Phytelephas macrocarpa Ruiz and Pav.

Palm, Talipot → Corypha umbraculifera L.

Palm, Thatch → Thrinax parviflora Sw.

Palm Wine → Arenga pinnata (Wurmb.) Merr., Caryota urens L., and Mauritia vinifera Mart.

Palma Ixtle → Samuela carnerosana Trel.

Palmaceae → Acanthorrhiza, Acrocomia, Ancistrophyllum, Areca, Arenga, Astrocaryum, Attalea, Bactris, Borassus, Brahea, Calamus, Calyptrogyna, Caryota, Ceroxylon, Chamaedorea, Chamaerops, Clinostigma, Coccothrinax, Cocos, Coelocccus, Copernica, Corypha, Daemonorops, Desmoncus, Elaeis, Eremospatha, Erythea, Eugeissona, Euterpe, Geonoma, Guilielma, Heterospathe, Hyphaene, Jessenia, Jubaea, Korthalsia, Leopoldina, Livistonia, Lodoicea, Manicaria, Mauritia, Maximilliana, Metroxylon, Nipa, Oenocarpus, Oncosperma, Orbignya, Oreodoxa, Phoenix, Pholidocarpus, Phytelephas, Pigafettia, Plectocomia, Raphia, Rhopalostylis, Roystonea, Sabal, Serenoa, Socratea, Thrinax, Trachycarpus, Washingtonia, Zalacca.

Palmarosa Oil → Cymbopogon Martini Stapf.

Palmate Butterbur → Petasites palmata Gray.

Palmate Violet → Viola palmata L.

Palmira Alstonia → Alstonia scholaris R. Br.

Palmita → Carludovica palmata Ruiz and Pav.

Palms, Useful → Palmaceae.

Palmyra Palm → Borassus flabellifer L.

Palo Blanco → Calycophyllum multiflorum Griseb.

Palo de Buba → Jacaranda Copaia (Aubl.) D. Don.

Palo Campechio → Haematoxylon campechianum L.

Palo Guate → Eysenhardtia polystachya (Ortega) Sarg.

Palo de Leche → Sapium pavonianum Muell. Arg.

Palo Muerto → Aextoxicon punctatum Ruiz and Pav.

Palo Negro → Lomatia hirsuta (Lam.) Diels.

Palo Pangue → Gunnera chilensis Lam.

Palo Santa → Bulnesia Sarmienti Lorentz.

Palo Verde → Cercidium Torreyanum (S.Wats) Sarg.

Palosápis Marsawa → Anisoptera thurifera Blanco.

Palosápis Oil → Anisoptera thurifera Blanco.

Palvaea Langsdorfii Berg. (Myrtaceae). - Tree. Brazil. Wood is used for general carpentry and agric. implements. Bark is considered astringent.

Pambotana Bark → Calliandra Houstoniana (Mill.) Standl.

Pampas Grass → Cortaderia argentea Stapf.

Panaeolus papilionaceus (Bull.) Quél. (Agaricaceae). - Basidiomycete. Fungus. Fruitbodies are used by some Indians of Central America as an intoxicating drug. Supposed to have an exhilarating effect.

Panama Hat Palm → Carludovica palmata Ruiz and Pav.

Panama Tree → Sterculia carthaginensis Cav.

Panau Resin → Dipterocarpus vernicifluus Blanco.

Panax Edgerleyi Hook. f. (syn. Nothopanax Edgerleyi Hook. f.). Raukawa. (Araliaceae). - Tree. New Zealand. Aromatic leaves are used by Maoris for making perfumed oils.

Panax fruticosum L. → Nothopanax fruticosum Miq.

Panax Murrayi F. v. Muell. (Araliaceae). - Tree. Australia. Source of a gum, resembling gum of Acacias, contains 85 % arabin. Recommended as a substitute for Gum Arabic.

Panax quinquefolia L. American Ginseng. (Araliaceae). - Herbaceous perennial. Eastern N. America. Cultivated in Wisconsin, Michigan, Oregon. Roots are exported to China. Favorite Chinese medicine. Stimulant, stomachic. Active principal is a glucosidal substance, panaquillon.

Pancheria obovata Brongn. and Gris. (Saxifragaceae). - Small tree. New Caledonia. Wood reddish-violet, fine-grained, durable; used for turnery; when it is polished it is very attractive.

Pancheria ternata Brongn. and Gris. (Saxifragaceae). - Tree. New Caledonia. Wood reddish-violet, fine-grained, very durable; used for cabinet work and turnery.

Panda oleosa Pierre. (syn. Porphyranthus Zenkeri Engl.). (Pandaceae). - Tree. Trop. Africa. Seeds are source of an oil used in cooking by the natives.

Pandaceae → Panda.

Pandanaceae → Freycinetia, Pandanus.

Pandanus amboinensis Warb. (Pandanaceae). - Tree. Mal. Archipelago. Leaves are used in the Moluccas for making mats.

Pandanus aquaticus F. v. Muell. (Pandanaceae). - Palm-like plant. Australia. Roasted seeds are eaten by the aborigenes of Queensland.

Pandanus atrocarpus Griff. (Pandanaceae). - Tall tree. Mal. Archipelago. Leaves are used on a large scale for manuf. of kadjangs, mats, hats etc. In Bangka they are used for sails of small vessels.

Pandanus Bidur Jungh. (Pandanaceae). - Tree. Malayan Archipelago. In Moluccas the fibres of brace-roots are used for ropes, strong twine and fishing-tackle.

Pandanus Copelandi Merr. (Pandanaceae). - Tree. N. Luzon to S. Mindanao (Philipp Islds.). Leaves are used in Philippines, for coarse mats and baskets.

Pandanus dubius Spreng. (Pandanaceae). - Tree. Amboina. Fibres from brace-roots made into cordage, used for seats of chairs, beds etc.

Pandanus edulis Thouars. (Pandanaceae). - Tree. Madagascar. Fruits eaten by natives.

Pandanus furcatus Roxb. (Pandanaceae). - Tree. E. India. Mal. Archipelago. Leaves used for mats.

Pandanus Heudelotianus Balf. f. (Pandanaceae). - Tree. Trop. Africa. A fibre is made from the brace roots. Leaves are made into mats.

Pandanus Houlletii Car. (Pandanaceae). - Tree. Screw-pine. Trop. Asia, esp. Malay. Penin. Fruits soft, sweet, pineapple flavor, eaten by natives.

Pandanus odoratissimus L. (syn. P. tectorus Soland.). Thatch Screw Pine. (Pandanaceae). - Mal. Archipelago, Pacif. Islds., Seychelles, Australia. Leaves are made into sugar bags, baskets, matting, cordage, hats and similar articles. Leaves are used in Hawai and other islands for thatching houses and floor coverings. Ripe phalanges, after pulp is decayed are employed for brushes. Male flowers give a powerful scent; they are source of an ess. oil, obtained by enfleurage. Seeds after careful preparation are consumed as food. Several varieties are in existance. Cultivated.

Pandanus odorus Thunb. (Pandanaceae). - Tree. Screw-pine. Trop. Asia, esp. Malaya. Young leaves used in cooking; in Java used in bean-curd.

Pandanus polycephalus Lam. (Pandanaceae). - Tree. Malaya. Young leaves are eaten raw by natives.

Pandanus simplex Merr. (Pandanaceae). - Tree-like plant. Philipp-Islds. Strips of leaves are much used in Philipp. Islds. for coarse and fine mats, bags, hats, baskets, picture frames, handbags, wall pockets and fancy slippers.

Pandanus togoensis Warb. (Pandanaceae). - Tree. Costal zone of Togo, (Afr.). Leaves are used by natives for making mats. Also P. Kerstingii Warb.

Pangium edule Reinw. (Samydaceae). - Java. Seeds contain prusic acid; are very poisonous when raw, after being cooked they are consumed as food, much sold in markets of Java. Oil from seeds is employed as illuminant. Crushed seeds are used to stupefy fish.

Panicularia fluitans Kuntze. → Glyceria fluitans (L.) R. Br.

Panicum ambiguum Trin. (Graminaceae). - Perennial grass. Trop. Asia, Pacif. Islds. Used as fodder for cattle.

Panicum barbinode Trin., Para Grass. (Graminaceae). - Perennial grass. S. America. Cultivated in warm countries; recommended for pasture and for hay.

Panicum Crus-Ardeae Willd. (syn. P. elongatum Poir., P. sulcatum Bert.). (Graminaceae). - Grass. S. America. Is considered an excellent forage grass in Brazil.

Panicum Crus-galli L. (syn. Oplismenus frumentaceus Kunth.) (Graminaceae). - Annual grass. Temp. zone. Used in Japan as a cereal, porridge, macaroni and in dumplings.

Panicum decompositum R. Br. (syn. P. laevinode Lindl., P. proliferum F. v. Muell.) Native Millet, Umbrella Grass. (Graminaceae). - Perennial herb. Australia. Grains are an excellent food for the aborigines. Pasture grass.

Panicum distachum L. → Brachiaria disticha (L.) Stapf.

Panicum flavidum Retz. (syn. P. brizoides Jacq.) Vandyke Grass. (Graminaceae). - Perennial grass. Australia. A good wintergrass, much liked by stock in Australia.

Panicum maximum Jacq., Guinea Grass. (Graminaceae). - Perennial grass. Tropics. Cultivated in warm countries, suitable as pasture grass.

Panicum microspermum Fourn. (syn. P. tricanthum Nees). (Graminaceae). - Perennial grass. Trop. America. The aromatic rhizome is considered in Brazil as an emollient, excitant and diuretic.

Panicum miliaceum L., Proso Millet, (Graminaceae). - Annual grass. India, Temp. zone. Cultivated in Old and New World. Source of a flour, used in bread. Also raised as food for cattle and hogs. Source of Braga or Busa, an alcoholic beverage; used in some parts of E. Europe, Balkans, Caucasia and Asia Minor. The beverage is also prepared from other cereals, among which oats, rye and barley.

Panicum obtusum H. B. K., Vine-Mesquite. (Graminaceae). - Perennial grass. Western United States and adj. Mexico. Seeds were ground with corn-meal and consumed as food by the Hopi Indians.

Panicum sanguinale L. (syn. Digitaria horizontalis Willd.). (Graminaceae). - Annual grass. Cosmopolitic. Used as fodder for live-stock.

Panicum sarmentosum Roxb. (Graminaceae). - Perennial grass. Trop. Asia. Mal. Archipelago. Used as food for live-stock.

Panicum stagninum Retz. (syn. P. Burgii Chev.). (Graminaceae). - Perennial grass. Trop. Africa, esp. Sudan, Centr. Africa. Source of a thick, sweet syrup used in confectionary by the natives, furnishes a beverage used by the Mussulmen in Timbuctoo.

Panicum texanum Buckl., Texas Millet, Colorado Grass. (Graminaceae). - Annual grass. Texas. Occasionally grown as food for live-stock.

Panicum turgidum Forsk. (Graminaceae). - Herbaceous grass. N. Africa, esp. Sahara region. Grains are used as food by different tribes in N. Africa.

Panopsis rubescens (Pohl) Pitt., Yolombo, Cedro Bordado. (Proteaceae). - Tree. Trop. S. America. Wood pinkish brown to reddish brown, moderately heavy, hard, straight-grained, easily worked; used locally for cabinet work, fancy articles.

Pansy → Viola tricolor L.

Panus rudis Fr. (syn. Lentinus rudis (Fr.) Henn.) (Agaricaceae). - Basidiomycete. Fungus. Cosmopolitian. Fruitbodies are used in Northern Caucasia for the preparation of a sheep cheese, called Airan. Young fruitbodies are occasionally eaten as food.

Páo d'Embira → Xylopia carminativa (Aruda) Fries.

Pão Manfin Oil → Agonandra brasiliensis Benth. and Hook.

Páo Rosa → Aniba rosaeodora Ducke., A. terminalis Ducke.

Paó Santo → Bulnesia Sarmienti Lorentz.

Papain → Carica Papaya L.

Papaver Rhoeas L., Corn Poppy, Red Poppy. (Papaveraceae). - Annual herb. Europe, temp. Asia; introd. in N. America. Petals are source of a red pigment; used for coloring esp. of wine and certain medicines. Flowers are employed medicinally since ancient times; used at the present as expectorant in home remedies.

Papaver somniferum L., Opium Poppy. (Papveraceae). - Annual herb. Mediterranean region to Iran. Cultivated in the Orient, Balkans, Iran, India, China etc. Source of Opium, Gum Opium or Crude Opium, an air-dried milky exudation from incised, unripe fruits. Extensively smoked as an intoxicant. Commercial products are called Turkey Opium, Persian Opium, Indian Opium, Chinese Opium and Egyptian Opium. Commercial varieties vary in appearance and quality, roughly classified as Soft or Shipping Opium, Druggists' and Manufacturer Opium. The Hadjikeuy is considered one of the best, followed by Malatia, Kharput and Sila. Druggists' Opium is largely used for manuf. of morphine, codeine, narcotine, laudenine, papaverine and many other alkaloids. Source of the toxic and extremely habit forming narcotic heroine or diamorphine; its manuf. is prohibited in some countries. Powdered Opium has been dried at not above 70°; used in Dover's Powder (powder of Ipecac and Opium). Various compounds are used in medicine as narcotic, antispasmodic, hypnotic, anodyne, analgestic, sedative, respiratory depression; to releave severe pain.

Seeds contain no Opium; they are used in baking and sprinkled on bread. They are source of a drying oil; used for manuf. of paints, varnishes and soaps; in foods, salad dressing. Oil cake is a good fodder for cattle. Sp. Gr. of oil 0.924—0.927; Sap. Val. 189—196; Iod. No. 132—140; Unsap. 0.5 %. Seeds are used for preparing emulsions for which the white var. is prefered; the bluish-black var. is generally used for baking. The capsules have been employed as fomentation in sprains, bruises, neuralgia, toothache etc.

Papaveraceae → Argemone, Bocconia, Chelidonium, Eschscholzia, Glaucium, Papaver, Sanguinaria.

Papaya → Carica Papaya L.

Papaya de Mico → Carica peltata Hook. a. Arn.

Papaya, Mountain → Carica candamarcensis Hook. f.

Papaya Orejona → Jacaratia mexicana A. DC.

Papaya de Terra Fria → Carica cestriflora Solms.

Paper Birch → Betula papyrifera Marsh.

Paper Mulberry → Broussonetia papyrifera Vent.

Papilionaceae → Leguminosaceae.

Papo Canary Tree → Canarium Schweinfurthii Engl.

Papoose Root → Caulophyllum thalictroides (L.) Michx.

Pappea capensis Eckl. and Zeyh. (Sapindaceae). - Shrub or small tree. S. Africa. Seeds are source of an oil; suitable as lubricant and for manuf. of soap.

Paprika → Capsicum frutescens L.

Papua Nutmeg → Myristica argentea Warb.

Papyrus Plant → Cyperus Papyrus L.

Para Arrowroot → Manihot esculenta Crantz.

Para Breadnut Tree → Brosimum paraense Hub.

Para Cress → Spilanthes acmella Murry.

Para Grass → Panicum barbinode Trin.

Para Ipecac → Cephaëlis Ipecacuanha (Brotero) Rich.

Para Nut → Bertholletia excelsa H. B. K.

Para Piassaba → Leopoldina Piassaba Wall.

Para Rhatany → Krameria argentea Mart.

Para Rubber → Hevea brasiliensis Muell. Arg.

Para Rubber Seed Oil → Hevea brasiliensis Muell. Arg.

Para Rubber Tree → Hevea brasiliensis Muell. Arg.

Para Sarsaparilla → Smilax Spruceana A. DC.

Parabarium Dindo Pierre. (Apocynaceae). - Woody vine. Indo-China. Has been mentioned as a source of rubber. Latex contains 35 % of crude rubber. Also P. Dinrang Pierre.

Paradaniella Oliveri Rolfe. Balsam Copaiva Tree. (Leguminosaceae). - Tree. Trop. W. Africa. Wood source of an oil and copal, Ako-Ogea, a hard fossilized gum; used for varnishes. Decoction of root is used for gonorrhoea and skin diseases.

Paradise Flower → Caesalpinia pulcherrima Swartz.

Paraguay Lignum → Bulnesia Sarmienti Lorentz.

Paraguay Palm → Acrocomia sclerocarpa Mart.

Paraguay Tea → Ilex paraguensis St. Hil.

Parameria barbata Schum. (Apocynaceae). - Tree. Burma. Source of a good rubber.

Parameria glandulifera Benth. (Apocynaceae). - Tree. S. E. Asia, Indo-China. Source of a good rubber. Latex coagulates by hot water.

Parameria philippensis Radlk. (Apocynaceae). - Woody vine. Philipp. Islds. Bark source of a fibre, used for manuf. of rope and for tying rice-bundles.

Parana Pine → Araucaria brasiliensis A. Rich.

Paratocarpus triandra Smith. (Moraceae). - Tree. Sumatra. Latex mixed with boiled rice, is used in Sumatra to kill rats.

Pareira root → Chondodendron tomentosum Ruiz and Pav.

Pareira, Yellow → Aristolochia glaucescens H. B. K.

Pereira, White → Abuta Candollei Triana and Planch.

Pariana lunata Nees. (Graminaceae). - Perennial grass. Colombia to Brazil. The broad leaves are used in the Chocco Intendancy of Colombia to wrap gold and platinum dust.

Parinarium annamense Hance. (Rosaceae). - Tree. China. Malaysia. Infusion of flowers is used in Siam as a cosmetic, applied after bathing.

Parinarium campestre Aubl. (Rosaceae). - Tree. Guiana. Fruits are edible, being of a pleasant flavor.

Parinarium curatellaefolium Planch. (Rosaceae). - Tree. Trop. Africa. Fruit edible, much esteemed by natives. Wood used for general carpentry.

Parinarium excelsum Sab. (Rosaceae). - Tree. Trop. Africa. Wood used for cabinet work, joinery, domestic utensils, construction work, furniture. Ashes of wood and bark are used in Congo for preparation of skins.

Parinarium macrophylla Sab., Ginger-bread Plum. (Rosaceae). - Tree. Trop. Africa. Fruits edible, much esteemed by the natives. Wood fairly hard, polishes well, light brown; used for boards, building purposes, canoes. Rind from fresh fruits imparts a pleasant scent to ointments. Decoction from bark and leaves is used as mouth wash. Lotion of macerated bark is employed for inflamed eyes.

Parinarium Mobola Oliv., Mobola Plum. (Rosaceae). - Tree. Trop. Africa. Fruit edible, has a strawberry flavor, one of the best native fruits in S. Africa.

Parinarium nonda F. v. Muell., Nonda Tree. (Rosaceae). - Tree. Australia, Fleshy drupes are consumed among the aborigenes.

Parinarium oblongifolium Hook. f. (Rosaceae). - Tree. Malacca. Wood orange to dark-brown, heavy, close-grained, very durable; is seldom attacked by termites; easily worked. Used for construction of houses, beams etc.

Parinarium polyandrum Benth. (Rosaceae). - Tree. Trop. Africa. Wood source of charcoal; used by native smiths. Lumber employed for building farm houses.

Parkia africana R. Br. (syn. P. biglobosa Benth.) African Locust. (Leguminosaceae). - Tree. Trop. West Africa. Fruits are edible, containing a sweet, yellow pulp, much esteemed by the natives. Seeds are used as condiment, also for colic. When parched they are used as coffee, called Café du Sudan. They are also made into cakes and used in sauces. Bark is used in Gambia as a specific for tooth ache. It is also used for tanning, giving a red color to leather. Wood is used in general carpentry.

Parkia filicoidea Welw. Fernleafed Nitta Tree, Westafrican Locust Bean. (Leguminosaceae). - Tree. Trop. Africa. Bark contains 12 to 14 % tannin; used in Sudan to produce a dark colored leather. Powdered seeds are used for flavoring native dishes and soups.

Parkia speciosa Hassk. (Leguminosaceae). - Tree. Malaysia. Cultivated in Malaya. Pods are used in food for flavoring, suggesting garlic. Likewise P. javanica Merr. Seeds are edible, consumed raw or popped in the shell. Sold in markets of Java.

Parkinsonia microphylla, Torr. (Leguminosaceae). - Small tree. S. Arizona, S. California and Sonora (Mex.). Seeds were eaten fresh or made into flour, also mixed with mesquite meal by the Indians of S. W. United States and Mexico.

Parmelia acebatulum Duby. (Parmeliaceae). - Lichen. Ireland. Employed in N. Ireland as a source of a brown-orange dye, to color home spuns, Harris Tweed.

Parmelia caperata (L.) Ach., Stone Crottle, Arcel, Wrinkled Lichen. (Parmeliaceae). - Lichen. Temp. zone. On stems and rocks. Source of a brown orange to lemon yellow dye; used in Isle of Man to stain woolens.

Parmelia conspersa (Ehrh.) Ach., Sprinkled Parmelia, Boulder Lichen. (Parmeliaceae). - Lichen. Temp. zone. On rocks. Used in England as a source of a red-brown dye, used to color woolens.

Parmelia centrifuga (L.) Ach., Ring Lichen. (Parmeliaceae). - Lichen. Temp. zone. On rocks. Used in Great Britain as a source of red-brown dye to color woolens.

Parmelia ciliaris Ach. → Anaptychia ciliaris (L.) Kbr.

Parmelia fahlunensis Ach. → Cetraria fahlunensis (L.) Schaer.

Parmelia furfuracea (L.) Ach. (syn. Evernia furfuracea E. Fr.). (Parmeliaceae). - Lichen. Temp. zone. On stems and wood. Source of an oleoresinous material, used in perfumery.

Parmelia kamtschadalis ? (Parmeliaceae). - Lichen. Trop. Asia. Source of a pale rose dye; used in India to print and perfume calico cloth.

Parmelia olivacea (L.) Ach., Bronze Shield Lichen. (Parmeliaceae). - Lichen. Temp. zone. On stems. Used in Great Britain as a source of a brown dye, to stain woolens.

Parmelia omphalodes (L.) Ach., Smoky Shield Lichen, Black Crottle, Corks. (Parmeliaceae). - Lichen. Temp. zone. On rocks. Used in Scandinavia, Ireland and Scotland as a source of a purple, crimson dye to stain woolens, wool for tweed. A much used dye lichen. Is easily fixed to yarns by simple mordants.

Parmelia physodes (L.) Ach., Dark Crottle, Puffed Shield Lichen. (Parmeliaceae). - Lichen. Temp. zone. On stems, rocks and wood. Used in Scotland and Scandinavia as a source of a brown dye to stain woolens. Was cooked and consumed in soup by the Indians of Wisconsin.

Parmelia prunastri Ach → Evernia prunastri (L.) Ach.

Parmelia saxatilis (L.) Ach. (Parmeliaceae). - Temp. zone. On rocks, stems and wood. Source of an orange, yellow and red-brown dye; used to stain woolens, esp. in Scotland and W. Ireland. Employed for Harris Tweed. The scent of this material is due to this dye. The lichens are usually collected in August when they are supposed to be richest in dye material.

Parmelia stygia (L.) Ach., Midnight Lichen. (Parmeliaceae). - Lichen. Temp. zone. On rocks Used in Great Britain as source of a dye to color woolens a brown color.

Parmelia, Sprinkled → Parmelia conspersa (Ehrh.) Ach.

Parmentiera alata Miers. (syn. Crescentia alata H. B. K.) (Bignoniaceae). - Tree. Mexico, Centr. America, Cuba, tropics of Old World. Wood used for wagons. In Nicaragua a cooling drink is made from seeds. Shells of fruits used for cups.

Parmentiera edulis DC. Food Candle Tree. (Bignoniaceae). - Tree. Mexico, Centr. America. Fruits eaten raw or cooked, of inferior quality.

Paronychia argentea Lam. Silver Nailroot. (Caryophyllaceae). - Perennial herb. Mediterranean region. An infusion of the herb is used in Morocco as diuretic and aphrodisiac.

Parosela lanata (Spreng.) Britt. (Leguminosaceae). - Perennial herb. Western N. America and adj. Mexico. Roots were consumed by Hopi Indians.

Parsley → Petroselinum hortense Hoffm.

Parsley, Cow → Chaerefolium sylvestre (L.) Shinz and Thell.

Parsley, Milk → Peucedanum palustre (L.) Moench.

Parsley, Turnip Rooted → Petroselinum hortense Hoffm.

Parsley, Water → Oenanthe sarmentosa Prel.

Parsnip → Pastinaca sativa L.

Parthenium argentatum Gray. Guayule. (Compositae). - Shrub. Semi-arid regions. S. W. of United States, adj. Mexico. Cultivated. Source of Guayule Rubber. Occasionally of commercial importance. Of value as an emergency rubber plant.

Parthenium Hysterophorus L. (Compositae). - Perennial herb. N. and S. America. Decoction of the boiled roots is used by the Koasati Indians to stop dysentery.

Parthenium integrifolium L. (Compositae). -Perenial herb. Eastern N. America. Fresh leaves were used by the Catawba Indians for burns.

Parthenocissus quinquefolia (L.) Planch. (syn. Ampelopsis quinquefolia Michx.) Virginia Creeper. (Vitaceae). Woody vine. N. America. Fruits are source of a pink dye, used by the Kiowa Indians (N. Mex.) for painting their skins and feathers to be used in war dances. Bark is employed as a tonic, alterative, expectorant; used for dropsy. Decoction of roots used by the Meskwaki for diarrhea.

Pasania cyclophora Gamble → Quercus cyclophora Endl.

Pasania densiflora Oerst. → Lithocarpus densiflora (Hook. and Arn.) Rehd.

Paspalum brevifolium Flueg. → Digitaria longiflora Pers.

Paspalum conjugatum Berg., Hilo Grass. (Graminaceae). - Perennial grass. Tropics. Plant produces a heavy growth, used as food for livestock.

Paspalum dilatatum Poir., Dallis Grass. (Graminaceae). - Perennial grass. S.America. Excellent pasture grass in S. of United States.

Paspalum mandiocanum Trin. (Graminaceae). - Perennial grass. Brazil. Cultivated. Considered an excellent fodder for live-stock in different parts of Brazil.

Paspalum notatum Fluegge., Bahia Grass. (Graminaceae). - Perennial grass. Mexico, West Indies to S. America. Cultivated as a source of fodder for live-stock.

Paspalum scrobiculatum L., Koda Millet. (Graminaceae). - Annual grass. Asia. Occasionally cultivated in Old and New World, as food for live-stock.

Paspalum Urvillei Steud. (syn. P. larranagai Arech.) Vasey Grass. (Graminaceae). - Perennial grass. S. America. Cultivated as a source of fodder for cattle.

Pasqueflower, Meadow → Anemone pratensis L.

Pasqueflower, Nodding → Anemone patens L.

Passerina filiformis L. (Thymelaeaceae). - Shrub. S. Africa, esp. Cape Peninsula to Transvaal. Bark is source of a cordage, used for tying down thatched roofs; also plaited into whipthongs; was formerly used for making slave-whips.

Passerina hirsuta L. (syn. (Thymelaea hirsuta Endl.). (Thymelaeaceae). - Shrub. Mediterranean region. Bark is source of a strong fibre.

Passiflora edulis Sims., Purple Grenadilla, Passion Fruit. (Passifloraceae). - Woody vine. Brazil. Cultivated in tropics and parts of subtropics of Old and New World. Fruits edible, pleasant, aromatic, rather acid, juicy pulp; eaten with sugar, for flavoring sherbets, icing cakes, confectionary, refreshing drinks, trifles (a sponge cake). In some countries fruit juice is manuf. on a commercial scale.

Passiflora incarnata L., May-Pop, Apricot Vine. (Passifloraceae). - Vine. Eastern N. America to Florida and Texas. Fruits edible, made into jelly and sherbet. Was cultivated by the Indians in Virginia. Dried flowering and fruit tops are used medicinally among eclectics in neuralgia, insomnia, diarrhea and dysmenorrhea.

Passiflora laurifolia L., Water-Lemon, Bell-Apple, Jamaica-Honeysuckle. (Passifloraceae). - Vine. Trop. America. Cultivated in some parts of West Indies. Fruits edible.

Passiflora ligularis Juss. Sweet Grenadilla. (Passifloraceae). - Woody vine. Trop. America. Fruit of a pleasant sweet flavor; used as breakfast fruit. Cultivated. Sold in markets.

Passiflora maliformis L., Curuba. (Passifloraceae). - Vine. Trop. America. Cultivated. Fruits edible, consumed in some parts of tropical America.

Passiflora mollissima (H. B. K.) Bailey., Tasco, Caruba de Castilla. (Passifloraceae). - Vine. N. and Centr. Ecuador, Colombia. Cultivated, esp. near Ambato and Ibarra (Ecuad.), also in Colombia. Fruits eaten raw, also in refreshing drinks, ice-creams and sherbets. They are made into Crema de Caruba.

Passiflora pinnatistipula Cav. Tasco, Gutupa. (Passifloraceae). - Vine. Ecuador and Colombia. Fruit greenish-yellow, round, 5 cm. in diam., pulp sweet, rather insipid; consumed by the inhabitants.

Passiflora Popenovii Killip., Grenadille de Quijos. (Passifloraceae). - Vine. Ecuador. Fruit edible, of delicious fragrance. Sold by Indians in markets of Ecuador.

Passiflora psilantha (Sodiro) Killip., Gullan. (Passifloraceae). - Vine. Ecuador. Sometimes cultivated. Fruits resembling those of P. mollissima. Used for similar purposes.

Passiflora quadrangularis L., Giant Granadilla, Barbardine, Granadilla Real. (Passifloraceae). - Strong vine. Trop. America. Widely cultivated in tropics. Fruits 15 to 20 cm. long, oval, smooth-skinned, yellow; pulp very palatable, eaten when raw also with wine; made into drinks and as a sauce. Unripe fruits are eaten as vegetable.

Passiflora salvadorensis Don Smith. (Passifloraceae). - Vine. El Salvador. Herb is used locally in El Salvador as a diuretic.

Passiflora tetrandra Banks and Soland → Tetraspathaea australis Raoul.

Passiflora tripartita (Juss.) Poir., Tasco. (Passifloraceae). - Vine. Ecuador. Cultivated. Fruit oblong, 7.5 cm long, deep yellow with red; pulp juicy, similar in flavor of P. mollissima. Sold in markets of Ecuador.

Passifloraceae → Adenia, Passiflora, Tetraspathaea.

Passion Fruit → Passiflora edulis Sims.

Passionflower, New Zealand → Tetraspathaea australis Raoul.

Pasta Althaea → Althaea officinalis L.

Pasteli → Ceratomia Siliqua L.

Pastinaca Opopanax L. → Opopanax Chironium (L.) Koch.

Pastinaca sativa L. (syn. Peucedanum sativum S. Wats.). Parsnip. (Umbelliferaceae). - Biennial herb. Europe, Asia, N. and S. America, Australia, New Zealand. Grown commercially. Roots are eaten as a vegetable when boiled, esp. for winter-use; also in soup, with meat and as salad. Varieties are: Hollow Grown, Guernsey, Long Dutch. Roots are also used as fodder for live-stock.

Pasture Grasses. → Agropyron, Agrostis, Alopecurus, Amphilophis, Arrhenaterum, Astrebla, Axonopus, Bouteloua, Brachiaria, Chloris, Cynodon, Cynosurus, Dactylis, Dichelachne, Eragrostis, Festuca, Haemarthria, Holcus, Lolium, Panicum, Paspalum, Phalaris, Poa, Sorghum.

Patagonula bahiensis Moric., Guayabi, Ope branco. (Boraginaceae). - Tree. S. America, esp. S. Brazil, Paraguay, Uruguay, Argentina. Heartwood used for fine furniture, interior finish of buildings, bentwood chairs, turnery. Sapwood is employed for agricultural implements, vehicles, yokes. Used for bows by Indians of Missiones.

Patashiti → Theobroma bicolor Humb. and Bonpl.

Patchouly → Pogostemon Cablin Benth.

Paterno → Inga Paterno Harms.

Patience Dock → Rumex Patientia L.

Patonia parvifolia Wight → Xylopia parvifolia Hook. f. and Thom.

Patridge Wood → Andira excelsa H. B. K.

Patti → Cannabis sativa L.

Pau Amarillo → Euxylophora paraensis Huber.

Pau-hoi → Machilus pauhoi Kanch.

Paullinia asiatica L. → Toddalia aculeata Pers.

Paullinia Cupana Kunth. Guarana. (Sapindaceae). - Shrub. S. America. Indians prepare an alcoholic beverage from the seeds with Cassava and water. Crushed seeds are medicinally used for chronic diarrhea. Contains guaranine, an alkaloid. Cultivated among the Indians. Seeds are made into Guarana Paste, sold in two grades, known as Bom and Poca, used as a stimulating drink, contains 4.88 % caffein.

Paullinia pinnata L. (syn. Serjania curassavica Radlk.) Barbasco, Nistamal. (Sapindaceae). - Tree. Mexico, Centr. America, West Indies. Plant is used to stupefy fish. Juice is employed for poisoning arrows. Negroes in the Antilles use the seeds for criminal poisoning. Stems are made into ropes.

Paulownia imperialis Sieb. and Zucc. (syn. P. tomentosa Koch.) (Scrophulariaceae). - Tree. Centr. and W. China. Wood made into charcoal; used for gun-powder. Wood was employed in China for the production of lute. Used in Japan for boxes, clogs, sandals, musical instruments and furniture.

Pausinystalia Yohimba Pierre. (Rubiaceae). - Tree. Trop. Africa. Bark used in Cameroon as aphrodisiac, also used in vet. medic. Contains two alkaloids, e. g. yohimbine and yohimbinine.

Pavia rubra L. → Aesculus Pavia L.

Pavonia Bojeri Baker. (Malvaceae). - Shrub. Madagascar. Bark is source of a fibre; used in Madagascar for cloth.

Pavonia dasypetala Turcz. (Malvaceae). - Herbaceous plant. Costa Rica to Venezuela. Bark is source of a strong fibre; used by the Indians.

Pavonia hirsuta Guill and Perr. (Malvaceae). - Tree. Trop. Africa. Mucilaginous root is used by the natives, in milk, to cause rapid production of butter after shaking.

Pavonia Schimperiana Hochst. (Malvaceae). - Half-shrub. Trop. Africa. Bark is source of a fibre; used for cordage by the natives. Also P. urens Cav.

Pawpaw → Asimina triloba (L.) Dunal.

Payena Leerii Kurz. (Sapotaceae). - Tree. Burma. Mal. Archipelago. Source of a Gutta Percha, of good quality, having a high contents of resin.

Payena obscura Burck. (Sapotaceae). - Large tree. Mal. Archipelago. Source of a Gutta Percha.

Pea, Asparagus → Psophocarpus tetragonolobus DC.

Pea, Capucine → Pisum arvense L.

Pea, Chick → Cicer arietinum L.

Pea, Field → Pisum arvense L.

Pea, Flat → Lathyrus sylvestris L.

Pea, Garden → Pisum sativum L.

Pea, Grass → Lathyrus sativus L.

Pea, Pigeon → Cajanus indicus Spreng.

Pea, Rosary → Abrus precatorius L.

Pea, Sweet → Lathyrus odoratus L.

Pea Tree → Caragana arborescens Lam.

Peanut → Arachis hypogaea L.

Peanut, Ground → Amphicarpa monoica (L.) Ell.

Peanut, Hog → Amphicarpa monoica (L.) Ell.

Peanut Oil → Arachis hypogaea L.

Peashrub, Siberian → Caragana arborenscens Lam.

Peavine, Grass → Lathyrus sativus L.

Peavine, Groundnut → Lathyrus tuberosus L.

Peach, Common → Amygdalus Persica L.

Peach Bitter, African → Sarcocephalus esculentus Afzel.

Peach Kernel Oil → Amygdalus Persica L.

Peach Palm → Bactris Gasipaës H. B. K.

Peach, Peento → Amygdalus Persica L.

Peachwood → Caesalpinia echinata Lam.

Peacock Flower → Caesalpinia pulcherrima Swartz.

Pear, Alligator → Persea americana Mill.

Pear, Balsam → Momordica Charantia L.

Pear, Common → Pyrus communis L.

Pear, Hard → Strychnos Henningsii Gilg.

Pear, Odara → Chrysophyllum africanum A. DC.

Pear, Red → Scolopia Mundtii Harv.

Pear, Thorn → Scolopia Zeyheri Arn.

Pear, White → Apodytes dimidiata E. Mey.

Pearl Fruit → Margyricarpus setosus Ruiz. and Pav.

Pearl Millet → Setaria glauca Beauv.

Pearl Moss → Chondrus crispus (L.) Stackh.

Pecan → Carya Pecan (Marsh.) Engl. and Graebn.

Pectinobacter amylophylum Makimov. (Bacteriaceae). - Bacil. Microorganism. An organism showing superior retting qualities of flax and hemp fibres.

Pectis papposa Harv., Fetid Marigold. (Compositae). - Herb. S. W. of United States. Flowers are used by the Indians of New Mexico for flavoring meat.

Pedaliaceae → Ceratotheca, Craniolaria, Pedalium, Rogera, Sesamum.

Pedalium Murex L. (Pedaliaceae). - Herbaceous plant. Trop. Africa. Leaves are consumed as a vegetable by the natives.

Pedicularis Langsdorffi Fisch. (syn. P. lanata Willd.). (Scrophulariaceae). - Perennial herb. Arctic Canada. Roots may be used as food, recommended in some parts of N. Canada.

Pedilanthus Pavonis (Klotzsch and Garcke) Boiss. (Euphorbiaceae). - Shrub. Mexico. Has emetic, purgative and emmenagogue properties. Source of a Candelilla Wax; used in the same way as Carnauba Wax.

Peento Peach → Amygdalus Persica L.

Peepul Tree → Ficus religiosa L.

Peganum Harmala L. (Zygophyllaceae). - Woody plant. Mediterranean region and Centr. Asia. Used in some parts of Egypt as alterative, aphrodisiac and lactogogue. Seeds are said to be narcotic. Seeds are source of a Turkish Red. They produce an oil, called Zit-el-Harmel.

Pegu Catechu → Acacia Catechu Willd.

Pelargonium antidysentericum Kostel. (Geraniaceae). - Woody plant. S. Africa. Tuberous roots, boiled in milk are used for dysentery in S. Africa.

Pelargonium capitatum Willd. (Geraniaceae). - Shrub. S. Africa. Source of an ess. oil; used in perfumery.

Pelargonium odoratissimum Ait. (Geraniaceae). Shrub. Trop. Africa. Cultivated. Source of an ess. oil; used in perfumery. The following species are occasionally considered in relation to their production of ess. oil: P. fulgidum Ait. butyric scented; P. capitatum var. rose unique, fine rose odor; P. graveolens L'Hér. rose and rue scented; P. quercifolium Baum. ladanum scented; P. glutinosum L'Hér. ladanum scented; P. vitifolium L'Hér. citronellal scented; P. crispum L'Hér. var. maximum citral scented; var. minima strawberry scented; P. exstipulatum Willd. pennyroyal scented; P. fragrans Willd. faint rose and tansy scented.

Pelargonium Radula L'Hér. (syn. P. roseum Willd.) (Geraniaceae). - Shrub S. Africa. Cultivated. Source of an ess. oil, Geranium Oil, used in perfumery.

Pelea anisata Mann. (Rutaceae). - Tree. Hawaii Islds. Anise scented fruits were strung into lei or garlands, used by Hawaiians.

Pelea fatuhivensis F. Br. (Rutaceae). - Tree. Fatuhiva, Marquesas. Leaves used by the natives of Marquesas for perfuming coconut oil.

Pelecyphora aselliformis Ehrenb., Peyotillo (Mex.) (Cactaceae). - Low growing, xerophytic plant. San Luis Potosi (Mex.) Used locally for fevers.

Pellitory → Anacyclus Pyrethrum DC.

Peltandra virginica (L.) Kunth., Virginian Tukkahoe, Green Arrow Arum. (Araceae). - Perennial herb. In swamps. Eastern United States. Starchy roasted corms were important source of food among the Indians in Virginia. Also boiled spadix and fruits were consumed.

Peltigera aphthosa (L.) Hoffm. (Peltigeraceae). - Lichen. Temp. zone. On soil. Source of dextro-mannose and dextro-galactose.

Peltigera canina (L.) Willd., Dog Lichen. (Peltigeraceae). - Lichen. Temp. zone. On rocks, between grass. Used in Europe as a source of dye to color woolens in iron-red color.

Peltiphyllum peltatum Engl. → Saxifraga peltata Torr.

Peltogyne paniculata Benth., Purple Heart. (Leguminosaceae). - Tree. Brazil, Guiana. Wood dullbrown, becoming violet or intense purple upon exposure; black when soaked in water; contains iron; heavy, hard, tough, medium to fine texture, durable, not difficult to work. Used for interior finish of rooms. Also P. densiflora Spruce, P. paradoxa Ducke, P. catingae Ducke, P. excelsa Ducke, P. gracilipes Ducke, P. Lecointei Ducke, P. maranhensis Ducke.

Peltophorum adnatum Griseb. (Leguminosaceae). - Tree. West Indies. Wood purplish; used for general carpentry.

Peltophorum Linnaei Benth., Brazil Wood. (Leguminosaceae). - Tree. Trop. America. Wood orange; used for cabinet work, wheel spokes.

Peltophorum pterocarpum Backer. (syn. Caesalpinia arborea Zoll.) (Leguminosaceae). - Tree. S. E. Asia to N. Australia. Used as shade-tree in coffee plantations. Bark is source of a yellow dye, employed in Batik-work in Java.

Peltophorum Vogelianum Walp. Cañafistula. (Leguminosaceae). - Tree. N. E. Argentina, Paraguay. Wood pinkish to red brown, medium texture, durable; used locally for carpentry, general construction work, turnery, furniture, vehicles.

Pemphis acidula Forster. (Lythraceae). - Shrub or small tree. Tropics of Old World, esp. near seacoast. Wood hard, heavy, difficult to split, easily polished; used for handles of tools, nails,

construction of vessels. Employed by the natives of Polynesia for making clubs, spears and fish hooks.

Penaea Sarcocolla L. (Penaeaceae). - Small shrub. S. and Centr. Africa. Source of Sarcocolla, a gum substance allied to Tragacanth. being composed of yellowish-red, roundish, small grains, having a liquorice taste. Also P. mucronata. L.

Penaeaceae → Penaea.

Pencil Cedar → Dysoxylon Fraseranum Benth.

Penicillin → Penicillium notatum Westling.

Penicillium brevicaule Sacc. (Aspergillaceae). - Fungus. Microorganism. This mold has been of importance to demonstrate the presence of arsenic in various compounds. Has also been used in criminal cases. It is characterized by developing the typical scent due to arsenic. This biological method has been considered much more sensitive than the chemical method.

Penicillium camemberti Thom. (Aspergillaceae). - Fungus. A mold that takes part in the manufacturing process of Camembert Cheese.

Penicillium citrinum Thom. (Aspergillaceae). - Fungus. A mold that has been suggested for manufacturing of citric acid as a result of fermentation. Also P. luteum Thom.

Penicillium notatum Westling (Aspergillaceae). - Fungus. A mold that has been widely used as an antibiotic or antibacterial to combate diseases of humans and animals. Source of Penicillin. Antibiotics form chemical substances that show a germicidal activity toward a number of microorganism in vivo or in vitro. This mold is grown in pure cultures. The organism is used for the prevention and treatment of infections caused by Gonococcus, Pneumococcus and a number of other Gram-positive organisms. It is ineffective against Gram-negative bacillary infections.

Penicillium roqueforti Thom. (Aspergillaceae). - Fungus. A mold that is used in the manufacturing of Roquefort Cheese in S. France. It is made principally of milk of ewes that are bred for their high milk-producing ability. The mold is inoculated with a dried bread product containing the spores of the mold. It is ground to a powder which is used as inoculum. The mottled and marbled appearance of the cheese is caused by the mold.

Pennantia corymbosa Forster. (Olacaceae). - Tree. New Zealand. Wood very hard, compact, durable. Used for handles of tools, ornamental turnery, cabinet work. Was used by Maoris to obtain fire by friction.

Pennisetum compressum R. Br. (Graminaceae). - Perennial grass. Australia. Considered a good fodder for live-stock.

Pennisetum glaucum (L.) R. Br. → Setaria glauca Beauv.

Pennisetum purpureum Schum., Napier Grass, Elephant Grass. (Graminaceae). - Perennial grass. Trop. Africa. Grown in warm countries, as pasture grass for farm animals. Source of a paper producing material, a substitute for esparto.

Penny Cress → Thlaspi arvense L.

Penny Flower → Lunaria annua L.

Pennyroyal → Pycnothamnus rigidus (Bart.) Small.

Pennyroyal, American → Hedeoma pulegioides (L.) Pers.

Pennyroyal, European → Mentha Pulegium L.

Pentaclethra filamentosa Benth., Iripil Bark Tree. (Leguminosaceae). - Tree. Trop. America. Wood light to dark brown, heavy, hard, tough, strong, straight-grained; used for furniture, floor beams and house frames.

Pentaclethra macrophylla Benth., Owala Oil Tree. (Leguminosaceae). - Trop. Africa. Seeds are source of Owala Butter or Oil, suitable for manuf. candles and soap. Seeds are used as food and made into bread by the natives. Lotion of bark is used for sores. Wood is hard; employed for turnery, wheel wrights' work and general carpentry.

Pentadesma butyracea Sabine., Butter Tree, Tallow Tree. (Guttiferaceae). - Tree. Trop. Africa. Seeds are source of an edible fat; used by the natives for cooking; it is called Lamy, Kanga or Sierra Leone Butter. Also employed for manuf. soap, margarine and candles. Sp. Gr. 0.857—0.860; Sap. Val. 186—199; Iod. No. 42—47; Unsap. 0.9—1.70 %. Fruits are sometimes used as an adulterant of Cola Nuts, Cola acuminata.

Pentstemon grandiflorus Nutt., Large Flowered Beard Tongue. (Scrophulariaceae). - Perennial herb. Prairies United States. Decoction of roots used by the Kiowa Indians (New Mex.) for tooth ache.

Pentzia virgata Less. (syn. P. incana O. Kuntze.). (Compositae). - Small shrub. S. Africa. An important source of fodder for livestock in S. Africa.

Peony, Chinese → Paeonia albiflora Pal.

Peony, Coral → Paeonia corallina Retz.

Pepino → Solanum muricatum Ait.

Pepino de Comer → Cyclanthera edulis Naud.

Pepper → Piper nigrum L.

Pepper, African → Xylopia aethiopica A. Rich.

Pepper, Black → Piper nigrum L.

Pepper, Cayenne → Capsicum frutescens L.

Pepper, Green → Capsicum frutescens L.

Pepper, Guinea → Xylopia aethiopica A. Rich.

Pepper, Indian Long → Piper longum L.

Pepper, Jaborandi → Piper longum L.

Pepper, Javanese Long → Piper retrofractum Vahl.

Pepper, Long → Piper longum L.

Pepper, Negro → Xylopia aethipica A. Rich.

Pepper Tree, California → Schinus Molle L.

Pepper Tree, Chilean → Schinus latifolius Engl.

Pepper Tree, Peru → Schinus dependens Ortez.

Pepper, White → Piper nigrum L.

Peppermint → Mentha piperita L.

Peppermint, Bastard → Tristania suaveolens Smith.

Peppermint Oil → Mentha piperita L.

Peppermint Tree → Eucalyptus amygdalina Lab.

Pepper Tree → Schinus Molle L.

Pepperweed, Desert → Lepidium Fremontii S. Wats.

Pepperidge → Nyssa multiflora Wang.

Peperomia leptostachya Hook. and Arn. (Piperaceae). - Succulent herb. Tropics. Juice of leaves used by natives of Pacif. Islds. for skin diseases, infection of eyes and for burns.

Peperomia vividispica Trel. (Piperaceae). - Perennial herb. Epiphyte. Centr. and S. America. Young leaves and stems ar consumed as salads by the natives.

Pera do Campo → Eugenia Klotzschiana Berg.

Perennial Lettuce → Lactuca perennis L.

Pereskia aculeata Mill. (syn. P. Pereskia (L.) Karst.) West Indian Gooseberry, Barbados Gooseberry, Lemon Vine. (Cactaceae). - Woody vine. Mexico. Cultivated. Fruit eaten raw or preserved. Leaves can be used as pot-herb.

Pereskia Bleo (H. B. K.) DC. (Cactaceae). - Shrub-like vine. S. America. Leaves are used as a vegetable in some parts of Colombia.

Pereskiopsis Porteri (Brandeg.) Britt. and Rose. (Cactaceae). - Xerophytic shrub. Mexico. Fruits edible, very acid.

Perezia multiflora Less. (Compositae). - Herbaceous plant. Argentina to Brazil. An infusion of the leaves is used in S. America as sudorific.

Perfumed Bdellium → Commiphora erythraea Engl. var. glabrescens Engl.

Perfumes. Lichen source of → Alectoria, Anaptychia, Cladonia, Evernia, Lobaria, Parmelia, Ramalina.

Perfumes, Plants source of → Abies, Acacia, Aglaia, Alectryon, Amburana, Amyris, Andropogon, Angelica, Aniba, Anisomeles, Annona, Anthemis, Aquilaria, Artemisia, Asperula, Backhousia, Baeckea, Boronia, Brachylaena, Bulnesia, Bursera, Canangium, Cedrus, Cheiranthus, Chimonanthus, Citrus, Clausenia, Combretum, Commiphora, Convallaria, Copaifera, Coriandrum, Corrigiola, Cuminum, Cupressus, Cydonia, Cymbopogon, Cyprus, Daniella, Daphne, Darwinia, Daucus, Dianthus, Dipteryx, Elettaria, Erythroxylon, Eucalyptus, Ferula, Fusanus, Gardenia, Gaultheria, Hedychium, Hibiscus, Hierochloë, Homoranthus, Hoslundia, Humulus, Hyacinthus, Hyssopus, Iris, Irvingia, Jasminum, Lavandula, Leptactina, Lippia, Liquidambar, Manilkara, Matricaria, Melissa, Mentha, Mesua, Myroxylon, Myrtus, Narcissus, Nardostachys, Ocimum, Opilia, Opopanax, Origanum, Panax, Pandanus, Pelargonium, Peucedanum, Philadelphus, Pogostemon, Polianthes, Reseda, Rhinacanthus, Rosa, Rosmarinus, Salvia, Santalum, Sassafras, Satureja, Saussurea, Sicana, Sindora, Spartium, Terminalia, Thuja, Unona, Valeriana, Vanilla, Vetiveria, Viola, Zingiber.

Perguetano → Moquilea utilis Hook.

Pergularia africana N. E. Br. (Asclepiadaceae). - Vine. Trop. Africa, esp. S. Leone to Old Calabar into Mozambique and Natal. Extract from plant is used to adulterate „Dragon's Blood" of commerce.

Pergularia minor Anders. → Telosma cordata Merr.

Periandra dulcis Mart. (Leguminosaceae). - Shrub. Brazil. Roots, Raiz Doce, are used as sustitute for Spanish and Russian Liquorice., containing 4.25 to 7.5 % glycyrhizin.

Pericopsis Mooniana Thw. (Leguminosaceae). Large tree. Ceylon. Wood valuable, brown, with dark shades. Used for furniture.

Perilla arguta Benth. (Labiaceae). - Annual herb. China, Japan. Cultivated. Young seeds are eaten in Japan when boiled or raw. Leaves and flower clusters are used as a condiment or eaten when salted. Leaves give a purplish red tint to the salted fruits of Prunus mume. Cotyledons of seedlings are used as a condiment. Var. Aoso or Shiroso has a strong flavor; is used as spice or preserved, by drying in salt.

Perilla frutescens (L.) Britt. (syn. P. ocimoides L.). (Labiaceae). - Herbaceous. India. Cultivated in India, Korea, Japan. Seeds source of Perilla Oil, a drying oil, resembling Linseed Oil. When dried, it is a hard, tough, brilliant film; used for water-proofing paper, manuf. cheaper lacquer varnishes; in printing ink, painting. In USA it is mixed with soy-bean oil for protective coatings. Sp. Gr. 0.930 - 937; Sap. Val. 188 - 197; Iod. No. 185—205; Unsap. 0.6—1.3 %. Leaves are used for flavoring dishes in India and China.

Perilla Oil → Perilla frutescens (L). Britt.

Periploca canescens Afz. (Asclepiadaceae). - Woody plant. Ivory Coast, Congo. Source of a black rubber of medium quality. Latex coagulates by lemon-juice and heating. Supposed to contain 88 % rubber.

Periploca graeca L. Grecian Silk Vine. (Asclepiadaceae). - Shrubby vine. S. Europe, W. Asia. Bark is source of periplocin, similar to digitalin.

Peristrophe bivalvis (L.) Merr. (Acanthaceae). - Shrub. Trop. Asia. Leaves and young stems are source of a yellow-orange to deep red-orange dye.

Pernambuco Cotton → Gossypium brasiliense Macf.

Pernambuco Jaborandi → Pilocarpus Jaborandi Holmes.

Pernambuco Wood → Caesalpinia echinata Lam.

Peroba Rosa → Aspidosperma polyneuron Muell. Arg.

Perobinha → Sweetia elegans Benth.

Perrottetia sandwicensis Gray. (Celastrinaceae). - Small tree. Hawaii Islds. Wood used by early Hawaiians to produce fire by friction.

Persea americana Mill. (syn. P. gratissima Gaertn.). Avocado Tree, Alligator Pear. (Lauraceae). - Tree. Mexico. Central America. Grown commercially in the tropics and the subtropics of the Old and the New World. Fruits known as Avocados or Alligator Pears, a salad fruit. Several commercial varieties among which I. Guatemalan Group: Taft, Taylor, Wagner, Linda. II. West Indian Group: Simmonds, Pollock, Trapp. Persea americana Mill. x P. drymifolia Cham. and Schlecht, a hybrid group contains the following commercial varieties: Fuerte and Lula. Seeds are source of Avocado Oil, a non-drying oil, being of mild and pleasant flavor. Has been recommended in cosmetics and is used in Hawaii as salad dressing. Sp. Gr. 0.9132; Sap. Val. 192.6; Iod. No. 94.4; Unsap. 1.6 %.

Persea Borbonia (L.) Raf., Red-Bay, Sweet-Bay. (Lauraceae). - S. and S. E. of United States, from Florida to Texas. Wood hard, heavy, very strong, close-grained, brittle, bright reddish-brown; sometimes used for interior finish of buildings, cabinet making; formerly used for ship and boat-building. Dried leaves are sometimes used as condiment for flavoring soup, especially crab-gumbo and stuffing roastfowl; resembling Laurus nobilis L.

Persea drymifolia Cham and Schlecht., Mexican Avocado. (Lauraceae). - Tree. Mexico. Plants are hardier than P. americana Mill. Fruits are smaller than those of the ordinary Avocado. Varieties are Northrop and Puebla.

Persea Lingue Nees. (Lauraceae). - Tree. S. America, Chile. Bark is used for tanning, called Cascara de Lingue.

Persea Meyeniana Nees. (Lauraceae). - Tree. S. America, Chile. Bark is used in Chile for tanning, source of Valdivia Leather.

Persea Schiedeana Nees. Coyo Avocado. (Lauraceae). - Tree. S. Mexico to Guatemala. Fruits are consumed in salads. Sold in markets of Centr. America.

Persianberry → Rhamnus infectorius L.

Persian Clover → Trifolium resupianatum L.

Persian Gum → Amygdalus leiocarpus Boiss.

Persian Insect Powder → Chrysanthemum coccineum Willd.

Persian Tragacanth → Astragalus pycnocladus Boiss. and Hassk.

Persimmon, Black → Diospyrus texana Scheele.

Persimmon, Common → Diospyros virginiana L.

Persimmon, Japanese → Diospyros Kaki L. f.

Persio → Roccella tinctoria DC.

Pertusaria corallina (L.) Th. Fr., White Crottle. (Pertusariaceae). - Lichen. Temp. zone. On rocks. Used in Scotland to give woolens a red-purple color.

Pertusaria pseudocorallina (Sw.) Arn. (Pertusariaceae). - Lichen. Temp., subarctic zones. Used in Norway and Sweden to give a red-purple color to woolens.

Peru Balsam → Myroxylon Pereirae Klotzsch.

Peru Balsam Tree → Myroxylon Pereirae Klotzsch.

Peru Peppertree → Schinus dependens Ortez.

Peruvian Bark → Cinchona officinalis L.

Peruvian Cotton Gossypium peruvianum Cav.

Peruvian Krameria → Krameria triandra Mart.

Peruvian Ratanhia → Krameria triandra Ruiz and Pav.

Petalostemon oligophyllum (Torr.) Rydb., Slender Prairie Clover. (Leguminosaceae). - Herbaceous perennial. Prairies of Western United States to Mexico. Roots eaten by the Kiowa Indians of New Mexico.

Petalostemon purpureum (Vent.) Purple Prairie Clover. Rydb. (Leguminosaceae). - Perennial herb. Wester N. and adj. Mexico. Leaves were source of tea. Roots were used for chewing by Indians of Missouri River regions.

Petasites frigidus (L.) Fries. (Compositae). - Perennial herb. Western N. America. Leaves used as greens by Eskimo in Alaska.

Petasites japonicus F. Schmidt. (Compositae). - Large perennial herb. Sachalin, Japan. Leaf-stalks when boiled in spring and summer are consumed as a vegetable in Japan, also when preserved in salt. Slightly bitter flower-buds are eaten when boiled; they are also used as a condiment.

Petasites palmata Gray. Palmate Butterbur, Sweet Coltsfoot. (Compositae). - Perennial herb. N. America. Ash derived from the plant used as a source of salt among some Indian tribes.

Petasites speciosa (Nutt.) Piper. (Compositae). - Perennial herb. British Columbia to California. Ashes from the plant furnished salt that was used by the Indians of the Pacific Coast regions.

Petitia Poeppigii Scheuer. (Verbenaceae). - Tree. West Indies. Wood strong, palisander-colored; used for navy construction.

Petiveria alliacea L., Gully-Root. (Phytolaccaceae). - Shrub. West Indies, Mexico to Centr. and S. America. Plant has a disagreeable skunk-like scent. Roots are placed between woolen

goods in West Indies as protection from insects. They are considered diuretic, expectorant, antispasmodic, sudorific, vermifuge, abortifacient, emmenagogue and used for nervous diseases.

Petroselinum hortense Hoffm. (syn. Petroselinum sativum Hoffm., Apium Petroselinum L.). Parsley. (Umbelliferaceae). - Biennial herb. S. Europe, Asia Minor. Widely cultivated as a vegetable in the Old and New World. Leaves used for flavoring soups, vegetables, meats, salads. Varieties are divided into those with plain-leaves and curled-leaves. Var. foliosum (Alef.) Thell.: Extra Double Curled, Moss Curled and Curled Dwarf. Turnip Rooted Pasley produces large roots, eaten when cooked. Var. tuberosum (Bernh.) Thell.: Hamburg. Dried fruits are used medicinally as stimulant, carminative, diuretic. Oil of Parsley is an ess. oil, mainly derived from the leaves; used for flavoring food products, sausages, canned soups, meats, vegetables, culinary sauces. Contains apiin, apiolin.

Pe-tsai → Brassica pekinensis Rupr.

Peucedanum araliaceum Benth. and Hook. f. (Umbelliferaceae). - Shrubby plant. Trop. Africa. Leaves are used by the natives for scenting clothes.

Peucedanum fraxinifolium Hiern. (Umbelliforaceae). - Shrubby plant. Trop. Africa. Leaves have scent of fennel; used for making lotions; infusion used as diuretic, depurative. In Sierra Leone leaves are used as vermifuge.

Peucedanum Galbanum Benth. (syn. Bubon Galbanum L.). (Umbelliferaceae). - Perennial herb. S. Africa. Decoction of the leaves is used among the farmers and Kaffirs of S. Africa as diuretic.

Peucedanum graveolens Benth. and Hook. (syn. P. sowa Kurz.). East Indian Dill. (Umbelliferaceae). - Herbaceous plant. Trop. Asia. Fruits are used for flavoring, source of apiol. Dried fruits are used medicinally.

Peucedanum officinale L. (syn. Selinum officinale Vest., P. altissimum Desf.) Hog's Fennel. (Umbelliferaceae). - Perennial herb. Europe, Asia. Root is used as sudorific, diuretic, emmenagogue, antiscorbutic. A resin, Gummi Peucedani is said to be similar to Gummi Ammoniacum.

Peucedanum Ostruthium (L.) Koch. (syn. Imperatoria Ostruthium L., Selenium Ostruthium Wallr.). Masterwort Hogfennel. (Umbelliferaceae). - Perennial herb. Europe, Asia. A gin is produced from the plant, used as stomachic. Is also used in certain herb-cheeses. Was formerly employed medicinally for chronic bronchial catarrh.

Peucedanum palustre (L.) Moench. (syn. Selenium palustre L., S. sylvestre L.). Milk Parsly. (Umbelliferaceae). - Biennial herb. Europe, Asia. In some Slavic countries the roots are used as a substitute for ginger. Was formerly used mecinally.

Peucedanum sativum S. Wats → Pastinaca sativa L.

Peumus Boldus Molina. (syn. Boldu Boldus (Mol.) Lyons., Boldea fragrans Gay.). (Monimiaceae). - Tree. Chile, Peru. Dried leaves are used medicinally as a mild diuretic, aromatic stimulant. Was formerly used in hepatic ailments. Contains an ess. oil; also boldine an alkaloid.

Peyote → Lophophora Williamsii (Lem.) Coulter.

Peyotillo → Pelecyphora aselliformis Ehrenb.

Phacelia tanacetifolia Benth. Tansy Phacelia. (Hydrophyllaceae). - Perennial herb. Western United States. Cultivated in some countries as a honey producing crop.

Phaeomeria hemisphaerica (Blume) Schum. (Zingiberaceae). - Perennial herb. Mal. Archipelago. Stems and leaves have been recommended for manuf. paper.

Phaeomeria magnifica Schum. (syn. Alpinia magnifica Rosc., Amomum magnifucum Benth., Alpinia speciosa D. Dietr.) (Zingiberaceae). - Perennial herb. Malaya. Cultivated. Used for seasoning food. Young flowering shoots are used in curries.

Phalaris arundinacea L., Red Canary Grass. (Graminaceae). - Perennial grass. Temp. Europe, Asia. Cultivated in Old and New World as pasture grass and for hay.

Phalaris canariensis L., Canary Grass. (Graminaceae). - Annual grass. Mediterranean region. Cultivated in Old and New World. Used as bird seed of commercial importance. Also used for human consumption.

Phallaceae → Dictyophora.

Phaseolus aconitifolius Jacq. Moth Bean. (Leguminosaceae). - Annual herb. Trop. Asia. Cultivated in India. Beans used as food, also as forage for livestock.

Phaseolus acutifolius Gray. var. latifolius Freem., Tepary Bean. (Leguminosaceae). - Annual vine. Arizona and adj. Mexico. Cultivated by different Indian tribes. Very drought resistant. Beans used as food, prepared in different ways. Many varieties and strains.

Phaseolus adenanthus E. Mey. (syn. P. rostratus Wall.) (Leguminosaceae). - Herbaceous vine. Tropics. Tuberous roots are eaten by Hindus.

Phaseolus angularis Wight., Adzuki Bean. (Leguminosaceae). - Annual herb. Cultivated for centuries in Japan and Korea, also in China and Manchuria. Beans are eaten boiled. Also made into a paste. Source of a bean meal, used for cakes and confectionary. Numerous varieties.

Phaseolus calcaratus Roxb., Rice Bean. (Leguminosaceae). - Annual herb. Trop. Asia. Cultivated in Japan, China, Korea, India, Mauritius, Java, Philipp. Islds. Seeds used as a food in many countries of Asia, eaten boiled, in soups

and in rice. Recommended as forage for animals and as a cover crop.

Phaseolus coccineus L. (syn. P. multiflorus Willd.). Scarlet Runner. (Leguminosaceae). - Annual to perennial vine. Centr. America. Cultivated. Young fruits eaten boiled. Several climbing and dwarf varieties.

Phaseolus diversifolius Pers. (Leguminosaceae). - Herbaceous plant. Southern United States. Boiled or mashed roots were eaten by Indians in Louisiana.

Phaseolus lunatus L. (syn. P. limensis Macf.). Lima Bean. (Leguminosaceae). - Annual to perennial herb. Trop. America. Cultivated. Beans consumed as food. Varieties are divided into Sieva and Lima groups. I. Lima Group. Pole varieties: Florida Butter Speckled, Large White Lime. Dwarf varieties: Fordhook Bush Lima, Burpee's Large Bush Lima. II. Sieva Group: Dwarf Sieva, Dwarf Carolina. Beans are also important in the canning industry.

Phaseolus Mungo L. (syn. P. aureus Roxb.) Mung Bean, Golden Gram, Black Gram, Urd. (Leguminosaceae). - Annual herb. Trop. Asia. Cultivated in S. India, Iran, Malaya, E. Africa, Greece. Grown in India since ancient times. Seeds are used as food. Straw is used as food for domestic animals. Also grown as a cover crops and green manure.

Phaseolus polystachyus (L.) B. S. P., Bean Vine, Wild Bean. (Leguminosaceae). - Herbaceous vine. Eastern and southern N. America. Seeds edible, consumed cooked, after being dried. Highly prized by North American Indians.

Phaseolus retusus Benth. (syn. P. Metcalfei Woot. and Standl.). Metcalf Bean. (Leguminosaceae). - Herbaceous plant. New Mexico, Arizona and adj. Mexico. Occasionally cultivated, Drought resistant. Recommended as a forage crop for dry regions.

Phaseolus vulgaris L. Common Bean, Kidney Bean, Haricot. (Leguminosaceae). - Annual plant. Introduced from S. America. Cultivated since the times of the Incas. Grown as a vegetable in the Old and New World. The whole plant is often used as forage for live-stock. Pole Beans develop into vines and need support; Bush or Dwarf Beans are low growers. There are in the different groups numerous varieties, adaptable to various regions, countries, market requirements, freezing, resistance to diseases etc. The many varieties are grown for their pods (snap beans); beans used in green-shelled condition (green-shell beans) and dry-shell beans or ripe seeds which are used in dry condition. Podded varieties are classified according to color of the fruits in green and yellow or wax podded. The varieties are divided in I. Dwarf-green podded e. g. Black Valentine, Bountiful, Burpee Stringless, Fordhook Favorite. II. Dwarf or Bush Wax podded varieties e. g. Black Wax Pencil Pod, Brittle Wax, Golden Wax. The varieties have either round or flat pods; round podded varieties are most desirable for home canning and freezing. III. Pole Green podded e. g. Kentucky Wonder, Horticultural. Kentucky Wonder is well adapted for canning purposes. IV. Pole Wax podded e. g. Golden Cluster and Kentucky Wonder Wax. Among the Snap Beans the strings and tough fibre of the pod has been bred out. As a result they break or snap easily. Field Beans are those grown for use in a dry state and include I. Kidney Beans, seeds more or less reniform to which belong White and Red Kidney, White Imperial. II. Marrow, thickness of the beans exceeding half the length, e. g. Yellow Eye, White and Red Marrow. III. Medium, thickness of seeds less than half the length, e. g. White Wonder, Burlingame Medium. IV. Pea, Navy Beans, seeds not reniform, 8 cm. or less in length, among which Boston Small Pea, Navy Bean, Snow Flake, Michigan Robust. Dried beans are much used in winter. Red Kidney and related varieties are in great demand either baked, in soups or in certain Spanish and Mexican dishes. Field Beans, Frijoles are a very important food among the peoples of many S. American countries.

Pheasant's Eye → Adonis aestivalis L.

Phellodendron amurense Rupr. Amur Cork Tree. (Rutaceae). - Tree. Japan, China. Bark of tree is used among the Ainu for skin diseases, especially for mizu-mushi. Berries used as expectorant.

Phellodendron sachalinense Sarg. Sachalin Cork Tree. (Rutaceae). - Tree. Japan, Korea, W. China. Wood heavy, hard, strong, close-grained, reddish-brown. Used in Japan for interior finish of houses, utensils, furniture, rail-road ties. Outer corky bark used by fishers for buoys of fishing-nets. Yellow bark source of a dye.

Phellopterus littoralis Benth. (Umbelliferaceae). - Herbaceous plant. Sachalin, Japan, China. Cultivated. Used as a condiment, resembling Angelica and Tarragon.

Phellopterus montanus Nutt. (Umbelliferaceae). - Herbaceous plant. New Mexico, Texas. Roots were baked, after being peeled, ground into meal and eaten by Indians in New Mexico.

Philadelphus mexicanus Schlecht. Mexican Mock Orange. (Saxifragaceae). - Shrub. Mexico. Scented water derived from destilled flowering branches was used in pre-Columbian times.

Philippine Gutta Percha → Palaquium Aherninanum Merr.

Phelypaea violacea Desf. (Orobanchaceae). - Herbaceous plant. N. Africa, Algerian Sahara. Roots are used as food by the Tuareg (N. Afr.), during times of distress. Also P. lutea Desf.

Phillyrea latifolia L. Tree. Phillyrea. (Oleaceae). - Tree. Mediterranean region. Wood fine-grained; used for turnery and for saddles in some parts of Greece. Also P. media L. Leaves are diuretic, emmenagogue and used in mouthwaters.

Philodendron Imbe Schott. (Araceae). - Perennial herb. Brazil. Aereal roots used for manuf. of rope and fibre.

Philodendron pertusum Kunth. and Bouché → Monstera deliciosa Liebm.

Philodendron radiatum Schott. (Araceae). - Perennial herb. Centr. America. Decoction of leaves used in baths for rheumatic pains and rikkets, in El Salvador. Also P. Warsewiczii.

Philodendron sagittifolium Liebm. (Araceae). - Herbaceous plant. S. Mexico. Aereal roots used for basketry by Chinantecs (Mex.). Also P. seguine Schott. and P. radiatum Schott.

Philodendron Selloum Koch. (Araceae). - Shrub. Brazil. Roots are drastic. Fruits are edible; used in compote.

Phleum alpinum L., Alpine Timothy. (Graminaceae). - Perennial grass. Alpine regions of Europe, Asia and N. America. Excellent forage grass in mountainous areas.

Phleum pratense L., Timothy Grass. (Graminaceae). - Perennial grass. Temp. Europe, Asia, N. America. Much cultivated as a forage crop, also for hay production.

Phoberos Mundtii Presl. → Scolopia Mundtii Harv.

Phoberos Zeyheri Presl. → Scolopia Zeyheri Arn.

Phoenix dactylifera L., Date Palm. (Palmaceae). - Tall palm. Probably from Arabia to India. Cultivated since time immemorial. Grown in tropical semi-arid desert regions of the Old and New World. Important staple food and dessert fruit. Berries are used in jams, cooking and alcoholic beverages. Fruits may be classified in (1) dry, containing high percentage of sugar; they are sun dried and keep indefinitely; (2) semi-dry are dried, packed loosely in boxes, they keep without fermentation and are much exported; (3) soft, contain a small amount of sugar, are eaten in fresh, ripe state. Leaves and the base of the leaves are source of fibre, used for ropes, baskets, cordage. Fibre from base of leaves mixed with camel hair is made into cloth; used for caravan tents. Blanched leaves are used in churches during Palm Sunday. Wood is employed for huts and houses. Varieties are I. Group. Dry sweet varieties: Deglet Noor, Amir Hadj, Braim, Dayri, Fara, Khalasa, Medjool, Daidy, Thoory. II. Group. Dry boiling varieties: Zehedi, Mubsli, Murdasing.

Phoenix pusilla Gaertn. (syn. P. zevlanica Trin.) Wild Date. (Palmaceae). - Palm. Damp, low areas. E. India. Ceylon. Leaves made into ornamental Kalutara baskets, mats and pouches.

Pholidocarpus Kingiana Ridl. (Palmaceae). - Tall palm. Trop. Asia, esp. Malay Penin. Midrib of leaves used for blow-pipe darts by Pagan races.

Pholiota cylindrica D. C. (syn. Agrocybe aegerita (Bruganti) Kuhner. (Agaricaceae). - Basidiomycete. Fungus. Mediterranean area. Fruitbodies are consumed as food. Occasionally cultivated. Sold in markets.

Pholiota marginata (Batsch) Quél. (syn. Agaricus marginatus Batsch.) (Agaricaceae). - Basidiomycete. Fungus. N. America. Fruitbodies are edible.

Pholiota mutabilis (Schäff.) Quél. (Agaricaceae). - Basidiomycete. Fungus. Fruitbodies are consumed as food. Sold in markets, esp. in Bavaria.

Pholiota nameko (T. Ito) S. Ito and Imai. (Agaricaceae). - Basidiomycete. Fungus. Fruitbodies are consumed as food in Japan. Sold in markets. Also P. terrestris Overholts and P. squarrosoides (Peck) Sacc.

Pholiota squanosa (Muell.) Karst. (Agaricaceae). - Basidiomycete. Fungus. N. America. The mushrooms are edible and can be eaten raw or cooked. They are of good quality.

Phormium tenax Forst. New Zealand Flax. (Liliaceae). - Perennial tall herb. New Zealand. Cultivated. Used for centuries by the Maories as fibre for clothing, cordage, string, fine white cloth, fishing nets. Also of commercial importance; fibre being strong is used for sheeting, toweling, sacking, table cloths. Ropes are said not to be of great strength.

Phoradendron californicum Nutt. Mesquite Mistletoe. (Loranthaceae). - Parasitic shrub on Acacia, Prosopis and Cercidium. Southwest of United States. Fruits are dried by the Papago Indians, stored and used as food.

Phoradendron juniperinum Engelm., Juniper Mistletoe. (Loranthaceae). - Shrub parasitic on Juniperus. SW of United States and Mexico. Seeds used by Hopi Indians as a substitute for coffee.

Phoradendron flavescens (Pursh) Nutt. American Christmas Mistletoe. (Loranthaceae). - Parasitic shrub. N. America. Plant is oxytoxic being efficacious in arresting post-partum hemorrhages; used by the Indians.

Photinia arbutifolia Lindl. Christmas Berry, California Holly. (Rosaceae). - Shrub or small tree. California to Mexico. Berries when raw or roasted were eaten by the Indians.

Photinia japonica Thunb. → Eriobotrya japonica Lindl.

Phragmites communis Trin., Reed Grass. (Graminaceae). - Perennial grass. Cosmopolit. Stems are made into mats, much used among florists for covering greenhouses and frames in winter to protect tender plants. Also used for covering roofs. Rootstocks are eaten by the Indians. A sweetish substance exudes which hardens into a gum as the result of punctures caused by insects which was consumed by the Indians. Indians in S. California collect plants, dry the rhizomes which are grinded into flour, which so-

metimes contains much sugar. It is heated over a fire and eaten as taffy.

Phrynium confertum (Benth. Schum. (Marantaceae). - Perennial herb. Trop. Africa. The stems are used among the natives for making mats anu basketry.

Phthirusa Theobromae (Willd.) Eichl. (Loranthaceae). - Parasitic shrub. Trop. America. Has been suggested as a source of rubber. Also P. pyrifolia (H. B. K.) Eichl.

Phulwara Butter → Madhuca butyracea Gm.

Phyllanthus corcovadensis Muell. Arg. (Euphorbiaceae). - Tree. Brazil. Root is used in some parts of Brazil for stones in the bladder. Employed in patent medicines. Also P. Niruri L. and P. acutifolius Mart.

Phyllanthus distichus (L.) Muell. Arg. (syn. P. acidus Skeels). Otaheite Gooseberry. (Euphorbiaceae). - Tree. Trop. India, Malaya. Cultivated in tne tropics. Fruits 2 to 3 cm. in diam., green, roundish; used in pickles and preserves.

Phyllanthus Emblica L., Emblic, Myrobolan. (Euphorbiaceae). - Shrub or small tree. Trop. Asia, India. Cultivated in tropics of Old and New World. Fruits very acid, occasionally eaten raw, usually consumed when preserved; source or a jelly. Bark was exported for its tannin.

Phyllanthus Llanosi Muell. Arg. → Glochidion Llanosi Muell. Arg.

Phyllanthus reticulatus Poir. (Euphorbiaceae). - Shrub. Tropics. Roots are source of a red dye. Leaves having cooling and diuretic properties.

Pnyllanthus Urinaria L. (Euphorbiaceae). - Herbaceous. Tropics. Used as diuretic. Likewise P. Niruri L.

Phyllitis Fascia Ktz. (Scytosiphonaceae). - Brown Alga. Sea Weed. Pac. Ocean and adj. waters. Plants are consumed as food among the farmers of the provinces Awa and Sagami, Japan. Sea weeds are dried in the sun.

Phyllocalyx edulis Berg. → Eugenia Selloi Berg.

Phyllocladus trichomanoides D. Don. (Podocarpaceae). - Tree. New Zealand. Bark astringent, contains tannic acid; also a red dye, used in preparation of kid for gloves. Wood very strong, dense, heavy, whitish. Used for mining wood, rail-road ties, marine piles.

Phyllophora rubens (L.) Grev. (Phyllophoraceae). - Red Alga. Baltic and Black Sea. Used for manuf. of agaroid in Southern Russia, esp. Odessa. Also source of iodine.

Phyllophoraceae (Red Algae) → Ahnfeltia, Gymnogongrus, Phyllophora.

Phyllostachys bambusoides Sieb. and Zucc. Giant Timber Bamboo. (Graminaceae). - Tall woody grass. China, Japan. Cultivated. Thick, woody stems are used for furniture, bridges, houses etc. Young buds are consumed as food in some parts of Japan and China.

Phyllostachys edulis (Rin.) Hairy Sheeth Edible Bamboo. (Graminaceae). - Tall woody, perennial grass. China, Japan. Cultivated in China and Japan. Very young buds are consumed in various dishes, soups etc. Often sold in markets.

Phyllostachys heteroclada Steud. (Graminaceae). - China, India. Woody grass bamboo. Source of a good paper; used for printing and writing also for papering windows.

Phyllostachys nigra Munro., Black Bamboo. (Graminaceae). - Woody grass. China, Japan. Young buds are consumed as vegetable in China and Japan.

Phyllostylon brasiliensis Cap. San Domingo Boxwood, West Indian Boxwood (U. S. trade name), Bois Blanc, Sabonero. (Ulmaceae). - Tree. Trop. America, West Indies to Argentina. Wood lemonyellow, fine texture, straight grained, takes high polish; used for knife scales; is stained black or „ebonized".

Physalis Alkekengi L., Alkekengi, Strawberry Tomato, Winter Cherry. (Solanaceae). - Perennial herb. Centr. and S. Europe, Balkan, Ural. Cultivated in Old and New World. Berries size of a cherry, occasionally eaten; care must be taken not to consume the surrounding calyx.

Physalis heterophylla Nees., Groundcherry. (Solanaceae). - Annual herb. N. America. Fruits were eaten raw or made into a sauce by Indians of different tribes. Also P. Fendleri Gray, P. lanceolata Michx., P. neomexicana Rydb., P. virginiana Mill., P. viscosa L. and others.

Physalis minima L., Sunberry. (Solanaceae). - Herbaceous plant. Tropics. Fruits are consumed as vegetable in different parts of the tropics.

Physalis pubescens L., Strawberry-Tomato, Downy Ground-Cherry, Dwarf Cape-Gooseberry. (Solanaceae). - Low annual herb. North America, West Indies, Mexico, Centr. and S. America, Europe, Asia. Berries are used in sauce and preserves. Occasionally cultivated.

Physcia pulverulenta (Schreb.) Nyl., Mealy Blister Lichen. (Physciaceae). - Lichen. Temp. zone. On stems. Source of a dye; used in Europe to give a yellow color to woolens.

Physic Nut → Caesalpinia Bonducella Flem. and Jatropha Curcas L.

Physocalymma scaberimum Pohl. (syn. P. floribundum Pohl., P. floridum Pohl). Brazilian Rosewood, Cego Maschado, Tulip Wood. (Lythraceae). - Tree. Peru, Brazil. Wood red with dark colored zones; hard, very dense, heavy, difficult to cut, easy to split and to polish; used for fancy articles, turnery, fine furniture.

Physostigma venenosum Balf. (Leguminosaceae). - Woody climber. Trop. Africa. Introd. in India and Brazil. Source of the Calabar Bean or Ordeal Bean. Dried, ripe seed used medicinally, made into physostigmine salicylate being myotic; stimulates glandular secretions and peristalsis. Also P. cylindrospermum.

Phytelephas macrocarpa Ruiz an Pav., Common Ivory Palm, Tagua Palm. (Palmaceae). - Tall palm. Trop. America. The very hard seeds are source of a vegetable ivory; used for manuf. buttons, inlays, chessman, knobs and similar material.

Phytelephas Seemanni Cook. Seemann Ivory Palm. (Palmaceae). - Tall palm. Centr. America. Very hard seeds are a source of a vegetable ivory; used for manuf. buttons and similar articles.

Phytolacca abyssinica Hoffm. (Phytolaccaceae). - Herbaceous perennial. Trop. E. Africa. Berries are source of a red dye. Young stems and leaves are eaten cooked in some parts of Africa.

Phytolacca acinosa Roxb. (Phytolaccaceae). - Perennial herb. Trop. Asia, China, Japan. Cultivated in some parts of India. Boiled leaves are consumed as a vegetable. Berries are used in Chinese pharmacy.

Phytolacca decandra L. (syn. Phytolacca americana L.) Common Pokeberry. (Phytolaccaceae). - Perennial herb. N. America. Berries are sometimes used as a source of red ink, also used for coloring wine and sugar confectionary. Dried root is used medicinally, being emetic, alterative, purgative. Contains a bitter saponin-like glucoside and an alkaloid, phytolaccine. Young shoots are sometimes consumed as a substitute for asparagus, or cooked as greens.

Phytolacca rivinoides Kunth and Bouché. Venezuela Pokeberry. (Phytolaccaceae). - Perennial herb. Centr. and S. America. Roots are sometimes used as a substitute for soap. Cooked leaves and shoots are eaten as vegetable in some parts of trop. America.

Phytolaccaceae → Gallesia, Petiveria, Phytolacca, Rivina, Stegnosperma.

Piaropus crassipes (Mart.) Britt. → Eichhornia crassipes Solms.

Piassave, Bahia → Attalea funifera Mart.

Piassava, Ceylon → Caryota urens L.

Piassava, Pará → Leopoldina Piassaba Wall.

Piauhy Rubber → Manihot piauhyensis Ule.

Picea alba Link. (syn. P. glauca (Moench.) Voss., P. canadensis (L.) B. S. P.). White Spruce. (Pinaceae). - Tree. N. America. Wood straightgrained, light yellow, not strong, light, soft; used for construction work, interior finish of buildings and wood pulp.

Picea asperata Mast. Dragon Spruce. (Pinaceae). - Tree. Western China. Wood used in China for general construction work.

Picea complanata Wils. (Pinaceae). - Tree. China. Wood used in China for general building purposes.

Picea Engelmanni (Parry) Engelm. Engelmann Spruce. (Pinaceae). - Tree. Western N. America. Wood close grained, not strong, soft, light pale yellow with red tinges. Used for construction of buildings, fuel and charcoal. Bark is sometimes used for tanning leather.

Picea excelsa Link. (syn. Abies excelsa Lam., Picea Abies (L.) Karst). Norway Spruce. (Pinaceae). - Tree. Europe, Balkans, N. Asia. Wood yellowish-white to brownish, medium hard, fairly elastic, easy to split, durable under water, easy to work; important lumber in many parts of Europe; used for general carpentry, shingles, telephone poles, boxes, bridges, boats, carriages, organs, musical instruments. Source of paperpulp and wood-shavings or excelsior. From the resin the Jura Turpentine is derived. Source of Burgundy Pitch; used medicinally as a stimulant and counter-irritant. Needles are source of an ess. oil, used in toilet preparations.

Picea Glehnii Mast. Sakhalin Spruce. (Pinaceae). - Tree. Isl Sachalin. Wood beautifully grained. Used in Japan for interior finish of buildings, sounding boards of violins, pianos and organs.

Picea jezoensis Carr. Yeddo Spruce, Yezo-matsu. (Pinaceae). - Tree. Eastern Asia. Wood light, straight grained, soft, white, lustrous, elastic, flexible. Used in Japan for general building material, interior finishing, furniture, shingles, matchboxes, paperpulp; for sounding boards of musical instruments. Resin is used for healing wounds among the Ainu.

Picea Mariana (Mill.) B. S. P. (syn. Picea nigra Ait.) Link. Black Spruce, Bog Spruce. (Pinaceae). - Small tree. Cold bogs of N. America. Wood used for paperpulp. In New England and Canada Spruce Gum, a resinous exudation is collected from the branches; used as masticatory. Best gum is derived from south-side of the trees. It is assorted into grades as to color and flavor. Spruce Beer is made by boiling the branches in water.

Picea purpurea Mast. Purple Cone Spruce. (Pinaceae). - Tree. Szechuan, Kansu (China). Wood brownish, resinous, close-grained; used in China for general construction work.

Picea rubra Link. (syn. P. rubens Sarg.). Red Spruce. (Pinaceae). - Tree. Eastern N. America. Wood soft, close-grained, light, not strong, light colored, slightly tinged with red; used for boxes, crates, planks, boards, furniture, sash, frames, doors, general millwork, sounding boards of musical instruments, paperpulp; was formlery used for ship-building. Source of spruce-gum.

Picea sitchensis (Bong.) Carr., Sitka Spruce. (Pinaceae). - Tree. Pacific Coast of N. America. Wood not strong, soft, light, straight-grained, light-brown, red tinged; used for interior finish of buildings, cooperage, boat-building, packing cases, wooden ware, fence posts, doors, blinds, sash, general millwork, refrigerators, boat construction, paperpulp.

Pichurim Bean → Nectandra Puchury-minor Nees. and Mart.

Pico de Gallo → Xylopia obtusifolia A. Rich.

Picotee → Dianthus Caryophyllus L.

Picralima Klaineana Pierre (syn. P. nitida Th. and Hel.). (Apocynaceae). - Tree. Trop. Africa. Bark is used by the natives as febrifuge and vermifuge. Seeds are employed by the inhabitants in place of quinine.

Picrasma excelsa (Swartz) Planch., Jamaica Quassia. (Simarubaceae). - Tree. West Indies. Wood white to yellowish white, soft, light, loosely-grained, easily to split, very bitter taste; used mainly medicinally; bitter tonic, anthelmintic; contains bitter quassin. Source of Lignum Quassiae Jamaicense.

Picrasma quassioides Benn. (syn. Nima quassioides Buch-Ham., Rhus ailanthoides Bunge). Nigaki, Shurni. (Anacardiaceae). - Small tree. Japan, Korea, China, Himalaya. Wood hard, yellow, takes polish. Used in Japan for mosaic work, utensils. Bark very bitter, used as an insecticide and febrifuge.

Picrorhiza Kurroa Royle. (Scrophulariaceae). - Herbaceous. Shensi (China). Used medicinally in China. Contains picrorhizin and cathartic acid.

Pictyothammus reticulatus (Chapm.) Small. → Asimina reticulata Chapm.

Pierardia sapida Roxb. → Baccaurea sapida Muell. Arg.

Pigafettia elata Wendl. (syn. Metroxylon elatum Mart.) (Palmaceae). - Tall palm. Celebes. Fibres from young leaves are made into thread, also manufactured into small rugs.

Pig Nut → Omphalea triandra L. and Simmondsia california Nutt.

Pignut Hickory → Carya glabra (Mill.) Sweet.

Pigeon Bean → Vicia Faba L.

Pigeon Pea → Cajanus indicus Spreng.

Pigeon Plum → Coccolobis laurifolia Jacq.

Pigments → Dyestuffs.

Pilgerodendron uviferum (Don.) Florin. (syn. Juniperus uvifera Don., Thuja tetragona Hook.) Ciprés de las Guaytecas. (Cupressaceae). - Tree. Isla de Chiloë, Patagonian Islds., Chile. Wood brownish, very durable; used in Chile for housebuilding, furniture, window sash, flooring, doors and telephone poles.

Pili Nut Oil → Canarium ovatum Engl.

Pilocarpus Jaborandi Holmes. (Rutaceae). - Tree. Brazil. Leaflets are used in medicine as a strong diaphoretic; in large doses they are emetic. Causes contraction of the eye. Source of philocarpine, an alkaloid. Also P. Selloanus.

Pilocarpus pinnatifolius Lem. (Rutaceae). - Tree. Trop. America. Source of Paraguay Jaborandi. Leaves are used in medicine; contain pilocarpine, an alkaloid.

Pilocereus lanatus Weber → Espostoa lanata (H. B. K.) Britt. and Rose.

Pimelodendron amboinicum Hassk. (syn. Carumbium amboinicum Miq.). (Euphorbiaceae). - Tree. Moluccas. Amboina. Bark used as purgative. Milky fluid from stem is used by Amboineese as a varnish for paintings, to resist moisture. When mixed with the juice of Alstonia scholaris R. Br. it is used to varnish wood.

Pimenta → Pimenta officinalis Lindl.

Pimenta acris (Swartz.) Kostel. (syn. Myrcia acris DC.). (Myrtacea). - Tree. East Indies. Cultivated. Source of Oil of Myrcia, Oil of Bay, Oleum Myrciae being distilled from the leaves. Used in soap industry, toilet waters, bay-rum. Formlery leaves were distilled with rum. Contains eugenol, myrcine, chavicol, methyl-eugenol.

Pimenta de Macaco → Xylopia carminativa (Aruda) Fries.

Pimenta officinalis Lindl. (syn. Eugenia Pimenta DC.). Allspice, Pimenta. (Myrtaceae). - Tree. Trop. America. Cultivated in Jamaica. Unripe dried berries used as a condiment in ketchups, sauces, sausages, pickles. Produces Oil of Pimenta, Oil of Allspice, an ess. oil used for flavoring. Also oil is obtained from the leaves. Dried, nearly ripe fruits are used medicinally as carminative, stimulant. Contain 65 to 80 % ess. oil, eugenol and a resin.

Pimiento → Capsicum frutescens L.

Pimpinella Anisum L. (syn. Anisum vulgare Gaertn., A. officinarum Moench.). Anise Plant. (Umbelliferaceae). - Annual herb. Probably from Orient. Introduced in Europe, Asia, N. America. Cultivated. Seeds used for flavoring foods, in confectionaries, bakery products, beverages, Anisette liqueur, anise milk. Ripe seeds used medicinally as a carminative, aromatic stimulant, carminative, diaphoretic Source of Oil of Anise, ess. oil, derived from S. Spain, S. Russia, Bulgaria. Contains 80 to 90 % anethol, methyl chavicol. Used in many home remedies.

Pimpinella Saxifraga L. (syn. Carum nigrum Baill.). Black Carroway. (Umbelliferaceae). - Herbaceous plant. N. Africa. Seeds are used as a condiment in some parts of the Mediterranean region.

Pimple Mallow → Callirrhoe pedata Gray.

Piña Fibre → Ananas comosus (L.) Merr.

Pinaceae → Abies, Agathis, Cedrus, Keteleria, Larix, Picea, Pinus, Pseudotsuga, Sequioa, Tsuga.

Pinckneya pubens Michx., Georgia Bark. (Rubiaceae) - Shrub or small tree S. Carolina to Florida. Bark has been sometimes used for intermittent fevers.

Pine, Aleppo → Pinus halepensis Mill.

Pine, Armand → Pinus Armandi Franch.

Pine, Austrian → Pinus nigra Arnold.

Pine, Aztec → Pinus Teocote Cham. and Schlecht.

Pine, Benguet → Pinus insularis Endl.

Pine, Black → Call:tris calcarata R. Br. and Podocarpus spicata R. Br.

Pine Bluegrass → Poa scabrella (Thurb.) Benth.

Pine, Brazilian → Araucaria brasiliensis A. Rich.

Pine, Bristle Cone → Pinus aristata Engelm.

Pine, Bull → Pinus Sabiniana Dougl.

Pine Bush → Hakea leucoptera R. Br.

Pine, Celery → Phyllocladus trichomanoides D. Don.

Pine, Chilghosa → Pinus Gerardiana Wall.

Pine, Chinese White → Pinus Armandi Franch.

Pine, Cluster → Pinus Pinaster Soland.

Pine, Digger → Pinus Sabiniana Dougl.

Pine, Foxtail → Pinus aristata Engelm.

Pine, Hickory → Pinus pungens Michx.

Pine, Hoop → Araucaria Cunninghamii Sweet.

Pine, Huon → Dacrydium Franklinii Hook.

Pine, Indian Blue → Pinus excelsa Wall.

Pine, Jack → Pinus Banksiana Lamb.

Pine, Japanese Stone → Pinus pumila Reg.

Pine, Korean → Pinus koraiensis Sieb. and Zucc.

Pine Lichen → Cetraria pinatri (Scop.) S. Gray.

Pine, Limber → Pinus flexilis James.

Pine, Loblolly → Pinus Taeda L.

Pine, Lodgepole → Pinus Murrayana Balf.

Pine, Long Leaved → Pinus palustris Mill.

Pine, Luzon → Pinus insularis Endl.

Pine, Masson → Pinus Massoniana D. Don.

Pine, Merkus → Pinus Merkusii Jungh. and De Vr.

Pine, Montezuma → Pinus Montezumae Lamb.

Pine, Moreton Bay → Araucaria Cunninghamii Sweet.

Pine, Mountain → Pinus montana Mill.

Pine, Murray → Callitris calcarata R. Br.

Pine Needle Oil → Pinus sylvetris L.

Pine, Neosia → Pinus Gerardiana Wall.

Pine, Nut → Pinus edulis Engelm.

Pine, Parana → Araucaria brasiliensis A. Rich.

Pine, Pitch → Pinus palustris Mill. and P. rigida Mill.

Pine, Pond → Pinus serotina Michx.

Pine, Red → Dacrydium cupressinum Solander. and Pinus resinosa Ait.

Pine, Rocky Mountain White → Pinus flexilis James.

Pine, Sand → Pinus clausa (Engelm.) Sarg.

Pine, Scotch → Pinus sylvestris L.

Pine, Screw → Pandanus spp.

Pine, She → Podocarpus elata R. Br.

Pine, Short Leaved → Pinus echinata Mill.

Pine, Slash → Pinus caribaea Morelet.

Pine, Southern → Pinus palustris Mill.

Pine, Southern Yellow → Pinus ponderosa Laws.

Pine, Sugar → Pinus Lambertiana Dougl.

Pine, Swiss Stone → Pinus Cembra L.

Pine, Table Mountain → Pinus pungens Michx.

Pine, Torrey's → Pinus Torreyana Carr.

Pine, Westland → Dacrydium Westlandicum T. Kirk.

Pine, White → Pinus albicaulis Engelm., P. monticola Dougl., P. Strobus L. and Podocarpus elata R. Br.

Pine, Yellow → Pinus arizonica Engelm.

Pine, Yellow Silver → Dacrydium intermedium T. Kirk.

Pineapple → Ananas comosus (L.) Merr.

Piney Varnish → Vateria indica L.

Pinguicula vulgaris Sm. Common Butterwort. (Lentibulariaceae). - Small perennial insectiverous plant. It has been claimed that the leaves act as coagulant of milk, used in some countries.

Pink Cedar → Acrocarpus fraxinifolius Wight and Arn.

Pink Root → Spigelia marilandica L.

Pink Spider Flower → Cleome integrifolia Torr. and Gray.

Piñon → Pinus cembroides Zucc. and P. edulis Englm.

Pinus albicaulis Engelm., White Bark Pine. (Pinaceae). - Tree. Pacific Coast of N. America. Large sweet seeds are consumed by the Indians.

Pinus aristata Engelm. Bristle Cone Pine, Foxtail Pine. (Pinaceae). - Tree. Western United States. Wood light, soft, not strong, light-red; sometimes used as mining-timber, and as fuel.

Pinus arizonica Engelm. Arizona Yellow Pine. (Pinaceae). - Tree. S. Arizona, Sonora, Chihuahua. Wood light, soft, not strong, brittle; occasionally used as coarse lumber in Arizona.

Pinus Armandi Franch. Armand Pine, Chinese White Pine. (Pinaceae). - Tree. Western and Centr. China to Formosa. Wood soft, close-grained, resinous; used in China for building purposes and cheaper grades of furniture.

Pinus Banksiana Lamb. (syn. P. divaricata (Ait.) Sudw. Jack Pine. (Pinaceae). - Tree. Eastern N. America. Wood fairly light, rather week in bending strength, lacking in stiffness, low in shock resistance, coarse; used for pulpwood, box lumber and fuel.

Pinus caribaea Morelet., Slash Pine, Swamp Pine. (Pinaceae). - Tree. Southeast of United States to Louisiana; Bahamas, Isle of Pines, highlands of Centr. America. Wood very hard, heavy, strong, coarse-grained, durable, dark orange; made into lumber; used for construction, rail-road ties. Source of turpentine, resin and naval stores, pulpwood.

Pinus Cembra L., Swiss Stone Pine. (Pinaceae). - Tree. Alps, Austria, Carpathians, N. W. Russia, N. Asia, Amur region. Wood soft, closely-grained, easily worked; used for furniture, turnery, for statues, containers of milk, alpine huts. Seeds used in certain pastries and milk-food, consumed in some parts of S. Germany, Russia and Norway. Leaves are source of Carpathian and Hungarian Turpentine.

Pinus cembroides Zucc. Nut Pine, Piñon, Mexican Piñon. (Pinaceae). - Tree. S. W. of United States, No. Mexico and Lower California. Large oily seeds are an important food among the natives. Sold in local markets.

Pinus clausa (Engelm.) Sarg., Sand Pine. (Pinaceae). - Tree. Florida, S. Alabama. Wood brittle, not strong, light, soft, yellow or red orange. Sometimes used for masts of small vessels.

Pinus Coulteri D. Don., Coulter Pine. (Pinaceae). - Tree. California. Seeds used as food by Indians in S. California.

Pinus echinata Mill., Short-leaved Pine, Yellow Pine. (Pinaceae). - Tree. Eastern and Southern N. America. Wood hard strong, heavy, coarse-grained, orange to yellow brown of very variable quality; manufactured into lumber, pulpwood.

Pinus edulis Engelm., Nut Pine, Piñon. (Pinaceae). - Tree. S. W. of United States and Northern Mexico. Seeds are an important food among the Indians and Mexicans. Frequently sold in markets. Wood brittle, soft, light, not strong; used as fuel, for fence posts and made into charcoal; sometimes used as timber.

Pinus excelsa Wall. Indian Blue Pine. (Pinaceae). - Tree. Temperate Himalaya to Kafiristan and Afghanistan; eastw. to Sikkim etc. Wood used for construction work in W. Himalaya. Produces turpentine and tar.

Pinus flexilis James. Limber Pine, Rocky Mountain White Pine. (Pinaceae). - Tree. Western N. America. Wood close-grained, soft, light, pale-yellow; becoming red on exposure; sometimes used for general carpentry.

Pinus Gerardiana Wall. Chilghoza Pine. (Pinaceae). - Tree. Himalayan region, Afghanistan etc. Seeds consumed as food in Afghanistan and India.

Pinus halepensis Mill., Aleppo Pine. (Pinaceae). - Tree. S. Europe, W. Asia. Stem is source of Greek Turpentine.

Pinus insularis Endl. Luzon Pine, Benguet Pine, Tree. Centr. and N. Luzon (Philipp-Islds.). Source of a turpentine.

Pinus Khasya Royle. Khasia Pine. (Pinaceae). - Tree. Khasia hills, Martaban, hills of Burma. Source of a valuable resin and turpentine.

Pinus koraiensis Sieb. and Zucc., Korean Pine. (Pinaceae). - Tree. Korea and Kamtchatka. Seeds are consumed by the natives.

Pinus Lambertiana Dougl., Sugar Pine. (Pinaceae). - Tree. Pacific Coast of United States. Wood light red-brown, straight-grained, light and soft; used for interior finish of buildings, shingles, different woodwork. A sweet substance is exuded from the wounds in the heartwood, when used in quantities, it has laxative properties.

Pinus longifolia Roxb. (Pinaceae). - Tree. N. and E. India. Wood useful in cold climates, not resistant to white ants. Charcoal used for Chinese fireworks. Trunk is source of Indian Turpentine.

Pinus Massoniana D. Don., Masson Pine. (Pinaceae). - Tree. S. China. Wood close-grained, durable, source of a valuable timber; used in China for various purposes. Also P. Henryi Mast.

Pinus Merkusii Jungh. Merkus Pine. (Pinaceae). - Tree. Burma to mountains N. Sumatra. Source of turpentine and colophony; an important commercial product. Wood is durable, does not easily rot in the soil; used for posts, beams and boards.

Pinus monophylla Torr. and Frem., Nut Pine, Piñon. (Pinaceae). - Tree. S. W. of United States to Baja California (Mex.). Wood used as fuel, also made into charcoal used for smelting. Seeds are an important food among the Indians.

Pinus montana Mill. (syn. Pinus Mugo Turra.). Mountain Pine. (Pinaceae). - Small tree or shrub Alpine regions of Europe. Wood used by mountainers for wooden slippers, Klozschuhe. Young twigs are source of an ess. oil, Oleum Pini Pumilionis. Source of Pine Needle Oil, obtained through steam distilliation, being colorless, of a pleasant, aromatic odor; used medicinally as a mild antiseptic, stimulant, inhalent with expectorant.

Pinus monticola Dougl. Western White Pine. (Pinaceae). - Tree. Western N. America. Wood not strong, light and soft, close and straight-grained, light-brown to red; used for construction work and interior finish of buildings.

Pinus Montezumae Lamb. (P. Russeliana Lindl., P. macrophylla Lindl.) Montezuma Pine. (Pinaceae). - Tree. Mexico, Centr. America. Resin is used in some parts of Mexico for healing open wounds.

Pinus Murrayana Balf. (syn. P. contorta var. Murrayana Engelm.). Lodgepole Pine. (Pinaceae). - Tree. Western N. America. Wood close and straight-grained, not strong, soft, light, easily worked, light yellow nearly white, not durable; used occasionally for rail-road ties, mining timber and as fuel.

Pinus palustris Mill. (syn. P. australis Michx. f.). Long-leaved Pine, Southern Pine. (Pinaceae). - Tree. Southern and southeastern United States. Important lumber tree. Wood known as Pitch Pine, Southern Pine, heavy, hard, strong and tough, durable, light red to orange, coarse-grained; used for general building, interior finish,

flooring, bridges, viaducts, railroad ties, fences, millwork, boxes, masts, spars, fuel, wood pulp, and charcoal. Much lumber is exported. Source of Turpentine, **Oleum Terebinthinae** or **Gum Turpentine**, obtained by steam distillation. Oil of Turpentine or Spirit of Turpentine, an ess. colorless oil, with typical odor and taste which increases with age, Spec. Gr. 0.854 to 0.868; distils at 68° to 77° C. Used as a solvent of waxes, paints, oils, resins; for manuf. of shoe, stove, furniture and other polishes. Oil of Turpentine rectified at Sp. Gr. 0.853 to 0.862 is used as anthelmintic, diuretic, mild antiseptic, carminative, rubefacient, expectorant; for gonorrhea, flatulence, cystitis, hiccup, tenia. Externally as ointment or liniment for rheumatism, arthritis, inhalation in bronchitis. Rosin, Colophony, Yellow Resin is left after distillation; is pale yellow to amber, brittle, Sp. Gr. 1.07 to 1.09. Soluble in alcohol, benzene, ether, carbon bisulph. Used in medicine for preparation of ointments and plasters. In industry for varnishes, printing inks, cement, soap, wood polishes, floor coverings, sealing wax, fire-works, plastics, sizes, rosin oil, waterproofing card board and walls. Rosin Oil, Retinol or Rosinol derived from dry distil. of rosin, is an oily liquid, yellow, fluorescent. Boils at 280° C. Used medicinally as ointment or liniment in skin diseases, gonorrhea. Industrially for carbon black in lithography, printing inks, retinol colors, brewer's pitch, axle greases, in varnishes.

Pinus nigra Arnold. (syn. P. Laricio Poir.). Austrian Pine. (Pinaceae). - Tree. Austria to Balkans. Source of Austrian Turpentine. Used in salves and plasters.

Pinus Parryana Engelm. Nut Pine, Piñon. (Pinaceae). - Tree. California and Baja California. Seeds are an important food among the Indians in Baja California (Mex.)

Pinus pentaphylla Carr. (syn. P. parviflora Sieb. and Zucc.). (Pinaceae). - Tree. Hokkaidô, Honshū (Jap.). Wood soft, light, straight-grained, easily worked. Used in Japan for house-building, ships, interior finish, cabinetwork, wooden wares, chips, shingles, matches, sculpture, wooden moulds for metal-casting, wooden pipes for water-works.

Pinus Pinaster (syn. P. maritima Lam.). Cluster Pine. (Pinaceae). - Tree. Mediterranean region. Source of Bordeaux Turpentine, French Oil Turpentine. Center of production is Bordeaux, France. Source of pulpwood.

Pinus ponderosa Laws., Western Yellow Pine, Ponderosa Pine. (Pinaceae). - Tree. Western N. America to N. Mexico and Lower California. Important lumber tree. Wood fine-grained, hard, strong, light-red; used for construction work, fence posts, rail-road ties and fuel.

Pinus pumila Engl., Japanese Stone Pine. (Pinaceae). - Small tree. Honshū (Jap.), Sachalin. Wood is used for charcoal.

Pinus pumilio Haenke. (syn. P. Mugo var. pumilio (Haenke) Zenari). (Pinaceae). - Centr. Europe to Balkan Penin. Source of Dwarf Pine Needle Oil, obtained by distillation. Used for toilet waters, bath salts, room sprays, soaps and other toilet preparations. It sterilizes the air when atomized. Has been recommended for ailments of the bronchii, lungs and bladder, diseases of the skin, scalp and acute rheumatism.

Pinus pungens Michx., Table Mountain Pine, Hickory Pine. (Pinaceae). - Tree. Eastern United States. Wood is made into charcoal in some parts of Pensylvania.

Pinus resinosa Ait., Red Pine. (Pinaceae). - Tree. Eastern America. Wood pale red, very close-grained, light and hard; used for buildings, bridges, piles, masts and spars. Bark employed for tanning leather.

Pinus rigida Mill., Pitch Pine. (Pinaceae). - Tree. Eastern and Southern America. Wood soft, light brown or red, not strong, coarse-grained, brittle, light, very durable; used for fuel, charcoal, occasionally made into lumber.

Pinus Sabiniana Dougl. Digger Pine, Bull Pine. (Pinaceae). - Tree. California. Produces upon distillation Abietine, a colorless aromatic liquid. Seeds are important food of Indians in California.

Pinus serotina Michx., Pond Pine. (Pinaceae). - Tree. N. Carolina to Florida. Wood heavy, brittle, soft, dark orange, coarse-grained, very resinous. Occasionally used as timber. Turpentine is produced along coastal region of N. Carolina.

Pinus Strobus L. Eastern White Pine. (Pinaceae). - Tree. Eastern United States. Wood straight-grained, light, not strong, easily worked; used for shingles, laths, cabinet-making, construction-work, woodenware, masts, matches, interior finish of buildings.

Pinus succinifera Conw. (Pinaceae). - Fossil tree. Reported from different parts of Europe, esp. Baltic region. Source of Amber, Baltic Amber, Succinite; used for beads, tobacco and cigarette pipes and other fancy articles. Amber of inferior quality is occasionally used in lacquers.

Pinus sylvestris L., Scotch Pine. (Pinaceae). - Tree. Centr. and N. Europe, N. Asia. Widely cultivated as forest tree in Europe. Wood soft, not strong, elastic, durable; used for furniture, general carpentry, rail-road ties, masts. Source of pulpwood, cellulose, tar, pitch and a fine soot. Needles are sometimes used as packing material. They are also source of a Pine Needle Oil, used in toilet and medicinal preparations.

Pinus Taeda L., Loblolly Pine. (Pinaceae). - Tree. Eastern and Southern America. Wood light brown, brittle, coarse-grained, not durable, weak; used for construction and interior finish of buildings.

Pinus Teocote Cham. and Schlecht., Aztec Pine, Jalocote, Ocote. (Pinaceae). - Tree. Mexico. Source of a terpentine, Trementina de Ocote, Ocotzol; used for different purposes, a balsamic stimulant. A tar is called Brea; used for torches and manuf. soap.

Pinus Torreyana Carr., Torrey's Pine. (Pinaceae). - Tree. California. The large seeds are used as food, eaten raw or roasted.

Pipe Gamboge → Garcinia Hanburyi Hook. f.

Piper aduncum L. (Piperaceae). - Shrub. Trop. America. Herb is astringent, diuretic and a stimulant.

Piper angustifolium Ruiz. and Pav. (Piperaceae). - Shrub. Mexico, West Indies, Centr. and S. America. Dried leaves are source of Matico. Used medicinally as antiseptic to the genetourinary tract; astringent, stimulant, styptic, vulnarary. Contains ess. oil in which is stearoptene matico camphor. Leaves are known in commerce as Matico, an export article.

Piper auritum H. B. K. (Piperaceae). - Shrub. Mexico and Centr. America. Leaves are used for seasoning tamales.

Piper bantamense Blume. (Piperaceae). - Vine. Mal. Archip. Bark is used among the Malayans and Javanese. Clothes are often washed with the bark to impart a pleasant scent.

Piper Betle L. (syn. Chavica auriculata Miq.) Betel, Sirih. (Piperaceae). - Vine, Malaysia, Indonesia etc. Widely cultivated and used as a masticatory in many Asiatic countries. For this purpose leaves are covered with lime and cutch. Slices of the nuts of the Betel-nut Palm, Areca Catechu L. in various stages of maturity, are placed on the leaf, to which are sometimes added tamarinds, cloves and other flavoring material, which is put in the mouth where it remains for some time, after which a flow of saliva is stimulated. The betel habit is supposed not to be harmful, it produces some stimulation and a feeling of well-being. It causes the teeth to become brown to red and finally black. Betel chewing has been practiced for centuries. Betel leaves are used in many household recepies under the natives among which a decoction is used for cleaning wounds.

Piper caducibracteum C. DC. (Piperaceae). - Vine. Mal. Archip. Young leaves are by some natives prefered to those of P. Betle L. in Betel chewing.

Piper capense L. f. (Piperaceae). - Shrub. S. Africa. Infusion of fruits was used by pioneers in Transvaal and Cape Colony as stomachic, stimulant and carminative for indigestion, colic and flatulency.

Piper Chaba Hunter. (syn. P. officinarum C. DC.). (Piperaceae). - Woody vine. Mal. Archipelago. Philipp. Islds. Seeds are source of a spice, sometimes called Long Pepper.

Piper Cubeba L., Cubeba. (Piperaceae). - Woody vine. Malaya. Cultivated. Source of Cubebs, Cubeb Berries. Much is derived from Java and Singapore. Used for flavoring certain cigarettes. Oil of Cubeb is employed for flavoring bitters; in food preparations; as lozenges for relief of conditions of the throat. Used medicinally as antiseptic, carminative, stimulating expectorant and diuretic. Also P. Clusii from W. Africa; P. sumatranum from Mal. Archip., P. borbonense from Bourbon and P. pedicellosum from Indo-Chine produce similar products.

Piper excelsum Forster. (Piperaceae). - Shrub. Australia, Pacif. Islds., Hawaii Islds. Uncooked flower-clusters are eaten by the natives of some of the Polynesian Islands. Leaves are used for leprosy. Decoction of leaves is used by the Maoris for toothache and rheumatic pains.

Piper fragile Benth. (Piperaceae). - Vine. Ternate and Moluccas. The bitter, fragile bark is used in native medicine for framboesia.

Piper hederaceum Cunningh. (Piperaceae). - Woody plant. New South Wales. Considered in Australia an important honey-yielder.

Piper longum L., Jaborandi Pepper, Long Pepper. (Piperaceae). - Woody vine. Himalayas to S. India. Cultivated. Source of Long Pepper; used for seasoning. Root is considered diuretic, stimulant and sudorific.

Piper medium Jacq. (Piperaceae). - Shrub. Trop. America. Herb is used in Costa Rica for snake bites.

Piper methysticum Forst; Kawa Pepper. (Piperaceae). - Woody plant. Pacific Islands. Cultivated. Roots are source of a beverage, called Kawa-Kawa, used by the natives of some of the Polynesian Islands. Dried rhizome and roots are used medicinally as antiseptic, diuretic, expectorant and genito-urinary stimulant. Contains methysticin, ω-methysticin.

Piper nigrum L., Pepper. (Piperaceae). - Woody vine. Cochin-China, India, Mal. Archip. Cultivated in tropics. Spice. Ground dried fruits are source of Black Pepper; when they are removed from the outer coating they are source of White Pepper. Much is derived from Java, Sumatra and Malaya. Used for flavoring. Important commercial grades are: Lampong Black Pepper, Tellicherry Black Pepper, Alleppy Pepper, Singore Pepper, Penang Black Pepper, Saigon Pepper, Madagascar Pepper, Muntok White Pepper, Madagascar White Pepper. Seeds are source of Oil of Pepper used for flavoring food-products, sausages, canned foods; in beverages; also employed in perfumery for adding bouquets of oriental types producing spicy notes. Used medicinally as a stimulant, febrifuge, tonic and irritant. Contains ess. oil in which are found dipentene, phellandrene, piperine (an alkaloid) and piperidine. Varieties are Balamcoota, Kallivalli, Cheridaki, Utharamvalli, Big Berry and Shortleaved.

Piper pseudonigrum C. DC. (Piperaceae). - Perennial vine. Tonkin-China. Seeds are occasionally used as a spice.

Piper retrofractum Vahl. Javanese Long Pepper. (Piperaceae). - Woody vine. Malaya. Cultivated. Used for seasoning in curries, preserves and pickles.

Piper sanctum (Miq.) Schlecht. (Piperaceae). - Shrub. S. Mexico. Aromatic leaves are used in some parts of Mexico for flavoring soups. Decoction of leaves used for indigestion and abdominal cramps.

Piper umbellatum L. (syn. Heckeria umbellata (L.) Kunth.). (Piperaceae). - Shrub. Trop. America. Decoction of roots is used by Mazatecs as febrifuge and diuretic. Tea from flowerclusters are used for coughs.

Piperaceae → Peperomia, Piper.

Piptadenia africana Hook. f. (Leguminosaceae). - Tree. Westcoast of Africa. Wood very hard, durable; used for planks in S. Nigeria.

Piptadenia Cebil Griseb. (Leguminosaceae). - Tree. Argentina. Source of Cebil Gum, contains 80 % arabin.

Piptadenia chrysostachys Benth. (Leguminosaceae). - Small tree. Centr. and E. Madagascar. Wood is used in Madagascar for certain musical instruments.

Piptadenia colubrina (Vel.) Benth. (syn. Mimosa colubrina Vell.). (Leguminosaceae). - Tree. S. America, esp. Brazil, Peru, Bolivia. Bark astringent. Seeds when finely ground, used as a snuff in Peru, is highly narcotic. Bark used in Brazil in tinctures, extracts, infusion or powders for diseases of the lungs. Used in home medicines in Brazil.

Piptadenia macrocarpa Benth., Cebil Colorado, Curupag. (Leguminosaceae). - Tree. Argentina, Chaco region. Bark source of tanning material.

Piptadenia peregrina (L.) Benth. Yoke, Cohoba Tree. (Leguminosaceae). West Indies, S. America. Source of Cohoba or Coxoba Snuff, being ground seeds, used with or without lime, causing drunkenness when inhaled, almost amounting to frenzy. Used by natives of Haiti and Santo Domingo in religious ceremonies. Wood very heavy, hard; used for rail-road ties.

Piptadenia rigida Benth. (syn. Acacia Angico Mart.). (Leguminosaceae). - Tree. Brazil. Source of Angico Gum; used in some parts of Brazil as mucilage.

Piptoporus Betulinus (Bull.) Karst. → Polyporus betulinus Bull.

Piptostegia Pisonis Mart. (Convolvulaceae). - Perennial vine. Brazil. Root is source of Brazilian Jalap; used medicinally as purgative; contains a resin.

Pipturus argenteus Wedd. (syn. Urtica argentea Forst., P. taitensis Wedd.). (Urticaceae). - Shrub or small tree. Mal. Archipelago, Pacif. Islands, Australia. Source of a fibre in Tahiti. Inner bark is used by the Samoans for mats, nets and fishing lines. Fibre is of fine texture, very strong, difficult to prepare. Bark is said to be source of a brown dye.

Pipturus Gaudichaudianus Wedd. (syn. P. albidus Gray.). (Urticaceae). - Shrub or tree. Hawaii Islands. Bark is source of a fibre, made into tapa or paper cloth; used by the natives of Hawaii.

Pipturus velutinus Wedd. (syn. P. incanus Wedd.). (Urticaceae). - Woody plant. Mal. Archipelago. New Caledonia. Bark is source of a fibre; used in New Caledonia for sails and nets.

Piquia Peroba → Aspidosperma tomentosum Mart.

Piratinera guianensis Aubl. → Brosimum Aubletii Poepp. and Endl.

Pirus → Pyrus.

Piscidia Erythrina L. (syn. Ichthyomethia piscipula (L.) Hitchc. Jamaica Fishfuddle Tree, Jamaica Dogwood. (Leguminosaceae). - Trop. America. Wood close-grained, very heavy, yellow brown; used for boat building, charcoal and fuel. Tree contains piscidin, an active principal, producing sleep. Powdered bark and roots were used by the Caribs in the West Indies to stupefy fish.

Pisolithus tinctorius (Pers.) Coker and Couch. (Sclerodermaticeae). - Basidiomycete. Fungus Fruitbodies are source of a brown dye which is used to color silk, esp. in southern France, Spain and Italy.

Pisonia alba Span. Maluko, Lettuce Tree. (Nyctaginaceae). - Small tree. Malaysia. Tender leaves when boiled, are eaten as a pot-herb among the natives.

Pisonia Brunioniana Endl. (Nyctacinaceae). - Tree. New Zealand, Pacif. Islds. Leaves used as diuretic: Roots are purgative.

Pisonia sylvestris Teijsm. and Binn. (Nyctaginaceae). - Tree. Moluccas. Cultivated in Indonesia. Leaves are eaten as a vegetable.

Pistache, Chinese → Pistacia chinensis Bunge.

Pistache, Egyptian → Pistacia Khinjuk Stocks.

Pistache, Lentisk → Pistacia lentiscus L.

Pistache, Mount Atlas → Pistacia atlantica Desf.

Pistache, Terebinth → Pistacia Terebinthus L.

Pistache, Turkish Terebinth → Pistacia mutica Fisch and May.

Pistachio, Common → Pistacia vera L.

Pistachio Galls → Pistacia lentiscus L.

Pistachio Nut → Pistacia vera L.

Pistachio Nut Oil → Pistacia vera L.

Pistacia atlantica Desf., Mount Atlas Pistache. (Anacardiaceae). - Tree. N. Africa. Wood is employed by the natives for different purposes.

Pistacia chinensis Bunge, Chinese Pistache. (Anacardiaceae). - Large tree. China. Wood used in China for rudder-posts. Young shoots eaten as vegetable; called huang-nu ya-tzu.

Pistacia Khinjuk Stocks. (Anacardiaceae). - Small tree. W. Himalaya. Source of hard galls; used occasionally for tanning and dyeing. Leaves used as fodder for camels and buffalos. Wood suitable for furniture and ornamental work. Fruits are used locally to impart a flavor to milk.

Pistacia lentiscus L., Lentisk Pistache. (Anacardiaceae). - Small tree. Mediterranean region. Cultivated. Source of Chios Mastic of commerce, derived from the bark of the trunk. Produces a pale varnish employed for coating paintings and metals, for lithographic processes and retouching negatives, as a constituent of incense, being a high grade and expensive comodity. It is also used as glue of beards of actors on the stage. Was used for embalming among ancient Egyptians. The mastic is chewed by Oriental women as breath sweetener. Sold in bazaars. Used medicinally as carminative, stimulant, to mucuous memberanes, protective for temporary teeth fillings, plasters. Contains 90 % resin, masticin, mastichic acid, also a bitter principle.

Pistacia mutica Fisch and Mey (syn. P. cabulica Stocks.) Turk Terebinth Pistache. (Anacardiaceae). - Tree. Centr. Asia. Trunk is source of a gum, known as Bombay Mastic, used in varnishes, also used for chewing in some parts of Iran. Much is derived from Afghanistan and Belutchistan.

Pistacia Terebinthus L., Terebinth Pistache. (Anacardiaceae). - Small tree. Mediterranean region. Formerly source of a turpentine. Plant produces Pistacia Calls, Carobbe di Guidea, caused by Pemphigus corniculatus. Source of tanning material, containing 50 to 60 % of tannin.

Pistacia vera L. Common Pistachio. (Anacardiaceae). - Small tree. Mediterranean region, Caucasia, W. Asia. Cultivated in subtropical areas of the Old and New World. Seeds called Pistachio Nuts are edible, 10 to 25 mm. long, of pleasant flavor, used in confectionary, icecreams etc. Several varieties e. g.: Alep, Tunis, Sicily. Source of Pistachio Nut Oil, a nondrying oil; used medicinally in E. India. Sp. Gr. 0.9179—0.9200; Sap. Val. 188—192; Iod. No. 85—98; Unsap. 1.0—3.1 %. Plant is source of Bokhara Calls, Gul-i-pista; much being derived from Iran; used as tanning material.

Pisum arvense L., Field Pea. (Leguminosaceae). Annual herb. Probably from Mediterranean region. Cultivated in different races and varieties; used for human consumption and as food for livestock. Also grown as green-manure. Var. elatius (Steven) Bieb. Bavarian Winter Pea, Black Podded Pea. Var. quadratum L. Capucine Pea, Smyrna Pea, Gray Konigsberger Pea. Var. arvense (L.) Gams. East Prussian Pea, Paluschke Pea, Sand Pea.

Pisum sativum L. Garden Pea. (Leguminosaceae). - Annual vine. Probably from Mediterranean region. Cultivated as a vegetable since ancient times. Grown as climbing or dwarf varieties. Seeds are consumed as vegetable, also in soup. They are sold fresh, canned, powdered, split and frozen pack. Of some varieties the entire pods are eaten. Varieties consumed as seeds with smooth seeds: Alaska, Creole, Mammouth; with wrinkled seeds: Surprise, American Wonder, Laxton's Progress, Graders, Alderman, Telephone. Varieties consumed as pods: Dwarf Green Sugar, Mammoth Melting Sugar.

Pitahaya → Acanthocereus pentagonus (L.) Britt. and Rose. Echinocactus stramineum (Engelm.) Ruempl and Lemaireocereus queretaroensis (Weber) Safford.

Pitahaya de Agosto → Echinocereus conglomeratus Forst.

Pitahaya Agria → Machaerocereus gummosus (Engelm.) Britt. and Rose.

Pitahaya Dulce → Lemaireocereus Thurberi (Engelm.) Britt. and Rose.

Pitahaya naranjada → Acanthocereus pentagonus (L.) Britt. and Rose.

Pitahaya Oregona → Hylocereus undatus (Haw.) Britt and Rose.

Pitanga → Eugenia uniflora L.

Pitanga tuba → Eugenia Selloi Berg.

Pitch, Burgundy → Picea excelsa Link.

Pitch, Canada → Tsuga canadensis (L.) Carr.

Pitch Pine → Pinus palustris Mill. and P. rigida Mill.

Pitcher Plants → Nepenthes distillatoria L. and Sarracenia purpurea L.

Pitchery → Duboisia Hopwoodii F. v. Muell.

Pithecolobium Auaremontemo Mart. (Leguminosaceae). - Tree. Brazil. Stem is source of a gum, said to resemble Gum Arabic.

Pithecolobium dulce Benth. Manila Tamarind. (Leguminosaceae). - Tree. Trop. America. Cultivated in tropics of Old and New World. Arillus of seeds pulpy, edible; used by poorer classes as a source of food, also made into a beverage. Bark produces a yellow dye; used for tanning skins. Fruits sold in markets.

Pithecolobium flexicaule (Benth.) Coult. (syn. Mimosa flexicaulis Benth., P. texense Coult.). (Leguminosaceae). - Shrub or small tree. Mexico, Texas. Seeds when roasted are consumed by the Mexicans. Green pods are eaten cooked. Thick seedcoat is used as substitute for coffee. Wood is durable; used for cabinet work and wagons.

Pithecolobium hymeneaefolia Benth. (Leguminosaceae). - Tree. Trop. America. Source of a gum, called Goma de Orore.

Pithecolobium lobatum Benth. (Leguminosaceae). - Tree. Burma, Mal. Archipelago. Cultivated. Leaves, fruits and flowers are eaten as a vegetable by the natives of Java.

Pithecolobium Saman Benth. Saman, Cow-Tamarind. (Leguminosaceae). - Tree. Trop. America. Wood durable, rich, dark when polished; excellent for furniture. Used as shade tree in coffee and cocoa plantations in Guiana.

Pithecolobium Unguis-Cati Benth., Cats' Claw. (Leguminosaceae). - Tree. Florida. West Indies and Centr. America. Fruits are edible, sometimes consumed as food.

Pithophora affinis Nordst. (Cladophoraceae). - Green Alga. Hawaii Islands. Is consumed among the Hawaiians with fresh water shrimps and salt. Also P. polymorpha.

Pitomba → Eugenia Luschnathiana Berg.

Pittosporaceae → Pittosporum.

Pittosporum crassifolium Soland., Karo Pittosporum. (Pittosporaceae). - Small tree or shrub. New Zealand. Wood very tough, whitish; used for inlay work.

Pittosporum phillyraeoides DC. (syn. P. angustifolium Lodd.). Butter-Bush. (Pittosporaceae). - Tree. Australia. Seeds very bitter, pounded into a flour, consumed by aborigines.

Pittosporum viridiflorum Sims. (Pittosporaceae). - Tree. S. Africa. Roasted bark from young trees is used by natives for dysentery.

Pituri → Duboisia Hopwoodii F. v. Muell.

Plagianthes betulinus Cunningh. Lowland Ribbon Wood. (Malvaceae). - Tree. New Zealand. Bark source of a fibre; used by Maoris for rope and twine, to make fishing-nets. Substitute for Raffia; used by gardeners for tying plants.

Plains Bamboo → Bambusa Balcooa Roxb.

Plaited Usnea → Usnea plicata Hoffm.

Planchonia valida Blume. (Lecythidaceae). - Tree. Mal. Archipelago. Young leaves are eaten as lalab with rice among the Javanese.

Plane Tree, Oriental → Platanus orientalis L.

Plantaginaceae → Plantago.

Plantago arenaria Waldst. and Kit. (syn. P. ramosa Gilib.) Asch. (Plantaginaceae). - Annual herb. S. and Centr. Europe, Caucasia, Siberia. Source of Spanish or French Psyllium Seed. Cultivated. Similar to P. Psyllium L.

Plantago Coronopus L., Crowfood Plantain, Buckthorn Plantain. (Plantaginaceae). - Annual herb. Europe, Asia, N. Africa. Introd. in Australia and New Zealand. Leaves are occasionally used in salads.

Plantago lanceolata L., Rib Grass. (Plantaginaceae). - Perennial herb. Europe, temp. Asia, introd. in N. America. Young leaves are sometimes consumed as vegetable, has been recommended as an emergency food during times of want. Also P. major L.

Plantago ovata Forsk. (syn. P. decumbens Forsk., P. Ispaghula Roxb.) (Plantaginaceae). - India, Centr. Asia, Mediterranean region. Annual herb. Cultivated in India. Source of Spogel, Ispaghul, Blond or Indian Plantago Seed. Similar to P. psyllium L. Used in Indian medicine. Considered beneficial in chronic dysentery.

Plantago Psyllium L., Psyllium. (Plantaginaceae). - Annual herb. Mediterranean region. Cultivated. Source of Psillium Seed, Plantain Seed or Flea Seed. Dried ripe seeds, used medicinally; being laxative, due to the swelling of the mucilagenous seeds. Contains mucilage, among which pentosanes and galactans.

Plantain, Banana → Musa paradisiaca L.

Plantain, Buckthorn → Plantago Coronopus L.

Plantain, Crawfood → Plantago Coronopus L.

Plantain Seed → Plantago Psyllium L.

Plantain, Water → Alisma Plantago L.

Platanthera bifolia (L.) Reichb. (Orchidaceae). - Perennial herb. Europe, Caucasia, Asia Minor, N. Africa. Dried roots source of Salep. → Orchis latifolia L.

Platanaceae → Platanus.

Platanus mexicana Moric., Mexican Sycomore. (Platanaceae). - Tree. Mexico. Wood used for general carpentry, dishes and spoons.

Platanus occidentalis L., American Svcomore, Buttonwood. (Platanaceae). - Tree. N. America. Wood tough, strong, difficult to split; used for manuf. boxes, crates, butcher's blocks, interior finish of houses and for furniture.

Platanus orientalis L., Oriental Sycomore, Oriental Plane, Doolb, Buna. (Platanaceae). - Tree. Mediterranean region, W. Asia. Wood brown, fairly hard, relatively heavy, very tough, difficult to split, not durable; used for wood pulp, inlay work.

Platea corniculata Becc. → Urandra corniculata Foxw.

Plathymenia reticulata Benth. Vinhatico. (Leguminosaceae). - Tree. Brazil, esp. lower Amazon to Sao Paulo. Wood lustrous yellow to light orange, medium texture, straight grained, easy to work, durable. Much esteemed in Brazil; common in markets of Pernambuco to Rio de Janeiro. Used for cabinet work, furniture, interior finish of buildings, parquet flooring, ship-building, general carpentry and construction work. Known in the trade as Brazilian Yellow Wood.

Platonia grandiflora Planch and Traina. Pahoorie. Ubacury. (Guttiferaceae). - Large tree. S. America, esp. Guianas, Brazil, Ecuador. Wood dull-yellow to orange-brown, hard, heavy, coarse to medium-grained, not difficult to work. Used for carpentry, ship-building, light cooperage, crating, general construction work.

Platonia insignis Mart., Bacury. (Guttiferaceae). - Tree. Brazil, esp. Amazon, Piauhy; Guiana. Fruits as large as an orange, yellow; flesh white, of pleasant flavor; used for compôtes, pastry and preserves. Sold in Brazil in local markets. Wood brownish-yellow, fairly heavy and dense, durable; used for flooring, fancy work and construction of buildings. Seeds are source of Bacury Kernel Oil, non-drying; used for manuf. candles and soap. Sap. Val. 191.8; Iod. No. 63.3; Melt. Pt. 51. 7° C; Unsap. 4.2 %.

Platycarya strobilacea Sieb. and Zucc. (Juglandaceae). - Tree. China, Japan. Bark is used in some parts of Japan for dyeing fishing-nets. Fruits are source of a black dye; used in China for dyeing cotton yarn and cotton goods.

Platylophus trifoliatus Don., White Aler (S. Afr.). (Cunoniaceae). - Tree. Cape Peninsula. Wood white, durable; used for furniture, boxes, picture frames, wagons.

Platymiscium dimosphandrum Don. Sm., Yama Rosewood, Yama Cocobolo. (Leguminosaceae). - Tree. Centr. America. Wood dark reddish-brown, dense, fine texture. Used for brush-backs, handles of kitchen knives, butts of billiard cues. Also P. Duckei Dug. from Nicaragua.

Platymiscium pinnatum (Jacq.) Dugand. (syn. P. polystachyum Benth.). Quira Macawood. (Leguminosaceae). - Tree. Trop. America, esp. N. of S. America, Trinidad, Atl. Centr. America. Wood bright-red to reddish or purplish brown, striped, tough, heavy, strong, not difficult to work, finishes smoothly; used locally for marimba bars, furniture, rail-road ties, bridges planking, for heavy durable construction work; exported to USA for manuf. of brush-backs, veneers, knife and tool nandles and turnery.

Platymiscium Ulei Harms. (syn. P. paraense Huber). Macacaúba. (Leguminosaceae). - Tree. Lower Amazon (Braz.). Wood easy to work; used for joinery and furniture.

Plectaneia microphylla Jum. and Perr. Mahavoahavana. (Apocynaceae). - Woody plant. Madagascar. Source of a rubber. Latex is often mixed with that of Landolphia.

Plectocomia Griffithi Becc. (Palmaceae). - Liane. Malay Penin. Canes used for chair-legs, miner's strong baskets.

Plectranthus esculentus N. E. Br., Kaffir Potato. (Labiaceae). - Perennial herb. Natal. Tubers are much esteemed by the natives of Natal as food.

Plectranthus floribundus N. E. Br. (Labiaceae). - Herbaceous perennial. Trop. Africa. Cultivated by natives of Yola, Rhodesia, Natal etc. Tuberous roots used as food, being of a pleasant taste; they are called Kaffir Potatoes or Umbondive.

Plectranthus rotundifolius Spreng. → Coleus rotundifolius Chev. and Perrot.

Plectranthus tuberosus Blume → Coleus rotundifolius Chev. and Perrot.

Plectridium pectinovorum Stoerer (Bacteriaceae). - A pectine dissolving anaerobic microorganism being of considerable value in the retting process of flax and hemp fibres.

Plectronia didyma Bedd. (syn. Canthium didymum Gaertn. f.). (Rubiaceae). - Tree. E. India, Ceylon. Wood is source of Ceylon Boxwood, being white, light brown; formerly used for cutlery.

Plectronia glabriflora Holland. (syn. Canthium glabriflorum Hiern.). (Rubiaceae). - Tree. Trop. Africa, esp. Old Calabar, Gold Coast, Cameroon. Wood fine-grained; like beech wood in hardiness; brownish white; used locally for carpentry.

Plectronia lanciflora Benth. and Hook. (syn. Canthium lanciflorum [Benth. and Hook.] Hiern.). (Rubiaceae). - Tree. S. Africa. Fruits size of a plum, light brown, edible; considered by Europeans as one of the best fruits of the region.

Plectronia ventosa L. (Rubiaceae). - Perennial plant. S. Africa. Herb is used by the Zulus for diarhea and dysentery.

Pleiogynium Solandri Engl. Burdekin Plum. (Anacardiaceae). - Tree. Queensland. Large seeds have a pleasant flavor, consumed in Australia.

Pleopeltis longissima Moore. (Polypodiaceae). - Perennial fern. Mal. Archipelago. Young shoots and leaves are consumed raw or boiled among the natives.

Pleurotus anas Van Overeem. (Agaricaceae). - Basidiomycete. Fungus. Fruitbodies are consumed in Indonesia where they are considered a delicacy. Also P. fissilis (Lev.) Sacc.

Pleurotus Eryngii (D. C.) Quél. (Agaricaceae). - Basidiomycete. Fungus. Fruitbodies are consumed in some parts of Italy. Known as Cardarellas. Sold in markets.

Pleurotus Opuntiae (Dur. and Lev.) Sacc. → Pleurotus ostreatus (Jacq.) Quél.

Pleurotus ostreatus (Jacq.) Quél. (syn. P. Opuntiae [Dur. and Lev.] Sacc., Agaricus ostreatus Jacq., P. Yuccae Mer.) (Agaricaceae). - Basidiomycete. Fungus. Mock Oyster. Cosmopolite. Fruitbodies are consumed in China. Collected material is often exported to China from eastern U.S.S.R. Cultivated on beach trunks in some parts of Germany. Sold in markets.

Pleurotus ulmarius → Clitocybe tessulata Bull.

Pleurotus Yuccae Mer. → Pleurotus ostreatus (Jacq.) Quél.

Plocarium alternifolium (Vahl.) Melch. (Theaceae). - Tree. Trop. Asia. Wood red, easy to bena, hard; used in Cochin-China for turnery and posts.

Pluchea indica Less. (Compositae). - Shrub. Trop. Asia, Australia. Leaves are eaten raw or cooked by the natives, also used as tea. Stimulates perspiration in fevers.

Pluchea sericea (Nutt.) Cov., Marsh Fleabane. (Compositae). - Northern Mexico and S. W. of United States. Straight stems are used by the Indians for shafts or arrows. Flowers are source of a honey.

Plukenetia conophora Muell. Arg. (Euphorbiaceae). - Woody vine. Trop. Africa. Cultivated. Seeds source of N'Ghat Oil, a drying oil; used by natives in food. Oil dries quicker than Linseed Oil. Sp. Gr. 0.936—0.939; Sap. Val. 190—192; Iod. No. 198—204; Unsap. 0.2 %.

Plum, Apricot → Prunus Simonii Carr.

Plum, Botoko → Flacourtia Ramontchi L'Hér.

Plum, Beach → Prunus maritima Wang.

Plum, Bullace → Prunus instititia L.

Plum, Burdekin → Pleiogynum Solandri Engl.

Plum, Cherry → Prunus cerasifera Ehrh.

Plum, Chickasaw → Prunus angustifolia Marsh.

Plum, Chinese → Prunus salicina Lindl.

Plum, Damson → Prunus instititia L.

Plum, Darling → Reynosia septentrionalis Urb.

Plum, Flatwoods → Prunus umbellata Ell.

Plum, Garden → Prunus domestica L.

Plum, Ginger Bread → Parinarium macrophylla Sab.

Plum, Governor's → Flacourtia Ramontchi L'Hér.

Plum, Hog → Prunus umbellata Ell., Spondias lutea L. and Symphonia globulifera L.

Plum, Hortulana → Prunus hortulana Bailey.

Plum, Icaco → Chrysobalanus Icaco L.

Plum, Japanese → Eriobotrya japonica Lindl. and Prunus salicina Lindl.

Plum, Java → Eugenia Jambolana Lam.

Plum, Jew → Spondias dulcis Forst. f.

Plum, Kaffir → Odina caffra Sond.

Plum, Klamath → Prunus subcordata Benth.

Plum, Marmelade → Calocarpum mammosum Pierre.

Plum of Martinique → Flacourtia inermis Roxb.

Plum, Mobola → Parinarium Mobola Oliv.

Plum, Natal → Carissa grandiflora A. DC.

Plum, Pacific → Prunus subcordata Benth.

Plum, Pigeon → Coccolobis laurifolia Jacq.

Plum, River → Prunus americana Marsh.

Plum, Runeala → Flacourtia cataphracta Roxb.

Plum, Sour → Owenia acidula F. v. Muell.

Plum, Tamarind → Dialium indum L.

Plum Tree → Prunus domestica L.

Plum, Wild Goose → Prunus Munsoniana Wight and Hedrick.

Plumbaginaceae → Plumbago, Statice.

Plumbago rosea L. (Plumbaginaceae). - Woody herb. Iran, India. Juice from plants is used by beggars to produce blisters on the body, causing pity. Plant is sudorific.

Plumbago scandens L. (Plumbaginaceae). - Herbaceous climber. Trop. America. Roots and leaves applied on skin cause blisters. Roots used for toothache. Beggars used leaves to cause sores on body, for purpose of creating pity.

Plumbago zeylanica L. (Plumbaginaceae). - Shrub. Trop. Asia, Africa. Leaves are used by natives for rheumatics. Powdered roots are employed for blistering; a decoction is used as antiscabious remedy. Said apt to cause abortion.

Plume Grass, Long Hair → Dichelachne crinita Hook. f.

Plume Grass, Short Hair → Dichelachne sciurea Hook. f.

Plumeria acutifolia Poir. (Apocynaceae). - Tree. Mexico. S. America. Supposed to produce a rubber.

Plumeria lancifolia Muell. Arg. (Apocynaceae). - Shrub. Brazil. Bark is antiastmatic, antisiphylitic, emmenagogue and purgative; used in Brazil. Exported to the United States and Europe.

Pluteus cervinus Quél. (syn. Hyporhodius cervinus [Schäff.] Henn.). (Agaricaceae). - Basidiomycete. Fungus. Cosmopolitan. Young fruitbodies are edible. They are tender, having a good flavor.

Pó da Bahia → Vetaireopsis speciosa Ducke.

Pó da Goa → Vataireopsis speciosa Ducke.

Poa abyssinica Juss. (syn. Eragrostis abyssinica Schrad.) Teff. (Graminaceae). - Annual grass. E. Africa, Abyssinia. Grown as a cereal in Abyssinia. There are varieties with white and black grains. Straw used as food for mules. It is mixed with clay, Tschickas, used for making huts.

Poa alpina L. (Graminaceae). - Perennial grass. Alpine and subalpine meadows. Europe, Asia Minor, Caucasia. A valuable pasture grass in moutainous regions.

Poa arachnifera Torr., Texas Bluegrass. (Graminaceae). - Perennial grass. Western United States. Occasionally cultivated as a winter pasture.

Poa compressa L., Canada Bluegrass. (Graminaceae). -Perennial grass. Europe, N. America. Cultivated in Old and New World. Used in pastures, recommended for race-horses.

Poa cynosuroides Retz. → Eragrostis cynosuroides Beauv.

Poa epilis Scribn., Skyline Bluegrass. (Graminaceae). - Perennial grass. Western N. America. A pasture grass in mountainous areas.

Poa Fendleriana (Steud.) Vasey., Mutton Grass. (Graminaceae). - Perennial grass. Western United States to N. Mexico. A pasture grass for mountainous regions.

Poa microstachya Link. → Eragrostis lugens Nees.

Poa nemoralis L. (syn. P. rigidula Koch.) Wood Bluegrass. (Graminaceae). - Perennial grass. Europe. temp. Asia, N. America. Grown as a pasture grass and for hay.

Poa palustris L. (syn. P. fertilis Host., P. serotina Ehrh.) Fowl Bluegrass. (Graminaceae). - Perennial grass. N. America, Europe, Asia. Grown as a pasture grass and for hay.

Poa pratensis L., Kentucky Bluegrass. (Graminaceae). - Perennial grass. Europe, N. America. Cultivated as a forage crop and for hay.

Poa rupicola Nash., Timberline Bluegrass. (Graminaceae). - Perennial grass. Western N. America. Used as pasture grass in rocky areas and mountain meadows.

Poa scabrella (Thurb.) Benth., Pine Bluegrass. (Graminaceae). - Perennial grass. Western part of the United States to Baja California (Mex.). Important forage grass for lower elevations in California.

Podgrass, Shore → Triglochin maritima L.

Podocarpaceae → Dacrydium, Phyllocladus, Podocarpus, Saxegothea.

Podocarpus amara Blume (Podocarpaceae). - Tree. Java. Wood easily worked; made into boards, posts and beams.

Podocarpus dacrydioides Rich. Kahika. (Podocarpaceae). - Tree. New Zealand. Wood light colored. Used for inside work, framing of houses, white wood furniture, panels, dado-work, covering-boards of boats, butter-boxes. Fruits eaten by the Maoris. Woodpulp for manuf. paper.

Podocarpus elata R. Br. White Pine, She Pine, Goongum. (Podocarpaceae). - Tree. New South Wales, Queensland. Wood tough, silky, fine-grained, easily worked. Used for cabinet-work; not readily attacked by teredo and termites.

Podocarpus elongata L'Hér., Quinteniqua Yellowwood. (Podocarpaceae). - Tree. S. Africa. Wood light, soft, moderately strong, durable, elastic, resinous, pale yellow-brown; used for beams, planks, rafters, furniture, rail-road ties.

Podocarpus ferrugineus Don. (Podocarpaceae). - Tree. New Zealand. Wood hard, tough, not easily worked. Used for cabinet work, turnery, frames of houses. Gum used in medicine.

Podocarpus Hallii Kirk. (Podocarpaceae). - Tree. New Zealand. Wood close-grained, firm, dull-red. Used for marine piles, resists teredo, also for ship-building, bridges, wharves.

Podocarpus madagascariensis Baker. (Podocarpaceae). - Tree. Madagascar. Wood is valuable in general carpentry and for house-building in Madagascar.

Podocarpus neriifolia D. Don. (Podocarpaceae). - Tree. Himalayas to S. W. China, New Guinea. Fleshy receptacle of fruits eaten by the natives of Nepal. Wood close-grained, light, fairly hard, yellowish; much used in Burma for general carpentry.

Podocarpus oleifolius D. Don. (Podocarpaceae). - Tree. From Bolivia to Costa Rica. Wood yellow fine uniform structure, easy to work. Used for cabinet work, carving, general carpentry. Also P. coriaceus Rich. West Ind. Venez. and Colombia.

Podocarpus Rumphii Blume. (Podocarpaceae). - Tree. Mal. Archipelago. Wood light yellow, fine-grained, not too hard, easily worked and polished, not attacked by borers; recommended for turnery, native vessels and boards.

Podocarpus spicata R. Br. Black Pine. (Podocarpaceae). - New Zealand. Wood of much commercial value, heavy, close-grained, easily worked, brown; used for floors of ballrooms, bridges, construction work, framing, joisting, rail-road ties. Bark is occasionally used for tanning.

Podocarpus taxifolia H. B. K. (syn. P. montana [Willd.] Lodd.). (Podocarpaceae). - Tree. S. America. Wood is used in Colombia for furniture and fine wood work. Also P. macrostachys Parl.

Podocarpus Thunbergii Hook., Yellow Wood. (Podocarpaceae). - Tree. S. Africa, esp. Cape Peninsula, Natal to Zululand. Wood bright yellow, evenly grained, very easy to work; used for furniture, coach and wagon work, general building purposes.

Podocarpus spinulosa R. Br. (Podocarpaceae). - Tree. Queensland. Fruits size of a plum; used in preserves.

Podocarpus Totara Cunningh. Totara. (Podocarpaceae). - Tree. New Zealand. Wood much esteemed, deep red, straight-grained, compact, very durable, strong, easily split, durable in water, resists teredo. Used by Maoris for canoes; also general carpentry, building purposes, joists, rafters, wheather-boards, rail-road ties, telegraph poles, bridges, wharves, construction work where large spans are required.

Podophyllum Emodi Wall. Himalayan Mayapple. (Berberidaceae). - Perennial herb. Himalaya, Sikkim to Kashmir. Used since time immemorial by Hindu physicians for its bile-expelling properties. Replaces P. peltatum L. during times of scarcity. Fruits are edible.

Podophyllum peltatum L. Common May Apple, Ground Lemon. (Berberidaceae). - Perennial herb. Easter N. America to Florida and Texas. Fruits edible, agreeable taste, must be eaten with moderation, causing colic. Dried rhizomes are collected in autumn, are used medicinally as drastic purgative, skin irritant; in billious and chronic constipation. Roots are derived from Virginia, N. Carolina and the Central States. Contains a purgative podophyllo-resin. Drug was known to the Indians.

Podaxaceae (Fungi) → Podaxis.

Podaxis carcinomalis (L.) Fr. (Podaxaceae). - Basidiomycete. Fungus. Africa. Spores are used by Hottentot women as a face powder. Fruitpodies are used among the natives for treating cancer tumors.

Podaxis pistillaris (L.) Fr. (Podaxaceae). - Basidiomycete. Fungus. Africa to Australia. Young fruitbodies are consumed as food.

Poet's Narcissus → Narcissus poeticus L.

Poga oleosa Pierre. Ovoga. (Rhizophoraceae). - Tree. Trop. Africa, esp. Gabon. Seeds are used as a condiment by the natives. Source of a pleasant flavored oil.

Pogogyne parviflora Benth. (Labiaceae). - Perennial herb. California. Herb was used for flavoring pinole by Indians of California.

Pogostemon Cablin Benth. (syn. P. Patchouli Pellet.). Patchouly. (Labiaceae). - Herbaceous. Malaya, Philipp. Islands. Cultivated in tropical countries. Source of Oil of Patchouly, an ess. oil obtained from leaves; used for soaps and perfumes. Employed for scenting carpets, shawls, woven material etc. Also P. Heyneanus Benth.

Pointleaf Manzanita → Arctostaphylos pungens H. B. K.

Poinxter Flower → Azalea nudiflora L.

Poison Hemlock → Conium maculatum L.

Poison Nut, Amazon → Strychnos Castelnaei Wedd.

Poison Nut, Curare → Strychnos toxifera Benth.

Poison Nut, St. Ignatius → Strychnos Ignatii Berg.

Pokeberry, Common → Phytolacca decandra L.

Pokeberry, India → Phytolacca acinosa Roxb.

Pokeberry, Venezuela → Phytolacca rivinoides Kunth. and Bouché.

Polakowskia Tacaco Pitt., Tacaco. (Cucurbitaceae). - Herbaceous vine. Costa Rica. Cultivated. Fruits are eaten boiled by the natives. Sold in local markets.

Polemoniaceae → Loeselia.

Polianthes tuberosa L., Tuberose (Amaryllidaceae). - Bulbous plant. Origin uncertain, Mexico? Cultivated as ornamental. Chinese in Java use flowers in vegetable soup. Flowers are source of Tuberose Flower Oil obtained by cold enfleurage; 1150 kg. flowers are source of 1 kg. tuberose extraction; used for very high grade perfumery. It has been suggested that gardenia perfume can be build up by blending with tuberose flower oil, ylang ylang, heliotropine and hydroxycitronellal. Many modern French perfumes contain some amounts of tuberose flower oil.

Poliomintha incana (Torr.) Gray. (syn. Hedeoma incana Torr.). (Labiaceae). - Shrub. Southwestern United States. Herb is eaten raw or boiled by the Hopi Indians. Flowers are used for seasoning foods.

Polish Wheat → Triticum polonicum L.

Polyadoa Elliottii Hook. (Apocynaceae). - Tree. S. Leone (Afr.). Wood used by natives for making combs.

Polyadoa umbellata Stapf. (Apocynaceae). - Tree. Trop. Africa, esp. Lagos, S. Nigeria, Cameroons. Wood very hard; used for different purposes.

Polyalthia Oliveri Engl. (Annonaceae). - Tree. Trop. W. Africa. Bark is used in Liberia as vermifuge. In Comeroons stems are employed as spear-shafts.

Polygala angulata DC. (Polygalaceae). - Perennial plant. Brazil. Source of White Ipecac or Poaya Blanca.

Polygala butyracea Heck. (Polygalaceae). - Herbaceous plant. Trop. Africa. Said not to be known in a wild state, probably relic of an ancient culture. Source of a fibre; used for cloth, nets, fishing lines, gunny bags. Seeds when parched and ground are used in soup.

Polygala costaricensis Chodat. (Polygalaceae). - Herb. Centr. America. Used as a substitute for Ipecacuanha.

Polygala javana DC. (syn. P. tinctoria Vahl.) (Polygalaceae). - Herbaceous plant. E. India, Malaysia. Herb is source of a blue dye.

Polygala Rugelii Shuttlw., Yellow Bachelor's Button. (Polygalaceae). - Perennial herb. Florida. Infusion of the herb was used by the Seminole Indians for snake-bites.

Polygala Senega L., Seneca, Snake Root. (Polygalaceae). - Perennial herb. Eastern N. America. Used by Seneca Indians for snake-bites. Dried root, dug in autumn; used medicinally as expectorant, irritant, stimulant; contains senegin, a glucosid.

Polygala theezans L. (Polygalaceae). - Woody plant. Java to Japan. Leaves are used as a tea in some parts of E. Asia, called Thé des Carolines.

Polygalaceae → Carpolobia, Polygala, Securidaca, Trigoniastrum, Xanthophyllum.

Polygonaceae → Antigonum, Coccolobis, Emex, Eriogonum, Fagopyrum, Muehlenbeckia, Polygonum, Rheum, Rumex, Triplaris.

Polygonatum biflorum (Walt.) Ell., Small Salomon's Seal. (Liliaceae). - Perennial herb. Eastern N. America, southward to Florida and Texas. Starchy rootstock was used as food by the Indians. Recommended in times of want.

Polygonatum giganteum A. Dietr. (syn. P. canaliculatum Pursh.) (Liliaceae). - Perennial herb. Western N. America, Temp. Asia, Japan. Rhizomes are source of a starch; used in Japan They are also eaten preserved in sugar or syrup, and consumed boiled or roasted by the Ainu.

Polygonatum officinale Moench. Solomon Seal. (Liliaceae). - Perennial herb. Europe. Very young shoots can be consumed in spring as asparagus. Recommended as an emergency food plant during times of want.

Polygonum aviculare L., Prostrate Knotweed, Knot Grass. (Polygonaceae). - Perennial herb. Temp. zone. Cosmopolitic. Herb is used in some parts of Europe as home remedy for lung complaints, hemorrhoids and rheumatism.

Polygonum barbatum L. (Polygonaceae). - Shrub. Tropics. Sap from pounded leaves is applied on wounds, is an effective cicatrizant.

Polygonum Bistorta L., Bistort, Snake Root, English Serpentary. (Polygonaceae). - Perennial herb. Europe, Asia, N. America. Dried rhizome is used medicinally as a tonic and astringent. Rhizomes were used in stews and in soups by the Cheyenne Indians.

Polygonum cuspidatum Sieb. and Zucc. Japanese Fleeceflower. (Polygonaceae). - Perennial herb. China, esp. Hupeh, Szechuan, Kiangsu; Japan. Cultivated in different countries. Used in Chinese Pharmacy. Contains cuspidatin and emodin. Bark of roots is source of a yellow dye.

Polygonum Douglasii Greene. Douglas Knotweed. (Polygonaceae). - Herbaceous perennial. Western N. America. Seeds were made into flour and used as food by the Klamath and other tribes of Indians in Oregon and Washington.

Polygonum fugax Small. (Polygonaceae). - Perennial herb. Arct. regions. Roots are consumed by the Eskimos. Also P. Bistortum L. and P. viviparum L.

Polygonum maritimum L. (Polygonaceae). - Perennial herb. N. Temp. Zone. Herb used for wounds, caused by burning.

Polygonum Maximowiczii Regel. (Polygonaceae). - Annual herb. Japan. Cultivated in Japan. Leaves of a green-leaved variety, having sharp, acrid taste, are used in cooking.

Polygonum Muehlenbergi (Meisn.) S. Wat. (Polygonaceae). - Herbaceous plant. N. America. Young shoots were consumed as food by the Sioux Indians.

Polygonum odoratum Lour. (Polygonaceae). - Annual herb. Cochin China. Used by the natives for seasoning meat and fish. Cultivated.

Polygonum sachalinense F. Schmidt., Sachalin Knotweed. (Polygonaceae). - Perennial herb. Sachalin. Young shoots are consumed as food by the Ainu.

Polygonum tinctorium Ait. (Polygonaceae). - Herbaceous plant. China. Source of a blue dye.

Polygonum Weyrichii F. Schmidt. (Polygonaceae). - Perennial herb. Sachalin and Japan. Fruits eaten after having been pounded in a mortar; used with millets by the Ainu.

Polymnia edulis Wedd. (Compositae). - Perennial plant. S. America. Root is edible; used as food in some parts of Colombia.

Polymnia Uvedalia L. Yellow Leafcup. (Compositae). - Perennial herb. Eastern N. America. Root is used by some Indian tribes as stimulant, anodyne and laxative.

Polynesian Chestnut → Inocarpus edulis Forster.

Polypodiaceae. → Acrostichum, Adiantum, Aspidium, Asplenium, Athyrium, Ceropteris, Cyclophorus, Diplazium, Drynaria, Drypteris, Lastrea, Mohria, Onoclea, Onychium, Pleopeltis, Polypodium, Pteridium, Sadlera, Scolopendrium, Stenochlaena.

Polypodium angustifolium Sw. (Polypodiaceae). - Fern. S. Mexico to Peru. Used in Peru for fever and dropsy; in Mexico as pectoral and diaphoretic.

Polypodium aureum L. (syn. Phlebodium aureum L.) J. Smith. (Polypodiaceae). - Fern. Tropics and subtropics of America. Rhizomes used by Mazatec Indians (Mex.) and other tribes as febrifuge, sudorific; also for coughs. Sold in local markets of Mexico, known as Calaguala.

Polypodium fimbriatum Max. (Polypodiaceae). - Perennial herb. Trop. America. Used in Colombia as diuretic and expectorant.

Polypodium furfuraceum Schlecht. (Polypodiaceae). - Fern. Trop. America. Used in El Salvador as analgesic.

Polypodium lanceolatum L. var. trichophorum Weatherb. (Polypodiaceae). - Fern. S. Mexico to Guatemala. Decoction of rhizomes used by Chinantecs for coughs and as gentle febrifuge.

Polypodium plebejum Schlecht. and Cham. (Polypodiaceae). - Fern. S. Mexico to S. America. Used by Mazatecs (Mex.) as an expectorant and purgative.

Polyporaceae (Fungi) → Boletus, Cryptosporus, Echinodontium, Favolus, Fistulina, Fomes, Ganoderma, Gyroporus, Haplopilus, Polyporus, Polystictus Poria, Trametes.

Polyporus alveolaris (D. C.) Bong and Sing. (syn. Hexagona Mori Poll.). (Polyporaceae). - Basidiomycete Fungus. Throughout Europe, Asia and America. Used for the manuf. of a brown dye for various textiles, esp. in Italy.

Polyporus anthelminticus Berk. (Polyporaceae). - Basidiomycete Fungus. India. Mycelium parasites on bamboo. Fruitbodies are used as anthelmintic for cattle in Burma.

Polyporus arcularius (Batsch) Fr. (Polyporaceae). - Basidiomycete. Fungus. Tropics. Fruitbodies are occasionally consumed in Malaysia.

Polyporus australiensis Wakefield. (Polypodiaceae). - Basidiomycete. Fungus. Young fruitbodies are source of a brown dye, used for coloring raffia fiber in Flinders Island, Australia.

Polyporus betulinus Bul. (syn. Piptoporus betulinus Bull.) Karst. (Polyporaceae). - Basidiomycete. Fungus. Fruitbodies were used for manufacturing of razor straps. Wood that has been attacked by the mycelium is employed for burnishing watches in Switzerland.

Polyporus Farlowii Lloyd. (Polyporaceae). - Basidiomycete. Fungus. S. W. of the United States. Fruitbodies of the fungus are consumed boiled, baked, also stored for winter use by the Indians of New Mexico.

Polyporus frondosus Dicks. (Polyporaceae). - Basidiomycete. Fungus. Temp. zone. Fruitbodies are consumed as food. Also P. ramosissimus Schäff.

Polyporus grammocephalus Berk. (Polyporaceae). - Basidiomycete. Fungus. Trop. Asia. Very young fruitbodies are consumed as food by the natives. Also P. udus Jungh. from Java and P. vibecinus Fr. from Pac. Islands.

Polyporus hispidus Bull. (syn. Inonotus hispidus [Bull.] Karst). (Polyporaceae). - Basidiomycete. Fungus. Temp. and dry-hot zones. Employed in some parts of Europe and Asia for dying wool, silk and cotton.

Polyporus nidulans Fr. → Hapalopilus nidulans (Fr.) Karst.

Polyporus officinalis (Vill.) Fr. (syn. Boletus Laricis Jacq.) (Polyporaceae). - Basidiomycete. Fungus. The fruitbodies were used in chirurgy in the same way as Boletus or Fungus Chirurgorum for absorbing blood and secretions of wounds. They are sometimes used as purgative and as a source of agaricic acid and agaricine.

Polyporus ovinus Schäff. (Polyporaceae). - Basidiomycete. Fungus. Temp. zone. Young fruitbodies are consumed as food. Sometimes sold in markets in Europe. Also P. Pes Caprae Pers. and P. cristata Pers.

Polyporus sanguineus (L.) Meyer. (syn. Pycnoporus sanguineus L.) Murr. (Polypodiaceae). - Basidiomycete. Fungus. Tropics. Fruitbodies are used in East India for intestinal troubles, venereal diseases and eczema. Sold in markets.

Polyporus squamosus (Huds.) Fr. (Polyporaceae). - Basidiomycete Fungus. Temp. zone. Fruitbodies are used for manuf. of razor straps. Young fruitbodies are sometimes consumed as food.

Polyporus sulfureus Bull. (Polyporaceae). - Basidiomycete. Fungus. Cosmopolite. Fruitbodies are cosumed when fresh or dried. Also used to manufacture a dye.

Polyporus tuckahoe (Gússow) Lloyd. (Polyporaceae). - Basidiomycete. Fungus. Northwestern North America. The sclerotia, tuckahoe, are used as food by the Indians.

Polyporus tunetanus (Pat.) Sacc. and Trott. (Polyporaceae). - Basidiomycete Fungus. Mediterranean region. The fruitbodies are consumed as food. They are found in burned over Cistus vegetation.

Polyporus udus Jungh. (Polyporaceae). - Basidiomycete. Fungus. Tropics. Young fruitbodies are consumed as food by the natives of Java and Halmahera.

Polyporus versipellis Fr. (syn. Leccinum aurantiacum [Bull.] Gray.) (Polyporaceae). - Basidiomycete. Fungus. Temp. zone. The fruitbodies are consumed as food in some countries of Europe. Sold in markets. They are pickled in Russia, known as krassny grib. They are exported to Germany where they are used instead of the Truffles from France.

Polyscias fruticosa Harms. → Nothopanax fruticosum Miq.

Polyscias Rumphiana Harms. → Nothopanax pinnatum Miq.

Polystictus sacer Fr. (Polyporaceae). - Basidiomycete. Fungus. Tropics of the Old World. The sclerotium is used as a remedy for consumption, asthma and colds in some parts of southern Asia and Africa.

Polytoca bracteta R. Br. (Graminaceae). - Perennial grass. Mal. Archipelago. Suitable as food for cattle.

Pomaderris zizyphoides Hook. and Arn. → Alphitonia zizyphoides Gray.

Pomegranate → Punica Granatum L.

Pometia eximea Bedd. (syn. P. tomentosa Teijsm and Binn.) (Sapindaceae). - Tall tree. Java. Wood strong, durable, brownish-red; used for building purposes and heavy bridges.

Pometia pinnata Forst. (Sapindaceae). - Large tree. Pacif. Islands. Fruits edible, roundish, smooth, 5 cm. or more in diam., aromatic, juicy, sweet, pleasant flavor. Seeds are consumed roasted or boiled. Wood is tough, hard, flexible, light red; used for cooperage, house building, agricultural implements and furniture.

Pompholyx sapidum Corda (Sclerodermataceae). - Basidiomycete. Fungus. Central and eastern Europe. The truffle-like fruitbodies are consumed as food and sold in markets.

Pompon Lily → Lilium pomponicum L.

Pompona Bova → Vanilla Pompona Schiede.

Poncirus trifoliata Raf. x Citrus sinensis Osbeck. Trifolate Orange x Sweet Orange. Citrange. (Rutaceae). - Tree. A hybrid group. Resistant to cold. Fruits used for drinks, marmelades. Varieties for home-use are: Rusk, Colman, Morton and Rustic.

(Poncirus trifoliata Raf. x Citrus sinensis Osbeck) x Fortunella margareta Swingle. (Trifoliate Orange x Sweet Orange) x Nagami Kumquat. Citrangequat. Tree. (Rutaceae). - Of hybria origine. Varieties: Sinton and Telfair. Resistant to cold. Fruits of little economic importance. Sometimes used in drinks.

Pond Apple → Annona glabra L.

Pondlily, Yellow → Nuphar advena Ait.

Pond Pine → Pinus serotina Michx.

Pongam Oil → Pongamia pinnata (L.) Merr.

Pongamia pinnata (L.) Merr. (syn. P. glabra Vent.). (Leguminosaceae). - Tree. Trop. Asia, Australia, Pacif. Islands. Seeds are source of Pongam or Hongay Oil; used in several countries of Asia as illuminant and treatment of skin diseases; has been recommended for manuf. soap and candles. Sp. Gr. 0.937; Sap. Val. 147—189; Iod. No. 83—84; Unsap. 2.4—9 %.

Pontederiaceae → Eichhornia, Monochoria.

Pontiak Manila → Agathis alba Foxw.

Pontianac Gutta → Dyera costulata (Miq.) Hook. f.

Pony Beebalm → Monarda pectinata Nutt.

Ponga Oil → Pongamia pinnata Merr.

Pop Corn → Zea Mays L.

Poplar, Arizona → Populus arizonica Sarg.

Poplar, Balsam → Populus balsamifera Muench.

Poplar, Black → Populus nigra L.

Poplar, Japanese → Populus Maximowiczii Henry.

Poplar, Silver Leaf → Populus alba L.

Poplar, White → Populus alba L.

Poplar, Yellow → Liriodendron tulipifera L.

Poppy, Californian → Eschscholtzia californica Cham.

Poppy, Corn → Papaver Rhoeas L.

Poppy, Horned → Glaucium flavum Crantz.

Poppy, Opium → Papaver somniferum L.

Poppy, Red → Papaver Rhoeas L.

Poppy Seed Oil → Papaver somniferum L.

Populus alba L., White Poplar, Silver-Leaf. Poplar. (Salicaceae). - Tree. Throughout Europe, Caucasia, W. Centr. Asia, Hamalaya. Cultivated. Wood yellowish, relatively coarse, very light and soft, elastic, not easy to bend; used for inner construction of rail-road cars, wooden shoes, matches, turnery, packing material; source of cellulose. Bark is used medicinally as tonic and febrifuge, being composed of flat pieces or quills. It contains salicin and glucosic populin.

Populus arizonica Sarg. Arizona Poplar. (Salicaceae). - Tree. S. W. of United States and adj. Mexico. Wood is used for watertroughs, carts and cartwheels.

Populus balsamifera Muench. (syn. P. Tacamahaca Mill.). Balsam Poplar. (Salicaceae). - Tree. N. America to Florida and Louisiana. Source of Takamahak, Balm of Gilead, Balsam Poplar Buds, much is derived from northern United States and Canada; being composed of air-dried, closed buds. Considered a stimulant and expectorant, contains balsamic resin, gallic acid, malic acid, populin, salicin and a light yellow ess. oil. Wood light brown; used for manuf. boxes, pails also for paperpulp and excelsior.

Populus grandidentata Michx., Largetooth Aspen, Great Aspen. (Salicaceae). - Tree. Eastern N. America. Wood used for manuf. of excelsior, paperpulp, matches, wooden-ware, veneer; occasionally employed as lumber. Cambium when boiled is consumed as food by serveral Indian tribes.

Populus heterophylla L., Swamp Cottonwood, Black Cottonwood. (Salicaceae). - Tree. N. America to Florida and Louisiana. Wood dull, grayish brown; manuf. in Mississippi Valley and Gulf States into lumber, known as Black Poplar; used for interior finish of buildings.

Populus Maximowiczii Henry. Japanese Poplar. (Salicaceae). - Tree. Pacific East Coast of Asia, Sachalin, Japan, Kamtschatka, Amurland, Manchuria, Korea. Wood dark-grey; used for matches, boxes, paperpulp, wooden wares.

Populus nigra L., Black Poplar. (Salicaceae). - Tree. Throughout Europe, temp. Asia. Cultivated. Wood coarse-grained, light-brown, very light, very soft, easy to split, not easy to bend, slightly elastic, not durable. Used for inner construction of rail-road cars, boxes, wood-carving, cigar-boxes, manuf. of cellulose. Buds are source of a resin, made into a salve, used in home-remedies; also considered an excellent medium for hair-growing.

Populus tremula L. European Aspen. (Salicaceae). - Shrub or tree. Throughout Europe, N. Africa, Caucasia, Siberia to Japan. Wood yellowish-white, very soft, light, elastic, coarse-grained, easy to split, durable indoors, not durable outdoors; used for interior of rail-road cars, cigar-boxes, wooden shoes, wood-carving, excelsior. Source of charcoal, used for gum-powder.

Populus tremuloides Michx., Trembling Aspen. (Salicaceae). - Tree. N. America and Mexico. Wood is much used for excelsior, woodpulp, matches, veneer and lumber.

Populus Wislizeni (S. Wats.) Sarg. (Salicaceae). - Tree. Southwestern N. America. Catkins or groups of male and female flowers were consumed by the Pueblos of New Mexico.

Poria cocos Wolf. (syn. Pachyma cocos Wolf.) Fr. (Polyporaceae). - Basidiomycete. Fungus. Temp. zone. Sclerotia growing on roots of trees are called Tuckahoe or Indian Bread. They are consumed as food by the Indians. Used in Chinese medicine.

Porlieria angustifolia (Engelm.) Gray., Soap Bush. (Zygophyllaceae). - Shrub or small tree. Mexico and S. W. Texas. Bark from roots is made into balls, sold as a kind of „amole"; used for washing woolen goods by the Indians.

Porlieria hygrometra Ruiz and Pav. (Zygophyllaceae). - Shrub. S. America, esp. Argentina, Peru, Chile. Herb is considered in some parts of S. America as vulnerary; used for rheumatism.

Porphyra leucosticta Thurot. (Bangiaceae). - Red Alga. Sea Weed. Pac. Ocean, Indian Ocean, Mediterranean and adj. waters. Considered a great delicacy among the Hawaiians. Eaten with limpet, a molusc and salt. It is called Limu Luan.

Porphyra nereocystis Anders. (Bangiaceae). - Red Alga. Sea Weed. Pac. Ocean and adj. waters. Is consumed as food in China.

Porphyra perforata J. Ag. (Bangiaceae). - Red Alga. Pacific Coast of North America, esp. California. Used as food by the Chinese, exported to China. Also eaten by the Indians of northern California.

Porphyra suborbiculata J. Ag. (Bangiaceae). - Red Alga. Sea Weed. Pac. Ocean. Consumed as a much prized food among the Chinese. Is much consumed during their New Year Feasts. It is called Tsu Choy. Cooked in soups, stewed meats, gravy; also made into pickles, preserves and sweet meats.

Porphyra tenera Kjellm. (Bangiaceae). - Red Alga. Coast of Eastern Asia, esp. around China and Japan. Used for the preparation of amanori, a relish much enjoyed in Japan. Is frequently cultivated, esp. in Tokyo Bay. Generally used in cooking in China and Japan, often consumed with soy. In Chinese restaurants in America often used in „seaweed soup". In Western Europe, esp. Great Britain and Ireland similar species are eaten when salted with pepper, vinegar and oil.; is also often eaten in stews; sprinkled with oatmeal or eaten with bacon during breakfast.

Porphyra umbilicalis (L.) J. Ag., Slack, Sloke. (Bangiaceae). - Red Alga. Temp. North Atlantic. Consumed as food in Great Britain, Ireland and some other European countries.

Porphyra vulgaris J. Ag. (Bangiaceae). - Red Alga. Pac. Ocean, Atl. Ocean and adj. waters. Was consumed as food by the Indians of California. Also eaten along the coast of Great Britain where it is considered a delicacy in some coastal towns. Is cultivated along the coast of Japan, where it is collected in spring; dried and used as food, called Asakusanori.

Portia Tree → Thespesia populnea Soland.

Portland Arrowroot → Colocasia antiquorum Schott.

Portuguese Ipecacuanha → Naregamia alata Wight and Arn.

Portuguese Oak → Quercus lusitanica Lam.

Portulaca lutea Forster. (Portulacaceae). - Herbaceous plant. Soc. Islds. Shoots are consumed as potherb by the natives.

Portulaca napiformis F. v. Muell. (Portulacaceae). - Herbaceous plant. Queensland, N. Australia. Tubers are used as food by the natives.

Portulaca oleracea L. Common Purslane, Pursley. (Portulacaceae). - Annual, low succulent herb. Europe; introd. in N. America. Cultivated in some countries as a vegetable. Eaten boiled, sometimes as greens in salad and soups. Also preserved for winter-use.

Portulaca quadrifida L. (Portulacaceae). - Herbaceous. Trop. Africa, Asia. Bruised leaves are used in Egypt as anticephalic.

Portulaca retusa Engelm. (Portulacaceae). - Annual herb. Western N. America. Plant is consumed as a vegetable by the Tewa and Hopi Indians in the S. W. of the United States.

Portulaca triangularis Jacq. → Talinum triangulare (Jacq.) Willd.

Portulacaceae → Calandrina, Claytonia, Lewisia, Portulaca, Portulacaria, Talinum.

Portulacaria afra Jacq. (Portulacaceae). - Succulent herb. S. Africa. Plant is a valuable stockfood in S. Africa in times of drought.

Posh,Té → Annona scleroderma Safford.

Posidonia australis Hook. f. (Potamogitonaceae). - Perennial waterplant. Australia. Source of Posidonia Fibre, Cellonia, Lanamar; used for sacks, coarse fabrics; also mixed with wool. Recommended for packing material and for stuffing.

Posidonia Fibre → Posidonia australis Hook. f.

Posoqueria latifolia Roem. and Schult. Brazilian Oak. (Rubiaceae). - Shrub or tree. E. Brazil. Source of Brazilian Oak Walking-Sticks.

Post Oak → Quercus obtusiloba Michx.

Pot Marjoram → Origanum vulgare L.

Potherb Fame Flower → Talinum triangulare Willd.

Potamogetonaceae → Posidonia, Zostera.

Potanin Sumach → Rhus Potanini Maxim.

Potato, Air → Dioscorea bulbifera L.

Potato Bean → Apios tuberosa Moench.

Potato, Chinese → Dioscorea Batatas Descne.

Potato, Common → Solanum tuberosum L.

Potato, Fendler → Solanum Fendleri Gray.

Potato, Garden → Solanum tuberosum L.

Potato, Hausa → Coleus rotundifolius Chev. and Perrot.

Potato, Kaffir → Plectranthus esculentus R. Br.

Potato, Sweet → Ipomoea Batatas Poir.

Potato, Wild → Solanum Fendleri Gray.

Potato Yam → Dioscorea esculenta (Lour.) Burk.

Potentilla Anserina L., Silverweed. (Rosaceae). - Perennial herb. Temp. zone. Source of Herba Anserinae, used as a tea or in wine for diarrhea, leucorrhoea, kidneystones, and arthritis. Also recommended for cramp. Roots are edible and have been recommended as emergency food during times of want.

Potentilla erecta (L.) Hampe. (syn. Tormentilla erecta L.) Tormentil. (Rosaceae). - Perennial herb. Europe, temp. Asia; introd. in N. America. Rhizome used as a tea; a tincture is employed for chronic intestinal catarrh and in brandy as a stomachic. Dried rhizome is used medicinally as tonic, astringent, contains chinovin, tormentillic acid and chinova acid.

Potentilla fruticosa L. Bush Cinquefoil. (Rosaceae). - Small shrub. Temp. to arct. zone N. Hemisphere. Dried leaves are used as a tea by the Eskimos in some parts of Alaska.

Potentilla glandulosa Lindl. (Rosaceae). - Perennial herb. Western N. America. Boiled leaves were source of a drink; used by the Indians of British Columbia.

Potentilla palustris Scop. → Comarum palustre L.

Potentilla rupestris L. (Rosaceae). - Temp. Europe, Asia. Leaves are used as a tea in some parts of Russia and Siberia, called Siberian Tea.

Potherb Mustard → Brassica japonica Sieb.

Pothoidium Lobbianum Schott. (Araceae). - Herbaceous plant. Philippine Islands. Source of a fibre, used in Philippines for fish-corrals.

Pothos cannaefolia Dryand. → Spathiphyllum candicans Poepp. and Endl.

Poui, Yellow → Tecoma serratifolia G. Don.

Poulard Wheat → Triticum turgidum L.

Poulsenia armata (Miq.) Standl. Tuma, Cucúa. (Moraceae). - Tree. S. Mexico, Centr. America to S. America. Bark used by aborigines for hammocks, mats, clothing, blankets.

Poupartia amazonica Ducke. (Anacardiaceae). - Tree. Brazil. Fruits are edible, consumed in different parts of Brazil.

Pourouma cecropiaefolia Mart. (Moraceae). - Tree. Brazil. Cultivated in Amazonbasin. Berries round, fleshy, flavor of a grape; consumed in different parts of Brazil.

Pouteria guyanensis Aubl. Jan Snijder. (Surinam). (Sapindaceae). - Tree. Guiana. Wood very hard, strong, durable, dark brown; used for furniture.

Pouzolzia occidentalis Wedd., Yaquilla. (Urticaceae). - Woody plant. Centr. America, Guiana, Venezuela. Fibre from the bark, called Yaquilla, is supposed to have similar properties as Ramie.

Pouzolzia pentandra Benn. (Urticaceae). - Woody plant. Trop. Asia. Fibres are used by the natives for manuf. sails.

Pouzolzia viminea Wedd. (Urticaceae). - Shrub. E. India, Malaysia. Fibres are used by the Javanese, since time immemoreal, for manuf. fishing nets and similar material.

Powdery Swiss Lichen → Nephroma parilis Ach.

Po-Yak Oil → Afrolicania elaeosperma Mildbr.

Pradosia lactescens Radlk. → Lucuma glycyphloea Mart. and Eichl.

Prainea Limpato Beumée → Artocarpus Limpato Miq.

Prairie Apple → Psoralia esculenta Push.

Premna arborea Roth → Gmelina arborea Roxb.

Premna integrifolia L., Headache Tree. (Verbenaceae). - Shrub or small tree. Malaysia. Wood very hard, close-grained, beautifully veined; used by the Javanese for handles of knives.

Premna nauseosa Blanco. (Verbenaceae). - Shrub. Trop. Asia. Leaves are occasionally used instead of those of Piper bettle for chewing with Arec Nuts.

Prickly Ash → Aralia spinosa L. and Zanthoxylum Clava-Herculis L.

Prickly Chaff Flower → Achyranthes aspera L.

Prickly Comfrey → Symphytum asperrinum Don.

Prickly Currant → Ribes lacustre Poir.

Prickly Gooseberry → Ribes Cynosbati L.

Prickly Juniper → Juniperus Oxycedrus L.

Prickly Leaved Elephant's Foot → Elephantopus scaber L.

Prickly Lettuce → Lactuca Scariola L.

Prickly Poppy, Mexican → Argemone mexicana L.

Prickly Poppy Seed Oil → Argemone mexicana L.

Prickly Turkey (gum). → Acacia Senegal (L.) Willd.

Primavera → Cybistax Donell-Smithii (Rose) Seib.

Primrose, Evening → Oenothera biennis L.

Primrose Malanga → Xanthosoma violaceum Schott.

Primula veris L. (syn. P. officinalis (L.) Hill.). Cowslip. (Primulaceae). - Perennial herb. Europe, temp. Asia. Flowers and roots are used in Europe as home-remedy for coughs and to cause perspiration. Leaves substitute for tea, also used to improve nervous condition. Contains primulin and cyclamin.

Primulaceae → Anagallis, Cyclamen, Dodecatheon, Glaux, Lysimachia, Primula, Samolus.

Princewood, Caribbean → Exostemma caribaeum Roem. and Schult.

Pringlea antiscorbutica R. Br., Kerguelen Cabbage. (Cruciferaceae). - Perennial herb. Antarctic regions. Eaten as cabbage by the Kerguelen. Leaves can be consumed as salad. Antiscorbutic.

Prioria copaifera Griseb. Cativo. Cautivo. (U. S. trade name). (Leguminosaceae). - Tree. Trop. America, esp. Colombia, Jamaica, Panama, Nicaragua, Costa Rica. Wood light brown, sometimes streaked, firm, strong, fine texture, easy to work, straight-grained, not durable, attacked by white ants. Used in Colombia for coarse furniture, boxes, crates, scaffolding and in U. S. A for veneer.

Pritchardia filifera Lindl. → Washingtonia filifera (Lindl.) Wendl.

Pritzelia sanguinea Kl. → Begonia sanguinea Raddi.

Privet, European → Ligustrum vulgare L.

Privet, Glossy → Ligustrum lucidum Ait.

Privet, Ibota → Ligustrum Ibota Sieb.

Privet, Japanese → Ligustrum japonicum Thunb.

Privet, Swamp → Forestiera acuminata (Michx.) Poir.

Propionibacterium Freudenreichii van Niel. (Bacteriaceae). - Bacil. Microorganism. Is important in the propionic acid fermentation. Also P. Jensenii, P. Peterssonii, P. Stermanii, P. pentosaccum and P. thoenii.

Proso Millet → Panicum miliaceum L.

Prosopis Algarobilla Griseb. (Leguminosaceae). - Tree. Brazil. Wood compact, little elastic; used for posts and construction work.

Prosopis juliflora DC., Mesquite. (Leguminosaceae). - Small tree. Trop. America. Wood close-grained, dark brown to red, heavy, resists influence of soil; used for railroad ties, fence-posts, fellies of wheels, fuel and charcoal. The ripe pods furnish an important food in some Mexican and Indian communities, also used as fodder for live-stock. Mesquite Gum forms an adhesive mucilage and may be used as emulsifying agent. The tears of this gum are smooth, light yellowish brown to dark brown. Roots contain 6 to 7 % tannin. Flowers are source of honey.

Prosopis nigra Hieron. (Leguminosaceae). - Tree. Argentina to Brazil. Wood is compact, resistant, durable; used for general carpentry. Fruits are used in some parts of Argentina for manuf. an alcoholic beverage.

Prosopis odorata Torr. and Frem. → Strombocarpa odorata (Torr. and Frem.) Torr.

Prosopis oblonga Benth. (Leguminosaceae). - Tree. Trop. Africa. Wood source of charcoal used by the Bongos and Jurs (Afr.) for iron smelting. Bark is used for tanning. Pods are employed to stupefy fish. Wood is hard, strong, very heavy, fine-grained, dark reddish brown, takes polish; used for carpentry, boat building and implements.

Prosopis pubescens Benth. (Leguminosaceae). - Shrub or tree. Mexico and Southwestern United States. Wood durable; used for tool handles. Fruits are edible, being source of an alcoholic beverage; used by the Indians of the Colorado River region.

Prostrate Amaranth → Amaranthus blitoides Wats.

Prostrate Knotweed → Polygonum aviculare L.

Protea mellifera L. Sugar Protea. (Proteaceae). - Shrub or small tree. S. Africa. Nectar from flowers is made by some farmers into a delicious syrup, Syrupus Protae; used for coughs and pulmonary affections. Flowers are source of an excellent honey.

Proteaceae → Banksia, Faurea, Gevuina, Grevillea, Hakea, Karmadecia, Knightia, Leucadendron, Leucospermum, Lomatia, Macadamia, Mimetes, Panopsis, Protea, Roupola, Stenocarpus.

Protium Aracouchini March. (syn. Icica Aracouchini Aubl.). (Burseraceae). - Tree. S. America. Source of a balsam; used for healing wounds.

Protium Copal L. (Burseraceae). - Small tree. Centr. America. Stem is source of a resin, used as incense in religious ceremonies among the Indians. Was already used for this purpose by the ancient Mayas.

Protium guianense March → Icica viridiflora Lam.

Protium heptaphyllum March. (syn. Icica heptaphylla Aubl.). (Burseraceae). - Tree. Trop. America. Source of an Elmira resin, called Hyawa Gum or Ronima Resin; used in varnishes and as an incense.

Protium Icicariba March. (syn. Icica Icicariba DC.). (Burseraceae). - Tree. Argentina, Paraguay, Uruguay. Stem is source of a resin, called Elemi Occidental.

Proustia pungens Poepp. (Compositae). - Shrub. Chile. Infusion of leaves and roots is used in some parts of Chile in baths for rheumatism and gout.

Provision Tree → Pachira aquatica Aubl.

Prunella grandiflora (L.) Jacq., Bigflower Selfheal. (Labiaceae). - Perennial Herb. Europe, temp. Asia. Herb is sometimes used in salads.

Prunus americana Marsh., American Wild Plum, River Plum. (Rosaceae). - Small tree. Eastern N. America to Florida, Texas and New Mexico, Rocky Mountain region and adj. states. Fruits are eaten raw or cooked, also made into preserves and jellies. Sold in local markets. Cultivated in some parts of Canada. Some varieties are: Cheney, Itasca, Aitkin, Oxford and Crimson.

Prunus Amygdalus Batsch. → Amygdalus communis L.

Prunus angustifolia Marsh., Chickasaw Plum, Mountain Cherry. (Rosaceae). - Small tree. Delaware to Florida and Mississippi. Fruits used in jellies and preserves. Sold in southern markets of the U.S.A. Occasionally cultivated. Varieties are: Caddo Chief, Ogeeche.

Prunus Armeniaca L. → Armeniaca vulgaris Lam.

Prunus avium L., Sweet Cherry, Mazard. (Rosaceae). - Tree. Europe, W. Asia. Cultivated in temp. zone of Old and New World. Fruits are consumed as dessert, canned, dehydrated, used in pies, source of beverages. Var. Juliana (L.) W. Koch. Heart Cherry, Geancherry Fruits heart-shaped with soft flesh of a dark color. Here belong: Black Tartarian, Black Republican, Early Rivers. Var. duracina (L.) W. Koch. Bigareaus. Hard Cherry, to which belong Yellow Spanish, Royal Ann, Esperen, Jaune de Buttner. Dukes are considered a hybrid of P. avium L. x P. Cerasus L. Cherry wood is hard, fairly heavy, difficult to split, elastic, easy to polish; used for turnery, furniture, instruments, wagons, inlaywork.

Prunus Besseyi Bailey., Western Sand Cherry. (Rosaceae). - Shrub. N. America. Stonefruits are purple black, sweet, edible.

Prunus brigantina Vill. → Armeniaca brigantina Pers.

Prunus cantabrigiensis Stapf. (syn. P. pseudocerasus Koids.). Chinese Sour Cherry. (Rosaceae). - Tree. Yangtze Valley (China). Cultivated in some parts of China. Tree produces a sour cherry; used as preserve.

Prunus Capollin Zucc., Capulin. (Rosaceae). - Tree. Mexico to Peru. Wood is used for general carpentry, cabinet work. Juice from the fruits is mixed with cornmeal to make a cake, called Cupultamal; consumed by the Indians in some parts of Latin America. Distilled water from the leaves is used as a cherry water. Fruits are eaten raw, stewed or preserved.

Prunus cerasifera Ehrh. (syn. P. Myrobalana Lois.). Cherry Plum. (Rosaceae). - Tree. W. Asia. Cultivated. Fruits subglobose, 2 to 3 cm. in diam, edible. Plant is used as stock of plum varieties.

Prunus Cerasus L., Sour Cherry. (Rosaceae). - Tree. S. E. Europe, W. Asia. Cultivated in temp. zones. Cherries are chiefly used for canning, as pitted cherries; used in pies, puddings; made into preserves, cherry cider, jelly, glacéfruits also put in brandy. Source of Kirschwasser, cherry brandy; the juice of Marasca, chiefly grown in Dalmatia, is an ingredient of a liqueur, called Marachino. Seeds are source of Cherry Kernel Oil, semi-drying oil, when refined used for cosmetics and as salad oil. Sp. Gr. 0.9222–0.929; Sap. Val. 192—198; Iod. No. 111—122. Stem is source of a gum; used in some countries in the cotton printing industry. Wood is fairly heavy, difficult to split, elastic, easy to polish; used for turnery, instruments and furniture. Varieties are: Montmorency, Early Richmond, Eugenie, English Morello. Leaves are used as substitute for tea.

Prunus domestica L., Plum Tree. Garden Plum. (Rosaceae). - Shrub or tree. Europe, W. Asia. Grown since antiquity. Used among Lake Dwellers. Cultivated in temp. zones. Includes most plums and prunes. Fruits consumed fresh, preserved, canned; made into jam, plum butter, juice, alcoholic beverages, liqueurs. Used in commercial canning. Prunes are derived from varieties with a high sugar content; they are dried artificially or in the sun and cured. Many prunes are derived from the United States, Jugoslavia and France. Some plum varieties are: Royale Hative, Tragedy, Yellow Egg, Columbia, Quackenbos, Victoria, Pond's Seedling, Coe's Golden Drop, Ickworth Imperatrice. Suitable prunes are: Prune d'Angen, Robe de Sergeant, Golden Prune, Silver Prune. Hybrids between different species are: Wickson, President, Gonzales, Excelsior, Golden. Wood reddish-brown, fairly hard, rather difficult to split, little durable; used for turnery and taps.

Prunus Grayana Maxim. Gray's Chokecherry. (Rosaceae). - Small tree. Hokkaidô, Shikoku (Jap.), W. Hupeh (China). Wood hard, compact, easily split, pale yellowish-brown. Used in Japan for printing blocks, engraving, turnery, handles of tools and implements, alpine-sticks, furniture, utensils. Salted flower buds and young fruits, having a pungent taste, are consumed in Japan.

Prunus hortulana Bailey., Hortulana Plum. (Rosaceae). - Tree. Central United States. Cultivated in So. States. Used in preserves and marmelades. Varieties are: Golden Beauty, Kanawha, Leptune, Moreman, Cumberland. Wood hard, strong, heavy; suitable for turnery.

Prunus instititia L., Bullace Plum, Damson Plum. (Rosaceae). - Shrub or small tree. Europe. Cultivated since ancient times. Var. syriaca (Borkh.) Koehne. Yellow Mirabella, with yellow, subglobose fruits, also Red Mirabelle, Mirabelle of Nancy, Early Mirabelle of Bergthold. Var. italica (Borkh.) Aschers and Graebn. Green Gage or Reine Claude, fruits green or greenish, also Small Greengage and Early Greengage.

Prunus Laurocerasus L. (syn. Laurocerasus officinalis Roem.). Cherry Laurel. (Rosaceae). - Shrub or small tree. S. E. Europe, Asia Minor. Cultivated as ornamental. Sometimes used as aromatic instead of bitter almond: small amount of leaves are employed for flavoring boiled milk. Poisonous in large quantities. Source of Cherry-Laurel Water, anodyne, antispasmodic, sedatve; used for coughs.

Prunus Mahaleb. L., Mahaleb Cherry, St. Lucie Cherry. (Rosaceae). - Tree. Europe, W. Asia. Wood called St. Luciewood, is hard, very heavy, difficult to split, polishes well; cumarin-scented; used for turnery, cigaret holders, tobacco pipes, walking sticks, snuff boxes. Decoction of leaves is used for flavoring milk; leaves are also used as a substitute for tobacco.

Prunus maritima Wang., Beach Plum. (Rosaceae). - Shrub. New Brunswick to Maine, Virginia. Fruits are made into an excellent jelly and preserves. Sold in local markets.

Prinus Maximowiczii Rupr. Miyana Cherry. (Rosaceae). - Tree. Hokkaidô, Honshū, Kyushu (Jap.), Korea, Manchuria, Amur. Wood very hard, heavy, close-grained, dark reddish-brown. Used in Japan for furniture, utensils, sporting goods, sculptures.

Prunus melanocarpa (A. Nels.) Rydb., Rocky Mountain Cherry. (Rosaceae). - Small tree. Western N. America. Fruits are used for jelly and jam. Favorite fruit among the Indians.

Prunus Mume Sieb. and Zucc. → Armeniaca Mume Sieb.

Prunus Munsoniana Wight and Hedrick., Wild Goose Plum. (Rosaceae). - Tree. N. America. Fruits edible, used for jellies, jams and preserves. Cultivated. Varieties are: Newman, Wildgoose, Milton, Osage, Texas Belle.

Prunus Padus L. (syn. Padus avium Mill.)., European Bird Cherry. (Rosaceae). - Tree. Europe, temp. Asia to Korea and Japan. Fruits in racemes, round, 6 to 8 mm. across, black; were formerly much used in jam since ancient times. Wood brownish, fine-grained, easy to work, heavy, hard, durable, polishes well; used for furniture, inside finish of houses, boat building, walking sticks.

Prunus paniculata Thunb. (syn. Prunus pseudo-cerasus Lindl.). (Rosaceae). - Tree. Japan. Cultivated in Japan. Flowers of the double flowering varieties that do not readily fall off, are salted and used as a tea. Variety: Yedozakura is especially esteemed for this purpose.

Prunus Persica (L.) Batsch. → Amygdalus Persica L.

Prunus Puddum Roxb. (syn. P. cerasoides D. Don.). (Rosaceae). - Tree. India, Himalayan region. Stem is source of a Cherry Gum; used for adulterating Tragacanth.

Prunus salicina Lindl. (syn. P. triflora Roxb.). Chinese or Japanese Plum. (Rosaceae). - Tree. China. Cultivated in California, Japan and China. Fruits edible. Varieties are Red June, Abundance, Satsuma and Kelsey.

Prunus serotina Ehrh., Wild Black Cherry, Rum Cherry. (Rosaceae). - Tree. N. America to Florida, Texas and New Mexico. Wood strong, hard light, close-grained, light brown; used for cabinet-making, interior finish of buildings. Fruits are used for flavoring alcoholic liqueurs. Bark is used medicinally; collected in autumn; used as a tonic and sedative. Much is derived from Michigan, Indiana, Virginia, N. Carolina. Also used in home-remedies as a syrup for coughs and bronchitis.

Prunus Simonii Carr., Apricot Plum. (Rosaceae). - Tree. N. China. Cultivated. Fruits are either bitter with an almond-like astringency while some have an agreeable taste.

Prunus spartoides (Spach.) Schneid. → Amygdalus spartoides Spach.

Prunus spinosa L. Blackthorn, Sloe. (Rosaceae). - Spiny shrub or small tree. Europe, Mediterranean region, W. Asia. Fruits globose, blue-black, with a bloom. 1 to 1.5 cm. across. Used for making liqueurs. Wood dark brown; used for turnery and walking sticks.

Prunus Ssiori F. Schmidt. (Rosaceae). - Tree. Hokkaidô, Honshū (Jap.), Manchuria, China. Wood heavy, strong, hard, tenaceous, close-grained, reddish-brown. Used in Japan for utensils, furniture, cases of instruments, engraving. Used for the shaft of implements that are used for collecting Laminaria (Brown-Algae).

Prunus subcordata Benth. Klamath Plum. Pacafic Plum. (Rosaceae). - Bush or small tree. California, Oregon. Fruits edible, eaten fresh or made into preserves. Wild fruits are gathered in large quantities.

Prunus tomentosa Thunb. Manchu Cherry. (Rosaceae). - Tree. Japan. Cherries are sweet, juicy, consumed in Japan.

Prunus umbellata Ell. Fladwoods Plum. Black Sloe, Hog Plum. (Rosaceae). - Small tree. Coastal Plain to S. Carolina and Florida. Fruits are much used for jelly and jam.

Prunus virginiana L. Common Choke Cherry. (Rosaceae). - Small tree. N. America. Bark used medicinally as a sedative, pectoral, tonic and astringent. Fruits are used for making pies and jellies. Occasionally sold in markets.

Psalliota arvensis (Schäff.) Fr. (syn. Agaricus arvensis Schäff., A. pratensis Scop.). Field Mushroom. (Agaricaceae). - Basidiomycete. Fungus. Temp. zone. Mushroom. Fruitbodies are edible and of fine quality.

Psalliota campestris (L.) Fr. (syn. Agaricus campestris L.) White Mushroom, Champignon. (Agaricaceae). - Basidiomycete. Fungus. The fruitbodies are an important food. This species is widely cultivated and is a considerable export article in many countries. The pure culture mycelium which is used for raising the fungus is called Spawn. The champignons are an important canning product.

Psamma arenaria R. and Sch. → Ammophila arundinacea Host.

Psathyrella Candolleana (Fr.) A. H. Smith. (syn. Hypholoma Candolleanum (Fr.) Quél., Agaricus appendiculatus Bull.) (Agaricaceae). - Basidiomycete. Fungus. Temp. zone. The fruitbodies are sometimes consumed in the U. S. A.

Pseudocedrela Kotschyi (Schroeinf.) Harms. (Meliaceae). - Tree. Trop. Africa, esp. Gold Coast. Source of a soluble gum, resembling that of Acacia seyal, though of inferior quality. Bark source of a brown dye used by the natives for dyeing cloth. Decoction of bark is used for fever and rheumatism.

Pseudocinchona africana Chev. (Rubiaceae). - Tree. Trop. Africa. It is claimed that bark has strong febrifugic properties.

Pseudocymopterus aletifolius Rydb. (Umbelliferaceae). - Herbaceous plant. New Mexico. Leaves cooked or as greens were eaten by Indians in New Mexico.

Pseudomonas ureae Berg. et al. (Bacteriaceae). - Bacil. Agriculturally a useful soil microorganism, being capable of converting urea into ammonium carbonate.

Pseudostachyum polymorphum Munro. (Graminaceae). - Large shrubby bamboo. E. India. Valuable for basket-work on tea estates. Culms. easy to split, flexible, durable.

Pseudotsuga taxifolia (Lam.) Britt. (syn. P. mucronata (Raf.) Sudw., P. Douglasii Carr.). Douglas Fir, Douglas Spruce, Red Fir. (Pinaceae). - Tree. Western N. America. Important lumber tree. Cultivated as forest tree in N. America and Europe. Wood variable, in density and quality, light yellow or red; made into lumber; used for construction, rail-road ties and fuel; source of wood pulp. Bark is used for tanning. Source of Oregon Balsam. A tea is occasionally made from the leaves.

Psiadia dodoneaefolia Steetz. (Compositae). - Shrub. Trop. Africa, Madagascar. Leaves are used by natives of Madagascar for annealing new water-pitchers.

Psidium Araca Raddi. (syn. P. guineense Sw.). Brazilian Guava, Araca. (Myrtaceae). - Shrub. Brazil. Occasionally cultivated. Fruits edible; used for jellies.

Psidium Cattleianum Sab., Strawberry Guava, Cattley Guava. (Myrtaceae). - Small bushy tree. Brazil. Introduced in tropics and subtropics. Fruits round, reddish-green, sweet, aromatic, eaten raw, excellent for jellies; recommended for custards, ices and drinks. Var. lucida, Chinese Strawberry Guava produces fruits of better quality, eaten raw, they are yellow.

Psidium cinereum Mart. (Myrtaceae). - Shrub. Brazil. Leaves are astringent. Fruits are used in some parts of Brazil for hemorrhages.

Psidium fluviatile Rich. (syn. P. guineense Pers). (Myrtaceae). - Shrub or small tree. Brazil, Guiana. Occasionally cultivated. Fruits are consumed by the natives.

Psidium Friedrichsthalianum Ndz., Costa Rican Guava. (Myrtaceae). - Small tree. Centr. America. Occasionally cultivated. Fruits acid; used for pies and made into jelly.

Psidium Guajava L. Common Guava. (Myrtaceae). - Shrub or small tree. West Indies, Mexico to Peru. Cultivated in the tropics and subtropics of the Old and New World. Often escaped of cultivation. Fruits eaten raw, made into jelly, preserves, pies, shortcakes. A thick jam, Guave Cheese is of commercial value in Florida and Cuba. Fruits are apple and pear shaped. There are few varieties among which Snow White, Persico, Guinea and Sour.

Psidium incanescens Mart. (Myrtaceae). - Woody plant. Brazil. Wood used in Brazil for agricultural tools, small objects. Leaves are used in medicine as astringent. Fruit is edible, has reputation of being antihemorrhagic.

Psidium molle Bertol., Guisaro. (Myrtaceae). - Shrub. S. Mexico, Centr. America. Sometimes cultivated. Fruits acid; used for jellies.

Psidium Oerstedeanum Berg., Arrayán. (Myrtaceae). - Small tree. Centr. America. Fruits are consumed by the natives; also made into jellies.

Psillium Seed → Plantago Psyllium L.

Psittacanthus calyculatus (DC.) Don. (syn. Loranthus calyculata DC.). (Loranthaceae). - Shrub parasitic in several species of trees. Mexico, Centr. America. Decoction of leaves and flowers is used for treating wounds, when distilled in water it is used as a cosmetic.

Psophocarpus palustris Desv. (syn. P. longepedunculatus Hassk.). (Leguminosaceae). - Perennial herb. Probably from trop. Africa. Cultivated. Young pods and rhizomes are consumed as food by the natives in some parts of Africa.

Psophocarpus tetragonolobus DC., Asparagus Pea, Goa Bean. (Leguminosaceae). - Perennial vine. Trop. Asia. Leaves, young sprouts and fruits are used in soups. Immature pods are eaten as vegetable. Ripe seeds when roasted are eaten with rice by the Javanese. Cultivated in the Old and New World.

Psoralea castorea S. Wats. Beaverbread Scurfpea. (Leguminosaceae). - Herbaceous plant. United States and adj. Mexico. Large roots were eaten cooked or boiled; also ground, made into mush or bread by Indians in Arizona and Nevada.

Psoralea esculenta Pursh., Common Breadroot. (Leguminosaceae). - Herbaceous perennial. Eastern and Southern United States. Turnip shaped roots are rich in starch, were consumed as food by the Indians and early settlers. Source of a starch made into cakes by Indians of several tribes. Has occasionally been recommended for cultivation.

Psoralea glandulosa L. (Leguminosaceae). - Herbaceous. Chile, Peru. Leaves are used as a tea in some parts of Chile.

Psoralea macrostachya DC. Leatherroot Scurfpea. (Leguminosaceae). - Perennial herb. Western United States. Inner bark is source of a fibre; used by the Indians for making a coarse thread.

Psoralea mephitica S. Wats. (Leguminosaceae). - Perennial herb. S. Utah, Arizona, adj. Mexico. Roots were eaten raw or cooked; also dried, ground into flour made into mush or bread by Indians of Utah. Roots are source of a yellow dye, used by the Luiseño Indians of California.

Psoralea pedunculata (Mill.) Vail. (Leguminosaceae). - Perennial herb. Eastern United States. Plant was used as aromatic bitter tonic by some Indian tribes.

Psorospermum febrifugum Spach. (Guttiferaceae). - Woody plant. Trop. Africa. Bark used as febrifuge and for leprosy by natives of Angola.

Psychotria Cooperi Standl. (Rubiaceae). - Shrub. S. America. Decoction of plant used for rheumatism in Chaco Intendancy of Colombia.

Psychotria emetica L. f. → Cephaelis emetica Pers.

Psychotria Ipecacuanha Stokes → Cephaëlis Ipecacuanha (Brotero) Rich.

Psychotria luzoniensis (Cham. and Schlecht.) Vill. (Rubiaceae). - Shrub. Philipp. Islds. Infusion of root is used as antidysenteric.

Psychotria rostrata Blume → Chasalia rostrata Miq.

Psychotria tomentosa Hemsl. (syn. Cephaelis tomentosa (Aubl.) Vahl.) (Rubiaceae). - Herbaceous. Colombia. Used as emmenagogue, also for relief of asthma.

Psyllium → Plantago Psyllium L.

Ptaeroxylon obliquum (Thunb.) Radlk., Cape Mahogony, Sneezewood. (Meliaceae). - Tree. S. Africa. Wood has a scent of pepper and causes sneezing; powder from wood is used by the natives as snuff for headache. Wood chips are used for moths.

Ptaeroxylon utile Eckl. and Zeyh., Sneezewood. (Meliaceae). - Tree. S. Africa. Wood hard, heavy, dense, close-grained, tough; used for gateposts, beams of bridges, harbour works. Employed by Kaffirs as tinder and to produce fire by friction.

Ptelea trifoliata L. Common Hop Tree. (Rutaceae). - Shrub or small tree. N. America. Bitter bark is used as a tonic. Fruits are occasionally used as a substitute for hops in making beer.

Pteridium aquilinum (L.) Kuhn. (syn. Pteris aquilina L.). Bracken, Eagle Fern. (Polypodiaceae). - Herbaceous plant. Fern. Cosmopolit. In spring tender stalks and fronds are boiled and eaten as a potherb in Japan. Rhizomes are consumed by some Indian races along the Pacific. Important food among the Maoris (New Zeal.). Has been used by Indians in N. America to destroy tapeworm.

Pteris aquilinia L. → Pteridium aquilinum (L.) Kuhn.

Pterocarpus Draco L. (syn. P. officinalis Jacq.) Dragonblood Padauk. (Leguminosaceae). - Tree. Guiana. Wood used in place of cork to keep fishing nets floating.

Pterocapus erinaceus Lam., African Rosewood, African Kino. (Leguminosaceae). - Tree. Trop. Africa. Wood valuable, fine-grained, hard, durable, dark red; used for cabinet work and turnery. Stem is source of Gum Kino, Sangue de Draco, a deep red, astringent resin; exported to Portugal and England. Is used by natives for treating wounds. Source of Camwood dye, derived from Zaria (Nigeria).

Pterocarpus lucens Guill. and Perr. (Leguminosaceae). - Tree. Trop. Africa. Wood hard, durable, yellowish-white; used for furniture and construction work.

Pterocarpus indicus Willd. (Leguminosaceae). - Tree. Trop. Asia. Wood durable, strong, beautifully colored; used for furniture.

Pterocarpus macrocarpus Kurz. (Leguminosaceae). - Tree. India: Burma. Wood source of Red Padauk or Brown Padauk, being reddish brown, hard, dense, relatively heavy; used for inlay work.

Pterocarpus Marsupium Roxb. Vengai Padauk. (Leguminosaceae). - Tree. E. India to Ceylon. Important lumber, durable, takes fine polish, seasons well; used for door and window frames, posts, furniture, agricultural implements, boat-building, carts, railway cariages, rail-road ties. Dried juice, source of Malabar or East Indian Kino is used medicinally, being astringent. Contains kinotannic acid.

Pterocarpus Osun Craib. Red Camwood. (Leguminosaceae). - Tree. Trop. Africa. Pounded heart-wood, bark and root are made into a paste, used by natives of So. Nigeria for coloring their skin. Source of Kino similar to P. erinaceus.

Pterocarpus santalinus L. f. Red Saunders. Sandalwood Padauk. (Leguminosaceae). - Tree. E. India, Ceylon to Philipp. Islands. Wood known as Red Sanderswood, Santalwood, Lignum Santali Rubrum is used for furniture, cabinet work. Source of a red color pigment, santalin. A colored powder is used for caste marks by the Hindus. Used as emetic since ancient times.

Pterocarpus Soyauxii Taub. African Padauk. Baywood, Red Wood. (Leguminosaceae). - Tree. West Trop. Africa. Source of a red to red-brown dye; used by natives of Ododobo, Afr. to color their bodies. Wood used for canoes and knife-handles.

Pterocarpus tinctorius Welw. (Leguminosaceae). - Tree. Trop. Africa. Wood and roots source of dye, resembling that of P. Soyauxii. Used in Pungo Adongo to paint newly-born children; also to color feet of stylish native women to imitate slippers and shoes.

Pterocarya fraxinifolia Spach. (syn. P. caucasica C. A. Mey). Caucasian Wing-nut. (Juglandaceae). - Tree. Caucasia to N. Iran. Wood light brown, soft, light, difficult to split; recommended for matches, wooden shoes.

Pterocarya rhoifolia Sieb. and Zucc., Japanese Wingnut. (Juglandaceae). - Tree. Japan. Wood light, soft, not easily cracked; used in Japan for chop sticks, wooden clogs and matches.

Pterocladia lucida (R. Br.) J. Ag. (Gelidiaceae). - Red Alga. Temperate S. Pacific, esp. Australia, Tasmania and New Zealand. Used as food in New Zealand. Other species are manufactured into agar in Japan and New Zealand.

Pterocelastrus variabilis Sond. (Celastraceae). - Tree. S. Africa. Leaves are source of tannin.

Pterogyne nitens Tul. (Leguminosaceae). - Tree. S. America, esp. Argentina, S. Paraguay, Brazil. Wood pinkish brown, becoming darker on exposure, resembles Swietenia, medium texture, heavy, tough, strong, hard; used in Argentina for cabinet work, furniture, interior finish of buildings, wheelwright work, rail-road ties, cooperage.

Pterolobium lacerans R. Br. (syn. Cantuffa exosa Gmel.) (Leguminosaceae). - Thorny tree. Trop. E. Africa. Leaves with dissolved iron rust are used for giving a black color to leather in some parts of E. Africa.

Pterorhachis Zenkeri Harms. (Meliaceae). - Shrub. Cameroon. Fresh bark has the taste of hazelnuts, used by the natives as aphrodisiac.

Pterospermum acerifolium Willd. (Sterculiaceae). - Tree. E. India. Mal. Archipelago. Wood strong, durable, resistant when under water, reddish-brown; used for houses, bridges, boats, canoes, the quality depends upon locality.

Pterospermum javanicum Jungh. (Sterculiaceae). - Tall tree. Java. Wood strong, durable, light, soft, close-grained; used for bridges, houses, boats and made into boards.

Ptychopetalum olacoides Benth. (Olacaceae). - Tree. Guiana, Brazil. Source of Muira Puama; used in medicine.

Ptychopetalum uncinatum Anselm. (Olacaeae). - Tree. Brazil. Plant is used in medicine, exported to Europe.

Puccinellia Nuttalliana (Schult.) Hitchc. Nuttall Alkali Grass. (Graminaceae). - Perennial grass. Western United States. An important forage grass, esp. in alkali soils.

Puchury Bean → Nectandra Puchury-minor Nees and Mart.

Pueraria Thunbergiana Benth. Thunberg Kudzu Vine. (Leguminosaceae). - Perennial to woody vine. China, Japan. Cultivated in mild countries of Old and New World. Grown as hay crop and green feeding for live-stock. Also for soiling. Recommended to prevent soil erosion. Roots are eaten cooked in some countries.

Puffball → Bovista nigrescens Pers., Calvatia lilacina (Berk. and Mont.) Lloyd and Lycoperdon spp.

Puffed Shield Lichen. → Parmelia physodes (L.) Ach.

Puget Balsamroot → Balsamorhiza deltoides Nutt.

Pugionum cornutum Gaertn. Sagai. (Cruciferaceae). - Annual to perennial herb. Salt prairies. Mongolia. Cultivated. Consumed as a vegetable in some parts of Mongolia.

Pulai → Alstonia scholaris R. Br.

Pulasan → Nephelium mutabile Blume.

Pulicaria crispa. Sch. (Compositae). - Herbaceous plant. Trop. Africa, Arabia, Egypt, India etc. Dried herb is used as vulnerary to bruises.

Pulmonaria officinalis L. Common Lungwort. (Boraginaceae). - Perennial herb. Europe, Centr. and S. Russia. Dried herb is used medicinally for treatment of bronchial affections.

Pulque → Agave atrovirens Karw., A. compluviata Trel., A. gracillispina Engelm., A. manisaga Trel. and A. melliflua Trel.

Pulsatilla → Anemone Pulsatilla L.

Pummelo → Citrus grandis Osbeck.

Pumpkin → Cucurbita Pepo L.

Pumpkin Seed Oil → Cucurbita Pepo L.

Punica Granatum L. Common Pomegranate. (Punicaceae). - Shrub or small tree. S. Europe. N. W. India. Widely cultivated in mild climates of Old and New World. Known since ancient times. Commercially grown in Mediterranean region, California. Fruits edible, roundish, size of an orange; skin tough, leathery. Fruits divided into unequal sections. Seeds surrounded by an edible, red, pink, sometimes whitish sub-acid pulp. Source of Grenadine, a refreshing drink. Several varieties, among which: Paper-Shell, Spanish Ruby, Wonderful, Jaffa, Granada Blanca, Tunsi, Galsi (white), Selimi or Bagdad (very large). Dried Pomegranate Bark or Granatum Bark, of which the wild is prefered, is anthelmintic, taenifuge; used for intermittent fever, diarrhea, night sweats. Contains the alkaloids: pelletierine and granatonine. Peel of fruits is source of tanning material, contains 26 % tannin. Produces a fine leather.

Punicaceae → Punica.

Purging Cassia → Cassia Fistula L.

Purging Flax → Linum catharticum L.

Purging Nut → Jatropha Curcas L.

Purple Cane Raspberry → Rubus parvifolius L.

Purple Cone Flower → Brauneria angustifolia (DC.) Heller.

Purple Cone Spruce → Picea purpurea Mast.

Purple Dead Nettle → Lamium purpurem L.

Purple Grenadilla → Passiflora edulis Sims.

Purple Heart → Peltogyne paniculata Benth.

Purple Loosestrife → Lythrum Salicaria L.

Purple Osierwillow → Salix purpurea L.

Purple Prairie Clover → Petalostemon purpureum (Vent.) Rydb.

Purple Tephrosia → Tephrosia purpurea Pers.

Purple Trillium → Trillium erectum L.

Purple Willow → Salix purpurea L.

Purretia coarctata Ruiz. and Pav. → Puya chilensis Mol. Sagg.

Purslane → Portulaca oleracea L.

Purslane, Suriname → Talinum quadrangulare Willd.

Purslane, Winter → Claytonia perfoliata Don.

Pursley → Portulaca oleracea L.

Pusa → Mouriria Pusa Gardn.

Puya chilensis Mol. Sagg. (syn. P. coarctata Fisch., Pourretia coarctata Ruiz. and Pav.) (Bromeliaceae). - Perennial herb. Chile. Leaves are source of a fibre; used for manuf. fishing nets. Also source of Chagual Gum or Maguey Gum.

Pycnanthemum albescens Torr. and Gray. (syn. Koellia albescens Ktze.). (Labiaceae). - Perennial herb. N. America. Decoction of boiled leaves is used by the Choctaw Indians to cause sweating as a relief from colds.

Pycnanthemum incanum (L.) Michx. (syn. Koellia incanum Ktze.). Atlantic Mountain Mint, Wild Basil. (Labiaceae). - Perennial herb. Eastern United States. Astringent decoction of the plant in water is used by the Koasati Indians to stop nosebleeding.

Pycnanthemum virginianum (L.) Durand and Jacks. (syn. Koellia virginiana (L.) MacM.) Virginiana Mountain Mint. (Labiaceae). - Eastern N. America. Perennial herb. Buds and flowers were used for flavoring meat and broth among the Chippewa Indians.

Pycnarrhena manilensis Vidal. Menispermanceae). - Vine. Philipp. Islands. Powdered roots used as tonic; very efficacious as a cicatrizant. Also used for snake-bites in Philipp. Islds.

Pycnocoma macrophylla Benth. (Euphorbiaceae). - Shrub. Trop. Africa. Fruits are used in Natal for tanning.

Pycnoporus sanguineus (L.) Murr. → Polyporus sanguineus (L.) Meyer.

Pycnothamnus rigidus (Bart.) Small. Pennyroyal. (Labiaceae). - Small shrub. Florida. Herb is used as a tea.

Pygeum africanum Hook. f., Red Stinkwood (Rosaceae). - Tree. Trop. Africa, especially Natal, Mozambique, Angola, Cameroons. Wood heavy, hard, strong, close-grained; used as wagon-wood in S. Africa.

Pyinma Andaman → Lagerstroemia hypoleuca Kurz.

Pyrethrum Balsamita (L.) Willd. → Chrysanthemum Balsamita L.

Pyrethrum roseum Lindl. → Chrysanthemum coccineum Willd.

Pyrethrum sinense DC. → Chrysanthemum sinense Sab.

Pyrola rotundifolia L. (Pyrolaceae). - Perennial herb. Europe, temp. Asia. Leaves used as a substitute for Chimaphila umbellata. Used in some parts of Central Europe as home-remedy for healing wounds. Contains arbutin, methylarbutin, ericolin, urson and gallic acid.

Pyrola secunda L. (Pyrolaceae). - Small perennial herb. Throughout Europe, W. Asia, Siberia, Japan, N. America, Mexico. Leaves, Herba Pirolae Secondae are uesd as substitute for leaves of Chimaphila umbellata. Also used occasionally as a tea.

Pyrolaceae → Chimaphila, Pyrola.

Pyrrhopappus carolinianus (Walt.) DC. (syn. Leontodon carolinianum Walt., Sitilias caroliniana Walt.) Raf.) Leafy-stemmed False Dendelion. (Compositae). - Herbaceous perennial. N. America. Roots were eaten in autumn by the Kiowa Indians.

Pyrularia pubera Michx. Oil Nut, Buffalo Nut. (Santalaceae). - Shrub. Eastern N. America. Fruits are edible known as Oil Nuts or Buffalo Nuts.

Pyrus angustifolia Ait. (syn. Malus angustifolia (Ait.) Michx.). Crab Apple. (Rosaceae). - Small tree. Eastern N. America to No. Florida and Louisiana. Wood hard, close-grained, light-brown tinged with red, heavy; used for tool-handles and levers. Fruits are used in jelly, pickles, preserves and cider.

Pyrus Aria Ehrh. → Sorbus Aria Crantz.

Pyrus Aucuparia Gaertn. → Sorbus Aucuparia L.

Pyrus baccata L. (syn. Malus baccata (L.) Borkh.) Chinese Crabapple. (Rosaceae). - Tree. N. China, Manchuria, Siberia. Fruits eaten by Siberians and Chinese when fresh, dried or preserved.

Pyrus chinensis Lindl., Chinese Pear. (Rosaceae). - Tree. China. Cultivated. Fruits eaten raw, with game or meat. Several varieties among which: Ya-kwam-li; Peking Pear or Pai-li; Pei-soo-li and Ta Suan Li.

Pyrus comunis L., Common Pear Tree. (Rosaceae). - Tree. Europe, W. Asia. Cultivated since early times. Grown in temp. zones, few varieties suitable in subtr. regions. Fruits eaten raw, also preserved, canned, dried, dehydrated, made into sweet pickles. Numerous varieties. Suitable for dessert are: Bartlett (syn. Bon Chrétien William) used for canning, Louise Bonne d'Avranche (L. B. of Jersey), Seigneur d'Esperen, Beurré Gris, Duchesse d'Angoulême, Nouveau Poiteau, Beurré Diel, Doyenné du Comice, Lectier, Passe Crasanne, Beurré Clairgeau, Calabas Bosc, Seckel, Kieffer. Suitable for culinary purposes are: Vicar, Pound and Catillac. Wood is reddish, heavy, little elastic, durable, takes polish; used for turnery, cutlery.

Pyrus coronaria L. (syn. Malus coronaria (L.) Mill.). Crab Apple, Fragrant Crap. (Rosaceae). - Tree. Eastern N. America. Wood close-grained, heavy, light-red, not strong; used for small domestic articles, handles of tools, levers. Fruits much used for jelly, cider and vinegar. Some varieties and hybrids are in existance. Early settlers buried the apples in the autumn, resulting in the loss of acidity of the fruits the following spring.

Pyrus Cydonia L. → Cydonia oblonga Mill.

Pyrus germanica Hook. f. → Mespilus germanica L.

Pyrus hupehensis Pamp. (syn. Malus hupehensis (Pamp.) Rehd., M. theifera Rhed.). Chinese Crab Apple. (Rosaceae). - Tree. China, Assam.

In China the leaves are used as a palatable, thirst quenching tea. Much exported from Shasi.

Pyrus Malus L. (syn. Malus communis DC., M. pumila Mill.). Common Apple Tree. (Rosaceae). - Europe, W. Asia. Cultivated in temperated zone of Old and New World. Source of the commercial apples. Some originated as hybrids with P. prunifolia Willd., P. baccata L. and others. Fruits used for dessert, also consumed after dried, cooked, made into apple sauce, apple butter, jam, jelly, marmelade; source of pectine, apple juice, apple syrup, cider; apple jack is an alcoholic beverage derived from cider. Fruits are made into canned „baked apple". Numerous varieties, some are suitable for certain regions and for certain industries. Grown in the United States are: Northern Spy, Gravenstein, McIntosh, Baldwin, Delicious, Rhode Island Greening, Wealthy, Stayman Winesap, Grims Golden, Jonathan, Yellow Newtown, Russet. Some important varieties grown in Europe are: Ripston Pippin, White Winter Calville, Red Winter Calville, Emperor Alexander, Beauty of Boskoop, Gravenstein, Blenheim Pippin, Yellow Transparent. Wood is hard, compact, close-grained, strong; used for turnery, rulers, handles of tools, knobs, canes, mallet heads.

Pyrus prunifolia Willd. (syn. Malus prunifolia (Willd.) Borkh.). Chinese Apple. (Rosaceae). - Tree. E. temp. China. Fruits eaten fresh or in preserves. Several varieties among which: Paiping-kua, a white apple from Peking; Hong-teng-ku; a red apple from Pautingfu.

Pyrus Ringo Wenzig. (Rosaceae). - Tree. China, Japan. Fruits 2 to 3 cm. long. Eaten in Japan fresh, also cut into slices and dried.

Pyrus rivularis Dougl. (syn. Malus rivularis Roem.) Oregon Crab Apple. (Rosaceae). - Tree. Western N. America, Alaska and Aleutian Islands. Fruits are used for jelly.

Pyrus Sieboldi Regel (syn. P. toringo Sieb.). (Rosaceae). - Tree. Japan. Bark is said to be source of a yellow dye; used in Japan.

Pyrus sinensis Lindl. (syn. P. ussuriensis Maxim.) (Rosaceae). - Tree. China. Cultivated. Fruits of a number of varieties are eaten in China among which Pel-li-tzu (White Pear) being apple shaped.

Pyrus torminalis Ehrh. → Sorbus torminalis (L.) Crantz.

Q

Quackgrass → Agropyron repens (L.) Beauv.

Quamo → Inga leptoloba Schlecht.

Quandong → Fusanus acuminatus R. Br.

Quararibea Fieldii Millsp. saha. (Bombaceae). - Tree. Yucatán (Mex.). Flowers are used in Mexico for flavoring chocolate.

Quassia africana Baill. (Simarubaceae). - Tree. Trop. Africa. Bark root and leaves are a very bitter tonic, also used as appetiser and as vermifuge.

Quassia amara L., Surinam Quassia, Bitter Wood. (Simarubaceae). - Tree. Guiana, Venezuela, N. Brazil. Cultivated in Colombia, Panama, West Indies. Wood used medicinally, bitter stomachic, stimulates gastric functions; anthelmintic against pinworms. Contains quassin, neoquassin. Also used as a poison in fly-paper. Often known a Lignum Quassiae Surinamense.

Quassia, Surinam → Quassia amara L.

Quassia Tree, Jamaica → Picrasma excelsa Planch.

Quebrachia Lorentzii Griseb. (syn. Schinopsis Lorentzii (Griseb.) Engl.). Quebracho Colorado. (Anacardiaceae). - Tree. Argentina, Paraguay, Uruguay. Bark and wood are source of an excellent tanning material; of much commercial value. Contains 18 to 20 % tannin. Also Q. Balansae Griseb.

Quebracho Blanco → Aspidosperma quebracho blanco Schlecht.

Quebracho Colorado → Schinopsis Balansae Engl. and Quebrachia Lorentzii Griseb.

Quebracho, Willowleaf Red → Schinopsis Balansae Engl.

Queen Crapemyrtle → Lagerstroemia Flos-Reginae Retz.

Queen-of-the-Meadow → Filipendula Ulmaria (L.) Maxim.

Queen's Root → Styllingia sylvatica L.

Queensland Arrowroot → Canna edulis Ker-Gawl.

Queensland Hemp → Sida rhombifolia L.

Queensland Nut → Macadamia ternifolia F. v. Muell.

Queensland Wild Lime → Microcitrus inodora Swingle.

Queso de Tuna → Opuntia megacantha Salm-Dyck.

Quercus alba L., White Oak. (Fagaceae). - Tree. Eastern N. America to Texas. Woody, heavy, close-grained, tough, strong, durable, light brown; used for construction, manuf. of agricultural implements, carriages, furniture, interior finish of buildings, rail-road ties, cooperage, fences, baskets, excellent fuel. White Oak Bark is medicinally the powdered bark, containing about 10 % tannin; used as astringent and tonic.

Quercus bicolor Willd. (syn. Q. platanoides (Lam.) Sudw.) Swamp White Oak. (Fagaceae). - Tree. Western United States. Wood strong, hard, heavy, light brown; used for construction, furniture, boat-building, carriages, interior finish, cooperage, fence-posts, rail-road ties and fuel.

Quercus Cerris L. European Turkey Oak. (syn. Q. Vallonea Kotschy.) (Fagaceae). - Tree. Europe, Orient. Source of a white substance, called Oak Manna, Gaz or Gazu. Much used in Iran. Sold in bazaars. Used for making a sweetmeat, called Gazenjubeen. When boiled with the leaves it forms a greenish cake used by Oriental ladies to offer to their guests. Also Q. persica Jaub. and Spach., Q. manifera Lindl. and Q. tauricola Klotszch. Q. Cerris is also source of Morea Galls; used as adulterant of Aleppo Galls, employed as tanning material.

Quercus chrysolepis Liebm., Canyon Live Oak, Maul Oak. (Fagaceae). - Tree. S. W. of United States and adj. Mexico. Wood very strong, tough, hard, close-grained, heavy, light brown; used for manuf. wagons and agric. implements.

Quercus coccinea Wangenh., Scarlet Oak. (Fagaceae). - Tree. Eastern N. America. Wood strong, coarse-grained, heavy, hard; used for similar purposes as Q. rubra L.

Quercus costata Blume (syn. Pasania costata Gamble). (Fagaceae). - Tree. Malaysia. Wood red-brown; used for building purposes.

Quercus crispula Blume. (Fagaceae). - Tree. Sachalin, Japan. Wood hard, easily worked, light brown. Used in Japan for furniture, cabinet-work, panelling, flooring, barrels, wagons, rail-road ties.

Quercus cuspidata Thunb. (Fagaceae). - Tree. Japan. Roasted acorns are consumed in some parts of Japan.

Quercus cyclophora Endl. (syn. Pasania cyclophora Gamble). (Fagaceae). - Tree. Malay Peninsula, Borneo. Wood hard, coarse; used in Sumatra for tool-handles and rice-pounders.

Quercus dendata Thunb. Daimyo Oak. (Fagaceae). - Tree. Hokkaidô, Honshū (Jap.), Formosa, Korea, Manchuria, China, E. Mongolia. Bark is used in Japan for tanning.

Quercus densiflora Sarg. → Lithocarpus densiflora (Hook. and Arn.) Rehd.

Quercus digitata (Marsh) Sudw. (sometimes united with Q. rubra L.). Spanish Oak, Red Oak. (Fagaceae). - N. America. Wood not durable, hard, strong, light red, coarse-grained; sometimes used for construction and for fuel. Bark rich in tannin. Also used medicinally, → Q. velutina Lam.

Quercus Emoryi Torr., Emory Oak. (Fagaceae). - Small tree. S. W. of United States and adj. Mexico. Acorns are consumed as food by Indians. Sold in local markets of New Mexico and adj. Arizona.

Quercus Faberi Hance. (Fagaceae). - Tree. China. Source of food of a silkworm Antheroea Pernyi. Also Q. serrata Car., Q. variabilis Blume and Q. aliena Carr.

Quercus Garryana Hook. Oregon White Oak. (Fagaceae). - Tree. Western N. America. Wood hard, strong, tough, close-grained, light-brown to yellow; used for manuf. of wagons, carriages, furniture, cooperage, ship-building, also for fuel.

Quercus glabra Thunb. (Fagaceae). - Tree. Japan. Roasted acorns are eaten in different parts of Japan.

Quercus glandulifera Blume. Glandbearing Oak. (Fagaceae). - Tree. Japan, Korea, China. Wood hard, strong, heavy, light redbrown. Used in Japan for beerbarrels, farm implements, sledges and wagons. Dead trunks are used for the cultivation of Sitake, Cortinellus Berkleyanum Ito and Imai, a mushroom.

Quercus glauca Thunb., Blue Japanese Oak. (Fagaceae). - Tree. Himalayan region, Japan. Wood is used for general construction. Acorns are consumed as food in different parts of India.

Quercus Ilex L. (syn. Q. Ballota Desf.)., Holly Oak. (Fagaceae). - Tree. Mediterranean region. Cultivated. Acorns are edible, of a pleasant flavor, consumed in Spain, Algeria and other Medit. countries. Leaves are source of Istrian Galls, caused by Cynips tinctoria Ol. var. nostra Destef., source of tanning material. Wood very strong, heavy, close-grained, hard, durable, elastic; used for furniture.

Quercus imbricaria Michx., Shingle Oak, Laurel Oak. (Fagaceae). - Tree. Eastern part of N. America. Wood coarse-grained, light brown tinged with red; hard; used for internal finishing of buildings, furniture, clap-boards, shingles. American Nutgalls are formed on this oak by Cynips aciculata.

Quercus lobata Née. California White Oak, Valley Oak, Western White Oak. (Fagaceae). - Tree. California. Acorns are used as a source of food, made into bread by the Digger Indians.

Quercus lusitanica Lam., Lusitanian Oak, Portuguese Oak. (Fagaceae). - Shrub or small tree. Mediterranean region. Source of Aleppo Galls, Gallae Turcicae, G. halepensis, Smyrna Galls; caused by Cynips gallae tinctoriae Oliv. Source of tannin and tanning material. Much is derived from Aleppo and Smyrna.

Quercus lyrata Walt., Overcup Oak, Water White Oak. (Fagaceae). - Tree. Eastern N. America. Wood hard, tough, strong, heavy, very durable, dark brown; used for the same purposes as Q. alba L. Decoction of astringent bark was used by the Cherokee and Creek Indians for dysentery; by the Chocatws for stomach ache.

Quercus macrocarpa Michx. Bur Oak, Mossy Cup Oak. (Fagaceae). - Tree. Eastern N. America. Wood hard, strong, tough, close-grained, very durable, heavy, light to dark brown; used

for construction, boat-building, furniture, carriages, cooperage, rail-road ties, fence posts, agricultural implements, baskets and fuel.

Quercus marylandica Du Roi. Black Jack Oak, Jack Oak. (Fagaceae). - Tree. Eastern United States. Wood used for the manuf. of charcoal.

Quercus mongolica Fischer. (syn. Q. crispula Blume, Q. grosserrata Blume). Mongolian Oak. (Fagaceae). - Tree. E. Siberia, N. China to Korea, Japan. Wood used in China for building purposes.

Quercus obtusiloba Michx. (syn. Quercus stellata Wang.) Post Oak, Iron Oak. (Fagaceae). - Tree. Eastern and southern United States. Wood close-grained, hard, durable when in contact with the soil, very heavy; used for rail-road ties, fencing, cooperage, sometimes in construction work and for fuel.

Quercus Pagoda Raf. (syn. Quercus pagodae-folia (Ell.) Ashe.) Spanish Oak, Swamp Spanish Oak. (Fagaceae). - Tree. Eastern United States. Wood light reddish brown, heavy, hard, strong; used for interior finishing, agricultural implements, furniture. Equal to that of Q. alba L.

Quercus palustris Du Roi. Swamp Spanish Oak, Pin Oak. (Fagaceae). - Tree. Eastern N. America. Wood strong, hard, heavy, light brown, coarse-grained; used for shingles, clapboards, sometimes for construction, interior finish and cooperage.

Quercus persica Jaub. and Spach. Manna Oak. (Fagaceae). - Tree. Iran. Acorns are used as food in some parts of Iran and Kurdistan.

Quercus Phellos L., Willow Oak. (Fagaceae). - Tree. Eastern N. America. Wood strong, not hard, heavy, light-brown, tinged with red, coarse-grained; used for fellies of wheels, clapboards, fuel, charcoal, sometimes for construction.

Quercus Prinus L., Chestnut Oak. (Fagaceae). - Tree. Eastern N. America. Wood strong, tough, hard, close-grained, heavy, durable when in contact with the soil; used for rail-road ties, fence posts and fuel. Bark is rich in tannin, used for tanning leather. Acorns are considered edible.

Quercus pubescens Willd. (syn. Q. lanuginosa Thuill., Q. apennina Zucc.). Pubescent Oak. (Fagaceae). - Tree. S. Europe, Asia Minor, Caucasia. Wood light brown, hard, close-grained, durable, also under water; not very elastic; used for the same purposes as Q. Robur.

Quercus reticulata H. B. K. Netleaf Oak. (Fagaceae). - Tree. Mexico. Acorns used as a substitute for coffee in San Luis Potosi (Mex.).

Quercus Robur L. (syn. Q. pedunculata Ehrh.) English Oak. (Fagaceae). - Tree. Europe, Caucasia, Asia Minor, N. Africa. Wood yellowish-brown, close-grained, heavy, hard, elastic, easy to split, very durable, resistant to water; used for general carpentry, furniture, wheels, rail-road ties, windows, frames, floors, shipbuilding, stair-cases. Source of acetic acid, tannic acid and charcoal. Acorns are used as food for swine, also used as substitute for coffee, Eichel Kaffee; sometimes eaten raw during times of distress. Bark is used for tanning. Leaves are source of Acorn Galls, Knoppern, caused by Cynips calices Burg. used as source of tannin and tanning material.

Quercus rubra L., Red Oak. (Fagaceae). - Tree. N. America. Important lumbertree. Wood hard, strong, close-grained, heavy, light reddish brown; used for interior finish of houses, construction work and furniture.

Quercus sessiliflora Salib., Durmast Oak. (Fagaceae). - Tree. Europe, Resembles Q. Robur L. and is used for the same purposes.

Quercus Suber L., Cork Oak. (Fagaceae). - Tree. S. W. Mediterranean region, esp. Spain, Algeria, Morocco. Cultivated in Portugal, Sardinia, S. France, Corsica, Istria, Dalmatia. Bark is a source of cork. It is occasionally also obtained from Q. pseudosuber Santi, Q. occidentalis Gay and Q. Fontanesii Trab.

Quercus sundaica Blume. (syn. Q. hystrix Korth.). (Fagaceae). - Tree. Malaysia. Wood heavy, hard; used for construction of houses, bridges and sheds.

Quercus tauricola Kotschy. (Fagaceae). - Tree. S. E. Europe, Asia Minor to Centr. Asia. Source of a gall, called Dead-Sea Apple, Rove, Mala Insana or Mala Sodomitica, caused by Cynips insana Westw., an insect. Source of a dye, Rouge d'Adrianople. Also valuable as tanning material and source of tannin. Also Q. lusitanica Lam. and other Oaks species.

Quercus texana Buckl. Texas Oak, Southern Red Oak. (Fagaceae). - Tree. Southern United States. Wood hard, heavy, light reddish, brown, close-grained; used in the Mississippi Valley for manuf. of lumber.

Quercus undulata Torr. (syn. Q. Cambelii Nutt.). Wavyleaf Oak. Shin Oak. (Fagaceae). - Tree. Western United States. Wood is used as fuel. Bark is occasionally used for tanning.

Quercus uruapanensis Trel. (Fagaceae). - Tree. Mexico. Source of an excellent timber.

Quercus variabilis Blume. Oriental Oak. (Fagaceae). - Tree. N. China, Korea, Japan. Burr-like capsules are source of a black dye; used in China for silk-yarn and fabrics.

Quercus velutina Lam. (syn. Q. tinctoria Bartr.). Black Oak. (Fagaceae). - Tree. N. America to Texas and Florida. Inner bark is source of a yellow dye, was formerly an important article of export. Used in printing calicos. Bark is also used for tanning. Source of Guercitron Extract.

Quercus virginiana Mill. Live Oak. (Fagaceae). - Tree. Southeast and South of the United States, Mexico, Centr. America and West Indies. Wood hard, strong, close-grained, tough, very heavy; formerly extensively used for ship-building.

Quihuicha → Amaranthus caudatus L.

Quillaja brasiliensis (St. Hil.) Mart. (syn. Fontanella brasiliensis St. Hil.). (Rosaceae). - Shrub or small tree. Trop. America, esp. S. Brazil and E. Argentina, Uruguay and Chile. Bark contains saponin and is used for cleaning fine textiles.

Quillaja Saponaria Mol. Soapbark Tree. Quillaja. (Rosaceae). - Tree. Chile and Peru. Dried inner bark is source of Soap Bark and Soap Tree Bark, an emulsifying agent, especially used for tars. It is employed instead of soap for washing clothes, giving a fine lustre to wool; is also used to wash the hair.

Quina do Campo → Strychnos Pseudoquina St. Hil.

Quina do Matto Bark → Cestrum Pseudoquina Mart.

Quince → Cydonia oblonga Mill.

Quince, Chinese → Cydonia sinensis Thouin.

Quinine → Cinchona.

Quinine Bush → Garrya ellipticus Dougl.

Quinoa → Amaranthus caudatus L. and Chenopodium Quinoa Willd.

Quinteniqua Yellowwood → Podocarpus elongata L'Hér.

Quira Macawood → Platymiscium polystachyum Benth.

Quisqualis indica L. Rangoon Creeper. (Combretaceae). - Woody vine. Trop. Asia. Mal. Archip. Cultivated. Roots and fruits are used as anthelmintic in some parts of Malaysia.

Quiza → Tuber Gennadii (Chatin) Pat.

R

Rabbit Bush → Chrysothamnus nauseosus Britt.

Radish → Raphanus sativus L.

Radish, Horse → Cochlearia Armoracea L.

Radish Tree, Horse → Moringa oleifera Juss.

Raffia Fibre → Raphia gigantea Chev., R. pedunculata Beauv.

Raffia Palm, Giant → Raphia gigantea Chev.

Raffia Palm, Ivory Coast → Raphia Hookeri Mann. and Wendl.

Raffia Palm, Madagascar → Raphia pedunculata Beauv.

Raffia Palm, Wine → Raphia vinifera Beauv.

Rafflesia patma Blume (Rafflesiaceae). - Java. (Parasite on Vitaceae). Flowerheads, patma sari, are uesd by the natives as astringent.

Rafflesiaceae → Cytinus, Rafflesia.

Rafnia perfoliata E. Mey. (syn. Vascoa perfoliaita DC.). (Leguminosaceae). - Perennial plant. S. Africa. Decoction of leaves is used by the farmers and Hottentots as diuretic.

Rage → Lobaria pulmonaria (L.) Hoffm.

Ragwort, Golden → Senecio aureus L.

Ragwort, Tansy → Senecio Jacobaea L.

Rain Tree, Manaca → Brunfelsia Hopeana Benth.

Raisin Tree, Japanese → Hovenia dulcis Thunb.

Raisins, Dried → Vitis vinifera L.

Rajapa eurhiza (Berk.) Sing. (syn. Collybia eurhiza Berk.) (Agaricaceae). - Basidiomycete Fungus. Trop. Asia. Mycelium grows in termite nests. Fruitbodies are consumed as food by the natives. Called in Malay „Djamoer rajap."

Ram's Horn → Martynia probocidea Glox.

Ramalina, Ivory-like → Ramalina scopulorum Ach.

Ramalina, Mealy → Ramalina farinacea (L.) Ach.

Ramalina calicaris (L.) Röhling. (syn. Lobaria calicaris Hoffm.). Common Twig Lichen. (Usneacee). - Lichen. Temp. zone. Used in Europe as a source of a yellow-red dye, to color woolens. Powdered plant was formerly used instead of starch for dyeing perukes and wigs.

Ramalina cuspidata (Ach.) Nyl. (Usneaceae). - Lichen. Temp. zone. Used in Europe as a source of a light brown dye for coloring woolens.

Ramalina farinacea (L.) Ach., Mealy Ramalina. (Usneaceae). - Lichen. Temp. zone. On stems. Used in Europe as a source of a light brown dye for coloring woolens.

Ramalina fraxinea (L.) Ach. (syn. Lobaria fraxinea Hoffm.) Ash Twig Lichen. (Usneaceae). - Lichen. Temp. zone, esp. Europe. On stems. Plant is used in perfumes and cosmetics. Also P. pollinaria (Westr.) Ach.

Ramalina scopulorum Ach., Ivory-like Ramalina. (Usneaceae). - Lichen. Temp. zone. Used in Scotland as a source of a yellow-brown to red-brown dye for coloring woolens. The lichens are boiled in water for one day and the wool is put next day in it. After boiling up, it is kept in the mixture until the desired color is obtained.

Rambutan → Nephelium lappaceum L.

Ramie → Boehmeria nivea (L.) Gaud.

Ramie Bukit → Alchornea villosa Muell. Arg.

Ramie Sengat → Abroma angusta L. f.

Ramon Breadnut Tree → Brosimum Alicastrum Swartz.

Rampion → Campanula Rapunculus L.

Ramtilla Oil → Guizotia abyssinica (L. f.) Cass.

Randia armata (Swartz) DC. → Basanacantha armata Hook. f.

Randia dumetorum Lam. (syn. R. floribunda DC.). (Rubiaceae). - Small tree. Trop. Asia. Berries globose, 2 to 3 cm. diam., yellow; edible when cooked; esteemed by the natives.

Randia maculata DC. (syn. Rothmannia longiflora Sal., Gardenia Stanleyana Paxt.). (Rubiaceae). - Shrub or small tree. Trop. Africa. Fruits source of a dye; used for tattooing the face among tribes of the interior of Africa.

Randia malleifera Benth. and Hook. (Rubiaceae). - Shrub or small tree. Trop. Africa. Sourec of an inky sap; used by the Mombutta and Niam-Niam tribes for dyeing the skin. Fruits produce an ink; used by the inhabitants of Bahr-el Ghazal, White Nile region.

Randia Ruiziana DC. (Rubiaceae). - Shrub. Trop. America. Pulp of fruits is edible; consumed by the natives.

Randonia africana Gosson. (Resedaceae). - Shrub. N. Africa, Sahara. A food for dromedaries and camels.

Rangoon Creeper → Quisqualis indica L.

Rangoon Rubber → Urceola esculenta Benth.

Ranunculaceae → Aconitum, Adonis, Anemone, Anemonella, Caltha, Cimicifuga, Clematis, Coptis, Delphinium, Helleborus, Hydrastis, Nigella, Paeonia, Ranunculus, Thalictrum, Xanthorrhiza.

Ranunculus Ficaria L., Lesser Celandrine. (Ranunculaceae). - Herbaceous perennial. Europe, temp. Asia. Bleached stems and leaves are occasionally eaten in some parts of Europe.

Ranunculus Pallasii Schlecht. (Ranunculaceae). - Perennial herb. Arct. regions. Rootstocks are consumed as food by the Eskimos in Alaska.

Ranunculus sceleratus L., Blister Buttercup. (Ranunculaceae). - Annual or perennial herb. Northern temp. hemisphere. Different parts of the plant cause blisters; used by beggers to attract sympathy.

Rapa → Brassica Napus L.

Rapa Bark → Ficus tinctoria Forster.

Rape, Bird → Brassica campestre L.

Rape Oil → Brassica Napus L.

Raphanus sativus L. (syn. R. Raphinastrum L. var. sativus L.) Domin. Radish. (Cruciferaceae). - Annual or perennial herb. Cultigen. Origin uncertain. Probably derived from R. Raphinastrum L. Roots are eaten as appetizer and for garnishing salads; occasionally the leaves are consumed as greens. They may be divided in: I Group. Spring Varieties: Scarlet Turnip, French Breakfast, Ne Plus Ultra, White Icicle. II Group. Summer Varieties: White Vienna, Golden Globe, Stuttgart. III Group. Winter Varieties. White Chinese, Chinese Rose, Long Black Spanish, Round Black Spanish.

Raphia gigantea Chev. Giant raffia Palm. (Palmaceae). - Palm. Trop. Africa. Young leaves are source of a Raffia used for tying. Piassave is derived from the petioles. Midribs are used for roofing of the huts; ladders an rafters. Stems yield a palmwine used as a beverage along the Gold Coast and Ivory Coast.

Raphia Hookeri Mann and Wendl. Ivory Coast Raffia Palm. Wine Palm. (Palmaceae). - Tall palm. Trop. Africa. Source of Mimbo, a palm wine, obtained from the sap of the young inflorescense. Fibre is made into cloth. Leaves are used for mats and roofing.

Raphia pedunculata Beauv. (syn. R. Ruffia Mart.) Madagascar Raffia Palm. (Palmaceae). - Palm. Madagascar. Leaves are source of a Raffia Fibre, used by nurserymen. Stem is source of a sweet beverage, Harafa; used by the natives. Shells of the fruits are made into snuff boxes and other small articles.

Raphia vinifera Beauv., Wine Raffia Palm. Bamboo Palm. (Palmaceae). - Palm. Trop. Africa. Base of the leaves are used as a source of Raffia. Also R. Hookeri Mann and Wendl., R. Welwitschi Wendl.

Raphiolepis japonica Sieb. and Zucc. (Rosaceae). - Tree. Japan. Bark is source of a brown dye; used in some parts of Japan for dying Tsumugi and other clothes.

Raphionacme Brownii Elliot. (Asclepiadaceae). - Herbaceous plant. Trop. Africa, esp. Sierra Leone. Turnip-like tubers are eaten raw or roasted by Fulani and Mandingo people.

Raphionacma utilis Brown and Stapf. (Asclepiadaceae). - Tree. Trop. Africa. Has been recommended as a source of rubber, called Ekanda Rubber.

Raputia alba Engl. (syn. Almeida alba St. Hil.). (Rutaceae). - Tree. Brazil. The bitter bark is excitant and febrifuge; also used in Brazil to stupefy fish.

Raputia aromatica Aubl. (syn. Siuris aromatica Gmel.). (Rutaceae). - Shrub. Guiana to Brazil. Bark is excitant, stomachic when used in small quantities; febrifuge when used in large amounts.

Raputia magnifica Engl. (Rutaceae). - Small tree. Brazil. Bark is used in S. America for intestinal worms. Contains chrysophanic acid.

Rasamala Resin → Altingia excelsa Noron.

Rasamala Wood Oil → Altingia excelsa Noron.

Rasp Pad → Flindersia australis R. Br.

Raspberry Acacia → Acacia acuminata Benth.

Raspberry, American → Rubus strigosus L.

Raspberry, Black → Rubus occidentalis L.

Raspberry, European → Rubus Idaeus L.

Raspberry, Purple Cane → Rubus parvifolius L.

Raspberry, Mauritius → Rubus rosaefolius Smith.

Raspberry, Wine → Rubus phoenicolasius Maxim.

Raspberry, Yellow Himalayan → Rubus ellipticus Smith.

Raspod, Zanzibar → Trachylobium Hornemannianum Hayne.

Rata, Northern → Metrosideros robusta Cunningh.

Rata Vine → Metrosideros scandens Soland.

Ratanhia Root → Krameria triandra Ruiz and Pav.

Ratanhia, Brazilian → Krameria argentea Mart.

Ratanhia, Para → Krameria argentea Mart.

Ratanhia, Savanilla → Krameria tomentosa St. Hil.

Ratany, Para → Krameria argentea Mart.

Ratomia apetala Griseb. (syn. Matayba apetala Radlk.). (Sapindaceae). - Tree. West Indies. Wood light colored; used for house-building, doors and windows.

Rattan → Calamus spp., Daemonorops spp.

Rattle Weed → Baptisia tinctoria (L.) R. Br.

Rattlesnake Master → Eryngium yuccifolium Michx.

Rattlesnake Weed → Brauneria angustifolia (DC.) Heller. and Daucus pusillus Michx.

Rauli → Nothofagus procera Oerst.

Rauwolfia canescens L., Trinidad Devilpepper. (Apocynaceae). - Tree. Trop. Africa. Juice from the fruits is used as a substitute for ink.

Rauwolfia heterophylla Roem. and Schutt. (Apocynaceae). - Shrub. S. Mexico. The latex has been reported to have emetic, cathartic, expectorant and diuretic properties. Used for treating dropsy. Juice from the fruits is used as a substitute for ink.

Rauwolfia perakensis King and Gamble. (Apocynaceae). - Roots have been used as adulterant of R. serpentina. Also R. densiflora (Wall.) Benth. ex Hook. f. and R. canescens L.

Rauwolfia serpentina (L.) Benth ex Kurz (Apocynaceae). - Erect evergreen subshrub. India, Ceylon, Andaman Islds., Burma, Siam, Indonesia. Root used in medicine as a tranquilizer in the treatment of hypertention and related ailments. Causes lowering of blood pressure, apparently without dangerous side-effects. The drugs have a sedative and tranquilizing action, said to result from a depressing effect on the hypothalamus. It is an ancient Indian medicine.

Ravenala madagascariensis Gmel. Madagascar Traveller's Tree. (Musaceae). - Large banana-like plant. Madagascar. Cultivated in the tropics of the Old and New World. It has been stated that water which is derived from the stem is used for drinking in time of scarcety of water. Stems are used by the natives for the construction of houses.

Ravensara aromatica Gmel. (syn. Agathophyllum aromaticum Willd.). Madagascar Nutmeg. (Lauraceae). - Tree. Madagascar. Seeds are source of Madagascar Nutmegs; used as spice. Tle strong aromatic bark is employed by the natives of Madagascar for manuf. rum.

Ravinson Oil → Brassica campestris L.

Recordoxylon amazonicum Ducke. (syn. Melanoxylon amazonicum Ducke.). (Leguminosaceae). - Tree. Amazonas (Braz.). Wood dark brown with yellowish hue, heavy, very hard, coarse texture, not easy to work, durable, resistant to insects; used for heavy and durable construction work.

Red Acaroid → Xanthorrhoea australis R. Br.

Red Alder → Alnus rubra Bong.

Red Ash → Fraxinus pubescens Lam.

Red Banana → Musa sapientum L.

Red Bark → Cinchona succirubra Pavon.

Red Bay → Persea Borbonia (L.) Raf.

Red Beech → Nothofagus fusca Oerst.

Red Beet → Beta vulgaris L.

Red Birch → Betula nigra L.

Red Box → Tristania conferta R. Br.

Red Buckeye → Aesculus Pavia L.

Red Box → Eucalyptus polyanthemos Scham.

Red Bud → Cercis canadensis L.

Red Canary Grass → Phalaris arundinacea L.

Red Cardinal → Erythrina arborea (Chapm.) Small.

Red Cedar → Acrocarpus fraxinifolius Wight. and Arn.

Red Cedar → Cedrela Toona Roxb.

Red Clover → Trifolium pratense L.

Red Currant → Ribes rubrum L.

Red Elm → Ulmus alata Michx.

Red Fescue → Festuca rubra L.

Red Fir → Abies magnifica Murr.

Red Flowered Mangrove → Lumnitzera coccinea Wight and Arn.

Red Fruited Lecidea → Mycoblastus sanguinarius (L.) Norm.

Red Gum → Eucalyptus rostrata Schlecht. and Liquidambar styraciflua L.

Red Ironwood → Reynosia septentrionalis Urb.

Red Ironbark → Eucalyptus leucoxylon F. v. Muell.

Red Mahogany → Khaya ivorensis Chev. and Eucalyptus resinifera Smith.

Red Mangrove → Rhizophora Mangle L. and others.

Red Maple → Acer rubrum L.

Red Mulberry → Morus rubra L.

Red Oak → Quercus rubra L.

Red Padauk → Pterocarpus macrocarpus Kurz.

Red Pear → Scolopia Mundtii Harv.

Red Peruvian Cotton → Gossypium microcarpum Tod.

Red Pine → Dacrydium cupressinum Solander, and Pinus resinosa Ait.

Red Poppy → Papaver Rhoeas L.

Red Root → Ceanothus americanus L.

Red Sandalwood → Adenanthera pavonina L.

Red Sanderswood → Pterocarpus santalinus L. f.

Red Sarsaparilla → Smilax Regelii Killip and Morton.

Red Spruce → Picea rubra Link.

Red Stinkwood → Pygeum africanum Hook. f.

Red Stoper → Eugenia confusa DC.

Redberry Bryony → Bryonia dioica Jacq.

Redberry Buckthorn → Rhamnus crocea Nutt.

Redroot Amaranth → Amaranthus retroflexus L.

Redshank Chamise → Adenostoma sparsifolium Torr.

Redtop → Agrostis alba L.

Redwater Tree → Erythrophleum guineense G. Don.

Redwood, Brazil → Brosimum paraense Hub. and Caesalpinia brasiliensis Sw.

Redwood → Sequioa sempervirens (Lamb.) Endl.

Redwood Indian → Soymida febrifugus Juss.

Reed, Giant → Arundo Donax L.

Reed, Giant Bur → Sparganium eurycarpum Engelm.

Reed Grass → Phragmitis communis Trin.

Reindeer Moss → Cladonia rangiferina (L.) Web.

Remijia pedunculata Flueck. (Rubiaceae). - Tree. S. America, esp. Colombia. Has been suggested as a source of quinine. Source of Cuprea Bark, being exported to Europe, esp. England.

Renealmia domingensis Horan. (syn. Alpinia aromatica Aubl.). (Zingiberaceae). - Perennial herb. Trop. America. Seeds are used in Brazil as emmenagogue. Juice was used by the Mayas for curing hemorrhoids.

Reniala Oil → Adansonia digitata L.

Rennet, Indian → Withamia coagulans Dunal.

Reptonia buxifolia A. DC. (Myrsinaceae). - Small tree. E. India. Drupe subglobose, 1 cm. across; pulp sweet; much esteemed by the natives. Sold in bazaars.

Rescue Grass → Bromus catharticus Vahl.

Reseda Luteola L. Wild Woad. (Resedaceae). - Biennal herb. Centr. Europe, Mediterranean region, Iran, Afghanistan. Formerly much cultivated. Source of an excellent deep yellow dye; used for coloring silk. Employed since neolithic times.

Reseda odorata L., Common Mignonette. (Resedaceae). - Annual herb. Mediterranean region. Cultivated as an ornamental. Source of an ess. oil; used in perfumery, was formerly of much importance. Oil is obtained by enfleurage and by the volatile solvent process. About 1150 to 1200 kg. flowers produce 1 kg. concrete, giving 350 gr. of alcohol soluble absolute.

Reseda Phyteuma L. (Resedaceae). - Herbaceous perennial. Mediterranean region, Asia Minor. Plants are occasionally used as a vegetable.

Resedaceae → Randonia, Reseda.

Resins. Plants source of. → Abies, Acacia, Agathis, Altingia, Amphimas, Amyris, Angophora, Araucaria, Balanites, Balanocarpus, Bertya, Boswellia, Calophyllum, Canarium, Cistus, Clusia, Convolvulus, Copaifera, Croton, Dacryodes, Daemonorops, Daniella, Dipterocarpus, Dorema, Dracaena, Entada, Eucalyptus, Ferula, Garcinia, Gardenia, Gossweilerodendron, Guiacum, Haronga, Hopea, Hymenaea, Libothamnus, Macaranga, Machaerium, Mammea, Melorrhoea, Pistacia, Populus, Protium, Rheedia, Schinus, Shorea, Spermolepis, Styrax, Symphonia, Tetraclinis, Trachylobium, Tsuga, Vateria, Vatica, Xanthorrhoea. See also under: frankincense, gums, kinos.

Resin, Abu Beka → Gardenia lutea Fresen.

Resin, Alriba → Canarium strictum Roxb.

Resin, Apitong → Dipterocarpus grandiflorus Blanco.

Resin, Baláu· → Dipterocarpus grandiflorus Blanco.

Resin, Black Dammar → Canarium bengalense Roxb.

Resin, Bulungu → Canarium edule Hook. f.

Resin, Combee →Gardenia gummifera L. f. and G. lucida Roxb.

Resin Dammar → Shorea Wiesneri Schiff.

Resin, Dikkamaly → Gardenia gummifera L. f.

Resin, Dragon's Blood → Dracaena Cinnabari Balf. f.

Resin of Fiji → Agathis vitiensis (Seem.) Warb.

Resin, Guaiac → Guaiacum sanctum L.

Resin, Hal-dummala → Vateria acuminata Hayne.

Resin, Levant Scammony → Convolvulus Scammonia L.

Resin, Maina → Calophyllum longifolium Willd.

Resin, Minyak Keruong → Dipterocarpus Kerrii King.

Resin, Pagsahingin → Canarium villosum Benth. and Hook.

Resin, Panau → Dipterocarpus vernicifluus Blanco.

Resin, Rasamal → Altingia excelsa Noron.

Resin, Ronima → Protium heptaphyllum March.

Resin, Sagapan → Ferula Szotitziana DC.

Resin, Sahing → Canarium villosum Benth. and Hook.

Resin, Saul → Shorea Talura Roxb.

Resin, Tabonico → Dacryodes hexandra Griseb.

Resin, Tacamahaca → Calophyllum tacamahaca Willd. Populus balsamifera Muench.

Resin, Thapsia → Thapsia garganica L.

Resin, Yellow → Pinus palustris Mill.

Resina de Cudpinole → Hymenaea Courbaril L.

Resina de Mamey → Mammea americana L.

Resina Lutea → Xanthorrhoea australis R. Br.

Resina Mani → Moronobea grandiflora Choisy.

Resina Ocuje → Calophyllum Calaba L.

Resina Papaman → Moronobea grandiflora Choisy.

Resina Scammoniae → Convolvulus Scammonia L.

Resina Thapsiae → Thapsia garganica L.

Restionaceae → Leptocarpus.

Retanilla Ephedra Brongn. (Rhamnaceae). - Shrub. Chile, Peru. Plant is used by the natives of Chile as astringent, carminative and for indigestion.

Retting of flax and hemp → Clostridium butyracum and Plectridium pectinovorum.

Revalenta Arabica → Ervum Lens L.

Rewarewa → Knightia excelsa R. Br.

Reynoldsia marchionensis F. Br. (Araliaceae). - Tree. Pacif. Islds. Leaves used for making fragrant coconutoil.

Reynosia septentrionalis Urb., Red Ironwood, Darling Plum. (Rhamnaceae). - Shrub or small tree. S. Florida, Keys, Bahamas, West Indies. Wood close-grained, very hard; used for cabinetwork. Fruits are edible. They are 1.5 to 2 cm. long, purple to black and of a pleasant flavor.

Rhabdoniaceae (Red Algae) → Catenella.

Rhamnaceae → Alphitonia, Ampelozizyphus, Ceanothus, Colletia, Colubrina, Condalia, Hovenia, Karwinskia, Retanilla, Reynosia, Rhamnus, Sageretia, Trevoa, Ventilago, Zizyphus.

Rhamnus cathartica L., Common Buckthorn. (Rhamnaceae). - Shrub or small tree. Throughout Europe, Asia, N. Africa. Wood hard, yellowish; used for small turnery. Berries used in medicine, usually derived from Hungary and Saxony; used for the preparation of Sirupus Rhamni Catharticae; being a purgative, cathartic; contains rhamnin and rhamnetin.

Rhamnus crocea Nutt., Redberry Buckthorn. (Rhamnaceae). - Shrub or small tree. S. W. of the United States and adj. Mexico. Fruits are consumed with meat by the Apache Indians.

Rhamnus dahurica Pall. (syn. R. utilis Dcne.). Dahuruan Buckthorn. (Rhamnaceae). - Shrub. Centr. and E. China. Leaves are used in China as a source of a green dye, called Chinese Green; used in painting and in the Orient for dyeing cotton and silk. Also R. tinctoria Waldst. and Kit.

Rhamnus Frangula L. (syn. Frangula Alnus Miller). Alder Buckthorn. (Rhamnaceae). - Shrub or small tree. Europe, Asia, N. Africa. Bark is used in medicine, contains frangulin, a glucosid. Used as home-remedy to „clean" the blood, also as a purgative; now largely replaced by R. Purshiana. Wood used for shoe-lasts, wooden-nails, veneer; also source of charcoal.

Rhamnus Humboldtiana Roem. and Schult. → Karwinskia Humboldtiana (Roem. and Schult.) Zucc.

Rhamnus infectorius L. (Rhamnaceae). - Shrub or small tree. S. Europe. Unripe berries called French Berries, Persian Berries, Yellow Berries, Graines d'Avignon, Fructus Rhamni, were source of a yellow dye. The trade in these berries was formerly of much importance in France, Hungary, Spain and Iran. Much comes from Turkey. Also are used: R. oleoides L. from Mediterr. Region; R. cathartica L. from Europe, N. Africa and Asia; R. graecus Boiss and Rent. from Greece and R. saxatilis L. from Centr. and S. Europe.

Rhamnus japonica Maxim. (syn. R. globosus Sieb. and Zucc.). Japanese Buckthorn. (Rhamnaceae), Small tree. Japan. Wood is close-grained; used in Japan for small furniture. Bark and fruit as laxative.

Rhamnus prinoides L'Hér. (Rhamnaceae). - Shrub. S. Africa. Leaves are used in Abyssinia as a stimulant; in honey wine, called Tetschen and in beer, called Tallas.

Rhamnus Purshiana DC. Cascara Buckthorn, Bearberry, Bearwood. (Rhamnaceae). - Shrub or small tree. Western N. America. Cascara Sagrada, the dried bark is used medicinally, being tonic, laxative, acts in the colon; contains emodin, purshinanin, a glucosied and cascarin.

Rhaponticum scariosum Lam. → Centaurea Rhaponticum L.

Rhatany, Brazilian → Krameria argentea Mart.

Rheedia acuminata Planch and Triana. (Gutti-feraceae). - Tree. Trop. America. Tree. is source of a greenish resin, called Maria Balsam.

Rheedia brasiliensis Planch and Triana., Baku-pari. (Guttiferaceae). - Tree. Brazil. Fruits ovate, orange-yellow, tough skinned, flesh sub-acid, white; very much prized by the inhabitants; used in jams. Sold in markets.

Rheedia edulis Planch and Triana. (Guttifera-ceae). - Small tree. Trop. America. Occasionally cultivated, esp. in Brazil. Fruits edible, acid, round, 2 cm. across, yellowish, good flavor; made into jams. Called Limao do Matto.

Rheedia lateriflora L. (Guttiferaceae). - Tree. West Indies, Trop. S. America. Tree is source of a hard wax.

Rheedia macrophylla Planch and Triana., Bacuru Pary. (Guttiferaceae). - Tree. Brazil. Cultivated. Fruits are edible, made into jams.

Rheedia Madruno (H.B.K.) Planch and Triana. (Guttiferaceae). - Tree. Trop. America. Fruits yellow, 6 to 7.5 cm. long, oval to elliptic, rough surface; pulp whitish, juicy, pleasant sub-acid, slightly aromatic; eaten raw, also used in preserves and drinks. Sold in local markets of Ecuador.

Rheum Emodi Wall., Himalaya Rhubarb. (Polygonaceae). - Perennial herb. Himalayan region. Source of Himalaya Rhubarb. Sold in bazaars as a drug.

Rheum hybridum Murray. Rhubarb. (Polygonaceae). - Large perennial herb. Is considered a hybrid R. Rhaponticum L. x R. palmatum L. Probably from Mongolia. Cultivated. Stewed leaf-stalks are used for dessert and eaten in pies. Rhubarb-wine is made from the juice. Varieties are: Victoria, Giant Crimson Winter, Linnaeus.

Rheum officinale Baill., Medicinal Rhubarb. (Polygonaceae). - Perennial herb. China and adj. territory. Dried rhizome and roots are used medicinally, being astringent, laxative, stomachic, purgative, tonic. Contains calcium oxalate, glu-cogallin and tetrarin. Cultivated. Much is exported from Shanghai. Compound Powder of Rhubarb or Gregory's Powder is powdered rhubarb with magn. oxid, and ginger.

Rheum palmatum L. Sorrel Rhubarb. (Polygonaceae). - Large perennial herb. Mongolia. Cultivated. Source of one of the finest Chinese Rhubarb. Used medicinally.

Rheum Rhaponticum L. Garden Rhubarb. (Polygonaceae). - Large perennial herb. S. Siberia. Cultivated in temp. zones of Old and New World. Stewed leaf-stalks are used as dessert. Stems are also used in pies. Juice is sometimes made into wine.

Rheum Ribes L. Currant Rhubarb. (Polygonaceae). - Perennial herb. Iran, Afghanistan, Turkestan. Roots are sold in markets of Teheran; used as vermifuge for horses.

Rheum undulatum L. (Polygonaceae). - Perennial herb. China. Leaves smaller than of R. hybridum Murray. Occasionally cultivated, esp. var. Rouge Hâtive de Tobolsk.

Rhinacanthus communis Nees. (Acanthaceae). - Shrub. Madagascar, India, Malaya, China. Seeds are used in Madagascar for scenting clothes.

Rhipogonum scandens Forst. Supplejack. (Liliaceae). - Liane. New Zealand. Stems made into baskets, occasionally made into rope-ladders.

Rhizobium leguminosarum Frank (syn. Bacterium radicicola Prazm.). (Bacteriaceae). - Bacil. Symbiotic nitrogen fixing microorganism, living in the tissues of the roots of leguminous plants. They increase the fertility of the soil. The organisms are grown in pure cultures which are sold commercially. Among the other species are R. trifoli Dang on clover; R. phaseoli Dang. on beans; R. melioti Dang. on Melilotus, R. lupini (Schroeter) Eckh. Baldw. and Feed. on lupines and others.

Rhizocarpon geographicum (L.) DC. f. atrovirens (L.) Mass. Map Lichen. (Lecideaceae). - Lichen. Temp. zone. On rocks. Used in Scandinavia as a source of a brown dye; used for staining woolens.

Rhizophora Mangle L., Red Mangrove. (Rhizophoraceae). - Tree. Coastal zone. Trop. America. Wood hard, close-grained, strong, heavy, dark reddish-brown, durable in water, is not attacked by mollusks and Teredo; used for wharf-piles, cabinet work. Bark is employed in dyeing and tanning leather; has been used as febrifuge and to stop hemorrhages. Flowers are source of a commercial honey.

Rhizophora mucronata Lam. (Rhizophoraceae). - Tree. Trop. Africa, Asia. Bark is used for tanning. Is often adulterated with other tanning barks.

Rhizophoraceae → Buguiera, Carallia, Ceriops, Kandelia, Poga, Rhizophora.

Rhizopogon rubescens Tul. (Hymenogastraceae). - Basidiomycete. Fungus. The subterranean fruitbodies are consumed as food in Japan.

Rhizopus japoncius Vuillemin. (Mucoraceae). - A mold that is used in converting starches into sugars. Also R. delemar.

Rhizopus nigricans Ehr. (Mucoraceae). - Fungus. A mold that has the ability to convert 40 to 50 % of the sugar consumed into fumaric acid. Other molds usually produce but small amounts or this substance.

Rhizopus oryzae Went and Prins Geerl. (Mucoraceae). - A mold having the ability to produce d-lactic acid from nutrient sugar media.

Rhode Island Bent → Agrostis vulgaris With.

Rhodes Grass → Chloris Gayana Kunth.

Rhodiola integrifolia Raf. → Sedum roseum Scop.

Rhododendron chrysanthum Pallas. (Ericaceae). - Shrub. Dahuria. Leaves are used as a tea in some parts of Centr. Asia, called Tartarian Tea.

Rhododendron maximum L., Great Laurel, Rose Bay. (Ericaceae). - Shrub to shrub-like tree. East of N. America. Wood brittle, hard, strong, heavy, close-grained, light brown; used for handles of tools, substitute for boxwood in engraving. Wood of the roots is made into tobacco pipes. Decoction of leaves is used as home remedy for rheumatism.

Rhododendron nudiflorum Torr. → Azalea nudiflora L.

Rhododendron, Singapore → Melastoma malabathricum L.

Rhodomelaceae (Red algae) → Acanthophora. Alsidium, Digenea, Laurencia.

Rhodomyrtus tomentosa Wight. Downy Rose Myrtle. (Myrtaceae). - Shrub. E. India, Malaysia. Cultivated in tropics and subtropics of Old and New World. Fruits occasionally used for pies.

Rhodopaxillus amethystinus (Berk. and Br.) van Overveen. (Agaricaceae). - Basidiomycete Fungus. Fruitbodies are sometimes consumed as food by the natives in the East Indies.

Rhodopaxillus Caffrorum (Kalckbr. and McOwan). (Agaricaceae). - Basidiomycete Fungus. S. Africa. The fruitbodies are consumed as food by the natives.

Rhodopaxillus nudus (Bull.) R. Maire (syn. Tricholoma nudum Bull.) Quél. (Agaricaceae). - Basidiomycete Fungus. Temp. zone. Fruitbodies are consumed as food in Europe and Japan. Sold in markets. This species is sometimes cultivated in cellars on dead leaves, esp. in France.

Rhodophyllaceae (Red Algae) → Cystoclonium.

Rhodosphaera rhodanthema Engl. (syn. Rhus rhodanthema F. v. Muell.). Yellow Cedar. (Anacardiaceae). - Tree. Australia. Wood fine-grained, soft, beautifully marked; used for cabinet work.

Rhodotachys argentina Baker → Bromelia serra Griseb.

Rhodymenia palmata (L.) Grev. (Rhodymeniaceae). - Red Alga. Pac. and Atl. Ocean. Eaten in some parts of Scotland and Ireland among the poorer classes of the coastal region. Consumed raw or cooked after having been dried. Sometimes sold in markets. It is abundant in New England where it is dried and consumed as a relish. The sea weed is also used as a condiment and is an esteemed ingredient of ragouts, giving it a red color and jelly-like consistance. Source of an alcoholic beverage among the natives of Kamschatka.

Rhodymeniaceae (Red Algae) → Rhodymenia.

Rhoicissus capensis (Burm. f.) Planch → Vitis capensis Burm. f.

Rhopalostylis sapida Wendl. and Drude. (Areca sapida Soland.) Nikau Palm. (Palmaceae). - Tall palm. New Zealand. Leaves used by Maoris of New Zealand for construction of huts.

Rhubarb → Rheum hybridum Murray. and others.

Rhubarb, Chinese →Centaurea Rhaponticum L.

Rhubarb, Currant → Rheum Ribes L.

Rhubarb, Garden → Rheum Rhaponticum L.

Rhubarb, Himalayan → Rheum Emodi Wall.

Rhubarb, Indian → Saxifraga peltata Torr.

Rhubarb, Medicinal → Rheum officinale Baill.

Rhubarb, Monk's → Rumex alpinus L.

Rhubarb, Sorrel → Rheum palmatum L.

Rhubarb Wine → Rheum hybridum Murray.

Rhus abyssinica Hochst. (Anacardiaceae). - Tree. E. Africa. Bark used by natives for tanning.

Rhus ailanthoides Bunge → Picrasma quassoides Benn.

Rhus albida Schousb. (syn. R. oxycanthoides Dum., R. dioica Willd.). (Anacardiaceae). - Shrub. N. Africa. Bark is used by the Tuareg (Afr.) for tanning sheepskins, an important article of export. Wood is source of good charcoal.

Rhus aromatica Ait. (syn. R. canadensis Marsh.). Fragrant Sumach, Sweet-scented Sumach. (Anacardiaceae). - Shrub. Eastern N. America to Florida and Texas. Split stems were used by Indians for making baskets. Soaked berries are source of a refreshing drink.

Rhus copallina L. (syn. Schmaltzia copallina Small.). Shining Sumach, Dwarf Sumach. (Anacardiaceae). - Shrub or small tree. Eastern N. America to Florida and Texas. Fruits were used in drinks by the Indians. Leaves are rich in tannin, used for tanning leather, also for dyeing. Decoction of roots was used by the Creek Indians for dysentery.

Rhus Coriaria L. Sicilian Sumach. (Anacardiaceae). - Shrub. Mediterranean region, Asia Minor. Leaves are source of an important tanning material. There are two recognized varieties: Mesculino with 25 to 35 % tannin and Feminella with less than 25 % tannin. Stems are sold as Gambazzo.

Rhus cotinoides (Nutt.) Britt. (syn. Cotinus americanus Nutt. Rhus americanus Nutt.). American Smoke Tree. (Anacardiaceae). - Shrub or small tree. Eastern N. America. Wood is locally used as a source of an orange dye.

Rhus Cotinus L. (syn. Cotinus coggyria Scop.). Common Smoke Tree. (Anacardiaceae). - Tree. Mediterranean region to China. Leaves are source of an important tanning material, called Ventilato. Powdered leaves are called Molito, containing 16 % tannin.

Rhus glabra L., Smooth Sumach, Red Sumach. (Anacardiaceae). - Shrub or small tree. Eastern N. America to Florida and Louisiana. Sumach fruits are considered astringent and tonic; used in drinks. Fruits are made by the Ojiba Indians into a cool beverage and during winter as a warm drink, mixed with maple sugar. They used decoction of flowers as astringent and refrigerant; used to gargle sore throat. Roots are source of an orange or yellow dye. Dried leaves mixed with tobacco were smoked by the Kiowa Indians.

Rhus Potanini Maxim., Potanin Sumach. (Anacardiaceae). - Tree. Centr. and W. China. Source of galls, called Ch'-pei-tzu, recommended for manuf. of indelible black ink.

Rhus rhodanthema F. v. Muell. → Rhodophaera rhodanthema Engl.

Rhus semialata Murr. (syn. R. chinensis Mill., R. javanica Thunb. not L.). Chinese Sumach. (Anacardiaceae). - Tree. China, Japan. Source of Japanese or Chinese galls, Wu-pei-tzu, caused by an insect Chermes Sp., used in China for dyeing blue silk. Also used for tanning.

Rhus simarubaefolia Gray. (syn. R. taitensis Guill.). (Anacardiaceae). - Tree. South Pacif. Islds. Wood used in Samoa for canoes.

Rhus succedanea L. Wax Tree. (Anacardiaceae). - Tree. China, Japan. Cultivated. Source of Vegetable Wax, Japan Tallow, Sumach Wax obtained from the mesocarp of the fruit, being pale yellow, flat cakes, substitute of beeswax; used in varnishes, floor waxes, wax varnishes, ointments, furniture polish, ingredients for plasters. Leaves are source of galls, caused by an Aphis; used for tanning material. Sold in bazaars of India.

Rhus sylvestris Sieb. and Zucc. (Anacardiaceae). - Tree. China and Japan. Plant is source of a lacquer. → R. vernicifera Stokes.

Rhus trilobata Nutt. (syn. Schmaltzia trilobata Nutt.). Small, S. Bakeri Greene.). Squaw Berry, Lemonade Sumach. (Anacardiaceae). - West of Ohio and Mississippi River into Mexico. Fruits used as food fresh or dried, by various Indian tribes. Also made into jam. When mashed in water, used as a beverage. Shoots were made into baskets. Leaves mixed with tobacco were used for smoking. Berries were also mixed with corn-meal and consumed as food by the Kiowa Indians (New Mex.).

Rhus sempervirens Scheele (syn. R. virens Lindh.) Evergreen Sumach. (Anacardiaceae). - Shrub or small tree. Mexico, S. W. of the United States. Leaves are smoked among the Indians of Texas, sometimes being mixed with tobacco.

Rhus typhina L. (syn. R. hirta (L.) Sudw.). Staghorn Sumach, Lemonade Tree. (Anacardiaceae). - Shrub or small tree. Eastern N. America. Acid fruits are made into drinks, known as Indian Lemonade.

Rhus vernicifera DC. (syn. Rhus verniciflua Stokes). Varnish Tree. (Anacardiaceae). - Tree. Japan, China. Source of a varnish, Japanese Lacquer which is obtained by cutting the bark and removing the exuding liquid. The emulsion is strained and when subjected to the air, it becomes finally hard and black, and forms the famous varnish. Plant is also source of an oil; used for candle making.

Rib Grass → Plantago lanceolata L.

Ribbon Eucalyptus → Eucalyptus viminalis Lab.

Ribbon Gum Kino → Eucalyptus amygdalina Lab.

Ribbon Wood, Lowland → Plagianthes betulinus Cunningh.

Ribes americanum Mill. (syn. R. floridum L'Hér.) American Wild Black Currant. (Grossulariaceae). - Shrub. Eastern N. America. Berries black, have a characteristic flavor; used for jellies.

Ribes aureum Pursh., Golden Current, Missouri Current, Buffalo Current. (Grossulariaceae). - Shrub. Western N. America. Currents are used for jellies and pies. Are of a characteristic flavor. Indians used dried fruits for pemmican, composed of pounded dried buffalo meat and fruits mixed with tallow which was made into cakes and loaves.

Ribes bracteosum Dougl. (Grossulariaceae). - Shrub. Western N. America. Currents are consumed fresh or boiled with salmon roe by the Eskimos of some parts of Alaska.

Ribes Cynosbati L. American Gooseberry, Prickly Gooseberry. (Grossulariaceae). - Shrub. Eastern N. America. Cultivated. Berries used for pies and preserves. Varieties: Downing, Pearl, Josselyn, Houghton.

Ribes Grossularia L., European Gooseberry. (Grossulariaceae). - Shrub. Europe, N. Africa, Caucasia, Manschuria, N. China etc. Cultivated in temp. zones. Fruits eaten raw, of a pleasant flavor, also made into wine. Unripe fruits when cooked are used in pies. Many large fruited varieties of red, green and yellow color, e. g.: Golden Drop, Industry, Green Ocean, Prince of Orange, Grosse Rouge Hâtive, Grosse Rouge Tardive etc. Can not be grown successfully in some parts of North America due to attacks of a Mildew.

Ribes lacustre Boir. Prickly Currant. (Grossulariaceae). - Shrub. N. America. Currants are consumed as food by Eskimos.

Ribes nigrum L., European Black Currant. (Grossulariaceae). - Shrub. Europe, Centr. Asia, Himalaya region. Cultivated. All parts of plants produce a characteristic scent when bruised. Berries deep black. Not eaten raw; made into jams and a liqueur, called Cassis. Decoction of dried leaves and twigs is a home-remedy for coughs. Varieties: Lee, Baldwin, Goliath, Westwick Choice.

Ribes oxycanthoides L., Smooth Gooseberry. (Grossulariaceae). - Shrub. Eastern N. America. Berries used for jams and pies.

Ribes rubrum L. (syn. R. silvestre Mert. and Koch). Red Currant. (Grossulariaceae). - Shrub. Europe, Asia. Widely cultivated. Fruits of commercial importance. Berries eaten fresh, in compote, as jam, sauce, fruit juice and wine. Barle-Muce Jelly was formerly derived from a whi-

te variety. There are several varieties, among which red fruited: Cumberland Red, Red Dutch, Victoria, Fay, Versailles Red; white fruited: Large White and White Grape.

Ribes triste Pall. Swamp Red Currant. (Grossulariaceae). - Small shrub. Eastern N. America, Siberia. Berries are suitable for pies and jellies.

Riboflavon → Candida Guillermondii (Cast.) Lang and Guerra and Eremothecium Ashby Schopf.

Ribseed → Aulospermum longipes (S. Wats.) Coult and Rose.

Rice → Oryza sativa L.

Rice Bean → Phaseolus calcaratus Roxb.

Rice, Indian → Zizania aquatica L.

Rice, Jungle → Echinochloa colonum (L.) Link.

Rice Paper → Fatsia papyrifera Hook.

Rice, Wild → Zizania aquatica L.

Richardsonia pilosa H. B. K. (syn. R. scabra St. Hil.) (Rubiaceae). - Shrub. Trop. and subtr. America. Source of Undulated or Farinaceous Ipecac.

Ricinodendron Rautanenii Schinz. (Euphorbiaceae). - Tree. Trop. S. W. Africa. Seeds source of Manketti Nut Oil, a drying oil; suggested for food, paints and varnishes. Sp. Gr. O.929—0.931; Sap. Val. 191,5—195; Iod. No. 129—137; Unsap. 0.85 %.

Ricinus communis L., Castor Bean, Castor-Oil Plant. (Euphorbiaceae). - Tree. Probably native to Africa. Seeds source of Castor-Oil. Cultivated. Much is derived from India, China, Manchuria, Mexico, Brazil. Keeps well. No. 1 oil is used for medical and technical purposes and as lubricant. No. 2 oil in crude stage employed for technical use. Much is converted into sulfonated castor oil, also called Turkey-Red Oil; used for dyeing cotton fabrics with alizarine. Manuf. into transparent soap, textile soap, typewriter inks, fly-paper, imitation leathers, lubricants; used in manuf. nitrocellulose, baking finishes, preparation of „perfume aromatics", sebacic acid, for synth. of nylon fibre. When dehydrated, used as drying oil in enamel, paint, varnish. Blown oil used for grinding lacquer paste colors, when hydrogenated and sulfonated employed for preparation of ointments. Sp. Gr. 0.958—0.968; Sap. Val. 177—187; Iod. No. 82—90; Unsap 0.3—0.7 %. Ground seed cakes are used as fertilizer, known as Castor Oil Pomace.

Ring Lichen → Parmelia centrifuga (L.) Ach.

Ringworm Senna → Cassia alata L.

Rinorea casteneaefolia (Spreng.) O. Ktze. (Violaceae). - Shrub. S. America. Leaves are consumed as a vegetable among the negro population in Brazil. Also R. physiphora (Mart. and Zucc.) O. Ktze.

Rio Rosewood → Dalbergia nigra Allem.

Rivea campanulata (L.) House. (Convolvulaceae). - Woody vine. Centr. America. Sap is used for coagulating latex of Castilla elastica.

River Bank Grape → Vitis vulpina L.

River Birch → Betula nigra L.

River Black Oak → Casuarina suberosa Otto and Dietr.

River Oak → Casuarina stricta Ait.

River Plum → Prunus americana Marsh.

Rivina humilis L., Rouge Plant. (Phytolacaceae). - Small shrub, Trop. America. Fruits are source of a red dye.

Rivulariaceae (Blue-green Algae) → Brachytrichia.

Robinia Pseudacacia L., Locust, Black Locust. (Leguminosaceae). - Tree. Eastern N. America. Wood strong, very hard, heavy, very durable, close-grained, brown; used for ship-building, turnery, construction, treenails, fuel. Seeds when cooked were consumed by the Indians. Plant has poisonous properties. Is said to be tonic, purgative and emetic. Flowers are source of a good honey.

Roble Colorado — Platymiscium pinnatum (Jacq.) Dugand.

Roblo Blanco → Tabebuia pentaphylla (L.) Hemsl.

Robusta Coffee → Coffea robusta Linden.

Roccella fuciformis (L.) Lam., Lima Weed, Angola Weed, Orcella. (Roccellaceae). - Lichen. Mediterranean region, Africa. Source of a purple-crimson or red-yellow dye; used in France and England for dyeing silk, woolens, carpet yarns; also for staining wood and marble. Source of Litmus, Orchil.

Roccella Montagnei Bél., Orcella Weed. (Roccellaceae). - Lichen. Africa, Asia, Australia. On trunk of trees. Used in Germany and Italy as a dye.

Roccella phycopsis Ach., Archil. (Roccellaceae). - Lichen. Asia. Source of a blue dye, used for coloring British Broadcloth. A tincture of alcohol is used for thermometers.

Roccella tinctoria DC., Orseille, Vulparaiso Weed. (Roccellaceae). - Lichen. Mediterranean region, Africa, Asia, Australia. Used as source of Litmus for testing alkalies giving a blue and for acids giving a red color. Employed for dyeing silks and woolens, also for coloring wine and liqueur, and used in laundry. Of commercial importance. Was employed before the time of Pliny. After the fall of the Roman Empire, its use was rediscovered about 1300 by Federigo, a Florentine, who became head of the Oricellari family; from which the name of the dye, Orseille was derived. The product is also called Persio. Most is derived from the Netherlands. The best lichen material is obtained from the Canary and Cape Verde Islds.

Rock Dammar → Hopea odorata Roxb.

Rock Elm → Ulmus Thomasii Sarg.

Rock Tripe → Gyrophora deusta (L.) Ach.

Rock Urceolaria → Urceolaria scruposa (Schreb.) Ach.

Rocket Larkspur → Delphinium Ajacis L.

Rocket Salad → Eruca sativa L.

Rocket, Yellow → Barbarea vulgaris R. Br.

Rocky Mountain Cherry → Prunus melanocarpa (A. Nels.) Rydb.

Rocky Mountain Fir → Abies lasiocarpa (Hook.) Nutt.

Rocky Mountain Flax → Linum Lewisii Pursh.

Rocky Mountain White Pine → Pinus flexilis James.

Rogersia adenophylla Gay. (Pedaliaceae). - Herbaceous plant. Red Sea region, Trop. Africa. Plant is source of a mucilagenous infusion, recommended for diarrhea.

Rohituka Oil → Amoora Rohituka Wight and Arn.

Rollinia deliciosa Safford., Biribá. (Annonaceae). - Medium sized tree. Brazil. Cultivated. Fruits known in markets of Rio de Janeiro as Fructa da Condessa. They are of excellent quality.

Rollinia longifolia St. Hil. (syn. R. dolabripetala (Raddi) St. Hil.). (Annonaceae). - Small tree. Brazil. Fruits edible, fleshy; occasionally cultivated.

Rollinia orthopetala A. DC. (Annonaceae). - Shrub or small tree. S. America. Fruits are edible, being sweet and of a pleasant flavor.

Rollinia Sieberi A. DC. Cachiman. (Annonaceae). - Small tree. Guiana. Fruits of the size of Annona squamosa, are supposed to have a good flavor; eaten by the natives.

Rolliniopsis discreta Safford. (Annonaceae). - Shrub or small tree. Brazil. Occasionally cultivated. Fruits edible, known as Fructa de Macaco, having taste like some species of Xvlopia.

Roman Chamomille → Anthemis nobilis L.

Roman Fennel → Foeniculum vulgare Mill.

Roman Wormwood → Artemisia pontica L.

Ronima Resin → Protium heptaphyllum March.

Roquette → Eruca sativa L.

Roripa Nasturtium-aquaticum (L.) v. Hayek → Nasturtium officinale R. Br.

Rosa Morada → Lonchocarpus hondurensis Benth.

Rosa Banksiae Ait. Banks' Rose. (Rosaceae). - Vine. China. Bark of roots used in W. Hupeh for tanning.

Rosa californica Cham and Schlecht. Californian Rose. (Rosaceae). - Shrub. Oregon to Baja California. Ripe fruits are eaten stewed or raw, after frost has sweetened them. Much esteemed among Spanish Californians who call them Macuatas.

Rosa canina L. Doghip, Dog-Rose. (Rosaceae). - Shrub. Europe, temp. Asia, N. Africa. Fresh fruits are used medicinally as a diuretic, refrigerant, mildly astringent. Leaves are used as a substitute for tea.

Rosa centifolia L. (syn. R. gallica L. var. centifolia Reg.). (Rosaceae). - Shrub. E. Caucasia. Cultivated. Petals distilled with water are a source of Aqua Rosae Fortior.

Rosa damascena Mill., Damask. Rose. (Rosaceae). - Shrub. Balkans, Asia Minor. Cultivated in Bulgaria, So. France and Turkey. Petals upon steam distillation are source of Otto of Roses, or Oil of Roses an ess. oil, important in perfumery; also as flavoring agent.

Rosa Eglanteria L. (syn. R. lutea Mill.). (Rosaceae). - Shrub. Iran, Afghanistan, Crimea, Asia Minor, Punjab. Cultivated. Flowers are used in Iran for colic and diarrhea. They are source of Gulangabin used in confectionary; being composed of petals and honey.

Rosa gallica L. French Rose. (Rosaceae). - Shrub. Europe, W. Asia. Cultivated. Dried petals are used medicinally as a mild astringent and tonic. Petals are collected before expansion of the flowers. Contain an ess. oil.

Rosa rugosa Thunb. Rugosa Rose. (Rosaceae). - Shrub. China, Japan. Rose hips are used as food by the Ainu.

Rosa pomifera Herm. (syn. R. villosa L. var pomifera (Herm.) Crép.) (Rosaceae). - Low shrub. Europe, W. Asia. Leaves are used as a substitute for tea. (deutscher Tee). Fruits, Rose-Hips are made into preserves and sauces; occasionally made into a beverage. Much esteemed in parts of Bavaria and Austria. Rosewine and Rosehoney were well known to the Romans. Also R. rugosa Thunb.

Rosaceae → Adenostoma, Afrolicania, Agrimonia, Alchemilla, Amelanchier, Amygdalus, Armeniaca, Cercocarpus, Chrysobalanus, Cliffortia, Comarum, Cotoneaster, Couepia, Cowania, Crataegus, Cydonia, Dryas, Eriobotrya, Ferolia, Filipendula, Fragaria, Geum, Gillenia, Hagenia, Licania, Margyricarpus, Mespilus, Neurada, Parinarium, Photinia, Potentilla, Prunus, Pygeum, Pyrus, Quillaja, Raphiolepis, Rosa, Rubus, Sanguisorba, Sorbus, Vauquelinia.

Rosary Pea → Abrus precatorius L.

Rose and Gold Lichen → Sticta aurata Ach.

Rose Apple → Eugenia Jambos L.

Rose Apple, Samarang → Eugenia javanica Lam.

Rose Apple, Watery → Eugenia aquea Burm. f.

Rose, Bank's → Rosa Banksiae Ait.

Rose, Californian → Rosa californica Cham. and Schlecht.

Rose, Christmas → Helleborus niger L.

Rose Dammar → Vatica Rassak Blume.

Rose, Dog → Rosa canina L.

Rose, Eglatine → Rosa Eglanteria L.

Rose Elf → Claytonia virginiana L.

Rose, French → Rosa gallica L.

Rose Gentian, Squarestem → Sabatia angularis (L.) Pursh.

Rose Hips → Rosa pomifera Herrm.

Rose Leak → Allium canadense L.

Rose, Stock → Sparmannia africana L. f.

Rosebay, East India → Ervatamia coronaria Stapf.

Rosella → Hibiscus Sabdariffa L.

Rosemary → Rosmarinus officinalis L.

Rosemary, Bog → Andromeda polifolia L.

Rosetta Rosewood → Dalbergia latifolia Roxb.

Rosewood → Amyris balsamifera L., Aniba rosaedora Ducke, Dysoxylon Fraserianum Benth., Eremophila Mitchelli Benth. and Synoum glandolosum Juss.

Rosewood, African → Pterocarpus erinaceus Lam.

Rosewood, Bahia → Dalbergia nigra Allem.

Rosewood, Black → Dalbergia latifolia Roxb.

Rosewood, Bombay → Dalbergia latifolia Roxb.

Rosewood, Brazilian → Dalbergia nigra Allem. and Physocallymma scaberimum Pohl.

Rosewood of British Honduras → Dalbergia Stevensoni Standl.

Rosewood, East Indian → Dalbergia latifolia Roxb.

Rosewood of Guatemala → Dalbergia cubilquitzensis (D. Sm.) Pitt.

Rosewood, Malabar → Dalbergia latifolia Roxb.

Rosewood, Nicaragua → Dalbergia retusa Hemsl.

Rosewood, Rio → Dalbergia nigra Allem.

Rosewood, Rosetta → Dalbergia latifolia Roxb.

Rosewood, Seychelles → Thespesia populnea Soland.

Rosewood of Southern India → Dalbergia latifolia Roxb.

Rosewood, White → Dalbergia nigra Allem.

Rosewood, Yama → Platymiscium dimosphandrum Don. Sm.

Rosha Grass → Cymbopogon Martini Stapf.

Rosin → Pinus palustris Mill.

Rosin Oil → Pinus palustris Mill.

Rosinol → Pinus palustris Mill.

Rosmarinus officinalis L., Rosemary. (Labiaceae). - Perennial plant. Mediterranean region, S. Europe. Cultivated in France, Spain, Dalmatian Islds., N. Africa. Herb used for flavoring sausages, also employed in perfumery, in Eau de Cologne, Hungarian Water. Flowers source of excellent honey. Oil of Rosemary, Oleum Rosemarini used medicinally, being carminative, stimulant; for liniments as a rubefacient. Also used for many home-remedies. Contains borneol.

Rostkovia grandiflora Hook. f. (syn. Marsippospermum grandiflorum Hook.). (Juncaceae). - Perennial herb. Chile, Antarctic Isles. Used for covering huts, manuf. baskets and similar articles.

Rothmannia longiflora Sal. → Randia maculata DC.

Rottboellia compressa L. f. (syn. Hemarthria compressa R. Br., H. guyanense Steud.). (Graminaceae). - Perennial grass. Tropics. Plant is considered an excellent forage grass.

Rottboellia exaltata L. f. (syn. Ophiurus appendiculatus Steud.) (Graminaceae). - Strong perennial grass. Tropics. Used as food for livestock.

Rottboellia fasciculata Desf. → Haemarthria fasciculata Kunth.

Rottboellia glandulosa Trin. (Graminaceae). - Perennial grass. Burma, Malaysia. Plants produce much leaf; used as fodder for cattle in Indonesia.

Rottlera discolor F. v. Muell. → Mallotus discolor F. v. Muell.

Rottlera japonica Spreng. (Euphorbiaceae). - Tree. Japan. Wood fine-grained, red; used in Japan for pillars of houses and for boxes.

Rouge d'Adrianople → Quercus tauricola Kotschy.

Rouge Plant → Rivina humilis L.

Rough Tongues → Aster macrophyllus L.

Round Cardamon → Amomum kepulaga Sprague.

Roupola brasiliensis Klotzsch. (Proteaceae). - Tree. Brazil. Wood durable, difficult to work; used for ship building, general carpentry. Also R. complicada H. B. K., R. heterophylla Pohl, and R. macrophylla Schott.

Roupala elegans Pohl. (Proteaceae). - Tree. Brazil. Wood is used for ship building and general carpentry.

Roupala montana Aubl., Trinidad Roupala. (Proteaceae). - Tree. Trinidad, Northern S.America. Infusion of bark is used locally in Trinidad as nerve stimulant.

Rourea glabra H. B. K. (Connaraceae). - Shrub. Tamaulipas to Tepec (Mex.), Centr. America, Venezuela, West Indies. Seeds are very poisonous to carnivorous animals. They are used for poisoning coyotes, also for criminal poisoning. Roots are source of a strong fibre, used for cordage. Bark used as a tanning material, gives a bright purple color to skins.

Rourea mimosoides Planch. (syn. Santaloides mimosoides Kuntze). (Connaraceae). - Woody climber. Nicobar Islds., Sumatra, Borneo. Decoction of roots is used for colic.

Rowan Tree → Sorbus Aucuparia L.

Roxburghiaceae → Stemona.

Royal Palm → Roystonia regia (H. B. K.) Cook.

Royal Salep → Allium Macleanii Baker.

Royena macrocalyx Guerke. (Ebenaceae). - Shrub. Trop. Africa. Roots and bark are source of a black dye.

Roystonia oleracea (Mart.) Cook. (syn. Oreodoxa oleracea Mart.). (Palmaceae). - Tall palm. West Indies. Very young leafbuds are consumed as a vegetable in some parts of the West Indies.

Roystonea regia (H. B. K.) Cook. (syn. Oreodoxa regia H. B. K.). Royal Palm. (Palmaceae). - Tall palm. S. Florida, West Indies, Centr. America. Cultivated. Trunks are used as wharf-piles and for construction. Fruits were consumed by the aborigines. Very young leafbuds are consumed in some parts of the West Indies.

Rozites caperata (Bers.) Karst. (Agaricaceae). - Basidiomycete. Fungus. Temp. zone. The fruitbodies are consumed as food. Sold in markets of Europe.

Rubber. Plants Source of → Alstonia, Apocynum, Asclepias, Bosquea, Carpidinus, Castilla, Chlorophora, Chonemorpha, Chrysothamnus, Clitandra, Cnidoscolus, Conopharyngia, Cryptostegia, Eucommia, Euphorbia, Ficus, Fockea, Fosteronia, Fouquieria, Funtumia, Glycine, Gonocrypta, Hancornia, Hevea, Holarrhena, Landolphia, Manihot, Marsdenia, Mascarenhasia, Micrechites, Omphalocarpum, Omphalogonus, Oncinotis, Parabarium, Parameria, Parthenium, Periploca, Phthirusa, Plectaneia, Plumeria, Raphionacme, Sapium, Scorzonera, Secamonopsis, Solidago, Struthanthus, Tabernaemontana, Taraxacum, Urceola, Voacanga, Willughbeia, Xylinabaria, Zschokkea.

Rubber Coagulants. Source of → Adansonia, Calonyction, Chrysophyllum, Conopharyngia, Costus, Ixonanthes, Orbignya, Rivea, Strophanthus.

Rubber, Assam → Ficus elastica Roxb.

Rubber, Bokalahy → Marsdenia verrucosa Decne.

Rubber, Camoto → Sapium taburu Ule.

Rubber, Caucho Blanco → Sapium pavonianum Muell. Arg.

Rubber, Ceara → Manihot Glaziovii Muell. Arg.

Rubber, Chupire → Euphorbia calyculata H. B. K..

Rubber, Colombia Virgin → Sapium Thomsoni God.

Rubber, Dahomey → Ficus Vogelii Miq.

Rubber, Ekanda →Raphionacma utilis N. E. Br.

Rubber, Fingotra → Landolphia crassipes Radl.

Rubber, Gaucho Blanco → Sapium Thomsoni God.

Rubber, Guayule → Parthenium argentatum Gray.

Rubber, Indian → Ficus elastica Roxb.

Rubber, Intisy → Euphorbia Intisy Drake del Cast.

Rubber, Jecquie → Manihot dichotoma Ule.

Rubber, Madagascar Rouge → Landolphia madagascariensis Benth. and Hook.

Rubber, Majunga Noir → Mascarenhasia arborescens A. DC.

Rubber, Majunga Rouge → Landolphia Perieri Jum.

Rubber, Mangabeira → Hancornia speciosa Gomez.

Rubber, Mozambique Blanc →Landolphia Kirkii Dyer.

Rubber, Mozambique Rouge → Landolphia Kirkii Dyer.

Rubber, Nyassa → Landolphia Kirkii Dyer.

Rubber, Orinoco Scrap → Sapium Jenmani Hemsl.

Rubber, Palay → Cryptostegia grandiflora R. Br.

Rubber, Para → Hevea brasiliensis Muell. Arg.

Rubber, Piauhy → Manihot piauhyensis Ule.

Rubber, Rangoon → Urceola Maingayi Hook. f.

Rubber, Rouge du Congo → Landolphia owariensis Beauv.

Rubber, Rouge du Kassai → Landolphia Gentilii de Wild.

Rubber, Rouge du Kassai → Landolphia owariensis Beauv.

Rubber, São Francisco → Manihot heptaphylla Ule.

Rubber, Serapat → Urceola acuto-acuminata Boerl.

Rubber, Sernamby → Sapium taburu Ule.

Rubber, Silk → Funtumia elastica Stapf.

Rubber, Tapuru → Sapium taburu Ule.

Rubber Tree, Male → Holarrhena Wulfsbergii Stapf.

Rubber Tree, Para → Hevea brasiliensis Muell. Arg.

Rubber, Vahimainty → Secamonopsis madagascariensis Jum.

Rubia cordifolia L., India Madder. (Rubiaceae). - Herbaceous perennial. Trop. and Temp. Asia, Africa. Used medicinally in China. Contains munjistin, a glucosid. Roots are source of a red dye. Much esteemed as lalab, a side dish with rice, among the Javanese.

Rubia khasiana Kurz. Khasia Madder. (Rubiaceae). - Perennial herb. Khasia. Roots are source of a red dye; used in Sikkim.

Rubia peregrina L. Levant Madder. (Rubiaceae). - Perennial herb. Mediterranean region. Decoction of flowers is used in N. Morocco as aphrodisiac. Powdered roots as emmenagogue and abortive. A decoction is used as diuretic.

Rubia sikkimensis Kurz. Sikkim Madder. (Rubiaceae). - Perennial herb. Himalayan region. Roots are source of a red dye; sold in the bazaars of Darjeeling.

Rubia tinctorum L., Common Madder. (Rubiaceae). - Perennial herb. S. Europe, Orient. Roots are source of a red dye, contains alizarin. Was much cultivated in former years, before the introduction of the analin dyes. Herb furnishes a good green-fodder when cut in second year.

Rubiaceae → Adina, Anthocephalus, Antirrhoea, Asperula, Basanacantha, Bikkia, Borreria, Calycophyllum, Canthium, Cephaelis, Cephalanthus, Chasalia, Chiococca, Cinchona, Coffea, Corynanthe, Craterispermum, Cremaspora, Crossopteryx, Danais, Diplospora, Esenbeckia, Euxylophora, Exostemma, Feretia, Gardenia, Genipa, Geophila, Grumilea, Guettarda, Heinsia, Hymenodictyon, Isertia, Ixora, Lecontea, Leptactina, Mitracarpum, Mitragyna, Morinda, Mussaenda, Oldenlandia, Oxanthus, Paederia, Pausinystalia, Pinkneya, Plectronia, Posoquera, Pseudocinchona, Psychotria, Randia, Remijia, Richardsonia, Rubia, Sarcocephalus, Schradera, Sickingia, Timonius, Uncaria, Vangueria.

Rubus adenotrichos Schlecht., Mora Común. (Rosaceae). - Shrub. Centr. and N. Ecuador, Mexico. Fruits edible, well flavored, of good quality; occasionally sold in local markets.

Rubus allegheniensis Porter. (syn. R. nigrobaccus Bailey.). Alleghany Blackberry, Mountain Blackberry. (Rosaceae). - Shrub. Eastern N. America. Berries are eaten and made into jam, preserves and compote.

Rubus amabilis Focke. (Rosaceae). - Small shrub. W. China. Fruits large, red, of good flavor; edible; consumed in China.

Rubus arcticus L. Arctic Bramble. (Rosaceae). - Perennial herb. Arctic regions. Leaves are used as a tea in some parts of Norway. Fruits yelloy, of a pleasant flavor, much esteemed by the Eskimos.

Rubus brasiliensis Mart. (Rosaceae). - Shrub. Brazil. Cultivated. Fruits are consumed by the inhabitants and are also used in drinks.

Rubus caesius L. European Dewberry. (Rosaceae). - Shrub. Europe. Fruits are consumed raw or used in jellies and preserves. Leaves are used as substitute for tea during times of lack of asiatic tea. Also other Rubus species are used for this purpose.

Rubus Chamaemorus L., Cloudberry, Yellow Berry, Salmonberry. (Rosaceae). - Low herbaceous plant. Northern temp. Hemisphere. Fruits edible, much prized in some localities, gathered from the woods in large quantities.

Rubus cochinchinensis Tratt. (Rosaceae). - Tall shrub. China, Cochin-China. Fruits edible; consumed by the natives.

Rubus ellipticus Smith., Yellow Himalayan Raspberry. (Rosaceae). - Shrub. E. India. Fruit edible, of good raspberry flavor; used for preserves. Naturalized in Jamaica. Introd. in Florida and California. Fruits are sometimes known as Golden Evergreen Raspberry.

Rubus Elmeri Focke. (Rosaceae). - Scrambling shrub. Mountains Luzon. (Philipp. Islds.). Fruits orange-yellow, edible well flavored.

Rubus floribundus H. B. K., Zarzamora. (Rosaceae.) - Shrub. Ecuador. Fruits edible, of good quality.

Rubus flagellaris Willd. (syn. R. villosus Ait.) Northern Dewberry. (Rosaceae). - Shrub. Eastern N. America. Variable species. Fruit edible. Cultivated. Eaten raw, in jams and pies, fruit juices, wines etc. With other species serves as an agreeable acidulous syrup, used in pharmaceutical preparations. Var. roribaccus Bailey is the Lucretia Dewberry.

Rubus geoides Sm. (Rosaceae). - Low straggling shrub. Magellan region, Falkland Islds. Fruits edible, red, juicy, large, delicious flavor.

Rubus glaucus Benth., Mora de Castilla. (Rosaceae). - Shrub. Ecuador and adj. territory. Cultivated. Fruit edible of excellent quality, pleasant, sub-acid, 2 to 3 cm. long, raspberry flavor; eaten fresh, in preserves and in a syrup called Jaropa de Mora. Some varieties have very delicate flavor.

Rubus ichangensis Hemsl. and Kuntze. (Rosaceae). - Shrub. Centr. and W. China. Fruits edible, small, red of good flavor. Eaten in some parts of China.

Rubus Idaeus L. European Red Raspberry. (Rosaceae). - Shrub. Europe, Asia. Widely cultivated. Of commercial importance. Fruits eaten raw, in compôte, jam, pastries, preserves, confectionary; used in beverages. Source of Raspberry Liqueur and a vinegar. Occasionally sold when dried. Numerous varieties with red and yellow fruits, among which Hornet, Superlative, Radboud, Malling Promise Preussen. Yellow Antwerp and Merveille de Quatre Saisons Jaune with yellow fruits. Dried leaves are used as a tea, esp. in times of emergency. Is closely related to R. strigosus Michx.

Rubus innominatus S. Moore. (Rosaceae). - Small shrub. Centr. and S. China. Fruits orange-red, edible; consumed in some parts of China.

Rubus macrocarpus Benth., Colombian Berry. (Rosaceae). - Shrub. Colombia, Ecuador. Cultivated. Fruits edible, very large, 5 cm. long, with a taste of the Loganberry.

Rubus microphyllus L. f. (syn. R. palmatus Thunb.). (Rosaceae). - Small shrub. China, Japan. Fruits edible, large, yellow.

Rubus moluccanus L. (syn. R. alcaeafolius Poiret.). (Rosaceae). - Shrub. Tonkin, Annam, Cambodia, Mal. Archip. Fruits are edible, consumed by the natives of several countries.

Rubus morifolius Sieb. (Rosaceae). - Shrub. Japan. Fruits edible, large, red; consumed in Japan.

Rubus neglectus Peck. Purple Cane Raspberry. (Rosaceae). - A hybrid group between R. strigosus and R. occidentalis. N. America, occasionally cultivated. Varieties are: Gladstone, Shaffer and Philadelphia.

Rubus occidentalis L., Black Raspberry, Black Cap. (Rosaceae). - Shrub. Eastern N. America to Colorado and Brit. Columbia. Cultivated in the United States. Fruits black, eaten raw, in sherbets, ice-creams, marmelade. Varieties are: Cumberland, Farmer, Gregg, Kansas. There are hybrids between R. occidentalis and R. strigosus.

Rubus parvifolius L. Japanese Raspberry. (Rosaceae). - Shrub. China, Japan. Fruits are edible; sold in the markets of China and Japan.

Rubus pectinellus Max. (Rosaceae). - Trainling shrub. Philipp. Islds., Japan. Fruits edible, bright red, sub-acid, juicy, fine flavored, of good quality.

Rubus phoenicolasius Maxim. Wine Raspberry, Wineberry. (Rosaceae). - Small shrub. Japan. Cultivated. Fruits small, edible, cherry red, acid, somewhat insipid; occasionally sold in markets of United States.

Rubus rosaefolius Smith., Bramble of the Cape, Mauritius Raspberry. (Rosaceae). - Shrub. Trop. Asia, introd. in other continents. Cultivated. Fruits edible, insipid; eaten fresh or cooked. Decoction of roots is used in Natal and Cape Colony for diarrhea.

Rubus roseus Poir., Mora de Rocoto, Huagra Mora. (Rosaceae). - Shrub. Peru, Ecuador. Fruits edible; resembling raspberries, they are made into refreshing drinks in Peru.

Rubus strigosus Michx. (syn. R. Idaeus L. var. strigosus Michx.) American Red Raspberry. (Rosaceae). - Shrub. Eastern N. America to New Mexico and Arizona. Cultivated in the United States and Canada. Fruits eaten raw, preserved, as compote, in drinks, sherbets, ice-cream, pastries, jams, confectionary. Important varieties are: Cuthbert, King, Marlboro, Herbert, Empire and Sunbeam. The often disputed, large fruited Loganberry, R. loganobaccus Bailey is probably a hybrid between R. Idaeus L. (or R. strigosus Mich.) Red Raspberry x R. ursinus Cham. and Schlecht, Californian Dewberry. The Mammoth Blackberry with large black fruits may be a hybrid of R. ursinus Cham. and Schlecht x a Blackberry. The Mayberry of Burbank is supposed to be a hybrid between R. palmatus Thunb. x R. strigosus Mich. var. Cuthberth. The Youngberry and Thornless Youngberry are derived from a cross between R. loganobaccus Bailey, Loganberry x R. flagellaris Willd. Dewberry. Also the Boysenberry and Thornless Boysenberry are of hybrid origin.

Rubus trivialis Michx., Southern Dewberry. (Rosaceae). - N. America. Shrub. Fruits edible; used for jams and preserves, also eaten raw. Infusion of the leaves is used by the Seminole Indians for stomach trouble.

Rue Anemone → Anemonella thalictroides (L.) Spach.

Rue, Columbine Meadow → Thalictrum aquifolium L.

Rue, Common → Ruta graveolens L.

Rue, European Goat's → Galega officinalis L.

Rugosa Rose → Rosa rugosa Thunb.

Rukam → Flacourtia Rukam Zoll. and Mor.

Rula Ixtle Fibre → Agave lophantha Schiede.

Rulingia pannosa R. Br. (syn. Commersonia dasyphylla Andr., Buettneria dasyphylla Gray.) (Sterculiaceae). - Tree. Australia, esp. Victoria, Queensland. Source of a very useful fibre.

Rum Cherry → Prunus serotina Ehrh.

Rumberry → Eugenia floribunda West.

Rumex abyssinicus Jacq. Spanish Rhubark Dock. (Polygonaceae). - Perennial herb. Abyssinia. Powdered rootstock is used to give a brick-red color to butter. Cultivated in Congo Basin. Also used by natives as pot-herb.

Rumex Acetosa L. Garden Sorrel. (Polygonaceae). - Perennial herb. Temp. Europe, Asia. Cultivated in temp. zones of Old and New World. Leaves eaten with spinach, also separately. Var. Oseille de Belleville.

Rumex alpinus L. Alpine Dock, Monk's Rhubarb. (Polygonaceae). - Perennial herb. Mountain regions. Centr. Europe, Balkans, Caucasia. Young leaves are eaten as salad and as spinach. Also employed as preservative of unsalted butter during summer months.

Rumex arcticus Trautv. (Polygonaceae). - Perennial herb. Arctic regions, Siberia, Alaska. Leaves are consumed when fresh, soured or in oil by the Eskimos of Alaska.

Rumex Berlandieri Meisn. (Polygonaceae). - Herbaceous perennial. Western N. America. Plant was consumed with the fruits of Opuntia spp. by the Indians of Arizona.

Rumex brasiliensis Link. (Polygonaceae). - Herbaceous perennial. Brazil. Leaves are used in Brazil as vegetable. Roots are tonic, febrifuge, diuretic.

Rumex crispus L., Curled Dock. (Polygonaceae). - Perennial herb. Europe, N. Asia; introd. in N. America, Mexico, Chile and New Zealand. Roots are used medicinally as alterative, tonic and laxative. Leaves are boiled and eaten as a potherb.

Rumex Ecklonianus Meisn. (Polygonaceae). - Perennial herb. S. Africa. Decoction of roots is used by the Kaffirs for tape-worm.

Rumex hymenosepalus Torr., Canaigre. (Polygonaceae). - Perennial herb. S. W. of United States and adj. Mexico. Tuberous roots, resembling those of Dahlias contain 35 to 60% tannin; used as tanning material, sometimes known as Raiz del India. Occasionally cultivated. Roots are source of a yellow dye; used by the Navahos for dyeing wool. Leaf-stalks are sometimes used for pies, instead of rhubarb. Leaves are eaten as greens. Roots are employed for colds and sore throat by the Hopi and Papago Indians.

Rumex obtusifolius L., Broad-leaved Dock, Bitter Dock. (Polygonaceae). - Perennial herb. Europe, temp. Asia. Young leaves are eaten as potherb. Cultivated.

Rumex mexicanus Meisn. (Polygonaceae). - Perennial herb. Western N. America and Mexico. Leaves were consumed as greens by several Indian tribes of the western states. Also R. occidentalis S. Wats.

Rumex Patienta L., Patience Dock. (Polygonaceae). - Perennial herb. Probably from S. Europe and W. Asia. Cultivated. Leaves are eaten like Sorrel.

Rumex paucifolius Nutt., Mountain Sorrel. (Polygonaceae). - Herbaceous perennial. Northwest of N. America. Leaves and stems are consumed as vegetable by the Klamath Indians in Oregon.

Rumex vescarius L. (Polygonaceae). - Perennial herb. N. Africa, Orient. Leaves are cooked and eaten by the Beduins in Arabia.

Runeala Plum → Flacourtia cataphracta Roxb.

Rush → Juncus glaucus Sibth.

Rush, Flowering → Butomus umbellatus L.

Russel River Lime → Microcitrus inodora Swingle.

Russet Buffaloberry → Shepherdia canadensis (L.) Nutt.

Russian Licorice → Glycyrrhiza glabra L.

Russian Thistle → Salsola Kali L.

Russian Vetch → Vicia villosa Roth.

Russula alutacea (Pers.) Fr. (Agaricaceae). - Basidiomycete. Fungus. Temp. zone. Fruitbodies are consumed as food. Sold in markets of Europe. Also R. Mariae Peck, R. vesca Fr., R. virescens (Schäff.) Fr., R. flava Rom. and several others.

Russula atrovirens Beeli (Agaricaceae). - Basidiomycete. Fungus. Trop. Africa, esp. Congo. Fruitbodies are consumed as food by the natives.

Ruta chalepensis L. (syn. R. angustifolia Pers.). Fringed Rue. (Rutaceae). - Shrub. Mediterranean region. Juice of the plant is used by the natives of Morocco for sore eyes. Also used as anthelmintic, emmenagogue and abortive. Sold in markets.

Ruta graveolens L., Common Rue. (Rutaceae). - Perennial herb. Mediterranean region. Cultivated in Old and New World. Leaves used as condiment, for flavoring sausages, sauces, beverages, in aromatic vinegar. Used medicinally as antispasmodic, emmenagogue; said to cause abortion. Used in several home remedies. Contains methyl nonul ketone, methyl anthranilate. Source of Oil of Rue, an ess. oil, being distilled from green parts of the plant. Algerian Oil is derived from R. montana L. and R. bracteosa DC.

Ruta tuberculata Forsk. (Rutaceae). - Herbaceous plant. N. Africa, Arabia, Iran. Source of an ess. oil; used by the natives for flavoring food.

Rutabaga → Brassica Napo-Brassica Mill.

Rutaceae → Acronychia, Aegle, Amyris, Atalantia, Balfourodendron, Barosma, Boronia, Calodendrum, Casimiroa, Chaetospermum, Citrus, Clausenia, Correa, Cusparia, Dendrosoma, Dictamnus, Eremocitrus, Evodia, Fagara, Feronia, Feroniella, Flindersia, Fortunella, Galipea, Galium, Melicope, Microcitrus, Monniera, Murraya, Pelea, Phellodendron, Pilocarpus, Poncirus, Ptelea, Raputia, Ruta, Toddalia, Triphasia, Zanthoxylum.

Rye → Secale cereale L.

Rye, Giant → Triticum polonicum L.

Rye, Giant Wild → Elymus condensatus Presl.

Rye Grass → Lolium perenne L.

Rye Grass, Italian → Lolium italicum R. Br.

Rye, Jerusalem → Triticum polonicum L.

Rye, Wild Beardless → Elymus triticoides Buckl.

Ryparosa caesia Blume. (Flacourtiaceae). - Small tree. Java. Wood durable, hard, heavy; used for beams, bridges, houses.

S

Sabal mexicana Mart. (syn. Inodes mexicana (Marst.) Standl.). (Palmaceae). - Centr. Mexico to Guatemala. Leaves source of important material for thatching, of commercial value.

Sabal Palmetto (Walt.) Todd., Cabbage Palmetto. (Palmaceae). - Tall palm. North Carolina to Florida, Gulf States and West Indies. Trunks are used for wharf-piles, resistant to water, also for construction of huts of Seminole Indians. Leaves are manuf. into baskets, mats; used for covering roofs of huts. Very young part of the bud is consumed as vegetable. Unfolded leaves are used for religious ceremonies during Easter. Stems are manuf. into stiff brushes. Polished cross-sections of the stems are employed for small tables. Fruits were used as food by the aboriginals. Flowers source of a honey.

Sabal texana Becc. (syn. Inodes texana Cook.) (Palmaceae). - Palm. Mexico. S. W. Texas. Leaves are used for thatching and chair seats.

Sabatia angularis (L.) Pursh. Squarestem Rosegentian, American Centaury. (Gentianaceae). - Biennial herb. Eastern N. America to Florida and Louisiana. Dried herb is used medicinally as a tonic. Contains erythrocentaurin, a bitter principle. Also S. Elliottii Steud. and E. campestris Nutt.

Sabiaceae → Meliosma.

Sabicu → Calliandra formosa Benth.

Saccharomycetaceae → Citromyces, Hansenia, Schizosaccharomyces, Saccharomyces, Torula and Torulopsis.

Saccharomyces anamensis Will. and Heinick. (Saccharomycetaceae). - Yeast. This microorganism is sometimes used in the process of alcoholic fermentation.

Saccharomyces apiculatus Reess. (Saccharomycetaceae). - Yeast. A microorganism causing alcohol fermentation of a number of different fruit juices.

Saccharomyces carlsbergensis Hansen. (Saccharomycetaceae). - Fungus. A bottom yeast which has been used in breweries of Copenhagen and other places. The fermentation process of bottom-fermenting yeasts, is usually carried out at 6 to 12° C. and is completed in 8 to 10 days, whereas that of top-fermenting yeast is completed in a shorter period, in 5 to 7 days and at a slightly higher temperature, 14 to 23° C. At this stage the beer is termed „green" or „young".

Saccharomyces cerevisiae Hansen. Beer Yeast, Bread Yeast. (Saccharomycetaceae). - Fungus. Important in the manufacturing of beer and baking bread. Many strains or varieties are characteristic for different types of beer. They are grown in pure cultures and sold commercially. Compressed yeast or Cerevisiae Fermentum Compressum is used as a laxative. It contains vitamin B.

Saccharomyces ellipsoides Hansen. Wine Yeast. (Saccharomycetaceae). - Fungus. This is the principal yeast used for the fermentation of must or grape juice. A number of strains are important in selecting the typical characteristics of a number of wines, among which the Burgundy and Tokay strains. They are grown in pure cultures and sold commercially. It has been found that the var. Steinberg and var. California Wine Yeast are important in the glycerol fermentation. Yeasts of this type that have been acclimated or „trained" to grow in an alkaline medium produce the highest amount of glycerol.

Saccharomyces Kefir. Beijerinck. (Saccharomycetaceae). - This yeast causes the fermentation of milk in the production of Kefir. Much used by the peoples of the Caucasion Mountains where it is extensively used as a food. It is prepared by inoculating the milk of mares, ewes, goats or cows with masses of the „seed", Kefir grains, having the resemblance of small cauliflowers which may be dried and preserved. The Kefir grains often contain also Lactobacillus caesi.

Saccharomyces monacensis Hansen. (Saccharomycetaceae). - This microorganism has been used as a bottom yeast in the manuf. of beer.

Saccharomyces Pasteurianus Hansen. Wild. Yeast. (Saccharomycetaceae). - Known as Wild Yeast, this organism causes turbidity of beer. Also S. turbitans Hansen.

Saccharomyces pyriformis (Saccharomycetaceae). - A yeast is used in the production of Ginger Beer. Also Bacterium vermiforme is associated with this process. There is apparently a symbiotic relationship between the two organism. It has been found that the two organism function best when in each others presence.

Saccharomyces Sake Yabe. (Saccharomycetaceae). - A yeast taking part in the fermentation of Sake, a widely used alcoholic beverage, yellow rice wine, among the Japanese. It contains 14 to 24 % alcohol. The methods of manufacturing vary. Steamed rice is inoculated with the spores of Aspergillus oryzae (Ahl) Cohn; incubated at 20° C. This causes starch to convert into fermentable sugars. Fermentation also takes place by S. tokyo and S. yeddo.

Saccharomyces secundus Groenew. (Saccharomycetaceae). - Yeast. Used for alcohol production in Java; causing a strong fermentation.

Saccharomyces theobromae Preyer. (Saccharomycetaceae). - A yeast taking part in the fermentation and curing of the cacao beans.

Saccharomyces Tuac Vorderm. (Saccharomycetaceae). - Yeast. This organism has been found in considerable numbers during the fermentation of Toewak, Palm Wine, derived from Cocos nucifera L., Borasus flabella L., Arenga pinnata Merr. and A. obtusifolia Mart. It was mainly found in Java.

Saccharomyces Vordermanii Went and Prins. Geerl. (Saccharomycetaceae). - Yeast is of importance in the production of Kustarak, an alcoholic beverage in Java.

Saccharomyces spec. (Saccharomycetaceae). - Fungus. A yeast designated as Saaz has been used as a bottom yeast in the production of beer.

Saccharum arundinaceum Retz. (Graminaceae). - Perennial grass. Trop. Asia, esp. India to China. Leaf sheaths source of Munj Fibre. Leaf blades employed for thatching houses and as paper material. Flower stems used for thatching boats, carts, also for making chairs, stools, tables, baskets, screens, etc.

Saccharum officinale L., Sugar Cane. (Graminaceae). - Tall perennial grass. Trop. Asia. Origin uncertain. Cultivated since ancient times. Grown in the tropics of the Old and New World, sometimes in subtr. regions. Stems are source of sugar cane, cane syrup. Molasses is used in cooking and candy making; also for rum, arrack, industrial alcohol used for explosives, synth. rubber, in combustion engines. Fresh stems are often chewed, mainly by the poorer classes. A mixture of bagasse or refuge canes and molasses called molascuit is given as food for cattle. Bagasse is also employed for manuf. paper, card board and as fuel. Wax from stems resembles Carnauba Wax. Sugar cane varieties are numerous, they are sometimes divided into the following races: Mauritius, Otaheite, Bourbon, Batavian, China, Singapore and Indian Cane.

Saccharum Sape St. Hil. → Imperata brasiliensis Trin.

Saccharum spontaneum L. (Graminaceae). - Perennial grass. Malaya. Young shoots are eaten boiled with rice in Java.

Saccopetalum Horsfieldii Benn. (Annonaceae). - Tree. Java. Wood tough, fine-grained; used for handles of lances.

Sachalin Agar → Ahnfeltia plicata (Huds.) Fries.

Sachalin Cork Tree → Phellodendron sachalinense Sarg.

Sachalin Knotweed → Polygonum sachalinense F. Schmidt.

Sachalin Spruce → Picea Glehnii Mast.

Sacred Datura → Datura meteloides DC.

Sadlera cyatheoides Kaulf. (Polypodiaceae). - Perennial fern. Hawaii. Islds. Source of a red dye in Hawaii. Starchy pith cooked, is eaten by the Hawaiians.

Safflower → Carthamus tinctoria L.

Saffron → Crocus sativus L.

Saffron Crocus → Crocus sativus L.

Saffron, Meadow → Colchicum autumnale L.

Saffron-yellow Solorina → Solorina crocea (L.) Ach.

Sagapan Resin → Ferula Szowitziana DC.

Sagapenum Gum → Ferula persica Willd.

Sage → Salvia officinalis L.

Sage, Black → Salvia mellifera Greene.

Sage, Blue → Salvia mellifera Greene.

Sage, Clary → Salvia Sclarea L.

Sage, Lindenleaf → Salvia tiliaefolia Vahl.

Sage, Lyre Leaved → Salvia lyrata L.

Sage, Meadow → Salvia pratensis L.

Sage, Thistle → Salvia carduacea Benth.

Sagebush → Artemisia tridentata Nutt.

Sagebrush Mariposa → Calochortus Gunnisonii S. Wats.

Sageretia Brandrethiana Aitch. (Rhamnaceae). - Shrub. Trop. Asia, esp. Orient, North of E. India. Fruits are consumed in some parts of Afghanistan.

Sageretia theezans (L.) Brong. (syn. Rhamnus theezans L.) (Rhamnaceae). - Shrub. E. India, Burma, China. Leaves are used as tea in Tonkin. They are often mixed with the leaves of Eugenia operculata Roxb.

Sagittaria latifolia Willd. (syn. S. variabilis Engelm.) Arrowleaf, Wapato, Duck Potato. (Alismaceae). - Perennial herb. North America, Mexico, Centr. America. Tuberous roots when boiled or roasted were used as food by the Indians, also consumed by the Chinese in California and in other states.

Sagittaria sagittifolia L., Arrowhead. (Alismaceae). - Perennial herb. Europe, Asia. Cultivated in China and Japan. Corns are consumed boiled as a vegetable.

Sago → Arenga, Caryota, Cycas, Mauritia, Metroxylon.

Sago Palm → Metroxylon Sagu Rottb.

Saguaragy Bark → Colubrina rufa Reiss.

Sahing Resin → Canarium villosum Benth. and Hook.

Sahuaro → Carnegiea gigantea (Engelm.) Britt. and Rose.

Saigon Cinnamon → Cinnamomum Loureirii Nees.

Saintfoin → Onobrychus viciaefolia Scop.

Saké → Oryza sativa L.

Sakulali, Bdellium → Commiphora Mukul Engl.

Salacia senegalensis DC. (syn. Hippocratea senegalensis Lam., H. verticillata Steud.). (Hippocrateaceae). - Shrub. Trop. Africa. Fruits are edible, consumed in some parts of Senegal.

Salai-gugul Gum → Boswellia serrata Roxb.

Salep → Anacamptis pyramidalis (L.) Rich., Orchis maculata L., O. militaris L., O. Morio L. and Platanthera bifolia (L.) Reichb.

Salep, Royal → Allium Macleanii Baker.

Salicaceae → Populus, Salix.

Salicornia australis Soland. (Chenopodiaceae). - Herbaceous. Salty situations. Australia. Young shoots are pickled and eaten by inhabitants.

Salicornia fruticosa L. Leadbush Glasswort. (Chenopodiaceae). - Mediterranean region. Small shrub. Used as fodder by camels. Also S. tetragona Del.

Salicornia herbacea L., Marsh Samphire. (Chenopodiaceae). - Annual to perennial herb. Coastal zones. Europe. In some parts of Europe the fleshy stems are consumed as spinach, also used as pickles in vinegar. Plants are occasionally

sold in local markets of France, Belgium and Holland.

Salix acutifolia Willd., Caspic Willow. (Salicaceae). - Shrub or tree. E. Russia, Siberia, Turkestan, Manchuria. Cultivated. Twigs are used for basketry.

Salix alba L. (syn. S. aurea Salisb.), White Willow. (Salicaceae). - Shrub or tree. Europe, temp. Asia, N. Africa. Introd. in N. America. Cultivated. Wood soft, easy to split and to bend, red to dark-brown; used for wooden shoes, esp. in Holland; boxes, general carpentry, sieves, toothpicks, handles of tools, boat-building, manuf. of paper-pulp. Twigs used for basketry, garden furniture, chairs etc. also used for tying twigs of plants against supports. Bark is used medicinally as astringent, contains 13 % tannin, 2,5 to 11.3 % salicin. Leaves are employed as substitute for tea.

Salix capensis Thunb. Cape Willow (Salicaceae). - Tree. S. Africa. Tender shoots containing salicin, a glucoside, are used by Hottentots for rheumatic fever.

Salix Caprea L. Goat Willow, Common Willow. (Salicaceae). - Shrub or small tree. Europe, N. Asia. Cultivated. Wood light, soft, easy to split, elastic; used in S .Germany for supports in vineyards. Source of a good charcoal, used for gunpowder.

Salix daphnoides Will. (syn. S. pulchra Wimner). (Salicaceae). - Small shrub. Temp. to arct. N. Hemisphere. Young shoots and catkins are eaten fresh or in seal oil by the Eskimos of Alaska. Cambium from inner bark is scraped off and consumed as food.

Salix fragilis L., Britle Willow. (Salicaceae). - Tree. Europe, introd. in N. America. Twigs are used for basketry. Wood is source of a charcoal used for gunpowder.

Salix Humboldtiana Willd. Humboldts' Willow. (Salicaceae). - Small tree. Central and South America. Widely used for basketry and similar work.

Salix irrorata Anders. Bluestem Willow. (Salicaceae). - Woody plant. S. W. of United States and adj. Mexico. Wood is source of a charcoal, used as body paint by Tewa Indians of New Mexico.

Salix jessoensis v. Seem. Yeddo Willow. (Salicaceae). - Tree. Hokkaidô (Jap.). Wood soft, light pale, brownish-white. Used in Japan for cheap wooden clogs, matches, tooth-picks, boxes, tailors boards.

Salix nigra Marsh. (syn. S. amygdaloides Anders.). Black Willow. (Salicaceae). - Tree. N. America. Wood soft, not strong, used for charcoal, fuel, excelsior, to a limited extend for paper pulp, with other willow species, also for artificial limbs. Bark is rich in tannin; used as home medicine for fevers.

Salix purpurea L. Purple Osierwillow. Purple Willow. (Salicaceae). - Shrub or small tree. Throughout Europe, temp. Asia, Africa. Twigs are used for basketry. Source of commercial salicin, being a bitter tonic, antiperiodic and antirheumatic.

Salix Sasaf Forsk. (Salicaceae). - Shrub. N. and Trop. Africa. Twigs are used by the natives for basket making. Leaves are source of a black dye; used for coloring mats.

Salix sitchensis Bong. Sitka Willow. (Salicaceae). - Shrub. Western N. America. Young supple twigs are made into baskets, frames for stretching skins etc. by the Eskimos of Alaska.

Salix triandra L. (syn. S. amygdalina L., S. auriculata Mill.) Almond Leaved Willow, French Willow. (Salicaceae). - Shrub or small tree. Europe, temp. Asia to China, Manchuria, Japan. Cultivated. Twigs are used for manuf. baskets.

Salix viminalis L. (syn. S. longifolia Lam., S. virescens Vill.) Basket Willow, Osier Willow, Common Osier. (Salicaceae). - Shrub or small tree. Europe, N. Asia. Cultivated. Much esteemed for manuf. baskets, chairs etc.

Sally, Black → Acacia melanoxylon R. Br.

Salmea Eupatoria DC. (syn. S. scandens (L.) DC., Bidens scandens L.). (Compositae). - Trailing shrub. Mexico. Roots when chewed deaden all sensation in the tongue. Remedy for toothache. Used as fish poison.

Salmonberry → Rubus Chamaemorus L.

Salomon's Seal, Small → Polygonatum biflorum (Walt.) Ell.

Salpichroa rhomboidea Miers. (Solanaceae). - Woody climber. Argentina. Fruits called Huevos de Gallo, are sometimes used as preserves. Sold in markets of Argentina and Paraguay.

Salsify → Tragopogon porrifolius L.

Salsify, Black → Scorzonera hispanica L.

Salsola aphylla L. (Chenopodiaceae). - Small shrub. S. Africa. Important source of food for live-stock in S. Africa.

Salsola asparagoides Miq. (Chenopodiaceae). - Annual herb. Coastal zone of China, Korea and Japan. Cultivated. Young boiled plants are consumed in Japan during spring and summer.

Salsola foetida Del. (Chenopodiaceae). - Herbaceous plant. Sahara region. Plants are used as food for camels.

Salsola Kali L., Russian Thistle. (Chenopodiaceae). - Annual. Temp. zone. Very young tender shoots when boiled are sometimes used as food. Recommended as an emergency food-plant.

Salsola Soda L. (Chenopodiaceae). - Annual herb. Sandy shores. Mediterranean Region, Asia. Cultivated in Japan. Leaves and stems when boiled are eaten in Japan as a vegetable. Is occasionally cultivated along coastal areas in Spain, S. France and Italy, where it is source of Barila, an impure carbonate of soda: was formerly used for the manuf. of soap and glass.

Salsola Zeyheri Shinz. (Chenopodiaceae). - Shrub. Alkali areas. S. Africa. Important source of fodder for livestock.

Saltbush → Atriplex canescens (Pursh) Nutt., Chenopodium auricomum Lindl., Salvadora persica Garc.

Saltbush, Fourwing → Atriplex canescens (Pursh) Nutt.

Saltbush, Mediterranean → Atriplex Halimus L.

Saltbush, Small → Altriplex campanulata Benth.

Saltwort → Batis maritima L.

Salvadora persica Garc., Salt Bush, Mustard Tree, Tooth-brush Tree. (Salvadoraceae). - Shrub or small tree. Trop. Africa, Asia. Shoots and leaves eaten as a salad, also as fodder for camels. Pungent, bitter fruits are used in medicine. Bark of root is acrid and vesicant. Twigs are used in Portug. E. Africa for cleaning teeth. Source of a vegetable salt, called Kegr, derived from the ash of the plant. Fat from seeds is manuf. into candles, also used for rubbing on the skin. Wood is used for different purposes.

Salvadoraceae → Dobera, Salvadora.

Salvia calycina Sibth. and Sm. (Labiaceae). - Perennial herb. Greece, Turkey. Infusion of herb is used as a tea in some parts of Greece. Also S. triloba L.

Salvia carduacea Benth., Thistle Sage. (Labiaceae). - Herbaceous plant. California, Arizona. Seeds when roasted were made into a flour; also used for a cool drink by the Indians of California.

Salvia chia Fern., Chia. (Labiaceae). - Herbaceous plant. Mexico. Cultivated. Seeds are source of a refreshing drink much used by the Mexicans. Chia Oil was used in Ancient Mexico for painting and also in native medicines.

Salvia Columbariae Benth. California Chia. (Labiaceae). - Herbaceous annual. Western N. America. Seeds were parched and ground into meal (pinole) and made into dark colored loaves and cakes, also mixed with wheat and corn meal, being an important food of the Indians. A popular drink is made from the seeds. Also S. apiana Jep.

Salvia lyrata L. Lyre-leaved Sage. (Labiaceae). - Perennial herb. E. of N. America to Florida and Texas. A salve is made from the astringent roots; used by the Catawba Indians for sores.

Salvia mellifera Greene; Black Sage, Blue Sage. (Labiaceae). - Herbaceous plant. California. Flowers are source of a rich and fine flavored honey; one of the best honey producers in the west of the United States. Also S. apiana Jep., S. carduacea Benth., S. columbariae Benth. and S. sonomensis Greene.

Salvia miltiorrhiza Bunge. (Labiaceae). - Perennial herb. Chihli, Kiangsi, Shantung (China). Used medicinally in China.

Salvia occidentalis Swartz. (Labiaceae). - Herbaceous plant. Tropical and subtropical America. Decoction from leaves used in parts of S. Mexico for dysentery and as a stomachic.

Salvia officinalis L. Garden Sage. (Labiaceae). - Small shrub or half-shrub. S. Europe, Mediterranean region, Balkans, Asia Minor, Syria. Cultivated as a kitchen herb in Old and New World. Used as a condiment for flavoring meats, stews, soups, fowl, sausage, canned foods, sage-milk. Source of Oil of Sage, ess. oil employed in flavoring and medicine, contains salvene, d-a-pinene, cineol, borneol, d-camphor. Dried leaves, collected during flowering time, are used medicinally, being a stimulant, carminative for dyspepsia. Used in several home-remedies.

Salvia pratensis L., Meadow Sage. (Labiaceae). - Perennial herb. Europe. Mucilage producing fruits are used in some countries for eye ailments.

Salvia Sclarea L., Clary Sage. (Labiaceae). - Perennial herb. Mediterranean region, Centr. Europe, Transcaucasia, Syria, Iran. Cultivated for oil production. Used as condiment. Oil of Clary is used in toilet waters, soaps, cosmetics, in high grade perfumes with lavender and bergamot oils in Eau de Cologne and in lavender waters, imparting a delicate note in perfumes. Used for European wines, Muscatel, Vermouth, liqueurs. Contains linalol. Powdered herb is used as snuff for headache.

Salvia tenella Sw. (syn. S. micrantha Vahl.) (Labiaceae). - Herbaceous perennial. S. Mexico. Herb used in Yucatan for ear-ache.

Salvia tiliaefolia Vahl. Lindenleaf Sage. (Labiaceae). - Herbaceous plant. Mexico and Centr. Americ. Seeds are sold in markets of Mexico for making a drink, sometimes it is mixed with barley water. It is said that the herb is used in Guatemala for killing head-lice.

Salvia viridis L. Bluebeard. (Labiaceae). - Annual to biennal herb. Mediterranean region to Iran. Ess. oil used for flavoring certain wines and beers. Occasionally cultivated. Flowers are a good source of honey.

Salviniaceae → Azolla.

Samadera indica Gaertn., Niepa Bark Tree. (Simarubaceae). - Small tree. India to Polynesia. Decoction of leaves is used to kill termites. Used in Indonesia as emetic, purgative and for bilious fevers.

Samarang Rose Apple → Eugenia javanica Lam.

Saman → Pithecolobium Saman Benth.

Sambucus australis Cham. and Schlecht. (Caprifoliaceae). - Shrub. Argentina, Chile, Brazil and adj. territory. Infusion of leaves is used in some parts of S. America for indigestion, as diuretic and sudorific. Berries are made into preserves.

Sambucus canadensis L. American Elder. (Caprifoliaceae). - Shrub or small tree. Eastern N. America to Florida and Texas. Black small fruits are made into jams, jellies, pies and wine. Dried flowers are used medicinally as a stimulant, carminative, diuretic, diaphoretic; contains an ess. oil. In home remedies it is used for sores and bruises. Constituent of eye lotions. Flowers are used in Elder Flower Water.

Sambucus coerulea Raf. (syn. S. glauca Nutt.). (Caprifoliaceae). - Shrub or small tree. Western N. America. Fruits when fresh or dried were used as food by several Indian tribes of the West. Wood was employed for making bows by the Luiseño Indians.

Sambucus Ebulus L. Dwarf Elder. (Caprifoliaceae). - Small shrub. Throughout Europe, W. Asia, N. Africa. Berries are source of a blue dye; used for coloring leather and yarn. Also used in some parts of Roumania to color wine. Leaves used in tea, among which Kneipp Tea.

Sambucus Gaudichaudiana DC. (Caprifoliaceae). - Shrub or tree. Australia. Fruits eaten by aborigines of Australia. Also S. xanthocarpa F. v. Muell.

Sambucus mexicana Presl., Mexican Elder. (Caprifoliaceae). - Tree. Southwestern United States to Centr. America. Fruits are consumed in jellies and pies.

Sambucus nigra L., European Elder. (Caprifoliaceae). - Shrub or small tree. Europe. Cultivated. Fruits are preserved with sugar; also made into wine. Syrup from fruits used in some countries as a purgative. Flowers, Flores Sambuci, are used as home-remedy for catarrh, as mouth wash and for gargling. Leaves and bark were formerly used in medicine. Pith from stems is used in microscopical technique for making slides.

Sambucus peruviana H.B.K. (Caprifoliaceae). - Medium sized tree. Bolivia, Peru, N. Argentina. Fruits are made into a syrup; used in Peru for throad troubles and for cure of ulcers. Leaves have excitant properties.

Samolus Valerandi L., Brookweed. (Primulaceae). - Perennial herb. Europe, temp. Asia to Japan, N. and S. America, S. W. Australia. Herb used as antiscorbutic. Young leaves are sometimes used in salad or eaten as spinach. Recommended as an emergency food plant.

Samphire → Crithmum maritimum L.

Samphire, Marsh →Salicornia herbacea L.

Samphire, Sea → Crithmum maritimum L.

Sam-rong → Sterculia lychnophora Hance.

Samuela carnerosana Trel. (Liliaceae). - Medium tree. Mexico. Trunk used for walls of houses. Flowers clusters eaten by cattle. Immature flower clusters boiled or roasted, are eaten by Indians. Leaves source of a fibre, Palma Ixtle.

Sandal Bead Tree → Adenanthera pavonina L.

Sandal Oil → Aptandra Spruceana Miers.

Sandalo Brasileiro → Aniba Canelilla (H. B. K.) Mez.

Sandalwood → Eremophila Mitchelli Benth., Pterocarpus santalinus L. f. and Santalum Yasi Seem.

Sandalwood, African → Baphia nitida Lodd.

Sandalwood, Australian → Santalum lanceolatum R. Br.

Sandalwood, Bastard → Eremophila Mitchelli Benth. and Myoporum platycarpum R. Br.

Sandalwood, East African → Osyris tenuifolia Engl.

Sandalwood, Fragrant → Fusanus spicatus R. Br.

Sandalwood, Hawaian → Santalum Freycinetianum Gaud.

Sandalwood, Lanceleaf → Santalum lanceolatum R. Br.

Sandalwood, New Caledonian → Santalum austro-caledonicum Vieil.

Sandalwood, Padauk → Pterocarpus santalinus L. f.

Sandalwood, Red → Adenanthera pavonina L.

Sandalwood, Scrub → Exocarpus latifolia R. Br.

Sandalwood, South Australian → Fusanus acuminatus R. Br.

Sandalwood, West Australian → Eucarya spicata Sprag. and Summ.

Sandalwood, West Indian → Amyris balsamifera L.

Sandalwood, White → Santalum album L.

Sandalwood, Yellow Fiji → Santalum Freycinetianum Gaud.

Sandarac → Tetraclinis articulata (Vahl.) Mast.

Sandbox Tree → Hura crepitans L.

Sanders, Yellow → Terminalia Hilariana Steud.

Sanderswood, Red → Pterocarpus santalinus L. f.

San Domingo Boxwood → Phyllostylon brasiliensis Cap.

San Domingo Mahogany → Swietenia Mahogony Jacq.

Sand Pine → Pinus clausa (Engelm.) Sarg.

Sando de Maranhão → Aptandra Spruceana Miers.

Sandoricum Koetjape Merr. (syn. S. indicum Cav.). Santol. (Meliaceae). - Medium sized tree. Malaya, Philipp. Islands. Fruits roundish, 6 cm. in diam., yellow-brown, whitish pulp; consumed by natives when fresh, with spices, dried or candied. Some varieties are seedless.

Sandpaper substitutes. Plants source of → Aphenanthe, Curatella, Dillenia, Equisetum, Ficus, Tetracera.

Sandpaper Tree → Curatella americana L.

Sandroot → Ammobroma sonorae Torr.

Sandwort, Seabeach → Arenaria peploides L.

Sang de Dragon → Dracaena Draco L.

Sanguinaria canadensis L., Blood Root. (Papaveraceae). - Perennial herb. Eastern N. America. Juice was used by the Indians to stain their faces. Also a remedy of early settlers for coughs. Dried rhizomes used medicinally as emetic, stimulating expectorant. Contains several alkaloids, e. g. sanguinarine, chelerythrine, and protopine; also a reddish resin.

Sanguis Draconis → Dracaena Draco L.

Sanguisorba officinalis L. Garden Burnet. (Rosaceae). - Perennial herb. Europe, temp. Asia, Japan, N. America. Young leaves are sometimes eaten with salads, giving a characteristic flavor. Also used as a vegetable in spring.

Sanicle, Wood → Sanicula europaea L.

Sanicula europaea L. European Wood Sanicle. (Umbelliferaceae). - Perennial herb. Europe. Asia Minor, Syria, Caucasia, temp. Asia. The herb was formerly used as an astringent, stomachic, resolvans; for diseases of the lungs and stomach.

Sansapote → Licania platypus (Hemsl.) Pitt.

Sansevieria abyssinica N. E. Br. (Liliaceae). - Perennial herb. E. Africa. Leaves are source of a fibre; used by the women of Bongoland for kilts, resembling horsetails.

Sansevieria senegambica Baker., African Bowstring Hemp. (Liliaceae). - Perennial herb. Trop. Africa. Leaves are source of a fibre, used for nets, fishing lines, bowstring; in leather work and shoes. Also S. liberica Ger. and Labr., S. trifasciata Prain.

Sansevieria thyrsiflora Thunb. (Liliaceae). - Perennial herb. S. Africa. Decoction of rootstock used by farmers and negroes in S. Africa for intestinal parasites.

Sansevieria zeylanica Willd., Ceylon Bowstrong Hemp (Liliaceae). - Perennial herb. Ceylon. Cultivated. Source of a fibre, known as Ceylon Bowstring Hemp; used for sails also for manuf. of paper.

Santalaceae → Acanthosyris, Arjona, Colpoon, Eucarya, Exocarpus, Fusanus, Jodina, Leptomeria, Osyris, Santalum.

Santalum album L. (syn. Sirium myrtifolium L.). White Sandalwood. (Santalaceae). - Tree. E. India. Cultivated. Wood heavy, hard, difficult to split, not attacked by termites; used for manuf. of incense. Oil of Santal or Oleum Santali obtained by steam distillation from the wood is used in medicine as a stimulant, expectorant, in bronchitis, gonorrhoea, gleet, chronic cystitis; as a urinary antiseptic. Also used in perfumery, has pleasant fragrance; blends well with patchouly, vetiver, geranium, musk etc.; is a powerful fixative. Powdered wood is used in cosmetics. Chinese use joss-sticks prepared from the wood. Sawdust kept in small bags is used for scenting clothes.

Santalum austro-caledonicum Vieill. New Caledonian Sandalwood. (Santalaceae). - Tree. New Caledonia. Wood yellow, scented, fine-grained, durable; used for boxes and small fancy articles. Wood was once an important article of export from N. Caledonia.

Santalum fernandezianum Phil. (Santalaceae). - Juan Fernandez. Sandalwood Tree of Juan Fernandez was once an important economic plant. It has been entirely eradicated.

Santalum Freycinetianum Gaud. (syn. S. insulare Bert.) Yellow Fiji Sandalwood, Hawaiian Sandalwood. (Santalaceae). - Shrub or tree. Hawaii. Islds. Wood was formerly highly valued in commerce.

Santalum lanceolatum R. Br., Lanceleaf Sandalwood, Australian Sandalwood. (Santalaceae). - Small tree. Trop. Australia. Wood yellowish, dense, close-grained, easy to polish; used as a Sandalwood. Other species that are source of Sandalwoods are: S. Preissianum Miq., S. Pilgeri Rock, S. paniculatum Hook. and Arn.; S. cuneatum (Hbd.) Rock., S. Cunninghamii Hook. from New Zealand; S. Hornei Seems from Isl. Eromanga; S. insulare Bertero from Isl. Tahiti.

Santalum marchionense Skottsb. (Santalaceae). - Tree. Pacif. Islds. Fragrant wood used for ceremonial purposes by natives. Source of Santalum scented wood oil, used for anointing the body and for embalming.

Santalum spicatum DC. → Eucarya spicata Sprag. and Summ.

Santalum Yasi Seem., Fiji Sandalwood Tree. (Santalaceae). - Tree. Pacific Islands. Wood source of Fiji Sandalwood; used for cabinet work. Produces Fiji Sandalwood Oil.

Santiria Griffithii Hook. f. (syn. Trigonochlamys Griffithii Hook. f.). (Burseraceae). - Tree. Trop. Asia., esp. Malaya. Wood scented, very hard, medium grained, yellowish white; used for house building and beams.

Sanwa Millet → Echinochloa frumentacea (Roxb.) Link.

São Francisco Rubber → Manihot heptaphylla Ule.

Sapelé, Heavy → Entandrophragma Candollei Harms.

Sapelé Mahogany → Entandophragma Candollei Harms.

Sapindaceae → Alectryon, Allophylus, Atalaya, Blighia, Chytranthus, Cupania, Diploglottis, Dodonaea, Erioglossum, Ganophyllum, Guioa, Harpulia, Heterodendrum, Koelreuteria, Lepidopetalum, Lepisanthes, Melicocca, Nephelium, Pappea, Paullinia, Pometia, Pouteria, Ratonia, Sapindus, Schleichera, Xerospermum.

Sapindus marginatus Willd. (syn. S. Drummondii Hook. and Arn.). Western Soapberry. (Sapindaceae). - Tree. South and southw. of United States and N. Mexico. Wood splits easily,

used for manuf. baskets, for harvesting cotton, frames of pack-saddles. Pulp of berries rubbed between fingers in water forms a lather; used as a substitute for soap.

Sapindus Mukorossi Gaertn. (syn. S. detergens Roxb.) Chinese Soapberry, Soapnut Tree. (Sapindaceae). - Tree. E. Asia, Himalaya. Fruits have been used by natives since immemorial times as detergens. Employed in Kashmir for washing shawls, silks etc. Used by Indian jewellers for restoration and brightening tarnished silverware. Saponaceous fruits are used in China and Korea as substitute for soap. Seeds are made into Buddists' rosaries.

Sapindus oahuensis Hillebr. (Sapindaceae). - Tree. Hawaii. Islds. Seeds are used by the natives as a cathartic.

Sapindus Rarak DC. (syn. Dittelasma Rarak Hook. f.). (Sapindaceae). - Tall tree. Malacca, Java. Fruits are used by the natives as a substitute for soap. Used for scabies and as insecticide. A 1 : 60 000 solution causes intoxication of fish.

Sapindus Saponaria L. Southern Soapberry. (Sapindaceae). - Tree. S. Florida, Trop. America. Seeds are used in Martinique and Guadeloupe for manuf. of oil.

Sapium Aucuparium Jacq. (syn. S. Hippomane Mey.). Milk Tree. Lecheros. (Euphorbiaceae). - Shrub or tree. Trop. America. Source of a good rubber. Chewed and coagulated sap is placed on twigs to catch small birds.

Sapium biloculare (S. Wats.) Pax. (Euphorbiaceae). - Shrub or tree. Mexico. Source of „jumping beans" much like those of Sebastiana pavoniana Muell. Arg. Finely chopped branches are used to stupefy fish.

Sapium Grahami Prain. (Euphorbiaceae). - Shrub. Trop. Africa, esp. Gold Coast, Togo, Nigeria. Crushed bark and leaves in water are used by native women for making red and orange marks on the face.

Sapium indicum Willd. (syn. Excoecaria indica Muell. Arg. (Euphorbiaceae). - Tree. Mal. Archipelago. Fruits are used to stupefy fish.

Sapium Jenmani Hemsl. (Euphorbiaceae). - Tree. Brit. Guiana. Occasionally cultivated. Source of a good rubber, called Orinoco Scrap.

Sapium madagascariensis Prain. (Euphorbiaceae). - Tree. Trop. Africa. Used in E. Africa as an arrow-poison.

Sapium pavonianum (Muell. Arg.) Hub., Palo de Leche. (Euphorbiaceae). - Tree. Colombia, Peru. Source of a rubber, called Gaucho Blanco.

Sapium sebiferum Roxb. Chinese Tallow Tree. (Euphorbiaceae). - Tree. Tropics. Cultivated. Pressed seeds are source of a fat, used for candles and soap.

Sapium stylare Muell. Arg., Gaucho Blanco, Caucho Mirado. (Euphorbiaceae). - Tall tree. S. America, esp. Venezuela, Ecuador. Source of a rubber.

Sapium taburu Ule., Tupuru, Seringeirana. (Euphorbiaceae). - Tree. Trop. S. America, esp. Peru, Ecuador, Brazil. Source of Tapuru, Sernamby, Camota Rubber. Is frequently mixed with Hevea-latex.

Sapium Thomsoni God. (syn. S. tolimense Muell. Arg.) (Euphorbiaceae). - Tree. Sub-equatorial Andean Region. Colombia, Ecuador. Source of a rubber, called Colombia Virgin or Scrap, Caucho Virgin, Caucho Blanco.

Sapodilla → Achras Sapota L.

Saponaria officinalis L., Soapwort, Soaproot. (Caryophyllaceae). - Perennial herb. Europe, Asia; introd. in N. America. Occasionally cultivated. Chopped roots form a lather in water; used for washing. Root is used in medicine as a laxative and alterative. Contains saponin, an alkaloid. Said to be used as fodder for sheep.

Sapotaceae → Achras, Anthocleista, Argania, Baillonella, Bassia, Bumelia, Butyrospermum, Callocarpum, Chrysophyllum, Dipholis, Dumoria, Ecclinusa, Lucuma, Madhuca, Manilkara, Micropholis, Mimusops, Omphalocarpum, Payena, Sideroxylon.

Sapote → Calocarpum mammosum Pierre.

Sapote, Black → Diospyros Ebenum Koen.

Sapote, Green → Calocarpum viride Pitt.

Sapote, Yellow → Lucuma salicifolia H. B. K.

Sapote, White → Casimiroa edulis La Llave.

Sappan Wood → Caesalpinia Sappan L.

Sapucaja Oil → Lecythis Ollaria L.

Saracura-mira → Ampelozizyphus amazonicus Ducke.

Sarcocephalus Diderrichii De Wild., West African Boxwood. (Rubiaceae). - Tree. W. Africa. Important lumber tree. Wood durable, hard, bright yellow, slightly veined; used for canoes, boards, tables, mortars, bridges.

Sarcocephalus esculentus Afzel. Doundaté. (Rubiaceae). - Shrub or tree. Trop. Africa. Fruit edible, 5 cm. across, deep red, rough, brownish; pulp juicy, sweet, much esteemed by the negroes and Europeans. Sold in markets of Senegambia and Sierra Leone. Root is used by the natives for fever, also as tonic and for indigestion; exported under the name of Peach Root. Base of African Peach Bitter. Roots are said to produce a yellow dye; used for the preparation of Kano or Morocco Leather. Flowerheads are consumed as vegetable by the natives.

Sarcochlamys pulcherrima Gaud. (Urticaceae). - Woody plant. Assam, Burma. Fibre is used as a substitute for Ramie. Sometimes it is called Duggal Fibre.

Sarcocolla Gum → Penaea Sarcocolla L.

Sarcolobus narcoticus Miq. (Asclepiadaceae). - Woody vine. Mal. Archipelago. Used by the Javanese for killing tigers and wild dogs. Also S. Spanoghei Miq.

Sarepta Mustard → Brassica Besseriana Adrz.

Sargassaceae (Brown Algae) → Sargassum.

Sargassum echinocarpum J. Ag. (Sargassaceae). - Brown Alga. Sea Weed. Pac. Ocean. Consumed as food, after having undergone a ripening process, by the Hawaiians. Usually the leaf-like parts are eaten. Also eaten fresh. Plants are sometimes broken into pieces, soaked in fresh water, stuffed with salmon before being roasted. Is also chopped with fish heads and salt. Consumed in meat gravies or stews. Called Limu Kala by the Hawaiians.

Sargassum enerve J. Ag. (Sargassaceae). - Brown Alga. Sea Weed. Pac. Ocean and adj. waters. Consumed as food among the Japanese.

Sargassum fusiforme (Harv.) Setch. (Sargassaceae). - Brown Alga. Coast of Eastern Asia, esp. China and Japan. Employed medicinally in China for glandular infections, lymphatic disorders and goiter. Young plants are used as vegetable and in soup; also as as cooling tea. Also consumed as food in Japan. Many species are used as manure and for manufacturing of iodine, potash and algin in Japan.

Sargassum Horneri J. Ag. (Sargassaceae). - Brown Alga. Sea Weed. Pac. Ocean and adj. waters. Used as an important fertilizer in the Amoy region, China. Also S. Thunbergii O. Kuntze and S. hemiphyllum (Turn.) J. Ag.

Sargassum siliquastrum J. Ag. (Sargassaceae). - Brown Alga. Pacif. Ocean. Plants when young are used as food in Japan.

Sargassum vulgare J. Ag. (Sargassaceae). - Brown Alga. Sea Weed. Pac. and Atl. Ocean. Consumed as food in some parts of the Philipp. Islds.

Sargent Lily → Lilium Sargentiae Wils.

Sarothamnus scoparius (L.) Wimmer. (syn. Spartium scoparium L., Sarothamnus vulgaris Wimmer). Broom. (Leguminosaceae). - Shrub. Europe, escaped in N. America. Plants used as a sand-binder. Twigs employed for manuf. brooms; a source of fibre, used as a substitute for jute. Bark produces a yellow and brown dye, used for coloring cloth and paper. Plant is a food for sheep and goats. Leaves and buds are kept in vinegar and salt, eaten in some parts of Germany, known as Brahm or Geiss Kappern. Twigs used sometimes to flavor beer. Flowers produce a good nectar for bees. Flowering tops collected before blooming, are used medicinally, being cardiac tonic, diuretic. Source of sparteine sulfate.

Sarracenia flava L., Trumpets (Sarraceniaceae). - Eastern N. America. Herbaceous. Occasionally used as diuretic and stomatic. Also S. purpurea L. Pitcherplant.

Sarraceniaceae → Sarracenia.

Sarsaparilla → Smilax spp., Sassafras albidum (Nutt.) Nees.

Sarsaparilla, Brown → Smilax Regelii Killip and Morton.

Sarsaparilla, Central American → Smilax Regelii Killip and Morton.

Sarsaparilla, Grey → Smilax aristolochiaefolia Mill.

Sarsaparilla, Honduras → Smilax Regelii Killip and Morton.

Sarsaparilla, Indian → Hemidesmus indicus R. Br.

Sarsaparilla, Jamaica → Smilax Regelii Killip and Morton.

Sarsaparilla, Mexican → Smilax aristolochiaefolia Mill.

Sarsaparilla, Para → Smilax Spruceana A. DC.

Sarsaparilla, Red → Smilax Regelii Killip and Morton.

Sarsaparilla, Tampico → Smilax aristolochiaefolia Mill.

Sarsaparilla, Vera Cruz → Smilax aristolochiaefolia Mill.

Sarsari → Silene macrosolen Steud.

Sarcina ureae (Beijerinck) Loehnis. (Coccoceae). - Bacil. Agriculturally a useful soil microorganism, being capable of converting urea into ammonium carbonate.

Sasah Bark → Aporosa frutescens Blume.

Sasanqua Camellia → Camellia Sasanqua Thunb.

Saskatoon Serviceberry → Amelanchier alnifolia Nutt.

Sassafras → Atherosperma moschatum Labill., Dorypha Sassafras Endl. and Sassafras albidum (Nutt.) Nees.

Sassafras, Black → Cinnamomum Oliveri Bailey.

Sassafras, Brazilian → Mespilodaphne Sassafras Meissn.

Sassafras, Chinese → Sassafras Tzumu Hemsl.

Sassafras albidum (Nutt.) Nees. (syn. S. officinale Nees and Eberm., S. varrifolium Kuntze.) Common Sassafras. (Lauraceae). - Tree. Eastern N. America to Texas and Florida. Wood weak, brittle, soft, durable in soil, coarse-grained, orange-brown; used for ox-yokes, cooperage, light boats, various articles that require lightness. Bark of Sassafras contains 5 to 9 % essential oil, principally safrol; used as aromatic, alterative, carminative, has antiseptic powers; used in dentistry to disinfect root canals. Also used as flavoring agent, employed in root-beer, sarsaparilla and other beverages; in chewing-gum, certain candies, semi-pharmaceuticals, patent-medicines, such as tooth-paste, mouth washes; for flavoring tobacco, in soaps, perfumes;

source of artificial heliotrope. Used in home-remedies as sudorific for colds, spring tonic to „thin the blood"; for high blood-pressure of old people, especially recommended by old country doctors.

Sassafras Goesianum Teijsm and Binn. → Massoia aromatica Bess.

Sassafras Tzumu Hemsl. Chinese Sassafras. (Lauraceae). - Tree. Hupeh, Ningpo, Chekiang (China). A valuable timber tree in China.

Sasswood → Erythrophleum guineense G. Don.

Sassy Bark → Erythrophleum guineense G. Don.

Satinwood → Chloroxylon Swietenia DC., Ferolia guyanensis Aubl. and Zanthoxylon flavum Vahl.

Satinwood, Brazilian → Euxylophora paraensis Huber.

Satinwood Tree → Chloroxylon Swietenia DC.

Satsuma → Citrus reticulata Blanco.

Satureja Acinos (L.) Scheele. (syn. Thymus Acinos L., Calamintha arvensis Lam., C. Acinos Clairv.). Spring Savory. (Labiaceae). - Annual to perennial herb. Europe, Asia; introd. in N. America. Herb is sometimes used medicinally as aromatic and carminative.

Satureja hortensis L., Summer Savory. (Labiaceae). - Annual herb. Mediterranean region, Centr. Europe, Dalmatia, Siberia. Introd. in many countries. Cultivated. Stems and leaves are used for flavoring. Source of Oil of Savory, an ess. oil, obtained by steam distillation, used for flavoring processed foods, in sausages, meats, catsups.

Sauerkraut → Brassica oleraceae L. var. capitata L.

Saul Resin → Shorea Talura Roxb.

Saurauia aspera Turc. (Dilleniaceae). - Tree. S. Mexico. Sweet and mucilaginous, fruits are eaten raw or cooked as food, among Indian tribes in Mexico.

Saurauia Conzattii Busc. (Dilleniaceae). - Tree. Endemic in Cuicatlán and Ixtlán (Mex.). Fruits used as food by Chinantecs and Guicatecs (Mex.). Also S. speluncicola Schult.

Saurauia Roxburghii Wall. (Dilleniaceae). - Tree. So. China, India-China, Burma. Mucilaginous matter derived from leaves is used as hair-pomade by Chinese women in Annam and Cambodia.

Sauropus albicans Blume (syn. S. androgynus Merr.). (Euphorbiaceae). - Shrub. Malaya. Young leaves are boiled with rice or used in soup in Java.

Saururaceae → Anemonopsis, Saururus.

Saururus cernuus L. Common Lizardtail. (Saururaceae). - Perennial herb. Eastern N. America. Mashed, boiled roots were applied to wounds as a poultice by the Choctow Indians. Medicinally it is a sedative, antispamodic and astringent.

Sausage Tree → Kigelia africana Benth.

Saussurea Lappa C. B. Clarke (syn. Aucklandia costus Falc.). Costus. (Compositae). - Perennial herb. Valley of Kashmir and adj. territory. Root has been used since time immemorial. Much is exported to China and the Red Sea area. Employed in perfumery, blends well with vetiver, rose geranium, sandal and patchouly, giving bouquets of the Oriental type. Odor of Costus Root Oil is heavy and lasting. Also used as incense. It is said that hairwash has reputation of gray hair turning black. Also used as fumigant, aphrodisiac and for skin diseases.

Savin → Juniperus Sabina L.

Savory, Spring. → Satureja Acinos (L.) Scheele.

Savory, Summer → Satureja hortensis L.

Saw Grass → Cladium effusum (Sw.) Torr.

Saw Greenbrier → Smilax Bona nox L.

Saw Palmetto → Serenoa serrulata (Michx.) Hook. f.

Sawarri Fat → Caryocar amygdaliferum Mutis.

Saxegothaea conspicua Lindl. Mañio. (Podocarpaceae). - Tree. Chile. Wood of excellent quality, similar to Podocarpus; easy to work, durable. Used in Chile for general carpentry.

Saxifraga erosa Pursh. (syn. S. micranthidifolia Steud.); Mountain Lettuce, Deer Tongue. (Saxifragaceae). - Perennial herb. Eastern United States. Herb is used in salads by mountain people of S. Pennsylvania.

Saxifraga peltata Torr. (syn. Peltiphyllum peltatum Engl.) Indian Rhubarb. (Saxifragaceae). - Perennial herb. California to Oregon. Thick fleshy leafstalks when peeled are eaten raw or boiled by the Indians.

Saxifraga punctata L. Dotted Saxifrage. (Saxifragaceae). - Perennial herb. Western N. America. Succulent leaves are consumed raw or after being kept in oil by Eskimos of Alaska.

Saxifragaceae → Astilbe, Callicoma, Ceratopetalum, Chrysosplenium, Deutzia, Hydrangea, Pancheria, Philadelphus, Saxifraga.

Scaevola Koenigii Vahl. (Goodeniaceae). - Shrub. Mal. Archipelago. Wood hard, durable in salt water, yellow to dark brown; used by the natives for wooden nails in boat building.

Scaevola spinescens R. Br. (Goodeniaceae). - Perennial plant. Australia. Fruits are consumed and much prized by aborigines of S. Australia.

Scammony → Convolvulus Scammonia L.

Scammony, Mexican → Ipomoea orizabensis Ledenois.

Scammony, Levant → Convolvulus Scammonia L.

Scandix Cerefolium L. → Anthriscus Cerefolium Hoffm.

Scandix grandiflora L. (Umbellifloraceae). - Perennial herb. Greece, Asia Minor. Young plants are eaten as salad, and are much esteemed in some parts of Greece.

Scandix odorata L. → Myrrhis odorata (L.) Scop.

Scandix Pecten-Veneris L., Venus' Comb, Lady's Comb. (Umbelliferaceae). - Annual herb. Temp. Europe, Asia, introd. in N. America. Young stem tops are occasionally used in salads.

Scarlet Oak → Quercus coccinea Wangenh.

Scarlet Runner → Phaseolus coccineus L.

Scented Mahogany → Carapa procera DC.

Scented Mahogany → Entandrophragma Candollei Harms.

Schaefferia frutescens Jacq. Florida Woodbox, Yellow Wood, Boxwood. (Celastraceae). - Tree. S. Florida, West Indies. Wood is sometimes used for engraving, in place of true Boxwood.

Schefflera digitata Forst. (Araliaceae). - Small tree. New Zealand. Wood soft; used by Maoris to obtain fire by friction.

Scheidweileria luxurians Kl. → Begonia luxurians Scheidw.

Schima Noronhae Reinw., Burma Guger Tree. (Theaceae). - Tall tree. Malaya. Wood heavy, hard, fine-grained, reddish-brown to dark-brown; used in Java for houses and bridges. Bark is used to stupefy fish.

Schima stellata Pierre. (Theaceae). - Tree. Cochin China. Wood reddish, beautifully marked; used for fancy articles.

Schima Wallichii (DC.) Choisy. (Theaceae). - Tree. Himalayan region to Sumatra. Wood is used as building material.

Schinopsis Balansae Engl. Willowleaf Red Quebracho, Quebracho Colorado. (Anacardiaceae). - Tree. W. Argentina and adj. territory. Source of tannin extracts used for tanning.

Schinopsis Lorentzii (Griseb.) Engl. → Quebrachia Lorentzii Griseb.

Schinus dependens Orteg. (syn. Amyris polygama Cav. Duvaua dependens DC.). Peru Peppertree. (Anacardiaceae). - Shrub. Chile, Brazil and adj. territory. Plant is used in some parts of Chile for rheumatism.

Schinus latifolius Engl. (syn. Litrea molle Griseb.). Chilean Peppertree. (Anacardiaceae). - Tree. S. America. It is said that the fruits are source of a fermented drink, used in some parts of Chile.

Schinus molle L. Brazil Pepper-Tree, California Pepper-Tree. (Anacardiaceae). - Tree. Andes of S. America. Cultivated. Much used in local medicine. Powdered bark is used as a purgative for domestic animals. In Mexico the fruit is ground, mixed with atole etc. to make a beverage. Seeds are used to adulterate pepper. A gum which exudes from the trunk is used for chewing, which is sometimes called American Mastic.

Schinus terebinthifolius Raddi. (syn. S. Aroeira Vell.). (Anacardiaceae). - Tree. Argentina, Parauay, Brazil and adj. territory. Cultivated. Aromatic plant is considered tonic and astringent. Stem is source of a resin, called Balsamo de Misiones.

Schizaceae → Lygodium.

Schizachyrium exile Stapf. (Graminaceae). - Annual grass. Nigeria, Nile region, India. Fodder grass, also used for thatching. Chopped up and mixed with clay, it is used in parts of Africa for building huts.

Schizoglossum shirense N. E. Brown. (Asclepiadaceae). - Woody plant. Trop. E. Africa. Used for dysentery; as a stomachic and aphrodisiac.

Schizosaccharomyces asporus Beijerinck. (Saccharomycetaceae). - Yeast. This species takes an important part in the alcohol fermentation and is probably the principal yeast in the production of Arak in Java.

Schizosaccharomyces pombe Lindner. (Saccharomycetaceae). - This yeast is sometimes used in the process of alcohol fermentation.

Schizosaccharomyces spec. (Saccharomycetaceae). - Fungus. Microorganism. A species of this genus is supposed to take part in the curing of the Candied Peel, Citrus medica L.

Schleichera oleosa Merr. Malay Lacktree. (Sapindaceae). - Tree. S. E. Asia, Mal. Archipelago. Wood close-grained, hard, heavy, resistant to moisture, reddish; used for small vessels, sugarmills; source of excellent charcoal. Bark is employed in tanning. Young leaves are eaten by natives with rice. Unripe fruits are pickled. Seeds are source of Macassar Oil, used in ointments, for candles, illumination and in Madura for Batik-work.

Schleichera trijuga Willd., Kussum Tree. (Sapindaceae). - Tree. India, Ceylon. Mal. Archipelago. Seeds source of Macassar or Kussum-Oil; non-drying. Used as illuminant; for manuf. soap, sometimes for food. Sap. Val. 227; Iod. No. 54.5.

Schmidelia africana DC. → Allophylus zeylanicus L.

Schoenocaulon officinale Gray. (syn. Sabadilla officinalis Brand and Ratzeb.) (Liliaceae). - Herbaceous plant. Centr. America. Seeds are used as insecticide, contain veratrin.

Schoutenia ovata Koch. (syn. Actinophora fragrans R. Br. (Tiliaceae). - Tree. Java. Wood very heavy, hard, close-grained, reddish-brown to dark red-brown, tough; used for handles of tools and wheels.

Schradera marginalis Standl. (Rubiaceae). - Woody vine. Colombia. Parts of the plant are chewed by the Citará Indians of Colombia to stain the teeth jet black.

Schweiser Tee → Dryas octopetala L.

Sciadopitys verticillata Sieb. and Zucc. (Taxodiaceae). - Tree. S. Japan. Wood reddish to yellowish - white, soft, elastic, resistant to moisture; used for waterworks and ship building.

Scirpodendron costatum Kurz. (Cyperaceae). - Tall sedge. Trop. Asia to Australia, Samoa. Cultivated in Sumatra; used for mat making. In the Philipp. Islds. used for hats.

Scirpus lacustris L. (syn. S. acutis L., S. validus Vahl.). Great Bulrush, Club Rush. (Cyperaceae). - Perennial herb. N. America. West Indies, Centr. America. Rhizomes were eaten raw or made into bread by the Indians. Young shoots were consumed in spring.

Scirpus nevadensis S. Wats. (Cyperaceae). - Perennial herb. Western N. America. Roots were consumed raw by the Cheyenne Indians.

Scirpus paludosus A. Nels. (Cyperaceae). - Perennial herb. N. America. Rhizomes were consumed raw or made into a flour for bread. Also pollen was mixed with bread by several Indians tribes of N. America.

Scirpus tuberosus Roxb. (Juncaceae). - Perennial herb. China, Japan. Cultivated in ricefields. Tubers are consumed as a vegetable in China and Japan. Tubers in China are source of a starch, called Batei-fun.

Sciuris aromatica Gmel. → Raputia aromatica Aubl.

Sclerocarya caffra Sond. Caffir Marvola Nut. (Anacardiaceae). - Large tree. S. Africa. Fruits size of plum, pleasant taste, excellent for jelly. Also made by natives of S. Africa into a fermented beverage.

Sclerocarya Schweinfurthii Schinz. (Anacardiaceae). - Tree. Trop. Africa. Fruits are used by the Ovambos, Afr. for the preparation of a beer-like beverage.

Sclerodermataceae (Fungi) → Pisolithus, Pompholyx.

Scolopendrium vulgare Sm., Hart's Tongue. (Polypodiaceae). - Perennial herb, fern. Temp. Zone. Fronds of the ferns are smoked by the Ainu; sometimes leaves are mixed with tobacco.

Scolopia Mundtii Harv. (syn. Phoberos Mundtii Presl.). Red Pear. (Flacourtiaceae). - Tree. Cape Peninsula to Natal. Wood hard, heavy, close-grained; used for felloes, wagons.

Scolopia Schreberi J. F. Gmel. (syn. S. Gaertneri Thw.). (Flacourtiaceae). - Medium sized tree. Ceylon. Wood hard, heavy, close-grained, dull-red; used for rafters, posts.

Scolopia Zeyheri Arn. (syn. Phoberos Zeyheri Presl.). Thorn Pear (S. Afr.). (Flacourtiaceae). - Thorny tree. S. Africa. Wood heavy, hard, close-grained; used for wagon-work, axles, felloes, spokes.

Scolymus hispanicus L., Spanish Oyster Plant. (Compositae). - Biennial herb. S. France, Mediterranean region. Roots when boiled are consumed as a vegetable in some parts of S. Europe. Centuries past the plant was more cultivated than at the present.

Scoparia dulcis L. Sweet Broom. (Scrophulariaceae). - Shrub. Trop. America. In some parts of the West Indies branches are placed in drinking wells to give the water a cool taste.

Scopolia → Scopolia carniolica Jacq.

Scopolia carniolica Jacq. Scopolia. (Solanaceae). - Perennial herb. Carpathian Mts., Croatia. Occasionally cultivated. Dried rhizome used medicinally. Resembles Belladonna, though it is more narcotic. Contains atropine, scopolamine and hyoscyamine. Scopolamine hydrobromide, a powerful hypnotic, causes sleep, resembling natural sleep.

Scorbute Grass → Cochlearia officinalis L.

Scorodocarpus borneensis (Baill.) Becc. (Olacaceae). - Tree. Borneo. Wood very heavy, close-grained, hard, dark-red to purple gray-brown; used for bridges, houses and boats. Known as „Koolim".

Scorzonera → Scorzonera hispanica L.

Scorzonera hispanica L. Scorzonera, Black Salsify. (Compositae). - Annual to perennial herb. Centr. Europe, Mediterranean region. Cultivated as a vegetable; in some countries of commercial importance. Roots when boiled are consumed as a vegetable. Contain much inulin. They are also used as substitute for coffee. Varieties: Russian Giant and Vulcan.

Scorzonera mollis Bieb. (syn. S. undulata Vahl.) (Compositae). - Perennial herb. Mediterranean region. Flowers are supposed to have the scent of chocolate; used by some peoples as salad.

Scorzonera Schweinfurthii Boiss. (Compositae). - Herbaceous plant. Egyptian-Arabic desert region. Roots are eaten as food by different tribes in N. Africa.

Scorzonera tau-saghys Lips. and Bosse., Tau-Saghys (Russ.) (Compositae). - Herbaceous. Plateau of Kara-Tau (Russia). Grown in some parts of Russia as a possible source of rubber. Latex contains 2 to 30 % rubber.

Scotch Broom → Sarothamnus scoparius (L.) Wimmer.

Scotch Maple → Acer Pseudo-Platanus L.

Scotch Pine → Pinus sylvestris L.

Scotch Thistle → Onopordon Acanthium L.

Screw Bean → Strombocarpa odorata (Torr. and Frem.) Torr.

Screw Pine → Panadanus spp.

Screwpine, Thatch → Pandanus odoratissima L.

Screwplant → Helicteres Isora L.

Scribbly Gum → Eucalyptus haemastoma Smith.

Scrophulariaceae → Alectra, Angelonia, Bonnaya, Buchnera, Calceolaria, Chelone, Conobea, Cordylanthes, Curanga, Digitalis, Euphrasia, Herpestis, Lathraea, Limnophila, Paulownia, Pedicularis, Penstemon, Picrorhiza, Scoparia, Verbascum, Veronica.

Scrub Sandalwood → Exocarpus latifolia R. Br.

Scurfpea, Beaverhead → Psoralea castorea S. Wats.

Scurfpea, Leatheroot → Psoralea macrostachya DC.

Scurvy Grass → Barbarea praecox R. Br.

Scurvy Grass → Cochlearia officinalis L.

Scutellaria baickalensis Georg. Baical Skullcap. (Labiaceae). - Perennial herb. Mongolia, Siberia; Chihli, Shantung (China). Used medicinally in China. Contains baicalin, scutellarin and an ess. oil.

Scutellaria lateriflora L. Sideflowering Skullcap. (Labiaceae). - Perennial herb. N. America, Mexico. Dried herb, collected during flowering time is used medicinally; being tonic, nervine, antispasmondic; used for nervous exhaustion, delirium tremens, neuralgia. Contains scutellarin, a glucoside.

Scytosiphonaceae (Brown Algae). → Hydroclathrus, Phyllitis.

Sea Bean → Mucuna gigantea DC.

Sea Buckthorn → Hippophae rhamnoides L.

Sea Girdles → Laminaria digitata (L.) Edmonson.

Sea Grape → Coccolobus univera (L.) Jacq.

Sea Holley → Acanthus ebracteatus Vahl.

Sea Island Cotton → Gossypium barbadense L.

Sea Kale → Crambe maritima L.

Sea Kale, Tatarian → Crambe tatarica Jacq.

Sea Milkwort → Glaux maritima L.

Sea Onion → Urginea Scilla Steinh.

Sea Rocket, American → Cakile edentula (Bigel) Hook.

Sea Samphire → Crithmum maritimum L.

Sea-Staff → Laminaria digitata (L.) Edmonson.

Seabeach Sandwort → Arenaria peploides L.

Seacoast Abronia → Abronia latifolia Esch.

Seatron → Nereocystis Luetkeana (Mert.) Post. & Rupr.

Sea Whistles → Ascophyllum nodosum (L.) Le Jolis.

Sebaca crassulaefolia Schlecht. (Gentianaceae). - Perennial herb. S. Africa. Used by natives for snake-bite, esp. of Bitis arietans.

Sebastiania Pavoniana Muell. Arg. (Euphorbiaceae). - Shrub. Mexico. Milky juice is used as arrow poison. Fruits source of „jumping beans." „Jumping" of seeds is caused by the larva of a small butterfly, Carpocapsa saltitans.

Secale africanum Stapf. (Graminaceae). - S. Africa. Straw is used by the natives for covering of huts.

Secale cereale L. Rye. (Graminaceae). - Annual grass. Probably from S. W. Asia. Widely cultivated in temperate zones of Old and New World. Important source of flour, made into Rye Bread, Black Bread; also made into alcohol, whisky, wodka. When roasted it is a substitute for coffee. In some countries rye is grown together with wheat. Also grown for hay, pasturage, green manure, as winter cover crop, to prevent erosion and as a sand-binder. Straw is used for manuf. paper, for packing, straw-products, bedding. Rye is more recent than wheat and barley. Some varieties are: Petkuser, Schlanstedt, Rosen, St. John, Abruzzi, South Georgia, Dean, Mammoth White, Virginia Winter, Rimpau.

Secale Cornuti → Claviceps purpurea (Fr.) Tul.

Secale montanum Guss. Mountain Rye. (Graminaceae). - S. Europe, N. Africa, Asia Minor and adj. territory. Grown as a grain crop and as fodder for animals. Varieties are divided into western and eastern groups. Some are perennial (var. Kuprijanovii Grossh.).

Secale sylvestre Host. (syn. S. fragile Marsch.) (Graminaceae). - Europe, trop. Asia. Grown in mountainous regions of China and Thibet. Grains are used as food.

Secamone myrtifolia Benth. (Asclepiadaceae). - Woody plant. Trop. Africa. Decoction of leaves is used by the natives as a purge.

Secamonopsis madagascariensis Jum. Vahimainty. (Asclepiadaceae). - Woody plant. Madagascar. Source of a good rubber.

Sechium edule Swartz. (syn. Chayota edulis Jacq.). Chayote, Guisquil. (Cucurbitaceae). - Perennial vine. Centr. America. Fruits and young shoots are consumed as a vegetable in Centr. America and also in S. of United States. Fruits are made in Centr. America into sweetmeat. Roots are consumed as food. Several varieties.

Sechsämtertropfen → Sorbus Aucuparia L.

Securidaca longipedunculata Fres. (Polygalaceae). - Small tree. Trop. Africa. Twigs are source of Buaze Fibre, supposed to be equal to flax. Leaves are used as purgative in Abyssinia. Powdered herb is employed as snuff in some parts of Togoland (Afr.). In Angola the plant forms one of the ingredients of the Ordeal Brew among the natives.

Securidaca philippensis Chodat. (syn. S. corymbosa Turcz.) (Polygalaceae). - Philipp. Islds. Bark contains saponin, is locally used in the Philipp. Islds. as a substitute for soap. Also S. Cumingii Hassk.

Sedum acre L. Goldmoss Stonecrop, Mossy Stonecrop. (Crassulaceae). - Perennial succulent herb. Europe, temp. Asia; introd. in N. America. Cultivated. Herb occasionally used as laxative.

Sedum album L. White Stonecrop, Wall Pepper. (Crassulaceae). - Perennial herb. Europe, temp. Asia, N. Africa. Herb is sometimes used in salads. Also S. reflexum L., S. Telephium L.

Sedum reflexum L. (syn. S. collinum Willd.). Jenny Stonecrop. (Crassulaceae). - Perennial, succulent herb. Leaves are sometimes consumed in salads or in soups. Also S. rupestre L.

Sedum roseum Scop. (syn. Rhodiola integrifolia Raf.). (Crassulaceae). - Herbaceous plant. Europe, N. America. Leaves are eaten fresh, soured or in oil by Eskimos of Alaska.

Seeds and grains used as food. → Abutilon, Acacia, Acanthosicyos, Adansonia, Aesculus, Agriophyllum, Amaranthus, Amphicarpa, Amygdalus, Anacardium, Arachis, Araucaria, Artemisia, Artocarpus, Arundinaria, Avena, Baillonella, Beckmannia, Bertholletia, Blepharis, Boscia, Brosimum, Buchanania, Cajanus, Canarium, Careya, Carnegiea, Carya, Castanea, Castanopsis, Ceratotheca, Cercidium, Chenopodium, Cicer, Corylus, Corynocarpus, Coula, Crescentia, Cucurbita, Digitaria, Dioon, Diplodiscus, Dipteryx, Durio, Elateriospermum, Eleusine, Enalus, Entada, Ervum, Erythrina, Eucarya, Euryale, Fagus, Gevuena, Ginkgo, Glyceria, Glycine, Gomphia, Gymnartocarpus, Hordeum, Inocarpus, Irvingia, Jubaea, Juglans, Latipes, Lecythis, Leucaena, Lupinus, Macadamia, Melocalamus, Mucuna, Nelumbium, Olneya, Omphalea, Oryza, Pachycereus, Palaquium, Pandanus, Pangium, Panicum, Parkia, Parkinsonia, Phaseolus, Pinus, Pistacia, Pisum, Pleiogynium, Psophocarpus, Quercus, Secale, Semecarpus, Simmondsia, Solidago, Sterculia, Terminalia, Torreya, Trapa, Treculia, Triticum, Tylostemon, Voandzeia, Yucca, Zea, Zizania.

Seepweed, Alkali → Suaeda fruticosa Forsk.

Seepweed, Desert → Suaeda suffrutescens Wats.

Sejal → Acacia tortilis Hayne.

Selaginella lepidophylla Spreng. (Selaginellaceae). - Perennial herb. Fern-ally. Mexico and adj. United States. Is used in some parts of Mexico as diuretic. Is sold in markets.

Selaginellaceae → Selaginella.

Selenicereus grandiflorus (L.) Britt. and Rose. (syn. Cereus grandiflorus Mill.). Night Blooming Cereus. (Cactaceae). - Xerophyte. Mexico. Cultivated. Source of a drug, similar to digitalin; used in rheumatism. Contains an alkaloid and glucoside. Drug is used in some parts of Trop. America for dropsy.

Selenipedium Chica Reichb. f., Vanilla Chica. (Orchidaceae). - Perennial herb. Centr. America, S. America. Was formerly used instead of ordinary vanilla.

Selinum officinale Vest. → Peucedanum officinale L.

Semba Gona Gum → Bauhinia variegata L.

Semecarpus Anacardium L. f. (syn. Anacardium orientale L.) Cashew Marking Nut Tree. Oriental Cashew-Nut. (Anacardiaceae). - Tree. Trop. Asia to Australia. Cultivated in the tropics. Juice with lime water is used as marking ink to write on cloth. Oil from seeds are used in India for protection against white ants and for floor dressing. Nuts are used for tanning.

Semecarpus Cassuvium Roxb. (Anacardiaceae). - Tree. Mal. Archipelago. Occasionally cultivated. Fleshy receptacle below the fruits is eaten by the natives. Sold in local markets of Banda, etc.

Semecarpus microcarpa Wall. (syn. S. cuneiformis Blanco.) (Anacardiaceae). - Shrub or small tree. Philip. Islds. Fruits are edible, resembling small cashew nuts.

Seminole Tea → Asimina reticulata Chapm.

Sempervivum tectorum L. Hen-and-Chickens, Roof Houseleek. (Crassulaceae). - Perennial succulent herb. Europe. Cultivated. Chewed leaves used as home-remedy for tooth-ache, also for stings of bees. Decoction of leaves is employed fo intestinal worms.

Senarr Gum → Acacia Seyal Delile.

Senarr, White → Acacia Senegal (L.) Willd.

Senebiera lepidioides Coss. (Cruciferaceae). - Herbaceous plant. Algerian Sahara. Used as food by the Tuareg (N. Afr.).

Seneca → Polygala Senega L.

Senecio Ambavilla Pers. (syn. Hubertia Ambavilla Bory). (Compositae). - Herbaceous plant. Madagascar, Bourbon. Used in Madagascar as diuretic, aromatic.

Senecio aureus L., Golden Ragwort, Life Root, Squaw Weed. (Compositae). - Perennial herb. N. America. Dried herb is used medicinally, being stimulant, emmenagogue, diuretic, uterine tonic. Contains an ess. oil; inuline and alkaloids senecine and senecifoline.

Senecio Jacobaea L., Tansy Ragwort. (Compositae). - Biennial to perennial herb. Throughout Europe, W. Siberia, Caucasia, Asia Minor, N. Africa, introd. in N. America. Leaves used as home-remedy for cramp, intestinal parasites, as tonic and stimulant.

Senecio Kaempferi DC. (Compositae). - Perennial herb. Japan. Petioles when boiled are eaten in Japan as a vegetable; also when salted.

Senecio nikoënsis Miq. (Compositae). - Herbaceous. Chihli, Shensi (China), Japan. Used medicinally in China.

Senecio palmatus Pall. (Compositae). - Perennial herb. Siberia, Sachalin, China, Japan. Used medicinally in China. Young leaves are used as food by the Ainu.

Senecio pteroneurus Sch. → Kleinia pteroneura DC.

Senecio rotundifolius Hook. f. (Compositae). - Shrub or small tree. New Zealand. Wood pale

brown, bright satiny luster, pretty silver grain. Used for inlaying, ornamental work, general turnery.

Senegal Ebony → Dalbergia melanoxylon Guill. and Perr.

Senegal Gum → Acacia Senegal (L.) Willd.

Senegal Khaya → Khaya senegalensis Juss.

Senegal Prickly Ash → Zanthoxylum senegalense DC.

Senna, Aden → Cassia holosericea Fresen.

Senna, Alexandrian → Cassia acutifolia Del.

Senna, American → Cassia marilandica L.

Senna, Arabian → Cassia angustifolia Vahl.

Senna, Coffee → Cassia occidentalis L.

Senna, Dog → Cassia obovata Collad.

Senn Gum → Bauhinia variegata L.

Senna, Italian → Cassia obovata Collad.

Senna, Mecca → Cassia angustifolia Vahl.

Senna, Ringworm → Cassia alata L.

Senna, Sickel → Cassia Thora L.

Senna, Smooth → Cassia laevigata Willd.

Senna, Tennevelley → Cassia angustifolia Vahl.

Sequoia gigantea Lindl. and Gord. (syn. S. Washingtoniana (Winsl.) Sudw., S. Wellingtoniana Seem.) Big Tree, Sequioa, Mammoth Tree. (Toxodiaceae). - California. Important lumber tree. Wood soft, not strong, brittle, very light, coarse-grained, becoming dark when exposed; used for lumber, construction work, shingles, fencing.

Sequioa sempervirens (Lamb.) Endl., Redwood. (Taxodiaceae). - Tree. California, Oregon. Important lumber tree. Wood soft, not strong, close-grained, light, easily worked, light, red, easily split, durable when in contact with soil; manufactured into lumber; used for building purposes, rail-road ties, shingles, fence-posts and wine butts.

Serapat Rubber → Urceola acuto-acuminata Boerl.

Serenoa serrulata (Michx.) Hook. f. (syn. S. repens (Bartr.) Small.) Saw Palmetto. (Palmaceae). - Dwarf-palm. South Carolina to Florida and Louisiana. Seeds were important food of the aborigines. Flowers are source of good honey. Dried, ripe berries are of medicinal importance as diuretic, anticatarrhal, sedative. They are collected in Aug. to Jan., sometimes they are dried to a prunelike substance. Used in pharmaceutical preparations. Contain dextrose, volatile and fixed oils. Fruits are supposed to have a stimulant action upon the mucuous membrane of the genito-urinary tracts, milder and less irritant than copaiba or cubeb.

Serianthes myriadena Planch. (syn. Acacia myriadena Bert.). (Leguminosaceae). - Tree. Pacif. Islds. Wood elastic, solid, straight-grained, resistant to water; used in New Caledonia for wagon making.

Seringeirana → Sapium taburu Ule.

Serjania curassavica Radlk. → Paullinia pinnata L.

Sernamby Rubber → Sapium taburu Ule.

Serradella → Ornithopus sativus Brot.

Service Berry → Amelanchie canadensis (L.) Medic.

Service Tree → Sorbus domestica L.

Service Berry, Saskatoon → Amelanchier alnifolia Nutt.

Service Berry, Western → Amelanchier alnifolia Nutt.

Service Berry, Shadblow → Amelanchoer canadensis (L.) Medic.

Sesame → Sesamum indicum L.

Sesamum angustifolium Engl. (Pedaliaceae). - Annual herb. Trop. Africa, esp. Tanganyika. Seeds have been suggested as a source of oil, suitable for manuf. of soap, and when refined for edible oil.

Sesamum calycinum Welw. (Pedaliaceae). - Herbaceous plant. Trop. Africa. Leaves are consumed as a vegetable by natives in E. Africa.

Sesamum indicum L. (syn. S. orientale L.) Oriental Sesame, Bene. (Pedaliaceae). - Annual herb. Trop. Asia. Cultivated in warm countries of Old and New World. Seeds source of Sesame, Bene, Teel or Gingli Oil. Semi-drying; used for edible purposes, iodized oils, manuf. soap, in salads, for cooking, margarine, shortenings, lard substitutes, cosmetics. Seed Oil Cake food for cattle and fertilizer. Many varieties with black, white brown, and dark-red seeds. Sp. Gr. of oil 0.928—0.926; Ac. Val. 1.4; Iod. No. 110.8; Sap. Val. 189.3; Unsap. 1.73 %. Seeds used for covering bread, cookies, cake, confectionary. Used medicinally for liniments, ointments; solvent of medicinal agents.

Sesbania aculeata Poir. (syn. Aeschynomene spinulosa Roxb.). (Leguminosaceae). - Tree. India, Ceylon, China, Trop. Africa. Stem is source of a fibre, known as Dundee Fibre; used for manuf. sails, fishing-nets. A substitute for Hemp in India.

Sesbania aegyptiaca Poir. (Leguminosaceae). - Shrub. E. Africa, India. Leaves are used as fodder for live-stock. Bark is source of a fibre, manuf. in India into ropes. Wood produces a charcoal. Decoction of leaves is used by the natives of Hausa for washing animals, as preventative for bite of tsetse fly. Also used as green manure.

Sesbania cinerascens Welw. (Leguminosaceae). - Woody plant. Trop. Africa. Seeds are recommended as fodder for live-stock in S. Africa.

Sesbania grandiflora (L.) Poir. (syn. Aeschynomene grandiflora L.). Agati Sesbania. (Leguminosaceae). - Tree. E. India to Australia. Cultivated in tropics of Old and New World. Flowers and green pods are used in some parts of so. Asia as salad and as pot-herb. Bitter bark is a

tonic and febrifuge. Leaves are diuretic and laxative.

Sesbania macrocarpa Muhl. (Leguminosaceae). - Annual. Southern and southwestern United States. Bark is source of a fibre, used by the Indians of the lower Colorado River for nets and fishing lines.

Sesbania tetraptera Hochst. (Leguminosaceae). - Herbaceous plant. Trop. Africa. Leaves are consumed by the natives as a pot-herb.

Seseli Carvi Lam. → Carum Carvi L.

Setaria glauca Beauv. (syn. Pennisetum glaucum (L.) R. Br.) Pearl Millet., Cat-tail Millet. (Graminaceae). - Annual grass. Trop. Africa. Cultivated in warm countries as a food for cattle.

Setaria italica L. (syn. Chaetochloa italica (L.) Scribn. Foxtail Mittel. (Graminaceae). - Annual grass. Europe, Asia. Cultivated in Old and New World, as haycrop, for pasture and as a cereal. Some varieties are: Hungarian, Golden Wonder, Siberian, Kursk.

Seven Bark → Hydrangea arborescens L.

Seville Orange → Citrus Aurantium L.

Seyal Acacia → Acacia Seyal Delile.

Shad Bush → Amelanchier canadensis (L.) Medic.

Shadblow Serviceberry → Amelanchier canadensis (L.) Medic.

Shaddock → Citrus grandis Osbeck.

Shagbark Hickory → Carya ovata (Mill.) Koch.

Shallon → Gaultheria Shallon Pursh.

Shallot → Allium ascalonicum L.

Shallu → Sorghum vulgare Pers.

Shama Millet → Echinochloa colonum (L.) Link.

She Pine → Podocarpus elata R. Br.

Shea Butter → Butyrospermum Parkii (Don.) Kotschy.

Shea Butter Tree → Butyrospermum Parkii (Don.) Kotschy.

Sheep Cheese → Panus rudis Fr.

Sheep's Fescue → Festuca ovina L.

Shepherd's Purse → Capsella Bursa pastoris (L.) Medic.

Shepherdia argentea Nutt. Silver Buffalo Berry. (Elaeagnaceae). - Shrub or small tree. N. America. Fruits used as jelly, also eaten dried with sugar. Indians dried fruits for winter use, often cooked with buffalo meat.

Shepherdia canadensis (L.) Nutt. (syn. Lepargyrea canadensis Greene). Russet Buffalo Berry. (Elaeagnaceae). - Shrub. N. America. Berries are pressed into cakes, dried or smoked which are consumed by the Eskimos of Alaska.

Shin Oak → Quercus undulata Torr.

Shingle Oak → Casuarina stricta Ait. and Quercus imbricaria Michx.

Shingle Tree → Acrocarpus fraxinifolius Wight. and Arn.

Shining Sumach → Rhus copallina L.

Shiny Asparagus → Asparagus lucidus Lindl.

Shir-Khist → Cotoneaster racemiflora (Desf.) Koch.

Shooting Star, Henderson → Dodecatheon Hendersonii A. Gray.

Shore Grindelia → Grindelia robusta Nutt.

Shore Lupine → Lupinus littoralis Dougl.

Shore Podgrass → Triglochin maritima L.

Shorea acuminata Dyer. (Dipterocarpaceae). - Tree. Mal. Penin. Wood heavy; used for house-building, bridges, boards.

Shorea aptera Burck. Borneo Shorea. (Dipterocarpaceae). - Tree. Mal. Archipelago, Borneo. Source of Borneo Tallow. Hard, brittle, solid, resembles Cacao-Butter. Used as substitute of Cacao-Butter. Sp. Gr. 0.52—0.860; Sap. Val. 189—200; Iod. No. 29—38; Unsap. 0.4—2 %.

Shorea Balangeran Burck. Yakal Shorea. (Dipterocarpaceae). - Tree. Mal. Archipelago. Source of a Dammar, Tangkawang Fat, used for candle making, manuf. soap and illumination. Also S. bracteolata Dyer, S. eximea Scheff. and S. laevis Ridl.

Shorea bracteolata Dyer. (Dipterocarpaceae). - Tree. Malaya. Wood yellow to brownish; used for boards and building vessels.

Shorea cochinchinensis Pierre. (Dipterocarpaceae). - Tree. Indo-China, Brit. Malaya to Kedah. Bark used in Cambodia as masticatory.

Shorea eximea Scheffer. (syn. Vatica sublacunosa Miq.). Almond Shorea. (Dipterocarpaceae). - Tree. India. Source of a resin, called Dammar Tubang or Dammar Klookoop; used by the natives for illumination.

Shorea glauca King. (Dipterocarpaceae). - Tree. Penang. Stem is source of Dammar Kumus.

Shorea hypochra Hance. Temal Shorea. (Dipterocarpaceae). - Tree. Cambodia. Source of Dammar Temak, resembling Dammar Penak from Balanocarpus Heimii King. suitable for varnishes.

Shorea Koordersii Brandis. (Dipterocarpaceae). - Tree, Celebes and Moluccas. Stem is source of a resin called Dammar Tenang; it is formed as result of pathological condition of the tree.

Shorea laevis Ridl. (Dipterocarpaceae). - Tree. Malaya. Wood hard; used by the natives for construction of houses, also source of an excellent charcoal. Stem produces a yellow white dammar, which is collected in large quantities. It is mixed with Perak Resin.

Shorea leprosula Miq. (Dipterocarpaceae). - Tree. Mal. Archipelago. Source of a resin, called Dammar Daging, used in native medicine. Wood easily worked, durable, attacked by insects; used by natives for flooring.

Shorea materialis Ridl. Balan. (Dipterocarpaceae). - Tree. Trop. Asia, esp. Terngganu to Jalore, eastcoast Mal. Penin. Wood very durable, used for different purposes. Exported. Used for bridges, wharves, beams and where great strength is required. Likewise S. platyclados Van Slooten, S. resinosa Foxw., S. Ridleyana King, S. utilis King.

Shorea selanica Blume. (Dipterocarpaceae). - Tree. Moluccas. Stem is source of a Dammar. Used by the natives for torches.

Shorea siamensis Miq. (syn. Pentacme siamensis Kurz.). (Dipterocarpaceae). - Tree. Burma, Siam, Indo-China. Wood very hard, heavy, very durable, strong; used for heavy construction; exported from Siam.

Shorea stenocarpa Burck. (Dipterocarpaceae). - Tree. Malaya. Cultivated for its seeds. Source of a Borneo Tallow.

Shorea Talura Roxb. (syn. S. Roxburghii G. Don., Vatica laccifera Wight). (Dipterocarpaceae). - Tree. Mal. Archipelago. Source of Saul or Sal Resin, formerly mistaken for Manila Kopal.

Shorea Tumbuggaia Roxb. (Dipterocarpaceae). - Tree. India. Stem is source of a resin.

Shorea vulgaris Pierre. (Dipterocarpaceae). - Tree. Trop. Asia, esp. Cochin China. Plant is source of Huile de Bois; used for making varnish.

Shorea Wiesneri Schiff. Gum Dammar. (Dipterocarpaceae). - Tree. Philip. Islds., Mal. Penin. Source of Resin Dammar, Gum Dammar; used for varnishes and plasters.

Short Buchu → Barosma betulina (Thunb.) Bartl. and Wendl. and B. crenulata (L.) Hook.

Short Hair Plume Grass → Dichelachne sciurea Hook. f.

Short Leaved Pine → Pinus echinata Mill.

Short Staple American Cotton → Gossypium herbaceum L.

Showy Milkweed → Asclepias speciosa Torr.

Shrub Lespedeza → Lespedeza bicolor Turcz.

Shrub Myrtle → Backhousia citridora F. v. Muel.

Siak Fat → Palaquium oleiferum Blanco.

Siam Benzoin → Styrax tonkinense Craib.

Siam Cardamon → Elettaria Cardamonum (L.) Maton.

Siam Gamboge → Garcinia Hanburyi Hook. f.

Siam Gamboge Tree → Garcinia Hanburyi Hook. f.

Siberian Filbert → Corylus heterophylla Fisch.

Siberian Motherwort → Leonurus sibericus L.

Siberian Peashrub → Caragana arborescens Lam.

Siberian Tea → Potentilla rupestris L.

Sicana odorifera Naud. Casa Banaya, Curuba. (Cucurbitaceae). - Vine. Peru, Brazil. Fruits are occasionally eaten as a vegetable in S. America, they are also used as preserves. The fragrant fruits are used to scent clothes in some parts of Latin America.

Sicilian Sumach → Rhus Coriaria L.

Sickingia Maxonii Standl. (Rubiaceae). - Tree. Costa Rica. Bark used in native medicine as a purgative and febrifuge.

Sickingia rubra (Mart.) Schum. (Rubiaceae). - Tree. Brazil, esp. Bahia to Rio de Janeiro. Bark source of a red dye. Wood used by aborigines for fancy articles, carpentry, inner construction.

Sickingia tinctoria (H. B. K.) Schum. (Rubiaceae). - Shrub. Brazil. Wood is source of a red dye.

Sickingia viridiflora Schum. (syn. Arariba viridiflora Allem.). (Rubiaceae). - Tree. Brazil. Bark is considered in some parts of Brazil a febrifuge.

Sickle Senna → Cassia Thora L.

Sickleleaf Acacia → Acacia harpophylla F. v. Muell.

Sida → Lagerstroemia parviflora Roxb.

Sida rhombifolia L. (syn. S. canariensis Willd.) Broomjute Sida. (Malvaceae). - Perennial herb. Tropics. Leaves are used as a tea in some parts of the Canary Islds. Stems is source of a good fibre, also S. acuta Murm. Fibre is sometimes called Queensland Hemp.

Siddhi → Cannabis sativa L.

Side-Oats Grama → Bouteloua curtipendula (Michx.) Torr.

Sideranthus spinulosus (Pursh) Sweet → Aplopappus spinulosus DC.

Sideritis theezans Boiss and Heldr. (Labiaceae). - Perennial herb. Greece. Leaves and flowering tops are used as an aromatic tea in Greece; sold in markets. Also S. peloponnensiaca Boiss and Heldr. and S. Roeseri Boiss. and Heldr.

Sideroxylon angustifolium Standl. (Sapotaceae). - Tree. Mexico. Bark used in Sinaloa for curdling of milk.

Sideroxylon cyrtobotryum Miq. (Sapotaceae). - Tree. Brazil. Trunk is source of Balata Rosada.

Sideroxylon dulcificum A. DC. (Sapotaceae). - Tree. Trop. Africa. Ripe, fresh fruits have the property of imparting a sweet taste to anything that is bitter, acid or sour, e. g. quinine, limejuice, unripe fruit, vinegar etc. when eaten immediately afterward. Used by natives to sweeten palm wine.

Sideroxylon glabrescens Miq. (Sapotaceae). - Tree. Mal. Archipelago, Bangka. Source of a chewing gum base.

Sideroxylon Mastichodendron Jacq. (syn. S. foetidissimum Jacq.) Mastic, Wild Olive. (Sapotaceae). - Tree. Florida, West Indies, Bahamas. Wood strong, hard, heavy, bright orange. Locally used for cabinet-work, boat-building.

Sideroxylon resiniferum Ducke. (Sapotaceae). - Tree. Brazil. Source of a Balata Rosada.

Siebold Walnut → Juglans Sieboldiana Maxim.

Siejas Wax → Langsdorffia hypogaea Mart.

Sierra Juniper → Juniperus occidentalis Hook.

Sierra Leone Copal → Copaifera Guibourtiana Benth.

Sierra Leone Butter → Pentadesmia butyracea Sabine.

Sikkim Madder → Rubia sikkimensis Kurz.

Silene macrosolen Steud. (Caryophyllaceae). - Perennial herb. E. Africa. Roots called Radix Ogkert or Sarsari; used in Abyssinia for tapeworm.

Siler montanum Crabtz → Laserpitium Siler L.

Silk Cotton Tree → Bombax buonopozence Beauv.

Silk Cotton Tree → Ceiba pentandra (L.) Gaertn.

Silk Oak → Grevilla robusta Cunnnigh.

Silk Rubber → Funtumia elastica Stapf.

Silk Rubber Tree → Funtumia elastica Sta f.

Silkworm food → Food for silkworms.

Silky Oak → Stenocarpus salignus R. Br.

Silphium of the Ancients → Ferula Narthex Boiss.

Silver Beech → Nothofagus Menziesii Oerst.

Silver Buffaloberry → Shepherdia argentea Nutt.

Silver Fir → Abies amabilis (Loud.) Forbes. and A. pectinata DC.

Silver Maple → Acer saccharinum L.

Silver Nailwood → Paronychia argentea Lam.

Silver Pine Needle Oil → Abies pectinata DC.

Silver Thistle → Onopordon Acanthium L.

Silver Wattle → Acacia dealbata Link.

Silverleaf Nightshade → Solanum elaeagnifolium Cav.

Silverleaf Poplar → Populus alba L.

Silverweed → Potentilla Anserina L.

Silybum Marianum Gaertn. (syn. Carduus Marianum L.). Holy Thistle. (Compositae). - Annual to perennial herb. Mediterranean region. Cultivated, often as an ornamental. Very young leaves are sometimes used in salads. Roasted seeds are employed as substitute for coffee.

Simaba Cedron Planch. (Simarubaceae). - Shrub or small tree. Centr. America to Colombia. Cotyledons from seeds are used for fevers and snake-bites in Centr. America.

Simaba ferruginea St. Hil. (Simarubaceae). - Tree. Brazil. Source of Calunga Bark used in Brazil for fever.

Simaba paraensis Ducke. (Simarubaceae). - Tree. Brazil. Bark is used as a substitute for Quassia.

Simaba salubris Engl. (Simarubaceae). - Shrub. Brazil. Bark is used in Brazil as febrifuge.

Simaba glandulifera Gardn. (Simarubaceae). - Shrub. Brazil. Bark is considered in Brazil a tonic and anthelmintic.

Simaruba amara Aubl. (syn. S. officinalis DC). (Simarubaceae). - Trop. America. Source of Orinoco Simaruba Bark, or Jamaica Bark used as bitter tonic.

Simaruba versicolor St. Hil. (Simarubaceae). - Tree. Brazil. Bark used in Brazil for intestinal worms and for snake-bites. When powdered it is used as an insecticide.

Simarubaceae → Brucea, Desbordesia, Hannoa, Harrisonia, Irvingia, Klaineodoxa, Odyendyea, Picrasma, Quassia, Samadera, Simaba, Simaruba, Soulamea.

Simmondsia californica Nutt. (syn. S. pabulosa Kellog. S. chinensis (Link.) Schneid.). Pignut, Goatnut. (Buxaceae). - Shrub or small tree. So. California and adj. Mexico. Seeds when raw or parched are consumed by the Indians, also used as a substitute for coffee. In no. Mexico and so. California seeds are roasted and ground with addition of yolk of hard-boiled egg, boiled with water, sugar and milk from which a beverage is made, being a substitute of coffee and chocolate. Seeds are source of Jojoba Oil, a non-drying oil; used by some chemists in melting point apparatus.

Simson Oil → Trianthema salsoloides Fenzl.

Sinapis alba L. (syn. Brassica alba (L.) Boiss.). White Mustard. (Cruciferaceae). - Annual herb. Europe. Cultivated in temp. zones of the Old and New World. Seeds are source of White Mustard, also of White Mustard Seed Oil, used as lubricant and illuminant. Sp. Gr. 0.9125—0.9160; Sap. Val. 171—177; Iod. No. 94—106.

Sinapis cernua Thunb. → Brassica nigra (L.) Koch.

Sinapis integrifolia Willd. → Brassica juncea Coss.

Sinapis juncea L. → Brassica juncea (L.) Cosson.

Sindora inermis Merr. (Leguminosaceae). - Tree. Philippine Islands, esp. S. Luzon, Mindanao. Trunk source of Kayu-gala oil; recommended for perfumery. Used in Philippines as illuminant and for skin-diseases.

Sindora leiocarpa Backer. (Leguminosaceae). - Tall tree. Sumatra. Wood soft, easily worked, not resistant to moisture, readish-gray; used for small household goods, spoons, trays, bowls, etc.

Sindora Supa Merr. (Leguminosaceae). - Tree. Philippine Islands. Trunk source of Supa-oil, non-drying. Used in Philippines as an illuminant and for skin-diseases, varnishes, paint, transparent paper and as an adulterant of other oils. Wood hard, heavy, dark bronze with age; used in Philippines for beams, rafters, joists, ship-wharf and bridge building, interior finish of buildings, cabinet-work and furniture.

Singapore Manila → Agathis alba Foxw.

Singapore Rhododendron → Melastoma malabathricum L.

Singhara Nut → Trapa bispinosa Roxb.

Sinkwa Towel Gourd. → Luffa acutangula Roxb.

Sintok Bark → Cinnamomum Sintok Blume.

Siparuna cujabana DC. (Monimiaceae). - Tree. Brazil. Bark used in Brazil as sudorific, occasionally as an abortive. Contains safrol. Also S. oligandra DC.

Sisal Agave → Agave sisalana Perrine.

Sisal Hemp → Agave sisalana Perrine.

Sisal, Yucatan → Agave fourcroydes Lemaire.

Sison ammi L. → Carum copticum (L.) Benth. and Hook.

Sissoo of India → Dalbergia Sissoo Roxb.

Sisymbrium Alliaria Scop. (syn. Alliaria officinalis Andr.) Hedge Garlic. (Cruciferaceae). - Perennial herb. Europe, temp. Asia. Leaves have the taste of garlic; they are used when mixed with other herbs for seasoning. Contains allylsulfid and rhodanallyl.

Sisymbrium canescens Nutt. (Cruciferaceae). - Herbaceous plant. N. America. Seeds are made into refreshing drinks with lime juice, claret and sirup; used in different parts of Mexico.

Sisymbrium officinale (L.) Scop. Hedge Mustard. (Cruciferaceae). - Annual herb. Temperated zone. The herb, Herba Erysimi, was formerly used as a stimulant, irritant, diuretic, resolvent and antiscorbutic.

Sisyrinchium acre Mann. (Iridaceae). - Perennial herb. Hawaii Islands. Source of a dye, used in Hawaii for ornamenting the body and for painless tattooing.

Sisyrinchium galaxoides Gomes → Trimeza lurida Salisb.

Sitilias caroliniana (Walt.) Raf. → Pyrrhopappus carolinianus (Walt.) DC.

Sitka Cypress → Chamaecyparis Nootkatensis (Lamb.) Sudw.

Sitka Spruce → Picea sitchensis (Bong.) Carr.

Sitka Willow → Salix sitchensis Bong.

Sium Sisarum L., Skirret, Chervin. (Umbelliferaceae). - Perennial herb. E. Asia, Mediterranean region. Was formerly cultivated. Roots were consumed boiled, also used in salads. Root is sometimes used as a substitute for coffee.

Six-rowed Barley → Hordeum hexastichon L.

Skeleton Weed. → Lygodesmia juncea (Pursh) D. Don.

Skirret → Sium Sisarum L.

Skullcap → Scutellaria lateriflora L.

Skullcap, Baikal → Scutellaria baicalensis Georg.

Shunk Cabbage → Spathyema foetida (L.) Raf.

Skyline, Bluegrass → Poa epilis Scribn.

Slack → Porphyra umbilicalis (L.) J. Ag.

Slash Pine → Pinus Caribaea Morelet.

Slender Prairie Clover → Petalostemon oligophyllum (Torr.) Rydb.

Slender Wheatgrass → Agropyron pauciflorum (Schwein.) Hitchc.

Slim Amaranth → Amaranthus hybridus L.

Slippery Elm → Fremontia californica Torr. and Ulmus fulva Michx.

Sloe → Prunus spinosa L.

Sloe, Black → Prunus umbellata Ell.

Sloe Blackthorn → Prunus spinosa L.

Sloke → Porphyra umbilicalis (L.) J. Ag.

Small Cranberry → Vaccinium Oxycoccus L.

Small Flowered Clover → Trifolium parviflorum Ehrh.

Small Galangal → Alpinia officinarum Hance.

Small Leaved Linden → Tilia cordata Mill.

Small Nettle → Urtica urens L.

Small Salmon's Seal → Polygonatum biflorum (Walt.) Ell.

Small Salt Bush → Atriplex campanulata Benth.

Small Soapweed → Yucca glauca Nutt.

Smilacina racemosa Desf. → Vagnera racemosa (L.) Morong.

Smilax aristolochiaefolia Mill. (syn. S. Medica Schlecht.). Mexican Sarsaparilla. (Liliaceae). - Vine. S. Mexico. One of the plants which is the source of commercial Sarsaparilla. Decoction of rhizomes used by some Indian tribes in Mexico as febrifuge, for digestive disorders, gonorrhoea, kidney troubles. Used in Mexico for scrofulus, skin diseases and rheumatism. Sold in markets. Cultivated. The product is called Mexican, Vera Cruz, Tampico and Gray Sarsaparilla.

Smilax auriculata Walt. (Liliaceae) - Perennial vine. Eastern N. America to Florida and Mississippi. The pounded rhizomes were source of a starch, consumed as food by the Seminole Indians. Also S. lanceolata L.

Smilax Beyrichii Kunth. (Liliaceae). - Perennial vine. Florida to Mississippi. Tuberous rootstock was made into bread, also eaten in soup by Indians of the southern states. Also S. glauca Walt., S. pseudo-china L. and S. rotundifolia L.

Smilax Bona-nox L. Saw Greenbrier, Bristly Greenbrier. (Liliaceae). - Perennial vine. Eastern and Southeastern United States to Texas. A flour was made from the large tuberous roots, which was made into bread and gruel, used as a food by the Indians.

Smilax calophylla Wall. (Liliaceae). - Wiry erect shrub. Malay Penin, Sumatra. Rhizome used as aphrodisiac and tonic.

Smilax China L. China Root, Greenbrier. (Liliaceae). - Vine. China, Japan. Dried rhizomes, pai-fu-ling (Chin.) used in China medicinally since ancient times.

Smilax, Costa Rica → Smilax Regelii Killip and Morton.

Smilax herbacea L. var nipponicum Miq. (Liliaceae). - Herb. Japan. Young leaves when boiled are eaten in Japan as a vegetable.

Smilax laurifolia L. (Liliaceae). - Perennial vine. Southeastern United States, esp. Florida and Georgia. Young shoots were consumed like Asparagus. The pounded rootstocks were source of a starch; used as food by the Seminoles and other Indians.

Smilax megacarpa DC. (Liliaceae). - Prickly climber. India to Java. Rhizome eaten by Jakuns. Fruits consumed in Java when preserved.

Smilax myosotiflora DC. (Liliaceae). - Vine. Siam, Mal. Archipelago, Java. Rhizome used among Malayas and jungle tribes as aphrodisiac.

Smilax pseudo-china L., Long-Stalked Greenbrier. (Liliaceae). - Perennial vine. Eastern N. America to Texas, West Indies. Starchy tubers when ground are source of a reddish flour used as food by the Indians.

Smilax oblongifolia Pohl. (Liliaceae). - Vine. Brazil. Roots are used in Brazil as tonic and purgative.

Smilax ornata Lam. (Liliaceae). - Herbaceous perennial. Centr. America. Dried root is source of a Sarsaparilla. Much is exported. It is sometimes called Jamaica Sarsaparilla of commerce. It is considerded tonic and alterative.

Smilax Regelii Killip and Morton. (Liliaceae). - Perennial herb. Centr. America. Source of Honduras or Brown Sarsaparilla. From related undescribed species are derived Jamaica, Costa Rica, Central America and Red Sarsaparilla.

Smilax Spruceana A. DC. (syn. S. papyracea Spruce). Para Sarsaparilla. (Liliaceae). - Perennial plant. Brazil. Was formerly source of Para Sarsaparilla; at present of little commercial value. Also called Sarsaparilla de Lisboa and S. de Marahaõ.

Smoke Tree → Rhus cotinoides (Nutt.) Britt.

Smoke Tree, Common → Rhus Cotinus L.

Smoky Shield Lichen → Parmelia omphalodes (L.) Ach.

Smooth Alder → Alnus rugosa (Du Roi) Spreng.

Smooth barked African Mahogany → Khaya anthotheca A. DC.

Smooth Brome → Bromus inermis Leyss.

Smooth Gooseberry → Ribes oxycanthoides L.

Smooth Holly → Hedycarya angustifolia Cunningh.

Smooth Senna → Cassia laevigata Willd.

Smyrna Galls → Quercus lusitanica Lam.

Smyrnium Olusatrum L., Maceron. (Umbelliferaceae). - Herbaceous plant. Mediterranean region, Caucasia, Syria, Canary Islands. Was cultivated for several centuries, until recent times when this species became gradually replaced by Celery.

Snail Clover → Medicago scutellata (L.) Willd.

Snake Gourd → Trichosanthes Anguinea L.

Snake Gourd, Japanese → Trichosanthes cucumeroides Max.

Snakeroot → Polygonum Bistorta L., Polygala Senega L. and Strychnos colubrina L.

Snakeroot, Black → Cimicifuga racemosa (L.) Butt.

Snakeroot, Button → Eryngium yuccifolium Michx.

Snakeroot, Texas → Aristolochia reticulata Nutt.

Snakeroot, Virginia → Aristolochia Serpentaria L.

Snakewood → Strychnos colubrina L.

Sneezeweed → Helenium tenuifolium Nutt.

Sneezewort → Achillea Ptarmica L.

Snow Lichen → Cetraria nivalis (L.) Ach.

Snow Trillium → Trillium grandiflorum (Michx.) Salisb.

Snowbread → Cyclamen europaeum L.

Snow-in-the-Mountain → Euphorbia marginata Pursh.

Snuff, Coxoba → Piptadenia peregrina Benth.

Soap Plant, Californian → Chenopodium californicum Wats.

Soap Pod → Acacia concinna DC.

Soapbark Tree → Quillaja Saponaria Mol.

Soapberry, Chinese → Sapindus Mukorossi Gaertn.

Soapberry, Southern → Sapindus Saponaria L.

Soapbush → Porliera angustifolia (Engelm.) Gray.

Soapnut Tree → Sapindus Mukorossi Gaertn.

Soaproot → Saponaria officinalis L.

Soaproot, Californian → Chlorogalum pomeridianum (Ker-Gawl) Kunth.

Soap substitutes. Plants source of → Agave, Albizzia, Ampelozizyphus, Chenopodium, Chlorogalum, Condalia, Cucurbita, Cyrtocarpa, Enterolobium, Fouquieria, Gallesia, Gleditsia, Gouania, Harpulia, Isertia, Melandrium, Momordica, Phytolacca, Quillaja, Sapindus, Saponaria, Securidaca, Stegnosperma, Trichosanthes, Vitis, Yucca, Zizyphus.

Soaptree Yucca → Yucca elata Engelm.

Soapweed, Small → Yucca glauca Nutt.

Soapwort → Saponaria officinalis L.

Socotra Aloë → Aloë Perryi Baker.

Socotra Dragon Blood → Dracaena schizantha Baker.

Socotro Dragonsblood Dracaena → Dracaena Cinnabari Balf.

Socratea durissima (Oerst.) Wendl. (syn. Iriartea durissima Oerst.) (Palmaceae). - Palm. Nicaragua to Panama. Stems used for thatching. Young buds are consumed boiled as a vegetable by the Indians.

Soilbacteria, Useful → Azotobacter chroococcum Beijerinck, Bacterium mycoides (Grotenf.) Migula, Clostridium Pasteurianum Winogr., Micrococcus urea Cohn, Nitrobacter Winogradskyii Buchanan, Rhizobium leguminosarum Frank and Sarcina urea Beijerinck.

Soja hispida Moench. → Glycina soja Sieb. and Zucc.

Sola Pith Plant → Aeschynomene aspera L.

Solanaceae → Acnistus, Atropa, Brunfelsia, Capsicum, Cestrum, Chamaedaracha, Cyphomandra, Datura, Duboisia, Fabiana, Hyoscyamus, Lycium, Lycopersicum, Mandragora, Nicotiana, Physalis, Salpichroa, Scopolia, Solanum, Withania.

Solanum aethiopicum L. (Solanaceae). - Herbaceous plant. Trop. Africa, Asia. Cultivated in Senegal, Sudan, Centr. Africa. Ripe fruits are consumed by the natives.

Solanum agrarium Sendt. (Solanaceae). - Shrub. Brazil. Fruits are consumed in some parts of Brazil. Decoction of leaves is used for gonorrhea.

Solanum alternato-pinnatum Steud. (syn. S. oleraceum Vell.). (Solanaceae). - Perennial plant. Argentina, Paraguay, Brazil and adj. territory. Leaves are said to be used as narcotic and diuretic in some parts of S. America.

Solanum andigenum Juz. and Buk. (Solanaceae). - Perennial herb. Peru to Colombia. Tubers are consumed as food in some parts of Colombia and Peru.

Solanum anomalum Thonn. (Solanaceae). - Herbaceous plant. Trop. Africa. Fruits called Children's Tomatos are used as a condiment in sauces and soups by the natives. They are also dried and preserved. Plant is occasionally cultivated.

Solanum aviculare Forst. (syn. S. vescum F. v. Muell., S. laciniatum Ait.). **Kangeroo-Apple**, Austr. (Solanaceae). - Herbaceous plant. Australia. Fully ripe fruits consumed raw, boiled or baked in some parts of Australia.

Solanum carolinense L. Coralina Horse Nettle. (Solanaceae). - Perennial herb. N. America. Air-dried ripe fruits are used medicinally, being sedative, antispasmodic.

Solanum Coronopus Dunal → Chamaesaracha Coronopus (Dunal) Gray.

Solanum dubium Fresn. (Solanaceae). - Herbaceous plant. Red Sea area, Arabia, E. Africa. Plants pounded and put in water, are used in Sudan to soak hides for removal of the hair.

Solanum Dulcamara L., Bittersweet. (Solanaceae). - Climbing shrub. Europe, Asia, N. Africa, N. America. Dried stems used medicinally, being sedative, diaphoretic, diuretic, hypnotic. Contains dulcamarin, a bitter glucosidal substance.

Solanum duplosinuatum Klotzsch. (Solanaceae). - Herbaceous plant. Trop. Africa. Yellow bitter fruits are used in soup. Occasionally cultivated. Employed for Strophanthus poisoning.

Solanum diversifolium Schlecht. (syn. S. Fendleri Van Heurck and Muell.) (Solanaceae). - Prickly shrub. Centr. America, West Indies. Young fruits are eaten with salt as a relish with codfish. Important food among negroes in the West Indies.

Solanum elaeagnifolium Cav. Silverleaf Nightshade. (Solanaceae). - Southern a. southwestern United States, Trop. America. Seeds used by the Navajo Indians (U. S. A.) to curdle milk. Used by the Kiowa Indians when mixed with braintissue for tanning buck skins. The plant contains an enzyme, similar to papain.

Solanum ellipticum R. Br. (Solanaceae). - Perennial plant. Australia. Fruits are consumed by the aborigenes of S. Australia.

Solanum Fendleri Gray. Findler Potato, Wild Potato. (Solanaceae). - Perennial herb. Southwestern United States and adj. Mexico to Panama. Tubers were eaten raw or cooked by Indians of New Mexico.

Solanum ferox L. (Solanaceae). - Small prickly plant. S. E. Asia, Malaysia. Fruits widely used in India, Siam, Malaysia as sour-relish in curries.

Solanum inaequilaterale Merr. (Solanaceae). - Shrub. Philipp. Islands. Leaves are used for smoking by the Moro-Subanus (Philipp Islds.).

Solanum incanum L. (Solanaceae). - Woody herb. E. Africa, Asia. Seeds are used in the Sudan for curdling milk.

Solanum indicum L. (Solanaceae). - Small shrub. S. E. Asia. Fruit eedible; when half ripe used in curries.

Solanum Jamesii Torr. (Solanaceae). - Perennial herb. Southwest of United States and adj. Mexico. Tubers were eaten boiled or raw by Indians of New Mexico and Arizona.

Solanum Lycopersicum L. → Lycopersicum esculentum Mill.

Solanum macrocarpon L. (Solanaceae). - Herbaceous plant. Origin unknown. Cultivated in French Guinea, Ivory Coast, Lower Dahomey, Mascar. Islands. Fruits are consumed as a vegetable by the natives.

Solanum Melongana L., Eggplant, Aubergine. (Solanaceae). - Annual to perennial or subshrub. Trop. Asia. Cultivated in warm countries of the Old and New World. Fruits are used as vegetable. Varieties are black, purple and white among which: Black Beauty, Early Long Purple, Florida High Bush, New Orleans Market, New York Improved.

Solanum muricatum Ait. Pepino. (Solanaceae). - Herbaceous plant. Subtropics. Origin probably Peru. Cultivated in Centr. and S. America. Often sold in markets. Fruits frequently seedless, round to long, greenish yellow with purplish-red blotches, juicy, argeeable flavor, somewhat aromatic, crisp; eaten raw.

Solanum nigrum L., Black Nightshade. (Solanaceae). - Herbaceous. Cosmopolit. Fruits called Wonderberries, some strains are eaten, used in pies. Supposed to have poisonous properties. Young shoots and leaves are sometimes cooked as pot-herb. Emergency food plant.

Solanum olivare Paill. (Solanaceae). - Herbaceous plant. Origin unknown. Cultivated along Ivory Coast, Dahomey, Congo. Fruits consumed by the natives.

Solanum paniculatum L. (Solanaceae). - Shrub. Brazil. Roots are used in Brazil as tincture or in wine for ailments of the liver, spleen and catarrh of the bladder. Used in many patent medicines in Brazil.

Solanum Pierreanum Bois., Olombe. (Solanaceae). - Perennial herb. Trop. Africa, esp. Gabon. Fruits are consumed by the natives.

Solanum piliferum Benth. (Solanaceae). - Herbaceous plant. Mexico. Fruits size of an egg, yellowish-green, scent of an apple; much esteemed in some parts of Mexico. Sold in markets.

Solanum quitoense Lam., Naranjillo. (Solanaceae). - Sub-shrub. Ecuador, Peru. Commercially grown in Baños, Tungurahua (Ecuad.). Fruits shipped to Quito. Fruits round, roundovate, 5 cm. diam bright orange; skin thick, leathery; pulp very juicy, refreshing, subacid; suitable for drinks and sherbets.

Solanum torvum Swartz. (Solanaceae). - Shrubby plant. Tropics. Young shoots are eaten raw or cooked in W. Java.

Solanum triflorum Nutt. (Solanaceae). - Herbaceous plant. Brit. Columbia to Mexico. Fruits when ripe were eaten raw, boiled or ground; also mixed with chili and salt, or made into a bread by the Zuni Indians.

Solanum tuberosum L., Potato. (Solanaceae). - Perennial herb. Temp. Andean region, S. America. Cultivated in Old and New World. Imported commercial crop. Tubers consumed as food, prepared in various ways. Some varieties are used for manufact. starch and alcohol. Numerous varieties: I. Cobbler Group: Irish Cobbler, Early Victor, Flourball. II. Triumpf Group: Triumpf (Bliss). II. Early Michigan Group: Early Albino, Early Michigan. IV. Rose Group: Early Rose, Early Durham, Rochester Rose. V. Early Ohio Group: Early Market, Early Ohio. VI. Hebron Group: Late Beauty of Hebron, Country Gentleman, Star of the East. VII. Burbank Group: Burbank, White Beauty, California Russet. VIII. Green Mountain Group: Green Mountain, Late Blightless, Star of Maine. IX. Rural Group: Late Victor, Ohio Wonder, Rural New Yorker, Prosperity. X. Pearl Group: Blue Victor, Pearl. XI. Peachblow Group: Improved Peachblow Dykman. Varieties used for manuf. starch and alcohol mainly grown in Europe are: Gertrud, Gastold, Prof. Wohltmann, New Imperator. Varieties for starch and animal fodder: Lech, Ferdinand Heine, President Krueger. The cultivated potato represents a great variety of forms which by some investigators have been recognized as species with several centers of origin. The center of origin of S. tuberosum L. is in Chile. Other species belong to the mountainous regions of the Andes. A great variety is found on the Peru - Bolivia tableland; among which S. ajanhuiri Juz. and Buk., S. Juzepczukii Buk., S. phureja Juiz and Buk. and S. mamilliferum Juz. and Buk. An isolated position is occupied by the areas of S. goniocalyx Juz. and Buk. comprising Central Peru; S. Rybini Juz and Buk. and S. boyacence Juz. and Buk. in Colombia. The widest area for distribution, Central Peru is taken by S. andigenum Juz. and Buk.

Solanum Uporo Dunal. (Solanaceae). - Herbaceous plant. Polynesia. Cultivated in Fidji. Fruits red, resembling tomatoes, eaten by the natives.

Solanum xanthocarpum Schrad. and Wendl. (Solanaceae). - Herbaceous plant. Tropics. Seeds are used as expectorant in asthma and catarrh.

Solenostemma Argel Hayne. (Asclepiadaceae). - Woody plant. Nubia. Pulverized leaves are used by the natives to heal sores of camels.

Solidago californica Nutt. Californica Goldenrod. (Compositae). - Perennial herb. California. Lotion of boiled leaves and stems is used in California for healing sores and cuts of man and domestic animals.

Solidago canadensis L. Canada Goldenrod. (Compositae). - Perennial herb. Eastern N. America. Seeds were eaten by Indians of several tribes. Also S. nana Nutt., S. spectabilis Gray and others. Recommended as an emergency food plant. S. canadensis is source of Canadian Golden Rod Oil.

Solidago Leavenworthii Torr. and Gray. Leavenworth Goldenrod. (Compositae). - Perennial herb. Southeast of the United States, esp. Florida to S. Carolina. Is source of a rubber. Improved strains may be valuable as rubber producers during times of emergency. Also S. altissima L., S. fistulosa Mill. and S. rigida L.

Solidago odora Ait., Sweet Goldenrod. (Compositae). - Perennial herb. Eastern N. America to Florida, Texas and Oklahoma. Crushed leaves are anise-scented; they are used as a tea. Source of an odorous ess. oil.

Solidago virgaurea L., European Goldenrod. (Compositae). - Perennial herb. Europe, Asia, Mediterranean region, introd. in N. America. Root and herb, Radix et Herba Virgae aurea occasionally used in medicine, being carminative, diuretic, nervine, exitans, digestive. Leaves are sometimes used as a substitute for tea. As home remedy it is used as tonic, astringent and has a reputation for healing wounds.

Solieriaceae (Red Algae) → Eucheuma, Meristotheca.

Solomon's Plumes → Vagnera racemosa (L.) Morong.

Solomon Seal → Polygonatum officinale Moench.

Solorina crocea (L.) Ach., Saffron-yellow Solorina. (Peltigeraceae). - Lichen. Temp. zone. On soil and rocks. Source of a yellow dye; used in Scotland for coloring woolens. The dye is abundantly formed in the thallus of the lichen.

Somali Gum → Acacia glaucophylla Steud.

Sonneratia acida L. f. (Sonneratiaceae). - Tree. Trop. Asia. The vertical roots or pneumatophores are used as substitute for cork.

Sonneratia alba Smith. (Sonneratiaceae). - Tree. Coastal zone, Trop. Asia, etc. Wood tough, hard, fairly heavy, light brown to dark chocolate brown; used for construction of houses, bridges. Leaves are eaten by the Malayans when cooked; also eaten in the Moluccas with fish.

Sonneratiaceae → Duabanga, Sonneratia.

Soncoya → Annona purpurea Moc. and Sassé.

Sophia halictorum Cock. (Cruciferaceae). - Herbaceous plant. Western N. America. Tender plants were cooked and used as food by Pueblo Indians of New Mexico.

Sophia incisa (Engelm.) Greene. (Cruciferaceae). - Perennial herb. Western N. America. Parshed and ground seeds were consumed as food by the Indians of Montana and Oregon.

Sophia parviflora (Lam.) Standl. (Cruciferaceae). - Herbaceous plant. Western N. America. Seeds were made into a mush or used in bread by Indian tribes of the western states.

Sophia pinnata (Walt.) Howell. (Cruciferaceae). - Herbaceous plant. N. America. Leaves when boiled or roasted between hot stones were eaten by Indians of several tribes. Seeds were cooked in water to a mush, and consumed with salt by the Indians.

Sophora chyrosphylla Seem. (syn. Edwardsia chrysophylla Slasib.). (Leguminosaceae). - Small tree. Hawaii Islands. Wood used for runners of the Hawaiian sled.

Sophora japonica L. Japanese Pagodatree. (Leguminosaceae). - Tree. N. China, Japan. Extract from leaves and fruits used in China to adulterate opium. A dye is prepared from pods, to color clothes yellow; with indigo becoming green.

Sophora secundiflora (Orteg.) Lag. Mescalbean Sophora. Lag. (Leguminosaceae). - Shrub or small tree. Southwestern United States and adj. Mexico. Red beans called Mescal Beans, used in necklaces, also used by tne Plain Indians as an intoxicant in the Red Bean Dance.

Sophora sericea Nutt., Silky Sophora. (Leguminosaceae). - Perennial herb. N. America to Mexico. Sweet roots were consumed as a delicacy by the Pueblos of New Mexico.

Sophora tetraptera Ait. Fourwing Sophora. (Leguminosaceae). - Shrub or small tree. New Zealand, Chile, Juan Fernandez. Wood pale brown, compact, dense, heavy, of great strength, tough, elastic, extemely durable. Used for bearings of shafts and machines, cabinet-work, ornamental turnery.

Sorbus americana Marsch. (syn. Pyrus americana DC.) Mountain Ash. (Rosaceae). - Tree. Eastern N. America. The astringent fruit is used in homeopathic remedies.

Sorbus Aria Crantz. (syn. Pyrus Aria Ehrh., Mespilus Aria Scop.). White Beam Tree. (Rosaceae). - Tree. Europe. Fruits made into brandy and vinegar, also occasionally baked in bread. Dried fruits used in some countries for coughs and catarrh.

Sorbus Aucuparia L. (syn. Pyrus Aucuparia Gaertn., Mespilus Aucuparia All.) Rowan Tree, European Mountain-Ash. (Rosaceae). - Tree. Europe, W. Siberia, Asia Minor. Fruits consumed when preserved, also used in jellies and compotes; substitute for coffee. Made into a liqueur after the first frost have made the fruits sweet, used in Germany with brandy, called Sechsämtertropfen. Ingredient of certain Russian Vodkas. Leaves and flowers used as adulterant of tea. Wood fine-grained, hard, fairly heavy, difficult to split, tough, elastic; used for turnery, wagons; source of cellulose.

Sorbus domestica L. Service Tree. (Rosaceae). - Tree. S. Europe, N. Africa, W. Asia. Cultivated. Fruits are eaten after touch of frost. Also made into wine. Var. pomifera Hayne has apple-shaped; var. pyrifera Hayne has pear-shaped fruits. Wood fine-grained, very heavy, difficult to split; used for furniture, wine-presses, screws and for material that need much friction. Bark is used for tanning, producing a beautiful brownish leather.

Sorbus torminalis (L.) Crantz. (syn. Pyrus torminalis Ehrh., Aria torminalis Beck.) Checker Tree. Service-Tree. (Rosaceae). - Tree. Europe, Caucasia, Asia Minor, Syria, N. Africa. Fruits used for making wine, brandy, called Aliziergeist in the Elsass, also made into vinegar.

Sorghum caudatum Stapf. (Graminaceae). - Herbaceous grass. W. Africa. var. colorans Snowd. is cultivated as a source of a red dye, derived from stems and leaf-sheaths. The dye is in red, yellow, black and black-purple colors.

Sorghum halepense (L.) Pers. (syn. Holcus halepensis L.). Johnson Grass. (Graminaceae). - Perennial grass. Mediterranean region. Cultivated in warm parts of Old and New World as fodder for cattle. Plant soon takes full possession of the land and is difficult to eradicate.

Sorghum margaritiferum Stapf. (Graminaceae). - Annual grass. French Congo, Middle Niger Area. Crains are used as food by the natives.

Sorghum vulgare Pers. (syn. Andropogon Sorghum (L.) Brot.) Sorghum. (Graminaceae). - Annual Grass. Cultivated since times immemorial in Asia, Africa, introd. in America. Several varieties or races are grown for different purposes, e. g.: I. Var. caffrorum (Retz.) Hubb. and Rehder. Kaffir. Food plant. Standard Black Hull; here belongs probably Hegari. II. Var. caudatum (Hack) A. F. Hill. Federita from Sudan, a grain crop. III. Var. cernuum (Ard.) Fiori and Paoli. White Durra, Egyptian Corn, Jerusalem Corn, a grain sorghum. N. Africa, S. W. Asia, Brit. India. Used for human consumption, also used as poultry feed. IV. Var. nervosum (Hack) Forbes and Hemsl. Kaoliang, Chinese Sorghum, source of grain, sugar and forage. Mukden White, Hansen Brown, Meyer Brown. V. Var Roxburghii (Hack) Haines. Shallu, from India, a grain crop. VI. Var. saccharatum (L.) Boerl. Sorgo, Sweet Sorghum, used for manuf. of syrup. Sumac, Gooseneck, Orange. VII. Var. subglabrescens (Steud.) A. F. Hill. Milo African. Juicy stems, sometimes poisonous to cattle when in green state. Grain used as food for cattle and poultry. VIII. Var. sudanensis (Piper) Hitchc. Sudan grass, used as pasture grass, also as hay. IX. Var. technica (Koern.) Fiori. Broomcorn used for manuf. brooms. Spanish, Evergreen Dwarf.

Sorghum → Sorghum vulgare Pers.

Sorghum, Sweet → Sorgum vulgare Pers.

Sorindeia madagascariensis DC. Crape Mango. (Anacardiaceae). - Tree. Madagascar. Fruits are consumed in some parts of Madagascar.

Sorindeia Warneckei Engl. (Anacardiaceae). - Tree. Trop. W. Africa. Juice from the fruits is source of a blue dye, used for tattooing and as a cosmetic.

Soroco → Espostoa lanata (H. B. K.) Britt. and Rose.

Sorrel → Rumex Acetosa L.

Sorrel, Jamaica → Hibiscus Sabdariffa L.

Sorrel, Mountain → Oxyria digyna (L.) Hill. and Rumex paucifolium Nutt.

Sorrel Rhubarb → Rheum palmatum L.

Sorrel, Wood → Oxalis Acetosella L.

Sotol → Dasylirion simplex Trel.

Sotol, Texas → Dasylirion texanum Scheele.

Sotol, Wheeler → Dasylirion Wheeleri S. Wats.

Soulamea amara Lam. (Simarubaceae). - Shrub or tree. Mal. Archipelago. Roots and fruits are locally used by natives for colic and pleurisy.

Sour Cherry → Prunus Cerasus L.

Sour Grass → Amphilophis pertusa Stapf.

Sour Gum → Nyssa multiflora Wang.

Sour Orange → Citrus Aurantium L.

Sour Plum → Owenia acidula F. v. Muell.

Soursop → Annona muricata L.

Soursop, Mountain → Annona montana Macf.

South African Doum Palm → Hyphaene crinita Gaertn.

South Australian Sandalwood → Fusanus acuminatus R. Br.

South Australian Sandalwood Oil → Fusanus acuminatus R. Br.

South Cameroon Ebony → Diospyros bipidensis Gurke.

Southern Cane → Arundinaria gigantea (Walt.) Chapm.

Southern Cypress → Taxodium distichum (L.) Rich.

Southern Dewberry → Rubus trivialis Michx.

Southern Magnolia → Magnolia grandiflora L.

Southern Red Cedar → Juniperus barbadensis L.

Southern Pine → Pinus palustris Mill.

Southern Red Oak → Quercus texana Buckl.

Southern Soapberry → Sapindus Saponaria L.

Southern White Cedar → Chamaecyparis tnvoides (L.) B. S. P.

Southern Wood → Artemisia Arbotanum L.

Southwestern Condalia → Condalia lycioides Gray.

Soy Bean → Glycine soja Sieb. and Zucc.

Soymida febrifuga Juss., Indian Red-Wood, Bastard Cedar. (Meliaceae). - Large tree. Trop. Asia, esp. Centr. and S. India. Bark source of strong red fibre, made into ropes in Chota Nagpur. Bark is also used for tanning. Wood suitable for house-building, carving, furniture, pestles, pounders, grain-mills.

Spanish Bajonet → Yucca aloifolia L.

Spanish Broom → Spartium junceum L.

Spanish Chestnut → Castanea sativa Mill.

Spanish Dagger → Yucca aloifolia L.

Spanish Esparcet → Hedysarum coronarium L.

Spanish Licorice → Glycyrrhiza glabra L.

Spanish Lime → Melicocca bijuga L.

Spanish Moss → Tillandsia usneoides L.

Spanish Oak → Quercus digitata (Marsh) Sudw.

Spanish Origanum Oil → Thymus capitatus Hoffm. and Link.

Spanish Oyster Plant → Scolymus hispanicus L.

Spanish Rhubarb Dock → Rumex abyssinica Jacq.

Spanish Turpethroot → Thapsia garganica L.

Sparassis crispa (Wulf.) Fr. (Clavariaceae). - Basidiomycete Fungus. Temp. zone. The fruitbodies are consumed as food. Sold in markets. Also: S. laminosa Fr.

Sparganiaceae → Sparganium.

Sparganium eurycarpum Engelm. Giant Bur Reed. (Sparganiaceae). - Perennial herb. N. America. Tubers were used as food by several Indians tribes.

Sparganophorus Vaill antii Crantz. (Compositae). - Herbaceous plant. Trop. Africa, esp. Nigeria, Gold Coast, Cameroon, Congo, Togo, Niam-Niam, Fernando Po, also W. Indies. Leaves are used as a condiment in soup.

Sparmannia africana L. f. Stock Rose. (Tiliaceae). - Shrub or tree. S. Africa. Bark is source of a fibre.

Spartium junceum L. Spanish Broom, Weavers Broom. (Leguminosaceae). - Shrub. Europe, Mediterranean region. Stems used for basketry. Bark employed in former times as a source of fibre; used for manuf. of mats and ropes, also for filling of matrasses and pillows. Occasionally used for making paper. Flowers are source of an ess. oil, recommended for perfumery, though is little used; it has been suggested to fit into the orange flower compounds; it blends well with ylang ylang. About 1200 kg. flowers are source of 1 kg. concrete, giving 300 to 350 gr. of absolute.

Spartium tinctorium Roth. → Genista tinctoria L.

Spathiphyllum candicans Poepp. and Endl. (syn. Pothos cannaefolia Durand, S. cannaefolium (Dryand) Schott.) (Araceae). - Perennial plant. Trop. America. Leaves are used by some Indians of Peru for flavoring tobacco.

Spathyema foetida (L.) Raf. (syn. Symplocarpus foetidus Salisb.). Skunk-Cabbage, Swamp-Cabbage. (Araceae). - Eastern N. America to Florida. Rootstock is emetic, antispasmotic and diuretic. Rhizomes were consumed after being dried or baked by the Indians as emergency food.

Spawn → Psalliota campestris (L.) Fr.

Spearleaf Grass Tree → Xanthorrhoea hastilis R. Br.

Spearmint → Mentha spicata L.

Spearmint Oil → Mentha spicata L.

Spear Wood → Acacia homalophylla Cunningh.

Speedwell, Drug → Veronica officinalis L.

Spelt → Triticum Spelta L.

Spergularia arvensis L. (syn. Spergula arvensis L.). Spurry. (Caryophyllaceae). - Annual herb. Probably indigenous to Europe. Cultivated in Europe as a food for live-stock.

Spermolepis gummifera Brongn. and Gris. (syn. Arillastrum gummiferum Panch.). (Myrtaceae). - Tree. New Caldonia. Wood reddish, hard, easily worked; used for interior work, has been recommended for ship-building. Stem is source of a resin, called Chene Gom. Bark is rich in tannin.

Sphaeralcea angustifolia (Cav.) G. Don., Narrowleaf Globemallow. (Malvaceae). - Perennial herb. Western United States and adj. Mexico. Stems were used for chewing among the Hopi Indians.

Sphaeranthus suaveolens DC. (Compositae). - Herbaceous. Trop. of Old World. Used in Abyssinia for scenting hair.

Sphaerococcus catilagineus J. Ag. (Sphaerococcaceae). - Red Alga. Sea Weed. Pac. Ocean and adj. waters. Used in Chinese medicine.

Sphaerococcus gelantinosus J. Ag. (Sphaerococcaceae). - Red Alga. Sea Weed. Pac. Ocean and adj. waters. Source of an Agar Agar in Malaya. Also S. Serra Kiz.

Sphaerococcaceae (Red Algae) → Gelidiopsis, Hypnea, Sphaerococcus.

Sphagnum cymbifolium Ehrh. Sphagnum Moss (Sphagnaceae). - Mossplant. Temp. zone. Bogs. Used for surgical dressings, as packing material and for growing of epiphytic greenhouse plants, e. g. orchids, bromeliads, aroids, ferns etc. in pots and hanging baskets. As turf, it is an excellent fuel; also useful as cover in stables and to make heavy soils more porous and loose.

Sphenocentrum Jollyanum Pierre. (Menispermaceae). - Woody plant. Trop. W. Africa. Roots are used by the natives as chewing sticks. They are acid at first, causing things eaten afterward to taste sweet. Also used for constipation and as a stomachic.

Sphenostylis Schweinfurthii Harms. Yam Bean. (Leguminosaceae). - Woody plant. Trop. W. Africa. Cultivated. Seeds and tubers are used as food by the natives.

Sphenostylis stenocarpa Hams. (Leguminosaceae). - Herbaceous vine. Trop. Africa, esp. Abyssinia, Angola, Ivory Coast, Dahomey, Togo. Cultivated. Seeds are consumed as food by the natives.

Spice Bush → Lindera Benzoin (L.) Blume.

Spices → Acrodiclidium, Amomum, Asarum, Calycanthus, Calyptranthes, Canella, Capsicum, Cinnamodendron, Cinnamomum, Drimys, Elettaria, Embelia, Eugenia, Ferula, Gastrochilus, Geum, Kaempfera, Laurelia, Laurus, Lindera, Lucuma, Monodora, Myristica, Myrtus, Nectandra, Piper, Ravensara, Thoningia, Unona, Xylopia, Zanthoxylum. See also Condiments, Flavorings, Essential Oils.

Spickled Alder → Alnus incana (L.) Moench.

Spicy Cedar → Tylostemon Mannii Stapf.

Spider Antelope Horn → Asclepiodophra decumbens (Nutt.) Gray.

Spider Flower, Pink → Cleome integrifolia Torr. and Gray.

Spigelia Anthelmia L. West Indian Spigelia. (Loganiaceae). - Herbaceous. Trop. America, esp. West Indies, Trinidad, N. of S. America.

Used as vermifuge. Also employed in criminal poisoning.

Spigelia Flemmingiana Cham. and Schlecht. (Loganiaceae). - Herbaceous plant. Brazil. Plant is considered anthelmintic and sudorific.

Spigelia glabrata Mart. (Loganiaceae). - Herbaceous plant. Brazil. Herb is considered anthelmintic, excitant, febrifuge and sudorific.

Spigelia marilandica L. Pinkroot. (Loganiaceae). - Perennial herb. Eastern N. America to Florida and Texas. Dried rhizome and roots, collected in the autumn, are used medicinally as an anthelmintic. Contains spigeline, resembling coniine and nicotine. Is often adulterated.

Spiked Loosestrife → Lythrum Salicaria L.

Spikenard → Nardostachys Jatamansi DC.

Spikenard, American → Aralia racemosa L.

Spikenard, Wild → Vagnera racemosa (L.) Morong.

Spilanthes acmella Murry., Para Cress. (Compositae). - Herbaceous. Tropics of Old and New World. Much used in parts of Trop. Asia for toothache; sold in Chinese herbalist's shops. Leaves are consumed boiled or in salads.

Spilanthes Mutisii H. B. K. (Compositae). - Perennial plant. S. America. A bitter tincture from the flowers is used in Colombia as a liver corrective.

Spinacea oleracea L., Spinach. (Chenopodiaceae). - Annual to perennial herb. Origin uncertain. Cultivated as a vegetable in temp. zones of Old and New World. Eaten as a pot-herb, also in soup. Preserved in cans. Several varieties, among which I. Savoy Group: Bloomsdale Savoy, Norfolk Savoy Leaved, Long Standing Savoy. II. Thickleaf Group: Giant Thick Leaf, Viroflay, Gaundry, Eskimo Giant. III. Long Standing Group: Victoria, King of Denmark. IV. Long Season Group: Long Season, Triumph, Juliana. V. Prickly Seeded Group: Prickly Winter, Amsterdam Giant, Hollandia.

Spinach → Spinacea oleracea L.

Spinach, New Zealand → Tetragonia expansa Murr.

Spindle Tree → Evonymus europaeus L.

Spiny Bamboo → Bambusa arundinacea Willd.

Spiny Vitis → Vitis Davidi (Roman) Foex.

Spiny Yam → Dioscorea spinosa Roxb.

Spiraea Ulmaria L. → Filipendula Ulmaria (L.) Maxim.

Spiraeanthemum samoense Gray. (Cunoniaceae). - Tall shrub or smal tree. Polynesia, Samoa. Wood used by natives for canoes and war-clubs.

Spirit of Turpentine → Pinus palustris Mill.

Spondianthus Preussii Engl. (syn. Megabarea Trillesii Pierre.). (Euphorbiaceae). - Perennial plant. Trop. Africa. Bark is used by the natives of Cameroon for poisoning rats.

Spondianthus ugandensis Hutch. (Euphorbiaceae). - Tree. Trop. W. Africa. Bark is boiled in rice, used in Liberia as a rat-poison.

Spondias dulcis Forst. f. (syn. S. cytherea Sonner.) Otaheite Apple, Ambarella. (Anacardiaceae). - Tree. Tropics. Occasionally cultivated in tropics of Old and New World. Fruits subacid of good quality, excellent for jellies, preserves, marmelade, in syrup with agar-agar; also used in pickles. Fruits are also called Jew Plum.

Spondias laosensis Pierre. (Anacardiaceae). - Tree. Trop. Asia. Fruits edible. Cultivated esp. in Tonkin.

Spondias lutea L., Yellow Mombin, Jobo, Hog-Plum. (Anacardiaceae). - Tree. Tropics. Cultivated. Fruits of good quality, usually eaten fresh, though less esteemed than other species.

Spondias mangifera Willd. (Anacardiaceae). - Tree. Trop. Asia. Fruits edible, astringent, acid; occasionally culticated in Indo-China. Flower-clusters are eaten as a vegetable or in salads.

Spondias purpurea L. (syn. S. Mombin L.). Red Mombin, Spanish Plum. (Anacardiaceae). - Small tree. Trop. America, Mexico, Centr. America. Fruits, called Ciruela (Span.), Jocote (Mex.) are eaten fresh, boiled, or dried. Flavor subacid and spicy. Sold in markets of Mexico and Guatemala.

Spondias tuberosa Arruda. Imbu, Hog Plum., Imbu. (Anacardiaceae). - Tree. N. E. Brazil. Seldom cultivated. Fruits of fine flavor, eaten fresh, also made into jelly. Imbuzada is a popular dessert in N. Brazil which is made from the fruit, boiled in sweet milk. Also source of a beverage.

Spondias Wirtgenii Hassk. (Anacardiaceae). - Tree. Java. Source of a gum, resembling Gum Arabic.

Sponge, Vegetable → Luffa aegyptiaca Mill.

Sporobolus cryptandrus (Torr.) Gray., Sand Dropseed. (Graminaceae). - Perennial grass. N. America. Seed was consumed by the Indians.

Sporolobus flexuosus (Thurb.) Rydb., Mesa Dropseed. (Graminaceae). - Perennial grass. Western N. America to Northern Mexico. Seeds were used as food by the Indians.

Sporobolus indicus R. Br. (syn. S. elongatus R. Br.). Sweet Grass. (Graminaceae). - Perennial grass. Tropics. Used for manuf. hats, characterized by its glossiness and golden-yellow color.

Sporobolus pallidus Lindl. (syn. S. Lindleyii Benth.). (Graminaceae). - Perennial grass. Australia. Seeds when ground are made into a paste and baked which is consumed by the aborigenes of Queensland.

Spoonwort → Cochlearia officinalis L.

Spotted Bur Clover → Medicago arabica (L.) All.

Spotted Gum → Eucalyptus goniocalyx F. v. Muell.

Spotted Wintergreen → Chimaphila umbellata Nutt.

Sprangletop, Green → Leptochloa dubia (H. B. K.) Nees.

Spindlewood Oxalis → Oxalis carnosa Molina.

Spreading Dogbane → Apoycynum androsaemifolium L.

Spreading Pasqueflower → Anemone patens L.

Spring Adonis → Adonis vernalis L.

Spring Beauty → Claytonia virginiana L.

Spring Grass → Anthoxanthum odoratum L.

Spring Savory → Satureja Acinos (L.) Scheele.

Sprinkled Parmelia → Parmelia conspersa (Ehrh.) Ach.

Sprout Lichen → Dermatocarpon miniatus (L.) Mann.

Spruce Beer → Picea Mariana (Mill.) B. S. P.

Spruce, Black → Picea Mariana (Mills.) B. S. P.

Spruce, Douglas → Pseudotsuga taxifolis (Lam.) Britt.

Spruce, Dragon → Picea asperata Mast.

Spruce, Engelmann → Picea Engelmanni (Parry) Engelm.

Spruce Gum → Picea Mariana (Mill.) B. S. P.

Spruce, Norway → Picea excelsa Link.

Spruce, Purple Cone → Picea purpurea Mast.

Spruce, Red → Picea rubra Link.

Spruce, Sakhalin → Picea Glehnii Mast.

Spruce, Sitka → Picea sitchensis (Bong.) Carr.

Spruce, White → Picea alba Link.

Spruce, Yeddo → Picea jezoensis Carr.

Spurge, Flowering → Euphorbia corollata L.

Spurge, Ipecac → Euphorbia Ipecacuanha L.

Squarestem Rosegentian → Sabatia angularis (L.) Pursh.

Squash → Cucurbita maxima Duch.

Squaw Berry → Rhus trilobata Nutt.

Squaw Grass → Elymus triticoides Buckl.

Squaw Root → Caulophyllum thalictroides (L.) Michx.

Squaw Weed → Senecio aureus L.

Squirrel Corn → Dicentra canadensis (Gold.) Walp.

Squirting Cucumber → Ecballium Elaterium A. Rich.

St. Ignatius Poison Nut → Strychnos Ignatii Berg.

St. John's Bread → Ceratonia Siliqua L.

St. Lucie Cherry → Prunus Mahaleb L.

St. Martha Wood → Caesalpinia echinata Lam.

St. Thomas Bean → Entada scandens Benth.

St. Vincent Arrowroot → Maranta arundinacea L.

Stachys californica Benth. (Labiaceae). - Perennial herb. California and adj. territory. Infusion of stems and leaves is used as a home remedy to wash sores and wounds. Soaked leaves are also used as poultice.

Stachys officinalis (L.) Trev., Common Betony. (Labiaceae). - Perennial herb. Europe to Caucasia. Used in various home remedies as tea, syrup, with honey or in wine for coughs, ailment of the stomach, kidneys, bladder and spleen.

Stachys palustris L., Marsh Woodwort. (Labiaceae). - Perennial herb. Europe, N. America. Tubers are sometimes consumed as food in some countries of Europe.

Stachys recta L. (syn. S. Betonica Crantz., S. bufonia Thuill.) (Labiaceae). - Perennial herb. Europe to Asia Minor. Decoction of herb is used among Slavic peoples to wash children to protect them from diseases.

Stachys Sieboldi Miq. (syn. S. affinus Bunge, S. tuberifera Naud.). Chinese, Japanese Artichoke. (Labiaceae). - Perennial herb. China, Japan. Cultivated in China, Belgium and France. Crisps, white tubers are boiled and consumed as a vegetable, sometimes they are eaten raw. In Japan they are eaten salted or when kept in plum vinegar. Contains the easily digestible stachyose and manneotetrose.

Stachytarpheta angustifolia Vahl. (Verbenaceae). - Herbaceous plant. Guiana. Roots are used as collyrium by the Arabs.

Stachytarpheta jamaicensis Vahl., Brazil Tea, Bastard Vervain. (Verbenaceae). - Herbaceous plant. Tropics of Old and New World. Sometimes used for dysentery.

Staghorn Sumach → Rhus typhina L.

Stanleya pinnatifida Nutt. (Cruciferaceae). - Herbaceous plant. Western United States. Stems and leaves were boiled in water and consumed by Indians of several tribes. Also S. albescens Jones and S. elata Jones.

Stanleyella Wrightii (Gray) Rydb. (Cruciferaceae). - Herbaceous plant. Southwestern United States and adj. Mexico. Source of a paint said to be used by the Tewa Indians for coloring pottery.

Star Anise → Illicium verum Hook. f.

Star Apple → Chrysophyllum Cainito L.

Star Apple, African → Chrysophyllum africanum A. DC.

Star Grass → Aletris farinosa L.

Star Tree, Ashleaf → Astronium fraxinifolium Schott.

Star Tree, Fetia → Astronium graveolens Jacq.

Starch. Plants source of. → Alsophila, Alstroemeria, Amorphophallus, Avena, Canna, Castanea, Castanospermum, Cicer, Colocasia, Corypha, Crinum, Curcuma, Dioscorea, Eleusine, Elymus, Encephalartos, Ephedra, Ervum, Erythronium, Euryale, Fagopyrum, Gleichenia, Hordeum, Lilium, Macrozamia, Manihot, Maranta, Marattia, Musa, Nelumbium, Oryza, Oryzopsis, Pachyrrhizus, Palaquium, Panicum, Phragmitis, Polygonatum, Psoralia, Salvia, Scirpus, Secale, Setaria, Smilax, Solanum, Sorghum, Tacca, Trapa, Treculia, Trichosanthes, Triticum, Zamia, Zea.

Starch, travancore → Curcuma angustifolia L.

Statice chilensis Phil. (Plumbaginaceae). - Perennial herb. Chile. Plant is used in Chile for treating ulcers, scrofula and dysentery.

Statice Limonium L. (Plumbaginaceae). - Perennial herb. Coastal region. Europe, Asia, N. America. Roots are used for tanning in some parts of Russia. Also S. latifolia Smith.

Statice tataricum L. (syn. Goniolimon tataricum (L.) Boiss.). (Plumbaginaceae). - Perennial plant. S. Europe, Caucasia, Siberia. Roots are used in Siberia for tanning.

Statice Thouini Viv. (Plumbaginaceae). - Perennial herb. Mediterranean region. Leaves are used by the natives in salads. Also S. ornata Ball.

Staudtia kamerunensis Warb. (Myristicaceae). - Tree. Cameroon. Source of a Bosé Wood, being dark red with black areas; used for different purposes.

Staudtia stipitata Warb. (Myristicaceae). - Tree. Trop. W. Africa. Wood hard, resistant to termites; orange or yellow-orange to red-brown; used for turnery, cabinet work, house building, walking-sticks, lead-pencils. Seeds are used for skin-diseases; also employed as bait for forest animals.

Stauntonia hexaphylla Decne. Japanese Staunton Vine. (Lardizabaliaceae). - Woody vine. Japan. Cultivated. Fruits white, pulpy, sweet, honeylike flavor; much esteemed in Japan.

Staurogyne elongata Kunze. (Acanthaceae). - Herbaceous plant. Mal. Archipelago. Young leaves are used with rice by the Javanese.

Stavisacre → Delphinium Staphisagria L.

Stegnosperma halimifolium Benth. (Phytolaccaceae). - Shrub. Mexico, West Indies, Central America. In Baja Calif. powdered roots are used as soap substitute.

Stelechocarpus Burakol Hook. f. (syn. Uvaria Burahol Blume). (Annonaceae). - Tree. Malaya. Fruits are edible, much esteemed by the natives.

Stellaria media (L.) Cyr., Common Chickweed. (Caryophyllaceae). - Annual herb. Sometimes uesd as a potherb. Emergency foodplant.

Stemmadenia Galeottiana (A. Rich.) Miers. (Apocynaceae). - Shrub. Cuba, S. E. Mexico, Guatemala. Latex is used by the Chinantex (Mex.) as a masticatory.

Stemona Burkelii Prain. (Roxburghiaceae). - Perenial. Malaya, Burma. Used by Burmese as insecticide. Also S. collinsae Craib.

Stemona tuberosa Lour. (Roxburghiaceae). - Woody plant. Trop. Asia. Roots have been recommended as an insecticide. Contain stemorin. Also S. sessilifolia Franch.

Stenocarpus daroides Brongn. and Gris. (Proteaceae). - Small tree. New Caledonia. Source of a fine wood; suitable for cabinet-making.

Stenocarpus lourinus Brongn. and Gris. (syn. S. laurifolius Panch. and Sebert.). (Proteaceae). - Tree. New Caledonia. Wood strong, dark colored, fine-grained, very difficult to work and to turn. Used for cabinet work.

Stenocarpus salignus R. Br. (syn. Hakea rubricaule Giord.). Silky Oak, Beef-Wood. (Proteaceae). - Tree. New South Wales. Wood hard, easy to split, close-grained, reddish. Used for furniture, walking-sticks, veneers, picture frames.

Stenochlaena palustris (L.) Mett. (Polypodiaceae). - Climbing fern. Mal. Archipelago. Young leaves are much esteemed as lalab with rice and in soup among the Javanese. Leaves are also used for braiding, anchor ropes and fish traps.

Stenocline incana Baker. (Compositae). - Shrub. Madagascar. Used in Madagascar as stomachic, tonic, astringent and aphrodisiac.

Sterculia carthaginensis Cav. (syn. S. apetala (Jacq.) Karst.). Panama Tree. (Sterculiaceae). - S. Mexico to S. America and West Indies. Wood is used for construction work.

Sterculia Chicha St. Hil. (Sterculiaceae). - Tree. Brazil, Guiana. Seeds called Castanha de Maranhão, Maranhão Nuts are edible; they are also source of an oil; used in paints and for lubricating watches.

Sterculia cordifolia Cav. (Sterculiaceae). - Tree. Trop. Africa. Pulp of fruits is edible; eaten by different tribes in Africa. Bark is source of a fibre; used as cordage.

Sterculia foetida L. (Sterculiaceae). - Tree. Tropics, esp. Indo-China, Malaya, Sumatra, Borneo, Africa, Australia. Seeds source of Kaloempang-Oil; non-drying. Sp. Gr. 0.922; Sap. Val. 188; Iod. No. 76.6. Seeds are eaten raw by the natives.

Sterculia lanceaefolia Roxb. (Sterculiaceae). - Tree. India, S. W. China. Seeds edible, also used in China for seasoning food.

Sterculia luzonica Warb. (Sterculiaceae). - Tree. Philippine Islands. Inner bark source of a fibre; used in the Philippines for the manuf. of rope.

Sterculia lychnophora Hance. (Sterculiaceae). - Tree. Trop. Asia, esp. Annam, Cochin-China. Seeds are source of a refreshing beverage, called Sam-rong in Cambodia.

Sterculia oblonga Mart. (syn. Eriobroma Klaineana Pierre). (Sterculiaceae). - Tree. Trop. Africa. Wood yellowish-gray, very dense, heavy, hard; used for boat building, railroad ties, turnery, tables. Wood is called Ekonge, Yellow Wood.

Sterculia plantanifolia L. f. → Firmiania simplex (L.) Wight.

Sterculia scaphigera Wall. (Sterculiaceae). - Tree. Trop. Asia. Seeds when macerated in water swell up to a gelatinous mass; used by the Siamese and Chinese as a delicacy after addition of some sugar. Also used as febrifuge.

Sterculia tomentosa Guill and Perr. (Sterculiaceae). - Tree. Trop. Africa. Source of a gum, called Icacia Chiche, Gomme de Sénégal used for dyeing fabrics, esp. in Europe. Also the following species are sources of gum S. hypochra Pierre from Cochin China; S. thorelii Pierre from Cochin China and S. rupestris Benth. from Australia.

Sterculia Tragacantha Lindl. African Tragacanth. (Sterculiaceae). - Tree. Trop. Africa, esp. Nigeria, Guinea. Source of a gum, resembles Tragacanth, Astragalus gummifer. Adulterant of Gum Arabic, Acacia Senegal, A. arabica. Wood used for posts, boards, construction work.

Sterculia urens Roxb. Kateeragum Sterculia. (Sterculiaceae). - Tree. Trop. Asia. Stem is source of an insoluble gum; used as substitute for tragacanth, called Kateera Gum, Karaya Gum.

Sterculiaceae → Abroma, Buettneria, Cola, Commersonia, Dombeya, Erythropsis, Firmiana, Fremontia, Cuazuma, Helicteres, Heritiera, Kleinhovia, Leptonychia, Melochia, Pterospermum, Rulingia, Sterculia, Tarrietia, Theobroma.

Stereocaulon paschale (L.) Hoffm. (syn. S. tomentosum Fr.). Eastern Lichen. (Cladoniaceae). - Lichen. Temp. zone. On sandy soils, on rocks. Source of an ash-green dye; used in some parts of Europe for dyeing woolens. Plant produces dextro-monnose and dextro-galactose.

Stereospermum chelonoides DC. (Bignoniaceae). - Tree. Tropics, Himalaya, S. China to Ceylon, Java. Wood hard, elastic, moderately durable, easy to work, grey; used for furniture, building purposes; in Assam used for tea-boxes, canoes.

Stevia Rebaudiana Bertoni → Eupatorium Rebaudianum Bertoni.

Sticta aurata Ach., Rose and Gold Lichen, Lungwort. (Stictaceae). - Lichen. Temp. and subarctic zone. Used in Great Britain and Scandinavia as a source of a dye for coloring woolens.

Sticta crocata (L.) Ach., Gold Edge Lichen. (Stictaceae). - Lichen. Temp. zone. Source of a brown dye, used for coloring woolens.

Sticta glomulerifera Del. (Stictaceae). - Lichen. Temp. zone. Plants were cooked and consumed as food by the Menominee and Ojibway Indians.

Sticta pulmonacea Ach., Oak Lung, Lungwort. (Stictaceae). - Lichen. Temp. zone. Used by the Herefordshire peasantry as a source of a pigment for dyeing stockings brown.

Stigeoclonium amoenum Kg. (Chaetophoraceae). - Green Alga. Hawaii Islds. Consumed by the Hawaiians with fresh shrimps and salt.

Stinging Nettle → Urtica dioica L.

Stinking Cedar → Torreya taxifolia Arn.

Stinkwood → Celtis Kraussiana Bernh.

Stinkwood, Black → Ocotea bullata E. Mey.

Stinkwood, Red → Pygeum africanum Hook. f.

Stipa comata Trin. and Rupr., Needle-and-Thread. (Graminaceae). - Perennial grass. Plains and prairies. N. America. An important forage grass for cattle in W. of United States.

Stipa hyalina Nees. (Graminaceae). - Perennial grass. S. America. Plant is used for forage in some parts of Brazil and Argentina.

Stipa Jarava Beauv. (syn. S. Ichu Kunth.) (Graminaceae). - Perennial grass. S. America to Mexico, esp. regions of the Andes. Considered a good fodder for live stock in dry areas.

Stipa tanacissima L. (syn. Macrochloa tanacissima (L.) Kunth.) Halfa, Esparta. (Graminaceae). - Herbaceous plant. Mediterranean region. Fibre is an important source in the paper industry along the Mediterranean region. Also used for ropes, sails, mats and similar comodities. Much is exported.

Stizolobium atterrinum Piper and Tracy → Mucuna atterrina (Piper and Tracy) Holland.

Stizolobium Deeringianum Bort. → Mucuna Deeringiana (Bort.) Small.

Stizolobium giganteum Spreng. → Mucuna gigantea DC.

Stizolobium niveum (DC.) Kuntze → Mucuna nivea DC.

Stock Rose → Sparmannia africana L. f.

Stoechas officinarum Moench. → Lavandula stoechas L.

Stoechas Oil → Lavandula stoechas L.

Stoneroot → Collinsonia canadensis L.

Stonecrop, Goldmoss → Sedum acre L.

Stonecrop, Mossy → Sedum acre L.

Stonecrop, White → Sedum album L.

Stone Crottle → Parmelia caperata (L.) Ach.

Stone Leek → Allium fistulosum L.

Stone Mint → Cunila origanoides (L.) Butt.

Storax, Levant → Liquidambar orientalis Mill.

Storax, Oriental → Liquidambar orientalis Mill.

Stouton (fermented drink). → Aubrya gabonensis Baill.

Stramonium → Datura Stramonium L.

Strassburg Turpentine → Abies pectinata DC.

Stratiotes aloides L., Crab's Claw. (Hydrocharitaceae). - Floating water plant. Centr. Europe, Caucasia, Siberia. Where plants are abundant, they are used as a manure.

Strawberry → Fragaria vesca L.

Strawberry Bush → Evonymus americanus L.

Strawberry Cactus → Echinacereus enneacanthus Engelm.

Strawberry, Chiloë → Fragaria chiloensis (L.) Duch.

Strawberry Guava → Psidium Cattleianum Sab.

Strawberry, Mexican → Echinocactus stramineus (Engelm.) Ruempl.

Strawberry Tomato → Physalis Alkekengi L.

Strawberry Tree → Arbutus Unedo L.

Strawberry, Virginian → Fragaria virginiana Duch.

Strebius asper Lour. (Urticaceae). - Shrub or small tree. Trop. Asia, esp. India, S. China to Philipp. Islds. Used for paper making in Siam. Decoction of bark used in India for dysentery, diarrhea, fevers. Old, rough course leaves used for polishing ivory. „It is one of the combustible substances that may be found as a packing in the thick cheroots smoked by the Burmese"

Streptomyces antibiotica (Waksman and Woodruff) Berg. et al. (Streptomycetaceae). - Fungus. Actinomycete used as an antibiotic. (→ Penicillium notatum Westling). Effective against Gram-negative bacteria. Also S. albus and S. griseus forming actinomycin, and S. aureofaciens forming aureomycin.

Strigelia acuminata Miers. → Styrax acuminatum Pohl.

Strigelia camporum Miers. → Stryrax camporum Pohl.

Stringy Bark → Eucalyptus obliqua L.'Hér.

Strobilanthes flaccidifolius Nees. (Acanthaceae). - Herbaceous plant. India, China, Siam. Cultivated, esp. in S. W. China to N. of Mal. Penin. Source of blue dye.

Strombocarpa odorata (Torr. and Frem.). Torr. (syn. Prosopis odorata Torr and Frem. S. pubescens (Benth.) Gray.). Screw Bean, Tornillo. (Leguminosaceae). - Shrub. Southeastern United States. Pods containing much sugar, were eaten by several Indian tribes.

Strombosia javanica Blume. (Olacaceae). - Tree. Mal. Archipelago. Wood strong, hard, close-grained, light, brown-yellow; used for fine carving, beams, boards and construction of houses.

Strophanthus Eminii Asch. and Pax. (Apocynaceae). - Woody plant. Trop. Africa. Decoction from plant is used as an arrow-poison.

Strophanthus gratus Franch. (Apocynaceae). - Vine. Trop. Africa, esp. Cameroons, Sierra Leone and Lagos. Juice is used by the natives as an arrow poison. Seeds have been recomended in medicine. Source of crystaline strophanthin.

Strophanthus hispidus DC. (Apocynaceae). - Perennial woody climber. S. Africa. Source of Brown Strophanthus. Used by natives as arrow poison. Used medicinally, → S. Kombé Oliv.

Strophanthus Kombé Oliv. (Apocynaceae). - Perennial woody climber. Trop. Africa. Source of Green Strophanthus; used by the natives of the vicinity of Lakes Tanganyika and Nyasa for arrow-poison. Dried, ripe seed used medicinally, being cardiac stimulant, diuretic. Contains strophanthin, a glucoside and kombic acid.

Strophanthus Preussii Engl. and Pax. (Apocynaceae). - Vine. Trop. Africa, esp. Nigeria, Old Calabar, Lagos, Fernando Po, Gold Coast. Juice is used along the Gold Coast for coagulating latex of Futumia elastica.

Strophanthus sarmentosus DC. (Apocynaceae). - Vine. Trop. Africa, esp. Nigeria, Senegambia to Lower Congo region. Used in Lagos as arrow-poison when mixed with „Isa". Source of commercial cortisone; used for treating rheumatoid arthritis and acute rheumatic fever.

Struthanthus syringifolius Mart. (Loranthaceae). - Parasitic shrub. Trop. America. Plant has been suggested as a source of rubber.

Strychni Semen → Strychnos nux-vomica L.

Strychnine Tree → Strychnos nux-vomica L.

Strychnos Atherstonei Harv. Cape Teak. (Loganiaceae). - Tree. S. Africa, esp. Cape Peninsula. Wood reddish brown, hard, heavy, tough; used for rural utensils, staves for coopers' work.

Strychnos Castelnaei Wedd., Amazon Poison Nut, Urari. (Loganiaceae). - Tree. S. America, esp. Orinoco region. Extract from bark is used as arrow poison; used by the Orinoco Indians. Source of a Curare. Causes paralysis of end of motor nerves of voluntary muscles. Anticonvulsive. Smallest amount in wounds is fatal.

Strychnos colubrina L. (Loganiaceae). - Liane. Western India, Ceylon. Supplies Snake root or Snake-wood; sold in bazaars of India. Wood used by natives for dyspepsia and malaria. Bark contains brucine and some strychnine.

Strychnos guineensis (Aubl.) Baill. (Loganiaceae). - Liane. Guiana. A source of curare, used in arrow poisons; used by Oajanas, Trios, Salvemas and Wamas Indians. Also S. toxifera and S. Mischerlichii.

Strychnos Henningsii Gilg. Hard Pear. (Loganiaceae). - Tree. Natal, Zululand. Wood durable, easily split; used for ax and pick handles, spokes.

Strychnos Ignatii Berg. (syn. Ignatia amara L.). St. Ignatius Poison Nut. (Loganiaceae). - Woody plant. Malaysia. Source of Ignatius Beans, Semen Ignatii, Faba St. Ignatii. Contain strychnin and brucin.

Strychnos Melinoniana Baill. (Loganiaceae). - Liane. Guiana. Used as aphrodisiac, by Indians and negroes in Dutch Guiana.

Strychnos nux-vomica L., Strychnine Tree. (Loganiaceae). - Small tree. East Indies, Ceylon, Coast of Malabar. Cultivated. Seeds known as Nux Vomica are used medicinally; source of strychnine, brucine; extremely poisonous. Sti-

mulant of nervous system, tonic, spinant, motor excitant. Much of the supply comes from the Madras Presidency. Was formerly used for poisoning animals. Was introduced into medicine in 1640.

Strychnos odorata Chev. (Loganiaceae). - Liane. Trop. W. Africa. Scented leaves are used by the native women to rub on the skin.

Strychnos Pseudoquina St. Hil. (Loganiaceae). - Shrub. Brazil. Bark, called Quina do Campo, is used in Brazil for intestinal worms, e. g. Ascaris, Oxyuris, Taenia, Dochmius.

Strychnos pubescens C. B. Clarke. (Loganiaceae). - Tall liane. Malay Penin. Used by wild tribes in dart poison.

Strychnos quadrangularis A. W. Hill. (Loganiaceae). - Liane. Malay Penin. Bark used by jungle tribes in dart poison.

Strychnos spinosa Lam. (syn. Brehmia spinosa Harv.). (Loganiaceae). - Small tree. Madagascar. Fruits edible, orange, size of an apple; pulp having an agreeable flavor.

Strychnos Tieute Lesch. (Loganiaceae). - Woody vine. Java. Root very bitter, used by the natives for the preparation of a strong poison, boiled with ginger, garlic etc.

Strychnos toxifera Benth. Urari (Braz.). Curare Poison Nut. (Loganiaceae). - Tree. Costa Rica to Brazil. A source of Curare, derived from the bark and roots; used by S. American Indians as arrow poison; esp. from blow-guns. Smallest amount entering the blood circulation, causes paralysis of motor nerves instantly. One of deadliest poisons.

Strychnos Unguacha A. Rich. (Loganiaceae). - Small tree. Trop. Africa, esp. Abyssinia to Schire Highlands, Nyasaland to origin of White Nile. The slimy, sweet fruits are consumed by the natives.

Stryphnodendron Barbatimam Mart. (syn. Mimosa, Barbatimam Vell.) Barbatimao. (Leguminosaceae). - Shrub. Brazil. Bark is antidiarrhetic, hemostatic; used for uterus hemorrhages. It is also an important source of commercial tanning material in local use, called Casea Virgindade. Leaves are said to be a tonic.

Stryphnodendron guianense Benth. (syn. Mimosa guianensis Aubl.). (Leguminosaceae). - Tree. Guiana. Wood very strong; used for furniture. Bark is source of tannin.

Styllingia sylvatica L., Queen's Root. (Euphorbiaceae). - Perennial herb. Coastal Plain to Florida and Texas and adj. territory. Dried root, collected in August; used medicinally; alterative antisyphilitic. Contains an ess. oil, an acrid resin and sylvacrol. Herb is said to be emetic and cathartic.

Styloceras laurifolium H. B. K. (Buxaceae). - Tree. Andes of Colombia, Ecuador, Peru, Bolivia. Wood yellowish-white, fine-grained, finishes easily; much esteemed locally for joinery.

Stylosanthes mucronata Willd., Wild Lucerne. (Leguminosaceae). - Annual herb. Trop. America and Africa. Plant has been recommended as a source of fodder for domestic animals.

Styphelia triflora Andr. (syn. S. glaucescens Sieb.) (Epacridaceae). - Shrub. New South Wales, Queensland. Fruits contain a sweetish pulp; consumed by the aborigines.

Styraceae → Styrax.

Styrax acumintum Pohl. (syn. Strigelia acuminata Miers.). (Styraceae). - Tree. Brazil. Wood is easily worked; used for carpentry. Also S. leprosum Hook. and Arn.

Styrax argenteum Presl. (Styraceae). - Tree. Mexico, Centr. America. Source of a gum; used in Costa Rica as incense in churches.

Styrax Benzoin Dryander. (Styraceae). - Tree. S. E. Asia, E. Indies. Cultivated. Source of Benzoin, Sumatra Benzoin. Used medicinally, being antiseptic, expectorant, diuretic and stimulant. Contains 75 % resinous substance, bensoresin.

Styrax camporum Pohl. (syn. Strigelia camporum Miers.). (Styraceae). - Shrub. Brazil. Stem is source of a resin; used for religious ceremonies in Brazil.

Styrax ferrugineum Nees and Mart. (Styraceae). - Tree. Argentina, Paraguay, Brazil and adj. territory. Stem and branches are source of an aromatic resin; used as incense in churches.

Styrax japonicum Sieb. and Zucc. (Styraceae). - Tree. Japan. Wood fine-grained, whitish; used in Japan for handles of umbrellas. Seeds are source of an oil.

Styrax tomentosum Humb. and Bonpl. (Styraceae). - Tree. S. America. Stem is source of a hard fragrant balsam.

Styrax tonkinense Craib. (Styraceae). - Tree. Malaya. Source of Siam Benzoin. Much is derived from Cochin-China. Used in perfumery and manuf. of soap. Employed in pharmacy, antiseptic, healing of wounds; made into tincture and balsams. Also mixed with chocolates; used as incense, sometimes mixed with myrrh.

Suaeda fruticosa Forsk. Alkali Seepweed. (Chenopodiaceae). - Shrub. N. temp. zone, Mediterranean region. Plant is used as fodder for camels.

Suaeda suffrutescens Wats. Desert Seepweed. (Chenopodiaceae). - Shrub. Southwestern United States. Plant is source of a black dye; used by the Indians.

Suakim Gum → Acacia Seyal Delile.

Suakwa Vegetable Sponge → Luffa aegyptiaca Mill.

Suari Fat → Caryocar amygdaliferum Mutis.

Suari Nuts → Caryocar nuciferum L.

Succinite → Pinus succinifera Conw.

Succory → Cichorium Intibus L.

Succus Conii → Conium maculantum L.

Sudan Grass → Sorghum vulgare Pers.

Sudan Teak → Cordia Myxa L.

Sugar. Plants, sources of → Acer, Arenga, Beta, Borassus, Caryota, Cocos, Caryphia, Nipa, Saccharum, Sorghum.

Sugar Apple → Annona squamosa L.

Sugar Beet → Beta vulgaris L.

Sugar Cane → Saccharum officinale L.

Sugar Grass → Glyceria fluitans (L.) R. Br.

Sugar Maple → Acer saccharum Marsh.

Sugar Palm → Arenga pinnata (Wurmb.) Merr.

Sugar Pine → Pinus Lambertiana Dougl.

Sugar Protea → Protea mellifera Thunb.

Sugarberry → Ehretia elliptica DC.

Sulla → Hedysarum coronarium L.

Sultanas → Vitis vinifera L.

Sumac → Sorghum vulgare Pers.

Sumach, Chinese → Rhus semialata Murr.

Sumach, Dwarf → Rhus copallina L.

Sumach, Lemonade → Rhus trilobata Nutt.

Sumach, Shining → Rhus copallina L.

Sumach, Sicilian → Rhus Coriaria L.

Sumach, Smooth → Rhus glabra L.

Sumach, Staghorn → Rhus typhina L.

Sumach, Sweet Scented → Rhus aromatica Ait.

Sumach Wax → Rhus succedanea L.

Sumatra Benzoin → Styrax Benzoin Dryander.

Summer Grape → Vitis aestivalis Michx.

Summer Savory → Satureja hortensis L.

Sun Berry → Physalis minima L.

Sundew → Drosera spp.

Sundri Tree → Heritiera minor Lam.

Sunflower → Helianthus annuus L.

Sunflower, Oblongleaf → Helianthus doronicoides Lam.

Sunflower, Oregon → Balsamorrhiza sagittata (Pursh.) Nutt.

Sunflower Seed Oil → Helianthus annuus L.

Sunn Hemp → Crotalaria juncea L.

Supa Oil → Sindora Supa Merr.

Supplejack → Rhipogonum scandens Forst.

Surinam Ant Tree → Triplaris surinamensis Cham.

Surinam Cherry → Eugenia uniflora L.

Surinam Greenheart → Tecoma leucoxylon (L.) Mart.

Surinam Purslane → Talinum triangulare Willd.

Surinam Quassia → Quassia amara L.

Sushi → Porphyra tenera Kjellm.

Suwari Nuts → Caryocar tomentosum Willd.

Swamp Bay → Magnolia glauca L.

Swamp Blueberry → Vaccinium corymbosum L.

Swamp Cabbage → Spathyema foetida (L.) Raf.

Swamp Cottonwood → Populus heterophylla L.

Swamp Ebony → Diospyros mespiliformis Hochst.

Swamp Mahogany → Eucalyptus botryoides Smith.

Swamp Milkweed → Asclepias incarnata L.

Swamp Oak → Casuarina equisetifolia L. and C. suberosa Otto and Dietr.

Swamp Privet → Forestiera acuminata (Michx.) Poir.

Swamp Red Currant → Ribes triste Pall.

Swamp Spanish Oak → Quercus Pagoda Raf. and Q. palustris Du Roi.

Swamp White Oak → Quercus bicolor Willd.

Swampwood → Dirca palustris L.

Swartzia madagascariensis Desv. (Leguminosaceae). - Tree. Madagascar, Trop. Africa. Wood very heavy, hard, durable, deep red; used for furniture manuf. pianos, heavy construction work.

Swatow Grass → Boehmeria nivea (L.) Gaud.

Swedish Shield Lichen → Cetraria fahlunensis (L.) Schaer.

Swedish Turnip → Brassica Napo-Brassica Mill.

Sweet Acacia → Acacia Farnesiana (L.) Willd.

Sweet Almond Oil → Amygdalus communis L.

Sweet Basil → Ocimum Basilicum L.

Sweet Bay → Magnolia glauca L. and Persea Borbonia (L.) Raf.

Sweet Bean → Cleditsia triacanthos L.

Sweet Blueberry → Vaccinium pennsylvanicum Lam.

Sweet Broom → Scoparia dulcis L.

Sweet Buckeye → Aesculus octandra Marsh.

Sweet Cassava → Manihot esculentus Crantz.

Sweet Cherry → Prunus avium L.

Sweet Cicily → Myrrhis odorata (L.) Scop.

Sweet Clover, White → Melilotus alba Desr.

Sweet Clover, Yellow → Melilotus officinalis (L.) Lam.

Sweet Coltsfoot → Petasites palmata Gray.

Sweet Corn Root → Calathea Alliquia (Aubl.) Lindl.

Sweet Elm → Ulmus fulva Michx.

Sweet Fennel → Foeniculum vulgare Mill.

Sweet Flag → Acorus Galamus L.

Sweet Gale → Myrica Gale L.

Sweet Goldenrod → Solidago odora Ait.

Sweet Grass → Hierochloe odorata (L.) Beauv. and Sporolobus indicum R. Br.

Sweet Grenadilla → Passiflora ligularis Juss.

Sweet Gum → Liquidambar styraciflua L.

Sweet Gum, Chinese → Liquidambar formosana Hance.

Sweet Leaf → Symplocos tinctoria (L.) L'Her.

Sweet Lemon → Citrus limetta Risso.

Sweet Majoram → Majorana hortensis Moench.

Sweet Marigold → Tagetes lucida Cav.

Sweet Orange → Citrus sinensis Osbeck.

Sweet Pea → Lathyrus odoratus L.

Sweet Potato → Ipomoea Batatas Poir.

Sweet Scented Sumach → Rhus aromatica Ait.

Sweet Sop → Annona squamosa L.

Sweet Sorghum → Sorghum vulgare Pers.

Sweet Trefoil → Trigonella coerulea (L.) Ser.

Sweet Vernalgrass → Anthoxanthum odoratum L.

Sweet Viburnum → Viburnum cassinoides L.

Sweet Violet → Viola odorata L.

Sweet Woodruff → Asperula odorata L.

Sweetia elegans Benth., Perobinha, Lapachillo. (Leguminosaceae). - Tree. S. Brazil, Argentina. Wood used for carpentry, joinery and manuf. of charcoal.

Sweetia panamensis Benth. Billyweb Sweetia. (Leguminosaceae). - Tree. Centr. America to N. Colombia, Venezuela. Bitter bark, Cascara Amarga, is used for scrofula. Wood strong, durable; used locally for heavy construction work, railroad ties, spokes, silks, implement frames. Also S. nitens (Vog.) Benth.

Swertia Chirata Buch.-Ham., Chirata. (Gentianaceae). - Annual herb. Himalayan region. Dried herb is used medicinally as bitter tonic, known as Chirata. Contains a bitter glucoside chiratin; ophelic acid and chiratogenin. Also other bitter plants from India are marketed as Chirata. Used by Hindoo and Mahommedan physicians. Sold in bazaars in India.

Swietenia Candollei Pitt., Venezuela Mahogany. (Meliaceae). - Tree. Guiana, Venezuela. Wood light reddish to brownish, hard, dense, easy to work; used especially in France for furniture and interior finish.

Swietenia cirrhata Blake., Venadillo. (Meliaceae). - Tree. Mexico. Wood much used in Sinaloa for carpentry.

Swietenia humilis Zucc. Mexican Mahogany. Palo Zopilote. (Meliaceae). - Tree. W. Mexico to El Salvador. Wood furnishes a mahogany; used for cabinet work and construction.

Swietenia macrophylla King. Honduras Mahogany. (Meliaceae). - Tree. Panama and adj. territory. Wood is source of a Mahogany.

Swietenia Mahagoni Jacq. Mahogany, Caoba Mahogany, Tabosa Mahogany, San Domingo Mahogany, West Indies Mahogany. (Meliaceae). - Tree. Trop. America. Wood hard, very durable, close-grained, strong, heavy, red brown, becoming darker on exposure; esteemed for cabinet making, interior finish of houses, railroad ties; formerly used for ship and boat building. Bitter bark is sometimes used in place of quinine for intermittent fevers.

Swiss Chard → Beta vulgaris L.

Swiss Stone Pine → Pinus Cembra L.

Switchsorrel → Dodonaea viscosa Jacq.

Sycomore → Platanus occidentalis L.

Sycomore Fig → Ficus Sycomorus L.

Sycomore, Mexican → Platanus mexicana Moric.

Sycomore, Oriental → Platanus orientalis L.

Sydney Peppermint Eucalyptus → Eucalyptus piperata Smith.

Symphonia clusioides Baker. (Cuttiferaceae). - Tree. Madagascar. Wood used for general carpentry. Stem is source of a gamboge-like resin.

Symphonia globulifera L., Hog Plum, Doctors Gum, Karamanni Wax. Guiana Symphonia. (Guttiferaceae). - Trop. Africa, America. Gum mixed with bees wax and powdered cnarcoal is used by the Indians of Brit. Guiana for cementing arrow heads and joining wood. Tree is source of a gum resin, called Karamanni Wax. Seeds are source of an oil used in Brazil for skin diseases; also employed in patent medicines.

Symphonia macrophylla Baker. (syn. Chrysopia macrophylla Camb.). (Guttiferaceae). - Shrub. Madagascar. Fruits are edible. Leaves are a source of food of a silkworm, Borocera madagascariensis.

Symphytum asperrinum Don., Prickly Gomfrey. (Boraginaceae). - Perennial herb. Caucasia to Armenia. N. Iran. Occasionally cultivated as food for hogs, goats and rabbits.

Symphytum officinale L., Common Confrey. (Boraginaceae). - Perennial herb. Europe, introd. in N. America. Young shoots can be consumed as asparagus; young leaves are eaten as spinach. Recommended as an emergency food crop.

Symplocarpus foetidus Salisb. → Spathyema foetida (L.) Raf.

Symplocaceae → Symplocos

Symplocos fasciculata Zoll. (syn. Dicalyx tinctorius Blume). (Symplocaceae). - Small tree. Malaya. Bark is source of a red dye; used in Java for the batik industry.

Symplocos tinctoria (L.) L'Hér. Common Sweet Leaf, Horse Sugar. (Symplocaceae). - Shrub or tree. Eastern N. America to Florida and Texas. Wood soft, close-grained, easily worked, brown or light red; used for turnery. Leaves and fruits are source of a yellow dye. Bitter, aromatic roots are used as a tonic.

Syncarpia laurifolia Ten. (syn. Metrosideros glomulifera Smith., M. procera Salisb.) Turpentine Tree, Boorea. (Myrtaceae). - Tree. New South Wales, Queensland. Wood durable, soft, dark brown, takes high polish, durable underground, difficult to burn. Used for ship-building.

Syndesmon thalictroides (L.) Hoffmz. → Anemonella thalictroides (L.) Spach.

Synoum glandulosum Juss. (syn. Trichilia glandulosa Smith.). Brush Bloodwood, Rosewood. (Meliaceae). - Tree. New South Wales, Queensland. Wood deep red, rose-scented, readily polished. Used for cabinet-work.

Syzygium Jambolanum DC. → Eugenia Jambolana Lam.

Syzygium multipetalum Brongn. and Gris. (Myrtaceae). - Tree. New Caledonia. Wood pale red, fine-grained, dense, hard; used for cabinet work.

Syzygium nodosum Miq. → Eugenia operculata Roxb.

Syzygium racemosum DC. → Eugenia jamboloides Koord. and Valet.

T

Taban Meran Fat → Palaquium oleiferum Blanco.

Tabebuia Donnell-Smithii Rose → Cybistax Donnell-Smithii (Rose) Seib.

Tabebuia Guayacum (Seem.) Hemsl. (Bignoniaceae). - Tree. Central America. Wood hard, dense, fine to medium grained, olive-brown to reddish-brown; used for tool-handles, boats and building purposes.

Tabebuia ipe (Mart.) Standl. Lapacho. (Bignoniaceae). - Tree. Trop. S. America, esp. Argentina, Paraguay, Brazil. Wood durable, strong.

Used for turnery, cabinet work, vehicles, carpentry. Yields a purple dye. Also T. serratifolia (Vahl.) Nich.

Tabebuia leucoxyla DC. (Bignoniaceae). - Tree. Trop. America, esp. Brazil. Wood used for wooden shoes. Infusion or tincture of bark is employed as a diuretic.

Tabebuia pentaphylla (L.) Hemsl. (syn. Bignonia pentaphylla L.) Zapatero, White Cedar, White Wood, Roblo Blanco. (Bignoniaceae). - Tree. Mexico, West Indies, Centr. America, Northern S. America. Wood light yellow to brown, very hard; used for cabinet work, house building, oars, wagons. Known in the trade as West Indian Boxwood. Used in Europe as substitute for Boxwood.

Tabernaemontana bovina Lour. (Apocynaceae). - Tree. Annam, Cochin China. Roots are source of a rubber.

Tabernaemontana citrifolia L., Cojón de Gato, Lecherillo. (Apocynaceae). - Shrub. Small tree. S. Mexico, West Indies, Centr. and S. America. Milky juice sometimes used to destroy warts.

Tabernaemontana coronaria Willd. → Ervatamia coronaria Stapf.

Tabernaemontana pachysiphon Stapf. → Conopharyngia pachysiphon Stapf.

Tabernaemontana Pandacaqui Poir. (Apocynaceae). - Tree. Philipp Islands to N. Guinea. Leaves are used as a bleaching agent in the Philipp Islds.

Tabernanthe Iboga Baill. (Apocynaceae). - Woody plant. Trop. Africa, esp. Gabun, Congo. The bitter Iboga Root is used by the natives as tonic, febrifuge and aphrodisiac.

Table Mountain Pine → Pinus pungens Michx.

Tablote → Guazuma ulmifolia L.

Tabonico Resin → Dacryodes hexandra Griseb.

Tabosa Mahogany → Swietenia Mahogany Jacq.

Tacaco → Polakowskia Tacaco Pitt.

Tacamahaca Gum → Bursera gummifera L.

Tacamahaca Gum (substitute) → Calophyllum Calaba L.

Tacamahaca Resin → Calophyllum tacamahaca Willd. and Populus balsamifera Muench.

Tacca hawaiiensis Limpr. f. (syn. T. oceanica Nutt.). Hawaiian Arrowroot. (Taccaceae). - Perennial herb. Hawaii Islds. Tubers are a source of food; was once of commercial importance.

Tacca pinnatifida Forst. (Taccaceae). - Perennial herb. Pacif. Islds., New Guinea. Cultivated. Tubers are source of Tahiti or Fiji Arrowroot.

Taccaceae → Tacca.

Tachia guianensis Aubl. (Gentianaceae). - Shrub. Brazil, Guiana. Roots used for fever in home remedies in some parts of Brazil.

Tasco → Passiflora mollissima (H. B. K.) Bailey. and P. tripartita (Juss.) Poir.

Taetsia fruticosa (L.) Merr. (Liliaceae). - Shrub. Pacif. Islands. Malaya. Leaves were used as garments for men and women in the islands of the Pacific. They were also used as collarettes and ankles for native dances. Cultivated among Samoans.

Tagetes minuta L. (syn. T. glandulifera Schrank.). (Compositae). - Annual herb. Trop. America. Herb is aromatic, bitter; an infusion is used as stomachic and carminative, diuretic and diaphoretic. Source of an ess. oil obtained by distillation, much obtained during seed formation:

suggested as fly and vermin repellent; also effective larvicide killing maggots in wounds.

Tagetes multifida DC. (syn. T. filicifolia Lag.) (Compositae). - Annual herb. Mexico, Centr. America. Herb is anise scented. Infusion of plants is used as diuretic in home medicines.

Tagetes lucida Cav. Sweet Marigold. (Compositae). - Annual herb. Mexico, Centr. America. Plants have tarragon-like flavor, used as a condiment by the natives.

Tagua Palm → Phytelephas macrocarpa Ruiz and Pav.

Tahiti Arrowroot → Tacca pinnatifida Forst.

Tahiti Chestnut → Inocarpus edulis Forster.

Takut Galls → Tamarix articulata Vahl.

Talauma mexicana (DC.) Don. → Magnolia mexicana DC.

Talca Gum → Acacia Seyal Delile.

Talha → Acacia tortilis Hayne.

Talha Gum → Acacia Seyal Delile.

Talinum aurantiacum Engelm. Orange Fame Flower. (Portulacaceae). - Perennial herb. Southwestern United States. Roots are cooked and consumed by the Indians of the Southwest of the United States.

Talinum triangulare Willd. (syn. Portulaca triangularia Jacq., T. racemosum Rohrb.) Potherb Fame Flower, Surinam Purslane. (Portulacaceae). - Perennial herb. Trop. America. Used as a vegetable, has the taste of purslane.

Talipot Palm → Corypha umbraculifera L.

Talki Gum → Acacia Seyal Delile.

Tall Albizia → Albizzia procera Benth.

Tall Meadow Oat Grass → Arrhenaterum avenaceum Beauv.

Tallow, Borneo → Shorea aptera Burck. and S. stenocarpa Burck.

Tallow, Jaboty → Erisma calcaratum Warm.

Tallow, Japan → Rhus succedanea L.

Tallow, Mafura → Trichilia emetica Vahl.

Tallow, Malabar → Vateria indica L.

Tallow, Minjak Tankawang → Isoptera borneensis Scheff.

Tallow, Mowrah Butter → Madhuca longifolia Gm.

Tallow Tree → Pentadesma butyracea Sabine.

Tallow Tree → Sapium sebiferum Roxb.

Tallow Wood → Ximenia americana L.

Tamarack → Larix americana Michx.

Tamarind → Tamarindus indica L.

Tamarind, Cow → Pithecolobium Saman Benth.

Tamarind, Manila → Pithecolobium dulce Benth.

Tamarind Plum → Dialium indum L.

Tamarindus indica L., Tamarind. (Leguminosaceae). - Large tree. Probably from S. Asia or Trop. Africa. Cultivated in tropics of Old and New World. Pods 6 to 15 cm. long, brownish when ripe; pulp soft, sweet mixed with a sour taste; used in drinks, preserves, syrup; also in chutnies and curries. Has antiscorbutic properties, was used by sailors instead of lime and lemon juice. Medicinally a mild laxative.

Tamaricaceae → Tamarix.

Tamarix articulata Vahl. (Tamariscaceae). - N. Africa, Arabia, Iran. Galls called Takut or Teggant are derived from the plant and used by the Arabs for tanning, producing fine qualities of sheep and goat skins, having a pink and purple color. Containing 40 % tannin. Infusion is used for enteritis and gastralgia. Wood is used for construction and as fuel in N. Africa.

Tamisan → Citrus longispina West.

Tampico Fibre → Agave falcata Engelm.

Tampico Sarsaparilla → Smilax aristolochiaefolia Mill.

Tampons → Aeschynomene uniflora E. Mey.

Tanacetum multiflorum Thunb. → Matricara multiflora Fenzl.

Tanacetum vulgare L. (syn. Chrysanthemum Tanacetum Karsch.). Common. Tansy. (Compositae). - Perennial herb. Northern temp. zone. Cultivated. Dried leaves and flowering tops are used medicinally, being stimulant, anthelmintic, emmenagogue. Contains ess. oil, borneol, thujone, camphor and resins. Home remedy for intestinal worms, insects, fleas and lice. Source of Oil of Tancy, an ess. oil. Contains an aromatic bitter principle; an overdose is dangerous.

Tang, Black → Fucus versiculosus L.

Tangelo → Citrus paradisi Macf. x C. reticulata Blanco var deliciosa.

Tangerine → Citrus reticulata Blanco.

Tangerine Oil → Citrus reticulata Blanco.

Tangle → Laminaria digitata (L.) Edmonson.

Tangle, Sweet → Laminaria saccharina (L.) Lamour.

Tanias → Xanthosoma sagittifolium Schott.

Tanning material. Plants source of → Acacia, Adansonia, Adenanthera, Adenopus, Albizzia, Alnus, Andromeda, Anogeissus, Arbutus, Arctostaphylos, Aspidosperma, Astrocaryum, Avicennia, Baloghia, Banksia, Blakea, Bruguiera, Bucklandia, Burkea, Caesalpinia, Calliandra, Calluna, Calotropis, Castanea, Casuarina, Catalpa, Cereops, Cistus, Colpoon, Comarum, Conocarpus, Copriaria, Cryptocarya, Curatella, Dacrydium, Dichrostachys, Dioscorea, Engelhardtia, Eremophila, Eucalyptus, Eucryphia, Ficus, Gordonia, Gunnera, Jatropha, Jodina, Kandelia, Laguncularia, Lecythis, Leucospermum, Lithocarpus, Loranthus, Lysiloma, Mahonia, Melanoxy-

lon, Mimetes, Myrica, Ocotea, Osyris, Oxalis, Pachycormus, Parkea, Persea, Phyllanthus, Picea, Pinus, Piptadenia, Pseudotsuga, Pterocelastrus, Punica, Pycnocoma, Quebrachia, Quercus, Rhizophora, Rhus, Rosa, Rourea, Schinopsis, Schleichera, Semecarpus, Sorbus, Soymida, Statice, Styphnodendron, Tamarix, Terminalia, Uncaria, Vitex, Weinmannia, Zizyphus, Zollernia.

Tansy → Tanacetum vulgare L.

Tansy Phacelia → Phacelia tanacetifolia Benth.

Tansy Ragwort → Senecio Jacobaea L.

Tapa Cloth → Broussonetia papyrifera Vent.

Tapana → Hieronima caribaea Muell. Arg.

Tape Grass → Vallisneria americana Michx.

Tapioca → Manihot esculenta Crantz.

Tapuru Rubber → Sapium taburu Ule.

Tara → Actinidia callosa Lindl.

Taraktogenos Kurzii King. (syn. Hydnocarpus heterophyllus Kurz.). Chaulmoogra. (Flacourtiaceae). - Tree. E. India. Seeds are source of Chaulmoogra Oil used for healing leprosy. Oil is externally a powerful irritant. Sp. Gr. 0.952; Sap. Val. 200.6; Iod. No. 101.5; Ac. Val. 2.6; Unsap. 0.29 %.

Taraxacum kok-saghyz Rodin., Kok-saghyz. (Compositae). - Herbaceous perennial. Turkestan. Roots are a source of rubber. Cultivated in some parts of Russia.

Taraxacum officinale Weber. (syn. Leontodon Taraxacum L.). Dandelion. (Compositae). - Perennial herb. Temp. zone. Cultivated as a vegetable; used in salads as greens or when bleached. Improved large leaved varieties are Coeur Plein, Vert de Montmagny, Ameliore Geant. Source of Dandelion Wine. Dried rhizome and roots used medicinally, being a bitter tonic, diuretic, aperient. Contains a bitter compound, also inulin and an ess. oil. Ground, roasted roots are sometimes used as substitute for coffee.

Tarchonanthus camphoratus L., Hottentot Tobacco. (Compositae). - Shrub. S. Africa. Wood close-grained, heavy; recommended for musical instruments, joiner's fancy work. Leaves have the taste fo camphor; they are chewed by the Mahommedans and smoked by Hottentots. Infusion is used for astma and as a diaphoretic.

Taramniya Gum → Combretum hypotilinum Diels.

Taro → Colocasia antiquorum Schott.

Tarragon → Artemisia Dracunculus L.

Tarragon, False → Artemisia dracunculoides Pursh.

Tarrietia javanica Blume. (Sterculiaceae). - Tree. Malaya, Philipp. Islands. Wood dark reddish-brown, fairly heavy, durable, easy to work; used for construction of houses, furniture, boxes, canoes; employed in place of Mahogany.

Tarrietia utilis Sprague. (Sterculiaceae). - Tree. Trop. W. Africa. Wood an excellent lumber, known as an African Mahogany; used for boards, canoes, general carpentry.

Tasmanian Blue Eucalyptus → Eucalyptus globulus Lab.

Tassel Grape Hyacinth → Muscari comosum Mil.

Tatarian Sea Kale → Crambe tatarica Jacq.

Tatary Buckwheat → Fagopyrum tataricum Gaertn.

Tauary → Couratari Tauari Berg.

Tau-Saghys → Scorzonera tau-saghys Lips.

Tawny Dailily → Hemerocallis fulva L.

Taxaceae → Taxus, Torreya.

Taxodiaceae → Crypomeria, Cunninghamia, Sciadopitys, Sequioa, Taxodium.

Taxodium distichum (L.) Rich., Southern Cypress, Bald Cypress. (Taxodiaceae). - Tree. Swampy areas of East, South East and South of the United States. Wood not strong, light, easily worked, light to dark brown; much used for construction, rail-road ties, posts, fences, cooperage, pumps; keeps well when in contact with moist soil.

Taxodium mucronatum Ten. (Taxodiaceae). - Mexico, esp. Sinaloa to Coahuila. Wood soft, light to dark brown; made into planks, used for furniture, general construction work, takes polishing. Acrid resin used in Mexico for wounds and ulcers. Bark is considered emmenagogue and diuretic.

Taxus baccata L., English Yew. (Taxaceae). - Tree. Europe, Algeria, Asia Minor, Caucasia, N. Iran. Wood elastic, dense, heavy, hard, durable, reddish to red-brown, easily to polish, bitter taste; used since time immemorial for handles of knives, bows, backs of combs, mountain sticks, wood-carving. Known as deutsches Ebenholz (German Ebony). In parts of Switzerland a decoction of leaves is used to combate troublesome insects of live-stock.

Taxus brevifolia Nutt., Pacific Yew. (Taxaceae). - Tree. British Columbia to California. Wood strong, bright red, heavy; used by the Indians of the Northwest for making spear-handles, paddles and bows.

Taxus canadensis Willd., American Yew, Ground Hemlock. (Taxaceae). - Shrub. Eastern N. America. Red fleshy disc around fruits is sometimes eaten; seeds are poisonous.

Taxus cuspidata Sieb. and Zucc. Japanese Yew. (Taxaceae). - Tree. China, Japan. Wood very fine-grained, takes beautiful polish. Used in Japan for furniture, utensils, interior finish, sculpture, marquetry, pencils, turnery. In olden times the shaku, a baton, was made from this wood. The Ainu use wood for arrows. Heartwood source of a brown dye, used for dyeing cloth.

Tchitscha (beverage) → Chenopodium Quinoa Willd.

Te del Monte → Clinopodium macrostemum (Benth.) Kuntze.

Tea → Camellia sinensis (L.) Kuntze.

Tea. Plants used for scenting → Aglaia, Chlorianthus, Jasminum, Nelumbium, Osmanthus.

Tea Substitutes → Acer, Achillea, Agapetes, Akebia, Althaea, Angraecum, Aniba, Aphloia, Aplopappus, Arctostaphylos, Artemisia, Aspalathus, Astragalus, Atherosperma, Athrixia, Baeckia, Borago, Calluna, Cassia, Catha, Ceanothus, Cedronella, Chamaedaphne, Chiogene, Chloranthus, Cistus, Cliffortia, Clinopodium, Coix, Corchorus, Correa, Crataegus, Croton, Cyclopia, Cymbopogon, Desmodium, Dictamnus, Diplospora, Dryas, Eheretia, Ephedra, Epilobium, Erythraea, Eupatorium, Eurya, Frankenia, Glycyrrhiza, Helichrysum, Hydrangea, Ilex, Ledum, Leptospermum, Leyssera, Ligustrum, Lindera, Lippia, Lithospermum, Litsea, Lysimachia, Matricaria, Melissa, Menyanthes, Miconia, Micromeria, Monarda, Myrtus, Neea, Petalostemon, Polygala, Potentilla, Primula, Pseudotsuga, Psoralea, Pycnothamnus, Pyrola, Pyrus, Rhododendron, Rosa, Rubus, Sageratia, Salvia, Sambucus, Sargassum, Sida, Sideritis, Solidago, Sorbus, Stachys, Teucrium, Tsuga, Tunica, Tussilago, Ulmus, Veronica, Viburnum, Vitex.

Tea, Arabian → Catha edulis Forsk.

Tea, Brazilian → Stachytarpheta jamaicensis Vahl.

Tea Bush, Wild → Leucas martinicensis R. Br.

Tea, Cape Barren → Correa alba Andr.

Tea, Caspa → Cyclopia subterenata Vog.

Tea, Chinese → Camellia sinensis (L.) Kuntze.

Tea, Hottentot → Helichrysum serpylifolium Lessing.

Tea, Kapporie → Epilobium angustifolium L.

Tea, Kapor → Epilobium angustifolium L.

Tea, Labrador → Ledum groenlandicum Oeder.

Tea, Mountain → Gaultheria procumbens L.

Tea, New Jersey → Ceanothus americanus L.

Tea, Oswego → Monarda didyma L.

Tea, Paraguay → Ilex paraguensis St. Hil.

Tea Seed Oil → Camellia Sasanqua Thunb.

Tea, Seminole → Asimina reticulata Chapm.

Tea, Siberian → Potentilla rupestris L.

Tea, Thorn → Cliffortia ilicifolia L.

Tea Tree, Broom → Leptospermum scoparium Forst.

Tea Tree, Heath → Leptospermum ericoides Rich.

Tea Tree, Yellow → Leptospermum flavescens Sm.

Teak, African → Oldfieldia africana Benth. and Hook.

Teak, Khartoum → Gordia Myxa L.

Teak, Malacca → Afzelia palembanica Baker.

Teak, Sudan → Cordia Myxa L.

Teak Wood → Tectona grandis L.

Teasel → Dipsacus fullonus L.

Teasel Clover → Trifolium parviflorum Ehrh.

Tecoma leucoxylon (L.) Mart. (Bignoniaceae). - Tree. Trop. America, esp. W. Indies, S. America. Source of Green Ebony, Greenheart Ebony, Surinam Greenheart, being yellow green to olive green, very hard, heavy, sinks in water; used for turnery, furniture, pianos, building purposes. The wood is not as resistant to the Teredo as Demarera Greenheart.

Tecoma ochracea Cham. (syn. T. Ipe Mart.). Lapacho. (Bignoniaceae). - Tree. S. America, esp. S. Brazil, Paraguay, Uruguay and Argentina. Wood is source of a yellow dye.

Tecoma serratifolia G. Don., Yellow Poui. (Bignoniaceae). - Tree. Trinidad. North of S. America. Wood hard, very durable, grey to greenish; suitable for outdoor work.

Tecoma stans (L.) H. B. K. (syn. Bignonia stans L.). Yellow Elder. (Bignoniaceae). - Shrub or small tree. Subtr., trop. America. Roots supposed to be powerful diuretic, vermifuge, tonic.

Tectona grandis L. Teak, Djati. (Verbenaceae). - Tree. E. India, Malaysia etc. Cultivated. Important timber tree. Wood very durable, strong, easily worked; used for general construction work, beams, bridges, boats, houses, furniture, rail-road ties, paving blocks. Bark source of a yellow dye; used for staining basket work etc. derived from Borassus flabelliformis.

Tee Coma → Albizzia procera Benth.

Teff → Poa abyssinica Juss.

Teggant Galls → Tamarix articulata Vahl.

Teinostachyum Dullooa Gamble. (Graminaceae). - Medium sized, tuffed bamboo. Trop. Asia, esp. N. E. Bengal. Stems used for umbrellas, small boxes, water-pails, baskets, mat-work, building purposes.

Teinostachyum Wightii Munro. (Graminaceae). - Tall, semi-scandent bamboo. Asia, esp. Slopes W. Ghats, N. Kanara to Cape Comorin. Used for construction of temporary bridges.

Telanthera polygonoides (R. Br.) Moq. (Amaranthaceae). - Perennial plant. Guiana, Venezuela. Used in some parts of Venzuela as a tonic, astringent and diuretic.

Telepathine → Banisteria Caapi Griseb.

Telfairia occidentalis Hook. f., Oyster Nut Tree, Krobonko. (Cucurbitaceae). - Perennial vine. Trop. Africa. Cultivated. Leaves are eaten as potherb. Seeds when cooked are consumed by the natives.

Telfairia pedata Hook. Zanzibar Oil Vine. (Cucurbitaceae). - Vine. Trop. Africa. Seeds edible, source of Castanha Oil; used for manuf. soap and candles.

Telitoxicum minutiflorum (Diels) Mold. (Menispermaceae). - Vine. Brazil. Root is used by the Tecuman Indians of Brazil for the preparation of curare, an arrowpoison. Also T. peruvianum Mold.

Telosma cordata Merr. (syn. T. odoratissima Cov., Pergularia minor Anders.) (Asclepiadaceae). - Woody climber. India, China. Flowers and leaves eaten by Siamese. Fleshy roots made into sweat-meat by Chinese in Java.

Telosma procumbens Merr. (Asclepiadaceae). - Woody climber. Philipp. Islands. Young fruits used in Philipp. Islds. as vegetable.

Temak Shorea → Shorea hypochra Hance.

Ten Months Yam → Dioscorea alata L.

Tenkwa-fun → Trichosanthes japonica Regel.

Tenio → Weinmannia trichosperma Cav.

Tennevelley Senna → Cassia angustifolia Vahl.

Teosinte → Euchlaena mexicana Schrad.

Tepary Bean → Phaseolus acutifolius Gray var. latifolius Freem.

Tephrosia candida DC. White Tephrosia. (Leguminosaceae). - Herbaceous plant. Trop. Asia. Plant has been recommended as a greenmanure in coffee and Hevea plantations.

Tephrosia cinerea Pers., Ashen Tephrosia. (Leguminosaceae). - Herbaceous plant. Trop. America. Plant has narcotic properties. Used in Guiana to stupefy fish.

Tephrosia nitens Benth. (Leguminosaceae). - Shrub. Trop. America. Herb is used in Brazil to stupefy fish.

Tephrosia periculosa Baker. (Leguminosaceae). - Perennial herb. Trop. E. Africa. Used to stupefy fish.

Tephrosia purpurea Pers. Purple Tephrosia. (Leguminosaceae). - Herbaceous plant. Tropics of Old and New World. Used in different countries to stupefy fish, ointment from roots used for elephantiasis. Juice applied for eruption of the skin. Decoction of roots used for indigestion, cough, liver and kidney affections. Decoction of plant was used by the Seminole Indians as a specific for nose bleeding. Roots are used by the natives of Kategum (Afr.) for flavoring milk. Has been recommended as a green manure and as cover crop in rubber plantations.

Tephrosia toxicaria (Sw.) Pers. (syn. Galega toxicaria Sw.) Fishdeath Tephrosia. (Leguminosaceae). - Shrub. Tropical and subtropical America. Roots used as fishpoison. Cultivated. Employed as insecticide.

Tephrosia virginiana (L.) Persoon (syn. Cracca virginiana L.). Goat's Rue, Virginia Tephrosia, Devil's Shoestrings. (Leguminosaceae). - Perennial herb. Eastern N. America to Florida and Louisiana. Decoction of boiled roots is used by the Cherokees as childs tonic. Roots are tonic, diaphoretic, anthelmintic and cathartic. Have been used by the Indians to stupefy fish.

Tephrosia Vogelii Hook. f. Vogel Tephrosia. (Leguminosaceae). - Perennial herb. Trop. Africa. Cultivated. Grown as green-manure and cover crop. Also used to stupefy fish. Has been suggested as an insecticide.

Terai Bamboo → Melocanna bambusoides Trin.

Terebinth Pistache → Pistacia Terebinthus L.

Terfaz → Terfezia Boudieri Chatin.

Terfezia Boudieri Chatin. Torfez, Terfaz. (Terfeziaceae). - Ascomycete. Fungus. N. Africa, Asia Minor. Fruitbodies of the mushroom are consumed as food among the Arabs. Also other species.

Terfezia Gennadi Chatin → Tuber Gennadi (Chatin) Pat.

Terfezia Magnusii Matt. → Choiromyces Magnusii (Matt.) Poal.

Terfeziaceae (Fungi) → Terfezia, Tirmania.

Terminalia Arjuna Wight and Arn., Arjan Terminalia. (Combretaceae). - Large tree. E. India, Malaya. Decoction of bark acts as cardiac stimulant. Wood is very hard, brown with dark streaks; used for house and boat building, carts and agricult. implements.

Terminalia Bellerica Roxb., Bellerica Terminalia. (Combretaceae). - Large tree. E. India, Malaya, Philipp. Islds. Fruits edible, dark red, smooth, 3 cm. across, like a small plum, pleasant flavor, subacid; used for preserves. They are called in commerce Baleric Myrobalans and are also used as a lotion for sore eyes. Leaves are used in India for tanning.

Terminalia Calamansanay (Blanco) Rolfe. (Combretaceae). - Tree. Throughout Philippine Islands, esp. N. Luzon. Bark is astringent, has lithotriptic qualities.

Terminalia Catappa L. Tropical Almond, Indian Almond, Myrobalan, Almendro. (Compretaceae). - Tree. Trop. Asia. Widely cultivated in tropics of Old and New World. Seeds eaten raw in many countries; contain a 50 % colorless oil, much esteemed in the Orient. Leaves are food of the Tasar Silkworm. Roots, bark and fruits are used for tanning. Fruits are source of a black dye; used in some parts of E. India for coloring teeth black.

Terminalia Chebula Retz. (Combretaceae). - Tree. Centr. Asia. Fruits are sold in Teheran; used medicinally as tonic, has aperient properties. Also used for tanning.

Terminalia comintana (Blanco) Merr. Binggas Terminalia. (Combretaceae). - Tree. Philipp. Islands. Fruit astringent; decoction is used in Philipp. Islds. for thrush and obstinate diarrhea.

Terminalia Hilariana Steud. (syn. T. capitata Sauv.). Yellow Sanders. (Combretaceae). - Tree. Trop. America. Wood strong, close-grained; used for posts, axles and general carpentry; takes a satiny finish. Known as Yellow Sanders.

Terminalia Kärnbachi Schum. (Combretaceae). - Tree. New Guinea. Fruits having the taste of almonds are consumed by the natives of N. Guinea.

Terminalia macroptera Guill. and Perr. Rebreb Terminalia. (Combretaceae). - Tree. Trop. Africa. Bark is source of a perfume; used by women of Sudan.

Terminalia mauritiana Lam. Mauritius Terminalia. (Combretaceae). - Tree. Mauritius, Réunion. Source of Jamrosa Bark; used for tanning. Contains 30 % tannin. Tanning takes place quickly and gives a light color to the leather.

Terminalia obovata Steud., Nargusta Terminalia, Yellow Olivier. (Combretaceae). - Tree. West Indies, S. America. Wood strong, moderately hard, takes a good polish; used for construction work and furniture.

Terminalia splendida Engl. and Diels. (Combretaceae). - Tree. Trop. Africa. Powdered bark is used as snuff by women of the Sudan.

Terminalia superba Engl. and Diels. Afara Terminalia. (Combretaceae). - Tree. Trop. W. Africa. Excellent lumber tree. Wood tough, strong, medium hard, splits easily; works easily; used shingles, paddles, frame work of mud-houses, house posts, door frames, coffins, canoes, boxes, bowls, also for construction work, bridges and general joinery. Better grades of wood are exported from Belg. Congo.

Ternstroemia elliptica Sw. (Theaceae). - Tree. West Indies. Astringent bark is used in Guadeloupe for diarrhea. Also T. obovalis A. Rich.

Ternstroemia japonica Thunb. (Theaceae). - Tree. Trop. Asia, esp. Ceylon and Burma to China and Japan, Mal. Penins. Wood is used for building material, ship building and manuf. of furniture. Also T. Wallichiana Ridl.

Ternstroemia Robinsonii Merr. (Theaceae). - Tree. Amboina. Crushed bark is used to stupefy fish. Wood brown; employed for handles of tools.

Terra Japonica → Acacia Catechu Willd.

Terra Japonica → Uncaria Gambier Roxb.

Tesota → Olneya Tesota Gray.

Tetracarpidium conophorum Hutch. and Dalz. (Euphorbiaceae). - Tree. Trop. Africa, esp. Sierra Leone, Cameroon, Nigeria. Seeds are consumed by the natives. Oil from seeds has been recommended for manuf. soap varnishes, and lacquer industry. In Sierra Leone leaves are eaten with rice.

Tetracera sarmentosa Vahl. (syn. Delima sarmentosa L., T. scandens Merr.). (Dilleniaceae). - Rambling shrub. Mal. Peninsula. Leaves are used by the natives as sand paper for smoothing their blow pipes and darts. In India they are used for polishing metal work. Decoction of leaves is used for dysentery also applied to boils.

Tetraclinis articulata (Vahl.) Mast. (syn. Callitris quadrivalvis Vent.). Arartree. (Cupressaceae). - Large shrub or tree. N. W. Africa. Source of a resin, called Sandarac, being composed of cylindrical, pale yellow tears, 5 to 30 mm. thick, of an aromatic, bitter taste. Is used for varnishes and medicinally as a mild stimulant. Resin is also used for preserving fine paintings, as lacqueur, spirit varnishes, for coating paper, labels, metal, bookbinder's and spirit varnishes, lacquers for photographic work, in dental cements, used as incense. Wood was highly valued in ancient times; used at present for bridges, house building.

Tetragonia expansa Murr., New Zealand Spinach. (Aizoaceae). - Annual herb. Australia, New Zealnad. Widely cultivated as a warm-weather crop. Leaves are consumed as spinach.

Tetrahit longiflorum Moench. → Galeopsis ochroleuca Lam.

Tetranthera polyantha Wall. → Litsea Cubeba Pers.

Tetrapleura Thonningii Benth. (Leguminosaceae). - Tree. Trop. W. Africa. Pods ground or roasted are used by natives in „black soup" in Old Calabar. Wood is used for doors, windows and benches.

Tetraspathaea australis Raoul. (syn. Passiflora tetrandra Banks. and Soland.). New Zealand Passionflower. (Passifloraceae). - Woody vine. New Zealand. Dried wood was used by the Maori as a slow match, to carry a spark from village to village.

Tetrastigma Harmandi Planch. (Vitaceae). - Woody vine. Trop. Asia, esp. Malaya. Berries are edible, flavor of a scuppernong grape; used for jellies.

Teucrium Chamaedrys L. (syn. T. officinale Lam., Chamaedrys officinalis Moench.). Chamaedrys Germander, Common Germander. (Labiaceae). - Halfshrub. Europe, N. Africa, W. Asia. Herb used in teas as stomachic, excitans, tonic, diuretic, febrifuge, antiscrophulosum; for ailments of the spleen, for dropsy and gout. Has been used medicinally since ancient times.

Teucrium Scordium L. (syn. Chamaedrys Scordium Moench.). Water Germander. (Labiaceae). - Perennial herb. Europe, Asia, N. Africa. Herb used as Extractum Scordii Dialysatum for lupus and actinomycose. In the Danube region plant is source of a yellow green dye; used for dyeing cloth.

Teucrium Scorodonia L., Wood Germander. (Labiaceae). - Perennial herb. Mediterranean region. Herb used in some parts of France as a home remedy for wounds and antihydroptic. Leaves are used to adulterate those of Digitalis.

Teucrium Thea Lour. (Labiaceae). - Perennial herb. China. Leaves are used in China as a source of tea.

Texas Black Walnut → Juglans rupestris Englm.

Texas Bluegrass → Poa arachnifera Torr.

Texas Millet → Panicum texanum Buckl.

Texas Oak → Quercus texana Buckl.

Texas Snakeroot → Aristolochia reticulata Nutt.

Texas Sotol → Dasylirion texanum Scheele.

Thalictrum aquifolium L., Columbine Meadowrue. (Ranunculaceae). - Perennial herb. Europe, N. Asia. Roots are eaten raw or roasted by the Ainu.

Thalictrum Hernandezii Tausch. (syn. T. lasiostylum Presl.). Alboquillo de Campo. (Ranunculaceae). - Perennial herb. S. America to Mexico. Root is diuretic, purgative. Decoction of plant is used in some parts of Argentina for rheumatics.

Thapsia garganica L., Gargan Deathcarrot. (Umbelliferaceae). - Herbaceous plant. Mediterranean region. Source of Spanish Turpethroot producing Thapsia Resin being epispastic, used as plasters.

Thapsia Resin → Thapsia garganica L.

Thatch Palm → Thrinax parviflora Sw.

Thatch Screwpine → Pandanus odoratissima L.

Thatching. Plants used for → Acanthorrhiza, Acrostichium, Amomum, Attalea, Brahea, Calamus, Calyptrogyna, Carludovica, Cocos, Copernica, Cymbopogon, Eucalyptus, Geonoma, Heliconia, Heteropogon, Hyparrhenia, Hyphaene, Imperata, Livistonia, Lodoicea, Manicaria, Maximilliana, Nipa, Nolina, Pandanus, Rostkovia, Sabal, Saccharum, Socratea, Washingtonia.

Thé de Bourbon → Angraecum fragrans Touars.

Thé des Carolines → Polygala theezans L.

Thea sinensis L. → Camellia sinensis (L.) Kuntze.

Theaceae → Adinandra, Anneslea, Camellia, Eurya, **Gordonia**, Laplacea, Plocarium, Schima, Ternstroemia, Visnea.

Thelepogon elegans Roth. (Graminaceae). - Annual grass. Trop. Africa, esp. Lagos, Katagum, Nigeria, Cameroons, Abyssinia, E. Africa, Nyasaland. Herb is given to horses as a bitter tonic.

Thelesperma gracile (Torz.) Gray. (Compositae). - Annual herb. Western United States. Decoction of leaves was used as a beverage by the Indians of New Mexico and Arizona. Also T. trifidum (Poir.) Britt.

Theloschistes flavicans (Swartz) Muell. Arg., Yellow Borrera. (Theloschistaceae). - Lichen. Source of a gamboge or yellow dye; used in Germany for dyeing woolens.

Theloschistes parietinus (L.) Norm. (Theloschistaceae). - Lichen. Temp. zone. On bark of trees. Source of a yellow dye; used in England and Sweden for dying woolens. Also used for painting Easter Eggs.

Thelygonaceae → Thelygonum.

Thelygonum Cynocrambe L. (Thelygonaceae). - Herbaceous plant. Mediterranean region. Young shoots are sometimes consumed as vegetable. Was formerly employed as a laxative.

Themeda triandra Forsk. (syn. Anthistiria imberbis Wood.). (Graminaceae). - Perennial grass. Africa, India, Australia. Used as fodder for live stock. Grains are used as a source of food by the natives during times of famine.

Theobroma albiflora Goud. (syn. Herrania albiflora Goud.). Cacao Montaras, Cacao Simarron. (Sterculiaceae). - Small tree. Colombia. Seeds are sometimes mixed with those of T. cacao L.

Theobroma bicolor Humb. and Bonpl. Nicaragua Cacao Tree, Cacao Blanco, Patashti. (Sterculiaceae). - Tree. Mexico, Centr. America, S. America. Seeds are used as a beverage. Has pleasant taste. Seeds contain small amount of theobromine and much cacao-butter. Cacao sometimes known as Tiger, Wariba and Patashti.

Theobroma Cacao L. Cacao Tree. (Sterculiaceae). - Small tree. Mexico, Centr. America. Cultivated in tropics of the Old and New World. Seeds have been a source of beverage among the Indians of Mexico since ancient times. Important commercial product. Seeds, Cacao Beans are source of cacao, chocolate, cacao-butter, widely used in confectionary, milk chocolates, cocoa nibs, powdered chocolate etc. Cacao Butter (45—50 % in seeds) is used in pharmaceutical preparations. Sp. Gr. 0.8823—0.8830; Sap. Val. 192—198; Iod. No. 32—40; Unsap. 0.3—0.8 %. The numerous varieties are divided in: I. Group Criollo: Amarillo Colorado. II. Forastero: Cundeamor verrugosa amarillo, Cundeamor verrugosa colorado, Amarillo, Colorado, Amelanedo marillo, Amelanedo colorado. III. Calabacillo: Amarillo and Colorado. Seeds have been used occasionally as money. Contain theobromine.

Theobroma glauca Karts. (Sterculiaceae). - Tree. Trop. America. Seeds source of a cacao, said to be of good quality.

Theobroma grandiflora Schum. → Guazuma grandiflora G. Don.

Theobroma Guazuma L. → Guazuma ulmifolia Lam.

Theobroma leiocarpa Bern., Cacao Calabacillo. (Sterculiaceae). - Tree. Trop. America. Seeds are source of an inferior cocoa.

Theobroma Mariae Schum. (syn. Abroma Marae Mart., Herrania Mariae Goud.) (Sterculiaceae). - Tree. Amazon Basin (Braz.). Beans are called Cacaoti and are sometimes mixed with those of T. cacao L. Seeds contain a fat, similar to that of Cacao Butter.

Theobroma Martiana Dietr. (Sterculiaceae). - Tree. N. Brazil. Seeds are sometimes mixed with those of T. cacao L., used as a beverage.

Theobroma microcarpa Mart. (Sterculiaceae). - Tree. Brazil. Cultivated in Bahia. Seeds are used as a substitute for cocoa.

Theobroma pentagona Bern., Cacao Lagarto. (Sterculiaceae). - Tree. Guatemala. Related to T. cacao L. Cultivated in some parts of Centr. America. Known as Cacao Lagarto.

Theobroma purpureum Pitt., Cacao de Mico. (Sterculiaceae). - Tree. Costa Rica to Panama. Seeds are the source of a bitter beverage, used by the Bribri Indians.

Theobroma speciosa Wild. (syn. T. angustifolium DC.) Cacao de Mico, Cacao Silvestre. (Sterculiaceae). - Small tree. Mexico, Centr. America, Brazil. Seeds are made into a beverage in some parts of Centr. America. Sometimes known as Cacao de Sonusco.

Theobroma subincanum Mart. (Sterculiaceae). - Tree. Trop. S. America, esp. Guiana, Brazil, Peru. Seeds are source of an inferior cocoa.

Thespesia populnea Soland., Portia Tree, Tulip Tree. (Malvaceae). - Small tree. Trop. Asia, Africa. Wood very hard, light brown with black streaks, keeps well under water; used for fancy work, recommended for boat building, wheel wrights, gun stocks. Bark is source of a strong fibre. Source of Seychelles Rosewood.

Thevetia Ahouai A. DC. (Apocynaceae). - Shrub. Trop. S. America, esp. Guiana, Brazil. Plant is emetic, cathartic, febrifuge; used for curing ulcers.

Thevetia nereifolia Juss., Exile Tree, Yellow Oleander. (Apocynaceae). - Shrub. Trop. America, West Indies. Cultivated in tropics of Old and New World. Bark powerful antiperiodic and febrifuge. Seeds source of Exile Oil in India. Seeds called Lucky Seeds or Lucky Beans also used as charms in some parts of West Indies.

Thingan Merawan → Hopea odorata Roxb.

Thistle, Canada → Cirsium arvense (L.) Scop.

Thistle, Carline → Carlina acaulis L.

Thistle, Holy → Silybum Marianum Gaertn.

Thistle, Russian → Salsola Kali L.

Thistle Sage → Salvia carduacea Benth.

Thistle, Scotch → Onopordon Acanthium L.

Thistle, Silver → Onopordon Acanthium L.

Thlapis arvense L., Penny Cress, Field Penny Cress. (Cruciferaceae). - Europe, N. Asia; introd. in N. America. Seeds were formerly used in medicine as Semen Thlaspeos for „cleaning" the blood; as diuretic and for rheumatism. They contain rhodonallyl and allylsulfid.

Thonningia sanguinea Vahl. (Balanophoraceae). - Parasitic on forest trees. Trop. Africa. Decoction of flowers is used in Yorubaland (Afr.) for sore throat, laryngitis. Aromatic root is used in Hausa as a spice.

Thornapple → Datura Stramonium L.

Thorn, Buffalo → Zizyphus mucronata Willd.

Thorn, Cape → Zizyphus mucronata Willd.

Thorn, Christ → Zizyphus Spina-Christi Willd.

Thorn, Cock Spur → Crataegus Crus-galli L.

Thorn Tea → Cliffortia ilicifolia L.

Thorn Pear → Scolopia Zeyheri Arn.

Thrinax parviflora Sw., Thatch Palm, Reef-Thatches. (Palmaceae). - Slender palm. Florida, West Indies. Leaves are extensively collected, cured and made into „artificial" palms; used for decoration in colder climates.

Thuja occidentalis L., White Cedar, Eastern Ar-bor-Vitae. (Cupressaceae). - Tree. Eastern N. America. Wood soft, brittle, very coarse-grained, durable, light, fragrant, pale yellow-brown; used for rail-road ties, fence posts, shingles. Leaves are source of Oil of White Cedar, Oleum Cedri Folii or Oil of Arbor Vitae, an essent. oil, being pale yellow, greenish yellow to colorless. Sp. Gr. 0.915—0.935. Used internally as expec-torant; antirheumatic, emmenagogue; external-ly for skin diseases; also used in perfumes. Wi-dely used as hedge plant and for wind-breaks.

Thuja plicata Don., Giant Arbor Vitae, Western Red Cedar. (Cupressaceae). - Tree. Western N. America. Fibre of inner bark is made by the Indians, into blankets, ropes, also as thatch for cabins. Wood not strong, brittle, soft and light, coarse-grained, easily split, dull brown; used for interior finish of buildings, sashes, doors, fences, cabinet-making, cooperage and shingles. Indians of the North West used to split planks for making their lodges, war canoes and carved totems to decorate their villages.

Thujopsis dolobrata Sieb. and Zucc. (syn. T. Hondai Henry). (Cupressaceae). - Tree. Japan. Wood straight grained, soft, durable, elastic. Used in Japan for building purposes, interior finish of houses, cabinet work, bridges, ship-building, rail-road ties, wooden-pipes for water-works, cooperage, sleighs, lacquered wares. Bark used for match-cord, for fillings between boards to prevent leaking of water.

Thurber Fescue → Festuca Thurberi Vasey.

Thyme → Thymus vulgaris L.

Thyme, Conehead → Thymus capitatus Hoffm. and Link.

Thyme, Creeping → Thymus Serpyllum L.

Thyme, Mastic → Thymus Mastichina L.

Thymelaea hirsuta Endl. → Passerina hirsuta L.

Thymelaeaceae → Aquilaria, Dais, Daphne, Daphnopsis, Dirca, Edgeworthia, Conystylus, Gyrinops, Lagetta, Lasiophora, Passerina, Wik-stroemia.

Thymus Acinos L. → Satureja Acinos (L.) Scheele.

Thymus capitatus Hoffm. and Link. (syn. Cori-dothymus capitatus Reichb.). Conehead Thyme. (Labiaceae). - Shrub. Mediterranean region. Source of an ess. oil derived from distillation;

used for flavoring. Sometimes called Spanish Origanum Oil being of the carvacrol type. Much is derived from Spain.

Thymus Chamissonis Benth. → Micromeria Chamissonis Greene.

Thymus hirtus Willd. (Labiaceae). - Herbace-ous perennial. N. Africa, Algarian Sahara, Spain. Used as a condiment and stomachic by diffe-rent tribes in N. Africa.

Thymus Mastichina L. Mastic Thyme. Spanish Marjoram. (Labiaceae). - Perennial herb. Me-diterranean regions. Source of Oil of Marjoram, Oil of Wild Marjoram; used for flavoring.

Thymus Serpyllum L. Creeping Thyme. (Labia-ceae). - Low prostrate half-shrub. Europe, temp. Asia, Africa, N. America. Occasionally cultiva-ted. Dried leaves and flowertops are used me-dicinally, being antispasmodic in whooping-cough. Contains carvacrol, thymol, cymene.

Thymus vulgaris L. Common Thyme. (Labia-ceae). - Half-shrub. Mediterranean region, Gree-ce, introd. in N. America. Cultivated. Used as a condiment, for flavoring meats, soups, sausa-ges. Dried leaves and flowering tops are used medicinally, being a stimulant, carminative. Source of Oil of Thyme, Oleum Thymi, a sti-mulant, antiseptic, antispasmodic; also applied as a liniment. Contains thymol and some carva-crol.

Thyrsostachys siamensis Gamble., Umbrella handle Bamboo, Monastry Bamboo. (Gramina-ceae). - Woody grass. Trop. Asia, esp. India, Burma to Tenasserim. Much cultivated in Mo-nastry Gardens. Used for umbrella handles.

Thysanolaena maxima Kuntze. (Graminaceae). - Perennial grass. Philipp. Islds. Panicles are used in the Philipp. Islds. for light dust brooms; used for highly polished hardwood floors.

Tiama Mahogany → Entandrophragma macro-phyllum Chev.

Tigridia Pavonia (L. f.) Ker-Gawl. Common Ti-gerflower. (Iridaceae). - Bulbous perennial. Centr. and S. Mexico, Centr. America. Roasted starchy bulbs are eaten as food by Mazatecs and other Indian tribes in Mexico; used since pre-hispanic times.

Tiger Lily → Lilium tigrinum Ker-Gawl.

Tigerflower, Common → Tigridia Pavonia (L. f.) Ker Gawl.

Tigerwood Lovoa → Lovoa Klaineana Pierre.

Tik → Curcuma angustifolia Roxb.

Tikoar → Curcuma angustifolia Roxb.

Tilia americana L. American Basswood, Linden. (Tiliaceae). - Tree. Eastern N. America to Flo-rida and Texas. Wood light, tough, soft, light brown tinged with red; used for excelsior, cheap furniture, wooden ware, carriages, inner soles of shoes, panels. Fibrous inner bark is made into mats, cordage an similar material. Flowers are source of a good honey.

Tilia cordata Mill. (syn. T. parvifolia Ehrh., T. ulmifolia Scop.) Small-Leaved Linden. (Tiliaceae). - Tree. Europe, Caucasia. Cultivated. Wood light colored, fairly coarse-grained, soft, easy to split, fairly elastic; used for making statues, wagons, tables, brakes, spoons, plates, kitchen-furniture, inlay-work, instruments, organs; source of excelsior and excellent charcoal; used for drawing and for gun-powder. Leaves are used to adulterate tobacco. Flowers employed in medicine and as a tea, causing perspiration, as stomachic, for cramp, also used as a mouthwash and gargle water. Flowers produce an excellent commercial honey. Also T. platyphyllos Scop. Bark produces a fibre used for manuf. Muscovite mats and other material.

Tilia japonica Simk. Japanese Linden. (Tiliaceae). - Tree. Japan, China. Wood light, soft, yellow white. Used in Japan for packing boxes, barrels, matches. Fibre of bark is used by Ainu for ropes.

Tilia Tuan Szyszyl. Tuan Linden. (Tiliaceae). - Tree. W. and Centr. China. Inner bark made into sandals; used by mountaineers in China.

Tiliaceae → Berria, Cistanthera, Columbia, Corchorus, Diplodiscus, Entelea, Grewia, Honkenua, Luehea, Muntingia, Schoutenia, Sparmannia, Tilia, Trichospermum, Triumfetta.

Tillandsia usneoides L. (syn. Dendropogon usneoides (L.) Raf.). Spanish Moss, Florida Moss, Wool-Crape. (Bromeliaceae). - Epiphyte. Southern United States, West Indies, South America to Argentina. Entire plants used as packing material. Cured stems employed in upholstery, known as Louisiana Moss, Vegetable Hair, Crin Végétal.

Tillandsia xiphoides Ker. (syn. T. odorata Gill., T. macrocnemis Griseb.). (Bromeliaceae). - Perennial herb. Argentina to Bolivia and adj. territory. The strongly scented flowers are used in the preparation of a medicine by the rural population for ailments of the chest.

Timberline Bluegrass → Poa rupicola Nash.

Timonius Rumphii DC. (Rubiaceae). - Tree. Malaysia, Queensland, S. Australia. Fruits of the shape of a wild apple, edible; esteemed by the natives.

Timor Chaste Tree → Vitex littoralis Cunningh.

Timothy, Alpine → Phleum alpinum L.

Timothy Grass → Pleum pratense L.

Tinder → Calvatia lilacina (Berk. and Mont.) Lloyd. and Fomes fomentarius (L.) Gill.

Tinospora Bakis Miers. (syn. Cocculus Bakis Guill. and Perr.). (Menispermaceae). - Liane. Trop. Africa. Bitter root is used in Senegal for fevers and as diuretic.

Tinospora cordifolia Miers. (syn. T. tuberculata Beumee.). Galancha Tinospora. (Menispermaceae). - Vine. Malaysia. Introd. in different parts of trop. Asia. Stems, laeves and roots are used in Hindu medicine for fevers. In E. India and Malaya it is used for washing sore eyes.

Tintero → Basanacantha armata Hook. f.

Tirmania africana Chat. Gros Terfaz Blanc. (Terfeziaceae). - Ascomycete. Fungus. N. Africa. Fruitbodies are consumed as food among the natives. They are collected in October.

Titi → Cliftonia monophylla (Lam.) Sarg.

Titi, Black → Cyrilla racemiflora L.

Tizon Orange → Citrus reticulata Blanco.

Tobacco, Aztec → Nicotina rustica L.

Tobacco, Common → Nicotiana Tabacum L.

Tobacco, Hottentot → Tarchonanthus camphoratus L.

Tobacco, Indian → Lobelia inflata L.

Tobacco, Tree → Nicotiana glauca Graham.

Tobacco, Tree Wild → Acnistus arborescens (L.) Schlecht.

Tobacco substitutes → Achillea, Arctostaphylos, Artemisia, Astilbe, Beta, Canabis, Cornus, Corylus, Crataegus, Daphniphyllum, Fagus, Laburnum, Myrrhis, Rhus, Scolopendrium, Tarchonanthes, Tilia.

Toddalia aculeata Pers. (syn. Paullinia asiatica L.). Wild Orange Tree. (Rutaceae). - Shrub. Subtr. Himalaya, Khasia Mts., China, Ceylon, Madagascar. Root called Lopez Root, is source of a yellow dye; used in India.

Toddalia lanceolata Lam. (Rutaceae). - Tree. Trop. E. Africa. Fruits are used in place of Cubeb, Piper Cubeba L. f. Also source of White Iron Wood.

Toddy → Arenga pinnata (Wurmb.) Merr.

Toledo Wood → Lecythis longipes Poit.

Tolu Balsam → Myroxylon toluiferum H. B. K.

Tolu Balsam Tree → Myroxylon toluiferum H. B. K.

Toluifera Balsamum L. → Myroxylon toluiferum H. B. K.

Toluifera Pereirae (Klotzsch.) Baill. → Myroxylon Pereirae Klotzsch.

Tomato → Lycopersicum esculentum Mill.

Tomato, Children's → Solanum anomalum Thonn.

Tomato, Strawberry → Physalis Alkekengi L.

Tomato, Tree → Cyphomandra betacea Sendt.

Tonka Beans → Dipteryx odorata Willd.

Tonka Bean (substitute) → Melilotus officinalis (L.) Lam.

Tonka, Dutch → Dipteryx odorata Willd.

Tonka Tree → Dipteryx odorata Willd.

Tootache Tree → Zanthoxylum Clava-Herculis L.

Tooth Brush Tree → Gouania lupuloides (L.) Urb.

Toothed Lavender → Lavandula dentata L.

Toothwort → Lathraea Squamaria L.

Toparejo → Inga Ruiziana G. Don.

Torchwood → Bikkia mariannensis Brogn.

Tordylium apulum L. (syn. Concylocarpus Apulus (L.) Hoffm.) (Umbelliferaceae). - Herbaceous perennial. Mediterranean region. Young plants are eaten as vegetable in some parts of Greece.

Torfez → Terfezia Boudieri Chatin.

Tormentil → Potentilla erecta (L.) Hampe.

Tormentilla erecta L. → Potentilla erecta (L.) Hampe.

Torresea cearensis Allem. → Amburana cearensis (Allem.) Sm.

Torrey's Pine → Pinus Torreyana Carr.

Torreya grandis Fort. Chinese Torreya. (Taxaceae). - Tree. China, esp. Hupeh and Szechuan. Roasted seeds are consumed in China; frequently sold in markets. Also used in medicine, called Fei Shu.

Torreya nucifera Sieb. and Zucc. Japanese Torreya. (Taxaceae). - Tree. Japan. Kernels are eaten in Japan raw or roasted, having an aromatic flavor. They are used in confectionary; also source of an oil. The var. Shibunashi-gaya is of the best quality, found the provinces Mino, Iga and Yoamato.

Torreya taxifolia Arn. (syn. Tumion taxifolia (Arn.) Greene. Florida Torreya. (Taxaceae). - Northwestern Florida. Wood is durable; sometimes used for fence posts.

Torula cremoris Hammer and Cordes. (Saccharomycetaceae). - A lactose forming yeast, being able to ferment all the lactose in a relatively short time.

Torula pulcherrima (Saccharomycetaceae). - A yeast that has been used commercially in the food and fodder yeast production for animal feeding. It is of value in converting waste, low cost carbchydrates and certain surplus materials into products used as food for cattle, hogs and other animals.

Torulopsis utilis (Hanneb.) Lodder (Saccharomycetaceae). - A yeast used in the manufacturing of animal fodder from waste, low-cost carbohydrates and surplus materials. It is valuable on account of its high protein content and vitamins of the B-complex. It produces little alcohol. Var. thermophila is able to grow under higher temperatures than most of the food-yeasts.

Totara → Podocarpus spinulosa R. Br.

Totum → Fillaeopsis discophora Harms.

Touch Wood → Amyris balsamifera L.

Touchardia latifolia Gaud., Olona. (Urticaceae). - Shrub. Hawaii Islands. Source of a strong fibre; used by the Hawaiians for fishing nets and fishing lines. Very resistant to fresh and salt water.

Touloucouna Oil → Carapa procera DC.

Towri → Caesalpinia digyna Rottl.

Toxicodendron capense Thunb. (Euphorbiaceae). - Tree. S. Africa. Fruits are used for poisoning hyanas.

Toxophoenix aculeatissima Schott. → Astrocaryum Ayri Mart.

Trachelospermum Stans Gray. (Apocynaceae). - Shrub. Mexico. Plant is used for poisoning cockroaches.

Trachycarpus excelsus Wendl. (Palmaceae). - Medium palm. N. E. India, China, Japan. Fibre from bark is used by Chinese for cordage, scrubbing-brushes; leaves made into hats and raincoats.

Trachylobium Hornemannianum Hayne. (syn. T. mossambicense Klotzch., Cynometra Spruceana Benth.). (Leguminosaceae). - Tree. Trop. Africa. Source of Anime or Zanzibar Copal; probably also of Mazambique Copal and Madagascar Copal.

Trachylobium verrucosum Oliv. (Leguminosaceae). - Tree. E. Africa, Mascar. Islds. Introd. in tropics. Root, trunk and fruits are source of a Zanzibar Copal, much is derived from semi fossil material dug from the soil. Also called Gum Copal of Madagascar. Related or identical to the previous species.

Tragacanth → Astragalus gummifera Lab.

Tragacanth, African → Sterculia Tragacantha Lindl.

Tragacanth, Indian → Astragalus heratensis Bunge.

Tragacanth, Morea → Astragalus cylleneus Boiss. and Heldr.

Tragacanth, Persian → Astragalus pycnocladus Boiss. and Haussk.

Tragopogon porrifolius L., Salsify, Vegetable Oyster, Oyster Plant. (Compositae). - Annual to biennial herb. Mediterranean region. Cultivated in Old and New World. Roots when boiled are eaten as a vegetable. Suitable for winter-use. A variety is Mammoth Sandwich Island. Latex from roots was used by Indians of Brit. Columbia for chewing.

Trametes lactina Berk. (Polyporaceae). - Basidiomycete. Fungus. Fruitbodies are used instead of pith for mounting different species of insects in collections.

Trapa bicornis L. (Onagrariaceae). - Annual waterplant. China. Cultivated in China, Korea and Japan. Seeds used as food in the Far East; consumeu in various dishes and preparations; preserved in honey and sugar. Eaten by the Chinese during Full Moon Festival. Seeds are boiled in water after which the kernels are removed and are eaten while still warm.

Trapa bispinosa Roxb. Singhara Nut. (Onagrariaceae). - Annual waterplant. Trop. Asia. Cultivated in India. Seeds are consumed by the natives. Also made into a flour which is much esteemed in some parts of India.

Trapa incisa Sieb. and Zucc. (Onagrariaceae). - Herbaceous waterplant, Japan. Seeds boiled or roasted, article of food among the Ainu.

Trapa natans L. Water Caltrop, Water Chestnut. (Onagrariaceae). - Annual waterplant. Europe, Mediterranean region, Asia, introd. in N. America, Australia. Seeds were used as food since Neolitic times. At present consumed in some parts of Centr. Europe. Sold in local markets of France. Seeds are also made into rosaries.

Travancore Starch → Curcuma angustifolia Roxb.

Traveller's Tree → Ravenala madagascariensis Gmel.

Travers Daisybush → Olearia Traversii F. v. Muell.

Trecul Yucca → Yucca treculeana Carr.

Treculia africana Decne., African Breadfruit Tree. (Moraceae). - Tree. Trop. Africa. Seeds are much esteemed as a food by the native negroes; they are sold in the markets of Angola, St. Thomé etc. Source of a flour in W. Africa. Fruits are a favorite food of elephants.

Tree Burrofat → Isomeris arborea L.

Tree Cotton → Gossypium arboreum L.

Tree Fern → Alsophila, Cibotium, Cyathea, Dicksonia.

Tree Fern, Black Stemmed → Cyathea medullaris Swartz.

Tree of Heaven → Ailanthus glandulosa Desf.

Tree Mallow → Lavatera plebeia Sims.

Tree Medic. → Medicago arborea L.

Tree Peony → Paeonia Moutan Sims.

Tree Tobacco → Nicotiana glauca Graham.

Tree, Wild Tobacco → Acnistus arborescens (L.) Schlecht.

Trefoil, Birds Foot → Lotus corniculatus L.

Trefoil, Sweet → Trigonella coerulea (L.) Ser.

Trema guineensis Ficalho. (Ulmaceae). - Shrub or small tree. W. and E. Africa, Madagascar. A tea obtained from small pieces of roasted wood is used in Lagos for dysentery. Wood soft, light; used for building purposes; in Angola it is used for the Samba Viola.

Trema orientale Blume. (Ulmaceae). - Tree. Mal. Archipelago. Though bark used for cordage, also source of a brown dye; used for treating fishing-nets.

Trembling Aspen → Populus tremuloides Michx.

Tremella fuciformis Berk. (Tremellaceae). - Basidiomycete, Fungus. Tropics and subtropics. Fruitbodies are consumed as food by the Chinese.

Tremellaceae (Fungi) → Tremella.

Trementina de Ocote (Turpentine) → Pinus Teocote Cham. and Schlecht.

Treversia moluccana Miq. → Boerlagiodendron palmatum Harms.

Trevoa trinervia Miers. (syn. Colletia Treba Bert.). Chilean Trevoa. (Rhamnaceae). - Shrub. Chile. Bark is used among the country people of Chile for wounds caused by burns.

Trianthema pentandra L. (Aizoaceae). - Prostrate herb. Trop. Africa, Asia. Plant is used in the Sudan as astringent in abdominal diseases. Leaves are consumed in some parts of India as a potherb during femines.

Trianthema salsoloides Fenzl. (Aizoaceae). - Herbaceous plant. Trop. Africa. Ashes of the herb with Simson Oil and lime are made into a soap in Sudan.

Tribulus terrestris L. (Zygophyllaceae). - Annual herb. Tropics of Old and New World. Fruits are used as diuretic in Sudan.

Trichilia capitata Klotzsch. (Meliaceae). - Tree. Trop. Africa. Roots are used in Portug. E. Africa by the natives for snake-bites.

Trichilia cedrata Chev. → Guarea cedrata (Chev.) Pel.

Trichilia emetica Vahl., Mafura Bitterwood, Cape Mahogany, Manubi Mahogany, Natal Mahogany. (Meliaceae). - Trop. Africa. Wood used for furniture, yokes, canoes. Seeds are source of a solid fat called Mafura Oil or Mafura Tallow, considered poisonous; used by the natives as ointment; commercially used for manuf. soap and candles. Sp. Gr. 0.931; Sap. Val. 195—202; Iod. No. 66—70; Unsap. 0.6—0.8 %.

Trichilia glabra L. (syn. T. havanensis Jacq.). (Meliaceae). - Tree. Trop. America. Wood compact, light yellow; used for handles of tools.

Trichilia glandulosa Smith → Synoum glandulosum Juss.

Trichilia quinquevalvis Montr. (Meliaceae). - Small tree. New Caledonia. Young wood yellowish white; old wood red, fine-grained, attractive; used for cabinet-making.

Trichilia spondioides Jacq. White Bitterwood. (Meliaceae). - Tree. West Indies. Wood has similar properties as T. glabra L.

Trichocereus chiloensis Colla (syn. Cereus Quisco Rem.). Cardon de Candelabro. (Cactaceae). - Xerophytic shrub. S. America, esp. Argentina and Chile. Fruits are edible, made into a brandy and a syrup.

Trichocereus pasacana (Weber) Britt. and Rose. (Cactaceae). - Xerophytic shrub. Argentina, Bolivia. Trunks are used by the natives of Bolivia for huts and goat corrals.

Tricholaena rosea Nees., Natal Grass. (Graminaceae). - Perennial grass. S. Africa. Cultivated in warm countries of Old and New World. Source of a hay.

Tricholaena sphacelata Benth. (Graminaceae). - Perennial grass. Trop. Africa, esp. Nigeria, Angola, Rhodesia, Brit. E. Africa. A good fodderplant for live-stock.

Tricholoma equestre (L.) Quél. (Agaricaceae). - Basidiomycete. Fungus. Temp. zone. Fruitbodies are consumed as food. Sold in markets of Europe and Japan. Also T. pessundatum (Fr.) Quél., T. columbetta (Fr.) Quél., T. flavobrunneum (Fr.) Quél. and other species.

Tricholoma Georgii (Fr.) Quél. (syn. T. gambosum (Fr.) Gill.). (Agaricaceae). - Basidiomycete. Fungus. Europe. Fruitbodies are consumed as food and are considered a delicacy. Sold in markets.

Tricholoma mongolica Imai (Agaricaceae). - Basidiomycete. Fungus. Eastern Asia, esp. Mongolia and China. Fruitbodies, called Pai ku mo, are consumed as food by the Mandshu and Mongolians. Sold in markets.

Tricholoma nudum Bull. → Rhodopaxillus nudus (Bull.) R. Maire.

Tricholoma rutinlans (Schäff.) Quél. (Agaricaceae). - Basidiomycete. Fungus. Temp. zone. Fruitbodies are sometimes consumed as food. Occasionally sold in markets.

Trichosanthes Anguina L. Edible Snake Gourd. (Cucurbitaceae). - Annual vine. Trop. Asia. The long fruits are cut in pieces, boiled and consumed by the natives.

Trichosanthes cucumeroides Max. Japanese Snake Gourd. (Cucurbitaceae). - Perennial vine. Japan, China. Roots are source of a starch, made in Japan. Dried fruits are used for washing, instead of soap.

Trichosanthes cucumerina L. (Cucurbitaceae). - Annual vine. Trop. Asia, Australia. Fruits are eaten cooked, they are much esteemed by the Hindus.

Trichosanthes japonica Regel. (Cucurbitaceae). - Perennial vine. Japan. Cultivated. Young fruits when salted or kept in soy, are consumed in Japan. Roots are source of a starch, called Tenkwa-fun.

Trichosanthes subvelutina F. v. Muell. (syn. T. palmata Benth.). (Cucurbitaceae). - Vine. Australia. Roots are roasted and consumed as food by the aborigines of Queensland.

Trichospermum Kurzii King. (Tiliaceae). - Tree. Nicobar Islands, Malayan Peninsula. Fibre used as rough cordage in S. Perak.

Trifolium alexandrinum L., Egyptian Clover. (Leguminosaceae). - Annual herb. Mediterranean region, N. Africa. Cultivated as a green manure and as fodder for live-stock.

Trifolium alpinum L. (Leguminosaceae). - Perennial herb. Alpine meadows in Europe and Caucasia. Considered excellent fodder for livestock in alpine regions.

Trifolium amabile H. B. K. Aztec Clover, Chicmu. (Leguminosaceae). - Herbaceous plant. Peru, Patagonia, Argentina to Mexico. Herb mixed with white maize and some other plants were used as a food, called Chucan, consumed by Indians of Peru.

Trifolium ciliatum Nutt. (Leguminosaceae). - Herbaceous plant. Western United States. Leaves and stems are eaten raw or cooked by the Luiseño Indians of California. Also T. gracilentum Torr. and Gray, T. microcephalum Pursh. and T. tridentatum Lindl.

Trifolum hybridum L., Alsike Clover. (Leguminosaceae). - Biennal to perennial herb. Temp. Europe, Caucasia, Asia Minor, N. Africa. Cultivated in Old and New World. Pasture and hay crop.

Trifolium incarnatum L., Crimson Clover. (Leguminosaceae). - Annual to biennal herb. Centr. and S. Europe, Balkans, N. Africa. Cultivated in Old and New World. Grown as cover-crop, green mannure; in pastures and as hay crop.

Trifolium Lupinaster L. (Leguminosaceae). - Perennial herb. Russia, Siberia, Poland, Korea, Amur region, Turkestan, Mongolia, Japan. Grown as a hay crop and as green fodder for cattle.

Trifolium pannonicum Jacq., Hungarian Clover. (Leguminosaceae). - Perennial herb. Centr. and S. Europe, Balkans, Caucasia, Ukrania. Occasionally cultivated as a hay crop in some parts of Centr. Europe.

Trifolium parviflorum Ehrh. Teasel Clover, Small flowered Clover. (Leguminosaceae). - Annual herb. S. and S. E. Europe. Occasionally cultivated in Hungary as green fodder for cattle, also as a hay crop.

Trifolium pratense L., Red Clover. (Leguminosaceae). - Perennial herb. Temp. Europe, Asia. Cultivated in both hemispheres, in pastures, for hay, silage often grown with Timothy grass. Dried inflorescence is used medicinally as alterative and sedative; contains a volatile oil, and coumaric acid. Flowers are source of a yellow dye; used with alum as a mordant.

Trifolium repens L., White Clover. (Leguminosaceae). - Perennial herb. Europe, temp. Asia, N. Africa. Cultivated in Old and New World. An important pasture plant and hay crop. Source of commercial honey.

Trifolium resupinatum L. (syn. T. suaveolens Willd.) Persian Clover. (Leguminosaceae). - Annual herb. Mediterranean region to Iran, Afghanistan and India. Cultivated as a source of fodder for live-stock.

Triglochin maritima L., Shore Podgrass. (Juncaginaceae). - Perennial herb. N. temp. zone. Very young leaves are eaten as vegetable. Has been recommended as an emergency food plant.

Triglochin procerum R. Br. (Juncaginaceae). - Perennial herb. Australia. Rhizomes are consumed by the natives in some parts of Australia.

Trigonella coerulea (L.) Ser., Sweet Trefoil. (Leguminosaceae). - Annual herb. Mediterranean region. Cultivated. Herb is used in Switzerland for flavoring green cheese.

Trigonella Foenum Graecum L., Fenugreek, Fenugrec. (Leguminosaceae). - Annual herb. Mediterranean region. Cultivated in S. Europe, N. Africa, India. Used as food for live-stock. Powdered seeds used medicinally as demulcent, emmolient; also used in vet. medic. Contains a mucilage and trigonelline.

Trigoniastrum hypoleucum Miq. (Polygalaceae). - Large tree. Sumatra, Malayan Penin. Wood hard, fine grained, pale lemon yellow. Used in Malaya to make tables.

Trilisa odoratissima (Walt.) Cass. Vanilla Trilisa, Deer's Tongue, Vanilla Plant. (Compositae). - Perennial herb. South and Southeast of the United States. Leaves are used as a flavoring agent in smoking tobacco. Contains coumarin.

Trillium erectum L. Purple Trillium, Bethroot. (Liliaceae). - Perennial herb. Eastern N. America. Dried rhizome and roots are used medicinally as a uterine stimulant. Contains trillin, a saponin.

Trillium grandiflorum (Michx.) Salisb. Snow Trillium, Wake Robin. (Liliaceae). - Perennial herb. Eastern N. America. Leaves have been recommended as greens when cooked; also other species of Trillium, as an emergency food-plant.

Trimeza lurida Salisb. (syn. Sisyrinchium galaxoides Gomes.). (Iridaceae). - Perennial herb. Trop. America. Plant is used in Colombia as a laxative.

Trinidad Devilpepper → Rauwolfia canescens L.

Trinidad Roupola → Roupola montana Aubl.

Triphasia aurantiola Lour. (syn. T. trifoliata DC.) (Rutaceae). - Shrub. Trop. Asia. Cultivated. Fruits are ovoid, red, fleshy, eaten raw, cooked or preserved.

Triplaris surinamensis Cham. Surinam Ant Tree. (Polygonaceae). - Tree. Guiana. Wood light, strong, tough, elastic, gray-brown, not very durable; used for interior finish and for boxes.

Tripsacum laxum Nash., Guatemala Grass. (Graminaceae). - Perennial grass. Centr. America. A good fodder grass for live stock in warm countries.

Tripterygium Wilfordii Hook. f. (Celastraceae). - Shrub. Chekiang (China), Japan. Powdered roots recommended as a powerful insecticide; used in some parts of China.

Tristachya leiostachya Nees. (Graminaceae). - Perennial grass. Trop. America. Considered an excellent fodder grass for live-stock in different parts of S. America.

Tristania conferta R. Br. (syn. T. subverticillata Wendl., Lophostemon arborescens Schott.) White Box, Brush Box, Red Box. (Myrtaceae). - Tree. Australia. Wood very strong, durable. Used for ship-building, bridges, wharves, not readily attacked by termites.

Tristania suaveolens Smith (syn. Melaleuca suaveolens Gaertn.). Bastard Peppermint, Broadleaved Water-Gum. (Myrtaceae). - Tree. Australia. Wood very strong, elastic, durable, close-grained, tough. Used for tool-handles, cogs of wheels, posts.

Tristania Whitiana Griff. (syn. T. sumatrana Miq.). (Myrtaceae). - Tree. Sumatra. Wood is hard, heavy; used by the natives for building houses.

Tristaniopsis Guillaini Vieill. (Myrtaceae). - Small tree. New Caledonia. Wood red, very hard, fine-grained and dense; used for turnery.

Triticum carthlicum Nevski (syn. T. persicum Vav.) Persian Wheat. (Graminaceae). - Cultivated as a grain crop in the turkish-caucasian region up to 2100 m. elevation.

Triticum compactum Host., Club Wheat. (Graminaceae). - Annual grass. Europe. Cultivated in Old and New World. Flour used for bread making, crackers, starchy breakfast foods.

Triticum dicoccum Schrank., Emmer. (Graminaceae). - Annual grass. Europe, Balkans, Caucasia, temp. Asia. Cultivated. Soarce of a flour, also used as feed for live-stock. Divided in Spring Emmer and Winter Emmer. Some varieties are: Heidelberger Emmer, Elsasser, Reisdinkel, Black Winter.

Triticum durum Desf., Durum Wheat, Macaroni Wheat. (Graminaceae). - Annual grass. S. Russia, Mediteranean region, Asia. Grown in Old and New World. Source of macaroni, semolina, spaghetti and similar products. Several varieties among which: Pentad, Kubanka, Akrona, Marouani.

Triticum Macha Dekr. and Men. (Graminaceae). - Related to T. spelta L. Cultivated in Georgia and other parts of S. Russia. Of little economic value.

Triticum monococcum L., Einkorn. (Graminaceae). - Annual grass. Europe. Cultivated. Supposed to be the most primitive wheat. Occasionally grown in Germany, Switzerland and Italy.

Triticum orientale Percival. Kharassan Wheat. (Graminaceae). - Grown as a grain crop in some of the Medit. countries, Abyssinia and countries of the Near East.

Triticum polonicum L., Polish Wheat, Astrakan Wheat, Giant Rye, Jerusalem Rye. (Graminaceae). - Annual grass. Centr. Europe. Flour suitable for macaroni, not adapted for bread making.

Triticum Spelta L., Spelt. (Graminaceae). - Annual grass. Europe. Cultivated for production of flour. Varieties are: White Spring, Alstroum, Red Winter.

Triticum sphaerococcus Percival. Indian Dwarf-wheat. (Graminaceae). - Grown as a grain crop in Pundjab and Central Provinces of India.

Triticum turgidum L., Poulard Wheat. (Graminaceae). - Annual grass. Cultivated. Flour made into macaroni, spaghetti, vermicelli; also used as feed for live-stock. Varieties: Alaska, Titanic.

Triticum vulgare Vill. (syn. T. sativum Lam.). Common Wheat. (Graminaceae). - Annual grass. Origin uncertain. Cultivated since ancient times in temp. and subtr. zones of the Old and New World. Flour of the hard wheat varieties are used in bread making; flour of the soft wheats are used for biscuits, cakes, crackers etc. Made into breakfast foods, bran flakes, puffed wheat, shredded wheat. Wheat Flour contains essentially starch and gluten; whole Wheat Flour or Graham Flour contains all constituents of cleaned grain. Source of alcoholic beverages, beer, industrial alcohol made into synth. rubber and explosives. Starch is used for sizing textiles. Straw is made into mats, seats of chairs, carpets, beehives, hats, baskets; used as packing material, for cattle to stand upon in stables, manuf. of paper material, stuffing of mattresses. The numerous varieties are divided into (1) Spring and Winter Wheats, (2) Hard and Soft Wheats, (3) Red and White Wheats, (4) Bearded and non-Bearded varieties.

Tritonia aurea Papp. → Crocosma aurea Planch.

Triumfetta Lappula L. (Tiliaceae). - Shrub. Mexico, West Indies, Centr. America, W. Africa. Bark source of a strong and fine fibre. An astringent mucilage is sometimes used for clarifying syrup.

Triumfetta rhomboidea Jacq. (Tiliaceae). - Woody plant. Tropics. Bark is source of a soft, glossy fibre; used for different purposes.

Triumfetta semitriloba Jacq. (Tiliaceae). - Shrub. Mexico, West Indies, Centr. America, S. America. Source of a tough fibre; used for rope and coarse cloth.

Trixus neriifolia Bonpl. → Libothamnus neriifolius Ernst.

Trochodendraceae → Trochodendron.

Trochodendron araloides Sieb. and Zucc. (Trochodendraceae). - Tree. Japan. Pounded bark is a source of bird-lime in Japan.

Troitzki Truffle → Choiromyces venosus (Fr.) Th. Fr.

Tropaeolaceae → Tropaeolum.

Tropaeolum brasiliense Casar. (Tropaeolaceae). - Vine. Brazil. Leaves are occasionally used in Brazil in salads; also as antiscorbutic.

Tropaeolum edule Paxt. (Tropaeolaceae). - Herbaceous plant. Chile, Peru. Tubers are consumed by the Indians in times of scarcity.

Tropaeolum majus L. Common Nasturtium. (Tropaeolaceae). - Herbaceous vine. S. America, esp. Peru, Bolivia, Colombia, Brazil. Cultivated as ornamental in Old and New World. Flower-buds and young fruits are used for flavoring vinegar, also used as capers.

Tropaeolum tuberosum Ruiz. and Pav. Tuber Nasturtium, Ysaño. (Tropaeolaceae). - Perennial vine. Peru. Cultivated in Peru, Chile and Bolivia. Tubers when boiled are consumed as vegetable. Also T. patagonicum Speg.

Tropical Almond → Terminalia Catappa L.

True Bay → Laurus nobilis L.

True Myall → Acacia pendula Cunningh.

Truffle → Terfezia spp. and Tuber spp.

Truffle, Troitzki → Choiromyces venosus (Fr.) Th. Fr.

Trumpet Lichen → Cladonia fimbriata (L.) Willd.

Trumpets → Sarracenia flava L.

Truxillo Coca → Erythroxylon nova-granatense (Morris) Hier.

Tsuga canadensis (L.) Carr. Eastern Hemlock. (Pinaceae). - Tree. Eastern N. America. Wood easy to work, light brown, red tinged; used as a coarse lumber, exterior finish of buildings, lumber, exterior finish of buildings, pulpwood. Bark used for tanning leather. Young branches produce, upon distillation Oil of Hemlock, Canada Pitch or Hemlock Pitch. An oleoresin is gathered from incisions in the trunk, or by boiling pieces of bark and wood, skinning off the oleoresin. Tea was made from the leaves; used by the Indians and by lumbermen of Maine and some parts of Canada.

Tsuga chinensis Pritz. Chinese Hemlock. (Pinaceae). - Tree. Hupeh, Szechuan, Shensi (China). Wood soft, durable; used in China for shingles.

Tsuga heterophylla (Raf.) Sarg., Western Hemlock. (Pinaceae). - Tree. Western N. America. Wood pale brown to almost white, easy to work, hard tough, moderately light, rather week in bending and compressive strength, used for building material, subflooring, joists, planking, rafters, boxes, crates, furniture, ladders, laundry appliances, refrigerators, sash, doors, general millwork, motor vehicles, Venetian blinds, pulpwood. Inner bark was used as food by the Indians of Alaska.

Tsuga yunnanensis (Franch.) Mast. Yunnan Hemlock. (Pinaceae). - Tree. W. China. Wood used for shingles and boards in China.

Tuan Linden → Tilia Tuan Szyszyl.

Tuart Eucalyptus → Eucalyptus gomphocephala DC.

Tuba (wine) → Corypha elata Roxb.

Tuber aestivum Vitt. (syn. T. cibarium Sowerb., T. albidum Fr.). Truffle, Truffe d'Été, Truffe Blanche. (Eutuberaceae). - Ascomycete. Fungus. Europe, Mediterranean Region. Fruitbodies are edible and of excellent quality. They have been consumed as food for centuries in many parts of Europe. They are eaten baked, with wine, in butter, oil, salads, and soups. They form an ingredient of certain liver-sausages, Trüffel-Leberwurst. They are consumed fresh when very young. They are sold in markets and are an important commercial commodity. Truffels are collected in summer and late autumn. They are gathered under trees by trained dogs and swine.

Tuber albidum Fr. → Tuber aestivum Vitt.

Tuber album With → Choiromyces venosus (Fr.) Th. Fr.

Tuber brumale Vitt. (syn. T. cibarium Fr.). Truffe Violette. (Eutuberacea). - Ascomycete. Fungus. Centr. and S. Europe, esp. N. Italy, France, Switzerland and S. Germany. Fruitbodies are consumed in various dishes and are of fine quality. They are sold in markets. Much is derived from France.

Tuber cibarium Bull. → Tuber melanosporum Vitt.

Tuber cibarium Fr. → Tuber brumale Vitt.

Tuber cibarium Sowerb. → Tuber aestivum Vitt.

Tuber excavatum Vittad. (syn. Vittadinion Montagnei Zobel). Truffe Jaune. (Eutuberaceae). - Ascomycete. Fungus. Centr. Europe, esp. N. Italy, France and England. Fruitbodies are consumed as food.

Tuber Gennadii (Chatin) Pat. (syn. T. lacunosum Matt., Terfezia Gennadii Chatin). (Eutuberaceae). - Ascomycete. Mediterranean region, Canary Islds. Fruitbodies are consumed as food, known as Quiza or Tartufi Bianchi.

Tuber griseum Fr. → Tuber magnatum (Pico) Vitt.

Tuber lacumosum Matt. → Tuber Gennadii (Chatin) Pat.

Tuber magnatum (Pico) Vitt. (syn. T. griseum Fr.) Truffe Grise, Truffe Blonde. (Eutuberaceae). - Ascomycete. Fungus. N. Italy, France. Fruitbodies are edible, being of a fine flavor. They are much esteemed in N. Italy and are of commercial importance.

Tuber melanospermum Vitt. (syn. T. cibarium Bull.). Truffe de France, Truffe Franche, Truffe Vraie. (Eutuberaceae). - Ascomycete. Fungus. N. Italy, Spain, France. Fruitbodies are edible and are of a fine flavor. Sold in markets. Truffles are collected in summer and autumn.

Tuber Nasturtium → Tropaeolum tuberosum Ruiz. and Pav.

Tuber uncinatum Chatin. Truffe de la Bourgogne. (Eutuberaceae). - Ascomycete. Fungus. Centr. and S. Europe. Fruitbodies are consumed as food.

Tuberose → Polianthes tuberosa L.

Tuberose Flower Oil → Polianthes tuberosa L.

Tuberous-rooted Chinese Mustard → Brassica napiformis Bailey.

Tubiflora squamosa (Jacq.) Kuntze. (syn. Elytraria tridentata Vahl. (Acanthaceae). - Annual. Centr. America. Used in El Salvador for dysentery.

Tu-chung → Eucommia ulmoides Hook.

Tuckahoe → Polyporus tuckahoe (Gussow) Lloyd.

Tuckahoe → Poria cocos Wolf.

Tuckahoe, Virginian → Peltandra virginica (L.) Kunth.

Tufted Bamboo → Oxytenanthera nigrocilata Munro.

Tulbaghia alliacea L. (Liliaceae). - Bulbous perennial, S. Africa. Decoction of bulbs is used in Cape Peninsula for intestinal worms. Also T. cepacea L. f. and T. violacea Harv.

Tule → Scirpus lacustris L.

Tulip, Cilician → Tulipa montana Lindl.

Tulip, Edible → Tulipa edulis Baker.

Tulip Tree → Liriodendron tulipifera L.

Tulip Wood → Harpullia pendula Planch.

Tulip Wood → Owenia venosa F. v. Muell.

Tulip Wood → Physocallymma scaberimum Pohl.

Tulipwood, Brazil → Dalbergia cearensis Ducke.

Tulipa edulis Baker. (syn. Orithia oxypetala Gray.) Edible Tulip. (Liliaceae). - Perennial herb. China, Japan. Bulb is source of a starch, made in some parts of Japan. Leaves are consumed as vegetable.

Tulipa Gesneriana L. Garden Tulip. (Liliaceae). - Bulbs of the various garden varieties have been used for making bread in the Netherlands during times of food shortage.

Tulipa montana Lindl. Cilician Tulip. (Liliaceae). - Bulbous perennial. Iran, Baluchistan, Afghanistan. Bulbs are consumed as food in Baluchistan.

Tulipastrum acuminata (L.) Small → Magnolia acuminata L.

Tuma → Poulsenia armata (Miq.) Standl.

Tuna → Opuntia.

Tuna, Miel de → Opuntia.

Tuna, Queso de → Opuntia.

Tunbridge Ware → Chlorosplenium aeruginosum (Oed.) De Not.

Tung Oil → Aleurites Fordii Hemsl.

Tung Oil, Japanese → Aleurites cordata Steud.

Tung Oil Tree → Aleurites Fordii Hemsl.

Tunica prolifera (L.) Scop. Little Leaf Tunic Flowers. (Caryophyllaceae). - Annual herb. S. and Centr. Europe, N. Africa, Caucasia. Flowers are sometimes used as a tea.

Tupelo, Gum → Nyssa aquatica L.

Tupelo, Water → Nyssa aquatica L. and N. biflora Walt.

Tupuru → Sapium taburu Ule.

Turbinaria ornata Kurz. (Fucaceae). - Brown Alga. Oceans and seas throughout Malaysia. Eaten raw by natives, also made into pickles. Likewise T. conoides J. Ag.

Turka Fibre → Apocynum venetum L.

Turkcap Lily → Lilium superbum L.

Turkey Corn → Dicentra canadensis (Gold.) Walp.

Turkey Red Oil → Agonandra brasiliensis Benth. and Hook. and Ricinus communis L.

Turkish Hazel → Corylus Colurna L.

Turkish Honey → Vitis vinifera L.

Turkish Red → Peganum Harmala L.

Turkish Terebinth Pistache → Pistacia mutica Fish and May.

Turmeric → Curcuma longa L.

Turnera diffusa Willd. (syn. T. aphrodisiaca Ward.). Damiana. (Turneraceae). - Small shrub. Trop. America. Dried leaves are used as laxative and stimulant.

Turneraceae → Turnera.

Turnip → Brassica Rapa L.

Turnip-rooted Celery → Apium graveolens L.

Turnip rooted Chervil → Chaerophyllum bulbosum L.

Turnip rooted Parsley → Petroselinum hortense Hoffm.

Turnip, Swedish → Brassica Napo-Brassica Mill.

Turnip Wood → Dysoxylon Muelleri Benth.

Turpentine → Pinus spp.

Turpentine, Bordeaux → Pinus Pinaster Soland.

Turpentine, Carpathian → Pinus Cembra L.

Turpentine, Greek → Pinus halepensis Mill.

Turpentine, Hungarian → Pinus Cembra L.

Turpentine, Indian → Pinus longifolia Roxb.

Turpentine, Japanese → Larix dahurica Turcz.

Turpentine, Jura → Picea excelsa Link.

Turpentine, Strassburg → Abies pectinata DC.

Turpentine, Trementina de Ocote → Pinus Teocote Cham. and Schlecht.

Turpeth Root → Ipomoea Turpethum R. Br.

Turpeth Root, Spanish → Thapsia garganica L.

Tussilago Farfara L. Common Coltsfoot. (Compositae). - Herbaceous perennial. Europe. Asia, N. Africa, N. America. Young leaves are occasionally used in soup, older leaves as a vegetable, and tea. Dried leaves are used medicinally, being demulcent, emolient, expectorant; used for colds, coughs, bronchial catarrh. Contains an acrid ess. oil, a bitter glucoside, resin and gallic acid.

Tussock Grass → Deschampsia caespitosa (L.) Beauv.

Tsubaki Oil → Camellia japonica L.

Twinflower Dolichos → Dolichos biflorus L.

Twinleaf → Jeffersonia diphylla (L.) Pers.

Twinvein Wattle → Acacia binervata DC.

Twisledleaf Garlic → Allium obliquum L.

Two-rowed Barley → Hordeum distichon L.

Tylophora brevipes (Turcz.) Vill. (Asclepiadaceae). - Shrub. Philipp. Islands. Roots used as substitute for Ipecacuanha; also considered specific for colic and emmenagogue.

Tylophora Perrottetiana Descne. (Asclepiadaceae). - Shrub. Phillip. Islands. Leaves are effective as vulnerary.

Tylostemon Mannii Stapf. Spicy Cedar. (Lauraceae). - Tree. Trop. W. Africa. Seeds are eaten raw or roasted by the natives; used in soup, as a vegetable and with rice. Sold in local markets. Fragrant flowers are used to flavor rice and other foods. Fruits are used for dysentery.

Typha angustifolia L. Narrowleaf Cat Tail. (Typhaceae). - Perennial herb. Cosmopolit. Rootstocks are sometimes consumed as food, during times of scarcity. Leaves are made into mats. Hairs of the fruits are used for stuffing mattresses and pillows.

Typha australis Schum. and Thonn. (Typhaceae). - Perennial herb. Africa. Rhizomes are eaten by the natives during times of scarcity. Silky florets are used for stuffing pillows.

Typhaceae → Typha.

Typhonium angustilobium F. v. Muell. (Araceae). - Herbaceous perennial. Australia. Corms after being roasted and pounded several times are used as food by the aborigines of N. Queensland.

Typhonium Brownii Schott. Merrin. (Araceae). - Perennial herb. Australia. Corms when roasted are eaten by the aborigines.

Typhonodorum Lindleyana Schott. (Araceae). - Tall herbaceous plant. Madagascar. Fruits, after long time boiling, are occasionally consumed as food by the natives of Madagascar.

U

Uapaca clusoides Baker. (Euphorbiaceae). - Small tree. Madagascar. Fruits small, edible, sold in native markets.

Uapaca guineensis Muell. Arg. (Euphorbiaceae). - Tree. Trop. Africa, esp. Upper Guinea, S. Leone, Nigeria, Upper Ubangi, Uganda. Fruit edible, has flavor of medlar. Wood a valuable timber; used for beams. Also U. Heudelotii Baill.

Uapaca sansibarica Pax. (Euphorbiaceae). - Tree. Trop. Africa, Zanzibar. Roots are source of a blue dye; used in Zanzibar for staining cloth.

Uapaca Staudtii Pax., Bihambi. (Euphorbiaceae). - Tree. Trop. Africa. Wood brown, relatively heavy and hard, durable, resistant to termites, easy to work; used for building purposes, furniture.

Ubacury → Platonia grandiflora Planch and Triana.

Ubussu → Manicaria saccifera Gaertn.

Ucuhuba Butter → Myristica surinamensis Roland.

Udo → Aralia cordata Thunb.

Uganda Aloë → Aloë ferox Mill.

Ugni Molinae Turcz. → Myrtus Ugni Mol.

Ulex europaeus L. Common Corse. (Leguminosaceae). - Shrub. Europe; introd. in N. America. Twigs when crushed between rollers are considered a food for cattle. The material is not poisonous before flowering.

Ulluco → Ullucus tuberosus Caldas.

Ullucus tuberosus Caldas., Ulluco. Basellaceae). - Perennial herb. S. America, esp. Bolivia, Peru, Colombia. Tubers having size of a nut, are consumed as food by the Indians.

Ulmaceae → Celtis, Chaetoptelea, Gironniera, Phyllostylon, Trema, Ulmus, Zelkova.

Ulmer Pipes → Acer campestre L.

Ulmo → Eucryphia cordifolia Cav.

Ulmus alata Michx., Winged Elm. Red Elm. (Ulmaceae). - Tree. Eastern N. America. Wood not strong, close-grained, difficult to split, heavy, hard, light brown; used for handles of tools and hubs of wheels. Inner bark is used for fastening covers of cotton-bales.

Ulmus americana L. American White Elm. (Ulmaceae). - Eastern N. America. Wood hard, strong, heavy, coarse-grained, light brown, hard to split; used for flooring, cooperage, hubs of wheels, boats and ship-building, saddle-trees and tool-handles.

Ulmus campestris L. (syn. U. glabra Mill.) English Elm. (Ulmaceae). - Tree. Centr. and S. Europe, N. Africa, Syria, N. Iran to Himalaya, Centr. China. Wood brown to reddish-brown, fairly hard, coarse-grained, elastic, difficult to split, durable, takes polish; used for wagons, pumps, ship-building, water-wheels, waterworks, furniture, tobacco-pipes, parts of rifles. Wood ash was source of potash. Leaves are sometimes used as tea, called Warsaw Tea.

Ulmus fulva Michx., Slippery Elm, Sweet Elm, Indian Elm. (Ulmaceae). - Eastern N. America. Wood very close-grained, hard, strong, heavy, easy to split, durable, dark-brown to red; used for sills of buildings, rail-road ties, agricultural implements, hubs of wheels, fence posts. Inner bark is used in medicine; it is collected in spring, contains a thick mucilage, it is demulcent in diarrhea, dysentery, inflamation of urinary tract, also poultice for abcesses. Indians of the Missouri River Valley cooked the bark with buffalo fat. Bark when dried and ground mixed with milk is said to be a food for infants and invalids.

Ulmus japonica Sarg., Japanese Elm. (Ulmaceae). - Tree. Japan, Korea, China, Manchuria, Amurland. Wood hard, heavy, light brown, not easily worked; used in Japan for hubs of wheels and turnery.

Ulmus mexicana (Liebm.) Planch. → Chaetoptelea mexicana Liebm.

Ulmus montana With., Mountain Elm. (Ulmaceae). - Tree. Centr. and S. Europe, Caucasia, Asia Minor, Centr. Asia, Manchuria, Japan. Wood resembling that of U. campestris; used for carriages, turnery, tough, less esteemed than U. campestris.

Ulmus Thomassii Sarg., Rock Elm. (Ulmaceae). - Tree. Eastern N. America. Wood very strong, hard, heavy, tough, light brown, tinged with red, close-grained; used for heavy agricultural implements, chairs, rail-road ties, sills of buildings, hubs of wheels, handles of tools.

Ulva fasciata Delile. (Ulvaceae). - Green Alga. Sea Weed. Consumed as food among the Hawaiians. It is boiled with squid forming a gelatinous mass when cold. Called Limu Lipahapaha.

Ulva Lactuca J. Ag. Lettuce Laver. (Ulvaceae). - Green Alga. Cosmopolitan. Consumed as food by Orientals also as salad and garnishing. Used as food for hogs and other domestic animals in various countries. Also U. latissima L., and other species.

Ulva nematoidea Kuetz. (Ulvaceae). - Green Alga. Plants are used as food by the natives of New Caledonia.

Ulva penniformis Mart. (Ulvaceae). - Green Alga. Sea Weed. Used medicinally in some parts of Siberia.

Ulva pertusa Kjellm. (Ulvaceae). - Green Alga. Sea Weed. Pacif. Ocean and adj. waters. Used as food in the Amoy region, China. It is cooked with oysters, meat or noodles. Also used in Chinese medicine for reducing high body temperatures.

Ulvaceae (Green Algae) → Enteromorpha, Ulva.

Umbellularia californica Nutt., Californian Laurel. (Lauraceae). - Oregon and California. Wood hard, close-grained, heavy, light brown, strong; used for furniture, interior finish of houses. Leaves produce upon steam distillation a pungent essential oil. Fruits contain umbellalic acid.

Umbelliferaceae → Aegopodium, Alepidia, Ammi, Ammodaucus, Anethum, Angelica, Anthriscus, Apium, Archangelica, Arctopus, Arracacia, Athamanta, Aulospermum, Bunium, Carum, Chaerefolium, Chaerophyllum, Cogswellia, Conium, Coriandrum, Crithmum, Cryptotaenia, Cuminum, Cymopterus, Daucus, Dorema, Eryngium, Ferula, Foeniculum, Heracleum, Heteromorpha, Hydrocotyle, Lagoecia, Laserpitium, Levisticum, Lichtensteinia, Ligusticum, Malabaila, Myrrhis, Oenanthe, Opopanax, Pastinaca, Petroselinum, Peucedanum, Phellopterus, Pimpinella, Pseudocymopterus, Sanicula, Scandix, Sium, Smyrnum, Thapsia, Tordylium.

Umbilicaria cylindrica Dub. → Cyrophora cylindrica (L.) Ach.

Umbilicaria pustulata (L.) Hoffm., Blistered Umbilicaria. (Gyrophoraceae). - Lichen. Temp. zone. On rocks. Formerly used in Norway and Germany as a source of a red, purple and brown dye; used for coloring woolens; upon treatment with ammonia. The plant material was made into a pulp with water and ammonia, allowed to ferment; taking 2 to 3 weeks. Also Evernia, Lecanora and Urceolaria spp. were treated in similar way.

Umbrella Grass → Panicum decompositum R. Br.

Umbrella Tree → Melia Azedarach L.

Umkokolo → Dovyalis caffra (Hook. f. and Harv.) Warb.

Uncaria Gambir Roxb., Bengal Gambir Plant. (Rubiaceae). - Vine. Trop. Asia. Leaves and twigs are source of Black Gambir, Cube Gambir, Terra Japonica; used in tanning, producing a soft porous leather, also used in dyeing, printing and for clearing beer. Leaves are used locally by the natives for chewing betel.

Uncaria guianensis Gmel. (Rubiaceae). - Vine. Guiana. Decoction of herb is used for dysentery.

Undaria pinnnatifida (Harv.) Suring. (syn. Ulopteryx pinnatifida Kjellm.) (Alariaceae). - Brown Alga. Eastern Coast of Asia, esp. China and Japan. Consumed as food in China and Japan, often with vinegar, boiled or roasted.

Undulated Ipecac → Richardsonia pilosa H. B. K.

Unicorn Plant → Martynia probocidea Glox.

Unicorn Root → Aletris farinosa L. and Chamaelirion luteum (L.) Gray.

Uniola racemiflora Trin. (Graminaceae). - Perennial grass. West Indies. Stems and leaves used for manuf. paper.

Unona carminativa Aruda → Xylopia carminativa (Aruda) Fries.

Unona concolor Willd. (Annonaceae). - Tree. Guiana. Fruits are used as a condiment.

Unona discolor Vahl. (syn. Desmos chinensis Lour.). (Annonaceae). - Tree. Trop. Asia. Decoction of roots is used for dysentery. Fruits are employed as condiment instead of pepper. The fragrant flowers are source of an ess. oil; used in perfumery.

Unona tripetalodea Moon. → Xylopia parvifolia Hook. f. and Thom.

Unona undulata Dunal. (Annonaceae). - Tree. Trop. Africa. Fruits used as a condiment.

Unona xylopioides Dun. (syn. Uvaria febrifuga Humb. and Bonpl.). (Annonaceae). - Tree. Trop. America. Leaves are used as febrifuge in some parts of Argentina and Parguay.

Unyoro Ebony → Dalbergia melanoxylon Guill and Perr.

Upas Tree → Antiaris toxicaria Lesch.

Uragoga Ipecacuanha Baill. → Cephaëlis Ipecacuanha (Brotero) Rich.

Uranda corniculata Foxw. (syn. Platea corniculata Becc.). (Icacinaceae). - Tall tree. W. Mal. Archipelago. Wood very heavy, does not split, has strong aromatic scent; used for heavy outdoor work, bridges, flag-poles.

Uraria crinita Desv. (Leguminosaceae). - Woody plant. Tropics of Old World. Decoction of roots used in Mal. Peninsul. for diarrhea. Crushed leaves used externally for lice. Plant is also used as greenmanure and cover crop in some parts of the tropics.

Uraria picta Desv. (Leguminosaceae). - Herbaceous. Trop. Africa. Pounded leaves are used in Lagos for gonorrhoea. In So. India supposed to be an antidote for bite of phúrsa snakes.

Urbanella procera Pierre → Lucuma procera Mart.

Urceola acuto-acuminata Boerl. (Apocynaceae). - Woody plant. Mal. Archipelago. Source of a rubber, known as Serapat Rubber.

Urceola brachysepala Hook. f. (Apocynaceae). - Woody vine. Malacca, Sumatra. Plant is source of a good rubber.

Urceola elastica Roxb. (Apocynaceae). - Tree. Trop. Asia, esp. Malacca, Borneo, Sumatra. Source of a rubber.

Urceola esculenta Benth. (Apocynaceae). - Woody plant. Trop. Asia, esp. E. India, Burma, Malaya. Source of Rangoon Rubber. Latex contains 55 to 83 % crude rubber.

Urceola Maingayi Hook. f. (Apocynaceae). - Woody plant. Mal. Archipelago, Sumatra, Borneo. Source of a rubber.

Urceola malaccensis Hook. f. (Apocynaceae). - Woody vine. Malacca. Source of a rubber.

Urceolaria calcarea Sommerf., Limestone Urceolaria. (Lecanoraceae). - Lichen. Temp. zone. On rocks. Used in Great Britain as a source of Cudbear, a red-crimson dye upon treatment with ammonia; employed for coloring woolens.

Urceolaria cinerea Ach., Greyish Urceolaria. (Lecanoraceae). - Lichen. Temp. zone. On rocks. Used in England as source of a red crimson dye to stain woolens.

Urceolaria scruposa (Schreb.) Ach., Rock Urceolaria. (Lecanoraceae). - Lichen. Temp. zone. On rocks. Source of a red brown dye; used in England for coloring woolens.

Urechites suberecta Muell. Arg. (Apocynaceae). - Woody plant. Trop. America. Source of Wooraia poison, an arrow poison, also used to stupefy fish. Employed medicinally for colic, gonorrhoe, psoriasis, warts and snake-bites.

Urena lobata L., Candillo, Aramina Plant. (Malvaceae). - Small shrub. Florida, Antilles, S. America, Africa, Asia. Cultivated. Stem is source of Aramina Fibre, being fine, lustrous, soft; comparable with jute; used for sacking cordage, coarse fabrics, ropes, hammocks, fishing tackle; said to resist termites and water.

Urena sinuata L. (syn. U. heterophylla Smith., U. Lappago Smith.). (Malvaceae). - Shrub. Tropics. Bark is source of a fibre; used in different parts of warm countries.

Urginea altissima Baker. (Liliaceae). - Perennial herb. Trop. Africa, esp. Nupe, Sierra Leone, Nyasaland, S. Africa. Bulbs are used in medicine instead of U. Scilla Steinh.

Urginea indica Kunth. India Drug Squill. (Liliaceae). - Perennial herb. Senegambia, Sierra Leone, Eritrea, Brit. E. Africa, India, etc. Bulbs are used as a substitute of the officinal Squill, U. Scilla Steinh.

Urginea Scilla Steinh. (syn. U. maritima (L.) Baker). Sea Onion. (Liliaceae). - Bulbous perenial. Mediterranean region. Used medicinally. Squills are derived from bulbs, collected in August. Cardiac stimulant, diuretic, emetic, expectorant, nauseant. Contain scillaren-A and scillaren-B glucosides. Red squills derived from a red variety as used as rat poison.

Urginea micrantha Solms. (Liliaceae). - Bulbous perennial. Trop. Africa, Red Sea area. Starch from the bulbs is used by the Hadendowas of Sudan to stiffen the hair.

Urtica argentea Forst. → Pipturus argenteus Wedd.

Urtica Breweri Wats. (Urticaceae). - Herbaceous perennial. Western N. America. Stems were source of a fibre; used by the Klamath Indians of Pac. Coast, for manuf. fishing-nets and cordage.

Urtica cannabina L. (Urticaceae). - Perennial herb. Siberia, Iran. Stems are source of a fibre; used for various purposes.

Urtica dioica L., Bigstring Nettle, Stinging Nettle. (Urticaceae). - Perennial herb. Temp. zone. Young tops of stems and leaves when boiled are consumed as a potherb and in soups among some country people of Europe. Bark is source of a fibre. Recommended as an emergency foodplant. Has been suggested as a possible source of commercial chlorophyll.

Urtica holosericea Nutt. (Urticaceae). - Perennial herb. Western N. America, Pacif. Islds. Stems are source of a fibre, used for manuf. cloth.

Urtica photiniphylla Cunningh. → Laportea photiniphylla Wedd.

Urtica Thunbergiana Sieb. and Zucc. (Urticaceae). - Perennial herb. Japan. Stems are source of a fibre. Young leaves and stems are eaten in Japan as a vegetable.

Urtica urens L., Dog Nettle, Small Nettle. (Urticaceae). - Annual herb. Cosmopolit. Stem is source of a fibre. Leaves and stem tops are sometimes eaten as a pot-herb, recommended as an emergency food-plant. Leaves and flowertops are medicinally a powerful diuretic.

Urticaceae → Aphananthe, Boehmeria, Cannabis, Cypholophus, Debregeasia, Girardinia, Humulus, Laportea, Pipturus, Pouzolzia, Sarcochlamys, Streblus, Touchardia, Urtica, Villebrunea.

Urari → Strychnos Castelnaei Wedd.

Urucurana → Hieronima alchorneoides Allem.

Usnea barbata Hoffm. Bearded Usnea, Old Man' Beard. (Usneaceae). - Lichen. Temp. zone. On stems of Conifers. Was during the 7th Century source of a Cyprus Powder, a toiled powder. Also other Usnea species were used. Lichens are source of an orange-red dye; used for staining woolens. Plants produce glucose upon hydrolysis of lichenin.

Usnea dasypoga (Ach.) Nyl. (Usneaceae). - Temp. zone. Used in medicine by the natives of Java.

Usnea florida (L.) Hoffm., Flowering Usnea. (Usneaceae). - Lichen. Temp. zone. On stems of Conifers. Used in Europe as a source of a green-yellow or red-brown dye, employed for coloring woolens.

Usnea plicata Hoffm., Plaited Usnea. (Usneaceae). - Lichen. Temp. zone. On stems of Conifers. Source of a green and yellow dye; used in Europe for dyeing woolens.

Utah Juniper → Juniperus utahensis (Engelm.) Lemm.

Uvalha → Eugenia Uvalha Gamb.

Uvaria Buesgenii Diels. Annonaceae). - Tree. Trop. Africa. Wood called Paddlewood, is dark orange red to orange, long fibred, hard, light, dense, resistant to termites; used for furniture, pencils, walking-sticks.

Uvaria Burakol Blume → Stelechocarpus Burakol Hook. f.

Uvaria Chamae Beauv. (Annonaceae). - Tree. Trop. Africa, esp. Nigeria, Senegal. Infusion of leaves is used by natives as eye-wash. Wood employed for oars on the river Casamance, Senegal. Used in Agbo, → Xylopia aethiopica.

Uvaria aethiopica Guill. and Poir. → Xylopia aethiopica A. Rich.

Uvaria febrifuga Humb. and Bonpl. → Unona xylopiodes Dun.

Uvaria monopetala Guill. and Perr. → Hexalobus senegalensis A. DC.

Uvaria rufa Blume. Calabao. (Annonaceae). - Shrubby vine. Trop. Asia, esp. Malaya, Philipp. Islds., Mal. Archipelago. Fruits edible, orange-yellow, granular, sweetish flesh. Sold in markets of Philippines. Much esteemed by the natives.

Uvularia sessiliflora L. (syn. Oakesia sessiliflora (L.) Wats.) Bellwort. (Liliaceae). - Perennial herb. Eastern N. America. Young shoots considered a substitute for Asparagus. Roots edible when cooked. Recommended as an emergency food-plant.

V

Vaccinium corymbosum L. Blueberry, Swamp Blueberry, High Bush Blueberry. (Ericaceae). - Shrub. Eastern N. America, southward to Georgia and Louisiana. Cultivated. Several improved large fruited varieties, e. g. Jersey, Rancocas, Dixy, Pioneer, Concord, Weymouth, Early Blue, Blue Ray, being eaten fresh; used in pies and pastries. They are often canned in water or in a syrup. Plants require an acid soil, pH 4.5—5.

Vaccinium erythrocarpum Michx. Dingleberry, Mountain Cranberry, Bear Berry. (Ericaceae). - Small shrub. Southeastern N. America. Fruits variable in flavor; used for jellies. Has been recommended for cultivation.

Vaccinium floribundum H. B. K. Colombian Blueberry, Andean Blueberry, Mortiño. (Ericaceae). - Shrub. Ecuador, Peru. Fruits edible, much esteemed by the natives; sold in markets of Andean villages.

Vaccinium leucanthum Schlecht. (Ericaceae). - Shrub. Mexico. Fruits edible. Occasionally sold in local markets of Mexico.

Vaccinium macrocarpon Ait. (syn. Oxycoccus macrocarpon Ait.) Pursh. Cranberry. (Ericaceae). - Low shrub. Eastern N. America. Cultivated in bogs. Fruits consumed cooked, as jelly, eaten with fowl. Varieties: Early Black, Howes, Centennial, McFarlin, Bennett, Searl, Prolific.

Vaccinium meridionale Sw. Billberry. (Ericaceae). - Shrub. Jamaica. Berries edible, made into jelly; used for tarts and pies.

Vaccinium Mortinia Benth. Mortiña. (Ericaceae). - Shrub. Colombia, Ecuador. Fruits edible, sold in local markets of Ecuador and Colombia.

Vaccinium Myrsinites Lam., Ground Blueberry. (Ericaceae). - Shrub. N. America. Fruits edible, eaten in compote and pies.

Vaccinium Myrtillis L. Blueberry, European Blueberry, Whortleberry, Billberry. (Ericaceae). - Small shrub. Europe. Temp. Asia, Western N. America. Fruits black, bloomy, 8 mm. across, sweet to sweetish acid. Important commercial product, esp. in Germany and Switzerland. Consumed fresh, cooked, in sugar, milk and wine. Used in pastries, compote, syrup and sauce. Was formerly also used in soup. Source of a wine, Heidelbeerwein, Heidelbeersekt and a distilled product Heidelbeergeist. Fruits are sometimes used to give wine a red color. As Fructus Myrtilli, Baccae myrtillorum they are used medicinally. Of less importance are the fruits of V. uliginosum L. Bog Billberry, which are occasionally used in brandy for stomach and intestinal catarrh, especially as home remedy.

Vaccinium myrtoides (Blume) Miq. (Ericaceae). - Shrub. Malaya. Fruits are used as preserves in the Philipp. Islds.

Vaccinium nitidum Andr., Blueberry. (Ericaceae). - Small shrub. Eastern N. America to Florida and Louisiana. Fruits edible; sold locally.

Vaccinium Oxycoccus L. (Oxycoccus palustris Pers.) Small Cranberry. (Ericaceae). - Low shrub. Temp. and arct. N. Hemisphere. Berries edible, sometimes consumed as Cranberries.

Vaccinium pennsylvanicum Lam., Low Sweet Blueberry, Early Sweet Blueberry. (Ericaceae). - Small shrub. Eastern N. America. Fruits used for pies, puddings etc., also consumed raw. Dried berries were much used by the Indians. Also V. vacillans Kalm.

Vaccinium Vitis-Idaea L., Cowberry, Foxberry. (Ericaceae). - Low shrub. Temp. Hemisphere. Berries red, edible, made into sauces and jellies, sometimes used as a substitute for Cranberries.

Vagnera racemosa (L.) Morong. (syn. Smilacina racemosa Desf.). Wild-Spikenards, Solomon's Plumes. (Liliaceae). - Perennial herb. N. America. Berries are occasionally consumed by country people.

Vahimainty Rubber → Secamonopsis madagascariensis Jum.

Valerian, African → Fedia Cornucopiae DC.

Valerian, Common → Valeriana officinalis L.

Valerian, Edible → Valeriana edulis Nutt.

Valeriana edulis Nutt. (Valerianaceae). - Perennial herb. N. America. Sometimes cultivated. Rootstocks are consumed as food when boiled by the Klamath and other Indian tribes.

Valeriana officinalis L., Valerian. (Valerianaceae). - Perennial herb, Europe, temp. Asia. Cultivated in Holland, England, Germany. Was used as a condiment during medieval times, and as a perfume during the XVI. Century. At present it is used as perfume in some Oriental countries. Dried rhizomes are employed medicinally as stomachic, nervine, antispasmodic, astringent of intestinal catarrh. Contains an ess. oil, bornyl valerate, bornyl formate. Cats are attracted by the herb.

Valeriana Wallichi DC. (Valerianaceae). - Perennial herb. Asia, esp. Himalayan region, India. Rhizomes are source of an oil, similar to that of V. officinalis L.

Valerianaceae → Fedia, Nardostachys, Valeriana, Valerianella.

Valerianella eriocarpa Desv., Italian Corn Salad. (Valerianaceae). - Annual herb. Mediterranean region. Cultivated. Leaves are used as a salad.

Valerianella olitoria Pollich. Corn Salad. (Valerianaceae). - Annual herb. Europe, Africa, Caucasia, introd. in N. America. Grown as a vegetable in some countries of Europe, occasionally in the United States. Leaves are used in salads, they are seldom cooked. Variety: Long Leaved.

Valley Oak → Quercus lobata Née.

Vallisneria americana Michx. Eel Grass, Tape Grass. (Hydrocharitaceae). - Perennial submersed herbaceous waterplant. Eastern N. America. A favorite food for ducks. Alsa V. spiralis L.

Vallisneria spiralis L. (Hydrocharitaceae). - Herbaceous, submersed waterplant. Europe, Asia, America. Young leaves are eaten in Annam in salads.

Vandyke Grass → Panicum flavidum Retz.

Vangueria esculenta S. Moore., Chirinda Medlar. (Rubiaceae). - Shrub. Trop. Africa. Fruits edible, having a bright gamboge color. Also

V. apiculata Schum, Small White Medlar and V. Munjiro S. Moore, Common Munjiro.

Vangueria madagascariensis J. F. Gmel. (syn. V. edulis Vahl.). Voavanga. (Rubiaceae). - Trop. Africa, Madagascar. Introduced in different parts of the tropics. Cultivated. Fruits edible, sweetish-acid, consumed by the natives. Fruits must be eaten when overripe.

Vaniera cochinchinensis Lour → Cudrania javanensis Trécul.

Vanilla Gardneri Rolf. (Orchidaceae). - Perennial vine. Brazil. Source of Vanilla of Brazil or Vanilla of Bahia, of minor commercial importance, used for flavoring.

Vanilla guianensis Splitgerb. (Orchidaceae). - Perennial vine. Brazil, Guayana. Pods resemble those V. Pompona Schiede. Used for flavoring.

Vanilla phaeantha Reich. (Orchidaceae). - Perennial herb. West Indies. Pods are occasionally collected and sold as Vanilla Beans.

Vanilla planifolia Andr., Vanilla. (Orchidaceae). - Perennial vine. Trop. America. Cultivated in Old and New World, among which in Mexico, Reunion, Bourbon, Mauritius, Tahiti, Venezuela, Brazil, Java etc. Source of Vanilla Beans or Vanilla Pods; used for flavoring in confectionary, perfumery, galenicals; in beverages. Contains 2—2.7 % vanillin, resin, vanillic acid and sugar. Some commercial grades are: Chica Fina, Saccata, Basura and Resecata.

Vanilla Pompona Schiede (syn. V. grandiflora Lindl.). (Orchidaceae). - Perennial vine. Trop. America. Occasionally cultivated in Martinique, Guadeloupe. Source of inferior vanilla beans, called Pompona Bova or Vanilla Bouffie.

Vanilla Beans → Vanilla planifolia Andr.

Vanilla Bouffie → Vanilla Pompona Schiede.

Vanilla Chica → Selenipedium Chica Reichb. f.

Vanilla of Bahia → Vanilla Gardneri Rolf.

Vanilla of Brazil → Vanilla Gardneri Rolf.

Vanilla Pod → Vanilla planifolia Andr.

Vanilla Trilisa → Trilisa odoratissima (Walt.) Cass.

Vanillosmopsis erythropappa Schultz-Bip. (syn. Albertinia Candolleana Gardn.). (Compositae). - Tree. Brazil. Wood durable; used for construction of boats, canoes, telegraph poles.

Variolaria orcina Ach. (Pertusariaceae). - Lichen. Source of a violet dye; used in France for dyeing woolens.

Varnish, Piney → Vateria indica L.

Varnish Tree → Rhus vernicifera DC.

Vascoa perfoliata DC. → Rafnia perfoliata E. Mey.

Vasconcella cestriflora A. DC. → Carica cestriflora Solms.

Vasey Grass → Paspalum Urvillei Steud.

Vatairea guianensis Aubl. (Leguminosaceae). - Tree. Guiana, Brazil. Wood yellowish, light brown to dark brown, planes fairly easily, not attacked by worms; recommended for decorative work, furniture, planking ships. Decoction of bark is used for dressing ulcers; sap is employed for ring-worm.

Vataireopsis speciosa Ducke., Angelim amarillo. (Leguminosaceae). - Tree. Brazil, esp. Central Amazon region. Bark source of Araroba powder, Pó da Goa, Pó da Bahia, used in medicine. Wood employed locally for carpentry and general construction.

Vateria acuminata Hayne. (Dipterocarpaceae). - Tree. E. India, Ceylon. Wood once extensively used for tea-boxes. Source of Hal-dummala, recommended for fine varnishes. Fruits eaten when made into Hal-pittu.

Vateria indica L., White Dammar, Indian Copal Tree, Malabar Tallow Tree. (Dipterocarpaceae). - E. India. Source of a resin called White Dammar, Piney Tallow, Dhupa Fat. Used in India for food, candles and for burning. Recommended medicinally instead of Pine Resin. Also employed in varnishes. Sp. Gr. 0.890; Sap. Val. 187—192; Iod. No. 36—41; Unsap. 1.2—2.5 %.

Vatica leucocarpa Foxw. (Dipteracarpaceae). - Tree. Borneo, Riouw Archipelago. Tree. Wood very heavy, hard, close-grained, durable, resists termites and decay, brown to almost black; used in general carpentry, house and boat-building.

Vatica papuana Dyer. (Dipterocarpaceae). - Tree. Borneo, Moluccas, New Guinea. Dead stem is source of a resin, called Damar Hiroe. High melting point. Considered a cheap copal. Some is exported.

Vatica Rassak Blume (syn. Retinodendron Rassak Korth.). (Dipterocarpaceae). - Tree. Borneo, Malaysia. Source of a resin, called Rose Dammar. Exported to Europe and America.

Vatica sumatrana V. Sl. (Dipterocarpaceae). - Tree. Sumatra. Stem is source of a resin; used among the natives for illumination.

Vatica Teijsmanniana Burck. (Dipterocarpaceae). - Tall tree. Sumatra, Bangka. Wood hard, easily worked, durable, resistant to termites and moisture; used for boards, beams of houses.

Vauquelinia corymbosa Correa. (Rosaceae). - Tree. Mexico, W. Texas. Bark source of a yellow dye used to color goat skins.

Vegetable Beefsteak → Fistulina hepatica (Huds.) Bull.

Vegetable Hair → Tillandsia usneoides L.

Vegetable Marrow → Cucurbita Pepo L.

Vegetable Oyster → Tragopogon porrifolius L.

Vegetable Sponge → Luffa aegyptiaca Mill.

Vegetable Wax → Rhus succedanea L.

Vegetables. Source of → Acanthopanax, Achillea, Achyranthes, Acrostichum, Adansonia, Adenophora, Aegopodium, Aerva, Algae, Alpinia, Agoseris, Agrostemma, Aizoon, Allium, Alocasia, Alpinia, Alsodeia, Alternanthera, Althaea, Amaranthus, Amorphophallus, Anemonella, Angelica, Anoectochilus, Antigonum, Apios, Apium, Aponogeton, Arachis, Aralia, Archangelica, Arctium, Areca, Arjona, Arracacia, Arthrocnemum, Arundinaria, Asclepias, Asparagus, Asphodelus, Aster, Athyrium, Atriplex, Attalea, Babiana, Balanites, Bambusa, Barbarea, Basella, Batis, Benincasa, Beta, Boerhaavia, Boltonia, Bomarea, Boussingaultia, Botrychium, Bowenia, Brasenia, Brassica, Bromelia, Bruguiera, Bunias, Bunium, Cajanus, Caladium, Calamus, Calandrina, Calathea, Calystegia, Campanula, Canavalia, Canna, Carica, Carum, Caryota, Celosia, Centaurea, Ceratopteris, Ceratotheca, Cercis, Chaerophyllum, Chamaedorea, Chamaerops, Chenopodium, Chrysanthemum, Cichorium, Claoxylon, Claytonia, Clematis, Clinocanthus, Cnicus, Coccinea, Cocos, Coleus, Colocasia, Corchorus, Corypha, Crambe, Crataeva, Crithmum, Cryptotaenia, Cucumis, Cucurbita, Cyamopsis, Cycas, Cyclanthera, Cymopterus, Cynara, Cynoglossum, Cyrtosperma, Daucus, Deeringia, Dendrocalamus, Dioscorea, Diplazium, Dolichos, Elaterium, Eleocharis, Emilia, Enhydra, Entada, Equisetum, Eremurus, Erucaria, Erythea, Euterpe, Foeniculum, Freycinetia, Gigantochloa, Glycine, Gnetum, Gundelia, Gunnera, Gynandropsis, Helianthus, Heliconia, Helwingia, Hibiscus, Humulus, Hydrocotyle, Ipomoea, Jacaratia, Lactuca, Lagenaria, Lasia, Lathyrus, Lepidium, Levisticum, Lilium, Limnocharis, Luffa, Lycium, Lycopersicum, Lycopus, Manihot, Martynia, Mesembryanthemum, Momordica, Muscari, Nasturtium, Neptunia, Nothopanax, Oenanthe, Oenothera, Olax, Oxalis, Pachyrrhizus, Pastinaca, Pedalium, Pereskia, Petasites, Phaseolus, Phyllostachys, Physalis, Phytolacca, Pisum, Plantago, Portulaca, Pteridium, Pueraria, Raphanus, Rheum, Rumex, Sagittaria, Salicornia, Salsaola, Sanguisorba, Scandix, Scorzonera, Sechium, Senecio, Sicana, Silybum, Sium, Smyrnium, Solanum, Spinaceae, Stachys, Taraxacum, Tetragonia, Thelygonum, Tordylium, Tragopogon, Trapa, Trichosanthes, Tropaeolum, Urtica, Valerianella, Veronica, Vicia, Vigna, Xanthosoma, Zea.

Velampisini Gum → Feronia limonia Swingle.

Velvet Bean → Mucuna nivea DC.

Velvet Grass → Holcus lanatus L.

Velvetleaf, Yellow → Limnocharis emarchinata Humb. and Bonpl.

Venezuela Mahogany → Swietenia Candollei Pitt.

Venezuela Pokeberry → Phytolacca rivinoides Kunth. and Bouché.

Vengai Padauk → Pterocarpus Marsupium Roxb.

Venidium arctotoives Less. (Compositae). - Perennial plant. S. Africa. Leaves are used by farmers and Kaffirs for healing wounds.

Ventilago maderaspatana Gaertn. (Rhamnaceae). - Woody vine. Trop. Asia, esp. India, Ceylon, Burma. Bark of root is source of a reddish dye, Ventilagin, used for coloring wool, cotton and silk. Roots are collected in Myore.

Ventilato → Rhus Cotinus L.

Venus' Comb → Scandix Pecten-Veneris L.

Vera Cruz Sarsaparilla → Smilax aristolochiaefolia Mill.

Veratrum album L. White False Hellebore, White Hellebore. (Liliaceae). - Perennial herb. Europe. Root medicinally used as Rhizoma Veratri or Radix Hellebori Albi; contains the alkaloids: jervin, pseudojervin, rubijervin; also a bitter glucosid: veratramarin.

Veratrum viride Ait., American False Hellebore, Indian Poke, American White-Hellebore. (Liliaceae). - Perennial herb. Eastern N. America. Powdered dried roots and rhizome used as insecticide. Occasionally used as cardiovascular and nerve sedative; slows heart, lowers blood pressure. Alkaloids: jervine, pseudo jervine, veratroidine. Much derived from North Carolina, Virginia, Illinois, Michigan. Collected in autumn. Medicinal properties were known to the Indians.

Verbascum Lychnitis L. (Scrophulariaceae). - Herbaceous plant. Europe, introd. in N. America. It has been claimed that flowers are sometimes used to destroy mice.

Verbascum phlomoides L. Clasping Mullein. (Scrophulariaceae). - Europe, introd. in N. America. Herbaceous plant. Ground capsules and seeds are used to stupefy fish. It has been claimed that fresh parts of the plant placed in cellars will drive away rats and mice.

Verbascum Thapsus L. Flannel Mullein. (Scrophulariaceae). - Biennial. Central and S. Europe, W. Asia. Introd. in N. America. Dried leaves are used medicinally, being demulcent, emolient; contain a bitter amorphic compound. Dried flowers are demulcent pectoral; contain an ess. oil, mucilage and a glucosidal coloring compound. Ground capsules have been used to stupefy fish, also V. phlomoides L., V. sinuatum L.

Verbena hastata L., American Blue Vervain, Wild Hyssop. (Verbenaceae). - Perennial herb. N. America. Dried herb, collected at flowering time, used medicinally as expectorant, diaphoretic. Contains verbenalin, a glucoside.

Verbenaceae → Avicennia, Callicarpa, Citherexylum, Clerodendron, Dicrastylis, Gmelina, Lantana, Lippia, Petitia, Premna, Stachytarpheta, Verbena, Vitex.

Verbesina virginiana L. White Crown Beard. (Compositae). - Perennial herb. Eastern United States. Pounded roots soaked in water were used by the Choctaw Indians for fever. Decoction of root is sudorific.

Vermouth Wine → Artemisia Absinthium L.

Vernonia amygdalina Delile., Bitter Leaf. (Compositae). - Shrub or small tree. Trop. Africa, esp. W. Africa, Congo, Abyssinia. Used as a bitter in S. Leone; as chewsticks in Abeokuta. Leaves are put in soup and in palaver souce in S. Leone.

Vernonia anthelmintica Willd. Kinka Oil Ironweed. (Compositae). - Perennial herb. Trop.Asia. Used in some parts of India for skin diseases, leprosy, and as abortive.

Vernonia Aschenborniana Schauer. (Compositae). - Shrub. Mexico, Centr. America. Herb is used in Guatemala for stomach complaints.

Vernonia Merana Baker. (Compositae). - Tree. Madagascar. Wood used in Madagascar for building houses.

Vernonia missurica Raf., Drummond's Ironweed. (Compositae). - Perennial herb. Prairies. Kansas, Arkansas to Texas and adj. territory. Flowers source of a purple dye. A watery decoction is used for dandruff.

Vernonia nigritiana Oliv. and Hiern., Nigerian Ironweed. (Compositae). - Woody plant. Trop. Africa. Decoction of roots is used by the natives for constipation, dysentery. Infusion is used as diuretic or emetic in Fr. Guinea. Sold in markets of Guinea, Senegal and Gambia. Herb is supposed to be febrifuge, antidysenteric, antihemorrhagic and has emetic properties. Contains vernonin, a glucoside, has been claimed to resemble digitaline.

Vernonia pectoralis Baker. (Compositae). - Shrub. Madagascar. Decoction of leaves is used as tonic.

Vernonia senegalensis Less., Bitters Tree of Gambia. (Compositae). - Tree. W. Africa, S. W. Africa. Bitter bark of trunk and root is used as a tonic, for fever and diarrhea in Angola. Astringent leaves are chewed in Gambia.

Vernonia Woodii Hoff. (Compositae). - Perennial plant. S. Africa. Infusion of bitter leaves is used by Kaffirs as stomachic.

Veronica Anagallis L. (Scrophulariaceae). - Perennial herb. Temp. zone. Cultivated. Leaves are consumed in salads in Japan.

Veronica Beccabunga L. Beccabunga Brooklime. (Scrophulariaceae). - Perennial herb. Europe, temp. Asia, Himalaya, Japan. Young shoots and stems are sometimes eaten as a vegetable.

Veronica officinalis L. Drug Speedwell. (Scrophulariaceae). - Perennial herb. Temp. Europe, Asia, N. America. Leaves are sometimes used as a tea in some countries of Europe.

Veronica salicifolia Forster. (Scrophulariaceae). - Shrub or small tree. New Zealand. Herb used by natives for diarrhea and dysentery.

Veronica virginica L. (syn. Leptandra virginica (L.) Nutt.). Culver's Root. (Scrophulariaceae). - Perennial herb. Eastern N. America to Georgia and Texas. Known to the Osage, Missouri and Delavare Indians as a violent purgative. Was also used by early settlers. Dried rhizome and roots used medicinally, being cathartic and emetic. Much is derived from Virginia, N. and S. Carolina.

Verpa bohemica (Krombh.) Schroet. (Helvellaceae). - Ascomycete. Fungus. Fruitbodies are consumed as food in Czechoslovakia and Austria. Resemble Morels.

Vervain, American Blue → Verbena hastata L.

Vervain, Bastard → Stachytarpheta jamaicensis Vahl.

Vetch, Bard → Vicia monantha Desf.

Vetch, Bitter → Vicia Ervillia (L.) Willd.

Vetch, Hungarian → Vicia pannonica Crantz.

Vetch, Canada Milk → Astragalus canadensis L.

Vetch, Chickling → Lathyrus sativus L.

Vetch, Common → Vicia sativa L.

Vetch, Hairy → Vicia villosa Roth.

Vetch, Gerard → Vicia Cracca L.

Vetch, Kidney → Anthyllis Vulneraria L.

Vetch, Milk → Astragalus glycyphyllos L.

Vetch, Monantha → Vicia monanthos Desf.

Vetch, Norbonne → Vicia narbonensis L.

Vetch, One Flowered → Vicia articulata Hornem.

Vetch, Russian → Vicia villosa Roth.

Vetch, Winter → Vicia villosa Roth.

Vetiver → Vetiveria zizanoides Stapf.

Vetiveria zizanoides Stapf. (syn. V. odorata Virey, Andropogon muricatus Retz.) Vetiver. (Graminaceae). - Perennial grass. Trop. Asia, esp. India, Burma, Ceylon. Cultivated. Roots made into aromatic scented mats, fans, ornamental baskets. Source of a volatile oil, used in perfumery, flavoring sherbets, cosmetics, soaps. Roots are also used for basketry.

Viburnum cassinoides L., Sweet Viburnum, Nanny Berry. (Caprifoliaceae). - Shrub. Eastern N. America. Fruits edible. Leaves are used as tea, called False Paraguay Tea.

Viburnum Opulus L., Cranberry Tree. (Caprifoliaceae). - Shrub or small tree. N. America, Europe, temp. Asia. Fruits have been recommended for jellies and as substitute for cranberries. Dried bark, High Bush Cranberry Bark, Cramp Bark, is used medicinally, being antispasmodic, much is derived from N. Carolina, Tennessee and Kentucky. Used by Indians as diuretic.

Viburnum pauciflorum La Pylaie. Moosebery Viburnum. (Caprifoliaceae). - Shrub. N. America. Fruits are eaten fresh or preserved for winter use by Eskimos of Alaska.

Viburnum prunifolium L., Black Haw, Stag Bush. (Caprifoliaceae). - Shrub or small tree. N. America. Fruits are edible, esp. when touched by frost. Bark of root contains viburnin, a bitter resinous principle; being astringent, tonic, uterine sedative, abortive, diuretic and nervine.

Viburnum theiferum Rehd. (syn. V. setigerum Hance). (Caprifoliaceae). - Shrub. Centr. and W. China. Leaves are source of a tea; used in some parts of China.

Vicia articulata Hornem. One-flowered Vetch. (Leguminosaceae). - Annual. Mediterranean region, Asia Minor, Madeira, Canary Islands. Occasionally cultivated as fodder for cattle and as green manure.

Vicia Cracca L. Gerard Vetch. (Leguminosaceae). - Perennial herb. Europe, Asia. Cultivated as fodder for cattle, as green manure and hay crop.

Vicia Faba L. (syn. Faba vulgaris Moench.). Broadbean. (Leguminosaceae). - Annual herb. Mediterranean region. Cultivated since ancient times. Grown in temp. zones of Old and New World. Var. major Harz. Broadbean. Beans are consumed as vegetable, eaten boiled. Varieties Mazagan, Windsor, Hangdown. Var. minor (Pieterm.) Harz. Pigeon Bean. Beans often fed to pigeons. Var. equina Pers. Horse Bean. Fed to domestic animals. Several varieties e. g. East Frisian, Rhineland, Holsteiner Horsebean. Seeds are sometimes eaten roasted. Also made into flour and mixed with wheat.

Vicia Ervilia (L.) Willd. (syn. Ervilia sativa Link., Ervum Ervilia L.). Bitter Vetch. Ervil. (Leguminosaceae).- - Annual herb. Mediterranean region. Cultivated. Seeds a source of fodder for cattle and pigeons. Used in some Mediterranean countries in soup. Considered sometimes as poisonous.

Vicia monanthos Desf. Bard Vetch, Monantha Vetch. (Leguminosaceae). - Annual or biennial herb. Mediterranean region. Occasionally grown as cover crop, green manure; also for hay or pasturage. Seeds are edible, similar to lentils.

Vicia narbonensis L., Narbonne Vetch. (Leguminosaceae). - Annual herb. Mediterranean region. Cultivated in S. Europe, Abyssinia as fodder for live-stock and green manure.

Vicia pannonica Crantz., Hungarian Vetch. (Leguminosaceae). - Annual. Mediterranean region, Balkans, Caucasia. Cultivated, used for pasturage, as hay crop and green manure.

Vicia sativa L., Common Vetch. (Leguminosaceae). - Annual to biennial herb. Mediterranean region, W. Asia. Widely grown in Old and New World, as green manure, green fodder for livestock, as hay crop and for winter pasturage.

Vicia villosa Roth., Hairy Vetch, Winter Vetch, Russian Vetch. (Leguminosaceae). - Annual to perennial herb. Mediterranean region, W. Asia, N. Africa. Grown as green manure, for pasture, as hay crop and silage.

Vigna Catjang Walp., Catjang, Hindu Cawpea. (Leguminosaceae). - Herbaceous vine. Tropics. Cultivated. Young pods are consumed as vegetable. Also raised as forage crop for live-stock. Numerous varieties.

Vigna sinensis Endl. (syn. Dolichos sinensis L.). Common Cowpea. (Leguminosaceae). - Herbaceous vine. Tropics. Cultivated in warm countries of Old and New World. Cultivated since ancient times. Seeds consumed when green or ripe, especially the Blackeye and White varieties. Also raised for pasturage, hay, ensilage, as green-manure. Numerous varieties among which: Brabham, Iron, Victor, Taylor, Early Bluff, Block. Young pods are consumed as French Beans.

Vigna triloba Walp. (Leguminosaceae). - Herbaceous vine. Trop. Africa. Pods are eaten as food in some parts of the Congo.

Vieillardia austro-caledonica Brogn. and Gris. (Nyctaginaceae). - Tree. New Caledonia. Wood coarse-grained, decays quickly; used for manuf. paper.

Villaresia Congonha Miers. (Icacinaceae). - Tree. S. America. Leaves are sometimes used as a tea in some parts of Brazil.

Villebrunea integrifolia Gaud. (Urticaceae). - Shrub or small tree. Trop. Asia, esp. E. Himalaya to Assam, Khusia hills, Sylhet, Manipur. Source of Mesakhi fibre; used for different purposes in India.

Vin d'Ananas → Ananas comosus (L.) Merr.

Vin de Cornoulle → Cornus Mas L.

Vine Maple → Acer circinatum Pursh.

Vine Mesquite → Panicum obtusum H. B. K.

Vine Spinach → Basella rubra L.

Vinegar → Acetobacter aceti (Kuntzing) Beijerinck.

Vinhatico → Plathymenia reticulata Benth.

Viola odorata L., Sweet Violet. (Violaceae). - Perennial herb. Europe, Asia. Cultivated. Tea from leaves is used for coughs. Var. Parma is mainly source in S. France of an ess. oil; used in perfumery and pomades. About 1000 to 1100 kg. leaves produce 400 gr. absolute. Flowers are mainly treated by the maceration method with hot fat producing pomades and extraits. About 100 kg. flowers are source of 31 gr. ess. oil. There is much competition with synthetic violet compounds, derived from orris and ionone compounds. Candied flowers are used in confectionary.

Viola palmata L., Palmate Violet. Early Blue Violet. (Violaceae). - Perennial herb. Eastern N. America to Florida. Plant is very mucilaginous; used by negroes for thickening soup, called „wild okra".

Viola tricolor L. Pansy. (Violaceae). - Annual herb. Europe, temp. Asia, introd. in N. and S. America. Herb occasionally used for diseases of children, esp. skin eruptions, rhachitis and scrophulosis. Decoction of leaves and flowers are used as expectorant in coughing; has sometimes harmful effects.

Violaceae → Alsodeia, Corynostylis, Ionidium, Melicytus, Rinorea, Viola.

Violet, Palmate → Viola palmata L.

Violet Root → Ferula Sumbul Hook. f.

Violet, Sweet → Viola odorata L.

Violet Wood → Copaifera bracteata Benth.

Violeta Wood → Dalbergia cearensis Ducke.

Virgilia capensis Lam. (Leguminosaceae). - Tree. S. Africa. Wood dark, light, soft; used for yokes, rafters, spars, fuel.

Virginia Creeper → Parthenocissus quinquefolia (L.) Planch.

Virginia Mountain Mint → Pycnanthemum virginianum (L.) Durand and Jacks.

Virginia Snakeroot → Aristolochia Serpentaria L.

Virginian Strawberry → Fragaria virginiana Duch.

Virginian Tuckahoe → Peltandra virginica (L.) Kunth.

Virola merendonis Pitt. Merendon Virola. (Myristicaceae). - Centr. America. Wood is used for boards in Guatemala.

Viscum album L. European Mistletoe. (Loranthaceae). - Parasitic shrub. Centr. and S. Europe, W. Iran. Viscid berries are a source of birdlime; also used since Roman times as home-remedy for epilepsy.

Vismia leonensis Hook. f. (Guttiferaceae). - Woody plant. Trop. Africa. An ointment is made from the bark; used in Africa for craw-craw disease; it contains a yellow resin.

Visnea Mocanera L. (Theaceae). - Shrub or tree. Canary Islands, Madeira. Fruits are source of a sirup, called Charcherquem or Lamedor de Moca; used by the natives for hemorrhages; also the astringent root is used. Berries called Mocanes are consumed by the inhabitants.

Vitaceae → Ampelocissus, Parthenocissus, Tetrastigma, Vitis.

Vitex Agnus Castus L. Chaste Tree. (Verbenaceae). - Tree. Mediterranean region. Cultivated in the Old and New World. Herb has been used for centuries as antiaphrodisiac. Source of a volatile oil. Young twigs are used for basketry. Fruits are employed as a substitute for pepper and are supposed to be antiaphrodisiac. Plant was regarded since antiquity as a symbol of chastity.

Vitex celebica Koord. (Verbenaceae). - Tall tree. Celebes. Wood hard, fine texture, heavy, elastic, resistant to moisture; used for house-building, bridges and boats.

Vitex Cienkowskii Kotschy and Perr. (Verbenaceae). - Tree. Trop. Africa. Fruits having the size of an olive; much esteemed by the natives. A gum is used in compounding Mallam's Ink; also considered an antidote for arrow poison. Leaves are used as subtitute for tea in Trop. Africa.

Vitex Cofassus Reinw. (syn. V. punctata Schau.) (Verbenaceae). - Tree. Mal. Archipelago. Celebes, Moluccas etc. Wood durable, resistant to sea-water and moist soil; used for building vessels.

Vitex divaricata Sw., Guiana Chaste Tree. (Verbenaceae). - Tree. West Indies, Guiana. Leaves are used for tanning, containing 14 % tannin. The wood is made into boards, and shingles.

Vitex littoralis Cunningh. (Verbenaceae). - Tree. New Zealand. Wood hard, dark-brown, heavy, very strong, durable, difficult to work. Used for bridges, construction work, ships' blocks, machine beds and bearings; where great strength and durability are necessary.

Vitex mollis H. B. K. (syn. V. lasiophylla Benth.). (Verbenaceae). - Tree. Mexico. Fruits are edible, about 1 to 2 cm. in diam; eaten raw. Sold in markets of Mexico. Leaves and fruits are used for diarrhoea.

Vitex parviflora Juss., Molave Chaste Tree. (Verbenaceae). - Mal. Archipelago, Philipp. Islands. Wood hard, strong, straw-yellow to light brown, easy to work; suitable for turnery and household goods.

Vitex peduncularis Wall. (Verbenaceae). - Tree. India, Burma. Recommended as febrifuge.

Vitex pubescens Vahl. (syn. V. latifolia Lam.). (Verbenaceae). - Tree. Trop. Asia. Wood hard, heavy, durable, fine-grained, yellow-brown to dark-brown; used for household goods, plows, handles of tools, furniture, building of houses.

Vitis aestivalis Michx., Summer Grape. (Vitaceae). - Woody vine. N. America. Berries edible. Occasionally cultivated.

Vitis aralioides Welw. (syn. Cissus aralioides Planch.). (Vitaceae). - Woody vine. Trop. Africa. Defoliated stems are used in the preparation of leather before tanning.

Vitis capensis Burm. f. (syn. Rhoicissus capensis (Burm. f.) Planch.). (Vitaceae). - Woody vine. S. Africa. Berries, edible, 2 cm. across, purplish black, reddish pulp; made into jelly in different parts of S. Africa.

Vitis caribaea DC. (Vitaceae). - Woody vine. N. America, West Indies, Centr. America. Fruits sour, edible, consumed in different regions.

Vitis Davidi (Roman) Foëx., Spiny Vitis. (Vitaceae). - Woody vine. W. China. Occasionally cultivated for its fruits. Berries black, globose, harsh flavor.

Vitis discolor Dalz. (syn. Cissus sicyoides L.). (Vitaceae). - Woody vine. Trop. America. Tough stems are used for baskets, when macerated in water, they will produce a lather, used for washing clothes.

Vitis Labrusca L., Fox Grape. (Vitaceae). - Woody vine. Eastern N. America. Cultivated. Berries are eaten raw, made into grape juice, sirup, jelly, jam, for flavoring candies and ice cream; manuf. into wine and non-alcoholic beverages. Seeds are source of oil, cream of tartar and tannin; pomace and press seed cake is recommended as food for live-stock. A jelly is made from the skins. Varieties are: Concord, Catawba, Isabella, Niagara, Delaware, Worden.

Vitis multistriata Baker. (syn. V. pentaphylla Guill. and Perr.). (Vitaceae). - Woody vine. Trop. Africa. Berries are edible, used fresh or preserved. Also V. Lecardii Car., V. Durandii Lec., V. Chantinii Lec., V. cornifolia Baker and V. palmatifida Baker.

Vitis pallida Wight and Arn. (syn. Cissus populnea Guill. and Perr.). (Vitaceae). - Woody vine. Trop. Africa, Asia. Fruits are edible, sold in markets of Abinsi, W. Afr. Viscid pericarp is used in soup. A viscid juice from the stem is employed to adulterate honey.

Vitis quadrangularis Wall. (syn. Cissus quadrangularis L.). (Vitaceae). - Woody vine. Trop. Africa, Asia. Juice is used by camel owners to cleanse and heal saddle-galls of camels.

Vitis rotundifolia Michx., Muscadine Grape, Southern Fox Grape. (Vitaceae). - Woody vine. Eastern United States. Cultivated in S. part of the United States. Fruits edible united into small bunches; thick skinned. Variety: Scuppernong. Sometimes sold in southern markets.

Vitis vinifera L. Common Grape, European Grape, Californian Grape, Grape Vine. (Vitaceae). - S. E. Europe, W. Asia. Cultivated in all continents. Grapes are consumed as dessert fruit; juice is mainly made into wine and champagne. Source of cognac, brandy, vermouth, liqueur, vinegar. Mistelle is mixture of must and alcohol. Is exported in large quantities from Algeria to France where it forms the foundation of vins apéritifs. Wines are classified as to origin: German, French, Californian, Chilean, Italian etc.; in sweet and dry; red and white; heavy, medium and light; natural and fortified; sparkling or still. Dry wines are used as table beverages with meals, like many Rhine, Mosel and Bordeaux wines, Claret, Chianti. Fortified sweet wines are consumed as appetizers or between meals, like cocktails and liqueurs, e. g. Madeira, Port, Tokay, Sherry, Angelica, Marsala, Muscatel. In dry wines the sugar is almost entirely fermented; in sweet wines fermentation is usually stopped before all sugar has been converted. To fortified wines of high alcoholic content, brandy or alcohol are added. To Medical Wines have been added quinine, iron compounds, cola, pepsin, rhubarb etc. They are recommended for sick or convalescent persons. Wine is used in religious ceremonies of the Roman Catholic and Jewish Faiths. The typical bouquet of wines originates after aging from four to five years or longer. Of importance are also climate, soil, region, variety of grape and nature of the yeast (→ Saccharomyces). Raisins are dried grapes among which Malagas and Muscatels are first grade table raisins; Lexias are suitable for cooking; Sultanas are small and oval. Californian raisins are divided into Layer, Seedless and Seeded. Currants are derived from a small fruited variety (var. apyrena Risso), grown in Greece. Waste products are used as food for live-stock, tannin, cream of tartar. There are numerous varieties, many are suitable to certain areas or for making certain vines or other products. A few groups and their wines are af follows: French Wines: White Bordeaux: Sauterne, Haute Saut., Barsac, Château Yquem, Château Latour-Blanche. Red Bordeaux is divided into First Crue: Château Margaux, C. Lafite, Second Crue: Château Bran-Cantenac, C. Lalande. Third Crue: Château Canbenac, C. d'Issan. Fourth Crue: Château Payet, C. Rochet. Fifth Crue: Château Belgrave. Burgundy wines have the name derived from the Province of Burgundy, France. They are heavier than Bordeaux. Red Burgundy: Romanée Conti, Chambertin, Richebourg, Clos de Vougeot. White Burgundy: Montrachet, Chablis and Meursault. Italian Wines: Chianti, Lacryma, Christi, Capri, Asti, Falerno, Barolo, Moscato di Stramboli. Spanish Wines: Malaga, Alicante, Rancio, Tarragona Port, Sherry. Portugeese Wines: Bucellas, Monsao, Calvel (Port). Madeira Wines: Madeira among which Verdeilho, Malmacy, Sercial, Bual. Rhine and Mosel Wines: Assmannshauser, Affenthaler, Geisenheimer, Bodenheimer, Zeltinger, Erdener, Oppenheimer. Austrian Wines: Gumpoldskircher, Luttenberger, Voslauer. Hungarian Wines: Tokay (Tokay Essentia, Tokay Ausbruch, Tokay Máslás), Szamarodni, Ruszti, Meneser, Adelsberger, Szegszarder. Bulgarian Wine: Euxenograd. Greek Wines: St. Elie, Mavrodaphne, Morea, Camerite, Nectar, Red Santorin. Red and white champagne are divided into sweet and dry and in non-Mousseux not effervescent; Crémant, moderately sparkling; Mousseux is enough effervescent to remove the cork with an audible report; Grand Mousseux is considered excessively effervescent. Red wines are derived from red grapes or are artificially colored; white wines are derived from white grapes. Zibebes are berries that were dried on the grapeplant. Grape Honey, Turkish Honey, Dips of the Arabs is an evaporated grapejuice used in the Orient as food and delicacy. A cake is made by mixing evaporated juice with flour and dried in the sun, a commercial product in the Orient. Seeds are source of Grape Seed Oil, a drying oil; used for manuf. soap, used in paints, as illuminant, when refined is also used in foods. Sp. Gr. 0.912—0.926; Sap. Val. 176—190; Iod. No. 125—157.

Vitis vulpina L., Frost Grape, River Bank Grape. (Vitaceae). - Woody vine. N. America. Fruits are edible, occasionally cultivated.

Vittadinion Montagnei Zobel → Tuber excavatum Vitt.

Voacanga africana Stapf. (Apocynaceae). - Shrub. Trop. W. Africa. Latex is used to adulterate rubber.

Voandzeia Poissoni Chev. (Leguminosaceae). - Herbaceous. Trop. Africa. Beans are eaten by natives of Dahomey (Africa).

Voandzeia subterranea Thon. Congo Goober, Bambara Ground Nut. (Leguminosaceae). - Annual herb. Origin uncertain, probably from Trop. Africa. Cultivated. Fruits ripen in the ground, like those of Arachis hypogaea L. Seeds are consumed as food by the natives of Africa.

Voavanga → Vangueria madagascariensis J. F. Smel.

Vochyaceae → Erisma.

Vogel Tephrosia → Tephrosia Vogelii Hook. f.

Volvaria volvacea (Bull.) Quél. (syn. V. esculenta Bres.) (Agaricaceae). - Basidiomycete. Fungus. Cosmopolite. Fruitbodies are consumed as food. Cultivated in Malaya, Burma on different waste material derived from plants, like that from sugar cane, coffee, rice etc. Sold in markets.

Vulparaiso Weed → Rocella tinctoria DC.

Vouacapoua americana Aubl. → Andira excelsa H. B. K.

W

Wake Robin → Trillium grandiflorum (Michx.) Salisb.

Walnut → Juglans regia L.

Walnut, African → Lovoa Klaineana Pierre and Sprague.

Walnut Black → Juglans nigra L.

Walnut, Black Bolivian → Juglans boliviana Dode.

Walnut, Cathay → Juglans cathayensis Maxim.

Walnut, Ecuador → Juglans Honorei Dode.

Walnut, English → Juglans regia L.

Walnut, Guatemalan → Juglans mollis Engelm.

Walnut, Mandschurian → Juglans mandschurica Maxim.

Walnut, Siebold → Juglans Sieboldiana Maxim.

Walnut, Texas Black → Juglans rupestris Engelm.

Wall Pepper → Sedum album L.

Wallaba → Eperua falcata Aubl.

Wallflower → Cheiranthus Cheiri L.

Wampi → Clausenia Lansium Skeels.

Wane → Ocotea rubra Mez.

Wapato → Sagittaria latifolia Willd.

Waras (dye) → Flemingia congesta Roxb.

Warty Leather Lichen → Lobaria scrobiculata (Scop.) DC.

Warty Olive → Olea verrucosa Link.

Washingtonia filifera (Lind.) Wendl. (syn. Neowashingtonia filifera Sudw., Pritchardia filifera Lindl.). Cañon Palm. (Palmaceae). - Tall palm. S. California and Baja California. Leaves are used by Indians for building huts. Fibres from leaves are employed for making basketry. Fruits are eaten dry or fresh, also ground into meal. Young central bud was roasted and consumed by the Indians.

Water Caltrop → Trapa natans L.

Water Chestnut → Eleocharis tuberosa Schultes. and Trapa natans L.

Water Chinquapin → Nelumbium luteum (Willd.) Pers.

Water Cress → Nasturtium officinale R. Br.

Water Dropwort → Oenanthe stolonifera Wall.

Water Germander → Teucrium Scordium L.

Water Gum → Nyssa biflora Walt.

Water Hyacinth → Eichhornia crassipes Solms.

Water Lemon → Passiflora laurifolia L.

Water Parsley → Oenanthe sarmentosa Presl.

Water Plantain → Alisma Plantago L.

Water Tree → Ilex capensis Harv.

Water Tupelo → Nyssa aquatica L. and N. biflora Walt.

Water Yam → Dioscorea alata L.

Waterleaf → Hydrophyllum appendiculatum Michx.

Waterlily, White → Nymphaea alba L.

Watermelon → Citrullus vulgaris Schrad.

Watery Rose Apple → Eugenia aquea Burm. f.

Wattle, Black → Acacia binervata DC. and A. decurrens Willd.

Wattle, Golden → Acacia pycnantha Benth.

Wattle Gum → Acacia pycnantha Benth.

Wattle, Silver → Acacia dealbata Link.

Wattle, Twinvein → Acacia binervata DC.

Wattung Urree → Banksia serrata L. f.

Wavyleaf Oak → Quercus undulata Torr.

Wax. Plants source of → Balanophora, Carthamus, Ceroxylon, Cinnamomum, Copernica, Euphorbia, Fraxinus, Langsdorffia, Ligustrum, Musa, Myoporum, Myrica, Pedilanthus, Rheedia, Rhus, Saccharum, Symphonia.

Wax, Candelilla → Pedilanthus Pavonis (Klotzsch and Garcke) Boiss.

Wax, Carnauba → Copernica cerifera Mart.

Wax Gourd → Benincasa cerifera Savi.

Wax, Karamanni → Symphonia globulifera L.

Wax Myrtle → Myrica cerifera L.

Wax, Siejas → Langsdorffia hypogaea Mart.

Wax, Sumach → Rhus succedanea L.

Wax Tree → Rhus succedanea L.

Wax, Vegetable → Rhus succedanea L.

Waxberry → Myricaria cordifolia L.

Wayaka Yam Bean → Pachyrrhizus angulatus Rich.

Weaversbroom → Spartium junceum L.

Weinmannia racemosa L., Kamahi. (Cunoniaceae). - Tree. New Zealand. Bark has been recommended as a source for tanning material. Also W. silvicola.

Weinmannia Selloi Engl. (Cunoniaceae). - Tree. S. America. Bark is a strong astringent, used in some parts of Brazil for treating wounds.

Weinmannia tinctoria Smith (syn. W. macrostachys DC.). (Cunoniaceae). - Tree. Mauritius, Bourbon. Bark is used for giving leather skins a red color.

Weinmannia trichosperma Cav. Tenio. (Cunoniaceae). - Tree. Chile. Bark is used for tanning in some parts of Chile.

Weld → Reseda Luteola L.

Welsh Onion → Allium fistulosum L.

Wendtia calycina Griseb. (Geraniaceae). - Halfshrub. Argentina. Plant is used in some parts of Argentina as carminative, for indigestion and dispepsia.

Werinnum Oil → Guizottia abyssinica (L. f.) Cass.

West African Boxwood → Sarcocephalus Diderrichii De Wild.

West African Locust Bean → Parkia filicoidea Welw.

West African Mahogany → Mitragyna macrophylla Hiern.

West Australian Sandalwood → Eucarya spicata Sprag. and Summ.

West Australian Sandalwood Oil → Eucalyptus spicata Sprague and Sum.

West Indian Boxwood → Tabebuia pentaphylla (L.) Hemsl.

West Indian Elemi → Bursera gummifera L.

West Indian Cedar → Cedrela odorata L.

West Indian Cherry → Malpighia punicifolia L.

West Indian Gherkin → Cucumis Anguria L.

West Indian Gooseberry → Pereskia cauleata Mill.

West Indian Mahogany → Swietenia Mahogany Jacq.

West Indian Sandalwood → Amyris balsamifera L.

West Indian Spigelia → Spigelia Anthelmia L.

Western Alder → Alnus rubra Bong.

Western Black Oak → Quercus Emoryi Torr.

Wester Buckeye → Aesculus arguta Buckl.

Western Catalpa → Catalpa speciosa Warder.

Western Flowering Dogwood → Cornus Nuttallii Aud.

Western Hackberry → Celtis reticulata Torr.

Western Hemlock → Tsuga heterophylla (Raf.) Sarg.

Western Larch → Larix occidentale Nutt.

Western Red Cedar → Thuya plicata Donn.

Western Sand Cherry → Prunus Besseyi Bailey.

Western Service Berry → Amelanchier alnifolia Nutt.

Western Soapberry → Sapindus marginatus Hook. and Arn.

Western Yellow Pine → Pinus ponderosa Laws.

Westland Pine → Dacrydium Westlandicum T. Kirk.

Wesy African Indigo → Lonchocarpus cyanescens Benth.

Wheat, Astrakan → Triticum polonicum L.

Wneat, Club → Triticum compactum Host.

Wheat, Common → Triticum vulgare Vill.

Wheat, Durum → Triticum durum Desf.

Wheat, Inca → Amaranthus caudatus L.

Wheat, Macaroni → Triticum durum Desf.

Wheat, Polish → Triticum polonicum L.

Wheat, Poulard → Triticum turgidum L.

Wheat, Wild → Elymus triticoides Buckl.

Wheatgrass, Slender → Agropyron pauciflorum (Schwein.) Hitchc.

Wheeler Sotol → Dasylirion Wheeleri S. Wats.

Wingleaf Prickly Ash → Zanthoxylon alatum Steud.

Whip Tree, Common → Luehea divaricata Mart.

Whipple Yucca → Yucca Whipplei Torr.

White Alder → Platylophus trifoliatus Don.

White Ash → Fraxinus americana L.

White Asparagus → Asparagus albus L.

White Beam Tree → Sorbus Aria Crantz.

White Bitterwood → Trichilia spondioides Jacq.

White Box → Eucalyptus hemiphloia F. v. Muell. and Tristania conferta R. Br.

White Brittlebush → Encelia farinosa Gray.

White Brown Beard → Verbesina virginiana L.

White Bryony → Bryonia alba L.

White Buttonwood → Laguncularia racemosa Gaertn. f.

White Campion → Melandrium album (Mill.) Garcke.

White Cedar → Thuya occidentalis L.

White Clover → Trifolium repens L.

White Crottle → Pertusaria corallina (L.) Th. Fr.

White Dammar → Vateria indica L.

White Dammarpine → Agathis alba Foxw.

White Dead Nettle → Lamium album L.

White Ebony → Diospyros Malacapai A. DC.

White Elm → Ulmus americana L.

White False Hellebore → Veratrum album L.

White Fir → Abies amabilis (Loud.) Forbes. and A. concolor Lindl. and Gord.

White Gentian → Laserpitium latifolium L.

White Gourd. → Benincasa cerifera Savi.

White Gum → Eucalyptus gomphocephala DC.

White Haiari → Lonchocarpus densiflorus Benth.

White Hellebore → Veratrum album L.

White Ipecac → Ionidium Ipecacuanha Vent.

White Ironbark → Eucalyptus leucoxylon F. v. Muell.

White Ironwood → Toddalia lanceolata Lam.

White Lupine → Lupinus albus L.

White Mahogany → Eucalyptus robusta Smith.

White Mangrove → Avicennia officinalis L. and Laguncularia racemosa Gaertn. f.

White Mahogany → Cybistax Donnel-Smithii (Rose) Seib. and Khaya anthotheca C. DC.

White Maple → Acer saccharinum L.

White Medlar → Vangueria esculenta S. Moore.

White Mulberry → Morus alba L.

White Mustard → Sinapis alba L.

White Mustard Seed Oil → Sinapis alba L.

White Oak → Quercus alba L.

White Pear → Apodytes dimidiata E. Mey.

White Pepper → Piper nigrum L.

White Pereira → Abuta Candollei Triana and Planch.

White Pine → Podocarpus elata R. Br. and Pinus albicaulis Engelm.

White Pine, Eastern → Pinus Strobus L.

White Pine, Western → Pinus monticola Doudl.

White Poplar → Populus alba L.

White Quebracho → Aspidosperma quebracho blanco Schlecht.

White Rosewood → Dalbergia nigra Allem.

White Sandalwood → Santalum album L.

White Sapote → Casimiroa edulis La Lave.

White Senaar → Acacia Senegal (L.) Willd.

White Spruce → Picea alba Link.

White Stonecrop → Sedum album L.

White Stopper → Eugenia axillaris (Sw.) Willd.

White Sweet Clover → Melilotus alba Desr.

White Tephrosia → Tephrosia candida DC.

White Waterlily → Nymphaea alba L.

White Willow → Salix alba L.

White Wood → Atalaya hemiglauca F. v. Muell. and Tabebuia pentaphylla (L.) Hemsl.

White Yam → Dioscorea alata L.

Whitespot Giant Arum → Amorphophallus campanulatus Blume.

Whitfildia longifolia T. And. (Acanthaceae). - Shrub. Trop. Africa, esp. Uganda, Kenya, Tanganyika, Nyasaland. N. Rhodesia. Juice from leaves is source of a black dye; used in Belg. Congo.

Whitfordiodendron erianthum Dunn. (syn. Adinobotrys erianthus Dunn., Milletia eriantha Benth.) (Leguminosaceae). - Woody climber. Malaya, Penin., Siam. Fruit very acid, eaten by natives when boiled.

Whitfordiodendron pubescens Burkill (syn. Adinobotrys atropurpureus Dunn., Padbruggea pubescens Craib.) (Leguminosaceae). - Trop. Asia. Wood heavy, loose-grained, brownish; used locally by Burmese and Talaings for beams, rafters of houses.

Widdringtonia cupressoides Endl. → Callitris cupressoides Schrad.

Wikstroemia japonica Miq. (Thymelaeaceae). - Shrub. Japan. Bark in Japan is source of a paper.

Wikstroemia sikokiana Franch. and Sav. (Thymelaeaceae). - Shrub. Japan. Plant is used for manuf. paper in some parts of Japan.

Wild Angelica → Angelica sylvestris L.

Wild Basil → Calamintha Clinopodium Benth.

Wild Bean → Phaseolus polystachyus (L.) B. S. P.

Wild Bergamot → Monarda fistulosa L.

Wild Black Cherry → Prunus serotina Ehrh.

Wild Black Currant → Ribes americanum Mill.

Wild Breadfruit → Artocarpus elastica Reinw.

Wild Cane → Gynerium sagittatum (Aubl.) Beauv.

Wild Cherimoya of Jalisco → Annona longiflora Wats.

Wild Cinnamon → Canella alba Murr.

Wild Durian → Cullenia excelsa Wight.

Wild Garlic → Allium canadense L.

Wild Ginger → Asarum canadense L.

Wild Goose Plum → Prunus Munsoniana Wight and Hedrick.

Wild Gourd → Cucurbita foetidissima H. B. K.

Wild Hyacinth → Camassia esculenta Lindl.

Wild Hyssop → Verbena hastata L.

Wild Leek → Allium tricoccum Ait.

Wild Lettuce → Lactuca canadensis L.

Wild Liquorice → Abrus precatorius L. and Glycyrrhiza sepidota (Nutt.) Pursh.

Wild Lucerne → Stylosanthes mucronata Willd.

Wild Mace → Myristica malabarica Lam.

Wild Mango → Irvingia gabonensis Baill.

Wild Oat → Avena fatua L.

Wild Olive → Sideroxylon Mastichodendron Jacq.

Wild Olive Seed Oil → Ximenia americana L.

Wild Orange Tree → Toddalia aculeata Pers.

Wild Plum → Prunus americana Marsh.

Wild Potato → Solanum Fendleri Gray.

Wild Rose → Zizania aquatica L.

Wild Rue → Anemonella thalictroides (L.) Spach.

Wild Spikenard → Vagnera racemosa (L.) Morong.

Wild Tea Bush → Leucas martinicensis R. Br.

Wild Wheat → Elymus triticoides Buckl.

Wild Yam → Dioscorea spinosa Roxb. and D. villosa L.

Willow, Almond Leaved → Salix triandra L.

Willow, Black → Salix nigra L.

Willow, Bluestem → Salix irrorata Anders.

Willow, Brittle → Salix fragilis L.

Willow, Caspic → Salix acutifolia Willd.

Willow, Common → Salix Caprea L.

Willow Fireweel Tree → Stenocarpus salignus R. Br.

Willow, Goat → Salix Caprea L.

Willow Oak → Quercus Phellos L.

Willow, Osier → Salix viminalis L.

Willow, Purple → Salix purpurea L.

Willow, Sitka → Salix sitchensis Bong.

Willow, White → Salix alba L.

Willowleaf Eucalyptus → Eucalyptus amydalina Lab.

Willowleaf Red Quebracho → Schonopsis Balansae Engl.

Willughbeia edulis Roxb. (Apocynaceae). - Woody vine. Cochin-China. Himalayan region, Malaya, India. Fruits edible, ovoid, size of a lemon, yellowish, of pleasant taste; eaten by the natives.

Willughbeia firma Blume. (Apocynaceae). - Woody vine. Mal. Archipelago, Borneo. Source of a rubber Getah Soesoe, Getah Borneo, Borneo Rubber. Latex coagulates by salt water.

Wine, Condurango → Marsdenia Reichenbachii Triana.

Wine, Dendelion → Taraxacum officinale Weber.

Wine, Gooseberry → Ribes Grossularia L.

Wine, Grape → Vitis vinifera L.

Wine, Raffia Palm → Raphia vinifera Beauv.

Wine Raspberry → Rubus phoenicolasius Maxim.

Wine, Rhubarb → Rheum hybridum Murray.

Wineberry → Rubus phoenicolasius Maxim.

Wineberry, Chilean → Aristotelia Maqui L'Hér.

Winepalm of Chile → Jubaea chilensis (Molina) Baill.

Wing Nut, Caucasian → Pterocarya fraxinifolia Spach.

Winged Elm → Ulmus alata Michx.

Winged Lophira → Lophira alata Banks.

Winter's Bark → Drimys Winteri Forst.

Winterberry → Ilex verticillata (L.) Gray.

Winter Cherry → Physalis Alkekengi L.

Winter Cress → Barbarea praecox R. Br.

Winter Crookneck → Cucurbita moschata Duch.

Winter Purslane → Claytonia perfoliata Don.

Winter Vetch → Vicia villosa Roth.

Winterbark Drimys → Drimys Winteri Forst.

Wintergreen → Gaultheria procumbens L.

Wintergreen, Spotted → Chimaphila umbellata Nutt.

Wintergreen, Indian → Gaultheria fragantissimum Wall.

Wissadula rostrata Planch. (Malvaceae). - Shrub. Tropics. Bark is source of a fibre, used for different purposes by natives of S. Africa. It is classed with jute.

Wissadula zeylancia Medic. (Malvaceae). - Half-shrub. Tropics. Source of an excellent fibre, very little used.

Wistaria brachybotrvs Sieb. and Zucc. (Leguminosaceae). - Woody vine. China, Japan. Cultivated. Bark is source of a fibre; used in Japan for threads and cloth.

Witch Hazel → Hamamelis virginiana L.

Witch Hazel Bark → Hamamelis virginiana L.

Withania coagulans Dunal, Cheesemaker, Indian Rennet. (Solanaceae). - Small shrub. E. India, Afghanistan. Fruit used by natives to coagulate milk.

Withania somnifera Dun. (Solanaceae). - Perennial plant. Africa, Mediterranean region, Orient. Decoction of the root and bark is used by Kaffirs for diseases of the rectum. Seeds are used in the Sudan to coagulate milk. They are considered diuretic and narcotic.

Witloof, Brussel → Cichorium Intibus L.

Woad → Isatis tinctoria L.

Wolfberry, Anderson → Lycium Andersonii Gray.

Wolfberry, Arabian → Lycium arabicum Schweinf.

Wood, Economic and Commercial. → Abies, Acacia, Acanthopanax, Acanthosyris, Acer, Acrocarpus, Acrodiclidium, Adenanthera, Adina, Adinandra, Aegiceras, Aeschynomena, Aesculus, Aetoxicon, Afzelia, Agathis, Ailanthus, Alangium, Albizzia, Alectryon, Alnus, Alphitonia, Alstonia, Altingia, Amburana, Amelanchier, Amerimnon, Amoora, Amyris, Anacardium, Andira, Anaba, Anisoptera, Annesleia, Anogeissus, Anthocephalus, Antirrhoea, Apodytes, Aporosa, Apuleia, Araucaria, Arbutus, Arenga, Argania, Aristotelea, Artocarpus, Aspidosperma, Astronium, Aubrya, Avicennia, Aydendron, Bactris, Balanocarpus, Balfourodendron, Bambusa, Banksia, Baphia, Barringtonia, Bassia, Behaimia, Beilschmiedia, Berlinia, Berria, Betula, Bischoffia, Bombacopsis, Bombax, Bowdichia, Brachystegia, Bridelia, Brosimum, Brya, Bucklandia, Buddleia, Bulnesia, Bursera, Butea, Buxus, Byrsonima, Cabralea, Caesalpinia, Calliandra, Callicoma, Callitris, Calodendron, Calophyllum, Calycophyllum, Caraipa, Carapa, Carinaria, Carpinus, Carpholobia, Carya, Caryocar, Casearia, Castanea, Castaniopsis, Casuarina, Catalpa, Cavanillesia, Cecropia, Cedrela, Cedrus, Ceiba, Celtis, Centrolobium, Cephalanthus, Ceratonia, Cercidiphyllum, Cercocarpus, Ceriops, Chaetocarpus, Chaetoptelea, Chamaecyparis, Chlorophora, Chloroxylon, Chrysophyllum, Cinnamomum, Cistanthera, Citerexylum, Cladrastus, Coccolobis, Cocos, Colletia, Colubrina, Copaifera, Cordia, Cornus, Corylus, Corynanthe, Coula, Crataegus, Crossopteryx, Cryptocarya, Cryptomeria, Ctenolophon, Cullenia, Cunninghamia, Cunonia, Cupania, Cuphocarpus, Cupressus, Curatella, Curtisia, Cyathocalyx, Cybistax, Cynometra, Dacrydium, Dalbergia, Daniella, Dendropanax, Detarium, Deutzia, Dialium, Dichrostachys, Didymopanax, Dillenia, Dimorphandra, Diospyros, Dipholis, Diplanthera, Diplotropis, Dipterocarpus, Distylium, Dodonaea, Dolichandrone, Doona, Dorypha, Drimys, Dryobalanops, Duabanga, Duguetia, Dumoria, Dysoxylon, Ekkebergia, Ehretia, Elaeocarpus, Elateriospermum, Enantia, Endospermum, Engelhardtia, Entandrophragma, Entelea, Enterolobium, Eperua, Erica, Erythrina, Erythrophleum, Eucalyptus, Euclea, Eucryphia, Eugenia, Eurya, Eusideroxylon, Euxylophora, Exocarpus, Exostemma, Eysenhardtia, Fagara, Fagraea, Fagus, Faurea, Ferolia, Ficus, Fillaeopsis, Firmiana, Fitzroya, Flacourtia, Flindersia, Forestiera, Fraxinus, Fusanus, Ganophyllum, Garcinia, Gardenia, Genipa, Gevuina, Girardinia, Gironneira, Gleditsia, Glochidion, Gluta, Gmelina, Gonioma, Gonystylus, Gordonia, Gossweilerodendron, Gossypiospermum, Goupia, Grevillea, Gewia, Griselinia, Guiacum, Guarea, Guazuma, Guettarda, Gymnocladus, Gyrinops, Gyrocarpus, Haematoxylon, Hakea, Haloxylon, Hannoa, Haploclathra, Haronga, Harpulia, Hedera, Hedicarya, Heinsia, Heritiera, Herminiera, Hibiscus, Hieronima, Hippomane, Hippophae, Holarrhena, Holocalyx, Homalium, Hopea, Humiria, Hura, Hydrangea, Hymenaea, Hymenodictyon, Ilex, Irvingia, Isoptera, Ixonanthes, Jacaranda, Juglans, Juniperus, Kalmia, Karwinskia, Karmadecia, Keteleria, Khaya, Kiggelaria, Klaineodoxa, Knightia, Laburnum, Lagerstroemia, Laplacea, Larix, Laurelia, Lecythis, Leitneria, Leptochlaena, Leptonychia, Leptospermum, Leucadendron, Leucopogon, Libocedrus, Licania, Ligustrum, Liquidambar, Liriodendron, Litsea, Lomatia, Lonchocarpus, Lophira, Lovoa, Loxopterygium, Luehea, Lumnitzera, Lysiloma, Maba, Machilus, Maclura, Maerua, Magnolia, Mallotus, Manilkara, Markhamia, Maxwellia, Maytenus, Melaleuca, Melanoxylon, Melia, Melocanna, Memecylon, Mesua, Metrosideros, Mezoneuron, Michelia, Microdesmia, Milletia, Mitragyna, Monotoca, Montrouziera, Morinda, Morus, Musanga, Myoporum, Myrianthus, Myristica, Myrocarpus, Myrsine, Nectandra, Neobaranonia, Nothofagus, Nuxia, Nyssa, Ochanostachys, Ochroma, Ochrosia, Ocotea, Odina, Oldfieldia, Olea, Olearia, Omphalocarpum, Oncosperma, Ormosia, Oroxylum, Osmanthus, Ostrya, Osyris, Owenia, Oxanthus, Palaquium, Palvaea, Pancheria, Panopsis, Paradaniella, Parinarium, Parkia, Parmentiera, Patagonula, Paulownia, Peltogyne, Peltophorum, Pemphis, Pennantia, Pentaclethra, Pericopsis, Persea, Petitia, Phellodendron, Phillyrea, Phoenix, Phyllocladus, Phyllostachys, Phyllostylon, Physocalymna, Picea, Picrasma, Pilgerodendron, Pinus, Piptadenia, Piscidia, Pistacia, Pithecolobium, Pittosporum, Platanus, Platonia, Plathymenia, Platylophus, Platymiscium, Plectronia, Ploearium, Podocarpus, Polyadoa, Pometia, Populus, Pouteria, Premna, Prioria, Prosopus, Prunus, Pseudotsuga, Psidium, Ptaeroxylon, Pterocarpus, Pterocarya, Pterogyne, Pterospermum, Pygeum, Pyrus, Quercus, Ratonia, Recordoxylon, Reynosia, Rhamnus, Rhizophora, Rhododendron, Rhodosphaera, Robinia, Rottlera, Roupola, Ryparosa, Saccopetalum, Salix, Santalum, Santiria, Sapindus, Sarcocephalus, Sassafras, Saxegothea, Scaevola, Schaeffera, Schima, Schleidera, Schoutenia, Sciadopitys, Scolopia, Scolymus, Scorodocarpus, Senecio, Sequioa, Serianthes, Shorea, Sickingia, Sideroxylon, Sindora, Sonneratia, Sophora, Sorbus, Soymida, Spermolepis, Staudtia, Stenocarpus, Sterculia, Stereospermum, Strombosia, Strychnos, Styphnodendron, Styloceras, Styrax, Swartzia, Sweetia, Swietenia, Symphonia, Symplocos, Synocarpia, Synoum, Syzygium, Tabebuia, Tamarix, Tarchonanthus, Tarrietia, Taxodium, Taxus, Tecoma, Terminalia, Ternstroemia, Tetraclinis, Tetraphleura, Tetraspathaea, Thespesia, Thuja, Thujopsis, Tilia, Toddalia, Torreya, Trema, Trichilia, Trigoniastrum, Tripla-

ris, Tristania, Tristaniopsis, Tsuga, Uapaca, Ulmus, Urandra, Uvaria, Vanillosmopsis, Vatairea, Vatairiopsis, Vatica, Vernonia, Virgilia, Vitex, Whitfordiodendron, Xerospermum, Ximenia, Xylia, Xylopia, Yucca, Zanthoxylum, Zelkova, Zollernia.

Wood Apple → Feronia limonia Swingle.

Wood Bluegrass → Poa nemoralis L.

Wood Crane's Bill → Geranium sylvaticum L.

Wood Germander → Teucrium Scorodonia L.

Wood Lily → Lilium philadelphicum L.

Wood Oil, Chinese → Aleurites Fordii Hemsl.

Wood Sanicle → Sanicula europaea L.

Wood Sorrel → Oxalis Acetosella L.

Wood Tar → Fagus ferruginea Dryand.

Woodbox, Florida → Schaefferia frutescens Jacq.

Woodland Angelica → Angelica sylvestris L.

Woodruff, Dyers' → Asperula tinctoria L.

Woodruff, Sweet → Asperula odorata L.

Woodfordia floribunda Salisb. (Lythraceae). - Large shrub. India, Himalaya, Trop. Africa. Source of a gum, Dhaura or Dham-ka-gond, mainly derived from Haranti and Mewar. Resembles Tragacanth-gum. Flowers are source of a dye.

Woodwort, Marsh → Stachys palustris L.

Wool Crape → Tillandsia usneoides L.

Woolly Buckthorn → Bumelia lanuginosa (Michx.) Persoon.

Woolly Milkweed → Asclepias eriocarpa Benth.

Woolly White Quebracho → Aspidosperma tomentosum Mart.

Wooraia Poison → Urechites suberecta Muell. Arg.

Worm Seed, American → Chenopodium ambrosioides L. var. anthelminticum L.

Worm Seed Plant, Levant → Artemisia Cina Berg.

Wormwood, Roman → Artemisia pontica L.

Woundwort → Laserpitium latifolium L.

Wrack, Bladder → Fucus vesiculosus L.

Wrack, Knobbed → Ascophyllum nodosum (L.) Le Jolis.

Wrack, Lady → Fucus vesiculosus L.

Wrack, Yellow → Ascophyllum nodosum (L.) Le Jolis.

Wrinkled Shield Lichen → Parmelia caperata (L.) Ach.

Wu-pei-tzu Galls → Rhus semialata Murr.

Wyethia amplexicaulis Nutt. (Compositae). - Perennial herb. Western part of the United States. Roots after being heated and undergone a fermentation were eaten by the Indians of Montana, Utah and Nevada. Also W. longicaulis Gray.

X

Xanthophyllum excelsum Miq. (Polygalaceae). - Tree. Java. Bark is used by Javanese for colic. Also source of a yellow dye.

Xanthophyllum lanceolatum J. J. Sm. Suir Tree. (Polygalaceae). - Tree. Sumatra. Seeds are source of an oil, used by the natives in the preparation of food. Used in Europe for manuf. of candles and soap.

Xanthophyllum glaucum Wall. (Polygalaceae). - Tree. Indo China, Burma. Leaves are used in Indo China to impart a bitter taste to beer.

Xanthoria candelaria (Ach.) Arn. (syn. Teloschistes candelarius (L.) Fink.). (Parmeliaceae). - Lichen. Temp. and subarctic zones. Source of yellow dye; used in Sweden for dyeing woolens.

Xanthorrhiza apiifolia L'Hér. (syn. X. simplicissima Marsh.). Yellow Root. (Ranunculaceae). - Small shrub. Eastern N. America to Florida and Tennessee. Roots are source of a yellow dye.

Xanthorrhoea arborea R. Br. Dackowar Grasstree. (Liliaceae). - Grass Tree. Australia. White tender base of leaves and extremities of young shoots are eaten as food by the arborigines of of N. Queensland.

Xanthorrhoea australis R. Br. Australian Grass Tree. (Liliaceae). - Tree-like plant. Australia. Source of Red Acaroid, Grass Tree Gum, Yakkagum, Resina Lutea. Also X. quadrangulata F. v. Muell. X. Preussii Endl.

Xanthorrhoea hastilis R. Br. Spearleaf Grass Tree. (Liliaceae). - Tree-like plant. Australia. Source of Botany Bay Gum, Grass Tree Gum, Resina Acaroides, Gummi Acaroides, Yellow Acaroid Resin. Used in the varnish industry.

Xanthorrhoea quadrangulata F. v. Muell. (Liliaceae). - Perennial plant. S. Australia. Gum from the stem is used by the aborigines of S. Australia for fixing weapons; also for attaching stone axes to the hafted handles.

Xanthosoma atrovirens C. Koch. Darkleaf Malange, Indian Kale. (Araceae). - Perennial herb. Northern S. America. The corms are sometimes consumed as food.

Xanthosoma brasiliense (Desf.) Engl. Belembe. (Araceae). - Tall perennial herb. S. America, West Indies. Cultivated. Leaves are cooked and consumed as spinach.

Xanthosoma Caracu Koch. and Bouché. Caracu. (Araceae). - Perennial herb. Trop. America. Cultivated in West Indies, among which Porto Rico, also in Centr. America. Corms are eaten boiled. Young, unfolded leaves are consumed as spinach, also used in stews and soups.

Xanthosoma Jacquinii Schott. Yautia Palma. (Araceae). - Large perennial herb. Venezuela, West Indies. Grown as a food for hogs in some parts of the West Indies.

Xanthosoma sagittifolium Schott. Yellow Yautia, Yautia Amarilla, Tanias. (Araceae). - Tall perennial herb. Trop. America. Corns are consumed cooked or roasted. Cultivated in Tropics of Old and New World. Young leaves are sometimes boiled and eaten as spinach.

Xanthosoma violaceum Schott. Primrose Malanga, Indian Kale, Otó. (Araceae). - Tall perennial herb. Trop. America. Cultivated in Tropics of Old and New World. Tubers are consumed cooked and eaten like potatoes. Young leaves when boiled and chopped are consumed as spinach.

Xerochlamys pilosa Baker. (Schizochlaenaceae). - Tree. Centr. Madagascar. Used in Madagascar for manuf. rum.

Xerophyllum tenax (Pursh.) Nutt. Common Bear-Grass. (Liliaceae). - Perennial herb. Western N. America. The tenaceous leaves were used by the Indians for making water-tight baskets. Also X. Douglasii Wats.

Xerospermum intermedium Radlk. (Sapindaceae). - Tree. Burma to Singapore. Wood hard, durable, light-brown; used for building purposes.

Ximenia americana L. Tallow Wood. (Olacaceae). - Tree. Trop. and Subtropics. Wood is sometimes used as a substitute for Sandalwood in Sudan. Fruits having the size of a plum are consumed in many countries, they are sometimes pickled. Seeds are source of Wild Olive Seed Oil, a non-drying oil, used in E. India as a substitute for Ghee. Sp. Gr. 0.9262; Sap. Val. 173.4; Iod. No. 82.5; Acid. Val. 2.3; Unsap. 1.7 %.

Xylaria obovata Berk. (syn. Coelorphopalon obovatum (Berk.) v. Overeem.) (Xylariaceae). - Ascomycete. Fungus. Malaya. Powdered fruit-bodies mixed with coconut oil are used among the natives in wounds that have been caused by burns.

Xylariaceae (Fungi) → Daldinia, Engleromyces, Xylaria.

Xylia dolabriformis Benth. (X. xylocarpa Taub.). (Leguminosaceae). - Tree. Malaysia, Siam, Burma. Wood very hard, difficult to saw, reddish brown; used for bridges, buffers, road pavements, ship building, rail-road ties, house posts.

Xylinabaria Reynaudi Jum. (Apocynaceae). - Woody plant. Tonkin. Source of a reddish rubber. Latex coagulates by heating.

Xylocarpus Granatum Koenig → Carapa moluccensis Lam.

Xylopia aethiopica (Dun.) Rich. (syn. Uvaria aethiopica Guill. an Poir.) (Annonaceae). - Tree. Trop. Africa. Wood light yellow, fairly heavy, elastic; used in Guinea for masts of small boats. Source of African Grains of Selim, Guinea or

Negro Pepper. Sold in markets of Senegal and other parts of Africa. Was formerly much used in Europe. This species was Piper aethiopicum of some old herbals.

Xylopia carminativa (Aruda) Fries. (syn. X. sericea St. Hil., Unonia carminativa Aruda.) Pao d'Embira, Pimenta de Macaco. (Annonaceae). - Tree. Brazil, esp. Minas Gereas to Brit. Guiana. Fruits have odor and taste of pepper, used as a condiment. Seeds employed in Brazilian medicine. Carminative. Bark is stringy and tenacious; used for cordage, especially for boat-cables.

Xylopia frustescens Aubl. (Annonaceae). - Tree. Trop. America. Fruits are used as a condiment. Bark is used for ropes.

Xylopia grandiflora St. Hil. Malagueto. (Annonaceae). - Tree. Trop. America, esp. Brazil to Panama and Cuba. Flowers used in pharmacy in Brazil as a tonic, carminative. Source of an ess. oil of a pepper-like flavor.

Xylopia obtusifolia A. Rich. (syn. Habzelia obtusifolia A. DC.) Guimba. (Annonaceae). - Shrub or small tree. Cuba. Wood much prized in Cuba for its yellow color, known as Pico de Gallo.

Xylopia parvifolia Hook. f. (syn. Unona tripetalodea Morn.) (Annonaceae). - Tree. Ceylon. Bark of root used by natives for ulcers. Flowers, fruits, bark of root are used in chewing betel in Ceylon.

Xylopia sericea St. Hil. (Annonaceae). - Tree. Brazil. Fruits are used as a condiment.

Xylopia striata Engl. (Annonaceae). - Tree. Cameroon. Source of Bosé Wood, being grey yellow; used for different purposes.

Xylopia undulata Beauv. (Annonaceae). - Tree. Trop. Africa. Fruits are used as a condiment.

Xyridaceae → Xyris.

Xyris communis Kunth. (syn. X. laxifolia Mart., X. indica Vell.). (Xyridaceae). - Perennial herb. Trop. America. Root is emetic, cathartic; used in some parts of Brazil for leprosy and eczema.

Xyris melanocephala Miq. (Xyridaceae). - Perennial herb. Java, Sumatra. Stems are manuf. into coarse mats; used by the natives.

Xysmalobium Heudelotianum Decne. (Asclepiadaceae). - Herbaceous perennial. Trop. Africa. Roots called Yakhop, are consumed as food by the natives of Senegambia and of other parts of Africa.

Y

Yage → Banisteria Caapi Griseb.

Yakal Shorea → Shorea Balangeran Burck.

Yakka Gum → Xanthorrhoea australis R. Br.

Yam, Affun → Dioscorea cayennensis Lam.

Yam, Akam → Dioscorea latifolia Benth.

Yam, Atlantic → Dioscorea villosa L.

Yam Bean → Pachyrrhizus erosus Urb., P. tuberosus Spreng.

Yam, Fancy → Dioscorea esculenta (Lour.) Burk.

Yam, Chinese → Dioscorea Batatas Decne.

Yam, Greater → Dioscorea alata L.

Yam, Guinea → Dioscorea cayennensis Lam.

Yam, Lesser Asiatic → Dioscorea esculenta (Lour.) Burk.

Yam, Long → Dioscorea transversa R. Br.

Yam, Negro → Dioscorea cayennensis Lam.

Yam, Spiny → Dioscorea spinosa Roxb.

Yam, Potato → Dioscorea esculenta (Lour.) Burk.

Yam, Ten Months → Dioscorea alata L.

Yam, Water → Dioscorea alata L.

Yam, White → Dioscorea alata L.

Yam, Wild → Dioscorea spinosa Roxb. and D. villosa L.

Yam, Yellow → Dioscorea cayennensis Lam.

Yama Cocobolo → Platymiscium dismophandrum Don. Sm.

Yama Rosewood → Platymiscium dimosphandrum Don. Sm.

Yamira piranga → Caesalpinia echinata Lam.

Yampa → Carum Gairdneri (Hook. and Arn.) Gray.

Yampi → Dioscorea trifida L. f.

Yangta → Actinidia chinensis Planch.

Yaragua Grass → Andropogon rufus Kunth.

Yarran, Curly → Acacia homalophylla Cunningh.

Yarrow → Achillea Millefolium L.

Yaruru → Aspidosperma excelsum Benth.

Yeast, Arak → Schizosaccharomyces asporus Beijerinnck.

Yeast, Beer → Saccharomyces cerevisiae Hansen.

Yeast, Bottom → Saccharomyces carlsbergensis Hansen.

Yeast, Bread → Saccharomyces cerevisiae Hansen.

Yeast, Palm Wine → Saccharomyces Tuac Vorderm.

Yeast, Wine → Saccharomyces ellipsoides Hansen.

Yaupon → Ilex vomitoria Ait.

Yautia Palma → Xanthosoma Jacquinii Schott.

Yautia, Yellow → Xanthosoma sagittifolium Schott.

Yaxci Fibre → Agave sisalana Perrine.

Yeddo Spruce → Picea jezoensis Carr.

Yeddo Willow → Salix jessoensis Seem.

Yellow Acaroid Resin → Xanthorrhoea hastilis R. Br.

Yellow Ash → Cladrastis lutea (Michx.) Koch.

Yellow Bachelor's Button → Polygala Rugelii Shuttlw.

Yellow Bark → Cinchona Calisaya Wedd.

Yellow Bark Oak → Quercus velutina Lam.

Yellow Berry → Rhamnus infectorius L.

Yellow Berry → Rubus Chamaemorus L.

Yellow Birch → Betula lutea Michx.

Yellow Borrera → Theloschistes flavicans (Swartz) Muell. Arg.

Yellow Box → Eucalyptus hemiphloia F. v. Muell.

Yellow Cedar → Rhodosphaera rhodanthema Engl.

Yellowbark Cinchona → Cinchona Calisaya Wedd.

Yellow Cirouaballi → Nectandra Pisi Miq.

Yellow Cypress → Chamaecyparis Nootkatensis (Lamb.) Sudw.

Yellow Fiji Sandalwood → Santalum Freycinetianum Gaud.

Yellow flowered Alfalfa → Medicago falcata L.

Yellow Gentian → Gentiana lutea L.

Yellow Himalayan Raspberry → Rubus ellipticus Smith.

Yellow, Indian → Mangifera indica L.

Yellow Jessamine → Gelsemium sempervirens (L.) Ait. f.

Yellow Lady's Slipper → Cypripedium parviflorum Salisb.

Yellow Lupine → Lupinus luteus L.

Yellow Mombin → Spondias lutea L.

Yellow Nut Grass → Cyperus esculentus L.

Yellow Oleander → Thevetia nereifolia Juss.

Yellow Olivier → Terminalia obovata Steud.

Yellow Pareira → Aristolochia glaucescens H. B. K.

Yellow Pine → Pinus arizonica Engelm.

Yellow Pondlily → Nuphar advena Ait.

Yellow Poplar → Liriodendron tulipifera L.

Yellow Poui → Tecoma serratifolia G. Don.

Yellow Resin → Pinus palustris Mill.

Yellow Rocket → Barbarea vulgaris R. Br.

Yellow Root → Xanthorrhiza apiifolia L'Hér.

Yellow Sanders → Terminalia Hilariana Steud.

Yellow Sand Verbena → Abroma latifolia Esch.

Yellow Sapote → Lucuma salicifolia H. B. K.

Yellow Silver Pine → Dacrydium intermedium T. Kirk.

Yellow Sweet Clover → Melilotus officinalis (L.) Lam.

Yellow Tea Tree → Leptospermum flavescens Sm.

Yellow Velvetleaf → Limnocharis emarchinata Humb. and Bonpl.

Yellow Wild Indigo → Baptisia tinctoria (L.) R. Br.

Yellow Yam → Dioscorea cayennensis Lam.

Yellow Yautia → Xanthosoma sagittifolium Schott.

Yellowheart Prickly Ash → Zanthoxylum flavum Vahl.

Yellowwood → Cladratis lutea (Michx.) Koch, Podocarpus Thunbergii Hook., Schaefferia frutescens Jacq., Sterculia oblonga Mart. and Zanthoxylum flavum Vahl.

Yellow Wood, Quintenigua → Podocarpus elongata L'Hér.

Yerba Maté → Ilex paraguensis St. Hil.

Yerba Santa → Eriodictyon californica (Hook. and Arn.) Torr.

Yew, American → Taxus canadensis Willd.

Yew, English → Taxus baccata L.

Yew, Pacific → Taxus brevifolia Nutt.

Ylang Ylang → Canangium odoratum L.

Yolombo → Panopsis rubescens (Pohl) Pitt.

Yoruba Indigo → Lonchocarpus cyanescens Benth.

Ysaño → Tropaeolum tuberosum Ruiz. and Pav.

Yucatan Sisal → Agave fourcroydes Lemaire.

Yuca dulce → Manihot dulce (J. F. Gmel.) Pax.

Yucca aloifolia L. Spanish Dagger, Spanish Bajonet. (Liliaceae). - Sparcely branched small tree. N. Carolina to Florida and Louisiana, West Indies. Fibres from leaves were used for making ropes by the early settlers. Also Y. filamentosa L., Y. gloriosa L. and Y. recurvifolia Salisb.

Yucca arborescens Trel. Joshua Tree. (Liliaceae). - Tree-like. Southwestern United States. Wood is cut into layers and made into boxes and small articles. Is also used as a wrapping material. Seeds are consumed as food by the Indians.

Yucca australis (Engelm.) Trel. (Liliaceae). - Large sparcely branched tree. Mexico. Leaves are source of a good fibre. Exported as Ixtle. Young stems and leaves produce an alcohol. Spongy material of trunk cut in strips, beaten flat and washed, is made into mats, used as pads (sudaderos) for pack animals.

Yucca baccata Torr. Dátil (Mex.) (Liliaceae). - Woody plant. Southeastern United States, Mexico. Fruits are consumed roasted, cooked or fresh by the Indians and Mexicans. Suitable for pies. Mexican Indians slice and dry the ripe fruits for winter use. Fresh flower buds are eaten by the natives. Fibres were used for basketry.

Yucca elata Engelm. Soap Tree Yucca. (Liliaceae). - Small almost unbranched tree. W. Texas to Arizona, Chihuahua (Mex.). Source of an excellent fibre; exported. Locally used for cordage; woven by Indians into mats and cloth. Roots have saponifying properties; used for washing clothes. Extract from roots is used to produce foam in beverages. Also other Yucca species are used for the same purpose.

Yucca elephanthipes Regel. Ozote. (Liliaceae). - Tall almost unbranched tree. Mexico to Centr. America. Cultivated. Flower buds consumed in soup. Sold in local markets.

Yucca Endlichiana Trel. (Liliaceae). - Tree. Mexico, esp. Coahuila. Leaves are source of an excellent fibre.

Yucca filamentosa L. Adam's Needle. (Liliaceae). - Perennial herb. Eastern N. America to Florida and Louisiana. Fruits are eaten by the Indians. Roots are used for cleaning.

Yucca glauca Nutt. (syn. Y. angustifolia Pursh). Small Soapweed. (Liliaceae). - Low shrub. Western United States and adj. Mexico. Leaves are manuf. into stable brooms. Fruits are cooked and consumed by the Indians, esp. during scarcity of food.

Yucca macrocarpa (Torr.) Cov. (Liliaceae). - Small almost unbranched tree. W. Texas to S. Arizona, also Chihuahua (Mex.). Leaves used by Indians in S. of New Mexico for baskets. Seeds were consumed as food by the natives.

Yucca mohavensis Sarg. (Liliaceae). - Tree-like plant. California and adj. territory. Green pods when roasted on coals were eaten by the Indians. Flowers were boiled and used as food. Leaves are source of a fibre, used by the Goahuilla Indians for weaving, saddle mats and sandals.

Yucca treculeana Carr. Trecul Yucca. (Liliaceae). - Small branched tree. Mexico to Texas. Leaves are source of a much used fibre.

Yucca Whipplei Torr. (syn. Herperoyucca Whipplei (Torr.) Baker.) Whipple Yucca. (Liliaceae). - Acaulescent plant. Mexico, California. Flowers eaten by Indians. Seeds when ground are eaten as a porridge.

Yunnan Hemlock → Tsuga yunnanensis (Franch.) Mast.

Z

Zacate → Nolina longifolia (Schult.) Hemsl.

Zakana → Boswellia Carterii Birdw.

Zakaton Grass → Epicampes macroura Benth.

Zalacca edulis Blume (syn. Salacca edulis Reinw.) Salak. (Palmaceae). - Malaysia, Sumatra, Java. Low, almost stemless palm. Fruits reddish brown, pleasant sweet when ripe. Young fruits are eaten pickled. Sold in local markets. Fruits preserved in cans with salt water and sugar, are eaten by Mahomedan pelgrims during their journey to Mecca.

Zamia furfuraceae Ait. (Cycadaceae). - Low palm-like plant. Centr. America. Roots are poisonous; used in some parts of Honduras for criminal poisoning.

Zamia integrifolia Ait. (syn. Z. floridana DC.). Coontie, Florida Arrowroot. (Cycadaceae). - Low palm-like plants with thick underground stem. Florida. Stem source of a starch; in former years sold as Florida Arrowroot. A staple food among the Seminole Indians of Florida.

Zamia latifolia Pren. (Cycadaceae). - Woody, palm-like plant. West Indies, Centr. America. Starch derived from stem is used in laundry work, also as food. Juice from plant is very poisonous.

Zamia Lindenii Regel. (Cycadaceae). - Small palm-like plant. Ecuador, Colombia and adj. territory. Seeds are used as food by the Choco Indians in Colombia.

Zamia spiralis R. Br. → Macrozamia spiralis Miq.

Zanthoxylum alatum Steud. (syn. Z. planispinum Sieb. and Zucc.). Wingleaf Prickly Ash. (Rutaceae). - Small tree. China. Aromatic seeds when powdered are used as stomachic in some parts of China.

Zanthoxylum americanum Mill. Northern Prickly Ash, Toothache Tree. (Rutaceae). - Shrub. Eastern N. America. Dried bark is used medicinally. → Z. Clava-Herculis L.

Zanthoxylum Bungei Planchon. (Rutaceae). - Small spiny shrub. China. Cultivated in Centr. and S. China. Seeds are source of Chinese Pepper, much used as a condiment in China.

Zanthoxylum Clava-Herculis L. Hercules Club Prickly Ash, Toothache Tree. (Rutaceae). - Shrub or tree. Southern and southeastern United States. Bark is used as a cure for toothache and rheumatism; is diaphoretic and alterative. Bark is much collected by negroes in the so. states. Berries are tonic, diaphoretic, mild stimulant and alterative.

Zanthoxylum emarginatum Sw. (syn. Z. coriaceum A. Rich, Fagara coriacea Kr. and Urb.) (Rutaceae). - Tree. West Indies. Wood yellow; used in general carpentry.

Zanthoxylum flavum Vahl. Yellow Heart Prickly Ash, Satin Wood, Yellow Wood. (Rutaceae). - Shrub or tree. Trop. America. Wood brittle, very hard, heavy, not strong, light orange; occasionally used for cabinet work, handles of tools and small objects.

Zanthoxylon nitidum DC. (syn. Fagara piperita Poureira). (Rutaceae). - Tree. China, Cochin-China, Tonkin. Seeds are used by the natives as condiment; also made into drinks.

Zanthoxylum piperatum DC. Japanese Prickly Ash. (Rutaceae). - Small spiny tree. China, Japan. Seeds are used as condiment in China and Japan.

Zanthoxylum Rhetsea DC. (Rutaceae). - Tree. India, Indo China. Fruits are used as a spice by the natives.

Zanthoxylum senegalense DC. Senegal Prickly Ash. (Rutaceae). - Tree. Trop. Africa, Nigeria. Bark sudorific, aromatic. Seeds when soaked in water are used by the natives for rheumatics.

Zanzibar Aloë → Aloë Perryi Baker.

Zanzibar Copal → Trachylobium Hornemannianum Hayne.

Zanzibar Ebony → Diospyros mespiliformis Hochst.

Zanzibar Raspod → Trachylobium Hornemannianum Hayne.

Zapatero → Tabebuia pentaphylla (L.) Hemsl.

Zapote Amarillo → Lucuma salacifolia H. B. K.

Zapote Blanco → Casimiroa edulis La Llave.

Zapote Cabillo → Licania platypus (Hemsl.) Pitt.

Zapote Negro → Diospyros Ebenum Koen.

Zapupe Fibre → Agave zapupe Trel.

Zarcemora → Rubus floribundus H. B. K.

Zea Mays L. Maize, Corn, Indian Corn. (Graminaceae). - Annual grass. Probably from American tropics. Grown since pre-Columbian times by the Indians. Cultivated in the Old and New World. Grains (fruits) are source of corn meal; starch made into baking powder; used for desserts, confectionary, cosmetics, adhesives, shoe polishes; source of alcoholic beverages, corn whisky, industrial alcohol, which can be made into explosives and synth. rubber; butyl alcohol, acetone; sirup, dextrose, maltose, dextrine; source of hominy, food for domestic animals, used for cloth sizing, paper sizing, laundry starching. Starch is used pharmaceutically as diluent for powders and as dusting powder; when swallowed as antidote of iodine poisoning and as base for suppositories. Cornsilks when green are used for manuf. of pharmaceutical preparations; diuretic. Green plants are used as fodder for live-stock. Germ of seeds is source of Corn Oil semi-drying; used for manuf. linoleum, in paints, varnishes, oil-cloth, soft soap, glycerine; used for edible purposes, in salads, lard substitutes. Sp. Gr. 0.921—0.927; Iod. No. 116—130; Sap. Val. 188—193; Unsap. 1.3—2.0 %. Press cake is used as food for live-stock. Z. Mays is divided in the following races, groups or subspecies: I. Group. Subsp. amylacea (Sturt.) Bailey. Soft Corn. No hard endosperm. II. Group. Subsp. indentata (Sturt.) Bailey. Dent Corn. White endosperm. Used for grain, fodder, and ensilage. III. Group. Subsp. indurata (Sturt.) Bailey. Flint Corn. Hard endosperm. IV. Group. Subsp. praecox Bonaf. Popcorn. Grains eaten when exposed to heat when „popped". Varieties: Golden Queen, White Rice, White Hulless. V. Group. Subsp. rugosa Bonaf. Sweet Corn. Used as vegetable when boiled. Varieties: Golden Bantam, Country Gentleman, Early Evergreen, Stowell's Evergreen. VI. Group. Subsp. tunica St. Hil. Pod Corn. Of no economic value.

Zebra Wood → Centrolobium robustum Mart. and Diospyros Kurzii Hiern.

Zebra Wood, Andaman → Diospyros Kurzii Hiern.

Zedoaria → Curcuma Zeodoaria Rosc.

Zedoary → Curcuma Zeodoaria Rosc.

Zelkova serrata (Thunb.) Mak. (syn. Z. Keaki Mayr.). Japanese Zelkova. (Ulmaceae). - Tree. Japan. Wood yellow white, tough, hard; used for house building, ships, furniture and handles of tools.

Zephyranthes Atamasco Herb. Atamasco Lily. (Amaryllidaceae). - Bulbous perennial. So. of the United States. Bulbs were eaten by the Creek Indians during times of scarcity.

Zerumbet Ginger → Zingiber Zerumbet (L.) Smith.

Zilla myagroides Forsk. (Cruciferaceae). - Herbaceous plant. Egypt, Arabia. Leaves when boiled are eaten as a vegetable by the Arabs.

Zingiber amaricans Noronha. (Zingiberaceae). - Perennial herb. Mal. Archip., Java. Rhizomes are used as a condiment by the natives, also as appetiser. Sold in markets of Java.

Zingiber Cassumunar Roxb. Cassumunar Ginger. (Zingiberaceae). - Perennial herb. Tonkin, Cochin China. Cultivated. Rhizomes are used as a condiment.

Zingiber Mioga (Thunb.) Rosc. Mioga Ginger. (Zingiberaceae). - Perennial herb. Japan. Cultivated in Japan. Rhizomes are source of Japanese Ginger, have a bergamot-like flavor. Young flowers, fruits and sprouts are eaten in Japan.

Zingiber officinale Rosc. Ginger. (Zingiberaceae). - Perennial herb. E. India. Cultivated in the tropics of the Old and New World. Washed and peeled rhizome is source of Ginger, a widely used spice. Rootstocks are sold dry, powdered, preserved in syrup and candied. Used in candies, pastry, cakes; also in ginger ale and ginger beer, wine and brandy. In perfumery it is used to impart scent of an oriental character. Known since antiquity. Commercially grown extensively in Jamaica. Some commercial grades are known as African Ginger, Cochin Ginger, Bleached Jamaica Ginger, Calcutta Ginger, Calicut Ginger. Contains ess. oil; chiefly terpenes, d-camphene, z-phellandrine, zingiberine, citral; borneol causing its flavor; a viscid oil, gingerol causes its pungency. Used in medicine as aromatic digestive stimulant and carminative.

Zingiber Zerumbet (L.) Smith. Zerumbet Ginger. (Zingiberaceae). - Perennial herb. Trop. Asia. Cultivated in Cochin China, Cambodia, Annam and Martinique. Rhizomes are source of Martinique Ginger.

Zingiberaceae → Aframomum, Amomum, Costus, Curcuma, Donax, Elettaria, Elettariopsis, Gastrochilus, Hedychium, Kaempfera, Phaeomeria, Renealmia, Zingiber.

Zizania aquatica L. Wild Rice, Indian Rice. (Graminaceae). - Perennial grass. Eastern N. America, southward to Florida and Texas. Grains were much consumed by the Indians. At present occasionally eaten in the United States and sold in markets.

Zizania latifolia Turcz. (Graminaceae). - Perennial grass. Manchuria, China, Korea, Japan. Cultivated in China, Japan and Korea. Very young shoots are eaten as a vegetable, sold in markets. Leaves are made into mats.

Zizyphora tenuior L. (Labiaceae). - Perennial herb. Iran, Baluchistan, Afghanistan and adj. territory. Herb used in Iran as cordial and stomachic.

Zizyphus Endlichii Loes. (Rhamnaceae). - Tree. Mexico. Fruits are edible. Bark is used by the natives for toothache.

Zizyphus Joazeiro Mart.(Rhamnaceae). - Shrub. Brazil. Used as fodder for livestock during severe dry spells. Fruits are sometimes consumed by the inhabitants.

Zizyphus Jujuba Mill. not Lam. (syn. Z. vulgaris Lam.). Common Jujub, Chinese Jujub. (Rhamnaceae). - Shrub or small tree. E. India, Malaysia. Cultivated in warm countries, esp. in China, Iran, Afghanistan, Arabia, Asia Minor, Metiterranean region. Fruits ovoid, thin dark skin; flesh whitish, mealy, sweet, pleasant flavor; eaten fresh, dried (like dates), boiled with millet or rice, stewed, baked, glacé fruit when boiled in honey or in syrup; also made into jujub-bread. Several varieties among which Yu, Lang, Mu-shing-hong. Leaves are source of food of a silk-worm, Epiphora bauhinae Cuer-Men.

Zizyphus Lotus Desf. Lotus fruit. (Rhamnaceae). - Small tree. N. Africa. Fruits edible. Is probably the edible Lotus of the Ancient Peoples of Lybia, called Lotophages.

Zizyphus lycioides Gray → Condalia lvcioides Gray.

Zizyphus mauritiana Lam. (syn. Z. Jujuba Lam. non Mill., Z. sosoria Roem. and Schult.). (Rhamnaceae). - Shrub or small tree. Trop. Africa, Asia. Fruits are edible, known as Tsa or Chinese Dates. Leaves are used for tanning.

Zizyphus mexicana Rose. (Rhamnaceae). - Tree. Mexico, esp. Colima to Oaxaca. Fruits are used as soap for washing clothing.

Zizyphus Mistol Griseb. Argentine Jujub. (Rhamnaceae). - Shrub or tree. Andean region. Berries edible, juicy, sweet; used in Bolivia for preparation of an alcoholic beverage, called Chicha.

Zizyphus mucronata Willd. Buffalo Thorn, Cape Thorn. (Rhamnaceae). - Small tree. Trop. Africa. Fruits edible. Decoction of root is used for lumbago. Seeds are made into rosaries by Musselmen.

Zizyphus sonorensis S. Wats. (Rhamnaceae). - Shrub or small tree. Mexico. Fruits used as substitute for soap, used for washing clothes.

Zizyphus Spina-Christi Willd. Christ Thorn. (Rhamnaceae). - Small tree. N. Africa, Arabia. Fruits are consumed among the Arabs, having taste of dried apples.

Zollernia ilicifolia Vog. Mocitahyba. (Leguminosaceae). - Tree. Brazil. Bark source of tannin. Wood used for rail-road ties, cabinet work and beams.

Zollernia paraensis Huber. Muirapinima Preta. (Leguminosaceae). - Tree. Brazil, esp. Lower Amazone. Wood highly prized; used locally for small cabinet work and turnery; in U. S. A. employed for handles of cutlery, brush-backs and butts of billiard cues.

Zornia diphylla (L.) Pers. (Leguminosaceae). - Perennial herb. Tropics. Herb used for dysentery. Used during dry seasons by Fulachs as food for cattle and horses.

Zostera marina L. Grass Weed, Grass Wack, Alva Marina. (Najadaceae). - Perennial herb, submersed in salt water. Europe. Dried leaves and stem material used as packing material, in pillows; also said to be manufactured into a street paving material.

Zschokkea Foxii Stapf. Minyadotana. (Apocynaceae). - Tree. Bolivia. Latex is used to adulterate Hevea and Castilla rubber.

Zuelania Roussoviana Pitt. (Flacourtiaceae). - Tree. Centr. America. Stem is source of a gum, said to be used as a vomitive.

Zumbic Tree → Adenanthera pavonina L.

Zygophyllaceae → Balanites, Bulnesia, Guiacum, Kallstroemia, Larrea, Nitraria, Peganum, Porliera, Tribulus, Zygophyllum.

Zygophyllum album L. (Zygophyllaceae). - Shrub. N. Africa, Canary Islds. The tips of the

flower clusters when dried have a pleasant scent of tea. An infusion is used as a toilet water.

Zygophyllum simplex L. (Zygophyllaceae). - Succulent herb. W. Asia, Trop. Africa, Red Sea area. Plant is used as fodder for camels.

Bibliography

PUBLICATIONS OF A GENERAL NATURE

BAILEY, L. H. Cyclopedia of American Agriculture. 4 vol. 4 ed. New York 1912. —, Manual of Cultivated Plants. 2 ed. New York 1949. —, Standard Cyclopedia of Horticulture. 3 vol. New York. Repr. 1952.

DIELS, L. Ersatzstoffe aus dem Pflanzenreich. Berlin 1912.

ENGLER, A. und K. PRANTL. Die natürlichen Pflanzenfamilien, nebst ihren Gattungen und wichtigeren Arten, insbesondere den Nutzpflanzen. Leipzig. 1 ed. 22 vol. 1887—1912; 2 ed. since 1924 in preparation.

GRAFE, V. Handbuch der organischen Warenkunde. 2 vol. Stuttgart 1927—1930.

HARTWICH, C. Die menschlichen Genussmittel. Leipzig 1911. — HEDRICK, U. P. Sturtevant's Notes on Edible Plants. New York Agric. Exp. Rept. 1919. — HILL, Albert F. Economic Plants. New York. 2 ed. 1952. — HOLLAND, J. H. Overseas Plant Products. London 1937.

KELSEY, Harlan R. and William A. DAYTON. Standardized Plant Names. 2 ed. Harrisburg 1942.

PEGLION, VITTORIO. Le nostre Piante Industriali. Bologna 1919.

SPRECHER v. BERNEGG, A. Tropische und subtropische Weltwirtschaftspflanzen. 5 vol. Stuttgart 1929—1936.

VANSTONE, J. H. Rawmaterials of Commerce. London 1929.

WEHMER, C. Die Pflanzenstoffe. 2 ed. Jena 1929. — WIESNER, J. von. Die Rohstoffe des Pflanzenreiches. 4 ed. Leipzig 1927—1928; 5 ed. unter Mitwirkung zahlreicher Fachgenossen herausgegebene Auflage von C. Regel. Weinheim 1960.

AGRICULTURAL CROPS

FESCA, Max. Der Pflanzenbau in den Tropen und Subtropen. 3 vol. Berlin 1904—1911.

KLAPP, E. Lehrbuch des Acker- und Pflanzenbaues. 4 ed. Berlin 1954.

ROEMER, Th. (Editor). Handbuch der Landwirtschaft. 5 vol. Berlin 1951—1954.

SCHIEMANN, Elisabeth. Weizen, Roggen, Gerste. Systematik, Geschichte und Verwendung. Jena 1948. — SCHLIPF's Praktisches Handbuch der Landwirtschaft. 31 ed. Berlin 1952. — SCHMIDT, G. A. und A. MARCUS. Handbuch der tropischen und subtropischen Landwirtschaft. 2 vol. Berlin 1943. — SEMLER, H. Die tropische Agrikultur. 4 vol. Wismar 1886. -- STEFFEREUD, Alfred. (Editor). Grass. Yearbook of Agriculture. U. S. Dept. of Agric. 1948. Washington, D. C. With 138 different contributions from several authors. —, Crops in Peace and in War. Yearbook of Agriculture. U. S. Dept. of Agric. 1950—

1951. Washington, D. C. With 137 different contributions from several authors.

WILCOX, E. V. Tropical Agriculture. New York 1916. — WOLFE, T. K. and M. S. KIPPS. The Production of Field Crops. 4 ed. New York 1953.

FRUITS AND FRUIT PRODUCTS

ARNOU, C. Les Industries de la Conservation des Fruits. Paris 1925.

BOIS, D. Phanérogames Fruitiers. Vol. II of: Les Plantes Alimentaires chez tous les Peuples et à travers les Ages. Paris 1929.

CHARLEY, V. L. S. The Commercial Production of Fruit Syrups. Fruit Prod. Journ. 17: 72—77, 83, 1937. GOURLEY, Joseph, H. and F. S. HOWLETT, Modern Fruit Production. New York 1946.

HEDRICK, U. P. Cyclopedia of Hardy Fruits. New York 1922 — HUME, H. The Cultivation of Citrus Fruits. New York 1945.

POPONOE, W. W. Manual of Tropical and Subtropical Fruits. New York 1921.

ROLET, A. Les Conserves des Fruits. Paris 1912.

SHOEMAKER, James S. Small Fruit Culture. 3 ed. New York 1955.

TANAKA, Tyôzaburô. Citrus Studies. Tokyo 1936. (in Japanese).

WICKSON, E. J. California Fruits and how to grow them. 9 ed. San Francisco 1921.

BEVERAGES

BABO, A. F. von und E. MACH. Handbuch des Weinbaues und der Kellerwirtschaft. 3 vol. Berlin 1910. — BOIS, D. Les Plantes à Boissons. Vol. IV of: Les Plantes Alimentaires chez tous les Peuples et à travers les Ages. Paris 1937.

CHENEY, R. H. Coffee. New York 1925. — CRUESS, W. V. Unfermented Fruit Juices. Circ. 220. Calif. Agric. Exp. Sta. 1920.

DUGAST, J. Vinification dans les Pays Chauds. Alger. 1930.

HALL, C. J. J. van. Cocoa. London 1914.

KROEMER, K. und G. KRUMHOLZ. Obst- und Beerenweine. Braunschweig 1932.

OTTAVI, Ottavio. Enologia Teorico-Pratica. 12 ed. Casale Montferrat 1931.

SANNINO, F. A. Tratado de Enologia. Barcelona 1925. — SCHANDERL, H. Mikrobiologie des Weines. Stuttgart 1950. — SCHOONMAKER, F. and T. MARVEL. The Complete Wine Book. New York 1934.

VENTRE, J. Traité de Vinification. 3 vol. Montpellier 1929.

WAGNER, P. American Wines and how to make them. New York 1933. — WHYMPER, R. Cocoa and Chocolate. Philadelphia 1921.

VEGETABLES AND VEGETABLE PRODUCTS

BECKER-DILLINGER, J. Handbuch des gesamten Gemüsebaues, einschl. der Gewürze und Küchenkräuter. 5. Aufl. Berlin 1950. — BERNADIN, Ch. 60 Champions Commestibles. Saint-Dié 1955. — BOIS, D. Phanérogames Léguminières. Vol. I of: Les Plantes Alimentaires chez tous les Peuples et à travers les Ages. Paris 1927.

ROLFS, P. H. Subtropical Vegetable-Gardening. New York 1929.

THOMAS, W. S. Field Book of Common Mushrooms. 3 ed. New York 1948. — THOMPSON, H. C. and William C. Kelly. Vegetable Crops. 5 ed. New York 1957.

SPICES AND HERBS

BOIS, D. Plantes à Épices, à Aromates, à Condiments. Vol. III of: Les Plantes Alimentaires chez tous les Peuples à travers les Ages. Paris 1934.

CLARKSON, Rosetta E. Herbs, Their Culture and Uses. New York 1948.

RIDLEY, H. N. Spices. London 1912.

SIEVERS, A. F. Production of Drug and Condiment Plants. U. S. Dept. Agric. Farm Bull. 1999. 1948.

FOOD AND FOOD TECHNOLOGY

CRUESS, W. V. Commercial Fruit and Vegetable Products. 3 ed. New York 1948.

PRESCOT, Samuel C. and Bernard E. PROCTOR. Food Technology. New York 1937.

SHERMAN, H. C. Food Products. 3 ed. New York 1936.

FOREST PRODUCTS

ANONYMOUS. An Index of the Minor Forest Products of the British Empire. Imp. Econ. Comm. Ser. H. M. Stat. Off. London 1936.

BROWN, H. P. and A. J. PANSHIN. Textbook of Wood Technology. 2 vol. New York 1949—1952.

CHOWDHURY, K. A. Sandalwood and its Indian Substitutes. Ind. Forester. 57: 431—433, 1931.

DESCH, H. E. Timber. Its Structure and Properties. 2 ed. London 1947.

HOWARD, Alexander L. A Manual of the Timbers of the World. London 1934.

KELLOG, R. S. Lumber and its Uses. 4 ed. New York 1931. — KOLLMANN, F. Technologie des Holzes. Berlin 1936.

MERVILLE, R. A List of True and False Mahoganies. Kew. Bull. 193. 1936. — MEYER, Hans. Buch der Holznamen. Hannover 1933—1936.

PATERSON, S. The Forest Area of the World and its potential Productivity. 1956.

STEFFEREUD, Alfred (Editor). Trees. Yearbook of Agriculture. U. S. Dept. of Agric. 1949. Washington, D. C. With 141 different contributions from several authors. — STONE, H. The Timbers of Commerce and Their Identification. London 1924.

TRENDELENBURG, R. Das Holz als Rohstoff. Seine Entstehung, stoffliche Beschaffenheit und chemische Verwertung. München 1939.

WISE, Luis (Editor). Wood Chemistry. New York 1944.

ZON, R. and W. N. Sparhawk. Forest Resources of the World. New York 1923.

MEDICINAL PLANTS

BROCQ-ROUSSEAU, D. et R. FABRE. Les Toxines Végétales Phytotoxines et Phytoagglutinines. Paris 1947.

CERUTI, O. Le Piante Medicinali. Torino 1945.

FREUDENBERG, G. und R. CAESAR. Arzneipflanzen, Anbau und Verwertung. Berlin 1954. — FOURNIER, P. Le Livre des Plantes Médicinales et Vénéneuses de France, 3 vol. Paris 1947—1948.

GATHERCOAL, Edmund N. and Elmer H. WIRTH. Pharmacognosy. 2 ed. Philadelphia 1949. — GESSNER, O. Die Gift- und Arzneipflanzen von Mitteleuropa. (Pharmakologie, Toxikologie, Therapie). 2. Aufl. Heidelberg 1953.

HERTWIG, H. Gesund durch Heilpflanzen. 6. Aufl. Berlin 1950. — HESSE, E. Die Rauch- und Genussgifte. Stuttgart 1938. — HOCKING, G. M. A Dictionary of Terms in Pharmacognosy and other Divisions of Economic Botany. Springfield, Ill. 1955. — HOWES, F. N. Fish Poison Plants. Bull. Misc. Inf. Kew. 129—152, 1930. — HUMPHRY, John. Drugs of Commerce. London 1921.

LEWIN, L. Phantastica, Narcotic and Stimulating Drugs. New York 1942. — LIMBACH, R. and K. BOSHART. Der Anbau von Heil-, Duft- und Gewürzpflanzen. Ein Ratgeber für den erwerbsmässigen Anbau. Berlin 1937.

MADAUS, Gerhard. Lehrbuch der Biologischen Heilmittel. 3 vol. Leipzig 1938. — MARZELL, H. Heil- und Nutzpflanzen der Heimat. 2. Aufl. Reutlingen 1947.

PAMMEL, J. H. A Manual of Poisonous Plants. Grand Rapids 1911. — PHARMACOPAEA INTERNATIONALIS. Welt-Gesundheits-Organisation. Genf. Since 1955. — PHARMACOPPOEIA of the different countries. — POTTER's Cyclopedia of Botanical Drugs and Preparations. 7 ed. London 1956.

TSCHIRCH, A. Handbuch der Pharmakognosie. 3 vol. in 6. Leipzig 1909—1925. 2 ed. Since 1931.

WEBER, U. Geschnittene Drogen. 1953. — WOODHOUSE, Roger P. Hayfever Plants. Waltham 1945.

OILS AND FATS

ELSDON, G. D. The Chemistry and Examination of Edible Oils and Fats. Their Substitutes and Adulterants. London 1926.

FRYER, P. J. Technical Handbook of Oils, Fats and Waxes. Cambridge 1917.

GROSSFELD, J. Fette und Oele, Lipoide, Wachse, Harze, ätherische Oele. Berlin 1939.

HEFTER, G. und H. SCHÖNFELD. Chemie und Technologie der Fette und Fettprodukte. 2. Aufl. Wien 1936—1939. — HILDITCH, T. P. The Industrial Chemistry of the Fats and Waxes. London 1927.

JEMIESON, George S. Vegetable Fats and Oils. New York 1943. — JUMELLE, H. Les Huiles Végétales. Paris 1921.

KOPPEL, C. van den en P. A. ROWAAN. Theezadolie. Ber. Afd. Handelsmus. Kol. Inst. Amsterdam. Nr. 121. 1938.

MENSIER, P.-H. Dictionaire des Huiles Végétales. Paris 1958.

SCHRADER, J. H. The Castor Oil Industry. Bull. 867. U. S. Dept. Agric. 1920.

UPHOF, J. C. Th. La Palma de Aceite como Industria Lucrativa. Bolet. Union. Panam. 84: 1—16, 1932.

ESSENTIAL OILS

FINNAMORE, H. The Essential Oils. London 1926.

GILDEMEISTER, E. und F. H. HOFFMANN. Die ätherischen Öle. 3 vol. Leipzig 1928—1931. — GUENTHER, Ernest S. Oil of Geranium. Soap. 14 (3); 26-29, (4): 28—31, (5): 28—32, (6): 30—33, 69—75, 1938. —, Oil of Ylan Ylang. American ePrfumer. 37 (8): 32—34, (9): 36—38, (10): 44—46, (11): 42—45, (11): 42—45, 70 (12): 42—44, 1938. —, Oil of Vetiver. Drug and Cosmetic Ind. 44: (6) 710—713, 728—732, 1939. —, Balsam Peru. Drug and Cosmetic Ind. 47 (1): 26-30, 47, 1940. —, Oil of Patchouly. American Perfumer. 41: (12) 27—32, 1941. —, Essential Oils. 5 vol. New York 1948.

LAZANNEREO, I. Manuel de Perfumerie. 2 ed. Paris 1928.

MAUNIER, E. Les Plantes à Perfums des Colonies Françaises. Marseille 1928.

PARRY, E. J. Cyclopedia of Perfumery. 2 vol. London 1925.

SPOON, W. Atjeh Patchouli-Olie. Ber. Afd. Handelsmus. Kol. Inst. Amsterdam, Nr. 71. 1932.

GUMS, RESINS AND WAXES

CORDEMOY, H. Jacob de. Les Plantes à Gommes et à Résines. Paris 1911.

HOWES, F. N. Sources of Vegetable Wax. Bull. Misc. Inf. Kew. 503, 526, 1936. — HOWES, F. N. Vegetable Gums and Resins. Waltham 1949.

PARRY, E. J. Gums and Resins. Their Occurance, Properties and Uses. London 1918.

TSCHIRCH, A. Die Harze und Harzbehälter. Berlin 1906. — TSCHIRCH, A. und Erich STOCK, Die Harze. 4 vol. Berlin 1933—1936.

VEITCH, F. P. and V. E. GROTLISCH. Turpentine, its Sources, Properties, Uses, Transportation and Marketing. Bull. 898. U.S. Depart. Agric. 1920.

WILLIAMSON, G. C. The Book of Amber. London 1932. — WOLFF, H. Die natürlichen Harze. Stuttgart 1928.

RUBBER

JUMELLE, H. L. Les Plantes à Caoutchouc et à Gutta. Paris 1903.

MEMLER, Karl. Handbuch der Kautschukwissenschaft. Leipzig 1930. — MOYLE, Alton. Bibliography and Collected Abstracts on Rubber Producing Plants (other than Species of Hevea). Circ. 99. Texas Agric. Exp. Station. 1942.

REINTGEN, Peter. Die Kautschukpflanzen. Eine wirtschaftsgeographische Studie. Beih. Tropenpfl. 6 Nr. 2, 3. 1905.

PECTIC SUBSTANCES, ALKALOIDS AND TANNINS

BRANFOOT, M. H. A Critical and Historical Study of the Pectic Substances of Plants. Dept. Sci. Ind. Brit. Food Invest. Bd. Spec. Reprt. 33, 1929.

DEKKER, J. Die Gerbstoffe, botan.-chem. Monographie der Tannide. Berlin 1913.

SUCHARIPA, R. Die Pektinstoffe. Braunschweig 1925. TRIER, G. und E. H. WINTERSTEIN. Die Alkaloide. Monographie der natürlichen Basen. 2. Aufl. Berlin 1931. — TRIMBLE, H. The Tannins. 2 vol. Philadelphia 1892.

DYESTUFFS

LEGGETT, F. Ancient and Medieval Dyes. 1944. MELL, C. D. Basic Dyes from Lichens. Textile Colourist. 57: 409—411, 1935.

PERKIN, A. G. and A. E. EVEREST, The Natural Organic Colouring Matters. London 1918.

RAWSON, C., W. M. GARDNER and W. F. LAYCOCK. A Dictionary of Dyes, Mordants and Other Compounds. London 1901.

FIBRES

DEWEY, Lyster H. Sisal and Henequen, plants yielding fiber for binder twine. Circ. 179. U. S. Dept. Agric. 1910. — DODGE, Chr. A Catalogue of the Useful Fibers of the World. Fibr. Ind. Reprt. 9. U. S. Dept. Agric. 1897.

GOULDING, E. and W. R. DUNSTAND. Cotton and Other Vegetable Fibres. London 1919.

HANNAN, W. I. Textile Fibres of Commerce. London 1902. — HENRY, Yves. Plantes à Fibres. Éléments d'Agriculture Coloniale. Paris 1924. — HERZOG, A. Mikrophotographischer Atlas der technisch wichtigen Pflanzenfasern. Berlin 1955.

ITERSON, J. van. Vezelstoffen. Vol. 12 of Onze Koloniale Landbouw. Haarlem 1917.

MICHOTTE, F. Les Sansevériés; Culture et Exploitation. Paris 1915. —, Les Bananiers Textiles, Culture et Exploitation. Paris 1911. —, Agave et Fourcroyas. 3 ed. Paris 1931. OAKLEY, F. I. Long Vegetable Fibres. London 1928. ROBINSON, Brittain B. Ramie Fiber Production. Circ. 585. U. S. Dept. Agric. 1940.

SCHILLING, Erw. Die Faserstoffe des Pflanzenreiches. Leipzig 1924.

TOBLER, F. Sesal und andere Agavefasern. Berlin-Charlottenburg 1931.

UPHOF, J. C. Th. El Arbol Kapok como Planta Fibrosa. Bol. Union PanAm. 64: 28-41, 1930. —, „Crin Végétal" uit Tillandsia usnoides. Tijdschr. Econ. Geogr. 23: 157-160, 1932.

WOOLMAN, Mary S. Textiles. New York 1943.

ECONOMIC MICROORGANISM

ARNULF, F. The New Wine Book. The Biology of Wine Fermentation and its Practical Importance. Whittier 1936.

BURKEY, L. A. Bulgarian Acidophilus Cultured Milks. Bur. Dairy Ind. U.S. Dept. Agric. 1947.

ELDER, A. L. The Commercial Production of Penicillin. Chem. Ind. 54: 501-504, 1944.

FISCHER, A. M. Yeast and Factor determining its Vitamin Potency. Brewers Digest. 13: 37 (nr. 10) 1938.

JONES, L. Meyer. Antibiotics in: Alfred Steffereud (edit.) Animal Diseases, Yearbook of Agriculture, U.S. Dept. Agric. 94—96. 1956.

KAYSER, E. The Races of Yeast and their Significance on the Bouquet of Wine. Chim. Ind. Spec. No. 610—625, May. 1924. — KÖHLER, H. Einführung in die Methoden der pflanzlichen Antibiotikaforschung. Berlin 1955.

LAFAR, F. Handbuch der technischen Mykologie. Jena 1904—1914.

NEUBRG, C. und F. WINDISCH. Über die Essiggärung und die chemischen Leistungen der Essigbakterien. Biochem. Zeit. 166: 454—460, 1925.

ORLA-JENSEN, S. The Lactic Acid Bacteria. Kgl. Danske Vidensk. Selsk. Skrifter, Naturvidensk. Math. Afd. Ser. 8—5: 51—196, 1919. —, Dairy Bacteriology. 2 ed. Philadelphia 1931.

PRESSCOTT, S. C. and C. G. DUNN. Industrial Microbiology. 3. ed. New York 1959.

RAMSBOTTOM, J. The Uses of Fungi. Brit. Ass. Adv. Sci. Ann. Report. 1936.

SLATOR, A. Yeast Crop and Factors which determine them. Journ. Chem. Soc. 119: 115—120, 1921.

ALGAE

BOTANICA MARINA. Intern Review of Seaweed Research and Utilization. Hamburg 1959.

CHAPMAN, V. J. Seaweeds and their Uses. New York 1950.

HENDRICK, James. The Value of Seaweeds as Raw Materials for Chemical Industry. Soc. Chem. Ind. Journ. London. 35: 565—574. 1916. — HUMM, Harold J. Agar Resources of the South Atlantic and East Atlantic Gulf States. Science. 100: 209—212, 1944.

MOORE, Lucy B. War-Time Interest in Marine Algae. Chron. Bot. 7: 406—409, 1943.

TRESSLER, D. K. Marine Products of Commerce. New York 1923.

LICHENS

PREZ-LLANO, George, Albert. Lichens, their Biological and Economic Significance. Bot. Rev. 10: 1—65, 1944.

ECONOMIC PLANTS IN RELATION TO THE CONTINENTS AND CERTAIN ISLAND GROUPS

EUROPE

ALLEN, H. The Wines of France. London 1924. — ARMSTRONG, S. F. British Grasses and their Employment in Agriculture. Cambridge 1937. — BORZA, Al. Etnobotanische Neuheiten aus Rumänien. Eine volkstümliche Apotheke. Bull. Gard. Sisal Museulin Bot. Univ. Cluj. 16: 17-27, 1937. (in Rumanian and German resumé). — BRANKE, J. W. and I. I. PARISHEW. Sweet Scented Plants of the Maritime Region and their Volatile Oils. Bull. Far. Eastern Branch. Acad. Sci. USSR. Nr. 23: 3—45, 1937. (in Russian and English resumé).

EDLIN, H. L. British Plants and their Uses. London 1951.

GISTL, R. und A. V. NOSTITZ. Handelspflanzen Deutschlands, österreichs und der Schweiz. Stuttgart 1932. — GRUSCHE, Herbert. Deutsche Gewürzpflanzen, ihre Bedeutung und Anwendung. Diss. Univ. Jena 1937.

HEDDREICH, Theodor von. Die Nutzpflanzen Griechenlands. Athen 1862. — HEGI, Gustav. Illustrierte Flora von Mitteleuropa. 13 vol. München 1905-1931. — HILL, Jason. Wild Foods of Britain. London 1939. — HOPPE, H. A. Europäische Drogen. 2 vol. 1948-1951.

ILJIN, M. M. Akademia Nauk SSSR. Botanicheskii institut. Rastitel'noe syr'e SSSR. Pod obshchei red. M. M. Il'ina. Moskva, 1950—57. 2 vol. Vol. 1: Technicheskie rasteniia. 1950. Vol. 2: Naturnye rasteniia. 1957 (in Russian). (Raw Materials of Plants of the USSR).

LA PUERTA, Don Gabriel de. Botánica descriptiva y determinación de las Plantas Indígenas y cultivadas en España de uso medicinal, alimenticio é industrial. Madrid 1891.

MAL'TSEW, A. I. The Use of Weeds and other Wild Plants. Bull. Appl. Bot. and Plant Breed. 13 (3): 85—89, 1922—1923 (1923). (in Russian). — MARZELL, H. Heil- u. Nutzpflanzen der Heimat. Reutlingen 1924. — MATTRIROLO, O. De un nuovo centro di produzione del Tartufo bianca del Piemonte. (T. magnatum Pico) in Istria. Ann. della R. decad. d'Agric. Torino. 75, 1932.

PANINI, F. Piante Medicinale d'Italia spontance e coltivate su vasta Scala. Roma 1925. — PERROT, E. Plantes Medicinales de France. 2 vol. Paris 1934.

REGEL, C. Beiträge zur Kenntnis von mitteleuropäischen Nutzpflanzen. Angew. Bot. 23: 137—151, 1941; 24: 278—302, 1942.

SADKOV, S. G. The Development of the Cultivation of Essential Oil Plants in USSR. Bull. Appl. Bot. Ser. A. Nr. 4: 77—84, 1932. (in Russian). — SINOVA, Elena S. Alges de la baie Novorossijok dans la Mer Noir et leur Utilization. Trav. Stat. Biol. Biol. Sébastepol. 4: 1—136, 1935. (in Russian with French résumé).

TOBLER, F. Deutsche Faserpflanzen und Pflanzenfasern. München 1938.

UPHOF, J. C. Th. De Rozenolieindustrie in Bulgarije. Tijdschr. Econ. Geograph. 29: 63—66, 1938.

VASILYEFF, P. V. The Treasures of the Soviet Woods (in Russion). Moscow 1949. — VAVILOV, N. F. Field Crops of Southeastern European Russia. Bull. Appl. Bot. and Plant. Breed. 13 (Suppl. 23): 1—228, 1922. (in Russian).

WERNECK, H. L. Bausteine zur Geschichte der Kulturpflanzen in den österreichischen Alpenländern. Angew. Bot. 20: 185—217, 1938.

ZHUKOVSKII, P. M. Wheat Crops of Georgia (West Transcaucasia). Sci. Papers. Sect. Tiflis. Bot. Gard. 3: 8—40, 1922—1923. (in Russian).

ASIA

AARONSOHN, Aaron. Agricultural and Botanical Explorations in Palestine. Bull. 180. Bur. Plant Ind. U.S. Dept. Agric. 1910. — ASAMI, Y. The Grab-Apples and Nectarines of Japan. Contributions of the system. Investigation of Fruit-Trees in Japan. Tokyo 1927.

BATCHELOR, John and KINGO MIYABE. Ainu Economic Plants. Trans. As. Soc. 21: 198—235, 1893.

BLASDALE, W. C. A Description of some Chinese Vegetable Food Material and their Nutritive and Economic Value. Bull. 68. Off. Exp. Sta. U.S. Dept. Agric. 1899. — BONAVIA, E. The Cultivated Oranges and Lemons etc. of India and Ceylon. 2 vol. London

1899. BROWN, W. H. and A. F. FISCHER. Philippine Bomboos. Philipp. Dept. Agric. and Nat. Res. Bur. For. Bull. 15. 1918. —, Philippine Fiber Plants. Philipp. Bur. For. Bull. 19. 1919. —, Minor Products of Philippine Forests. Dept. Agric. and Nat. Res. Bur. For. Philipp Islds. Bull. 22. 1920. — BRUCK, Werner, Friedrich. Der Faserbau in Holländisch-Indien und auf den Philippinen. Beih. Tropenfl. Bd 13 Nr. 5-6. 1912. — BURKILL, I. H. Dictionary of the Economic Products of the Malay Peninsula. 2 vol. Oxford 1935.

CHAN, L. A Brief History of Chinese Herbs and Medicine. Bull. Torrey Bot. Club. 66: 563—568, 1939. — CHEMA, G. S. Commercial Fruits of India. Bombay 1954. — CHOPRA, Ram Nath. The Medical and Economic Aspects of some Indian Medical Plants. Patna 1929—1930. — CHUN, Woon Young. Chinese Economic Trees. Shanghai 1922. — CREVOST, Ch. et Ch. Lemarie. Catalogue des Produits de l'Indochine. 5 vol. Hanoi 1917—1935.

DASTUR, J. Useful Plants in India and Pakistan. 1954. — DESCH, H. E. Manual of Malayan Timbers. 2 vol. Singapore 1954. — DRURY, Heber. The Useful Plants of India. Madras 1858.

EBERT, F. Beiträge zur Kenntniss des chinesischen Arzneischatzes. Früchte und Samen. Diss. Univ. Zürich 1907.

FAIRCHILD, D. G. Japanese Bamboos and their Introduction into America. Bur. Plant Ind. Bull. 43. U.S. Dept. Agric. 1903. — FOXWORTHY, F. W. Commercial Timber Trees of the Malay Peninsula. Singapore 1927. —, Forest Products of the Malay Peninsula. Mal. For. Rec. Singapore. Nr. 2. 1922.

GAMBLE, J. S. A Manual of Indian Timbers. London 1902. — GRANT, J. W. and A. N. P. WILLIAMS. Burma Fruits and their Cultivation. Dept. Agric. Burma. Bull. 30. 1936. — GRESSHOFF, M. Beschrijving der Giftige en Bedwelmende Planten bij de Vischvangst in gebruik. Med.'s Lands Plantentuin. Batavia 1895. — GUERRERO, Leon Maria. Medicinal Uses of Philippine Plants. Bull. 22. Dept. Agric. and Nat. Res. Bur. For. Philipp. Islds. 1921. — GUENTHER, Ernest S. East Indian Oil of Sandalwood. Drug and Cosmetic Ind. 48 (5): 534—540, (6): 668—669, 1940. — HEYNE, H. De Nuttige Planten van Nederlandsch Indië. 3 vol. 2. ed. Batavia 1927. — HOOPER, David and Henry FIELD, Useful Plants and Drugs of Iran and Iraq. Field Mus. Nat. Hist. Bot. Ser. 9 (3). 1937.

ICHIMURA, T. Important medicinal Plants of Japan. (in Japanese and English). Kanagawa 1932.

KING, F. H. Farmers for Centuries or Permanent Agriculture in China, Korea and Japan. 2. ed. New York 1927. — KIRTIKAR, K. R. and B. D. BASU. Indian Medicinal Plants. 2 vol. Allahabad 1918. — KLAUTKE, P. Nutzpflanzen und Nutztiere Chinas. Hannover 1922. — KOEKICHIKNINSHO, T. Oeconomic Plants of Eastern Asia. 9 vol. - Supplements by Chikinshofuroku 9 vol. (in Japanese) Tokyo. — KOKIN, A. J. New Resources of Wild Food Plants of Asia Media. Bull. Appl. Bot. Ser. A. Nr. 7: 137—149, 1933 (in Russian).

LAFONT, M. le, Les Cultures de l'Archipel des Comores. Bibl. Agric. Col. Paris 1902. — LEWIS, Frederick, The Vegetable Products of Ceylon. Colombo 1934.

McCLURE, F. A. Notes on Bamboo Culture with special reference to Southern China. Hong Kong Nat. 9: 4—18, 1938. — MEYER, Frank N. Agricultural Explorations in the Fruit and Nut Orchards of China. Bull.

204. Bur. Plant Ind. U. S. Dept. Agric. 1911. — MULLER, T. Industrial Fiber Plants of the Philippines. Philipp. Bur. Ed. Bull. 9. 1913. — MURRAG, James A. The Plants and Drugs of Sind. London 1881.

NEKRASSOVA, V. L. Les Plantes Textiles chez les Ainos, Ghiliaks et Goldes. Sovietskaja Botanika. 6: 125—141, 1934. (in Russian).

OCHSE, J. J. Fruits and Fruit Culture in the Dutch East Indies. Batavia. 1931. —, Vegetables of the Dutch East Indies. Buitenzorg 1931. — OKAMURA, Kintaro. Uses of Algae in Japan. Proc. Pacif. Sci. Congr. 5th. Canada. 1933. 4: 3153—3161, 1934.

PEARSON, R. S. and H. P. BROWN. Commercial Timbers of India, Supplies, Anatomical Structure, Physical and Chemical Properties and Uses. Calcutta 1932. — PERROT, Em. et C. L. GATIN. Les Algues Marines Utiles et en particulier les Algues Alimentaires d'Extrême-Orient. Inst. Oceanogr. Ann. 3 Fasc. 1: 1—101, 1911. — PETELOT, Alfred. Les Plantes Médciinales du Cambodge, du Laos, et du Viêt-nam. Arch. Agron. Cambodge. 4 vol. 1958—1959.

RAIZADA, B. et B. S. VARMA. Plantes des Indes reputées ichthyotoxiques. Rev. Bot. Appl. 17: 752—757, 1937. — READ, B. C. and JUCH'IANG, Liu. Plantae Medicinalis Sinensis. 2 ed. Peking 1927. —, Chinese Medicinal Plants from the Pen Ts'ao Kang Mu. 3 ed. Peking 1936. — RICHARDSON, H. K. Native Oil Production in Western China. Chem. Met. Eng. 27 (21): 1032, 1920. — ROWAAN, P. A. Nederlandsch Indië als Leverancier op de Wereldmarkt van Vetzaden en Plantaardige Vetten. Ber. Afd. Handelsmus. Kol. Inst. Amsterdam. Nr. 88. 1934. —, De Aetherische Oliën van Nederlandsch Indië. Kol. Inst. Amsterdam. 1938.

SATOW, E. M. The Cultivation of Bamboos in Japan. Trans. Asiat. Soc. Japan. 27 pt. 3. 1899. — SHIH, C. Y. Studies in Chinese Economic Botany. Thesis Soochow Univ. Peking 1940. — SERGEYER, M. A. On the Utilization of Plant Resources of Kamchatka. Bull. Far Eastern Branch. Acad. Sci. USSR. 28 Nr. 1: 109—112, 1938 (in Russian). — SHIBATA, Kieta. Shijen shokubutsee jiten. (A Cyclopedia of useful plants and plant products). Tokyo 1949. (in Japanese). — SINSKAIA, E. N. Field Crops of Altai. Bull. Appl. Bot. and Plant. Breed. 14 (1): 359—376, 1925 (in Russian). — SMITH, H. M. Seaweed Industries in Japan. Bull. 24. U. S. Bur. Fisheries. 1904. — SPOON, W. Atjeh Terpentijn, tegenwoordige kwaliteit en nieuwe beoordeelingen. Ber. Afd. Handelsmus. Kol. Inst. Amsterdam. Nr. 57. 1930. — SPOON, W. en P. A. ROWAAN. Sumatra-terpentijn en -colophonium in Nederland. Ber. Afd. Handelsmus. Kol. Inst. Amsterdam. Nr. 173. 1941.

TSENG, Cheng-Kwei. Gloiopeltis and other Economic Seaweeds of Anoy, China. Lingnan Sci. Journ. 12: 43—63, 1933. —, Economic Seaweeds of Kwantung Province, S. China. Lingnan Sci. Journ. 14: 93—104, 1935. — TSCHIRCH, A. Indische Heil- und Nutzpflanzen und deren Cultur. Berlin 1892.

WALKER, Egbert H. The Plants of China and their Usefulness to Man. Smithon. Reprt. Washington 325-363, 1943. — WATT, G. A Dictionary of the Economic Products of India. 6 vol. Calcutta 1889—1896. —, The Commercial Products of India. London 1908. — WEST, A. P. and W. H. BROWN, Philippine Resins, Gums, Seed Oils and Essential Oils. Philipp. Bur. For. Bull. 20. 1920. — WESTER, P. J. The Food Plants of the Philippines. Bull. 39. Dept. Agric. and Nat. Res. Bur. Agric. Philipp. 1925.

AFRICA

AINSLIE, J. R. A List of Plants used in Native Medicine in Nigeria. Inst. Paper Imp. Forestry Inst. Oxford. no. 7. 1937.

BALLY, P. R. O. Native Medicinal and Poisonous Plants of East Africa. Kew Bull. 10—26, 1937. — BOCQUILLON-LIMOUSIN, Henri. Les Plantes Utiles de la Tunisie. Monde des Plantes. 4: 241—244, 260, 276—279, 288—289, 305—312, 1895. — BRAUN, Alexander. Beitrag zur Kenntnis der Abyssinischen Culturpflanzen. Flora 31: 89—98, 1848. — BRAUN, Karl. Gewürze und Aromatika der Völker des früheren Deutsch-Ostafrika. Heil- u. Gewürz-Pflanz. 11: 55—86, 113—141, 1928. — BUSSE, Walter. Ueber Heil- und Nutzpflanzen Deutsch-Ostafrikas. Ber. Deut. Pharm. Gesellsch. 14: 187—207, 1904.

CHEVALIER, Aug. Les Végétaux Utiles de l'Afrique Tropicale. Paris 1905. —, Nos Connaissances Actuelles sur la Géographie Botanique et la Flore Économique du Sénégal et du Soudan. 1910. —, Énumération des Plantes Cultivées par les Indigènes en Afrique Tropicale et des Espèces Naturalisées dans le même Pays et ayant probablement été cultivées à une Époque plus ou moins reculée. Bul. Soc. Natl. Acclim. France. 59: 65—79, 104—110, 133—138, 239—242, 312—318, 341—346, 386—392, 1912. —, Énumération des Plantes Cultivées par les Indigènes en Afrique Tropicale. Paris 1912. —, Les Productions Végétales du Sahara et de ses Confins Nord et Sud. Rev. Bot. Appl. 12: 669—924, 1932. —, Une Enquête sur les Plantes Medicinales de l'Afrique Occidentale. Rev. Bot. Appl. et d'Agric. Trop. 17: 165—175, 1937. —, La Somalie Française. Sa Flore et ses Productions Végétales. Rev. Bot. Appl. 19: 663—687, 1939. — CHIOVENDA, Emilio. Vegetali utilizzati nella medicina indigena dell' Eritrea, Somalia e regione vicine. Atti Cong. Studi Colon. 17: 351—376, 1931. — CORDEMOY, Jacob de. Flore de l'Ile de la Réunion avec l'Indication des Proprietés Économiques et Industrielles des Plantes. Paris 1895.

DALZIEL, J. M. Vegetable Products of Kontagora Province, Northern Nigeria. Bull. Imp. Inst. 5: 255—266, 1907. —, The Useful Plants of West Tropical Africa. London 1937. — DUBARD, Marcel et Victor CAYLA. Liste de quelques Plantes Utiles du Moroc d'après les Documents rapportés par M. de Gironcourt. Agric. Prat. Pays Chauds. 9: 95—106, 1909.

ENGLER, A. Die Pflanzenwelt Ost-Afrikas und Nachbargebiete. Theil C. Die Nutzpflanzen Ost-Afrikas. Berlin 1895. — ESDORN, Ilse. Afrikanische Kopale, besonders Kongo-Kopal. Kolonialforstl. Merkbl. Reihe 2. Nr. 1, 1—13, 1942.

FICALHO, Conde de. Noticia de algunas productos vegetaes importantes on puco conhecidas da Africa Portugueza. Jorn. Sci. Math., Phys., e Nat. Lisboa 1878. —, Plantas Uteis da Africa Portugueza. Lisboa 1884.

GOESTER, L. E. De Geneesmiddelen van Groot Nederland. Overzicht der in Zuid-Afrika meest gebruikte Plantaardige Geneesmiddelen. Pharm. Weekbl. Nederland. 51: 1019—1024, 1035—1041, 1107—1117, 1129-1140, 1200—1204, 1240—1249, 1265—1275, 1914. — GREENWAY, P. J. Dyeing and Tanning Plants in East Africa. Great Britt. Imp. Inst. Bull. 39: 222—245, 1941.

HECKEL, Ed. Les Plantes Utiles de Madagascar. Marseille 1910. — HEMSLEY, W. B. On the Vegetable Productions of Abyssinia. Journ. Trav. and Nat. Hist. 1; 309—318, 1868. — HOLLAND, J. H. The Useful Plants of Nigeria. Kew Bull. Ad. Ser. IX. London 1922.

ISSAC, W. E. Agar from South African Seaweeds. Nature 151: 532, 1943.

LELY, H. V. The Useful Woods of Northern Nigeria. London 1925.

PAILLIEUX, Auguste et Désiré BOIS. De Quelques Plantes Alimentaires de l'Abyssinie. Rev. Sci. Nat. Appl. Soc. Nat. Acclim. France. 37: 803—809, 1890. — PAOLI, Guido. Cenno sulle Piante Utili e Utilizzabili Raccolte dalla Missione. In Stefanini, Giuseppe e Paoli, G. Ricerche Idrogeologiche, Botaniche ed Entomoliche fatte nella Somalia Italiana Meridionale. Relaz. e Monog. Agric. Colon. Ist. Agric. Colon. Ital. 7: 225—250, 1916. — PELLEGRIN, François. De Quelques Bois Utiles du Gabon. Bul. Soc. Bot. France. 84: 639—645, 1938. — PERROT, Emile. Sur les Productions Végétales du Moroc, la Constitution du Sol Marocain et les Influences Climatologiques. Paris 1921. —, Sur les Productions Végétales Indigènes ou Cultivées de l'Afrique Occidentale Française. (Sahara, Soudan Nigérien, Haute-Volta, Guinée). Trav. Off. Natl. Mat. Prem. Vég. (Paris) Notice 31. 1929. — PERROT, Emilie et Louis GENTIL. Sur les Productions Végétales du Moroc. Minst. Comm. et Industr. Not. 10. 1921. — PHILIPPS, E. P. Economic Plants of South Africa. Off. Yearb. Union S. Afr. 8 (1910—25): 47—53, 1927. — POBÉGUIN, Henri. Essai sur la Flore de la Guinée Française. Produits Forestiers, Agricoles et Industriels. Paris 1906. — PORTO da CRUZ, Visconde do. A Flora Madeirense na Medicina Polular. Brotéria. Cién. Nat. 4: 35—46, 71—78, 139—144, 145—154, 1935.

RANCON, André. La Flore Utile du Bassin de la Gambie. Bull. Soc. Géogr. Com. Bordeaux. II, 18: 324—338, 353—382, 385—410, 417—442, 467—477, 496-508, 518—541, 545—558, 1895. — REIN, G. K. Die im englischen Sudan, in Uganda und dem nördlichen Kongostaate wild und halbwild wachsenden Nutzpflanzen. Tropenpfl. 13: 374—379, 532—539; 15: 217—220, 387—393, 1911.

SANTESSON, C. G. Einige Drogen aus dem Kamerun-Gebiete und ihre einheimische Verwendung. Ark. Bot. 20 A nr. 8. 1926. — SCHWEINFURTH, Georg. Vegetationscharakter und Nutzpflanzen der Niamniam- und Monbutto-Länder. Zeitschr. f. Erdk. Berlin, 6: 234—248, 1871. —, Sur l'origine africaine des Plantes cultivées en Egypte. Bull. de l'Inst. Egypte. Nr. 12: 200—206, 1873. —, La Piante utile dell'Eritrea. Bol. Soc. Soc. Afr. Italia. 10: 233—286, 1891. — SÉBIRE, R. P. A. Les Plantes Utiles du Sénégal. Paris 1899. — SHABETAI, J. R. Agricultural Crops and Economic Plants of Egypt. Bot. and Plant. Breed. Sect. 10. Min. Agric. Egypt. Cairo 1933. — SIM, T. P. Forest Flora and Forest Resources of Portuguese East Africa. Aberdeen 1909.

TROTTER, Alesandro. Flora Economica della Libia. Roma 1915.

VERDOORN. I. C. Edible Wild Fruits of the Transvaal. Bull. Dept. Agric. and For. Union. S. Afr. Nr. 185 (Plant Ind. Ser. 29) 1938. — VOLKENS, Georg. Die Nutzpflanzen Togos. Notizbl. K. Bot. Gart. Berlin. App. 22. 1909—1910.

WAIBEL, Leo. Die Rohstoffgebiete des tropischen Afrika. Leipzig 1937. — WARBURG, O. Die Kulturpflanzen Usambaras. Mitt. Deutsch. Schutzgeb. 7 (2) 1894. — WATT, J. M. and M. G. BREYER-BRANDWIJK. The Medical and Poisonous Plants of Southern Africa, being an Account of their Medical Uses, Chemical Composition, Pharmacological Effects and Toxicology in Man and Animal. Edinburgh 1932. —, The Medical and Poisonous Plants of South Africa. Edinburgh 1939. — WILDEMAN, Émile de. Notices sur les Plantes Utiles ou intéressantes de la Flore du Congo. Bruxelles 1903—1906. —, Notices sur les Plantes Utiles ou Intéressantes de la Flore du Congo. Publ. Et. Ind. du Congo. Bruxelles 1903-1908. —, Les Forêts Congolaises et leurs principales Essences Économiques. Bruxelles 1926. —, A Propos de Médicaments Indigènes Congolais. Mem. Sect. Sci. Nat. de Med. Inst. Roy. Colon. Belge Collect. v. 3 fasc. 3. 1935. —, Sur des Plantes Médicinales ou Utiles du Mayumbe. (Congo Belge) d'après des Notes du R. P. Wellens (1891—1924). Mem. Sect. Sci. Nat. et Med. Inst. Roy. Colon. Belge Collect. v. 6, fasc. 4. 1938. — WILLIAMS, P. O. Useful and ornamental Plants of Zanzibar and Pemba. 1949.

AUSTRALIA

BAILEY, F. M. Queensland Woods, with a brief popular description of the trees, their distribution, qualities, uses of timber etc. 3 ed. London 1899. — BENSON, A. H. Fruits of Queensland. Dept. Agric. and Stock. Queensland. 1914. — BOAS, I. H. The Commercial Timbers of Australia. 1947.

CHELANE, J. B. and T. H. JOHNSON. Aboriginal Names and Uses of Plants in the Northern Flinders Ranges. Trans. Roy. Soc. So. Austr. 63: 172—179, 1939. — CLELAND, J. B. and T. H. JOHNSTON. Aboriginal Names and Uses of Plants at the Granites, Central Australia. Trans. Roy. Soc. So. Austr. 63: 22—26, 1939. — COLENSO, William. On the Vegetable Food of the Ancient New Zealanders before Cook's Visit. Trans. and Proc. New Zeal. Inst. 13: 3—38, 1881.

FITZGERALD, W. V. Trees of Western Australia, with notes on their use and distribution. Journ. and Proc. Mueller Bot. Soc. West Austral. 1: 1—78, 1903. GUENTHER, Ernest S. Australian Eucalyptus Oils. Drug and Cosmetic Ind. 51 (2): 160—167, (3): 281—285, (4): 404—409, 1942.

KARNBACH, Ludw. Über die Nutzpflanzen der Eingeborenen in Kaiser-Wilhelmsland. Bot. Jahrb. Engl. (Beih.) 37: 10—19, 1892.

MAIDEN, J. H. Some reputed Medicinal Plants of New South Wales. Proc. Linn. Soc. N. S. Wales. II, 3: 355—393, 1888. —, Useful Plants of Australia. Sydney 1889.

RUMSEY, H. J. Australian Nuts and Nut Growing in Australia. 1927.

WARBURG, O. Das Pflanzenkleid und die Nutzpflanzen Neu-Guineas. In Krieger. Neu-Guinea 36—72, 1899.

PACIFIC ISLANDS

CHUNG, H. L. and J. C. RIPPERTON. Utilization and Composition of Oriental Vegetables in Hawaii. Bull. 60. Hawaii Agric. Exp. Sta. 1924.

JOUAN, Henri. Les Plantes Alimentaires de l'Océanie. Mem. Soc. Nat. Sci. Nat. Cherbourg. 19: 33—83, 1875.

KAAIAKAMANU, D. M. and J. K. AKINA. Hawaiian Herbs of Medicinal Value. Honolulu 1922.

McLELLAND, C. K. Crasses and Forage Plants of Hawaii. Bull. 36. Hawaii Agric. Exp. Sta. 1915. — MERRIL, E. Emergency Food Plants and Poisonous Plants of the Islands of the Pacific. U. S. War Dept. Techn. Man. 1943. — MILLER, C. D. Fruits of Hawaii. Honolulu. 1956.

NADEAUD, Jean. Plantes Usuelles des Tahitiens. Montpellier 1864.

REED, Minnie. The Economic Seaweeds of Hawaii and their Food Value. Hawaii. Agric. Exp. Sta. Ann. Reprt. 1906. — REINKE, Franz. Die Nutzpflanzen Samoas und ihre Verwendung. Jahresb. Schl. Ges. Vaterl. Cult. 73 Abt. II (Naturw. C.): 22—46, 1896. — RIPPERTON, J. C. The Hawaiian Tree Fern as a Commercial Source of Starch. Bull. 53. Hawaii Agric. Exp. Sta. 1924. — RIPPERTON, J. C. and N. A. RUSSELL. Hawaiian Vegetables and their Function in the Diet. Ext. Bull. 9. Hawaii Agric. Exp. Sta. 1926. SAFFORD, W. E. The Useful Plants of the Island Guam. Contrib. U. S. Nat. Herb. vol. 9. 1905. — SETCHELL, William Albert. American Samoa. Pt. II. Ethnobotany of the Samoans. Publ. 341. Carneg. Inst. Washington 1924.

VIEILLARD, Eug. Plantes Utiles de la Nouvelle-Caledonie. Ann. Sci. Nat. 4. Bot. 16: 28—76, 1862.

WILDER, Gerrit, P. Fruits of the Hawaiian Islands. Honolulu 1911.

NORTH AMERICA

ANDERSON, J. P. Plants used by the Eskimos of the Northern Bering Sea and Arctic Regions of Alaska. Amer. Journ. Bot. 26: 714—716, 1939.

BOISEN, Anton T. and J. A. NEWLIN. The Commercial Hickories. Bull. 80 For. Ser. U. S. Dept. Agric. 1910. — BRYAN, N. G. Navajo Natural Dyes. Indian Handc. nr. 2. Chiloces. 1940. — BURD, J. S. Economic Value of Pacific Coast Kelps. Bull. 248. Univ. Calif. Agric. Exp. Sta, 1915.

CASTETTER, Edw. The Ethnobiology of the Chiricahua and Mescat Apache. Univ. New Mex. Bull. 297. Biol. Ser. 4. nr. 5. 1—63, 1936. — COVILLE, F. V. Notes on the Plants used by the Klamath Indians of Oregon. Contrib. U. S. Nat. Herb. 5. Pt. 2. 87—108, 1897. — CROCKETH, W. H. Vermont Maple Sugar. Bull. 38. Comm. Agric. Vermont. Bur. Publ. 1929.

DAYTON, W. A. et al. Range Plant Handbook. For Serv. U. S. Dept. Agric. 1937.

FRASER, Melville J. The Irish Moss Industry of Massachusetts. Fishery Markets News. 4 (3): 25—28, 1942. GALLOWAY, B. T. Bamboos: Their Culture and Uses in the United States. Dept. Bull. 1329. U. S. Dept. Agric. 1925. — GILMORE, M. R. Uses of Plants by the Indians of the Missouri River Region. 33 rd Ann. Rept. Bur. Ethnol. 1911—1912. 41—154, 1919. — GRIFFITH, David, George L. BIDWEL and Charles E. GOODRICH. Native Pasture Grasses of the United States. Bull. 201. U. S. Dept. Agric. 1915. — GORMAN, M. W. Economic Botany of Southeastern Alaska. Pittonia 3: 64—85, 1896.

HALL, William L. and H. MAXWELL. Uses of Commercial Woods of the United States. II Pines. Bull. 99. For. Ser. U. S. Dept. Agric. 1911. — HENKEL, Alice. American Medicinal Barks. Bull. 139. Bur. Plant Ind. U. S. Dept. Agric. 1909. —, American Medicinal Leaves and Herbes. Bull. 219. Bur. Plant Int. U. S. Dept. Agric. 1911. —, American Medicinal Flowers, Fruits and Seeds. Bull. 26. U. S. Dept. Agric. 1913. — HOUGH, R. B. The American Woods. Lowville 1888—1928.

LEECHMAN, Douglas. Vegetables Dyes from North American Plants. Saint Paul 1945.

MAXWELL, Hu. Uses of Commercial Woods of the United States. Beech, Birches and Maples. Bull. 12. U. S. Dept. Agric. 1913. — McELHANNEY, T. A. Canadian Woods: Their Properties and Uses. Ottawa 1935. — MEDSGER, Oliver Perry. Edible Wild Plants.

New York 1943. — MUENSCHER, Walter Conrad. Poisonous Plants of the United States. New York 1944.

OERTEL, Everett. Honey and Pollen Plants of the United States. Circ. 554. U. S. Dept. Agric. 1939.

PATTERSON, Flora W. and Vera K. CHARLES. Mushrooms and other Common Fungi. Bull. 175. U. S. Dept. Agric. 1915.

RABAK, Frank. Wild Volatile Oil Plants and their Economic Importance. Bur. Plant Ind. Bull. 235. U.S. Dept. Agric. 1912. — RECORD, S. J. The Identification of the Timbers of Temperate North America. New York 1934.

SAUNDERS, C. F. Useful Plants of the United States and Canada. 3. ed. New York 1934. — STEEDMAN, E. V. The Ethnobotany of the Thompson Indians of British Columbia. 45th Ann. Rept. Bur. Am. Ethnol. Smithon. Inst. 443—522, 1927—1928 (1930). — STEVENSON, M. C. Ethnobotany of the Zuñi Indians. 30th. Ann. Reprt. Bur. Amer. Ethnol. Smithon. Inst. 31—102, 1915. — STURROCK, David. Tropical Fruits for Southern Florida and Cuba and their Uses. Publ. Atk. Inst. Harv. Univ. Jamaica Plain. 1940.

TAYLOR, Lyda Averill. Plants used as Curatives by certain Southeastern Tribes. Bot. Mus. Harv. Univ. 1940. — TRAIN, Percy, JAMES R. HENRICHS and W. ANDREW. Medical Uses of Plants by Indian Tribes of Nevada. Rev. ed. Contrib. Flora Nevada. 45: 1—139, 1957.

UPHOF, J. C. Th. Het Gebruik der Cactusplanten in Noord-Amerika. Tijdschr. Econ. Geogr. 25: 149—158, 1934.

WELLS, Sidney D. and John de RUE. The Suitability of American Woods for Paper pulp. Dept. Bull. 1485. U.S. Dept. Agric. 1927. — WOOD, Horatio C. and Arthur OSOL. The Dispensatory of the United States of America. 23 rd ed. 1943. — WOOTON, E. O. Certain Desert Plants as Emergency Stock Feed. Bull. 728. U.S. Dept. Agric. 1918.

YANOVSKY, Elias. Food Plants of the North American Indians. Misc. Publ. 237. U.S. Dept. Agric. 1936.

CENTRAL AMERICA AND MEXICO

BENEDICT, F. G. and M. STEGGERDA. Food of the present day Maya Indians of Yucatan. Carnegie Inst. Publ. 436: 155—188, 1936. — BUKASOV, S. M. The Cultivated Plants of Mexico, Guatemala and Colombia. Bull. Appl. Bot. Gen. Plant Breed. Suppl. 47. (in Russian and English). 1930.

DIESELDORFF, E. P. Las Plantas Medicinales del Departemento de Alto Verapaz. Anal. Soc. Geogr. e Hist. Guatemala. 16: 33—46, 1939.

ENDLICH, Rud. Der Ixtle und seine Stammpflanzen. Beih. Tropenpfl. Bd. 9 Nr. 5. 1908.

GRIFFITH, David. The Prickly Pear and other Cacti as Food for Stock. Bull. 74. U.S. Dept. Agric. 1905. — GRIFFITH, David and R. F. HARE. The Tuna as Food for Man. Bull. 116 Bureau Plant Ind. U.S. Dept. Agric. 1907. — GUENTHER, Ernest S. Mexican Linaloe Oil. Drug and Cosmetic Ind. 49 (2) 146—150, 161, 1941. — GUZMAN, David, J. Especies Utiles de la Flora Salvadoreña. ed. 2. San Salvador 1950.

HAGEN, V. W. von. Mexican Paper-Making Plants. Journ. N. Y. Bot. Gard. 44: 1—10, 1943. — HARPER, R. M. Useful Plants of Yucatan. Bull. Torr. Bot. Club. 59: 279—288, 1932.

LUNDELL, C. L. Plants probably utilized by the Old Empire Mayas of Petén and adjacent Lowlands. Pap. Mich. Ac. Sci. 24: 37—56, 1939.

MARTINEZ, Maximo. Plantas Utiles de Mexico. 2. ed. Mexico 1936. —, Plantas Medicinales de Mexico. 3. ed. Mexico 1944.

NORIEGA, J. M. Las Plantas Mexicanas y algunas exoticas productoras de materias colorantes. Publ. Sec. Ind. Com. y Trab. Dept. de Aprovisionam. General. 1919.

PITTIER, H. Ensayo sobre las Plantas Usuales de Costa Rica. Washington 1908. —, The Middle American Species of the Genus Inga. Journ. Dept. Agric. Porto Rico. 13: 117—177, 1920. — POPENOE, Wilson. The Useful Plants of Copan. Amer. Anthrop. 21: 125—138, 1919.

ROQUÉ, José Maria. Flora medico-guatemalteca apuntes para la Materia Medica de la Republica de Guatemala. Guatemala 1941. — ROSE, J. N. Notes on useful Plants of Mexico. Contrib. U.S. Nat. Herb. 5 pt. 4. 1899. — ROYS, L. The Ethnobotany of the Maya. Tulane Univ. Middle. Am. Res. Ser. Publ. 2. 1—359, 1931.

STANDLEY, Paul C. Trees and Shrubs of Mexico. Contrib. U.S. Nat. Herb. 23: 1—1721, 1920—1926. —, Flora of the Panama Canal Cone. Contrib. U.S. Nat. Herb. 27: 1—416, 1928. —, Flora of Costa Rica. Field Mus. Nat. Hist. Bot. Ser. 1938.

WEST INDIES

BARRETT, O. W. The Food Plants of Porto Rico. Journ. Dept. Agric. Porto Rico. 9: 61—208, 1925. — BECKWITH, Martha Warren. Notes on Jamaican Ethnobotany. Poughkeepsie. N. Y. 1927. — BERK, L. H. van. Bijdrage tot de Kennis der West Indische Volksgeneeskruiden. Dis. Utrecht 1930.

COOK, O. F. and G. N. COLLINS. Economic Plants of Porto Rico. Contrib. U.S. Nat. Herb. 8 pt. 2: 57—269, 1903.

FREEMAN, W. G. and R. O. WILLIAMS. The Useful and Ornamental Plants of Trinidad and Tobago. Mem. Dept. Agric. Trin. and Tob. 1928.

HARRIS, W. Notes on Fruits and Vegetables of Jamaica. Jamaica 1913. — HODGES, W. H. Plants used by the Dominican Caribs. Journ. N. Y. Bot. Gard. 43: 189—201, 1942. — MAZA, Manual G. de la, y JIMINEZ y Juan T ROIG y MESA. Flora de Cuba. Bol. 22. Estic. Exp. Agron. Cuba 1914.

ROIG Y MESA, J. T. Catalogo de Maderas Cubanas. Bol. 57. Estac. Agric. Cuba 1935. —, Plantas Medicinales, Aromaticas o Venenosas de Cuba. Habana 1945. — ROWAAN, P. A. Curaçoa-Aloë. Ber. Afd. Handelsmus. Kol. Inst. Amsterdam. Nr. 101. 1936.

SOUTH AMERICA

ARBELAEZ, Enrique, Perez. Plantas Utiles de Colombia. Bogota 1936. — BENOVIST, R. Quelques Plantes toxiques utilisées par les Indiens de l'Equateur. Bull. Mus. Nat. Hist. Nat. 7: 145—147, 1935. — BOERGER, Alberto. Investigaciones Agronómicas. 3 vol. Montevideo 1943. — BROCADET, A. P. Plantes Utiles do Brésil. Paris 1921. — BUKASOV, S. M. Potatoes of South America and their breeding possibilities. Lenin Acad. USSR. Inst. Plant Ind. Bull. (Suppl.) 58: 1—192, 1933.

CARVALHO, J. B. de. Notas sobre a Industria de Oleos Vegetaes no Brisil. Rio de Janeiro 1924. — CASTRO, E. B. Las Maderas Argentinas. Su Importancia Industrial. Rosario 1918. — CORREA, Pio. Diccionario das Plantas Uteis do Brasil e das Exoticas Cultivadas. Minist. da Agric. Vol. 1 (A—Cap). Vol. 2 (Car—E). Rio de Janeiro 1926, 1931. —

DEVEZ, G. Les Plantes Utiles et les Bois Industriels de la Guyane. Paris 1932. — DOMINGUEZ, J. A. Contribuciones a la Materia Médica Argentina. (Primera Contribución). Trab. Inst. Bot. y Farm. Fac. Cien. Med. Buenos Aires 1928.

FONSECA, E. T. da. Oleos Vegetaes Brasileiros. Rio de Janeiro 1922. — FREISE, Frederico W. Plantas Medicinales Brasileiras. Sao Paulo 1934.

GARDENAS, Martin. Contribuciones a la Flora Economica de Bolivia. Publ. Univ. Autonoma „Simon Bolivar" 1941. — GONGGRIJP, J. W. De Beteekenis van de Exploitatie van Rozenhoutoilie, afkomstig van Aniba rosaeodora Ducke. Ber. Afd. Handelsmus. Kol. Inst. Amsterdam. Nr. 138. 1939. — GRESHOFF, M. De Nuttige Planten van Fransch Guyana in Verband met Suriname beschouwd. Bul. Kolon. Mus. Haarlem. 25: 23—45, 1901.

HARRISON, J. B. and C. K. BANCROFT. Food Plants of British Guiana. Journ. Bd. Agric. Brit. Guian. 10: 143—177, 1917. — HERING, C. J. Overzicht van de Cultuurgewassen en Boschproducten in Verband met Nijverheid en Handel in de Kolonie Suriname, Nederlandsch Guiana. Paramaribo. 1902—1903. — HERRERA, F. L. Plantas Endemicas domesticadas por los antiguos peruanos. Rev. Mus. Nac. Lima. 11: 25—30, 1942. — HOEHNE, F. C. Plantas e Substancias Vegetais Toxicas e Medicinais. São Paulo. 1939. — HOHENKERK, L. S. British Guiana Timbers. Journ. Board. Agric. Brit. Guiana. 12: 152—185, 1919.

KRUKOFF, B. A. and H. N. MOLDENKE. Studies of American Menispermaceae, with special Reference to Species used in the Preparation of Arrow-Poison. Brittonia 3: 1—74, 1938.

LATCHAM, Ricardo, Eduardo. La Agricultura Precolombiana en Chile y los Países Vecinos. Ed. Univ. Chile. Santiago 1936.

MOLFINA, J. F. Notas Sobre las Nyctaginaceas usades de la Flora Argentina. Buenos Aires, Minist. Agric. Secc. Propag. e Inf. nr. 755. 1928. —, Plantas usuales de la Flora Argentina. Almanaque Min. Agric. Argentina. 9: 457—466, 1934.

NAVARRO de ANDRADE, W. Les Bois Indigenes de Sao Paulo. Sao Paulo 1916.

PARDAL, Ramon. Medicina Aborigen Americana. Buenos Aires 1937. — PARODI, Domingo. Notas sobre algunas Plantas usuales del Paraguay, de Corrientes y de Missiones. Buenos Aires 1886. — PARODI, L. R. Las Plantas Indigenas no Alimentacias Cultivadas en la Argentina. Rev. Argent. Agron. 1: 165—212, 1912. — Relaciones de la agricultura prehispanica con la agricultura Argentina actual. Observ. gen. sobre la domest. de las Plantas. An. Acad. Nac. Agron. y Veter. Buenos Aires 1: 115—167, 1935. — PENNA, Meira. Diccionario Brasileiro de Plantas Medicinais. Descriçao das Plantas Medicinais Indigenas e das Exóticas Aclimadas no Brasil. Rio de Janeiro 1941. — PEREIRA, Huascar. Diccionario das Plantas Uteis do Estado de S. Paulo. Sao Paulo 1929. — PÉREZ, ARBELÁEZ, Enrique. Plantas Medicinales más usades en Bogotá. Bol. Agric. Colombia. Supl. no. 32. 1934. — POPENOE, Wilson. Economic Fruit Bearing Plants of Ecuador. Contrib. U. S. Nat. Herb. 24 pt. 5: 100—134, 1924. — PFEIFFER, J. Th. De Houtsoorten van Suriname. Afd. Handelsmuseum. nr. 6. Kon. Ver. Kol. Inst. Amsterdam 1927. — PITTIER, H. Manual de las Plantas Usuales de Venezuela. Caracas 1926. Suppl. 1939.

RANGEL, Jose, Luiz, e Haya S. SCHNEIDER. Copaés do Brasil. Inst. Nac. de Techn. (Minist. d. Trabalho, Indust. e Comm.) Rio de Janeiro 1936. — RECORD, S. J. and C. D. MELL. Timbers of Tropical America. New Haven 1924. — REICHE, K. F. Los Productos Vejetales Indigenas de Chile. Bol. Soc. Fom. Fabril Santiago (Chile). 32: 481—486, 679—684, 776—784, 1915. — RODRIGUEZ, P. M. Plantas Medicinales del Paraguay. Asunción 1915. — ROWAAN, P. A. Divi-Divi van Curaçao. Ber. Afd. Handelsmus. Kol. Inst. Amsterdam. Nr. 144. 1940.

SANTA CRUZ, Alcibíades. Plantas Medicinales de la Región de Concepción. Rev. Chilena Hist. Nat. 25: 241—252, 1923. — SEMLER, Henrique. Plantas Uteis do Deserto. Serv. de Inf. e Devulg. Minist. da Agric. Ind. e Comm. Rio de Janeiro 1913. — STAHEL, G. The Nuttige Planten van Suriname. Surin. Dept. Landb. Bull. 57. 1942. — STORNI, Julio. Vegetales que utilizaban nuestros Indigenas para su Alimentación. Tucuman 1937.

WILLIAMS, L. Woods of Northeastern Peru. Chicago 1936. —, Maderas Economicas de Venezuela. Caracas 1939.

WORKS OF HISTORICAL VALUE

CANDOLLE, A. de. Origine des Plantes Cultivées. 3 ed. Paris 1884. (transl. in English and German). — CHABRAEUS, Dominicus. Stirpium Icones et Sciagraphia. Genevae 1666. — CLUSIUS, Carolus (l'Écluse, Charles de). Rariorum Plantarum Historia. Antwerpiae 1601. — CULPEPER, Nicholas. A Physicall Directory or A translation of the London Dispensatory Made by the Colledge of Physicians in London. London 1649.

DODONAEUS, Rembertus. (Dodoens, Rembert). Cruydeboeck. Antwerpen 1554.

FERRARII, Bapt. Hesperides s. de Malorum Aureorum. Romae 1646. — FISCHER, Hermann. Mittelalterliche Pflanzenkunde. München 1929. — FUCHS, Leonard. De Historia Stirpium Commentarii. Basil. 1542.

GERAD, John (Gerarde, John). The Herball or Generall Historie of Plantes. London 1597.

HEER, Oswald. Die Pflanzen der Pfahlbauten. Zürich 1865. — HERNANDEZ, Francisco. Nova Plantarum, Animalum et Mineralium Mexicanorum Historia. Madrid 1651.

KNOOP, Jean, Herman. Pomologie ou Description des Meilleures Sortes de Pommes et de Poires. Amsterdam 1771 (also in Dutch).

LOBELIUS, Mathias (de l'Obel, Mathias). Stirpium Adversaria Nova. Londini 1570.

MAGNUS, Albertus. Liber aggregationis seu liber secretorum Albertii magni de virtutibus herbarum. Per Johannem de Annunciata de Augusta. 1478. — MATTHIOLUS, Petrus Andreas (Mattioli, Pierandrea). Compendium De Plantis Omnibus, ... de quibus scripsit suis in commentariis in Dioscoridem editis. Venetiis 1571. — MOLDENKE, H. N. und A. L. Plants of the Bible. Waltham 1952. — MUNTING, Abrahamus. Waare Oeffening der Planten. Amsterdam 1672. —, Nauwkeurige Beschrijving der Aard-Gewassen. 2 vol. Leyden 1696.

ORTUS SANITATIS (Hortus Sanitatis). Moguntia 1491. QUINTENYE, Joannes de la. Instruction pour les Jardins Fruitiers et Potagers, avec un Traité des Orangers, suivi de quelques reflexions sur l'Agriculture. Paris 1690.

REEDE TOT DRAKENSTEIN, H. A. Hortus Malabari-
cus. 12 vol. Amstelodami 1678—1703. — RUMPHIUS,
G. E. Herbarium Amboinense. 6 vol. Amstelodami
1741—1755.

SERRES, Oliver. Théatre d'Agriculture et Ménage des
Champs. Paris 1600.

THEOPHRASTUS. Theophrasti de historia et causis
plantarum, Libri Quindecim. Theodor Gaza inter-
prete. Excussum Luteciae, in aedibus Christiani We-
chel. 1529. — TURNER, Robert. The British Physi-
cian; or The Nature and Vertues of English Plants.
London 1664.

VAVILOV, N. I. Studies on the Origin of Cultivated
Plants. Bull. Appl. Bot., Gen. and Plant Breed. Lenin-
grad 1926. — VOLCKAMER, J. C. Nürnbergische
Hesperides oder Beschreibung der Citronat-, Citro-
nen- und Pomeranzen-Früchte. 2 vol. Nürnberg 1708—
1714.

WOENIG, Franz, Die Pflanzen im alten Aegypten.
2. Aufl. Leipzig 1897.

JOURNALS

ANGEWANDTE BOTANIK. (Bis 1918: Jahresbericht
d. Ver. f. Angewandte Botanik.) Since 1903. BULLE-
TIN Applied Botany and Plant Breeding. Moscow.
Since 1920.

ECONOMIC BOTANY. New York. Since 1947.

REVUE de Botanique Appliquée et d'Agriculture Coloniale. Paris. Since 1921.

Hafner Books on Botany

BOWER, F. O.
  The Origin of a Landflora. A Theory based upon the Facts of
  Alternation. Illus. (London 1900) Reprint 1959 $ 10.00

BOWER, F. O.
  Primitive Land Plants, also known as the Archegoniatae. Illus.
  (London 1935) Reprint 1959 $ 10.00

CANDOLLE, A. DE.
  Origin of Cultivated Plants. (London 1886) Reprint 1959 $ 5.00

DIOSCORIDES.
  The Greek Herbal. Illus. by a Byzantine A. D. 512: Englished
  by John Goodyear A. D. 1655: Edited and first printed A. D.
  1933, by Robert T. Gunther. (Oxford 1934) Reprint 1959 $ 15.00

BULLER, A. REGINALD.
  Researches on Fungi. Originally published 1909—1934. 6 vols.
  Reprint 1958. Single volumes   ea. $ 15.00   set $ 90.00

CARPENTER, RICHARD J.
  An Ecological Glossary. Defining nearly 3000 terms. 368 pp.,
  maps. Originally published 1938. Reprint 1956. $ 5.75

CLEMENTS, FREDERIC E.
  Dynamics of Vegetation. Selections from the Writings of
  F. E. Clements, 292 pp. 70 plates, glossary. 1949. $ 4.50

CLEMENTS, FREDERIC E.
  Plant Succession and Indicators. A definitive edition of Plant
  Succession and Plant Indicators. 469 pp., 44 plates, bibliography.
  1928. $ 7.50

CLEMENTS, FREDERIC E. and SHEAR, CORNELIUS, L.
  The Genera of Fungi ill. by Edith S. Clements. 496 pp., 58 plates.
  Glossary of Latin and English terms. 34-page index. Reprint 1957 $ 15.00

GAEUMANN, ERNST
  The Fungi. A Description of their Morphological Features and
  of their Evolutionary Development. Translated from the German
  and Arranged as an American text by F. L. Wynd. 420 pp., 440
  ill., bibliog. 1952 $ 10.00

FUNDER, SIGURD
  Practical Mycology. Manual for Identification of Fungi, 146 pp.,
  num. ill., table of classification, 1953 $ 6.50

SEWARD, A. C.
  Plant Life Trough the Ages. Originally published in 1933. Re-
  printed 1959 $ 12.50

HAFNER PUBLISHING COMPANY

31 East 10th Street   New York 3, N. Y.

An Interesting New Publication:

»Fertilizer Use«
Nutrition and Manuring of Tropical Crops
by **Prof. Dr. A. Jacob** and **Dr. H. v. Uexküll,**

491 pages, illustrated, published by the Verlagsgesellschaft für Ackerbau,
Hannover 1958.

Tropical and sub-tropical agriculture has grown considerably in importance during
the last decades. On the one hand technical progress and improvement in the means
of communication have opened up wide territories for the intensification of agricultural production. On the other hand, the increasing population is steadily accen-
tuating the need for more food. Consequently the measures such as manuring for
increasing the crop yields have become profitable in practically all parts of the
tropics and sub-tropics and have frequently proved to be a real necessity.

In contrast to the regions of intensive agriculture in the temperate zones our knowledge of the manuring of tropical crops is still very imcomplete in certain respects.
It is, therefore, particularly welcome that an attempt has now been made to sum-
marise in a single volume the complex field of the manuring of the different tropical
crops.

It is a long time since a book of this kind was published and it appears at a time
when the problems of manuring in tropical agriculture have become especially acute.
The book is divided into two parts, a special one dealing with the use of manures
for the individual crops, and a general one devoted to the principles of plant nutrition and manuring.

In the special part the reader is made conversant with the present position and the
particular problems of the use of manures for the different important tropical and
sub-tropical crops. This part is based on a revised summary of numerous widely
distributed individual publications. The concise, clear presentation considerably
facilitates the reading of the book and it will certainly soon become a popular
handbook in science, advisory work and practice.

The crops treated in detail are as follows:

    I   Cereals (Rice, Wheat, Maize, Barley, Millet)
    II   Tuber Crops (Potatoes, Cassava, Yams, Sweet Potatoes)
   III Sugar-cane
   IV Fodder Crops (Lucerne, Lespedeza, Kudzu)

V	Oil-Seeds (Soya-bean, Groundnut, Olive, Tung Tree, Ricinus)
VI	Fibre Plants (Cotton, Kenaf, Sisal, Ramie, Manila Hemp, Hemp, Flax)
VII	Stimulants (Tobacco, Tea, Coffee, Cocoa)
VIII	Rubber
IX	Fruits (Citrus Varieties, Avocado, Papaya, Mango, Banana, Pineapple, Grape-vine, Apples, Pears).

An exhaustive bibliography is appended to each section dealing with a crop and undoubtedly this increases the value of the book as a basis for future work. In addition, there are numerous tables and illustrations distributed throughout the text. The general part comprises an introduction to the principles of plant nutrition and manuring. In the first section of this part the effects and functions of the individual plant nutrients (including the trace elements) in plant metabolism are described. All that is necessary to understand the effects of the nutrients in the fertilizers is concisely represented; for instance, the behaviour of the nutrients in the soil, the form in which they are absorbed by the plant, the functions of the nutrients in metabolism as well as the symptoms of deficiency or excess. In this connection the most recent discoveries have been taken into consideration; for example chlorine and sodium are not treated amongst the unnecessary elements but as vitally important trace elements which can be absorbed by most plants in larger quantities without injury.

The second section deals with the more important fertilizers and their properties. Here the organic fertilizers are compared with the inorganic ones and emphasis is laid upon the point that the organic and inorganic fertilizers are supplementary in character and that it is only by a combination of both kinds of fertilizers that one can provide the ideal nutrient conditions for the plant. In dealing with the inorganic fertilizers, besides considering these individually, particular attention is given to the proups of fertilizers having comparable properties. As regards mixed and complete manures the most recent developments have been discussed.

The third section deals with the mutual relationships between manuring and the other yield factors such as soil, climate, plant and cultivation by Man. It is shown clearly that manuring represents only one method of increasing agricultural production and this can only achieve complete success when all the other yield factors are equally favourable and brought into a correct relationship with the use of fertilizers.

Particular attention should also be given to the extensive appendix of tables containing much important and useful information regarding volumes, weights, conversion factors and statistics. Since this book supplies numerous diverse data on the theme „Nutrition and Manuring of the Most Important Crops of the Tropics and Sub-Tropics", doubtless it will be used constantly as a reference book. In order to meet this purely external requirement, the octavo size, (5" × 8"), has been chosen which is particularly handy for the writing table; in addition, the book is provided with a flexible binding and several book-marks.

Qualitas Plantarum et Materiae Vegetabiles

Organ of the Confoederatio Internationalis ad Qualitates Plantarum Edulium Perquirendas (CIQ) and of the International Commission for Plant Raw Materials

EDITORS:

A. Th. Czaja, Aachen - L. Genevois, Bordeaux - C. Regel, Graz
W. Schuphan, Geisenheim

This journal publishes original contributions on genetics, biochemistry, physiology, ecology, culture and preparation of edible and of medicinally and technically used plants with special regard to their qualifying substances forming the "nutritional" or the "technical" value.

The double volume III/IV contains the papers and discussions of the "Deuxième Colloque International sur la qualité végétale", Paris 1957 (President E. F. Terroine). Every paper is accompagnied of extensive summaries in english, french and german.

Subscription per volume of 400 pages: dutch guilders 48.—

VEGETATIO

ACTA GEOBOTANICA

International Review of Plant sociology, Ecology and Plantgeography

EDITORS:

J. Braun-Blanquet, Montpellier - R. Tüxen, Stolzenau/Weser

Vegetatio publishes original papers on plantsociology, contributions from the important fields of ecology, history of the flora (including analysis of pollen) and plantgeography; papers on new problems, views and methods as well as those which deal with less known subjects; papers on applied plantsociology; **a personal column and reviews of new books and pamphlets.**

Subscription per volume   dutch guilders 40.—

Bestimmungstabellen der Blattminen von Europa einschließlich des Mittelmeerbeckens und der Kanarischen Inseln

by

Prof. Dr. Erich Martin Hering

All the known leaf mines of Europe, including the Mediterranean Basin and the Canaries, are set out in dichotomic identification keys, arranged in alphabetical order of the plant genera. **They make possible the determination of the mine producer from the form of the mine.** For each species additional characteristics of the mine, time of occurrence, geographical distribution (with a reference to those species also occurring in North America) are included.

1957. 2 volumes and atlas. 1406 pages with 725 figures. cloth. dutch guilders **194.-**

Dr. W. JUNK, PUBLISHERS, THE HAGUE, THE NETHERLANDS

PAREYS BLUMENGÄRTNEREI

(Parey's gardening of flowers)

Description, culture and use of horticultural ornamental plants

Second, revised edition

Under contribution of many specialists

edited by

Fritz Encke

Director of the PALMENGARTEN, Frankfort-on-the-Main.

Two volumes with about 1760 pages, about 1000 pictures accompanying the text, and 40 coloured plates.

PAREYS BLUMENGÄRTNEREI gives a general and complete view on all kinds of plants that are nowadays used or might be used for horticultural purposes, both professional and private. You will find in the book not only flowers and ornamental plants, but just everything that might be in a flourishing and verdant condition either in gardens, parks, glass-houses, or chambers. It is thus the most voluminous German publication dealing with all kinds of ornamental plants, a book that will be just as indispensable for the gardener to whom horticulture is a means of living as for the gardener by passion. All important kinds are shown by particularly beautiful, impressive, and typical pictures which facilitate determinating and recognizing them. Forty multi-coloured plates are the special floral ornaments of the book. In the same way like the originals do, they let you see modern cultivations of choice descendants, but also rare and interesting appearances.

„Sixteen expertes and high specialists have been contributors to this outstanding publication. The descriptions of cultivations bring about reliable advices and hints for all plants, whereas the more than 1000 pictures which accompany the text represent an intuitive material that cannot be equalled. This makes PAREYS BLUMENGÄRTNEREI an indispensable work of reference."

Schweizerisches Gartenbau-Blatt

The two volumes of the work have been published since the middle of 1957 by totally about twenty-two monthly deliveries. Each delivery comprises eighty pages and two coloured plates. The reduced price of subscription is DM 12.40. When subscribing you will engage yourself to purchase the total of the work.

Paul Parey, Publisher / Berlin and Hamburg

Handbuch der Pflanzenzüchtung

(Manual of Plant Breeding)

founded by TH. ROEMER and W. RUDORF

Second Revised Edition. In Six Volumes

With the collaboration of numerous contributors in Germany and foreign countries and with the cooperation of Prof. Dr. H. K. Hayes, St. Paul/USA, and Prof. Dr. A. Müntzing, Lund/Sweden

edited by

Prof. Dr. H. KAPPERT, Berlin, and Prof. Dr. W. RUDORF, Köln

Volume I — Fundamentals of Plant Breeding
872 pages with 175 illustrations. Clothbound. By subscription DM 160.—
Single DM 192.—

Volume II — Breeding of Grain Species
631 pages with 94 illustrations. Clothbd. By subscr. DM 122.—

Volume III — Breeding of Tubers and Root Crop Plants
331 pages with 59 illustrations. Clothbound. By subscr. DM 73.—

Volume IV — Breeding of Forage Plants

Volume V — Breeding of Special Cultivated Plants

Volume VI — Breeding of Fruits and Legumes, Viniculture and Silviculture

The reprint of this international standard work brings a modern and exhaustive performance of the actual knowledge in the field of plant growing for the purpose of getting bigger quantities and better qualities of foodstuffs for human beings; it shows the actual and possible capacities in modern cultivations and so doing animates both labourers and research stations to go on working with sucess. The work has been published since the middle of 1955 with totally about forty monthly deliveries of five printed sheets each. The reduced subscription price of each delivery is DM 13.50.

Zeitschrift für Acker- und Pflanzenbau

(Periodical for Agriculture and Plant Growing)

edited by

Prof. Dr. W. BROUWER, Stuttgart, and Prof. Dr. O. TORNAU, Göttingen

The periodical which is rich of traditions is a central publishing agent of the German institutes for agriculture and the cultivation of plants. It comprehends the entire proceeds of scientific work in this field by original articles which are prior to summaries in German and English.

Every year there are published eight to ten parts, each of which has about eight printing sheets, at indefinite periods. Four parts form one volume. The price of each printing sheet is about DM 3.50.

Zeitschrift für Pflanzenzüchtung

(Periodical for the Cultivation of Plants)

Edited by

Prof. Dr. E. ÅKERBERG, Svalöf, Prof. Dr. Dr. h. c. H. KAPPERT, Münster, Prof. Dr. H. KUCKUCK, Hannover, Prof. Dr. W. RUDORF, Köln, Prof. Dr. Dr. h. c. H. STUBBE, Gatersleben, and Prof. Dr. Dr. h. c. E. v. TSCHERMAK, Wien

The periodical for the cultivation of plants publishes in German, English, and French original works, collective reports, book reviews, and compositions of single persons on the doctrine of heredity that has been applied to plants, and on the practical cultivation of useful plants in agriculture, horticulture, and sylviculture.

Every year there are published six to eight parts, each of which has about eight printing sheets, at indefinite periods. Four parts form one volume. The price of each printing sheet is about DM 3.50.

Paul Parey, Publisher / Berlin and Hamburg

BOTANICA MARINA

International Review for Seaweed Research and Utilization

Editorial Committee:

R. Biebl, Österreich

L. R. Blinks, U.S.A.

T. Braarud, Norge

V. J. Chapman, New Zealand

M. R. Droop, Great Britain

J. Feldmann, France

C. Hoffmann, Deutschland

H. A. Hoppe, Deutschland

T. Levring, Sverige

G. F. Papenfuss, U.S.A.

M. Parke, Great Britain

L. Provasoli, U.S.A.

O. Schmid, Deutschland

E. Steemann Nielsen, Danmark

A. R. A. Taylor, U.S.A.

W. R. Taylor, U.S.A.

Y. Yamada, Japan

E. G. Young, Canada

Editorial Office:

Studiengesellschaft zur Erforschung von Meeresalgen e. V., Hamburg

The BOTANICA MARINA will contain original articles covering: oecology, botany, chemistry, physiology, plankton-algae, seaweed and seaweed products technology. It is intended to publish a Special Issue once a year devoted to technological and industrial problems.

Subscriptions: DM 28.— per annum, postage extra
(4 issues annually; 32 pages, format 17 × 24 cm). Single copies DM 7.—.

HEINZ A. HOPPE · DROGENKUNDE

MANUAL OF VEGETABLE AND ANIMAL RAW MATERIALS

Seventh edition, amended and extended 1958, 17,5 × 25 cm, 1231 pages
linen bound DM 78.—

Contents: Part I and II deal with vegetable and animal raw materials in 1838 paragraphs (derivation, origin, trade names, ingredients and uses). Part III gives a table of countries of origin. Part IV indicates possibilities for the use in various industries. Part V gives the important terms in foreign languages (English, French, Portuguese, Spanish). The index contains more than 20 000 catchwords.

CRAM, DE GRUYTER & CO. HAMBURG 1

In der SAMMLUNG GÖSCHEN erschienen:

Date Due